CONDENSED PYRIDAZINES INCLUDING CINNOLINES AND PHTHALAZINES

This is the twenty-seventh volume in the series
THE CHEMISTRY OF HETEROCYCLIC COMPOUNDS

THE CHEMISTRY OF HETEROCYCLIC COMPOUNDS

A SERIES OF MONOGRAPHS

ARNOLD WEISSBERGER and EDWARD C. TAYLOR

Editors

CONDENSED PYRIDAZINES INCLUDING CINNOLINES AND PHTHALAZINES

Edited by

Raymond N. Castle

DEPARTMENT OF CHEMISTRY
BRIGHAM YOUNG UNIVERSITY
PROVO, UTAH

AN INTERSCIENCE® PUBLICATION

JOHN WILEY & SONS,
NEW YORK · LONDON · SYDNEY · TORONTO

An Interscience® Publications

Library of Congress Cataloging in Publication Data:

Castle, Raymond Nielson, 1916–
Condensed pyridazines including cinnolines and phthalazines.

(The Chemistry of heterocyclic compounds, v. 27)
Includes bibliographical references.
1. Pyridazine. I. Singerman, Gary M. II. Title.

QD401.C34 547'.593 72-6304

ISBN 0-471-38211-6

Printed in the United States of America

10 9 8 7 6 5 4 3 2 1

Contributors

Natu R. Patel, *Gulf Research and Development Company, Merriam, Kansas*
Gary M. Singerman, *Gulf Research and Development Company, Pittsburgh, Pennsylvania*
B. Stanovnik, *Department of Chemistry, University of Ljubljana, Ljubljana, Yugoslavia*
M. Tišler, *Department of Chemistry, University of Ljubljana, Ljubljana, Yugoslavia*

The Chemistry of Heterocyclic Compounds

The chemistry of heterocyclic compounds is one of the most complex branches of organic chemistry. It is equally interesting for its theoretical implications, for the diversity of its synthetic procedures, and for the physiological and industrial significance of heterocyclic compounds.

A field of such importance and intrinsic difficulty should be made as readily accessible as possible, and the lack of a modern detailed and comprehensive presentation of heterocyclic chemistry is therefore keenly felt. It is the intention of the present series to fill this gap by expert presentations of the various branches of heterocyclic chemistry. The subdivisions have been designed to cover the field in its entirety by monographs which reflect the importance and the interrelations of the various compounds, and accommodate the specific interests of the authors.

In order to continue to make heterocyclic chemistry as readily accessible as possible new editions are planned for those areas where the respective volumes in the first edition have become obsolete by overwhelming progress. If, however, the changes are not too great so that the first editions can be brought up-to-date by supplementary volumes, supplements to the respective volumes will be published in the first edition.

ARNOLD WEISSBERGER

Research Laboratories
Eastman Kodak Company
Rochester, New York

EDWARD C. TAYLOR

Princeton University
Princeton, New Jersey

Preface

The subject matter in this book was originally intended for inclusion in a single volume on pyridazines, condensed pyridazines, and cinnolines and phthalazines. It became apparent, however, that the topics should be divided, simple pyridazines comprising one volume and condensed pyridazines, which are limited for the most part to two rings, making up a separate volume.

This volume is organized into three sections that deal with the two benzo-pyridazines, cinnolines and phthalazines, and condensed pyridazines in which the second ring contains one or more heteroatoms.

The literature on cinnolines and phthalazines covered here continues that reviewed by J. C. E. Simpsom in *Condensed Pyridazine and Pyrazine Rings* (*Cinnolines, Phthalazines and Quinoxalines*), Interscience Publishers, Inc., New York (1953), with only relatively small areas of overlap caused by an effort to provide continuity. The literature is discussed up to mid-1971 using *Chemical Abstracts* as the guide to the original literature.

The field of condensed pyridazines containing heteroatoms in both rings has in many instances experienced comparatively little research and, therefore, these rings provide many fruitful research areas. Furthermore, a number of possible condensed pyridazine rings have not appeared in the literature; they present opportunities for research in the synthesis of new condensed pyridazine rings.

I hope that this volume will stimulate research in heterocyclic chemistry by alerting chemists to the fascinating and challenging problems that await solution in these ring systems.

I am indebted to the four authors, Professor Tišler, Dr. Stanovnik, Dr. Patel, and Dr. Singerman, for their cooperation and understanding in the preparation of this volume.

RAYMOND N. CASTLE

Provo, Utah
April 1972

Contents

CHAPTER 1

Cinnolines

GARY M. SINGERMAN

Gulf Research and Development Company, Pittsburgh, Pennsylvania

Part A. Cinnoline

I. Physical Properties

Cinnoline (1,2-diazanaphthalene) (**1**) is a pale yellow solid which is soluble in water and organic solvents. It rapidly liquifies and turns green on standing in air (1), apparently with little decomposition (2). It may be safely stored as the yellow solid under nitrogen at 0° C. When crystallized from ether, it forms a colorless etherate complex which melts at 24–25° C (3). The solvent-free base melts at 40–41° C (2) [37–38° C (1)] and has a boiling point of 114° C at 0.35 mm Hg (2). The cinnoline ring system was first prepared in 1883 (4), and unsubstituted cinnoline was subsequently made in 1897 (3). Cinnoline itself is toxic and shows antibacterial action against *Escherichia coli* (3). Neither it nor its derivatives have been found in nature. It is numbered according to IUPAC nomenclature as indicated in structure **1**.

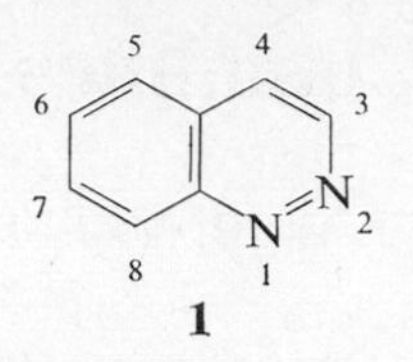

1

Cinnoline, with a pK_a equal to 2.51 in 50% aqueous ethanol at 21–22° C (5) [2.70 (6) or 2.29 (7) in water at 20° C], is a weak base compared to quinoline [pK_a 4.94 in water at 20° C (6)] and isoquinoline [pK_a 5.40 in water at 20° C (7)].

Cinnoline has a dipole moment of 4.14 D in benzene solution compared to values of 4.32 D for pyridazine (8), 2.18 D for quinoline, and 2.52 D for isoquinoline (9), all in benzene solution.

The first and second ionization potentials of cinnoline have been determined by photoelectron spectroscopy to be 8.51 and 9.03 eV (electron volts), respectively. The first ionization potential corresponds to loss of nonbonding electrons (the "lone pair" electrons) from nitrogen, and the second is a π ionization. For comparison, the first ionization potential of naphthalene (π ionization) is 8.11 eV, that of quinoline (π ionization) is 8.62 eV, that of quinazoline (possibly π ionization, but uncertain) is 9.02 eV, and that of phthalazine (lone pair, nonbonding electron ionization) is 8.68 eV (10). Additional information concerning the ionization potentials of the azabenzenes and azanaphthalenes is given in an excellent review of photoelectron spectroscopy by Worley (11).

The heat of atomization of cinnoline, although not determined experimentally, has been calculated by a self-consistent field molecular orbital treatment to be 79.167 eV. This is similar to the calculated heats of atomization of phthalazine (79.215 eV), quinazoline (80.306 eV), and quinoxaline (79.739 eV) (12).

Several molecular orbital calculations of the π-electron density distribution in cinnoline have been made by the Hückel method (13–16). The results of three of these calculations are given in structures **2**, **3**, and **4**. The π-electron distribution in structure **4a** was calculated by the complete neglect of differential overlap (CNDO) method (35). Although the electron density assignments for these structures are not in complete agreement with each other, all four locate the highest electron density for the ring carbon atoms at positions 5 and 8, indicating that electrophilic substitution should occur preferentially at these sites. This is borne out experimentally, at least for

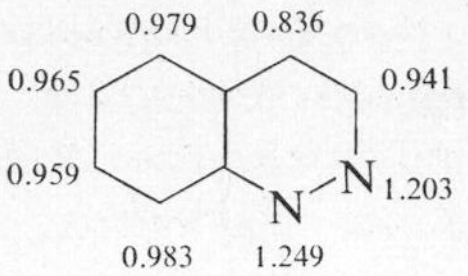

0.979 0.836 0.965 0.941 0.959 N N 1.203 0.983 1.249

2

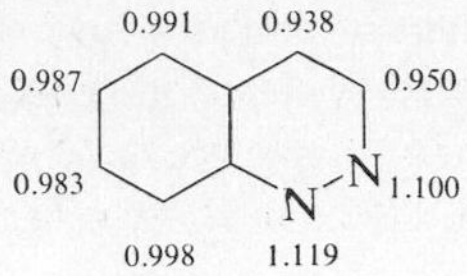

0.991 0.938 0.987 0.950 0.983 N N 1.100 0.998 1.119

3

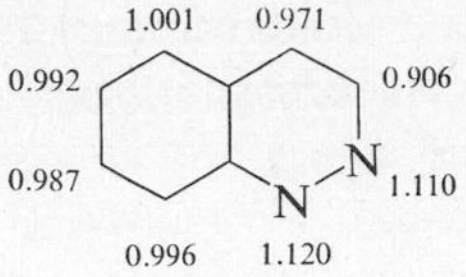

1.001 0.971 0.992 0.906 0.987 N N 1.110 0.996 1.120

4

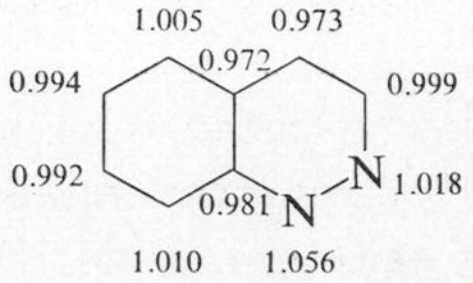

1.005 0.973 0.972 0.994 0.999 0.992 N 1.018 0.981 N 1.010 1.056

4a

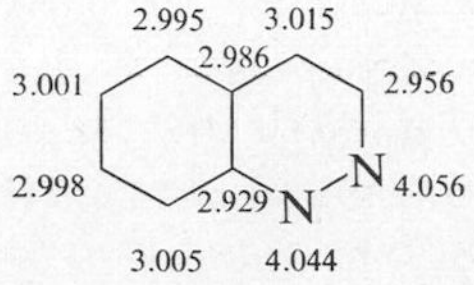

2.995 3.015 2.986 3.001 2.956 2.998 N 4.056 2.929 N 3.005 4.044

4b

simple electrophilic substitution reactions such as nitration (2, 17, 18), and concurs with the results of calculations by Dewar (19, 24). A higher π-electron density is assigned to N-1 than to N-2 in structures **2**, **3**, **4**, and **4a**, whereas a higher σ-electron density is assigned to N-2 than to N-1 in structure **4b** (35). Experimentally it is known that cinnoline undergoes *N*-oxidation (20, 21), protonation (22), and alkylation (23) preferentially at N-2. The 2-cinnolinium ion is calculated to be slightly more stable than the 1-cinnolinium ion (22). Recent molecular orbital calculations by Palmer and co-workers (24a) indicate that the electron densities are essentially equal at *N*-1 and *N*-2 for cinnoline, 4-methylcinnoline, 3-methylcinnoline, and 3,4-dimethylcinnoline, leading Palmer to conclude that preferential *N*-2 protonation is simply a result of steric hindrance to *N*-1 protonation by the *peri* C-8 proton. This correlates with experimental work by Palmer and McIntyre (24b) where such a steric effect was claimed to be balanced by a substituent in the 3-position.

The IR (infrared) absorption spectrum of cinnoline (Table 1A-1) has been recorded and absorption modes have been assigned to the bands when possible (25, 26). The Raman spectrum of cinnoline also has been recorded (26).

The UV (ultraviolet) absorption spectrum of cinnoline has been recorded in several different solvents. The spectral parameters thus obtained are given in Table 1A-2. In addition, several theoretical calculations of transition energies and band intensities have been made (27–29). The spectrum of cinnoline is reported to display from three to six absorption maxima in various solvents in the range 200–380 mμ (millimicrons) which are attributable to π-π^* transitions and an n-π^* absorption of low intensity at 390 mμ (in ethanol) due to the nonbonding electrons of the ring nitrogen atoms.

This n-π^* band is observed to occur at shorter wavelengths in both diazabenzenes and diazanaphthalenes when the two nitrogen atoms are nonvicinal (30). Thus the n-π^* band of quinazoline (1,3-diazanaphthalene) is reported to occur as a shoulder at 330 mμ (31). The UV absorption spectrum of cinnoline has been compared to those of naphthalene and phthalazine (32), and a very fine discussion of the spectrum of cinnoline in the vapor state is given by Wait and Grogan (33), supplemented by Glass, Robertson, and Merritt (34).

The pmr (proton magnetic resonance) spectrum of cinnoline is given in Table 1A-3. A study of the carbon-13 magnetic resonance spectrum of cinnoline has been published (35).

Mass spectral studies show that cinnoline fragments upon electron impact to lose first nitrogen and then acetylene (36, 37). The structure of the C_8H_6 cation resulting from initial loss of N_2 is unknown, even though 3- and 4-deuteriocinnoline and 3,4-dideuteriocinnoline were prepared and subjected to mass spectral analysis in an attempt to elucidate this cation's structure. Incorporation of deuterium in the acetylene arising by fragmentation of the C_8H_6 cation is completely random (37).

II. Methods of Preparation

A. Decarboxylation of Cinnoline-4-carboxylic Acid

The best method now used to prepare cinnoline is the thermal decarboxylation of cinnoline-4-carboxylic acid (**5**) in benzophenone at 155–165°, which gives cinnoline (**1**) in 72% yield (1) and 4,4′-dicinnolinyl (**6**) in about 5.6% yield (2). The decarboxylation of **5** is regarded as proceeding through

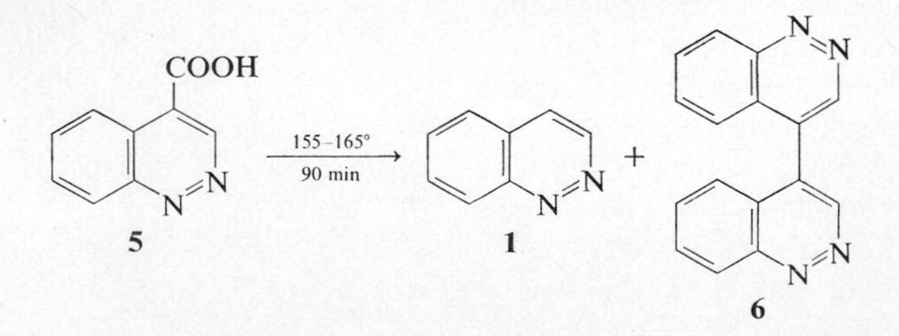

the protonated cinnolinyl anion (**7**) from which both cinnoline and 4,4′-dicinnolinyl are thought to be produced, the latter by attack of the anion **7** on cinnoline (2). However, since 4,4′-dicinnolinyl is formed easily from cinnoline by action of the free radical initiator, *N*-nitrosoacetanilide, the possibility exists that decarboxylation may proceed at least in part via a free radical pathway (38) (see Section 1A-III-B).

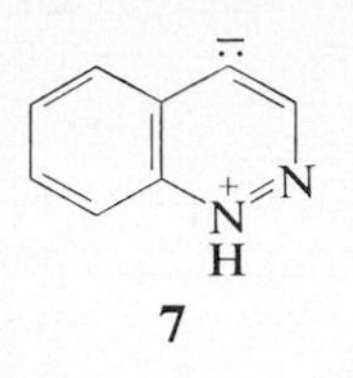

B. Removal of a 3- or 4-Halo Group

Cinnoline was first prepared by the chemical reduction of 4-chlorocinnoline (**8**) with iron and 15% sulfuric acid to give 1,4-dihydrocinnoline (**9**), which was then oxidized to cinnoline with mercuric oxide (3). The

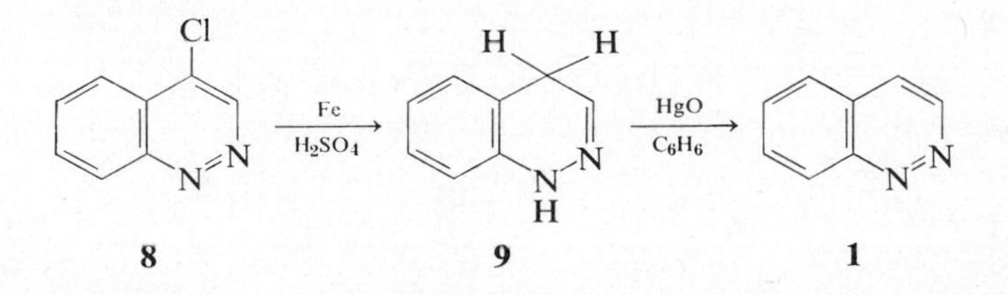

reduced cinnoline **9** was originally assigned a 1,2-dihydro structure but has since been shown by pmr to have a 1,4-dihydro structure (39). Chemical reduction of 4-chlorocinnoline with lithium aluminum hydride in ether solution gives only 4,4′-dicinnolinyl (**6**) (40), whereas catalytic reduction of 4-chlorocinnoline in methanol with palladium on calcium carbonate gives only a trace of cinnoline, the main product also being 4,4′-dicinnolinyl (2).

Cinnoline also has been prepared by treating 4-chlorocinnoline with toluene-*p*-sulfonylhydrazide and decomposing the resulting 4-(toluene-*p*-sulfonylhydrazino)cinnoline (**10**) with aqueous sodium carbonate (17). The yield of cinnoline by this route is good.

Treatment of 4-chlorocinnoline with hydrazine followed by oxidation of the resultant 4-hydrazinocinnoline (**11**) with aqueous copper sulfate gives cinnoline in a yield of about 56% (41).

Reduction of 3-bromocinnoline by hydrazine over palladium-charcoal in methanolic potassium hydroxide gives cinnoline in approximately 56% yield (42).

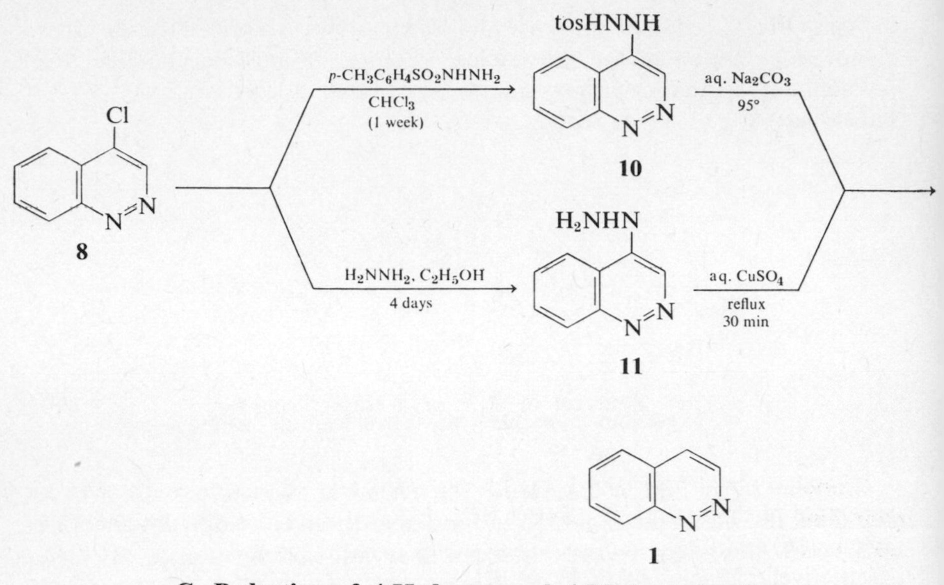

C. Reduction of 4-Hydroxy- and 4-Methoxycinnoline

Direct reduction of 4-hydroxycinnoline (**12**) with lithium aluminum hydride in refluxing tetrahydrofuran for 8 hours followed by gentle oxidation of the resulting partially reduced cinnoline with mercuric oxide in refluxing benzene gives cinnoline in 74% yield (43). Reduction of **12** with lithium aluminum hydride in refluxing 1,2-dimethoxyethane for 3 hours without

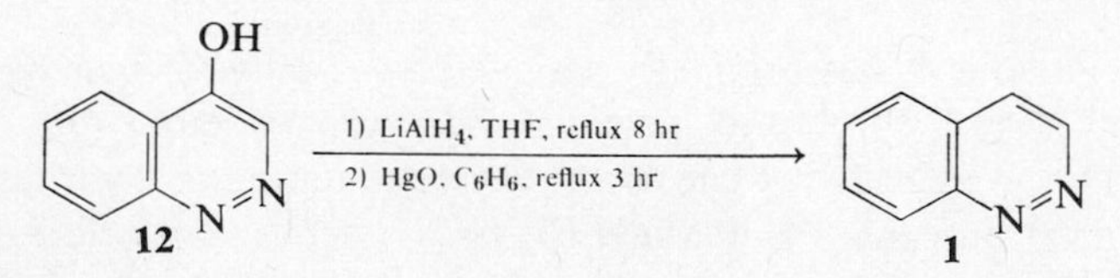

subsequent treatment with mercuric oxide yields a mixture of cinnoline and 1,2,3,4-tetrahydrocinnoline (**13**), while a similar reduction of 4-methoxycinnoline (**14**) in a benzene-ether solution also gives **1** and **13**.

Cinnoline may be prepared from 4-hydroxycinnoline by polarographically reducing the hydroxycinnoline in acid solution, making the solution slightly alkaline, and then oxidizing the intermediate 1,4-dihydrocinnoline anodically. No isolation of the intermediate is necessary, and the overall yield of cinnoline is 70–80% (44).

Only 1,2,3,4-tetrahydrocinnoline is isolated when 3-hydroxycinnoline is reduced by lithium aluminum hydride in refluxing 1,2-dimethoxyethane for 3 hours (40).

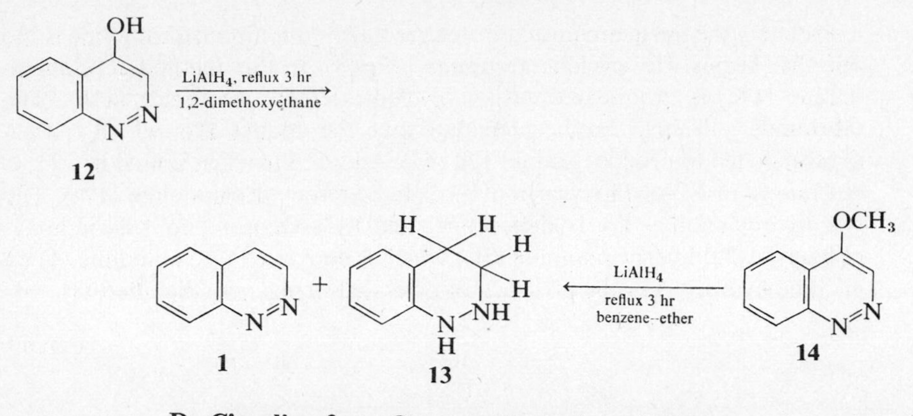

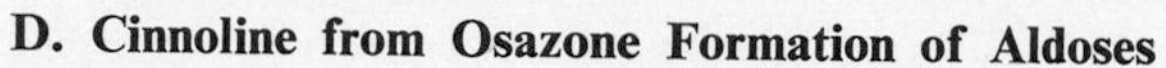

D. Cinnoline from Osazone Formation of Aldoses

When D-glucose (**15**) is heated with aqueous hydrochloric acid and an excess of phenylhydrazine, there is obtained a mixture of D-glucose phenylosazone (**16**) and 1-(3-cinnolinyl)-D-*arabino*tetritol (**17**) in yields of 35 and 20%, respectively. Treatment of **17** with UV light in aqueous sodium hydroxide for 8 hours then gives cinnoline (**1**). Under the same conditions D-mannose gives 25% of D-glucose phenylosazone (**16**) and 16% of the cinnoline derivative **17** (45).

CHO | CHOH | $(CHOH)_3$ | CH_2OH (**15**) + $C_6H_5NHNH_2$ —(aq. HCl, steam bath, 2–3 hr)→

$CH{=}NNHC_6H_5$ | $C{=}NNHC_6H_5$ | $(CHOH)_3$ | CH_2OH (**16**) + 3-cinnolinyl–$(CH(OH))_3$–CH_2OH (**17**) —(UV light, aq. NaOH, 8 hr)→ cinnoline (**1**)

E. Miscellaneous

Desulfurization of 4-mercaptocinnoline with wet Raney nickel in ethanol gives cinnoline, isolated in 30% yield as its picrate (46).

Probably the most circuitous route ever taken to synthesize cinnoline is the one that begins with cycloöctatetraene (47, 48). In this method, cycloöctatetrene (**17a**) is brominated to give cycloöctatetrene dibromide (**17b**). The dibromide and ethyl azodicarboxylate give the adduct **17c**, which is then debrominated to produce adduct **17d** (47). Adduct **17d** when heated to 350° C rearranges to 1,2-diethoxycarbonyl-1,2,4*a*,8*a*-tetrahydrocinnoline (**17e**). The tetrahydrocinnoline **17e** is dehydrogenated by *o*-chloranil to 1,2-diethoxycarbonyl-1,2-dihydrocinnoline (**17f**), which is converted into cinnoline (**1**) by alkaline hydrolysis in the presence of activated manganese dioxide (48).

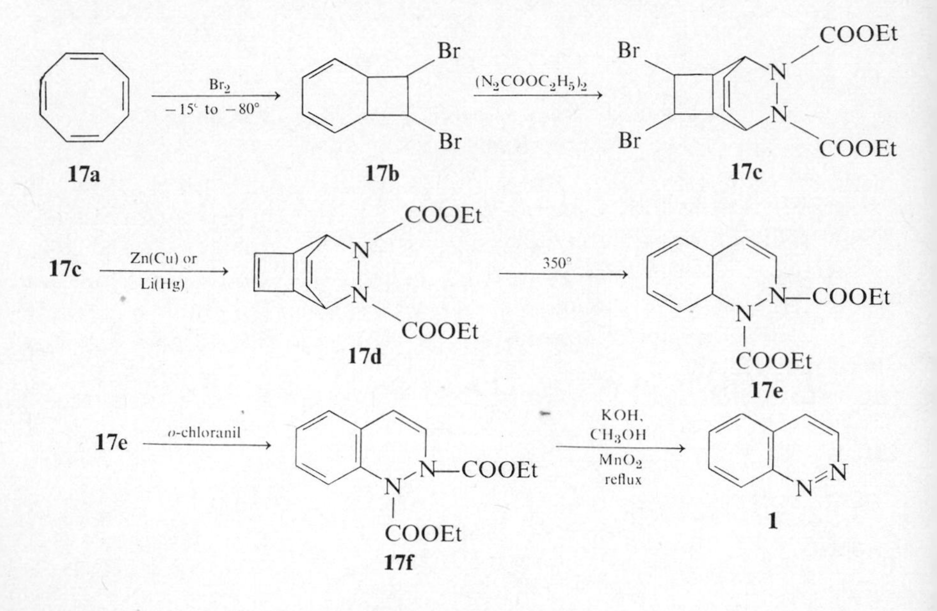

III. Reactions

A. Electrophilic Substitution

Very little experimental work has been done on electrophilic substitution reactions of cinnoline. Dewar (19) determined on the basis of molecular orbital calculations that the relative order of reactivities at different positions of the cinnoline ring system toward simple electrophilic substitution is $5 = 8 > 6 = 7 > 3 \gg 4$. That is, the 5- and 8-positions should be most reactive, and the 4-position should be least reactive toward electrophilic substitution. This is confirmed experimentally in the case of nitration in sulfuric acid, which results in 33% of 5-nitrocinnoline (**18**) and 28% of

8-nitrocinnoline (**19**) as the sole nitration products (2, 17). The species nitrated is not cinnoline itself but the protonated 2-cinnolinium cation. At 80°, in the acidity range 76–83% sulfuric acid, this cation is nitrated 287 times more slowly than the isoquinolinium cation. This gives some measure of the deactivating power of the unprotonated N-1 atom on the nitration of the 2-cinnolinium ion (18).

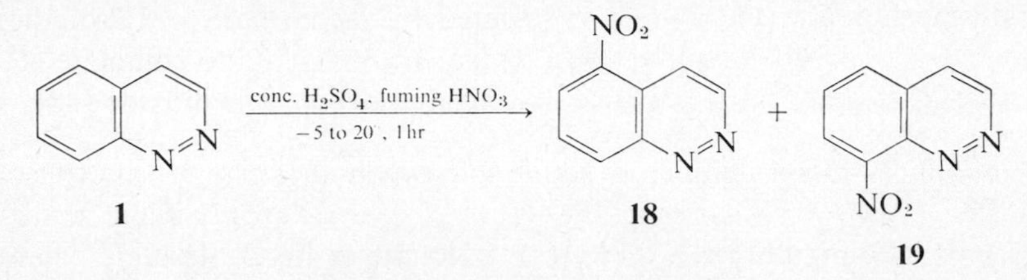

B. Free Radical Phenylation

Interaction of cinnoline with *N*-nitrosoacetanilide at 55–60° for 3 hours yields a complex mixture of products, 4,4′-dicinnolinyl (**6**) being the principal

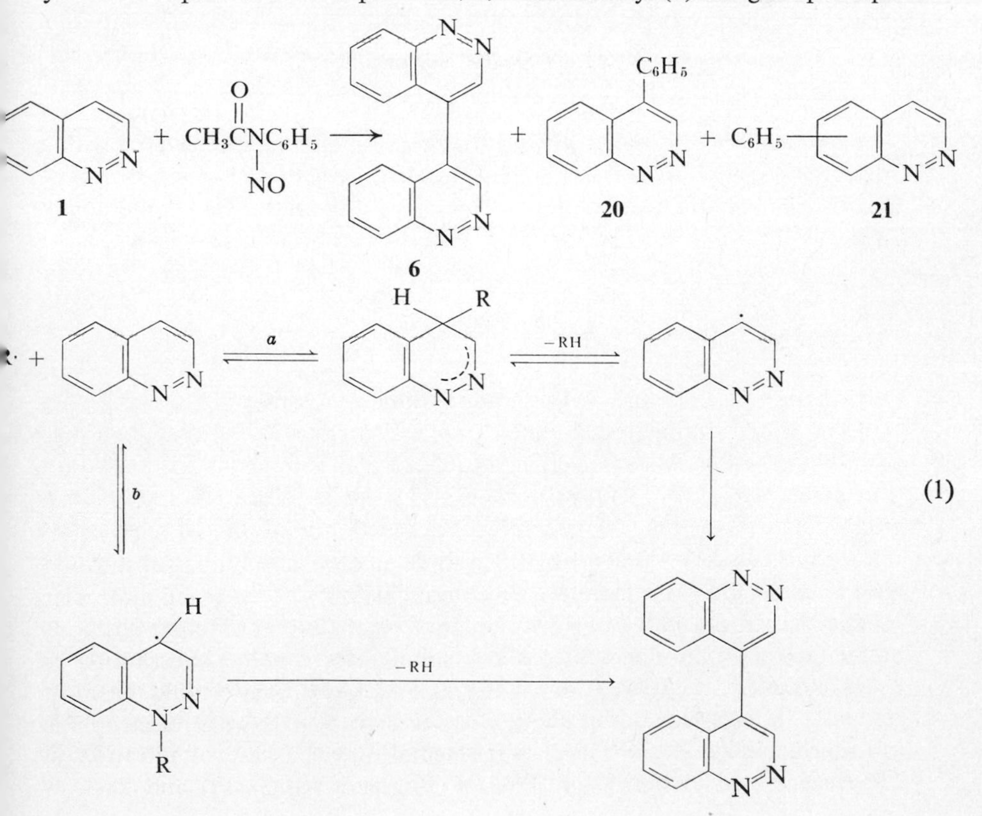

product with a concentration in the product mixture 10 times that of all other products combined. Equal amounts of 4-phenylcinnoline (**20**) and 5 (or 8)-phenylcinnoline (**21**) (mp 131°; picrate, mp 208°) together with lesser amounts of four unidentified products are also obtained (38). Formation of 4,4′-dicinnolinyl is thought to arise by one of two routes, path *a* or path *b* shown in Eq. 1 (38).

C. Salt Formation

Cinnoline forms a number of stable salts, including a hydrochloride, mp 156–160° C (3), a picrate, mp 196–196.5° (1), a chloroplatinate, mp 280° (dec.) (3), an aurichloride, mp 146° (3), and many *N*-alkylcinnolinium halides from the interaction of cinnoline with an alkyl halide. It has been believed for many years that quaternization of cinnoline takes place at N-1, but recently Ames (22, 23, 49) showed on the basis of chemical and spectroscopic evidence that protonation and alkylation of cinnoline both occur at N-2.

D. *N*-Oxide Formation

When cinnoline is treated with hydrogen peroxide in acetic acid or with a percarboxylic acid, there is obtained a mixture of cinnoline 1-oxide, cinnoline 2-oxide, and a small amount of cinnoline 1,2-dioxide. The predominant isomer is the 2-oxide. This reaction is discussed in Section 1J.

E. Reduction

Reduction of cinnoline with lithium aluminum hydride in refluxing ether solution gives 1,4-dihydrocinnoline (**9**) in 51% yield (39), whereas treatment of cinnoline with amalgamated zinc in refluxing 33% aqueous acetic acid for 2 hr yields indole (**22**) in approximately 57% yield. When this reduction is stopped as soon as the yellow color of the reaction mixture is discharged (4 min), then 1,4-dihydrocinnoline (**9**) is isolated and can be further reduced with amalgamated zinc in acetic acid to give indole (50).

A study of the catalytic hydrogenation of cinnoline in ethanol or in ethanolic hydrochloric acid at both low and high pressure has been performed using five different catalysts (51). Low-pressure hydrogenations were carried out at 27° C and 60 psi, while high-pressure hydrogenations were performed at 122–195° and 2230–2950 psi. Catalysts used were 5% rhodium on alumina, 5% rhodium on carbon, 5% palladium on carbon, ruthenium oxide, and

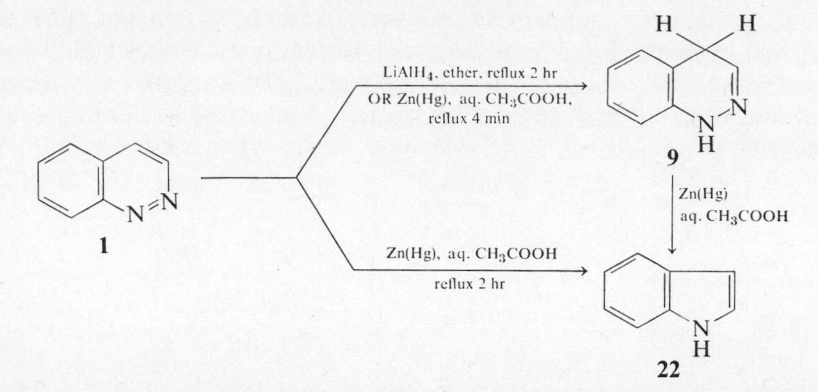

platinum oxide. Seven compounds were isolated from these reductions. Six were positively identified as 1,4-dihydrocinnoline (**9**), 1,2,3,4-tetrahydrocinnoline (**13**), indole (**22**), 2,3-dihydroindole (**23**), *cis*-octahydroindole (**24**), and *o*-aminophenethylamine (**25**). The seventh compound was not identified, but is proposed to be 1,1′,4,4′-tetrahydro-4,4′-dicinnolinyl (**26**). Each individual hydrogenation gives from one to five of the foregoing products, depending upon the reaction conditions and time. Specific results are summarized in Table 1A-4.

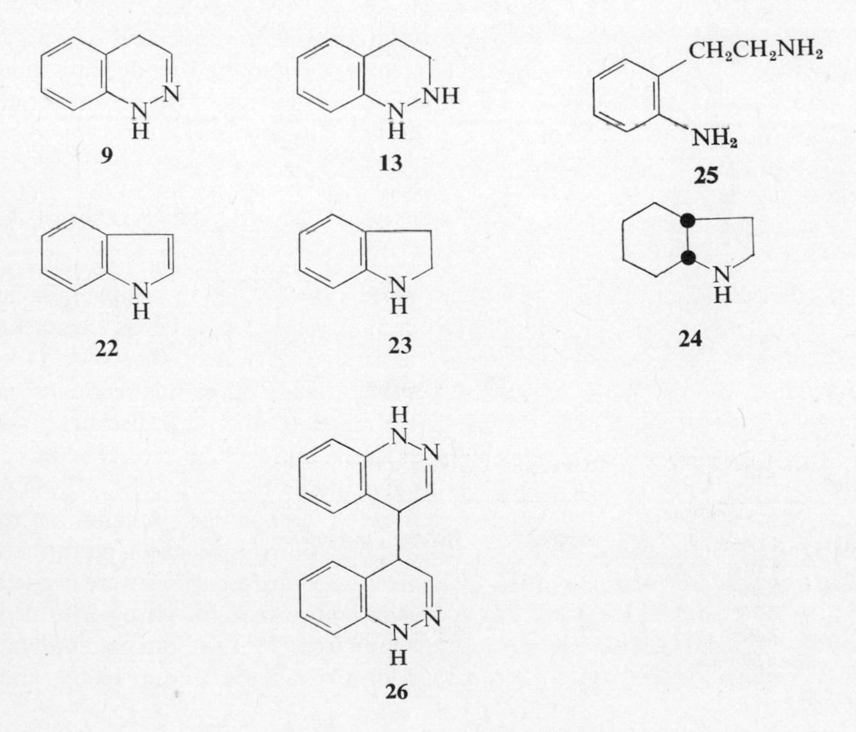

Cinnoline undergoes a reductive formylation when heated with formic acid and formamide to give 1-formamidoindole (**27**) in 55% yield. Under the same conditions 4-cinnolinecarboxylic acid also gives **27**, being decarboxylated during the reaction, while 4-methylcinnoline gives 1-formamido-3-methylindole (52).

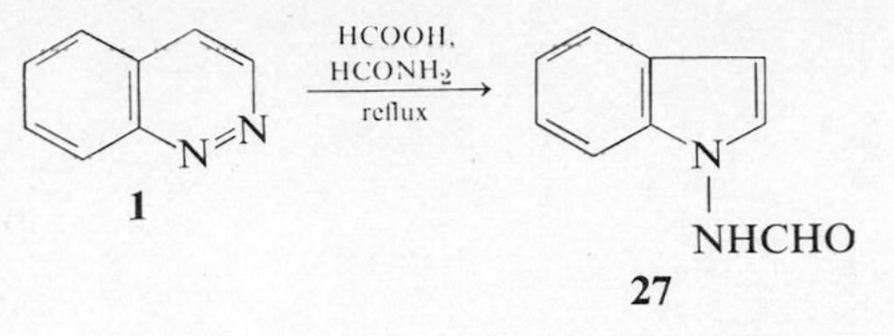

F. Reaction with Dimethylketene

Cinnoline reacts with two equivalents of dimethylketene in ether solution to give 4-isopropylidene-1,1-dimethyl-4*H*[1,3,4]oxadiazino[4,3-a]cinnolin-2(1*H*)one (**28**). While this adduct is thermally stable and can be sublimed *in vacuo*, it is easily hydrolyzed in alkaline solution to the amido acid **29** (53).

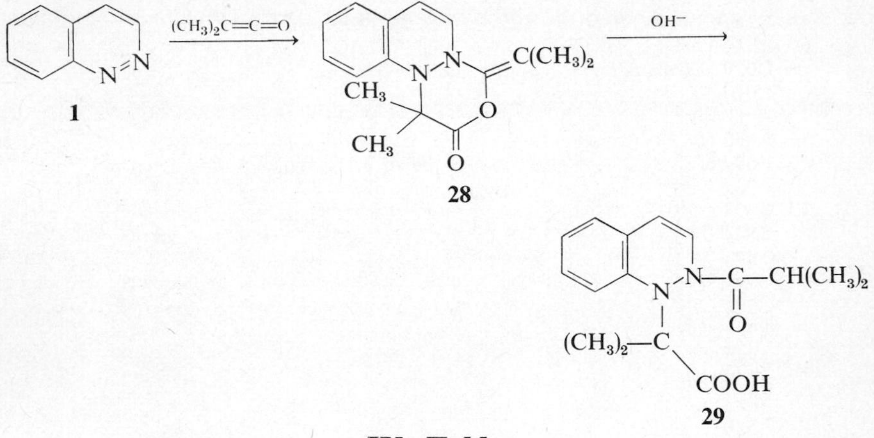

IV. Tables

TABLE 1A-1. Infrared Spectral Data for Cinnoline[a]

Absorption[b] (cm^{-1})	Intensity[c]	Absorption mode[d]
3,054	—	—
3,016 sh.	—	CH stretching
2,990 sh.	—	CH stretching
2,966 sh.	—	—
2,925	—	—

TABLE 1A-I. (*continued*)

Absorption[b] (cm^{-1})	Intensity[c]	Absorption mode[d]
1,663 sh.	vw	—
1,638 sh.	w	—
1,620 (1,623)	m	Skeletal stretching
1,593	w	Skeletal stretching
1,580 (1,581)	vs	Skeletal stretching
1,550 (1,553)	w-m	Skeletal stretching
1,538	w-m	—
1,491 (1,494)	s	Skeletal stretching
— (1,477)	w	Skeletal stretching
1,461	vw	—
1,440 (1,441)	m-s	Skeletal stretching
1,416 (1,417)	m-s	CH bending
1,410 sh.	vw	Skeletal stretching
1,392 (1,393)	s	Skeletal stretching
1,334	vw	—
1,291 (1,293)	m-s	Skeletal stretching (CH bending?)
1,258 (1,259)	m	CH bending
1,251	m	CH bending
1,221	vw	—
1,179 (1,182)	m-s	CH bending
1,170 sh.	w	—
1,158 (1,160)	w-m	CH bending
1,138 (1,139)	m-s	(Skeletal bending?)
1,117	vw	—
1,090 (1,091)	vs	Skeletal distortion
1,070	vw	—
— (1,041)	w	—
1,030 (1,029)	w	—
1,007 (1,008)	w-m	Skeletal distortion (skeletal stretching)
962 (964)	m	CH bending
875 (875)	m	CH bending
844 (843)	vs	CH bending
821 (823)	m	Skeletal breathing (CH bending)
794 (794)	m	CH bending
774 (775)	m	Skeletal distortion (CH bending)
748 (748)	vs	Skeletal distortion
716	vw	CH bending
650 (650)	m-s	Skeletal distortion
633	m-s	Skeletal distortion
530	m	Skeletal distortion
512	m	Skeletal distortion
470	s	Skeletal bending
373	s	Skeletal distortion

[a] Data taken from reference 26 except those in parentheses, which are from reference 25; the spectral data are those of cinnoline as a liquid.

[b] Abbreviation "sh." means that the absorption is a shoulder on another peak.

[c] Symbols used to designate intensity are vs = very strong, s = strong, m = medium, w = weak, and vw = very weak.

[d] A question mark indicates uncertainty of assignment; a dash indicates no assignment was made.

TABLE 1A-2. Ultraviolet Absorption Spectrum of Cinnoline

Solvent	λ_{max}(mμ)	log E_{max}	Ref.
Ethanol	276	3.45	54
	286	3.42	
	308.5	3.29	
	317	3.25	
	322.5	3.32	
	390	2.42	
Cyclohexane	222	4.66	27
	276	3.52	
	287	3.48	
	310	3.35	
	318	3.30	
	322	3.40	
	392.5	2.45	
Methanol	225	4.60	27
	280	3.40	
	320	3.41	
H_2O $pH = 7$	226 (225)	4.64 (4.63)	7, 55
	283 (285)	3.38 (3.37)	
	290 (—)	3.38 (—)	
	321 (320)	3.44 (3.43)	
H_2O $pH = 0.3$	237	4.59	7
	294	3.31	
	305	3.32	
	353	3.40	
n-hexane	199	4.48	56
	219.5	4.72	
	275	3.49	
	322	3.40	

TABLE 1A-3. Proton Magnetic Resonance Spectrum of Cinnoline[a]

Parameter[b]	Solvent	
	Acetone	CCl_4[c]
τ_3	0.678	0.783
τ_4	1.917	2.24 ± 0.05
τ_5	1.988	
τ_6	2.163	
τ_7	2.068	
τ_8	1.510	1.523
$J_{3,4}$	5.75	5.80
$J_{4,8}$	0.83	—
$J_{5,6}$	7.87	—
$J_{5,7}$	1.57	—
$J_{5,8}$	0.85	—
$J_{6,7}$	6.94	—
$J_{6,8}$	1.34	—
$J_{7,8}$	8.64	—

[a] Data taken from reference 57; Spectra are recorded relative to tetramethylsilane as an internal reference.

[b] Chemical shift value is given in τ (tau) units in ppm; coupling constants are given as J in cps.

[c] Spectrum in carbon tetrachloride is not analyzed fully because of the near coincidence of the chemical shifts of protons 4, 5, 6, and 7. Besford, Allen, and Bruce (39) report the spectrum of cinnoline in carbon tetrachloride as a doublet at 0.78τ (J = 6.0 cps), a multiplet at 1.5–1.7τ, and a multiplet at 2.1–2.4τ using tetramethylsilane as an internal reference.

TABLE 1A-4. Catalytic Hydrogenation of Cinnoline[a]

Catalyst	Solvent	Temp. (°C)	Pressure (psi)	Time (hr)	Products[b]	Yield (%)
5% Rh/Al_2O_3	Ethanol	27	60	9	**9**	70
					26	30
		125	2,610	5	**22**	38
					13	5
					23	10
					25	48
		195	2,950	2	**24**	100
	Ethanol-HCl	27	60	12	**9**	100
		122	2,450	8	**22**	21
					24	79
5% Rh/C	Ethanol	27	60	12	**9**	52
					26	48
		122	2,230	3	**22**	25
					23	16
					24	59
RuO_2	Ethanol	27	60	25	**9**	56
					26	44
		123	2,500	5	**24**	100
	Ethanol-HCl	27	60	2	NR	—
5% Pd/C	Ethanol	27	60	3	**9**	37
					26	63
		124	2,530	5	**22**	14
					13	19
					23	21
					24	3
					25	43
	Ethanol-HCl	27	60.5	1	**9**	100
		125	2,400	6	**24**	86
					25	14
PtO_2	Ethanol	27	60	0.5	**26**	—
		122	2,500	8	**22**	1
					13	81
					25	18
	Ethanol-HCl	27	61.5	0.5	**22**	4
					13	65
					23	7
					24	2
					25	23
		27	60	3	**13**	100
		125	2,350	6	**22**	16
					13	36
					23	21
					25	27

[a] Data taken from reference 51.

[b] Products are assigned numbers within this column according to their assignments within the text of this chapter (Section 1A-III-E).

References

1. T. L. Jacobs, S. Winstein, R. E. Henderson, and E. C. Spaeth, *J. Am. Chem. Soc.*, **68,** 1310 (1946).
2. J. S. Morley, *J. Chem. Soc.*, **1951,** 1971.
3. M. Busch and A. Rast, *Ber.*, **30,** 521 (1897).
4. V. von Richter, *Ber.*, **16,** 677 (1883).
5. J. R. Keneford, J. S. Morley, J. C. E. Simpson, and P. H. Wright, *J. Chem. Soc.*, **1949,** 1356.
6. A. Albert, R. Goldacre, and J. Phillips, *J. Chem. Soc.*, **1948,** 2240.
7. A. R. Osborn, K. Schofield, and L. N. Short, *J. Chem. Soc.*, **1956,** 4191.
8. M. T. Rogers and T. W. Campbell, *J. Am. Chem. Soc.*, **75,** 1209 (1953).
9. R. J. W. LeFevre and J. W. Smith, *J. Chem. Soc.*, **1932,** 2810.
10. M. J. S. Dewar and S. D. Worley, *J. Chem. Phys.*, **51,** 263 (1969).
11. S. D. Worley, *Chem. Revs.*, **71,** 295 (1971).
12. M. J. S. Dewar and T. Morita, *J. Am. Chem. Soc.*, **91,** 769 (1969).
13. A. Pullman, *Rev. Sci.*, **86,** 219 (1948); *Chem. Abstr.*, **43,** 2095 (1949).
14. H. C. Longuet-Higgins and C. A. Coulson, *J. Chem. Soc.*, **1949,** 971.
15. S. C. Wait, Jr. and J. W. Wesley, *J. Mol. Spectrosc.*, **19,** 25 (1966).
16. O. W. Adams and P. C. Lykos, quoted by A. H. Gawer and B. P. Dailey, *J. Chem. Phys.*, **42,** 2658 (1965).
17. E. J. Alford and K. Schofield, *J. Chem. Soc.*, **1953,** 609.
18. R. B. Moodie, E. A. Qureshi, K. Schofield, and J. T. Gleghorn, *J. Chem. Soc.* (*B*), **1968,** 312.
19. M. J. S. Dewar and P. M. Maitlis, *J. Chem. Soc.*, **1957,** 2521.
20. M. Ogata, H. Kano, and K. Tori, *Chem. Pharm. Bull.* (*Tokyo*), **11,** 1527 (1963).
21. I. Suzuki, M. Nakadate, T. Nakashima, and N. Nagasawa, *Tetrahedron Lett.*, **1966,** 2899.
22. D. E. Ames, G. V. Boyd, A. W. Ellis, and A. C. Lovesey, *Chem. Ind.* (*London*), **1966,** 458.
23. D. E. Ames and H. Z. Kucharska, *J. Chem. Soc.*, **1964,** 283.
24. M. J. S. Dewar and C. C. Thompson, Jr., *J. Am. Chem. Soc.*, **87,** 4414 (1965).

24a. M. H. Palmer, A. J. Gaskell, P. S. McIntyre, and D. W. W. Anderson, *Tetrahedron*, **27,** 2921 (1971).

24b. M. H. Palmer and P. S. McIntyre, *Tetrahedron*, **27,** 2913 (1971).

25. W. F. L. Armarego, G. B. Barlin, and E. Spinner, *Spectrochim. Acta*, **22,** 117 (1966).
26. R. W. Mitchell, R. W. Glass, and J. A. Merritt, *J. Mol. Spectrosc.*, **36,** 310 (1970).
27. G. Favini, S. Carra, V. Pierpaoli, S. Polezzo, and M. Simonetta, *Il Nuovo Cimento*, **8,** 60 (1958).
28. G. Favini, I. Vandoni, and M. Simonetta, *Theor. Chim. Acta* (*Berlin*), **3,** 45, 418 (1965).
29. L. Goodman and R. W. Harrell, *J. Chem. Phys.*, **30,** 1131 (1959).
30. R. C. Hirt, F. T. King, and J. C. Cavagnol, *J. Chem. Phys.*, **25,** 574 (1956).
31. S. F. Mason, *J. Chem. Soc.*, **1962,** 493.
32. E. D. Amstutz, *J. Org. Chem.*, **17,** 1508 (1952).
33. S. C. Wait, Jr., and F. M. Grogan, *J. Mol. Spectrosc.*, **24,** 383 (1967).
34. R. W. Glass, L. C. Robertson, and J. A. Merritt, *J. Chem. Phys.*, **53,** 3857 (1970).
35. R. J. Pugmire, D. M. Grant, M. J. Robins, and R. K. Robins, *J. Am. Chem. Soc.*, **91,** 6381 (1969).

36. J. R. Elkins and E. V. Brown, *J. Heterocyclic Chem.*, **5**, 639 (1968).
37. M. H. Palmer, E. R. R. Russell, and W. A. Wolstenholme, *Org. Mass Spectrom.*, **2**, 1265 (1969).
38. C. M. Atkinson and C. J. Sharpe, *J. Chem. Soc.*, **1959**, 3040.
39. L. S. Besford, G. Allen, and J. M. Bruce, *J. Chem. Soc.*, **1963**, 2867.
40. D. E. Ames and H. Z. Kucharska, *J. Chem. Soc.*, **1962**, 1509.
41. K. Schofield and T. Swain, *J. Chem. Soc.*, **1950**, 392.
42. E. J. Alford and K. Schofield, *J. Chem. Soc.*, **1953**, 1811.
43. C. M. Atkinson and C. J. Sharpe, *J. Chem. Soc.*, **1959**, 2858.
44. H. Lund, *Acta Chem. Scand.*, **21**, 2525 (1967).
45. H. J. Haas and A. Seeliger, *Ber.*, **96**, 2427 (1963).
46. H. J. Barber and E. Lunt, *J. Chem. Soc.* (*C*), **1968**, 1156.
47. R. Askani, *Chem. Ber.*, **102**, 3304 (1969).
48. G. Maier, *Chem. Ber.*, **102**, 3310 (1969).
49. D. E. Ames, G. V. Boyd, R. F. Chapman, A. W. Ellis, A. C. Lovesey, and D. Waite, *J. Chem. Soc.* (*B*), **1967**, 748.
50. L. S. Besford and J. M. Bruce, *J. Chem. Soc.*, **1964**, 4037.
51. J. D. Westover, Ph.D. Dissertation (Brigham Young University, Provo, Utah), August, 1965; University Microfilms (Ann Arbor, Mich.), Order No. 66-2137.
52. D. E. Ames and B. Novitt, *J. Chem. Soc.* (*C*), **1970**, 1700.
53. M. A. Shah and G. A. Taylor, *J. Chem. Soc.* (*C*), **1970**, 1642.
54. J. M. Hearn, R. A. Morton, and J. C. E. Simpson, *J. Chem. Soc.*, **1951**, 3318.
55. G. Favini, *Rend. Ist. Lombardo Sci.*, *pt. I*, *Classe Sci. Mat. e Nat.*, **94A**, 331 (1960).
56. G. Favini and I. R. Bellobono, *Rend. Ist. Lombardo Sci. Lett.*, **A99**, 380 (1965).
57. P. J. Black and M. L. Heffernan, *Aust. J. Chem.*, **18**, 707 (1965).

Part B. Alkyl- and Arylcinnolines

I. Methods of Preparation

A. Widman-Stoermer Synthesis

The most widely used method to prepare cinnolines which have an alkyl, aryl, or heteroaryl group at the 4-position or at both the 3- and 4-positions is the Widman-Stoermer synthesis. By this method, a diazotized *o*-aminoarylethylene (**1**, R_1 = alkyl, aryl, or heteroaryl; R_2 = hydrogen, alkyl, aryl, or heteroaryl) cyclizes upon standing to give the cinnoline (**3**). Table 1B-1 lists most of the *o*-aminoarylethylenes that have been successfully diazotized

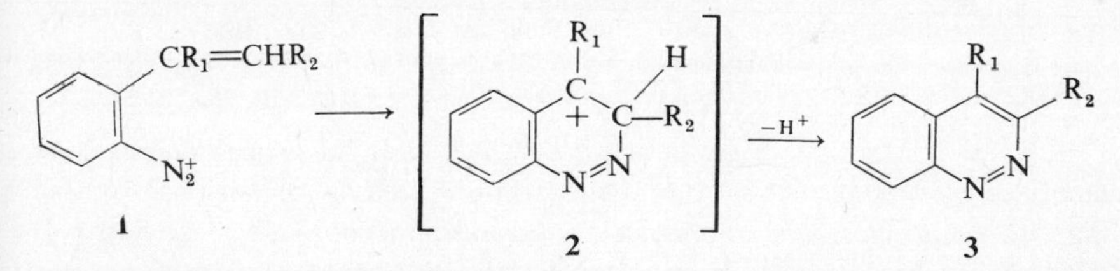

and cyclized to cinnolines. Inspection of this table shows that the α-carbon of the ethylene moiety is always substituted with an alkyl, aryl, or heteroaryl group (designated as R_1 in the table and in structures **1**, **2**, and **3**). This appears to be necessary because all attempts to prepare unsubstituted cinnoline or cinnolines substituted only in the benzenoid ring or only at the 3-position have failed. That is, all cinnolines prepared by the Widman-Stoermer method are substituted at the 4-position, or at both the 3- and 4-positions with alkyl, aryl, or heteroaryl groups. Attempts to prepare cinnoline-4-carboxylic acids by diazotization of *o*-aminoarylethylenes in which R_1 is the electron-attracting carboxyl group have met with failure.

These results are explicable when one considers that the success of the reaction depends upon the stability of the intermediate benzylic carbonium ion **2**, or, to be more precise, it depends upon the energy difference between the diazonium ion **1** and the transition state that leads to **2**. In the most successful reactions, R_1 is an electron-donating group such as the *p*-methoxyphenyl group which can stabilize **2** by charge delocalization, hence lowering the energy of the transition state between **1** and **2** and increasing the rate of the reaction. Even when R_1 is a heteroaryl group such as 2-pyridinyl, charge delocalization in **2** by the pyridinyl ring can occur and the reaction is successful. Apparently, when R_1 is hydrogen or carboxyl, the energy difference between **1** and the transition state leading to **2** is high enough to be essentially insurmountable.

B. Borsche Synthesis

The Borsche synthesis is described in Section 1C-I-A-3, as a method to prepare 4-hydroxycinnolines (**5**) by the diazotization and cyclization of 2-aminoacetophenones (**4**). Little more need be added here except that at least 30 4-hydroxycinnolines substituted at the 3-, 6-, 7-, or 8-position with an alkyl or aryl group (**5**, R = alkyl, aryl; R′ = 3-, 6-, 7-, or 8-alkyl or aryl) have been prepared by this method.

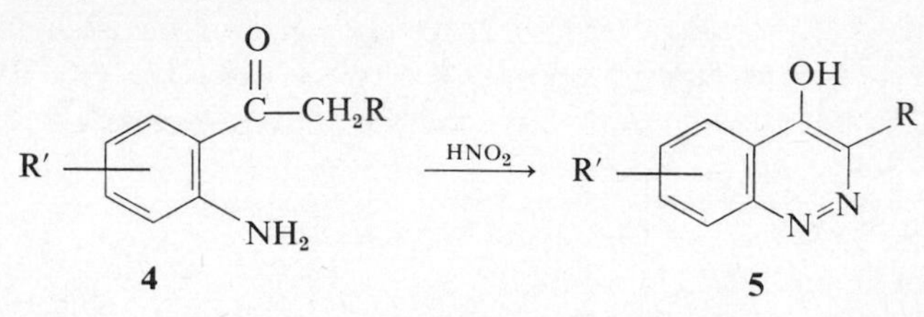

Table 1B-2 lists some 2-aminoacetophenones that have been successfully diazotized and cyclized to the corresponding alkyl- or aryl-substituted 4-hydroxycinnolines. It is interesting that apparently no 5-alkyl- or 5-aryl-4-hydroxycinnolines have yet been prepared by the Borsche synthesis.

An ingenious modification of the Borsche synthesis was devised by Baumgarten (1), who coupled diazotized 2-aminoacetophenone (**6**) with nitromethane in a dilute, basic solution to give nitroformaldehyde 2-acetylphenylhydrazone (**7**) in yields up to 98%. Cyclization of **7** in the presence of aluminum oxide then gives 4-methyl-3-nitrocinnoline (**8**) in a yield of 59%.

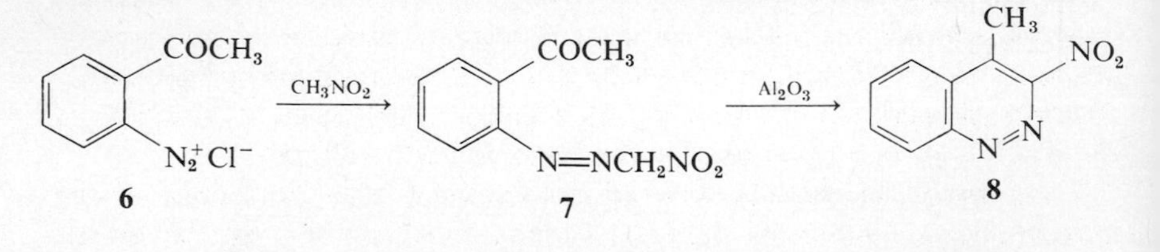

C. 4-Alkylcinnolines from 4-Chloro- and 4-Methylsulfonylcinnolines

Several 4-alkylcinnolines have been prepared by the nucleophilic displacement of chloride ion from 4-chlorocinnoline (see Table 1B-5). For example, the condensation of 4-chlorocinnoline (**9**) with the sodio derivative of phenylacetonitrile in benzene solution yields α-(4-cinnolinyl)phenylacetonitrile (**10**) in 94% yield (2). This is easily converted to α-(4-cinnolinyl)-

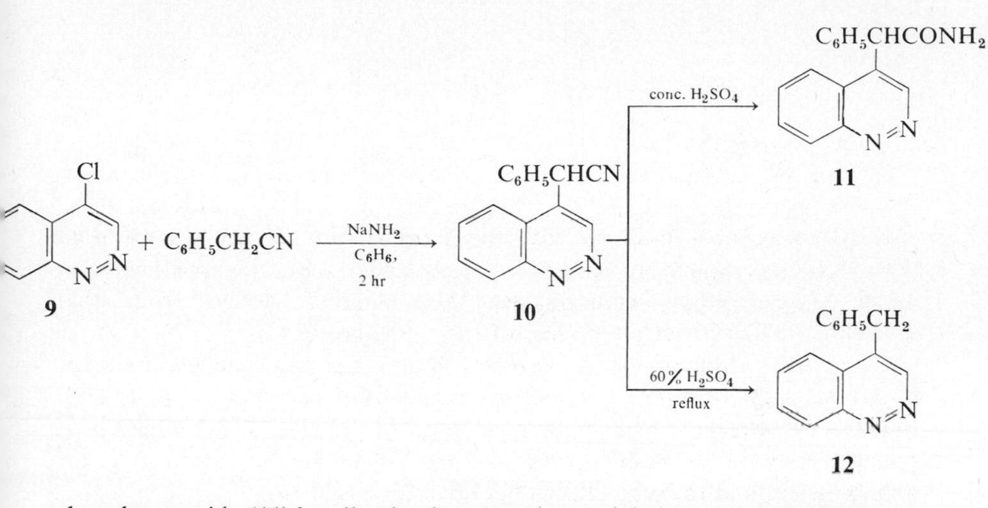

phenylacetamide (**11**) by allowing it to stand overnight in concentrated sulfuric acid at room temperature. Alternatively, **10** is readily converted to 4-benzylcinnoline (**12**) by refluxing it for 1 hr in 60% sulfuric acid (2).

A number of α-(ω-dialkylaminoalkyl)-α-phenyl-α-(4-cinnolinyl)acetonitriles (**13**) have been prepared similarly from α-(ω-dialkylaminoalkyl)-phenylacetonitriles and 4-chlorocinnoline. These may be hydrolyzed and decarboxylated in refluxing 60% sulfuric acid to the corresponding 4-[(1-phenyl-ω-dialkylamino)alkyl]cinnolines (**14**) (3). Specific examples are listed in Table 1B-5. Likewise the condensation of 3-benzyl-4-chlorocinnoline with phenylacetonitrile in benzene solution, using sodium amide as the condensing agent, followed by treatment of the product with refluxing aqueous sulfuric acid yields 3,4-dibenzylcinnoline (**15**) (4).

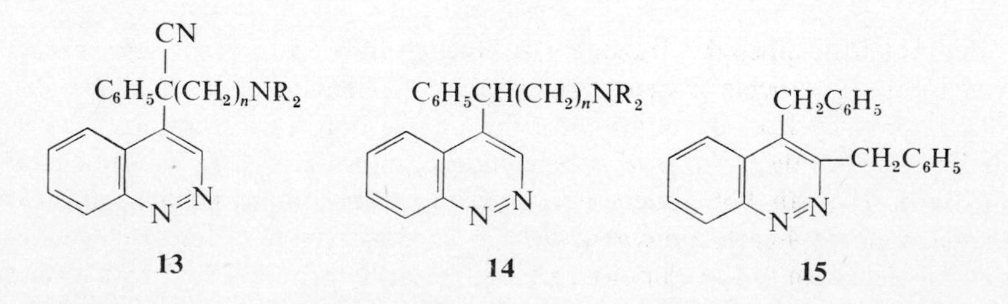

An indirect preparation of 4-methylcinnoline (**17**) from 4-chlorocinnoline has been executed by condensation of 4-chlorocinnoline with ethyl cyano-acetate, followed by hydrolysis of the resulting ethyl 4-cinnolinylcyanoacetate (**16**) in hot aqueous hydrochloric acid (5).

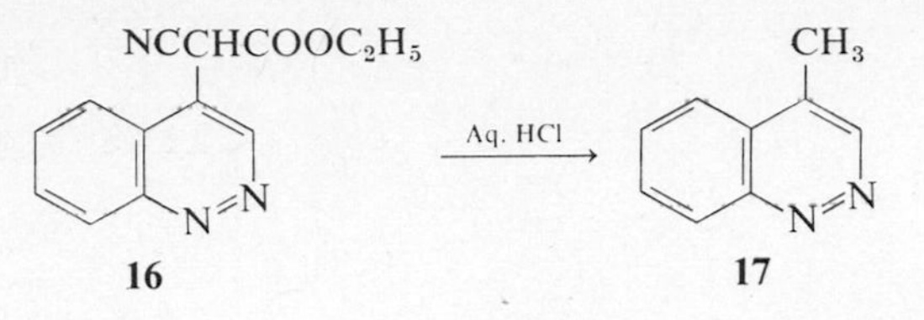

Ethyl acetoacetate does not give the expected product when condensed with 4-chlorocinnoline in benzene solution under the influence of sodium amide. Instead, ethyl 4-cinnolinylacetate is obtained, derived from ethyl α-(4-cinnolinyl)acetoacetate by loss of the acetyl group (5).

Generally, sodium amide in benzene solution is a good condensing agent to convert 4-chlorocinnolines to 4-alkylcinnolines of the types just described, but when this reagent gives poor results, as in the case of 3,4-dimethoxy-phenylacetonitrile or *p*-aminophenylacetonitrile, potassium amide in liquid ammonia is found to be an effective condensing agent (6).

Hayashi and Watanabe (6a–6d) have studied the reaction of 4-methyl-sulfonylcinnoline with ketones under the influence of basic condensing agents to give 4-alkylcinnolines. This reaction is discussed in Section 1F-III, and the 4-alkylcinnolines prepared by this method are listed in Table 1B-5. Like 4-chlorocinnoline, 4-methylsulfonylcinnoline also reacts under basic conditions with esters and nitriles having labile α-hydrogen atoms to give 4-alkylcinnolines (6e).

D. Cyclization of Phenylhydrazone Derivatives

1. *Benzaldehyde Phenylhydrazones* (*Stolle-Becker Synthesis*)

Benzaldehyde phenylhydrazone (**18**), when allowed to react with excess oxalyl chloride, yields *N*-benzylideneamino-*N*-phenyloxamyl chloride (**19**), which can be cyclized by aluminum chloride in chloroform solution (7) or in methylene chloride (8) to give *N*-benzylideneaminoisatin (**20**). Treatment of the isatin **20** with hot aqueous sodium or potassium hydroxide gives 3-phenylcinnoline-4-carboxylic acid (**21**) in 75–85% yields (7, 8). Rearrangement of the isatin to the cinnoline almost certainly proceeds first by alkaline hydrolysis of the amide linkage in the isatin to open the ring. This would be followed by a recyclization and aromatization to give the cinnoline. The carboxyl group of **21** can be removed thermally to give 3-phenylcinnoline (**22**) in 51–74% yields (7).

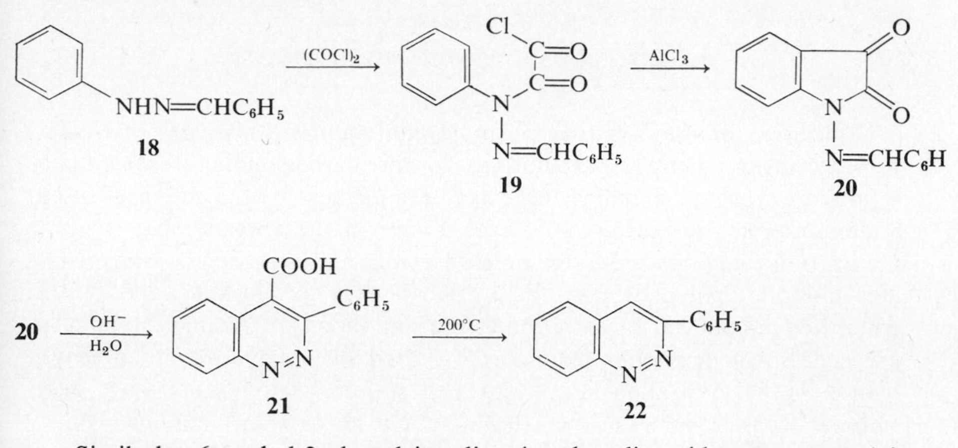

Similarly, 6-methyl-3-phenylcinnoline-4-carboxylic acid was prepared in 53% overall yield from benzaldehyde *p*-tolylhydrazone. Thermal decarboxylation then gave 6-methyl-3-phenylcinnoline. This same sequence, however, failed with benzaldehyde *p*-chlorophenyl-, *p*-anisyl-, and 1-naphthylhydrazone, so that the reaction does not appear to represent a general synthesis of 3-arylcinnolines (7). In addition, substituents attached to the benzene ring of the phenylhydrazine portion of the benzaldehyde phenylhydrazone may be limited to those allowing mild conditions for the Friedel-Crafts cyclization step (8).

Several attempts to use acetaldehyde phenylhydrazone (to give 3-methylcinnoline-4-carboxylic acid), phenylacetaldehyde phenylhydrazone (to give 3-benzylcinnoline-4-carboxylic acid), or isonicotinaldehyde phenylhydrazone [to give 3-(4-picolinyl)cinnoline-4-carboxylic acid] have been unsuccessful (8).

2. *Benzil Monophenylhydrazone*

Benzil monophenylhydrazone (**23**) may be cyclized in 75–80% sulfuric acid solution to 3,4-diphenylcinnoline (**24**) in 75% yield (9). The use of concentrated sulfuric acid lowers the yield of **24** and produces sulfonated benzil monophenylhydrazones (10).

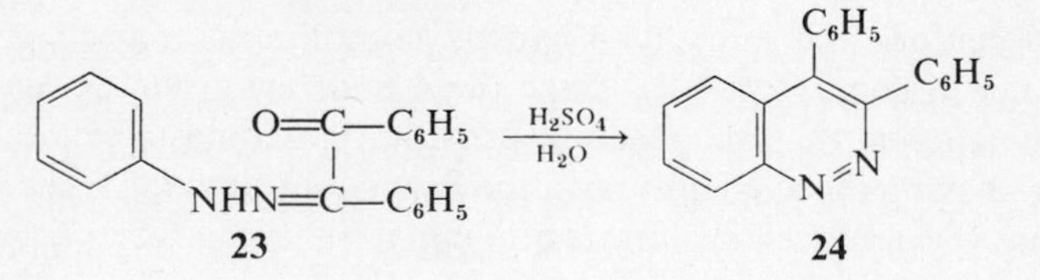

The use of substituted benzil monophenylhydrazones in this reaction has not been reported.

3. *Phenylglyoxal Monophenylhydrazones*

Cyclization of the *o*-hydroxyphenylglyoxal monophenylhydrazones (**25**, R = hydrogen, methyl, or chlorine) to the corresponding 6-substituted 4-(*o*-hydroxyphenyl)cinnolines (**26**) has been effected by the action of fused aluminum chloride at 180–190° C for 5 min in the absence of a solvent. Cyclization could not be effected in the presence of inert solvents. Anhydrous zinc chloride, polyphosphoric acid, phosphorus oxychloride, and boron trifluoride did not cause cyclization to the cinnoline (11). A single attempt to prepare 4-phenylcinnoline from unsubstituted phenylglyoxal monophenylhydrazone by cyclization in concentrated sulfuric acid solution gave only sulfonic acid derivatives of the hydrazone, but no 4-phenylcinnoline (10).

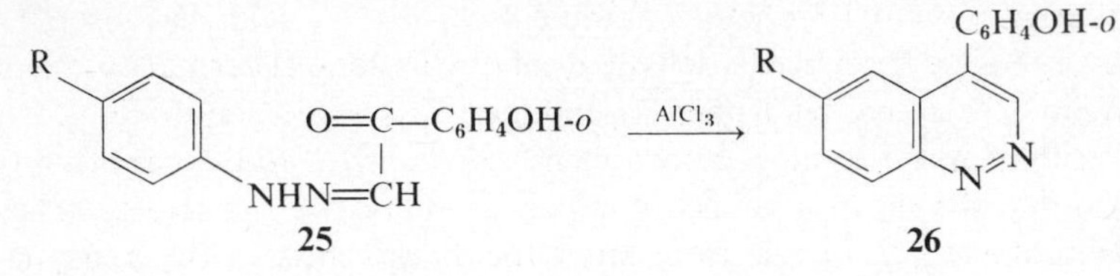

4. *Mesoxalyl Chloride Phenylhydrazones*

The titanium tetrachloride-catalyzed cyclization of mesoxalyl chloride phenylhydrazone (**27**) or one of its substituted derivatives is especially useful for the preparation of 4-hydroxycinnoline-3-carboxylic acid (**28**) or 4-hydroxycinnoline (**29**) or their derivatives which are substituted in the benzenoid ring. This reaction is discussed in detail in Section 1C-I-A-4.

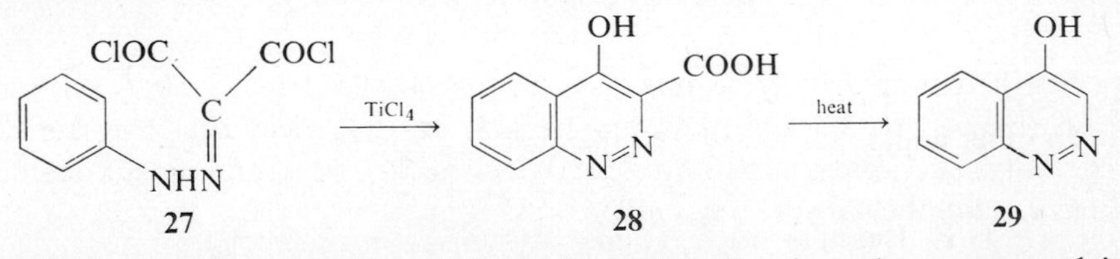

Both 6-methyl- and 8-methyl-4-hydroxycinnoline have been prepared in this manner in good yield (12). Since the 4-hydroxy group can be removed from the cinnoline ring, the reaction provides a method to synthesize alkylcinnolines having the alkyl groups in the benzenoid ring only. The 4-hydroxy group may be removed by first refluxing it in a mixture of phosphorus pentachloride and phosphorus oxychloride to convert it to the 4-chlorocinnoline. This is then converted to the 4-toluene-*p*-sulfonylhydrazone derivative, which will decompose in hot aqueous sodium carbonate solution to give a cinnoline with a hydrogen atom at the 4-position (13–15).

E. 2-Phenylisatogen to 3-Phenylcinnoline

The reaction of 2-phenylisatogen (**30**) with ethanolic ammonia at 140–145° C in an autoclave for 6 hr gives 26% of 4-hydroxy-3-phenylcinnoline 1-oxide (**31**). A possible mechanism for this rearrangement is discussed in Section 1J-I-B. Catalytic hydrogenation of **31** over Raney nickel then gives 4-hydroxy-3-phenylcinnoline (**32**) in 45% yield (16). Reduction of **32** with lithium aluminum hydride followed by gentle oxidation of the product with mercuric oxide then gives 3-phenylcinnoline (**22**) (17).

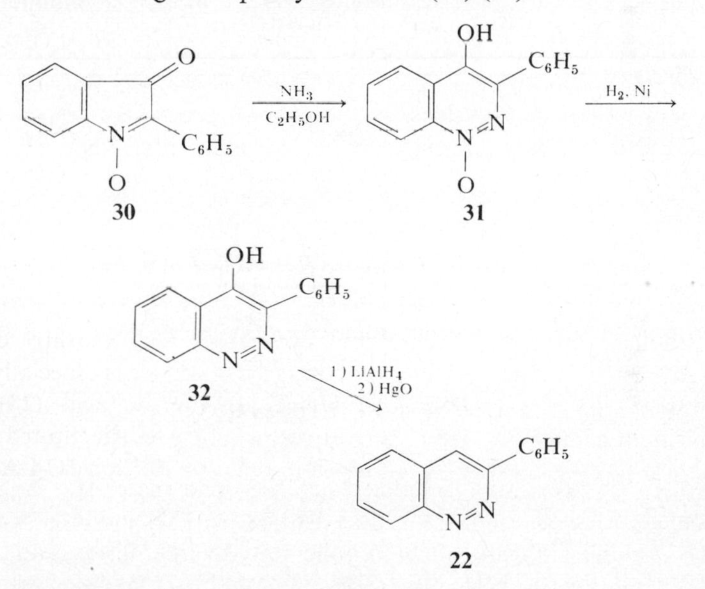

No other cinnolines have been prepared by this route.

F. Isatin to 4-Alkyl- and 4-Aryl-3-hydroxycinnolines

The reaction of isatin (**33**) with Grignard reagents yields 3-alkyl- or 3-aryldioxindoles. In the example shown, phenylmagnesium bromide and isatin give 3-phenyldioxindole (**34**) in 55% yield (18). The dioxindole can be cleaved in aqueous alkaline solution to give 2-aminomandelic acid (**35**), which upon diazotization and reduction is converted to 1-amino-3-phenyldioxindole (**37**) by spontaneous cyclization of the hydrazine **36**. If **37** is hydrolyzed back to the hydrazine **36**, and the reaction mixture is then carefully neutralized, 3-hydroxy-4-phenylcinnoline (**38**) is obtained in 46% yield (18).

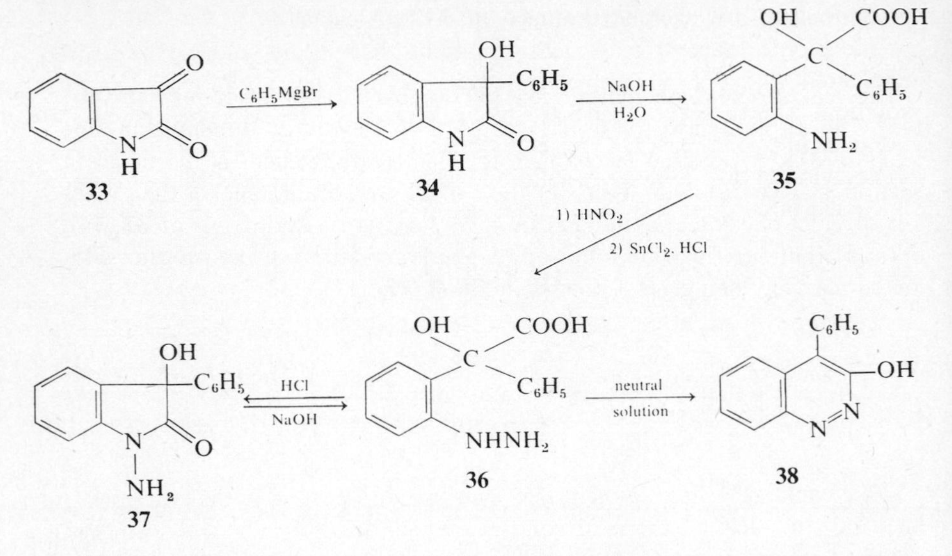

By this same route, 4-benzyl-3-hydroxycinnoline has been synthesized from isatin and benzylmagnesium chloride (18), but apparently no other alkyl- or arylcinnolines have been prepared in this way.

II. Physical Properties

Generally, the alkyl- and arylcinnolines are solid materials, ranging in color from colorless to orange-red. Most are yellow. They are basic materials that form a number of salts, including picrates, hydrochlorides, and methiodides. Very little has been recorded about their actual basic strengths, but the alkylcinnolines are expected to be somewhat more basic than cinnoline itself, in accordance with the usual slight base-strengthening effect of alkyl groups in heterocyclic bases. Like cinnoline itself (and unlike 4-aminocinnoline), the basic center of 3-, 4-, and 8-methylcinnoline is at N-2. This is known by molecular orbital calculations (19) and by experimentation (19). The experimental proof was obtained by showing that the UV spectra of protonated 3-, 4-, and 8-methylcinnoline are very similar to the spectra of the corresponding 2,3-, 2,4- and 2,8-dimethylcinnolinium perchlorates. The spectra of the 1,3-, 1,4-, and 1,8-dimethylcinnolinium perchlorates are distinctly different (19).

The infrared spectra of a few alkyl- and arylcinnolines have been recorded (7, 20), as have the pmr (proton magnetic resonance) spectra (Table 1B-3). The mass spectra of a number of alkyl- and arylcinnolines have been reported (20a–20c). Fragmentation patterns can be used to distinguish between arylcinnolines and arylquinoxalines (20b).

The dipole moment of 4-methylcinnoline in benzene solution is 4.53 D compared with the moment of cinnoline itself, which is 4.14 D in benzene solution. The moment is directed along a perpendicular line to the N—N bond, as in pyridazine (21). For comparative purposes, the dipole moment of 4-methylquinoline in benzene solution is 2.52 D (21), whereas that of quinoline itself in benzene is 2.18 D (22).

III. Reactions

A. Reduction

1. *Catalytic*

Maycock (23) has studied the hydrogenation of 4-methylcinnoline in neutral and acidic ethanolic solutions at 2000–3000 psig, and at 60 psig over 5% rhodium on alumina, ruthenium oxide, 5% palladium on activated charcoal, 5% rhodium on activated charcoal, and platinum oxide. With these catalysts, seven products have been obtained, six of which have been identified. In order of decreasing yield, these are *o*-amino-β-methylphenethylamine (**39**), octahydroskatole (**40**), 1-ethyloctahydroskatole (**41**), 2,3-dihydroskatole (**42**), 1,4-dihydro-4-methylcinnoline (**43**), and 4-methyl-1,2,3,4-tetrahydrocinnoline (**44**). The dihydrocinnoline (**43**) is listed as 3,4-dihydro-4-methylcinnoline in reference 23, but it is almost certain to be the 1,4-dihydro isomer instead, and is given as such here. The particular product or products obtained in any one case depends upon the catalyst and the conditions. In neutral ethanolic solution, for example, the reduction of 4-methylcinnoline over 5% rhodium on charcoal, or over platinum oxide, favors formation of *o*-amino-β-methylphenethylamine (**39**), whereas in acidic ethanolic solution the main product is octahydroskatole (**40**). Over ruthenium oxide in neutral ethanolic solution, 1-ethyloctahydroskatole (**41**) is obtained.

An interesting application of the catalytic reduction of 4-methylcinnoline was devised by Elslager, Worth, and Pericone (24), who converted 4-methylcinnoline (**17**) into 1,3,4,5-tetrahydro-5-methyl-2*H*-1,3-benzodiazepine-2-thione (**45**) by subjecting 4-methylcinnoline to a reductive scission over

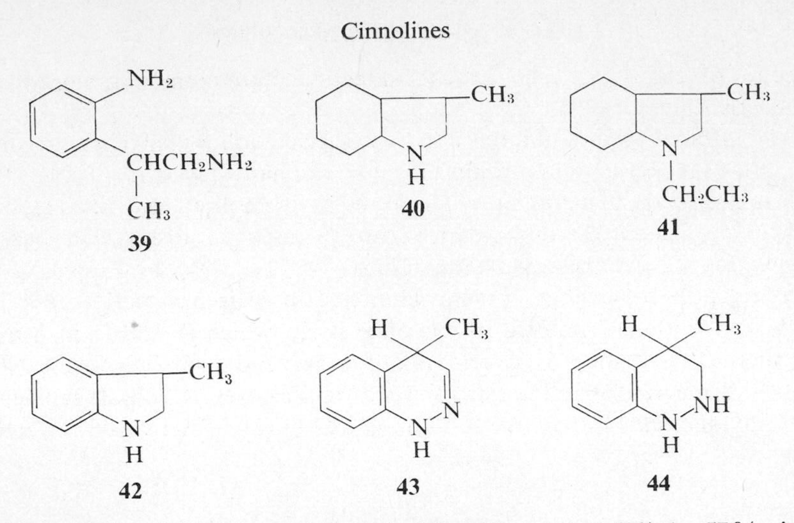

Raney nickel to give *o*-amino-*β*-methylphenethylamine (**39**) in 57% yield, then treating **39** with carbon disulfide to give the diazepine **45** in 65% yield. Several derivatives of the diazepine **45** were then prepared, none of which possessed appreciable biological activity against a broad spectrum of helminths in mice and in the inhibition of ADP-induced thrombocyte aggregation *in vitro* (24).

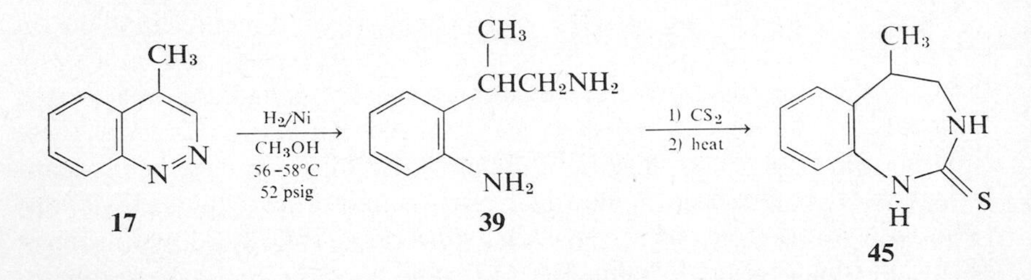

Catalytic reduction of 3,4-diphenylcinnoline and its 7-methyl homolog over Adams platinum catalyst in absolute ethanol yields 1,4-dihydro-3,4-diphenylcinnoline and 1,4-dihydro-3,4-diphenyl-7-methylcinnoline, respectively (10). 4-Phenylcinnoline is reduced almost quantitatively to 1,4-dihydro-4-phenylcinnoline by hydrogenation over palladium on activated aluminum oxide in acetic acid solution (24a).

2. *Chemical*

Like cinnoline itself, 4-methylcinnoline, 4-phenylcinnoline, 3-methyl-4-phenylcinnoline, and 3,4-dimethylcinnoline are reduced by lithium aluminum hydride in refluxing ether solution to their 1,4-dihydro derivatives and not to their 1,2-dihydro derivatives, as previously thought (25, 26). The 1,4-dihydro derivatives are stable enough to be isolated. They are in equilibrium with the corresponding 1-aminoindoles in hot, dilute, aqueous hydrochloric acid solution (26). On the other hand, reduction of these same four alkyl- and arylcinnolines with sodium in ethanolic solution yields in each case a mixture of the corresponding 1,4-dihydro derivative (**46**, R′ = methyl or phenyl; R = hydrogen or methyl) and the corresponding indole (**47**). No 1-amino-indole was detected (26).

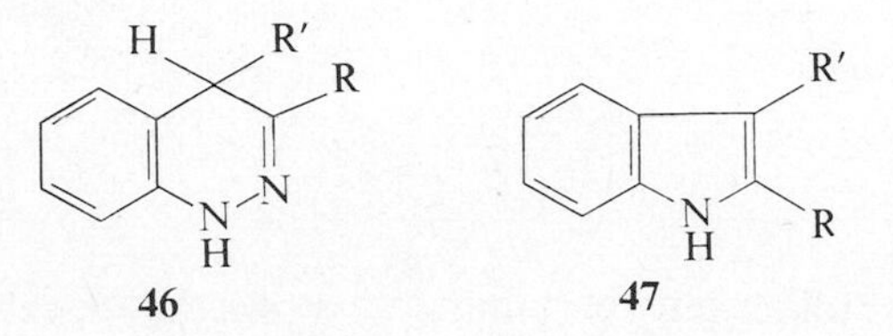

Reduction of 4-methylcinnoline (**17**) with amalgamated zinc in refluxing aqueous acetic acid solution for 2 hr gives 82% of skatole (**48**), but allowing the reaction mixture to reflux for only 8 min affords principally 1,4-dihydro-4-methylcinnoline (**43**) together with a small amount of skatole (27).

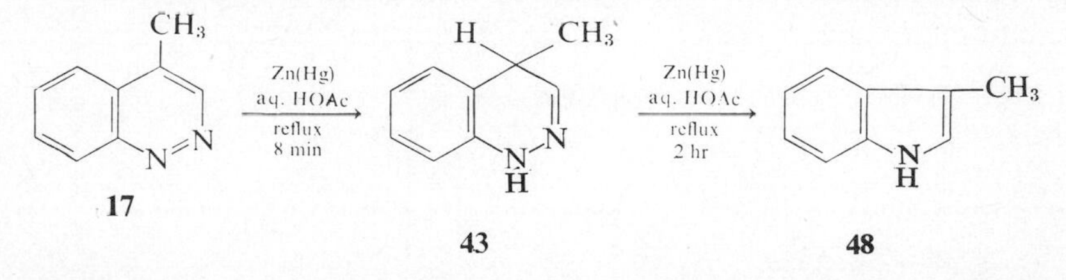

Similarly, when 4-phenylcinnoline is reduced with amalgamated zinc in refluxing aqueous acetic acid solution for 2 hr, 3-phenylindole is obtained in 88% yield. Other 4-arylcinnolines also give the corresponding 3-substituted indoles under these conditions (28). When the reaction time is only 9 min, however, 4-phenylcinnoline yields principally 1,4-dihydro-4-phenylcinnoline together with some 3-phenylindole (27). Since both 1,4-dihydro-4-phenyl-cinnoline and 1,4-dihydro-4-methylcinnoline can be further reduced with amalgamated zinc in refluxing aqueous acetic acid to the corresponding indoles, it is likely that the dihydrocinnolines are intermediates in the reduction of cinnolines to indoles. The nitrogen atom eliminated in the reductive contraction to an indole was shown to be N-2 by reduction of 4-phenyl[2-^{15}N]cinnoline (27).

Reduction of 3-phenylcinnoline with zinc dust and barium hydroxide in refluxing aqueous ethanolic solution for 4 hr yields 1,4-dihydro-3-phenyl-cinnoline (8).

A reductive formylation of 4-methylcinnoline (**17**) may be effected by heating it in a mixture of formic acid and formamide. The product, obtained in 74% yield, is 1-formamido-3-methylindole (**48a**). In the same way, 3,4-dimethylcinnoline is converted to 1-formamido-2,3-dimethylindole (28a).

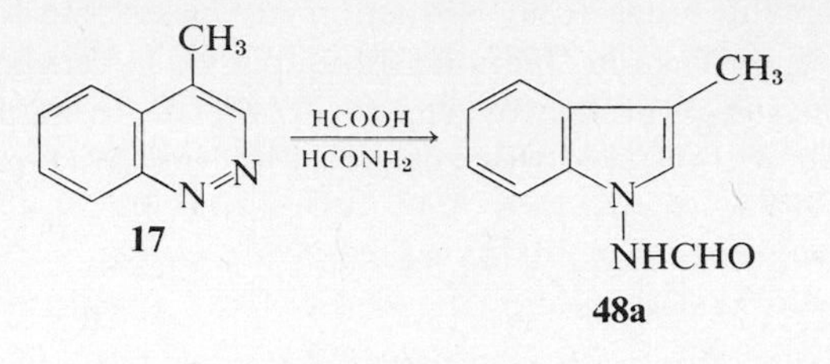

3. *Electrolytic*

A fine polarographic study by Lund (29) shows that the electrolytic reduction of 4-methylcinnoline (**17**) at −0.4 V (SCE) in 1*N* aqueous hydrochloric acid solution consumes two electrons per molecule to give 1,4-dihydro-4-methylcinnoline (**43**). Under these same conditions, but at −1.0 V (SCE), the reduction consumes four electrons per molecule to give skatole (**48**). The dihydrocinnoline **43** is also reduced in acid solution to skatole.

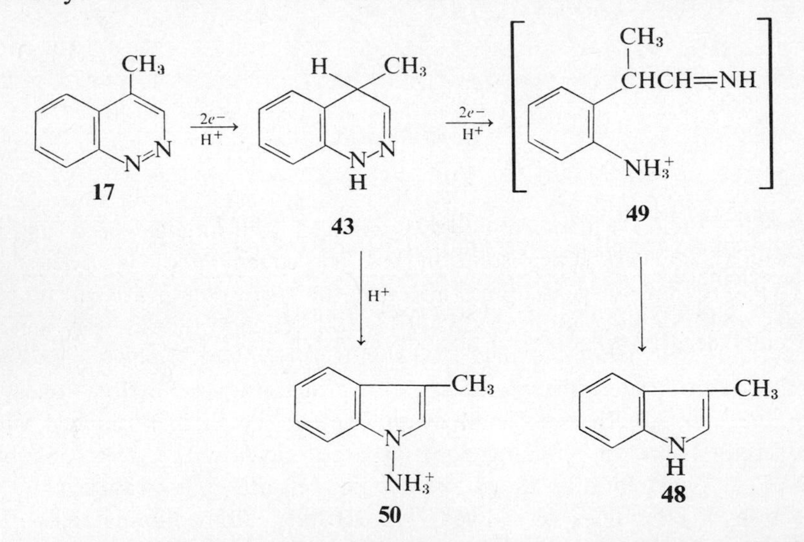

In alkaline solution, however, 4-methylcinnoline is reduced to 1,4-dihydro-4-methylcinnoline (**43**), which cannot be further reduced and is not oxidized at the anode.

The formation of skatole from 1,4-dihydro-4-methylcinnoline could follow two routes. 1,4-Dihydrocinnolines are known to rearrange to *N*-aminoindoles in acid solution, so that *N*-aminoskatole (**50**) is a possible intermediate. However, **50** cannot be reduced under these conditions to skatole (**48**) and therefore is not an intermediate in the reduction of **43** to **48**. On the other hand, 1,4-dihydrocinnolines are cyclic phenylhydrazones, and the first step in the reduction of most phenylhydrazones is a cleavage of the nitrogen-nitrogen bond. Therefore, the aldimine **49** is a likely intermediate, which undergoes ring closure with loss of the imino nitrogen atom to give skatole (29). This is in agreement with the fact that the nitrogen atom lost during the reductive contraction of 4-phenylcinnoline to 3-phenylindole, using amalgamated zinc in refluxing aqueous acetic acid, is N-2 (27).

In 2*N* hydrochloric acid solution at −0.5 V (SCE), 3-phenylcinnoline (**22**) is electrolytically reduced to 1,4-dihydro-3-phenylcinnoline (**51**). Further reduction at −1.2 V (SCE) cleaves the nitrogen-nitrogen bond and the intermediate aldimine is further reduced to the only isolable product, 1-(2′-aminophenyl)-2-phenyl-2-aminoethane (**52**). No 2-phenylindole was obtained (29).

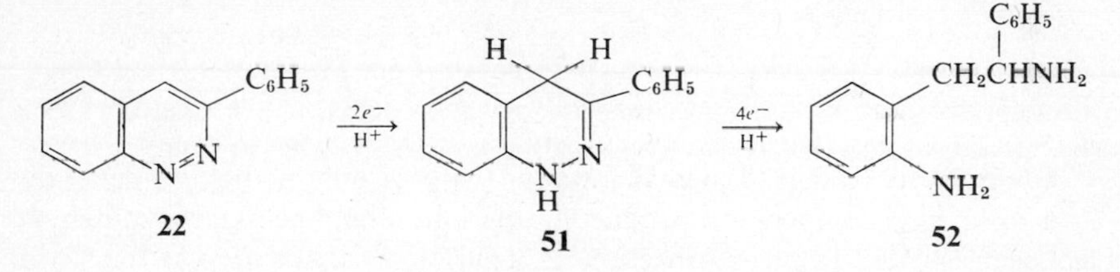

B. Oxidation

In general, the nitrogen-containing ring of the alkyl- and arylcinnolines is more stable than the benzenoid ring to oxidation, so that oxidation of 4-phenylcinnoline (**53**) with hot, aqueous potassium permanganate yields 90% of 5-phenylpyridazine-3,4-dicarboxylic acid (**54**), which can then be thermally decarboxylated to 4-phenylpyridazine (30). This does not mean

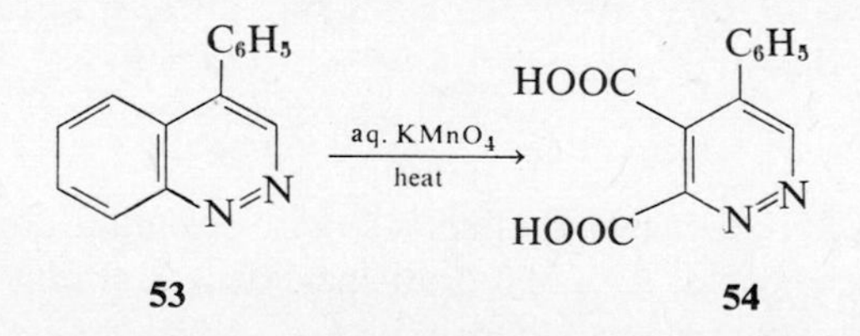

that the nitrogen-containing ring cannot be oxidized. Oxidation of 3-phenylcinnoline with hydrogen peroxide in glacial acetic acid for 2 hr on a steam bath yields, in addition to the expected 3-phenylcinnoline 1- and 2-oxides, small amounts of indazole and benzoic acid. The benzoic acid could only have been derived from the 3-phenyl group and the 3-carbon atom of the cinnoline nucleus (31).

Because the cinnoline ring system is so easily oxidized, at least by potassium permanganate or alkaline potassium ferricyanide, it is not feasible to prepare cinnoline-4-carboxylic acid (**56**) from 4-methylcinnoline (**17**) by direct oxidation. In fact, oxidation of 4-methylcinnoline with aqueous potassium permanganate at 28° C produces only a small amount of a material which does not precipitate when its alkaline solution is acidified. The material yields a picrate derivative (32). No other products were isolated from this oxidation. For this reason, one takes advantage of the acidity of the methyl group in 4-methylcinnoline to prepare 4-styrylcinnoline (**55**), which can then be gently oxidized in pyridine solution to cinnoline-4-carboxylic acid, using aqueous potassium permanganate as the oxidizing agent (33, 34).

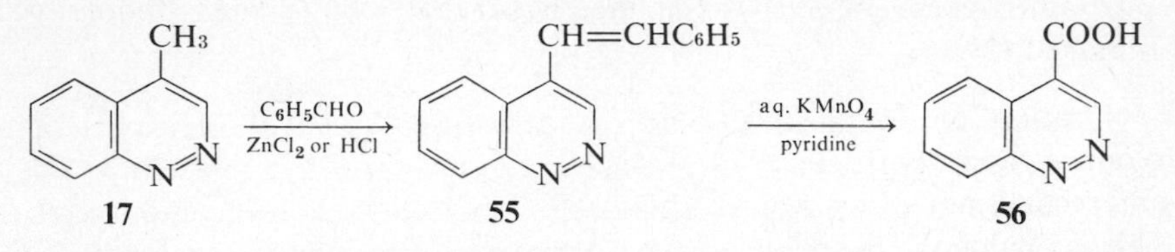

Attempts to oxidize 4-methylcinnoline with selenium dioxide generally lead to tarry decomposition products, although cinnoline-4-carboxaldehyde has been isolated from this reaction as its semicarbazone derivative (**58**) in 7% yield (35). Oxidation of a material claimed to be 4-hydroxymethyl-1,2-dihydrocinnoline (probably 4-hydroxymethyl-1,4-dihydrocinnoline, **57**) with selenium dioxide in dioxane solution, followed by treatment with semi-carbazone hydrochloride also yields the semicarbazone **58**, but in a yield of 58% (35).

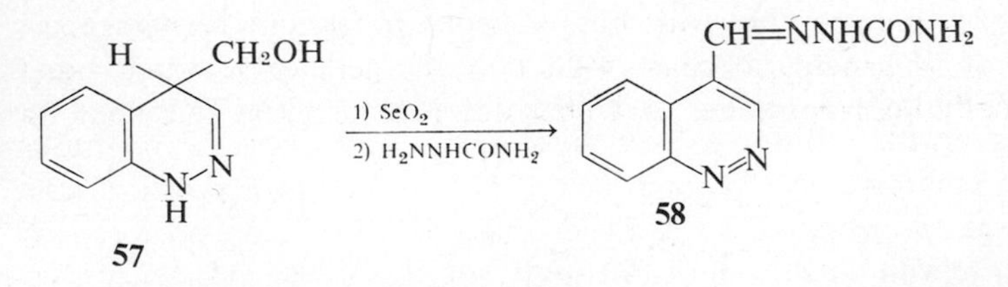

Oxidation of 3,4-dibenzylcinnoline (**15**) with selenium dioxide in refluxing acetic acid solution yields 45% of 3,4-dibenzylidene-3,4-dihydrocinnoline (**59**) and 15% of 3,4-dibenzoylcinnoline (**60**) (4).

Oxidation of alkyl- and arylcinnolines with hydrogen peroxide or with peracids such as 3-chloroperbenzoic acid yields principally the corresponding cinnoline 1- and 2-oxides. Since this particular type of oxidation is discussed in detail in Section 1J, it will not be discussed further here.

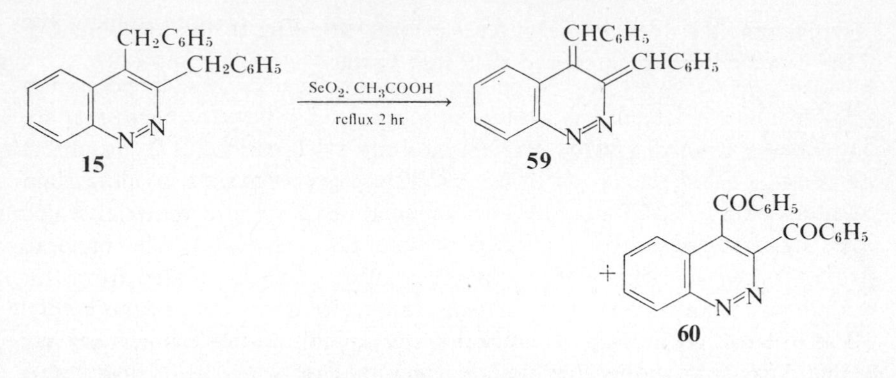

C. Nitration

Nitration of 4-methylcinnoline in concentrated sulfuric acid solution produces approximately 28% of 4-methyl-8-nitrocinnoline and about 13% of a second mononitro isomer, thought to be 4-methyl-5-nitrocinnoline (36). This is analogous to the behavior of cinnoline itself, which is converted by nitration into a mixture of 5- and 8-nitrocinnoline. In contrast, 4-methylcinnoline 2-oxide yields a mixture of its 6- and 8-nitro derivatives. Nitration of 4-ethyl-3-methylcinnoline in concentrated sulfuric acid solution gives about 85% of its 8-nitro derivative. No other mononitro isomer could be isolated (37).

D. Reactions of the Methyl Group in 4-Methylcinnoline

The methyl substituent of 4-methylcinnoline is sufficiently acidic that it will condense with refluxing benzaldehyde in the presence of zinc chloride to give 4-styrylcinnoline (**55**) in good yield (33). Certain substituted benzaldehydes such as *p*-methoxybenzaldehyde and 3,4-dimethoxybenzaldehyde also condense with 4-methylcinnoline in the presence of anhydrous zinc chloride, although the yield of the styrylcinnoline is rather low in the latter case. Condensation of 2-thiophenealdehyde with 4-methylcinnoline under the same conditions proceeds similarly, but again the yield is low (10–12%) and large amounts of tarry materials are produced. Salicylaldehyde, furfural, *p*-nitrobenzaldehyde, and phenylacetaldehyde are reported to produce only tars under the same conditions using zinc chloride as the catalyst (38). Chloral reacts with 4-methylcinnoline in pyridine solution to give 4-(3,3,3-trichloro-2-hydroxypropyl)cinnoline (**61**). Under the same conditions,

3-methylcinnoline does not react with chloral, attesting to the lower acidity of the 3-methyl group compared with that of the 4-methyl group (39).

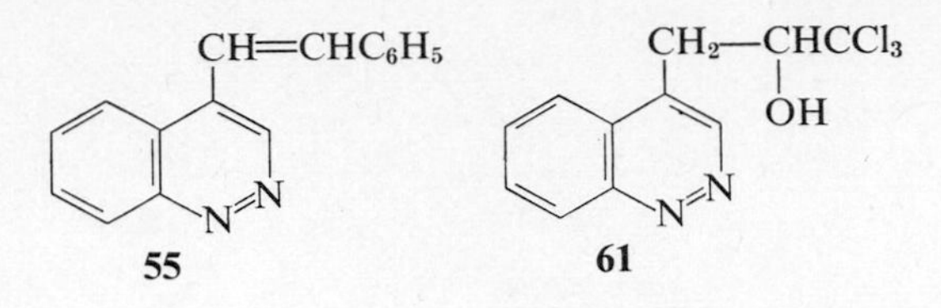

The 3-methyl group may be sufficiently activated by *N*-alkylation, however, so that 3-methylcinnoline methiodide will condense with *p*-dimethylamino-benzaldehyde in refluxing acetic anhydride to give a small yield of 3-(*p*-dimethylaminostyryl)cinnoline methiodide (39). Whether the *N*-methyl group is at N-1 or at N-2 of the cinnoline ring is unknown in this case. *N*-alkylation also enhances the reactivity of the 4-methyl group, so that 4-methylcinnoline ethiodide will condense with *p*-dimethylaminobenzaldehyde to give the 4-styryl derivative in the absence of any catalyst (32). Such activation of a 3- or 4-methyl substituent by quaternization would also be operative in the condensation of 8-hydroxy-4-methylcinnoline hydrochloride with benzaldehyde where an almost quantitative yield of 8-hydroxy-4-styrylcinnoline is produced (34). It would be interesting to apply the conditions of this latter reaction to cases where zinc chloride is an ineffective catalyst or leads to decomposition.

On nitrosation with ethyl nitrite in ethanolic hydrochloric acid, 4-methylcinnoline (**17**) affords the oxime derivative of 4-cinnolinecarboxaldehyde (**61a**) in 80% yield (39a).

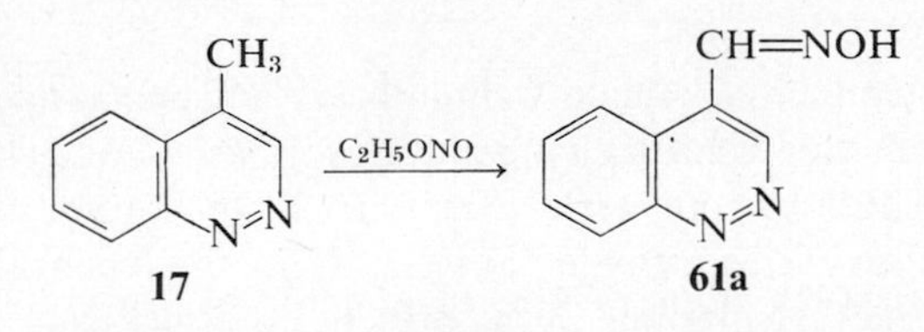

A few cyanine dyes containing the cinnoline nucleus have been prepared (40–41a), such as the dye **62** from condensation of 4-methylcinnoline ethiodide and ethyl orthoformate in refluxing acetic anhydride and the dye **63** from 4-anilinovinylcinnoline methiodide and lepidine methiodide in acetie anhydride containing triethylamine. The 4-anilinovinylcinnoline methiodide **64** is prepared from 4-methylcinnoline methiodide and diphenylformamidine

(40, 41). In none of these cases is it known whether the *N*-alkyl group is at N-1 or N-2, although the materials are claimed to be N-1 salts.

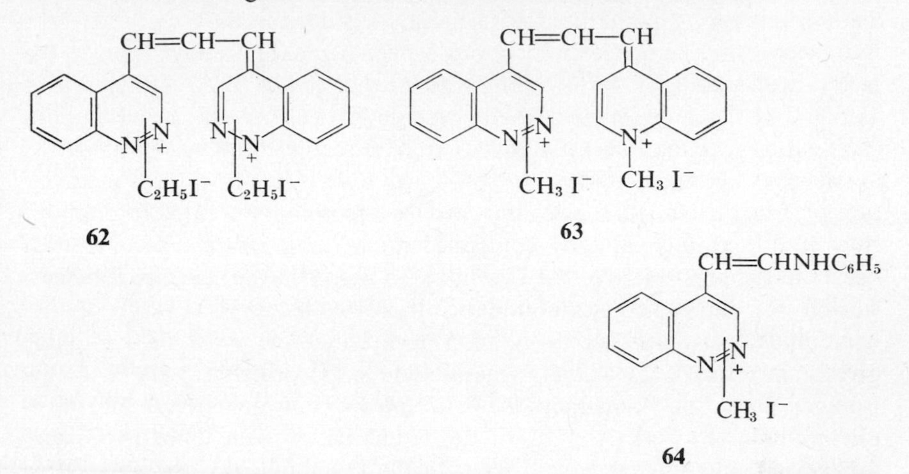

Conversion of 4-methylcinnoline into 4-cinnolinylmethylpyridinium iodide (**65**) can be effected in 70% yield by treating it with iodine in refluxing pyridine solution (35). Compound **65** can then be transformed into cinnoline-4-carboxaldehyde in good yield as discussed in Section 1I-II. Finally, 4-methylcinnoline reacts with *p*-nitrosodimethylaniline in ethanol in the presence of anhydrous sodium carbonate to give the anil **66** in low yield (42).

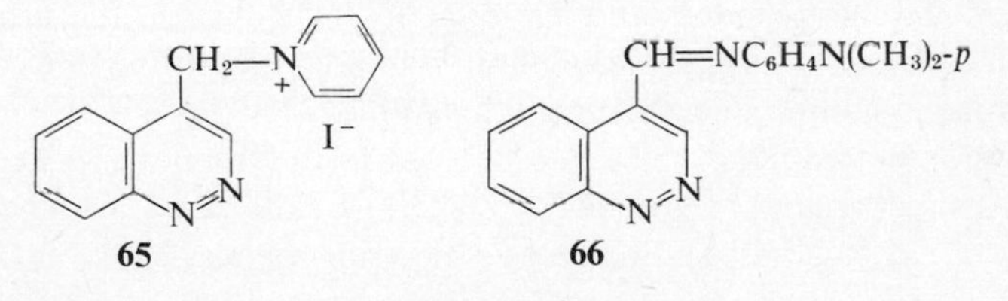

E. Quaternization

The fact that the ethiodide of 4-methylcinnoline is so highly reactive that it condenses with *p*-dimethylaminobenzaldehyde in the absence of a catalyst was earlier interpreted to mean that quaternary salt formation occurs at N-1 in 4-methylcinnoline (32). In fact, quaternization of 4-methylcinnoline with simple alkyl halides occurs principally at N-2, and the cinnoline ethiodide which condensed with *p*-dimethylaminobenzaldehyde was almost certainly the 2-ethiodide. In general, alkylation of the alkyl- and arylcinnolines with alkyl halides or sulfates, such as dimethyl sulfate, occurs predominantly at

N-2 unless the steric effect of a group in the 3-position of the cinnoline ring shifts more of the alkylation to the N-1 position. For example, when 4-methylcinnoline is quaternized with methyl iodide, the ratio of the N-2 to N-1 methyl salts obtained is 10:1, but 3-phenylcinnoline yields 78% of the N-1 salt and 22% of the N-2 salt with methyl iodide (29). Alkylation of 4-ethyl-3-methylcinnoline with methyl iodide gives the 1- and 2-methiodides in equal proportions (37). Methylation of 3-methylcinnoline with methyl iodide gives two methiodides, N-1 (31%) and N-2 (46%), whereas methylation of 8-methylcinnoline gives only the N-2 methiodide (13). Protonation, since it is thermodynamically controlled rather than kinetically controlled, occurs at the most basic of the two nitrogen atoms in the cinnoline nucleus. Thus 3-, 4-, and 8-methylcinnoline are all protonated at N-2.

As mentioned, quaternization of 3-phenylcinnoline with methyl iodide gives a mixture of its 1- and 2-methiodide salts, principally the 1-methyl isomer. Only the 2-methiodide is isolated from quaternization of 4-phenylcinnoline (42a). A study of the reduction of such *N*-methyl-3- and 4-phenylcinnolinium salts, together with the reduction of certain *N*-methyl alkylcinnolinium salts, has been performed under a variety of reduction conditions (42a). This study showed that the reduction of *N*-methylcinnolinium salts with amalgamated zinc and hydrochloric acid can be used to determine the site of quaternization. For example, reduction of 2,3-dimethylcinnolinium iodide with amalgamated zinc and hydrochloric acid yields 2-methylindole and 2-methylindoline. The same reduction of 1,3-dimethylcinnolinium iodide produces 1,2-dimethylindole and 1,2-dimethylindoline. On the other hand, reduction of 1-methyl-3-phenylcinnolinium iodide under the same conditions gives 2-*o*-methylaminophenyl-1-phenethylamine.

Polarographic reduction may also be used to differentiate between cinnolinium salts methylated at N-1 or N-2. The N-1 salts show two one-electron waves, whereas N-2 salts show one two-electron wave (42a).

F. Pharmacology

Very few alkyl- and arylcinnolines have been investigated for pharmacological activity, but some 3-(*p*-dialkylaminoalkoxyphenyl)cinnolines such as 3-[4-(2-diethylaminoethoxy)phenyl]cinnoline (**67**) are claimed to be antiulcer agents, as demonstrated by their inhibition of ulceration in the Shay rat. They are also antiinflammatory agents, showing a phenylbutazone-like effect on edematous conditions, and they are said to possess antibiotic activity against organisms such as the protozoa *Tetrahymena gelleii* and the alga *Chlorella vulgaris*. They have also demonstrated activity as appetite inhibitors and central nervous system stimulants (43).

A group of 4-aryl-3-heteroarylcinnolines such as 4-(4-methoxyphenyl)-3-(2-pyridinyl)cinnoline (**68**) and 4-(4-methoxyphenyl)-3-(2-quinolinyl)cinnoline (**69**) has been shown to possess very little antibacterial activity (42).

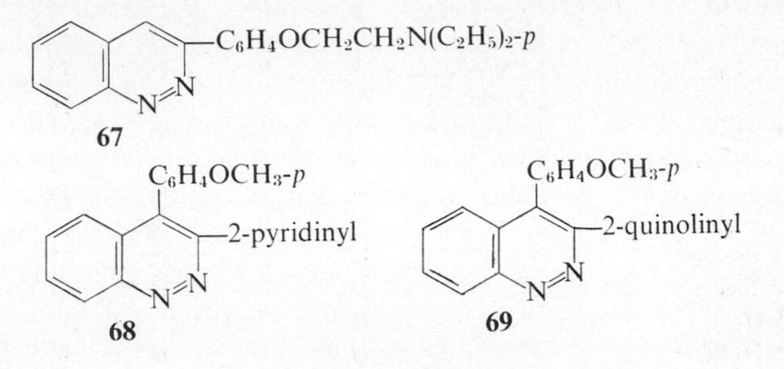

IV. Tables

TABLE 1B-1. *o*-Aminoarylethylenes Which Have Been Diazotized and Cyclized to Cinnolines (Widman-Stoermer Synthesis)

R_3, $CR_1{=}CHR_2$, R_4, NH_2, R_5

R_1	R_2	R_3	R_4	R_5	Reference
CH_3	H	H	H	CH_3O	14, 34
C_6H_5	H	H	H	CH_3O	14
CH_3	CH_3	H	H	H	26
C_6H_5	H	H	H	H	27, 28
CH_3	H	CH_3O	CH_3O	H	44
2,5-$(CH_3O)_2C_6H_3$	H	H	H	H	28
2,3-$(CH_3O)_2C_6H_3$	H	H	H	H	28
3,5-$(CH_3O)_2C_6H_3$	H	H	H	H	28
p-$CH_3OC_6H_4$	H	H	H	H	28
C_2H_5	CH_3	H	H	H	37
CH_3	H	H	H	H	37
CH_3	H	CH_3	H	H	37
CH_3	H	H	H	CH_3	37
CH_3	H	Br	H	H	37
$(CH_3)_3C$	H	H	H	H	37
C_6H_5	C_6H_5	H	H	H	46
C_6H_5	1-Naphthyl	H	H	H	46
p-$CH_3OC_6H_4$	2-Pyridinyl	H	H	H	42, 45
p-$CH_3OC_6H_4$	2-Quinolinyl	H	H	H	42, 45

TABLE 1B-1. (*continued*)

R_1	R_2	R_3	R_4	R_5	Reference
C_6H_5	2-Pyridinyl	H	H	H	42, 45
2-Pyridinyl	H	H	H	H	42, 45
p-$CH_3C_6H_4$	2-Pyridinyl	H	H	H	42
p-$CH_3C_6H_4$	2-Quinolinyl	H	H	H	42
C_6H_5	2-Quinolinyl	H	H	H	42
C_6H_5	2-Pyridinyl	Cl	H	H	42
C_6H_5	2-Quinolinyl	Cl	H	H	42
p-$CH_3OC_6H_4$	3-Pyridinyl	H	H	H	42
2-Pyridinyl	CH_3	H	H	H	42
3-Pyridinyl	H	H	H	H	42
4-Pyridinyl	H	H	H	H	42
2-Thienyl	H	H	H	H	42
CH_3	H	H	COOH	H	47
p-HOC_6H_4	H	H	H	H	48
p-$CH_3C_6H_4$	H	H	H	H	30
C_6H_5	H	Br	H	H	49
p-HOC_6H_4	H	Cl	H	H	49
2-HO-5-ClC_6H_3	H	Cl	H	H	49
C_6H_5	CH_3	H	H	H	30, 50
p-$CH_3OC_6H_4$	CH_3	H	H	H	50
C_6H_5	$C_6H_5CH_2$	H	H	H	46
p-$CH_3OC_6H_4$	C_6H_5	H	H	H	50
CH_3	H	H	H	Cl	51
CH_3	H	H	H	NO_2	51
CH_3	H	H	Cl	H	32
CH_3	H	Cl	H	H	32
CH_3	$(CH_3)_2CH$	H	H	H	51a
CH_3	C_2H_5	H	H	H	51a
C_2H_5	CH_3	H	H	H	51a

TABLE 1B-2. 2-Aminoacetophenones Which Have Been Diazotized and Cyclized to Alkyl- or Aryl-Substituted 4-Hydroxycinnolines (Borsche Synthesis)

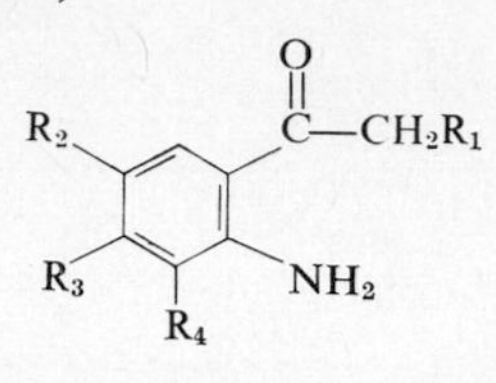

R_1	R_2	R_3	R_4	Reference
C_6H_5	H	H	H	52
C_6H_5	CH_3	H	H	52
1-Naphthyl	H	H	H	52
C_6H_5	NO_2	H	H	52
H	H	C_6H_5	H	17
H	C_6H_5	H	H	17
H	H	CH_3	H	53
H	NO_2	CH_3	H	53
$C_6H_5CH_2$	H	H	H	4
CH_3	H	H	H	54, 58
C_2H_5	H	H	H	54
$(CH_2)_3COOH$	H	H	H	54
H	H	CH_3	Cl	55
H	Cl	CH_3	H	55
H	Br	CH_3	H	55
H	CH_3	CH_3	H	59
H	—$(CH_2)_3$—		H	59
H	—$(CH_2)_4$—		H	59
Cl	—$(CH_2)_3$—		H	59
Cl	—$(CH_2)_4$—		H	59
H	H	—$(CH_2)_3$—		59
H	H	—$(CH_2)_4$—		59
Cl	CH_3	H	H	59
Cl	CH_3	CH_3	H	59
CH_3	H	H	NO_2	54
CH_3	Cl	H	H	54
CH_3	Br	H	H	54, 58
CH_3	NO_2	H	H	54, 58
$CH_2COOC_2H_5$	H	$COOC_2H_5$	H	56
$CH_2COOC_2C_5$	CH_3O	CH_3O	H	57
H	H	H	CH_3	53

TABLE 1B-3. Proton Magnetic Resonance Spectra of Alkyl- and Arylcinnolines[a]

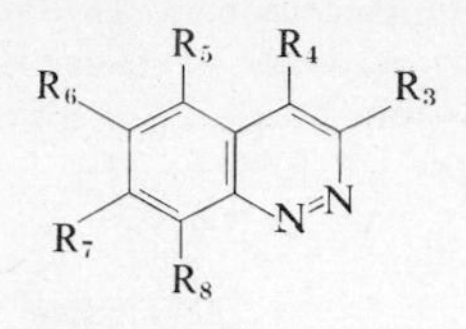

R_3	R_4	R_5	R_6	R_7	R_8	Solvent	Ref.
H	CH_3	H	H	H	H	CCl_4	37
1.00(Q)[b]	7.42	←——2.25(M)——→			1.6		
1.05(S)	7.49(S)	←——2.0–2.5(M), 1.5–1.8(M)——→				CCl_4	25
H	CH_3	H	CH_3	H	H	CCl_4	37
0.95(Q)[b]	7.35	2.40(S)	7.40	2.40(D)	1.65(D)		
				$J_{7,8} = 9$			
H	CH_3	H	H	H	CH_3	CCl_4	37
1.00(Q)[b]	7.38	←——2.45(M)——→			7.08		
CH_3	CH_3	H	H	H	H	CCl_4	37
7.12	7.42	←——2.25(M)——→			1.65		
H	CH_3	H	Br	H	H	CCl_4	37
0.95(Q)[b]	7.35	1.95(D)	—	2.18(D)	1.65(D)		
		$J = 2$		$J_1 = 9$	$J = 9$		
				$J_2 = 2$			
CH_3	CH_2CH_3	H	H	H	H	CCl_4	37
7.12	7.0(Q)	←——2.35(M)——→			1.72		
	8.75(T)						
	$J = 7$						
H	$(CH_3)_3C$	H	H	H	H	CCl_4	37
0.8	8.35	1.55(M)	2.25(M)	2.25(M)	1.55(M)		
H	CH_3	H	H	H	NO_2	$CDCl_3$	37
0.7(Q)[b]	7.24	←——2.0(M)——→			—		
CH_3	CH_2CH_3	H	H	H	NO_2	$CDCl_3$	37
7.02	6.85(Q)	1.95(D)	2.15(T)	1.65(D)	—		
	8.70(T)	$J_1 = 8$	$J = 8$	$J = 2$			
	$J = 7$	$J_2 = 2$					
CH_3	CH_3	H	H	H	NO_2	$CDCl_3$	37
7.02	7.31	2.00(D)	2.25(T)	1.77(D)	—		
		$J_1 = 8$	$J = 8$	$J_1 = 8$			
		$J_2 = 2$		$J_2 = 2$			
H	$(CH_3)_3C$	H	H	H	NO_2	CF_3COOH	37
2.0	8.35	←——2.0(M)——→			—		
NO_2	CH_3	H	H	H	H	$CDCl_3$	37
—	7.15	←——1.9(M)——→			1.35		
H	C_6H_5	H	H	H	H	CCl_4	25
0.97(S)	2.55	←——1.5–1.7(M), 2.0–2.6(M)——→					
H	$2,3\text{-}(CH_3O)_2C_6H_3$	H	H	H	H	CCl_4	25
0.93(S)	CH_3O at 6.08(S)	←1.4–1.6(M), 2.2–2.4(M), 2.8–3.3(M)→					
	and 6.54(S)						
H	$3,4\text{-}(CH_3O)_2C_6H_3$	H	H	H	H	$CDCl_3$	25
0.75(S)	CH_3O at 6.02(S)	←1.3–1.5(M), 2.1–2.3(M), 2.90(S)→					
	and 6.04(S)						

TABLE 1B-3. (*continued*)

R_3	R_4	R_5	R_6	R_7	R_8	Solvent	Ref.
C_6H_5	C_6H_5	H	H	H	H	$CDCl_3$	25
←1.3–1.5(M), 2.2–2.9(M)→							
CH_3 7.17	CH_3O 5.94	H 2.38	CH_3O 5.94 or 5.99	CH_3O 5.99 or 5.94	H 2.88	$CDCl_3$	62
CH_3 7.06	Cl —	H 2.34	CH_3O 5.92	CH_3O 5.92	H 2.82	$CDCl_3$	62
CH_2COOCH_3 CH_2 at 5.58 CH_3 at 6.24	Cl —	H 2.26	CH_3O 5.88	CH_3O 5.88	H 2.72	$CDCl_3$	62
C_6H_5 Not given	H 1.83	H Not given	H Not given	H Not given	H 1.28–1.53	$CDCl_3$	31
4-ClC_6H_4 1.78(D) 2.47(D) $J = 9$	H 1.87(S)	H ←2.03–2.25(M)→	H	H	H 1.28–1.55	$CDCl_3$	31
4-$CH_3O_6H_4$ CH_3O at 6.13(S) Ar at 1.80(D), 2.95(D) $J_{Ar} = 9$	H 1.98(S)	H ←2.13–2.37(M)→	H	H	H 1.28–1.62	$CDCl_3$	31
C_6H_5 Not given	H 2.03(S)	H 2.53(S)	CH_3 7.48(S)	H 2.47(D)	H 1.63(D) $J_{7,8} = 10$	$CDCl_3$	31
4-ClC_6H_4 1.87(D) 2.53(D) $J = 9$	H 2.03(S)	H 2.33(S)	CH_3 7.45(S)	H 2.53(D)	H 1.60(D) $J_{7,8} = 10$	$CDCl_3$	31
C_6H_5 1.83–2.83(M)	$COOC_2H_5$ Not given	H ←1.83–2.83(M)→	H	H	H 1.20–1.50	$CDCl_3$	31
H 0.70(S)	$C_6H_5CH{=}CH$ 2.65(M)	H 2.03(M)	H ←2.65(M)→	H	H 1.68(M)	$CDCl_3$	20c

[a] Chemical shift values are given immediately below the proton(s) to which they refer and are in τ units; the multiplicity of each absorption, when stated, is given in parentheses following the τ values as S = singlet, D = doublet, T = triplet, Q = quartet, M = multiplet; J = coupling constant in cps. Chemical shift positions refer to tetramethylsilane as an internal standard.

[b] Poorly defined quartet, collapsing to a singlet upon irradiation of the 4-methyl signal.

TABLE 1B-4. Ultraviolet Absorption Spectra of Alkyl- and Arylcinnolines

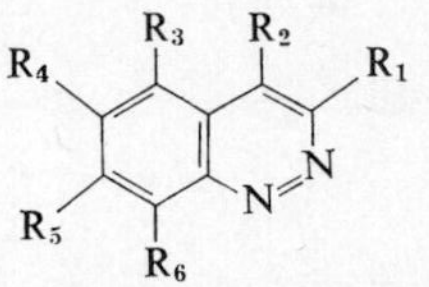

R_1	R_2	R_3	R_4	R_5	R_6	Solvent	λ_{max} (mμ)[a]	log ε_{max}	Ref.
H	CH_3	H	H	H	H	Water pH = 7.0	225	4.68	60
							283	3.50	
							292	3.52	
							308	*3.42*	
							320	3.54	
							330	*3.47*	
						Water pH = 0.3	235	4.58	60
							300	*3.40*	
							306	3.44	
							349	3.52	
H	H	H	H	H	CH_3	Ethanolic perchloric acid	243	4.40	19
							295	3.15	
							307	*2.94*	
							375	3.36	

H	C_6H_5	H	H	H	H	Dioxane	227	4.62	28
							298	3.86	
							327	3.77	
H	*p*-$CH_3OC_6H_4$	H	H	H	H	Dioxane	226	4.60	28
							315	*3.96*	
							330	4.02	
H	*p*-HOC_6H_4	H	H	H	H	Dioxane	226	4.60	28
							315	3.97	
							331	4.04	
p-HOC_6H_4	H	H	H	H	H	Methanol	266	4.47	61
							298	4.27	
p-$CH_3OC_6H_4$	H	H	H	H	H	Methanol	265	4.50	61
							297	4.27	
p-$(C_2H_5)_2NCH_2CH_2OC_6H_4$	H	H	H	H	H	Methanol	264	4.52	61
							295	4.30	
p-$HOOCCH_2OC_6H_4$	H	H	H	H	H	Methanol	265	4.50	61
							296	4.27	
$CH(CH_3)_2$	CH_3	H	H	H	H	Ethanol	226	4.81	51a
							281	3.54	
							292	3.51	
							329	3.55	

[a] Italics refer to shoulders or inflections.

TABLE 1B-5. Alkyl- and Arylcinnolines

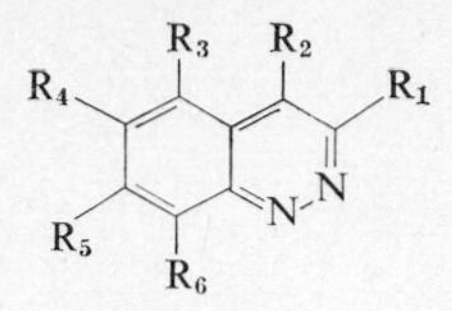

R_1	R_2	R_3	R_4	R_5
CH_3	H	H	H	H
p-$(CH_3)_2NC_6H_4CH{=}CH-$	H	H	H	H
H	$CCl_3CH(OH)CH_2-$	H	H	H
C_6H_5	C_6H_5	H	H	H
H	C_6H_5	H	H	H
C_6H_5	C_6H_5	H	H	CH_3
C_6H_5	C_6H_5	H	H	H
C_6H_5	H	C_6H_5	H	H
C_6H_5	H	C_6H_5	CH_3	H
C_6H_5	H	C_6H_5	H	CH_3
H	H	H	C_6H_5	H
H	H	H	H	C_6H_5
C_6H_5	H	H	H	H
H	C_6H_5CHCN (attached via CH)	H	H	H
H	m-$CH_3OC_6H_4CHCN$ (attached via CH)	H	H	H
H	p-$CH_3OC_6H_4CHCN$ (attached via CH)	H	H	H

R_6	Prep.[a]	MP (°C)	Remarks[b]	Ref.
H	A	58.5–61	Yellow plates (ether-ligroin); picrate deriv., mp 168–170°, or 175–176.5; perchlorate deriv., mp 156–158°; 1-methiodide deriv., mp 159–161°; 2-methiodide deriv., mp 200–202°	13, 15, 19 39, 65
H	B	335–340 (dec.)	Impure; identity questionable	39
H	C	165–166 (dec.)	Silver leaflets (ethanol)	39
H	D, F	149–150, 151–152	Almost colorless crystals; methiodide deriv., mp 246° (dec.)	9, 10
H	E, H	67–67.5 65.5–66	Yellow crystals; picrate deriv., mp 158°, perchlorate, mp 202–204; 2-methiodide, mp 224–226°	30, 63, 28, 42a
H	F	178	Yellow crystals; picrate deriv., mp 155°	10
CH_3	F	158	Yellow crystals	10
H	G	123	Almost colorless crystals (ethanol); structure uncertain; picrate deriv., mp 164–165°	10
H	G	183	Almost colorless crystals; structure uncertain; picrate deriv., mp 181–182°	10
H	G	170	Almost colorless crystals (ethanol); structure uncertain; picrate deriv., mp 194–195°	10
H	I	111–111.5	Yellow needles (ligroin); picrate deriv., mp 176°	17
H	I	116	Yellow needles (aq. ethanol); picrate deriv., mp 177°	17
H	I, O	118–119, 120.5–121.5	Yellow needles (ligroin); picrate deriv., mp 151°; 1-methiodide deriv., mp 205–207°; 2-methiodide deriv., mp 176–178°	7, 8, 17, 42a
H	J, EE	197.5–198.5, 200–200.5	Yellow plates (aq. ethanol)	2, 5, 6e
H	J	194.5–195	Orange-red plates (aq. ethanol)	2
H	J	183–185	Orange-red plates (aq. ethanol)	2

TABLE 1B-5. (*continued*)

R_1	R_2	R_3	R_4	R_5
H	$C_6H_5\underset{\mid}{C}HCN$	H	CH_3O	CH_3O
H	$C_6H_5\underset{\mid}{C}HCONH_2$	H	H	H
H	$C_6H_5\underset{\mid}{C}HCONH_2$	H	CH_3O	CH_3O
H	$C_6H_5CH_2$	H	H	H
p-HOC_6H_4	H	H	H	H
$C_6H_4OCH_2COOH$-p	H	H	H	H
CH_3—(pyrrolidine N)—CH_3, N—CH_2CH_2O—C_6H_4—	H	H	H	H
p-$(C_2H_5)_2NCH_2CH_2OC_6H_4$	H	H	H	H
p-$CH_3OC_6H_4$	H	H	H	H
p-ClC_6H_4	H	H	H	H
C_6H_5	H	H	CH_3	H
p-ClC_6H_4	H	H	CH_3	H
$\underset{OH}{CH_2}-\left(\underset{OH}{CH}-\right)_3$	H	H	H	H
$\underset{OAc}{CH_2}-\left(\underset{OAc}{CH}-\right)_3$	H	H	H	H
$\underset{OH}{CH_2}-\left(\underset{OH}{CH}-\right)_3$	H	H	H	H
$\underset{OH}{CH_2}-\left(\underset{OH}{CH}-\right)_2$	H	H	H	H
H	CH_3	H	H	H
H	$C_5H_5N^+$—CH_2— I^- (pyridinium)	H	H	H
H	p-$(CH_3)_2NC_6H_4$—$\underset{O}{N}$=CH—	H	H	H

R_6	Prep.[a]	MP (°C)	Remarks[b]	Ref.
H	J	220–221	Orange-red plates (aq. ethanol)	2
H	K	248–249	Yellow crystals	2
H	K	234.5–236.5	Opaque white crystals (aq. ethanol)	2
H	L	104.5	Nearly colorless crystals (aq. ethanol)	2
H	M	235–237	(Methanol-benzene)	43, 61
H	N	230–232	(Aq. ethanol)	61
H	N	102–103	(Ligroin)	43, 61
H	N	67–68	(Ligroin or hexane)	43, 61
H	O	111–112	(Ether)	8, 64
H	O	146.5–147.5	(Ligroin)	8, 64
H	O	139.6–140.3, 138.5–139.5	Yellow solid (ligroin or methanol)	7, 8
H	O	185–186	(Methanol)	8, 64
H	P	205–206	(Water) R_1 = D-*arabo*-tetrahydroxybutyl	65
H	Q	131–132	Yellow needles (ethanol) R_1 = tetracetyl-D-*arabo*-tetrahydroxybutyl	65
H	P	159–160	(Methanol) R_1 = D-*lyxo*-tetrahydroxybutyl	65
H	P	149–150	(Ethanol) R_1 = D-*threo*-trihydroxypropyl	65
H	H, HH	72	Picrate deriv., mp 177–179°; ethiodide deriv., mp 155–157°; 2-methiodide deriv., mp 201–203°; perchlorate deriv., mp 148–149°	19, 35, 37, 40, 41, 60 6a,
H	R	186–187 (dec.)	Yellow prisms (ethanol)	35
H	S	191–192 (dec.)	Brown needles (ethanol)	35

TABLE 1B-5 (*continued*)

R_1	R_2	R_3	R_4	R_5
H	NO_2—CH═CH—	H	H	H
H	C_6H_5CH═CH—	H	H	H
H	*p*-$CH_3OC_6H_4$CH═CH—	H	H	H
H	3,4-$(CH_3O)_2C_6H$ ═CH—	H	H	H
H	(thiophene ring, S) —CH═CH—	H	H	H
H	C_6H_5—N (triazole ring, N═, N═) —CH═CH—	H	H	H
2-Pyridyl	*p*-$CH_3OC_6H_4$	H	H	H
2-Pyridyl	C_6H_5	H	H	H
2-Quinolyl	*p*-$CH_3OC_6H_4$	H	H	H
2-Quinolyl	C_6H_5	H	H	H
2-Pyridyl	*p*-$CH_3C_6H_4$	H	H	H
2-Quinolyl	*p*-$CH_3C_6H_4$	H	H	H
3-Pyridyl	*p*-$CH_3OC_6H_4$	H	H	H
2-Pyridyl	*p*-HOC_6H_4	H	H	H
H	2-Pyridyl	H	H	H
CH_3	2-Pyridyl	H	H	H
H	3-Pyridyl	H	H	H
CH_3	3-Pyridyl	H	H	H
H	4-Pyridyl	H	H	H
H	2-Thienyl	H	H	H
H	5-Br-2-$CH_3OC_6H_3$	H	H	H
H	*p*-$CH_3OC_6H_4$	H	H	H
H	*p*-HOC_6H_4	H	H	H
H	2,5-$(CH_3O)_2C_6H_3$	H	H	H
H	2,5-$(HO)_2C_6H_3$	H	H	H
H	3,4-$(CH_3O)_2C_6H_3$	H	H	H
H	3,4-$(HO)_2C_6H_3$	H	H	H
H	2,3-$(CH_3O)_2C_6H_3$	H	H	H
H	2,3-$(HO)_2C_6H_3$	H	H	H
H	*p*-$(CH_3)_2NC_6H_4$CH═CH—	H	H	H
H	C_6H_5NHCH═CH—	H	H	H
H	(quinoline ring, N—CH_3) ═CH—CH═CH—	H	H	H

R6	Prep.[a]	MP (°C)	Remarks[b]	Ref.
H	T	141 (dec.)	Yellow–brown granules (ethanol)	35
H	C	121.8–122.6	Green–gold plates (aq. ethanol)	38
H	C	112.2–112.6	Golden plates (methanol)	38
H	C	193–194	Yellow needles (aq. ethanol)	38
H	C	113–113.6	Yellow needles (methanol)	38
H	C	205.2–205.6	Yellow needles (methanol)	38
H	H	157–158	Yellow tablets (aq. methanol)	42, 45
H	H	145–146	Yellow crystals (elute in benzene over alumina); picrate deriv., mp 194–196°	42, 45
H	H	151–152	Yellow tablets (methanol)	42, 45
H	H	162–163	Yellow prisms (aq. ethanol)	42
H	H	164–165	Yellow needles (aq. ethanol)	42
H	H	153–154	Yellow needles (aq. ethanol or benzene-ligroin)	42
H	H	145–146	Yellow needles (aq. ethanol)	42
H	M	265–266	Yellow prisms (ethanol)	42
H	H	125–126, 128–129	Prisms (ligroin-ethyl acetate); picrate deriv., mp 201–203°C	42, 45
H	H	155–156	Yellow prisms (benzene-ligroin)	42
H	H	141–142	Yellow needles (benzene-ligroin)	42
H	H	192–193	Yellow prisms (ethanol)	42
H	H	—	Picrate deriv., mp 276–278°	42
H	H	85–86	Yellow prisms (ether)	42
H	H	138–138.5	Yellow needles (ligroin)	28
H	H	85	Yellow needles (ligroin)	28
H	M	237–238	Orange-yellow crystals (butanol)	28
H	H	130	Greenish-yellow needles (ligroin)	28
H	M	285–287	Yellow needles by sublimation	28
H	H	152.5	Yellow blades (ligroin)	28
H	M	266–268 (dec.)	Yellow needles (acetic acid)	28
H	H	120.5–121	Yellow prisms (ligroin)	28
H	M	277–281 (dec.)	Yellow needles by sublimation	28
H	C	201	Red crystals (ethyl acetate)	40
H	U	—	Methiodide deriv., mp 236–237°; ethiodide deriv., mp 227–228°	40
H	V	—	Methiodide deriv., mp 263–264° (dec)	40

TABLE 1B-5 (*continued*)

R_1	R_2	R_3	R_4	R_5
H	N $=CH—CH=CH—$ C_2H_5	H	H	H
H	$CH—CH=CH—$ N CH_3	H	H	H
H	$CH—CH=CH—$ N C_2H_5	H	H	H
H	$CH—CH=CH—$ N N CH_3	H	H	H
H	$CH—CH=CH—$ N N C_2H_5	H	H	H
H	S $=CH—CH=CH—$ N C_2H_5	H	H	H
H	O $=CH—CH=CH—$ N C_2H_5	H	H	H
H	4-Cinnolinyl	H	H	H

R_6	Prep.[a]	MP (°C)	Remarks[b]	Ref.
H	V	—	Ethiodide deriv., mp 263°	40, 41
H	V	—	Methiodide deriv., mp 250°	40, 41
H	V	—	Ethiodide deriv., mp 237–238°	40
H	W	—	Methiodide deriv., mp 283–284° (dec); structure of R_2 uncertain (CH_3 may be at N-2)	40
H	W	—	Ethiodide deriv., mp 253–255°; structure of R_2 uncertain (C_2H_5 may be at N-2)	40, 41
H	X	—	Ethiodide deriv., mp 270–271°	40
H	Y	—	Ethiodide deriv., mp 185–186°	40
H	E, AA	232–234, 238	Yellow needles (methanol, then benzene)	63, 66

TABLE 1B-5 (*continued*)

R_1	R_2	R_3	R_4	R_5
H	$o\text{-}HOC_6H_4$	H	H	H
H	$o\text{-}HOC_6H_4$	H	CH_3	H
H	$p\text{-}(CH_3)_2NC_6H_4N{=}CH{-}$	H	H	H
H	$C_6H_5-\overset{\vert}{\underset{CN}{C}}-CH_2CH_2N(CH_3)_2$	H	H	H
H	$C_6H_5-\overset{\vert}{\underset{CN}{C}}-CH_2CH_2N(C_2H_5)_2$	H	H	H
H	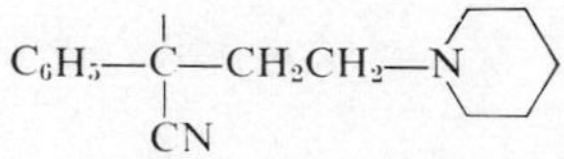 $C_6H_5-\overset{\vert}{\underset{CN}{C}}-CH_2CH_2-N\bigcirc$	H	H	H
H	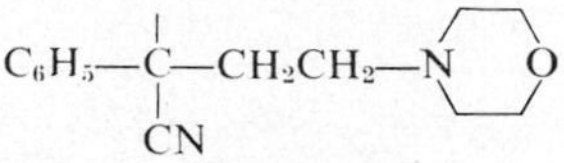 $C_6H_5-\overset{\vert}{\underset{CN}{C}}-CH_2CH_2-N\bigcirc O$	H	H	H
H	$C_6H_5-\overset{\vert}{\underset{CN}{C}}-(CH_2)_3-N(CH_3)_2$	H	H	H
H	$C_6H_5-\overset{\vert}{\underset{CN}{C}}-(CH_2)_3-N(C_2H_5)_2$	H	H	H
H	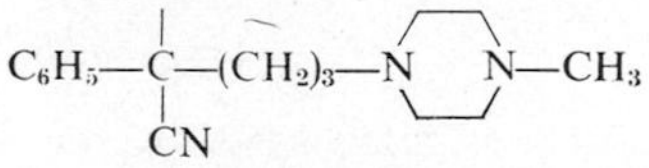 $C_6H_5-\overset{\vert}{\underset{CN}{C}}-(CH_2)_3-N\bigcirc N-CH_3$	H	H	H
H	$C_6H_5\overset{\vert}{C}HCH_2CH_2N(CH_3)_2$	H	H	H
H	$C_6H_5\underset{\vert}{C}HCH_2CH_2N(C_2H_5)_2$	H	H	H
H	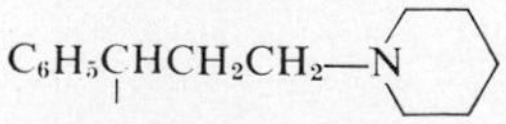 $C_6H_5\underset{\vert}{C}HCH_2CH_2-N\bigcirc$	H	H	H
H	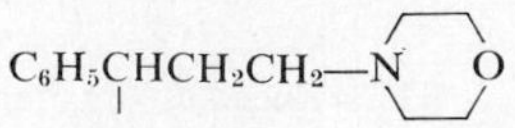 $C_6H_5\underset{\vert}{C}HCH_2CH_2-N\bigcirc O$	H	H	H

R_6	Prep.[a]	MP (°C)	Remarks[b]	Ref.
H	Z	188	Yellow needles (ether-hexane); *O*-acetyl deriv., mp 103–104°; *O*-tosylate deriv., mp 110–11°	11
H	Z	170–171	Yellow needles; *O*-tosylate deriv., mp 115–116°	11
H	BB	196–197	Red leaflets (ethanol)	42
H	J	—	Red syrup; bp 198–201° at 0.015 mm Hg	3
H	J	—	Bp 200–204° at 0.005 mm Hg; acidic succinate deriv., mp 147–148°	3
H	J	171–172	Yellow granules (ethanol)	3
H	J	—	Red syrup; bp 235–238° at 0.02 mm Hg	3
H	J	—	Red syrup; bp 216–220° at 0.025 mm Hg	3
H	J	—	Red syrup; bp 216–219° at 0.025 mm Hg	3
H	J	—	Red syrup; bp 248–252° at 0.025 mm Hg	3
H	L	—	Red syrup; bp 176–179° at 0.015 mm Hg	3
H	L	—	Orange syrup; bp 178–181° at 0.015 mm Hg	3
H	L	—	Red syrup; bp 210–212° at 0.02 mm mm Hg	3
H	L	—	Red syrup; bp 222–226° at 0.015 mm Hg	3

TABLE 1B-5 (*continued*)

R_1	R_2	R_3	R_4	R_5
H	$C_6H_5CH(CH_2)_3N(CH_3)_2$	H	H	H
H	$C_6H_5CH(CH_2)_3N(C_2H_5)_2$	H	H	H
H	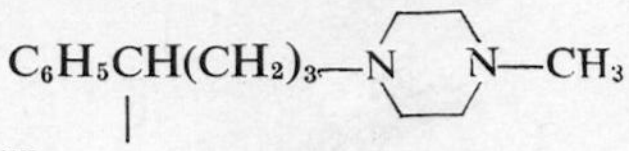 $C_6H_5CH(CH_2)_3—N\quad N—CH_3$	H	H	H
H	H	H	H	H
H	$NCCHCOOC_2H_5$	H	H	H
H	$—CH(COOC_2H_5)_2$	H	H	H
H	$—CH_2COOC_2H_5$	H	H	H
H	*p*-$NH_2C_6H_4CH_2$—	H	H	H
$C_6H_5CH_2$	$C_6H_5CH_2$	H	H	H
$C_6H_5CH_2$	H	H	H	H
CH_3	C_2H_5	H	H	H
H	$(CH_3)_3C$	H	H	H
H	CH_3	H	CH_3	H
H	CH_3	H	H	H
CH_3	CH_3	H	H	H
H	*p*-FC_6H_4CHCN	H	H	H
H	*p*-ClC_6H_4CHCN	H	H	H
H	*p*-BrC_6H_4CHCN	H	H	H
H	*p*-IC_6H_4CHCN	H	H	H
H	*o*-ClC_6H_4CHCN	H	H	H
H	2,4-$Cl_2C_6H_3CHCN$	H	H	H
H	3,4-$Cl_2C_6H_3CHCN$	H	H	H
H	3,4-$(CH_3O)_2C_6H_3CHCN$	H	H	H
H	*p*-$NH_2C_6H_4CHCN$	H	H	H
H	$(C_6H_5)_2CCN$	H	H	H
H	1-Naphthyl-CHCN	H	H	H

R_6	Prep.[a]	MP (°C)	Remarks[b]	Ref.
H	L	—	Red syrup; bp 190–194° at 0.02 mm Hg	3
H	L	—	Red syrup; bp 200–204° at 0.025 mm Hg	3
H	L	—	Red syrup; bp 216–220° at 0.015 mm Hg	3
CH_3	CC	—	Oil, bp 84° at 0.1 mm Hg; picrate deriv., mp 185–186°; 2-methiodide deriv., mp 175–177°; perchlorate deriv., mp 146–147° (dec.)	13, 19
H	J, EE	230–233 (dec.), 226.5–227	—	5, 6e
H	J, EE	48	Dipicrate deriv., mp 102–103°	5, 6e
H	J, EE	72–72.5	Picrate deriv., mp 110°	5, 6e
H	L	176–177	Yellow needles (benzene)	6
H	L	114–115	(Ether-ligroin)	4
H	CC	74–75	Yellow crystals (ether-ligroin)	4
H	H	78 63–65	(Ligroin), picrate deriv., mp 164–166° (dec.)	37, 51a
H	H	—	Oil	37
H	H	75–76	—	37
CH_3	H	90–91	—	37
H	H	119–120, 118–120	(Ligroin)	26, 37
H	J	170–172.8	Orange-red needles (benzene)	6
H	J	177.6–178.4	Orange needles (benzene)	6
H	J	188.2–188.8	Orange needles (benzene)	6
H	J	185.6–186.4	Orange-red needles (benzene)	6
H	J	200.4–201.2	Yellow needles (aq. ethanol)	6
H	J	188–188.8	Golden needles (aq. ethanol)	6
H	J	236.6–237.6	Orange-gold platelets (ethanol)	6
H	J	180.6–181.2	Orange needles (ethanol)	6
H	J	236.4–237.4	Red needles (ethanol)	6
H	J	186.2–186.8	Yellow crystals (aq. ethanol)	6
H	J	224–224.8	Orange crystals (ethanol)	6

TABLE 1B-5 (*continued*)

R_1	R_2	R_3	R_4	R_5
H	2-Naphthyl-CHCN	H	H	H
H	CH_3 CH_3— N N—$CH(CH_3)_2$ CHCN	H	H	H
CH_3	C_6H_5	H	H	H
H	C_6H_5	H	CH_3	H
H	C_6H_5	H	CH_3	CH_3
H	C_6H_5	H	CH_3	H
H	2,5-$(CH_3)_2C_6H_3$	H	H	H
H	*p*-ClC_6H_4	H	CH_3	H
H	C_6H_5	H	Cl	H
H	C_6H_5	H	Cl	H
H	*p*-$CH_3C_6H_4$	H	Cl	H
H	—CH=NOH	H	H	H
$(CH_3)_2CH$	CH_3	H	H	H
C_2H_5	CH_3	H	H	H
H	$C_6H_5COCH_2$	H	H	H
H	$C_6H_5COCHCH_3$	H	H	H
H	$C_6H_5COCHC_2H_5$	H	H	H
H	$C_6H_5COCH(CH_2)_2CH_3$	H	H	H
H	$C_6H_5COCHCH(CH_3)_2$	H	H	H
H	$C_6H_5COC(CH_3)_2$	H	H	H
H	CH_3COCH_2	H	H	H
H	$C_2H_5COCHCH_3$	H	H	H
H	$C_2H_5COCH_2$	H	H	H
H	$CH_3COCHCH_3$	H	H	H
H	$CH_3(CH_2)_2COCH_2$	H	H	H
H	$(CH_3)_2CHCH_2COCH_2$	H	H	H
H	$(CH_3)_2CHCOCH_2$	H	H	H
H	$(CH_2)_5COOH$	H	H	H
H	CH_3 O $CH(CH_3)_2$	H	H	H

R_6	Prep.[a]	MP (°C)	Remarks[b]	Ref.
H	J	248.2–249	Golden needles (aq. ethanol)	6
H	J	207–208	Yellow needles (aq. ethanol)	6
H	H	135–136	(Ligroin)	24a
H	H	125	(Ligroin)	24a
H	H	82	(Benzene)	24a
CH_3	H	—	Oily liquid	24a
H	H	—	Oily liquid	24a
H	H	116	(Ethanol-water)	24a
CH_3	H	85	(Ligroin)	24a
H	H	139	(Ethanol-water)	24a
H	H	143	(Ethanol)	24a
H	DD	195–198	—	39a
H	H	100–101	(Ligroin), bp 140°/1 mm	51a
H	H	—	BP 110°/0.5 mm	51a
H	EE	163–164	(Benzene-ligroin)	6a, 6b, 6c
H	EE	135–135.5	—	6a, 6c
H	EE	98–98.5	—	6a
H	EE	94–94.5	—	6a
H	EE	105.5–106	—	6a
H	EE	120–121	—	6a
H	EE	94	—	6a, 6b, 6c
H	EE	—	Oil	6a
H	EE	112–112.5	—	6a, 6c
H	EE	94–94.5	—	6a, 6c
H	EE	85.5	—	6a
H	EE	91–92.5	—	6a
H	EE	79	—	6a
H	FF	141–142	—	6a, 6b, 6c
H	EE	129	—	6a

TABLE 1B-5 (*continued*)

R_1	R_2	R_3	R_4	R_5
H	C_5H_8=C(COOH)—$(CH_2)_3$—	H	H	H
H	C_2H_5	H	H	H
H	$CH_3CH_2CH_2$	H	H	H
H	$CH_3(CH_2)_3$—	H	H	H
H	$(CH_3)_2CHCH_2$	H	H	H
H	$(CH_3)_2CH$	H	H	H
H	4-Cinnolinyl-$CHCH_3$ |	H	H	H
H	4-Cinnolinyl-CHC_2H_5 |	H	H	H
H	4-Cinnolinyl-$CHCOCH(CH_3)_2$ |	H	H	H
H	4-Cinnolinyl-$CHCOCH_2CH(CH_3)_2$	H	H	H
H	2-Cyclohexanone	H	H	H
CH_3	CN	H	H	H
C_2H_5	CN	H	H	H
H	4-Cinnolinyl-CH_2—	H	H	H
H	—$CH(CN)_2$	H	H	H
H	$C_6H_5C{\equiv}C$—	H	H	H

[a] A, from 4-chloro-3-methylcinnoline by conversion to its 4-toluene-*p*-sulfonylhydrazino derivative and decomposition of this in hot, aqueous sodium carbonate solution or by the Wolff-Kishner reduction of cinnoline-3-carboxaldehyde. B, from 3-methylcinnoline methiodide and *p*-dimethylaminobenzaldehyde in acetic anhydride solution at 160° for 2 hr. C, by condensation of the appropriate aldehyde with 4-methylcinnoline. D, by cyclization of benzil monophenylhydrazone in 75–80% sulfuric acid solution. E, by treatment of cinnoline with *N*-nitrosoacetanilide at 55–60° for 3 hr. F, by dehydrogenation of the appropriately substituted 5,6,7,8-tetrahydrocinnoline over palladium on charcoal. G, a rearranged product of uncertain structure, obtained by dehydrogenation of the corresponding methyl-substituted 3,4-diphenyl-5,6,7,8-tetrahydrocinnoline over Pd/C. H, by diazotization and cyclization of the appropriate *o*-aminoarylethylene. I, by reduction of the corresponding 4-hydroxycinnoline with lithium aluminum hydride in tetrahydrofuran. J, by condensation of the appropriate 4-chlorocinnoline with the appropriate phenylacetonitrile, or other active hydrogen compound, using sodium amide in dry benzene or potassium amide in liquid ammonia as the condensing catalyst. K, by hydrolysis of the corresponding nitrile in sulfuric acid at room temperature to give the amide listed. L, by hydrolysis and decarboxylation of the corresponding 4-cinnolineacetonitrile in refluxing 50–60% sulfuric acid solution. M, by demethylation of the corresponding methoxyphenylcinnoline in refluxing 48% hydrobromic acid solution. N, from 3-(4-hydroxyphenyl)-cinnoline and chloroacetic acid or the appropriate alkyl halide in aqueous basic solution, depending on the product. O, by thermal decarboxylation of the corresponding cinnoline-4-carboxylic acid. P, from the interaction of phenylhydrazine with D-glucose (or D-mannose), D-galactose, or D-xylose in hot aqueous hydrochloric acid solution, depending on the product. Q, by acetylation of 3-(D-*arabino*-tetrahydroxybutyl)cinnoline with acetic anhydride. R, from 4-methylcinnoline, iodine, and pyridine, reflux 3 hr. S, from 4-cinnolinylmethylpyridinium iodide and *p*-nitrosodimethylaniline in aqueous sodium hydroxide solution. T, from 4-cinnolinecarboxaldehyde and nitromethane in aqueous-ethanolic sodium hydroxide solution. U, from 4-methylcinnoline methiodide or ethiodide and

R_6	Prep.[a]	MP (°C)	Remarks[b]	Ref.
H	GG	131–131.5	—	6a, 6b
H	HH	21	Picrate deriv., mp 145°	6a
H	HH	—	Picrate deriv., mp 155°	6a
H	HH	—	Picrate deriv., mp 123°	6a
H	HH	—	Picrate deriv., mp 97°	6a
H	HH	—	Picrate deriv., mp 14.25° (?); mp may be misprint in *Chem. Abstr.* reference	6a
H	II	214	—	6b, 6c
H	II	—	—	6b
H	EE	—	—	6b
H	EE	—	—	6b
H	EE	135–136	—	6c
H	JJ	197.5–198	—	6c, 6d
H	KK	160–161.5	—	6c
H	LL	—	—	6d
H	EE	286.5	—	6e
H	EE	107.5–108	—	6e

N,N′-diphenylformamidine. V, from 4-anilinovinylcinnoline methiodide (or ethiodide), and quinoline methiodide (or ethiodide) or lepidine methiodide (or ethiodide) in acetic anhydride solution containing triethylamine, depending on the product. W, from 4-methylcinnoline methiodide (or ethiodide), and ethyl orthoformate in refluxing acetic anhydride solution. X, from 4-methylcinnoline ethiodide and 2-(β-acetanilinovinyl)benzothiazole ethiodide in molten sodium acetate at 130° for 30 min. Y, from 4-methylcinnoline ethiodide and 2-(β-acetanilinovinyl)benzoxazole ethiodide, acetic anhydride, and sodium acetate. Z, by heating the appropriate *o*-hydroxyphenylglyoxal-2-phenylhydrazone with fused, anhydrous aluminum chloride at 180–190° for 5 min. AA, by reduction (catalytic or with lithium aluminum hydride) of 4-chlorocinnoline. BB, by condensation of 4-methylcinnoline with *p*-nitrosodimethylaniline in refluxing ethanol solution containing anhydrous sodium carbonate. CC, by decomposition of the corresponding 4-(toluene-*p*-sulfonylhydrazino)-cinnoline in hot aqueous sodium carbonate solution. The hydrazinocinnoline was obtained from the 4-chlorocinnoline and toluene-*p*-sulfonylhydrazide. DD, by nitrosation of 4-methylcinnoline with ethyl nitrite. EE, by condensation of 4-methylsulfonylcinnoline with the appropriate ketone or other active hydrogen compound in the presence of a basic condensing agent. FF, by condensation of 4-methylsulfonylcinnoline with cyclohexanone as in EE, except with concomitant hydrolytic scission of the cyclohexanone ring. GG, by condensation of 4-methylsulfonylcinnoline with cyclopentanone in the presence of a basic condensing agent. HH, by alkaline hydrolysis of 2-(4-cinnolinyl)acetophenone or the appropriate 2-alkyl-2-(4-cinnolinyl)acetophenone. II, same as EE except alkaline hydrolysis of an acyl group from the original ketone occurs concomitantly. JJ, by reaction of 4-methylsulfonylcinnoline with acetophenone using potassium cyanide in dimethyl sulfoxide as base. KK, same as JJ except using propiophenone in place of acetophenone. LL, by reaction of 4-cinnolinecarbonitrile with acetophenone in dimethyl sulfoxide in the presence of potassium carbonate.

[b] Solvent employed in recrystallization is enclosed in parentheses.

References

1. H. E. Baumgarten and M. R. DeBrunner, *J. Am. Chem. Soc.*, **76**, 3489 (1954).
2. R. N. Castle and F. H. Kruse, *J. Org. Chem.*, **17**, 1571 (1952).
3. R. N. Castle and M. Onda, *Chem. Pharm. Bull.* (*Tokyo*), **9**, 1008 (1961).
4. D. E. Ames, H. R. Ansari, and A. W. Ellis, *J. Chem. Soc.* (*C*), **1969**, 1795.
5. Y. Mizuno, K. Adachi, and K. Ikeda, *Pharm. Bull.* (*Japan*), **2**, 225 (1954).
6. R. N. Castle and D. B. Cox, *J. Org. Chem.*, **19**, 1117 (1954).

6a. E. Hayashi and T. Watanabe, *Yakugaku Zasshi*, **88**, 593 (1968); *Chem. Abstr.* **69**, 96627 (1968).

6b. E. Hayashi and T. Watanabe, *Yakugaku Zasshi*, **88**, 742 (1968); *Chem. Abstr.*, **69**, 77200 (1968).

6c. E. Hayashi and T. Watanabe, Japan Pat., 7019,908, July 7, 1970; *Chem. Abstr.*, **73**, 87934 (1970).

6d. E. Hayashi and T. Watanabe, *Yakugaku Zasshi*, **89**, 1092 (1969); *Chem. Abstr.*, **72**, 3451 (1970).

6e. E. Hayashi and T. Watanabe, *Yakugaku Zasshi*, **88**, 94 (1968); *Chem. Abstr.*, **69**, 19106 (1968).

7. H. E. Baumgarten and J. L. Furnas, *J. Org. Chem.*, **26**, 1536 (1961).
8. H. S. Lowrie, *J. Med. Chem.*, **9**, 664 (1966).
9. B. P. Moore, *Nature*, **163**, 918 (1949).
10. C. F. H. Allen and J. A. Van Allen, *J. Am. Chem. Soc.*, **73**, 5850 (1951).
11. V. V. Bhat and J. L. Bose, *Chem. Ind.* (*London*), **1963**, 1930; **1965**, 1655.
12. H. J. Barber, K. Washbourn, W. R. Wragg, and E. Lunt, *J. Chem. Soc.*, **1961**, 2828.
13. D. E. Ames, R. F. Chapman, and D. Waite, *J. Chem. Soc.* (*C*), **1966**, 470.
14. E. J. Alford, H. Irving, H. S. Marsh, and K. Schofield, *J. Chem. Soc.*, **1952**, 3009.
15. E. J. Alford and K. Schofield, *J. Chem. Soc.*, **1953**, 609.
16. W. E. Noland and D. A. Jones, *J. Org. Chem.*, **27**, 341 (1962).
17. C. M. Atkinson and C. J. Sharpe, *J. Chem. Soc.*, **1959**, 2858.
18. H. E. Baumgarten and P. L. Creger, *J. Am. Chem. Soc.*, **82**, 4634 (1960).
19. D. E. Ames, G. V. Boyd, R. F. Chapman, A. W. Ellis, A. C. Lovesey, and D. Waite, *J. Chem. Soc.* (*B*), **1967**, 748.
20. R. N. Castle, D. B. Cox and J. F. Suttle, *J. Am. Pharm. Assoc.*, **48**, 135 (1959).

20a. M. H. Palmer and E. R. R. Russell, *Org. Mass. Spectrom.*, **2**, 1265 (1969).

20b. S. N. Bannore, J. L. Bose, K. G. Das, and V. N. Gogte, *Indian J. Chem.*, **7**, 654 (1969).

20c. J. R. Elkins and E. V. Brown, *J. Heterocyclic Chem.*, **5**, 639 (1968).

21. M. T. Rogers and T. W. Campbell, *J. Am. Chem. Soc.*, **75**, 1209 (1953).
22. R. J. W. LeFevre and J. W. Smith, *J. Chem. Soc.*, **1932**, 2810.
23. W. E. Maycock, M. S. Thesis, Brigham Young University, Provo, Utah, September, 1964.
24. E. F. Elslager, D. F. Worth, and S. C. Pericone, *J. Heterocyclic Chem.*, **6**, 491 (1969).

24a. F. Schatz and T. Wagner-Jauregg, *Helv. Chim. Acta*, **51**, 1919 (1968).

25. L. S. Besford, G. Allen, and J. M. Bruce, *J. Chem. Soc.*, **1963**, 2867.
26. D. I. Haddlesey, P. A. Mayor, and S. S. Szinai, *J. Chem. Soc.*, **1964**, 5269.
27. L. S. Besford and J. M. Bruce, *J. Chem. Soc.*, **1964**, 4037.
28. J. M. Bruce, *J. Chem. Soc.*, **1959**, 2366.

28a. D. E. Ames and B. Novitt, *J. Chem. Soc.* (*C*), **1970**, 1700.

29. H. Lund, *Acta Chem. Scand.*, **21**, 2525 (1967).
30. R. Stoermer and H. Fincke, *Ber.*, **42**, 3115 (1909).

31. H. S. Lowrie, *J. Med. Chem.*, **9**, 670 (1966).
32. C. M. Atkinson and J. C. E. Simpson, *J. Chem. Soc.*, **1947**, 808.
33. T. L. Jacobs, S. Winstein, R. B. Henderson, and E. C. Spaeth, *J. Am. Chem. Soc.*, **68**, 1310 (1946).
34. A. Albert and A. Hampton, *J. Chem. Soc.*, **1952**, 4985.
35. R. N. Castle and M. Onda, *J. Org. Chem.*, **26**, 4465 (1961).
36. E. A. Hobday, M. Tomlinson, and H. Irving, *J. Chem. Soc.*, **1962**, 4914.
37. M. H. Palmer and E. R. R. Russell, *J. Chem. Soc.* (*C*), **1968**, 2621.
38. R. N. Castle and D. B. Cox, *J. Org. Chem.*, **18**, 1706 (1953).
39. E. J. Alford and K. Schofield, *J. Chem. Soc.*, **1953**, 1811.
39a. H. Bredereck, G. Simchen, and P. Speh, *Ann. Chem.*, **737**, 39 (1970).
40. A. B. Lal, *Chem. Ber.*, **94**, 1723 (1961).
41. A. B. Lal, *J. Indian Chem. Soc.*, **36**, 64 (1959).
41a. A. K. Misra, A. Nayak, and M. K. Rout, *J. Indian Chem. Soc.*, **48**, 165, 1971.
42. A. J. Nunn and K. Schofield, *J. Chem. Soc.*, **1953**, 3700.
42a. D. E. Ames, B. Novitt, and D. Waite, *J. Chem. Soc.* (*C*), **1969**, 796.
43. H. S. Lowrie, U.S. Pat. 3,239,523, March 8, 1966.
44. J. M. Bruce and P. Knowles, *J. Chem. Soc.*, **1964**, 4046.
45. K. Schofield, *J. Chem. Soc.*, **1949**, 2408.
46. J. C. E. Simpson, *J. Chem. Soc.*, **1943**, 447.
47. O. Widman, *Ber.*, **17**, 722 (1884).
48. C. M. Atkinson and J. C. E. Simpson, *J. Chem. Soc.*, **1947**, 1649.
49. J. C. E. Simpson and O. Stephenson, *J. Chem. Soc.*, **1942**, 353.
50. J. C. E. Simpson, *J. Chem. Soc.*, **1946**, 673.
51. K. Schofield and T. Swain, *J. Chem. Soc.*, **1949**, 1367.
51a. M. A. Shah and G. A. Taylor, *J. Chem. Soc.* (*C*), **1970**, 1642.
52. D. W. Ockenden and K. Schofield, *J. Chem. Soc.*, **1953**, 3706.
53. J. R. Keneford, J. S. Morley, and J. C. E. Simpson, *J. Chem. Soc.*, **1948**, 1702.
54. J. R. Keneford and J. C. E. Simpson, *J. Chem. Soc.*, **1948**, 354.
55. J. R. Keneford and J. C. E. Simpson, *J. Chem. Soc.*, **1947**, 227.
56. C. F. Koelsch, *J. Org. Chem.*, **8**, 295 (1943).
57. K. Schofield and J. C. E. Simpson, *J. Chem. Soc.*, **1945**, 520.
58. N. J. Leonard and S. N. Boyd, *J. Org. Chem.*, **11**, 419 (1946).
59. K. Schofield, T. Swain, and R. S. Theobald, *J. Chem. Soc.*, **1949**, 2399.
60. A. R. Osborn, K. Schofield, and L. N. Short, *J. Chem. Soc.*, **1956**, 4191.
61. H. S. Lowrie, *J. Med. Chem.*, **9**, 784 (1966).
62. A. W. Ellis and A. C. Lovesey, *J. Chem. Soc.* (*B*), **1967**, 1285.
63. C. M. Atkinson and C. J. Sharpe, *J. Chem. Soc.*, **1959**, 3040.
64. H. S. Lowrie (to G. D. Searle and Company), U.S. Pat. 3,265,693, Aug. 9, 1966.
65. H. J. Haas and A. Seeliger, *Chem. Ber.*, **96**, 2427 (1963).
66. D. E. Ames and H. Z. Kucharska, *J. Chem. Soc.*, **1962**, 1509.

Part C. Hydroxycinnolines

I. Methods of Preparation

A. 4-Hydroxycinnolines

1. *Richter Synthesis*

This synthesis was developed by Richter (1) in 1883 when he showed that *o*-aminophenylpropiolic acid (**1**) may be diazotized and cyclized to give 4-hydroxycinnoline-3-carboxylic acid (**2**). The reaction was not used until

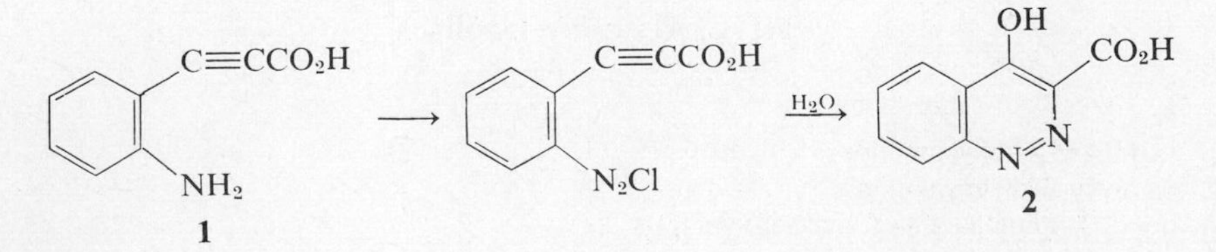

1945 when it was further refined by Schofield and Simpson (2) and extended to include the preparation of 4-hydroxycinnolines (**4**) from diazotized *o*-aminophenylacetylenes (**3**; R = H or phenyl). The cyclization of **3** fails when R is a 2-pyridyl group (3).

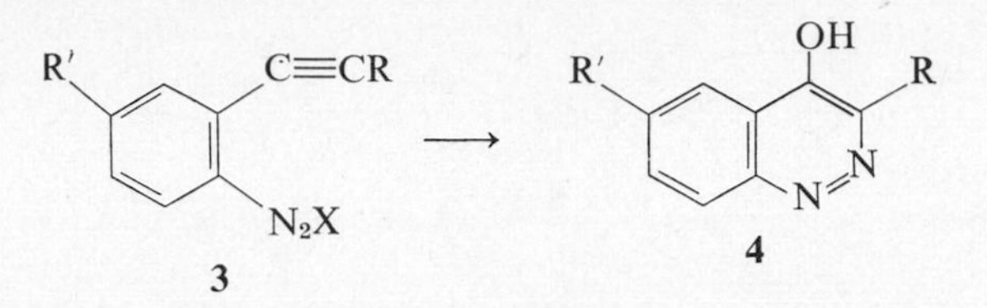

Nevertheless, the Richter synthesis has found only limited application. The 4-hydroxycinnolines that do not contain a 3-carboxylic acid group are more conveniently prepared via the Borsche synthesis (Section 1C-I-A-3) and those that contain a 3-carboxylic acid group may be obtained from the Friedel-Crafts cyclization of mesoxalyl chloride phenylhydrazones (Section 1C-I-A-4). Discussions of the mechanism of the Richter synthesis may be found in the reviews by Simpson (4) and by Jacobs (5).

2. *Pfannstiel and Janecke Synthesis*

A low yield of 5-chloro-3-phenyl-4-hydroxycinnoline (**6**) is realized when 6-chloro-2-hydrazinobenzoic acid (**5**) is refluxed with benzaldehyde. The principal product in this reaction is 4-chloroindazolone (**7**) (6).

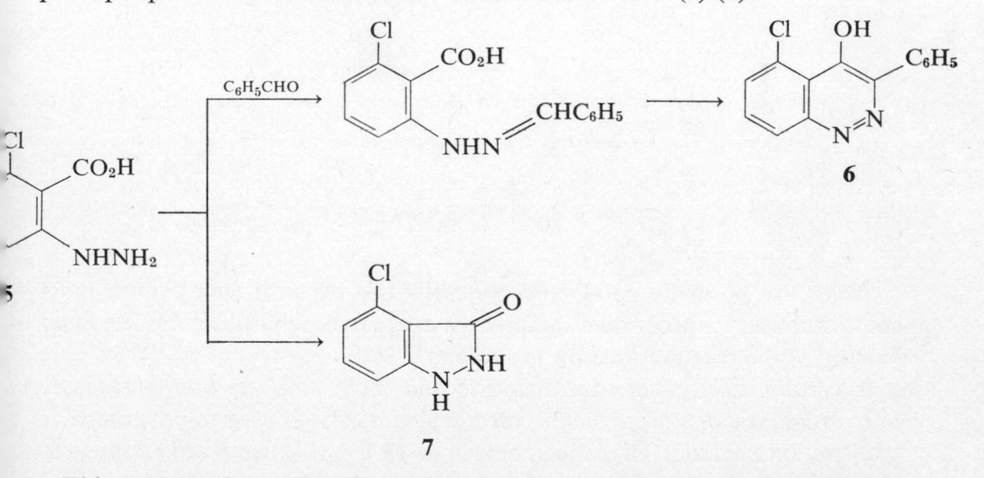

This preparation of compound **6** by the cyclization of benzaldehyde 2-carboxy-3-chlorophenylhydrazone is the only successful example of the preparation of a 4-hydroxycinnoline by this method. An attempt to extend this synthesis to the cyclization of 2-carboxy-5-chlorophenylhydrazones (**8**; X = $COOC_2H_5$, $COCH_3$, H) was met by a notable lack of success (7).

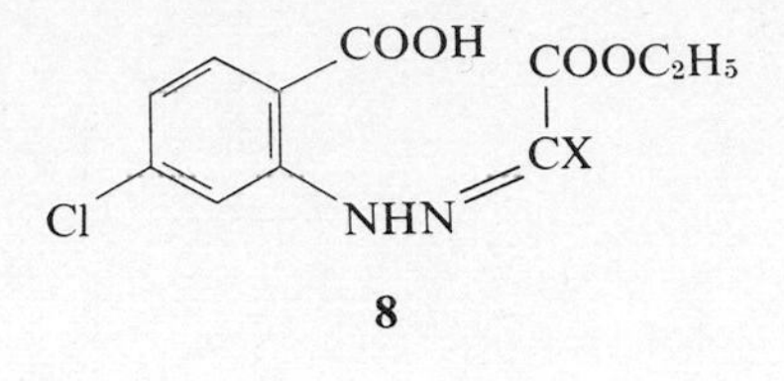

3. *Borsche Synthesis*

This synthesis, discovered in 1941 by Borsche and Herbert (8), is at present the most versatile and widely employed method for the preparation of 3-, 5-, 6-, 7-, and 8-substituted 4-hydroxycinnolines. Initially, the Borsche synthesis involved the diazotization of substituted 2-aminoacetophenones (**9**) by interaction with sodium nitrite in dilute or concentrated hydrochloric or sulfuric acid, followed by the spontaneous cyclization of the diazonium salt **10** to give the 4-hydroxycinnoline **11**.

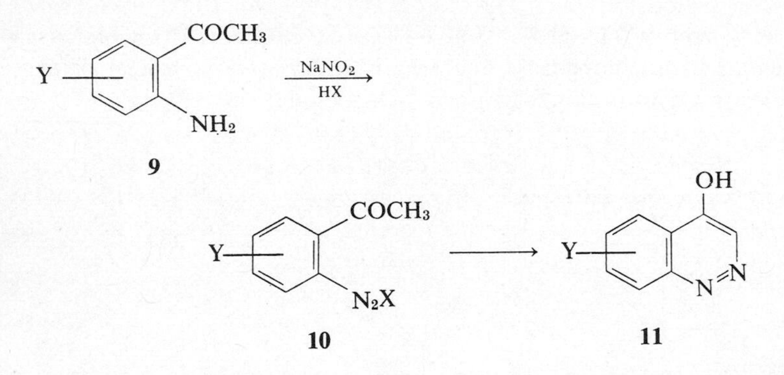

Yields are normally good, and generally the reaction seems to be most successful when concentrated rather than dilute hydrochloric acid is used and heating of the reaction mixture is avoided.

In certain cases, where diazotization and cyclization are not satisfactory in hydrochloric or sulfuric acids, formic acid has been used to advantage as the reaction medium. Thus the azobenzene **12** is not satisfactorily converted into 4-hydroxy-6-phenylazocinnoline (**13**) in hydrochloric or sulfuric acids of various strengths, but this conversion occurs smoothly in formic acid (9).

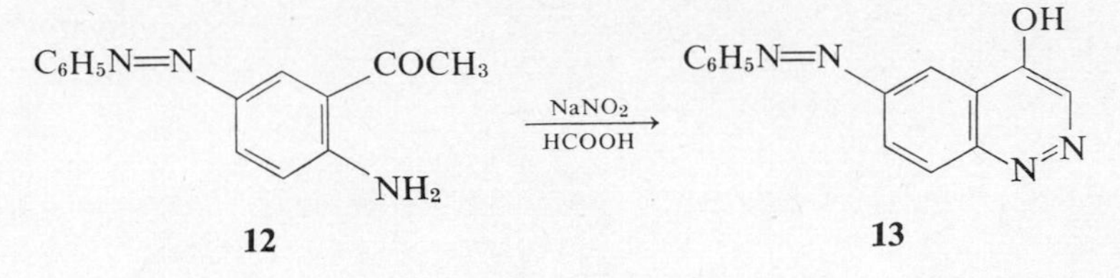

In some cases, group exchange has been observed. For example, diazotization and cyclization of 4-chloro-5-nitro-2-aminoacetophenone in hydrochloric acid yields 6,7-dichloro-4-hydroxycinnoline (10).

The Borsche reaction has been extended to include the diazotization and cyclization of ω-halogeno-*o*-aminoacetophenones and other *o*-aminoaryl ketones (**14**; R = halogen, alkyl, or aryl) to give 3-substituted 4-hydroxycinnolines (**15**) (11, 12).

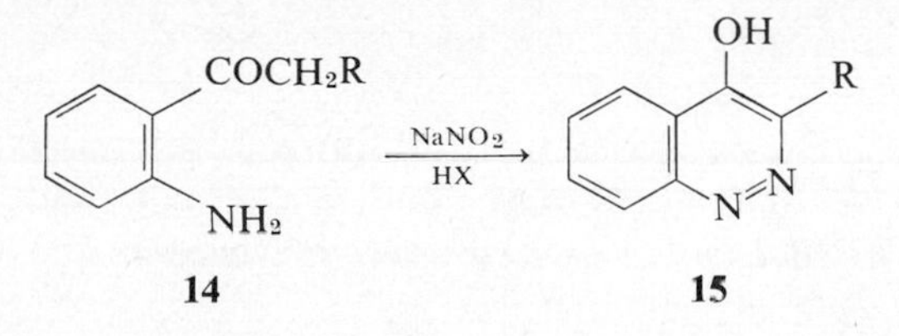

In the diazotization and cyclization of ω-halogeno-*o*-aminoacetophenones (**14**; R = halogen) an exchange of halogen atoms may occur if the reaction is carried out in a halogen acid different from that corresponding to the original ω-halogen atom of **14** (12).

When *o*-aminophenyl benzyl ketones (**16**; R = H, CH_3, NO_2) are diazotized in concentrated hydrochloric acid, subsequent cyclization may occur in two ways, one path leading to a 4-hydroxy-3-phenylcinnoline (**17**), the second path leading via a Pschorr reaction to a 9-phenanthrol (**18**). Under a

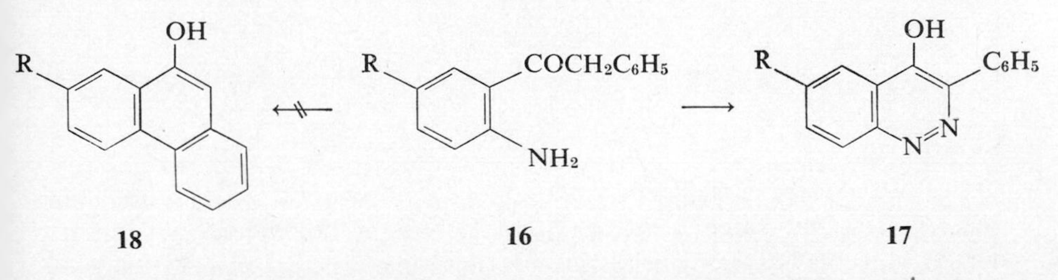

variety of conditions appropriate to the Pschorr reaction, the cinnoline is the only isolable product, being obtained in yields of 61 to 90% (11).

The mechanism of the Borsche synthesis, which is thought to proceed through the enol form of the ketone, is discussed in the review by Simpson (4), and 4-hydroxycinnolines synthesized by this reaction before 1949 are tabulated in this review. The 4-hydroxycinnolines which have been prepared in the same manner since 1949 are listed in Table 1C-5.

4. *Friedel-Crafts Cyclization of Mesoxalyl Chloride Phenylhydrazones*

This method, first described by Barber and co-workers in 1956 (13), starts from readily accessible diethyl mesoxalate phenylhydrazones (**19**; Y = CH_3, OCH_3, NO_2, Cl, Br, F); the products are 4-hydroxycinnoline-3-carboxylic acids (**22**), which on decarboxylation give 4-hydroxycinnolines (**11**).

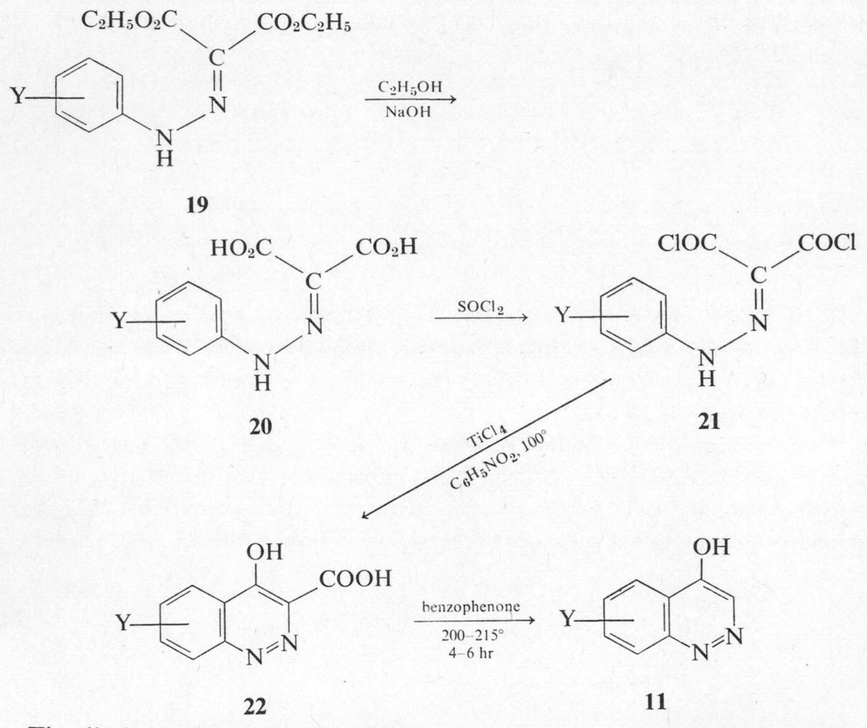

The diethyl mesoxalate phenylhydrazones **19** may be obtained by coupling diazotized aromatic amines with diethyl malonate in the presence of sodium acetate, and they are assigned a phenylhydrazone structure rather than the tautomeric azo structure on the basis of chemical and spectroscopic evidence (13).

The diesters **19** are hydrolyzed to the diacids **20** in boiling ethanolic sodium hydroxide. In certain cases hydrolysis may be followed by decarboxylation. Thus two diesters (**19**; Y = *p*-OH and *p*-$C_6H_5CH_2O$—) undergo substantial decarboxylation during hydrolysis, even at 25°, and three diacids (**20**; Y = *p*-CH_3O—, *p*-CH_3CONH—, and *p*-C_2H_5OOCNH—) show similar instability on attempted recrystallization.

The mesoxalyl chloride phenylhydrazones **21** are moderately stable crystalline materials which may be prepared by heating the diacids **20** with thionyl chloride in an inert solvent. Phosphorus pentachloride is effective in certain cases where reaction with thionyl chloride is sluggish.

Cyclization of the acid chlorides **21** to the cinnolines **22** may be effected by heating the acid chlorides with titanium tetrachloride in nitrobenzene at 100° and hydrolyzing the products in dilute alkali. The 4-hydroxycinnoline-3-carboxylic acids **22** may then be decarboxylated in benzophenone or Dowtherm at 200–215° to give the corresponding 4-hydroxycinnolines (**11**).

In this manner a variety of substituted 4-hydroxycinnoline-3-carboxylic acids (**22**) and 4-hydroxycinnolines (**11**) have been prepared. Overall yields vary widely, 5,8-dichloro-3-carboxy-4-hydroxycinnoline being obtained in only 11% yield whereas 8-methyl-3-carboxy-4-hydroxycinnoline is obtained in 92% yield (13).

When the mesoxalyl chloride phenylhydrazone carries a substituent in the 2- or 4-position, cyclization will yield only one 4-hydroxycinnoline (Eqs. 1 and 2), but when the substituent is in the 3-position, two isomeric 4-hydroxycinnolines may theoretically be obtained (Eq. 3).

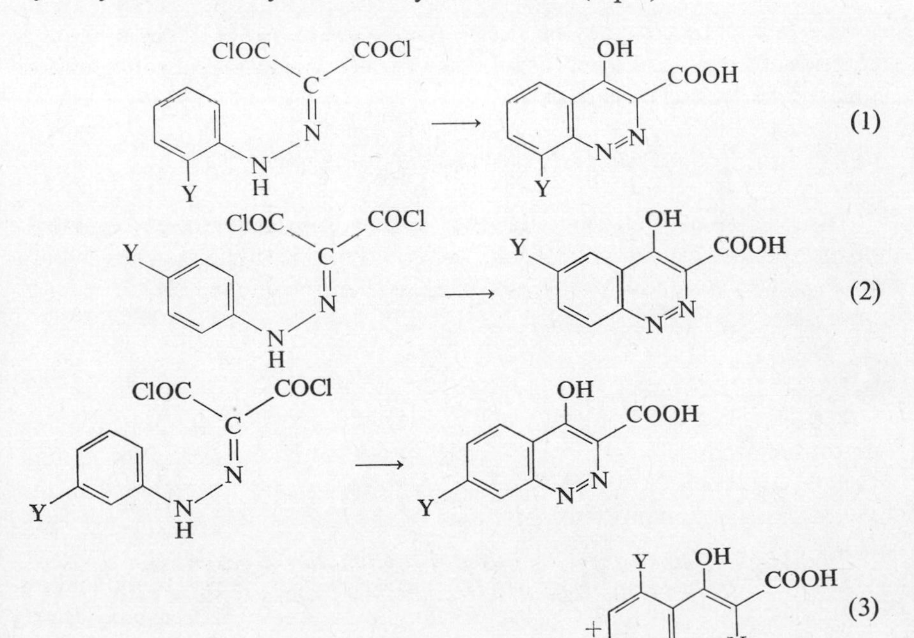

This imposes a limitation on the use of this reaction for the preparation of 5- or 7-substituted 4-hydroxycinnolines, since the reaction product may be a mixture of both isomers; in fact, Shoup and Castle (14) demonstrated that a mixture of 5-chloro-4-hydroxy- and 7-chloro-4-hydroxycinnoline is obtained from the cyclization of mesoxalyl chloride 3-chlorophenylhydrazone. They similarly obtained a mixture of 5-bromo-4-hydroxy- and 7-bromo-4-hydroxycinnoline from mesoxalyl chloride 3-bromophenylhydrazone, but only 7-fluoro-4-hydroxycinnoline is obtained from the cyclization of mesoxalyl chloride 3-fluorophenylhydrazone.

Attempts to extend the synthesis to the cyclization of glyoxylyl chlorides (**23**; Z = H, CN, NO_2, $CO_2C_2H_5$; Y = H, *p*-Cl, *p*-OCH_3, *p*-CH_3, *p*-$C_6H_5CH_2O$, *p*-CH_3CONH) were unsuccessful, except in the case of **23** (Z = $CO_2C_2H_5$, Y = *p*-CH_3), which gave 4-hydroxy-6-methylcinnoline-3-carboxylic acid instead of the expected 4-hydroxy-3-carbethoxy-6-methylcinnoline (13).

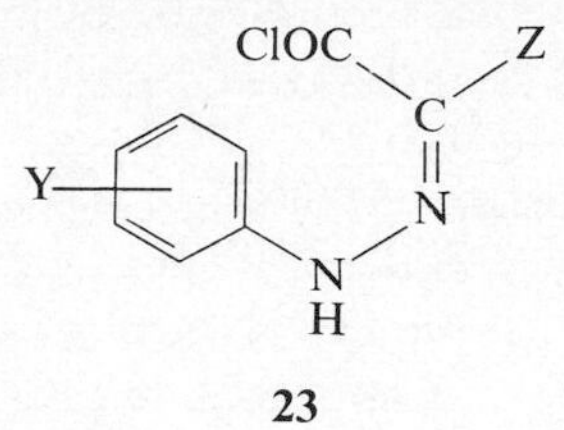

23

This lack of success may be due in part to the existence of the monoacid chlorides **23** in *cis* and *trans* forms, one of which would have a configuration unfavorable to ring closure (13).

5. *Hydrolysis of 4-Amino-3-nitrocinnolines*

The reaction of 3-nitrocinnoline (**24**) with hydroxylamine gives 4-amino-3-nitrocinnoline (**25**) in 50–62% yield. When warmed in dilute aqueous sodium hydroxide solution, this produces 4-hydroxy-3-nitrocinnoline (**26**) in 75–83% yield (15).

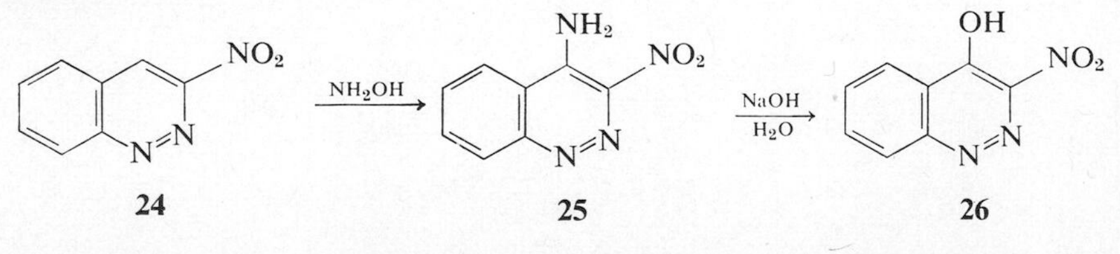

24 **25** **26**

6. *High-Temperature Reaction of a Cinnoline-4-carboxylic Acid with Potassium Hydroxide and Cupric Oxide*

Fusion of the potassium salt of 3-phenylcinnoline-4-carboxylic acid in a manner similar to the preparation of phenols from sulfonic acids gives only a

5% yield of 3-phenyl-4-cinnolinol, but when 3-phenylcinnoline-4-carboxylic acid (**27**) is heated to 200° with potassium hydroxide and cupric oxide in mineral oil or other inert solvent, the distillate, collected above 200°, contains 62% of 4-hydroxy-3-phenylcinnoline (**28**) and 6% of 3-phenylcinnoline (**29**) (16).

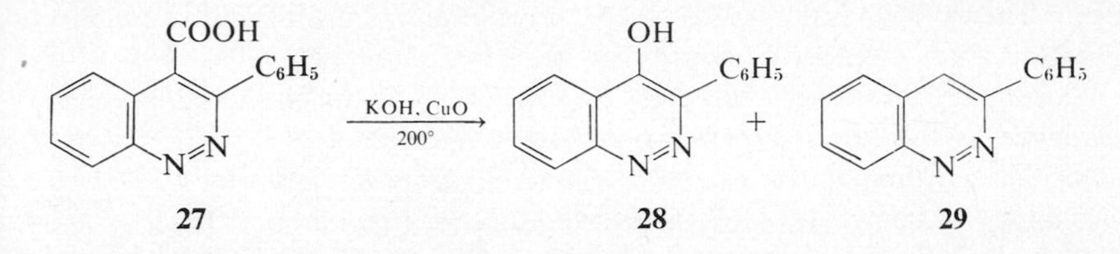

When 3-phenylcinnoline itself is subjected to the same conditions, only 9% of 4-hydroxy-3-phenylcinnoline is obtained, with a 74% recovery of 3-phenylcinnoline. This suggests that initial decarboxylation is not the principal route in the conversion of the cinnoline-4-carboxylic acid (**27**) to the 4-hydroxycinnoline (**28**) (16).

This reaction is severely limited by the strenuous conditions employed, which may affect other substituent groups. Thus 3-(4-chlorophenyl)cinnoline-4-carboxylic acid (**30**) is converted into 4-hydroxy-3-(4-hydroxyphenyl)-cinnoline (**31**) in low yield, and 3-(4-methoxyphenyl)cinnoline-4-carboxylic acid (**32**) is converted into a low-yield mixture of compound (**31**) and 3-(4-methoxyphenyl)cinnoline (**33**) (16).

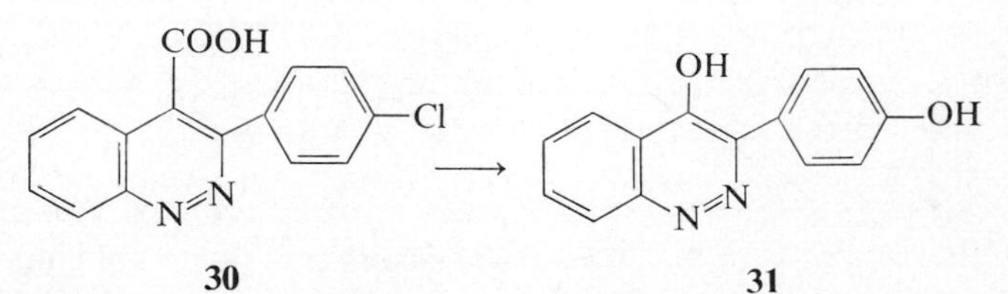

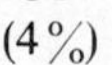

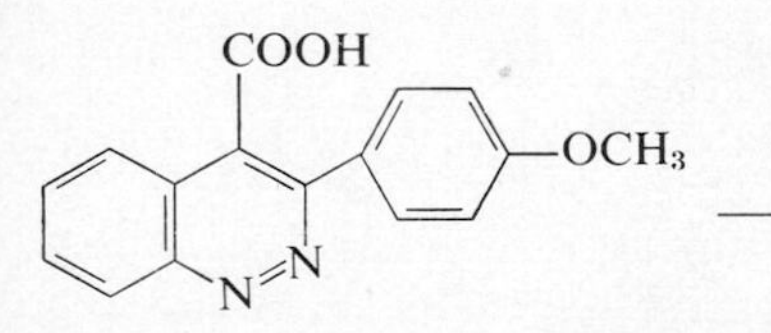

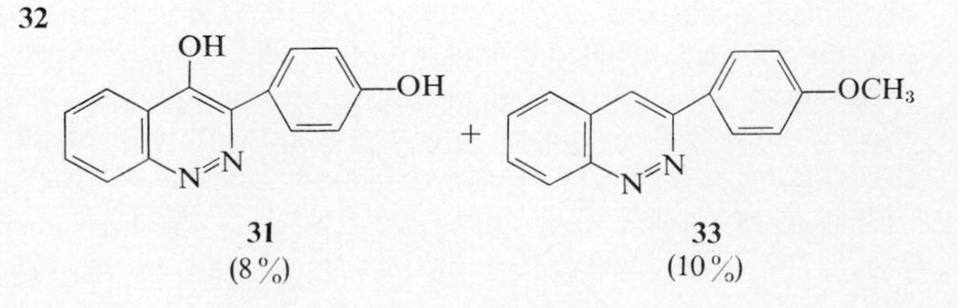

7. *Reaction of 3-Phenylcinnoline with t-Butyl Hypochlorite or Phosphorus Oxychloride-Dimethylformamide*

When a solution of 3-phenylcinnoline (**29**), *t*-butyl hypochlorite, and methylene chloride is allowed to stand at room temperature for 20 hr there is obtained a 2.5% yield of 4-hydroxy-3-phenylcinnoline (**28**) (16).

During an unsuccessful attempt to prepare 3-phenylcinnoline-4-carboxaldehyde by allowing a solution of 3-phenylcinnoline, phosphorus oxychloride, and dimethylformamide to stand at room temperature for 3 days, 4-hydroxy-3-phenylcinnoline was produced in a small amount (16).

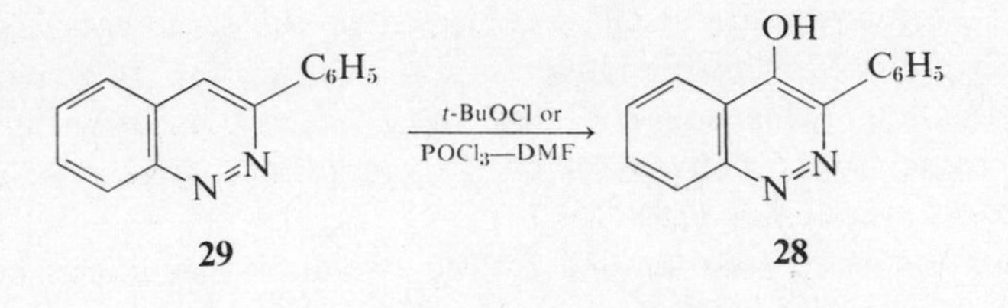

8. *Hydrolysis of 4-Chloro-, 4-Methylsulfonyl-, and 4-Methylsulfinylcinnolines*

4-Chloro-, 4-methylsulfonyl-, (16a, 16b) and 4-methylsulfinylcinnolines (16c) are readily hydrolyzed to 4-hydroxycinnolines in water, dilute acid, or base. Hydrolysis of 4-methylsulfinylcinnoline, however, is accompanied by formation of 4-methylthiocinnoline in significant amount.

B. 3-Hydroxycinnolines

1. *Neber-Bossel Synthesis*

The first preparation of 3-hydroxycinnoline was described by Bossel (17) in his Inaugural Dissertation at Tübingen in 1925. The compound was mentioned again in 1929 by Neber (18), who described its reduction by phosphorus and hydriodic acid to oxindole, citing Bossel's work for its preparation. The Neber-Bossel synthesis apparently was not used until 1952 when Alford and Schofield (19), working with an abstract of Bossel's dissertation, showed the reaction to be a practical method for preparing 3-hydroxycinnoline.

In the Neber-Bossel synthesis an aqueous solution of sodium *o*-aminomandelate (**34**) and sodium nitrite is added to concentrated hydrochloric acid at 0°, and the resulting diazonium salt (**35**) is then reduced to *o*-hydrazinomandelic acid (**36**) by stannous chloride. The hydrazine **36** is then cyclized by boiling in aqueous hydrochloric acid to give 3-hydroxycinnoline (**37**) in 59–62% yields (19).

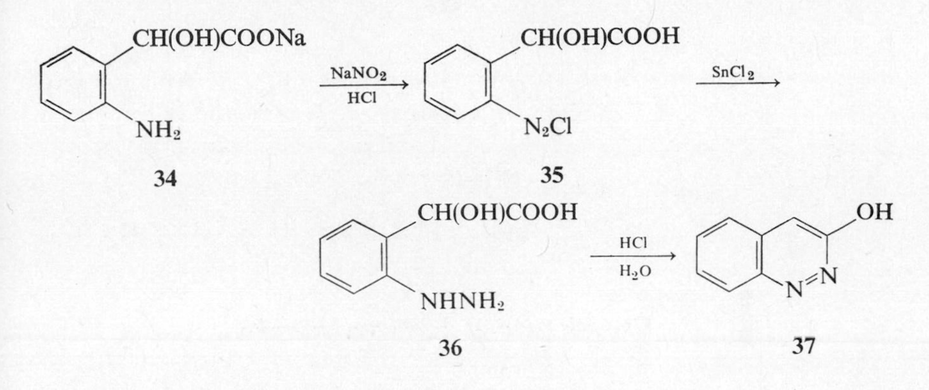

o-Aminomandelic acid itself cannot be diazotized directly in acid solution because it is rapidly converted into dioxindole (**38**).

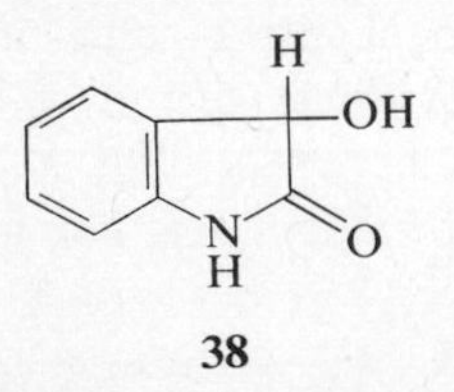

A low yield of 6-chloro-3-hydroxycinnoline is obtained in this same manner from sodium 2-amino-5-chloromandelate, and 3,6-dihydroxycinnoline is obtained in approximately 20% yield from sodium 2-amino-5-methoxymandelate, demethylation of the methoxy group occurring concomitantly with cyclization (19).

When α-substituted potassium *o*-aminomandelates (**39**; R = C_6H_5 or $C_6H_5CH_2$) are diazotized in hydrochloric acid at 0° and then reduced by stannous chloride in hydrochloric acid at 0°, 3-substituted 1-aminodioxindoles (**40**) are produced on warming the reaction mixture to 5–10°. The 1-aminodioxindoles may be hydrolyzed in hot, dilute, aqueous base to give the *o*-hydrazinomandelate derivatives (**41**), which will cyclize to the 4-substituted 3-hydroxycinnolines (**42**; R = C_6H_5 or $C_6H_5CH_2$) upon neutralization of the reaction mixture with hydrochloric acid (20).

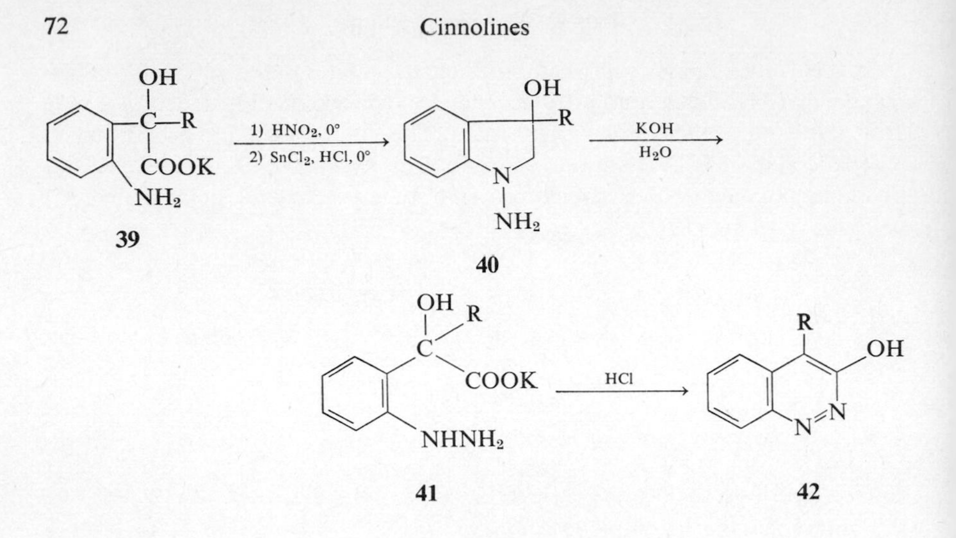

2. *Diazotization of 3-Aminocinnolines*

Diazotization of 3-aminocinnolines (**43**; R = H or CH_3; Y = H, 6-Cl or 7-Cl) in a dilute mineral acid solution gives 3-hydroxycinnolines (**44**) in 40–86% yields. When the diazotization is performed in concentrated hydrochloric or hydrobromic acid, 3-chloro- or 3-bromocinnolines (**45**; X = Cl or Br) are formed (18–53%) along with the 3-hydroxycinnolines (29–40%) (21).

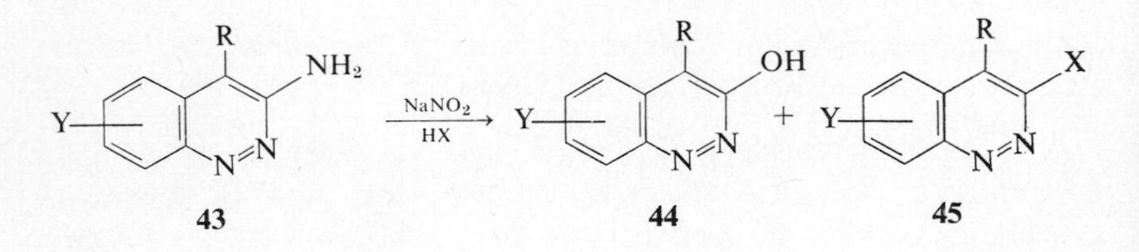

3. *Oxidation of 1-Aminoöxindoles*

Treatment of 1-aminoöxindole (**46**) in benzene with lead tetraacetate at room temperature for 2 hr gives 3-hydroxycinnoline (**37**) in 63% yield (22), while the interaction of equimolar amounts of the oxindole and *t*-butyl hypochlorite in dry benzene for 25 min at room temperature gives the cinnolinol in nearly quantitative yield (23, 23a).

It is important that an excess of *t*-butyl hypochlorite is not used in the

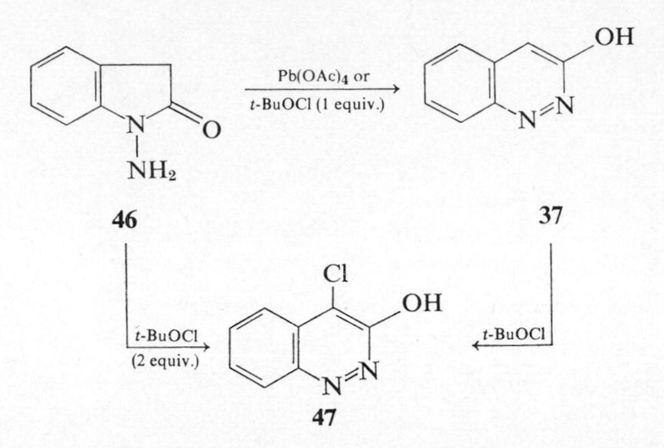

conversion of **46** to **37** since an excess of this reagent causes chlorination of the cinnoline ring in the 4-position. Thus 3-hydroxycinnoline is converted into 4-chloro-3-hydroxycinnoline (**47**) in good yield by treatment with *t*-butyl chloride in benzene, while similar treatment of one equivalent of 1-aminoöxindole itself with two equivalents of *t*-butyl hypochlorite results in a high yield of 4-chloro-3-hydroxycinnoline (**47**) (23, 23a).

Molecular chlorine also may be used to convert 1-aminoöxindoles into 3-hydroxycinnolines, as demonstrated by the conversion of 4-chloro- and 6-chloro-1-aminoöxindole into the corresponding 5-chloro- and 7-chloro-3-hydroxycinnolines by bubbling chlorine gas into a benzene solution of the 1-aminoöxindoles at room temperature. The reaction with molecular chlorine, however, is not as clean and as convenient as the reaction with *t*-butyl hypochlorite (23).

A mechanism for the oxidative rearrangement of 1-aminoöxindoles involving a nitrene intermediate has been proposed (22, 23). The 1-aminoöxindole **46** is first oxidized to a nitrene (**48**) which rearranges via an electron shift (path *a* or *b*) to the 3-cinnolone **49**. The cinnolone may then tautomerize to the 3-cinnolinol **37**. Baumgarten (23a), however, cautions that there is no evidence for the existence of the nitrene **48** and that oxidation of 1-amino-öxindole with *t*-butyl hypochlorite or chlorine might proceed through 1-(*N*-chloroamino)oxindole or 1-(*N*,*N*-dichloroamino)oxindole, both of which could lose chlorine simultaneously with the rearrangement of the oxindole to the cinnolinol. Postulation of a nitrene intermediate is then unnecessary. This same argument, but involving a lead triacetate group bonded through the lead atom to the 1-amino substituent of the oxindole, applies to the oxidative rearrangement caused by lead tetraacetate.

Whether or not this reaction can be developed into a general synthesis of 3-hydroxycinnolines appears to depend upon the availability of substituted 1-aminoöxindoles which at present are relatively inaccessible compounds.

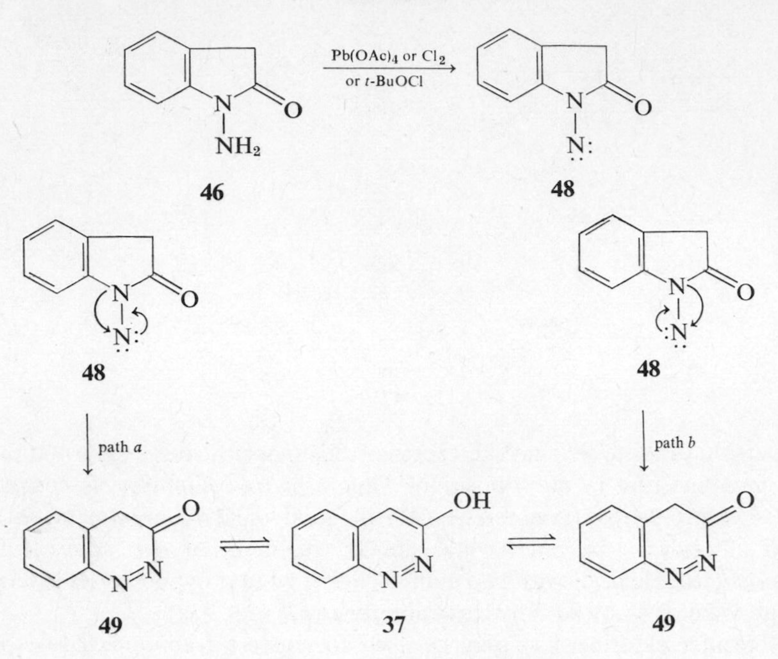

C. 5-, 6-, 7-, and 8-Hydroxycinnolines

All of these compounds have been prepared via the Borsche synthesis, starting with a 2-aminoacetophenone (**50**) which is substituted with a methoxy group in the 3-, 4-, 5-, or 6-position (24, 25). The aminoacetophenone is diazotized and cyclized by interaction with sodium nitrite in concentrated hydrochloric acid to give the Borsche product, 5-, 6-, 7-, or 8-methoxy-4-hydroxycinnoline (**51**), which is then converted into the 4-chlorocinnoline **52** by treatment with refluxing phosphorus oxychloride. The 4-chloro derivatives are then converted to the 4-toluene-*p*-sulfonylhydrazino derivatives **53** by replacement of the chloro group with toluene-*p*-sulfonylhydrazide in chloroform. Alkaline decomposition of **53** gives the methoxycinnolines **54**. The 5-, 6-, and 7-methoxycinnolines are stable crystalline solids, whereas the 8-methoxy isomer quickly liquefies and becomes green in air. All are demethylated by hydrobromic acid to give the 5-, 6-, 7-, and 8-hydroxycinnolines **55**. Overall yields are good. The 5-, 6-, and 7-hydroxycinnolines are crystalline solids, melting significantly higher than 8-hydroxycinnoline. The 8-hydroxy isomer and the 5-hydroxy isomer give a red dye with benzenediazonium chloride, but diazo coupling is not observed with the 6- and 7-isomers (25).

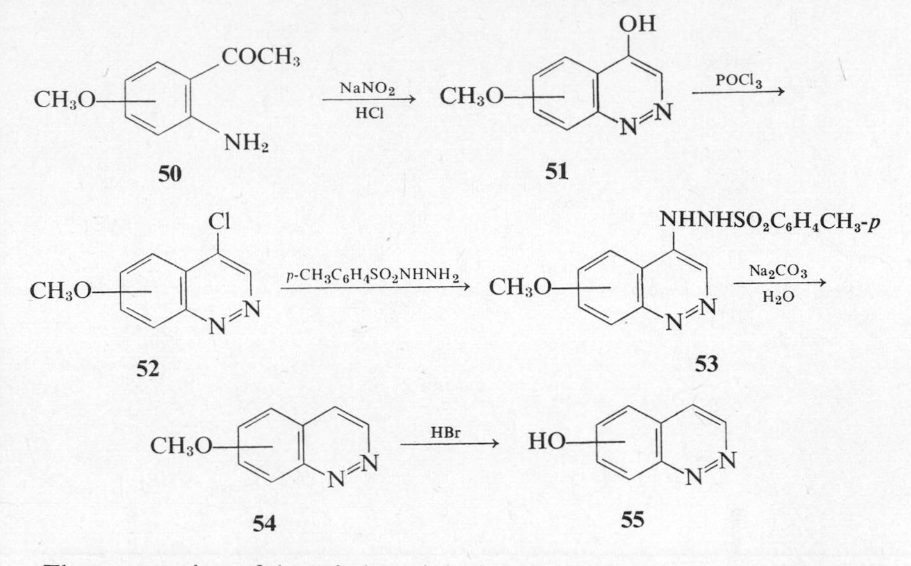

The preparation of 4-methyl- and 4-phenyl-8-hydroxycinnolines has been accomplished by causing 2-amino-3-methoxyacetophenone (**56**) to react with methylmagnesium iodide or phenylmagnesium bromide to give the carbinols **57** ($R = CH_3$ or C_6H_5), which are readily dehydrated. Diazotization and cyclization of the ethylenes **58** give the 4-substituted 8-methoxycinnolines **59**, which can be converted by hydrobromic acid into the corresponding 8-hydroxy derivatives (**60**; $R = CH_3$ or C_6H_5) (24).

Alternatively, 8-hydroxy-4-methylcinnoline is obtained in good yield from methyl 3-methoxyanthranilate (**61**) (26) by a series of transformations which are quite similar to those just described for the conversion of the acetophenone (**56**) to 8-hydroxy-4-methylcinnoline (**60**; $R = CH_3$). This alternative sequence, shown in Eq. 4, is obviously unsuitable for the preparation of the 4-phenyl derivative. The 4-methyl group may then be removed by condensation of the hydrochloride of 8-hydroxy-4-methylcinnoline (**62**) with benzaldehyde in the presence of dry gaseous hydrochloric acid, oxidation of the resulting styryl intermediate (**63**) by potassium permanganate in pyridine to give the 4-carboxylic acid derivative (**64**) and then thermal decarboxylation of the acid to give 8-hydroxycinnoline (**65**) (26).

II. Physical Properties and Structure

All of the hydroxycinnolines that have been prepared up to the present time are high-melting crystalline compounds. The 4-hydroxycinnolines

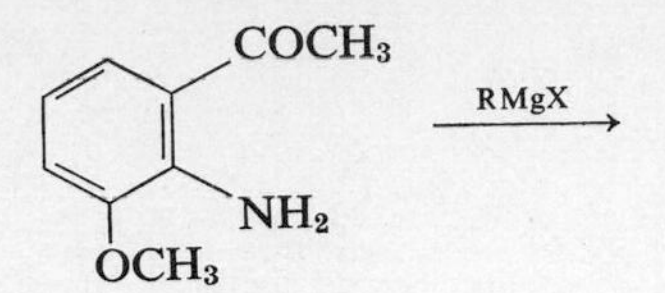

$COCH_3$
RMgX
NH_2
OCH_3

56

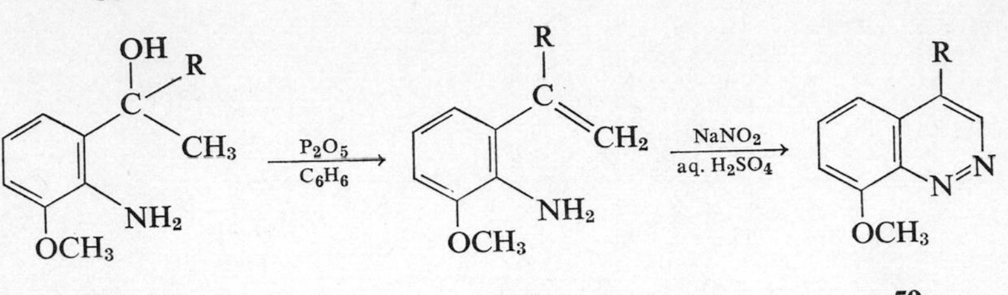

OH
R
C
CH_3
P_2O_5
C_6H_6
NH_2
OCH_3
57
R
C
CH_2
$NaNO_2$
aq. H_2SO_4
NH_2
OCH_3
58

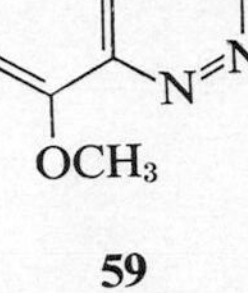

R
N=N
OCH_3
59

HBr

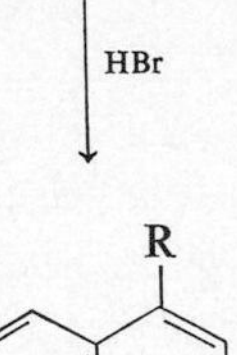

R
N=N
OH

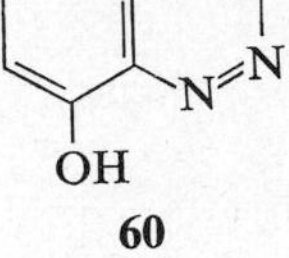

60

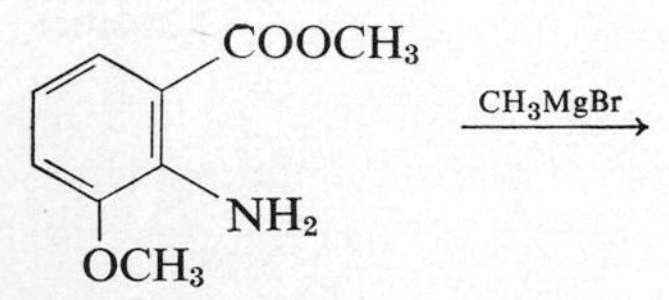

$COOCH_3$
CH_3MgBr
NH_2
OCH_3
61

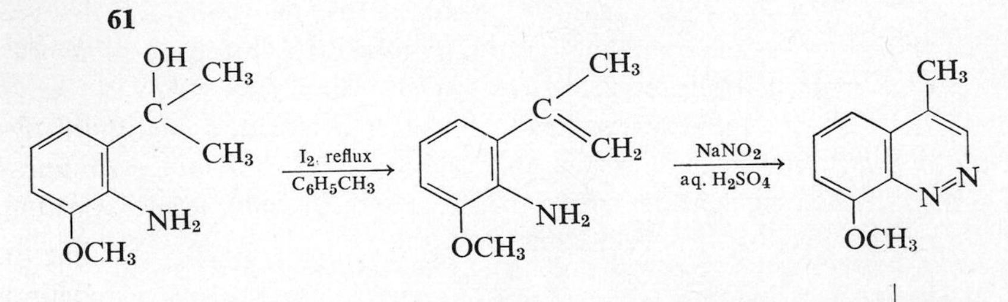

OH
CH_3
C
CH_3
I_2, reflux
$C_6H_5CH_3$
NH_2
OCH_3
CH_3
C
CH_2
$NaNO_2$
aq. H_2SO_4
NH_2
OCH_3
CH_3
N=N
OCH_3

HBr

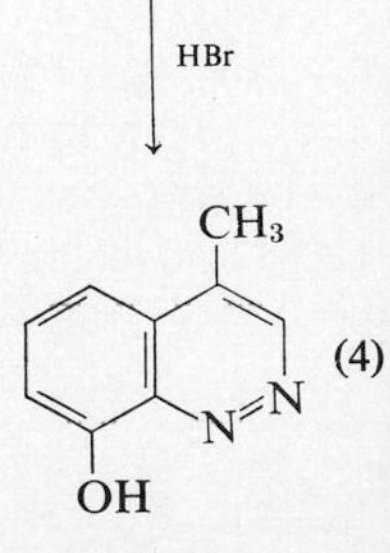

CH_3
N=N
OH

(4)

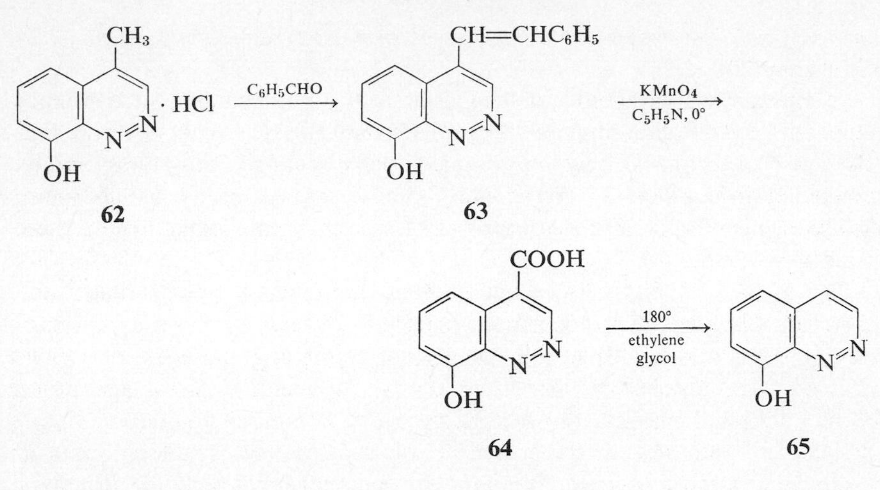

characterized before 1949 are listed by Simpson (4) and those characterized from 1949 to the present time are given in Table 1C-5; Tables 1C-6 and 1C-7 list the melting points and preparations of the 3-, 5-, 6-, 7-, and 8-hydroxy-cinnolines.

The hydroxycinnolines show acidic properties, dissolving in aqueous sodium hydroxide solution, except that 5-chloro-3-phenyl-4-hydroxycinnoline is reported to be insoluble in this reagent (6). Strongly electron-attracting groups, such as nitro or cyano, when placed in the 6-position of the 4-hydroxycinnoline ring, cause these compounds to be soluble in aqueous sodium carbonate (2, 27).

The order of acidity for hydroxycinnolines which are unsubstituted except for the hydroxyl group itself is as follows: 5- > 6- > 7- > 8- > 3- > 4-hydroxycinnoline. Exact pK_a values are listed in Table 1C-4. Thus it is seen that 5-hydroxycinnoline (pK_a 7.40 in water) is the strongest acid in the series and that 4-hydroxycinnoline (pK_a 9.27 in water) is the weakest acid. All are therefore stronger acids than phenol (pK_a 9.98) and the naphthols (28), and each is a stronger acid than the corresponding hydroxyquinoline except that 3-hydroxyquinoline is reported to be a somewhat stronger acid than 3-hydroxycinnoline (29).

Basic strengths of the hydroxycinnolines (Table 1C-3) reveal the following order of basicity for the compounds which are unsubstituted except for the hydroxyl group: 6- > 7- > 8- > 5- > 3- > 4-hydroxycinnoline. The strongest base is 6-hydroxycinnoline (pK_a 3.65 in water) and the weakest base is 4-hydroxycinnoline (pK_a −0.35 in water). The introduction of one additional cyclic nitrogen atom into a hydroxyquinoline is invariably base-weakening (30), and each of the hydroxycinnolines is a weaker base than the corresponding hydroxyquinoline. For comparative purposes 6-hydroxyquinoline has pK_a 5.17 (in water) as a base and 4-hydroxyquinoline has pK_a 2.27 (in water) (29).

In considering the lability of ring protons, a proton magnetic resonance study of 4-hydroxycinnoline in 98% dideuterium sulfate at 160° C shows that the 8-proton is most rapidly exchanged, the exchange being substantially completed in less than 15 hours. The 5- and 6-protons are exchanged more slowly, while the 3- and 7-protons are essentially unaffected under these conditions (30a).

The 5-, 6-, 7-, and 8-hydroxycinnolines are phenolic in character and therefore the ultraviolet absorption spectrum of each of these compounds closely resembles the spectrum of its corresponding methoxy derivative (24, 28). The absorption curves for the 6-, 7-, and 8-hydroxy derivatives show additional maxima above 400 mμ when compared to their methoxy derivatives, and these are ascribed to the existence of small amounts of zwitterionic structures in equilibrium with the uncharged phenolic structures (Eqs. 5–7) (28).

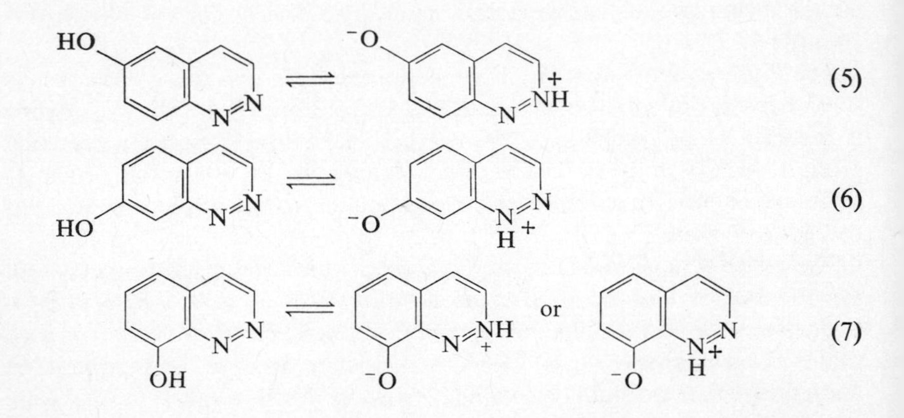

The site that accepts the proton in the zwitterionic form of 6-hydroxycinnoline (Eq. 5) is N-2, while the site in 7-hydroxycinnoline (Eq. 6) is N-1, these assignments arising from a study of dissociation constants (28). The site of protonation of 8-hydroxycinnoline (Eq. 7) is uncertain (28), although it is claimed that N-1 is the proton-accepting site on the basis of IR absorption data (31). In the case of 5-hydroxycinnoline, a study of UV spectra and dissociation constants indicates that it is very unlikely that appreciable zwitterion formation occurs (28).

Both 4-hydroxycinnoline (**66**) and 3-hydroxycinnoline (**37**) possess principally the amide structure **66a** $\leftrightarrow$ **66b** and **37a** $\leftrightarrow$ **37b** respectively, both in the solid state and in solution (28, 32, 33).

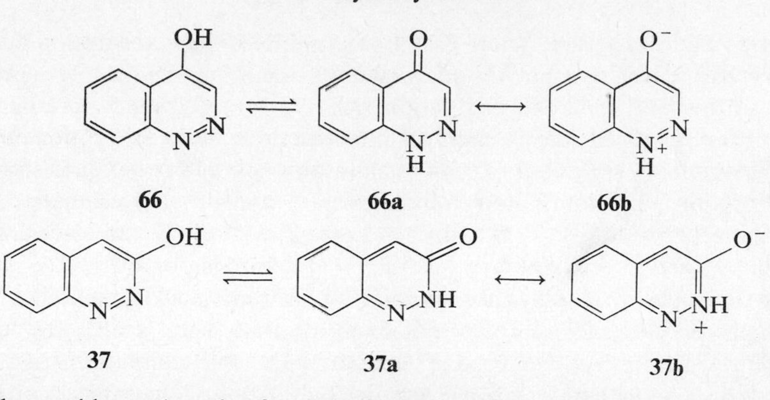

The amide:enol ratio for 4-hydroxycinnoline in aqueous solution is calculated to be 3600:1 (29) or 4000:1 (34) compared with 24,000:1 for 4-hydroxyquinoline (29). Recently Katritzky and co-workers (34a) determined that the amide:enol ratio for 3-hydroxycinnoline is 380:1. They attributed this relatively low value to destabilization of the amide form (**37a**) by its *ortho*-quinonoid structure in which the benzenoid character of the homocyclic ring is partially disrupted.

When heterocyclic compounds are substituted with a hydroxyl group α or γ to a ring-nitrogen atom, their UV absorption spectra are closely similar to those of their *N*-methyl derivatives and are different from the spectra of their *O*-methyl derivatives. This shows that these hydroxy compounds exist predominantly in the amide form (35). Thus the 4-hydroxycinnoline spectrum shows bands at about 285 and 295 mμ, which are characteristic of 1-alkyl-4-cinnolones (36), and the spectrum of 3-hydroxycinnoline resembles that of 2-methyl-3-cinnolone (32), showing in 95% ethanol an intense maximum near 230 mμ and a second weaker band near 400 mμ (21). Table 1C-2 lists the UV spectral parameters of several hydroxycinnolines.

Studies of ionization constants and the ultraviolet spectra of all the ionic forms of 4-hydroxycinnoline show clearly that the tautomeric form **66a** ↔ **66b** with hydrogen at N-1 is preferred to structure **67** with hydrogen at N-2 (37, 38). This is noteworthy because it is known that the principal product resulting from the alkylation of 4-hydroxycinnoline is the N-2 isomer **68** (Section 1C-III-E).

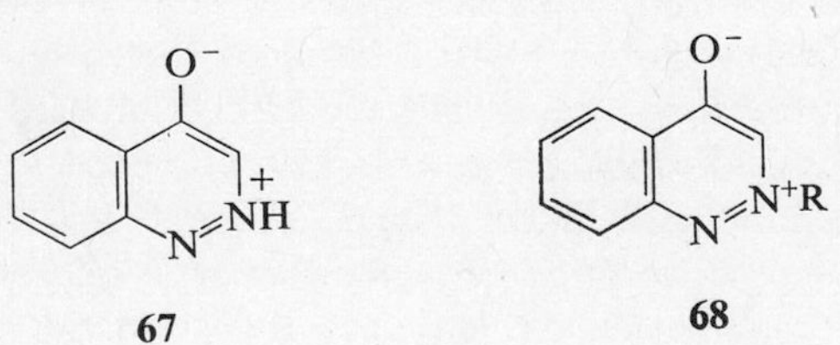

As would be expected, 3- and 4-hydroxycinnolines show absorption due to N—H and C=O, but not O—H, stretching vibrations in their IR spectra both in the solid state and in solution (33, 39, 40), whereas 5-, 6-, 7-, and 8-hydroxycinnolines, having mainly enolic structures, show absorption due to O—H stretching vibrations but fail to show absorption due to C=O stretching vibrations (33). An IR study indicates that 3- and 4-hydroxycinnolines are intermolecularly N—H · · · O hydrogen-bonded, while 5-, 6-, and 7-hydroxycinnolines are intermolecularly O—H · · · O hydrogen-bonded. The same study indicates that 8-hydroxycinnoline is intramolecularly O—H · · · N hydrogen-bonded, although UV data indicate that there is little hydrogen bonding between the 8-hydroxyl group and N-1 in this compound (24).

Although 5-, 6-, and 7-hydroxycinnoline fail to give a noticeable color with ferric chloride, 8-hydroxycinnoline is known to form chelate complexes with ferric ions and many other metal ions; however, these are generally less stable than the corresponding 8-hydroxyquinoline (oxine) complexes. The use of 8-hydroxycinnoline as a complexing reagent does not appear to offer any analytical advantage over oxine (30, 41, 42).

A mass spectral study (42a) has shown that fragmentation of 4-hydroxycinnoline occurs with the consecutive loss of two hydrogen cyanide molecules, carbon monoxide, and a hydrogen atom to give ions at m/e 119, 92, 64, and 63. In addition, 4-hydroxycinnoline fragments by an alternate pathway with an initial loss of a hydrogen atom.

III. Reactions

A. Metatheses

1. *Hydroxy- to Chlorocinnolines*

When 4-hydroxycinnolines are heated in phosphorus oxychloride or in a mixture of phosphorus oxychloride and phosphorus pentachloride, the corresponding 4-chloro derivatives are produced, usually in good yields (24, 25, 43, 44). In certain cases, the effectiveness with which phosphorus oxychloride causes this transformation seems to depend upon its condition, "aged" phosphorus oxychloride—which has been stored in contact with the atmosphere for some time—being more active than the freshly distilled material. Albert and Gledhill attributed this difference in activity to the presence of phosphoric acid in the "aged" reagent (45) and Baumgarten showed that in one case, addition of a small amount of phosphoric acid to fresh phosphorus oxychloride did enhance the reactivity of this reagent.

However, in a limited series of experiments, he found that the addition of phosphoric acid to fresh phosphorus oxychloride appeared to have no generally beneficial effect (15).

Nuclear substituents may occasionally suffer replacement during this reaction. Thus when 6-bromo-4-hydroxycinnoline (**69**) is heated with phosphorus oxychloride alone, the expected product, 6-bromo-4-chloro-cinnoline (**70**) is obtained, but when **69** is heated with a mixture of phosphorus oxychloride and phosphorus pentachloride, 4,6-dichlorocinnoline (**71**) is obtained (46). Similarly, 4-hydroxy-3-nitrocinnoline (**72**) yields the expected 4-chloro-3-nitrocinnoline **73** when treated with phosphorus oxychloride alone, but it yields 3,4-dichlorocinnoline (**74**) when treated with a mixture of phosphorus oxychloride and phosphorus pentachloride (15).

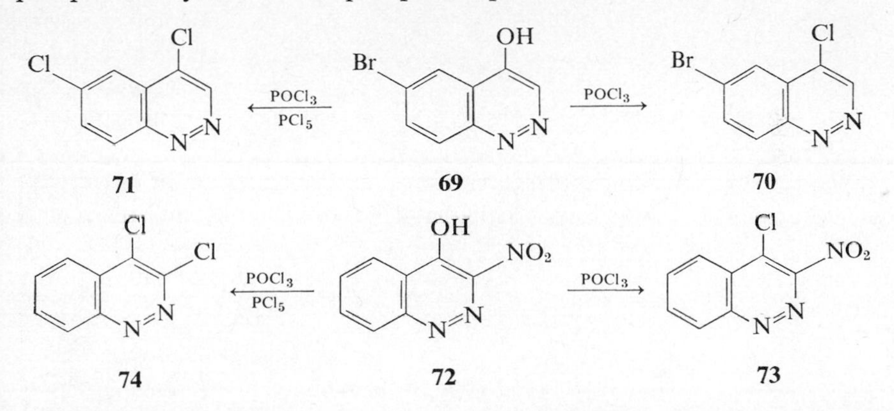

In addition to effecting the conversion of 4-hydroxycinnolines to their 4-chloro derivatives, phosphorus oxychloride also effects the conversion of 3-hydroxycinnoline to 3-chlorocinnoline in a low yield (32).

2. *Hydroxy- to Mercaptocinnolines*

Thiation of 4-hydroxycinnoline (**66**) may be carried out by allowing it to react with phosphorus pentasulfide in a refluxing pyridine solution to give 4-mercaptocinnoline (**75**) in nearly quantitative yield. The 6,7-dimethoxy derivative of 4-hydroxycinnoline gives 6,7-dimethoxy-4-mercaptocinnoline in somewhat poorer yield in the same manner (47).

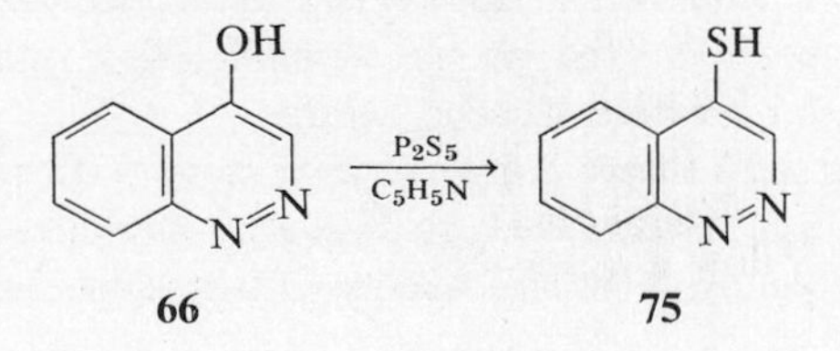

Treatment of 6-bromo- and 6-chloro-4-hydroxycinnoline (**76**; X = Br or Cl) with phosphorus pentasulfide in refluxing pyridine for 3 hr gives high yields of 4,6-dimercaptocinnoline (**77**). Subjecting 6-fluoro-4-hydroxycinnoline to these same conditions results in degradation of the starting material and only a small amount of a mixture of 4,6-dimercaptocinnoline and 6-fluoro-4-mercaptocinnoline is isolated. When pyridine is replaced by dry toluene as the solvent, a mixture of the monomercapto compound (**78**; X = Br or Cl) and 4,6-dimercaptocinnoline is obtained. Under these conditions a good yield of 6-fluoro-4-mercaptocinnoline is obtained from 6-fluoro-4-hydroxycinnoline.

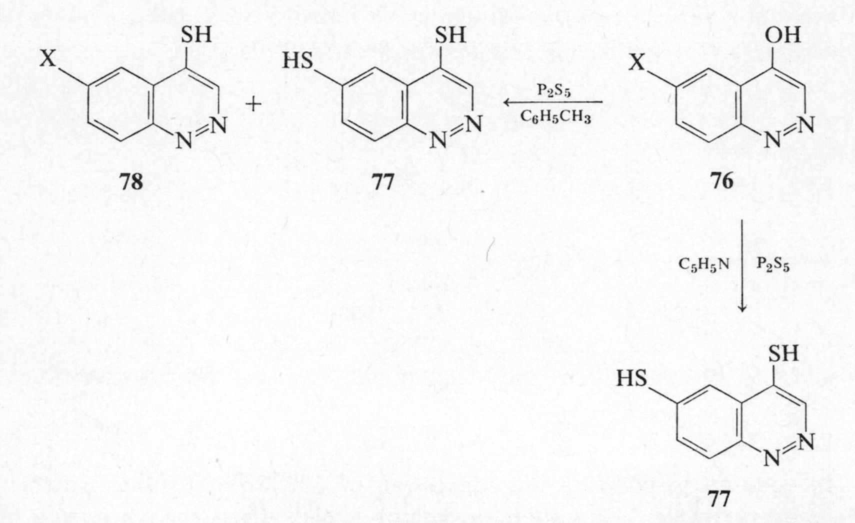

Displacement of a halogen atom from the 8-position of 4-hydroxycinnoline has not been observed, so that 8-fluoro-, 8-chloro-, and 8-bromo-4-mercaptocinnoline are obtained when the corresponding 8-halo-4-hydroxycinnolines are treated with phosphorus pentasulfide in boiling pyridine solution. Generally 7-halo substituents are also inert toward displacement, although a small amount of 4,7-dimercaptocinnoline was isolated in one instance from the reaction of 7-fluoro-4-hydroxycinnoline with phosphorus pentasulfide. Similar treatment of 5-halo-4-hydroxycinnolines in refluxing pyridine results in complete degradation of the organic material. However, 5-chloro-4-mercaptocinnoline is obtained in good yield from 5-chloro-4-hydroxycinnoline when dry toluene is used as the solvent (48). A number of other 4-hydroxycinnolines have been thiated (48a).

Although attempts to thiate 3-hydroxycinnoline with phosphorus pentasulfide have been unsuccessful (48a), its N-2 alkyl derivatives may be thiated. Thus the intensely yellow 2-(tetra-*O*-acetyl-*β*-D-glucopyranosyl)-3-cinnolone

79 is converted to violet-colored 2-(tetra-*O*-acetyl-*β*-D-glucopyranosyl)-3-thiocinnolone (**80**) by interaction of **79** with phosphorus pentasulfide in boiling pyridine (49), and 2-methyl-3-cinnolone is likewise converted to 2-methyl-3-thiocinnolone in refluxing benzene (51).

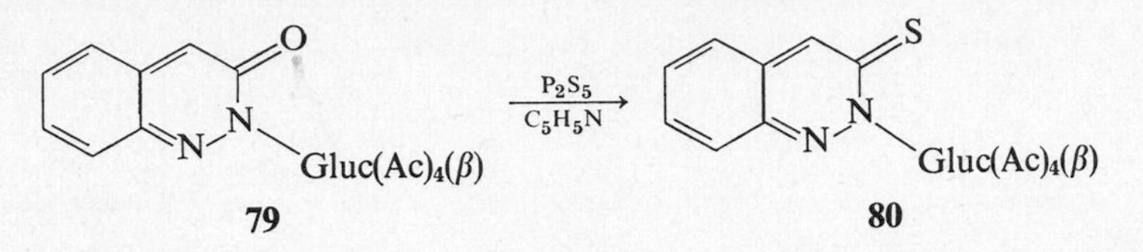

Both the N-1 and N-2 alkyl derivatives of 4-hydroxycinnoline are similarly converted to their 4-mercapto analogs. For example, 1,4-dihydro-1-methyl-4-cinnolone (**81**) is converted to 1,4-dihydro-1-methyl-4-thiocinnolone (**82**) the anhydro-base of 4-hydroxy-2-methylcinnolinium hydroxide (**83**) is while converted to its sulfur analog **84** (36, 37).

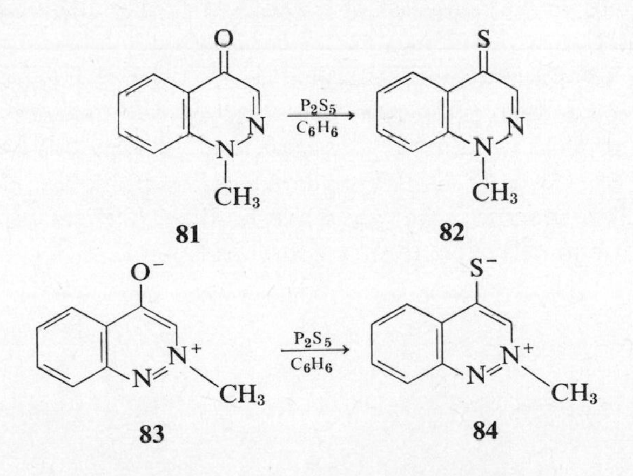

3. *Hydroxy- to Aminocinnolines*

Although 3-hydroxyquinoline is converted into 3-aminoquinoline by interaction with ammonium sulfite and aqueous ammonia at 130–140°, 3-hydroxycinnoline is unchanged by this process (32). The reaction is successful with 5-, 6-, 7-, and 8-hydroxycinnoline, converting these materials into their corresponding amino derivatives in fair yields (Eq. 8) (24, 25).

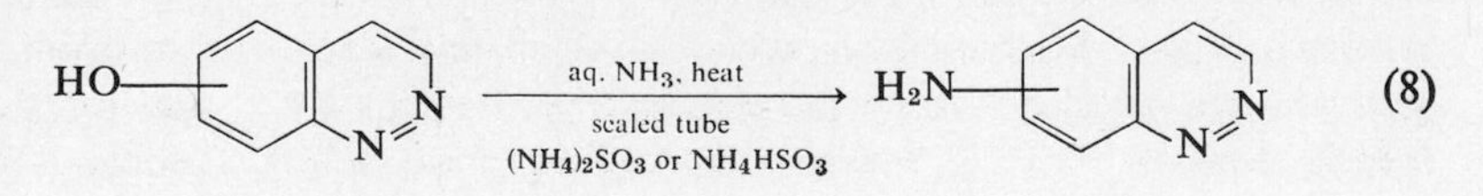

B. Electrophilic Substitution

1. *Halogenation*

Both sulfuryl chloride and bromine in acetic acid will directly halogenate 4-hydroxy-, 6-chloro-4-hydroxy-, and 6-bromo-4-hydroxycinnoline to give the corresponding 3-halogeno-4-hydroxycinnolines in about 20% yield (50). Ames (36) reasoned that bromination in acetic acid is impeded by protonation of the cinnoline by the hydrobromic acid which is formed during the reaction. Accordingly, he demonstrated that the bromination of 4-hydroxycinnoline (**66**) can be effected almost quantitatively by addition of bromine to an alkaline solution of 4-hydroxycinnoline. The product is 3-bromo-4-hydroxycinnoline (**85**). A similar yield of 3-bromo-4-hydroxy-8-nitrocinnoline is obtained from 4-hydroxy-8-nitrocinnoline in this same manner (51), and 3-bromo-4-hydroxy-6-nitrocinnoline is obtained similarly from 4-hydroxy-6-nitrocinnoline (38).

Iodination of 4,8-dihydroxycinnoline in an acidic aqueous mixture of potassium iodide and potassium iodate gives 4,8-dihydroxy-5,7-diiodocinnoline (30a) while 4,5-dihydroxycinnoline gives its 6,8-diiodo derivative in a yield of 86% (51a). 4,5-Dihydroxycinnoline-8-sulfonic acid gives a 6-iodo derivative under the same conditions while 4,8-dihydroxycinnoline-5-sulfonic acid gives a 7-iodo derivative in very good yield (51a).

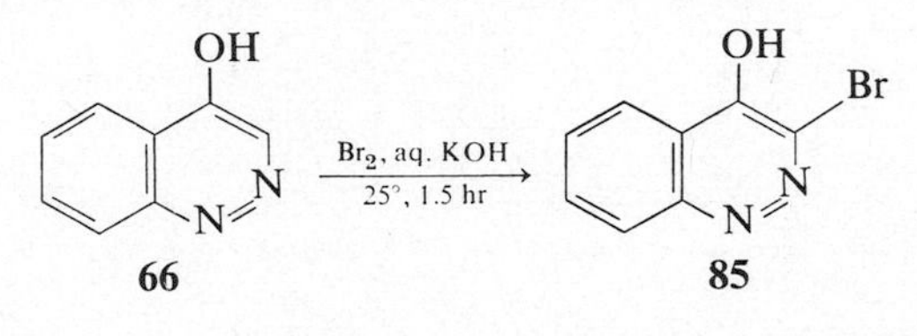

2. *Nitration*

Schofield and Simpson (2, 4) found that the nitration of 4-hydroxycinnoline yields 6-nitro-4-hydroxycinnoline as the principal product, together with small amounts of the 8-nitro isomer, and a third compound to which they tentatively assigned the 3-nitro structure. Subsequently, Baumgarten (15) confirmed that the third isomer is indeed 3-nitro-4-hydroxycinnoline. Recently, Schofield and co-workers (51b) studied the kinetics and products of the nitration of 4-hydroxycinnoline in sulfuric acid solution. They reported the products from the reaction at 25° in 84.9% sulfuric acid as 0.96% 3-nitro-, 0.38% 5-nitro-, 58.4% 6-nitro-, 0.36% 7-nitro-, 39.9%

8-nitro-4-hydroxycinnoline and 5% of an unidentified compound. The 3-nitro isomer was postulated to arise by nitration of the free base, whereas the remaining mononitro isomers were produced by nitration of the protonated 4-hydroxycinnolinium cation. UV spectral data from 4-hydroxycinnoline, 1-methyl-4-cinnolinone, and its 2-methyl isomer in sulfuric acid solution indicated that the predominant 4-hydroxycinnolinium cation in the system was the one protonated at *N*-1. This contrasts with alkylation of 4-hydroxycinnoline, which occurs principally at *N*-2 (section 1C-III-E). The 8-nitro derivative is obtained from 4-hydroxy-7-methylcinnoline (52), while nitration of 4-hydroxy-8-methylcinnoline gives the 5-nitro derivative (38). Nitration of 1-methyl-4-cinnolone gives 1-methyl-6-nitro-4-cinnolone (38).

The nitration of a number of 5-, 6-, 7-, and 8-chloro-4-hydroxycinnolines has been studied (51a). The mononitrated cinnolines so obtained are listed in Table 1C-5. Apparently, nitrations of 3-, 5-, 6-, 7-, and 8-hydroxycinnolines have not been reported.

3. *Sulfonation*

The sulfonation of 8-chloro-4-hydroxycinnoline in fuming sulfuric acid at 170–180° gives 8-chloro-4-hydroxycinnoline-5-sulfonic acid in a yield of 51% (30a, 51a). Sulfonation of 5-chloro-4-hydroxycinnoline in fuming sulfuric acid at 95° gives the 8-sulfonic acid derivative in 46–55% yields (51a).

C. Reduction

Direct reduction of 4-hydroxycinnoline (**66**) with lithium aluminum hydride in refluxing tetrahydrofuran for 8 hr followed by gentle oxidation of the reduction product with red mercuric oxide gives cinnoline (**86**) in 74% yield (44). In a similar manner 6- and 7-phenylcinnoline are obtained in lower

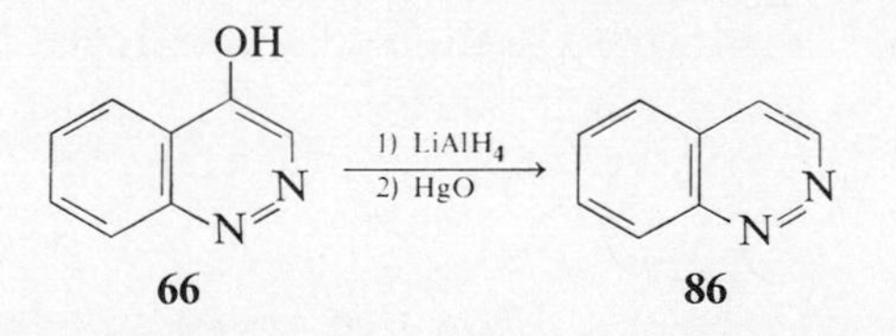

yields from 4-hydroxy-6-phenyl- and 4-hydroxy-7-phenylcinnoline, respectively, but 4-hydroxy-3-phenylcinnoline yields by the same procedure a red material thought to be a complex of 3-phenylcinnoline and a dihydro-3-phenylcinnoline. Two picrate derivatives are isolated from this material and each gives 3-phenylcinnoline on alkaline decomposition. Picric acid evidently acts here as an oxidizing agent (44). Reduction of 6,7-dimethoxy-4-hydroxycinnoline with lithium aluminum hydride followed by oxidation with mercuric oxide gives 6,7-dimethoxycinnoline (53).

When 4-hydroxycinnoline (**66**) is reduced with lithium aluminum hydride in refluxing 1,2-dimethoxyethane and the reaction mixture is examined without prior oxidation, there is obtained a mixture of cinnoline (**86**) and 1,2,3,4-tetrahydrocinnoline (**87**). When 3-hydroxycinnoline (**37**) is reduced similarly, 1,2,3,4-tetrahydrocinnoline (**87**) is isolated in about 45% yield (54). This same procedure converts 3-hydroxy-4-phenylcinnoline (**88**) into 1,4-dihydro-4-phenylcinnoline (**89**) (55).

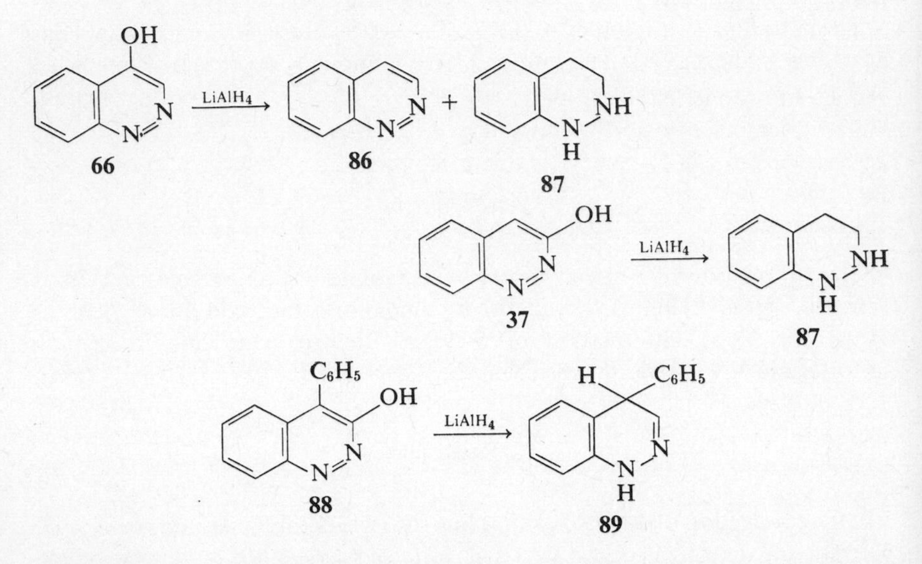

Reduction of 3-hydroxycinnoline (**37**) with zinc dust and sulfuric acid gives 1-aminoöxindole (**90**) in 50–76% yields (22), a transformation that is postulated to occur by reduction of the cinnolinol to 1,2,3,4-tetrahydro-3-oxocinnoline which then suffers hydrolytic acyl-nitrogen cleavage followed by recyclization to give 1-aminoöxindole (55a). Reduction of 3-hydroxycinnoline with red phosphorus and hydriodic acid gives oxindole (**91**) and ammonia (18).

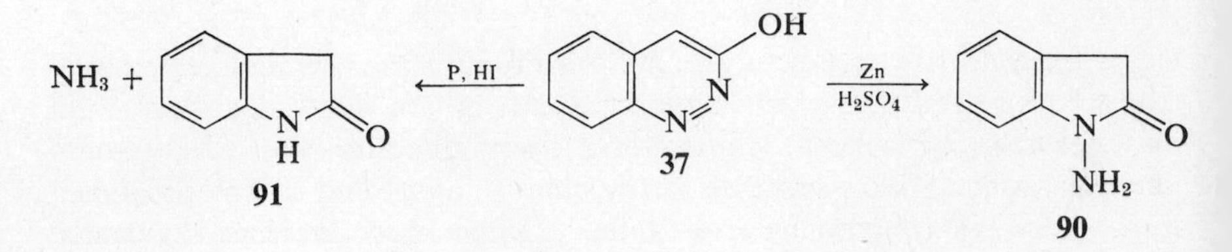

A polarographic study shows that in acidic or neutral solution, 4-hydroxycinnoline is reduced in a two-electron reduction to 4-keto-1,2,3,4-tetrahydrocinnoline, which may be further reduced in acidic or neutral solution in a two-electron step to give 4-hydroxy-1,2,3,4-tetrahydrocinnoline. But in alkaline solution, 4-keto-1,2,3,4-tetrahydrocinnoline is reduced in a one-electron step to a pinacol system in which the two cinnoline rings are dimerized at the C-4 position. Anodic oxidation of this then apparently yields 4,4′-dicinnolinyl. An interesting variation is that 4-hydroxycinnoline may be reduced polarographically in acid solution at the potential of the second wave; the solution is then made slightly alkaline and the product anodically oxidized to cinnoline in 70–80% yield (55b).

In slightly acid solution 3-hydroxycinnoline is reduced polarographically to 3-keto-1,2,3,4-tetrahydrocinnoline, but in strongly acid solution the main reduction product is 1-aminoöxindole (55b).

Treatment of 3-hydroxycinnoline with hydroxylamine-*O*-sulfonic acid in aqueous alkali at 60° gives oxindole (32%). At lower temperature the intermediate in this reaction, 2-aminocinnolin-3-one (mp 130°), can be isolated. It decomposes at its melting point or in boiling toluene to oxindole (55c).

D. 4-Hydroxycinnoline-3-carboxylic Acid with Pyridine and Acetic Anhydride

The product formed by warming together 4-hydroxycinnoline-3-carboxylic acid with pyridine and acetic anhydride is shown by Morley (56) to be the anhydro base of 5-acetyl-5:13-dihydro-12-hydroxy-13-oxocinnolino[2,3-*c*]-pyrido[1,2-*a*]imidazolinium hydroxide (**92**), and not the 4-(2-pyridyl)-cinnoline derivative (**93**) reported earlier (4, 57).

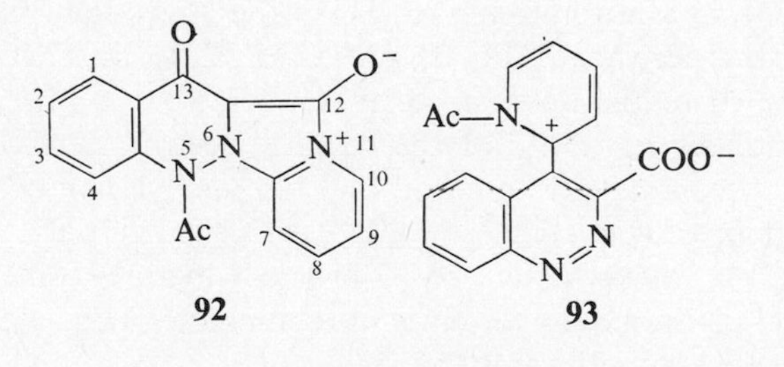

E. Alkylation

Ames and Kucharska (58) found that the methylation of 4-hydroxycinnoline (**66**) with dimethyl sulfate, diazomethane, or methyl iodide (36) gives principally the anhydro base of 2-methyl-4-hydroxycinnolinium hydroxide

(**83**), together with a small amount of 1-methyl-4-cinnolone (**81**). This is in contrast to earlier work in which it was erroneously assumed that the principal product from these reactions is 1-methyl-4-cinnolone (2, 4).

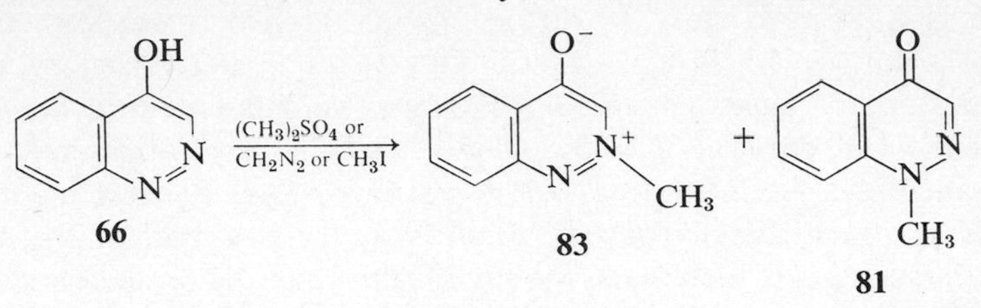

A variety of other alkylating agents, such as diethyl sulfate, ethyl iodide, isopropyl bromide, benzyl chloride, and 2-bromoethanol have been used to alkylate 4-hydroxycinnoline (36, 59). In all of these cases the principal product is the 2-alkyl derivative. Methylation of 6-chloro-4-hydroxycinnoline with dimethyl sulfate or methyl iodide gives a mixture of the 1-methyl and 2-methyl derivatives, the principal product being the 2-methyl isomer (60), and 4-hydroxy-6-nitrocinnoline is similarly methylated to give a mixture of its 1-methyl and 2-methyl derivatives (38). The 2-methyl isomer was wrongly formulated earlier (2, 4) as the methyl nitronate **94**.

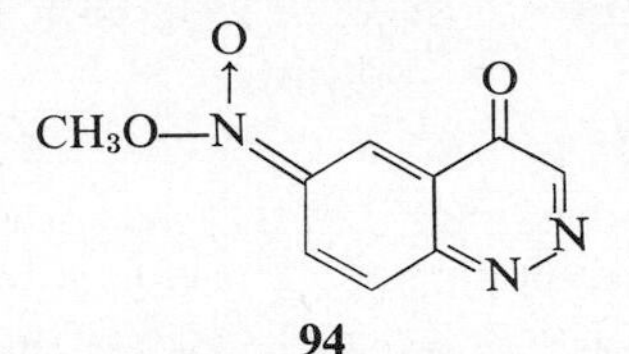

On the other hand, alkylation of 4-hydroxycinnolines which have a substituent group such as a phenyl (61, 61c), methyl (38), bromo (36), carbethoxy (55), or a nitro group (15, 51) in the 3-position of the ring gives principally 1-alkyl derivatives (Eq. 9; $Y = C_6H_5$, CH_3, Br, $COOC_2H_5$, NO_2). This is attributed to the dominance of steric factors in the alkylation of 4-hydroxycinnolines regardless of the relative electron densities at N-1 or N-2 (36, 38). Accordingly, it is found that the methylation of 8-substituted 4-hydroxycinnolines (Eq. 10; $Y = CH_3$, NO_2) gives only the 2-methyl derivatives (38). Methylation of 3-bromo-4-hydroxy-8-methylcinnoline (Eq. 11) and 3-bromo-4-hydroxy-8-nitrocinnoline (Eq. 12) gives the 2-methyl derivative in both cases (38, 51).

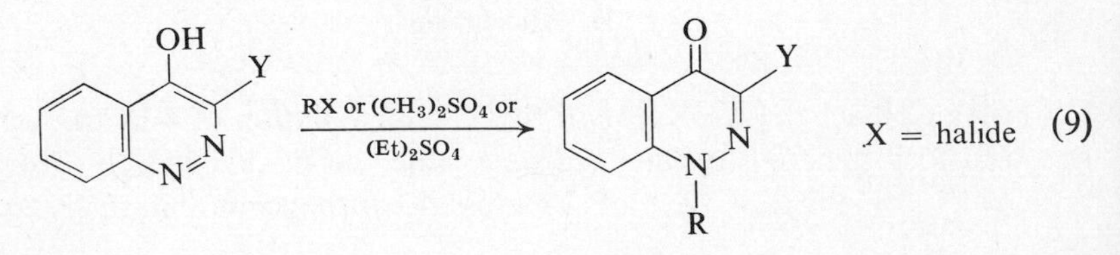

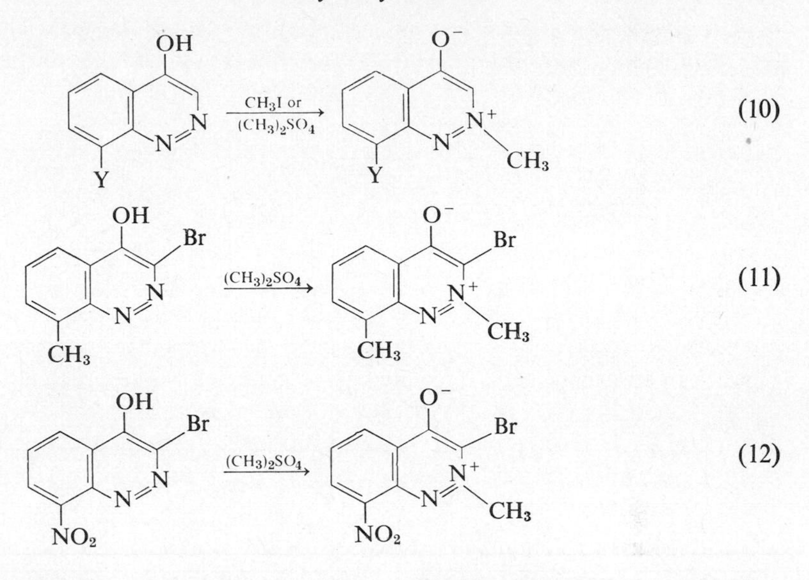

In exception to the preceding pattern, the cyanoethylation of 4-hydroxycinnoline with acrylonitrile in the presence of benzyltrimethylammonium hydroxide (Triton B), gives exclusively 1-cyanoethyl-4-cinnolone (**95**). This result has been explained on the basis of the reversibility of the base-catalyzed cyanoethylation which will lead to the formation of the most stable product, despite the steric factors favoring attack at N-2 (36). Wagner and Heller found that the reaction of 4-hydroxycinnoline with 2,3,4,6-tetra-*O*-acetyl-α-D-glucopyranosyl bromide in alkaline aqueous acetone gives a mixture of 4-(tetra-*O*-acetyl-β-D-glucopyranosyloxy)cinnoline (**96**) and 1-(tetra-*O*-acetyl-β-D-glucopyranosyl)-4-cinnolone (**97**) (62, 63). Compound **96** rearranges to **97** when treated with mercuric bromide in boiling toluene (62).

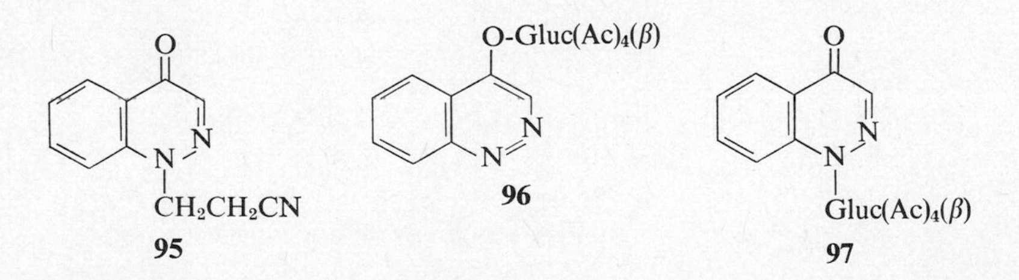

In contrast with the results of alkylation, 4-hydroxycinnoline and its 6,7-dimethoxy derivative are acetylated at the 1-position by acetic anhydride (64). Earlier work erroneously ascribed the 4-acetoxycinnoline structure (*O*-acylation) to 1-acetyl-4-cinnolone (4).

Methylation of 3-hydroxycinnoline (**37**) with ethereal diazomethane or with methyl sulfate and a deficiency of sodium hydroxide gives the bright yellow 2-methyl-3-cinnolone **98** (32, 58).

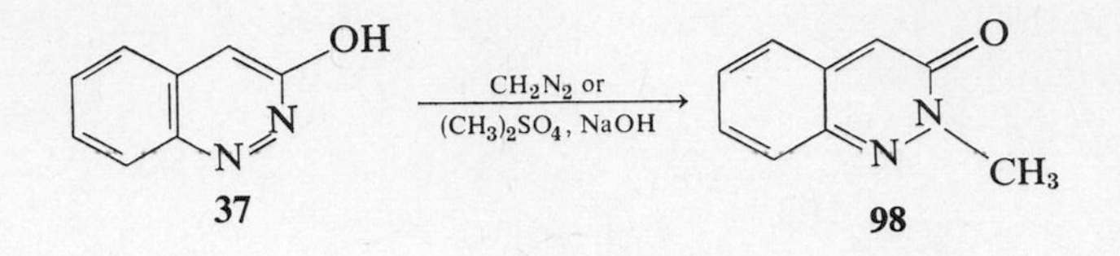

When 3-hydroxycinnoline is treated with methyl sulfate and an excess of sodium hydroxide, apparently a tarry material is formed from which a small amount of an orange solid is isolated. This material was regarded tentatively as the 1-methyl isomer **99**, even though its elemental analysis was unsatisfactory (32); in a repetition of this experiment by other workers, the 2-methyl derivative **98** was obtained, but none of the 1-methyl isomer **99** could be detected (58).

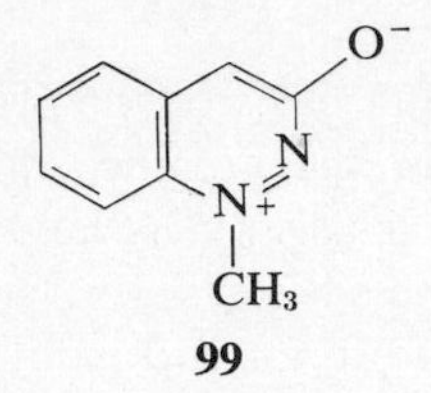

Benzylation of 3-hydroxycinnoline with benzyl chloride gives the 2-benzyl derivative (59), and cyanoethylation of 3-hydroxycinnoline with acrylonitrile in the presence of benzyltrimethylammonium hydroxide likewise results in 2-alkylation to give 2-cyanoethyl-3-cinnolone (36). Treatment of 3-hydroxycinnoline with tetra-*O*-acetyl-α-D-glucopyranosyl bromide in alkaline aqueous acetone gives 2-(tetra-*O*-acetyl-β-D-glucopyranosyl)-3-cinnolone (**100**), while interaction of this same bromide with the silver salt of 3-hydroxycinnoline in refluxing toluene gives 3-(tetra-*O*-acetyl-β-D-glucopyranosyloxy)-cinnoline (**101**) together with a lesser amount of **100**. Compound **101** may be rearranged to the cinnolone **100** by treating it with mercuric bromide in boiling toluene (49, 65). Methylation of 3-hydroxy-4-phenylcinnoline with dimethyl sulfate gives only 2-methyl-4-phenyl-3-cinnolone (61c).

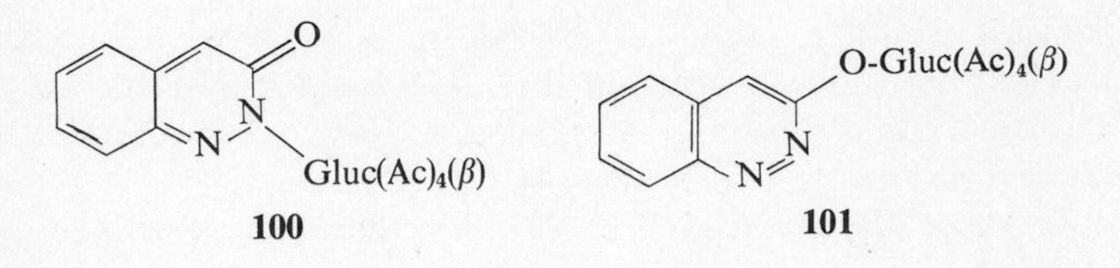

When 8-hydroxycinnoline is dissolved in ethereal diazomethane, a brilliant emerald-green color is produced; this then changes to deep blue. A dark blue material precipitates and it has been suggested that this possesses a structure similar to the brick-red product **102** which is formed under the same conditions from oxine (24).

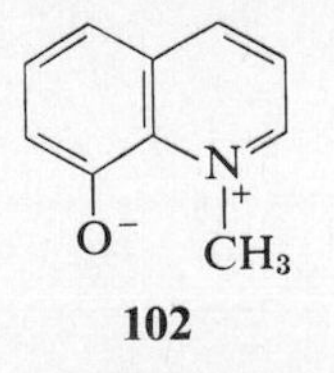

102

Certain differences in spectral parameters will often allow one to decide whether the product resulting from alkylation of a 4-hydroxycinnoline is a 1-alkyl or a 2-alkyl derivative. For example, 4-hydroxycinnolines and their 1-alkyl derivatives have similar ultraviolet absorption spectra, showing characteristic bands at about 285 and 295 mμ, whereas the 2-alkyl and *O*-alkyl derivatives show distinctly different spectra (35, 36, 38, 61). The infrared spectra of 1-methyl-4-cinnolones show great similarity to those of the parent compounds (61). One might expect the 2-methyl derivatives to exhibit somewhat different spectra, especially in the carbonyl stretching absorption region. A comparison of the pmr spectra of several 1-methyl-4-cinnolones with the spectra of the corresponding 2-methyl derivatives in deuterochloroform reveals that the 1-methyl group consistently shows higher field absorption than the 2-methyl group. The difference in τ values for the two series is found to be about 0.18–0.23τ. This difference is attributed to the greater polarization of the 2-methyl anhydro base than of 1-methyl-4-cinnolone. In addition, the C-8 proton resonance of any given 1-methyl-4-cinnolone occurs at higher field than that of its corresponding 2-methyl isomer. On changing to a more polar solvent, the absorption positions of the C-8 proton in both the N-1- and N-2-methyl cinnolones shifts to lower field, but to a greater extent in the N-1-methyl isomer (38, 61, 61a, 61b). The pmr spectra of a number of N-1- and N-2-methyl-4-cinnolones are given in Table 1C-1a.

IV. Tables

TABLE 1C-1. Proton Magnetic Resonance Spectra of 4-Hydroxycinnolines[a]

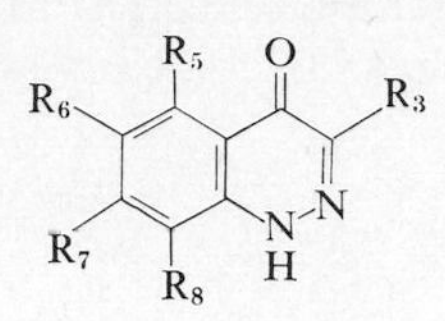

R_3	R_5	R_6	R_7	R_8	Ref.
H	H	H	H	H	64
2.12(S)	1.83(D)	←——2.17–2.7(M)——→		2.57(Q)	
	$J = 8.5$			$J_1 = 7, J_2 = 2$	
2.17	1.89	2.52	2.16	2.34	30a
	$J_{5,6} = 8.3$	$J_{6,7} = 7.1$	$J_{7,8} = 8.4$		
	$J_{5,7} = 1.6$	$J_{6,8} = 1.0$			
	$J_{5,8} = 0.5$				
COOH	H	H	H	H	64
—	1.69(D)	←—— 1.9–2.5(M) ——→		2.31(Q)	
	$J = 9$			$J_1 = 6$	
				$J_2 = 2$	
—	1.80	2.32	2.01	2.15	30a
	$J_{5,6} = 8.2$	$J_{6,7} = 7.0$	$J_{7,8} = 8.6$		
	$J_{5,7} = 1.4$	$J_{6,8} = 1.0$			
	$J_{5,8} = 0.5$				
H	H	OCH_3	OCH_3	H	64
2.23(S)	2.54(S)	{5.97(S)	6.02(S)}	2.94(S)	
		or			
		{6.02(S)	5.97(S)}		
H	Cl	H	H	H	30a
2.33	—	2.63	2.32	2.50	
		$J_{6,7} = 7.6$	$J_{7,8} = 8.6$		
		$J_{6,8} = 1.0$			
COOH	Cl	H	H	H	30a
—	—	2.40	2.16	2.27	
		$J_{6,7} = 7.7$	$J_{7,8} = 8.6$		
		$J_{6,8} = 1.1$			
H	H	Cl	H	H	30a
2.22	2.03	—	2.20	2.30	
	$J_{5,7} = 2.5$		$J_{7,8} = 9.1$		
	$J_{5,8} = 0.3$				
H	H	H	Cl	H	30a
2.23	1.99	2.59	—	2.40	
	$J_{5,6} = 8.8$	$J_{6,8} = 1.9$			
	$J_{5,8} = 0.5$				

TABLE 1C-1. (*continued*)

R_3	R_5	R_6	R_7	R_8	Ref.
H 2.10	H 1.98 $J_{5,6} = 8.1$ $J_{5,7} = 1.3$	H 2.58 $J_{6,7} = 7.7$	H 2.07	Cl —	30a
COOH —	H 1.85 $J_{5,6} = 8.2$ $J_{5,7} = 1.2$	H 2.40 $J_{6,7} = 7.7$	H 1.92	Cl —	30a
H 2.29	Cl —	Cl —	H 2.09 $J_{7,8} = 9.0$	H 2.47	30a
H 2.22	Cl —	H 2.64 $J_{6,7} = 8.7$	H 2.13	Cl —	30a
H 2.24	H 1.88 $J_{5,8} = 0.9$	Cl —	Cl —	H 2.19	30a
NO_2 —	H 1.70 $J_{5,8} = 0.5$	Cl —	Cl —	H 2.03	30a
Cl —	H 2.04 $J_{5,7} = 2.4$ $J_{5,8} = 0.5$	Cl —	H 2.20 $J_{7,8} = 9.0$	H 2.36	30a
H 2.12	H 1.90 $J_{5,7} = 2.3$	Cl —	H 2.07	Cl —	30a
COOH —	H 1.76 $J_{5,7} = 2.5$	Cl —	H 1.91	Cl —	30a
H 2.12	H 2.01 $J_{5,6} = 8.7$	H 2.43	Cl —	Cl —	30a
COOH —	H 1.94 $J_{5,6} = 8.8$	H 2.25	Cl —	Cl —	30a
H 2.19	OH —	H 3.32 $J_{6,7} = 7.9$ $J_{6,8} = 0.9$	H 2.33 $J_{7,8} = 8.4$	H 2.98	30a
H 2.24	H 2.74 $J_{5,6} = 8.1$ $J_{5,7} = 1.1$	H 2.50 $J_{6,7} = 7.7$	H 2.93	OH —	30a
COOH —	H 1.42 $J_{5,6} = 8.1$ $J_{5,7} = 1.4$	H 2.27 $J_{6,7} = 8.1$	H 1.23	NO_2 —	30a

TABLE 1C-1. *(continued)*

R_3	R_5	R_6	R_7	R_8	Ref.
H	Cl	NO_2	H	H	30a
2.26	—	—	1.79	2.39	
			$J_{7,8} = 9.5$		
H	Cl	H	H	NO_2	30a
2.11	—	2.47	2.43	—	
		$J_{6,7} = 8.7$			
H	NO_2	Cl	H	H	30a
2.13	—	—	1.95	2.18	
			$J_{7,8} = 9.2$		
H	H	Cl	H	NO_2	30a
2.00	1.62	—	1.29	—	
	$J_{5,7} = 2.5$				
COOH	H	Cl	H	NO_2	30a
—	1.53	—	1.24	—	
	$J_{5,7} = 2.5$				
COOH	H	NO_2	H	Cl	30a
—	1.29	—	1.29	—	
H	H	NO_2	Cl	Cl	30a
2.05	1.47	—	—	—	
COOH	H	NO_2	Cl	Cl	30a
—	1.33	—	—	—	
H	Cl	H	H	SO_3H	30a
2.23	—	2.59	2.03	—	
		$J_{6,7} = 8.1$			
H	SO_3H	H	H	Cl	30a
2.16	—	2.00	1.94	—	
		$J_{6,7} = 8.1$			
H	OH	H	H	SO_3H	30a
2.05	—	3.26	2.00	—	
		$J_{6,7} = 8.3$			
H	SO_3H	H	H	OH	30a
2.38	—	2.10	2.98	—	
		$J_{6,7} = 8.0$			
H	Br	CH_3	H	H	30a
2.34	—	—	2.56	2.34	
			$J_{7,8} = 8.5$		
COOH	Br	CH_3	H	H	30a
—	—	—	2.46	2.30	
			$J_{7,8} = 8.6$		
H	I	H	I	OH	30a
1.92	—	1.61	—	—	
H	OH	I	H	SO_3K	30a
1.94	—	—	1.74	—	
H	SO_3Na	H	I	OH	30a
1.30	—	1.65	—	—	

[a] Chemical shift values are given in τ units; the multiplicity of an absorption, when this was indicated in the original paper, is given in parentheses following the τ value as S = singlet, D = doublet, Q = quartet, and M = multiplet. J = coupling constant in cps. In all cases, the solvent used is dimethyl sulfoxide, and the reference standard is tetramethylsilane or tetramethylammonium sulfate.

TABLE 1C-1a. Proton Magnetic Resonance Spectra of *N*-Methyl-4-cinnolones[a]

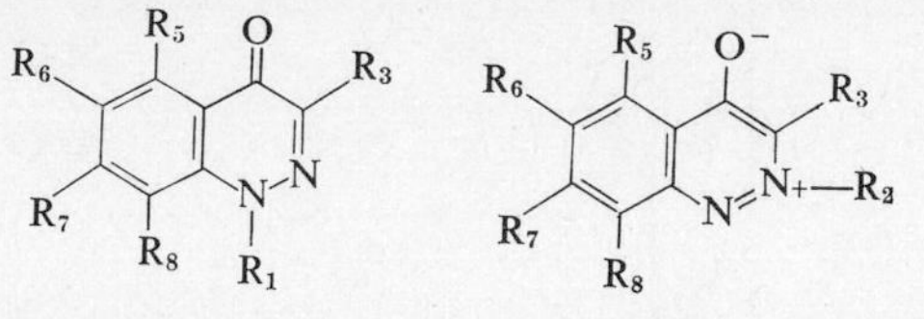

R_1	R_2	R_3	R_5	R_6	R_7	R_8	Solvent	Ref.
CH_3	—	H	H	H	H	H		
5.92		2.23	1.70	—	—	2.60	$CDCl_3$	61b
5.86		2.21	1.80	—	—	2.46	$(CD_3)_2SO$	61b
5.19		1.00	1.30	—	—	1.85	CF_3COOH	61b
—	CH_3	H	H	H	H	H		
	5.70	1.99	1.61	—	—	2.48	$CDCl_3$	61b
	5.61	2.20	1.82	—	—	2.41	$(CD_3)_2SO$	61b
CH_3	—	CH_3	H	H	H	H		
5.99		7.62	1.70	—	—	—	$CDCl_3$	61b
—	CH_3	CH_3	H	H	H	H		
	5.80	7.43	1.78	—	—	—	$CDCl_3$	61b
CH_3	—	Br	H	H	H	H		
5.87		—	1.63	—	—	2.70	$CDCl_3$	61b
—	CH_3	Br	H	H	H	H		
	5.41	—	1.66	—	—	2.46	$CDCl_3$	61b
CH_3	—	H	H	NO_2	H	H		
5.86		2.18	0.88	—	1.47	2.47	—	61a
			$J_{5,7} = 3.0$			$J_{7,8} = 10.2$		
—	CH_3	H	H	NO_2	H	H		
	5.62	2.02	0.81	—	1.53	2.11	Not given	61a
			$J_{5,7} = 4.0$			$J_{7,8} = 10.0$		
CH_3	—	H	H	CH_3O	CH_3O	H		
6.00		2.23	2.39	5.92	5.92	3.30	$CDCl_3$	61a
—	CH_3	H	H	CH_3O	CH_3O	H		
	5.72	2.16	2.39	5.95 or 5.99	5.95 or 5.99	2.93	$CDCl_3$	61a

[a] Chemical shifts are given in τ units, immediately below the proton(s) to which they refer. J = coupling constant in cps. Where a chemical shift value is not assigned, it means that the absorption for that proton is too complex to assign a value for the center of its multiplet.

TABLE 1C-2. Ultraviolet Absorption Spectra of Hydroxycinnolines

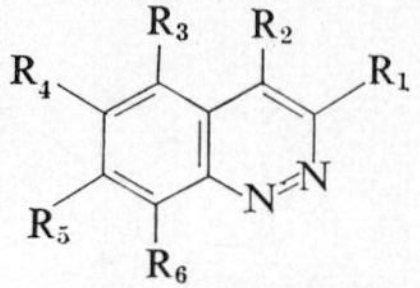

R_1	R_2	R_3	R_4	R_5	R_6	Solvent	$\lambda_{max}(m\mu)^a$	$\log E_{max}$	Ref.
OH	H	H	H	H	H	95% methanol	300	2.84	32
							312	2.76	
							400	3.47	
						Water	226	4.67	
							300	2.93	
							312	2.89	
							394	3.51	
						0.01*N* HCl	300	2.88	
							398	3.46	
						0.01*N* NaOH	385	3.62	
						95% ethanol	230	4.66	21
							282	*3.41*	
							312	*3.20*	
							400	3.52	
OH	CH_3	H	H	H	H	95% ethanol	235	4.65	21
							280	*3.40*	
							393	3.72	
OH	$C_6H_5CH_2$	H	H	H	H	95% ethanol	234	4.67	21
							288	*3.41*	
							298	3.73	

OH	C_6H_5	H	H	H	H	95% ethanol	233	4.62	21
							322	*3.35*	
							413	3.81	
OH	H	H	Cl	H	H	95% ethanol	237	4.67	21
							299	3.26	
							310	3.24	
							323	3.28	
							390	3.45	
OH	H	H	H	Cl	H	95% ethanol	236	4.70	21
							282	3.21	
							390	3.44	
H	OH	H	H	H	H	Water pH = 5.0	207	4.49	34, 37
						(neutral species)	227	4.12	
							234	4.11	
							253	3.98	
							261	*4.49*	
							284	3.35	
							337	4.09	
							352	4.01	
						Water pH = 11.5	211	4.48	
						(anionic species)	240	4.17	
							336	4.02	
							345	*3.99*	
						H_2SO_4 pH = −3.6	205	4.25	
						(cationic species)	233	4.47	
							249	4.08	
							294	*3.37*	
							305	3.54	
							338	3.76	

TABLE 1C-2. (*continued*)

R_1	R_2	R_3	R_4	R_5	R_6	Solvent	$\lambda_{max}(m\mu)^a$	log E_{max}	Ref.
						Methanol	211	4.32	14
							231	4.03	
							238	4.06	
							253	3.94	
							283	3.48	
							294	3.50	
							338	4.15	
							352	4.09	
						Ethanol	228	4.01	36
							236	4.04	
							251	3.92	
							260	*3.80*	
							282	3.45	
							292	3.48	
							336	4.10	
							352	4.07	
H	OH	Cl	H	H	H	Methanol	214	4.36	14
							232	3.90	
							241	3.86	
							259	3.90	
							266	3.84	
							289	3.62	
							301	3.72	
							342	4.09	

H	OH	Br	H	H	H	Methanol	216	4.33	14
							234	3.96	
							265	3.89	
							270	3.86	
							292	3.61	
							304	3.69	
							346	4.09	
H	OH	H	F	H	H	Methanol	210	4.17	14
							234	4.11	
							241	4.12	
							283	3.54	
							294	3.60	
							342	4.15	
							357	4.14	
H	OH	H	Cl	H	H	Methanol	214	4.14	14
							240	4.23	
							245	4.23	
							288	3.68	
							299	3.77	
							345	4.09	
							360	4.09	
H	OH	H	Br	H	H	Methanol	215	4.40	14
							243	4.31	
							249	4.31	
							290	3.74	
							301	3.81	
							348	4.10	
							363	4.10	

TABLE 1C-2. *(continued)*

R_1	R_2	R_3	R_4	R_5	R_6	Solvent	$\lambda_{max}(m\mu)^a$	log E_{max}	Ref.
H	OH	H	H	F	H	Methanol	208	4.34	14
							225	4.02	
							235	4.08	
							248	4.06	
							256	3.96	
							280	3.42	
							288	3.39	
							331	4.14	
							345	4.11	
H	OH	H	H	Cl	H	Methanol	214	4.47	14
							240	4.14	
							253	4.20	
							262	4.13	
							285	3.38	
							295	3.34	
							336	4.12	
							350	4.13	
H	OH	H	H	Br	H	Methanol	217	4.36	
							243	4.11	
							257	4.19	
							265	4.16	
							289	3.41	
							339	4.09	
							352	4.10	
H	OH	H	H	H	F	Methanol	209	4.29	14
							228	4.05	
							255	3.93	

							284	3.63	
							294	3.68	
							336	4.12	
H	OH	H	H	H	Cl	Methanol	212	4.35	14
							229	3.99	
							258	3.89	
							266	3.84	
							285	3.55	
							294	3.52	
							342	4.14	
H	OH	H	H	H	Br	Methanol	213	4.38	14
							234	3.84	
							261	3.91	
							267	3.84	
							286	3.53	
							294	3.49	
							344	4.17	
C_6H_5	OH	H	H	H	H	Methanol	262	4.22	61
							310	4.07	
							352	4.08	
CH_3	OH	H	H	H	H	Ethanol	208	4.47	38
							230	*4.10*	
							238	4.13	
							250	4.02	
							282	3.56	
							292	3.54	
							343	4.10	
							357	4.09	

TABLE 1C-2. (*continued*)

R_1	R_2	R_3	R_4	R_5	R_6	Solvent	$\lambda_{max}(m\mu)^a$	log E_{max}	Ref.
H	OH	H	H	H	CH_3	Ethanol	211	4.40	38
							229	4.02	
							236	3.99	
							258	3.90	
							267	3.86	
							288	3.43	
							297	3.42	
							340	4.04	
							354	*3.95*	
Br	OH	H	H	H	CH_3	Ethanol	214	4.38	38
							235	4.09	
							243	4.13	
							261	3.98	
							271	3.95	
							293	3.59	
							302	3.57	
							346	4.08	
							362	3.97	
H	OH	H	H	H	NO_2	Ethanol	230	4.16	38
							255	*3.99*	
							280	*3.79*	
							388	4.01	
H	OH	H	NO_2	H	H	Ethanol	236	4.14	38
							258	*3.89*	
							266	3.86	
							275	*3.76*	

							322	3.85	
							365	3.99	
Br	OH	H	NO_2	H	H	Ethanol	210	4.22	38
							238	4.16	
							270	*4.00*	
							279	*3.92*	
							324	3.93	
							370	4.11	
H	OH	NO_2	H	H	CH_3	Ethanol	210	4.38	38
							230	3.93	
							257	3.83	
							266	*3.76*	
							290	3.51	
							300	3.57	
							340	3.94	
							351	*3.90*	
H	OH	H	NO_2	H	CH_3	Ethanol	240	4.19	38
							273	3.86	
							282	*3.83*	
							314	*3.77*	
							327	3.91	
							360	4.03	
NO_2	OH	H	Cl	H	H	Ethanol	222	4.42	38
							343	4.11	
							353	*4.05*	
CH_2CO_2H	OH	H	OCH_3	OCH_3	H	Ethanol	227	4.54	66
							257	4.43	
							265	4.45	

TABLE 1C-2. (*continued*)

R_1	R_2	R_3	R_4	R_5	R_6	Solvent	$\lambda_{max}(m\mu)^a$	log E_{max}	Ref.
CH_2CO_2H	OH	H	OCH_3	OCH_3	H	Ethanol			
							285	*3.34*	
							343	4.26	
							357	4.25	
$CH_2CO_2CH_3$	OH	H	OCH_3	OCH_3	H	Ethanol	228	4.42	66
							257	4.28	
							266	4.30	
							290	*3.00*	
							342	4.08	
							357	4.07	
CH_3	OH	H	OCH_3	OCH_3	H	Ethanol	226	4.26	66
							256	4.20	
							264	4.22	
							286	*3.40*	
							344	4.00	
							359	4.01	
H	H	OH	H	H	H	H_2SO_4 pH = −0.25	257	4.47	28
							310	3.16	
							328	*3.05*	
							410	3.28	
						Water pH = 4.52	245	4.53	
							300	*3.06*	
							357	3.39	
						Water pH = 11.0	234	4.23	
							262	4.42	
							338	3.48	
							385	*3.19*	
							425	3.11	

						Water pH = 4.66 (isoelectric value)	244	4.51	35
							305	*3.08*	
							356	3.39	
						Ethanol	245	4.52	
							305	*3.09*	
							359	3.39	
H	H	H	OH	H	H	0.1*N* HCl	255	4.42	28
							310	*3.40*	
							324	3.53	
							367	3.80	
						Water pH = 5.43	240	4.50	
							280	3.54	
							310	*3.60*	
							320	3.66	
							414	3.03	
						Water pH = 10.0	259	4.45	
							311	3.57	
							376	3.82	
						Ethanol	240	4.61	
							259	*3.41*	
							308	*3.61*	
							325	3.68	
							340	*3.59*	
							423	*2.04*	
						Water pH = 5.58 (isoelectric value)	238	4.38	35
							281	3.57	
							320	3.67	
							402	3.08	

TABLE 1C-2. (*continued*)

R_1	R_2	R_3	R_4	R_5	R_6	Solvent	$\lambda_{max}(m\mu)^a$	log E_{max}	Ref.
H	H	H	OH	H	H	Ethanol	241	4.61	
							326	3.69	
H	H	H	H	OH	H	0.1*N* HCl	250	4.55	28
							307	3.53	
							325	*2.93*	
							388	3.44	
						Water pH = 5.43	236	4.58	
							267	3.66	
							358	3.47	
							449	2.26	
						Water pH = 10.0	253	4.62	
							286	3.82	
							402	3.54	
							428	*3.39*	
						Ethanol	237	4.63	
							264	*4.00*	
							284	*3.68*	
							296	*3.43*	
							306	*3.07*	
							358	3.59	
						Water pH = 5.44	235	4.63	35
						(isoelectric value)	267	3.71	
							354	3.47	
							448	2.45	
						Ethanol	237	4.68	
							275	*3.79*	
							355	3.60	

H	H	H	H	H	OH	95% methanol	245	4.53	24
							303	3.03	
							315	3.04	
							360	3.43	
						0.01*N* HCl	252	4.62	
							304	3.07	
							430	3.32	
						0.01*N* NaOH	256	4.40	
							330	3.43	
							425	3.49	
						Water pH = 5.47	242	4.51	35
						(isoelectric value)	296	3.09	
							359	3.42	
							525	1.32	
						Ethanol	246	4.53	
							303	3.05	
							362	3.43	
H	CH_3	H	H	H	OH	95% methanol	246	4.42	24
							310	3.09	
							360	3.50	
						0.01*N* HCl	250	4.38	
							308	3.14	
							420	3.43	
						0.01*N* NaOH	255	4.40	
							330	3.46	
							415	3.60	

[a] Italics refer to shoulders or inflections.

TABLE 1C-3. Basic Strengths of Hydroxycinnolines

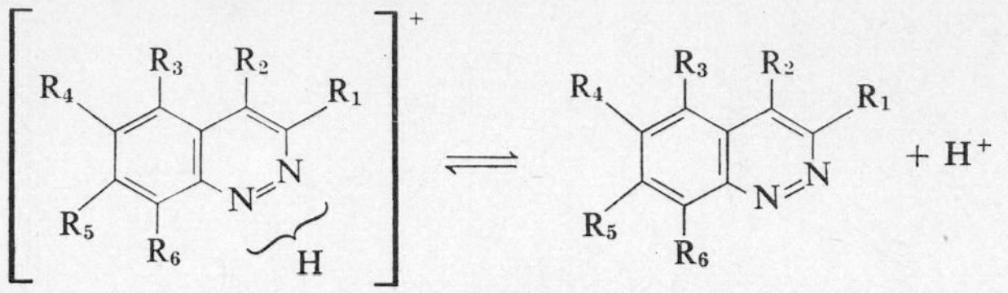

R_1	R_2	R_3	R_4	R_5	R_6	Basic pK_a[a]	Solvent	Temp. (°C)	Ref.
OH	H	H	H	H	H	0.21 ± 0.04	Water	20	29
H	OH	H	H	H	H	-0.35 ± 0.05	Water	20	29
H	H	OH	H	H	H	1.92 ± 0.06	Water	20	28, 29
H	H	H	OH	H	H	3.65 ± 0.01	Water	20	28, 29
H	H	H	H	OH	H	3.31 ± 0.02	Water	20	28, 29
H	H	H	H	H	OH	2.74 ± 0.02	Water	20	30
						2.86 ± 0.06	Water	14 ± 3	24
						2.56 ± 0.06	Water	20	28
H	CH_3	H	H	H	OH	3.18 ± 0.02	Water	20	30
						3.28 ± 0.02	Water	14 ± 3	24

[a] pK_a = negative logarithm of the ionization constant.

TABLE 1C-4 Acidic Strengths of Hydroxycinnolines

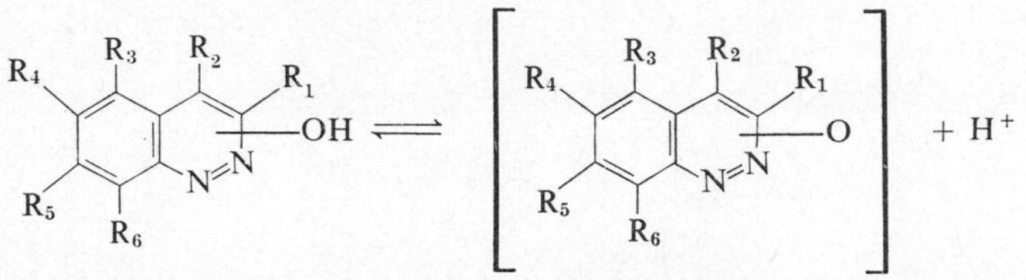

R_1	R_2	R_3	R_4	R_5	R_6	Acidic pK_a[a]	Solvent	Temp. (°C)	Ref.
OH	H	H	H	H	H	8.62 ± 0.02	Water	20 ± 0.05	21
						9.08 ± 0.03	50% ethanol	20 ± 0.05	21
						8.64	Water	20 ± 0.1	32
						8.61 ± 0.02	Water	20	28
OH	H	H	Cl	H	H	8.41 ± 0.01	50% ethanol	20 ± 0.05	21
OH	H	H	H	Cl	H	8.48 ± 0.02	50% ethanol	20 ± 0.05	21
H	OH	H	H	H	H	9.27 ± 0.03	Water	20	29
H	H	OH	H	H	H	7.40 ± 0.02	Water	20	28, 29
H	H	H	OH	H	H	7.52 ± 0.01	Water	20	28, 29
H	H	H	H	OH	H	7.56 ± 0.02	Water	20	28, 29
H	H	H	H	H	OH	8.20 ± 0.04	Water	20	30
						8.13 ± 0.07	Water	14 ± 3	24
						8.17	Water	20	28
H	CH_3	H	H	H	OH	8.34 ± 0.05	Water	20	30
						8.67 ± 0.04	Water	14 ± 3	24

[a] pK_a = negative logarithm of the ionization constant.

TABLE 1C-5. 4-Hydroxycinnolines

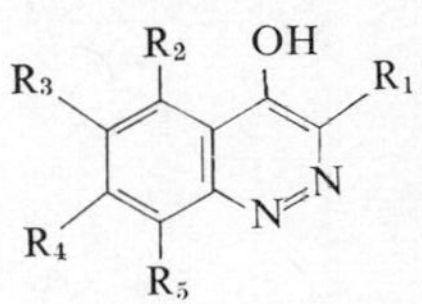

R_1	R_2	R_3	R_4	R_5	Prep.[a]	MP (°C)	Remarks[b]	Ref.
H	H	H	H	I	A	261–262	Yellow leaflets (ethanol)	67
H	H	H	H	OCH_3	A	164–165 (monohydrate)	White needles of monohydrate (water)	24
C_6H_5	H	H	H	H	B	265–267, 268–270	White leaflets (methanol)	11, 16
C_6H_5	H	CH_3	H	H	B	310–312	Cream-colored plates	11
C_6H_5	H	NO_2	H	H	B	347–348	Yellow needles	11
α-Naphthyl	H	H	H	H	C	285–286	Small plates	11
H	CH_3O	H	H	H	A	275 (dec.)	Yellow needles (water); a polymorphic form, m.p. 245° (dec.) observed, which changed on standing into the higher-melting form	25
H	H	H	CH_3O	H	A	255–257	Needles (water)	25
NO_2	H	H	H	H	D	284.5–285.5	Yellow needles (ethanol)	15
H	H	H	C_6H_5	H	A	323–325	Colorless needles (ethanol or methanol)	44
H	H	C_6H_5	H	H	A	294–296	Colorless needles (ethanol)	44

TABLE 1C-5 (*continued*)

R_1	R_2	R_3	R_4	R_5	Prep.[a]	MP (°C)	Remarks[b]	Ref.
COOH	H	H	H	CH_3	E	263–264 (dec.)	Colorless needles (acetic acid)	4
COOH	H	CH_3	H	H	E	269 (dec.)	Fawn-colored needles (dimethylformamide)	13
COOH	H	H	H	Cl	E	247–248 (dec.)	Fawn-colored plates (acetic acid)	13
COOH	Cl	H	H	H	E	263–264 (dec.)	Silvery plates (dimethylformamide)	13
COOH	H	Cl	H	H	E	267 (dec.), 263–264 (dec.)	Colorless needles (dimethylformamide)	3, 13
COOH	H	H	Cl	Cl	E	249–250 (dec.)	Colorless prisms (dimethylformamide)	13
COOH	H	Cl	H	Cl	E	250–251 (dec.)	Fawn-colored needles (acetic acid)	13
COOH	Cl	H	H	Cl	E	248–249 (dec.)	Fawn-colored needles (acetic acid)	13
COOH	Cl	Cl	H	H	E	268 (dec.)	Fawn-colored needles (acetic acid)	13
COOH	H	Br	H	H	E	256–258 (dec.) 264 (dec.)	Yellow needles (dimethylformamide)	3, 13
COOH	H	H	H	OCH_3	E	274–275 (dec.)	Pink needles (acetic acid)	13
COOH	{Br / H}	CH_3 or CH_3	{H / Br}	H	E	262–263 (dec.)	Fawn-colored needles (dimethylformamide)	13
COOH	H	OCH_3	H	H	E	258–259 (dec.) 268 (dec.)	Fawn-colored needles (acetic acid)	2, 13
COOH	H	H	H	p-$CH_3C_6H_4SO_3$	E	255 (dec.)	Colorless needles (acetic acid)	13

COOH	H	H	H	NO_2	E	253–254 (dec.)	Pale green plates (acetic acid)	13
COOH	H	NO_2	H	H	E	275 (dec.)	Yellow needles (acetic acid)	13
H	H	CH_3	H	H	F	271	Fawn-colored needles (acetic acid)	13
H	Cl	H	H	H	F	330–332	Fawn-colored needles (dimethylformamide)	13
H	H	H	Cl	Cl	F	261–262 253–254	Colorless needles (acetic acid)	13, 68
H	H	Cl	H	Cl	F	221–223	Grey needles (acetic acid)	13
H	Cl	H	H	H	F	222–224	Colorless needles (acetic acid)	13
H	Cl	Cl	H	H	F	336–337	Fawn-colored needles (acetic acid)	13
H	Br or H	CH_3 CH_3	H Br	H	F	288–289	Colorless needles (2-ethoxyethanol)	13
H	H	H	H	OCH_3	F	162–163	Colorless needles (water)	13
p-HOC_6H_4	H	H	H	H	G	256–258	(Ethanol–toluene)	16
p-ClC_6H_4	H	H	H	H	H	329–330	White plates (butanone); sublimes 250–330° at 0.05 mm	16
H	H	OCH_3	OCH_3	H	A	271–272 305–306	White powder 1-Acetyl deriv., mp 215°	43, 64
COOH	H	H	H	F	E	254–255	(Glacial acetic acid)	48
COOH	H	H	H	Br	E	256–257	(Glacial acetic acid) *S*-benzylthiouronium salt deriv., mp 198°	48, 69

TABLE 1C-5 (*continued*)

R_1	R_2	R_3	R_4	R_5	Prep.[a]	MP (°C)	Remarks	Ref.
COOH	H	H	F	H	E	267–268	(Glacial acetic acid)	48
COOH	H	F	H	H	E	264	(Glacial acetic acid)	48
H	H	H	H	F	F	262–263	(Methanol)	48
H	H	H	H	Br	F	190–192	(Methanol)	48
H	H	H	F	H	F	226–227	(Methanol)	48
H	Br	H	H	H	F	325–327	(Methanol)	48
H	H	H	Br	H	F	285–286	(Methanol)	48
H	H	F	H	H	A, F	267	White needles (methanol)	48
Br	H	H	H	H	I	272–275 276.5–277.5	(Ethanol)	12, 36
$CH_2CO_2CH_3$	H	OCH_3	OCH_3	H	J	315–316 (dec.)	(Methanol)	66
CH_3	H	OCH_3	OCH_3	H	K	338–339	(2-Methoxyethanol)	66
Br	H	H	H	CH_3	I	285–287	(Ethanol)	38
Br	H	NO_2	H	H	I	300–301	(Ethanol)	38
COOH	NO_2	H	H	CH_3	E	249–250 (dec.)	(2-Methoxyethanol)	38
COOH	H	NO_2	H	CH_3	E	254–256	(Acetic acid)	38
H	NO_2	H	H	CH_3	F, L	255–256	(Acetic acid)	38
H	H	NO_2	H	CH_3	F	274–276	(2-Methoxyethanol)	38
Br	H	H	H	NO_2	I	235–237	(Ethanol)	51
H	H	$C_6H_5N{:}N$—	H	H	M	289–300	Yellow needles (acetic acid); acetate deriv., mp 179–180°	9
H	H	*m*-$CH_3COC_6H_4N{:}N$—	H	H	M	279–289 (dec.)	Red needles (acetic acid); acetate deriv., yellow needles, mp 178–180°	9

H	H	OH N:N—	H	H	N	>320	Impure material; no elemental analysis reported; insol. in organic solvents; assumed diacetyl deriv., mp >300°	9
Cl	H	Br	H	H	O	309–310	—	50
Br	H	Br	H	H	P	315–316	White needles (ethanol); acetyl deriv., mp 179–180°	50
C_2H_5COO—	H	Cl	H	H	Q	228–230	(Ethanol)	55
H	H	Cl	H	H	R	296–298	—	48a
F	F	F	F	F	R	226–228	—	70
Br	H	CH_3O	CH_3O	H	I	318–319	—	71
CH_3	H	H	H	NO_2	S	235–236	—	72
H	H	—O—CH_2—O—		H	A	—	—	73
Br	H	—O—CH_2—O—		H	I	—	—	73
CN	H	—O—CH_2—O—		H	S	—	—	73
COOH	H	Cl	H	NO_2	L	217–218	(Methanol)	51a
H	H	Cl	H	NO_2	F	159–160	(Ethanol)	51a
H	NO_2	Cl	H	H	F, L	337–340	(Acetic acid)	51a
H	H	Cl	H	NH_2	U	277–278 (dec.)	(Acetic acid)	51a
NO_2	H	Cl	H	H	L	298–299 (dec.)	(Acetic acid)	51a
NH_2	H	Cl	H	H	U	335–337 (dec.)	(Acetic acid)	51a
$COOCH_3$	H	H	H	H	V	229–230 (dec.)	Yellow microprisms (methanol)	51a
$COOC_2H_5$	H	H	H	H	V	194–195 (dec.)	—	51a
$COOCH_3$	H	Cl	H	H	V	254–255 (dec.)	—	51a

TABLE 1C-5 (*continued*)

R_1	R_2	R_3	R_4	R_5	Prep.[a]	MP (°C)	Remarks[b]	Ref.
$COOC_2H_5$	H	Cl	H	H	V	233–235 (dec.)	—	51a
$CONHNH_2$	H	H	H	H	W	>275 (dec.)	(Dimethylformamide)	51a
$CONHNH_2$	H	Cl	H	H	W	>290 (dec.)	(Dimethylformamide)	51a
$CONH_2$	H	H	H	H	W	354–355 (dec.)	(Acetic acid)	51a
Cl	H	Cl	H	H	R	307–308	(Ethanol)	51a
COOH	H	NO_2	H	Cl	L	244–245 (dec.)	(Acetic acid)	51a
H	H	NO_2	H	Cl	F	248–249	(Acetic acid)	51a
H	H	NH_2	H	Cl	U	310–311 (dec.)	(Nitrobenzene)	51a
H	Cl	NO_2	H	H	L	306–310 (dec.)	(Nitrobenzene)	51a
H	Cl	H	H	NO_2	L	220–221	(Acetic acid)	51a
H	OH	NO_2	H	H	X	255–257 (dec.)	Brown prisms (acetic acid)	51a
NO_2	H	Cl	Cl	H	L	298–299 (dec.)	Acetic acid	51a
COOH	H	NO_2	Cl	Cl	L	252–253 (dec.)	(Acetic acid)	51a
H	H	NO_2	Cl	Cl	F	238–240	(Acetic acid)	51a
H	Cl	H	H	SO_3H	Y	>360	(Water)	51a
H	OH	H	H	SO_3H	Z	—	Sodium salt, mp >360°	51a
H	SO_3H	H	H	OH	Z	—	Potassium salt, mp >360°	51a
H	OH	I	H	SO_3H	AA	—	Sodium salt, mp >360°	51a
H	OH	I	H	I	AA	265–267 (dec.)	Yellow needles (dimethylformamide)	51a
H	SO_3H	H	I	OH	AA	—	Potassium salt, mp 205–210° (dec.)	51a
H	SO_3H	H	H	Cl	Y	>360	(Water)	51a
H	OH	H	H	H	BB	216–218	Yellow feathery needles (water)	51a
H	H	H	H	OH	CC	340–345 (dec.)	(Acetic acid)	51a

[a] A, by diazotization and cyclization of the appropriately substituted 2-aminoacetophenone with sodium nitrite in hydrochloric acid. B, by diazotization and cyclization of the appropriately substituted 2-aminophenyl benzyl ketone with sodium nitrite in concentrated hydrochloric acid. C, by diazotization and cyclization of *o*-aminophenyl α-naphthylmethyl ketone with sodium nitrite in concentrated hydrochloric acid. D, by treatment of 4-amino-3-nitrocinnoline with dilute aqueous sodium hydroxide. E, by Friedel-Crafts cyclization of the appropriately substituted mesoxalyl chloride phenylhydrazone with titanium chloride in nitrobenzene. F, by decarboxylation of the corresponding cinnoline-3-carboxylic acid in benzophenone or Dowtherm at 190–215° C. G, by treatment of 3-(4-chlorophenyl)cinnoline-4-carboxylic acid or 3-(4-methoxyphenyl)cinnoline-4-carboxylic acid with potassium hydroxide and cupric oxide at 200–300° C. H, by refluxing 3-(4-chlorophenyl)cinnoline 1-oxide in thionyl chloride for 1 hr, followed by refluxing 1 hr in aqueous-ethanolic potassium hydroxide. I, by treatment of the appropriate 4-hydroxycinnoline with bromine in aqueous potassium hydroxide. J. by esterification of 4-hydroxy-6,7-dimethoxycinnoline-3-acetic acid with methanol or by diazotization and cyclization of methyl-*β*-(4-amino-4,5-dimethoxybenzoyl)propionate with sodium nitrite in hydrochloric acid. K, by decarboxylation of 4-hydroxy-6,7-dimethoxycinnoline-3-acetic acid in sulfuric acid at 200°, or by diazotization and cyclization of 2-amino-4,5-dimethoxypropiophenone with sodium nitrite in concentrated hydrochloric acid at 0°. L, by nitration of the corresponding 4-hydroxycinnoline. M, by diazotization and cyclization of the appropriately substituted 2-aminoacetophenone with sodium nitrite in formic acid. N, by diazotization and cyclization of 3,3′-diacetyl-4,4′-diaminoazobenzene with sodium nitrite in formic acid. O, by chlorination of 6-bromo-4-hydroxycinnoline with sulfuryl chloride in acetic acid. P, by bromination of 6-bromo-4-hydroxycinnoline with bromine in acetic acid. Q, by esterification of 6-chloro-4-hydroxycinnoline-3-carboxylic acid in ethanol using boron trifluoride etherate as catalyst. R, by hydrolysis of the corresponding 4-chlorocinnoline or hexafluorocinnoline. S, by peroxide oxidation of 4-ethyl-3-methyl-8-nitrocinnoline. T, by treatment of the corresponding 3-bromocinnoline with cuprous cyanide in refluxing dimethylformamide. U, by reduction of the corresponding nitrocinnoline. V, from the corresponding 4-hydroxycinnoline-3-carboxylic acid. W, from the methyl or ethyl ester of the corresponding carboxylic acid. X, by hydrolysis of 5-chloro-4-hydroxy-6-nitrocinnoline. Y, by sulfonation of the corresponding cinnoline in fuming sulfuric acid. Z, by treatment of the corresponding 5- or 8-chlorocinnoline with concentrated potassium hydroxide solution at 200°. AA, by iodination of the corresponding cinnoline. BB, by boiling 4,5-dihydroxycinnoline-8-sulfonic acid in aqueous sulfuric acid. CC, by boiling 4,8-dihydroxycinnoline-5-sulfonic acid in aqueous sulfuric acid.

[b] Solvent employed in recrystallization is enclosed in parentheses.

TABLE 1C-6. 3-Hydroxycinnolines

R_1	R_2	R_3	R_4	R_5	Prep.[a]	MP (°C)	Remarks[b]	Ref.
H	H	H	H	H	A, B, C	201–203 200–201 203.5–204.5	Yellow needles (water, then benzene); benzoyl deriv., mp 148–149°	19, 21, 22, 23a
H	H	Cl	H	H	A, C	262–265 304–305 (dec.)	Yellow needles (ethanol)	19, 21
H	H	OH	H	H	C	>300	Yellow nodules (hydrochloric acid); 6(3?)-benzoyloxy-3(6?)-hydroxycinnoline deriv., mp 233–235°	19
C_6H_5	H	H	H	H	D	300–302 (dec.)	Yellow blades (ethanol)	20

$C_6H_5CH_2$	H	H	H	H	E	222–225 (dec.)	Yellow needles (benzene-ligroin)	20
CH_3	H	H	H	H	A	232–234 (dec.)	Crystals (dry benzene); this compound has mp 204–205° (dec.) when recrystallized from aqueous ethanol	21
H	H	H	Cl	H	A, F	263–264.5 (dec.)	Yellow crystals (dry benzene); this compound has mp 255° when recrystallized from ethanol	21, 23, 23a
H	Cl	H	H	H	G	268 (dec.)	Yellow crystals (ethanol)	23, 23a
Cl	H	H	H	H	H	220	Yellow crystals (benzene or ethanol)	23, 23a

[a] A, by diazotization of the appropriately substituted 3-aminocinnoline in aqueous mineral acid. B, by oxidation of 1-aminoöxindole with lead tetraacetate. C, by diazotization of the appropriately substituted sodium *o*-aminomandelate with sodium nitrite in concentrated hydrochloric acid, reduction of the resulting diazonium salt to the *o*-hydrazinomandelic acid derivative with stannous chloride, and cyclization of the hydrazino compound to the cinnoline in boiling hydrochloric acid solution. The 3,6-dihydroxycinnoline was prepared from 2-amino-5-methoxymandelic acid, demethylation of the methoxy group accompanying the cyclization of the intermediate hydrazino compound. D, by hydrolysis of 1-amino-3-phenyldioxindole in aqueous potassium hydroxide. The resulting potassium α-phenyl-2-hydrazinomandelate was then cyclized to the 3-cinnolinol by careful neutralization with hydrochloric acid. E, by hydrolysis of 1-amino-3-benzyldioxindole in aqueous potassium hydroxide. The resulting potassium α-benzyl-2-hydrazinomandelate was then cyclized to the 3-cinnolinol by careful neutralization with hydrochloric acid. F, by treatment of one equivalent of 6-chloro-1-aminoöxindole in dry benzene with one equivalent of *t*-butyl hypochlorite. G, by treatment of one equivalent of 4-chloro-1-aminoöxindole in dry benzene with one equivalent of *t*-butyl hypochlorite. H, by treatment of one equivalent of 1-aminoöxindole in dry benzene with two equivalents of *t*-butyl hypochlorite, or by treatment of one equivalent of 3-hydroxycinnoline with one equivalent of *t*-butyl hypochlorite.

[b] Solvent employed in recrystallization is enclosed in parentheses.

TABLE 1C-7. 5-, 6-, 7-, and 8-Hydroxycinnolines

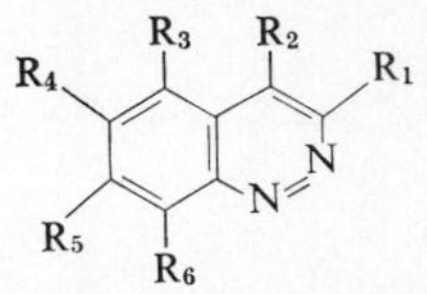

R_1	R_2	R_3	R_4	R_5	R_6	Prep.[a]	MP (°C)	Remarks[b]	Ref.
H	H	OH	H	H	H	A	285 (dec.)	Brownish-yellow cubes (ethanol)	25
H	H	H	OH	H	H	A	>300	Cream-colored needles (water)	25
H	H	H	H	OH	H	A	>300	Needles (water, then ethanol)	25
OH	H	H	OH	H	H	B	>300	Yellow nodules (hydrochloric acid)	19
H	H	H	H	H	OH	A	185–186 186–187	Yellow needles (benzene); sublimes 115–120° at 0.05 mm	24, 26
H	CH_3	H	H	H	OH	A	177–178.5 176.5–177.5	Yellow prisms (benzene); sublimes 150–160° at 0.1 mm	24, 26
H	C_6H_5	H	H	H	OH	A	142–143.5	Fawn-colored needles (benzene-ligroin)	24
H	COOH	H	H	H	OH	C	200 (dec.)	Red needles (water)	26
H	$C_6H_5CH{:}CH$	H	H	H	OH	D	200–201	Yellow needles (benzene or methanol); 8-benzoyl deriv., mp 212–213°	26
H	CH_3NH	H	OH	H	H	A	324–326 (dec.)	(Water)	48a

[a] A, by demethylation of the corresponding 5-, 6-, 7-, or 8-methoxycinnoline in refluxing hydrobromic acid. B, by diazotization of sodium 2-amino-5-methoxymandelic acid with sodium nitrite in concentrated hydrochloric acid, reduction of the resulting diazonium salt to 2-hydrazino-5-methoxymandelic acid, and cyclization to the cinnoline in boiling hydrochloric acid solution. Demethylation of the methoxy group accompanied by cyclization. C, by oxidation of 8-benzoyloxy-4-styrylcinnoline with potassium permanganate in pyridine. The 8-benzoyl group was removed during work-up. D, by treatment of 8-hydroxy-4-methylcinnoline monohydrochloride with benzaldehyde and dry gaseous hydrochloric acid for 1.5 hr at 155–160°.

[b] Solvent employed in recrystallization is enclosed in parentheses.

References

1. V. von Richter, *Ber.*, **16**, 677 (1883).
2. K. Schofield and J. C. E. Simpson, *J. Chem. Soc.*, **1945**, 512.
3. K. Schofield and T. Swain, *J. Chem. Soc.*, **1949**, 2393.
4. J. C. E. Simpson, in *The Chemistry of Heterocyclic Compounds*, Vol. V, A. Weissberger, Ed., Interscience, New York-London, 1953.
5. T. L. Jacobs, in *Heterocyclic Compounds*, Vol. VI, R. C. Elderfield, Ed., Wiley, New York, 1957, p. 143.
6. K. Pfannstiel and J. Janecke, *Ber.*, **75**, 1096 (1942).
7. N. J. Leonard, S. N. Boyd, and H. F. Herbrandson, *J. Org. Chem.*, **12**, 47 (1947).
8. W. Borsche and A. Herbert, *Ann.*, **546**, 293 (1941).
9. J. McIntyre and J. C. E. Simpson, *J. Chem. Soc.*, **1952**, 2606.
10. C. M. Atkinson and J. C. E. Simpson, *J. Chem. Soc.*, **1947**, 232.
11. D. W. Ockenden and K. Schofield, *J. Chem. Soc.*, **1953**, 3706.
12. K. Schofield and J. C. E. Simpson, *J. Chem. Soc.*, **1948**, 1170.
13. H. J. Barber, K. Washbourn, W. R. Wragg, and E. Lunt, *J. Chem. Soc.*, **1961**, 2828; Br. Pat. 762,184, Nov. 28, 1956.
14. R. R. Shoup and R. N. Castle, *J. Heterocyclic Chem.*, **2**, 63 (1965).
15. H. E. Baumgarten, *J. Am. Chem. Soc.*, **77**, 5109 (1955).
16. H. S. Lowrie, *J. Med. Chem.*, **9**, 670 (1966).

16a. E. Hayashi and T. Watanabe, *Yakugaku Zasshi*, **88**, 94 (1968).

16b. G. B. Barlin and W. V. Brown, *J. Chem. Soc.* (*C*), **1967**, 2473.

16c. G. B. Barlin and W. V. Brown, *J. Chem. Soc.* (*C*), **1969**, 921.

17. G. Bossel, *Inaugural Dissertation*, Tübingen, May 1925; E. J. Alford and K. Schofield, *J. Chem. Soc.*, **1952**, 2102.
18. P. W. Neber, G. Knoller, K. Herbst, and A. Trissler, *Ann.*, **471**, 113 (1929).
19. E. J. Alford and K. Schofield, *J. Chem. Soc.*, **1952**, 2102.
20. H. E. Baumgarten and P. L. Creger, *J. Am. Chem. Soc.*, **82**, 4634 (1960).
21. H. E. Baumgarten, W. F. Murdock, and J. E. Dirks, *J. Org. Chem.*, **26**, 803 (1961).
22. H. E. Baumgarten, P. L. Creger, and R. L. Zey, *J. Am. Chem. Soc.*, **82**, 3977 (1960).
23. W. F. Wittman, Ph.D. Dissertation, University of Nebraska, Lincoln, 1965; University Microfilms (Ann Arbor, Mich.) Order No. 65-11,435.

23a. H. E. Baumgarten, W. F. Whittman, and G. J. Lehmann, *J. Heterocyclic Chem.*, **6**, 333 (1969).

24. E. J. Alford, H. Irving, H. S. Marsh, and K. Schofield, *J. Chem. Soc.*, **1952**, 3009.
25. A. R. Osborn and K. Schofield, *J. Chem. Soc.*, **1955**, 2100.
26. A. Albert and A. Hampton, *J. Chem. Soc.*, **1952**, 4985.
27. K. Schofield and J. C. E. Simpson, *J. Chem. Soc.*, **1945**, 520.
28. A. R. Osborn and K. Schofield, *J. Chem. Soc.*, **1956**, 4207.
29. A. Albert and J. N. Phillips, *J. Chem. Soc.*, **1956**, 1294.
30. A. Albert and A. Hampton, *J. Chem. Soc.*, **1954**, 505.

30a. A. R. Katritzky, E. Lunt, B. Ternai, and G. J. T. Tiddy, *J. Chem. Soc.*, (B), 1243.

31. F. J. C. Rossotti and H. S. Rossotti, *J. Chem. Soc.*, **1958**, 1304.
32. E. J. Alford and K. Schofield, *J. Chem. Soc.*, **1953**, 1811.
33. S. F. Mason, *J. Chem. Soc.*, **1957**, 4874.
34. A. Albert and G. B. Barlin, *J. Chem. Soc.*, **1962**, 3129.

34a. A. J. Boulton, I. J. Fletcher, and A. R. Katritzky, *J. Chem. Soc.* (*B*), **1971**, 2344.

35. S. F. Mason, *J. Chem. Soc.*, **1957**, 5010.

36. D. E. Ames, R. F. Chapman, H. Z. Kucharska, and D. Waite, *J. Chem. Soc.*, **1965,** 5391.
37. G. B. Barlin, *J. Chem. Soc.*, **1965,** 2260.
38. D. E. Ames, R. F. Chapman, and D. Waite, *J. Chem. Soc.* (*C*), **1966,** 470.
39. R. N. Castle, D. B. Cox, and J. F. Suttle, *J. Am. Pharm. Assoc.*, **48,** 135 (1959).
40. R. R. Shoup and R. N. Castle, *J. Heterocyclic Chem.*, **1,** 221 (1964).
41. H. Irving and H. S. Rossotti, *Analyst*, **80,** 245 (1955).
42. H. Irving and H. S. Rossotti, *J. Chem. Soc.*, **1954,** 2910.
42a. J. R. Elkins and E. V. Brown, *J. Heterocyclic Chem.*, **5,** 639 (1968).
43. R. N. Castle and F. H. Kruse, *J. Org. Chem.*, **17,** 1571 (1952).
44. C. M. Atkinson and C. J. Sharpe, *J. Chem. Soc.*, **1959,** 2858.
45. A. Albert and W. Gledhill, *J. Soc. Chem. Ind.* (*London*), **64,** 169 (1945).
46. N. J. Leonard and S. N. Boyd, *J. Org. Chem.*, **11,** 419 (1946).
47. R. N. Castle, H. Ward, N. White, and K. Adachi, *J. Org. Chem.*, **25,** 570 (1960).
48. R. N. Castle, R. R. Shoup, K. Adachi, and D. L. Aldous, *J. Heterocyclic Chem.*, **1,** 98 (1964).
48a. H. J. Barber and E. Lunt, *J. Chem. Soc.* (*C*), **1968,** 1156.
49. G. Wagner and D. Heller, *Z. Chem.*, **4,** 349 (1964).
50. K. Schofield and T. Swain, *J. Chem. Soc.*, **1950,** 384.
51. D. E. Ames and R. F. Chapman, *J. Chem. Soc.* (*C*), **1967,** 40.
51a. E. Lunt, K. Washbourn, and W. R. Wragg, *J. Chem. Soc.* (*C*), **1968,** 687.
51b. R. B. Moodie, J. R. Penton, and K. Schofield, *J. Chem. Soc.* (*B*), **1971,** 1493.
52. J. R. Keneford, J. S. Morley, and J. C. E. Simpson, *J. Chem. Soc.*, **1948,** 1702.
53. J. M. Bruce and P. Knowles, *J. Chem. Soc.*, **1964,** 4046.
54. D. E. Ames and H. Z. Kucharska, *J. Chem. Soc.*, **1962,** 1509.
55. D. E. Ames, R. F. Chapman, and H. Z. Kucharska, *J. Chem. Soc.*, Suppl. No. 1, **1964,** 5659.
55a. L. S. Besford and J. M. Bruce, *J. Chem. Soc.*, **1964,** 4037.
55b. H. Lund, *Acta Chem. Scand.*, **21,** 2525 (1967).
55c. C. W. Rees and A. A. Sale, *J. Chem. Soc.* (*D*), **1971,** 531.
56. J. S. Morley, *J. Chem. Soc.*, **1959,** 2280.
57. K. Schofield and J. C. E. Simpson, *J. Chem. Soc.*, **1946,** 472.
58. D. E. Ames and H. Z. Kucharska, *J. Chem. Soc.*, **1963,** 4924.
59. D. E. Ames and H. Z. Kucharska, *J. Chem. Soc.*, **1964,** 283.
60. D. E. Ames, *J. Chem. Soc.*, **1964,** 1763.
61. H. S. Lowrie, *J. Med. Chem.*, **9,** 784 (1966).
61a. A. W. Ellis and A. C. Lovesey, *J. Chem. Soc.* (*B*), **1967,** 1285.
61b. A. W. Ellis and A. C. Lovesey, *J. Chem. Soc.* (*B*), **1968,** 1898.
61c. D. E. Ames, B. Novitt, and D. Waite, *J. Chem. Soc.* (*C*), **1969,** 796.
62. G. Wagner and D. Heller, *Z. Chem.*, **4,** 386 (1964).
63. G. Wagner, private communication.
64. J. M. Bruce, P. Knowles, and L. S. Besford, *J. Chem. Soc.*, **1964,** 4044.
65. G. Wagner, *Z. Chem.*, **6,** 367 (1966).
66. D. E. Ames and A. C. Lovesey, *J. Chem. Soc.*, **1965,** 6036.
67. K. Schofield and R. S. Theobald, *J. Chem. Soc.*, **1950,** 395.
68. J. R. Keneford and J. C. E. Simpson, *J. Chem. Soc.*, **1947,** 227.
69. A. Jart, A. J. Bigler, and V. Bitsch, *Anal. Chim. Acta*, **31,** 472 (1964).
70. R. D. Chambers, J. A. MacBride, and W. K. R. Musgrave, *J. Chem. Soc.* (*D*), **1970,** 739.
71. A. N. Kaushal and K. S. Narang, *Indian J. Chem.*, **6,** 350 (1968).
72. M. H. Palmer and E. R. R. Russell, *J. Chem. Soc.* (*C*), **1968,** 2621.
73. W. A. White, German Pat. 2,005,104, August 6, 1970; *Chem. Abstr.*, **73,** 77269 (1970).

Part D. Halocinnolines

I. Methods of Preparation

A. 4-Halocinnolines

4-Chlorocinnoline (**2**) has been known since 1892 (1). The most commonly employed method to obtain it or a substituted 4-chlorocinnoline is by treatment of 4-hydroxycinnoline (**1**) or a substituted derivative of 4-hydroxycinnoline with phosphorus oxychloride, the reaction conditions varying from a gentle warming of the mixture for five min to refluxing it for more than one hr (2–10). Phosphorus oxychloride containing a small amount of *N*,*N*-dimethylaniline (11) or mixtures of phosphorus oxychloride and

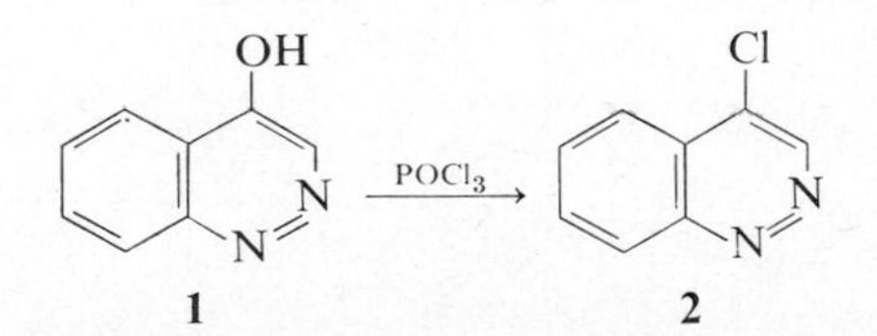

phosphorus pentachloride (10–13) have been used successfully in cases where the use of phosphorus oxychloride alone has been unsatisfactory. One disadvantage in the use of mixtures of phosphorus oxychloride and phosphorus pentachloride is that metathesis of other nuclear substituents may occur, 4-hydroxy-6-nitrocinnoline giving, for example, both 4-chloro-6-nitro- and 4,6-dichlorocinnoline with this mixture (14). Further examples of such metatheses were given in Section 1C-III-A-1. The problem of group exchange may be largely avoided by use of phosphorus oxychloride alone or by the use of thionyl chloride containing a catalytic amount of phosphorus pentachloride in an inert solvent such as chlorobenzene or *o*-dichlorobenzene (14).

A much less frequently used preparation of 4-chlorocinnolines is by treatment of a cinnoline 1-oxide (**3**; Y = H, C_6H_5, *p*-ClC_6H_4, *p*-$CH_3OC_6H_4$, and *p*-$CH_3C_6H_4$) or cinnoline 2-oxide (**4**; Y = C_6H_5, *p*-ClC_6H_4, *p*-$CH_3OC_6H_4$, and *p*-$CH_3C_6H_4$) with phosphorus oxychloride in chloroform solution. Oxygen is removed from the ring nitrogen atom during the course of the reaction. Good yields of the 4-chloro derivative are generally realized (15–17).

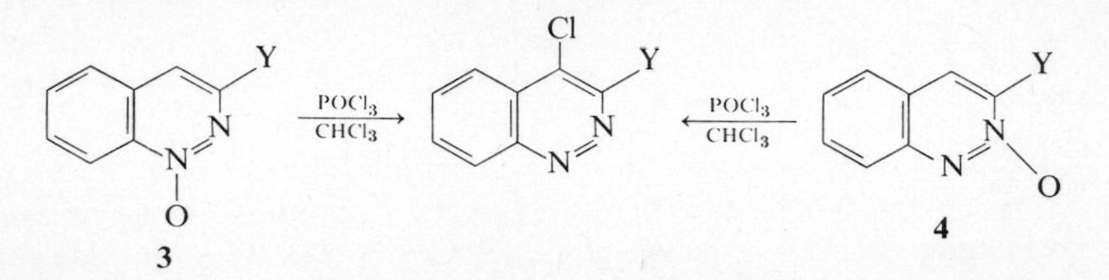

Minor methods used to prepare 4-chlorocinnolines involve treatment of 1-aminoöxindole (**5**) with two equivalents of *t*-butyl hypochlorite to give 93% of 4-chloro-3-hydroxycinnoline (**6**), treatment of 3-hydroxycinnoline (**7**) with an equimolar amount of *t*-butyl hypochlorite to give 77% of 4-chloro-3-hydroxycinnoline (**6**) (18, 18a) (see Section 1C-I-B-3), and displacement of the 4-nitro group from 3-methoxy-4-nitrocinnoline 1-oxide by heating this compound in concentrated hydrochloric acid at 100° C for 1 hr to give 4-chloro-3-methoxycinnoline 1-oxide (19). It should then be possible to remove the 1-oxide function by treating the cinnoline 1-oxide with phosphorus trichloride in chloroform solution.

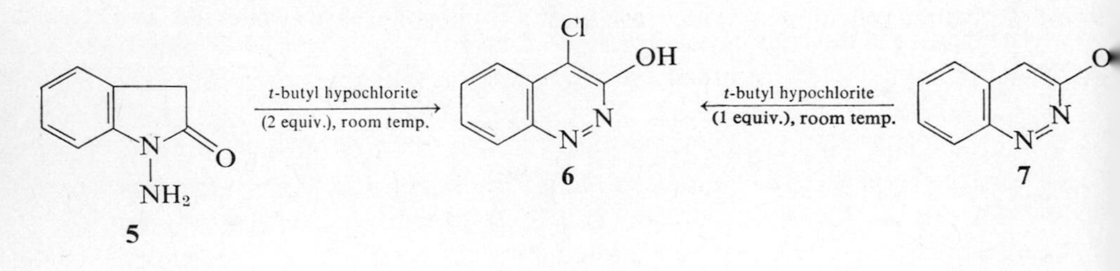

B. 3-Halocinnolines

In a manner analogous to the principal preparation of 4-chlorocinnoline, 3-chlorocinnoline (**8**) may be prepared by chlorination of 3-hydroxycinnoline (**7**) with phosphorus oxychloride, although in this case the yield of the chloro compound is rather low (20). A second method to introduce chlorine in the 3-position of the cinnoline ring is by direct halogenation of a 4-hydroxycinnoline (**1**) with sulfuryl chloride in acetic acid to give a 3-chloro-4-hydroxycinnoline (**9**) (21, 21a).

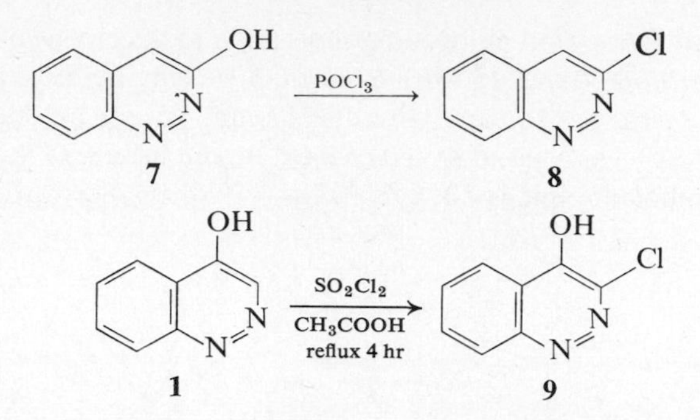

A third way to introduce a chlorine atom into the 3-position of the cinnoline ring is by metathesis of a 3-nitro group. For example, when 4-chloro-3-nitrocinnoline (**10**) is heated with a mixture of phosphorus oxychloride and phosphorus pentachloride, 87% of 3,4-dichlorocinnoline (**11**) is obtained. Chlorination of 4-hydroxy-3-nitrocinnoline with this same chlorinating mixture yields two products, 4-chloro-3-nitrocinnoline and 3,4-dichlorocinnoline (7). Refluxing 6-chloro-4-hydroxy-3-nitrocinnoline with phosphorus oxychloride gives 3,4,6-trichlorocinnoline together with 4,6-dichloro-3-nitrocinnoline (21b).

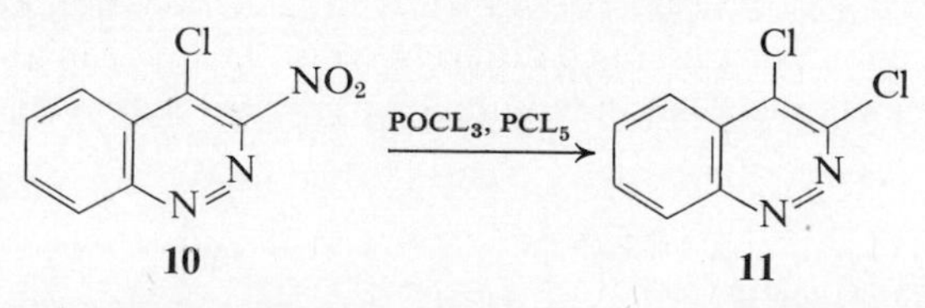

An effective, almost quantitative method for the preparation of 3-bromo-4-hydroxycinnolines (**13**) involves bromination of a 4-hydroxycinnoline (**12**; Y = H, 6-nitro, 8-nitro, 8-methyl, 6,7-dimethoxy, and 6,7-methylenedioxy) with bromine in an aqueous alkaline medium (10, 22–23b). Bromination of 4-hydroxycinnolines in acetic acid solution is much less effective, giving only low yields of the 3-bromocinnolines. The reaction probably is impeded when the cinnoline ring nitrogens are protonated by the hydrobromic acid which is formed during the bromination (22).

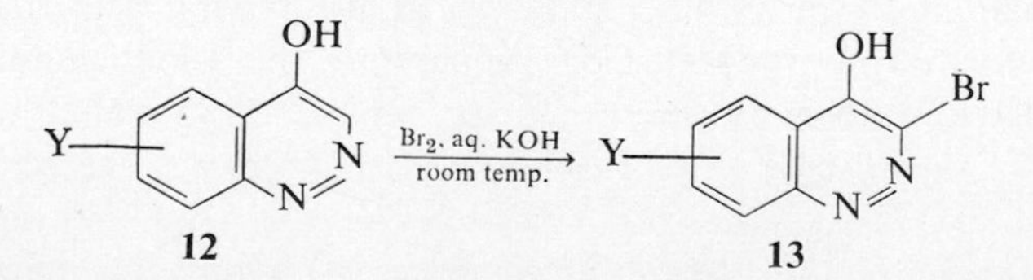

The parent compound, 3-bromocinnoline, otherwise unsubstituted, has been obtained in this manner by bromination of 4-hydroxycinnoline to give 3-bromo-4-hydroxycinnoline (**14**), conversion of this with phosphorus oxychloride to 3-bromo-4-chlorocinnoline (**15**), replacement of the 4-chloro substituent in compound **15** with toluene-*p*-sulfonylhydrazide to give the 4-hydrazino derivative **16**, and, finally, decomposition of this in refluxing aqueous sodium carbonate to give 3-bromocinnoline (**17**). The overall yield of 3-bromocinnoline is 80% from the 4-hydroxy derivative (9).

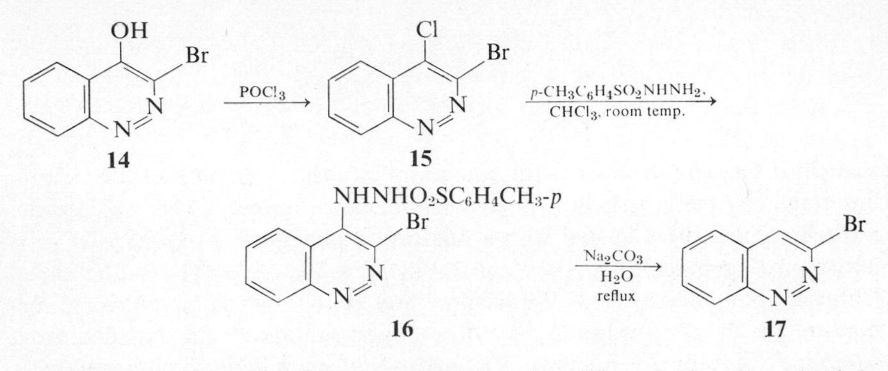

ω-Halo-*o*-aminoacetophenones (**18**; X = Br, Cl) may be diazotized in sulfuric acid to give 3-halo-4-hydroxycinnolines (**19**; X = Br, Cl) via the Borsche synthesis (see Section 1C-I-A-3). If the diazotization is effected with a halogen acid containing halogen different from the halogen atom contained in the *o*-aminoacetophenone, then halogen exchange may occur (21, 24).

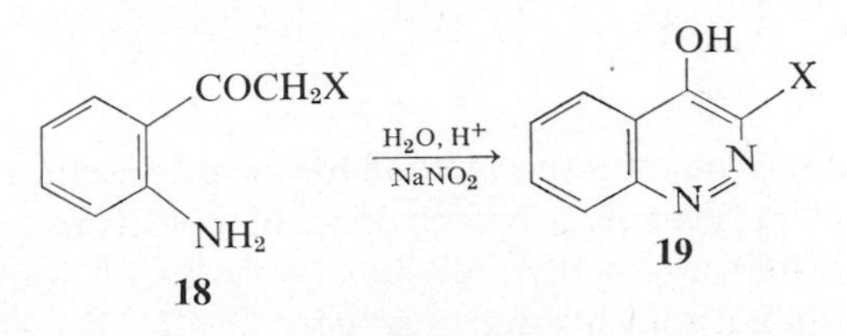

Finally, a 3-halo substituent can be introduced into the cinnoline ring by diazotization of a 3-amino group in concentrated hydrochloric or hydrobromic acid. In this case, the 3-aminocinnoline (**20**; R = H, CH_3) gives the 3-halocinnoline **21** (X = Br, Cl) in 18–53% yields together with the 3-hydroxycinnoline **22** in 29–40% yield. In dilute mineral acid, only the 3-hydroxy derivative is produced (25). Apparently, no 3-iodocinnolines have been described in the literature.

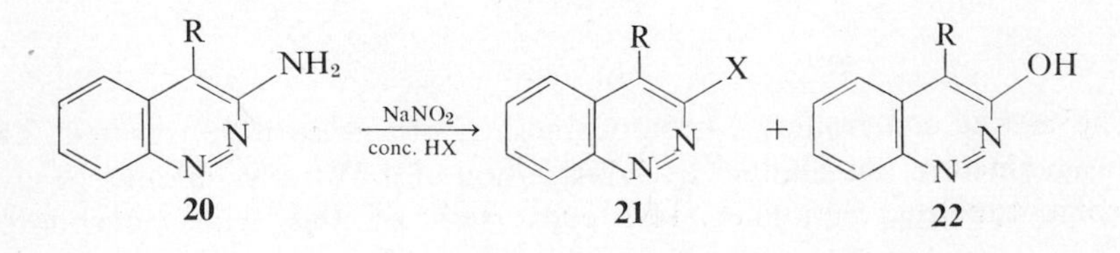

C. 5-, 6-, 7-, and 8-Halocinnolines

Of several different ways in which cinnolines bearing a halogen atom in the 5-, 6-, 7-, and 8-positions have been prepared, six utilize the diazotization of an aromatic amine.

In the first of the methods involving diazotization, the diazotization and spontaneous cyclization of halogenated *o*-aminophenylethylenes such as **23** has been used to prepare cinnolines **24**, which are substituted with halogen in the benzenoid ring (26). This is an example of the Widman-Stoermer synthesis (Section 1B-I-A).

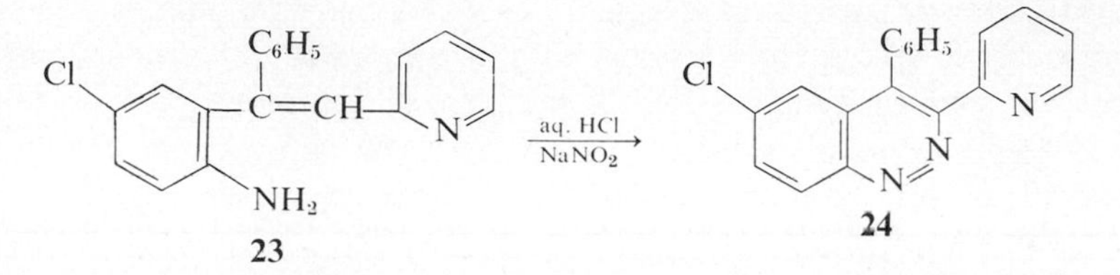

Second, the Borsche synthesis (Section 1C-I-A-3), which involves diazotization and cyclization of an *o*-aminoacetophenone, has been used successfully, 8-iodo-4-hydroxycinnoline (**26**) being prepared from 2-amino-3-iodoacetophenone (**25**) in this way (27), and 6-fluoro-4-hydroxycinnoline (**28**) being similarly obtained from 2-amino-5-fluoroacetophenone (**27**) (28).

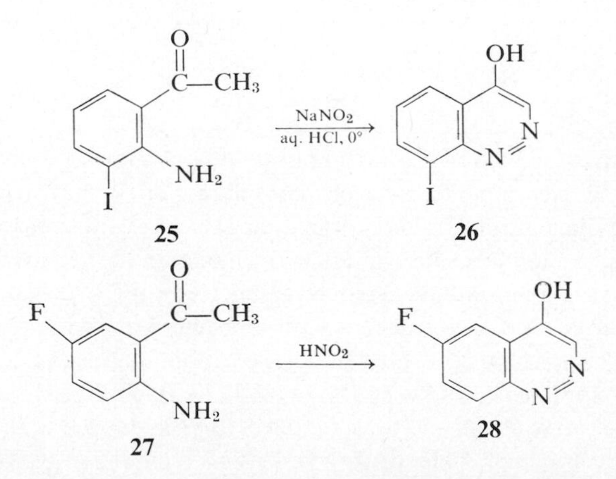

Third, diazotization of 8-amino-4-methylcinnoline (**29**), followed by treatment of the resultant diazonium salt with cuprous chloride, gives 8-chloro-4-methylcinnoline (**30**) (29). In the same way, 5-amino-6-chloro-4-hydroxycinnoline is converted to 5,6-dichloro-4-hydroxycinnoline, while

6-amino-8-chloro-4-hydroxycinnoline is converted to 6,8-dichloro-4-hydroxycinnoline (21b).

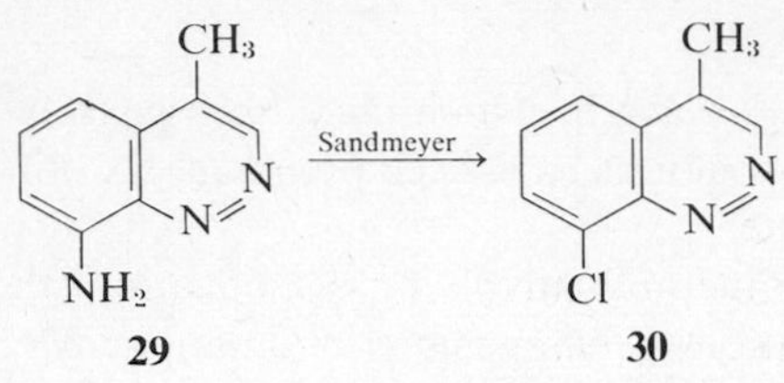

In a fourth sequence utilizing the diazotization reaction, *o*-hydrazinomandelic acids are prepared and then subjected to a Neber-Bossel cyclization (Section 1C-I-B-1) to give 3-hydroxycinnolines which are substituted with halogen in the benzenoid ring (30, 31). Yields by this method have been low. As an example of this procedure, sodium 2-amino-5-chloromandelate (**31**) is diazotized, and the diazonium salt is reduced with stannous chloride to give 5-chloro-2-hydrazinomandelic acid (**32**), which is then cyclized in refluxing aqueous acid solution to give the chlorocinnoline **33**.

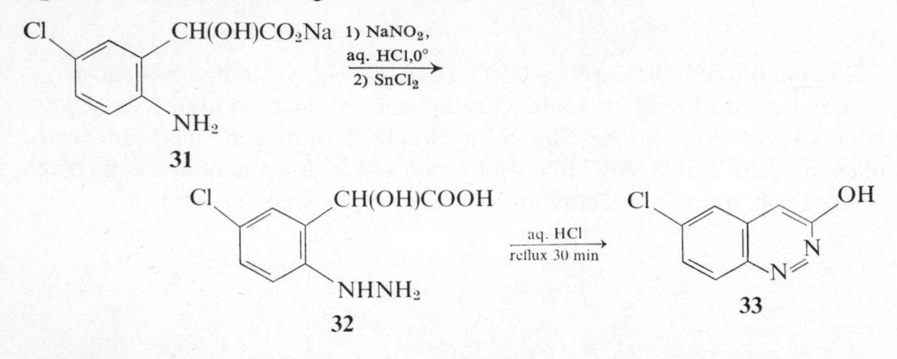

A fifth procedure involves diazotization of an appropriately substituted *o*-aminobenzaldehyde. The *o*-aminobenzaldehyde is diazotized and the intermediate diazonium salt is then treated in one of the following ways to give various cinnolines substituted with halogen in the benzenoid ring (32, 33). First, the *o*-aminobenzaldehyde **34** (X = 4-Cl, 5-Cl) is diazotized and allowed to react with acetoacetic acid to give the pyruvaldehyde x-halo-2-formylphenylhydrazone **35**, which undergoes spontaneous cyclization to give the 3-acetylcinnoline derivative **36** (X = 6-Cl, 7-Cl). The diazonium salt of **34** could alternatively be allowed to react with ethyl hydrogen malonate to give the ethyl glyoxylate phenylhydrazone **37**, which likewise cyclizes spontaneously, yielding the 3-carbethoxycinnoline compound **38** (X = 6-Cl, 7-Cl). With nitromethane, the diazonium salt of **34** gives the nitroformaldehyde phenylhydrazone **39**. This material will cyclize in aqueous potassium hydroxide to give a 3-nitrocinnoline derivative (**40**; X = 6-Cl, 7-Cl).

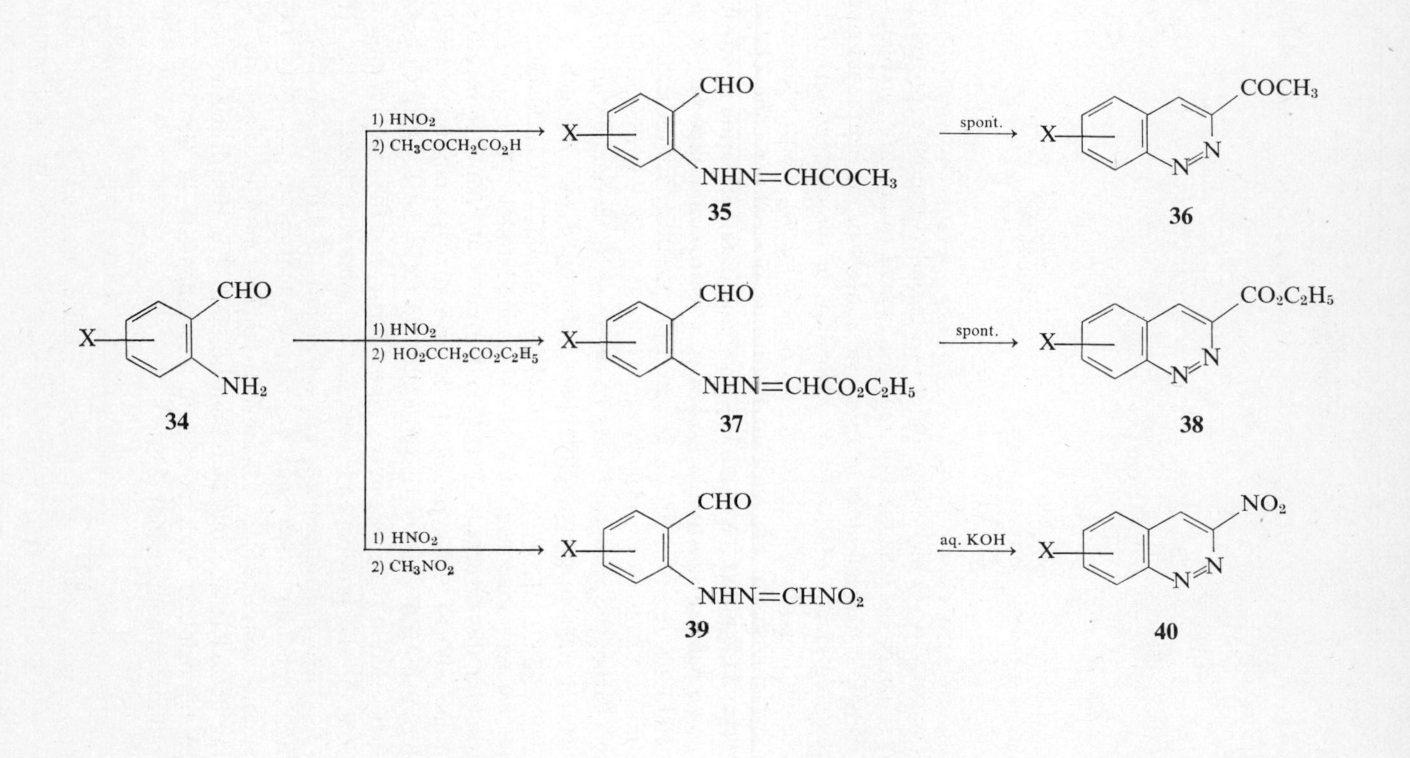

CHO
NH2
X
34
1) HNO2
2) CH3COCH2CO2H
CHO
NHN=CHCOCH3
35
spont.
COCH3
36
1) HNO2
2) HO2CCH2CO2C2H5
CHO
NHN=CHCO2C2H5
37
spont.
CO2C2H5
38
1) HNO2
2) CH3NO2
CHO
NHN=CHNO2
39
aq. KOH
NO2
40

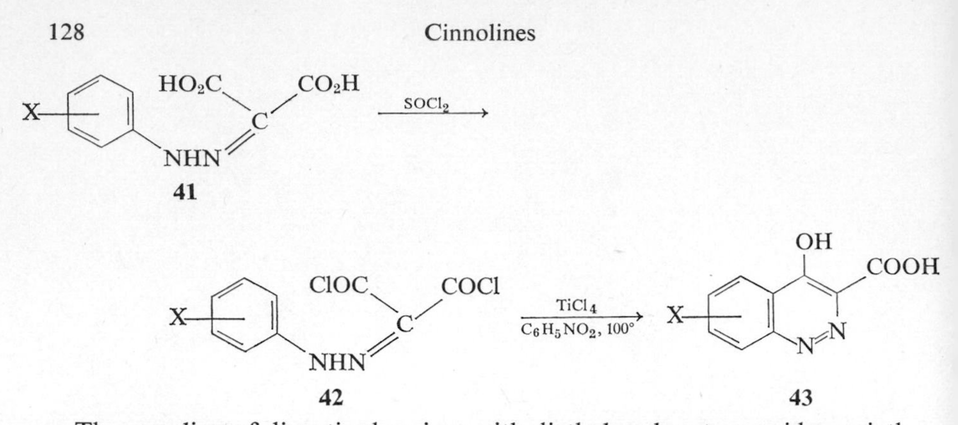

The coupling of diazotized amines with diethyl malonate provides a sixth way in which the diazotization reaction is used to prepare cinnolines substituted with a halogen atom or other substituent in the benzenoid ring. Such a coupling reaction followed by hydrolysis of the ester groups will yield a mesoxalic acid phenylhydrazone (**41**; X = halogen), which is converted with thionyl chloride to the mesoxalyl chloride phenylhydrazone **42**. The acid chloride **42** can then be cyclized with titanium tetrachloride in nitrobenzene to give a halogen-substituted 3-carboxy-4-hydroxycinnoline (**43**; X = 5-, 6-, 7-, and 8-Cl; 6- and 8-Br; 6-, 7-, and 8-F; 5,6-diCl; 5,8-diCl; 6,8-diCl; and 7,8-diCl). Yields vary widely, from 11 to 96% (5, 13) (Section 1C-I-A-4). The 3-carboxylic acid group may be removed by thermal decarboxylation.

Several ways to prepare cinnolines substituted with halogen in the benzenoid ring do not involve diazotization. The first of these methods involves oxidative rearrangement of a halo-1-aminoöxindole (**44**; X = 4-Cl, 6-Cl), a reaction that proceeds cleanly and rapidly at room temperature in dry benzene with *t*-butyl hypochlorite to give almost quantitative yields of the halogen-substituted 3-hydroxycinnoline **45** (X = 5-Cl, 7-Cl). Chlorine itself may be used in place of *t*-butyl hypochlorite, but the reaction appears to be less clean, and products are obtained in somewhat lower yields (18, 18a). In a related rearrangement, both 1-amino-2,3-dihydro-3-hydroxy-4,5,6,7-tetrafluoroindole (**45a**) and 1-amino-4,5,6,7-tetrafluoroindole (**45b**) are smoothly converted to 1,4-dihydro-5,6,7,8-tetrafluorocinnoline (**45c**) by being heated in aqueous hydrochloric acid. The dihydrocinnoline **45c** can then be oxidized with potassium permanganate in acetone to 5,6,7,8-tetrafluorocinnoline (**45d**) (33a).

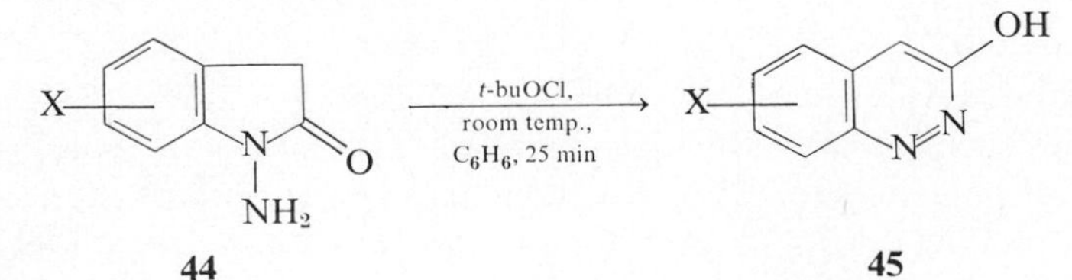

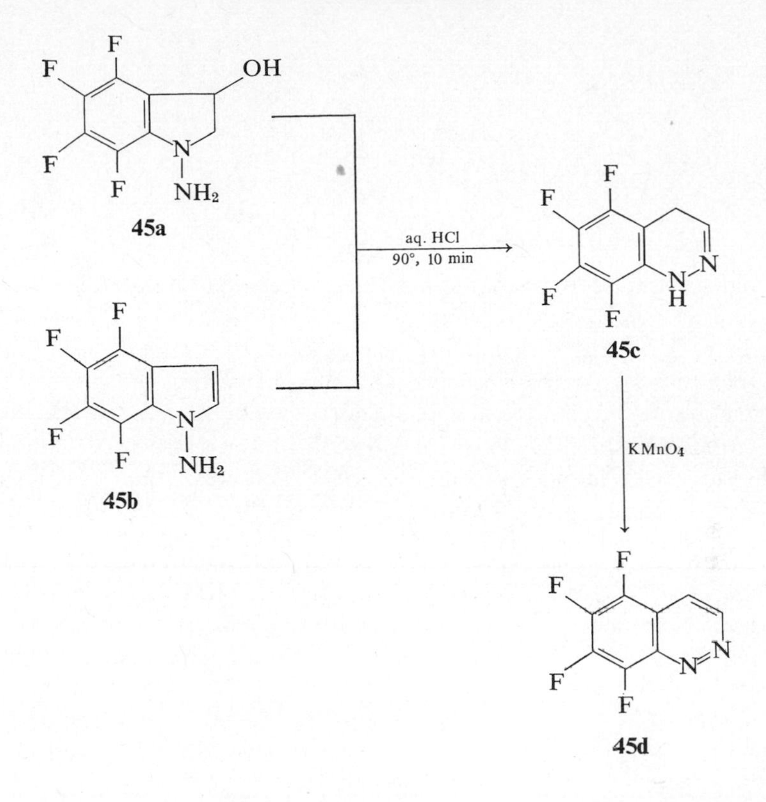

4,6-Dichlorocinnoline (**48**) has been prepared by metathesis of a 6-nitro group while attempting to convert 4-hydroxy-6-nitrocinnoline (**46**) into 4-chloro-6-nitrocinnoline (**47**) with a mixture of phosphorus oxychloride and phosphorus pentachloride. In this case the hydroxy derivative **46** yields a mixture of **47** and **48** (14).

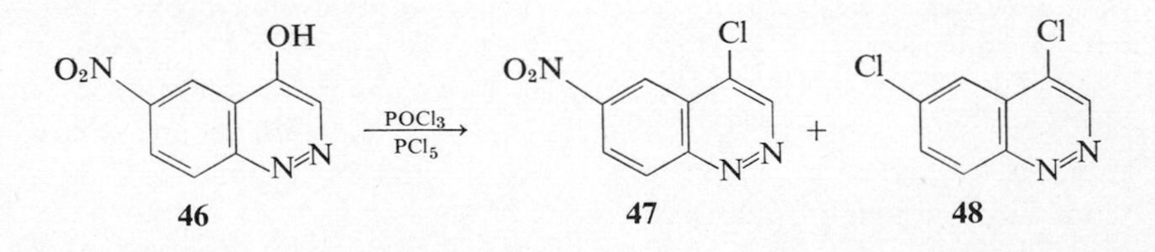

Several iodo-substituted cinnolines have been prepared by treating a cinnoline such as 4,5-dihydroxycinnoline-8-sulfonic acid (**48a**) with a mixture of potassium iodide and potassium iodate in acidic medium. In this case, the product is 4,5-dihydroxy-6-iodocinnoline-8-sulfonic acid (**48b**) (21b).

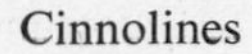

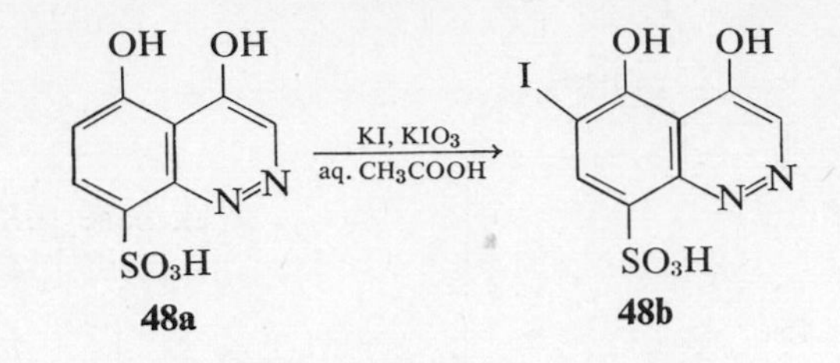

Finally, hexafluorocinnoline has been prepared from 7,8-dichloro-4-hydroxycinnoline (**48c**). In this transformation **48c** is converted first to 4-hydroxy-3,7,8-trichlorocinnoline (**48d**) by interaction with sulfuryl chloride, and then to 3,4,7,8-tetrachlorocinnoline (**48e**) by heating **48d** with a mixture of phosphorus oxychloride and phosphorus pentachloride. Chlorination of the tetrachlorocinnoline **48e** with chlorine in the presence of aluminum trichloride gives hexachlorocinnoline (**48f**), and this, upon reaction with potassium fluoride, gives hexafluorocinnoline (**48g**). Hexafluorocinnoline is unstable inasmuch as it is rapidly hydrolyzed in moist air to 4-hydroxy-3,5,6,7,8-pentafluorocinnoline. Upon irradiation with a mercury lamp at 100°, **48g** isomerizes in low yield to hexafluoroquinazoline (**48h**) (21a).

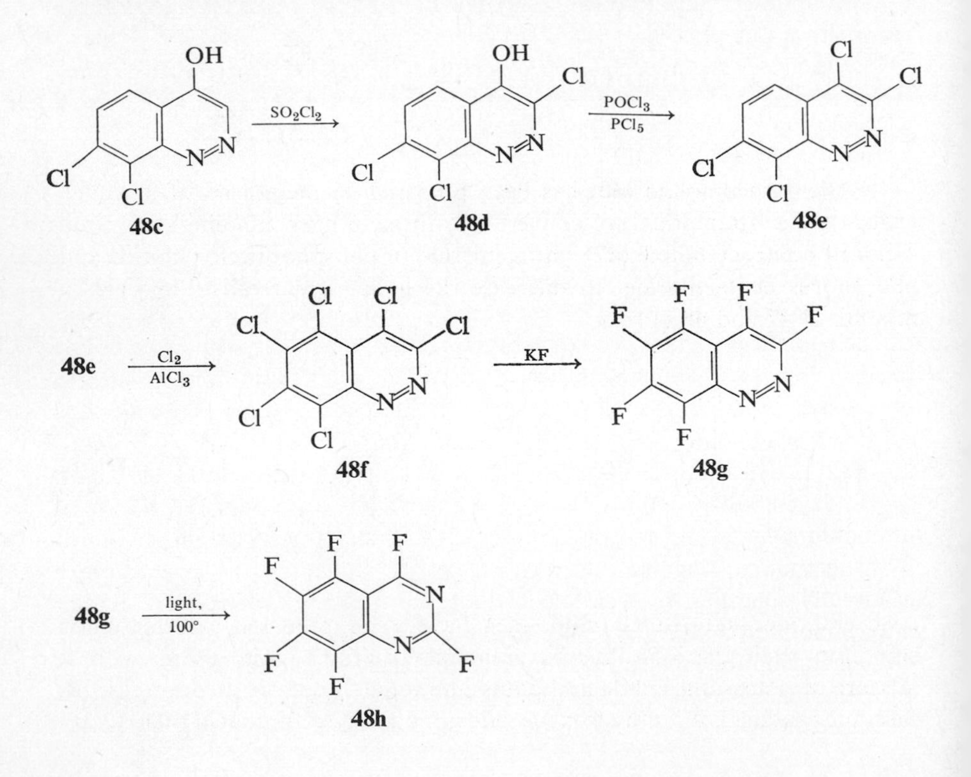

II. Physical Properties

The skin irritant (34) 4-chlorocinnoline is a weak base [pK_a 2.08 in 50% aqueous ethanol (35)], even weaker than cinnoline itself [pK_a 2.51 in 50% aqueous ethanol (35)].

The infrared spectrum of 4-chlorocinnoline disubstituted with methoxy groups in the 6- and 7-positions displays a band at 12.57 μ (801 cm^{-1}) which is attributed to C-Cl stretching absorption (36), while 4-mercapto- and 4-hydroxycinnolines substituted with halogens at the 5-, 6-, 7-, and 8-positions show absorption in the 8.3–10.0 μ (1200–1000 cm^{-1}) region which is due again to vibrations involving the halogen atom (37).

The ultraviolet absorption spectrum of 3-chlorocinnoline shows absorption maxima in 95% methanol at 285 mμ (log E_{max} 3.33) and 330 mμ (log E_{max} 3.43); in 0.01N hydrochloric acid at 285 mμ (log E_{max} 3.28) and 335 mμ (log E_{max} 3.42); and in 0.01N sodium hydroxide at 285 mμ (log E_{max} 3.40) and 338 mμ (log E_{max} 2.54) (20). The spectral characteristics of the UV absorption spectra of 3-, 4-, 6-, 7-, and 8-chlorocinnolines are discussed by Osborn and Schofield (38), although the actual absorption maxima are not recorded in this paper.

The pmr spectrum of 4-chlorocinnoline in carbon tetrachloride using tetramethylsilane as a reference standard displays a singlet at 0.98 τ and multiplets at 1.4–1.8 τ and 2.0–2.5 τ in a ratio of 1:1:3 (39).

III. Reactions

The most useful property of 4-chlorocinnolines is the ease with which the 4-chloro substituent undergoes nucleophilic displacement by a great variety of nucleophilic reagents. These reactions are summarized for 4-chlorocinnoline itself in Fig. 1D-1. Thus 4-chlorocinnoline (**2**) reacts with boiling water to give 4-hydroxycinnoline (**1**) (40) and with alkoxide ions or aryloxide ions to give a wide variety of 4-alkoxy- or 4-aryloxycinnolines (**49**; R = alkyl, aryl) (2, 4, 10, 11, 13–15, 22, 41–45). An alternative method to obtain 4-phenoxycinnoline (**50**) is to heat the 4-chloro derivative **2** with a mixture of ammonium carbonate and phenol (11, 43), although in the case of 4-chloro-3-nitrocinnoline, 4-amino-3-nitrocinnoline is obtained by this procedure in essentially quantitative yield instead of the 4-phenoxy derivative. Other nitrocinnolines do give the 4-phenoxy derivatives (7). Once formed, the 4-phenoxycinnoline **50** can be converted to 4-aminocinnoline (**51**) by heating with ammonium acetate at 160°, and in fact 4-phenoxy derivatives often have been prepared for just this purpose. The 4-aminocinnolines **51**, however,

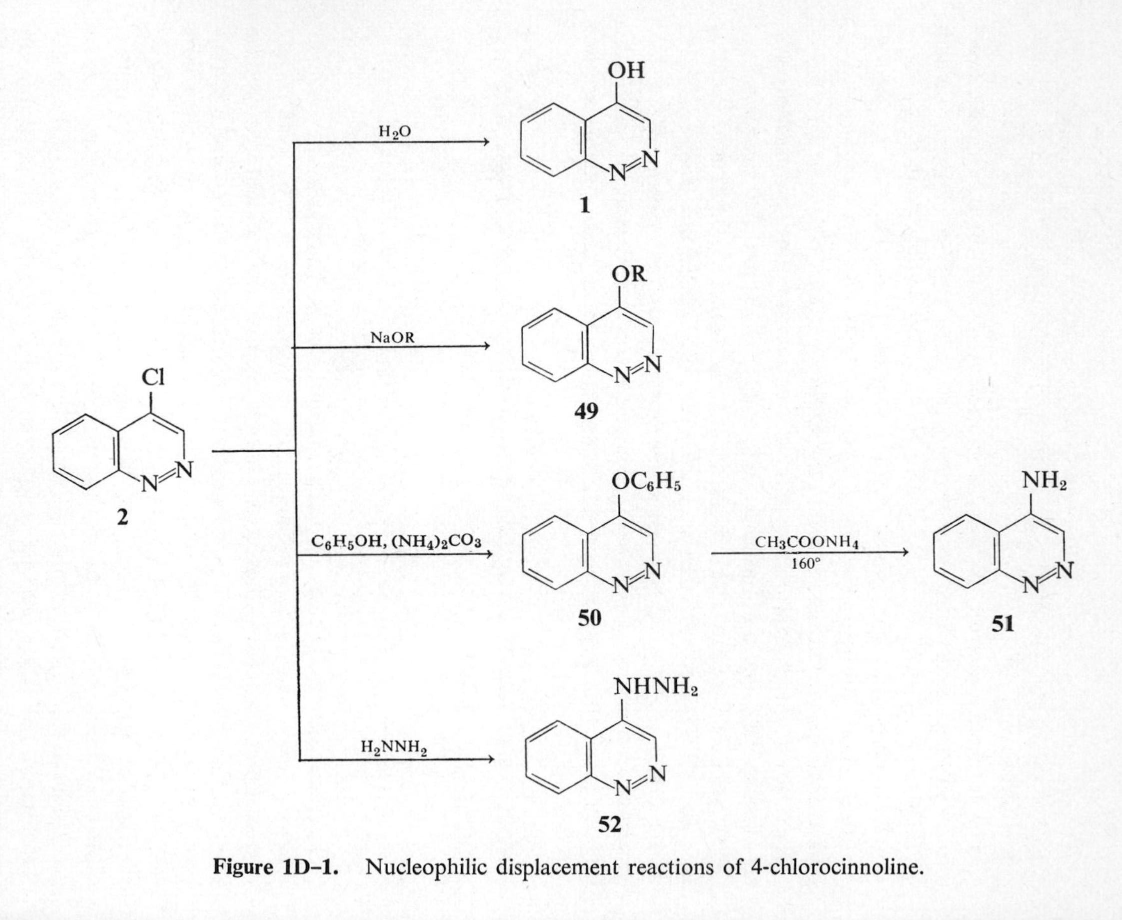

Figure 1D–1. Nucleophilic displacement reactions of 4-chlorocinnoline.

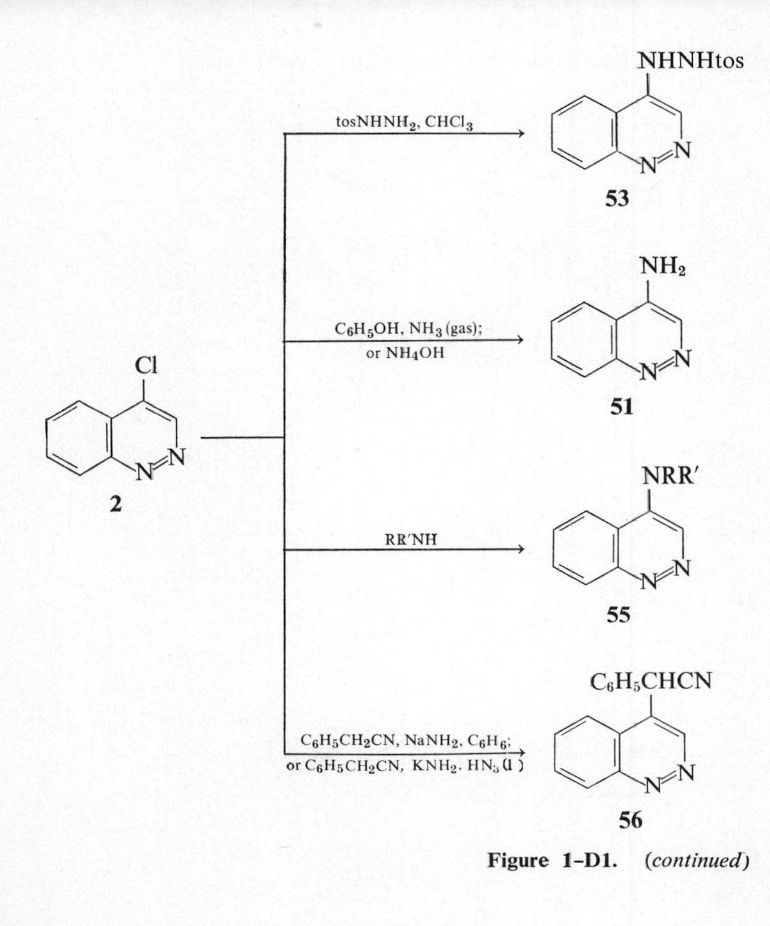

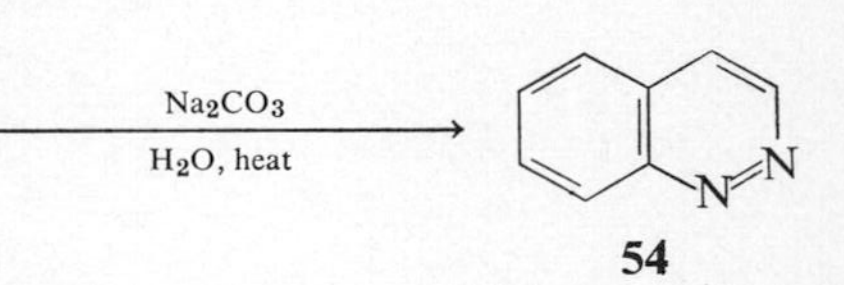

Figure 1–D1. *(continued)*

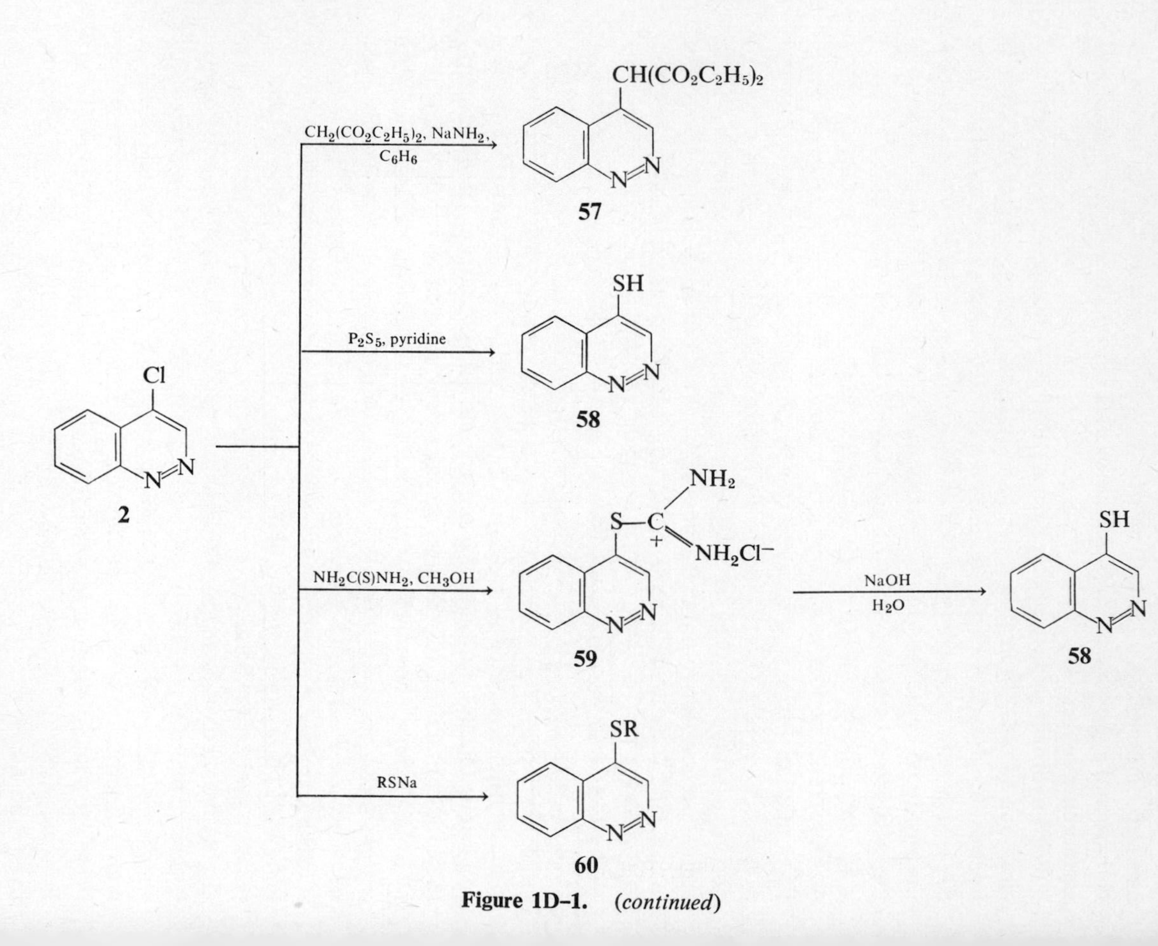

Figure 1D–1. *(continued)*

can also be prepared directly from the 4-chloro compound **2**, by treating **2** with phenol and gaseous ammonia (14, 45–47) or by allowing it to react with ammonium hydroxide (36).

With hydrazine, the 4-chloro derivative **2** gives the expected 4-hydrazinocinnoline **52** (48, 49) and with the substituted hydrazine, toluene-*p*-sulfonylhydrazide, gives the corresponding 4-toluene-*p*-sulfonylhydrazinocinnoline **53**, which can be decomposed in hot aqueous sodium carbonate to give unsubstituted cinnoline (**54**). This constitutes an excellent method to remove the 4-chloro substituent from the cinnoline ring (3, 6, 9, 38, 49).

Primary and secondary alkyl- and arylamines will displace chlorine from 4-chlorocinnoline to give the 4-amino derivatives **55** (R, R′ = alkyl; R = H, R′ = alkyl or aryl). This reaction can be carried out without solvent in the presence of excess amine, but better yields of product are reported when dimethylsulfoxide is used as a solvent (7, 16, 17). In a variation of the amination reaction, 4-methylaminocinnoline is obtained in good yield by treatment of 4-chlorocinnoline with phenol and methylamine at 140° for 40 min (2).

Sodio derivatives of active hydrogen compounds effectively displace the 4-chloro substituent from the cinnoline nucleus. For example, phenylacetonitrile in refluxing dry benzene with sodium amide gives 81% of α-(4-cinnolinyl)phenylacetonitrile (**56**) (12, 34, 50, 51), and under the same conditions diethyl malonate gives 44% of diethyl (4-cinnolinyl)malonate (**57**) (50). Examples of similar displacement reactions with other active hydrogen compounds such as ethyl acetoacetate, ethyl cyanoacetate, and α-substituted phenylacetonitriles are also recorded in the literature (12, 34, 50, 51).

A 70% yield of 4-cinnolinethiol (**58**) is obtained by treating 4-chlorocinnoline with phosphorus pentasulfide in refluxing pyridine solution (52), a reaction which is general for activated halogen-substituted heterocycles. An alternative method to prepare 4-cinnolinethiol is to warm 4-chlorocinnoline with a methanolic solution of thiourea. The resultant thiouronium salt (**59**), which precipitates from the solution, is readily converted in good yield into 4-cinnolinethiol (**58**) by heating it in an aqueous sodium hydroxide solution (41). The usual procedure to obtain 4-alkylthio- or 4-arylthiocinnolines (**60**; R = alkyl, aryl) is by reaction of a 4-chlorocinnoline with the sodio derivative of an alkyl- or arylmercaptan (5, 41, 52a). Alternatively, 4-alkylthiocinnolines may be obtained by the interaction of 4-cinnolinethiol with an alkyl halide. This procedure obviously will not work with an aryl halide.

A kinetic study (53) indicates that the nucleophilic displacement of chloride ion from 4-chlorocinnoline occurs by a one-stage bimolecular mechanism in the same fashion as the displacement of activated halide from similar nitrogen-containing heterocycles. There is no kinetic evidence for the existence of a stable intermediate such as **61**, although this structure is considered to be a good guide to the nature of the transition state. Neither is there any kinetic evidence for a "hetaryne" type of elimination-addition mechanism, as is

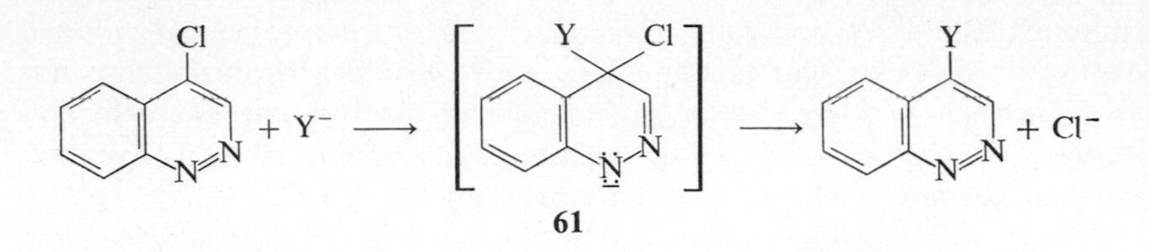

known to occur in the case of 3-chloroquinoline with lithium piperidide in piperidine to give a mixture of 3- and 4-piperidinoquinoline via a 3,4-quinolyne (54).

When the reactivity of 4-chlorocinnoline toward ethoxide ion is compared to the reactivity of other chloro-substituted heterocycles toward this same nucleophile, it is found that 4-chloroquinazoline is more reactive than 4-chlorocinnoline, which has approximately the same reactivity as 1-chlorophthalazine, and all three of these are much more reactive than 4-chloroquinoline. More specifically, the rate coefficient k at 20° C for the reaction of ethoxide ion with 4-chlorocinnoline is 4.74×10^{-3} to 4.90×10^{-3} liter/(mole)(sec), with 1-chlorophthalazine it is 1.82×10^{-3} to 1.90×10^{-3} liter/(mole)(sec), and with 4-chloroquinoline, k is 6.5×10^{-7} liter/(mole)(sec). Accordingly, the enthalpy of activation, $\Delta H^{\ddagger}$ —the extra energy required to go over the energy barrier (transition state) between reactants and products, which is associated with bonding forces—is found to be greater for 4-chloroquinoline ($\Delta H^{\ddagger}$ 21,800 cal/mole) than for 4-chlorocinnoline ($\Delta H^{\ddagger}$ 15,200 cal/mole). The entropy of activation, $\Delta S^{\ddagger}$—associated with the freedom of atomic motions—is found to be -12.3 cal/(mole)(deg) for 4-chloroquinoline and -17.3 cal/(mole)(deg) for 4-chlorocinnoline (53).

Qualitatively, it is found that 4-chloroquinazoline reacts much more rapidly with hydrazine in ethanol to give the 4-hydrazino derivative than does 4-chlorocinnoline under similar conditions (48), and in reaction with the sodium salt of active hydrogen compounds such as phenylacetonitrile or ethyl cyanoacetate, 4-chloroquinazoline is again found to be more reactive than 4-chlorocinnoline. In this case, 4-chlorocinnoline is found to be significantly more reactive than 1-chlorophthalazine, which in turn is more reactive than 2-chloroquinazoline, 1-chloroisoquinoline, and 2-chloroquinoline (50).

A halogen atom located at the 3-position of the cinnoline nucleus is much less reactive in nucleophilic displacement reactions than is a halogen atom at the 4-position, as is evidenced by the conversion of 3,4-dichlorocinnoline

(**11**) to 4-amino-3-chlorocinnoline (**62**) (48) and by the hydrolysis of 3,4,6-trichlorocinnoline (**62a**) to 3,6-dichloro-4-hydroxycinnoline (**62b**) (21b). A

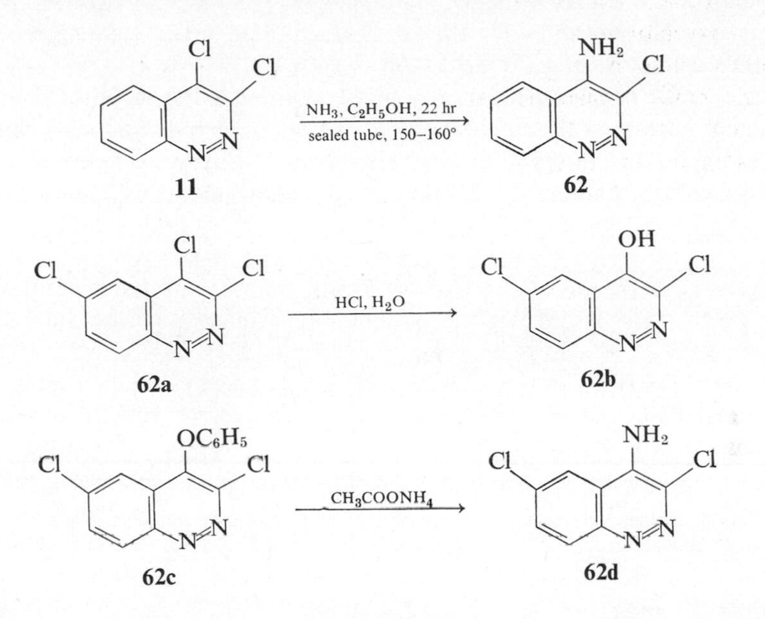

4-phenoxy group is also more easily displaced than a halogen atom at the 3-position, as shown by the conversion of 3,6-dichloro-4-phenoxycinnoline (**62c**) to 3,6-dichloro-4-aminocinnoline (**62d**) (21b). Nevertheless, a 3-halo substituent can be displaced from the cinnoline ring by a nucleophilic reagent when reaction conditions are sufficiently stringent. For example, interaction of 3-bromocinnoline with sodium methoxide in methanol at 110–120° C in a sealed tube gives 3-methoxycinnoline in good yield, while treatment of 3-bromocinnoline with aqueous ammonia at 130–140° C in the presence of copper sulfate gives 3-aminocinnoline (20). 3-Bromocinnoline and anhydrous dimethylamine in a sealed tube at 65° for 24 hr give 3-dimethylaminocinnoline (20a). Reaction of 2-aminothiophenol with 3-bromo-4-chloro-6,7-dimethoxycinnoline (**62e**) or 3,4-dichloro-6,7-dimethoxycinnoline gives 2,3-dimethoxy-7*H*-cinnolino[4,3-*b*]-1,4-benzothiazine (**62f**) (23b). Refluxing 3-bromo-4-(*o*-hydroxyanilino)-6,7-dimethoxycinnoline (**62g**) with phosphorus pentasulfide in pyridine solution gives 2,3-dimethoxy-12*H*-cinnolino[3,4-*b*]-1,4-benzothiazine (**62h**) (23b). Both **62f** and **62h** are produced in yields of 60–80% by these reactions.

Halogen exchange may occur at the 3-position of the cinnoline ring as illustrated by the metathesis of the 3-bromo substituent of 3-bromo-4-chlorocinnoline (**15**) in phosphorus oxychloride solution at 95° C to give

3,4-dichlorocinnoline (**11**) (21) or by the conversion of 3-bromocinnoline itself to 3-chlorocinnoline under similar conditions (20). Treatment of 3-bromo-6,7-dimethoxy-4-hydroxycinnoline (**62i**) witn a mixture of phosphorus oxychloride and phosphorus pentachloride gives 3,4-dichloro-6,7-dimethoxycinnoline in good yield (23b).

Nucleophilic displacement of a halogen atom which is substituted in the benzenoid portion of the cinnoline ring has been observed in a few instances. For example, the attempt to convert 6-bromo-4-hydroxycinnoline to 6-bromo-4-chlorocinnoline in a mixture of phosphorus oxychloride and

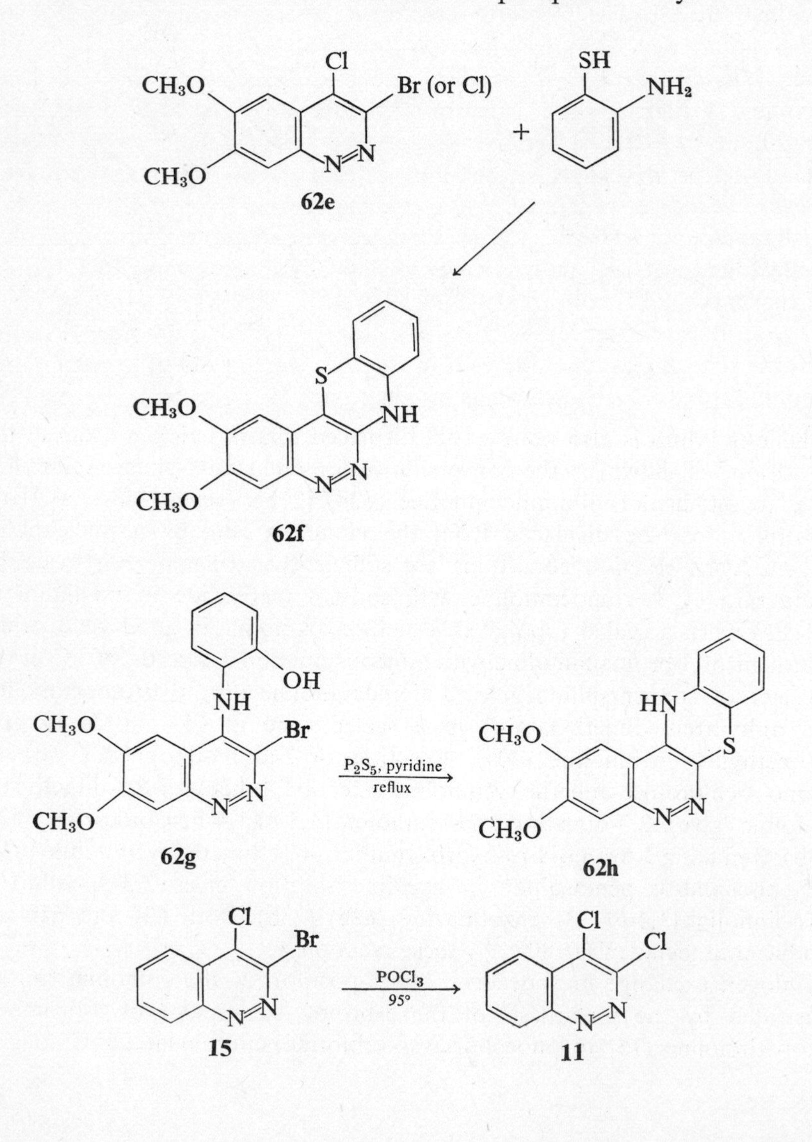

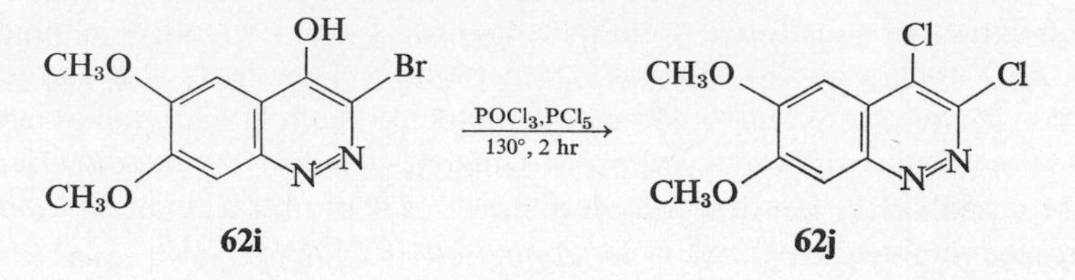

phosphorus pentachloride results in partial replacement of the 6-bromo substituent to give a mixture of 4,6-dichlorocinnoline and 6-bromo-4-chloro-cinnoline (5, 21). The 5-chloro substituent in both 5-chloro-4-hydroxy-6-nitrocinnoline and 5-chloro-4-hydroxy-8-nitrocinnoline is labile enough to undergo hydrolysis and to be displaced by amines (21b). Treatment of 6-bromo-, 6-chloro-, and 6-fluoro-4-hydroxycinnoline with phosphorus pentasulfide in refluxing pyridine solution all give 4,6-dimercaptocinnoline in high yields (5). Under the same conditions, 7- and 8-halo substituents appear to be inert to phosphorus pentasulfide, whereas 5-chloro and 5-bromo-4-hydroxycinnoline are completely degraded. With phosphorus pentasulfide in refluxing toluene solution, however, the 5-chloro substituent in 5-chloro-4-hydroxycinnoline remains intact (5).

The product obtained from the reduction of 4-chlorocinnoline depends upon the reducing agent employed. In the first preparation of unsubstituted cinnoline, iron in aqueous sulfuric acid was used as the reducing agent. This resulted in 1,4-dihydrocinnoline, which was converted to cinnoline by gentle oxidation with mercuric oxide (55). Cinnoline (**54**) is also obtained when 4-chlorocinnoline (**2**) is treated with toluene-*p*-sulfonylhydrazide and then decomposing the resulting 4-(toluene-*p*-sulfonylhydrazino)cinnoline (**53**) by heating it in aqueous sodium carbonate. Generally, this method gives good yields, and it has been employed not only with 4-chlorocinnoline but also

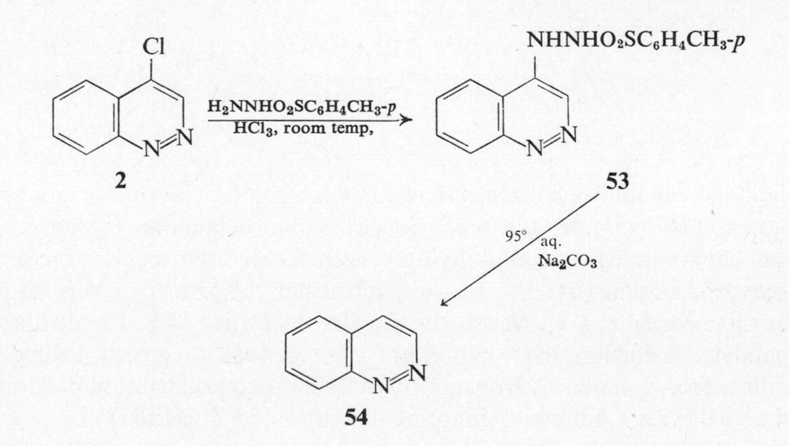

with a variety of substituted 4-chlorocinnolines such as 8-methyl-, 6-chloro-, 7-chloro-, 8-chloro, 5-methoxy-, 6-methoxy-, 7-methoxy-, 8-methoxy-, 5-nitro-, 7-nitro, and 8-nitro-4-chlorocinnoline (3, 6, 9, 10, 38). An exception is 4-chloro-6-nitrocinnoline, whose 4-(toluene-*p*-sulfonylhydrazino) derivative reportedly does not decompose in aqueous sodium carbonate or sodium hydroxide to give the desired 6-nitrocinnoline (9, 49).

Catalytic reduction of 4-chlorocinnoline (**2**) with palladium on calcium carbonate in methanol solution, on the other hand, yields principally 4,4′-dicinnolinyl (**63**) and only a trace of cinnoline (49). Chemical reduction by lithium aluminum hydride yields only 4,4′-dicinnolinyl (**63**) (56), and a polarographic study has shown that 4-chlorocinnoline is reduced electrolytically to 4,4′-dicinnolinyl also (57).

Zinc dust reduction of 4-chloro-3-hydroxycinnoline (**6**) in acidic medium gives 73% of 1-aminoöxindole (**5**), while the same reduction of 5- and 7-chloro-3-hydroxycinnoline gives 4-chloro-1-aminoöxindole and 6-chloro-1-aminoöxindole, respectively (18a).

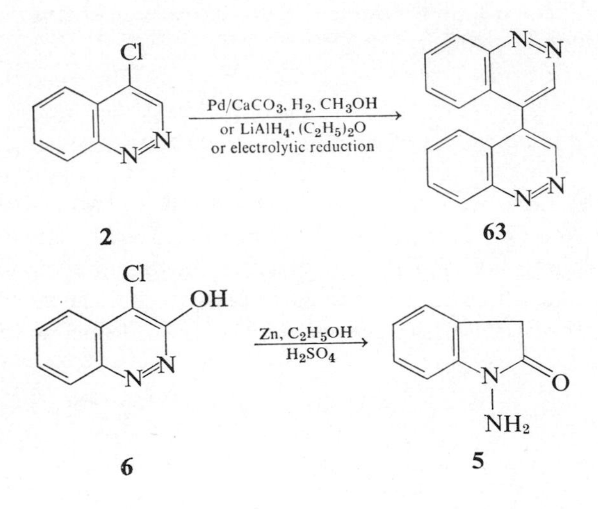

The 3-halocinnolines behave differently from their 4-chloro isomers upon reduction. For example, the reduction of 3-chlorocinnoline (**8**) by excess lithium aluminum hydride in refluxing benzene-ether solution gives 3-chloro-1,4-dihydrocinnoline (**64**), the 3-chloro substituent being inert to this reagent under these conditions. However, the dihydro derivative **64** will lose chlorine by catalytic reduction over palladium on charcoal to give 1,4-dihydrocinnoline (**65**). A similar hydrogenation of 3-chlorocinnoline using palladium on charcoal gives 1,4-dihydrocinnoline directly in 55% yield (58).

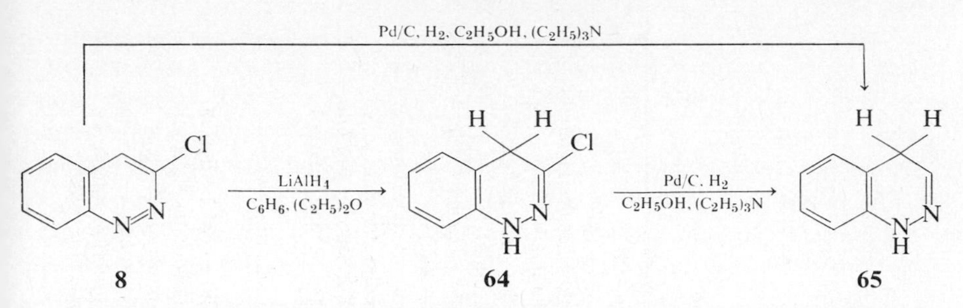

Under somewhat different conditions, 3-bromocinnoline (**17**) is reduced directly to cinnoline (**54**) over palladium on charcoal in methanolic potassium hydroxide solution containing a catalytic amount of hydrazine hydrate (20).

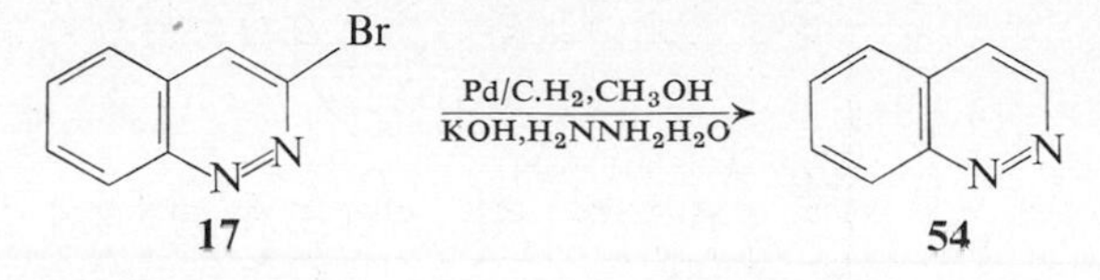

Similarly, the 3-bromo group may be reductively removed by palladium on charcoal from the 1-alkyl derivatives **66** (R = methyl, ethyl, *n*-propyl, *i*-propyl, and benzyl) and 2-alkyl derivatives **68** (R = CH_3) of 3-bromo-4-hydroxycinnolines to yield the corresponding 1-alkyl **67** and 2-alkyl **69** derivatives of 4-hydroxycinnoline (22, 23).

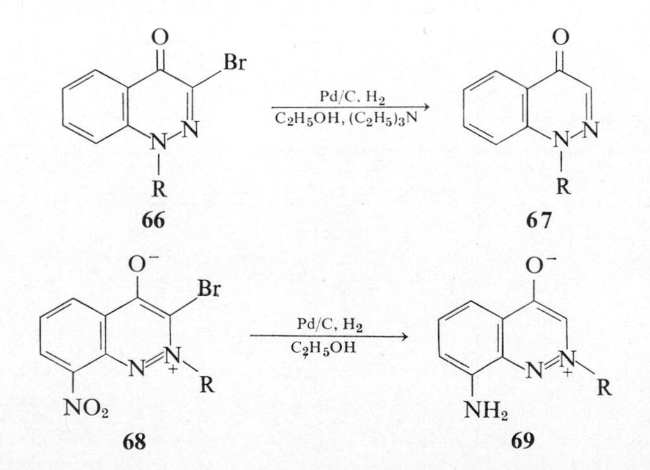

Under similar conditions with palladium over charcoal in ethanol, the 6-chloro group of the 1- and 2-methyl derivatives of 6-chloro-4-hydroxycinnoline is removed to give the corresponding *N*-methyl derivatives of 4-hydroxycinnoline (59).

IV. Tables

TABLE 1D-1. Halocinnolines[a]

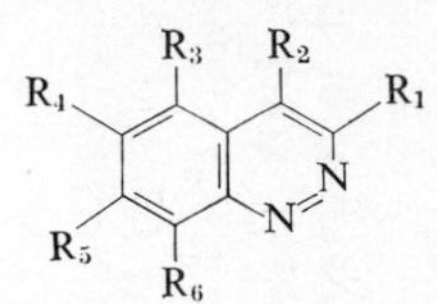

R_1	R_2	R_3	R_4	R_5	R_6	Prep.[b]	MP (°C)	Remarks[c]	Ref.
H	Cl	H	H	H	OCH_3	A	142–143	Yellow needles (acetone)	3
$—CH_2CO_2CH_3$	Cl	H	OCH_3	OCH_3	H	A	151–152	(Methanol)	4
CH_3	Cl	H	OCH_3	OCH_3	H	A	222–223	Yellow crystals (ethyl acetate)	4
H	Cl	H	F	H	H	A	89	Yellow needles (cyclohexane)	5
H	Cl	OCH_3	H	H	H	A	141–142	White needles (ligroin)	6
H	Cl	H	H	OCH_3	H	A	178–179	Needles (ligroin)	6
NO_2	Cl	H	H	H	H	A	169–170	Yellow needles (ligroin)	7
Cl	Cl	H	H	H	H	A, B	128–129, 126–127	White needles (ligroin)	7, 21
Cl	OC_6H_5	H	H	H	H	C	127–128 120–121	Crystals (ligroin)	7, 21
H	Cl	H	C_6H_5	H	H	A	151–152	Yellow needles (ligroin)	8
H	Cl	H	H	C_6H_5	H	A	124–125	Yellow plates (ligroin)	8
Br	Cl	H	H	H	H	A	153–154	Lemon-yellow blades (acetone)	9
Br	H	H	H	H	H	D	92–93	(Ligroin) picrate deriv., mp 118–120°	9

H	Cl	H	$C_6H_5N{=}N—$	H	H	E	161–162	(Ethyl acetate)	11
H	Cl	H	*m*-$CH_3COC_6H_4N{=}N—$	H	H	E	194–196	Orange needles (benzene)	11
H	Cl	H	Cl, N=N—, N=N (ring structure)	H	H	E	>300	—	11
H	Cl	H	OCH_3	OCH_3	H	E	195–196	Cream needles (ethanol)	12
H	OC_6H_5	Cl	H	H	H	C	118–119	Brown prismatic needles (ligroin)	13
H	$CH_3C(=O)—O—$	Cl	Cl	H	H	F	178–179	Colorless prisms (ligroin)	13
H	OC_6H_5	Cl	Cl	H	H	C	189	Colorless needles (Benzene-ligroin)	13
H	Cl	Cl	Cl	H	H	E	Not stated	—	13
H	Cl	H	NO_2	H	H	G	138–139	—	14
p-$ClC_6H_4—$	Cl	H	H	H	H	H, K	138–140	(Acetone)	16, 17
p-$CH_3OC_6H_4—$	Cl	H	H	H	H	I, L	130–131	(Ligroin or hexane)	16, 17
C_6H_5	Cl	H	H	H	H	E, J	120–121	Yellow needles (ligroin)	16
p-$CH_3C_6H_4—$	Cl	H	H	H	H	M	Not stated	—	17
p-$ClC_6H_4—$	Cl	H	CH_3	H	H	N	Not stated	—	17
Cl	H	H	H	H	H	D, O, P	90–91, 91–92	Colorless needles (ligroin); sublimes at 80° and 0.2 mm	9, 20, 25
Cl	Cl	H	Cl	H	H	A	141–142	Yellow needles (ligroin)	21

TABLE 1D-1. (*continued*)

R_1	R_2	R_3	R_4	R_5	R_6	Prep.[b]	MP (°C)	Remarks[c]	Ref.
Br	H	H	H	H	H	Q	92–93	Colorless needles (ligroin); sublimes at 80° and 0.2 mm	25
Cl	CH_3	H	H	H	H	P	145–146	Crystals (ligroin); sublimes at 120° and 0.2 mm	25
Br	CH_3	H	H	H	H	Q	158.5–159	From ligroin, then sublimed at 150° and 0.2 mm	25
2-Pyridyl	C_6H_5	H	Cl	H	H	R	143–144	Yellow prisms (aq. ethanol)	26
2-Quinolyl	C_6H_5	H	Cl	H	H	S	205–206	Yellow needles (ethanol)	26
H	CH_3	H	H	H	Cl	T	124–125, 126–127	Yellow needles (ligroin); picrate deriv., mp 177–178° and 179–180°	29, 60
$—C(=O)CH_3$	H	H	Cl	H	H	U	206–207	(Ligroin)	32
$—COOC_2H_5$	H	H	Cl	H	H	V	152.5–153	Yellow needles (ligroin)	32
$—C(=O)CH_3$	H	H	H	Cl	H	W	211–212	Yellow needles (ligroin)	32
$—COOC_2H_5$	H	H	H	Cl	H	X	200–201	Yellow needles (ligroin)	32
NO_2	H	H	H	Cl	H	Y	165.5–166	Yellow needles (acetone or ethyl acetate)	33

NO_2	H	H	Cl	H	H	Y	227–228	Yellow plates (ethyl acetate)	33
H	H	H	Cl	H	H	D, Z	131–131.5, 119–120	Yellow needles (ligroin)	38, 48
H	H	H	H	Cl	H	D, AA	89–90, 92–93	Needles (ligroin or water)	38, 49
H	H	H	H	H	Cl	D	88–89	Yellow crystals (ligroin)	38
H	H	H	Br	H	H	Z	129–130	Buff-colored needles (ether-ligroin)	48
Br	Cl	H	OCH_3	OCH_3	H	A	219	Yellow needles (ethanol)	23b
Cl	Cl	H	OCH_3	OCH_3	H	BB	231	Green-yellow solid (acetic acid)	23b
H	H	F	F	F	F	CC	107–108	(Hexane)	33a
H	Cl	H	CL	H	H	E, G	112–113	(Ligroin); hydrochloride deriv., mp 149–151°	59a
H	Cl	H	Cl	Cl	H	E, G	140–141	Hydrochloride deriv., mp 164–167°	59a
H	Cl	H	H	Cl	Cl	E, G	223–225	Hydrochloride deriv., mp 225 (dec.)	59a
H	Cl	H	H	H	Cl	E, G	146–147	—	59a
H	OC_6H_5	H	Cl	H	H	C	129–130	(Ligroin)	59a
H	OC_6H_5	H	Cl	Cl	H	C	161–162	—	59a
H	OC_6H_5	H	H	Cl	Cl	C	221–222	—	59a
H	OC_6H_5	H	H	H	Cl	C	151–152	—	59a
Cl	Cl	H	H	Cl	Cl	E	253–254.5	—	21a
Cl	Cl	Cl	Cl	Cl	Cl	DD	188–190	—	21a
F	F	F	F	F	F	EE	100–102	—	21a

TABLE 1-D1. (*continued*)

R_1	R_2	R_3	R_4	R_5	R_6	Prep.[a]	MP (°C)	Remarks[b]	Ref.
F	F	Cl	F	F	F	EE	98–100	—	21a
NO_2	Cl	H	Cl	H	H	A	—	—	21b
NO_2	OC_6H_5	H	Cl	H	H	C	111–112	(Ligroin)	21b
Cl	OC_6H_5	H	Cl	H	H	C	—	Not isolated	21b
H	Cl	Cl	NO_2	H	H	E	—	Hydrochloride deriv., mp 195–200° (dec.)	21b

[a] Halocinnolines which bear a hydroxy, mercapto, or amino substituent are not given in this table. Hydroxy-substituted halocinnolines may be found in Tables 1C-2 and Table 1C-3, while halocinnolines bearing a mercapto, alkylthio, or arylthio group may be found in Table 1F-3 and amino-substituted halocinnolines are given in Table 1H-3.

[b] A, by treatment of the corresponding 4-hydroxycinnoline with phosphorus oxychloride. B, by treatment of 4-chloro-3-nitrocinnoline with a mixture of phosphorus oxychloride and phosphorus pentachloride. C, by interaction of the corresponding 4-chlorocinnoline with potassium or sodium phenoxide. D, by treatment of the corresponding 4-chlorocinnoline with toluene-*p*-sulfonylhydrazide followed by decomposition of the resulting 4-(toluene-*p*-sulfonylhydrazino)cinnoline in hot aqueous sodium carbonate solution. E, by treatment of the corresponding 4-hydroxycinnoline with a mixture of phosphorus oxychloride and phosphorus pentachloride. F, by interaction of 5,6-dichloro-4-hydroxycinnoline with acetic anhydride. G, by refluxing the corresponding 4-hydroxycinnoline with thionyl chloride containing a small amount of phosphorus pentachloride in an *o*-dichlorobenzene or chlorobenzene solution. H, by refluxing a solution of 3-(4-chlorophenyl)cinnoline 1-oxide in phosphorus oxychloride for 1.5 hr. I, by refluxing of 3-(4-methoxyphenyl)cinnoline-1-oxide in phosphorus oxychloride for 1.5 hr. J, by refluxing solutions of either 3-phenylcinnoline-1-oxide in phosphorus oxychloride or 3-phenylcinnoline-2-oxide in phosphorus oxychloride for 1.5 hr. K, by refluxing a

mixture of 3-(4-chlorophenyl)cinnoline 1-oxide and 2-oxide in phosphorus oxychloride for 3 hr. L, by refluxing a mixture of 3-(4-methoxyphenyl)-cinnoline 1-oxide and 2-oxide in phosphorus oxychloride for 3 hr. M, by refluxing a mixture of 3-(4-tolyl)cinnoline 1-oxide and 2-oxide in phosphorus oxychloride for 3 hr. N, by refluxing a mixture of 3-(4-chlorophenyl)-6-methylcinnoline 1-oxide and 2-oxide in phosphorus oxychloride for 3 hr. O, by treatment of 3-hydroxycinnoline with phosphorus oxychloride. P, by diazotization of the corresponding 3-aminocinnoline in concentrated hydrochloric acid. Q, by diazotization of the corresponding 3-aminocinnoline in 48% hydrobromic acid. R, by diazotization and cyclization of 1-(2-amino-5-chlorophenyl)-2-(2-pyridinyl)ethylene. S, by diazotization and cyclization of 1-(2-amino-5-chlorophenyl)-1-phenyl-2-(2-quinolinyl)ethylene. T, by diazotization of 8-amino-4-methylcinnoline with sodium nitrite in aqueous hydrochloric acid, and then treatment of the resulting diazonium salt with cuprous chloride. U, by diazotization of 2-amino-5-chlorobenzaldehyde followed by treatment of the resultant diazonium salt with acetoacetic acid to give pyruvaldehyde 4-chloro-2-formylphenylhydrazone, which cyclizes spontaneously to give the cinnoline derivative. V, by diazotization of 2-amino-5-chlorobenzaldehyde followed by treatment of the resultant diazonium salt with ethyl hydrogen malonate to give ethyl glyoxalate 4-chloro-2-formylphenylhydrazone, which cyclizes spontaneously to give the cinnoline derivative. W, by diazotization of 2-amino-4-chlorobenzaldehyde followed by treatment of the intermediate diazonium salt with acetoacetic acid to give pyruvaldehyde 5-chloro-2-formylphenylhydrazone, which is then cyclized to the cinnoline in concentrated sulfuric acid. X, by diazotization of 2-amino-4-chlorobenzaldehyde followed by treatment of the intermediate diazonium salt with ethyl hydrogen malonate. The resultant ethyl glyoxalate 5-chloro-2-formylphenylhydrazone partially cyclized spontaneously to give a small amount of the cinnoline derivative. Y, by diazotization of the appropriately substituted 2-aminobenzaldehyde, treatment of the diazonium salt with nitromethane to give a nitroformaldehyde 2-formylphenylhydrazone, which was then cyclized in aqueous potassium hydroxide to give the 3-nitrocinnoline derivative. Z, by oxidation of the appropriately substituted 4-hydrazinocinnoline in refluxing aqueous 10% cupric sulfate solution. AA, by catalytic hydrogenation of 4,7-dichlorocinnoline in methanol solution over palladium hydroxide on calcium carbonate at room temperature and 4–5 atm pressure for 15 min. BB, by reaction of the corresponding 3-bromo-4-hydroxycinnoline with a mixture of phosphorus oxychloride and phosphorus pentachloride. CC, by heating 1-amino-4,5,6,7-tetrafluoroindole or 1-amino-2,3-dihydro-3-hydroxy-1,2,3,4-tetrafluoroindole in hydrochloric acid solution. DD, by chlorination of 3,4,7,8-tetrachlorocinnoline with chlorine and aluminum chloride. EE, by fluorination of hexafluorocinnoline with potassium fluoride.

[c] Solvent employed in recrystallization is enclosed in parentheses.

References

1. M. Busch and M. Klett, *Ber.*, **25**, 2847 (1892).
2. C. M. Atkinson and A. Taylor, *J. Chem. Soc.*, **1955**, 4236.
3. E. J. Alford, H. Irving, H. S. Marsh, and K. Schofield, *J. Chem. Soc.*, **1952**, 3009.
4. D. E. Ames and A. C. Lovesey, *J. Chem. Soc.*, **1965**, 6036.
5. R. N. Castle, R. R. Shoup, K. Adachi, and D. L. Aldous, *J. Heterocyclic Chem.*, **1**, 98 (1964).
6. A. R. Osborne and K. Schofield, *J. Chem. Soc.*, **1955**, 2100.
7. H. E. Baumgarten, *J. Am. Chem. Soc.*, **77**, 5109 (1955).
8. C. M. Atkinson and C. J. Sharpe, *J. Chem. Soc.*, **1959**, 2858.
9. E. J. Alford and K. Schofield, *J. Chem. Soc.*, **1953**, 609.
10. D. E. Ames, R. F. Chapman, and D. Waite, *J. Chem. Soc.*, (*C*), **1966**, 470.
11. J. McIntyre and J. C. E. Simpson, *J. Chem. Soc.*, **1952**, 2606.
12. R. N. Castle and F. H. Kruse, *J. Org. Chem.*, **17**, 1571 (1952).
13. H. J. Barber, K. Washbourn, W. R. Wragg, and E. Lunt, *J. Chem. Soc.*, **1961**, 2828.
14. H. J. Barber, E. Lunt, K. Washbourn, and W. R. Wragg, *J. Chem. Soc.* (*C*), **1967**, 1657.
15. I. Suzuki, T. Nakashima, N. Nagasawa, and T. Itai, *Chem. Pharm. Bull.* (*Tokyo*), **12**, 1090 (1964).
16. H. S. Lowrie, *J. Med. Chem.*, **9**, 670 (1966).
17. H. S. Lowrie, U.S. Pat. 3,265,693 (Cl 260-247.5), Aug. 9, 1966.
18. W. F. Wittman, Ph.D. Dissertation, University of Nebraska, Lincoln (1965); University Microfilms (Ann Arbor, Mich.), Order No. 65-11,435.

18a. H. E. Baumgarten, W. F. Wittman, and G. J. Lehmann, *J. Heterocyclic Chem.*, **6**, 333 (1969).

19. M. Ogata, H. Kano, and K. Tori, *Chem. Pharm. Bull.* (*Tokyo*), **11**, 1527 (1963).
20. E. J. Alford and K. Schofield, *J. Chem. Soc.*, **1953**, 1811.

20a. A. J. Boulton, I. J. Fletcher, and A. R. Katritzky, *J. Chem. Soc.* (*B*), **1971**, 2344.

21. K. Schofield and T. Swain, *J. Chem. Soc.*, **1950**, 384.

21a. R. D. Chambers, J. A. H. MacBride, and W. K. R. Musgrave, *J. Chem. Soc* (*D*), **1970**, 739.

21b. E. Lunt, K. Washbourn, and W. R. Wragg, *J. Chem. Soc* (*C*), **1968**, 687.

22. D. E. Ames, R. F. Chapman, H. Z. Kucharska, and D. Waite, *J. Chem. Soc.*, **1965**, 5391.
23. D. E. Ames and R. F. Chapman, *J. Chem. Soc.* (*C*), **1967**, 40.

23a. W. A. White, German Pat., 2,005,104, Aug. 6, 1970; *Chem. Abstr.*, **73**, 77269 (1970).

23b. A. N. Kaushal and K. S. Narrang, *Indian J. Chem.*, **6**, 350 (1968).

24. K. Schofield and J. C. E. Simpson, *J. Chem. Soc.*, **1948**, 1170.
25. H. E. Baumgarten, W. F. Murdock, and J. E. Dirks, *J. Org. Chem.*, **26**, 803 (1961).
26. A. J. Nunn and K. Schofield, *J. Chem. Soc.*, **1953**, 3700.
27. K. Schofield and R. S. Theobald, *J. Chem. Soc.*, **1950**, 395.
28. R. N. Castle, Proc. 3rd Int. Congr. Chemotherapy, Stuttgart, July, 1963. Thieme Verlag, Stuttgart, 1964, pp. 1064–1068.
29. L. McKenzie and C. S. Hamilton, *J. Org. Chem.*, **16**, 1414 (1951).
30. E. J. Alford and K. Schofield, *J. Chem. Soc.*, **1952**, 2102.
31. H. E. Baumgarten and P. L. Creger, *J. Am. Chem. Soc.*, **82**, 4634 (1960).
32. H. E. Baumgarten and C. H. Anderson, *J. Am. Chem. Soc.*, **80**, 1981 (1958).

33. H. E. Baumgarten, D. L. Pedersen, and M. W. Hunt, *J. Am. Chem. Soc.*, **80,** 1977 (1958).
33a. V. P. Petrov and V. A. Barkhash, *Khim. Geterotsikl. Soedin.*, **1970** (3), 381; *Chem. Abstr.*, **73,** 25229 (1970).
34. R. N. Castle and D. B. Cox, *J. Org. Chem.*, **19,** 1117 (1954).
35. J. R. Keneford, J. S. Morley, J. C. E. Simpson, and P. H. Wright, *J. Chem. Soc.*, **1949,** 1356.
36. R. N. Castle, D. B. Cox, and J. F. Suttle, *J. Am. Pharm. Assoc.*, **48,** 135 (1959).
37. R. R. Shoup and R. N. Castle, *J. Heterocyclic Chem.*, **1,** 221 (1964).
38. A. R. Osborn and K. Schofield, *J. Chem. Soc.*, **1956,** 4207.
39. J. M. Bruce, P. Knowles, and L. S. Besford, *J. Chem. Soc.*, **1964,** 4044.
40. J. R. Keneford, J. S. Morley, J. C. E. Simpson, and P. H. Wright, *J. Chem. Soc.*, **1950,** 1104.
41. R. N. Castle, H. Ward, N. White, and K. Adachi, *J. Org. Chem.*, **25,** 570 (1960).
42. H. S. Lowrie, *J. Med. Chem.*, **9,** 784 (1966).
43. C. M. Atkinson, J. C. E. Simpson, and A. Taylor, *J. Chem. Soc.*, **1954,** 1381.
44. R. N. Castle and M. Onda, *J. Org. Chem.*, **26,** 2374 (1961).
45. E. M. Lourie, J. S. Morley, J. C. E. Simpson, and J. M. Walker, Br. Patent 702,664, January 20, 1954; Appl. July 5, 1950.
46. W. C. Austin, L. H. C. Lunts, M. D. Potter, and E. P. Taylor, *J. Pharm. Pharmacol.*, **11,** 80 (1959).
47. J. R. Keneford, E. M. Lourie, J. S. Morley, J. C. E. Simpson, J. Williamson, and P. H. Wright, *J. Chem. Soc.*, **1952,** 2595.
48. K. Schofield and T. Swain, *J. Chem. Soc.*, **1950,** 392.
49. J. Morley, *J. Chem. Soc.*, **1951,** 1971.
50. Y. Mizuno, K. Adachi, and K. Ikeda, *Pharm. Bull.* (*Tokyo*), **2,** 225 (1954).
51. R. N. Castle and M. Onda, *Chem. Pharm. Bull.* (*Tokyo*), **9,** 1008 (1961).
52. R. N. Castle, K. Kaji, G. A. Gerhardt, W. D. Guither, C. Weber, M. P. Malm, R. R. Shoup, and W. D. Rhoads, *J. Heterocyclic Chem.*, **3,** 79 (1966).
52a. S. M. Yarnal and V. V. Badiger, *J. Med. Chem.*, **11,** 1270 (1968).
53. N. B. Chapman and D. Q. Russell-Hill, *J. Chem. Soc.*, **1956,** 1563.
54. H. J. Den Hertog and H. C. Van der Plas, in *Advances in Heterocyclic Chemistry*, Vol. 4, A. R. Katritzky, Ed., Academic Press, New York and London, 1965, pp. 121–142.
55. M. Busch and A. Rast, *Ber.*, **30,** 521 (1897).
56. D. E. Ames and H. Z. Kucharska, *J. Chem. Soc.*, **1962,** 1509.
57. H. Lund, *Acta Chem. Scand.*, **21,** 2525 (1967).
58. D. E. Ames, R. F. Chapman, and H. Z. Kucharska, *J. Chem. Soc.*, Suppl. 1, **1964,** 5659.
59. D. E. Ames, *J. Chem. Soc.*, **1964,** 1763.
59a. E. Lunt, K. Washbourn, and W. R. Wragg, *J. Chem. Soc.* (*C*), **1968,** 1152.
60. K. Schofield and T. Swain, *J. Chem. Soc.* **1949,** 1367.

Part E. Alkoxy- and Aryloxycinnolines

I. Methods of Preparation

A. 3- and 4-Alkoxy- and Aryloxycinnolines

By far the most prevalent method of preparing 4-alkoxycinnolines (**2**; R = alkyl) is the interaction of a 4-chlorocinnoline (**1**) in alcohol solution with the sodium or potassium salt of the same alcohol, conditions varying from a brief warming to a reflux period of 2 to 3 hr (1–10). Alternatively, the 4-chlorocinnoline has been successfully made to react with sodium or potassium alkoxides in benzene or toluene solution at room temperature for 48 hr or at reflux temperature for 4 or 5 hr (3, 5, 11).

The 4-phenoxycinnolines (**2**; R = phenyl or substituted phenyl) are similarly prepared by treatment of the appropriate 4-chlorocinnoline with a sodium or potassium phenoxide in the same phenol (12–14), with potassium carbonate in phenol (15), or by gentle warming of the 4-chlorocinnoline with ammonium carbonate in phenol (12, 14, 16). Further heating of the 4-phenoxycinnoline with excess ammonium carbonate in phenol results in the conversion of the 4-phenoxy derivative to a 4-aminocinnoline (15). The 4-alkoxy- and 4-phenoxycinnolines prepared as just described, too numerous to discuss in detail here, are listed in Table 1E-1.

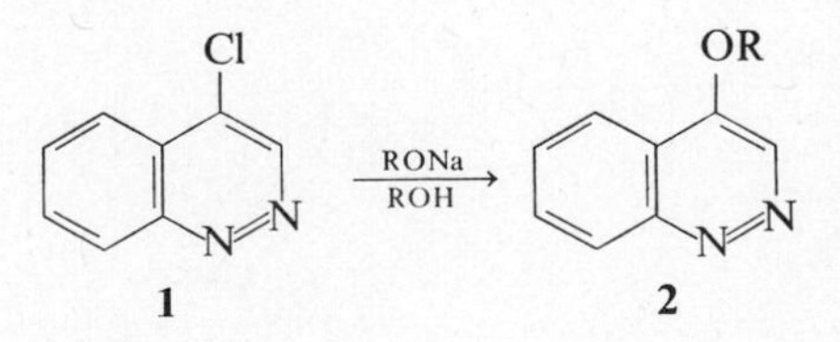

Less frequently, 4-alkoxycinnolines have been synthesized by other means. For example, 4-methoxycinnoline 1-oxide (**4**) may be prepared by displacement of the nitro group of 4-nitrocinnoline 1-oxide (**3**) by sodium methoxide in methanol (10), and 3-methoxy-4-nitrocinnoline 1-oxide similarly yields 3,4-dimethoxycinnoline 1-oxide by treatment with sodium methoxide in methanol (17). Interaction of 4-hydroxycinnoline 2-oxide (**5**) with a methanolic solution of methyl iodide and silver oxide in a sealed tube for 3 hr at 100° C gives 50% of 4-methoxycinnoline 2-oxide (**6**) (10). Under these same conditions, however, 4-hydroxycinnoline 1-oxide (**7**) yields principally the 1-methoxycinnolone (**8**) and only a small amount of 4-methoxycinnoline 1-oxide (**4**) (10). The oxygen atom may then be removed from the ring nitrogen atom by treatment of the cinnoline *N*-oxide with phosphorus trichloride in chloroform solution. Alkylation of 1-methyl-4-cinnolinone with triethyloxonium fluoroborate gives 4-ethoxy-1-methylcinnolinium fluoroborate while the anhydro base of 4-hydroxy-2-methylcinnolinium hydroxide similarly gives 4-ethoxy-2-methylcinnolinium fluoroborate (17a).

With the preceding exceptions, 4-alkoxy- or 3-alkoxycinnolines cannot be prepared from 4-hydroxy- or 3-hydroxycinnolines by reaction with alkyl

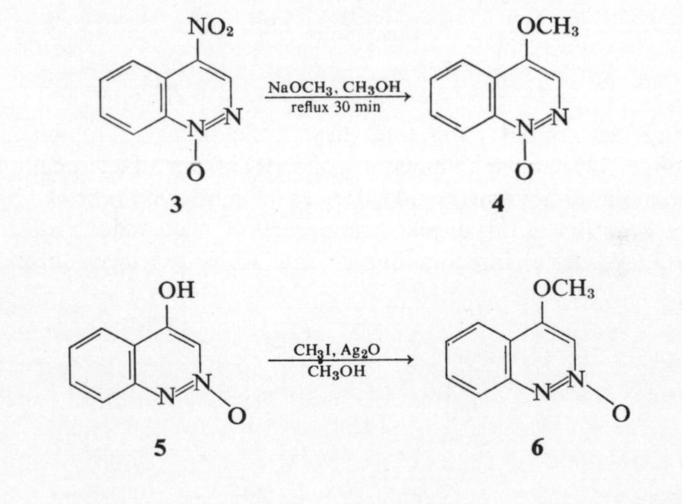

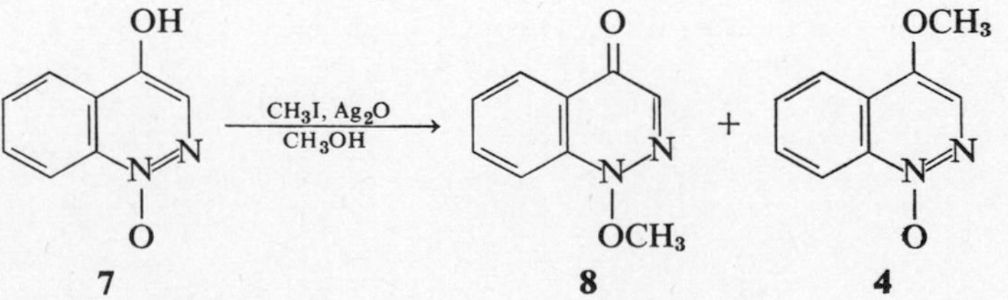

halides, dimethyl sulfate, diethyl sulfate, or diazomethane, since these reactions yield *N*-alkyl- rather than *O*-alkylcinnolines (3, 5, 18–21).

4-Methoxycinnoline (**10**) has been prepared in good yield by displacement of the 4-methylsulfonyl group from 4-methylsulfonylcinnoline (**9**) by sodium methoxide in methanol (22, 22a). Methylsulfinyl, methylthio (22b), and nitrile (22c) groups have also been displaced from the 4-position of the cinnoline ring by methoxide ion.

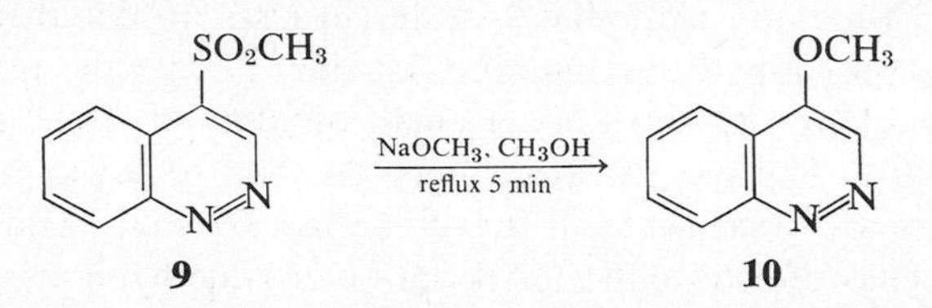

The only other preparation of a 4-alkoxycinnoline to appear in the literature is an inadvertent one, in which the attempted catalytic reduction of 4-chloro-6-phenylcinnoline in aqueous ethanolic sodium hydroxide solution over a palladium-charcoal catalyst gave only 4-ethoxy-6-phenylcinnoline (23).

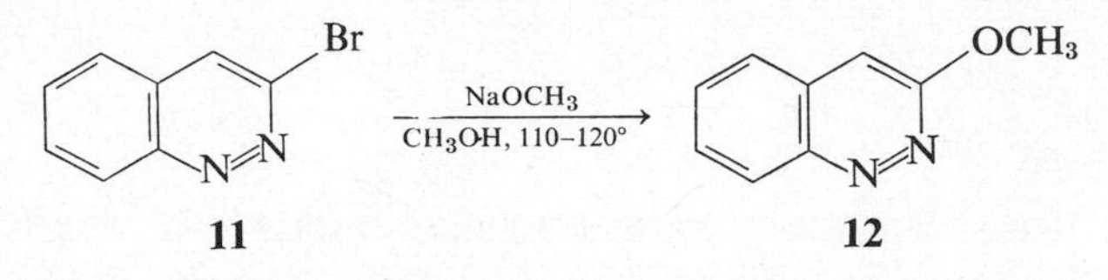

Although few 3-alkoxycinnolines have been prepared to date, 3-methoxycinnoline (**12**) may be obtained in good yield from 3-bromocinnoline (**11**) and methanolic sodium methoxide after 17 hr in a sealed tube at 110–120° C (21). In a variation of this displacement reaction, the 3-chloro substituent of 3-chloro-5,6,7,8-tetrahydrocinnoline-1-oxide (**13**) has been displaced by

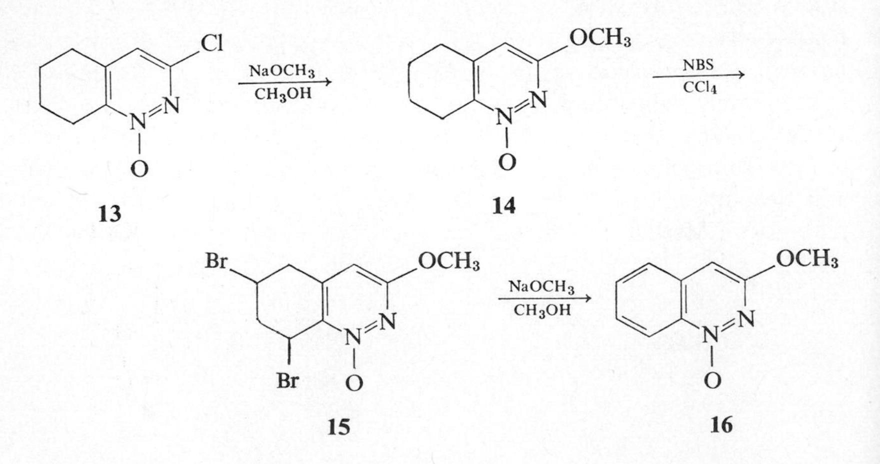

methoxide ion in methanolic sodium methoxide to give 3-methoxy-5,6,7,8-tetrahydrocinnoline 1-oxide (**14**). Compound **14** can then be dibrominated using *N*-bromosuccinimide in carbon tetrachloride to give 6,8-dibromo-3-methoxy-5,6,7,8-tetrahydrocinnoline 1-oxide (**15**), and this then is dehydrobrominated with sodium methoxide in methanol to yield 3-methoxycinnoline 1-oxide (**16**) (17, 24). This material could probably be deoxygenated with phosphorus trichloride in chloroform solution to give 3-methoxycinnoline itself.

Like quinoline-1-oxide, cinnoline-1-oxide (**17**) can be nitrated by benzoyl nitrate in chloroform solution. By this procedure, **17** gives 3-nitrocinnoline 1-oxide (**18**) in 71% yield, and interaction of **18** with sodium methoxide in methanol solution then produces 3-methoxycinnoline 1-oxide (**16**) in 95% yield (9, 25).

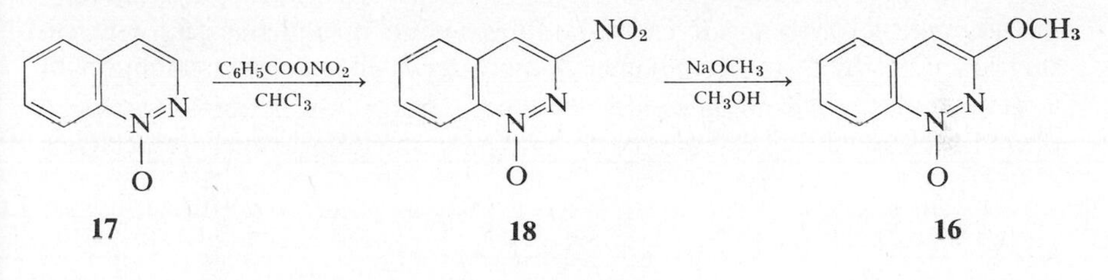

B. 5-, 6-, 7-, and 8-Alkoxycinnolines

Unlike 3- and 4-alkoxycinnolines, which are usually prepared by nucleophilic displacement by alkoxide ion of some other substituent in these same positions of the cinnoline ring, all of the 5-, 6-, 7-, and 8-alkoxycinnolines described to date have been prepared by building the alkoxy groups into the system before formation of the cinnoline ring itself. Thus 5-, 6-, 7-, and 8-methoxy-4-hydroxycinnolines (**20**) and 6,7-dimethoxy-4-hydroxycinnoline have all been prepared via the Borsche diazotization and cyclization of the appropriately substituted methoxy-2-aminoacetophenone (**19**) (see Section 1C-I-A-3) (26–29). The 4-hydroxy group may then be removed by conversion to the 4-chlorocinnoline, this to the 4-(toluene-*p*-sulfonylhydrazino)cinnoline and decomposition of this by heating it in aqueous sodium carbonate (28, 29). Methyl 6,7-dimethoxy-4-hydroxycinnoline-3-acetate (**22**) is

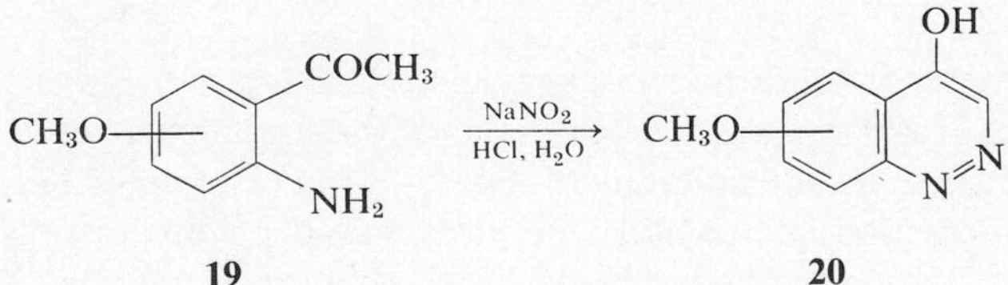

prepared in a similar manner through diazotization and cyclization of methyl β-(2-amino-4,5-dimethoxybenzoyl)propionate (**21**) (6).

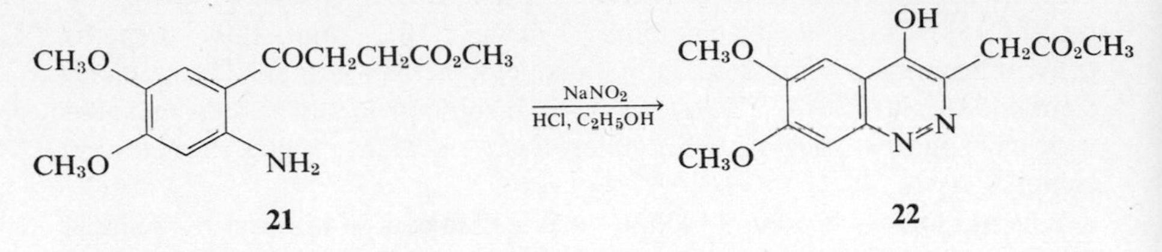

The Widman-Stoermer synthesis (see Section 1B-I-A) has been used to prepare 4-methyl- and 4-phenylcinnolines substituted with methoxy groups in the benzenoid portion of the ring system. Thus 8-methoxy-4-methylcinnoline (**24**) is obtained by diazotization and cyclization of 2-(2-amino-3-methoxyphenyl)propene (**23**) with sodium nitrite in aqueous sulfuric acid solution (29, 30). In like manner 8-methoxy-4-phenylcinnoline and 6,7-dimethoxy-4-methylcinnoline are obtained from α-(2-amino-3-methoxyphenyl)styrene and 2-(2-amino-4,5-dimethoxyphenyl)propene, respectively (29, 31).

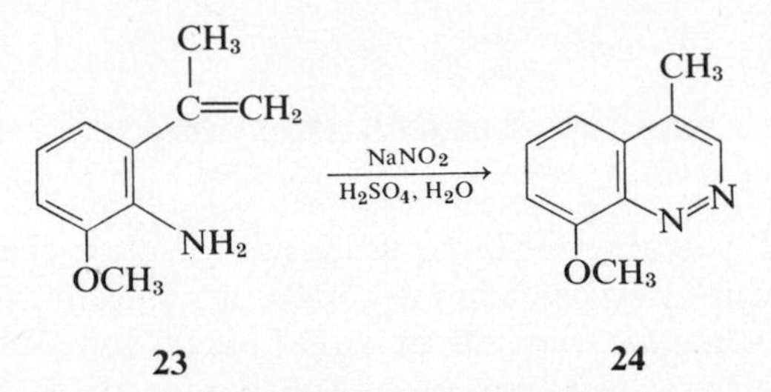

Cyclization of mesoxalyl chloride *p*-methoxyphenylhydrazone (**25**) in dry nitrobenzene at 95° C in the presence of titanium tetrachloride yields 4-hydroxy-6-methoxycinnoline-3-carboxylic acid (**26**) in 76% yield (32).

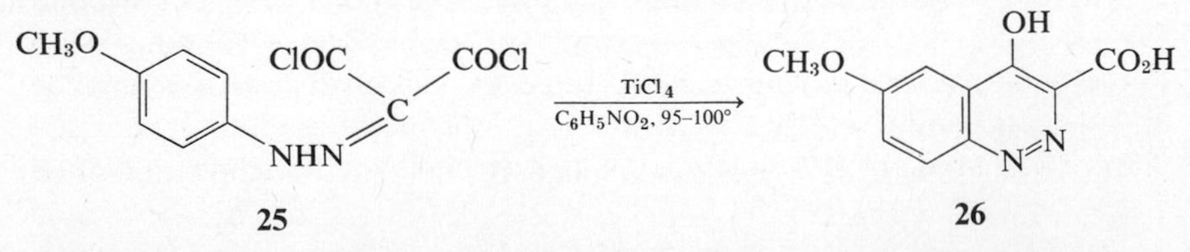

A low yield of 6-methoxy-3-phenylcinnoline-4-carboxylic acid (**29**) is obtained by rearrangement of 5-methoxy-*N*-benzylideneaminoisatin (**28**) in refluxing aqueous potassium hydroxide solution. The isatin **28** is obtained from the interaction of benzaldehyde *p*-methoxyphenylhydrazone (**27**) with oxalyl chloride followed by cyclization with aluminum chloride (33).

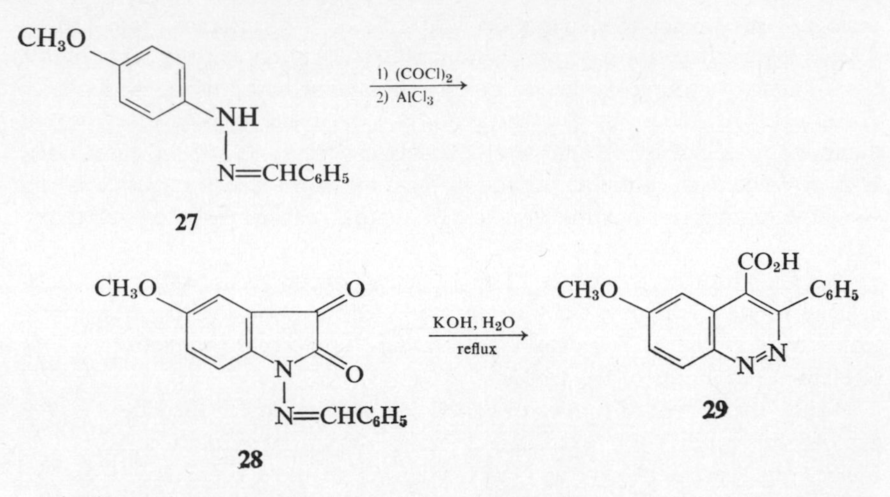

Finally, reductive cyclization of the alkoxy-substituted 2,β-dinitrostyrenes **29a** with lithium aluminum hydride in refluxing tetrahydrofuran produces a mixture of the corresponding alkoxycinnolines **29b** and indoles **29c**, both in poor yield (33a).

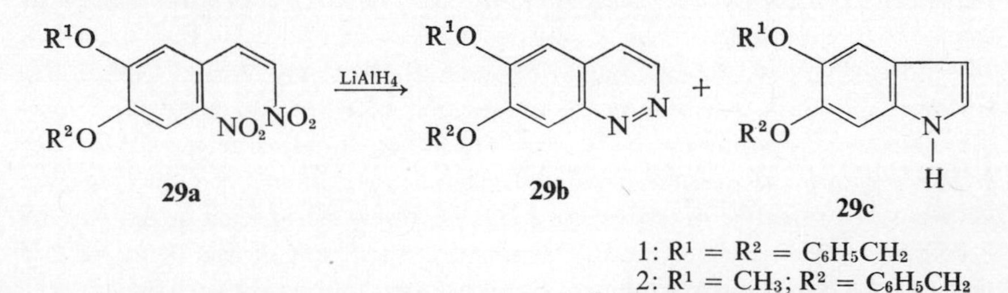

II. Physical Properties

Like the 4-chlorocinnolines, the 4-alkoxy- and 4-aryloxycinnolines are generally lower-melting and much more soluble in organic solvents than the corresponding 4-hydroxycinnolines.

The basic strength of 4-methoxycinnoline is such that 4-methoxyquinoline [pK_a 5.35 in 50% ethanol (34), 6.65 in water (35)] is a stronger base than is 4-methoxycinnoline [pK_a 2.71 in 50% ethanol (34), 3.21 in water (36)], which shows approximately the same basic strength as 4-methoxyquinazoline [pK_a 2.73 in 50% ethanol (34), 3.13 in water (36)]. Compared with the unsubstituted compounds, the effect of a 4-methoxy group is base-strengthening in the quinoline and cinnoline systems but base-weakening in the quinazoline system. Thus we see that cinnoline [pK_a 2.51 in 50% ethanol (34), 2.70 in water (36)] is a weaker base than is 4-methoxycinnoline. On the other hand, the 4-phenoxy group is found to be base-weakening, and hence 4-phenoxycinnoline [pK_a 2.27 in 50% ethanol (34)] is an even weaker base than is cinnoline itself. As expected, 4-methylthiocinnoline [pK_a 3.13 in water (20)] is more basic than cinnoline, although the increase in basicity caused by the 4-methylthio group is somewhat less than that caused by the 4-methoxy group.

The IR absorption spectra of 6,7-dimethoxycinnolines which are substituted with a 4-hydroxy, 4-chloro, or 4-alkyl group show five to nine absorption bands in the 6.17–9.48 μ region which have been assigned to C—O stretching vibrations of the methoxy group (37).

During the course of studies designed to shed light upon the extent of the tautomeric equilibrium exhibited by the 4-hydroxycinnolines, the UV absorption spectra of several 4-alkoxycinnolines have been recorded. The spectral parameters so obtained are given in Table 1E-2, together with the parameters for other alkoxycinnolines. Briefly, the spectra of the 4-hydroxycinnolines resemble those of their *N*-alkyl derivatives and are unlike those of the 4-alkoxycinnolines, pointing to a 4-cinnolone structure as the preferred tautomer for the 4-hydroxycinnolines. In fact, the UV absorption spectra of the 4-alkoxycinnolines are quite distinct from those of the *N*-alkylcinnolines (6) and can be used to distinguish between *O*-alkyl- and *N*-alkylcinnolines. Further, UV data can be used to distinguish between *N*-1-alkyl- and *N*-2-alkyl-4-hydroxycinnolines (5–7). On the other hand, one would expect hydroxy groups substituted in the benzenoid portion of the cinnoline ring system to be phenolic in nature, and this is confirmed at least in the case of 8-hydroxycinnoline, since the UV absorption spectrum of this compound is found to be very similar to that of 8-methoxycinnoline (29).

A methoxy group attached to the cinnoline ring system, at least at the 4-, 6-, and 7-positions, absorbs in the 5.70–6.12τ region of the pmr spectrum (7, 38), somewhat downfield from the absorption position of the methoxyl protons of anisole (6.22τ) or 6-methoxyquinoline (6.13 τ). The pmr absorption spectra of several methoxycinnolines are given in Table 1E-3.

III. Reactions

Molecular orbital calculations, using a self-consistent version of the Hückel method, predict that the preferred site of protonation of 4-methoxycinnoline is at N-1, but 3-, 5-, 6-, and 8-methoxycinnolines and 6,7-dimethoxycinnoline should all be protonated at N-2. No prediction could be made for 7-methoxycinnoline because the calculated π-electron energy in the transition state (localization index) for protonation at N-1 is too similar to this same quantity calculated for protonation at N-2 (39).

Like the 4-hydroxycinnolines, which are alkylated predominantly at N-2 in the absence of a group having large steric requirements at position 3, quaternization of 4-phenoxycinnoline with methyl iodide in ethanol is found experimentally to result in methylation at N-2 only. Similarly, 4-methoxycinnoline is methylated by methyl iodide principally at N-2, but in this case a small amount of the N-1 isomer is also obtained (5, 40). At first glance this may seem to contradict the results of the molecular orbital calculations concerning protonation, but the problem of kinetically versus thermodynamically controlled reactions must be considered here. Apparently, little work has been done on the alkylation of alkoxycinnolines other than 4-alkoxycinnolines.

N-Oxidation of 4-methoxycinnoline with phthalic monoperacid in ether solution gives principally the 2-oxide (33%) together with a smaller amount (18%) of the 1-oxide (10).

The phenoxy group of 4-phenoxycinnoline is readily cleaved in aqueous acid solution to give 4-hydroxycinnoline, as are the phenoxy group of 2-methyl-4-phenoxycinnolines and the methoxy group of 4-methoxy-2-methylcinnolines, which in refluxing 48% hydrobromic acid solution give the corresponding 2-methyl-4-hydroxycinnoline derivatives (40, 41). In addition, simply heating 4-methoxy-2-methylcinnolinium iodide at 150–160° C 1 hr yields the anhydro base of 4-hydroxy-2-methylcinnolinium hydroxide (40). The 8-methoxy group of 8-methoxycinnoline derivatives may also be cleaved in refluxing 48% hydrobromic acid solution to give the corresponding 8-hydroxycinnolines (29, 30).

The methoxy groups of both 4-methoxycinnoline 1- and 2-oxide are easily displaced nucleophilically by treating them with aqueous 5% sodium hydroxide solution on a steam bath, resulting in 57–87% yields of the 4-hydroxycinnoline-*N*-oxides (10).

An interesting feature of certain 4-alkoxy-3-phenylcinnolines [**30**; R = CH_3 or $CH_2CH_2N(C_2H_5)_2$] is that upon attempted distillation, these materials rearrange to the 1-alkyl-4-cinnolones (**31**) (3), a reaction also observed for 4-alkoxyquinolines.

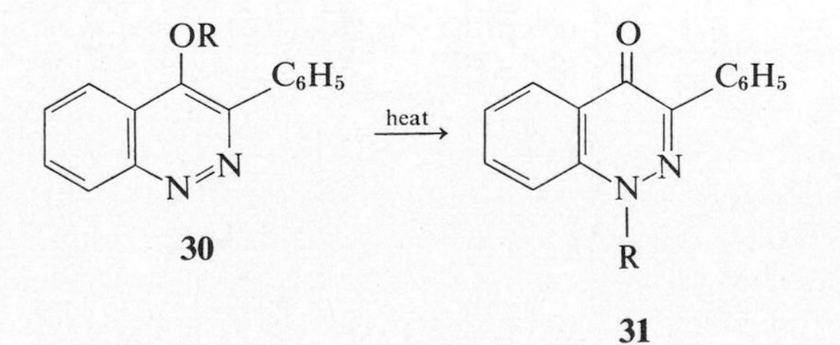

IV. Tables

TABLE 1E-1 Alkoxy- and Aryloxycinnolines[a]

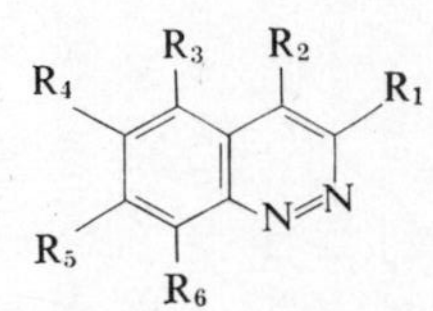

R_1	R_2	R_3	R_4	R_5	R_6	Prep.[b]	MP (°C)	Remarks[c]	Ref.
H	OCH_3	H	H	H	H	A	127–128	—	1, 2
H	OC_2H_5	H	H	H	H	A	Not stated	Picrate deriv., mp 174°; 2-methiodide deriv., mp 142–144° (benzene–ligroin)	1, 17a
H	$OCH(CH_3)_2$	H	H	H	H	A	100–101	Picrate deriv., mp 163°	1, 5
C_6H_5	$OCH_2CH_2N(C_2H_5)_2$	H	H	H	H	A	Not stated	Orange oil	3
C_6H_5	OCH_3	H	H	H	H	A	106–108	Yellow needles (ligroin)	3
H	OCH_3	H	OCH_3	OCH_3	H	A	210 (dec.)	(Methanol)	4
H	OC_2H_5	H	OCH_3	OCH_3	H	A	185–187	(Ethanol)	4
H	$OCH_2C_6H_5$	H	H	H	H	A	104–106	(Benzene–ligroin)	5
H	$OCH_2CH_2CH_3$	H	H	H	H	A	49–50	(Benzene–ligroin)	5
CH_2COOCH_3	Cl	H	OCH_3	OCH_3	H	B	151–152	(Methanol)	6
CH_3	Cl	H	OCH_3	OCH_3	H	B	222–223	Yellow crystals (ethyl acetate)	6
CH_3	OCH_3	H	OCH_3	OCH_3	H	A	182–183	Yellow needles (ethyl acetate)	6
CH_3	OCH_3	H	H	H	H	A	77–77.5	(Benzene–ligroin)	7
H	OCH_3	H	H	H	CH_3	A	104–105	(Benzene–ligroin)	7
H	OCH_3	H	H	NO_2	H	A	200	Yellow flat blades (acetone)	8

H	$OCH_2CH_2N(CH_3)_2$	H	H		H	H	A	76–77	BP 162–167 at 0.05 mm; acidic *d*-tartrate deriv., mp 129–131°	11
H	$OCH_2CH_2N(C_2H_5)_2$	H	H		H	H	A	—	Red syrup, bp 154–160° at 0.08 mm; acidic *d*-tartrate deriv., mp 131–133°	11
H	OCH_2CH_2—N (piperidine ring)	H	H		H	H	A	—	Red syrup, bp 170–178° at 0.065 mm; acidic *d*-tartrate deriv., mp 128–131°	11
H	OCH_2CH_2—N (morpholine ring, O)	H	H		H	H	A	—	Red syrup, bp 205–[illegible]0° at 0.02 mm; acidic *d*-tartrate deriv., mp 170–171°	11
H	OCH_2CH_2—N (piperazine ring) N—CH_3	H	H		H	H	A	—	Red syrup, bp 205–208° at 0.04 mm; dipicrate deriv., mp 230–232°	11
H	$O(CH_2)_3N(CH_3)_2$	H	H		H	H	A	—	Red syrup, bp 151–153° at 0.01 mm; dipicrate deriv., mp 160–162°	11
H	$O(CH_2)_3N(C_2H_5)_2$	H	H		H	H	A	—	Red syrup, bp 164–169° at 0.05 mm; acidic *d*-tartrate deriv., mp 190–193° (dec.)	11
H	$O(CH_2)_3$—N (piperidine ring)	H	H		H	H	A	—	Semisolid, bp 177–181° at 0.02 mm; acidic *d*-tartrate deriv., mp 166–168°	11

TABLE 1E-1 (*continued*

R_1	R_2	R_3	R_4	R_5	R_6	Prep.[b]	MP (°C)	Remarks[c]	Ref.
H	$O(CH_2)_3$—N N—CH_3	H	H	H	H	A	—	Red syrup, bp 195–198° at 0.01 mm; tripicrate deriv., mp 243–245° (dec.)	11
H	$OCH_2CH(CH_3)CH_2$—N N—CH_3	H	H	H	H	A	—	Red syrup, bp 193–196° at 0.02 mm; dipicrate deriv., mp 213–215°	11
H	O— N—CH_3	H	H	H	H	A	88–90	BP 178–181° at 0.02 mm; dipicrate deriv., mp 192–194°	11
H	OC_6H_5	H	$C_6H_5N{=}N$—	H	H	C	167–169	Red needles (ethyl acetate)	12
H	OC_6H_5	H	OC_6H_5 N=N— N N	H	H	A	254–258 (dec.)	Orange-red needles (benzene)	12, 13
Cl	OC_6H_5	H	H	H	H	A	127–128, 120–121	Crystals (ligroin)	15, 44
NO_2	OC_6H_5	H	H	H	H	A	144.5–145	Tan plates (ligroin)	15
H	OC_6H_5	H	H	NH_2	H	D	179–180	Greenish-yellow plates (benzene)	16
H	OC_6H_5	H	H	H	NH_2	D	130	Rhomboid crystals (benzene–ligroin)	16
H	OC_6H_5	NH_2	H	H	H	D	199–200 (dec.)	Yellow needles (benzene)	16
H	OC_6H_5	NO_2	H	H	H	C	123	Colorless needles (ligroin)	16

H	OC_6H_5	H	OC_6H_5 N=N NHCNH— ‖ $NCOC_6H_5$	H	H	E	244–245	Colorless needles (2-ethoxyethanol); trihydrochloride deriv., mp >350°	16
OCH_3	H	H	H	H	H	F	40–42	Impure; picrate deriv., mp 155–157.5°	21
H	OC_2H_5	H	C_6H_5	H	H	G	176	Colorless needles (ligroin or aq. alcohol); picrate deriv., mp 168°	23
H	$C_6H_5\underset{\vert}{C}HCN$	H	OCH_3	OCH_3	H	H	220–221	Orange-red plates (aqueous ethanol)	26
H	$C_6H_5\underset{\vert}{C}HCONH_2$	H	OCH_3	OCH_3	H	I	234.5–236.5	White crystals (aqueous ethanol)	26
H	Cl	H	OCH_3	OCH_3	H	A	195–196	Cream needles (ethanol)	26
H	$p\text{-}IC_6H_4\underset{\vert}{C}HCN$	H	OCH_3	OCH_3	H	H	252.6–253.0 (dec.)	Orange-red crystals (benzene)	27
H	$p\text{-}ClC_6H_4\underset{\vert}{C}HCN$	H	OCH_3	OCH_3	H	H	224.4–225.2 (dec.)	Orange-red crystals (benzene)	27
H	$p\text{-}BrC_6H_4\underset{\vert}{C}HCN$	H	OCH_3	OCH_3	H	H	241.4–241.8 (dec.)	Orange-red powder (benzene)	27
H	$m,p\text{-}Cl_2C_6H_3\underset{\vert}{C}HCN$	H	OCH_3	OCH_3	H	H	275.2–276.2 (dec.)	Orange needles (ethanol)	27
H	$m\text{-}CH_3OC_6H_4\underset{\vert}{C}HCN$	H	OCH_3	OCH_3	H	H	222.2–223.2 (dec.)	Red-gold needles (ethanol)	27
H	$p\text{-}NH_2C_6H_4\underset{\vert}{C}HCN$	H	OCH_3	OCH_3	H	J	206–207	Red needles (aqueous ethanol)	27
H	$m,p\text{-}(CH_3O)_2C_6H_3\underset{\vert}{C}HCN$	H	OCH_3	OCH_3	H	J	250–250.8	Orange needles (ethanol)	27
H	$(C_6H_5)_2CCN$	H	OCH_3	OCH_3	H	H	238.6–239.4	Yellow crystals (ethanol)	27

TABLE 1E-1 (*continued*)

R_1	R_2	R_3	R_4	R_5	R_6	Prep.[b]	MP (°C)	Remarks[c]	Ref.
H	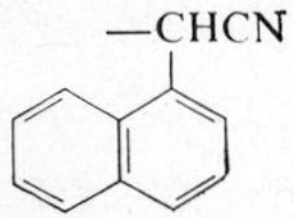 —CHCN	H	OCH_3	OCH_3	H	H	247.2–248.0	Orange-gold crystals (ethanol)	27
H	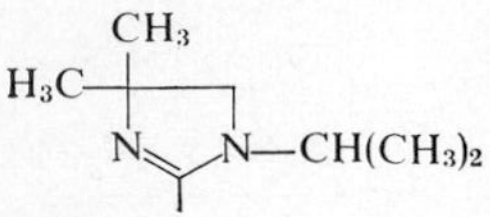 CH_3, H_3C, N, N—$CH(CH_3)_2$	H	OCH_3	OCH_3	H	H	242–243 (dec.)	Yellow crystals (Al_2O_3 chromatography, ethanol)	27
H	Cl	OCH_3	H	H	H	B	141–142	White needles (ligroin)	28
H	p-$CH_3C_6H_4SO_2NHNH$—	OCH_3	H	H	H	K	Not stated	Hydrochloride deriv., mp 221° (dec.)	28
H	p-$CH_3C_6H_4SO_2NHNH$—	H	OCH_3	H	H	K	Not stated	Monohydrate deriv., mp 199–201° (dec.)	28
H	p-$CH_3C_6H_4SO_2NHNH$—	H	H	OCH_3	H	K	Not stated	Hydrochloride deriv., mp 169–172° (dec.)	28
H	H	OCH_3	H	H	H	L	92–93.5	Cream blades (ligroin)	28
H	H	H	OCH_3	H	H	L	87–88	Monohydrate, colorless needles (ligroin)	28
H	H	H	H	OCH_3	H	L	109–110	Monohydrate, needles (ligroin)	28
H	Cl	H	H	OCH_3	H	B	178–179	Needles (ligroin)	28
H	Cl	H	H	H	OCH_3	B	142–143	Yellow needles (acetone)	29
H	OC_6H_5	H	H	H	OCH_3	A	123–124	Colorless needles or plates (aq. acetone)	29

H	p-$CH_3C_6H_4SO_2NHNH$—	H	H	H	OCH_3	K	Not stated	Hydrochloride deriv., mp 169–172 (dec.)	29
H	H	H	H	H	OCH_3	L	67–70.5	Colorless needles (benzene–ligroin); picrate deriv., mp 189–191.5°	29
H	CH_3	H	H	H	OCH_3	M	130–132, 131–132	Orange needles (benzene–ligroin); yellow prisms (ligroin)	29, 30
H	C_6H_5	H	H	H	OCH_3	N	85–89	Yellow plates (chloroform–ligroin)	29
H	CH_3	H	OCH_3	OCH_3	H	O	180–182	Yellow crystals, sublime 135° at 0.005 mm	31
H	H	H	OCH_3	OCH_3	H	P	123–124.5	Yellow crystals (cyclohexane)	31
H	OC_6H_5	Cl	H	H	H	A	118–119	Brown prismatic needles (ligroin)	32
H	OC_6H_5	Cl	Cl	H	H	A	189	Colorless needles (benzene–ligroin)	32
C_6H_5	COOH	H	OCH_3	H	H	Q	229–230	(Acetic acid)	33
NO_2	OC_6H_5	H	Cl	H	H	A	111–112	(Ligroin)	46
H	OC_6H_5	H	Cl	H	H	A	129–130	(Ligroin)	47
H	H	H	$C_6H_5CH_2O$	C_6H_5-CH_2O	H	R	156–157	(Benzene–ligroin)	33a
H	H	H	OCH_3	C_6H_5-CH_2O	H	R	165	(Benzene–ligroin)	33a
H	H	H	OCH_3	OCH_3	H	R	122–124	(Cyclohexane)	33a
Br	Cl	H	OCH_3	OCH_3	H	B	219	Yellow needles (ethanol)	48

TABLE 1E-1 (*continued*)

R_1	R_2	R_3	R_4	R_5	R_6	Prep.[b]	MP (°C)	Remarks[c]	Ref.
Cl	Cl	H	OCH_3	OCH_3	H	S	231	Green-yellow solid (acetic acid)	48
Br	*o*-HOC_6H_4NH	H	OCH_3	OCH_3	H	T	256	Yellow needles (dioxane)	48
H	C_6H_5NH	H	OCH_3	OCH_3	H	T	223	Yellow prisms (ethanol)	48

[a] Alkoxy- and aryloxycinnolines which also bear a hydroxy or mercapto substituent are not given in this table. Hydroxy-substituted alkoxy- and aryloxycinnolines may be found in Tables 1C-5 and 1C-6, while alkoxy- and aryloxycinnolines bearing a mercapto, alkylthio, or arylthio group may be found in Table 1F-3.

[b] A, by treatment of the appropriate 4-chlorocinnoline with the sodium or potassium salt of the appropriate alcohol or phenol. B, by treatment of the corresponding 4-hydroxycinnoline with phosphorus oxychloride. C, by heating the appropriate 4-chlorocinnoline with phenol and ammonium carbonate, or by passing dry ammonia through a warm suspension of the 4-chlorocinnoline in phenol. D, by reduction of the corresponding nitrocinnoline with stannous chloride in glacial acetic acid. E, by interaction of 7-amino-4-phenoxycinnoline with dichloromethylenebenzamide ($C_6H_5N{=}CCl_2$) in nitromethane at room temperature. F, from 3-bromocinnoline and methanolic sodium methoxide in a sealed tube at 110–120° C 17 hr. G, inadvertently, during the attempted reduction of 4-chloro-6-phenylcinnoline over palladium-charcoal in aqueous alcoholic sodium hydroxide solution. H, by condensation of the appropriately substituted phenylacetonitrile with 4-chloro-6,7-dimethoxycinnoline in dry benzene in the presence of sodium amide. I, by hydrolysis of α-(6,7-dimethoxy-4-cinnolinyl)phenylacetonitrile in concentrated sulfuric acid solution. J, by condensation of the appropriately substituted phenylacetonitrile with 4-chloro-6,7-dimethoxycinnoline in liquid ammonia in the presence of potassium amide. K, by interaction of the appropriately substituted 4-chlorocinnoline with toluene-*p*-sulfonylhydrazide in chloroform. L, by decomposition of the appropriately substituted 4-toluene-*p*-sulfonylhydrazinocinnoline by heating in an aqueous sodium carbonate solution. M, by diazotization and cyclization of 2-(2-amino-3-methoxyphenyl)propene with sodium nitrite in aqueous sulfuric acid. N, by diazotization and cyclization of α-(2-amino-3-methoxyphenyl)styrene with sodium nitrite in aqueous sulfuric acid. O, by diazotization and cyclization of 2-(2-amino-4,5-dimethoxyphenyl)propene with sodium nitrite in aqueous hydrochloric acid. P, by the action of lithium aluminum hydride in refluxing tetrahydrofuran on the corresponding 4-hydroxycinnoline, followed by gentle oxidation of the product with red mercuric oxide. Q, by rearrangement of 5-methoxy-*N*-benzylideneaminoisatin by refluxing it in 50% aqueous potassium hydroxide. R, by reductive cyclization of the corresponding alkoxy-substituted 2,β-dinitrostyrene with lithium aluminum hydride. S, by treatment of the corresponding 3-bromo-4-hydroxycinnoline with a mixture of phosphorus oxychloride and phosphorus pentachloride. T, by reaction of the appropriately substituted 4-chlorocinnoline with aniline or a substituted aniline.

[c] Solvent employed in recrystallization is enclosed in parentheses.

TABLE 1E-2. Ultraviolet Absorption Spectra of Alkoxycinnolines[a]

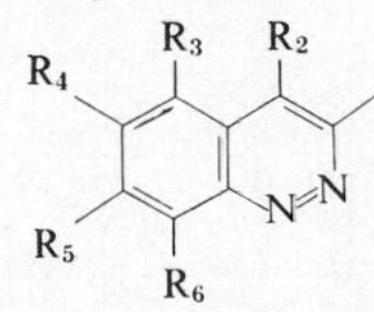

R_1	R_2	R_3	R_4	R_5	R_6	Solvent	$\lambda_{max}(m\mu)^b$	log E_{max}	Ref.
H	H	H	H	H	OCH_3	95% methanol	242	4.64	29
							295	3.32	
							350	3.64	
						0.01*N* HCl	247	4.34	
							300	3.08	
							422	3.28	
						0.01*N* NaOH	242	4.38	
							295	3.12	
							358	3.44	
CH_3	OCH_3	H	OCH_3	OCH_3	H	Ethanol	243	4.65	6
							304	3.87	
H	OCH_3	H	H	H	H	Ethanol	226	4.60	5
							284	*3.72*	
							292	3.75	
							317	3.65	
						Water pH = 7.0 (neutral species)	*211*	*4.30*	2, 20
							227	4.59	
							293	3.71	
							319	3.69	
							328	*3.65*	

TABLE 1E-2. (*continued*)

R_1	R_2	R_3	R_4	R_5	R_6	Solvent	$\lambda_{max}(m\mu)^b$	log E_{max}	Ref.
						Water pH = 1.0	233	4.50	2, 20
						(cationic species)	254	3.97	
							296	*3.53*	
							307	3.71	
							338	3.79	
C_6H_5	OCH_3	H	H	H	H	Methanol	248	4.59	3
							288	3.83	
							332	3.54	
C_6H_5	$—O(CH_2)_2NEt_2$	H	H	H	H	Methanol	248	4.54	3
							288	3.79	
							330	3.54	
H	H	OCH_3	H	H	H	Water pH = 7.0	245	4.53	45
							305	*3.15*	
							353	3.44	
H	H	H	OCH_3	H	H	Water pH = 7.0	240	4.56	45
							316	3.72	
H	H	H	H	OCH_3	H	Water pH = 7.0	237	4.66	45
							269	3.46	
							346	3.50	
CH_3	OCH_3	H	H	H	H	Ethanol	225	4.75	7
							281	3.53	
							292	3.53	
							326	3.61	

H	OCH_3	H	H	H	CH_3	Ethanol	219	4.67	7
							286	3.68	
							295	3.72	
							323	3.71	
H	OCH_3	H	NO_2	H	H	Ethanol	218	4.38	7
							235	4.36	
							255	4.18	
							290	3.79	
							299	3.76	
							349	3.67	
H	OCH_3	H	H	H	NO_2	Ethanol	221	4.32	7
							293	3.71	
							321	3.63	
							390	3.63	

[a] Ultraviolet absorption data for alkoxycinnolines which are substituted with a hydroxy group are not given here. These may be found in Table 1C-2.

[b] Italics refer to shoulders or inflections.

TABLE 1E-3. Proton Magnetic Resonance Spectra of Alkoxycinnolines[a]

R_3	R_4	R_5	R_6	R_7	R_8	Ref.
H	CH_3	H	OCH_3	OCH_3	H	31
1.02(S)	7.42(S)	2.98(S)	5.91(S)	5.91(S)	2.29(S)	
H	H	H	OCH_3	OCH_3	H	31
0.83(D)	2.29(Q)	3.01(S)	{5.88(S)	5.93(S)}	2.24(S)	
			or			
$J = 6$	$J_1 = 6$, $J_2 = 0.5$		{5.93(S)	5.88(S)}	(also part of H_4)	
CH_3	OCH_3	H	OCH_3	OCH_3	H	38
7.17	5.94	2.38	{5.94	5.99}	2.88	
			or			
			{5.99	5.94}		
CH_3	Cl	H	OCH_3	OCH_3	H	38
7.06	—	2.34	5.92	5.92	2.82	
H	OCH_3	H	OCH_3	OCH_3	H	38
1.15	5.90	2.32	5.94	5.94	2.71	
CH_2COOCH_3	Cl	H	OCH_3	OCH_3	H	38
CH_2 at 5.58; CH_3 at 6.24	—	2.26	5.88	5.88	2.72	
H	OCH_3	H	Br	H	H	38
0.95	5.82	1.57	—	1.67	2.09	
		$J_{5,7} = 3.5$		$J_{7,8} = 9.4$		

[a] Chemical shift values are in τ units; the multiplicity of each absorption when stated is given in parentheses following the τ value as S = singlet, D = doublet, Q = quartet; J = coupling constant in cps, given below the proton to which it refers. The pmr spectra of alkoxycinnolines which bear a hydroxy group may be found in Table 1C-1. In all cases deuteriochloroform was used as the solvent, and tetramethylsilane was used as the reference standard.

Often 4-phenoxycinnolines have been prepared as intermediates in the synthesis of 4-aminocinnolines. When a 4-phenoxycinnoline is heated at 175° C in phenol in the presence of ammonium chloride and dry ammonia (12, 13) or heated at 137° C with molten ammonium acetate (16), good yields of the 4-amino compounds are obtained. Substituted amines can also be prepared in this manner; dry methylamine, ammonium chloride, phenol, and a 4-phenoxycinnoline at 135° C, for example, give the 4-methylaminocinnoline (13).

Reduction of 4-methoxycinnoline by lithium aluminum hydride in refluxing benzene solution for 3 hr yields a mixture of cinnoline and 1,2,3,4-tetrahydrocinnoline (42). A polarographic study has shown that the electrolytic reduction of 4-methoxycinnoline in an aqueous citric acid buffer solution consumes two electrons per molecule, and the product is 4-keto-1,2,3,4-tetrahydrocinnoline. The intermediate product in this reduction is thought to be 1,2-dihydro-4-methoxycinnoline, a type of enolic ether which suffers loss of the methoxy group by hydrolysis to give the final product (43).

Reduction of 4-phenoxycinnoline by zinc in ammonia or by hydrogenation over palladium on alumina gives unstable 1,4-dihydrocinnoline and phenol (43a).

References

1. K. Adachi, *J. Pharm. Soc. Japan*, **75,** 1426 (1955).
2. A. Albert and G. B. Barlin, *J. Chem. Soc.*, **1962,** 3129.
3. H. S. Lowrie, *J. Med. Chem.*, **9,** 784 (1966).
4. R. N. Castle, H. Ward, N. White, and K. Adachi, *J. Org. Chem.*, **25,** 570 (1960).
5. D. E. Ames, R. F. Chapman, H. Z. Kucharska, and D. Waite, *J. Chem. Soc.*, **1965,** 5391.
6. D. E. Ames and A. C. Lovesey, *J. Chem. Soc.*, **1965,** 6036.
7. D. E. Ames, R. F. Chapman, and D. Waite, *J. Chem. Soc.* (*C*), **1966,** 470.
8. C. M. Atkinson and A. Taylor, *J. Chem. Soc.*, **1955,** 4236.
9. I. Suzuki, T. Nakashima, N. Nagasawa, and T. Itai, *Chem. Pharm. Bull.* (*Tokyo*), **12,** 1090 (1964).
10. I. Suzuki and T. Nakashima, *Chem. Pharm. Bull.* (*Tokyo*), **12,** 619 (1964).
11. R. N. Castle and M. Onda, *J. Org. Chem.*, **26,** 2374 (1961).
12. J. McIntyre and J. C. E. Simpson, *J. Chem. Soc.*, **1952,** 2606.
13. E. M. Lourie, J. S. Morley, J. C. E. Simpson, and J. M. Walker, Br. Patent 702,664, Jan. 20, 1954; Appl. July 5, 1950.
14. H. J. Barber E. Lunt, K. Washbourn, and W. R. Wragg, *J. Chem. Soc.* (*C*), **1967,** 1657.
15. H. E. Baumgarten, *J. Am. Chem. Soc.*, **77,** 5109 (1955).
16. C. M. Atkinson, J. C. E. Simpson, and A. Taylor, *J. Chem. Soc.*, **1954,** 1381.
17. M. Ogata, H. Kano, and K. Tori, *Chem. Pharm. Bull.* (*Tokyo*), **11,** 1527 (1963).

17a. D. E. Ames and B. Novitt, *J. Chem. Soc.* (*C*), **1969,** 2355.

18. D. E. Ames and H. Z. Kucharska, *J. Chem. Soc.*, **1963,** 4924.
19. D. E. Ames and R. F. Chapman, *J. Chem. Soc.* (*C*), **1967,** 40.
20. G. B. Barlin, *J. Chem. Soc.*, **1965,** 2260.
21. E. J. Alford and K. Schofield, *J. Chem. Soc.*, **1953,** 1811.
22. G. B. Barlin and W. V. Brown, *J. Chem. Soc.* (*B*), **1967,** 736.

22a. E. Hayashi and T. Watanabe, *Yakugaku Zasshi*, **88,** 94 (1968); *Chem. Abstr.*, **69,** 19106 (1968).

22b. G. B. Barlin and W. V. Brown, *J. Chem. Soc.* (*B*), **1968,** 1435.

22c. T. Watanabe, *Yakugaku Zasshi*, **89,** 1167 (1969); *Chem. Abstr.*, **72,** 3452 (1970).

23. C. M. Atkinson and C. J. Sharpe, *J. Chem. Soc.*, **1959,** 2858.
24. M. Ogata, H. Kano, and K. Tori, *Chem. Pharm. Bull.* (*Tokyo*), **10,** 1123 (1962).
25. I. Suzuki, T. Nakashima, and T. Itai, *Chem. Pharm. Bull.* (*Tokyo*) **11,** 268 (1963).

26. R. N. Castle and F. H. Kruse, *J. Org. Chem.*, **17,** 1571 (1952).
27. R. N. Castle and D. B. Cox, *J. Org. Chem.*, **19,** 1117 (1954).
28. A. R. Osborn and K. Schofield, *J. Chem. Soc.*, **1955,** 2100.
29. E. J. Alford, H. Irving, H. S. Marsh, and K. Schofield, *J. Chem. Soc.*, **1952,** 3009.
30. A. Albert and A. Hampton, *J. Chem. Soc.*, **1952,** 4985.
31. J. M. Bruce and P. Knowles, *J. Chem. Soc.*, **1964,** 4046.
32. H. J. Barber, K. Washbourn, W. R. Wragg, and E. Lunt, *J. Chem. Soc.*, **1961,** 2828.
33. H. S. Lowrie, *J. Med. Chem.*, **9,** 664 (1966).
33a. I. Baxter and G. A. Swan, *J. Chem. Soc. (C)*, **1968,** 468.
34. J. R. Keneford, J. S. Morley, J. C. E. Simpson, and P. H. Wright, *J. Chem. Soc.*, **1949,** 1356.
35. A. Albert and J. N. Phillips, *J. Chem. Soc.*, **1956,** 1294.
36. A. Albert and A. Hampton, *J. Chem. Soc.*, **1954,** 505.
37. R. N. Castle, D. B. Cox, and J. F. Suttle, *J. Am. Pharm. Assoc.*, **48,** 135 (1959).
38. A. W. Ellis and A. C. Lovesey, *J. Chem. Soc. (B)*, **1967,** 1285.
39. D. E. Ames, G. V. Boyd, R. F. Chapman, A. W. Ellis, A. C. Lovesey, and D. Waite, *J. Chem. Soc. (B)*, **1967,** 748.
40. D. E. Ames, R. F. Chapman, and H. Z. Kucharska, *J. Chem. Soc.*, Suppl. 1, **1964,** 5659.
41. D. E. Ames, *J. Chem. Soc.*, **1964,** 1763.
42. D. E. Ames and H. Z. Kucharska, *J. Chem. Soc.*, **1962,** 1509.
43. H. Lund, *Acta Chem. Scand.*, **21,** 2525 (1967).
43a. F. Schatz and T. Wagner-Jauregg, *Helv. Chim. Acta*, **51,** 1919 (1968).
44. K. Schofield and T. Swain, *J. Chem. Soc.*, **1950,** 384.
45. A. R. Osborn and K. Schofield, *J. Chem. Soc.*, **1956,** 4207.
46. E. Lunt, K. Washbourn, and W. R. Wragg, *J. Chem. Soc. (C)*, **1968,** 687.
47. E. Lunt, K. Washbourn, and W. R. Wragg, *J. Chem. Soc. (C)*, **1968,** 1152.
48. A. N. Kaushal and K. S. Narang, *Indian J. Chem.*, **6,** 350 (1968).

Part F. Mercapto- and Alkylthiocinnolines

I. Methods of Preparation

A. Hydroxy- to Mercaptocinnolines

The conversion of hydroxycinnolines into their mercapto analogs by interaction with phosphorus pentasulfide in refluxing pyridine, toluene, or benzene has been discussed in detail in Section 1C-III-A-2 and need not be discussed further here.

B. Halo- to Mercaptocinnolines

When 4-chlorocinnoline (**1**) is treated with phosphorus pentasulfide in refluxing pyridine for 1 hr, a 70% yield of 4-mercaptocinnoline (**2**) is obtained. Generally, nitrogen heterocycles which have a halogen atom attached to the ring at a position where it is active toward nucleophilic substitution undergo this reaction, halogen being replaced by the mercapto group (1). A second method that has been used to prepare 4-mercaptocinnolines is the action of thiourea on 4-chlorocinnolines. The intermediate thiouronium salt **3** is readily hydrolyzed to the 4-mercaptocinnoline in good yield by heating it in an aqueous alkaline solution (2).

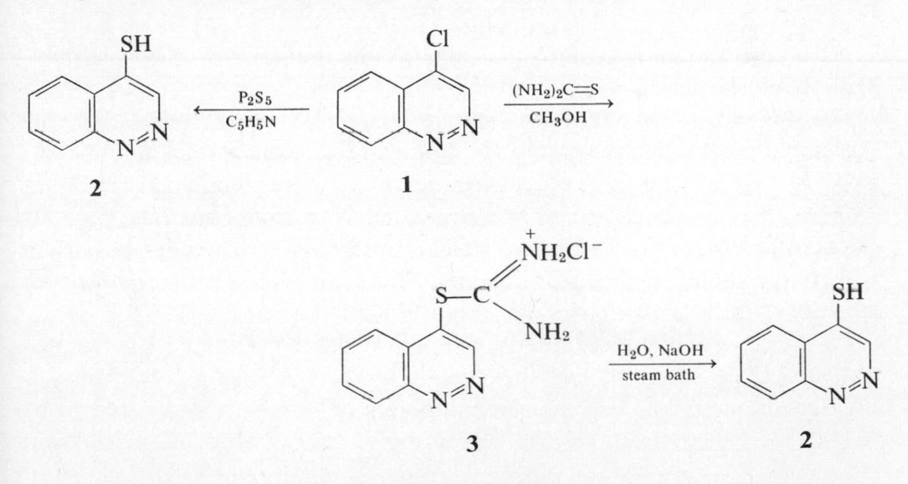

With few exceptions, the mercaptocinnolines reported in the literature to date are 4-mercaptocinnolines. These are listed in Table 1F-3. The few other mercaptocinnolines which have been prepared are discussed in Section 1C-III-A-2.

C. Alkylthio- and Arylthiocinnolines

Two general methods are available for the preparation of 4-alkylthio- and 4-arylthiocinnolines. In the first, 4-chlorocinnoline (**1**) is allowed to react with a potassium or sodium alkylmercaptide or arylmercaptide to give the 4-alkylthiocinnoline or 4-arylthiocinnoline **4** (3–6). In the second procedure,

4-mercaptocinnoline (**2**) is allowed to react with an alkyl halide or dimethyl sulfate in the presence of a base to give the alkylthio derivative **4** (2–4, 7); in this method, halogen-substituted nitrogen heterocycles such as 2-chloroquinoxaline have been used successfully in place of simple alkyl halides (2).

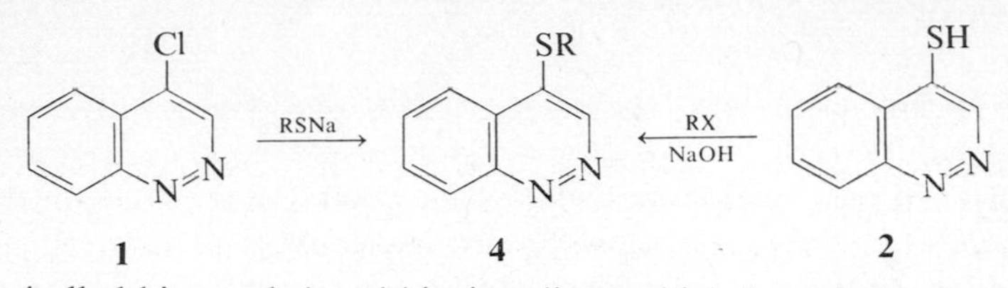

The 4-alkylthio- and 4-arylthiocinnolines which have been prepared by these methods are listed in Table 1F-3.

II. Physical Properties and Structure

All of the known mercapto- and alkylthiocinnolines are solid, relatively high-melting materials.

The 4-mercaptocinnolines are stronger acids and weaker bases than the corresponding 4-hydroxy derivatives. For example, 4-mercaptocinnoline has a pK_a of 7.09 (as an acid in water), whereas 4-hydroxycinnoline has a pK_a of 9.27 (in water). In comparing basic strengths, 4-mercaptocinnoline, possessing a pK_a of -1.83 (as a base in water), is seen to be a weaker base than 4-hydroxycinnoline (pK_a -0.35 in water). The *S*-methyl derivatives, however, are only slightly weaker bases than their *O*-methyl analogs (8).

On the basis of UV absorption data and ionization constants, 4-mercaptocinnoline (**2**) is shown to exist principally as the tautomeric 1,4-dihydro-4-cinnolinethione **5** (9). The ratio of tautomers (NH:SH) is calculated to be 90,000:1 in aqueous solution (8). Accordingly, the UV absorption spectrum

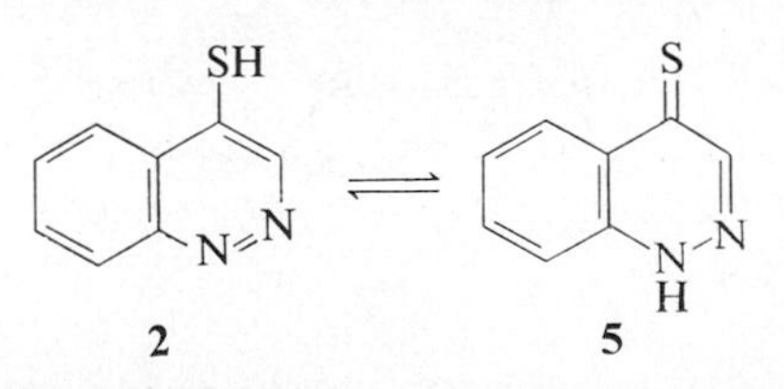

of 4-mercaptocinnoline resembles that of its *N*-alkyl derivatives rather than that of 4-methylthiocinnoline (9, 10), displaying a high-intensity band near 420 mμ which is associated with the thiocarbonyl group (11). Tables 1F-1 and 1F-2 give the UV spectral parameters for some mercapto- and alkylthiocinnolines. Both 4-mercaptocinnoline and its ring-substituted derivatives display a broad N—H stretching vibration absorption in their infrared spectra in the solid state, indicating intermolecular N—H · · · S hydrogen bonding. A band occurring near 1171 cm^{-1} which is absent from the spectra of the corresponding 4-hydroxycinnolines is attributed to absorption by the thiocarbonyl group. Absorption bands (near 2600 cm^{-1}) due to S—H stretching vibrations are absent (12).

III. Reactions

Unlike 4-hydroxycinnoline, which undergoes *N*-alkylation (Section 1C-III-E), 4-mercaptocinnoline undergoes almost exclusively *S*-alkylation (Eq. 1) (2, 3, 10). In only one case has a trace of the 1-alkylated product been reported (4). The 1- and 2-methyl derivatives of 4-mercaptocinnoline and 2-methylcinnoline-3-thione are likewise methylated on sulfur rather than on the second nitrogen atom (Eqs. 2, 3, 4, respectively) (13). For this reason

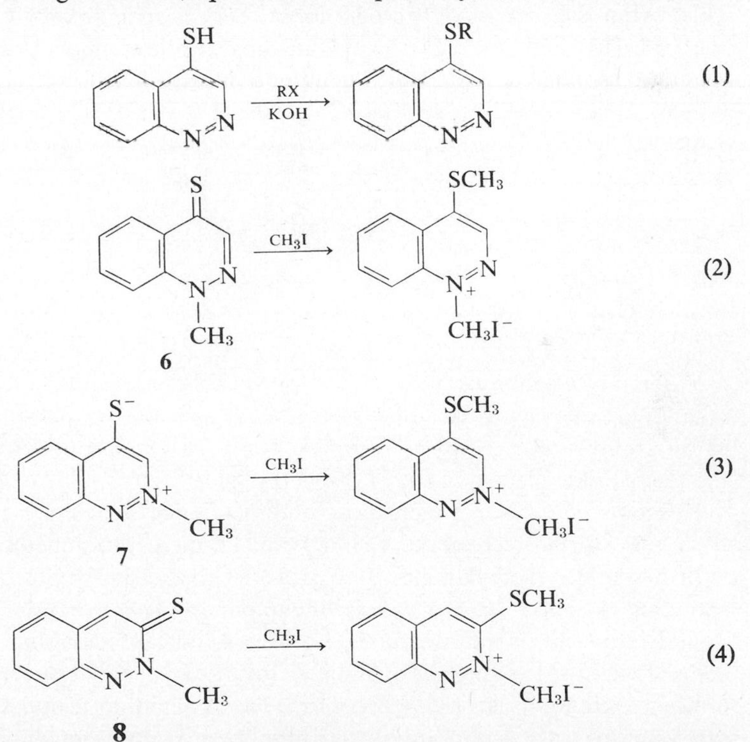

1,4-dihydro-1-methyl-4-thiocinnolone (**6**), the anhydro- base of 4-mercapto-2-methylcinnolinium hydroxide (**7**), and 2,3-dihydro-2-methyl-3-thiocinnolone (**8**) must be obtained from the reaction of the corresponding *N*-methyl hydroxycinnolines with phosphorus pentasulfide (see Section 1C-III-A-2).

Methylation of 4-methylthiocinnoline (**9**) with methyl iodide in refluxing benzene solution yields a mixture of 2-methyl-4-methylthiocinnolinium iodide (**10**) (59%) and 1-methyl-4-methylthiocinnolinium iodide (**11**) (13%) (13).

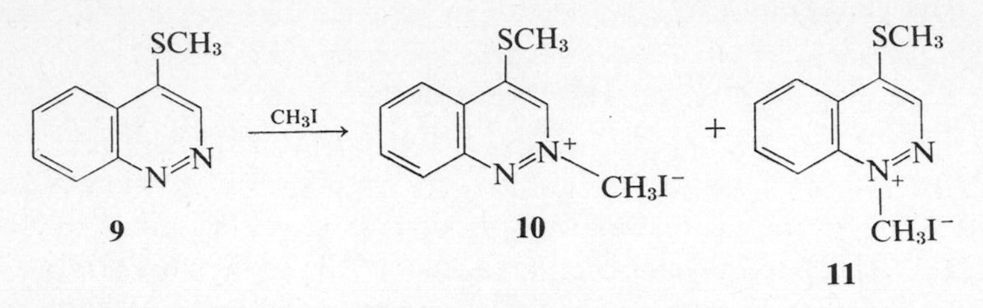

Wagner (4, 14, 15) demonstrated that it is possible to cause *S*-alkyl cinnolines to rearrange to the *N*-alkyl isomers. Thus the tetra-*O*-acetyl-*β*-D-glucopyranoside of 4-mercaptocinnoline (**12**) rearranges to a mixture of the N-1 (**13**) and N-2 (**14**) acetyl glucopyranosides upon treatment with mercuric bromide in refluxing toluene or xylene.

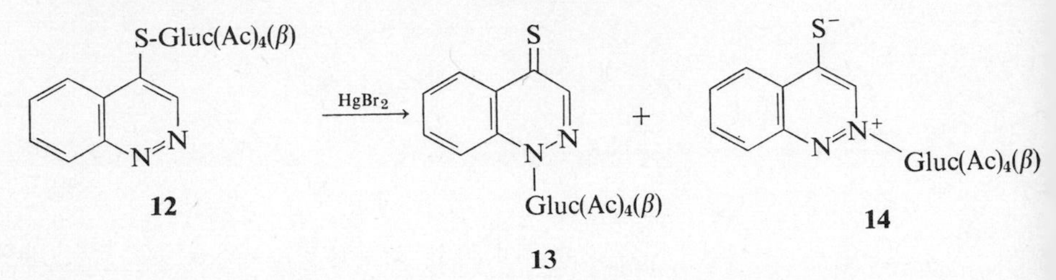

When dilute solutions of 5-, 6-, 7-, or 8-halo-4-mercaptocinnoline in 95% ethanol are allowed to stand for several days at room temperature, the mercapto group is replaced by a hydroxyl group. This reaction was observed by the appropriate change in the UV spectra (11).

Reaction of certain 4-mercaptocinnolines, such as 4-mercaptocinnoline itself, 6-chloro-4-mercaptocinnoline, and 4-mercapto-6-methoxycinnoline with ammonia, methylamine, or hydrazine gives good yields of the corresponding 4-amino- and 4-hydrazinocinnolines. This reaction failed with 4-mercapto-6-nitrocinnoline and 3-amino-4-mercaptocinnoline (7). With sodium methoxide in methanol at 66° C for 3 hr, 4-methylthiocinnoline gives 4-methoxycinnoline in 71% yield (16). The methylthio group is less easily displaced from the 4-position of the cinnoline ring than is a chloride ion or a methylsulfonyl group. Thus the ratio of reactivity of 4-chloro- to 4-methylthiocinnoline (Cl/SCH_3) with methoxide ion at 30° C is 1.23×10^2, whereas the ratio of reactivity of 4-methylsulfonyl- to 4-methylthiocinnoline (SO_2CH_3/SCH_3) with methoxide ion is 5.11×10^3 under the same conditions (16).

Desulfurization of 4-mercaptocinnoline with wet Raney nickel in ethanol at room temperature produces cinnoline in about 30% yield. Desulfurization of 6-chloro-4-mercaptocinnoline under similar conditions likewise produces cinnoline, the 6-chloro substituent being removed by hydrogenolysis. Under more drastic conditions (tenfold excess of Raney nickel, prolonged refluxing in ethanol) the reaction gives unidentified products (7).

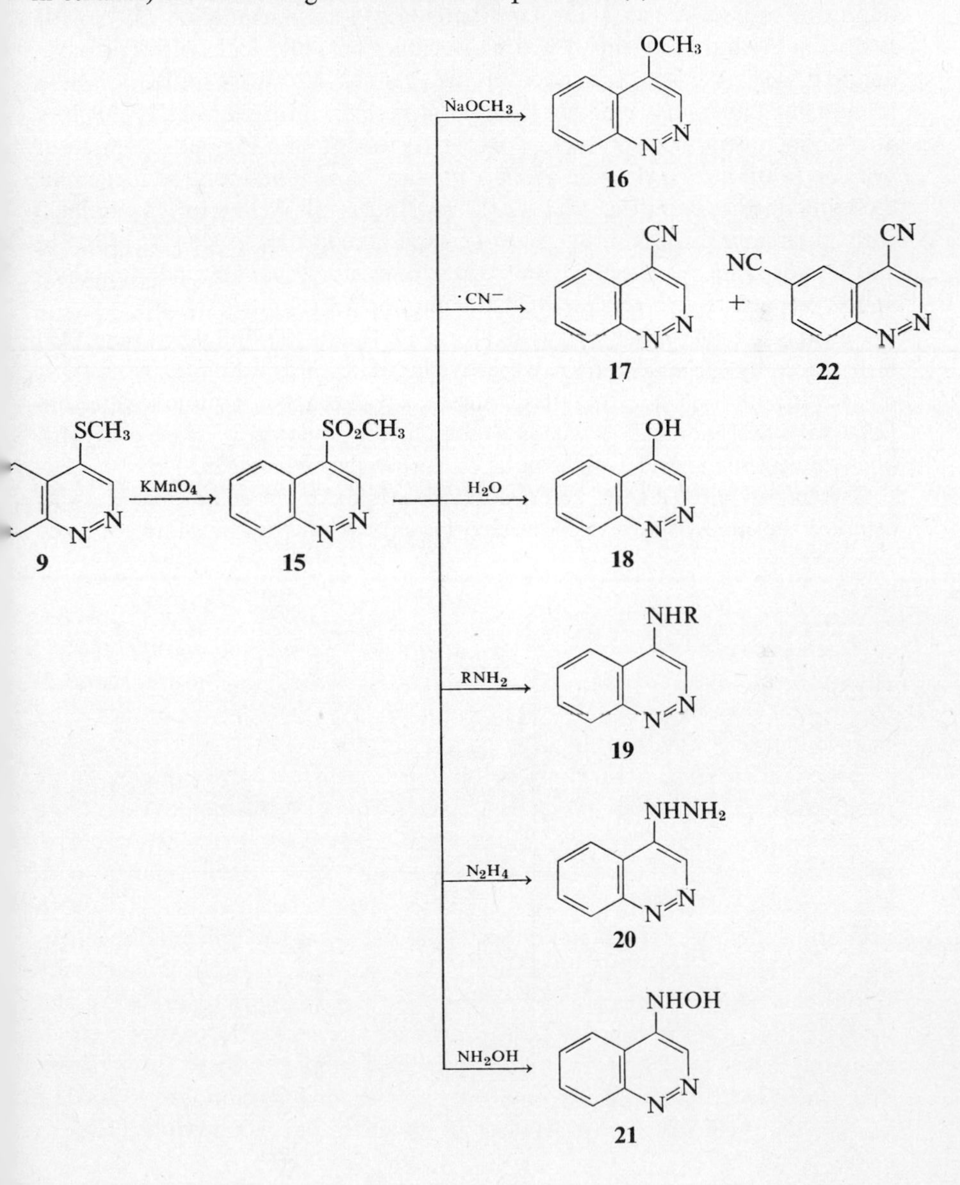

Oxidation of 4-methylthiocinnoline (**9**) with potassium permanganate in aqueous acetic acid yields 4-methylsulfonylcinnoline (**15**). This compound undergoes nucleophilic displacement reactions with extreme ease, even more readily than does 4-chlorocinnoline. Thus with sodium methoxide in methanol, it yields 4-methoxycinnoline (**16**); with water, dilute hydrochloric acid or aqueous sodium hydroxide it yields 4-hydroxycinnoline (**18**); with methylamine or aniline it yields the corresponding 4-aminocinnoline (**19**); with hydrazine hydrate it yields 4-hydrazinocinnoline (**20**); and with hydroxylamine it yields 4-hydroxyaminocinnoline (**21**) (17–19). With sodium cyanide in refluxing dimethylformamide (5 min), 4-methylsulfonylcinnoline (**15**) gives 4-cinnolinecarbonitrile (**17**) (18), while treatment of (**15**) with potassium cyanide in dimethyl sulfoxide gives a mixture of 4-cinnolinecarbonitrile and 4,6-cinnolinedicarbonitrile (**22**) (20). Treatment of 4-cinnolinecarbonitrile with potassium cyanide also gives 4,6-cinnolinedicarbonitrile (20, 21). At 100° C for 1 hr in dimethyl sulfoxide, potassium cyanide and 4-methylsulfonylcinnoline give principally 4,6-cinnolinedicarbonitrile (21).

As expected, the methylsulfonyl group of 4-methylsulfonylcinnoline (**15**) is displaced by carbanions to give 4-alkylcinnolines. For example, interaction of **15** with ethyl acetate in a basic medium yields ethyl 4-cinnolinylacetate (**23**), with malononitrile it yields 4-cinnolinylmalononitrile (**24**), and with phenylacetylene it gives 1-(4-cinnolinyl)-2-phenylacetylene (**25**) (19).

Carbanions derived from ketones having labile α-hydrogen atoms likewise displace the methylsulfonyl group from 4-methylsulfonylcinnoline, but the

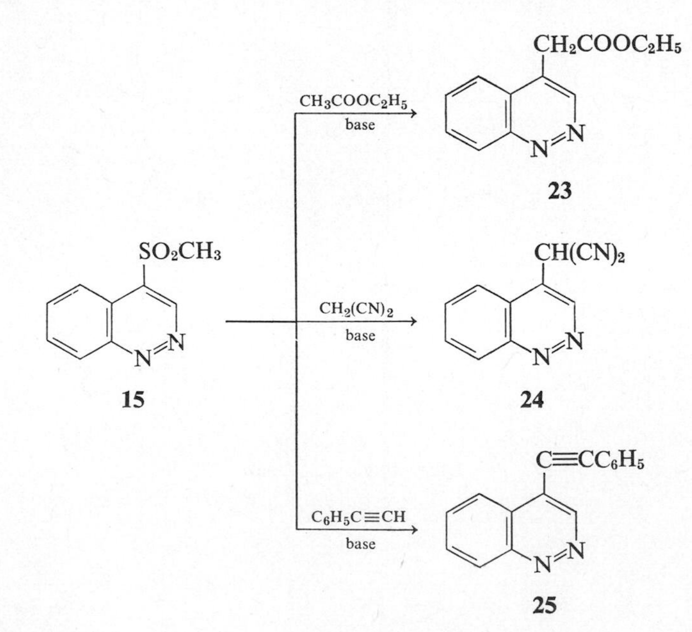

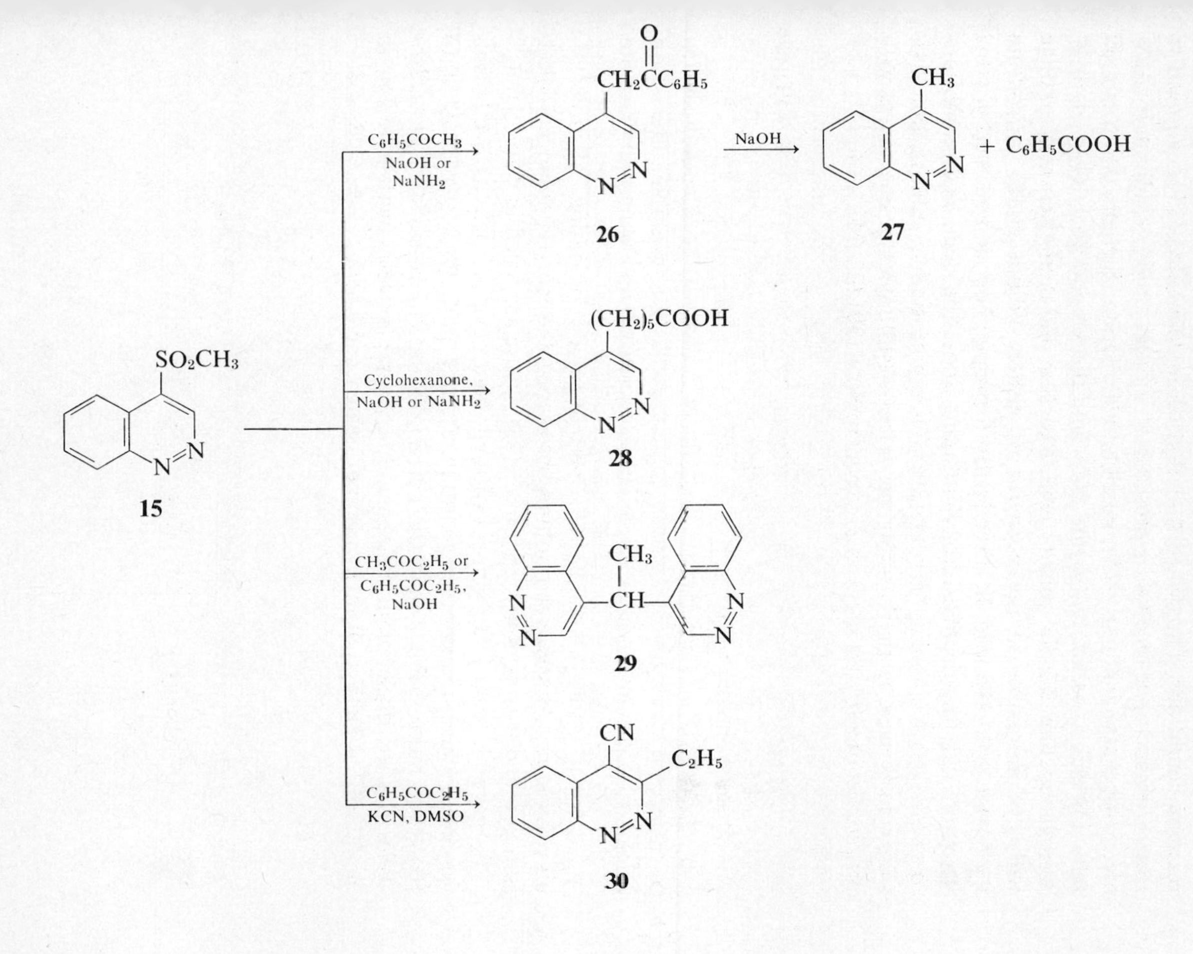
SO_2CH_3
15
$C_6H_5COCH_3$
NaOH or
$NaNH_2$
O
$CH_2CC_6H_5$
26
NaOH
CH_3
$+ C_6H_5COOH$
27
Cyclohexanone,
NaOH or $NaNH_2$
$(CH_2)_5COOH$
28
$CH_3COC_2H_5$ or
$C_6H_5COC_2H_5$,
NaOH
CH_3
CH
29
$C_6H_5COC_2H_5$
KCN, DMSO
CN
C_2H_5
30

type of product obtained depends upon the structure of the ketone and upon reaction conditions. Thus when 4-methylsulfonylcinnoline (**15**) is allowed to react with acetophenone in the presence of sodium amide or sodium hydroxide, 4-phenacylcinnoline (**26**) is obtained. Upon further treatment in refluxing 30% sodium hydroxide solution, the ketone **26** is cleaved to 4-methylcinnoline (**27**) and benzoic acid. Reaction of 4-methylsulfonylcinnoline with cyclohexanone in the presence of either sodium amide or sodium hydroxide gives the acid **28** directly. Both propiophenone and methyl ethyl ketone react with 4-methylsulfonylcinnoline under the same basic conditions to give 1,1-bis(4-cinnolinyl)ethane (**29**). When potassium cyanide in dimethylsulfoxide is used as the basic agent, 4-methylsulfonylcinnoline reacts with propiophenone to give 4-cyano-3-ethylcinnoline (**30**) (22–24).

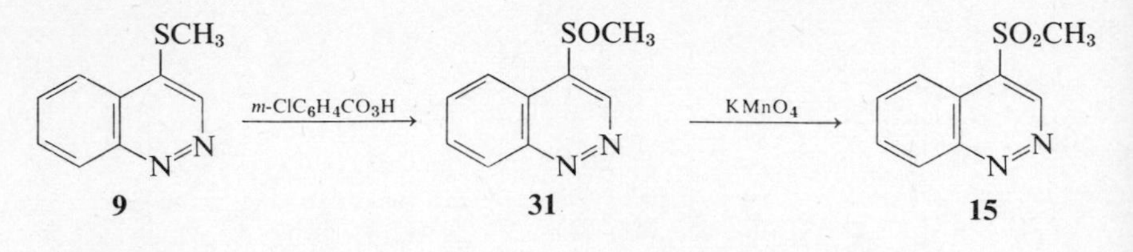

As with 4-methylthiocinnoline, permanganate oxidation of other 4-alkylthio- and 4-arylthiocinnolines produces the corresponding 4-alkylsulfonyl- and 4-arylsulfonylcinnolines (25). These are listed in Table 1F-4.

In general, methylsulfonyl-substituted heterocycles such as methylsulfonylquinolines, -quinoxalines, -cinnolines, and -phthalazines are from 40 to 100 times more reactive than the corresponding chloro-substituted heterocycles toward nucleophilic reagents. For example, with methoxide ion in methanol, 4-methylsulfonylcinnoline (rate coefficient, $k = 5.43 \times 10^{-2}$

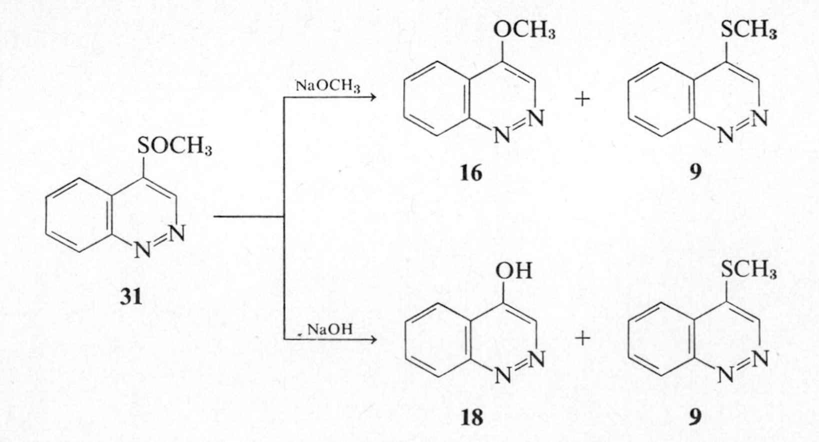

IV. Tables

TABLE 1F-1. Ultraviolet Absorption Spectra of 4-Mercaptocinnolines

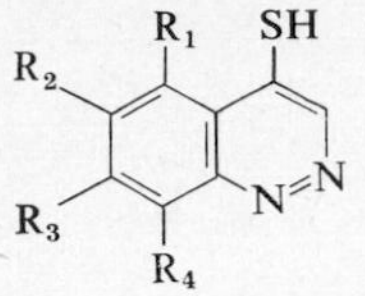

R_1	R_2	R_3	R_4	Solvent	λ_{max}(mμ)[a]	log E_{max}	Ref.
H	H	H	H	Methanol	224	4.58	11
					264	3.82	
					275	3.75	
					308	3.39	
					422	4.30	
				Water pH = 4.0 (neutral species)	*217*	*4.46*	8, 9
					222	4.56	
					252	3.77	
					271	3.63	
					305	3.25	
					417	4.27	
				Water pH = 9.5 (anionic species)	221	4.52	8, 9
					247	3.82	
					275	3.29	
					380	4.13	
				H_2SO_4 pH = −4.0 (cationic species)	211	4.10	8, 9
					238	4.40	
					248	*4.24*	
					288	*3.36*	
					295	3.37	
					333	3.82	
					369	4.02	
Cl	H	H	H	Methanol	231	4.48	11
					272	3.53	
					281	3.53	
					328	3.51	
					421	4.28	
					430	4.26	
H	F	H	H	Methanol	221	4.55	11
					267	3.83	
					313	3.33	
					424	4.25	
H	Cl	H	H	Methanol	229	4.63	11
					271	3.94	
					281	3.85	
					322	3.37	
					430	4.26	

TABLE 1F-1. (*continued*)

R_1	R_2	R_3	R_4	Solvent	$\lambda_{max}(m\mu)$[a]	log E_{max}	Ref.
H	Br	H	H	Methanol	231	4.56	11
					273	3.95	
					284	3.85	
					299	3.38	
					340	3.41	
					361	3.53	
					429	4.28	
H	H	F	H	Methanol	221	4.53	11
					270	3.87	
					306	3.41	
					420	4.28	
H	H	Cl	H	Methanol	227	4.60	11
					267	3.84	
					278	3.77	
					296	3.50	
					427	4.23	
H	H	H	F	Methanol	220	4.50	11
					266	3.59	
					276	3.56	
					316	3.40	
					417	4.31	
H	H	H	Cl	Methanol	227	4.50	11
					262	3.64	
					269	3.63	
					279	3.60	
					315	3.32	
					425	4.30	
H	H	H	Br	Methanol	227	4.46	11
					263	3.66	
					271	3.65	
					281	3.61	
					320	3.36	
					426	4.27	

[a] Italics refer to shoulders or inflections.

liter/(mole)(sec) at 5.0° C) is more than 50 times as reactive as 4-chlorocinnoline with methoxide ion under identical conditions ($k = 9.55 \times 10^{-4}$ liter/(mole)(sec) at 5.0° C) (17).

Careful oxidation of 4-methylthiocinnoline (**9**) with *m*-chloroperbenzoic acid gives 4-methylsulfinylcinnoline (**31**) (mp 157° C) (16). Further oxidation of **31** with potassium permanganate gives 4-methylsulfonylcinnoline (**15**) (26).

The methylsulfinyl group has the same order of reactivity in displacement reactions as the methylsulfonyl group, but the reaction of 4-methylsulfinylcinnoline (**31**) with methoxide ion is anomalous in that it not only gives the

TABLE 1F-2. Ultraviolet Absorption Spectra and Basic Strengths of Methylated 4-Mercaptocinnolines

Compound	Solvent	$\lambda_{max}(m\mu)$[a]	log E_{max}	Basic pK_a[b,c]	Ref.
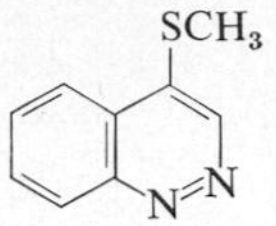 SCH_3, N=N	Water pH = 6.0	226	4.41	3.13 ± 0.03	8, 9
	(neutral species)	*246*	*4.03*		
		271	*3.28*		
		349	4.04		
	Water pH = 0.0	219	4.26		
	(cationic species)	240	4.36		
		256	*3.90*		
		281	*3.23*		
		385	4.18		
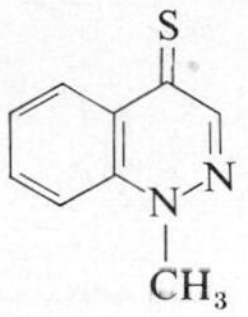 S, N, N, CH_3	Water pH = 1.0	223	4.57	−1.29 ± 0.1	9
	(neutral species)	252	3.83		
		264	*3.76*		
		275	3.69		
		308	3.30		
		427	4.36		
	H_2SO_4 pH = −3.29	211	4.11		
	(cationic species)	239	4.36		
		250	*4.23*		
		300	3.42		
		336	3.82		
		372	4.08		
		386	*3.99*		
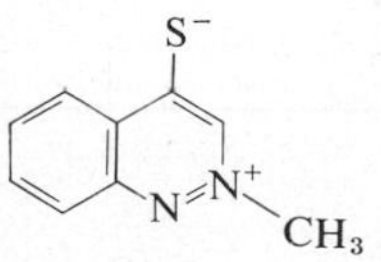 S^-, $N=N^+$, CH_3	Water pH = 7.0	227	4.51	−0.80 ± 0.05	8, 9
	(neutral species)	265	3.91		
		274	*3.79*		
		317	3.45		
		424	4.10		
	H_2SO_4 pH = −3.0	215	4.09		
	(cationic species)	244	4.49		
		285	3.40		
		333	3.68		
		367	*3.93*		
		377	3.95		
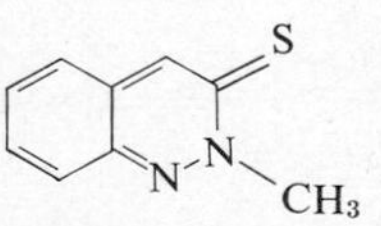 S, N, N, CH_3	Not stated	224	4.30	—	13
		255	*4.08*		
		277	4.36		
		320	4.18		
		329	4.20		
		433	3.30		

[a] Italics refer to shoulders or inflections.

[b] pK_a = negative logarithm of the ionization constant.

[c] In water at 20°C.

TABLE 1F-3. 4-Mercapto- and 4-Alkylthiocinnolines

R	R_1	R_2	R_3	R_4	Prep.[a]	MP (°C)	Remarks[b]	Ref.
H	H	H	H	H	A, B, C	204–205, 205–207	Orange needles (dissolution in aqueous base then precipitation by acidification)	1, 2, 7, 8
H	H	OCH_3	OCH_3	H	A, B	216–217	Yellow crystals (acetic acid)	2
CH_3	H	H	H	H	D	98, 96.5–97	Yellow needles (cyclohexane)	2, 8
CH_3	H	OCH_3	OCH_3	H	D	215–217	Yellow crystals (benzene)	2
N=N	H	H	H	H	D	180–181	(Ethanol)	2
CH_3O, CH_3O, N=N	H	OCH_3	OCH_3	H	D	220–225	(Ethanol)	2
N=N	H	OCH_3	OCH_3	H	D	193	(Ethanol)	2

	H	H	H	H	D	153–154	Yellow needles (ethanol)	2
	H	OCH_3	OCH_3	H	D	210	(Ethanol)	2
H	H	H	H	F	B	232	Orange solid (methanol)	3
H	H	H	H	Cl	B	228, 234–235	Blue solid (methanol or toluene)	3, 7
H	H	H	H	Br	B	270–271	Violet-blue solid (methanol)	3
H	Cl	H	H	H	B	192–193	Red solid (methanol)	3
H	H	H	F	H	B	241–242	Orange-red solid (methanol)	3
H	H	H	Cl	H	B	208–209	Orange solid (methanol)	3
H	H	F	H	H	B	179–180	Orange solid (methanol)	3
H	H	Cl	H	H	B	181–182, 191–192	Red solid (methanol or anisole)	3, 7
H	H	Br	H	H	B	187–189	Purple solid (methanol)	3
C_2H_5	H	H	H	H	D	98	(Cyclohexane)	3
$C_6H_5CH_2$	H	H	H	H	D	141	(Ethanol)	3
o-$ClC_6H_4CH_2$	H	H	H	H	D	136	(Benzene)	3
p-$ClC_6H_4CH_2$	H	H	H	H	D	178	(Benzene)	3
2,4-$Cl_2C_6H_3CH_2$	H	H	H	H	D	153	(Benzene)	3
2,4-$Cl_2C_6H_3CH_2$	H	H	H	F	D	151	(Benzene)	3
2,4-$Cl_2C_6H_3CH_2$	H	H	H	Br	D	194	(Benzene)	3
CH_3	H	CH_3S	H	H	E	162	(Benzene–cyclohexane)	3
$C_6H_5CH_2$	H	$C_6H_5CH_2S$	H	H	E	128	(Benzene–cyclohexane)	3
o-$ClC_6H_4CH_2$	H	*o*-$ClC_6H_4CH_2S$	H	H	E	148	(Benzene–cyclohexane)	3
p-$ClC_6H_4CH_2$	H	*p*-$ClC_6H_4CH_2S$	H	H	E	162	(Benzene–cyclohexane)	3

TABLE 1F-3. (*continued*)

R	R_1	R_2	R_3	R_4	Prep.[a]	MP (°C)	Remarks[b]	Ref.
2,4-$Cl_2C_6H_3CH_2$	H	2,4-$Cl_2C_6H_3CH_2S$	H	H	E	145	(Benzene–cyclohexane)	3
3,4-$Cl_2C_6H_3CH_2$	H	3,4-$Cl_2C_6H_3CH_2S$	H	H	E	128–130	(Benzene–cyclohexane)	3
H	H	SH	H	H	F	245 (dec.)	Orange-brown powder (ethanol or methanol)	3
H	H	H	SH	H	G	231–233	Orange powder (methanol)	3
$C_6H_5CH_2$	H	F	H	H	H	79	(Methanol)	3
o-$ClC_6H_4CH_2$	H	F	H	H	H	114	(Benzene–cyclohexane)	3
p-$ClC_6H_4CH_2$	H	F	H	H	H	162	(Ethanol)	3
3,4-$Cl_2C_6H_3CH_2$	H	F	H	H	H	178	(Ethanol)	3
[pyranose ring: CH_2OAc, O, OAc, AcO, OAc]	H	H	H	H	D	129–131	Colorless solid	4
[pyranose ring: CH_2OH, O, OH, HO, OH]	H	H	H	H	I	164–167 (dec.)	Colorless solid	4
C_6H_5	H	OCH_3	OCH_3	H	H	165	(Ethanol–water)	6
p-$CH_3C_6H_4$	H	OCH_3	OCH_3	H	H	181–182	(Ethanol–water)	6
o-ClC_6H_4	H	OCH_3	OCH_3	H	H	152	(Ethanol–water)	6

m-ClC_6H_4	H	OCH_3	OCH_3	H	H	176–177	(Ethanol–water)	6
p-ClC_6H_4	H	OCH_3	OCH_3	H	H	199	(Ethanol–water)	6
2,5-$Cl_2C_6H_3$	H	OCH_3	OCH_3	H	H	200–201	(Ethanol–water)	6
3,5-$Cl_2C_6H_3$	H	OCH_3	OCH_3	H	H	177–178	(Ethanol–water)	6
3,4-$Cl_2C_6H_3$	H	OCH_3	OCH_3	H	H	192	(Ethanol–water)	6
2,4-$Cl_2C_6H_3$	H	OCH_3	OCH_3	H	H	179	—	5
2,3-$Cl_2C_6H_3$	H	OCH_3	OCH_3	H	H	184	—	5
3,6,4-$Cl_2(HO)C_6H_2$	H	OCH_3	OCH_3	H	H	238	—	5
5-Cl-2-$CH_3C_6H_3$	H	OCH_3	OCH_3	H	H	161	—	5
H	H	OCH_3	H	H	B	210–212	(Ethanol–water)	7
H	H	NO_2	H	H	B	189–192	(Acetic acid)	7
CH_3	H	Cl	H	H	D	190–191	(Ligroin)	7
CH_3	H	NO_2	H	H	D	185–187 (dec.)	(Ligroin)	7
CH_2COOH	H	Cl	H	H	D	214–215 (dec.)	(Acetic acid or dimethylformamide)	7
$(CH_3)_2NC(O)$—	H	Cl	H	H	J	159–161	Yellow solid (ligroin)	7

[a] A, by reaction of the appropriately substituted 4-chlorocinnoline with thiourea followed by alkaline decomposition of the thiouronium salt. B, by interaction of the appropriately substituted 4-hydroxycinnoline with phosphorus pentasulfide in refluxing pyridine or toluene solution. C, by interaction of 4-chlorocinnoline with phosphorus pentasulfide in refluxing pyridine solution. D, by reaction of the appropriately substituted sodium cinnoline-4-mercaptide with either an alkyl halide, dimethyl sulfate, or a heterocyclic aryl halide in alcoholic or aqueous solution. E, by interaction of 4,6-dimercaptocinnoline with the appropriate alkyl halide in aqueous-ethanolic sodium hydroxide. F, by reaction of 6-chloro- or 6-bromo-4-hydroxycinnoline with phosphorus pentasulfide in refluxing pyridine solution. G, by reaction of 7-fluoro-4-hydroxycinnoline with phosphorus pentasulfide in refluxing toluene solution. H, by reaction of the appropriate sodium alkylmercaptide with the appropriately substituted 4-chlorocinnoline in refluxing benzene solution. I, by deacylation of 4-(tetra-*O*-acetyl-β-D-glucopyranosylmercapto)-cinnoline with sodium methoxide. J, by reaction of 6-chloro-4-mercaptocinnoline with dimethylcarbamoyl chloride.

[b] Solvent employed in recrystallization is enclosed in parentheses.

TABLE 1F-4. 4-Alkylsulfonyl- and 4-Arysulfonylcinnolines

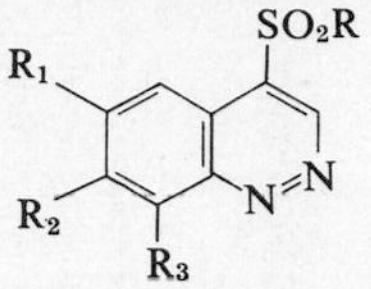

R	R_1	R_2	R_3	Prep.[a]	MP (°C)	Ref.
CH_3	H	H	H	A	178, 183–184, 188–189	17, 20, 25
C_2H_5	H	H	H	A	122–123	25
CH_3	OCH_3	OCH_3	H	A	239–240	25
CH_3	H	H	OCH_3	A	236–237.5	25
CH_3	Cl	H	H	A	198.5–200	25
CH_3	H	H	Cl	A	204–205.5	25
CH_3	H	Cl	Cl	A	229–230.5	25
CH_3	CH_3	H	H	A	213–214	25
CH_3	H	H	NO_2	A	213–215	25
CH_3	CN	H	H	A	—	25
$CH_3(CH_2)_3$	H	H	H	A	58–59	25
$C_6H_5CH_2$	Cl	H	H	A	190.5–191	25
p-$NO_2C_6H_4$	Cl	H	H	A	—	25
C_6H_5	Cl	H	H	A	—	25
p-ClC_6H_4	OCH_3	OCH_3	H	B	—	5

[a] A, by permanganate oxidation of the corresponding 4-alkylthiocinnoline in aqueous acetic acid. B, by reaction of 4-chloro-6,7-dimethoxycinnoline with *p*-$ClC_6H_4SO_2Na$ in dimethyl sulfoxide.

expected 4-methoxycinnoline (**16**) (60%) but also gives 4-methylthiocinnoline (**9**) in 20% yield (16). In a similar manner, reaction of **31** with aqueous sodium hydroxide gives 4-hydroxycinnoline (**18**) and 4-methylthiocinnoline (**9**) in approximately equal amounts (26). With *n*-butylamine or hydrazine, 4-methylsulfinylcinnoline gives 4-*n*-butylaminocinnoline and 4-hydrazinocinnoline, respectively, both in good yield (26). No methylthiocinnoline was reported in the product mixture in these cases.

Several 4-alkylthiocinnolines, including 4-benzylthiocinnoline, 4-(2,4-dichlorobenzylthio)cinnoline, 4-(3,4-dichlorobenzylthio)cinnoline, 8-chloro-4-(2,4-dichlorobenzylthio)cinnoline, 4,6-bis-(2,4-dichlorobenzylthio)cinnoline, and 4-(4-chlorobenzylthio)-6,7-dimethoxycinnoline exhibit antitumor activity against experimental rodent tumor or in cell culture or both (3, 27). The 4-alkylsulfonyl- and 4-arylsulfonylcinnolines are claimed to be useful as bactericides (25).

References

1. R. N. Castle, K. Kaji, G. A. Gerhardt, W. D. Guither, C. Weber, M. P. Malm, R. R. Shoup, and W. D. Rhoads, *J. Heterocyclic Chem.*, **3,** 79 (1966).
2. R. N. Castle, H. Ward, N. White, and K. Adachi, *J. Org. Chem.*, **25,** 570 (1960).
3. R. N. Castle, R. R. Shoup, K. Adachi, and D. L. Aldous, *J. Heterocyclic Chem.*, **1,** 98 (1964).
4. G. Wagner and D. Heller, *Z. Chem.*, **4,** 386 (1964).
5. S. M. Yarnal and V. V. Badiger, *Arch. Pharm.* (*Weinheim*), **303,** (7), 560 (1970); *Chem. Abstr.*, **73,** 77174Z (1970).
6. S. M. Yarnal and V. V. Badiger, *J. Med. Chem.*, **11,** 1270 (1968).
7. H. J. Barber and E. Lunt, *J. Chem. Soc.* (*C*), **1968,** 1156.
8. A. Albert and G. B. Barlin, *J. Chem. Soc.*, **1962,** 3129.
9. G. B. Barlin, *J. Chem. Soc.*, **1965,** 2260.
10. D. E. Ames, R. F. Chapman, H. Z. Kucharska, and D. Waite, *J. Chem. Soc.*, **1965,** 5391.
11. R. R. Shoup and R. N. Castle, *J. Heterocyclic Chem.*, **2,** 63 (1965).
12. R. R. Shoup and R. N. Castle, *J. Heterocyclic Chem.*, **1,** 221 (1964).
13. D. E. Ames and R. F. Chapman, *J. Chem. Soc.* (*C*), **1967,** 40.
14. G. Wagner, *Z. Chem.*, **6,** 367 (1966).
15. G. Wagner, private communication.
16. G. B. Barlin and W. V. Brown, *J. Chem. Soc.* (*B*), **1968,** 1435.
17. G. B. Barlin and W. V. Brown, *J. Chem. Soc.* (*B*), **1967,** 736.
18. G. B. Barlin and W. V. Brown, *J. Chem. Soc.* (*C*), **1967,** 2473.
19. E. Hayashi and T. Watanabe, *Yakugaku Zasshi*, **88,** 94 (1968); *Chem. Abstr.*, **69,** 19106K (1968).
20. E. Hayashi, Y. Akahori, and T. Watanabe, *Yakugaku Zasshi*, **87,** 1115 (1967); *Chem. Abstr.*, **68,** 49538 (1968).
21. E. Hayashi, Japanese Pat. 7010,349, April 14, 1970; *Chem. Abstr.*, **73,** 45529C (1970).
22. E. Hayashi and T. Watanabe, *Yakugaku Zasshi*, **88,** 742 (1968); *Chem. Abstr.*, **69,** 77200X (1968).
23. E. Hayashi and T. Watanabe, Japanese Pat. 7019,908, July 7, 1970; *Chem. Abstr.*, **73,** 87934M (1970).
24. E. Hayashi and T. Watanabe, *Yakugaku Zasshi*, **88,** 593 (1968); *Chem. Abstr.*, **69,** 96627h (1968).
25. E. Hayashi, H. Tani, H. Ishihama and T. Mizutani, Japanese Pat. 7025,511, August 24, 1970; *Chem. Abstr.*, **73,** 131020M (1970).
26. G. B. Barlin and W. V. Brown, *J. Chem. Soc.* (*C*), **1969,** 921.
27. R. N. Castle, *Proc. 3rd Int. Congr. Chemotherapy* Stuttgart, July, 1963. Thieme Verlag, Stuttgart, 1964, pp. 1064–1068.

Part G. Nitrocinnolines

I. Methods of Preparation

A. Direct Nitration

Molecular orbital calculations predict that the direct nitration of unsubstituted cinnoline (**1**) will take place at the 5- and 8-positions of the ring (1). Experimentally, the nitration of cinnoline with a mixture of nitric acid and sulfuric acid is found to give 5-nitrocinnoline (**2**) and 8-nitrocinnoline (**3**) in nearly equal proportions (2, 3). Since nitration of 2-methylcinnolinium perchlorate occurs at a rate similar to that for the nitration of cinnoline and also is nitrated equally at the 5- and 8-positions, the nitration of cinnoline probably involves nitration of the 2-cinnolinium cation rather than unprotonated cinnoline. The 2-cinnolinium cation is nitrated 287 times more slowly than the isoquinolinium cation at 80°, a fact which gives a measure of the deactivating effect of the unprotonated N-1 atom on the nitration of cinnoline (3).

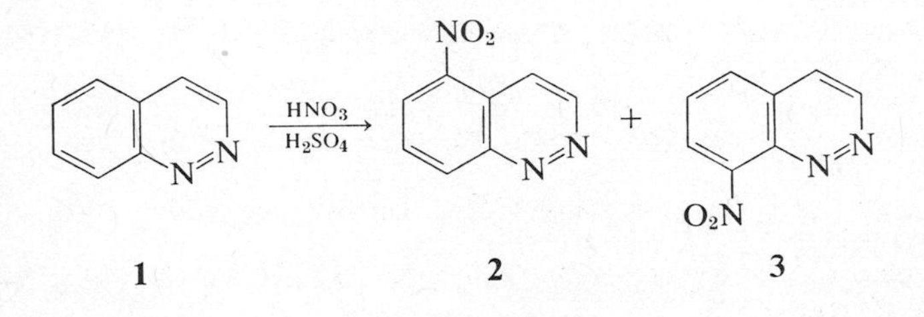

The electron distribution within the ring changes with the introduction of a 4-hydroxy group so that the direct nitration of 4-hydroxycinnoline (**4**) yields principally the 6-nitro derivative **5**. Smaller amounts of 4-hydroxy-8-nitrocinnoline (**6**) and 4-hydroxy-3-nitrocinnoline (**7**) are also isolated (4–7).

Nitrations of several 4-hydroxycinnolines which bear additional substituents have been carried out, and in these cases the position of the nitro group in the product is determined not only by the hydroxy group but also by the additional substituent(s). These nitrations are summarized in Table 1G-1.

Just as 4-hydroxycinnoline is nitrated principally in the 6-position, so also is 4-aminocinnoline (**8**), which in fuming nitric acid and concentrated sulfuric acid at 0–5° C gives 4-amino-6-nitrocinnoline (**9**) in good yield (12).

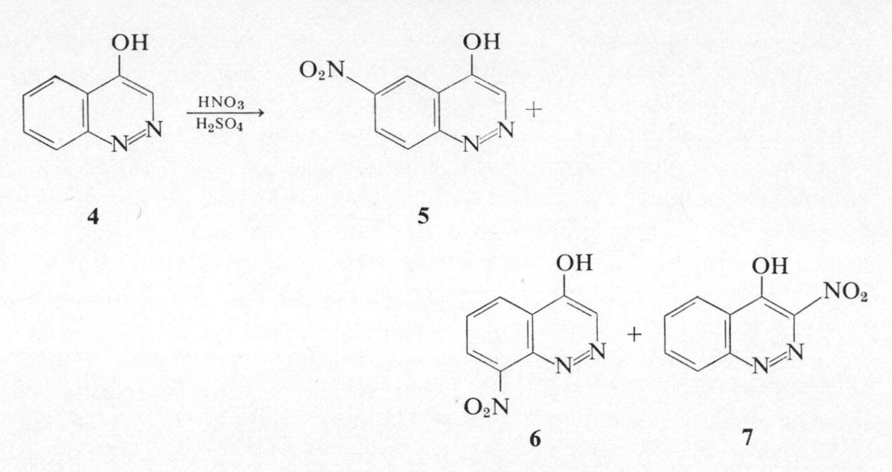

This route to **9** avoids the preparation of unstable 4-chloro-6-nitrocinnoline. On the other hand, 4-methylcinnoline (**10**), other 4-alkyl- and 3,4-dialkylcinnolines, and 4-chlorocinnoline (**12**) are nitrated in the 8-position to give **11** and **13**, respectively (5, 8, 13, 14).

As mentioned earlier, nitration of unsubstituted cinnoline yields the 5- and 8-nitro derivatives. The positions at which nitration takes place are changed when cinnoline is *N*-oxidized. For example, nitration of cinnoline

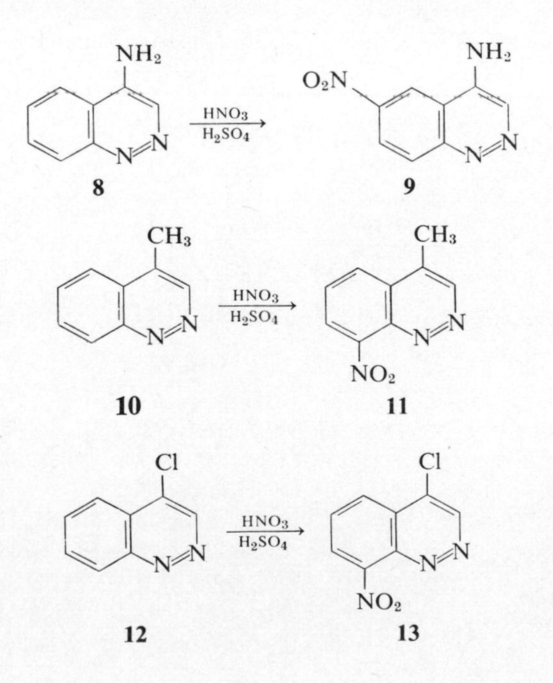

1-oxide with a mixture of nitric acid and sulfuric acid affords either 4-nitrocinnoline 1-oxide or 4,5-dinitrocinnoline 1-oxide, or both, depending upon the reaction conditions. Nitration of cinnoline 1-oxide with benzoyl nitrate in chloroform yields the 3-nitro derivative in 71% yield (15, 16). Nitration of cinnoline 2-oxide gives still different results. In a nitric acid and sulfuric acid mixture, the 2-oxide yields a mixture of 5-, 6-, and 8-nitrocinnoline 2-oxide. The proportion of isomers formed depends upon the acidity (percentage of sulfuric acid) of the nitrating medium, the amounts of 5- and 8-nitro isomers increasing with increasing acidity and the amount of 6-nitrocinnoline 2-oxide decreasing with increasing acidity (17). Schofield and co-workers (17) suggest that the 5- and 8-nitro isomers are formed by nitration of protonated cinnoline 2-oxide, whereas the 6-nitro isomer arises by nitration of the unprotonated 2-oxide. Benzoyl nitrate converts cinnoline 2-oxide to its 5-nitro derivative in low yield (18, 19). A more complete discussion of the nitration of cinnoline *N*-oxides is given in Section 1J-III-A.

B. Borsche Synthesis

The Borsche synthesis is discussed in Section 1C-I-A-3. It has been used to prepare 4-hydroxycinnolines (**15**) which are substituted with a nitro group in the 5-, 6-, 7-, or 8-position by diazotization and cyclization of the appropriate nitro-2-aminoacetophenone (**14**). Yields of the cinnoline are generally

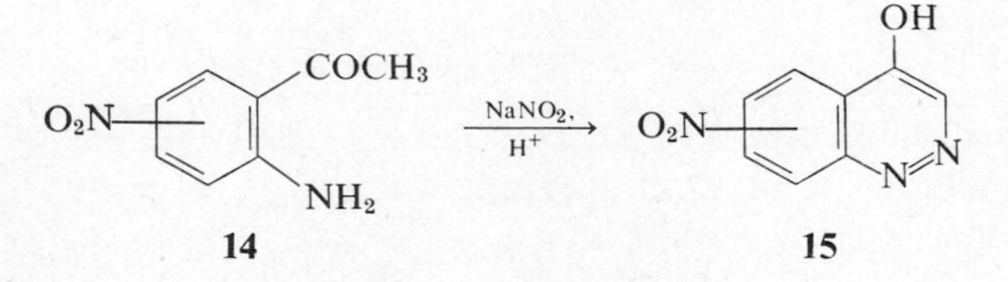

good. Examples of nitro-substituted 4-hydroxycinnolines which have been prepared in this way are listed in Table 1G-2.

C. Friedel-Crafts Cyclization of Mesoxalyl Chloride Phenylhydrazones

The cyclization of mesoxalyl chloride phenylhydrazones has been used in a few instances to obtain 4-hydroxycinnolines which are substituted with nitro groups in the benzenoid ring. The mesoxalyl chloride phenylhydrazones, moderately stable crystalline materials, are prepared from the corresponding diacids with excess thionyl chloride or with phosphorus pentachloride in an inert solvent. The diacid is obtained from the interaction of a diazotized aniline with diethyl malonate followed by hydrolysis of the ester functions (see Section 1C-I-A-4).

For example, mesoxalyl chloride *p*-nitrophenylhydrazone (**16**) may be cyclized with titanium tetrachloride in nitrobenzene at 95° C for 4 hr to give 20% of 4-hydroxy-6-nitrocinnoline-3-carboxylic acid (**17**), which is easily thermally decarboxylated at 180–210° to give 4-hydroxy-6-nitrocinnoline (**5**) (24). The yield of **17** may be increased to 35% if the phenylhydrazone is heated with excess titanium tetrachloride in nitrobenzene for 42 hr, but it is doubtful if high yields of nitro-substituted cinnolines will ever be realized by this method because of the deactivating effect the nitro group exerts upon the Friedel-Crafts cyclization step. Stannic chloride is reported to be as effective as titanium tetrachloride in effecting the cyclization of the acid chloride, but the work-up of the reaction mixture is more difficult (7).

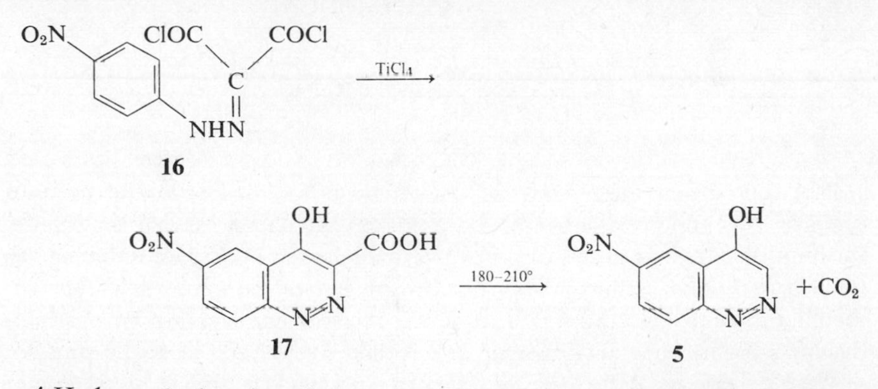

4-Hydroxy-8-nitrocinnoline-3-carboxylic acid (24) and 8-methyl-4-hydroxy-6-nitrocinnoline-3-carboxylic acid (9) have been prepared in a similar fashion from the appropriately substituted mesoxalyl chloride phenylhydrazone.

D. Cyclization of Nitroformaldehyde *o*-Carbonylphenylhydrazones

Nitromethane may be coupled with diazotized *o*-aminobenzaldehyde to give nitroformaldehyde *o*-formylphenylhydrazone (**18**). Cyclization of **18** in the presence of aluminum oxide or sodium hydroxide yields 3-nitrocinnoline (**19**). Aluminum oxide is more effective than sodium hydroxide in effecting the cyclization of **18**, but even with aluminum oxide overall yields from *o*-aminobenzaldehyde are only in the range of 22–29% (25). Beginning with the appropriately substituted *o*-aminobenzaldehyde and proceeding in this same manner, 7-chloro-3-nitrocinnoline, 6-chloro-3-nitrocinnoline, and 6,7-methylenedioxy-3-nitrocinnoline have been prepared in overall yields of

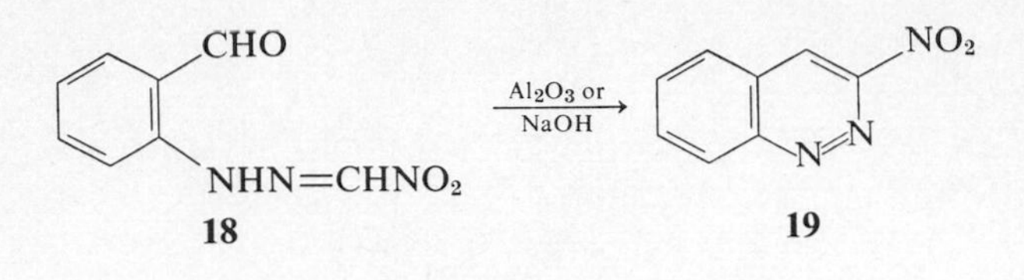

10–17% (26). It is inadvisable to use acetone as a solvent for the cyclization reaction since it will react with the formyl group of the nitroformaldehyde *o*-formylphenylhydrazone. For example, the cyclization of nitroformaldehyde 5-chloro-2-formylphenylhydrazone in acetone in the presence of base gives rise to nitroformaldehyde 5-chloro-2-(3-oxo-1-butenyl)phenylhydrazone (**20**) in 10–26% yields by interaction of the phenylhydrazone with acetone (26).

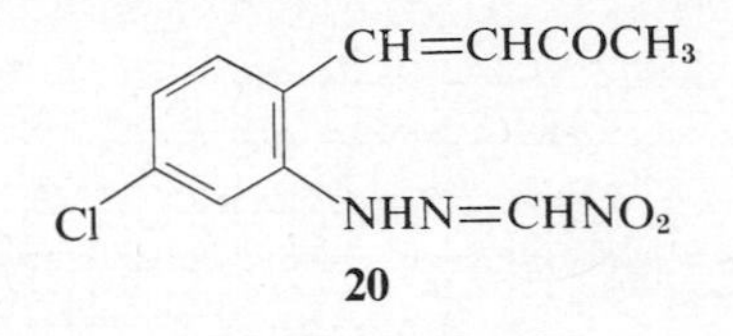

Probably the greatest drawback to this method has been the need to prepare *o*-aminobenzaldehydes. Baumgarten managed to avoid this by forming the ethylene acetal of *o*-nitroformaldehyde (**21**) before reducing the nitro group to an amino group, diazotizing it, and coupling it with nitromethane. The product from this sequence, the ethylene acetal of nitroformaldehyde *o*-formylphenylhydrazone (**22**), is then hydrolyzed in acidic solution to the aldehyde **18**. A further improvement in yield is obtained by the use of an anion exchange resin rather than aluminum oxide or sodium hydroxide to cyclize the aldehyde to 3-nitrocinnoline (**19**). In this way an overall yield of 3-nitrocinnoline from *o*-nitrobenzaldehyde of 46% is realized (26).

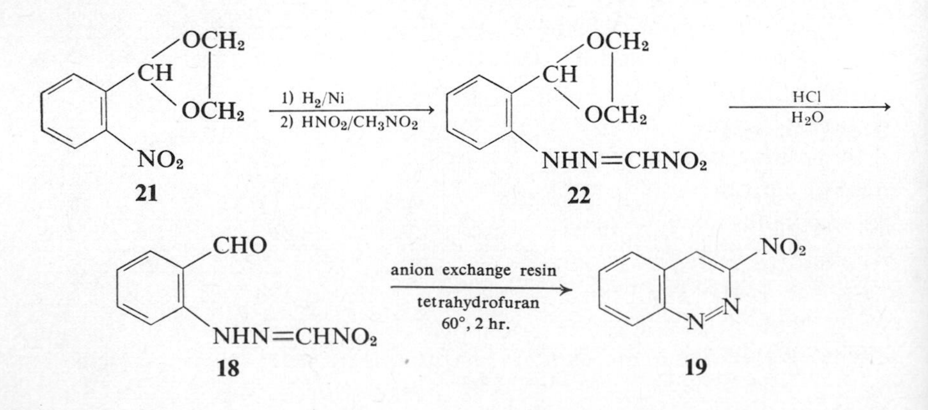

Diazotization of *o*-aminoacetophenone (**23**) and coupling with nitromethane gives nitroformaldehyde *o*-acetylphenylhydrazone (**24**), which can be cyclized to 4-methyl-3-nitrocinnoline (**25**) over aluminum oxide in 59% yield. Dilute aqueous sodium hydroxide in place of aluminum oxide produces only a trace of the cinnoline (25). Attempts to prepare 4-hydroxy-3-nitro-

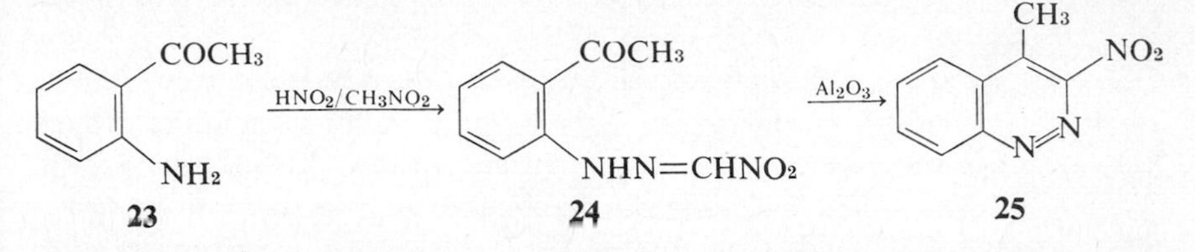

cinnoline from anthranilic acid or from its methyl ester by this route have been unsuccessful (25).

The pmr spectra of 3-, 5-, and 8-nitrocinnoline have been reported (27) and are given in Table 1G-5. The pmr spectra of nitrocinnolines bearing additional substituents may be found in other tables throughout this work. Those having a 4-hydroxy group, for example, are recorded in Table 1C-1, while those having alkyl substituents are given in Table 1B-3.

A mass spectra study (27) has shown that 3-, 5-, and 8-nitrocinnoline all undergo initial fragmentation with cleavage of the nitro group, resulting in loss of NO_2 and to a lesser extent NO. After loss of NO, all three cinnolines then suffer further fragmentation resulting in the consecutive loss of N_2 (or CO) and CO (or N_2). After initial loss of NO_2, however, 3-nitrocinnoline consecutively loses two HCN groups while the initial ion from 5- and 8-nitrocinnoline (from loss of NO_2) fragments with the consecutive loss of N_2 and C_2H_2. In addition, the ion resulting from 3- and 8-nitrocinnoline by initial loss of NO_2 also fragments with the loss of C_4H_3.

II. Reactions

The most obvious and useful reaction of a nitro group in an aromatic system is its reduction to an amino group. Nitrocinnolines which have been reduced to aminocinnolines are given in Table 1G-3.

In addition to the nitrocinnolines listed in Table 1G-3, the N-1 and N-2 methyl derivatives of several nitrocinnolines have been reduced to the corresponding aminocinnolines without loss of the *N*-methyl substituent with iron powder in water (30) or over palladium-charcoal in ethanol (12).

Several nitrocinnoline 1-oxides and 2-oxides have also been reduced to aminocinnolines, but in these cases one obtains either the deoxygenated aminocinnolines or a mixture of the aminocinnoline and the aminocinnoline

N-oxide, depending upon the catalyst system. For example, 4-nitrocinnoline 1-oxide is reduced with hydrogen over Raney nickel in methanolic solution to give 75% of 4-aminocinnoline (15, 16); the reduction of 6- and 8-nitrocinnoline 2-oxides over palladium-charcoal in ethanol gives principally 6- and 8-aminocinnoline 2-oxides as the products, but reduction of these same two nitrocinnoline 2-oxides over Adams platinum catalyst yields a mixture of the aminocinnolines and the aminocinnoline 2-oxides (18, 19).

Group exchange has been observed in the case of the 6-nitro group where 7-chloro-4-hydroxy-6-nitrocinnoline gives 4,6,7-trichlorocinnoline when heated with a mixture of phosphorus oxychloride and phosphorus pentachloride (8) and where 4-hydroxy-6-nitrocinnoline gives both 4-chloro-6-nitrocinnoline and 4,6-dichlorocinnoline upon treatment with phosphorus oxychloride and phosphorus pentachloride. In the latter case, replacement of the 6-nitro group by a 6-chloro substituent can be suppressed by the use of phosphorus oxychloride alone; a better approach is the use of thionyl chloride containing a small amount of phosphorus pentachloride in an inert solvent such as chlorobenzene (7).

One case of the replacement of a 5-nitro group has been reported, wherein 6-chloro-4-hydroxy-5-nitrocinnoline was heated at reflux temperature in pyrrolidine for 12 hr. The product from this reaction was 6-chloro-4-hydroxy-5-pyrrolidinocinnoline (11).

The 3-nitro group in 4-hydroxy-3-nitrocinnoline is also sufficiently labile to be replaced by chlorine when the hydroxy compound is heated with a mixture of phosphorus oxychloride and phosphorus pentachloride. Both 4-chloro-3-nitrocinnoline and 3,4-dichlorocinnoline are obtained from this reaction. Again, the use of phosphorus oxychloride alone will minimize the replacement of the 3-nitro group (6). Treatment of 6-chloro-4-hydroxy-3-nitrocinnoline with refluxing phosphorus oxychloride for 3 hr yields a mixture of 4,6-dichloro-3-nitrocinnoline and 3,4,6-trichlorocinnoline (11).

The nitro groups in both 3-nitrocinnoline 1-oxide and 4-nitrocinnoline 1-oxide will undergo nucleophilic displacement reactions. For example, 4-nitrocinnoline 1-oxide is converted to 4-methoxycinnoline 1-oxide in 65% yield by heating the nitro compound in methanolic sodium methoxide solution on a steam bath for 30 min. Heating 4-nitrocinnoline 1-oxide in acetyl chloride at 50° C for 30 min produces 4-chlorocinnoline 1-oxide in 74% yield (31), while treatment of 4,5-dinitrocinnoline 1-oxide with concentrated hydrochloric acid at 100° C for 1 hr gives 4-chloro-5-nitrocinnoline 1-oxide in 94% yield (15). Treatment of 3-nitrocinnoline 1-oxide with refluxing methanolic sodium methoxide for 1 hr gives 3-methoxycinnoline 1-oxide in 95% yield (15).

Both 5- and 8-nitrocinnoline are unstable in warm, dilute aqueous sodium hydroxide solution but stable in hot 4*N* ammonium hydroxide and in mineral acids. When 8-nitrocinnoline is dissolved in warm aqueous sodium hydroxide, the color of the solution changes immediately from yellow to deep red to deep olive green. Acidification of the solution precipitates a dark, amorphous solid of unknown structure and of melting point greater than 300° C; 5-nitrocinnoline behaves similarly (2). Apparently, 4-hydroxy-6-nitrocinnoline is also unstable in aqueous alkali solutions (4).

At this point perhaps a brief discussion of the *N*-methylation of nitro-substituted 4-hydroxy- and 4-aminocinnolines is in order, since a certain amount of confusion in the early literature exists concerning the structures of the products from this reaction. This confusion has only recently been clarified.

First, methylation of 4-hydroxy-6-nitrocinnoline (**5**) with dimethyl sulfate yields both an N-1 methyl derivative (**26**) and an N-2 methyl derivative (**27**) (9, 12); originally the N-2 methyl derivative **27** was incorrectly assigned the methyl nitronate structure **28** (4).

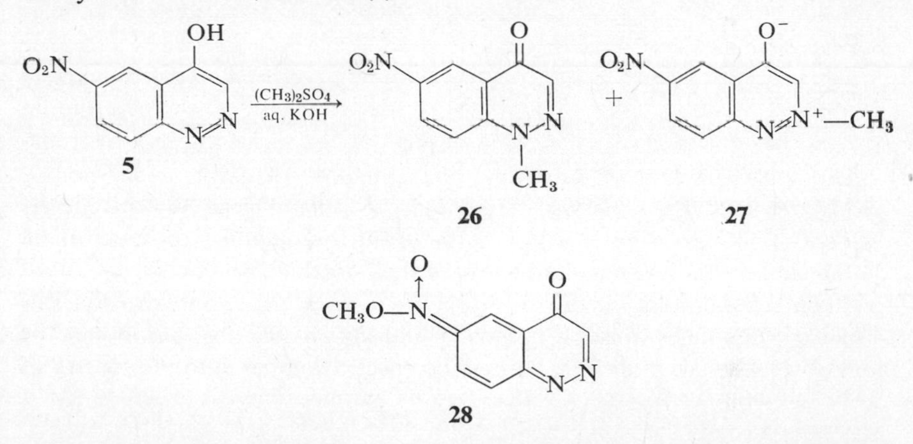

In a similar case, methylation of 3-methyl-4-hydroxy-6-nitrocinnoline (**29**) yields an N-1 methyl derivative (**30**) and an N-2 methyl derivative (**31**) (12) but not a methyl nitronate (**32**), the structure which had been earlier assigned to compound (**31**) (32).

In general, whether methylation of nitro-substituted 4-hydroxycinnolines occurs at N-1 or at N-2 depends primarily upon steric effects rather than electronic effects. Thus 4-hydroxy-3-nitrocinnoline is methylated only at N-1, the least sterically hindered nitrogen atom (6. 12), whereas 4-hydroxy-8-nitrocinnoline is methylated only at N-2 (9). Where the nitro group cannot exert a steric influence, as in 4-hydroxy-6-nitrocinnoline, a mixture of the N-1 and N-2 methyl derivatives is obtained. In the case of 3-bromo-4-hydroxy-8-nitrocinnoline where steric interference to quaternization exists at both nitrogen atoms, only the N-2 methyl isomer is obtained (12).

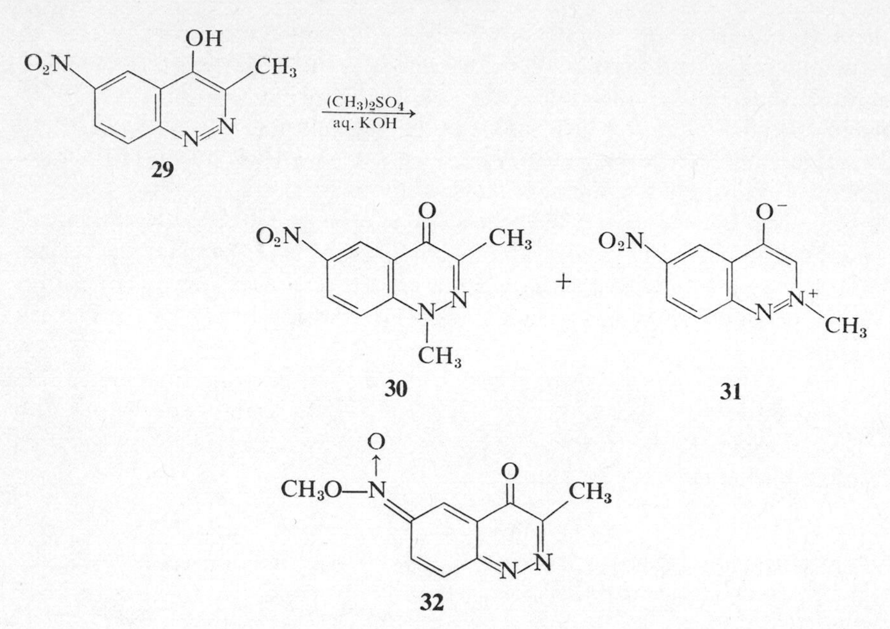

The quaternization of 4-amino-6-nitrocinnoline (**9**) with methyl iodide yields two salts, the 1-methiodide **33** and the 2-methiodide **34** (12). Earlier work (33) had erroneously assigned the *aci*-nitro 1-methiodide structure **36** to the isomer which is now known to be the 2-methiodide salt **34**.

It is interesting that the presence of a nitro group in the anhydro bases of 4-hydroxycinnolines causes a bathochromic shift of the absorption band at longest wavelength in the UV absorption spectra compared to the spectra of

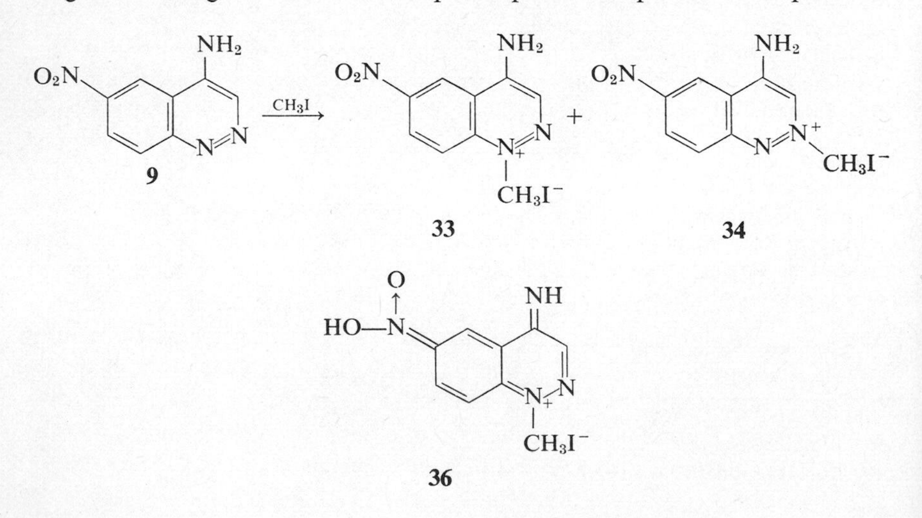

the corresponding anhydro bases which do not bear a nitro group. This is especially marked in the case of the 6-nitro derivative **27**, where the bathochromic shift may be due to the existence of the resonance-contributing **37**. Such a bathochromic shift is not observed in the UV absorption spectrum of 1-methyl-6-nitro-4-cinnolone where a resonance-contributing form analogous to **37** is not possible (9).

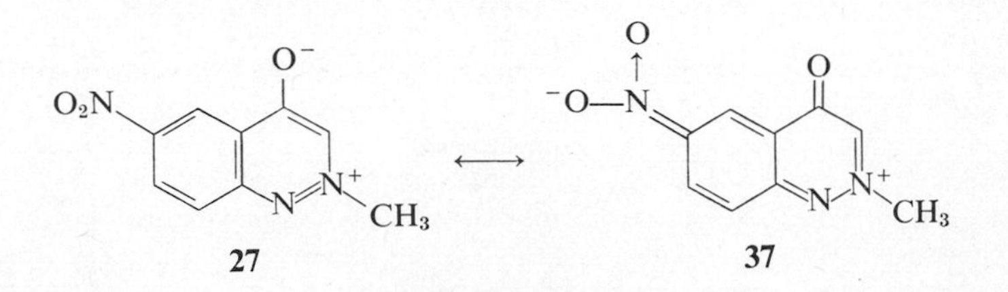

III. Tables

TABLE 1G-1. Principal Mononitration Product(s) from Direct Nitration of Substituted 4-Hydroxycinnolines

Substituted 4-Hydroxycinnoline	Principal Mononitration Product(s)	Ref.
4-Hydroxy-8-methylcinnoline	4-Hydroxy-8-methyl-5-nitrocinnoline	8, 9
4-Hydroxy-7-methylcinnoline	4-Hydroxy-7-methyl-8-nitrocinnoline[a]	8
4-Hydroxycinnoline-3-carboxylic acid	4-Hydroxy-6-nitrocinnoline-3-carboxylic acid	7
5-Chloro-4-hydroxycinnoline	5-Chloro-4-hydroxy-6-nitrocinnoline 5-Chloro-4-hydroxy-8-nitrocinnoline	10, 11
6,7-Dichloro-4-hydroxycinnoline	6,7-Dichloro-4-hydroxy-3-nitrocinnoline	10
6-Chloro-4-hydroxycinnoline-3-carboxylic acid	6-Chloro-4-hydroxy-8-nitrocinnoline-3-carboxylic acid	10, 11
7,8-Dichloro-4-hydroxycinnoline-3-carboxylic acid	7,8-Dichloro-4-hydroxy-6-nitrocinnoline-3-carboxylic acid	10, 11
7,8-Dichloro-4-hydroxycinnoline	7,8-Dichloro-4-hydroxy-6-nitrocinnoline	10
1-Methyl-4-cinnoline	1-Methyl-6-nitro-4-cinnolone	9
6-Chloro-1-methyl-4-cinnolone	6-Chloro-1-methyl-3-nitro-4-cinnolone	9
Anhydro base of 4-hydroxy-2-methylcinnolinium hydroxide	Anhydro base of 4-hydroxy-2-methyl-8-nitrocinnolinium hydroxide	5, 9
Anhydro base of 4-hydroxy-2,8-dimethylcinnolinium hydroxide	Anhydro base of 4-hydroxy-2,8-dimethyl-5-nitrocinnolinium hydroxide	9
6-Chloro-4-hydroxycinnoline	6-Chloro-4-hydroxy-5-nitrocinnoline[a] 6-Chloro-4-hydroxy-3-nitrocinnoline[b]	11
8-Chloro-4-hydroxycinnoline-3-carboxylic acid	8-Chloro-4-hydroxy-6-nitrocinnoline-3-carboxylic acid	11

[a] Principal product when the nitrating medium is a mixture of nitric acid and concentrated sulfuric acid.

[b] Principal product when the nitrating medium is nitric acid alone.

TABLE 1G-2. Borsche Synthesis of Nitro-Substituted 4-Hydroxycinnolines

Acetophenone	Diazotization Medium	Product	Ref.
2-Amino-6-nitroaceto-phenone	Acetic acid-sulfuric acid	4-Hydroxy-5-nitro-cinnoline	20
2-Amino-5-nitroaceto-phenone	Acetic acid-sulfuric acid	4-Hydroxy-6-nitro-cinnoline	7, 21
2-Amino-4-methyl-5-nitroacetophenone	Hydrochloric acid	4-Hydroxy-7-methyl-6-nitrocinnoline	8
2-Amino-4-nitroaceto-phenone	Acetic acid-sulfuric acid or aqueous formic acid[a]	4-Hydroxy-7-nitro-cinnoline	20, 22
2-Amino-3-nitro-acetophenone[b]	Acetic acid-sulfuric acid	4-Hydroxy-8-nitro-cinnoline	23

[a] If the diazotization of 2-amino-4-nitroacetophenone is carried out in aqueous formic acid at 0° C, then one obtains 4-hydroxy-7-nitrocinnoline; but if gentle warming is used to help dissolve the acetophenone, then the only product is 2-formamido-4-nitroacetophenone (see reference 22).

[b] If the diazotization of 2-amino-3-nitroacetophenone is carried out in concentrated hydrochloric acid solution at 70–80° C, then group exchange occurs and one obtains 45% of 8-chloro-4-hydroxycinnoline (see reference 8).

TABLE 1G-3. Reduction of Nitrocinnolines to Aminocinnolines

Nitrocinnoline	Reducing Medium	Product	Yield (%)	Ref.
3-Nitrocinnoline	$SnCl_2$, HCl	3-Aminocinnoline	31	25
	Fe, CH_3COOH		70	26
4-Amino-3-nitrocinnoline	$SnCl_2$, HCl	3,4-Diaminocinnoline	76	6
5-Nitrocinnoline	Pd-C, C_2H_5OH	5-Aminocinnoline	37	18, 19
5-Nitro-4-phenoxycinnoline	$SnCl_2$, CH_3COOH, $(CH_3CO)_2O$	5-Amino-4-phenoxy-cinnoline	79	28
6-Nitrocinnoline	Pd-C, C_2H_5OH	6-Aminocinnoline	40	18, 19
4-Amino-6-nitrocinnoline	$SnCl_2$, HCl	4,6-Diaminocinnoline	90	29, 30
	$SnCl_2$, HCl, CH_3COOH		84	7
	Ni, H_2, DMF		86	7
4-Amino-3-methyl-6-nitrocinnoline	$SnCl_2$, HCl, CH_3COOH	4,6-Diamino-3-methylcinnoline	—	30
4-Anilino-6-nitrocinnoline	$SnCl_2$, HCl, CH_3COOH	6-Amino-4-anilino-cinnoline	—	30
4-Hydroxy-6-nitrocinnoline	Ni, H_2, DMF	6-Amino-4-hydroxy-cinnoline	55	7
	Fe, HCl, or CH_3COOH		—	4
4-Hydroxy-6-nitrocinnoline-3-carboxylic acid	Ni, H_2, aq. NaOH	6-Amino-4-hydroxy-cinnoline-3-carboxylic acid	58	7
7-Nitro-4-phenoxycinnoline	$SnCl_2$, CH_3COOH, $(CH_3CO)_2O$	7-Amino-4-phenoxy-cinnoline	86	28
4-Hydroxy-7-nitrocinnoline	Fe, CH_3COOH	7-Amino-4-hydroxy-cinnoline	—	20
4-Amino-7-nitrocinnoline	Zn, $(CH_3CO)_2O$, CH_3COONa	4,7-Diacetamido-cinnoline	—	20
8-Nitrocinnoline	Pd-C, C_2H_5OH	8-Aminocinnoline	28	18, 19
	Zn, $(CH_3CO)_2O$, CH_3COONa	8-Acetamidocinnoline	—	2
4-Hydroxy-8-nitrocinnoline	Fe, CH_3COOH	8-Amino-4-hydroxy-cinnoline	—	20
4-Amino-8-nitrocinnoline	Fe, H_2O, $FeSO_4$	4,8-Diaminocinnoline	—	20
8-Nitro-4-phenoxycinnoline	$SnCl_2$, CH_3COOH, $(CH_3CO)_2O$	8-Amino-4-phenoxy-cinnoline	—	28
4-Methyl-8-nitrocinnoline	$SnCl_2$, HCl	8-Amino-4-methyl-cinnoline	—	5
6-Chloro-4-hydroxy-3-nitrocinnoline	Fe, CH_3COOH	3-Amino-6-chloro-4-hydroxycinnoline	74	11
4-Amino-6-chloro-3-nitrocinnoline	$SnCl_2$, HCl	3,4-Diamino-6-chloro-cinnoline	86	11
8-Chloro-4-hydroxy-6-nitrocinnoline	$SnCl_2$, CH_3COOH, $(CH_3CO)_2$, HCl	6-Amino-8-chloro-4-hydroxycinnoline	42	11

TABLE 1G-4. Nitrocinnolines

R_1	R_2	R_3	R_4	R_5	R_6	Prep.[a]	MP (°C)	Remarks[b]	Ref.
H	H	H	H	H	NO_2	A, S	136–138, 132–132.5	Yellow needles (benzene or water)	2, 23
H	H	NO_2	H	H	H	A, S	151–152, 147–148.5	Yellow needles (ethanol or water)	2, 23
H	NH_2NH	H	NO_2	H	H	B	>330	Maroon solid; hydrochloride deriv., mp 215° (dec.)	2
H	$(CH_3)_2C{=}NNH$	H	NO_2	H	H	C	179–180	Orange needles (aq. acetone)	2
H	H	H	NO_2	H	H	D	205–206	Yellow needles (ethanol)	2
H	*p*-$CH_3C_6H_4SO_2NHNH$	H	NO_2	H	H	E	212–213 (dec.) 190–192 (dec.)	Orange-red needles (aq. acetic acid)	2, 23
H	OH	H	NO_2	H	H	A, M	330–331 335–337	Red-brown prismatic needles (acetic acid)	4, 7
						F	336	Yellow prisms (acetic acid)	24
H	Cl	H	NO_2	H	H	G	135–137 138–139	Decomp. on standing a few hours	4, 7
H	C_6H_5NH	H	NO_2	H	H	H	228.5–229.5	Orange needles (aq. ethanol)	4
H	OCH_3	H	NO_2	H	H	I	194–194.5	Yellow needles (ethanol)	4
H	CH_3	H	H	H	NO_2	A	138–139 (dec.) 137–138 (dec.)	Yellow plates (methanol)	5, 13, 14

NO_2	NH_2	H	H	H	H	J	308–308.5	Yellow cottony solid (ethanol)	6
NO_2	OH	H	H	H	H	K	284.5–285.5 (dec.)	Yellow needles (ethanol)	6
NO_2	Cl	H	H	H	H	G	169–170	Yellow needles (ligroin)	6
NO_2	OC_6H_5	H	H	H	H	L	144.5–145	Tan plates (ligroin)	6
NO_2	C_6H_5NH	H	H	H	H	H	187–188	Yellow needles (ethanol)	6
COOH	OH	H	NO_2	H	H	A, N	275–276 (dec.)	Yellow needles (acetic acid)	7, 24
H	OC_6H_5	H	NO_2	H	H	L	188–191	—	7
H	NH_2	H	NO_2	H	H	O, A	288–290 (dec.) 290–291 297–298 (dec.) 289–291	(Precipitation from aq. HOAc by addition of base or from 2-methoxyethanol); Methiodide salt, mp 269–270° (dec.) or 273–275° (dec.); ethiodide salt, mp 180–181°	7, 29, 30 12
H	CH_3NH	H	NO_2	H	H	P, H	345–346 (dec.) >360	Yellow needles (2-propanol); methiodide salt, mp 229–230° (dec.); ethiodide salt, mp 216°	7, 33 34
H	$CH_3CH_2CH_2NH$	H	NO_2	H	H	P	287 (dec.)	Methiodide salt, mp 202–203° (dec.)	7
H	6-O_2N-cinnolin-4-yl-$NH(CH_2)_6NH$—	H	NO_2	H	H	P	>360	Dihydrochloride, mp 309–310° (dec.); dimethomethylsulfate deriv., mp 230–231° (dec.); dimethiodide deriv., mp 243° (dec.)	7

TABLE 1G-4 (*continued*)

R_1	R_2	R_3	R_4	R_5	R_6	Prep.[a]	MP (°C)	Remarks[b]	Ref.
Br	OH	H	NO_2	H	H	Q	300–301	(Ethanol)	9
COOH	OH	NO_2	H	H	CH_3	N	249–250 (dec.)	(2-Methoxyethanol)	9
COOH	OH	H	NO_2	H	CH_3	N	254–256	(Acetic acid)	9
H	OH	NO_2	H	H	CH_3	F, A	255–256	Colorless needles (acetic acid)	8, 9
H	OH	H	NO_2	H	CH_3	F	274–276	(2-Methoxyethanol)	9
H	OH	H	H	CH_3	NO_2	A	243–244 (dec.)	Lemon-yellow blades (acetic acid)	8
H	Cl	H	H	CH_3	NO_2	G	210–211 (dec.)	Green blades (benzene)	8
H	C_6H_5NH	H	H	CH_3	NO_2	H	262–263 (dec.)	Greenish-yellow needles (ethanol)	8
H	OC_6H_5	H	H	CH_3	NO_2	L	172–173	Cream needles (alcohol)	8
H	OH	H	NO_2	CH_3	H	M	250–251	Brown rhombohedra (acetic acid)	8
H	OC_6H_5	H	H	H	NO_2	L	166–167	Yellow or green needles (benzene)	8, 20
H	NH_2	H	H	H	NO_2	O	242–243 (dec.) 235–236 (dec.)	Yellow or rust-red needles (Ethanol); 4-acetal deriv., mp 291–292°	8, 20, 33
H	OC_6H_5	NO_2	H	H	CH_3	L	160–161	Colorless rods (benzene-ligroin)	8
H	Cl	NO_2	H	H	CH_3	G	130–131	Yellow-green prisms	8
H	NH_2	NO_2	H	H	CH_3	O	242–243 (dec.)	Orange-red needles (precipitation from acetic acid by addition of base)	8
H	C_6H_5NH	NO_2	H	H	CH_3	H	166–168	Orange-red needles (aqueous ethanol)	8
Br	OH	H	H	H	NO_2	Q	235–237	(Ethanol)	12

H	OCH_3	NO_2	H	H	H	I, R	180–182	Yellow powder (benzene-hexane)	15
H	OH	NO_2	H	H	H	M	304–305	Colorless needles (ethanol); acetyl deriv., mp 185–186°	20
H	OH	H	H	NO_2	H	M	295–296, 300	Yellow needles (ethanol); acetyl deriv., mp 140–141°	20, 28
H	Cl	NO_2	H	H	H	G	170–171	Yellow crystals (ether-ligroin)	20
H	Cl	H	H	NO_2	H	G	148–149	Yellow needles (ether-ligroin)	20
H	Cl	H	H	H	NO_2	G	167–169	gold leaflets (ether-ligroin)	20
H	OC_6H_5	H	H	NO_2	H	L	172–173	Yellow needles (benzene-ligroin)	20
H	NH_2	H	H	NO_2	H	O	300–301 (dec.)	Yellow leaflets (aq. ethanol); 4-acetyl deriv., mp 235°	20, 28
H	H	H	H	NO_2	H	S	153–154.5	Yellow crystals (water)	23
H	OH	H	H	H	NO_2	M	183–185	Yellow needles (ethanol)	23
H	p-$CH_3C_6H_4SO_2NHNH$	H	H	H	NO_2	E	195–196 (dec.)	Brown needles (acetic acid)	23
H	p-$CH_3C_6H_4SO_2NHNH$	H	H	NO_2	H	E	195–196 (dec.)	Orange crystals (acetic acid)	23
COOH	OH	H	H	H	NO_2	N	253–254 (dec.)	Green plates (acetic acid)	24
NO_2	H	H	H	H	H	T	205.5–206.5, 204–205	Cream needles (methanol or acetone)	25, 26
NO_2	CH_3	H	H	H	H	U	188–189	Brown needles (ethanol)	25
NO_2	H	H	H	Cl	H	T	165.5–166	Yellow needles (ethyl acetate)	26
NO_2	H	H	Cl	H	H	T	227–228	Yellow plates (ethyl acetate)	26

TABLE 1G-4 (*continued*)

R_1	R_2	R_3	R_4	R_5	R_6	Prep.[a]	MP (°C)	Remarks[a]	Ref.
NO_2	H	H	—O—CH_2—O—		H	T	>250	Cream needles (ethyl acetate)	26
H	OC_6H_5	NO_2	H	H	H	L	122–123	Colorless needles (ligroin)	28
CH_3	NH_2	H	NO_2	H	H	H	320 (dec.)	Orange-yellow solid	30
H	OCH_3	H	H	NO_2	H	I	200	Yellow blades (acetone)	33
CH_3	C_2H_5	H	H	H	NO_2	A	102–103	—	14
CH_3	CH_3	H	H	H	NO_2	A	150–151	—	14
H	$C(CH_3)_3$	H	H	H	NO_2	A	195	—	14
H	SO_2CH_3	H	H	H	NO_2	V	213–215	—	35
COOH	OH	H	Cl	H	NO_2	A	217–218 (dec.)	(Methanol)	11
H	OH	H	Cl	H	NO_2	F	157–160	(Ethanol)	11
H	OH	NO_2	Cl	H	H	A	334–336	(Acetic acid, ethanol)	11
NO_2	OH	H	Cl	H	H	A	298–299 (dec.)	(Acetic acid)	11
NO_2	OC_6H_5	H	Cl	H	H	L	111–112	(Ligroin)	11
NO_2	NH_2	H	Cl	H	H	O	337–338 (dec.) 341 (dec.)	(2-Ethoxyethanol)	11
NO_2	Cl	H	Cl	H	H	G	128–130	—	11
COOH	OH	H	NO_2	H	Cl	A	244–245 (dec.)	(Acetic acid)	11
H	OH	H	NO_2	H	Cl	F	248–249	(Acetic acid)	11
H	OH	Cl	NO_2	H	H	A	306–310 (dec.)	(Nitrobenzene)	11
H	OH	Cl	H	H	NO_2	A	220–221	(Nitrobenzene, acetic acid)	11
H	OH	OH	NO_2	H	H	W	255–257 (dec.)	Brown prisms (acetic acid)	11
H	CH_3NH	CH_3NH	NO_2	H	H	X	—	Acetic deriv., mp 268–269° (dec.)	11
COOH	OH	H	NO_2	Cl	Cl	A	252–253 (dec.)	(Acetic acid)	11

H	OH	H	NO_2	Cl	Cl	F	238–240	(Acetic acid)	11
H	OH	Piperidino (N—)	NO_2	H	H	Y	222–223 (dec.)	Orange-red solid (toluene)	11
H	OH	Pyrrolidino (N—)	H	H	NO_2	Y	202–204	(Toluene)	11

[a] A, by direct nitration of cinnoline or a substituted cinnoline. B, by interaction of the corresponding 4-chlorocinnoline with hydrazine. C, by interaction of the corresponding 4-hydrazinocinnoline with refluxing acetone. D, by decomposition of 6-nitro-4-isopropylidenehydrazinocinnoline in refluxing aqueous cupric sulfate solution. E, from the corresponding 4-chlorocinnoline and toluene-*p*-sulfonylhydrazide in chloroform solution. F, by decarboxylation of the corresponding cinnoline-3-carboxylic acid. G, by heating the corresponding 4-hydroxycinnoline with phosphorus oxychloride alone, with a mixture of phosphorus oxychloride and phosphorus pentachloride, with a mixture of phosphorus oxychloride and dimethylaniline, or with thionyl chloride containing a small amount of phosphorus pentachloride. H, from the interaction of the corresponding 4-chlorocinnoline with the appropriate amine. I, from the interaction of the corresponding 4-chlorocinnoline with methanolic sodium methoxide. J, by treatment of 3-nitrocinnoline with ethanolic hydroxylamine. K, by hydrolysis of the corresponding 4-aminocinnoline with aqueous sodium hydroxide. L, by heating the corresponding 4-chlorocinnoline with potassium or ammonium carbonate and phenol or with sodium phenoxide in phenol. M, by diazotization and cyclization of the appropriately substituted 2-aminoacetophenone. N, by cyclization of the appropriately substituted mesoxalyl chloride phenylhydrazone with titanium tetrachloride in nitrobenzene (see text). O, by fusion of corresponding 4-phenoxycinnoline with ammonium acetate or by heating the corresponding 4-chlorocinnoline with ammonia, ammonium carbonate, or ammonium acetate in phenol. P, from the 4-phenoxycinnoline with the appropriate amine in phenol or dimethylformamide. Q, from the corresponding 4-hydroxycinnoline with bromine in aqueous potassium hydroxide solution. R, from 4-methoxy-5-nitrocinnoline 1-oxide with phosphorus trichloride in chloroform solution. S, by decomposition of the corresponding 4-(toluene-*p*-sulfonylhydrazino)cinnoline in hot aqueous sodium carbonate solution. T, by cyclization of the appropriately substituted nitroformaldehyde 2-formylphenylhydrazone with potassium hydroxide, aluminum oxide, or an anion exchange resin (see text). U, by cyclization of nitroformaldehyde 2-acetylphenylhydrazone with aluminum oxide. V, by permanganate oxidation of the corresponding 4-methylthiocinnoline. W, by hydrolysis of the corresponding 5-chlorocinnoline. X, by interaction of 4,5-dichloro-6-nitrocinnoline with methylamine. Y, by interaction of the corresponding 5-chlorocinnoline with the appropriate amine.

[b] Solvent employed in recrystallization is enclosed in parentheses.

TABLE 1G-5. Proton Magnetic Resonance Spectra of Nitrocinnolines[a]

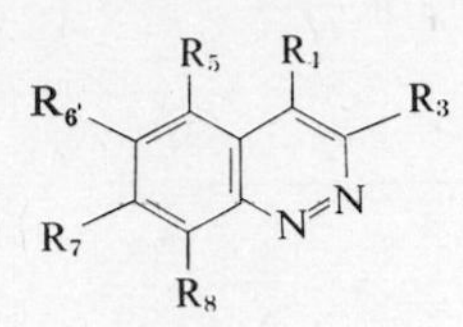

R_3	R_4	R_5	R_6	R_7	R_8	Solvent	Ref.
NO_2	H	H	H	H	H	DMSO-d_6	27
—	0.77(D) $J = 1$	1.70(M)	←—	~1.93(D)	—→ 1.40(M)		
H	H	NO_2	H	H	H	$CDCl_3$	27
0.60(D) $J = 6$	1.20(DD) $J = 6, 1$	—	1.17(D) $J = 8$	2.10(T) $J = 8$	1.38(DD) $J = 8, 1$		
H	H	H	H	H	NO_2	$CDCl_3$	27
0.6(D) $J = 6$	~2.05(M)	2.80(D) $J = 5$	←—	~2.05(M)	—→ —		

[a] Chemical shift values are given immediately below the proton to which they refer and are in τ units; the multiplicity of each absorption is given in parentheses following the τ value as D = doublet, DD = doublet of doublets, M = multiplet, and T = triplet; J = coupling constant in cps. Chemical shifts (τ) are recorded relative to tetramethylsilane.

References

1. M. J. S. Dewar and P. M. Maitlis, *J. Chem. Soc.*, **1957**, 2521.
2. J. S. Morley, *J. Chem. Soc.*, **1951**, 1971.
3. R. B. Moodie, E. A. Qureshi, K. Schofield, and J. T. Gleghorn, *J. Chem. Soc. (B)*, **1968**, 312.
4. K. Schofield and J. C. E. Simpson, *J. Chem. Soc.*, **1945**, 512.
5. K. Schofield and T. Swain, *J. Chem. Soc.*, **1949**, 1367.
6. H. E. Baumgarten, *J. Am. Chem. Soc.*, **77**, 5109 (1955).
7. H. J. Barber, E. Lunt, K. Washbourn, and W. R. Wragg, *J. Chem. Soc. (C)*, **1967**, 1657.
8. J. R. Keneford, J. S. Morley, and J. C. E. Simpson, *J. Chem. Soc.*, **1948**, 1702.
9. D. E. Ames, R. F. Chapman, and D. Waite, *J. Chem. Soc. (C)*, **1966**, 470.
10. A. R. Katritzky, E. Lunt, B. Ternai, and G. J. T. Tiddy, *J. Chem. Soc. (B)*, **1967**, 1243.
11. E. Lunt, K. Washbourn, and W. R. Wragg, *J. Chem. Soc. (C)*, **1968**, 687.
12. D. E. Ames and R. F. Chapman, *J. Chem. Soc. (C)*, **1967**, 40.
13. L. McKenzie and C. S. Hamilton, *J. Org. Chem.*, **16**, 1414 (1951).
14. M. H. Palmer and E. R. R. Russell, *J. Chem. Soc. (C)*, **1968**, 2621.
15. I. Suzuki, T. Nakashima, N. Nagasawa, and T. Itai, *Chem. Pharm. Bull. (Tokyo)*, **12**, 1090 (1964).
16. I. Suzuki, T. Nakashima, and T. Itai, *Chem. Pharm. Bull. (Tokyo)*, **11**, 268 (1963).
17. J. T. Gleghorn, R. B. Moodie, E. A. Qureshi, and K. Schofield, *J. Chem. Soc. (B)*, **1968**, 316.

18. E. J. Alford and K. Schofield, *J. Chem. Soc.*, **1953,** 1811.
19. I. Suzuki, T. Nakashima, and N. Nagasawa, *Chem. Pharm. Bull.* (*Tokyo*), **11,** 1326 (1963).
20. K. Schofield and R. S. Theobald, *J. Chem. Soc.*, **1949,** 2404.
21. K. Schofield and J. C. E. Simpson, *J. Chem. Soc.*, **1945,** 520.
22. C. M. Atkinson, J. C. E. Simpson, and A. Taylor, *J. Chem. Soc.*, **1954,** 1381.
23. E. J. Alford and K. Schofield, *J. Chem. Soc.*, **1953,** 609.
24. H. J. Barber, K. Washbourn, W. R. Wragg, and E. Lunt, *J. Chem. Soc.*, **1961,** 2828.
25. H. E. Baumgarten and M. R. DeBrunner, *J. Am. Chem. Soc.*, **76,** 3489 (1954).
26. H. E. Baumgarten, D. L. Pedersen, and M. W. Hunt, *J. Am. Chem. Soc.*, **80,** 1977 (1958).
27. J. R. Elkins and E. V. Brown, *J. Heterocyclic Chem.*, **5,** 639 (1968).
28. C. M. Atkinson, J. C. E. Simpson, and A. Taylor, *J. Chem. Soc.*, **1954,** 1381.
29. E. M. Lourie, J. S. Morley, J. C. E. Simpson, and J. M. Walker, Br. Pat. 702,664, Jan. 20, 1954.
30. J. R. Keneford, E. M. Lourie, J. S. Morley, J. C. E. Simpson, J. Williamson, and P. H. Wright, *J. Chem. Soc.*, **1952,** 2595.
31. I. Suzuki and T. Nakashima, *Chem. Pharm. Bull.* (*Tokyo*), **12,** 619 (1964).
32. J. R. Keneford, J. S. Morley, J. C. E. Simpson, and P. H. Wright, *J. Chem. Soc.*, **1950,** 1104.
33. C. M. Atkinson and A. Taylor, *J. Chem. Soc.*, **1955,** 4236.
34. W. Hepworth and F. H. S. Curd, U.S. Pat. 2,585,935 (Cl. 260-250), Feb. 19, 1952.
35. E. Hayashi, H. Tani, H. Ishihama and T. Mizutani, Japanese Pat., 7025,511, Aug. 24, 1970; *Chem. Abstr.*, **73,** 131020 (1970).

Part H. Aminocinnolines

I. Methods of Preparation

A. Reduction of Nitrocinnolines

All of the possible aminocinnolines (3-, 4-, 5-, 6-, 7-, and 8-aminocinnoline) have been prepared by the chemical or catalytic reduction of the appropriate nitrocinnoline. This method of preparation is discussed as a reaction of the nitrocinnolines in Section 1G-II and will not be discussed further here. The aminocinnolines prepared by reduction and by other methods are given in Table 1H-4.

B. Nucleophilic Displacement

Halide, phenoxide, methylsulfonyl (CH_3SO_2-), methylsulfinyl (CH_3SO-), and even nitrile (1a) and mercapto (1b) groups are labile in the 4-position of the cinnoline ring and are easily displaced by ammonia and by amines to give the corresponding 4-aminocinnolines.

The conversion of 4-chlorocinnolines to 4-aminocinnolines is discussed in Section 1D-III. Generally, experimentalists have claimed that 4-aminocinnoline itself is prepared in better yields from 4-phenoxycinnoline and ammonia than from 4-chlorocinnoline and ammonia, but a wide variety of 4-alkylamino- and 4-arylaminocinnolines have been prepared from 4-chlorocinnoline and the appropriate amine in very good yield, both in the absence of solvent and in dimethyl sulfoxide solution (1–4).

Heating 4-phenoxycinnoline with ammonium acetate at 137–140 °C (5, 6) or with dry ammonia in phenol in the presence of ammonium chloride at 175° C (7, 8) yields 4-aminocinnoline in good yields. Similar procedures, using methylamine instead of ammonia, give the 4-methylamino derivative from the appropriate 4-phenoxycinnoline (9, 10).

The methylsulfonyl group is displaced from the 4-position of the cinnoline ring with extreme ease, even more easily than the chloride anion. Prepared by the oxidation of 4-methylthiocinnoline with potassium permanganate in aqueous acetic acid, 4-methylsulfonylcinnoline reacts with aniline, ammonia, methylamine, *n*-propylamine, or hydrazine to give the corresponding 4-amino-, 4-alkylamino-, or 4-hydrazinocinnoline (11, 11a). Similarly, 4-methylsulfinylcinnoline gives 4-*n*-butylaminocinnoline or 4-hydrazinocinnoline when allowed to react with *n*-butylamine or hydrazine hydrate, respectively (12).

All of the nucleophilic displacements discussed occur at the 4-position of the cinnoline ring; substituents at other positions apparently are not labile enough to be displaced by amines with the following known exceptions: 3-bromocinnoline treated with aqueous ammonia at 130–140° C in the presence of copper sulfate gives 3-aminocinnoline in good yield (13). In contrast, 3-bromo-4-hydroxycinnoline gives no recognizable product when heated with aqueous ammonia in the presence of copper sulfate (14). 3-Bromocinnoline and anhydrous dimethylamine in a sealed tube at 65° for 24 hr give 3-dimethylaminocinnoline in 40% yield (14a).

C. Hydroxy- to Aminocinnolines

The heating of an aromatic compound which bears a phenolic hydroxyl group with aqueous ammonium sulfite or bisulfite will convert the hydroxyl group to an amino group. This reaction has been utilized to convert 5-, 6-, 7-, and 8-hydroxycinnoline to the corresponding 5-, 6-, 7-, and 8-aminocinnoline (15, 16) (Eq. 1). Whereas 3-hydroxyquinoline

$$HO\text{—(cinnoline, N=N)} \longrightarrow H_2N\text{—(cinnoline, N=N)} \qquad (1)$$

can be converted to 3-aminoquinoline, 3-hydroxycinnnoline is claimed to be unchanged by this process (13).

D. Schmidt Rearrangement (Acetyl- to Aminocinnolines)

Treatment of 7-acetyl-4-hydroxycinnoline (**1**) with sodium azide and concentrated sulfuric acid at room temperature for 3 hr gives—on a small scale—about 55% of 7-acetamido-4-hydroxycinnoline (**2**), which can then be hydrolized in refluxing aqueous hydrochloric acid to 7-amino-4-hydroxy-

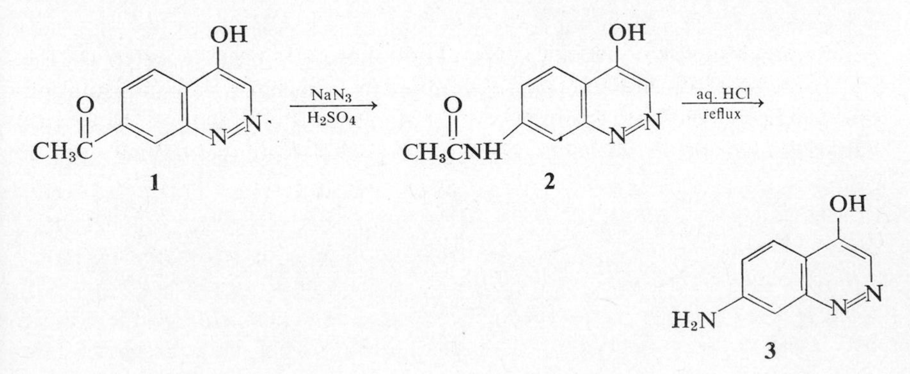

cinnoline (**3**) (6). In a similar manner, 3-acetylcinnoline is converted to 3-aminocinnoline, but in only 10–15% yields (17).

E. Beckmann Rearrangement of Oximes

The reaction of 7-acetyl-4-phenoxycinnoline (**4**) with hydroxylamine gives 7-acetyl-4-hydroxylaminocinnoline oxime (**5**). The oxime is rearranged by hydrogen chloride in an acetic acid-acetic anhydride mixture to either 7-acetamido-4-hydroxylaminocinnoline (**6**) or 4-hydroxylaminocinnoline-7-carboxymethylamide (**7**); it is not clear which (6).

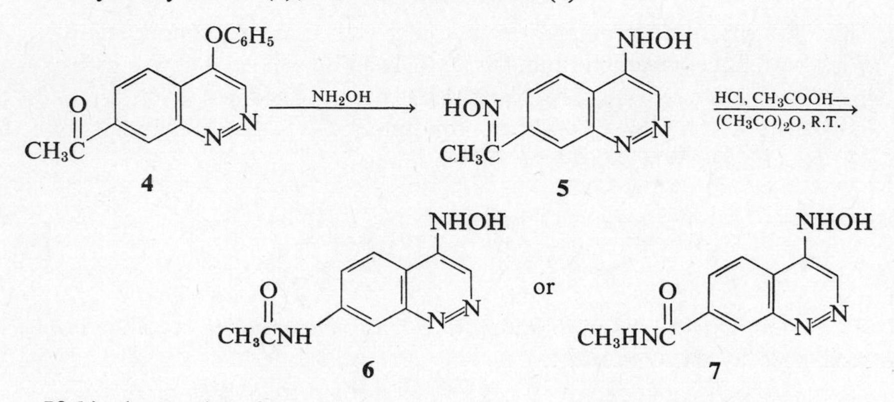

If **6** is the product from this reaction, then the 7-aminocinnoline derivative could be obtained from it by hydrolysis. It is also unknown whether the product from the Beckmann rearrangement of 7-acetyl-4-hydroxycinnoline oxime is the 7-acetamido or the 7-carboxymethylamide derivative (6). Obviously, the reaction needs closer attention.

F. Miscellaneous

Two preparations of specific aminocinnolines have been reported. In the first, 3-nitrocinnoline (**8**) and hydroxylamine in an ethanol-methanol solution give 50–62% yields of 4-amino-3-nitrocinnoline (**9**). Reduction of **9** with stannous chloride in hydrochloric acid gives 3,4-diaminocinnoline (**10**) in

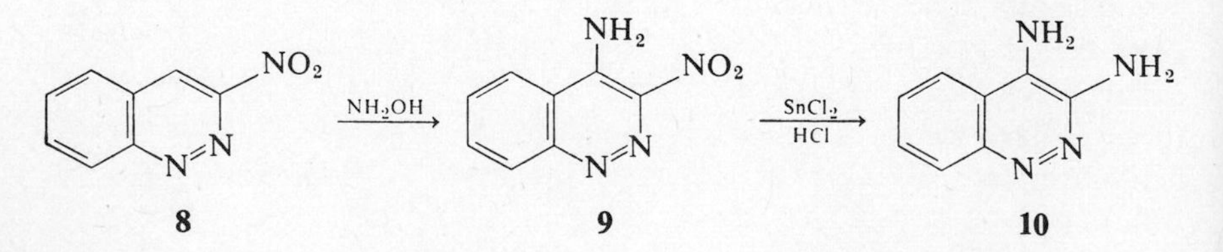

76% yield (18). This is similar to the previously known reaction of 3-nitroquinoline with hydroxylamine to give 4-amino-3-nitroquinoline.

Finally, the reduced cinnoline, 1,2-dicarbomethoxy-4-methoxy-4-(1,2-dicarbomethoxy)hydrazino-1,2,3,4-tetrahydrocinnoline (**13**), when refluxed with hydrazine hydrate gives 4-hydrazinocinnoline (**14**) in 55–60% yield (19). The reduced cinnoline **13** is prepared by the condensation of dimethyl azoformate (**11**) and α-methoxystyrene (**12**).

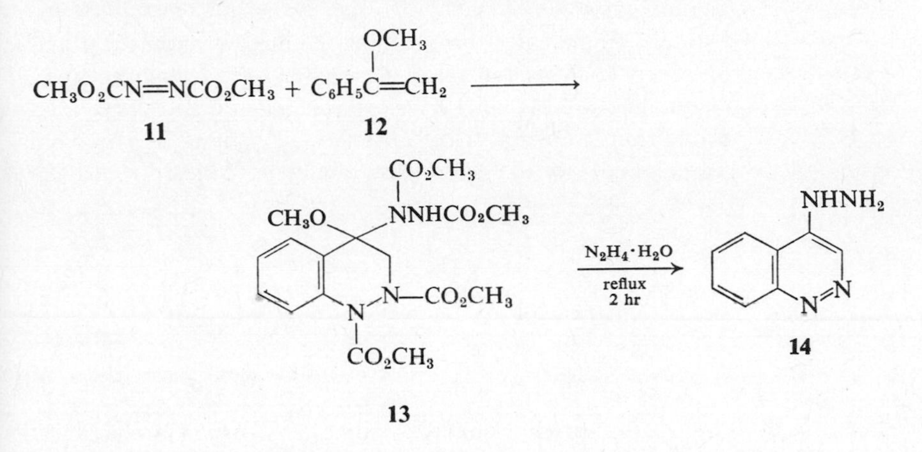

II. Physical Properties and Structure

All of the monoaminocinnolines, otherwise unsubstituted, are yellow crystalline materials melting in the range 157–218° C except 8-aminocinnoline, a yellow solid which melts at 91–92° C (20, 21). Its picrate derivative is jet black in color (16).

Aqueous acidic solutions of 5-aminocinnoline are violet-red in color, whereas 6- and 7-aminocinnoline under the same conditions form yellow solutions (15). The color of 8-aminocinnoline in aqueous acidic solution is reported as violet-red (15) and as deep purple (16). Aqueous solutions of 3-aminocinnoline are greenish-yellow with a greenish-blue fluorescence (compared with the violet fluorescence of 3-aminoquinoline) (13).

There are apparently two crystalline modifications of 4-hydrazinocinnoline. The first appears as orange-red needles or plates, mp 301° C (dec.) (19) [296–297° C (dec.) (14)]. This form may be converted to a second modification, deep yellow irridescent plates, mp 229° C (dec.) (19) [226–227° C (dec.) (14)] by refluxing the first form in ethanol. Both give the same benzal derivative, mp 307° C (dec.) (19) and the same hydrochloride, mp 244–245° C (14).

It is well known that an amino substituent in most *N*-heteroaromatic systems exists in the amino form (—NH_2) rather than the imino form (=NH). Infrared studies of the amino derivatives of pyridine, quinoline, pyrimidine, and related systems in which tautomerism is possible are offered as evidence for the preference of the amino over the imino form (22–24). The amino groups of 5-, 6-, 7- and 8-aminocinnoline most certainly exist in the amino form, but published reports present a confusing picture for the structure of 4-aminocinnoline.

One infrared study (25) indicates that 4-aminocinnoline is anomalous and exists in the imino form **15**. This conclusion is reached as follows. The study reports 4-aminocinnoline to have an IR absorption band at 3311 cm^{-1} and at 1675 cm^{-1} in addition to several other absorption bands (25). The band at 3311 cm^{-1} corresponds to the absorption band at 3325 cm^{-1} in the

spectrum of 1,2-dihydro-2-imino-1-methylpyridine (**17**), which is characteristic of the imino structure (22). Compounds that bear the amino substituent in the amino (—NH_2) form would be expected to show this absorption band at 100–200 cm^{-1} higher frequency. Thus 2- and 4-aminopyridine both show a pair of frequencies at 3400 and 3500 cm^{-1}, characteristic of the (—NH_2) structure (23).

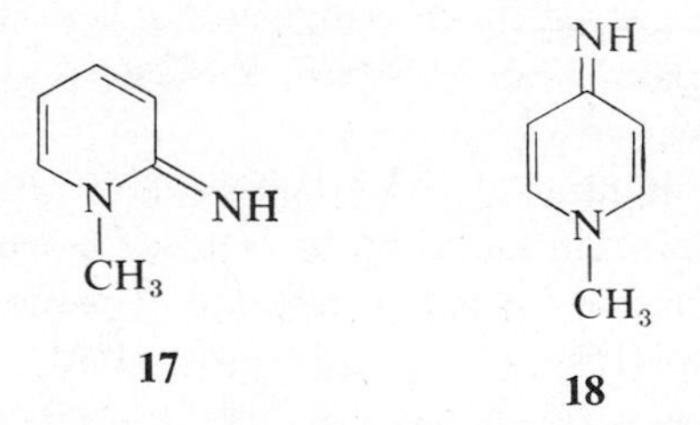

Also indicative of the imino structure is the strong deformation band reported to be at 1675 cm^{-1} in the spectrum of 4-aminocinnoline (25). The (C=NH) deformation band in the spectrum of the imino compound **17** and in the spectrum of 1,4-dihydro-4-imino-1-methylpyridine (**18**) is a strong absorption occurring at 1650–1660 cm^{-1}. On the other hand, the (—NH_2) deposition band for 2- and 4-aminopyridine and for 2- and 4-aminoquinoline—all known to exist in the (—NH_2) form—occurs at significantly lower frequencies of 1619–1629 cm^{-1} (22).

Although these comparisons seem to indicate that 4-aminocinnoline exists in the imino form **15** rather than the amino form **16**, a second infrared study (26) indicates just the opposite by reporting absorption bands in the IR spectrum of 4-aminocinnoline at 3421 and 3505 cm^{-1} instead of the single band at 3311 cm^{-1} discussed above. Unfortunately, this second study does not give the complete IR spectrum of 4-aminocinnoline and the absorption bands (if any) in the 1620–1660 cm^{-1} range are unrecorded. Even so, the two bands at 3421 and 3505 cm^{-1} indicate the amino structure **16** for 4-aminocinnoline. A very recent study by Katritzky and co-workers (14a) shows on the basis of UV spectra and basicity measurements that the amino: imino ratio for 4-aminocinnoline is 12,880:1, and that this same ratio for 3-aminocinnoline is 2×10^8:1. This high value for the amino:imino ratio in 3-aminocinnoline is attributed to the fact that the imino form of 3-aminocinnoline possesses an *ortho*-quinonoid system in the homocyclic ring.

A pmr study (1) has shown that the 4-alkylamino-3-phenylcinnolines (**19**; R=alkyl) exist in the amino form as shown and not in the imino form. The pmr spectra of 3-, 4-, 5-, and 8-aminocinnoline are given in Table 1H-4.

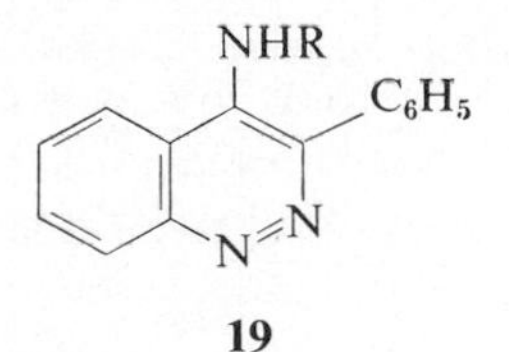

19

Substitution of an amino group into the cinnoline ring system has a base-strengthening effect; hence all of the monoaminocinnolines are stronger bases than cinnoline itself, with the possible exception of 5-aminocinnoline. Even so, each monoaminocinnoline, having two ring nitrogen atoms, is a weaker base than the corresponding aminoquinoline. That is, 3-aminocinnoline is a weaker base than 3-aminoquinoline; 4-aminocinnoline is a weaker base than 4-aminoquinoline, and so on. Basic strengths of the aminoquinolines are recorded in reference 27, while the basic strengths of the aminocinnolines are given in Table 1H-2. From this table it can be seen that the order of basic strengths for the monoaminocinnolines (in order of decreasing basicity) is: $4 > 6 > 7 > 3$, $8 > 5$-aminocinnoline. Thus 4-aminocinnoline (pK_a 6.26–6.85) is the strongest base and 5-aminocinnoline (pK_a 2.70) is the weakest base. This order of basicities is borne out by the values of the force constants for the ($—NH_2$) stretching frequencies in the IR spectra. The values of the force constants follow the order of basicity with the most basic 4-aminocinnoline having the highest force constant (6.64×10^5 D/cm) and the least basic 5-aminocinnoline having the lowest force constant (6.56×10^5 D/cm). Basicity and force constant values are correlated through the fact that the more interaction the amino substituent has with a ring nitrogen atom—to stabilize the protonated cinnolinium ion and hence increase basicity—the higher will be the force constant (26).

There are three places where the protonation and alkylation of the monoaminocinnolines might occur: at the amino group, at N-1, or at N-2. Of these, we can eliminate the amino group on the basis of UV absorption curves and IR data. The UV absorption curves of the *N*-heteroaromatic bases such as quinoline or cinnoline generally show a marked bathochromic shift when the base is protonated. This is also true for *N*-heteroaromatic amines, including the monoaminocinnolines, when they are converted into their monocations. Typical aromatic amines do not show this bathochromic shift upon protonation, indicating very strongly that the *N*-heteroaromatic amines accept the proton on the ring nitrogen atom (13, 26, 28). In the case of 3-aminocinnoline, quaternization with methyl iodide yields two salts, both of which show characteristic (NH_2) absorption bands in their IR spectra. Therefore methylation does not occur at the amino group; the two products are N-1 methyl and N-2 methyl salts (28).

As to the question of which of the two ring nitrogen atoms is most basic, we cannot use the site of quaternization with an alkyl halide because this reaction is subject to kinetic control, but protonation is thermodynamically controlled so that the most basic of the two ring nitrogen atoms of the monoaminocinnolines would be indicated by the principal site of protonation. Knowledge of this fact allowed Ames and his co-workers in England to show experimentally that the most basic ring nitrogen atom in 3-aminocinnoline is N-2, and that the most basic ring nitrogen atom in 4-aminocinnoline is N-1. This was done by comparing the UV absorption spectra of 3- and 4-aminocinnolinium perchlorate with the spectra of the 1- and 2-methyl salts. The results correlated with those obtained by molecular orbital calculations (28). Similar work has not been done for the remaining monoaminocinnolines.

One can utilize the observed basicities to predict the site of protonation (the basic center) of the monoaminocinnolines. Let us first assume that for 5-, 6-, 7-, and 8-aminocinnoline the amino group will have a negligible inductive effect upon the basicity of the molecule. The significantly greater basicity of the monoaminocinnolines (except 5-aminocinnoline) compared to cinnoline itself must therefore be due to a resonance interaction between the amino group and the protonated ring nitrogen atom. This resonance interaction increases basicity by delocalizing the positive charge and hence increasing the stability of the aminocinnolinium cation. Only one of the two ring nitrogen atoms can interact in this way with an amino substituent at any given position of the cinnoline ring. Using this approach we can see that the

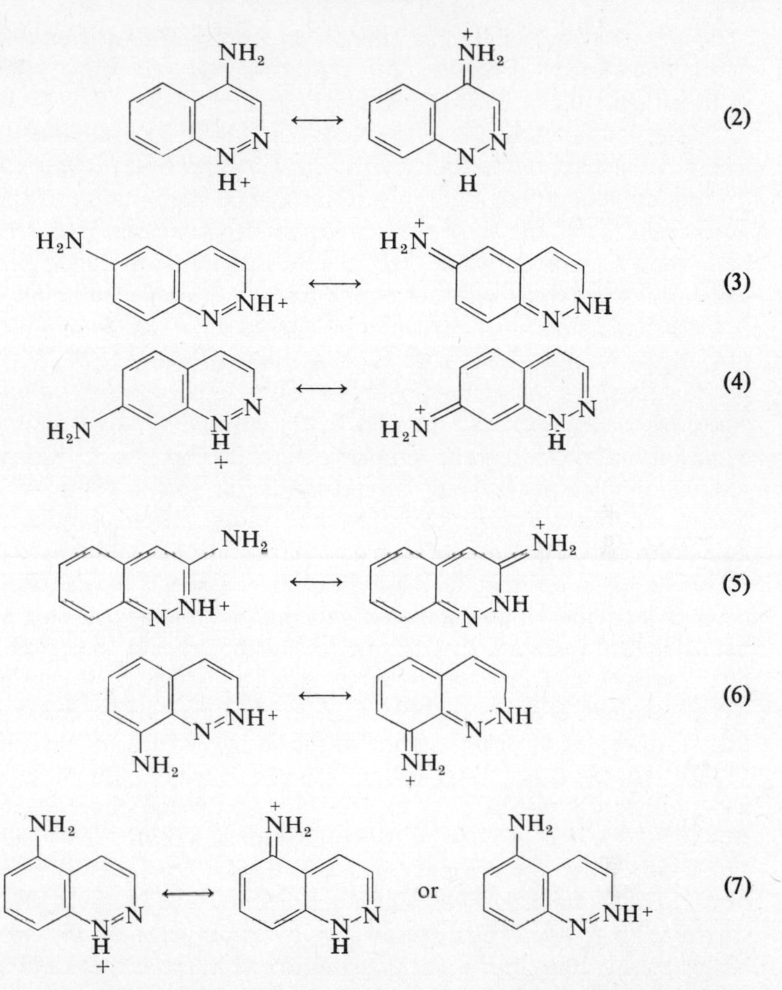

most basic site in 4-aminocinnoline is N-1 (Eq. 2). This has been confirmed experimentally, as mentioned earlier. The most basic site in 6-aminocinnoline would be N-2 (Eq. 3); in 7-aminocinnoline it would be N-1 (Eq. 4). The most basic site in 3-aminocinnoline has been shown experimentally to be N-2 (Eq. 5), and in 8-aminocinnoline the basic center is predicted to be N-2 (Eq. 6). An exception is 5-aminocinnoline (Eq. 7), whose basicity (pK_a 2.70) is not sufficiently different from that of cinnoline itself (pK_a 2.70 or 2.29) to indicate any great amount of resonance interaction, even though such an interaction would dictate protonation at N-1. In addition, the force constant for (—NH_2) stretching absorption in the IR spectrum of 5-aminocinnoline, as discussed earlier, is the lowest for all of the monoaminocinnolines, indicating the least amount of interaction with the protonated ring nitrogen atom. Therefore, 5-aminocinnoline may be protonated at N-2, the same site at which cinnoline itself is protonated.

To substantiate this theory, we note that 3-aminoisoquinoline (pK_a 5.05), where resonance stabilization of the cation is possible, is more basic than 3-aminoquinoline (pK_a 4.95), where resonance stabilization of the cation is not possible. For the same reason 6-aminoisoquinoline (pK_a 7.17) is more basic than 6-aminoquinoline (pK_a 5.63); 7-aminoisoquinoline (pK_a 6.20) is less basic than 7-aminoquinoline (pK_a 6.65); 8-aminoisoquinoline (pK_a 6.06) is more basic than 8-aminoquinoline (pK_a 3.99); and 4-aminoisoquinoline (pK_a 6.28) is much less basic than 4-aminoquinoline (pK_a 9.17). The 5-position is again anomalous because one would expect 5-aminoquinoline, where resonance stabilization of the cation is possible, to be more basic than 5-aminoisoquinoline, where resonance stabilization is not possible. In fact, just the opposite is observed: 5-aminoquinoline (pK_a 5.46) is less basic than 5-aminoisoquinoline (pK_a 5.59). [All pK_a values for the quinolines and isoquinolines in this paragraph refer to water solutions at 20° C (26, 27)].

One can also use resonance stabilization as a means to predict the relative order of basicities of the monoaminocinnolines. Comparison of Eqs. 2 to 7 indicates that 4-aminocinnoline (pK_a 6.26–6.85) should be the most basic of all the monoaminocinnolines because resonance stabilization does not disrupt the aromaticity of the benzonoid ring. In all other cases, benzonoid aromaticity is disrupted by delocalization of the positive charge to the amino group. In addition, resonance stablilization of the 4-aminocinnolinium cation involves a *p*-quinonoid structure (Eq. 2). This is more stable than an *o*-quinonoid structure. The next most basic is 6-aminocinnoline (pK_a 5.04). Inspection of Eq. 3 shows that resonance stabilization of its cation involves a symmetrical *p*-quinonoid structure. Following 6-aminocinnoline in basicity is 7-aminocinnoline (pK_a 4.85) whose cation also has a *p*-quinonoid resonance-contributing form, although it is not as symmetrical as the corresponding form for 6-aminocinnoline. The remaining aminocinnolines all have a less stable *o*-quinonoid structure contributing to the stabilization of their cations. Thus 3-aminocinnoline (pK_a 3.63–3.70) (Eq. 5) and 8-aminocinnoline (pK_a 3.68) (Eq. 6) are less basic than 7-aminocinnoline, because of *o*-quinonoid resonance and because both 3- and 8-aminocinnoline may destabilize their cation inductively. This inductive effect would be a direct effect in the case of 3-aminocinnoline and an indirect one in the case of 8-aminocinnoline. Hydrogen-bonding between the 8-amino group and N-1 would place a partial positive charge at N-1. N-1 could then destabilize the cation by draining electron density from N-2, the site of protonation. One would predict that 5-aminocinnoline (Eq. 7) could have about the same basicity as 3- and 8-aminocinnoline, or even to be somewhat more basic because inductive destabilization of its cation is not as likely as it is for the 3- and 8-isomers. In fact, the basicity of 5-aminocinnoline (pK_a 2.70) is lower than the basicity of 3- and 8-aminocinnoline.

In summary, the basic center of 4-aminocinnoline is known to be at N-1. The basic center of 3-aminocinnoline is known to be at N-2. We predict the basic center of 6- and 8-aminocinnoline to be at N-2, and the basic center of 7-aminocinnoline to be at N-1. The site of the basic center in 5-aminocinnoline is not established.

A mass spectral study of 3-, 4-, 5-, and 8-aminocinnoline shows that initial fragmentation of all of these aminocinnolines occurs with loss of molecular nitrogen to given an *m/e* 117 ion which then loses HCN and H. In addition, 4-aminocinnoline fragments by an alternate pathway with initial loss of HCN to give an *m/e* 118 ion. The mass spectra of 5- and 8-aminocinnoline are essentially identical (28a).

III. Reactions

A. Diazotization

An amino substituent at any position of the cinnoline ring may be diazotized, although the diazonium salt of 3-aminocinnoline (and probably also 4-aminocinnoline) is considerably less stable than the diazonium salts of 5-, 6-, 7-, and 8-aminocinnoline. Thus the 5-, 6-, 7-, and 8-aminocinnolines, which give colored aqueous acidic solutions, all form colorless solutions upon diazotization; they then produce red dyes by coupling with β-naphthol in alkaline solution (15). On the other hand, attempts to couple the diazonium salt, prepared in dilute aqueous acidic solution, of 3-aminocinnoline with either α- or β-naphthol have not been successful (29). Most likely, the cinnoline-3-diazonium salt reacts with the solvent system immediately after its formation to give 3-hydroxycinnoline. This reaction occurs before the coupling reaction with α- or β-naphthol can take place. In fact, when solutions of 3-aminocinnoline (**20**) in dilute hydrochloric acid or sulfuric acid are treated with sodium nitrate in the cold, nitrogen is evolved immediately and

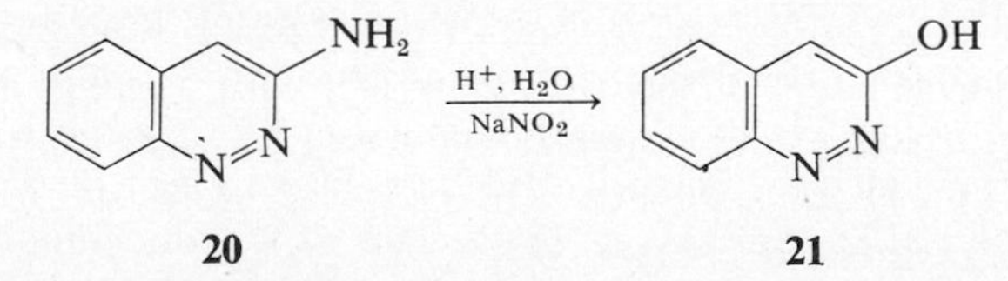

3-hydroxycinnoline (**21**) is obtained (29). This is also true for the 1- and 2-methyl derivatives of 3-aminocinnoline (28, 30). If the diazotization of 3-aminocinnoline or a substituted 3-aminocinnoline is carried out in concentrated hydrochloric acid or 48% hydrobromic acid, then a mixture of the

3-hydroxycinnoline and the 3-chloro- or 3-bromocinnoline is obtained (29, 30).

Treatment of 4-amino-2-methyl-6-nitrocinnolinium iodide in 40% sulfuric acid with sodium nitrite produces the expected anhydro base of 4-hydroxy-2-methyl-6-nitrocinnolinium hydroxide (30).

Diazotization of 8-amino-4-methylcinnoline (**22**) in hydrochloric acid solution followed by treatment of the diazonium salt with cuprous chloride (Sandmeyer reaction) yields 8-chloro-4-methylcinnoline (**23**) (31, 32), but

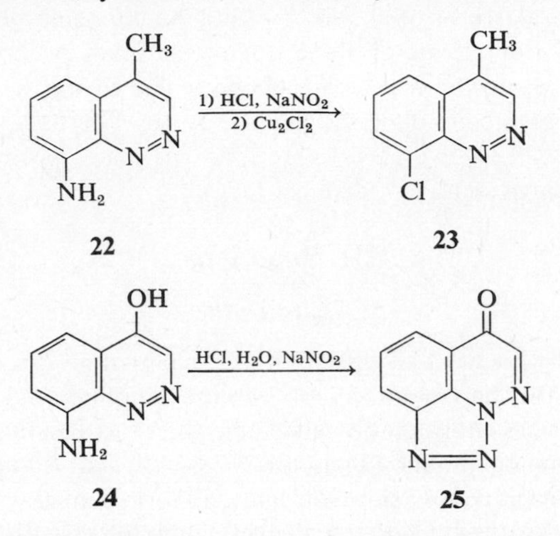

diazotization of 8-amino-4-hydroxycinnoline (**24**) in aqueous hydrochloric acid gives 1,8-azo-1,4-dihydrocinnolin-4-one (**25**) (6). The anhydro base of 6-amino-4-hydroxy-2-methylcinnolinium hydroxide can be diazotized and treated with cuprous chloride to give the 6-chloro derivative (30).

6-Chloro-4-hydroxy-5-nitrocinnoline has been reduced, diazotized, and treated with cuprous chloride to give 5,6-dichloro-4-hydroxycinnoline. This same sequence was used to convert 5-chloro-4-hydroxy-6-nitrocinnoline into 5,6-dichloro-4-hydroxycinnoline (32a).

Several substituted 6-amino-4-hydroxycinnoline 1- and 2-methyl salts, and 8-amino-4-hydroxycinnolinium hydroxide have been deaminated by diazotization and reduction of the diazonium salt with hypophosphorus acid (30). Attempts to deaminate 3-aminocinnoline and 3-amino-4-methylcinnoline similarly with hypophosphorus acid, however, were not successful and gave only the 3-hydroxy derivative (29).

B. Oxidative Deamination of 4-Hydrazinocinnolines

Oxidative deamination of 4-hydrazinocinnoline (**14**) may be effected by a refluxing aqueous cupric sulfate solution. The product is cinnoline (**26**).

Similarly, 6-chloro-4-hydrazino- and 6-bromo-4-hydrazinocinnoline are deaminated, but 3-chloro-4-hydrazinocinnoline undergoes violent decomposition under the same conditions to give a tarry, intractible product (14). On the other hand, attempts to deaminate 4-hydrazino-6-nitrocinnoline by

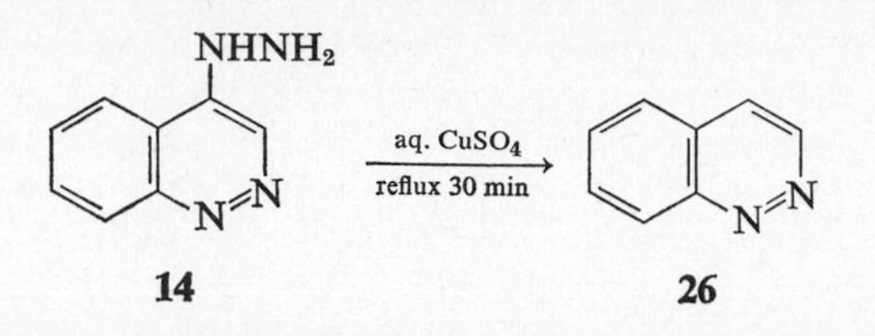

refluxing aqueous cupric sulfate have not been successful. This material does give a hydrazone with acetone, however, which is readily converted by aqueous cupric sulfate into 6-nitrocinnoline (33).

C. Hydrolysis

The presence of a nitro group in the cinnoline ring system, or methylation at N-1 or N-2, further enhances the ability of the cinnoline ring system to take part in nucleophilic displacement reactions at the 4-position. Thus 4-amino-3-nitrocinnoline (**9**) undergoes alkaline hydrolysis simply by being warmed with dilute aqueous potassium hydroxide to give 4-hydroxy-3-nitrocinnoline (**27**) (18). Both 4-amino-1-methylcinnolinium iodide (**28**) and 4-amino-2-methylcinnolinium iodide (**30**) are hydrolyzed in aqueous base to give the corresponding 4-hydroxy derivatives, **29** and **31**, respectively (34). Certain other 4-aminocinnolines, such as the 6-chloro- and 6-nitro

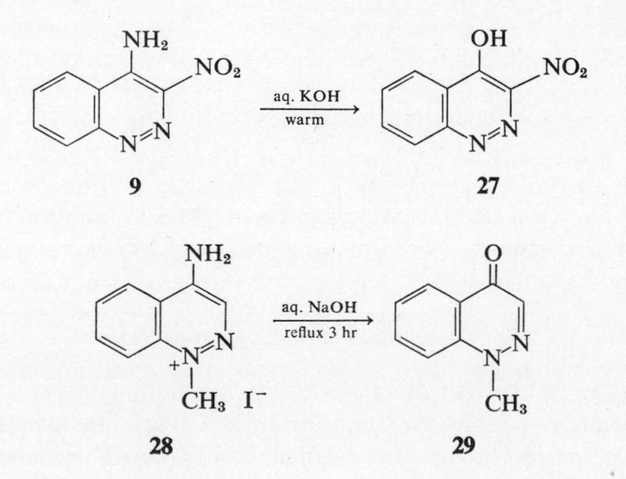

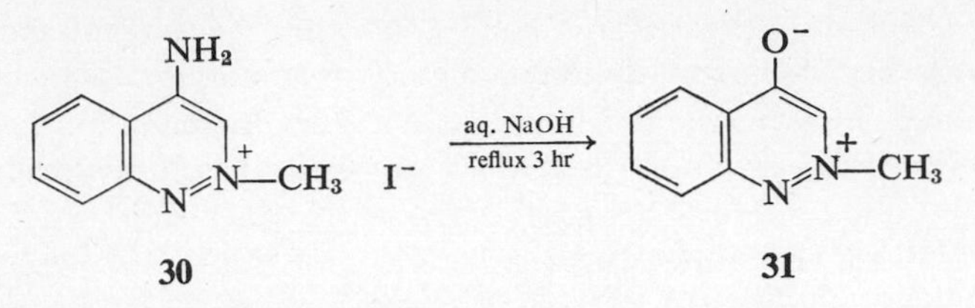

quaternary salt derivatives, and unquaternized 4-amino-6-nitrocinnoline also undergo ready hydrolysis to give the 4-hydroxycinnolines (35, 36). The 4-methylamino group is displaced in the same manner as the 4-amino group from the cinnoline quaternary salts (36, 37).

D. Skraup Reaction

Like certain other primary aromatic amines, 8-amino-4-methylcinnoline (**22**) undergoes the Skraup reaction. With glycerol or with acrolein it gives 1-methyl-3,4,5-triazaphenanthrene (**32**), and with crotonaldehyde diacetate

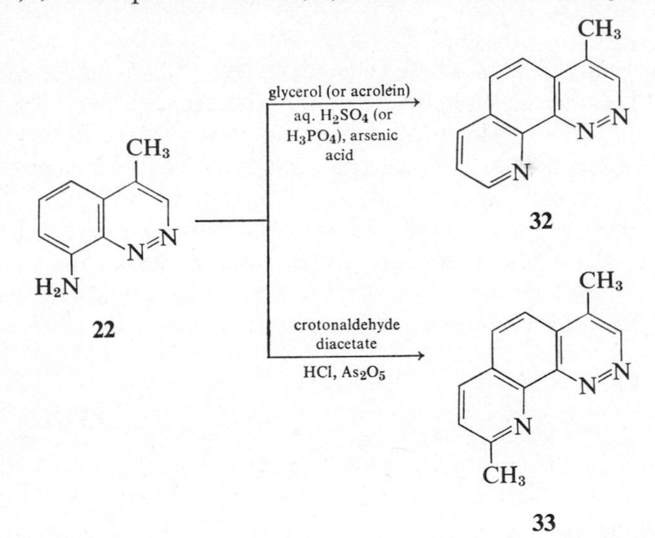

it gives 1,6-dimethyl-3,4,5-triazaphenanthrene (**33**) (38, 39). Both **32** and **33** form colored complexes with metal ions such as copper(I), copper(II), iron(II), and cobalt(II) (39).

E. Protonation and Quaternization

Protonation of 3-aminocinnoline occurs predominantly on N-2, but 4-aminocinnoline is protonated mainly on N-1 (28). The position of protonation for the remaining monoaminocinnolines is not known, although predictions for the most likely site have been given earlier in this chapter.

Methylation of 3-aminocinnoline with methyl iodide results in an N-1 methiodide and an N-2 methiodide in approximately equal yields, the higher melting salt being the 2-methyl isomer (28); 4-aminocinnoline also gives both an N-1 and an N-2 methiodide, although in this case twice as much of the N-2 methiodide is obtained than the N-1 methiodide (34, 40). Similarly, 4-acetamidocinnoline gives 37% of the 2-methiodide and only 13% of the 1-methiodide upon quaternization with methyl iodide (40). The 6-chloro- and 6-nitro- derivatives of 4-aminocinnoline are also known to be methylated at both ring nitrogen atoms (30, 35).

F. Schiff Base Formation

The aminocinnolines and hydrazinocinnolines, like other primary amines, yield Schiff bases by interaction with aldehydes and ketones. For example, when 3,4-diaminocinnoline (**10**) and phenanthroquinone (**34**) are heated together in glacial acetic acid, bright orange needles of 9,10,11,16-tetraazatribenz[*a*,*c*,*h*]anthracene (**35**) are obtained (18). In the same way, 6-chloro-3,4-diaminocinnoline reacts with diacetyl to give 6-chloro-2,3-dimethyl-1,4,9,10-tetrazaphenanthrene. Treatment of the diamine with

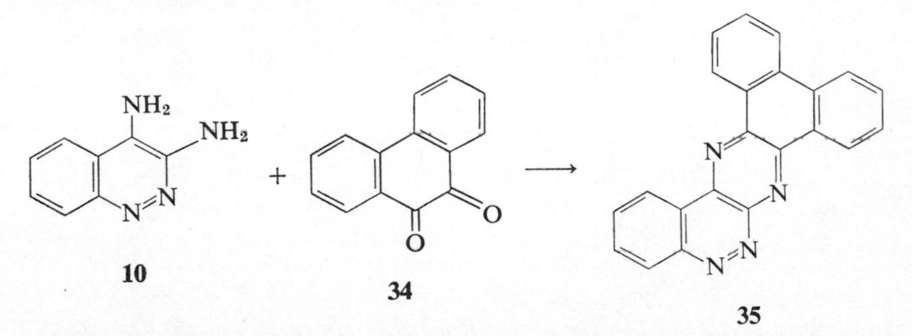

formic acid at 95° produces 8-chloroimidazo[4,5-c]cinnoline, while diazotization gives 8-chloro-1,2,3-triazolo[4,5-c]cinnoline (32a). 4-Hydrazinocinnoline gives a benzal derivative (**36**) with benzaldehyde (19), and 4-hydrazino-6-nitrocinnoline forms orange, hairlike needles of acetone 6-nitro-4-cinnolinylhydrazone (**37**) when refluxed in acetone for 10 min (33).

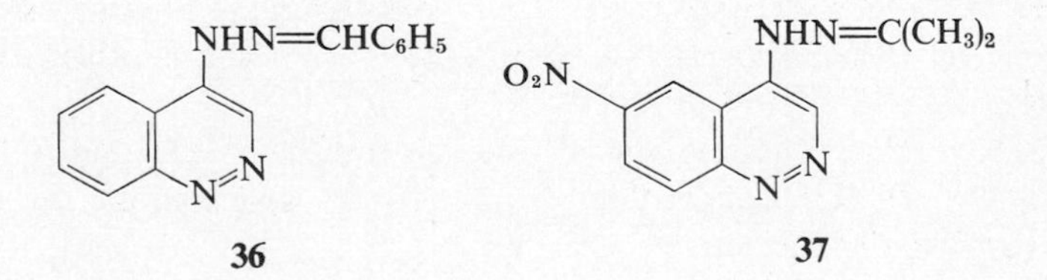

G. Nitration of 4-Aminocinnoline

Treatment of 4-aminocinnoline (**16**) with fuming nitric acid and concentrated sulfuric acid at 0–5° C for 2 hr gives a fair yield of 4-amino-6-nitrocinnoline (**38**) (30). 4-Hydroxycinnoline is also nitrated principally at the 6-position. This contrasts with the nitration of unsubstituted cinnoline, where 5- and 8-nitrocinnoline are obtained as products.

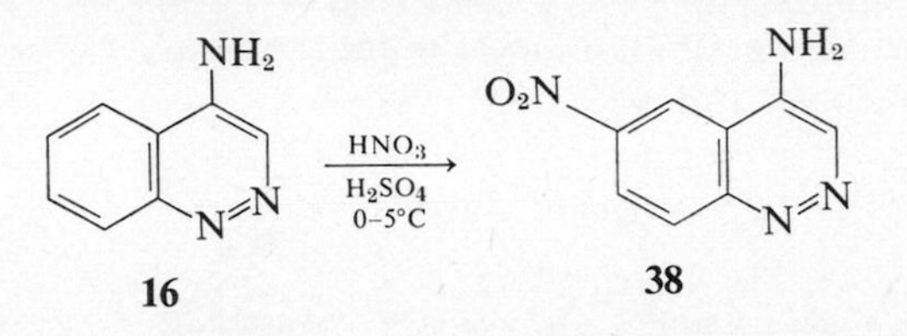

H. Formation of a Pyrido[3,2-*h*]cinnoline

The interaction of 8-amino-4-methylcinnoline with diethyl ethoxymethylenemalonate yields 8-(2,2-dicarbethoxyvinylamino)-4-methylcinnoline (**39**).

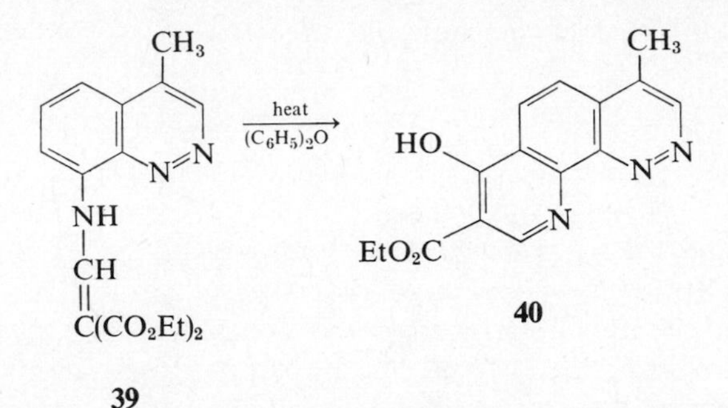

This material can be cyclized in hot diphenyl ether to form 3-carbethoxy-4-hydroxy-7-methylpyrido[3,2-*h*]cinnoline (**40**) (32).

I. Biological Activity

Of all the substituted cinnolines that have been prepared, the aminocinnolines and their quaternary salts are the only ones found to have any significant biological activity, although this may simply be because other cinnolines have been less thoroughly screened.

The sulfanilamido derivatives of 4-aminocinnoline (**41**) and 3-amino-4-methylcinnoline (**42**) display antibacterial activity, even though not as much as many other sulfonamides (41).

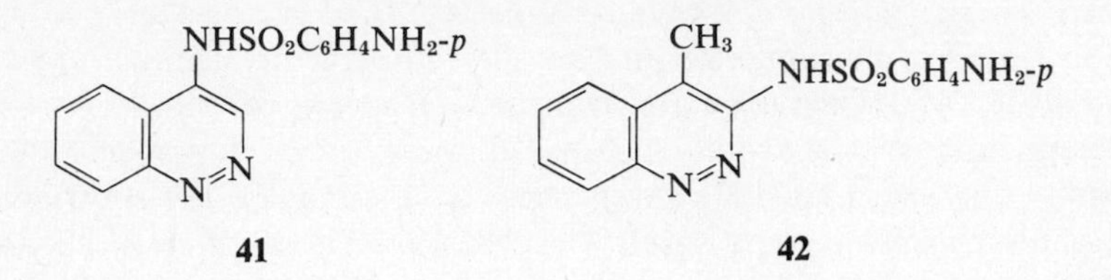

41 42

4-Hydrazinocinnoline is a skin-sensitizing agent (42), while *N*,*N*-dimethyl-*N*′-benzyl-*N*′-(4-cinnolinyl)ethylenediamine (**43**) shows good antihistamine activity (43).

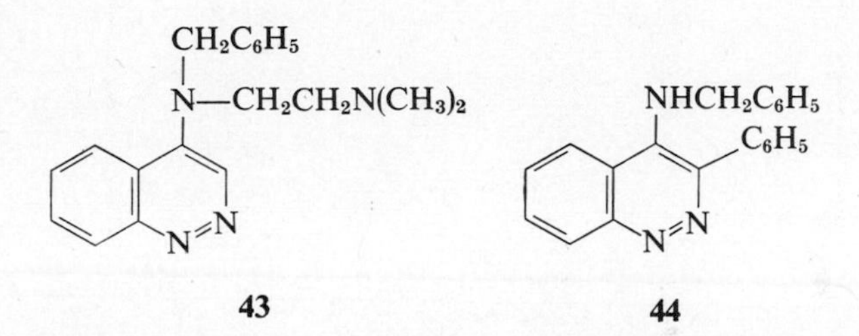

43 44

Certain 4-alkylamino-3-arylcinnolines, such as 4-benzylamino-3-phenylcinnoline (**44**), 4-benzylamino-3-(4-chlorophenyl)cinnoline, and 4-(2-hydroxyethylamino)-3-phenylcinnoline are claimed to possess antiulcer activity, to be pepsin inhibitors, and to be antialgal agents (2). The closely related 3-aryl-4-dialkylaminoalkylaminocinnolines such as 4-(2-dimethylaminoethylamino)-3-phenylcinnoline (**45**) and 4-(2-piperidinoethylamino)-3-phenylcinnoline are claimed to show antibiotic action against bacteria (*Diplococcus pneumoniae*), protozoa (*Tetrahymena gelleii*), fungi (*Candida albicans*), and algae (*Chlorella vulgaris*). These materials also inhibit the germination of seeds of *Trifolium* (4). Similar antibiotic activity is claimed for the 3-phenyl-4-piperazinylcinnolines, such as 4-(4-methyl-1-piperazinyl)-3-phenylcinnoline (**46**) and 4-(4-benzyl-1-piperazinyl)-3-phenylcinnoline. In addition, these piperazinylcinnolines are claimed to be diuretic agents (3).

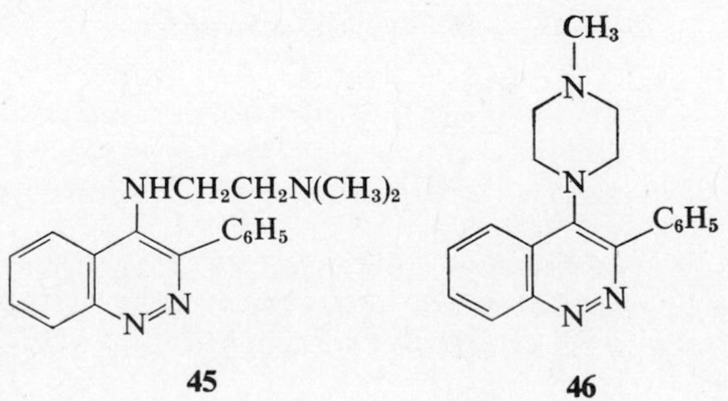

45 46

The polymethylene *bis*-cinnolinium salts (**47**; R = H or acetyl, *n* = 4-22), derived from 4-aminocinnoline and an alkyl diiodide are antimicrobial and antiparasitic (44). When there are 20 methylene groups (**47**; R = H, *n* = 20) the compound shows activity against filariasis (45), a parasitic disease transmitted by mosquitos. However, these compounds do not show activity against *Hymenolepii nana*, a parasitic worm (46). The position of alkylation in **47**, N-1 or N-2, is not known with certainty. For this reason, the methylene groups in the accompanying structure **47** are not shown as a specific attachment to N-1 or N-2.

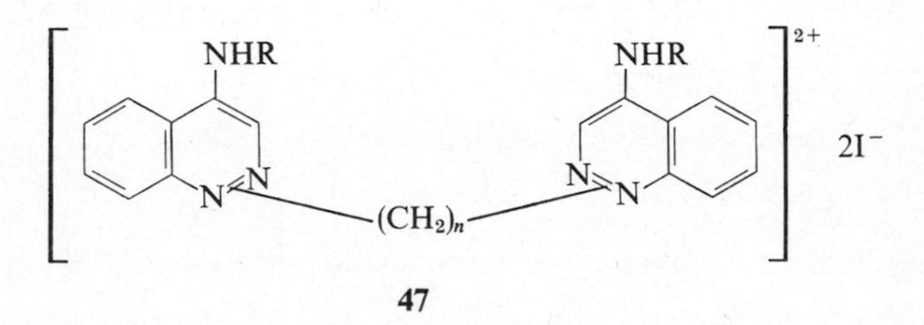

47

The dimethiodide salts of certain pyrimidinylaminocinnolines such as the dimethiodide of 4-amino-6-(2′-amino-6′-methylpyrimidinyl-4′-amino)cinnoline (**48**) and substituted derivatives are claimed to have powerful trypanocidal activity (9, 47). Compound **48** is prepared from 2-amino-4-chloro-6-methylpyrimidine 1-methiodide and 4,6-diaminocinnoline methiodide. The cinnoline is claimed to be a 1-methyl salt but may actually be the 2-methyl isomer. As expected, the 6-amino group and not the 4-amino group of the cinnoline system acts as the nucleophilic center, displacing chloride from the 4-position of the pyrimidine ring to give **48**.

The quaternary salts of several derivatives of 4-amino-6-nitrocinnoline and of 4-amino-6-nitrocinnoline itself were tested for trypanocidal activity,

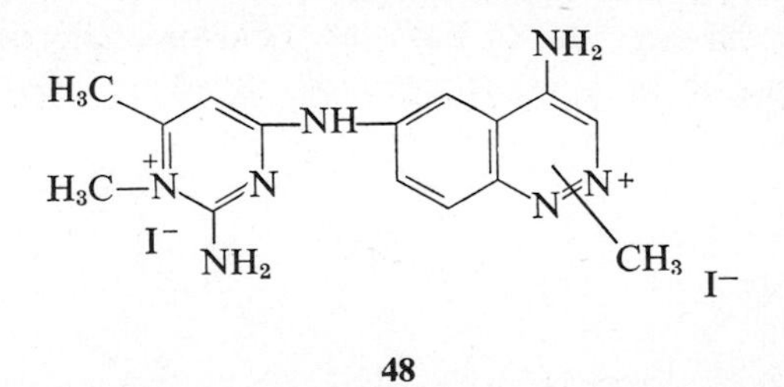

48

primarily against *T. congolense*. Except for the weak trypanocidal activity of 4-methylamino-6-nitrocinnoline methiodide, none of these compounds showed any useful level of activity, although several were active at near toxic doses (48).

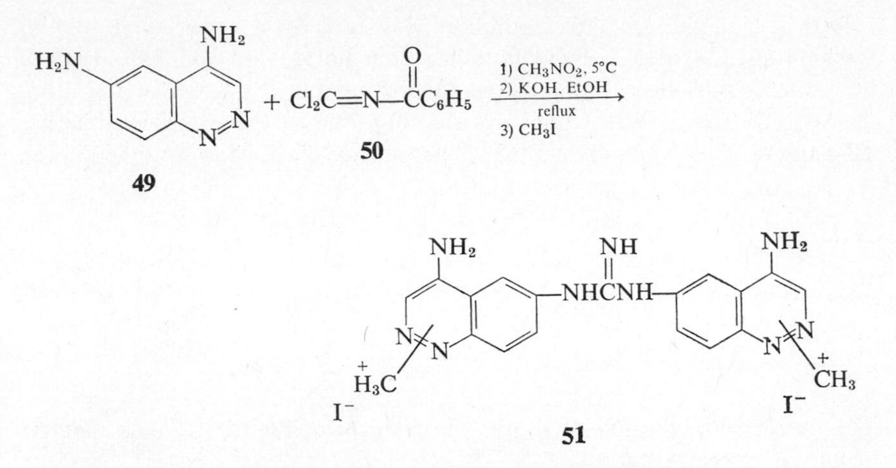

N^1,N^3-*Bis*(4-amino-6-cinnolinyl)guanidine dimethiodide (**51**), known as "cinnoline 528," is a powerful trypanocidal agent against *T. congolense*, a parasitic protozoan which is the cause of an infection in cattle.

Compound **51** is prepared from 4,6-diaminocinnoline (**49**) and dichloromethylenebenzamide (**50**) in nitromethane solution, followed by hydrolysis

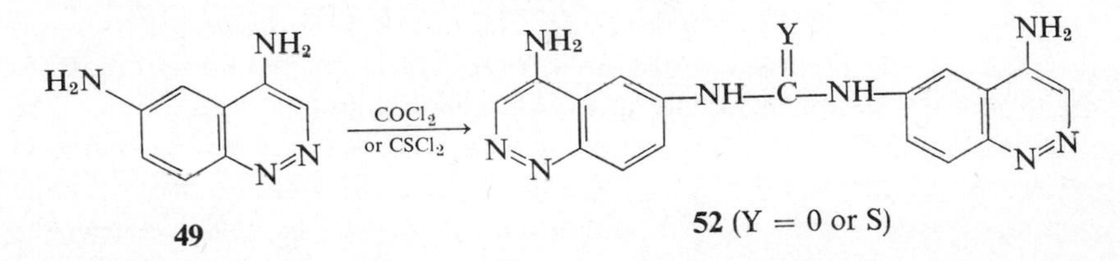

of the initial condensation product with alcoholic potassium hydroxide (5). Similar compounds (**52**; Y = O or S) are prepared from 4,6-diaminocinnoline (**49**) and carbonyl chloride (phosgene) or thiocarbonyl chloride to give the urea or thiourea derivative, respectively (49). The methiodide salt of the urea compound **52** (Y = O) is only slightly active against *T. congolense*, and the thiourea analog **52** (Y = S) is completely inactive, as are their

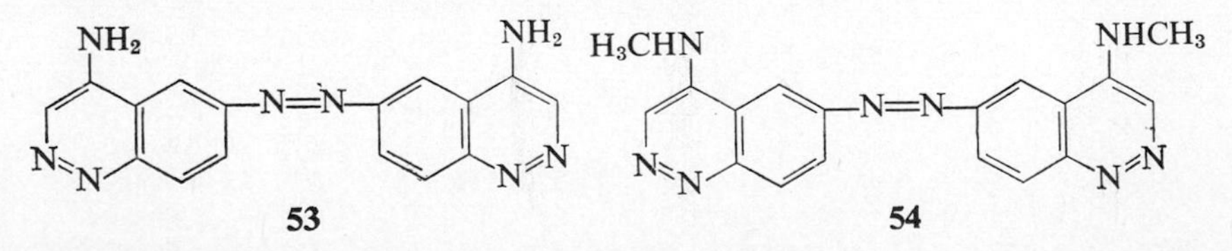

3,3′-dimethyl derivatives. This is interesting because the 3,3′-dimethyl derivative of the only active guanadine compound (**51**) is also inactive (7, 49–51).

Both 4,4′-diamino-6,6′-azocinnoline (**53**) and 4,4′-*bis*-methylamino-6,6′-azocinnoline (**54**) yield two isomeric dimethochlorides. In each case, both of the isomeric salts show trypanocidal activity against *T. congolense* but differ markedly in the extent of their action; only one of the isomeric dimethochlorides of **53** and only one of the dimethochlorides of **54** is curative (8, 10).

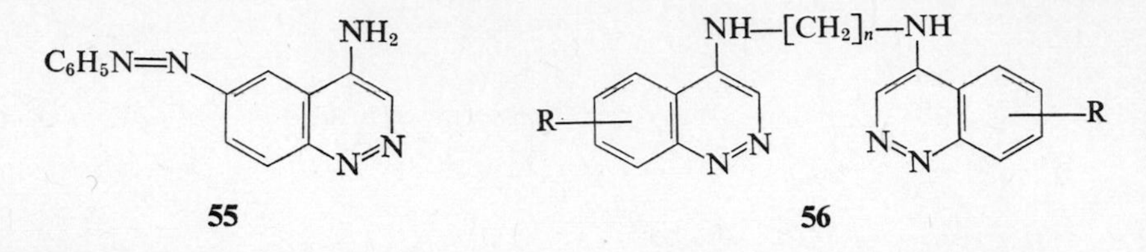

The monomethochloride of 4-amino-6-phenylazocinnoline (**55**) is inactive against *T. congolense* (8).

Several di(cinnolin-4-ylamino)alkanes (**56**; R = 6-Cl, 6,7-Cl_2, and 7,8-Cl_2; n = 2–10) have been prepared by reaction of a 4-phenoxycinnoline with a diamine. None showed trypanocidal or amebicidal activity (51a).

IV. Tables

TABLE 1H-1. Ultraviolet Absorption Spectra of Aminocinnolines

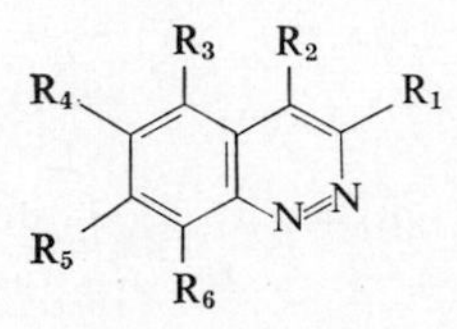

R_1	R_2	R_3	R_4	R_5	R_6	Solvent	λ_{max} (mμ)[a]	log E_{max}	Ref.
NH_2	H	H	H	H	H	95% ethanol	237	4.65	54
							384	3.56	
						95% methanol	237	4.59	13
							385	3.43	
						0.01*N* HCl	235	4.59	13
							310	2.88	
							315	2.83	
							405	3.45	
						0.01*N* NaOH	233	4.53	13
							372	3.36	

TABLE 1H-1. (*continued*)

R_1	R_2	R_3	R_4	R_5	R_6	Solvent	λ_{max} (mμ)[a]	log E_{max}	Ref.
						Water (pH = 7.0)	211	4.30	26
							234	4.61	
							283	*3.29*	
							304	*2.93*	
							374	3.44	
						Water (pH = 2.02)	211	4.08	26
							237	4.61	
							314	2.87	
							327	2.71	
							405	3.47	
NH_2	H	H	Cl	H	H	95% ethanol	220	4.31	29
							244	4.61	
							388	3.48	
NH_2	H	H	H	Cl	H	95% ethanol	*220*	*3.89*	29
							243	4.69	
							276	*3.72*	
							395	3.48	
H	H	H	NH_2	H	H	Water (pH = 9.22)	206	4.36	26
							256	4.50	
							315	3.58	
							362	3.74	
						Water (pH = 2.02)	220	4.36	26
							276	4.34	
							326	3.51	
							338	3.47	
							416	4.02	
H	H	H	H	NH_2	H	Water (pH = 9.22)	206	4.43	26
							249	4.60	
							283	3.77	
							385	3.43	
						Water (pH = 2.02)	210	4.36	26
							268	4.61	
							292	*3.63*	
							318	*3.23*	
							446	3.67	
H	H	H	H	H	NH_2	Water (pH = 9.22)	251	4.41	26
							327	3.21	
							395	3.40	
						Water (pH = 0.3)	*245*	*4.18*	26
							260	4.22	
							344	3.44	
							520	3.48	

TABLE 1H-1. (*continued*)

R_1	R_2	R_3	R_4	R_5	R_6	Solvent	λ_{max} (mμ)[a]	log E_{max}	Ref.
H	NH_2	H	H	H	H	Water (pH = 9.21)	212	4.52	26
							238	4.18	
							255	*3.77*	
							310	*3.71*	
							338	4.03	
						Water (pH = 4.35)	211	4.45	26
							236	4.16	
							262	3.83	
							295	3.24	
							345	4.12	
							360	*4.03*	
						Not stated	240	4.10	55
							325	*3.95*	
							345	4.06	
H	H	NH_2	H	H	H	Water (pH = 7.0)	*235*	*4.14*	26
							254	4.36	
							334	3.28	
							382	3.34	
						Water (pH = 0.3)	236	4.25	26
							280	4.00	
							348	3.32	
							490	3.00	
H	$NHCOCH_3$	H	H	H	H	Not stated	226	4.63	55
							251	3.64	
							303.5	3.90	
							328.5	3.87	
H	$NHCOCH_3$	H	NO_2	H	H	Not stated	229	4.70	55
							260	4.23	
							305	3.86	
							362	3.86	
H	C_6H_5NH	H	H	H	H	Not stated	248	4.16	55
							364	4.18	
H	C_6H_5NH	H	NO_2	H	H	Not stated	247	4.39	55
							278	4.07	
							343.5	3.79	
							418	4.07	
H	NH_2	H	NO_2	H	H	Not stated	255.2	4.20	55
							272	4.17	
							326.5	3.72	
							402	3.91	

[a] Italics refer to shoulders or inflections.

TABLE 1H-2. Basic Strengths of Aminocinnolines

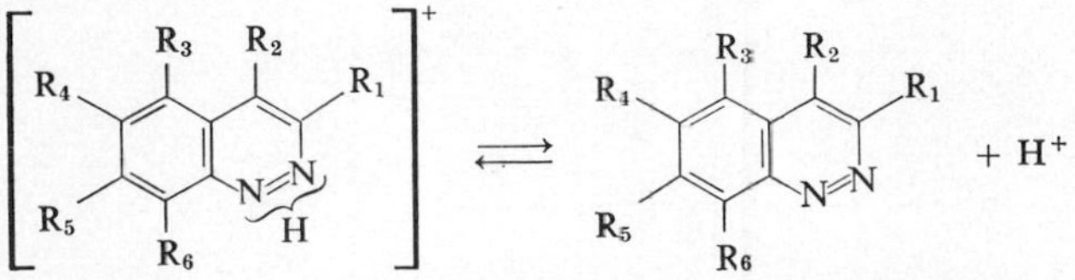

R_1	R_2	R_3	R_4	R_5	R_6	pK_a^a	Solvent	Temp (°C)	Ref.
H	NH_2	H	H	H	H	6.84	Water	20	27
						6.85	Water	20	26
						6.26	50% aq. ethanol	25	52
						6.25	50% aq. ethanol	24–25	53
H	NH_2	H	NO_2	H	H	5.1	50% aq. ethanol	20	53
						5.08	50% aq. ethanol	25	52
H	NH_2	H	Cl	H	H	5.4	50% aq. ethanol	28	53
H	NH_2	H	NH_2	H	H	6.86	50% aq. ethanol	25	52
H	C_6H_5NH	H	H	H	H	5.31	50% aq. ethanol	21–22	52
H	$C_6H_5CH_2NCH_2CH_2N(CH_3)_2$	H	H	H	H	pK_a' 7.33[b] pK_a'' 3.96	50% aq. ethanol	20 ± 0.2	43
H	p-$CH_3OC_6H_4CH_2NCH_2CH_2N(CH_3)_2$	H	H	H	H	pK_a' 7.43 pK_a'' 4.11	50% aq. ethanol	20 ± 0.2	43
NH_2	H	H	H	H	H	3.70	Water	20	26, 29
						3.63	Water	20	13
NH_2	H	H	Cl	H	H	3.24	Water	20	29
NH_2	H	H	H	Cl	H	3.28	Water	20	29
H	H	NH_2	H	H	H	2.70	Water	20	26
H	H	H	NH_2	H	H	5.04	Water	20	26
H	H	H	H	NH_2	H	4.85	Water	20	26
H	H	H	H	H	NH_2	3.68	Water	20	26

[a] pK_a = negative logarithm of the ionization constant.

[b] pK_a' refers to the basicity of a nitrogen atom in the aliphatic substituent; pK_a'' refers to the basicity of a ring nitrogen atom.

TABLE 1H-3. Aminocinnolines

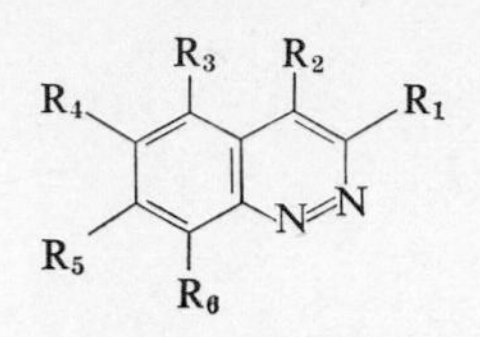

R_1	R_2	R_3	R_4	R_5
C_6H_5	$C_6H_5CH_2NH$—	H	H	H
C_6H_5	$C_6H_5CH_2CH_2NH$—	H	H	H
C_6H_5	N—	H	H	H
C_6H_5	$HOCH_2CH_2NH$—	H	H	H
C_6H_5	*p*-$CH_3OC_6H_4NH$—	H	H	H
C_6H_5	—$NHC_6H_4COOCH_3$-*o*	H	H	H
C_6H_5	—$NHCH_2CH_2OCH_2CH_2OH$	H	H	H
C_6H_5	NH_2NH—	H	H	H
C_6H_5	$(C_2H_5)_2NCH_2CH_2NH$—	H	H	H
C_6H_5	$(C_2H_5)_2N(CH_2)_3CHNH$— (with CH_3 on CH)	H	H	H
C_6H_5	O N—$(CH_2)_3NH$—	H	H	H
C_6H_5	$(CH_3)_2N(CH_2)_3NH$—	H	H	H
C_6H_5	$(CH_3)_2NCH_2CH_2NH$—	H	H	H
C_6H_5	$(CH_3)_2NCH_2CH_2$— N—	H	H	H
p-ClC_6H_4	$(CH_3)_2N(CH_2)_3NH$—	H	H	H
C_6H_5	N—CH_2CH_2NH—	H	H	H
C_6H_5	$(CH_3)_2N(CH_2)_3N$— (with CH_3 on N)	H	H	H
C_6H_5	$(CH_3)_2NCH_2CH_2N$— (with CH_3 on N)	H	H	H
C_6H_5	$H_2NCH_2CH_2NH$—	H	H	H

R_6	Prep.[a]	MP (°C)	Remarks[b]	Ref.
H	A	141–142	(Ether)	1, 2
H	A	145–146	(Benzene)	1, 2
H	A	166.5–167.5	(Methanol)	1
H	A	144–145	White plates (methanol-benzene)	1, 2
H	A	150–151	(Ether-ligroin or butanone)	1
H	A	163–165	(Methanol)	1
H	A	112–113	Pink plates (benzene)	1, 2
H	A	—	Hydrochloride deriv., mp 209–210°	1
H	A	82–84	Yellow blades (pentane)	1, 4
H	A	57–59	(Ligroin-acetone)	1, 4
H	A	145–146	(Benzene-ligroin)	1, 4
H	A	—	Dihydrochloride deriv., mp 204–206°	1, 4
H	A	105–107	White needles (hexane)	1, 4
H	A	103–105	(Ligroin–butanone)	1
H	A	118–119	(Ether–ligroin or butanone)	1, 4
	A	138–140	(Ether–ligroin or butanone)	1, 4
	A	—	Dihydrochloride deriv., mp 215–218°	1
	A	65–67	(Ligroin–acetone); dihydrochloride deriv., mp 195–197°	1, 4
	A	—	Dihydrochloride deriv., mp 255–257°	1, 4

TABLE 1H-3. (*continued*)

R_1	R_2	R_3	R_4	R_5
C_6H_5	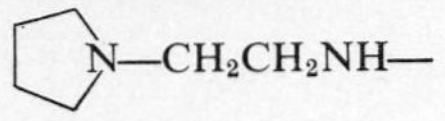 N—CH_2CH_2NH—	H	H	H
C_6H_5	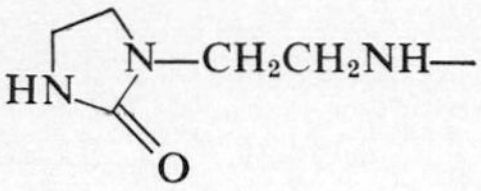 HN N—CH_2CH_2NH— (O)	H	H	H
C_6H_5	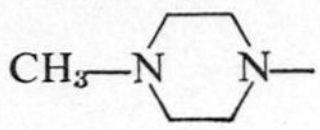 CH_3—N N—	H	H	H
C_6H_5	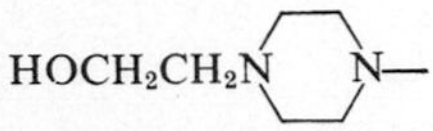 $HOCH_2CH_2$N N—	H	H	H
C_6H_5	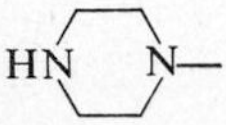 HN N—	H	H	H
C_6H_5	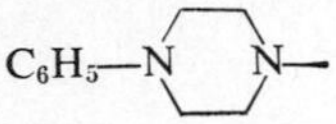 C_6H_5—N N—	H	H	H
C_6H_5	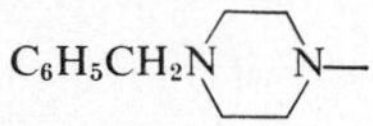 $C_6H_5CH_2$N N—	H	H	H
C_6H_5	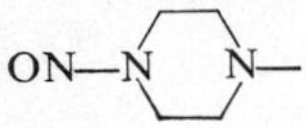 ON—N N—	H	H	H
C_6H_5	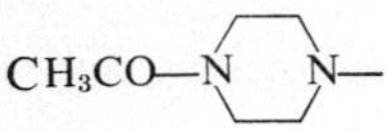 CH_3CO—N N—	H	H	H
C_6H_5	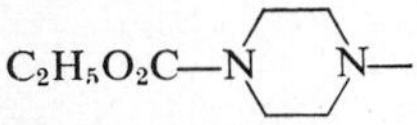 $C_2H_5O_2C$—N N—	H	H	H
p-ClC_6H_4	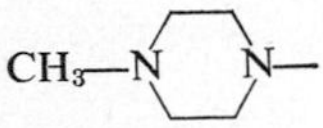 CH_3—N N—	H	H	H
p-$CH_3OC_6H_4$	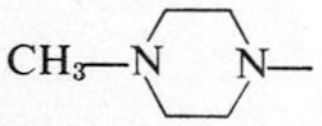 CH_3—N N—	H	H	H
C_6H_5	 *p*-ClC_6H_4CON N—	H	H	H
C_6H_5	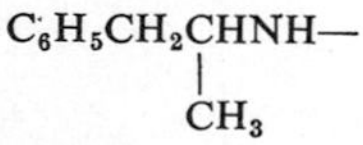 $C_6H_5CH_2CH(CH_3)NH$—	H	H	H
H	C_6H_5O—	H	H	NH_2
H	NH_2	H	H	NH_2
H	NH_2	H	H	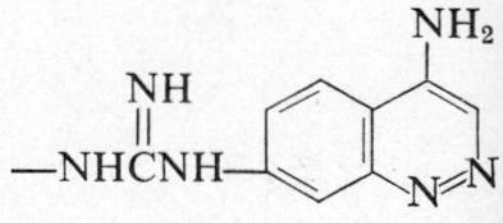 —NHC(=NH)NH— (NH_2, N=N)

Prep.[a]	MP (°C)	Remarks[b]	Ref.
A	82–84	Yellow prisms (hexane)	1, 4
A	168–170	(Benzene)	1
A	128–129	Yellow prisms (hexane)	1, 3
A	170–173 174–176	Brown needles (benzene)	1, 3
A	171–173	Yellow clusters (benzene)	1, 3
A	181–182	(Methanol)	1, 3
A	110–111	Yellow clusters (hexane)	1, 3
A	204–205	(Methanol)	1
B	209–210	Yellow clusters (acetone–hexane)	1, 3
A	128–129	(Ethyl acetate)	1, 3
A	151–153	(Ether–ligroin or butanone)	1, 3
A	156–157	(Ether)	1, 3
C	193–196	(Ether)	1
A	152.5–153.5	Yellow needles (ether)	2
D	179–180	Greenish-yellow plates (benzene)	5
E	250 (dec.)	Yellow needles (water); methiodide deriv., mp 262–263°	5
F	238–240	Identity uncertain; trihydrochloride deriv., mp 330° (dec.)	5

TABLE 1H-3. (*continued*)

R_1	R_2	R_3	R_4	R_5
H	$NHCOCH_3$	H	H	NO_2
H	NH_2	H	NH ‖ —NHCNH— NH_2 N=N	H
H	$C_6H_5O—$	H	H	H
H	$C_6H_5O—$	NH_2	H	H
H	OH	H	H	NH_2
H	$CH_3COO—$	H	H	$CH_3CONH—$
H	OH	H	H	H
H	$CH_3COO—$	H	H	H
H	NH_2	H	H	NO_2
H	NH_2	H	H	H
H	$CH_3CONH—$	H	H	$CH_3CONH—$
H	NH_2	H	H	H
H	$CH_3CONH—$	H	H	H
H	OH	H	H	$CH_3CONH—$
H	—NHOH	H	H	NOH ‖ $CH_3C—$
H	—NHOH	H	H	$CH_3CONH—$
H	$CH_3NH—$	H	$NHCH_3$ —N=N— N=N	H
H	NH_2	H	NH_2 —N=N— N=N	H
H	NH_2	H	NO_2	H

R_6	Prep.[a]	MP (°C)	Remarks[b]	Ref.
H	G	235	(Ethanol); methiodide deriv., mp 188–189°	5
H	F, H	250–255 245–250	Trihydrochloride deriv., mp 335; (dec.)° dimethiodide deriv., mp 270° (dec.)	5, 7, 49
NH_2	D	130	Rhombohedral crystals (benzene–ligroin)	5
H	D	199–200 (dec.	Yellow needles (benzene)	5
H	D	276–277	Fawn-colored needles (aq. ethanol)	6
H	G	>330	Colorless needles (aq. ethanol)	6
NH_2	D	290–291	Yellow needles (aq. acetic acid)	6
CH_3CONH—	G	282–283	Tan needles (aq. ethanol)	6
H	E	300–301 (dec.)	Yellow leaflets (aq. ethanol); methiodide deriv., mp 236–238° (dec.)	5, 6
NO_2	E, I	235–236 (dec.) 242–243 (dec.)	Yellow needles (aq. ethanol) or rust-red prismatic needles (ethanol)	6, 59
H	J	312 (dec.)	Colorless needles (aq. ethanol)	6
NH_2	D	167–168	Buff solid (ether–ligroin)	6
CH_3CONH—	G	299–300	Greenish-yellow needles (aq. ethanol)	6
H	K	>330	Colorless crystals (ethanol)	6
H	L	264–265 (dec.)	Yellow crystals (ethanol)	6
H	M	229–230	Yellow needles (ethanol)	6
H	N	>320	Scarlet solid; dihydrochloride deriv., mp >320°; dimethiodide, mp 274–278° (dec.); dimethochloride has two isomers, mp 296–300° (dec.) and mp 291–294° (dec.)	7, 10
H	N	>320	Red needles; dihydrochloride, mp >320°; dimethiodide, mp 284–288° (dec.); dimethochloride has two isomers, mp 282–285° (dec.) and mp 312–314° (dec.)	7, 8
H	I, O, E	288–290 (dec.)	1-Methiodide, deriv., mp 207° (dec.)	7, 9, 30, 48, 51
		290–291 297–298 (dec.)	1-Methomethylsulfate, mp 234–236°	

TABLE 1H-3. (*continued*)

R_1	R_2	R_3	R_4	R_5
H	NH_2	H	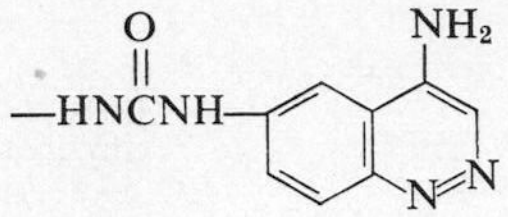 O ‖ —HNCNH— NH₂ N=N	H
H	NH_2	H	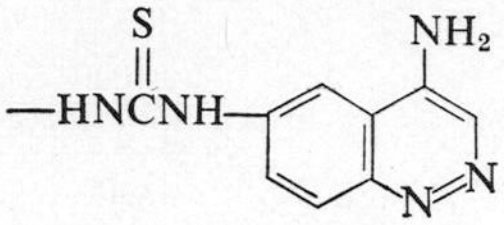 S ‖ —HNCNH— NH₂ N=N	H
H	NH_2	H	$C_6H_5N{=}N$—	H
H	CH_3CONH—	H	$C_6H_5N{=}N$—	H
H	NH_2	H	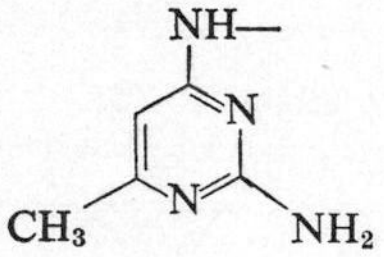 NH— N N CH_3 NH_2	H
H	NH_2	H	NH_2	H
H	NH_2	H	CH_3CONH—	H
H	NH_2	H	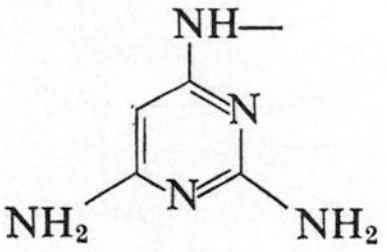 NH— N N NH_2 NH_2	H
H	NH_2	H	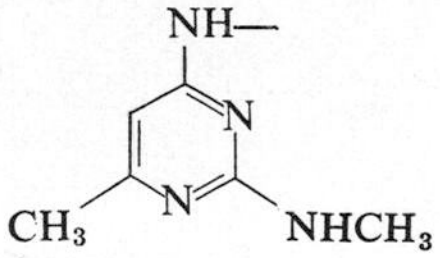 NH— N N CH_3 $NHCH_3$	H
CH_3	NH_2	H	NH_2	H
H	CH_3NH	H	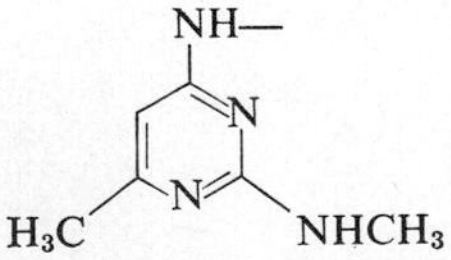 NH— N N H_3C $NHCH_3$	H
H	CH_3NH	H	NO_2	H
H	CH_3NH	H	NH_2	H

R_6	Prep.[a]	MP (°C)	Remarks[b]	Ref.
I	P	—	Yellow needles (aq. acetone); dimethochloride, mp 304–305° (dec.) and 295–297° (dec.)	7, 49
I	Q	—	Dihydrochloride deriv., mp 240–245° (dec.); dimethochloride, mp 205–280°; dimethiodide, mp 270–275° (dec.)	7, 49
I	E	300–301 (dec.)	Golden needles (2-ethoxyethanol); methiodide deriv., mp 253–255° (dec.); methochloride deriv., mp 250–255° (dec.)	8
I	G	271–275 (dec.)	Red leaflets (2-ethoxyethanol)	8
I	R	320 (dec.)	(Aq. ethanol); dimethiodide deriv., mp 305° (dec.); dimethochloride deriv., mp 298° (dec.); dihydrochloride monohydrate deriv., mp >370°	9, 47
I	D	262–264 (dec.) 266–267 268–270	Diacetyl deriv., mp 275–277°; 1-methiodide deriv., mp 284°	9, 48, 51
I	G	361 (dec.)	(Water)	9
I	G	—	3′-Methiodide hydroiodide deriv., mp 342° (dec.)	9
I	R	—	1,1′-Dimethiodide deriv., mp 310° (dec.)	9
I	D	270 (dec.) 272–274	Hydrochloride deriv., mp 325° (dec.); diacetyl deriv., mp 297–298° (dec.)	9, 51
I	R	—	1,1′-Dimethiodide deriv., mp 272–274°; 1′-methiodide trihydrate deriv., mp 296° (dec.)	9
I	A, S	>360 345–346 (dec.)	Yellow needles (isopropanol)	9, 36, 48
I	D	300	(Water); 1-methiodide deriv., mp 276–278°	9

TABLE 1H-3. (*continued*)

R_1	R_2	R_3	R_4	R_5
H	NH_2	H	NH— N N $NHCH_3$	H
H	NH_2	H	NH_2 N N NH—	H
H	NH_2	H	NH— N N C_2H_5 $NHCH_3$	H
H	NH_2	H	NH— N N CH_3 $NHCH(CH_3)_2$	H
CH_3	NH_2	H	NH— N N H_3C NH_2	H
H	CH_3NH	H	H	H
H	$CH_3CH_2CH_2NH$	H	H	H
H	NH_2	H	H	H
H	$NH_2NH—$	H	H	H
H	$CH_3(CH_2)_3NH—$	H	H	H
NH_2	H	H	H	H
$CH_3CONH—$	H	H	H	H
Cl	$NH_2NH—$	H	H	H
H	$NH_2NH—$	H	Cl	H
H	$NH_2NH—$	H	Br	H
Cl	$C_6H_5NHNH—$	H	H	H

R_6	Prep.[a]	MP (°C)	Remarks[b]	Ref.
I	R	—	1,1′-Dimethiodide deriv., mp 274–278°	9
I	T	260 (as dihydrate)	Dihydrochloride deriv., mp >360°; 1,1′-dimethiodide deriv., 322° (dec.)	9, 47
I	R	—	1,1′-Dimethiodide deriv., mp 320° (dec.)	9
I	R	—	1′-Methiodide deriv., mp 270° (dec.); 1,1′-dimethiodide deriv., mp 320° (dec.)	9, 47
I	R	—	1′-Methiodide dihydrate deriv., mp 270°; (dec.); 1,1′-dimethiodide dihydrate deriv., mp 284° (dec.)	9, 47
I	A, U	228–230	Colorless needles (water); hydrochloride deriv., mp 284°; picrate deriv., mp 253–255°	11, 36
I	U	169–170	(Benzene-cyclohexane)	11
I	A, D, U	208–209 213–214 215.5–218	Yellow crystals (water); picrate deriv., mp 280°; perchlorate deriv., mp 215–217°; acetyl deriv., mp 272–273° or 277° (dec.); benzoyl deriv., mp 222–224°; 2-methiodide deriv., mp 256–258°; 1-methiodide deriv., mp 206–208°	11, 25, 28, 34, 45, 56, 57
I	A, U, V, EE, S, FF	298	Exists in two forms: first, mp 296–297° (dec.) or 287–289° (dec.); second form, mp 226–227° or 229°. Both give same hydrochloride, mp 244–245° and benzal deriv., mp 307°	11, 12, 14, 19, 1a, 1b, 11a
I	V	154–155	Picrate deriv., mp 134–135°	12
I	D, W, X	163–164.5 165–166.5	Yellow needles (benzene or ethyl acetate); perchlorate deriv., mp 183–184°	13, 17, 28, 54, 60, 14a
I	G	225–226	Silky needles (water)	13
I	A	>300	Orange leaflets (ethanol)	14
I	A, S, FF	>320	Orange-yellow crystals (ethanol)	14, 1b
I	A	>300	Orange crystals (ethanol)	14
I	A	134–135	Brown leaflets (aq. ethanol)	14

TABLE 1H-3. (*continued*)

R_1	R_2	R_3	R_4	R_5
Cl	NH_2	H	H	H
H	p-$CH_3C_6H_4SO_2NHNH$—	CH_3O	H	H
H	H	NH_2	H	H
H	p-$CH_3C_6H_4SO_2NHNH$—	H	CH_3O	H
H	H	H	NH_2	H
H	p-$CH_3C_6H_4SO_2NHNH$—	H	H	CH_3O
H	H	H	H	NH_2
H	p-$CH_3C_6H_4SO_2NHNH$—	H	H	H
H	H	H	H	H
H	$(CH_3)_2N$	H	H	H
$(CH_3)N$	H	H	H	H
Br	NH_2	H	H	H
H	H	H	H	H
NO_2	NH_2	H	H	H
NH_2	NH_2	H	H	H
NO_2	C_6H_5NH—	H	H	H
H	CH_3CONH—	H	H	H
H	CH_3	H	H	H
H	CH_3	H	H	H
H	NH_2NH—	H	NO_2	H
H	$(CH_3)_2C{=}NNH$—	H	NO_2	H
H	CH_3CONH—	H	H	H
H	CH_3	H	H	H
H	CH_3	H	H	H
H	CH_3	NH_2	H	H
H	$(CH_3)_2NCH_2CH_2NCH_2C_6H_5$ (N bond below)	H	H	H
H	$(CH_3)_2NCH_2CH_2NCH_2C_6H_4OCH_3$-$p$ (N bond below)	H	H	H
H	$(CH_3)_2NCH_2CH_2NC_2H_5$ (N bond below)	H	H	H

R_6	Prep.[a]	MP (°C)	Remarks[b]	Ref.
H	A, E	228–229	White needles (ethanol)	14
H	A	—	Hydrochloride deriv., mp 221° (dec.)	15
H	D, Y	157–159 160–161	Yellow prisms (benzene–cyclohexane) or yellow-green needles (chloroform)	15, 20, 21
H	A	199–201 (dec.) (as hydrate)	(Ethanol)	15
H	D, Y	201–202 203–204	Yellow prisms (benzene) or yellow scales (benzene–ethanol)	15, 20, 21
H	A	—	Hydrochloride deriv., mp 169–172° (dec.)	15
H	Y	191–192	Yellow needles (ethyl acetate)	15
CH_3O	A	—	Hydrochloride deriv., mp 169–172° (dec.)	16
NH_2	D, Y	91–92	Yellow needles (ligroin); picrate deriv., mp 236–238°	16, 20, 21
H	—	—	Brown oil, solidifying below 30°; picrate deriv., mp 209–210°	14a
H	—	107–108	Yellow plates (Light petroleum)	14a
H	—	246–247	White needles (ethanol); methiodide deriv., mp 287–288°	14a
CH_3CONH—	G, J	173–175 177–178	Yellow prisms (benzene–ligroin); picrate deriv., mp 238–239; (dec.)	16, 33
H	Z	308–308.5	Yellow, cottonlike solid (ethanol)	18
H	D	220–220.5 (dec.)	(Water); monohydrochloride deriv., mp 316–317° (dec.)	18
H	A	187–188	Yellow needles (ethanol)	18
H	G	277 (dec.)	(Ethanol); 2-methiodide deriv., mp 248–250°	25, 40
NH_2	D	126–127 128–133 133–134	Orange prisms (ether–ligroin)	31, 32, 39
-NHCH=C-$(COOC_2H_5)_2$	AA	155–156	(Ethanol)	32
H	A	>330	Maroon solid; hydrochloride deriv., mp 215° (dec.)	33
H	BB	179–180	Orange, hairlike needles (aq. acetone)	33
NO_2	G	291–292	Colorless needles (methyl ethyl ketone)	36
CH_3CONH—	G	158	Needles (ethanol)	39
C_6H_5CONH—	CC	170	Prisms (ethanol)	39
H	D	165–166	Yellow solid; orange solid after exposure to moist air and then melts at 174–176°	39
	A	—	Boiling point 190–195° at 0.2 mm; picrate deriv., mp 202–204; (dec.)	43
	A	—	BP 220–230° at 0.02 mm; picrate deriv., mp 137–139°	43
	A	—	BP 140–148° at 0.05 mm; picrate deriv., mp 205.5–208° (dec.)	43

TABLE 1H-3. (*continued*)

R_1	R_2	R_3	R_4	R_5
H	NH_2	H	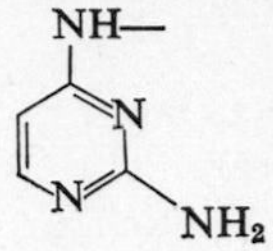 NH— N N NH_2	H
H	$CH_3CH_2CH_2NH—$	H	NO_2	H
H	OH	H	NH_2	H
COOH	OH	H	NH_2	H
H	$—HN(CH_2)_6NH$ 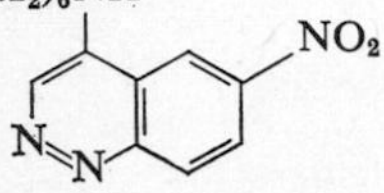 NO_2 N=N	H	NO_2	H
CH_3	NH_2	H	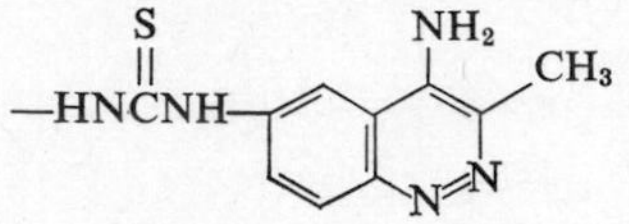 S ‖ —HNCNH— NH_2 CH_3 N=N	H
CH_3	NH_2	H	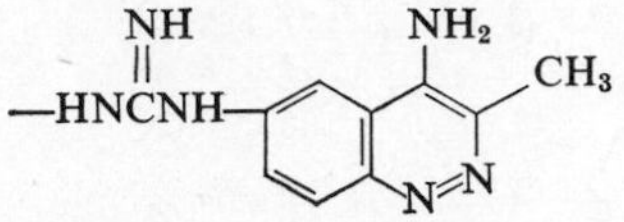 NH ‖ —HNCNH— NH_2 CH_3 N=N	H
CH_3	NH_2	H	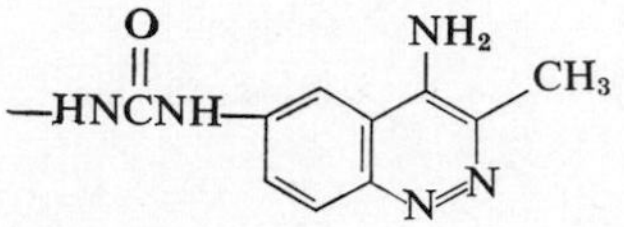 O ‖ —HNCNH— NH_2 CH_3 N=N	H
CH_3	NH_2	H	NO_2	H
H	$C_6H_5NH—$	H	NH_2	H
H	$C_6H_5NH—$	H	$CH_3CONH—$	H
NH_2	CH_3	H	H	H
H	*p*-$CH_3C_6H_4SO_2NHNH—$	H	H	H
CH_3	*p*-$CH_3C_6H_4SO_2NHNH—$	H	H	H
Cl	*p*-$CH_3C_6H_4SO_2NHNH—$	H	H	H
Br	*p*-$CH_3C_6H_4SO_2NHNH—$	H	H	H
H	*p*-$CH_3C_6H_4SO_2NHNH—$	H	H	H
H	*p*-$CH_3C_6H_4SO_2NHNH—$	H	NO_2	H
H	*p*-$CH_3C_6H_4SO_2NHNH—$	H	H	NO_2
H	$C_6H_5NH—$	H	H	CH_3
H	NH_2	H	H	CH_3
H	NH_2	H	H	H
H	$C_6H_5NH—$	H	H	H

6	Prep.[a]	MP (°C)	Remarks[b]	Ref.
	R	—	1-Methiodide monohydrate deriv., mp 320° (dec.)	47
	S	287 (dec.)		48
	D	265–268 269–270	Yellow needles (water)	48, 61
	D	299–300 (dec.)	(Aq. hydrochloric acid)	48
	S	>360	Dihydrochloride deriv., mp 309–310° (dec.)	48
	Q	—	Dihydrochloride deriv., mp 290° (dec.); dimethochloride deriv., mp 313–314° (dec.)	49
	F	240–270 (dec.)	Yellow needles (aq. ethanolic ammonia); trihydrochloride deriv., mp 315° (dec.); dimethiodide deriv., mp 251–253 and 280°	49
	P	—	Dimethochloride deriv., mp 288–290° (dec.)	49
	I	320 (dec.)	Orange-yellow solid	51
	D	259–261 (dec.)	Yellow gelatinous mass; hydrochloride deriv., mp 301–302° (dec.)	51
	G	289–290 (dec.)	Light brown needles	51
	D	159.5–160	Yellow needles (benzene)	54
	A	—	Hydrochloride deriv., mp 224–226;	58
	A	—	Hydrochloride deriv., mp 186–187° (dec.)	58
	A	167–169	Colorless crystals (acetic acid)	58
	A	187–189 (dec.)	Pink solid, impure	58
O_2	A	195–196 (dec.)	Brown needles (acetic acid)	58
	A	190–192 (dec.) 212–213 (dec.)	Red prisms (acetic acid) or orange-red needles (aq. acetic acid)	33, 58
	A	195–196 (dec.)	Orange crystals (acetic acid)	58
O_2	A	262–263 (dec.)	Greenish-yellow prismatic needles (ethanol)	59
O_2	—	300 (dec.)	Sandy microprisms (ethanol)	59
H_2	—	145–152	Colorless blades (water)	59
H_3	A	231–232	Lemon-yellow needles (ethanol)	59

TABLE 1H-3. (*continued*)

R_1	R_2	R_3	R_4	R_5
NH_2	H	H	H	Cl
NH_2	H	H	Cl	H
Br	*o*-HOC_6H_4NH	H	CH_3O	CH_3O
H	C_6H_5NH	H	CH_3O	CH_3O
F	NH_2	F	F	F
H	C_6H_5NH	H	H	H
H	—NHOH	H	H	H
H	CH_3NH	H	Cl	H
H	NH_2	H	Cl	H
H	CH_3NH	H	CH_3O	H
H	CH_3NH	H	OH	H
p-$BrC_6H_4N{=}N$—	OH	H	Cl	H
NH_2	OH	H	Cl	H
NO_2	NH_2	H	Cl	H
Cl	NH_2	H	Cl	H
NH_2	NH_2	H	Cl	H
H	OH	H	NH_2	H
H	OH	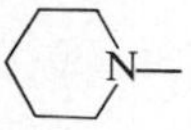 N—	NO_2	H
H	CH_3NH	CH_3NH	NO_2	H
H	OH	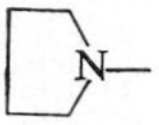 N—	H	H
H	$CH_3CH_2CH_2NH$	H	Cl	H
H	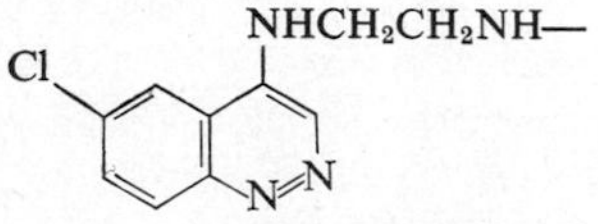 $NHCH_2CH_2NH$— Cl N=N	H	Cl	H
H	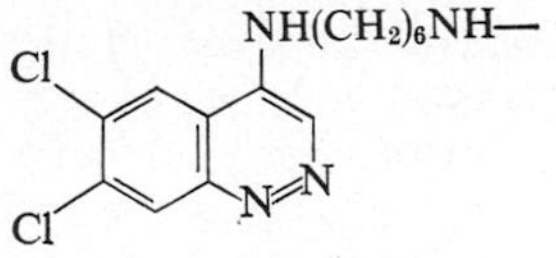 $NH(CH_2)_6NH$— Cl Cl N=N	H	Cl	Cl
H	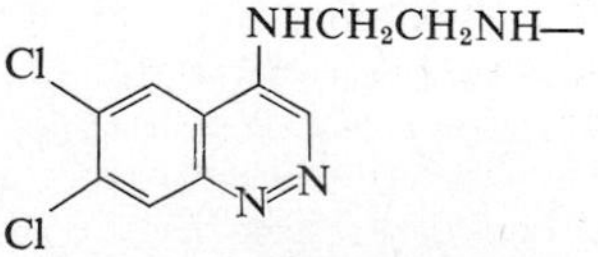 $NHCH_2CH_2NH$— Cl Cl N=N	H	Cl	Cl
H	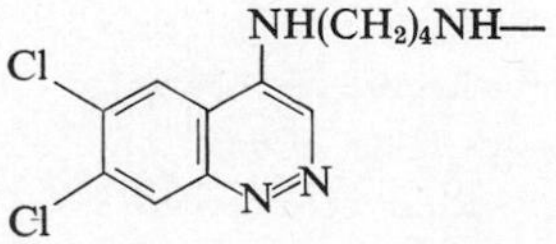 $NH(CH_2)_4NH$— Cl Cl N=N	H	Cl	Cl

Prep.[a]	MP (°C)	Remarks[b]	Ref.
D	202 (dec.)	Yellow plates (benzene)	60
D	215 (dec.)	Yellow needles (benzene)	60
A	256	Yellow needles (dioxane)	62
A	223	Yellow prisms (ethanol)	62
DD	179–181	—	63
EE, U	240	—	1a, 11a
U	210–210.5	—	11a
FF, S	283–285	(Ethanol); 2-methiodide deriv., mp 243–244°	1b, 51a
FF, E	260–264, 277	—	1b, 51a
FF, S	235–237	—	1b
GG	324–326 (dec.)	(Water)	1b
HH	307–309 (dec.)	(Acetic acid)	32a
D	335–337 (dec.)	(Acetic acid)	32a
E, A	337–338, 341 (dec.)	(2-Ethoxyethanol)	32a
E	301–302 (dec.)	(Acetic acid)	32a
D	>293 (dec.)	(Dimethylformamide)	32a
D	310–311 (dec.)	(Nitrobenzene)	32a
II	222–223 (dec.)	Orange-red solid (toluene)	32a
II	—	Acetate deriv., mp 268–269 (dec.)	32a
II	202–204	(Toluene)	32a
S	234–235 (dec.)	(Ethanol)	51a
S	>360	Dihydrochloride deriv., mp >320 (dec.)	51a
S	>360	(Dimethylformamide)	51a
S	>340	—	51a
S	>290	—	51a

TABLE 1H-3. (*continued*)

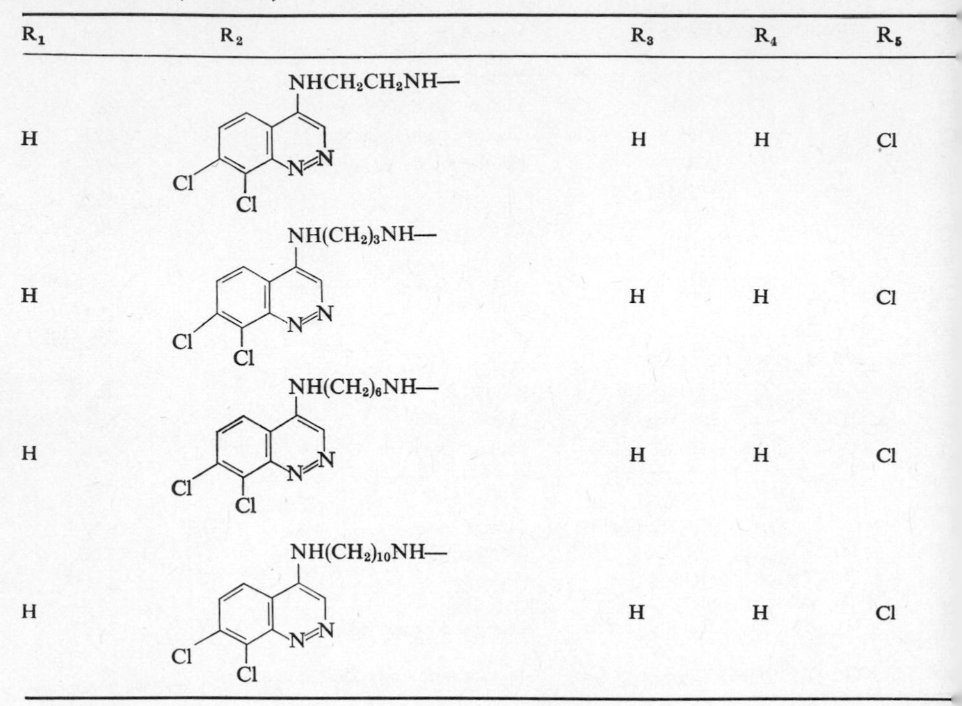

R_1	R_2	R_3	R_4	R_5
H	$NHCH_2CH_2NH$—	H	H	Cl
H	$NH(CH_2)_3NH$—	H	H	Cl
H	$NH(CH_2)_6NH$—	H	H	Cl
H	$NH(CH_2)_{10}NH$—	H	H	Cl

[a] A, by the interaction of the appropriately substituted 4-chlorocinnoline derivative with an ami
from 4-(*N*-piperizinyl)cinnoline and *p*-chlorobenzoyl chloride. D, by chemical or catalytic reduction of
4-aminonitrocinnoline to give the diaminocinnoline. E, by heating the appropriately substituted 4-pheno
of the corresponding diaminocinnoline with thiocarbonyl chloride followed by treatment of the result
and acetic anhydride. H, by hydrolysis of the condensation product from 4,6-diaminocinnoline and
ammonia, ammonium carbonate, or ammonium acetate, all in phenol solution. J, by reductive acylation
anhydride, and sodium acetate. K, by the Schmidt rearrangement of 7-acetyl-4-hydroxycinnoline w
amine. M, by the Beckmann rearrangement of 7-acetyl-4-hydroxylaminocinnoline oxime in a mixt
6,6′-azocinnoline with methylamine in phenol (or with ammonia in phenol, depending on the produ
chloride in acetone or water. Q, by condensation of the appropriate diaminocinnoline with thiocarbo
chloropyrimidine. S, by heating the appropriate 4-phenoxycinnoline with an amine in phenol or
responding 1-methiodide salts. U, from 4-methylsulfonylcinnoline and the appropriate amine. V, fr
aqueous ammonia in the presence of cupric sulfate. X, from 3-acetylcinnoline and sodium azide in c
sponding hydroxycinnoline, ammonium hydrogen sulfite (or ammonium sulfite), and ammonia, and
ethoxymethylenemalonate in benzene. BB, from 4-hydrazino-6-nitrocinnoline and acetone. CC,
cinnoline with ammonia. EE, by reaction of 4-cinnolinecarbonitrile with an amine or hydrazine. FF, fr
cinnoline in refluxing hyriodic acid. HH, from reaction of 6-chloro-4-hydroxycinnoline with *p*-bron
amine.

[b] Solvent employed in recrystallization is enclosed in parentheses.

R6	Prep.[a]	MP (°C)	Remarks[b]	Ref.
Cl	S	>355	—	51a
Cl	S	>360	—	51a
Cl	S	>360	—	51a
Cl	S	280–285 (chars)	—	51a

ydrazine, or *p*-toluenesulfonylhydrazide. B, from 4-(*N*-piperizinyl)cinnoline and acetic anhydride. C, orresponding nitrocinnoline derivative to give the monoaminocinnoline, or by reduction of the appropriate innoline with ammonium acetate or with ammonia and ammonium chloride in phenol. F, by condensation cinnolinylthiourea with mercuric oxide and methanolic ammonia. G, from the appropriate aminocinnoline hloromethylenebenzamide ($Cl_2C = NCOC_6H_5$). I, by heating the 4-chlorocinnoline derivative with -amino-7-nitrocinnoline (or of 8-nitrocinnoline, depending upon the product) with zinc dust, acetic odium azide in concentrated sulfuric acid. L, by the treatment of 7-acetyl-4-phenoxycinnoline with hydroxyl- f acetic acid, hydrochloric acid, and acetic anhydride (see text). N, from the interaction of 4,4′-diphenoxy- , by nitration of 4-aminocinnoline. P, from the appropriate 4,6-diaminocinnoline derivative and carbonyl hloride. R, from the corresponding 6-aminocinnoline derivative and the appropriately substituted 4- imethylformamide. T, from 4,6-diaminocinnoline and 4-amino-2-chloropyrimidine,or from their cor- methylsulfinylcinnoline and the appropriate amine. W, by heating 3-chloro- or 3-bromocinnoline with ntrated sulfuric acid followed by hydrolysis of the resultant 3-acetamidocinnoline. Y, From the corre- om 3-nitrocinnoline and hydroxylamine in ethanol. AA, from 8-amino-4-methylcinnoline and diethyl enzoylation (Schotten-Bauman) of 8-amino-4 methylcinnoline. DD, by treatment of hexafluoro- 4-mercaptocinnoline and the appropriate amine. GG, by demethylation of the corresponding 6-methoxy- enzenediazonium chloride. II, by treatment of the corresponding 5-chlorocinnoline with the appropriate

TABLE 1H-4. Proton Magnetic Resonance Spectra of Aminocinnolines[a]

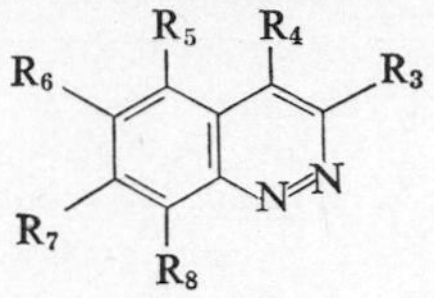

R_3	R_4	R_5	R_6	R_7	R_8	Solvent	Ref.
NH_2	H	H	H	H	H	DMSO-d_6	28a
—	3.00(S)	←——	2.47(M)	——→	1.83(D) $J = 8$		
H	NH_2	H	H	H	H	DMSO-d_6	28a
1.27(S)	2.63(S)	←—— 2.35(M)	——→	←—— 1.83(M)	——→		
H	H	NH_2	H	H	H	DMSO-d_6	28a
0.87(D) $J = 6$	1.75(D) $J = 6$	—	3.18(Q) $J = 4$	2.48(D) $J = 4$	2.48(D) $J = 4$		
H	H	H	H	H	NH_2	$CDCl_3$	28a
0.97(D) $J = 6$	3.15(M)	2.45(D) $J = 6$	3.15(M)	2.70(D) $J = 7$	4.62(S)		

[a] Chemical shift values are given immediately below the proton(s) to which they refer and are in τ units. The multiplicity of each absorption, when stated, is given in parentheses following the τ values as S = singlet, D = doublet, Q = quartet, M = multiplet; J = coupling constant in cps. Chemical shift positions refer to tetramethylsilane as standard.

References

1. H. S. Lowrie, *J. Med. Chem.*, **9**, 670 (1966).
1a. T. Watanabe, *Yakugaku Zasshi*, **89**, 1167 (1969); *Chem. Abstr.*, **72**, 3452 (1970).
1b. H. J. Barber and E. Lunt, *J. Chem. Soc.* (*C*), **1968**, 1156.
2. H. S. Lowrie (to G. D. Searle & Co.), U.S. Pat. 3,239,526 (Cl. 260-250), Mar. 8, 1966.
3. H. S. Lowrie (to G. D. Searle & Co.), U.S. Pat. 3,272,818 (Cl. 260-268), Sept. 13, 1966.
4. H. S. Lowrie (to G. D. Searle & Co.), U.S. Pat. 3,265,693 (Cl. 260-247.5), Aug. 9, 1966.
5. C. M. Atkinson, J. C. E. Simpson, and A. Taylor, *J. Chem. Soc.*, **1954**, 1381.
6. K. Schofield and R. S. Theobald, *J. Chem. Soc.*, **1949**, 2404.
7. E. M. Lourie, J. S. Morley, J. C. E. Simpson, and J. M. Walker, Br. Pat. 702,664, Jan. 20, 1954.
8. J. McIntyre and J. C. E. Simpson, *J. Chem. Soc.*, **1952**, 2606.
9. W. Hepworth and F. H. S. Curd (to Imperial Chemical Industries, Ltd.), U.S. Pat. 2,585,935 (Cl. 260-250), Feb. 19, 1952.
10. J. McIntyre and J. C. E. Simpson, *J. Chem. Soc.*, **1952**, 2615.
11. G. B. Barlin and W. V. Brown, *J. Chem. Soc.* (*C*), **1967**, 2473.
11a. E. Hayashi and T. Watanabe, *Yakugaku Zasshi*, **88**, 94 (1968).
12. G. B. Barlin and W. V. Brown, *J. Chem. Soc.* (*C*), **1969**, 921.

13. E. J. Alford and K. Schofield, *J. Chem. Soc.*, **1953**, 1811.
14. K. Schofield and T. Swain, *J. Chem. Soc.*, **1950**, 392.
14a. A. J. Boulton, I. J. Fletcher, and A. R. Katritzky, *J. Chem. Soc.*, (*B*), **1971**, 2344.
15. A. R. Osborn and K. Schofield, *J. Chem. Soc.*, **1955**, 2100.
16. E. J. Alford, H. Irving, H. S. Marsh, and K. Schofield, *J. Chem. Soc.*, **1952**, 3009.
17. H. E. Baumgarten and C. H. Anderson, *J. Am. Chem. Soc.*, **80**, 1981 (1958).
18. H. E. Baumgarten, *J. Am. Chem. Soc.*, **77**, 5109 (1955).
19. K. Alder and H. Niklas, *Ann.*, **585**, 97 (1954).
20. I. Suzuki, T. Nakashima, and N. Nagasawa, *Chem. Pharm. Bull.* (*Tokyo*), **11**, 1326 (1963).
21. I. Suzuki, T. Nakashima, and N. Nagasawa, *Chem. Pharm. Bull.* (*Tokyo*), **14**, 816 (1966).
22. C. L. Angyal and R. L. Werner, *J. Chem. Soc.*, **1952**, 2911.
23. J. D. S. Goulden, *J. Chem. Soc.*, **1952**, 2939.
24. S. F. Mason, *J. Chem. Soc.*, **1958**, 3619.
25. R. N. Castle, D. B. Cox, and J. F. Suttle, *J. Am. Pharm. Ass.*, **48**, 135 (1959).
26. A. R. Osborn, K. Schofield, and L. N. Short, *J. Chem. Soc.*, **1956**, 4191.
27. A. Albert, R. Goldacre, and J. Phillips, *J. Chem. Soc.*, **1948**, 2240.
28. D. E. Ames, G. V. Boyd, R. F. Chapman, A. W. Ellis, A. C. Lovesey, and D. Waite, *J. Chem. Soc.* (*B*), **1967**, 748.
28a. J. R. Elkins and E. V. Brown, *J. Heterocyclic Chem.*, **5**, 639 (1968).
29. H. E. Baumgarten, W. F. Murdock, and J. E. Dirks, *J. Org. Chem.*, **26**, 803 (1961).
30. D. E. Ames and R. F. Chapman, *J. Chem. Soc.* (*C*), **1967**, 40.
31. K. Schofield and T. Swain, *J. Chem. Soc.*, **1949**, 1367.
32. L. McKenzie and C. S. Hamilton, *J. Org. Chem.*, **16**, 1414 (1951).
32a. E. Lunt, K. Washbourn, and W. R. Wragg, *J. Chem. Soc.* (*C*), **1968**, 687.
33. J. S. Morley, *J. Chem. Soc.*, **1951**, 1971.
34. D. E. Ames, R. F. Chapman, and H. Z. Kucharska, *J. Chem. Soc.*, Suppl. 1, **1964**, 5659.
35. D. E. Ames, *J. Chem. Soc.*, **1964**, 1763.
36. C. M. Atkinson and A. Taylor, *J. Chem. Soc.*, **1955**, 4236.
37. H. J. Barber and E. Lunt, *J. Chem. Soc.*, **1965**, 1468.
38. F. H. Case and J. A. Brennan, *J. Am. Chem. Soc.*, **81**, 6297 (1959).
39. E. A. Hobday, M. Tomlinson, and H. Irving, *J. Chem. Soc.*, **1962**, 4914.
40. D. E. Ames, R. F. Chapman, H. Z. Kucharska, and D. Waite, *J. Chem. Soc.*, **1965**, 5391.
41. S. Suzuki, K. Ueno, and K. Mori, *Yakugaku Kenkyu*, **34**, 224 (1962) (Japanese).
42. R. L. Mayer, P. C. Eisman, and D. Jaconia, *J. Invest. Dermatol.*, **24**, 281 (1955).
43. N. B. Chapman, K. Clarke, and K. Wilson, *J. Chem. Soc.*, **1963**, 2256.
44. E. P. Taylor, M. D. Potter, H. O. J. Collier, and W. C. Austin, Br. Pat. 812,994 (C07d), May 6, 1959.
45. W. C. Austin, L. H. C. Lunts, M. D. Potter, and E. P. Taylor, *J. Pharm. Pharmacol.*, **11**, 80 (1959).
46. B. K. Bhattacharya and A. B. Sen, *Br. J. Pharmacol.*, **24**, 240 (1965).
47. W. Hepworth and F. H. S. Curd (to Imperial Chemical Industries, Ltd.), U.S. Pat. 2,585,936, Feb. 19, 1952.
48. H. J. Barber, E. Lunt, K. Washbourn, and W. R. Wragg, *J. Chem. Soc.* (*C*), **1967**, 1657.
49. J. S. Morley and J. C. E. Simpson, *J. Chem. Soc.* **1952**, 2617.
50. E. M. Lourie, J. S. Morley, J. C. E. Simpson, and J. M. Walker, *Br. J. Pharmacol. Chemother.*, **6**, 643 (1951).

51. J. R. Keneford, E. M. Lourie, J. S. Morley, J. C. E. Simpson, J. Williamson, and P. H. Wright, *J. Chem. Soc.*, **1952,** 2595.
51a. E. Lunt, K. Washbourn, and W. R. Wragg, *J. Chem. Soc.* (*C*), **1968,** 1152.
52. J. R. Keneford, J. S. Morely, J. C. E. Simpson, and P. H. Wright, *J. Chem. Soc.*, **1949,** 1356.
53. J. S. Morley and J. C. E. Simpson, *J. Chem. Soc.*, **1949,** 1014.
54. H. E. Baumgarten and M. R. DeBrunner, *J. Am. Chem. Soc.*, **76,** 3489 (1954).
55. J. M. Hearn, R. A. Morton, and J. C. E. Simpson, *J. Chem. Soc.*, **1951,** 3318.
56. I. Suzuki, T. Nakashima, and T. Itai, *Chem. Pharm. Bull.* (*Tokyo*), **11,** 268 (1963).
57. I. Suzuki, T. Nakashima, N. Nagasawa, and T. Itai, *Chem. Pharm. Bull.* (*Tokyo*), **12,** 1090 (1964).
58. E. J. Alford and K. Schofield, *J. Chem. Soc.*, **1953,** 609.
59. J. R. Keneford, J. S. Morley, and J. C. E. Simpson, *J. Chem. Soc.*, **1948,** 1702.
60. H. E. Baumgarten, D. L. Pedersen, and M. W. Hunt, *J. Am. Chem. Soc.*, **80,** 1977 (1958).
61. K. Schofield and J. C. E. Simpson, *J. Chem. Soc.*, **1945,** 512.
62. A. N. Kaushal and K. S. Narang, *Indian J. Chem.*, **6,** 350 (1968).
63. R. D. Chambers, J. A. MacBride, and W. K. R. Musgrave, *J. Chem. Soc.* (*D*), **1970,** 739.

Part I. Cinnolinecarboxylic Acids, Aldehydes, Ketones, and Nitriles

I. Cinnolinecarboxylic Acids and Derivatives

A. Acids

There are five serviceable preparations of cinnolinecarboxylic acids; two produce 4-cinnolinecarboxylic acids, two produce 3-cinnolinecarboxylic acids, and one has yielded a 7-cinnolinecarboxylic acid. Apparently, no cinnolines bearing the carboxy group at positions other than the 3-, 4-, and 7-positions have been reported. All five of the methods used to prepare cinnolinecarboxylic acids are described in detail in other chapters, so that only a brief summary of each is given here.

The first preparation of 4-cinnolinecarboxylic acids involves conversion of a 4-methylcinnoline to a 4-styrylcinnoline, which can be gently oxidized to the 4-cinnolinecarboxylic acid (see Sections 1B-III-B and 1B-III-D). The second preparation of 4-cinnolinecarboxylic acids involves the alkaline cleavage of an *N*-benzilideneaminoisatin, which then recyclizes to give a 3-aryl-4-cinnolinecarboxylic acid (the Stolle-Becker synthesis; see Section 1B-I-D-1).

Of the two preparations of 3-cinnolinecarboxylic acids, the first is the Richter synthesis, in which an *o*-aminophenylpropiolic acid is diazotized and cyclized to give a 4-hydroxy-3-cinnolinecarboxylic acid (see Section 1C-I-A-1). The second synthesis of 3-cinnolinecarboxylic acids involves the Friedel-Crafts cyclization of mesoxalyl chloride phenylhydrazones. This procedure also yields a 4-hydroxy-3-cinnolinecarboxylic acid (see Section 1C-I-A-4).

Finally, 4-methyl-7-cinnolinecarboxylic acid (1) has been prepared by the Widman-Stoermer diazotization and cyclization of 3-amino-4-isopropenylbenzoic acid (see Section 1B-I-A). The cinnolinecarboxylic acids prepared by the methods outlined are listed in Table 1I-1.

Both 3- and 4-cinnolinecarboxylic acids are readily decarboxylated at 150–200° C (2–12). Decarboxylation of 4-cinnolinecarboxylic acid is the most useful method to obtain unsubstituted cinnoline, as described in Section 1A-II-A. Decarboxylation of a 3-cinnolinecarboxylic acid has been used to control the position of *N*-methylation of 4-hydroxycinnolines (3, 11). Methylation of 4-hydroxycinnoline (**1**) with methyl iodide occurs predominantly at N-2 to give the anhydro base of 2-methyl-4-hydroxycinnolinium hydroxide (**2**), but ethyl 6-chloro-4-hydroxy-3-cinnolinecarboxylate (**3**) is methylated predominantly at N-1 because of the steric influence of the 3-carbethoxy group. The resulting 1-methyl-4-cinnolone derivative (**4**) when subjected to ester hydrolysis and then thermal decarboxylation yields 6-chloro-1-methyl-4-cinnolone (**5**) (3, 11).

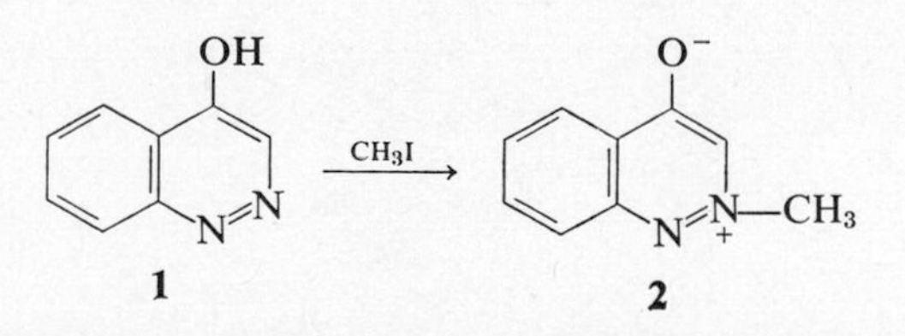

Simple thermal decarboxylation in the absence of a solvent, or in an inert solvent, results principally in the replacement of the carboxy group with a hydrogen atom, but when 3-phenyl-4-cinnolinecarboxylic acid (**6**) is heated at 200° C in the presence of potassium hydroxide, cupric oxide, and metallic copper, the reaction yields 62% of 4-hydroxy-3-phenylcinnoline (**7**) together with a little 3-phenylcinnoline (13, 14).

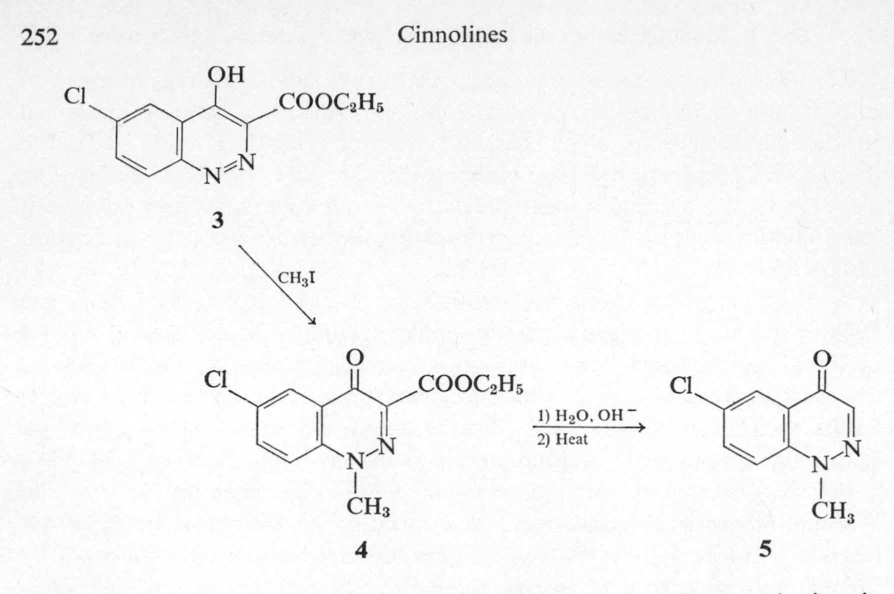

Reduction of 4-cinnolinecarboxylic acid (**8**) by amalgamated zinc in refluxing aqueous acetic acid occurs stepwise to give first 1,4-dihydrocinnoline (**9**) and then indole (**10**) in 77% yield. Unsubstituted cinnoline also

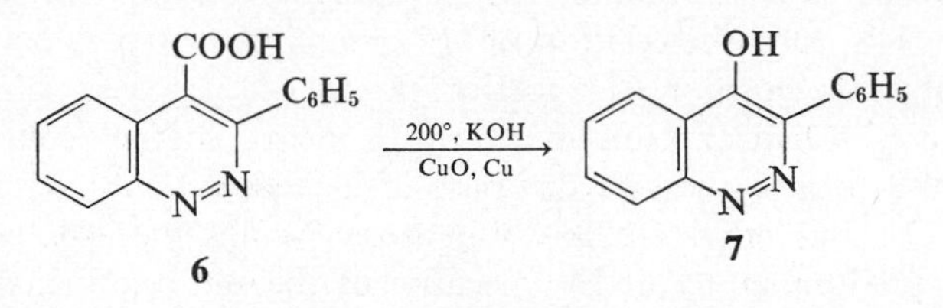

gives **9** and then **10** under these same conditions. The decarboxylation of **8** probably occurs at the dihydro stage, since 4-cinnolinecarboxylic acid is only slowly decarboxylated in refluxing aqueous acetic acid solution (15).

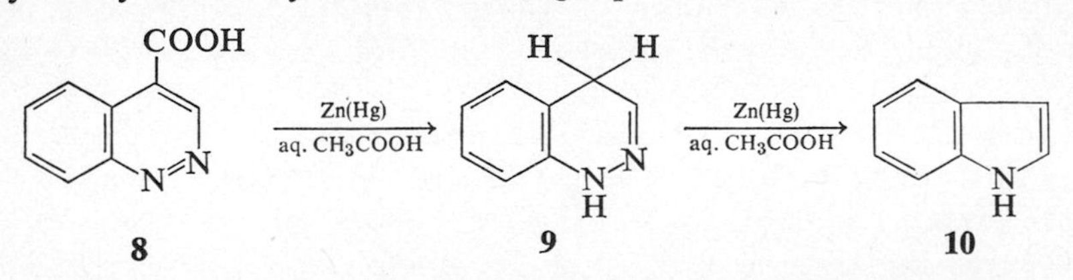

Reductive formylation of 4-cinnolinecarboxylic acid (**8**), by heating it with formic acid and formamide, causes it to undergo decarboxylation and ring contraction to give 1-formamidoindole (**11**) in 32% yield, the same product being obtained from cinnoline itself (16).

The polarographic reduction of 3-phenyl-4-cinnolinecarboxylic acid (**6**) in both acidic and basic solutions consumes two electrons per molecule and

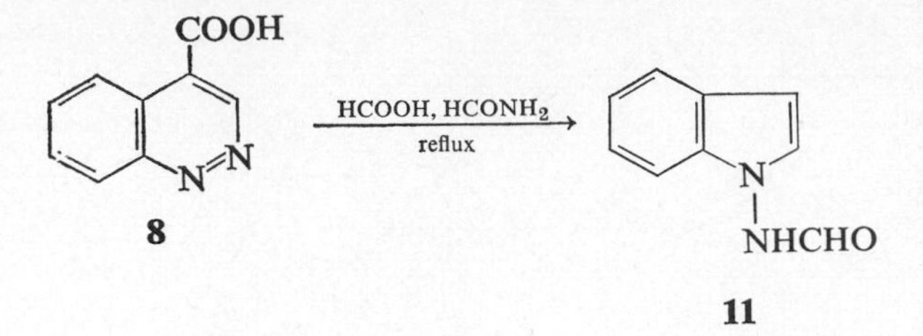

yields 1,4-dihydro-3-phenylcinnoline (**12**). In acidic solution, and at a more negative potential, the dihydro compound suffers hydrogenolysis of the nitrogen-nitrogen bond in a four-electron reduction to give 1-(*o*-aminophenyl)-2-phenyl-2-aminoethane (**13**). Reduction of 3-phenylcinnoline similarly gives **12** and then **13** under these same conditions (17).

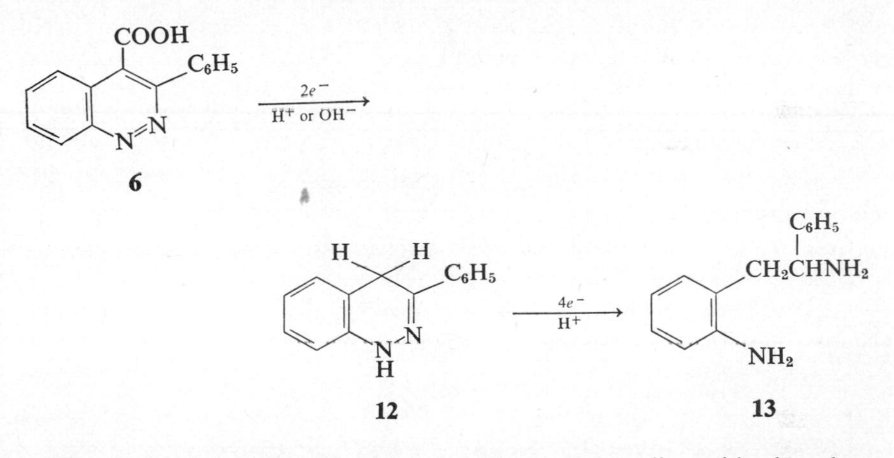

Nitration of certain 4-hydroxycinnoline-3-carboxylic acids has been examined and found to require more drastic conditions than nitration of the corresponding 4-hydroxycinnolines. Nitration of 6-chloro-4-hydroxycinnoline-3-carboxylic acid in a mixture of nitric acid and concentrated sulfuric acid at 80–85° C gives 6-chloro-4-hydroxy-8-nitrocinnoline-3-carboxylic acid as the main product (30–45%), together with a small amount of the 5-nitro isomer. Under the same conditions, 8-chloro-4-hydroxycinnoline-3-carboxylic acid is converted to its 6-nitro derivative in 40–55% yields. Nitration of 7,8-dichloro-4-hydroxycinnoline-3-carboxylic acid also occurs at the 6-position, but in this case a higher reaction temperature is required and yields are poor (18).

The *N*-oxidation of 4-cinnolinecarboxylic acid by hydrogen peroxide in acetic acid has been examined, and the only product isolated was 4-carboxycinnoline-2-oxide (19). This is unlike 4-methylcinnoline, which under the same conditions gives a mixture of its 1- and 2-oxides and a small amount of its 1,2-dioxide derivative (19).

The pmr spectrum of 4-cinnolinecarboxylic acid has been recorded (20).

B. Esters

Cinnolinecarboxylic acids may be esterified in the usual ways. Ethyl 4-cinnolinecarboxylate has been prepared from 4-cinnolinecarboxylic acid in 77% yield by refluxing a solution of the acid and absolute ethanol in the presence of sulfuric acid (2). A number of esters of 4-cinnolinecarboxylic acids have also been prepared by first converting the carboxy group to a carbonyl chloride group, then reacting this with the appropriate alcohol (6, 21, 22). The 4-cinnolinecarbonyl chloride may be prepared by the interaction of a 4-cinnolinecarboxylic acid with thionyl chloride (6), or from potassium 4-cinnolinecarboxylate and oxalyl chloride (21, 22). Ethyl 6-chloro-4-hydroxy-3-cinnolinecarboxylate has been prepared from the correspondingly substituted 3-cinnolinecarboxylic acid and ethanol, using ethereal boron trifluoride as the catalyst (11). Other 3-cinnolinecarboxylic acid esters have been prepared in like manner and with gaseous hydrogen chloride as catalyst (18).

Certain esters of cinnolinecarboxylic acids have been prepared by the cyclization of noncinnoline compounds. Thus ethyl β-(2-amino-4-carbethoxybenzoyl)propionate (**14**) has been diazotized and cyclized to ethyl 7-carbethoxy-4-hydroxy-3-cinnolinylacetate (**15**) (23).

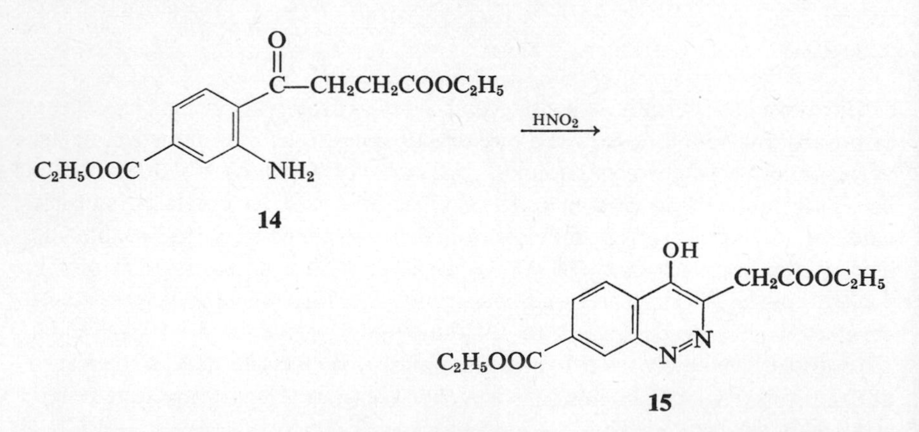

Ethyl 3-cinnolinecarboxylate (**18**) (and its 6- and 7-chloro analogs) has been prepared from *o*-aminobenzaldehyde (**16**) by diazotizing **16** and then coupling it with ethyl hydrogen malonate to give the intermediate ethyl glyoxylate *o*-formylphenylhydrazone (**17**), which cyclizes spontaneously to the cinnoline **18** (24). Unfortunately, yields by this method have been low (1–12%).

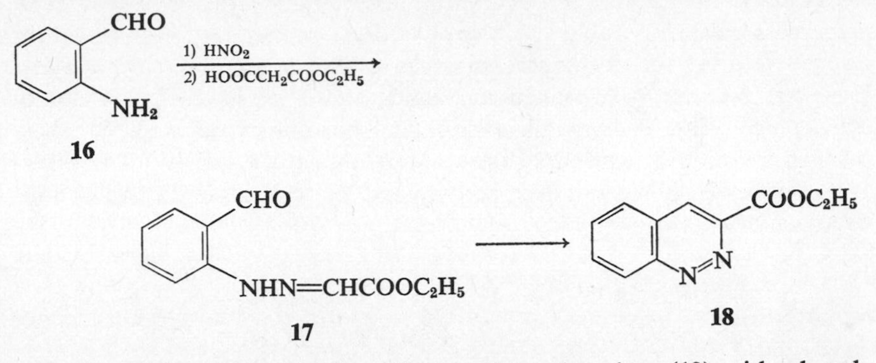

The reaction of ethyl 3-phenyl-4-cinnolinecarboxylate (**19**) with phenylmagnesium bromide is interesting in that it does not give the expected Grignard product. Instead, this reaction produces low yields of the 1-phenyl ester **20**, and the 1,4-dihydro ester **21** (6). In addition, the cinnoline **19** does not give the expected hydrazide with methylhydrazine in refluxing butanol solution. The methylhydrazine acts as a reducing agent instead, so that the only product isolated is the 1,4-dihydro butyl ester **22** in low yield (6).

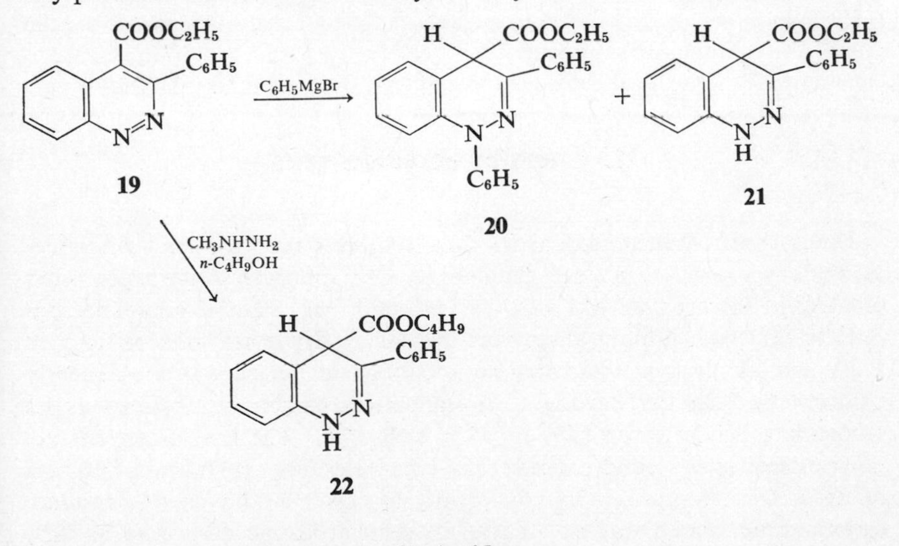

C. Amides

With the exception of 4-hydroxycinnoline-3-carboxamide, obtained from the reaction of the corresponding ethyl ester with ammonia (18), the only cinnolinecarboxamides apparently reported in the literature are 4-cinnolinecarboxamides. All of these have been prepared from a 4-cinnolinecarbonyl chloride and the appropriate amine (6, 22). The cinnolinecarboxamides so prepared are listed in Table 1I-1. Many of these amides have been tested for pharmacological activity, where most have shown little, if any, activity. However, 3-phenyl-4-cinnolinecarboxamide shows slight antiinflammatory activity (6). The *N*-aminoalkylcinnolinecarboxamides **23** and **24** show hypotensive activity while **25** shows antiinflammatory activity. The piperazine amide **26** shows antiulcer activity and the hydrazide **27** shows anti-inflammatory activity (6).

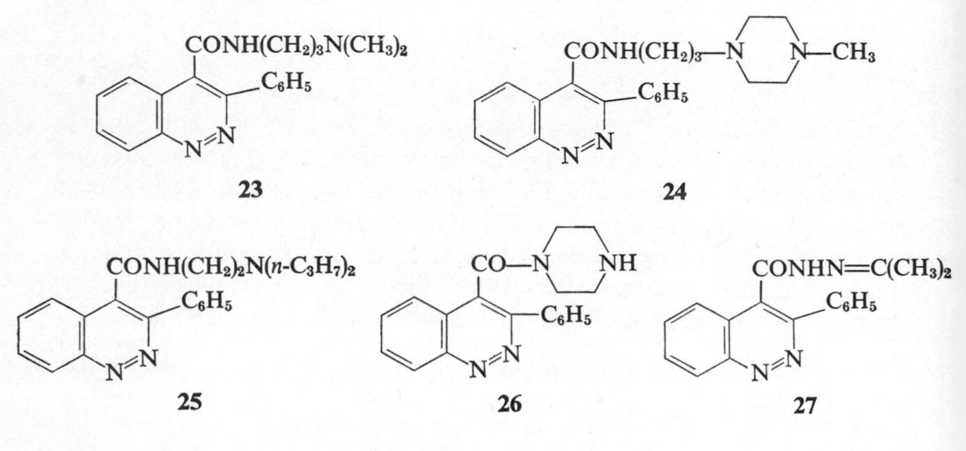

II. Cinnolinecarbaldehydes

The only cinnolinecarbaldehydes that have been reported are 4-cinnolinecarbaldehyde and 3-cinnolinecarbaldehyde. One approach to the preparation of 4-cinnolinecarbaldehyde was by reduction of ethyl 4-cinnolinecarboxylate (**28**) with lithium aluminum hydride in dry ether solution to give a dihydro-4-hydroxymethylcinnoline in 57% yield, which was subsequently oxidized by selenium dioxide to 4-cinnolinecarbaldehyde, isolated as its semicarbazone derivative (**29**) in 58% yield (21). The free aldehyde was not isolated. In a second procedure, 4-cinnolinecarbonyl chloride (**30**) was reduced by tri-*t*-butoxyaluminohydride to give a dihydro-4-cinnolinecarbaldehyde, which was isolated as its semicarbazone derivative in 32% yield. Oxidation of this with selenium dioxide gave the semicarbazone of 4-cinnolinecarbaldehyde (**29**). Again the free aldehyde was not isolated (21). Direct oxidation of 4-methylcinnoline with selenium dioxide followed by treatment with semicarbazide also gives the semicarbazone **29**, but in only

7% yield (21). Nitrosation of 4-methylcinnoline (**31**) with ethyl nitrite in ethanolic hydrochloric acid solution gives 4-cinnolinecarbaldehyde oxime (**32**) in 80% yield (25). No attempt was made to hydrolyze **32** to the free aldehyde.

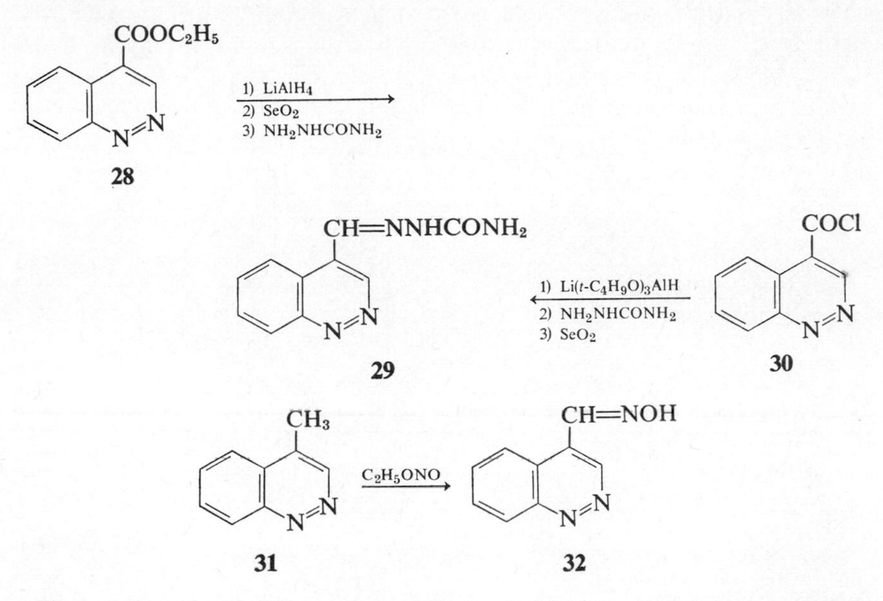

The preparation of free 4-cinnolinecarbaldehyde (**36**) from 4-methylcinnoline was achieved in a four-step sequence by first converting 4-methylcinnoline into 4-cinnolinylmethylpyridinium iodide (**33**) in 70% yield by heating it in a solution of pyridine and iodine. Heating **33** with aqueous *p*-nitrosodimethylaniline gave *N*,*N*-dimethyl-*N*′-(4-cinnolinylmethylene)-*p*-phenylenediamine *N*′-oxide (**34**) in 92% yield. Cleavage of **34** with aqueous hydrochloric acid followed by treatment with hydrazine gave the hydrazone derivative **35**. Interaction of the hydrazone **35** with nitrous acid than gave the free aldehyde **36** in 66% yield (21). The overall yield of **36** from 4-methylcinnoline is thus about 40%. After recrystallization from benzene-petroleum ether, 4-cinnolinecarbaldehyde is obtained in the form of yellow needles; it is not stable on standing (21).

In an interesting and rather surprising transformation, the natural sugars D-glucose, D-mannose, D-galactose, and D-xylose have been converted into 3-cinnolinecarbaldehyde. When these sugars are treated with phenylhydrazine in aqueous acetic acid at 115–120° for 3 hr, there are obtained not only the expected phenylosazones, but also certain 3-substituted cinnolines in 16–25% yields. The 3-substituted cinnoline produced from both D-glucose and D-mannose in this reaction is 3-(D-*arabo*-tetrahydroxybutyl)cinnoline (**37**).

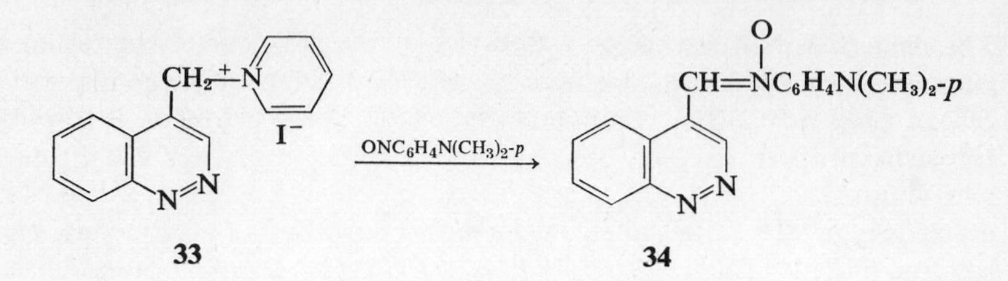

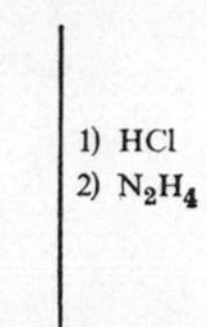

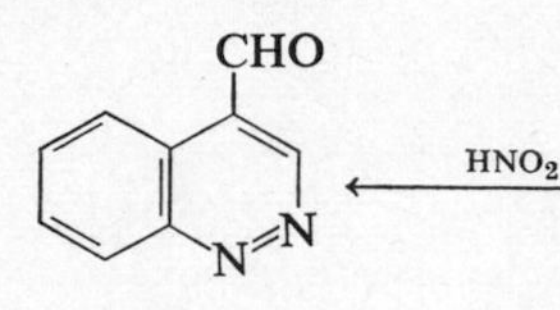

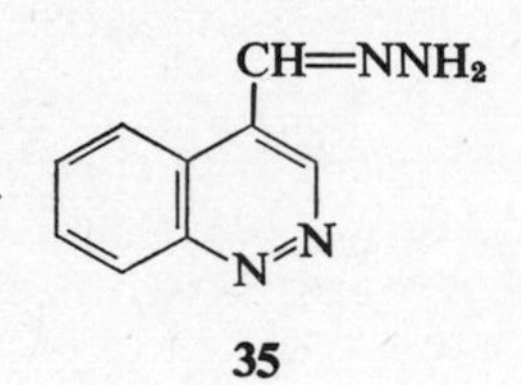

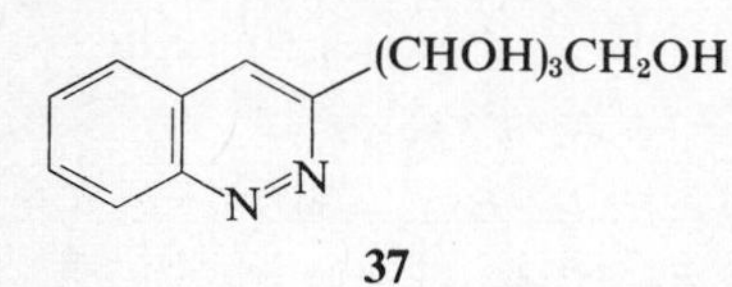

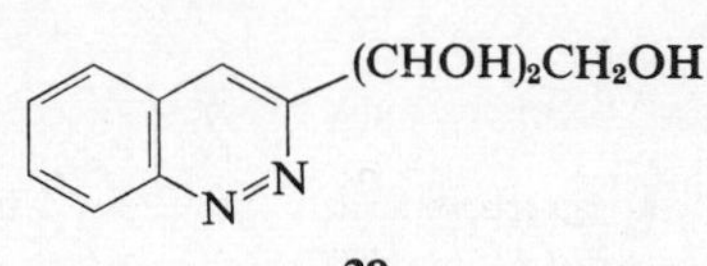

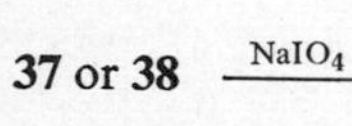

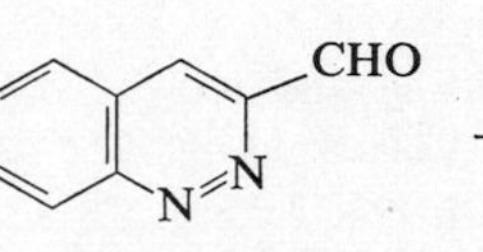

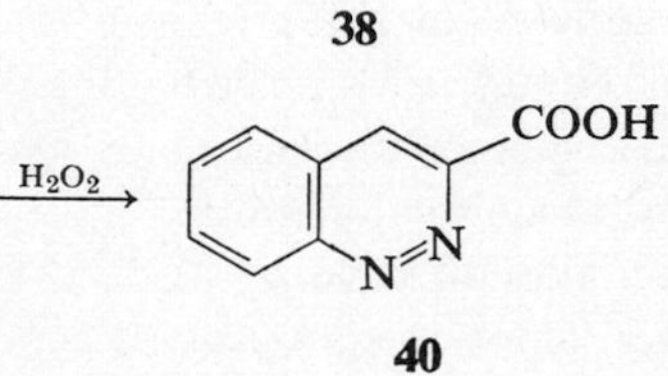

UV Light

Wolff-Kishner reduction

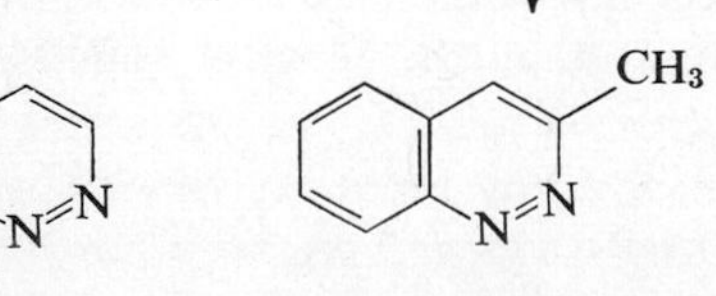

The cinnoline produced from D-galactose is 3-(D-*lyxo*-tetrahydroxybutyl)-cinnoline, which has the same formula as **37** but differs in the configuration of the side chain. The cinnoline produced from D-xylose is 3-(D-*threo*-trihydroxypropyl)cinnoline (**38**). All three of these cinnolines when treated with aqueous sodium periodate in the dark give 50–87% yields of 3-cinnoline-carbaldehyde (**39**) (26). The aldehyde is easily oxidized in an aqueous, basic solution of hydrogen peroxide at room temperature to 3-cinnolinecarboxylic acid (**40**) or reduced by hydrazine hydrate to 3-methylcinnoline (**41**). In the presence of UV light, the side chain is cleaved from 3-(D-*arabo*-tetrahydroxy-butyl)cinnoline to give unsubstituted cinnoline (**42**) (26).

III. Cinnolinyl Ketones

Condensation of ethyl 4-cinnolinecarboxylate (**28**) with ethyl acetate gives the keto ester **43**. Acidic hydrolysis of **43** then yields 4-cinnolinyl methyl ketone (**44**). The overall yield of **44** is 61%. It may be oxidized by potassium hypochlorite to 4-cinnolinecarboxylic acid in 77% yield (2).

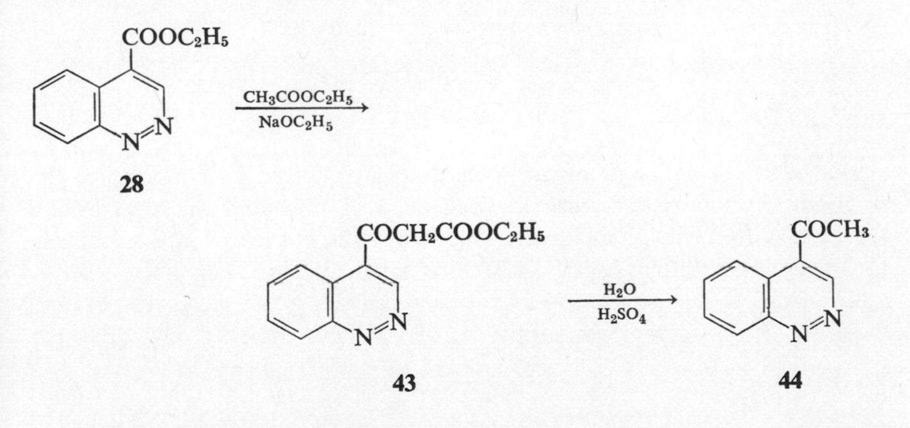

The measured dipole moment of 4-cinnolinyl methyl ketone, 2.52 D, is considerably lower than the calculated value, 3.42 D, indicating that the acyl group has a greater interaction with the cinnoline ring than with the benzene ring. This is also true for ethyl 4-cinnolinecarboxylate (**28**), which has a measured dipole moment of 3.62 D and a calculated moment of 4.00 D (27). In the case of 4-cinnolinyl methyl ketone, this means that the contribution of structure **45** to the ground state of the molecule would be greater than would be the contribution of a similar structure to the ground state of phenyl methyl ketone (27).

The preparation of 4-cinnolinyl chloromethyl ketone (**47**) was accomplished by treating 4-cinnolinecarbonyl chloride (**30**) with diazomethane, then treating the resultant diazoketone **46** with dry ethereal hydrogen chloride.

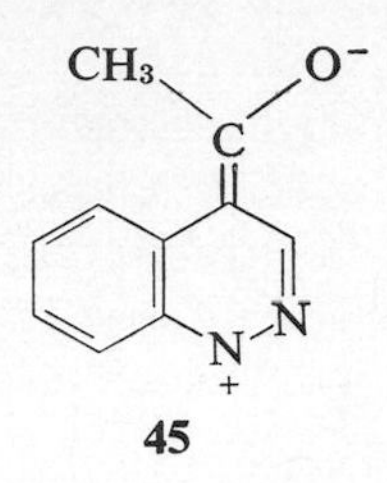

The hydrochloride salt of the ketone **47** is a hygroscopic red solid which becomes a dark oil on exposure to air. Whereas the acid chloride **30** can be converted to the hydrochloride of **47** in nearly quantitative yield, conversion of the hydrochloride to the free base **47** could be accomplished in only 29% yield. The free base is quite unstable and decomposes on standing (2).

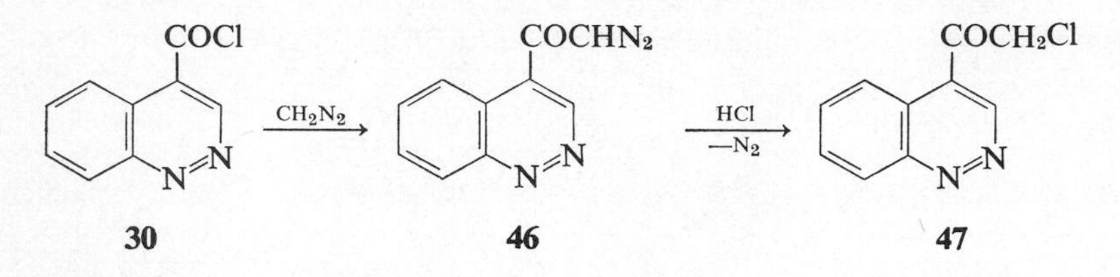

Diazotization of *o*-aminobenzaldehyde (**16**) followed by coupling of the resultant diazonium salt with acetoacetic acid produces unstable pyruvaldehyde 1-(*o*-formylphenyl)hydrazone (**48**) which cyclizes spontaneously to 3-cinnolinyl methyl ketone (**49**). This same reaction, beginning with 2-amino-4-chloro- or 2-amino-5-chlorobenzaldehyde, yields the corresponding 7-chloro- and 6-chloro-3-cinnolinyl methyl ketones, respectively, although

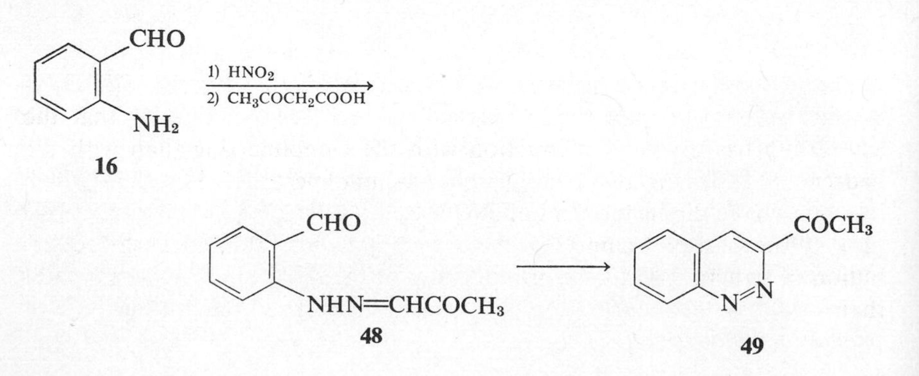

in lower yields, and the cyclization step is not always spontaneous and complete. The presence of a strong acid appears to aid cyclization (24). Treatment of 3-cinnolinyl methyl ketone (**49**) in concentrated sulfuric acid with sodium azide (the Schmidt reaction) produces 15% of 3-aminocinnoline (24).

One reported example of a cinnolinyl ketone in which the acyl group is not attached to the 3- or 4-position of the cinnoline ring is that of 4-hydroxy-7-cinnolinyl methyl ketone (**51**). This compound was prepared by diazotization and cyclization of 2-amino-1,4-diacetylbenzene (**50**). Like 3-cinnolinyl methyl ketone, **51** also undergoes the Schmidt reaction to give 7-amino-4-hydroxycinnoline (28).

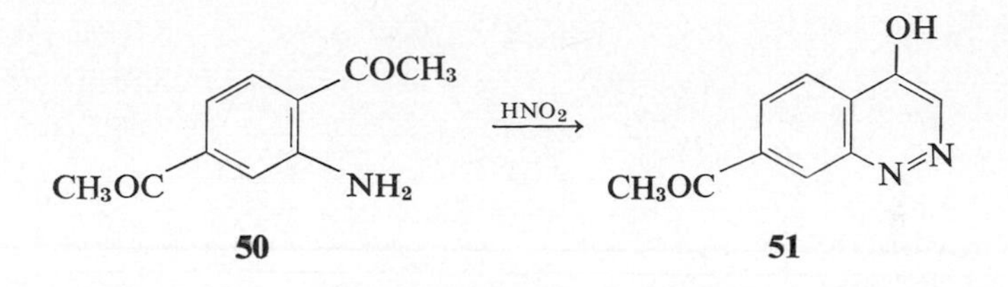

The polynuclear cinnolinyl ketone, 11-indeno[1,2-*c*]cinnolin-11-one (**53**), is prepared in about 75% yield from the Friedel-Crafts cyclization of 3-phenyl-4-cinnolinecarbonyl chloride (**52**) in carbon disulfide solution, using aluminum chloride as the catalyst (6).

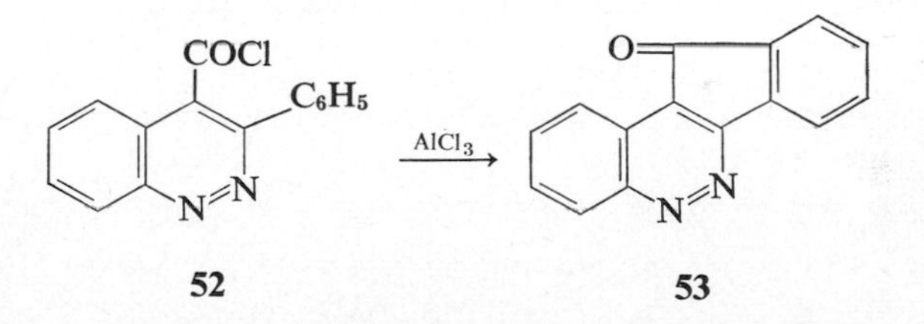

The diketone 3,4-dibenzoylcinnoline **55** was first synthesized in 1908 by oxidation of 1,3-diphenyl-2*H*-pyrrolo[3,4-*c*]cinnoline (**54**) with nitric acid in acetic acid solution (29). The structure of the diketone **55** was confirmed in 1969 when 3,4-dibenzylcinnoline (**57**) was oxidized with selenium dioxide in refluxing acetic acid solution to give 15% of **55** and 45% of 3,4-dibenzylidene-3,4-dihydrocinnoline (**58**) (30). An early report (31) claimed that the interaction of 3,4-dibenzoylcinnoline (**55**) with ethanolic ammonium hydrogen sulfide produced the dithione analog of **55**, but a reinvestigation of this reaction showed that in fact an almost quantitative yield of 3,4-dibenzoyl-1,4-dihydrocinnoline (**56**) is obtained (30). None of the dithione analog of **55** was detected.

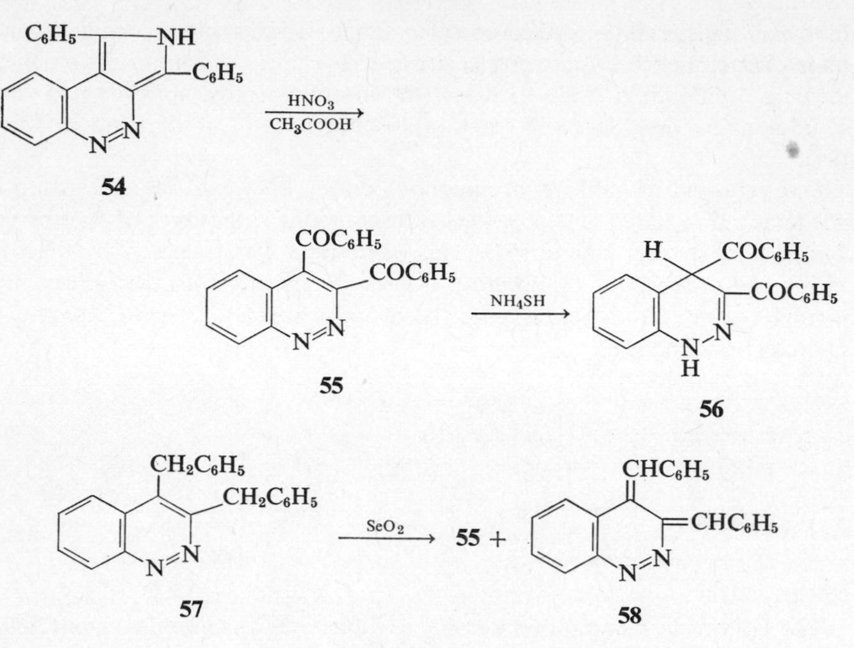

The Wolff-Kishner reduction of 3,4-dibenzoylcinnoline (**55**) with hydrazine hydrate in hot diethylene glycol gives only the pyrrolocinnoline **54**, not the expected 3,4-dibenzylcinnoline **57**. Reduction of **55** with zinc and acetic acid yields 1,3-diphenylfuro[3,4-*c*]cinnoline (**59**) (30).

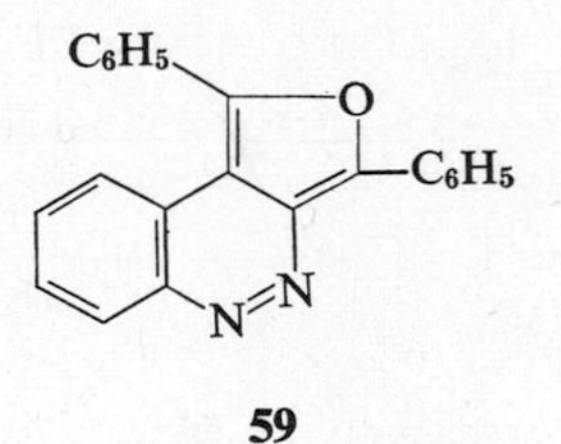

IV. Cinnolinecarbonitriles

Treatment of 4-methylsulfonylcinnoline (**60**) with an excess of potassium cyanide in dimethyl sulfoxide solution yields a mixture of 4-cinnolinecarbonitrile (**61**) and 4,6-cinnolinedicarbonitrile (**62**). Treatment of **61** with potassium cyanide under the same conditions gives **62** (32). However, interaction of 4-methylsulfonylcinnoline with only a slight excess of sodium cyanide in dimethylformamide solution gave 68% of 4-cinnolinecarbonitrile

(**61**), and no 4,6-cinnolinedicarbonitrile (**62**) was isolated (33). The dinitrile **62** is claimed to have antibacterial activity (34).

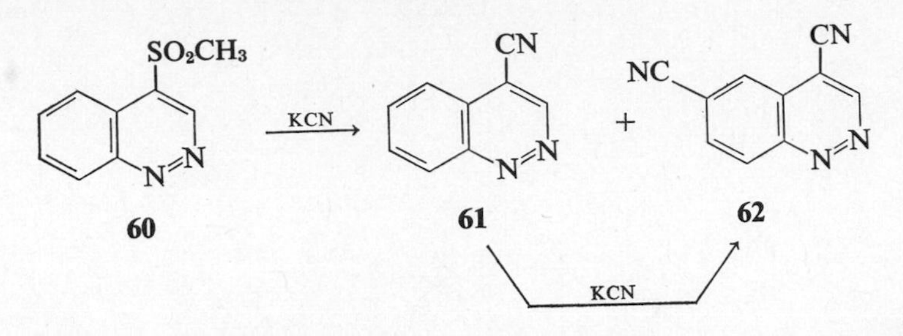

Other than these two cinnolinecarbonitriles, only a few other cinnolinecarbonitriles are described in the literature, these being 4-hydroxy-6-cinnolinecarbonitrile (**64**), obtained by the diazotization and cyclization of 2-amino-5-cyanoacetophenone (**63**) (35), 4-hydroxy-6,7-methylenedioxycinnoline-3-carbonitrile (**66**), and a number of its 1-alkyl derivatives. The nitrile **66** was obtained by treatment of the corresponding 3-bromocinnoline **65** with cuprous cyanide (36).

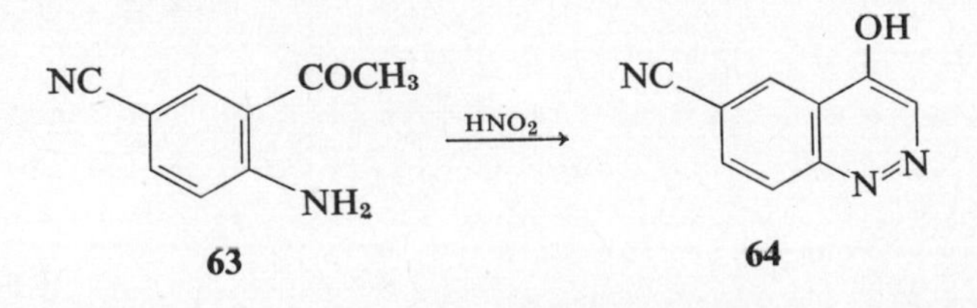

The nitrile group of 4-cinnolinecarbonitrile (**61**) undergoes the usual reactions of this group, such as hydrolysis to the acid **8**, reaction with alkaline hydrogen peroxide to give 4-cinnolinecarboxamide (**67**), and reaction with hydroxylamine to give 4-cinnolinecarboxamide oxime (**68**) (37)

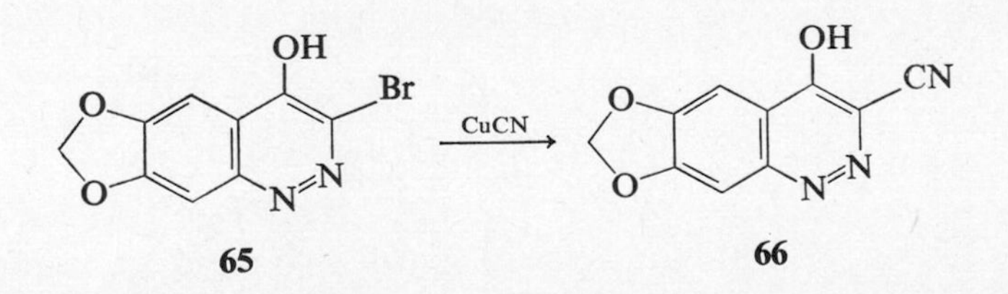

In addition, the nitrile group of 4-cinnolinecarbonitrile (**61**) is readily displaced by certain nucleophilic reagents such as methoxide ion, aniline, and hydrazine to give 4-methoxycinnoline (**69**), 4-anilinocinnoline (**70**), and 4-hydrazinocinnoline (**71**), respectively (37).

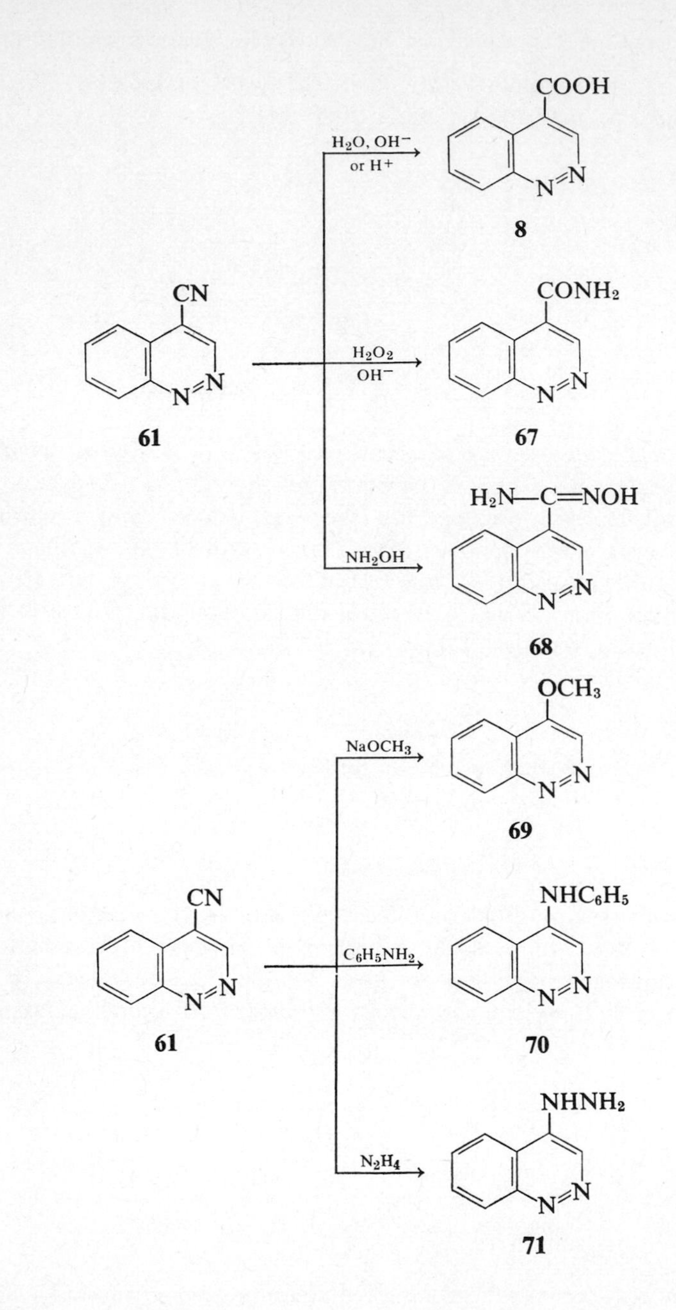
COOH
H_2O, OH^-
or H^+
N=N
8
CN
N=N
61
H_2O_2
OH^-
$CONH_2$
N=N
67
$H_2N—C=NOH$
NH_2OH
N=N
68
OCH_3
$NaOCH_3$
N=N
69
CN
N=N
61
NHC_6H_5
$C_6H_5NH_2$
N=N
70
$NHNH_2$
N_2H_4
N=N
71

V. Tables

TABLE 1I-1. Cinnolinecarboxylic Acids and Derivatives, Aldehydes, Ketones, and Nitriles[a]

R_1	R_2	R_3	R_4	R_5	R_6	Prep.[b]	MP (°C)	Remarks[c]	Ref.
H	CN	H	H	H	H	A	139–140, 146.5	Yellow crystals or orange solid (benzene-ligroin)	32, 33
H	CN	H	CN	H	H	A	179	Yellow crystals	32
$—COC_6H_5$	$—COC_6H_5$	H	H	H	H	B	162–163	Yellow needles (ether-ligroin)	30
H	Cl	H	H	$—COCH_3$	H	D	147–148	Yellow needles (ether-ligroin)	28
H	CHO	H	H	H	H	E	147–149	Yellow needles (benzene-ligroin); semicarbazone deriv., mp 234–235° (dec.); hydrazone deriv., mp 297° (dec.); 2,4-dinitrophenylhydrazone deriv., mp 316–317° (dec.); oxime deriv., mp 223° (dec.)	21
p-FC_6H_4	COOH	H	H	H	H	J	216–217	(Butanone)	6
C_6H_5	COOH	H	CH_3O	H	H	J	229–230	(Acetic acid)	6

TABLE 1I-1 (*continued*)

R_1	R_2	R_3	R_4	R_5	R_6	Prep.[b]	MP (°C)	Remarks[c]	Ref.
C_6H_5	COCl	H	H	H	H	N	139–142	Yellow needles (ligroin)	6
H	COCl	H	H	H	H	N	—	—	2
C_6H_5	—$COOC_2H_5$	H	H	H	H	C	92–93	Yellow prisms (methanol)	6
C_6H_5	—$CONH_2$	H	H	H	H	M	272–273	Yellow prisms (ethanol)	6
p-ClC_6H_4	—$CONH_2$	H	H	H	H	M	224–228	(Ethanol)	6
p-FC_6H_4	—$CONH_2$	H	H	H	H	M	236–237	(Ethanol)	6
p-$CH_3C_6H_4$	—$CONH_2$	H	H	H	H	M	266–267	(Ethanol)	6
p-$CH_3OC_6H_4$	—$CONH_2$	H	H	H	H	M	248–250	(Ethanol)	6
p-ClC_6H_4	—$CONH_2$	H	CH_3	H	H	M	236–237	(Ethanol)	6
C_6H_5	—$CONHCH_3$	H	H	H	H	M	209–212	(Methanol)	6
C_6H_5	—CO—N(pyrrolidine ring)	H	H	H	H	M	160–162	(Ether–ligroin)	6
C_6H_5	—CO—N(piperidine ring)	H	H	H	H	M	173–175	(Methanol)	6
C_6H_5	—$CONHCH_2C_6H_5$	H	H	H	H	M	234–235	White needles (chloroform)	6
C_6H_5	—$CONHCH(C_6H_5)_2$	H	H	H	H	M	214–215	(Methanol)	6
CHO	H	H	H	H	H	F	119–120	Yellow solid (ligroin); semicarbazone deriv., mp 253–254°	26
COOH	H	H	H	H	H	G	206 (dec.)	(Methanol)	26
H	COOH	H	H	H	H	H	195–196 (dec.)	(Methanol)	2
p-HOC_6H_4	COOH	H	H	H	H	I	239–240	(Ethanol)	13
p-ClC_6H_4	COOH	H	H	H	H	J	210–211	(Acetone)	6, 10
p-$CH_3OC_6H_4$	COOH	H	H	H	H	J	249–250	(Acetic acid)	6, 10
p-$CH_3C_6H_4$	COOH	H	H	H	H	J	207–208	(Acetone)	6, 10

p-ClC_6H_4	$COOH$	H	CH_3	H	H	J	211–212	(Butanone)	6, 10
C_6H_5	$COOH$	H	H	H	H	J	224–225	Yellow prisms (ethanol)	6, 10
C_6H_5	$COOH$	H	CH_3	H	H	J	222–223, 229–229.5	Yellow solid (ethanol)	6, 10
$—COCH_3$	H	H	H	H	H	K	155–156	Yellow needles (ligroin)	24
$—COCH_3$	H	H	Cl	H	H	K	206–207	(Ligroin)	24
$—COCH_3$	H	H	H	Cl	H	K	211–212	Yellow needles (ligroin)	24
$—COOC_2H_5$	H	H	H	H	H	L	97–97.5	Yellow needles (ligroin)	24
$—COOC_2H_5$	H	H	Cl	H	H	L	152.5–153	Yellow needles (ligroin)	24
$—COOC_2H_5$	H	H	H	Cl	H	L	200–201	Yellow needles (ligroin)	24
H	$—COOC_2H_5$	H	H	H	H	C	48.5–49.5	Yellow plates (hexane)	2
H	$—COOCH_2CH_2N(CH_3)_2$	H	H	H	H	C	—	Monopicrate deriv., mp 194–196°	22
H	$—COOCH_2CH_2N(C_2H_5)_2$	H	H	H	H	C	—	Acidic *d*-tartrate deriv., mp 110–112°	22
H	$—COOCH_2CH_2—N\bigcirc$	H	H	H	H	C	—	Acidic *d*-tartrate deriv., mp 119–121°	22
H	$—COOCH_2CH_2—N\bigcirc O$	H	H	H	H	C	—	Acidic *d*-tartrate deriv., mp 126–128°	22
H	$—COO(CH_2)_3N(CH_3)_2$	H	H	H	H	C	—	Acidic *d*-tartrate deriv., mp 142–144°	22
H	$—COO(CH_2)_3N(C_2H_5)_2$	H	H	H	H	C	—	Acidic *d*-tartrate deriv., mp 114–116°	22
H	$—COO—\bigcirc N—CH_3$	H	H	H	H	C	—	Acidic *d*-tartrate deriv., mp 193–195° (dec.)	22
H	$—CONH(CH_2)_3N(CH_3)_2$	H	H	H	H	M	—	BP 202–205° at 0.07 mm Hg; monopicrate deriv., mp 174–176°	22

TABLE 1I-1 (*continued*)

R_1	R_2	R_3	R_4	R_5	R_6	Prep.[b]	MP (°C)	Remarks[c]	Ref.
H	$—CONH(CH_2)_3N(C_2H_5)_2$	H	H	H	H	M	—	BP 210–212° at 0.07 mm Hg; monopicrate deriv., mp 140–142°	22
H	$—CONH(CH_2)_3—N$ (piperidino ring)	H	H	H	H	M	—	BP 224–227° at 0.02 mm Hg; monopicrate deriv., mp 196–198°	22
H	$—CONH(CH_2)_3—N$ (morpholino ring, O)	H	H	H	H	M	—	BP 245–248° at 0.025 mm Hg; monopicrate deriv., mp 181–184°	22
H	$—CONHCH_2CH_2N(C_2H_5)_2$	H	H	H	H	M	—	BP 210–214° at 0.02 mm Hg; monopicrate deriv., mp 148–150°	22
C_6H_5	$—CONH(CH_2)_3N(CH_3)_2$	H	H	H	H	M	127–128	Yellow flakes (methylene chloride–ligroin)	6
C_6H_5	$—CONH(CH_2)_3—N\quad N—CH_3$ (piperazine ring)	H	H	H	H	M	179–180	(Acetone)	6
C_6H_5	$—CONHCH_2CH_2N(CH_3)_2$	H	H	H	H	M	137–139	(Ether–ligroin)	6
C_6H_5	$—CONHCH_2CH_2N(n\text{-}C_3H_7)_2$	H	H	H	H	M	109–110	(Ligroin)	6
C_6H_5	$—CONHCH_2CH_2—N\quad O$ (morpholino ring)	H	H	H	H	M	130–131	(Acetone–ligroin)	6
C_6H_5	$—CO—N\quad NH$ (piperazine ring)	H	H	H	H	M	—	Maleic acid deriv., mp 186–187°	6

C_6H_5	—CO—N(piperazine)N—CH_3	H	H	H	H	M	184–185	(Methylene chloride–ligroin)	6
C_6H_5	—CO—N(piperazine)N—CH_2CH_2OH	H	H	H	H	M	125–127	(Ether–methanol)	6
C_6H_5	—CO—N(piperazine)N—NH_2	H	H	H	H	M	164–167	(Ether–methanol)	6
C_6H_5	—CO—N(piperazine)N—$NHCOCH_3$	H	H	H	H	M	181–182	(Ether–benzene)	6
C_6H_5	—$CONHNH_2$	H	H	H	H	M	198–200	(Benzene)	6
C_6H_5	—$CONHN{=}C(CH_3)_2$	H	H	H	H	M	201–203	(Acetone)	6
C_6H_5	—$CONHN(CH_3)_2$	H	H	H	H	M	237–240	(Methanol)	6
C_6H_5	—$CON(CH_3)N(CH_3)_2$	H	H	H	H	M	191–194	(Methanol)	6
C_6H_5	—CONHNH—(piperidin-4-yl)N—CH_3	H	H	H	H	M	223–226	(Acetone)	6
C_6H_5	—CONH—N(piperazine)N—CH_3	H	H	H	H	M	216–218	(Acetone–ligroin)	6
C_6H_5	—$COO(CH_2)_3N(CH_3)_2$	H	H	H	H	C	—	Maleic acid deriv., mp 153–155°	6
C_6H_5	—$COOCH_2CH_2$—N(piperazine)N—CH_3	H	H	H	H	C	—	Maleic acid deriv., mp 150–151°	6
H	—$COCH_3$	H	H	H	H	O	100–101	Yellow crystals (hexane); oxime deriv., mp 165–165.5°	3

TABLE 1I-1 (*continued*)

R_1	R_2	R_3	R_4	R_5	R_6	Prep.[b]	MP (°C)	Remarks[c]	Ref.
H	$—COCH_2COOC_2H_5$	H	H	H	H	P	81.5–82	—	2
H	$—CO(CH_2)_5NHCOC_6H_5$	H	H	H	H	Q	115.5–116.5	(Methanol)	2
H	$—COCH_2Cl$	H	H	H	H	R	95–100 (dec.)	Yellow crystals (benzene–hexane); decomposes on standing	2

[a] Those cinnolinecarboxylic acids and derivatives which also bear a hydroxyl substituent are not listed here. Instead, these are given in Table 1C-5.

[b] A, from 4-methylsulfonylcinnoline and potassium or sodium cyanide in dimethyl sulfoxide or dimethylformamide solution. B, by oxidation of 3,4-dibenzylcinnoline with selenium dioxide. C, by esterification of the corresponding 3-cinnolinecarboxylic acid or the appropriate 4-cinnolinecarbonyl chloride. D, from the corresponding 4-hydroxycinnoline and phosphorus oxychloride. E, by converting 4-methylcinnoline to 4-cinnolinecarbaldehyde hydrazone which is then cleaved to 4-cinnolinecarbaldehyde by nitrous acid (see text). F, by sodium periodate oxidation of 3-(D-*arabo*-tetrahydroxybutyl)cinnoline, or 3-(D-*lyxo*-tetrahydroxybutyl)cinnoline, or 3-(D-*threo*-trihydroxypropyl)cinnoline, which are the reaction products from phenylhydrazine and D-glucose, D-galactose, or D-xylose, respectively (see text). G, by oxidation of 3-cinnolinecarbaldehyde with alkaline hydrogen peroxide. H, by oxidation of 4-styrylcinnoline with potassium permanganate in aqueous pyridine solution. I, from the demethylation of 3-(4-methoxyphenyl)-4-cinnolinecarboxylic acid in refluxing 48% hydrobromic acid. J, by treatment of the appropriate *n*-benzilideneaminoisatin with hot, aqueous sodium or potassium hydroxide solution to give the corresponding 3-phenyl-4-cinnolinecarboxylic acid. K, by diazotization of the appropriately substituted *o*-aminobenzaldehyde followed by coupling with acetoacetic acid. The resulting pyruvaldehyde 2-formylphenylhydrazone then cyclizes to the 3-cinnolinyl methyl ketone. L, by diazotization of the appropriately substituted *o*-aminobenzaldehyde followed by coupling with ethyl hydrogen malonate. The resulting ethyl glyoxalate 2-formylphenylhydrazone then cyclizes to the corresponding ethyl 3-cinnolinecarboxylate. M, from the appropriate 4-cinnolinecarbonyl chloride and an amine, ammonium hydroxide, or hydrazine derivative, depending on the product. N, by refluxing a solution of the appropriately substituted 4-cinnolinecarboxylic acid and thionyl chloride. O, from the condensation of ethyl 4-cinnolinecarboxylate with ethyl acetate followed by acidic hydrolysis and decarboxylation of the intermediate keto ester. P, from the condensation of ethyl 4-cinnolinecarboxylate with ethyl acetate. Q, from ethyl 4-cinnolinecarboxylate by condensation with ethyl 6-benzamidohexanoate followed by hydrolysis. R, by treatment of 4-cinnolinecarbonyl chloride with diazomethane, then hydrochloric acid.

[c] Solvent employed in recrystallization is enclosed in parentheses.

References

1. O. Widman, *Ber.*, **17,** 722 (1884).
2. T. L. Jacobs, S. Winstein, R. B. Henderson, and E. C. Spaeth, *J. Am. Chem. Soc.*, **68,** 1310 (1946).
3. D. E. Ames, R. F. Chapman, H. Z. Kucharska, and D. Waite, *J. Chem. Soc.*, **1965,** 5391.
4. K. Schofield and J. C. E. Simpson, *J. Chem. Soc.*, **1945,** 512.
5. R. N. Castle, R. R. Shoup, K. Adachi, and D. L. Aldous, *J. Heterocyclic Chem.*, **1,** 98 (1964).
6. H. S. Lowrie, *J. Med. Chem.*, **9,** 664 (1966).
7. H. E. Baumgarten and J. L. Furnas, *J. Org. Chem.*, **26,** 1536 (1961).
8. H. J. Barber, E. Lunt, K. Washbourn, and W. R. Wragg, *J. Chem. Soc. (C)*, **1967,** 1657.
9. H. J. Barber, K. Washbourn, W. R. Wragg, and E. Lunt, *J. Chem. Soc.*, **1961,** 2828.
10. H. S. Lowrie (to G. D. Searle & Co.), U.S. Pat. 3,265,693, Aug. 9, 1966.
11. D. E. Ames, R. F. Chapmán, and H. Z. Kucharska, *J. Chem. Soc.*, Suppl. 1, **1964,** 5659.
12. J. S. Morley, *J. Chem. Soc.*, **1951,** 1971.
13. H. S. Lowrie, *J. Med. Chem.*, **9,** 670 (1966).
14. H. S. Lowrie (to G. D. Searle & Co.), U.S. Pat. 3,167,552, Jan. 26, 1965.
15. L. S. Besford and J. M. Bruce, *J. Chem. Soc.*, **1964,** 4037.
16. D. E. Ames and B. Novitt, *J. Chem. Soc. (C)*, **1970,** 1700.
17. H. Lund, *Acta Chem. Scand.*, **21,** 2525 (1967).
18. E. Lunt, K. Washbourn, and W. R. Wragg, *J. Chem. Soc. (C)*, **1968,** 687.
19. M. H. Palmer and E. R. R. Russell, *J. Chem. Soc. (C)*, **1968,** 2621.
20. J. R. Elkins and E. V. Brown, *J. Heterocyclic Chem.*, **5,** 639 (1968).
21. R. N. Castle and M. Onda, *J. Org. Chem.*, **26,** 4465 (1961).
22. R. N. Castle and M. Onda, *J. Org. Chem.*, **26,** 2374 (1961).
23. C. F. Koelsch, *J. Org. Chem.*, **8,** 295 (1943).
24. H. E. Baumgarten and C. H. Anderson, *J. Am. Chem. Soc.*, **80,** 1981 (1958).
25. H. Bredereck, G. Simchen, and P. Speh, *Ann. Chem.*, **737,** 39 (1970).
26. H. J. Haas and A. Seeliger, *Ber.*, **96,** 2427 (1963).
27. M. T. Rogers and T. W. Campbell, *J. Am. Chem. Soc.*, **75,** 1209 (1953).
28. K. Schofield and R. S. Theobald, *J. Chem. Soc.*, **1949,** 2404.
29. F. Angelico, *Atti. acad. Lincei*, **17,** II, 655 (1908); *Chem. Abstr.*, **4,** 1618 (1910).
30. D. E. Ames, H. R. Ansari, and A. W. Ellis, *J. Chem. Soc. (C)*, **1969,** 1795.
31. F. Angelico and C. Labisi, *Gazz. Chim. Ital.*, **40,** I, 411 (1910); *Chem. Abstr.*, **5,** 1092 (1911).
32. E. Hayashi, Y. Akahori and T. Watanabe, *Yakugaku Zasshi*, **87,** 1115 (1967); *Chem. Abstr.*, **68,** 49538 (1968).
33. G. B. Barlin and W. V. Brown, *J. Chem. Soc. (C)*, **1967,** 2473.
34. E. Hayashi, Japanese Pat. 7010,349, April 14, 1970; *Chem. Abstr.*, **73,** 45529 (1970).
35. K. Schofield and J. C. E. Simpson, *J. Chem. Soc.*, **1945,** 523.
36. W. A. White, German Pat. 2,005,104, August 6, 1970; *Chem. Abstr.*, **73,** 77269 (1970).
37. T. Watanabe, *Yakugaku Zasshi*, **89,** 1167 (1969); *Chem. Abstr.*, **72,** 3452 (1970).

Part J. Cinnoline *N*-Oxides

I. Methods of Preparation

A. *N*-Oxidation of Cinnolines

The first recorded preparation of cinnoline *N*-oxides is that of Atkinson and Simpson (1) in 1947; they found that the treatment of certain 3,4-disubstituted cinnolines with hydrogen peroxide in acetic acid produces the corresponding mono *N*-oxides in 80–90% yields. Thus were prepared the *N*-oxides of 4-phenyl-3-methylcinnoline, 4-phenyl-3-benzylcinnoline, 3-phenyl-4-*p*-anisylcinnoline, 4-*p*-anisyl-3-methylcinnoline, and 3,4-diphenylcinnoline, all of which were formulated as the 1-oxides **1**. This assignment was made on the erroneous assumption (2) that the basic center of 4-substituted cinnolines is N_1 and was not proven experimentally. Subsequent work

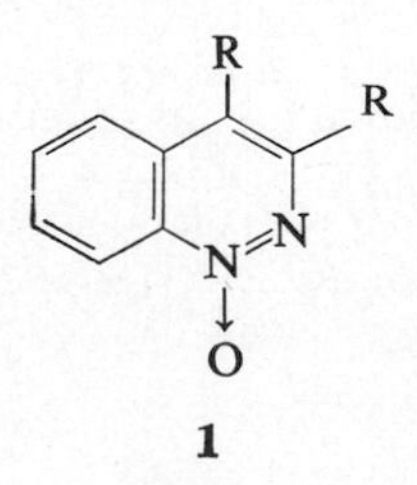

1

has shown that in general the presence of a bulky 3-substituent in the cinnoline ring (in the absence of an 8-substituent) causes the 1-oxide to be the principal isomer. When no 3-substituent is present, the 2-oxide is usually the predominant isomer.

When Ogata, Kano, and Tori (3, 4) treated cinnoline (**2**) with hydrogen peroxide in acetic acid at 70° for 6 hr there was obtained a mixture of cinnoline 1-oxide (**3**) and cinnoline 2-oxide (**4**), the ratio of **3** to **4** being 1:1.4. When this reaction was repeated under similar conditions by Suzuki and co-workers, compounds **3**, **4**, cinnoline 1,2-dioxide [**5**, colorless needles, mp 235° (dec.)] and indazole (**6**) were isolated in 25.9, 49.4, 0.3, and 3.0% yields, respectively (5, 5a).

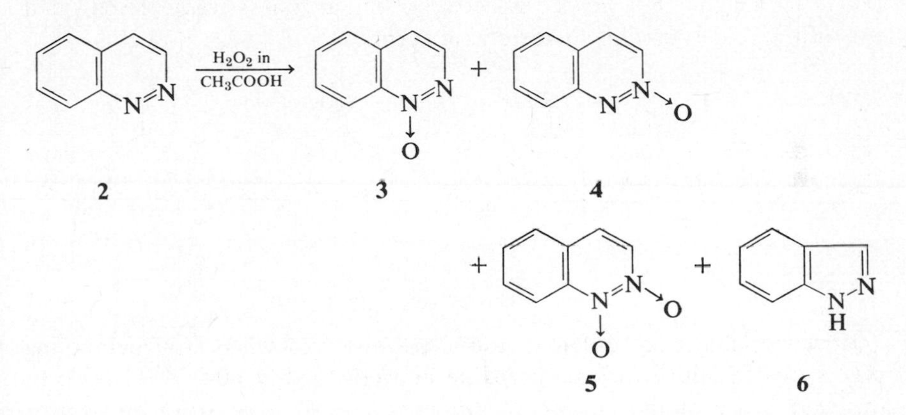

When cinnoline (**2**) is heated with hydrogen peroxide in acetic acid at 110–120° for 8 hr, the yield of the dioxide **3** is increased to 13%. A mechanism to explain the appearance of the ring-contracted product **6** has not yet been advanced. However, Lowrie (5b) observed that the oxidation of 3-phenylcinnoline with hydrogen peroxide in acetic acid gives a mixture of the 1- and 2-oxides (principally the 1-isomer), indazole, and benzoic acid arising from the 3-phenyl group of 3-phenylcinnoline. No 3-phenylindazole was found in the reaction mixture. These facts suggest that loss of the 3-carbon from the cinnoline ring is a possible pathway in its conversion to indazole. Similar results are observed when 3-(4-chlorophenyl)cinnoline is oxidized with hydrogen peroxide in acetic acid, when a product mixture consisting of the 1- and 2-oxides, indazole, and *p*-chlorobenzoic acid is obtained. No 3-(4-chlorophenyl)indazole was observed in the reaction mixture.

Treatment of the mono *N*-oxides **3** and **4** with hydrogen peroxide in acetic acid at 110–120° gives the dioxide **5** in 25 and 4% yields, respectively (5, 5a).

When 4-methylcinnoline is allowed to react with hydrogen peroxide in acetic acid at 70° for 6 hr there is produced a mixture in good yield of both

4-methylcinnoline 1-oxide and 4-methylcinnoline 2-oxide in a ratio of 1/2 (3, 4) together with a small amount (about 4%) of 4-methylcinnoline 1,2-dioxide as yellow needles, mp 168–169° (or mp 172° on a Köfler block) (6). On the other hand, 3-methyl-4-ethylcinnoline and 3,4-dimethylcinnoline give mixtures of their 2-oxide and 1,2-dioxide derivatives in a ratio of 2-oxide/1,2-dioxide of 2/1. No 1-oxide was obtained. Cinnoline-4-carboxylic acid gives only a 2-oxide derivative, whereas 4-methyl-3-nitrocinnoline gives only 4-methyl-3-nitrocinnoline 1-oxide (6a).

If 5-nitrocinnoline (**7**) is heated with hydrogen peroxide in acetic acid at 60–70° for 8 hr, there is obtained a mixture of 5-nitrocinnoline 1-oxide (**8**), 5-nitrocinnoline 2-oxide (**9**), and the ring-contracted product 4-nitroindazole (**10**) in 18, 43, and 16% yields, respectively. Similar results are observed when persulfuric acid is used as the oxidizing agent (7).

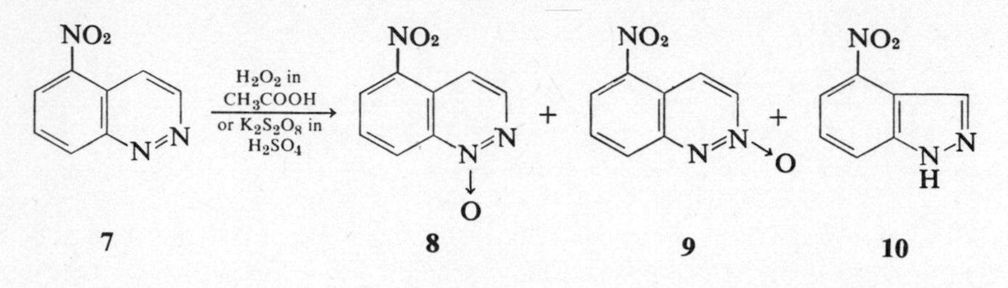

However, somewhat different results are observed when 8-nitrocinnoline (**11**) is treated with hydrogen peroxide in acetic acid at 60–70° for 8 hr. In this case, none of the 1-oxide is obtained. Instead, a mixture of 8-nitro-cinnoline 2-oxide (**12**), 7-nitroindazole (**13**), and 8-nitro-4-cinnolinol (**14**) in 20, 45, and 3% yields, respectively, is realized (7).

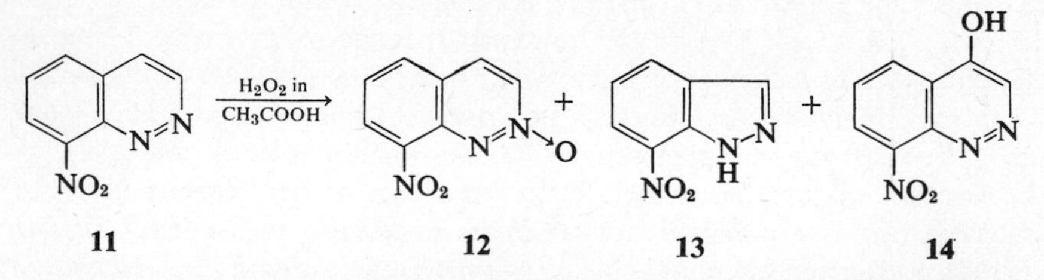

Other oxidizing agents such as ethereal phthalic monoperacid, hydrogen peroxide in sulfuric acid, and persulfuric acid in sulfuric acid also give mixtures of **12**, **13**, and **14** in various proportions from **11**, while chromic acid in acetic acid gives only the 4-hydroxy compound **14** in a yield of 15% with recovery of **11** (8). In general, an 8-nitro substituent appears to promote oxidation of the 4-position of the cinnoline ring. For example, 4-methyl-8-nitrocinnoline and 3,4-dimethyl-8-nitrocinnoline are oxidized by hydrogen peroxide in acetic acid to 8-nitro-4-hydroxycinnoline and 3-methyl-8-nitro-4-hydroxycinnoline, respectively. In contrast, 4-methyl-8-nitroquinoline is converted under the same conditions to 2-amino-3-nitroacetophenone (6a).

Both 4-methoxycinnoline (**15a**) and 4-chlorocinnoline (**15b**) when allowed to stand in an ethereal phthalic monoperacid solution for 2 weeks give mixtures of the corresponding 1-oxides and 2-oxides. Thus **15a** gives the 1-oxide **16a** and the 2-oxide **17a** in 18 and 33% yields, respectively, while **15b** gives the 1-oxide **16b** and the 2-oxide **17b** in 28 and 43% yields, respectively (8).

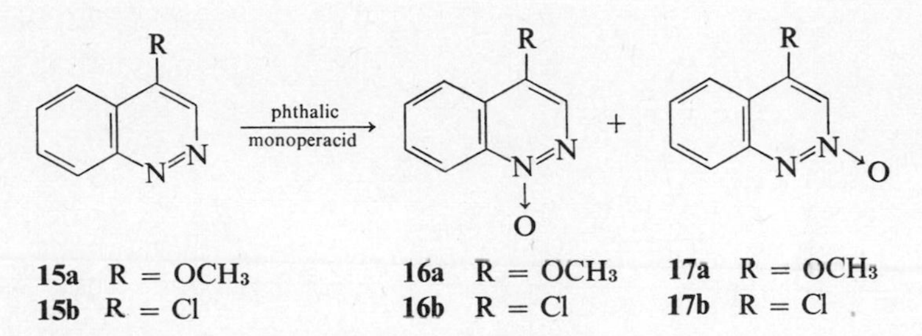

In contrast, only 3-chlorocinnoline 1-oxide, but no 3-chlorocinnoline 2-oxide, is observed in the product mixture when 3-chlorocinnoline is treated with hydrogen peroxide in acetic acid at 70° for 6 hr (4).

B. Ring Expansion by Nitrogen Insertion into 2-Phenylisatogen

In 1962, Noland and Jones (9) found that 2-phenylisatogen (**18**) will undergo ring expansion and nitrogen insertion when treated with ethanolic ammonia in an autoclave at 140–145° for 6 hr to give 3-phenyl-4-cinnolinol 1-oxide (**19**) in 26% yield.

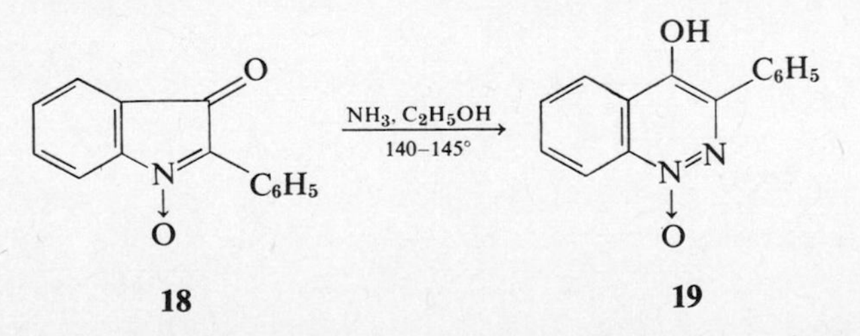

These workers postulated that this transformation is initiated by the attack of ammonia on the isatogen **18** to give the addition product **20**, which then undergoes ring cleavage to give the transient *o*-nitrosophenyl enamine

21. Compound **21** then recyclizes to give the dihydrocinnoline *N*-oxide **22**, which undergoes dehydrogenation, possibly by air oxidation, to give the product 3-phenyl-4-cinnolinol 1-oxide **19**.

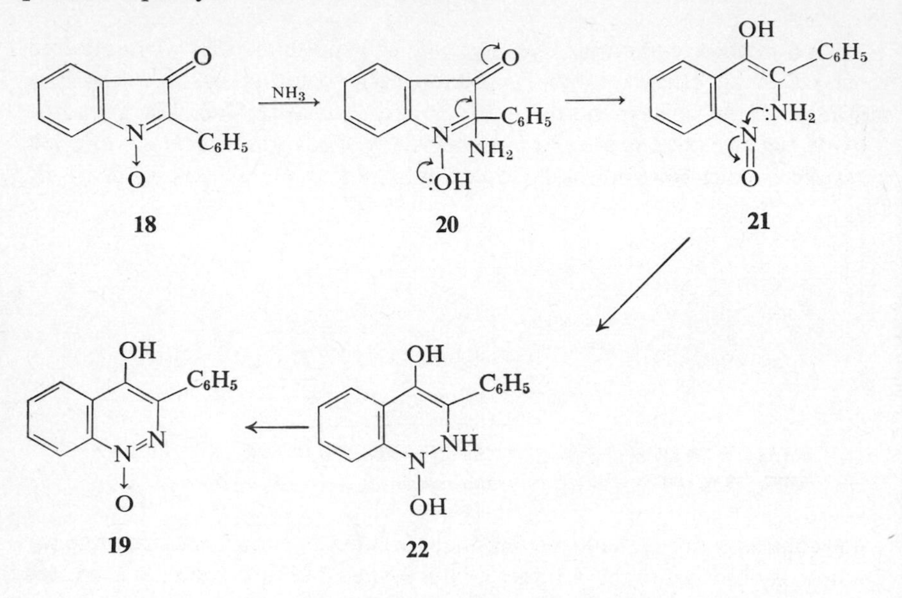

C. Cyclization of a *o*-Nitrophenyl Enamine

In 1954 Kröhnke and Vogt (10) observed that when an ethanolic solution of the enamine **23**, α-(*p*-nitrophenyl)-β-(*o*-nitrophenyl)vinylamine, is exposed to sunlight for several hours, a reaction occurs and the product, in the form of yellow needles, precipitates from the solution. They considered this material to be 3-(*p*-nitrophenyl)cinnoline 1-oxide (**24**) on the basis of an elemental analysis. It was obtained in a yield of 75%.

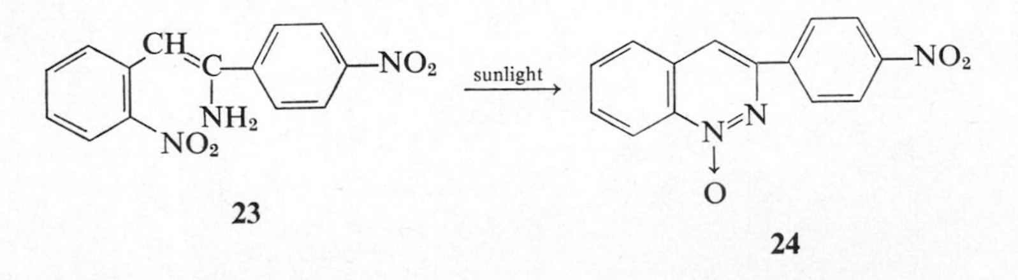

Compound **23** may be prepared in good yield by the interaction of 1-(*p*-nitrobenzyl)pyridinium bromide with *o*-nitrobenzaldehyde, as shown in Eq. 1, to give 1-[*o*-nitro-α-(*p*-nitrophenyl)styryl]pyridinium bromide, which is then converted to **23** by heating briefly in a mixture of pyridine, piperidine, and water.

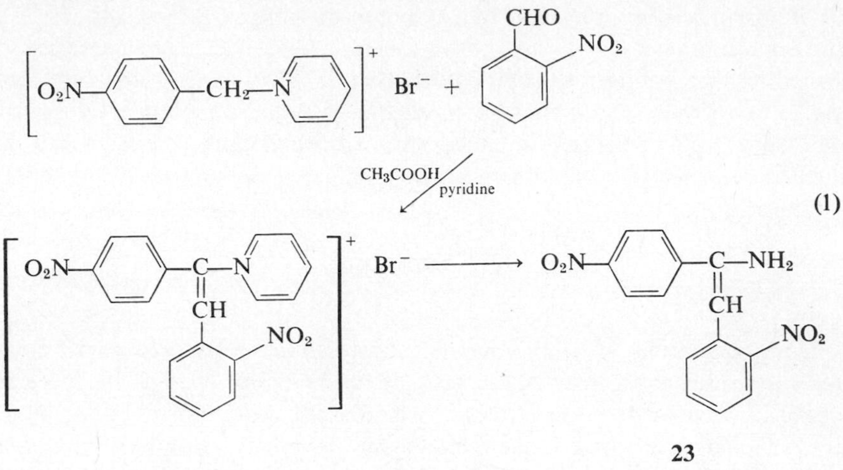

It is not yet known whether this reaction can be developed into one of a general nature, since the only attempt to prepare a cinnoline 1-oxide by this procedure is that which has just been discussed for the preparation of **24**.

D. Aromatization of 5,6,7,8-Tetrahydrocinnoline 1-Oxides

During the course of an investigation designed to show whether *N*-oxidized cinnoline derivatives, obtained by the interaction of the cinnoline derivatives with hydrogen peroxide in acetic acid, are 1-oxides or 2-oxides, Ogata, Kano, and Tori (3, 4) prepared cinnoline 1-oxide and four substituted cinnoline 1-oxides by bromination of their corresponding 5,6,7,8-tetrahydro derivatives with *N*-bromosuccinimide followed by dehydrobromination with sodium methoxide, as shown in Eq. 2.

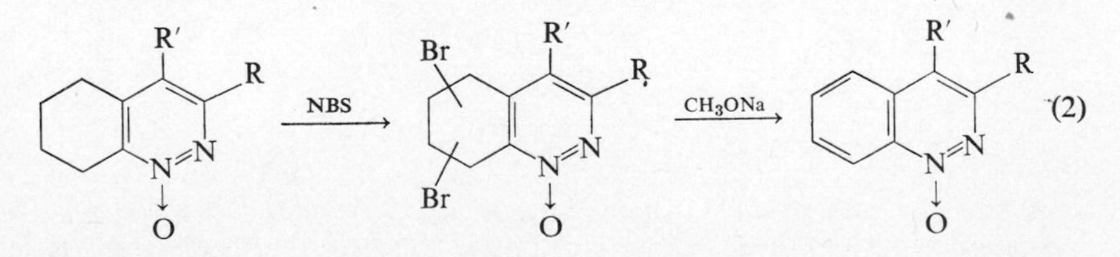

Thus are prepared cinnoline 1-oxide, 3-chlorocinnoline 1-oxide, 3-methylthiocinnoline 1-oxide, and 4-chloro-3-methoxycinnoline 1-oxide in low yields, while 3-methoxycinnoline 1-oxide is prepared in a yield of approximately 28%. Bromination of both 5,6,7,8-tetrahydrocinnoline 1-oxide and its 3-chloro analog gives first the monobromo derivatives, which are then further brominated to the 5,8-dibromo derivatives, whereas bromination of 3-methoxy-5,6,7,8-tetrahydrocinnoline 1-oxide gives a 6,8-dibromo derivative in a single step. The positions of the bromine atoms in the dibromo derivatives of 3-methylthio-5,6,7,8-tetrahydrocinnoline 1-oxide and of 4-chloro-3-methoxy-5,6,7,8-tetrahydrocinnoline 1-oxide have not been established.

II. Structure

Upon oxidation of cinnoline or substituted cinnolines to their mono *N*-oxides, one is faced with the task of determining whether the product obtained is a 1-oxide or a 2-oxide, or which product in a mixture of products is the 1-oxide and which is the 2-oxide. This problem was solved by Ogata, Kano, and Tori (3, 4), who used the following approach.

From previous work with pyridazines it was known that the *N*-oxidation of 3-chloro-5-methylpyridazine (**25**) (11) and 3-chloro-6-methylpyridazine (**27**) (12, 13) by hydrogen peroxide in acetic acid gives the corresponding 1-oxides, **26** and **28**, respectively. Ogata, Kano, and Tori were thus reasonably confi-

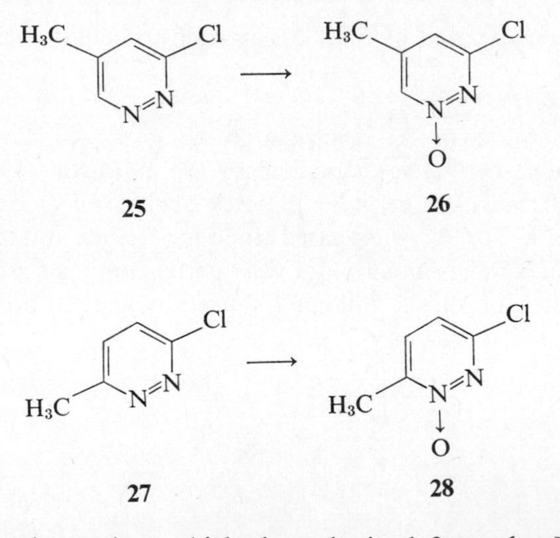

dent that the sole product which they obtained from the *N*-oxidation of 3-chloro-5,6,7,8-tetrahydrocinnoline (**29**) in hydrogen peroxide and acetic acid was the 1-oxide **30** from which chlorine was removed by hydrogenation over palladium-carbon to give the unsubstituted tetrahydrocinnoline 1-oxide

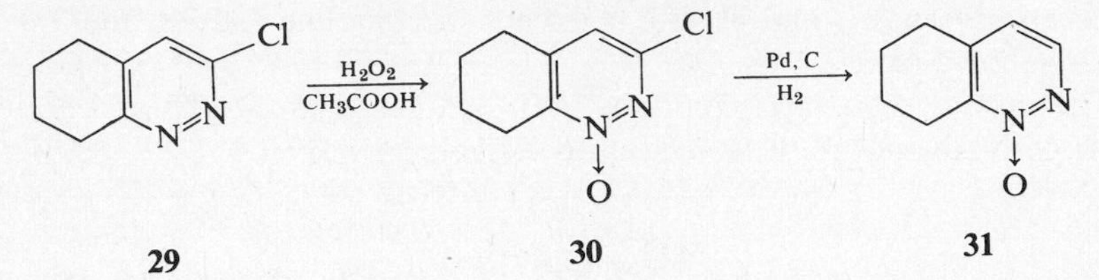

(**31**). Compound **31** was then shown to be identical with one of the two isomeric *N*-oxides that were obtained when 5,6,7,8-tetrahydrocinnoline (**32**) was *N*-oxidized by hydrogen peroxide in acetic acid. Compound **31** was

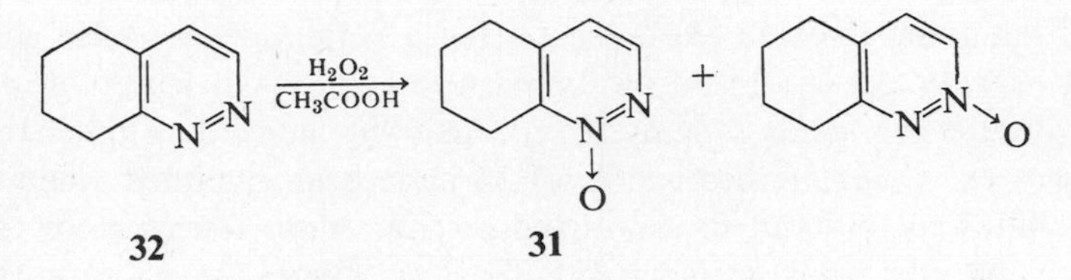

brominated in the 5- and 8-positions by *N*-bromosuccinimide and was then dehydrobrominated in methanolic sodium methoxide to give cinnoline 1-oxide (**3**), thus establishing its structure. In the course of this work, several

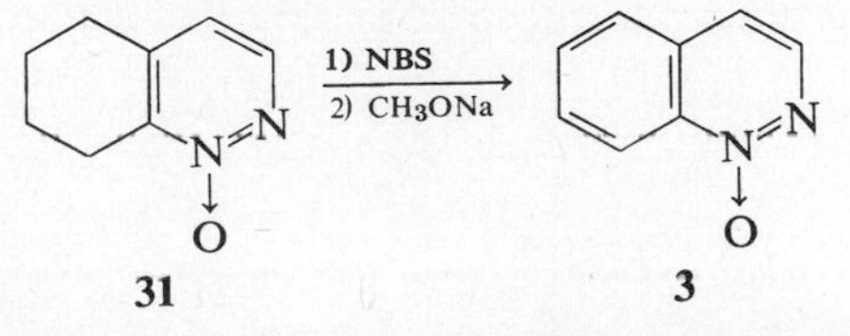

substituted 5,6,7,8-tetrahydrocinnoline 1-oxides were prepared and are listed in Table 1J-1.

As stated in Section 1J-A, *N*-oxidation of 4-methylcinnoline with hydrogen peroxide in acetic acid yields both 4-methylcinnoline 1-oxide and 4-methylcinnoline 2-oxide, whose structures could now be assigned by a comparison of their UV and pmr spectra to those of cinnoline 1-oxide and cinnoline 2-oxide (4). Ultraviolet absorption spectral data which have been recorded in the literature for cinnoline *N*-oxides are given in Table 1J-2.

Proton magnetic resonance studies of several cinnoline 1-oxides and 2-oxides have been performed (3, 4, 14) and the spectral parameters are listed in Table 1J-3. It is notable that the H_8 proton signal of the 1-oxides appears at a lower field than the other protons of the benzenoid ring, although this same proton in the 2-oxides remains buried in the multiplet signal due to the remaining benzenoid ring protons. Ogata, Kano, and Tori attribute this downfield shift of the H_8 proton, which is at the *peri*-position to the *N*-oxide group of cinnoline 1-oxides, to a magnetic anisotropy effect of the *N*-oxide group. It is also of interest that the coupling constant ($J_{3,4}$)

between protons H_3 and H_4 is 6.0–6.2 cps in cinnoline 1-oxide and 7.0 cps in cinnoline 2-oxide.

The mass spectra of cinnoline 1- and 2-oxides, 1,2-dioxide, and various alkyl derivatives of these have been investigated (14a).

III. Reactions

A. Nitration

When cinnoline 1-oxide (**3**) is treated with a mixture of nitric acid and sulfuric acid, or potassium nitrate in sulfuric acid, 4-nitrocinnoline 1-oxide (**32**) is obtained in yields ranging from 3 to 64%, depending upon reaction conditions (15, 16). The best yields of **32** have been obtained when the *N*-oxide is allowed to stand in the mixed acid at room temperature for 8 hr and then the mixture is heated at 50° for 1 hr. When the mixture of nitric

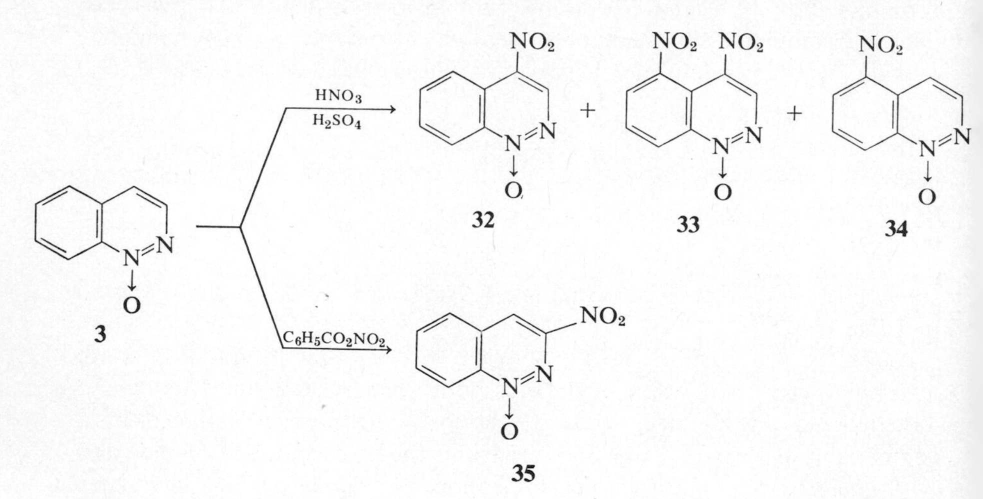

acid and sulfuric acid is replaced with a mixture of fuming nitric acid and sulfuric acid, only small yields of **32** are realized, but 44–50% yields of 4,5-dinitrocinnoline 1-oxide (**33**) are then obtained. In one experiment, a small amount (2%) of 5-nitrocinnoline 1-oxide (**34**) was detected.

Both 4-nitrocinnoline 1-oxide (**32**) and the 5-nitro isomer (**34**) when treated with fuming nitric acid in sulfuric acid give 4,5-dinitrocinnoline 1-oxide (**33**).

On the other hand, when cinnoline 1-oxide is treated with freshly prepared benzoyl nitrate in chloroform, 3-nitrocinnoline 1-oxide (**35**) is obtained in a 71% yield (15, 16).

Treatment of 3-methoxycinnoline 1-oxide with concentrated nitric acid in acetic acid gives 3-methoxy-4-nitrocinnoline 1-oxide (4).

Nitration of cinnoline 2-oxide gives different results from nitration of cinnoline 1-oxide. Thus when cinnoline 2-oxide (**4**) is allowed to react with a mixture of nitric acid and sulfuric acid, or with potassium nitrate in sulfuric acid, there is obtained a mixture of 8-nitrocinnoline 2-oxide (**12**), 6-nitrocinnoline 2-oxide (**36**), and 5-nitrocinnoline 2-oxide (**9**), the mixture of nitrated products being obtained in yields ranging from 0.6 to 97% depending upon the reaction conditions (17, 18).

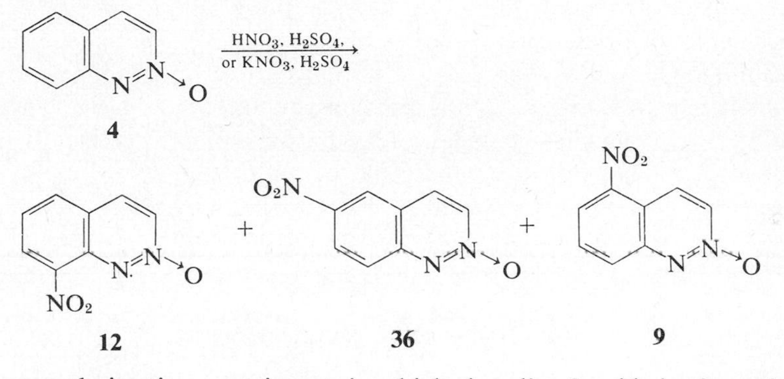

In several nitration experiments in which cinnoline 2-oxide is nitrated with potassium nitrate in sulfuric acid, there are obtained individual yields of up to 72% of the 8-nitro isomer **12**, 25% of the 6-nitro isomer **36**, and 21% of the 5-nitro isomer **9**, depending upon reaction conditions. With one exception, where the total yield of the three nitrated products was only 2.1%, the principal product in each individual experiment is the 8-nitro isomer. Suzuki, Nakashima, and Nagasawa rationalized this result by suggesting that the orienting effect of the 2-oxide function will direct an attacking electrophile to the 6- and 8-positions of the carbocyclic ring, as shown by the accompanying resonance-contributing structures, while the effect of the cinnoline ring itself is to direct such an incoming group to the 5- and 8-positions. Thus both effects are cumulative at the 8-position, and 8-nitrocinnoline 2-oxide (**12**) is expected to be the principal product.

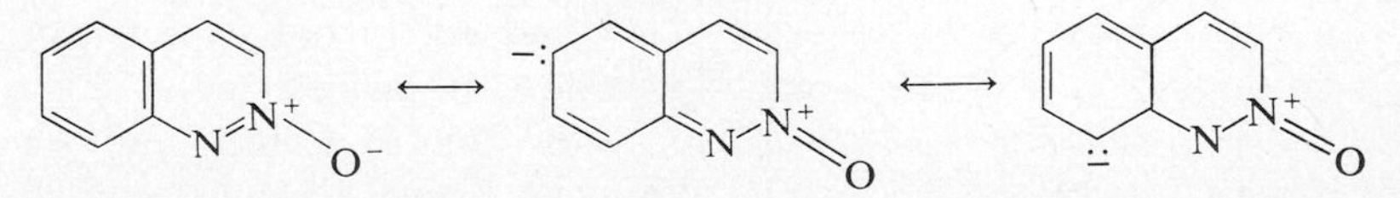

A kinetic study of the nitration of cinnoline 2-oxide in sulfuric acid showed that the amounts of 5- and 8-nitrocinnoline 2-oxides increase with increasing acidity (in the acidity range 64–90% sulfuric acid), whereas the amount of 6-nitrocinnoline 2-oxide decreases. It was suggested that nitration of the protonated cinnoline 2-oxide cation gives the 5- and 8-nitro derivatives while nitration of unprotonated cinnoline 2-oxide produces the 6-nitro isomer (18a).

When cinnoline 2-oxide (**4**) is treated with benzoyl nitrate in chloroform for one week at room temperature, a low yield (1.5%) of 5-nitrocinnoline 2-oxide (**9**) is obtained, and 91% of **4** is recovered (17, 18).

Several nitrated cinnoline *N*-oxides have been found to display antifungal, antibacterial, and anticancer activity. Thus it is claimed that 3-methoxy-4-nitrocinnoline 1-oxide is useful as an antifungal agent (19), while 3-nitrocinnoline 1-oxide and 6-nitrocinnoline 2-oxide show strong activity against Ehrlich tumor cells and certain species of bacteria (20). The preparation of nitrocinnoline *N*-oxides by nitration of the corresponding cinnoline *N*-oxides is outlined in a patent which claims that compounds such as 4-nitrocinnoline 1-oxide, 5-nitrocinnoline 1-oxide, 4,5-dinitrocinnoline 1-oxide, 5-nitrocinnoline 2-oxide, 3-nitrocinnoline 2-oxide, 6-nitrocinnoline 2-oxide, and 3-nitrocinnoline 1-oxide are useful as bactericides, fungicides, and cancer remedies (21). Another patent (21a) claims 4-nitrocinnoline 1-oxide and 5- or 8-nitrocinnoline 2-oxides as bactericides.

B. Nucleophilic Substitution

Just as the nitro group of 4-nitroquinoline 1-oxide may be displaced by a variety of nucleophiles, so also may the nitro group of 4-nitrocinnoline

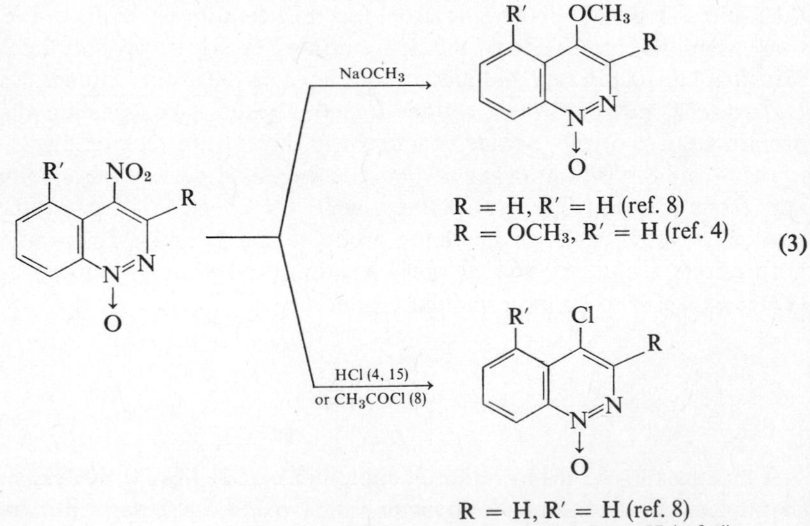

1-oxides, as indicated in Eq. 3 where the nitro group is displaced by a methoxy group, through interaction with sodium methoxide, or by a chloro group through interaction with acetyl chloride or concentrated hydrochloric acid.

One instance of nucleophilic substitution of a nitro group in the 3-position of the cinnoline 1-oxide ring has been reported (15, 16). Thus 3-nitrocinnoline 1-oxide (**35**) interacts with sodium methoxide in methanol to give 3-methoxycinnoline 1-oxide (**37**) in a yield of 95%.

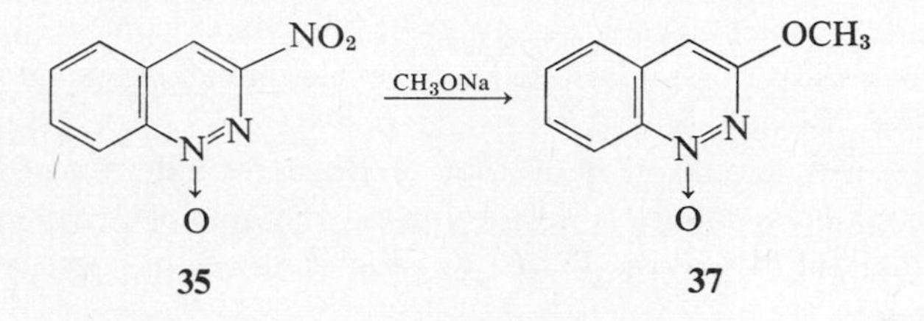

In a similar manner, chloro and methoxy groups in the 4-position of the cinnoline 1-oxide ring are found to be replaceable by nucleophilic reagents such as the hydroxide anion and the methoxide anion as shown in Eq. 4.

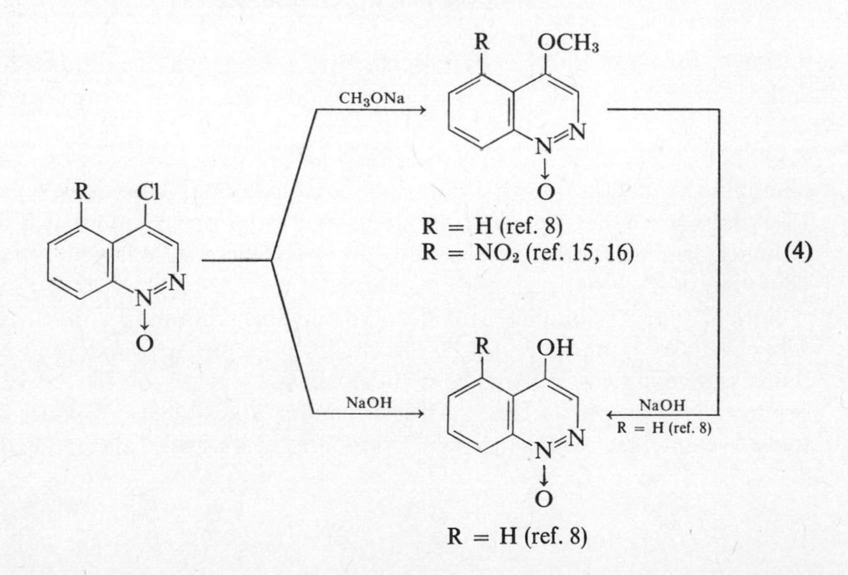

The chlorine atom in 4-chlorocinnoline 2-oxide (**17b**) is similarly labile to nucleophilic reagents, being replaced by methoxy, ethoxy, or hydroxy groups when **17b** is allowed to interact with the corresponding sodium alkoxide or hydroxide (8).

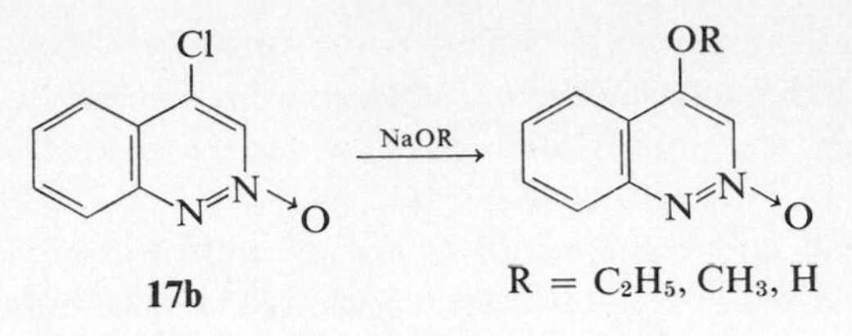

The reaction of cinnoline 1-oxide (**3**) with phosphorus oxychloride affords 4-chlorocinnoline in good yield (15, 16), whereas 4,5-dinitrocinnoline 1-oxide is reported to be inert to this reagent (15). The conversion of **3** to 4-chlorocinnoline is most probably initiated by a nucleophilic attack by the oxygen atom of the *N*-oxide function on the phosphorus atom of phosphorus oxychloride, with expulsion of a chloride ion. It would then be terminated by a nucleophilic attack by chloride ion on the adduct (at the 4-position) with concomitant loss of the $PO_2Cl_2^-$ anion from the nitrogen atom of the ring.

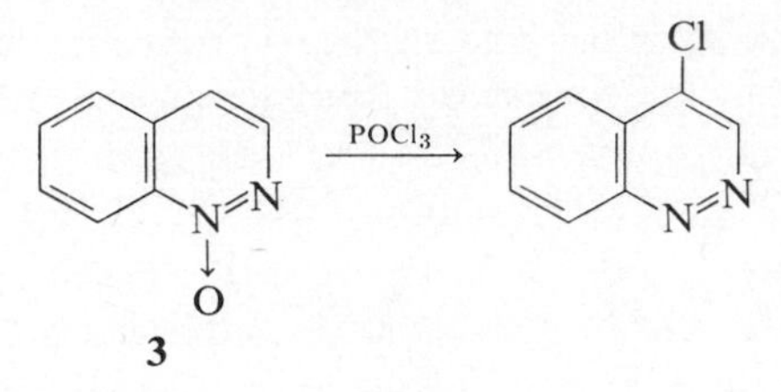

Cinnoline 2-oxides also may be converted in some cases into 4-chlorocinnolines by treatment with phosphorus oxychloride. Thus it is reported (5b) that when either 3-phenylcinnoline 1-oxide or 3-phenylcinnoline 2-oxide is boiled in phosphorus oxychloride for $1\frac{1}{2}$ hr, there is obtained 4-chlorocinnoline as the product in both instances.

Both 4-hydroxycinnoline 1-oxide and 4-hydroxycinnoline 2-oxide themselves behave as nucleophiles in the presence of methyl iodide and silver oxide, attacking methyl iodide with either the oxygen of the 4-hydroxy group or the oxygen of the *N*-oxide function. Thus when a mixture of 4-hydroxycinnoline 1-oxide (**38**), methyl iodide, silver oxide, and methanol is

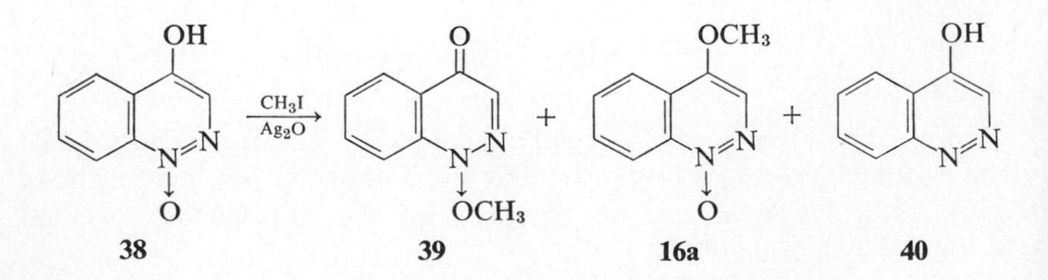

heated in a sealed tube on a steam bath for 3 hr, the major product is 1-methoxy-4(1H)cinnolinone (**39**) (28%; yellow needles, mp 82–83°) accompanied by a small amount (4%) of 4-methoxycinnoline 1-oxide (**16a**) and some 4-hydroxycinnoline (**40**) (18%) (8).

When 4-hydroxycinnoline 2-oxide (**41**) is allowed to react with methyl iodide and silver oxide under the same conditions described for the 1-oxide **38**, the major product (50%) is 4-methoxycinnoline 2-oxide (**42**), accompanied by 4-hydroxycinnoline (**40**) (8).

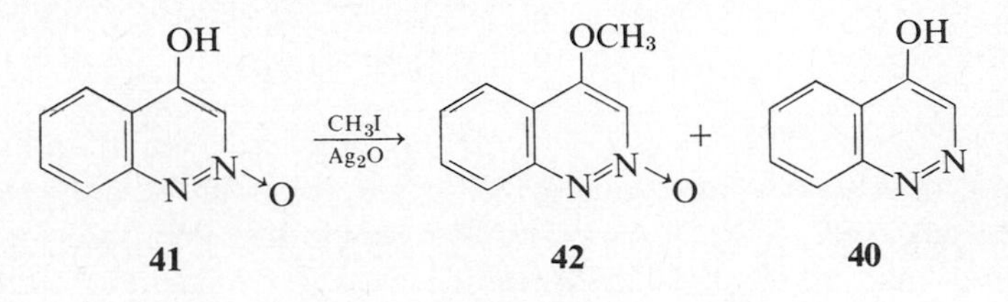

C. Reduction

1. *Catalytic Reduction*

Catalytic reductions of unsubstituted cinnoline 1-oxide and cinnoline 2-oxide are apparently not recorded in the literature, but such reductions of several substituted cinnoline *N*-oxides have been reported. For example, 3-phenyl-4-cinnolinol 1-oxide (**19**) is hydrogenated at 2 atm in ethanol over Raney nickel to give 3-phenyl-4-cinnolinol (**43**) in a 45% yield (9).

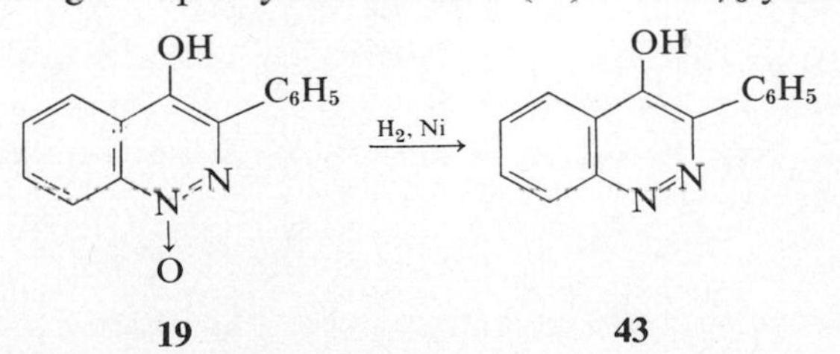

Hydrogenation of 4-nitrocinnoline 1-oxide (**32**) over Raney nickel causes removal of the *N*-oxide function and reduces the nitro group to give 4-aminocinnoline (**44**) in 75% yield (15, 16).

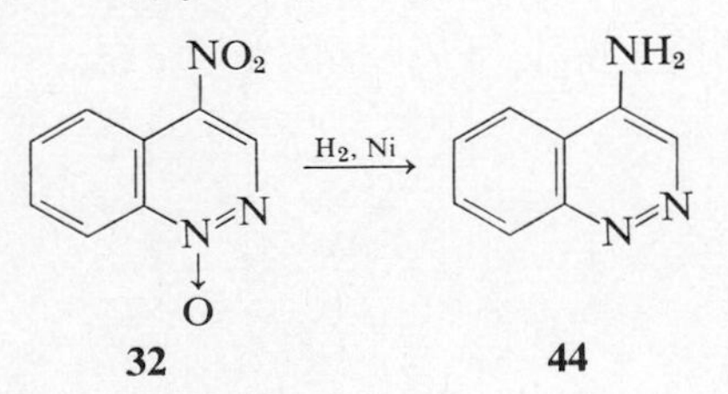

When either 8-nitrocinnoline 2-oxide (**12**) or 6-nitrocinnoline 2-oxide (**36**) is hydrogenated over palladium-charcoal in ethanol, the corresponding aminocinnolines are obtained in yields of 53 and 70%, respectively, but hydrogenation of **12** and **36** over Adams platinum catalyst in ethanol gives mixtures of the corresponding aminocinnolines and aminocinnoline 2-oxides, as shown for **36** in Eq. 5.

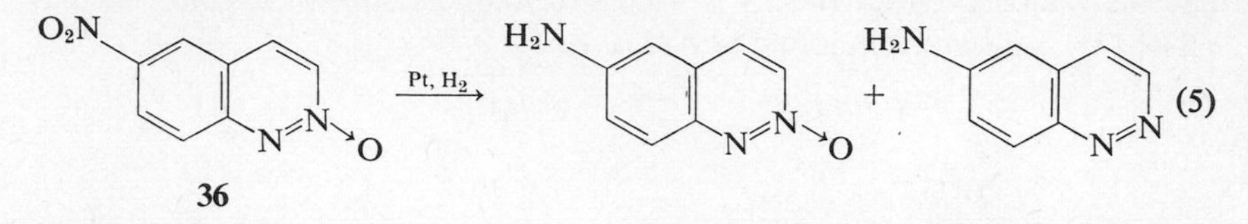

Hydrogenation of 5-nitrocinnoline 2-oxide (**8**) over Adams platinum catalyst, however, gives a 5-aminodihydrocinnoline (**45**) in 36% yield, which may have either a 1,2-dihydro or a 1,4-dihydro structure, and a small amount (5%) of 5-aminocinnoline (**46**) (17, 18). Compound **45** is most

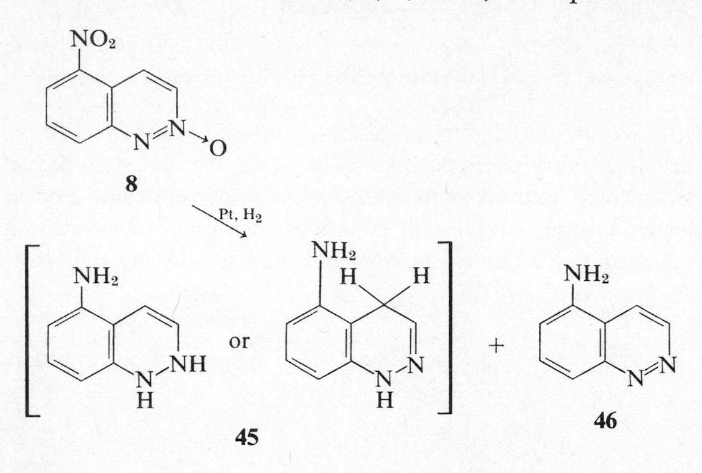

likely to have a 1,4-dihydro structure, since it has been shown (22) that dihydrocinnoline itself and its 4-methyl and 4-phenyl derivatives all exist as the 1,4-dihydro isomers.

Both 4-chlorocinnoline 1-oxide and 4-chlorocinnoline 2-oxide, when subjected to hydrogenation over palladium-charcoal in ethanol containing a small amount of 10% sodium hydroxide, suffer loss of the chlorine atom

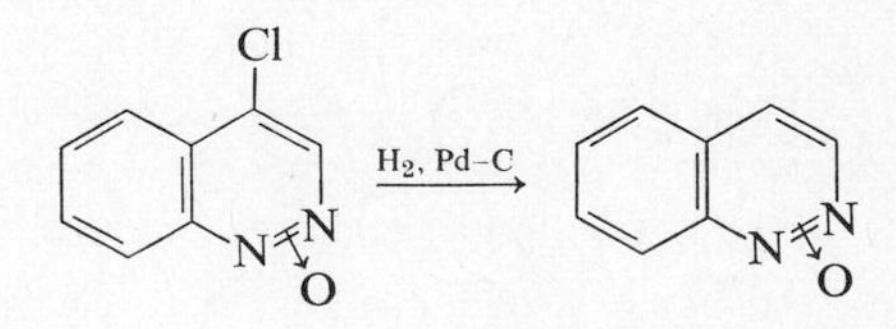

while the *N*-oxide function remains intact. Thus are obtained the corresponding cinnoline 1-oxide or cinnoline 2-oxide as indicated in Eq. 6. In the case of 4-chlorocinnoline 2-oxide, the reduction is also accompanied by the formation of 4-ethoxycinnoline 2-oxide (**47**) in 19% yield (8).

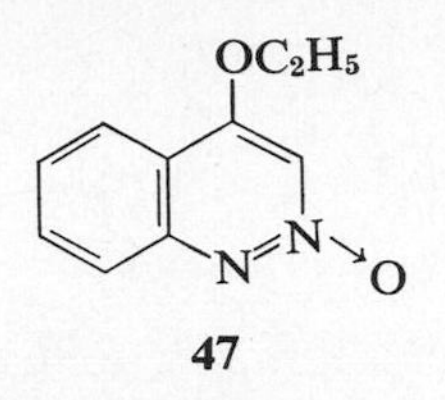

When cinnoline 1,2-dioxide (**5**) is subjected to hydrogenation over palladium-charcoal, and three molar equivalents of hydrogen are taken up by the reaction, there is obtained a dihydrocinnoline, mp 83° (5, 5a), which is most likely 1,4-dihydrocinnoline (**48**) (22); but when the hydrogenation of **5** is stopped after the uptake of only one molar equivalent of hydrogen, a mixture of the dihydro compound **48**, cinnoline, cinnoline 1-oxide, and cinnoline 2-oxide is obtained in 3.4, 3.9, 17.5, and 64.6% yields, respectively (5, 5a).

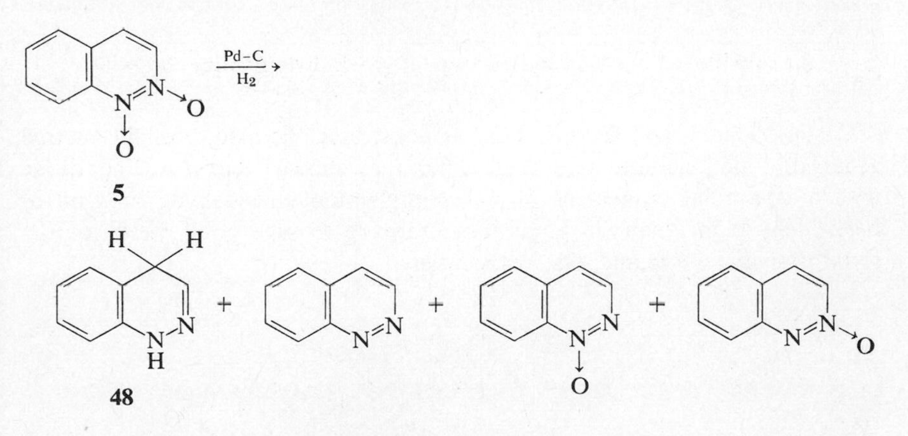

2. *Reduction of the* N—*O Bond by Phosphorus Trichloride*

Treatment of cinnoline 1-oxide (**3**) and cinnoline 2-oxide (**4**) with phosphorus trichloride in refluxing chloroform for 2 hr results in partial deoxygenation to give cinnoline (**2**) in 27.4 and 12.6% yields, respectively, while **3** and **4** are recovered in 58 and 40% yields (5a). A similar treatment of cinnoline 1,2-dioxide gives a small amount (10.1%) of cinnoline 2-oxide with recovery of the dioxide in a 72.7% yield (5, 5a). Phosphorus trichloride in chloroform will also deoxygenate 4-methoxy-5-nitrocinnoline 1-oxide to give 4-methoxy-5-nitrocinnoline (15).

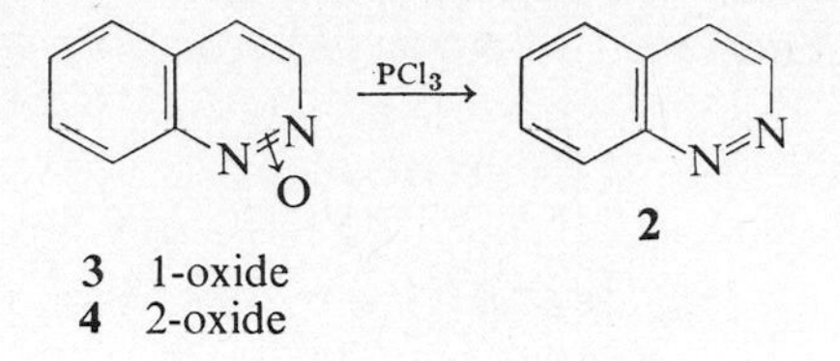

3. *Polarographic Reduction*

A polarographic reduction study has revealed that in aqueous solution, cinnoline 1,2-dioxide is more easily reduced than cinnoline itself, which is in turn more easily reduced than the cinnoline mono *N*-oxides. Of the two mono *N*-oxides, cinnoline 1-oxide is somewhat more easily reduced than cinnoline 2-oxide, and this study suggests that cinnoline 1,2-dioxide is first reduced to the 2-oxide and then rapidly reduced to the dihydrocinnoline **48** (5, 5a).

D. Acidity of the Methyl Group of 4-Methylcinnoline 2-Oxide

Castle, Adachi, and Guither (23) demónstrated the acidity of the methyl group of 4-methylcinnoline 2-oxide when they showed that it will condense with a one-molar equivalent of 4-dimethylaminobenzaldehyde or 4-nitrobenzaldehyde in ethanolic potassium ethoxide to give good yields of the styryl derivatives **49a** and **49b**, respectively.

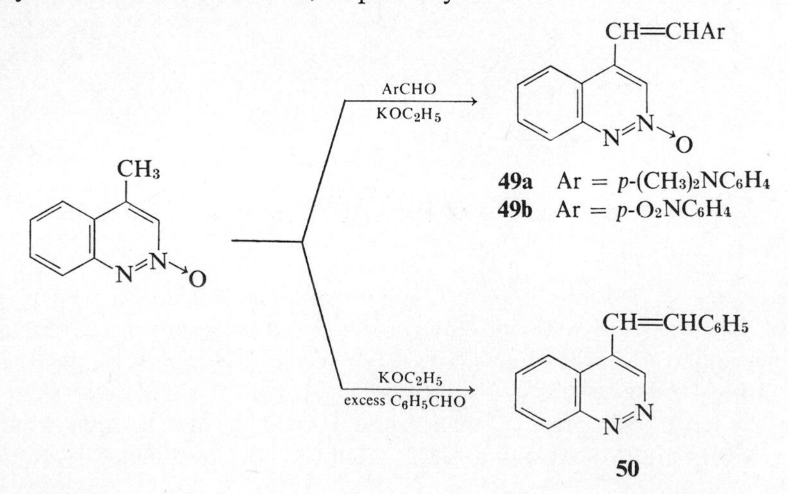

When 4-methlycinnoline 2-oxide is heated under reflux with excess benzaldehyde in the presence of potassium ethoxide, deoxygenation of the ring nitrogen occurs, and 4-styrylcinnoline (**50**) is obtained as the product.

Experiments that might show the acidity of methyl groups in the 3-position of cinnoline 2-oxide and in the 3- and 4-positions of cinnoline 1-oxide have not yet been reported in the literature.

E. Reaction of Cinnoline 2-Oxides with Phenylmagnesium Bromide

The reaction of cinnoline 2-oxide with phenylmagnesium bromide gives phenanthrene (**51**, R = H), *trans*-stilbene (**52**, R = H), *cis*-stilbene (**53**, R = H), 1,2-dihydro-2,3-diphenylcinnoline (**54**, R = H), and 2-styrylazobenzene (**55**, R = H), all in yields of 1–15%. The same products (except R = CH_3) are obtained from 4-methylcinnoline 2-oxide and phenylmagnesium bromide in similar yields. A pathway for the formation of these products has been proposed wherein initial reaction of the cinnoline 2-oxide with phenylmagnesium bromide gives the adduct **56**. Adduct **56** then undergoes cleavage to give **57**, which further reacts to give the diazonium ion **58**. This ion then undergoes loss of nitrogen, further reaction with phenylmagnesium bromide, or cyclization to produce the products **51**, **52**, **53**, **54**, and **55** (24).

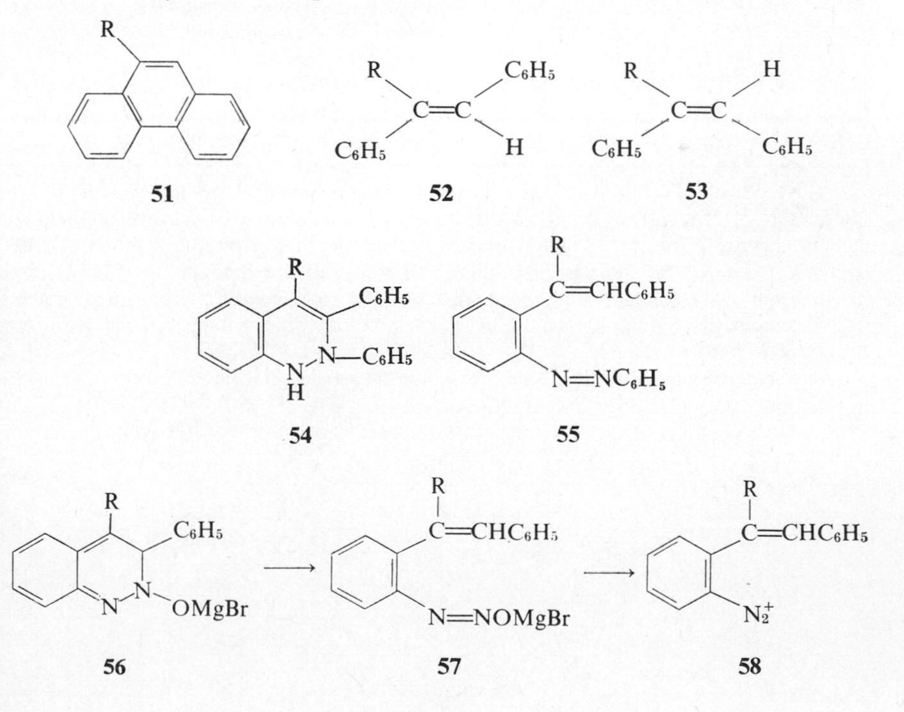

TABLE 1J-1. 5,6,7,8-Tetrahydrocinnoline 1-Oxides

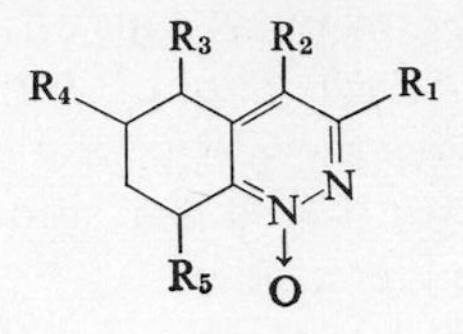

R_1	R_2	R_3	R_4	R_5	Prep.[a]	MP (°C)	Remarks[b]	Ref.
H	H	H	H	H	A, C	100–100.5	Needles (benzene)	4, 3
Cl	H	H	H	H	A	133–134	Needles (benzene–cyclohexane)	4, 3
CH_3O	H	H	H	H	B	101–102	Needles (benzene-cyclohexane)	4, 3
CH_3S	H	H	H	H	B	124–125.5	Needles (benzene-cyclohexanol)	4
H	H	H	H	Br	D	146–147, 145–146	Prisms (methanol)	4, 3
Cl	H	H	H	Br	D	93–94	Needles (ethanol)	4, 3
H	H	Br	H	Br	E	149–150	Needles (ethanol)	4, 3
Cl	H	Br	H	Br	E	137–138	Prisms (ethanol)	4, 3
CH_3O	H	H	Br	Br	D	151–152	Needles (ethanol)	4, 3
Cl	H	H	CH_3	H	A	133–135.5	Needles (benzene-cyclohexane)	4
H	H	H	CH_3	H	C	120–122	Needles	4
H	H	H	CH_3	Br	D	144–145	Prisms	4
CH_3O	H	H	CH_3	H	B	168–169	Plates	4
CH_3O	Cl	H	H	H	A	132–133	Needles (cyclohexane)	4

[a] A, by heating the tetrahydrocinnoline in hydrogen peroxide and acetic acid. B, by interaction of the corresponding 3-chlorotetrahydrocinnoline 1-oxide with methanolic sodium methoxide or with sodium methyl sulfide, depending on the product. C, by catalytic reduction of the corresponding 3-chlorotetrahydrocinnoline 1-oxide. D, by bromination of the corresponding tetrahydrocinnoline 1-oxide with *N*-bromosuccinimide. E, by bromination of the corresponding 8-bromotetrahydrocinnoline 1-oxide with *N*-bromosuccinimide.

[b] All compounds listed in this table are reported as being colorless. Solvent employed in recrystallization is enclosed in parentheses.

TABLE 1J-2. Ultraviolet Absorption Spectra of Cinnoline *N*-Oxides

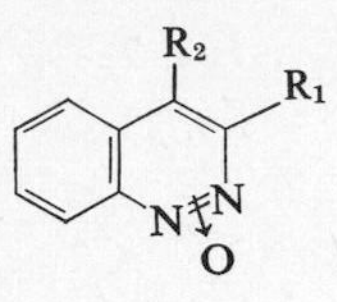

R_1	R_2	Position of Oxygen	Solvent	λ_{max} (mμ)	log E_{max}	Ref.[a]
H	H	1	Ethanol	219 (—)	4.41 (—)	3, 4, (5a)
			(Ethanol)	229.5 (230)	4.40 (3.70)	
				303 (–)	3.71 (–)	
				315 (315)	3.71 (4.03)	
				351 (351)	3.90 (3.89)	
				367 (367)	3.88 (3.87)	
H	H	2	Ethanol	218 (–)	4.48 (–)	3, 4, (5a)
			(Ethanol)	262 (262)	4.44 (4.47)	
				(267, shoulder)		
				308 (310)	3.76 (3.68)	
				351 (352)	3.70 (3.69)	
				360 (361)	3.70 (3.69)	
H	CH_3	1	Ethanol	219 (220)	4.42 (4.36)	3, 4, (6)
			(Ethanol)	229.5 (—)	4.37 (—)	
				253 (263)	3.93 (4.12)	
				308 (—)	3.70 (—)	
				318 (315)	3.72 (3.58)	
				357 (360)	3.94 (3.75)	
				369 (—)	3.90 (—)	
H	CH_3	2	Ethanol	222 (224)	4.33 (4.42)	3, 4, (6)
			(Ethanol)	261 (262)	4.37 (4.57)	
				307 (—)	3.80 (—)	
				348.5 (—)	3.70 (—)	
				357 (357)	3.68 (3.54)	
C_6H_5	OH	1	Ethanol	225	4.39	9
				246 (infl.)	4.09	
				269	4.09	
				330	3.95	
H	H	1,2	Ethanol	234	4.26	5, 5a
				258	4.35	
				273	4.55	
				340	3.81	
H	CH_3	1,2	Ethanol	236	4.42	6
				275	4.60	
				348	3.85	

[a] References in parentheses refer to absorption values and solvents that are also enclosed in parentheses.

TABLE 1J-3. Proton Magnetic Resonance Spectra of Cinnoline *N*-Oxides[a]

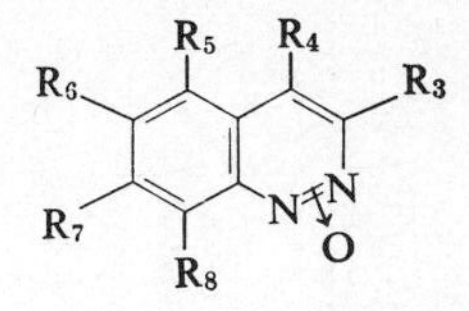

Position of Oxygen	R_3	R_4	R_5	R_6	R_7	R_8	Solvent[b]	Ref.
1	H	H	H	H	H	H		
	1.67	2.50	—	—	—	1.33	$CDCl_3$	3, 4, 14
	$J_{3,4} = 6.2$	$J_{4,8} = 0.9$						
	1.75	2.61	—	—	—	1.41	$CDCl_3$	5, 5a
	$J_{3,4} = 6.0$	$J_{4,8} = 1.0$				$J_{7,8} = 6.0$		
	1.8	2.53	←	2.3(M)	→	1.33	—	6a
1	H	CH_3	H	H	H	H		
	1.87	7.42	←	2.1(M)	→	1.35	$CDCl_3$	3, 4, 6, 14
	$J_{3,4} = 1$							
	1.83	7.42	←	2.3(M)	→	1.33	—	6a
1	H	CH_3	H	CH_3	H	H		
	1.90	7.40	2.40	7.44	2.40	1.45	—	6a
1	NO_2	CH_3	H	H	H	H		
	—	7.0	←	1.70(M)	→	1.10	CF_3COOH	6a
1	OCH_3	H	H	H	H	H		
	5.93	3.08	—	—	—	1.52	$CDCl_3$	4, 14
		$J_{4,8} = 1.0$						
1	C_6H_5	H	H	H	H	H		
	2.75–1.83(M)	2.23	←	2.75–1.83(M)	→	1.55–1.22	$CDCl_3$	5b

1	p-ClC_6H_4	H	H	H	H	H		
	1.98(D), 2.55(D)	2.17	←——	2.50–2.00(M)	——→	1.53–1.22	$CDCl_3$	5b
1	p-$CH_3OC_6H_4$	H	H	H	H	H		
	2.05(D) 3.05(D), 6.18(S)	←——	——	2.58–2.33(M)	——	——→ 1.62–1.32	$CDCl_3$	5b
1	C_6H_5	$COOC_2H_5$	H	H	H	H		
	2.67–1.67(M)	—	←——	2.67–1.67(M)	——→	1.38–1.08	$CDCl_3$	5b
2	H	H	H	H	H	H		
	1.79	1.94	—	—	—	—	$CDCl_3$	3, 4, 14
	$J_{3,4} = 7.0$	$J_{4,8} = 1.0$						
	1.85	2.05	—	—	—	—	$CDCl_3$	5, 5a
	1.80	1.90	←——	——	2.3(M)	——→	—	6a
2	H	COOH	H	H	H	H		
	1.40	—	←——	——	2.10(M)	——→	$(CH_3)_2SO$	6a
2	H	CH_3	H	H	H	H		
	1.90	7.37	←——	——	2.2(M)	——→	$CDCl_3$	3, 4, 6, 6a, 14
	$J_{3,4} = 1$							
2	H	CH_3	H	CH_3	H	H		
	1.90	7.40	2.35	7.42	2.35	2.20	—	6a
2	CH_3	CH_3	H	H	H	H		
	7.40	7.40	←——	——	2.37(M)	——→	—	6a
2	CH_3	C_2H_5	H	H	H	H		
	7.35	6.88, 8.35	←——	——	2.30(M)	——→	—	6a
2	C_6H_5	H	H	H	H	H		
	2.60–2.00(M)	1.90	←——	——	2.60–2.00(M)	——→	$CDCl_3$	5b
1,2	H	H	H	H	H	H		
	1.92	2.55	—	—	—	1.66	$CDCl_3$	5, 5a
	$J_{3,4} = 7.5$	$J_{4,8} = 1.0$				$J_{7,8} = 9.6$		

TABLE 1J-3. (continued)

Position of Oxygen	R_3	R_4	R_5	R_6	R_7	R_8	Solvent[b]	Ref.
1,2	H	CH_3	H	H	H	H		
	1.90	7.40	←	2.20(M)	→	1.70	$CDCl_3$	6, 6a
	$J_{3,4} = 1.2$							
1,2	CH_3	C_2H_5	H	H	H	H		
	7.32	6.95, 8.70	←	2.20(M)	→	1.60	—	6a
1,2	CH_3	CH_3	H	H	H	H		
	7.36	7.36	←	2.30(M)	→	1.70	—	6a

[a] Chemical shift values, given in τ units, are recorded relative to the reference standard tetramethylsilane, and are placed immediately below the proton(s) to which they refer. A dash below a proton indicates that a chemical shift was not assigned to that proton. The multiplicity of an absorption, when indicated, is given in parentheses following the τ value as S = singlet, D = doublet, and M = multiplet or center of multiplet. J = coupling constant in cps.

[b] A dash indicates that solvent employed was not identified in original paper.

TABLE 1J-4. Cinnoline 1-Oxides

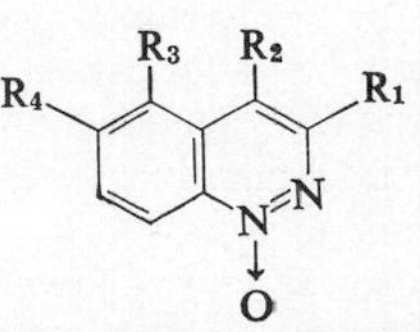

R_1	R_2	R_3	R_4	Prep.[a]	MP (°C)	Remarks[b]	Ref.
H	H	H	H	A, B, I	110.5–111.5, 111–112	Nearly colorless plates (benzene-petroleum benzin)	4, 8
H	CH_3	H	H	A	94–95	Pale yellow needles (benzene-petroleum benzin)	4
Cl	H	H	H	A, B	168–169	Pale yellow needles (benzene or benzene-cyclohexane)	4
CH_3O	H	H	H	B, S	94–95, 92.5–93.5	Yellow needles (benzene-petroleum benzin, or diisopropyl ether)	4, 15, 16
CH_3S	H	H	H	B	119–120	Yellow needles (benzene-petroleum benzin)	4
CH_3O	NO_2	H	H	C	154–155	Yellow needles (ethanol)	4
CH_3O	Cl	H	H	D, B	169–170	Yellow needles (ethanol)	4
CH_3O	CH_3O	H	H	D	116–117	Yellow prisms (benzene-cyclohexane)	4
C_6H_5	OH	H	H	E	220–227	Yellow-green needles (methanol-water); sublimes to white solid above 198°	9
p-$NO_2C_6H_4$	H	H	H	F	267–268	Yellow needles (dioxane-alcohol)	10
H	H	NO_2	H	A, G, N	182–183	Yellow needles (acetone)	7, 15
H	Cl	H	H	H, J	94–94.5	Colorless needles (ether)	8
H	CH_3O	H	H	K, L	107–108	Yellow needles (diisopropyl ether)	8
H	OH	H	H	K, M	153 (dec.)	Yellow needles (methanol)	8
H	NO_2	H	H	N, O	161–162	Yellow needles (acetone)	15, 16
H	H	NO_2	H	N	182–183	Eluted from alumina with benzene	15

TABLE 1J-4. (*continued*)

R_1	R_2	R_3	R_4	Prep.[a]	MP (°C)	Remarks[b]	Ref.
H	NO_2	NO_2	H	N, P	191–192	Yellow needles (acetone)	15
H	Cl	NO_2	H	Q	238–240	Pale yellow needles (methanol)	15
H	CH_3O	NO_2	H	K	222–224	Yellow powder (benzene-hexane)	15
NO_2	H	H	H	R	214–215	Yellow needles (chloroform)	15, 16
C_6H_5	H	H	H	A	138–139	Yellow needles (methanol)	5b
p-ClC_6H_4	H	H	H	A, T	184–186	Yellow needles (benzene)	5b
p-$CH_3OC_6H_4$	H	H	H	T	165–169	Crystals (methanol)	5b
C_6H_5	$CO_2C_2H_5$	H	H	A	90–92	Crystals (ligroin)	5b
H	CH_3	H	CH_3	A	160–161	—	6a
H	CH_3	H	H	A	94–95	—	6a
H	H	H	H	A	107	—	6a
NO_2	CH_3	H	H	A	170	—	6a

[a] A, by heating the cinnoline in hydrogen peroxide and acetic acid. B, by bromination of the 5,6,7,8-tetrahydrocinnoline 1-oxide derivative with *N*-bromosuccinimide, followed by dehydrohalogenation by sodium methoxide. C, by nitration of 3-methoxycinnoline 1-oxide. D, by displacement of the nitro group of 3-methoxy-4-nitrocinnoline 1-oxide by chloride ion in concentrated hydrochloric acid, or by the methoxide ion from sodium methoxide, depending on the product. E, by reaction of 2-phenylisatogen with ammonia in ethanol at 140–145° (see text). F, by photochemical cyclization of a *o*-nitrophenyl enamine (see text). G, by oxidation of the cinnoline with persulfuric acid in sulfuric acid. H, by oxidation of the cinnoline in ethereal phthalic monoperacid. I, by catalytic hydrogenation of 4-chlorocinnoline 1-oxide. J, by reaction of 4-nitrocinnoline 1-oxide with acetyl chloride. K, by treatment of 4-chlorocinnoline 1-oxide with methanolic sodium methoxide, or with sodium hydroxide, depending on the product. L, by reaction of 4-nitrocinnoline 1-oxide with methanolic sodium methoxide. M, by reaction of 4-methoxycinnoline 1-oxide with sodium hydroxide. N, by nitration of cinnoline 1-oxide in nitric acid and sulfuric acid. O, by nitration of cinnoline 1-oxide with potassium nitrate in sulfuric acid. P, by nitration of 4-nitro- or 5-nitrocinnoline 1-oxide in nitric acid and sulfuric acid. Q, by treatment of 4,5-dinitrocinnoline 1-oxide with concentrated hydrochloric acid. R, by treatment of cinnoline 1-oxide with benzoyl nitrate. S, by treatment of 3-nitrocinnoline 1-oxide with methanolic sodium methoxide. T, by treatment of the cinnoline with *m*-chloroperbenzoic acid in benzene.

[b] Solvent employed in recrystallization is enclosed in parentheses.

TABLE 1J-5. Cinnoline 2-Oxides

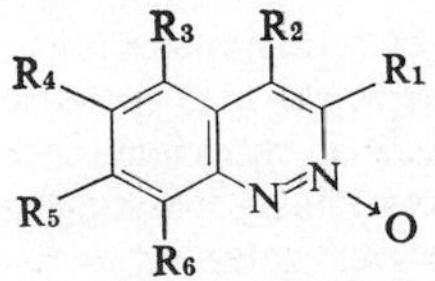

R_1	R_2	R_3	R_4	R_5	R_6	Prep.[a]	MP (°C)	Remarks[b]	Ref.
H	H	H	H	H	H	A, E	125–126, 122–123	Colorless plates (benzene), colorless needles (acetone)	4, 8
H	CH_3	H	H	H	H	A	151–152	Pale yellow needles (benzene)	4
H	H	NO_2	H	H	H	A, B, I, J	215–217	Yellow needles (ethanol)	7, 17, 18
H	H	H	H	H	NO_2	A, B, I	228 (dec.)	Yellow needles (ethanol)	7, 17, 18
H	Cl	H	H	H	H	C	150–151	Yellow needles (benzene)	8
H	C_2H_5O	H	H	H	H	D	190–191	Colorless needles (benzene)	8
H	CH_3O	H	H	H	H	D, C	176–177	Colorless needles (benzene)	8
H	OH	H	H	H	H	F, G	257 (dec.)	Colorless needles (methanol)	8
H	*p*-$(CH_3)_2NC_6H_4CH{:}CH$	H	H	H	H	H	247–248	Eluted from alumina with chloroform	23
H	*p*-$NO_2C_6H_4CH{:}CH$	H	H	H	H	H	287–288	Crystalline solid (dimethylformamide)	23
H	$C_6H_5CH{:}CH$	H	H	H	H	C	205–207	Yellow needles (ethanol)	23
H	H	H	NO_2	H	H	I	212–213	Pale yellow needles (ethanol)	17, 18
H	H	H	NH_2	H	H	K	247 (dec.)	Yellow needles	17, 18
H	H	H	H	H	NH_2	K	212 (dec.)	Yellow scales	17, 18
NO_2	H	H	H	H	H	I	227–228	—	21
C_6H_5	H	H	H	H	H	A	181–182	White flakes (methylene chloride-ligroin)	5b
H	H	H	H	H	H	A	126	—	6a

TABLE 1J-5. (*continued*)

R_1	R_2	R_3	R_4	R_5	R_6	Prep.[a]	MP.[a] °C	Remarks[b]	Ref.
H	CH_3	H	H	H	H	A	147–148	—	6a
H	CH_3	H	CH_3	H	H	A	230–231	—	6a
CH_3	C_2H_5	H	H	H	H	A	98–99	—	6a
CH_3	CH_3	H	H	H	H	A	160–161	—	6a
H	COOH	H	H	H	H	A	242	—	6a
H	2,4-$Cl_2C_6H_3SO_2$—	H	OCH_3	OCH_3	H	L	227	—	25
H	C_6H_5	H	OCH_3	OCH_3	H	L	239 (dec.)	—	25
H	*p*-$CH_3C_6H_4$	H	OCH_3	OCH_3	H	L	268 (dec.)	—	25
H	*o*-ClC_6H_4	H	OCH_3	OCH_3	H	L	217 (dec.)	—	25
H	m-ClC_6H_4	H	OCH_3	OCH_3	H	L	233	—	25
H	*p*-ClC_6H_4	H	OCH_3	OCH_3	H	L	264 (dec.)	—	25
H	2,5-$Cl_2C_6H_3$	H	OCH_3	OCH_3	H	L	254 (dec.)	—	25
H	3,5-$Cl_2C_6H_3$	H	OCH_3	OCH_3	H	L	220 (dec.)	—	25
H	3,4-$Cl_2C_6H_3$	H	OCH_3	OCH_3	H	L	226 (dec.)	—	25
H	2,3-$Cl_2C_6H_3$	H	OCH_3	OCH_3	H	L	233 (dec.)		25
H	3,6,4-$Cl_2(HO)C_6H_2$	H	OCH_3	OCH_3	H	L	240 (dec.)	—	25
H	5,2-$Cl(CH_3)C_6H_3$	H	OCH_3	OCH_3	H	L	245 (dec.)	—	25

[a] A, by heating the cinnoline in hydrogen peroxide and acetic acid. B, by oxidation of the cinnoline with persulfuric acid in sulfuric acid, with ethereal phthalic monoperacid, or with hydrogen peroxide alone or in sulfuric acid. C, by oxidation of the cinnoline in ethereal phthalic monoperacid. D, by reaction of 4-chlorocinnoline 2-oxide with ethanolic sodium ethoxide, or with methanolic sodium methoxide, depending on the product. E, by catalytic hydrogenation of 4-chlorocinnoline 2-oxide. F, by reaction of 4-chlorocinnoline 2-oxide with sodium hydroxide. G, by reaction of 4-methoxycinnoline 2-oxide with sodium hydroxide. H, by reaction of 4-methylcinnoline 2-oxide with the corresponding 4-substituted benzaldehyde. I, by nitration of cinnoline 2-oxide with nitric acid or with potassium nitrate in sulfuric acid. J, by reaction of cinnoline 2-oxide with benzoyl nitrate in chloroform. K, by catalytic hydrogenation of the corresponding nitrocinnoline 2-oxide. L, by oxidation of the corresponding 4-arylthiocinnoline with hydrogen peroxide in acetic acid.

[b] Solvent employed in recrystallization is enclosed in parentheses.

References

1. C. M. Atkinson and J. C. E. Simpson, *J. Chem. Soc.*, **1947**, 1649; J. C. E. Simpson, in *The Chemistry of Heterocyclic Compounds*, Vol. V, A. Weissberger, Ed., Interscience, New York–London, 1953, p. 9.
2. C. M. Atkinson and J. C. E. Simpson, *J. Chem. Soc.*, **1947**, 808.
3. M. Ogata, H. Kano, and K. Tori, *Chem. Pharm. Bull.* (*Tokyo*), **10**, 1123 (1962).
4. M. Ogata, H. Kano, and K. Tori, *Chem. Pharm. Bull.* (*Tokyo*), **11**, 1527 (1963).
5. I. Suzuki, M. Nakadate, T. Nakashima, and N. Nagasawa, *Tetrahedron Lett.*, **1966**, 2899.

5a. I. Suzuki, M. Nakadate, T. Nakashima, and N. Nagasawa, *Chem. Pharm. Bull.* (*Tokyo*), **15**, 1088 (1967).

5b. H. S. Lowrie, *J. Med. Chem.*, **9**, 670 (1966).

6. M. H. Palmer and E. R. R. Russell, *Chem. Ind.* (*London*), **1966**, 157.

6a. M. H. Palmer and E. R. R. Russell, *J. Chem. Soc.* (*C*), **1968**, 2621.

7. I. Suzuki, T. Nakashima, and N. Nagasawa, *Chem. Pharm. Bull.* (*Tokyo*), **13**, 713 (1965).
8. I. Suzuki and T. Nakashima, *Chem. Pharm. Bull.* (*Tokyo*), **12**, 619 (1964).
9. W. E. Noland and D. A. Jones, *J. Org. Chem.*, **27**, 341 (1962).
10. F. Kröhnke and I. Vogt, *Ann.*, **589**, 26 (1954).
11. M. Ogata and H. Kano, *Chem. Pharm. Bull.* (*Tokyo*), **11**, 35 (1963).
12. M. Ogata and H. Kano, *Chem. Pharm. Bull.* (*Tokyo*), **11**, 29 (1963).
13. T. Nakagome, *Yakugaku Zasshi*, **81**, 1048 (1961); **82**, 250 (1962).
14. K. Tori, M. Ogata, and H. Kano, *Chem. Pharm. Bull.* (*Tokyo*), **11**, 681 (1963).

14a. M. H. Palmer and E. R. R. Russell, *Org. Mass. Spectrom.*, **2**, 1265 (1969).

15. I. Suzuki, T. Nakashima, N. Nagasawa, and T. Itai, *Chem. Pharm. Bull.* (*Tokyo*), **12**, 1090 (1964).
16. I. Suzuki, T. Nakashima, and T. Itai, *Chem. Pharm. Bull.* (*Tokyo*), **11**, 268 (1963).
17. I. Suzuki, T. Nakashima, and N. Nagasawa, *Chem. Pharm. Bull.* (*Tokyo*), **14**, 816 (1966).
18. I. Suzuki, T. Nakashima, and N. Nagasawa, *Chem. Pharm. Bull.* (*Tokyo*), **11**, 1326 (1963).

18a. J. T. Gleghorn, R. B. Moodie, E. A. Qureshi, and K. Schofield, *J. Chem. Soc.* (*B*), **1968**, 316.

19. H. Kano and M. Ogata, Japanese Pat. 22,008 (1964); *Chem. Abstr.*, **63**, 5658 (1965).
20. F. Miyazawa, T. Hashimoto, S. Iwahara, T. Itai, I. Suzuki, S. Sako, S. Kamiya, S. Natsume, T. Nakashima, and G. Okusa, *Bull. Nat. Inst. Hyg. Sci.* (*Japan*), **81**, 98 (1963); *Chem. Abstr.*, **63**, 3477 (1965).
21. T. Itai, I. Suzuki, and T. Nakajima, Japanese Pat. 1183 (1967) Cl. 16E46; *Chem. Abstr.*, **66**, 95067 (1967).

21a. K. Koshinuma, C. Ishizeki, S. Iwahara, T. Itai, T. Nakashima, S. Sueyoshi, K. Nato, S. Kamiya, and I. Suzuki, *Eisei Shikenjo Hokoku*, **1968**, (86), 157; *Chem. Abstr.*, **71**, 110108q (1969).

22. L. S. Besford, G. Allen, and J. M. Bruce, *J. Chem. Soc.*, **1963**, 2867.
23. R. N. Castle, K. Adachi, and W. D. Guither, *J. Heterocyclic Chem.*, **2**, 459 (1965).
24. H. Igeta, T. Tsuchiya, T. Nakai, G. Okusa, M. Kumagai, J. Miyoshi, and T. Itai, *Chem. Pharm. Bull.* (*Tokyo*), **18**, 1497 (1970).
25. S. M. Yarnal and V. V. Badiger, *Arch. Pharm.* (*Weinheim*), **303**, 560 (1970); *Chem. Abstr.*, **73**, 77174z (1970).

Part K. Reduced Cinnolines

I. Reduced Cinnolines from Reduction of Cinnolines

The reduction of cinnolines, discussed in detail in previous sections, is considered here in more general terms for convenience. It should be noted that in all such reductions, whether chemical, catalytic, or electrolytic, the benzenoid ring usually is left intact, whereas the hetero ring is reduced in varying degrees depending upon the reduction conditions or upon the particular cinnoline being reduced. Thus the system may be reduced to a 1,4-dihydro form or a 1,2,3,4-tetrahydro form; the hetero ring may be cleaved and rearranged to an indole or it may be simply cleaved to give an alkyl-substituted benzene.

A. Chemical Reduction

Reduction of cinnoline, 4-methylcinnoline, 4-phenylcinnoline, 3-methyl-4-phenylcinnoline, and 3,4-dimethylcinnoline by lithium aluminum hydride in refluxing ether solution produces the corresponding 1,4-dihydro derivatives and not the 1,2-dihydro derivatives as proposed in the early literature. These are stable enough to be isolated and their pmr spectra have been recorded (1,2). They are in equilibrium with the corresponding 1-aminoindoles in hot, dilute, aqueous hydrochloric acid solution (2). An early report (3) that 1,4-dihydro-4-phenylcinnoline disproportionates into 4-phenylcinnoline and 1,2,3,4-tetrahydro-4-phenylcinnoline in the presence of concentrated hydrochloric acid could not be duplicated. Instead, the corresponding 1-aminoindole was obtained (2). Further reduction of 1,4-dihydro-4-methylcinnoline with Adams platinum catalyst in glacial acetic acid at atmospheric pressure gives 4-methyl-1,2,3,4-tetrahydrocinnoline (4).

With diethyl butylmalonate in the presence of sodium ethoxide, 1,4-dihydro-4-phenylcinnoline gives a 1,2-butylmalonyl derivative (**1**), by rearrangement of the 1,4-dihydrocinnoline to its 1,2-dihydro form (5), as in the preparation of the corresponding phthaloyl derivative (**2**; R = CH_3 or C_6H_5) from phthalic anhydride in pyridine solution (1). Several other alkyl- and aryl-substituted 1,2-malonyl-1,2-dihydrocinnolines similar to **1** have been prepared and are claimed to be antiinflammatory, analgesic, and antipyretic agents (6, 6a). It is interesting that these compounds which show biological activity are structurally related to the drug Phenylbutazone (**3**) which is also an antiinflammatory, analgesic, and antipyretic agent.

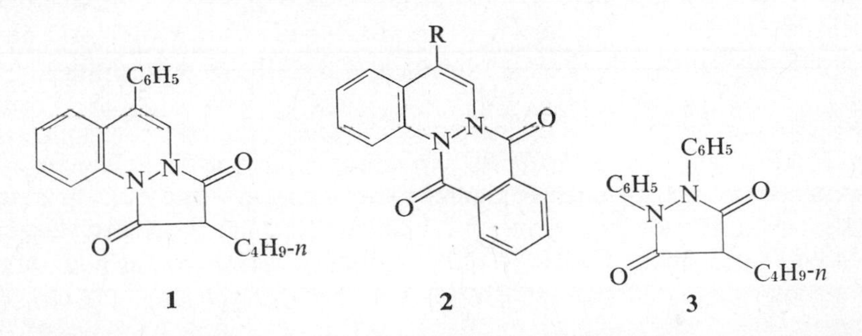

In other reductions with lithium aluminum hydride, both 4-hydroxycinnoline and 3-hydroxycinnoline are reduced in refluxing 1,2-dimethoxyethane to 1,2,3,4-tetrahydrocinnoline (7, 8), whereas 3-hydroxy-4-phenylcinnoline is reduced to 1,4-dihydro-4-phenylcinnoline under the same conditions (5). Both 2-methyl-3-cinnolinone and the anhydro base of 4-hydroxy-2-methylcinnolinium hydroxide are reduced by lithium aluminum hydride to 2-methyl-1,2,3,4-tetrahydrocinnoline. In the same way, 1-methyl-4-cinnolinone is reduced to 1-methyl-1,2,3,4-tetrahydrocinnoline (8). Reduction of 4-methoxycinnoline by lithium aluminum hydride in refluxing benzene solution yields a mixture of cinnoline and 1,2,3,4-tetrahydrocinnoline (7). Reduction of 4-chlorocinnoline by lithium aluminum hydride gives 4,4′-dicinnolinyl (7), but with iron powder in aqueous sulfuric acid solution it gives 1,4-dihydrocinnoline, which may be oxidized to cinnoline with mercuric oxide (9). Unlike the 4-chloro isomer, 3-chlorocinnoline is reduced by lithium aluminum hydride in a refluxing benzene-ether solution to 3-chloro-1,4-dihydrocinnoline, which can be dechlorinated over palladium on charcoal in ethanol solution to give 1,4-dihydrocinnoline. Hydrogenation of 3-chlorocinnoline over palladium on charcoal in ethanol gives 1,4-dihydrocinnoline directly (5).

Treatment of cinnoline with amalgamated zinc in refluxing aqueous acetic acid for 2 hr yields indole in approximately 57% yield. When this reaction is stopped as soon as the yellow color of the reaction mixture is discharged (4 min), then 1,4-dihydrocinnoline is isolated and can be further reduced with amalgamated zinc in acetic acid to indole (10). In the same way, 4-methylcinnoline and 4-phenylcinnoline are reduced by amalgamated zinc in refluxing aqueous acetic acid first to their 1,4-dihydro derivatives, and then to skatole and 3-phenylindole, respectively. The 1,4-dihydrocinnolines are probably intermediates in the production of the indoles (10). Under the same conditions with amalgamated zinc and acetic acid, 4-cinnolinecarboxylic acid is reduced and decarboxylated first to 1,4-dihydrocinnoline and then to indole (10). Reduction of 3-phenylcinnoline with zinc dust and barium hydroxide in refluxing aqueous ethanol, however, yields only 1,4-dihydro-3-phenylcinnoline (11). Treatment of 1-methyl-4-cinnolinone with zinc dust in acetic acid opens the hetero ring to give 2-methylaminoacetophenone, while the anhydro base of 2-methyl-4-hydroxycinnolinium hydroxide with zinc and ammonia gives 2-aminoacetophenone (8). 1-Methyl-3-phenylcinnolinium iodide is reduced by sodium borohydride (or electrolytically) to 1,4-dihydro-1-methyl-3-phenylcinnoline (11a).

Reduction of 4-methyl-, 4-phenyl-, 3,4-dimethyl-, and 3-methyl-4-phenylcinnoline with sodium in ethanol solution yields in each case a mixture of the corresponding 1,4-dihydro derivative and the corresponding indole (2).

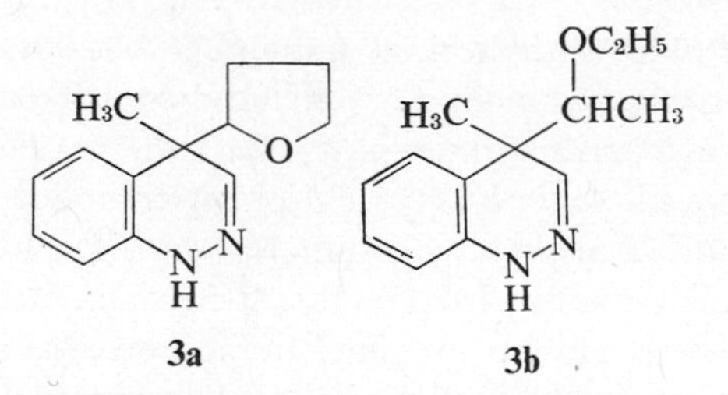

Reduction of 2-methyl-3-cinnolinone with phosphorus and hydriodic acid gives methylamine and oxindole (8).

4-Methylcinnoline, on irradiation in ethers, produces low yields of photoproducts such as 1,4-dihydro-4-methyl-4-(2-tetrahydrofuryl)cinnoline (**3a**) and 1,4-dihydro-4-methyl-4-(α-ethoxyethyl)cinnoline (**3b**) by 1,4-addition of the ethers (11b).

Treatment of 4-hydroxy-6,7-methylenedioxycinnoline-3-carbonitrile (**3c**) with sodium hydride in dimethylformamide followed by addition of an alkyl halide gives 1-alkyl-6,7-methylenedioxy-4(1*H*)oxocinnoline-3-carbonitriles (**3d**). Acidic hydrolysis of the nitriles **3d** gives the corresponding 3-carboxylic acids which, together with their salts having pK_a greater than 7.5, show antimicrobial activity (11c).

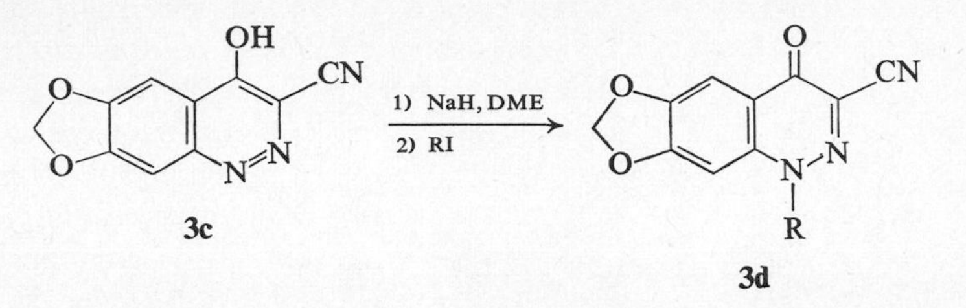

1,3-Dipolar cycloaddition of dimethyl acetylenedicarboxylate to the anhydro base of 4-hydroxy-2-methylcinnolinium hydroxide (**3e**) gives the adduct **3f** (11d).

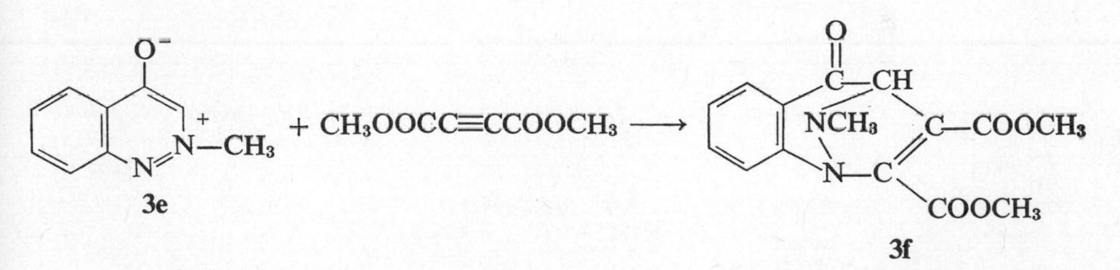

A well-executed investigation by Shah and Taylor (11e) outlines the reaction of cinnoline and various alkylcinnolines with dimethylketene. Dimethylketene forms two types of adduct with cinnolines, both having an oxadiazinocinnolinone structure. Cinnolines lacking a substituent in the 3-position form 1,3,4-oxadiazino[4,3-*a*]cinnolinones resulting from nucleophilic attack on the ketene by N-2. For example, unsubstituted cinnoline and two moles of dimethylketene give 4-isopropylidene-1,1-dimethyl-4*H*-[1,3,4]oxadiazino[4,3-*a*]cinnolin-2(1*H*)-one (**3g**), which undergoes alkaline hydrolysis to give the amido acid **3h**. 3-Alkylcinnolines form 1,3,4-oxadiazino[3,4-*a*]cinnolinones predominantly. These arise from nucleophilic attack on the ketene by N-1. For example, 3-isopropyl-4-methylcinnoline and dimethylketene give 6-isopropyl-1-isopropylidene-4,4,7-trimethyl-1*H*-[1,3,4]-oxadiazino[3,4-*a*]cinnolin-3(4*H*)-one (**3i**), which can be hydrolyzed to the amido acid **3j**. Hydrogenation of amido acids possessing the structure of **3h** in ethanol or acetic acid over platinum oxide causes reduction of the double bond in the hetero ring. The product from this reduction, when treated with polyphosphoric acid, unexpectedly gives 2-isobutyryl-1,2,3,4-tetrahydrocinnolines such as **3k**.

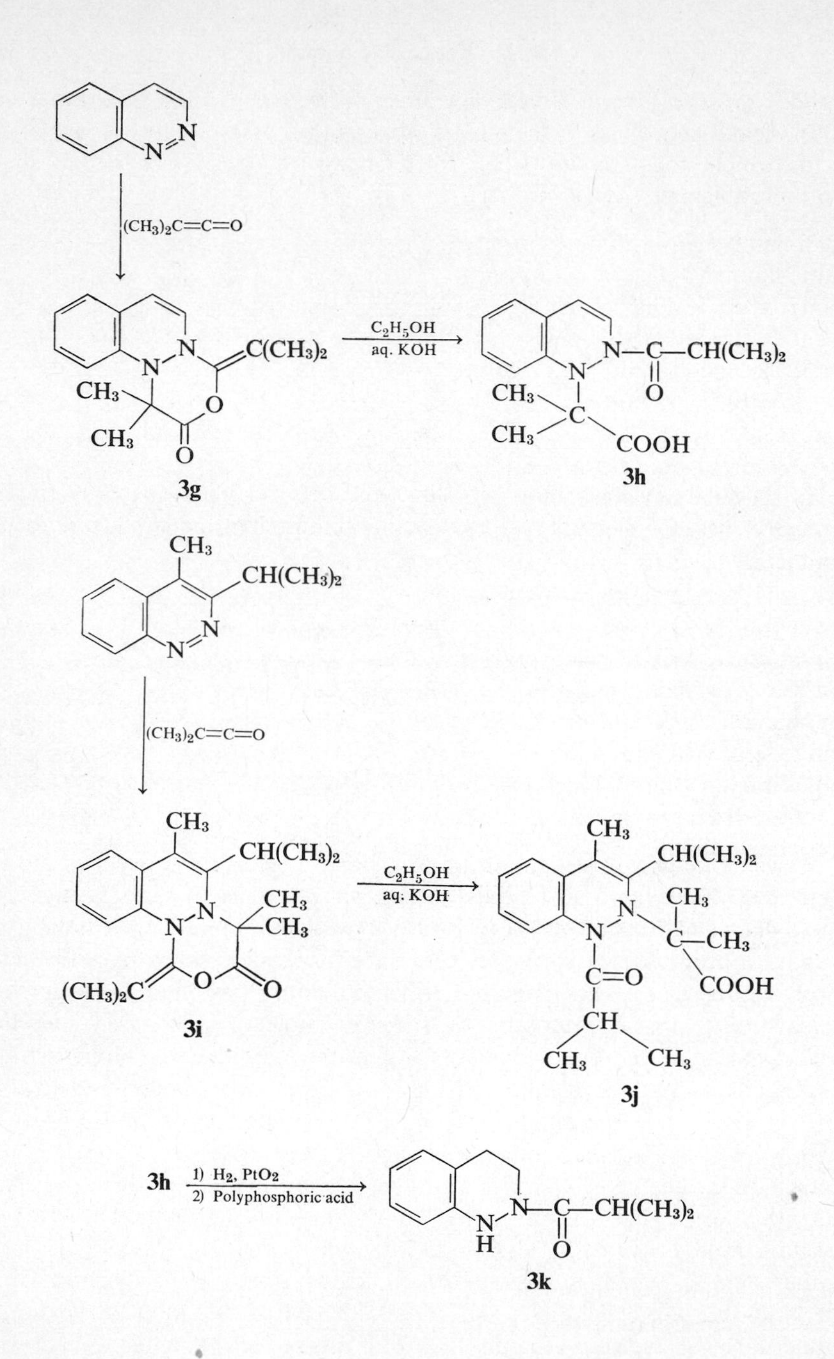
(CH3)2C=C=O
C2H5OH
aq. KOH
3g
3h
3i
3j
3h
1) H2, PtO2
2) Polyphosphoric acid
3k

B. Catalytic Reduction

Catalytic hydrogenation of both cinnoline and 4-methylcinnoline over such catalysts as rhodium on alumina or charcoal, palladium on charcoal, ruthenium oxide, and platinum oxide produces in each case at least seven compounds, among which are the corresponding 1,4-dihydrocinnolines and 1,2,3,4-tetrahydrocinnolines, the corresponding indoles, 2,3-dihydroindoles, octahydroindoles and *o*-aminophenethylamine (from cinnoline) or *o*-amino-β-methylphenethylamine (from 4-methylcinnoline) (see Sections 1A-III-E and 1B-III-A-1). 4-Phenylcinnoline is reduced almost quantitatively to 1,4-dihydro-4-phenylcinnoline by hydrogenation in acetic acid over palladium on activated alumina or charcoal (6a).

Both 3,4-diphenylcinnoline and 3,4-diphenyl-7-methylcinnoline are easily converted to their 1,4-dihydro derivatives in absolute ethanol by hydrogenating over Adams platinum catalyst at 70° for 12 min (12). Catalytic reduction of 3-chlorocinnoline over palladium on charcoal in ethanol containing a small amount of triethylamine gives 1,4-dihydrocinnoline (5). On the other hand, reduction of 4-chlorocinnoline over palladium-charcoal on calcium carbonate in methyl alcohol yields principally 4,4′-dicinnolinyl and a trace of cinnoline (13). Reduction of cinnoline 1,2-dioxide over palladium on charcoal also yields 1,4-dihydrocinnoline. Interruption of this reaction before completion produces a mixture of 1,4-dihydrocinnoline, cinnoline, cinnoline 1-oxide, and cinnoline 2-oxide (14, 15).

C. Electrolytic Reduction

Electrolytic reduction of 4-methylcinnoline in aqueous hydrochloric acid solution produces 1,4-dihydro-4-methylcinnoline. Further reduction under the same conditions but at a higher potential gives skatole. In alkaline solution, however, the dihydro compound cannot be further reduced to skatole. In aqueous hydrochloric acid solution, 3-phenylcinnoline is reduced to 1,4-dihydro-3-phenylcinnoline, and further reduction under the same conditions produces 1-(2′-aminophenyl)-2-phenyl-2-aminoethane. 4-Chlorocinnoline is reduced electrolytically to 4,4′-dicinnolinyl. Electrolytic reduction of 4-methoxycinnoline in aqueous citric acid solution produces 4-keto-1,2,3,4-tetrahydrocinnoline. In both acidic and neutral solution, 4-hydroxycinnoline is first reduced to 4-keto-1,2,3,4-tetrahydrocinnoline and then to 4-hydroxy-1,2,3,4-tetrahydrocinnoline. In slightly acidic solution, 3-hydroxycinnoline is reduced electrolytically to 3-keto-1,2,3,4-tetrahydrocinnoline, but in strongly acidic solution the main reduction product is

1-aminoöxindole. Cinnoline 1-oxide, 2-oxide, and 1,2-dioxide are all reduced to 1,4-dihydrocinnoline. Cinnoline 1-oxide is more easily reduced than cinnoline 2-oxide, so that cinnoline 1,2-dioxide appears to be first reduced to the 2-oxide and then rapidly to 1,4-dihydrocinnoline. Reduction of 3-phenyl-4-cinnolinecarboxylic acid in both acidic and basic solutions yields 1,4-dihydro-3-phenylcinnoline. Further reduction in acidic solution causes hydrogenolysis of the nitrogen-nitrogen bond to give 1-(2′-aminophenyl)-2-phenyl-2-aminoethane (16).

II. Reduced Cinnolines from Cyclization Reactions

A. From 2-Oxocyclohexylacetones, Acetophenones, or Acetaldehydes

Interaction of 2-oxocyclohexylacetone (**4**) with hydrazine in ethanolic solution gives 3-methyl-4,4*a*,5,6,7,8-hexahydrocinnoline (**5**), which is unstable in air. Aromatization of the crude hexahydrocinnoline by hydrogen exchange with cyclohexene over palladium on charcoal produces 3-methyl-5,6,7,8-tetrahydrocinnoline (**6**) in 56% yield (17).

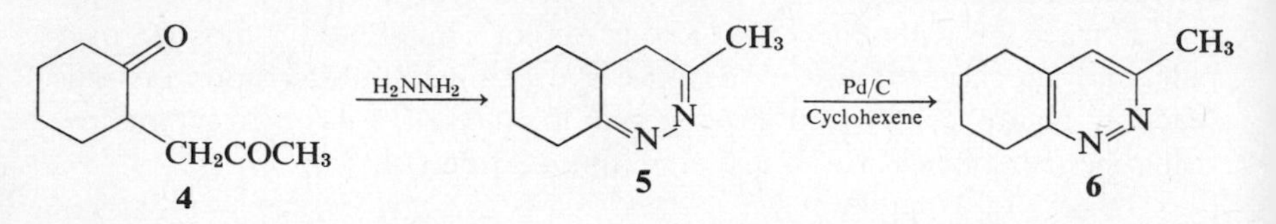

Similarly, α-(2-oxocyclohexyl)acetophenone gives 3-phenyl-5,6,7,8-tetrahydrocinnoline (**7**) in 88% yield (17); catalytic dehydrogenation of this over palladium on charcoal in 1 atm of carbon dioxide does not yield the expected 3-phenylcinnoline. Instead, 39% of 2-phenylindole (**8**) is obtained (18).

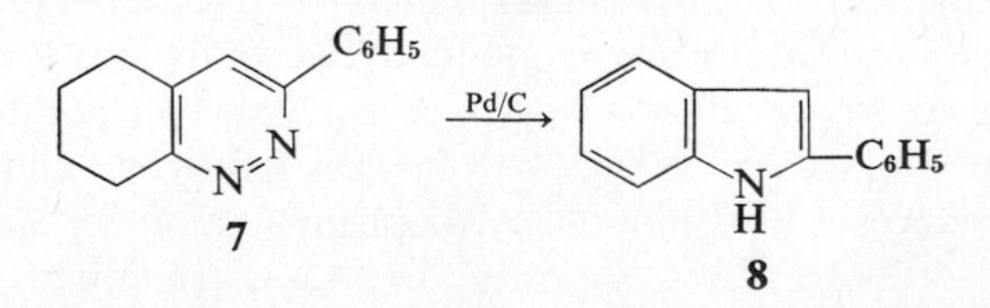

When 2-(hydroxydesyl)-4(or 5, or 6)-methylcyclohexanones (**9**) are treated with hydrazine, water is eliminated and 6(or 7, or 8)-methyl-5,6,7,8-tetrahydrocinnolines (**10**) are formed in 80–90% yields. When these tetrahydrocinnolines are dehydrogenated, using palladium on charcoal, ammonia is evolved and a mixture of substances results, the composition of which depends upon the reaction conditions, but the principal product in each case is a diphenylindole (**11**) (12).

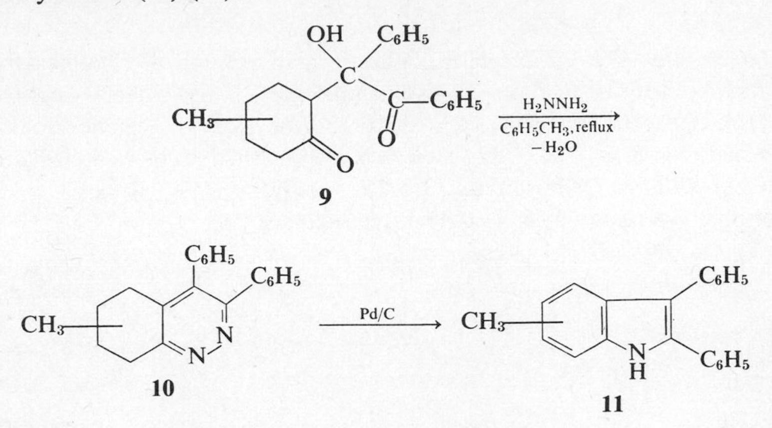

Similarly, α-(1-hydroxy-2-oxocyclohexyl)acetone (**12**) reacts with hydrazine to give 3-methyl-5,6,7,8-tetrahydrocinnoline (**6**) and both α-methyl-α-(1-hydroxy-2-oxocyclohexyl)acetone (**13**) and α-methyl-α-hydroxy-α-(2-oxocyclohexyl)acetone (**15**) with hydrazine give 3,4-dimethyl-5,6,7,8-tetrahydrocinnoline (**14**) (19, 20).

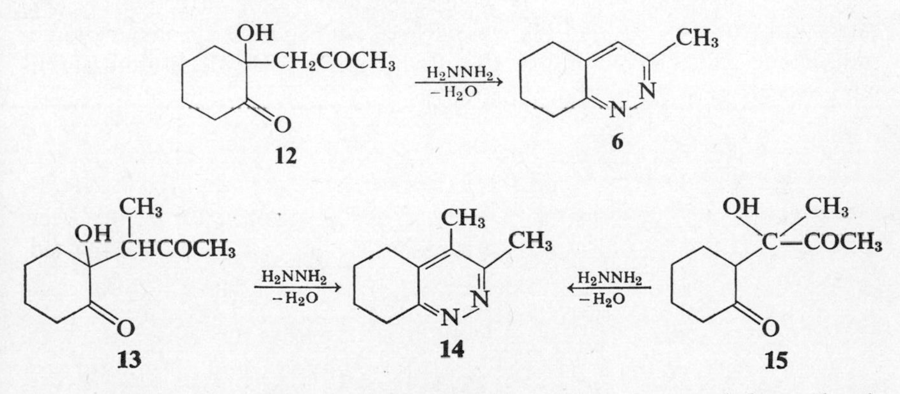

Unsubstituted 5,6,7,8-tetrahydrocinnoline (**17**) is prepared from the 4-oxoaldehyde **16** obtained from the condensation of glyoxal and potassium cyclohexanone-2-carboxylate (20).

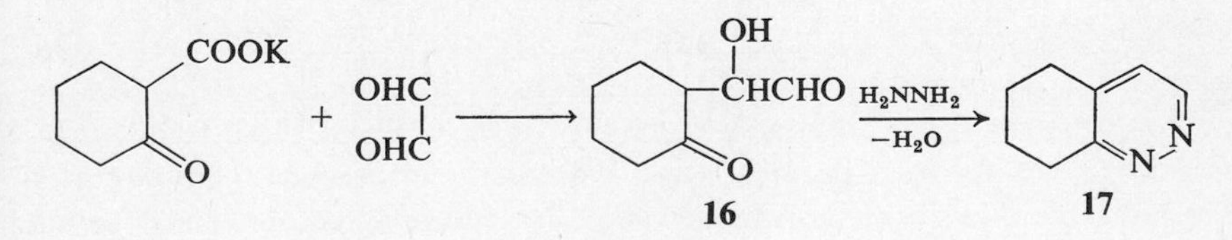

B. From 2-Oxocyclohexylacetic Acids and Esters

Ring closure of a 2-oxocyclohexylacetic acid or one of its esters with hydrazine produces a reduced 3-oxocinnoline. Thus when 1-methyl-2-oxocyclohexylacetic acid (**18**; R = CH_3) (21) or 2-oxo-1-phenylcyclohexylacetic acid (**18**; R = C_6H_5) (22) is treated with hydrazine, there are obtained 4*a*-methyl-4,4*a*,5,6,7,8-hexahydro-3(2*H*)cinnolinone (**19**; R = CH_3) and the 4*a*-phenyl analog **19** (R = C_6H_5), respectively.

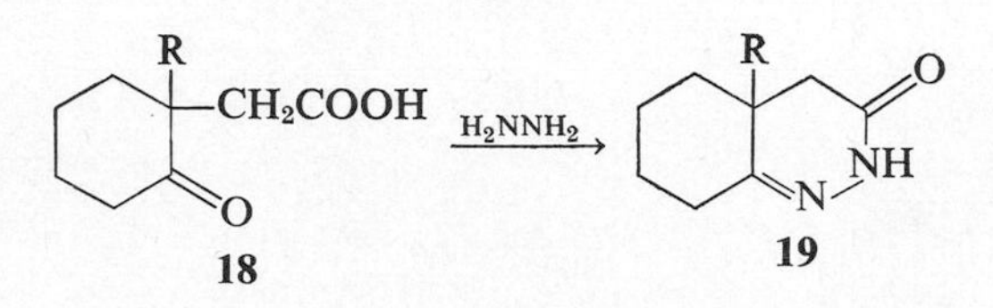

This reaction was used advantageously by a group of Japanese workers (22a, 22b) in the synthesis of a number of azamorphinans such as **19e**. As an example, 1-(3-methoxyphenyl)-2-oxocyclohexylacetic acid (**19a**) was cyclized with hydrazine to give 2,3,4,4*a*,5,6,7,8-octahydro-4*a*-(3-methoxyphenyl)-3-cinnolinone (**19b**). The cinnolinone **19b** was then further reduced catalytically to give **19c**, and **19c** was cyclized with sodium hydroxymethanesulfonate to give the cinnolinone **19d**. Reduction of **19d** with lithium aluminum hydride gave the azamorphinan **19e**.

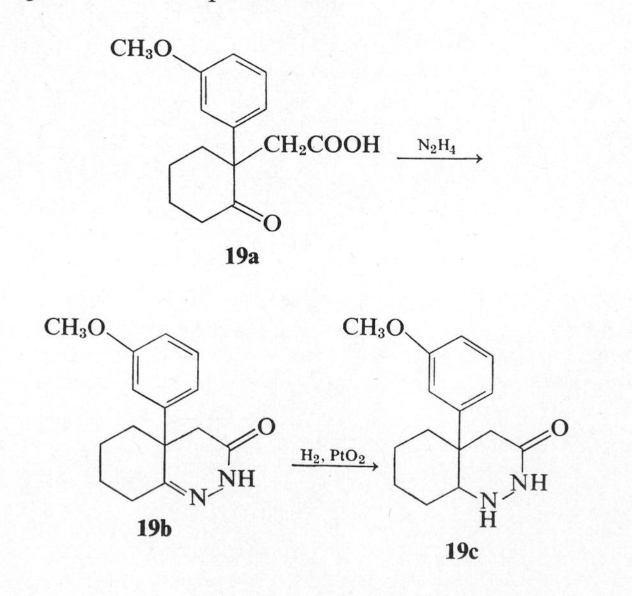

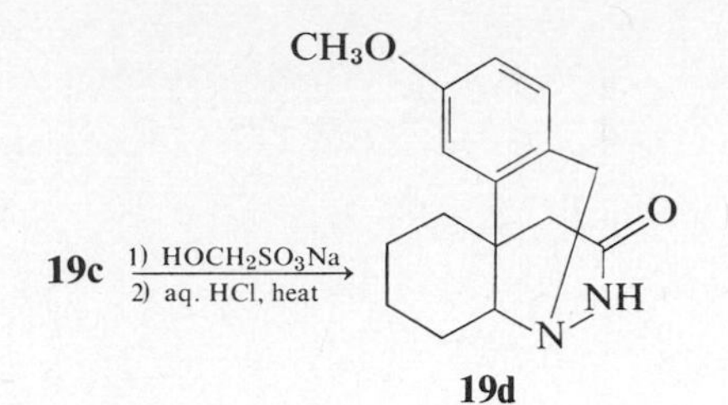

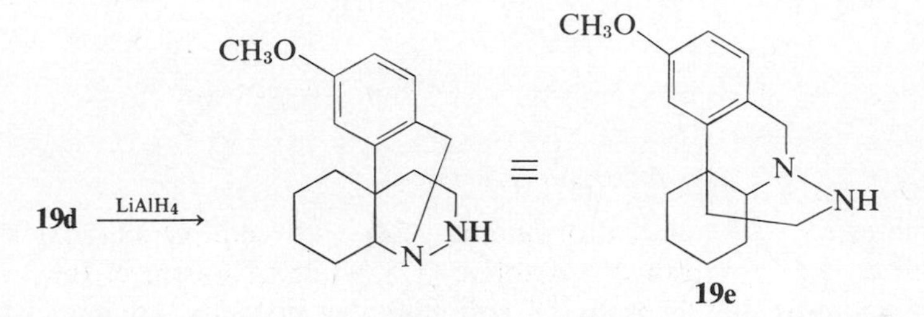

Condensation of ethyl 2-oxocyclohexylacetate (**20**) with hydrazine yields 50–75% of 4,4*a*,5,6,7,8-hexahydro-3(2*H*)cinnolinone (**21**) (17, 23, 24). With phenylhydrazine, the 4-oxo ester **20** gives the 2-phenyl analog of **21** (24). The reduced cinnoline **21** is also obtained by condensation of diethyl 2-oxocyclohexylmalonate (**22**) with hydrazine (25).

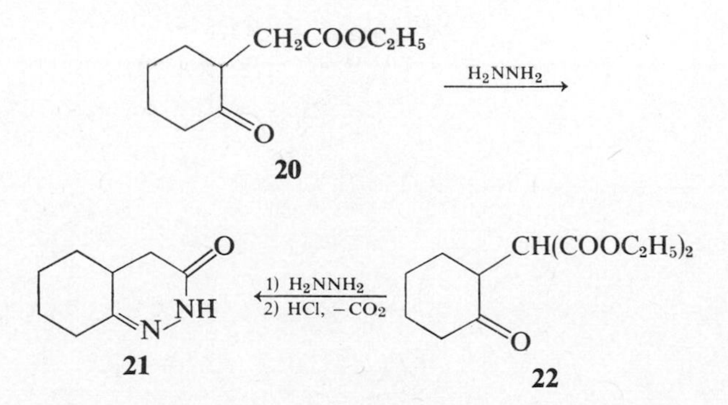

Aromatization of **21** with bromine in warm acetic acid gives 51–82% of 5,6,7,8-tetrahydro-3(2*H*)cinnoline (**23**), which with phosphorus oxychloride gives 65–83% of 3-chloro-5,6,7,8-tetrahydrocinnoline (**24**) (17, 23). Reaction

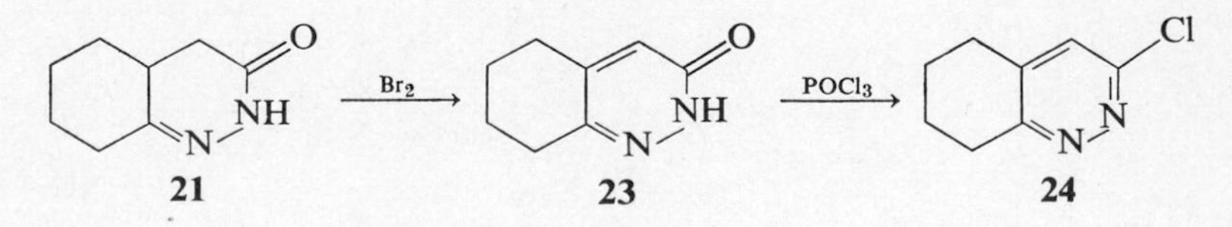

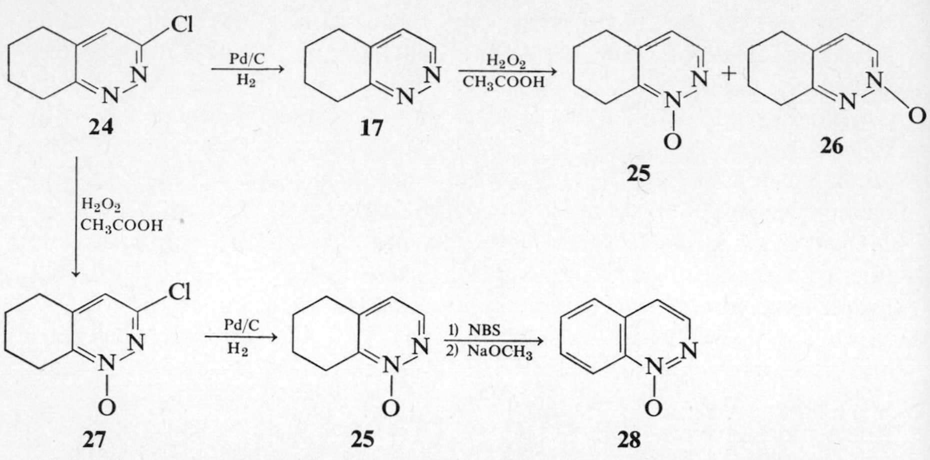

of 3-chloro-5,6,7,8-tetrahydrocinnoline (**24**) with sodium methoxide gives 3-methoxy-5,6,7,8-tetrahydrocinnoline (17), while treatment of the 3-chloro compound with red phosphorus and refluxing constant boiling hydriodic acid gives 75% of 3-iodo-5,6,7,8-tetrahydrocinnoline (23).

Hydrogenation of 3-chloro-5,6,7,8-tetrahydrocinnoline (**24**) over palladium on charcoal gives good yields of 5,6,7,8-tetrahydrocinnoline (**17**) (26), which with hydrogen peroxide in acetic acid gives a mixture of its N-1 and

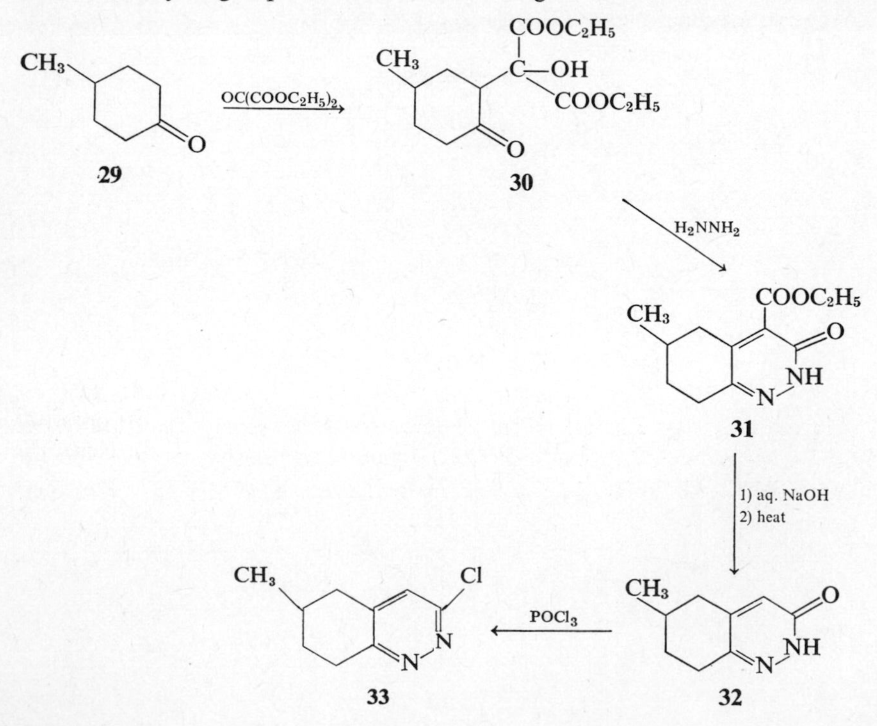

N-2 oxides (**25**, **26**). On the other hand, this same *N*-oxidation of 3-chloro-5,6,7,8-tetrahydrocinnoline yields only the 3-chloro 1-oxide **27**, which can be hydrogenated over palladium on charcoal to give pure 5,6,7,8-tetrahydrocinnoline 1-oxide (**25**). Treatment of **25** with *N*-bromosuccinimide, then with sodium methoxide, produces the fully aromatic cinnoline 1-oxide (**28**). Similarly, treatment of **27** with *N*-bromosuccinimide and then with sodium methoxide gives 3-chlorocinnoline 1-oxide (26).

Heating a mixture of 4-methylcyclohexanone (**29**) and diethyl oxomalonate (diethyl mesoxalate) gives the addition product **30**. Treatment of **30** with hydrazine hydrate in ethanol gives ethyl 6-methyl-3(2*H*)oxo-5,6,7,8-tetrahydro-4-cinnolinecarboxylate (**31**), which can be hydrolyzed and thermally decarboxylated to 6-methyl-5,6,7,8-tetrahydro-3(2*H*)cinnolinone (**32**). Chlorination of **32** with phosphorus oxychloride then gives the corresponding 3-chloro derivative **33** (26).

C. From 2,2-Dihydroxycyclohexylideneacetic Acid γ-Lactones

Lactonization of 1-hydroxy-2-oxocyclohexylacetic acid (**34**) followed by hydrolysis of the resulting γ-acetoxylactone (**35**) gives *cis*-2,2-dihydroxycyclohexylideneacetic acid γ-lactone (**36**) (27). With hydrazine, the lactone

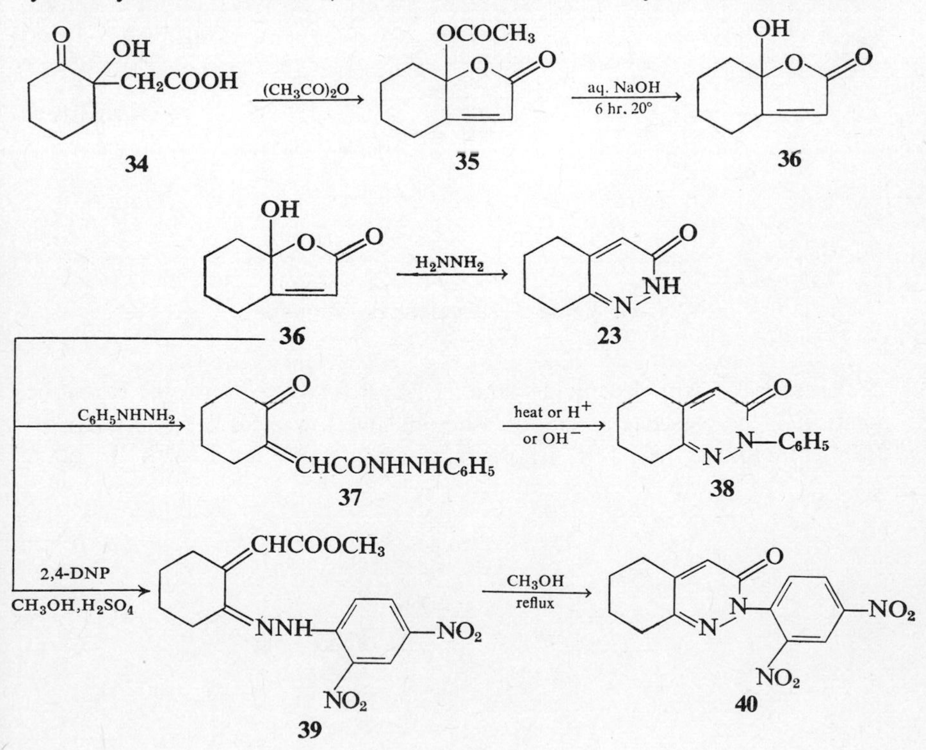

36 gives 5,6,7,8-tetrahydro-3(2*H*)cinnolinone (**23**) (28). With phenylhydrazine, 2-oxo-*cis*-cyclohexylideneacetic acid phenylhydrazide (**37**) is formed from **36**; on heating only, or on treatment with acid or alkali, **37** cyclizes and rearranges to 2-phenyl-5,6,7,8-tetrahydro-3(2*H*)cinnolinone (**38**) (28). With 2,4-dinitrophenylhydrazine in a methanol-sulfuric acid solution, the lactone **36** produces the 2,4-dinitrophenylhydrazone **39**, which in refluxing methanol cyclizes to 2-(2,4-dinitrophenyl)-5,6,7,8-tetrahydro-3(2*H*)cinnolinone (**40**) (27).

Menthofuran (**41**), a terpenoid substance present in peppermint oil, undergoes autoxidation (or oxidation by chromic acid or by hydrogen peroxide in acetic acid) to give the lactone **42**, which when boiled in ethanol with *p*-tolylhydrazine gives the reduced cinnoline **43** (29).

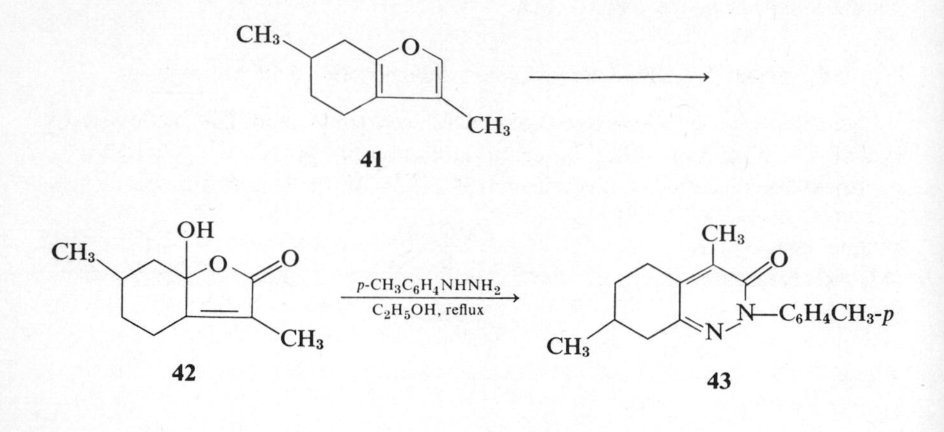

D. From 1,2-Cyclohexanediones

Cyanoacetic acid hydrazide and 1,2-cyclohexeanedione in ethanolic solution in the presence of sodium ethoxide give 4-cyano-5,6,7,8-tetrahydro-3(2*H*)cinnolinone (**44**) (30, 31), which when methylated with dimethyl

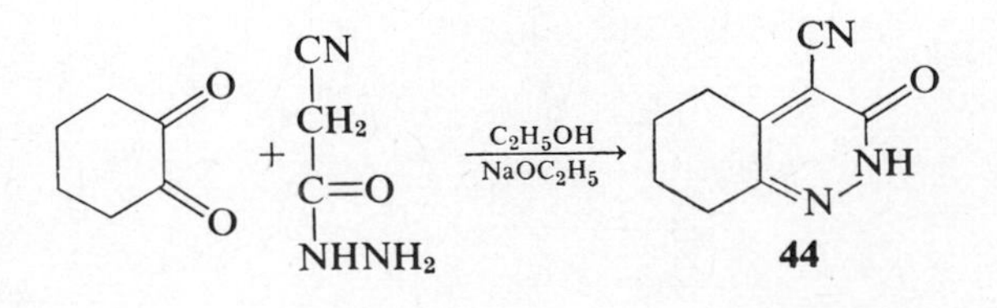

sulfate yields 4-cyano-2-methyl-5,6,7,8-tetrahydro-3(2*H*)cinnolinone, claimed to be useful as an analgesic (32, 33).

The monohydrazone **45** resulting from the condensation of 1,2-cyclohexanedione and *N*-methyl-4-piperidylhydrazine, when allowed to react with ethyl cyanoacetate in the presence of ammonium acetate and acetic acid in refluxing benzene solution, gives 4-cyano-2-(1-methyl-4-piperidyl)-5,6,7,8-tetrahydro-3(2*H*)cinnolinone (**46**). This material is said to be a mild antipyretic and analgesic (34, 35).

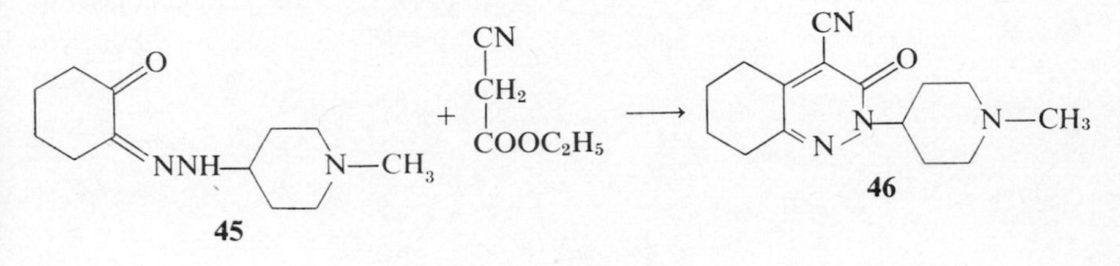

E. From Styrenes or 1-Vinylcyclohexenes

The Diels-Alder addition of diethyl azodicarboxylate to 1-vinylcyclohexene gives diethyl 1,2,3,5,6,7,8,8*a*-octahydrocinnoline-1,2-dicarboxylate (**47**). Hydrogenation of **47** over Adams platinum catalyst gives diethyl decahydrocinnoline-1,2-dicarboxylate (**48**), which in refluxing 48% hydrobromic acid is converted to decahydrocinnoline (**49**), an almost colorless oil which darkens rapidly in air. It forms a phthaloyl derivative and a dibenzoyl derivative (36).

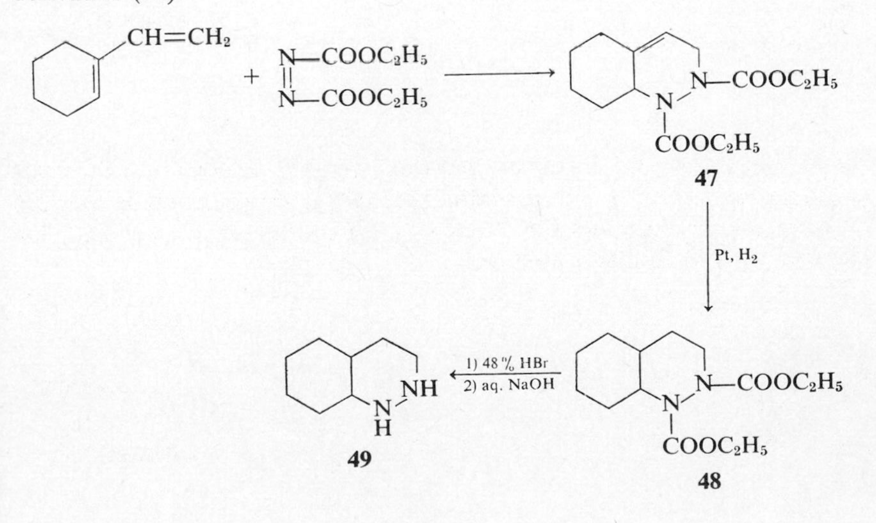

Interaction of styrene with diethyl azodicarboxylate gives tetraethyl 4-(1,2-dicarboxyhydrazino)-1,2,3,4-tetrahydro-1,2-cinnolinedicarboxylate (**51**) (37), probably by way of the Diels-Alder adduct **50**.

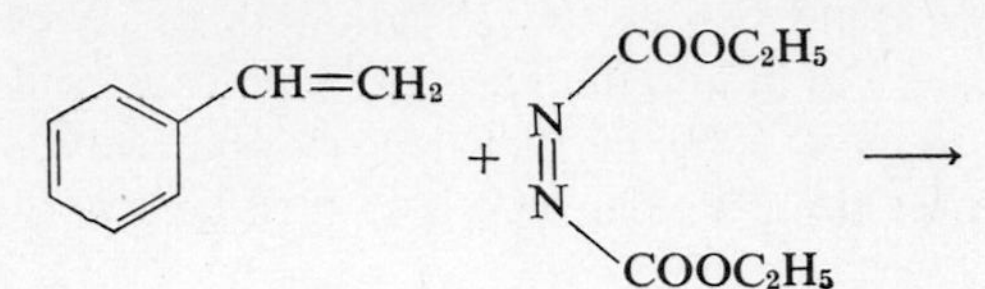

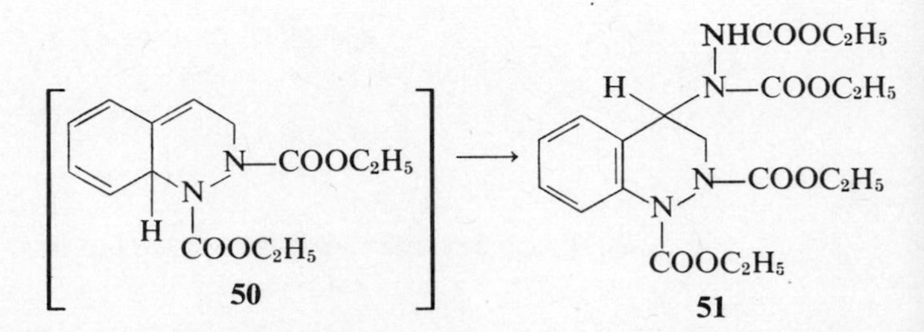

Similarly, certain α-substituted styrenes (**52**; R = H, OCH_3, or C_6H_5) interact with dimethyl azodicarboxylate to yield the adducts **53** (R = H, OCH_3, or C_6H_5) (38). When 1,2-dicarbomethoxy-4-methoxy-4-(1,2-dicarbo-

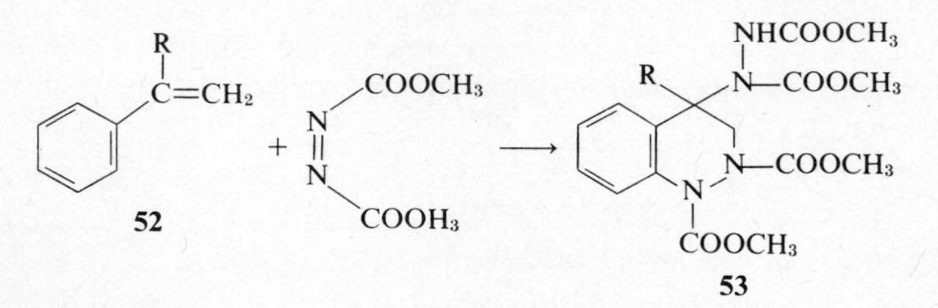

methoxy)hydrazino-1,2,3,4-tetrahydrocinnoline (**53**; R = OCH_3) is boiled in hydrazine hydrate, 4-hydrazinocinnoline (**54**) is obtained, while with 2*N*-hydrochloric acid at 70° for 30 min, 1,2-dicarbomethoxy-4-oxo-1,2,3,4-tetrahydrocinnoline (**55**) is obtained.

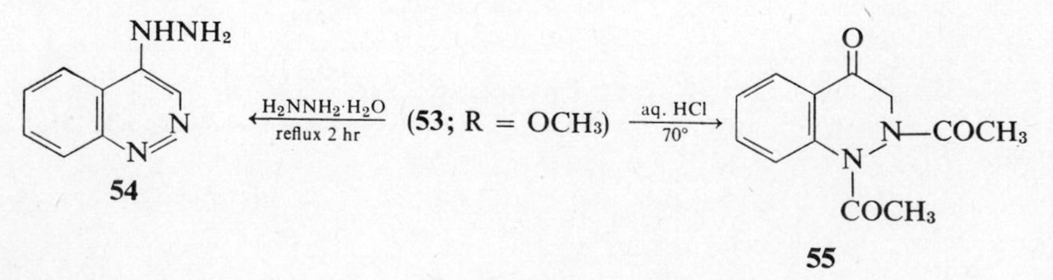

However, when the R group in **53** is phenyl (**53**; R = C_6H_5), reaction with hydrazine hydrate under the same conditions gives the indole **56**. When **53** (R = C_6H_5) is treated with hydrazine hydrate under more stringent conditions (sealed tube, 180°, 7 hr) 1-amino-3-phenylindole (**57**) is obtained and may be converted to 3-phenylindole (**58**) by hydrogenation over Raney nickel.

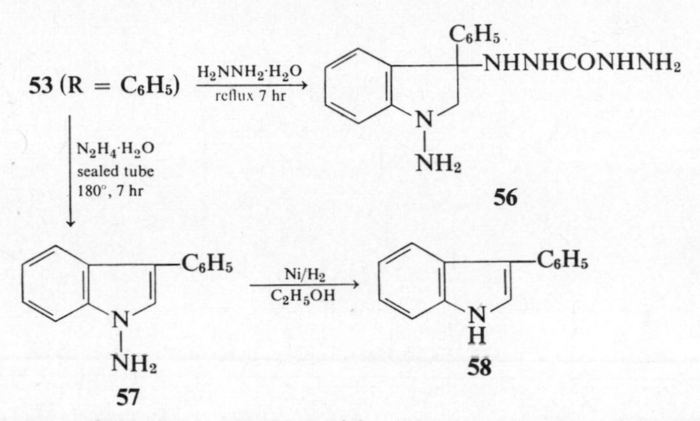

Treatment of **53** (R = H) first with aqueous potassium hydroxide then with aqueous acetic acid yields the tetrahydrocinnoline **59** (38).

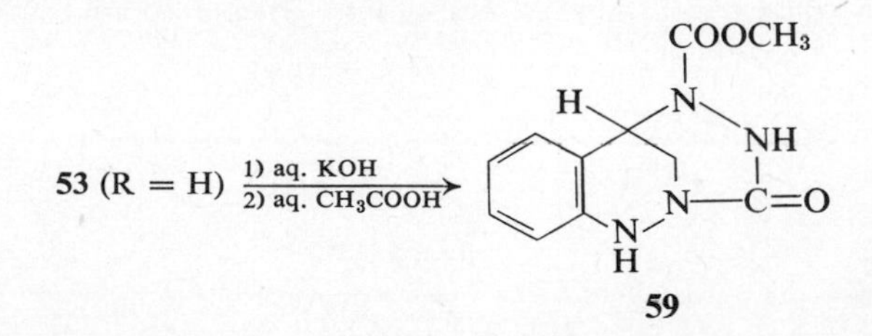

Addition of styrene to an electrochemical oxidation mixture of 1,1-diphenylhydrazine and perchloric acid gives 1,4-diphenyl-1,2,3,4-tetrahydrocinnoline, which may be oxidized in neutral solution to 1,4-dihydro-1,4-diphenylcinnoline (38a).

F. From Isoxazole Hydrazides

The isoxazole-3-carboxylic acid hydrazides **59a** (R = H, CH_3) are converted in good yield to 4-amino-5,6,7,8-tetrahydrocinnolin-2*H*(3)ones (**59c**; R = H, CH_3) by hydrogenation over Raney nickel in ethanol. The cinnolinones arise from the intermediates **59b** (R = H, CH_3) resulting from cleavage of the isoxazole ring (38b).

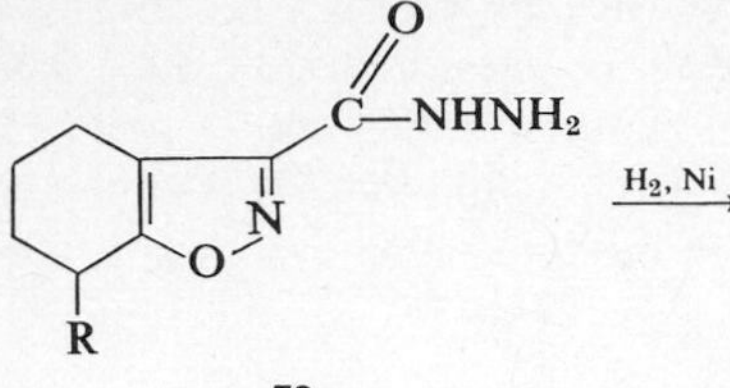

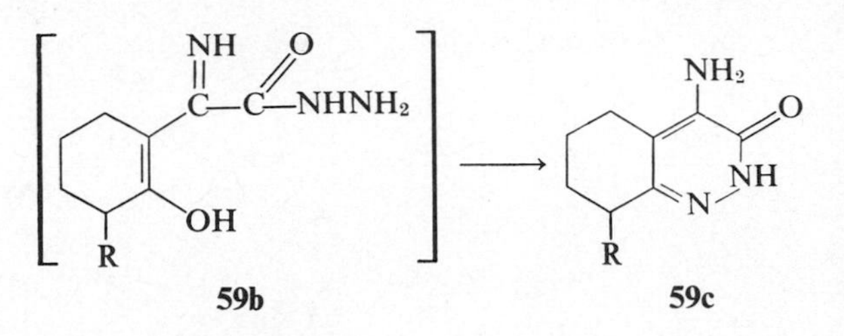

G. Miscellaneous

The sulfide **60** with excess hydrazine hydrate in hot acetic acid gives the thiadiazepine **61**, which is smoothly decomposed by heat in boiling ethylene glycol to 3-phenyl-5,6,7,8-tetrahydrocinnoline (**7**) with evolution of hydrogen sulfide (39).

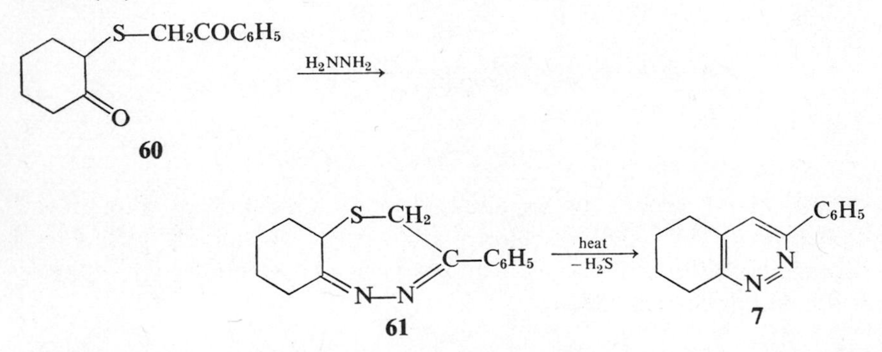

Condensation of diethyl mesoxalate with α-alkyl- or α-aryl-substituted phenylhydrazine **62** (R = CH_3, C_2H_5, or C_6H_5) yields a diethyl mesoxalate *N*-alkyl(or *N*-aryl)phenylhydrazone **63**. Partial hydrolysis of the hydrazone to its half-ester **64** followed by treatment with thionyl chloride gives an α-carbethoxyglyoxylyl chloride *N*-substituted phenylhydrazone **65**. Treatment of **65** under Friedel-Crafts conditions with titanium tetrachloride in nitrobenzene at 100° C gives the 1-alkyl(or aryl)-1,4-dihydro-4-cinnolinone-3-carboxylic acid **66** (R = CH_3, C_2H_5 or C_6H_5) (40, 41).

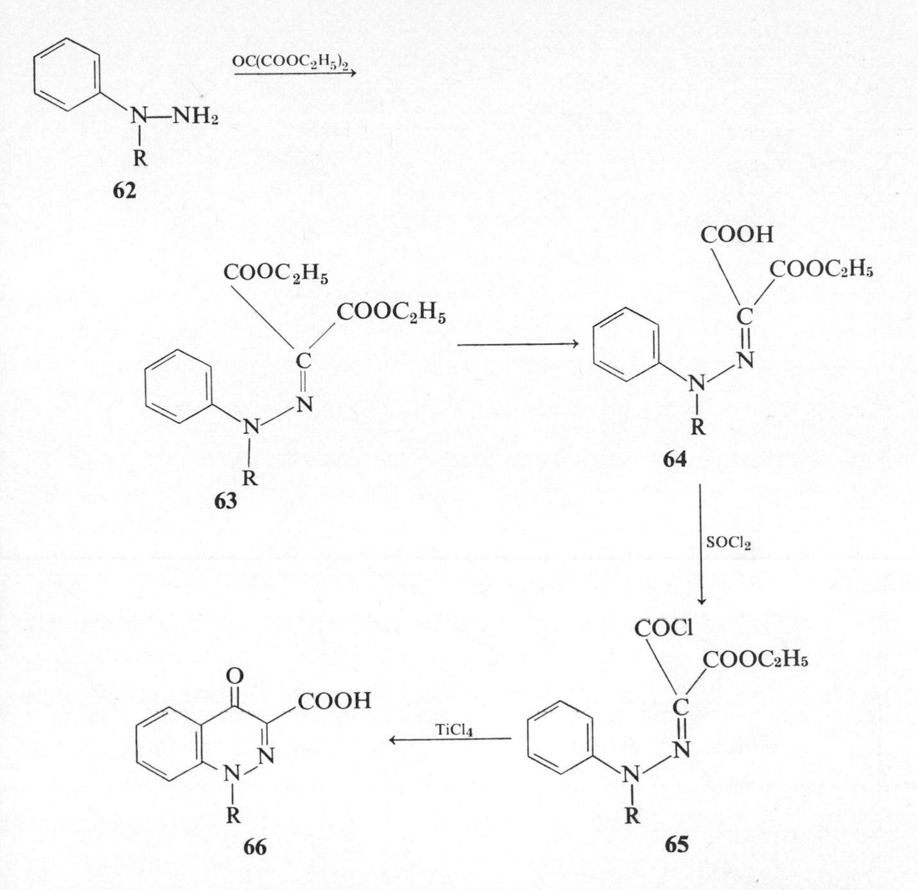

A related reaction involves the cyclization of phenylhydrazides such as **66a** in the presence of sulfuric acid to give 3-oxo-4,4-diaryl-1,2,3,4-tetrahydrocinnolines(**66b**; R = aryl, R′ = alkyl or halogen) (41a–41d). The pmr spectra of several such 1-alkyl-4-cinnolinones (without the 3-carboxy group) together with the spectra of the corresponding anhydro bases of 4-hydroxy-2-alkylcinnolinium hydroxides have been recorded in a number of different solvents, and the effect of the *N*-alkyl group on the spectra has been examined.

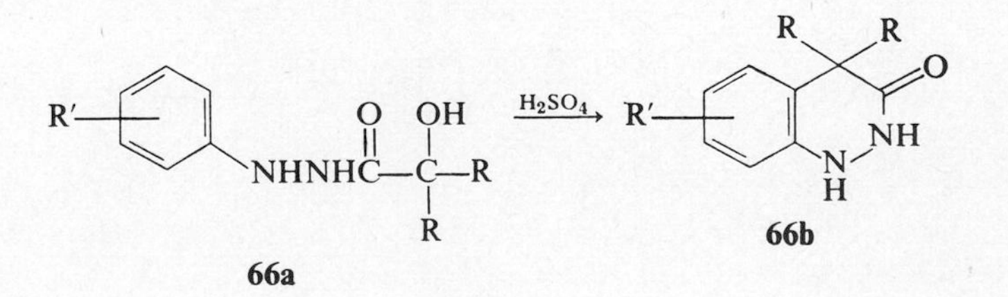

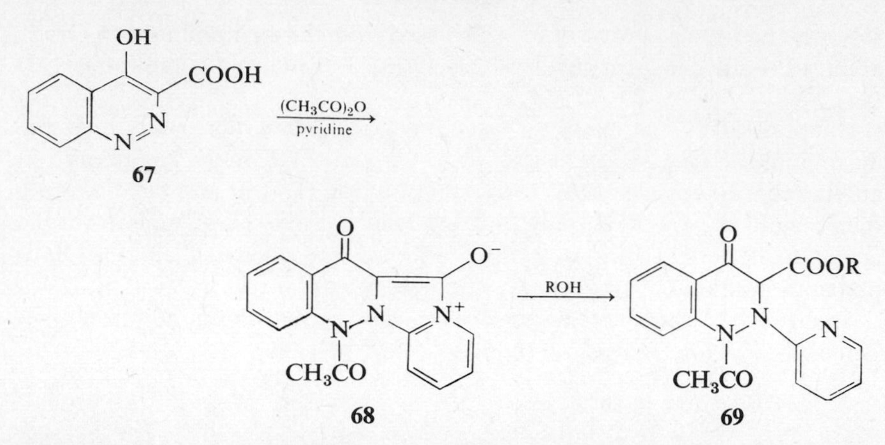

For example, the *N*-1-methyl resonance of a given 4-cinnolinone is always observed at a somewhat higher field than the methyl resonance of its *N*-2-methyl analog because of the more polar nature of the anhydro base (42).

Warming 4-hydroxy-3-cinnolinecarboxylic acid (**67**) with pyridine and acetic anhydride gives compound **68**, which is converted to the reduced 4-cinnolinone **69** (R = CH_3, C_2H_5 or *i*-C_3H_7) by the action of boiling lower alcohols (43).

Ethyl pentafluorobenzoylacetate (**70**) couples with benzenediazonium chloride at 0° to give the phenylazo derivative **71**. Treatment of **71** with aqueous sodium hydroxide solution gives 1,4-dihydro-3-carbethoxy-1-phenyl-5,6,7,8-tetrafluoro-4-cinnolinone (**72**). With aniline in refluxing

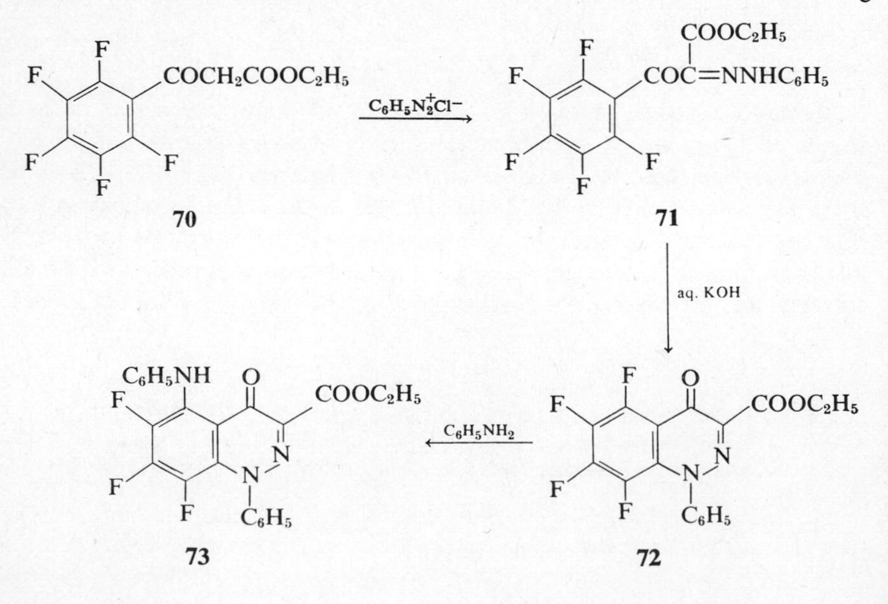

toluene, the 5-fluoro group of **72** is displaced from the 4-cinnolinone to give 5-anilino-1,4-dihydro-3-carbethoxy-1-phenyl-6,7,8-trifluoro-4-cinnolinone **(73)** (44).

Coupling of 4-hydroxy-5,6,7,8-tetrahydrocoumarin **(74)** with benzenediazonium sulfate gives the 3-phenylazo derivative **75**. When **75** is boiled in an ethanolic-aqueous sodium hydroxide solution for 1 hr and the mixture is then acidified, 1,4,5,6,7,8-hexahydro-1-phenyl-4-oxo-3-cinnolinecarboxylic acid **(76)** is obtained in good yield. The 1-*p*-nitrophenyl analog of **76** is prepared in the same way (45).

Treatment of the cyclohexene **77** with benzenediazonium solution gives 1-phenyl-5,6,7,8-tetrahydro-4(1*H*)cinnolinone **(78)** (46).

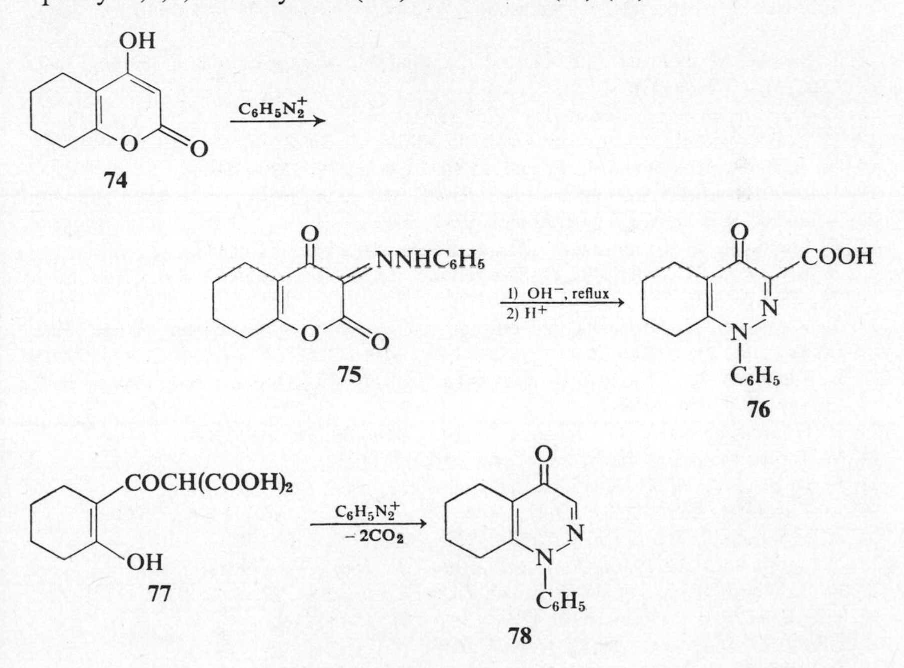

References

1. L. S. Besford, G. Allen, and J. M. Bruce, *J. Chem. Soc.*, **1963**, 2867.
2. D. I. Haddlesey, P. A. Mayor, and S. S. Szinai, *J. Chem. Soc.*, **1964**, 5269.
3. P. W. Neber, G. Knoller, K. Herbst, and A. Trissler, *Ann.*, **471**, 113 (1929).
4. R. N. Castle and M. Onda, *J. Org. Chem.*, **26**, 4465 (1961).
5. D. E. Ames, R. F. Chapman, and H. Z. Kucharska, *J. Chem. Soc.*, Suppl. 1, **1964,** 5659.
6. A.-G. Siegfried, French Pat. 1,393,596, Feb. 15, 1965.

6a. F. Schatz and T. Wagner-Jauregg, *Helv. Chim. Acta*, **51,** 1919 (1968).

7. D. E. Ames and H. Z. Kucharska, *J. Chem. Soc.*, **1962,** 1509.
8. D. E. Ames and H. Z. Kucharska, *J. Chem. Soc.*, **1963,** 4924.
9. M. Busch and A. Rast, *Chem. Ber.*, **30,** 521 (1897).
10. L. S. Besford and J. M. Bruce, *J. Chem. Soc.*, **1964,** 4037.
11. H. S. Lowrie, *J. Med. Chem.*, **9,** 664 (1966).
11a. D. E. Ames, B. Novitt, and D. Waite, *J. Chem. Soc.* (*C*), **1969,** 796.
11b. T. T. Chen, W. Dorscheln, H. Goth, M. Hesse, and H. Schmid, *Helv. Chim. Acta*, **51,** 632 (1968).
11c. W. A. White, German Pat. 2,005,104, Aug. 6, 1970; *Chem. Abstr.*, **73,** 77269 (1970).
11d. D. E. Ames and B. Novitt, *J. Chem. Soc.* (*C*), **1969,** 2355.
11e. M. A. Shah and G. A. Taylor, *J. Chem. Soc.* (*C*), **1970,** 1642.
12. C. F. H. Allen and J. A. Van Allan, *J. Am. Chem. Soc.*, **73,** 5850 (1951).
13. J. Morley, *J. Chem. Soc.*, **1951,** 1971.
14. I. Suzuki, M. Nakadate, T. Nakashima, and N. Nagasawa, *Tetrahedron Lett.*, **1966,** 2899.
15. I. Suzuki, M. Nakadate, T. Nakashima, and N. Nagasawa, *Chem. Pharm. Bull.* (*Tokyo*), **15,** 1088 (1967).
16. H. Lund, *Acta Chem. Scand.*, **21,** 2525 (1967).
17. H. E. Baumgarten, P. L. Creger, and C. E. Villars, *J. Am. Chem. Soc.*, **80,** 6609 (1958).
18. H. E. Baumgarten and J. L. Furnas, *J. Org. Chem.*, **26,** 1536 (1961).
19. J. Levisalles and P. Baranger, *Compt. Rend.*, **242,** 1336 (1956).
20. J. Levisalles, *Bull. Soc. Chim. France*, **1957,** 1009.
21. E. Buchta, G. Wolfrum, and H. Ziener, *Chem. Ber.*, **91,** 1552 (1958).
22. E. B. Carton, H. Lederle, L. H. Schartzman, and G. F. Woods, *J. Am. Chem. Soc.*, **74,** 5126 (1952).
22a. T. Kametani, K. Kigasawa, M. Hiiragi, and N. Wagatsuma, *Chem. Pharm. Bull.* (*Tokyo*), **16,** 296 (1968).
22b. T. Kametani, K. Kigasawa, K. Wakisaka, and N. Wagatsuma, *Chem. Pharm. Bull.* (*Tokyo*), **17,** 1096 (1969).
23. R. H. Horning and E. D. Amstutz, *J. Org. Chem.*, **20,** 707 (1955).
24. W. Reid and A. Draisbach, *Chem. Ber.*, **92,** 949 (1959).
25. G. R. Schultze, M. H. Korgami, and F. Boberg, *Ann.*, **688,** 122 (1965).
26. M. Ogata, H. Kano, and K. Tori, *Chem. Pharm. Bull.* (*Tokyo*), **11,** 1527 (1963).
27. A. Mondon, H. Menz, and J. Zander, *Chem. Ber.*, **96,** 826 (1963).
28. Wang Yu and Huang Jing-Jain, *Acta Chim. Sin.* **28**(6), 351 (1962).
29. R. B. Woodward and R. H. Eastman, *J. Am. Chem. Soc.*, **72,** 399 (1950).
30. P. Schmidt and J. Druey, *Helv. Chim. Acta.*, **37,** 134 (1954).
31. Ciba Ltd., Br. Patent, 788,393, Jan. 2, 1958.
32. P. Schmidt and J. Druey, *Helv. Chim. Acta*, **40,** 1749 (1957).
33. J. Druey and P. Schmidt (to Ciba Pharmaceutical Products, Inc.), U.S. Patent, 2,835,672, May 20, 1958.
34. E. Jucker and R. Suess, *Helv. Chim. Acta*, **42,** 2506 (1959).
35. E. Jucker, *Angew. Chem.*, **71,** 321 (1959).
36. J. M. Bruce and P. Knowles, *J. Chem. Soc.*, **1964,** 4046.
37. L. Horner and W. Naumann, *Ann.*, **587,** 81 (1954).
38. K. Alder and H. Niklas, *Ann.*, **585,** 97 (1954).
38a. G. Cauquis and M. Genies, *Tetrahedron Lett.*, **1970,** 3403.
38b. M. Ruccia and N. Vivona, *J. Heterocyclic Chem.*, **8,** 289 (1971).
39. J. D. Loudon, L. B. Young, and A. A. Robertson, *J. Chem. Soc.*, **1964,** 591.
40. H. J. Barber and E. Lunt, *J. Chem. Soc.*, **1965,** 1468.

41. E. Lunt and T. L. Threlfall, *Chem. Ind.*, **1964**, 1805.
41a. I. S. Berdinskii, T. M. Dimitrieva, P. A. Petyunin, O. O. Makeeva, and G. N. Pershin, *Khim.-Farm. Zh.*, **3**, 15 (1969); *Chem. Abstr.*, **71**, 49491 (1969).
41b. I. S. Berdinskii, *Khim. Geterotsikl. Soedin.*, **1969** (5) 885; *Chem. Abstr.*, **72**, 111398 (1970).
41c. I. S. Berdinskii, E. Yu. Posyagina, *Zh. Org. Khim.*, **6**, 151 (1970); *Chem. Abstr.*, **72**, 90394 (1970).
41d. I. S. Berdinskii, E. Yu. Posyagina, *Khim. Geterotsikl. Soedin.*, **1970** (6), 844; *Chem. Abstr.*, **73**, 120578 (1970).
42. A. W. Ellis and A. C. Lovesey, *J. Chem. Soc* (*B*), **1968**, 1393.
43. J. S. Morley, *J. Chem. Soc.*, **1959**, 2280.
44. A. T. Prudchenko, G. S. Shchegoleva, V. A. Barkhash, and N. N. Vorozhtsov, Jr., *Zh. Obshch. Khim.*, **37** (11), 2487 (1967); *Chem. Abstr.*, **69**, 36059 (1968).
45. E. Ziegler and E. Noelken, *Monatsh. Chem.*, **91**, 850 (1960).
46. E. Noelken and E. Ziegler, *Monatsh. Chem.*, **91**, 1162 (1960).

CHAPTER II

Phthalazines

NATU R. PATEL

Gulf Research & Development Co., Merriam, Kansas

Part A. Phthalazine

I. Phthalazine

A. Physical Properties

Phthalazine (**1**) (benzo[*d*]pyridazine, β-phenodiazine) forms white, hard prisms, mp 90–91° (1). It boils at 315–317° at atmospheric pressure with decomposition but can be distilled at 189°/29 mm; 175°/17 mm (2, 3). It sublimes at 125°/0.3 mm, forming colorless crystals, mp 91.5–92.0° (4). Phthalazine is easily soluble in water and most organic solvents except ligroin. Neither phthalazine nor any of its derivatives has been isolated from nature.

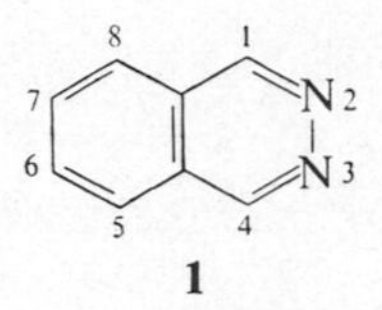

1

Phthalazine, with a pK_a value of 3.5 in water at 20° C (5) [3.47 (6); 3.47 in 1% ethanol at 20° C (7)], has a higher base strength than its parent diazine, pyridazine (pK_a 2.3). This is explained simply by high single-bond character of N—N bond in phthalazine (5). Phthalazine is a stronger base than cinnoline (pK_a 2.3) but weaker base than isoquinoline (pK_a 5.4).

The ionization potential of phthalazine has been calculated by the self-consistent molecular field method to be 9.57 ev (8). The theoretically calculated dipole moment for phthalazine is 4.56 D (9). The experimental value is found to be 4.88 D in *p*-xylene (10) compared with isoquinoline (2.38 D in *p*-xylene) and pyridazine (4.56 D in benzene).

B. Theoretical Aspects

Theoretical calculations of the π-electron distribution in phthalazine by Longuet-Higgins and Coulson (11) gave the charge densities as shown in **2**. Recent calculations, by Gawer and Dailey (12) and Wait and Wesley (13) gave the π-electron density distribution as shown in **3** and **4**, respectively,

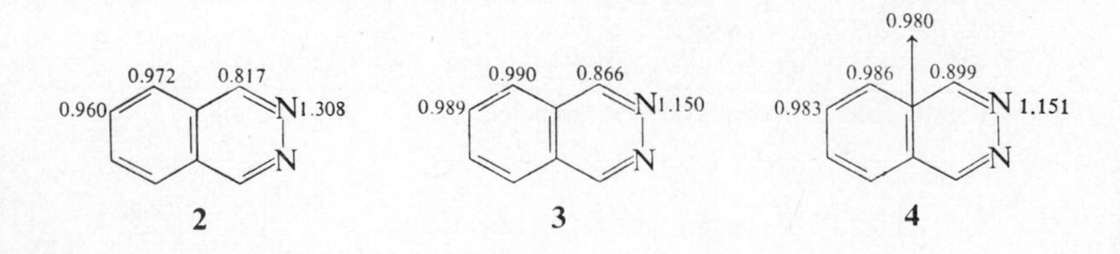

2 **3** **4**

which are in close agreement with each other. Recently Pugmire and coworkers (61), using extended Hückel theory and complete neglect of differential overlap wave function, calculated the σ and π charge densities and the bond orders as shown below:

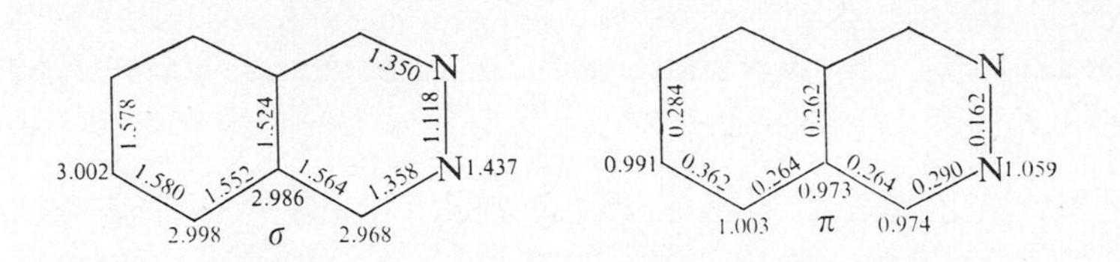

The σ and π charges are outside the rings and the bond orders inside the rings.

The electron densities are closely correlated with the chemical behavior of phthalazine which on electrophilic substitution, for example nitration gives 5-nitrophthalazine. In cases of nucleophilic substitution, leaving groups, for example halogen, mercapto and its derivatives substituted at C_1 or C_4, are easily displaced.

C. Spectra

1. *Infrared Spectra*

The IR absorption spectrum of phthalazine has been recorded in the literature (Fig. 2A-1) but no attempt has been made for assignments of various bands.

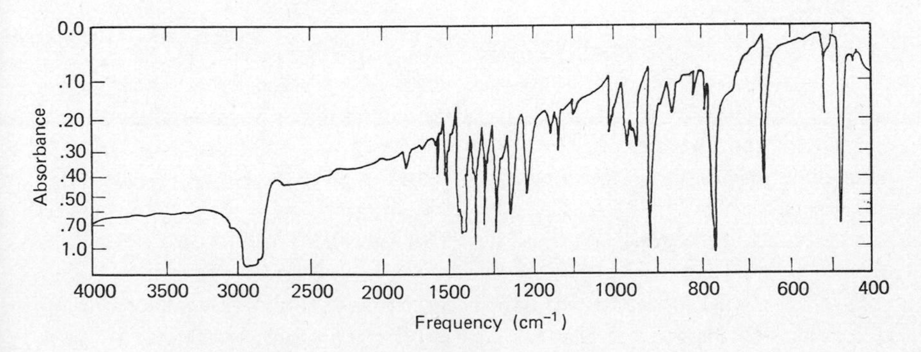

Fig. 2A-1. Infrared spectrum of phthalazine in nujol (1).

2. *Ultraviolet Spectra*

The UV spectrum of phthalazine has been studied in detail. A number of theoretical calculations of transition energies and band intensities have been reported (14–18). The absorption spectra of a phthalazine single crystal and of mixed crystals with naphthalene at 4.2° K have been reported and analyzed (19).

The UV spectrum of phthalazine has been recorded in various solvents, including cyclohexane (20), methylcyclohexane (21), 95% ethanol (20), methanol (22), water (5, 20), and in presence of acid (5, 22). The band position and their intensities are recorded in Table 2A-1. The UV spectra of phthalazine and naphthalene are very similar, with the latter showing slightly greater absorbancy. The strong bands of phthalazine occur at somewhat shorter wavelengths (21). The weak bands have been observed to show "blue shift" in more polar solvents (20).

3. *Phosphorescence Spectra*

Absorbance and phosphorescence spectra of phthalazine were determined in polar, nonpolar, and acidic solvents and compared with the corresponding spectra of naphthalene (23). Phthalazine shows characteristic n-π absorbance band and upon excitation of this band it shows the same π-π^* phosphorescence band as naphthalene. Recently, phosphorescence, excitation, and absorption spectra of phthalazine were obtained (62). The excitation spectrum corresponds to the absorption spectrum, indicating the phosphorescence originates from phthalazine itself.

4. *Proton Magnetic Resonance Spectra*

The pmr spectrum of phthalazine is measured at 100 Mc/sec, and assignment of the protons has been made (24) and is shown along with the computed spectrum in Fig. 2A-2. The aromatic ring protons (5, 6, 7, 8) appear as two halves of an AA′BB′ multiplet. The low-field half was assigned to protons 5 and 8, and the high-field half to protons 6 and 7. The small difference in chemical shifts between the two groups of protons has been rationalized as due to the greater distance of the nitrogen atoms from the aromatic ring protons (24).

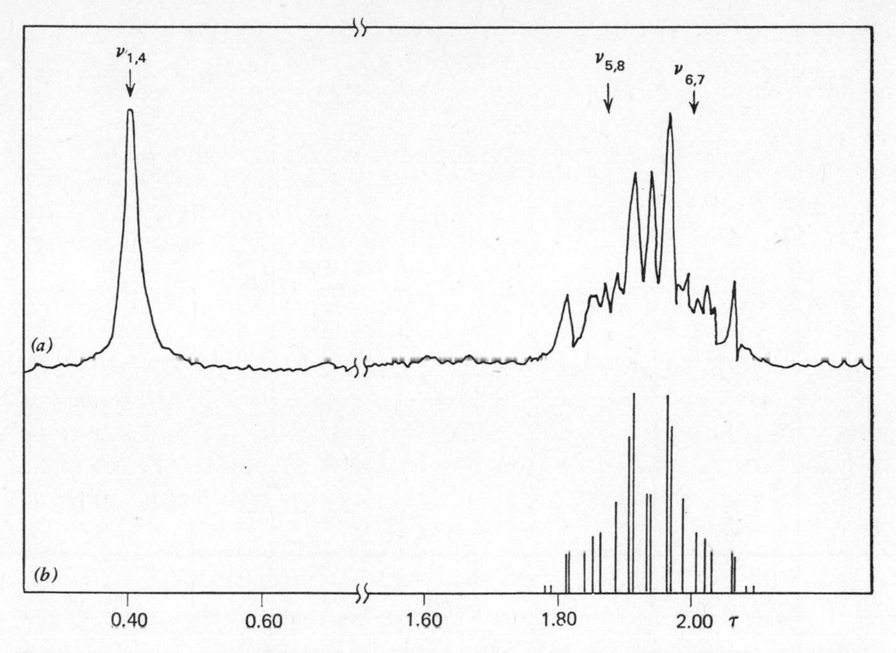

Fig. 2A-2. (*a*) Experimental and (*b*) calculated spectrum of phthalazine ($\simeq$0.01 g/ml in acetone) at 100 Mc/sec.

Assignment of Protons	Coupling Constants
$\nu_{1,4} = 0.4\tau$	$J_{4,8} = 0.4$ cps
$\nu_{5,8} = 1.87\tau$	$J_{5,8} = 8.17$ cps
$\nu_{7,8} = 2.00\tau$	$J_{5,7} = 1.24$ cps
	$J_{5,8} = 0.57$ cps
	$J_{6,7} = 6.76$ cps

Carbon-13 proton spin coupling constant for C_1 and C_4 in phthalazine has been observed to be 181 cps in deuteriochloroform and has been found to be in agreement with the calculated value of 180.5 cps (25). Recently C^{13} chemical shifts of phthalazine have been reported (61).

The chemical shifts of phthalazine have been compared with the π electron densities (12, 26). Only a rough correlation was observed. The chemical shifts of protons bonded to carbon atoms adjacent to nitrogen atoms seems particularly anomalous (12).

5. *Mass Spectra*

The mass spectrum of phthalazine (Fig. 2A-3) and several of its C_1 and C_4 substituted derivatives has been reported (27). Some ions in the spectrum

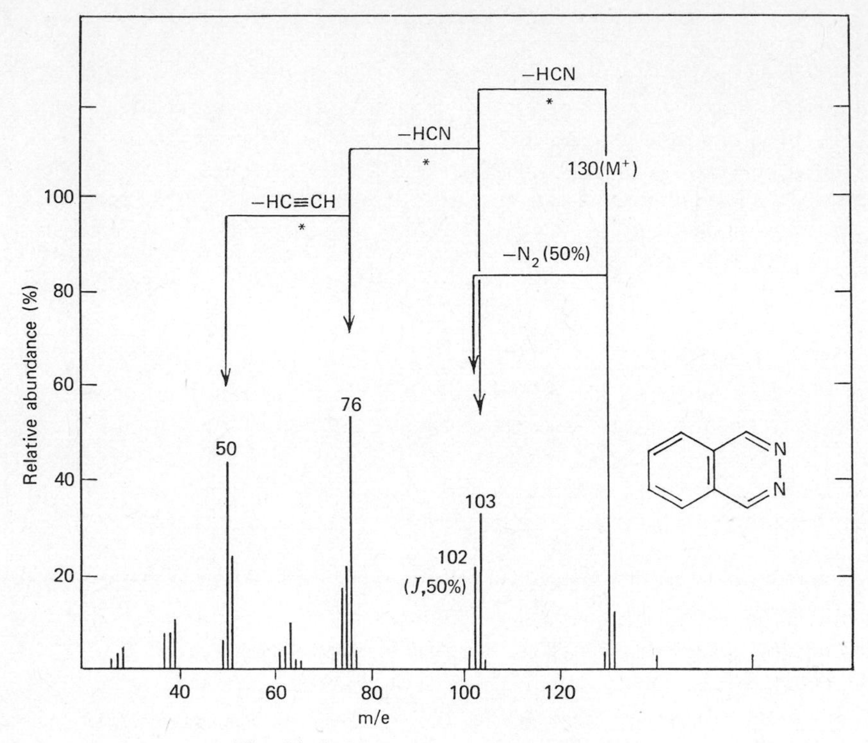

Fig. 2A-3. Mass spectrum of phthalazine.

have the following composition:

m/e	Composition
103	C_7H_5N
102	C_7H_4N (50%)
	C_8H_6 (50%)
75	C_6H_3 (80%)
	C_5HN (20%)
50	C_4H_4

The two adjacent nitrogen atoms in phthalazine can be detected by the presence of an M—N_2 peak. However, the main fragmentation scheme is M—HCN—HCN—HC≡CH— to form $[HC{\equiv}C{-}C{\equiv}CH]^+$ *m/e* 50. This is explained by quantum mechanical calculations showing phthalazine to exist preferentially in the C═N form. When amino, chloro, methyl, or methoxy groups are present either at C_1 and/or at C_4 of the phthalazine nucleus, the processes M—N_2 and M—N_2H predominate and M—HCN ions are small or absent (27).

6. *Electron Spin Resonance Spectra*

Electron spin resonance measurements of the phthalazine anion radical in dimethyl sulfoxide and complete assignments of hyperfine coupling constants have been reported by selective deuteration (28). The observed and calculated spectra are in good agreement. Several open-shell SCF—LCAO—MO calculations of the π electron spin densities of the phthalazine anion have been reported (29–31). The calculations are in good agreement with the experimental results.

D. Polarography

The polarographic reduction waves of phthalazine in aqueous solution at various *p*H values at 25° are recorded (7). The $E_{0.5}$ value varied linearly, -0.83 to -1.61 (against saturated calomel electrode), for *p*H 1–12.7, with $E_{0.5}/p\text{H}$ of 69 mV. The electrochemical reduction in DMF showed that the protonation of the mononegative ions is the first step in the dimerization of phthalazine (63). Lund and Jensen (64) have investigated the polarographic behavior of phthalazine and some of its derivatives and discussed the controlled potential electrolysis of these compounds. The reduction of phthalazine in acid solution furnished a mixture of α,α'-diamino-*o*-xylene and isoindoline. The composition of the mixture depended primarily on the *p*H. The reduction in alkaline medium may yield a dimeric product, 1,2-dihydrophthalazine or 1,2,3,4-tetrahydrophthalazine depending primarily on the cathode potential.

E. Adsorption Chromatography

Measurements by linear elution adsorption chromatography on alumina and calculations by the theory of intramolecular electronic effects showed that the adsorption energy of phthalazine (and other nitrogen heterocycles) is affected by steric and electronic factors (32). Similar conclusions have been drawn in a thin layer chromatographic study of phthalazine on alumina and the authors have discussed the geometry of adsorption (33). The Rf value of phthalazine has been reported to be 0.24 with 2-butanone as the mobile phase.

F. Methods of Preparation

1. *Condensation*

Phthalazine (**1**) was first prepared by Gabriel and Pinkus (2) in 1893 from $\alpha,\alpha,\alpha',\alpha'$-tetrachloro-*o*-xylene (**5**) and aqueous hydrazine by heating under pressure at 150° for 2 hr. Recently this method has been modified to manufacture **1** by heating **5** with hydrazine sulfate in >90% sulfuric acid (yield

86%) (34) (Eq. 1). Similarly, phthalazine has been prepared by condensation

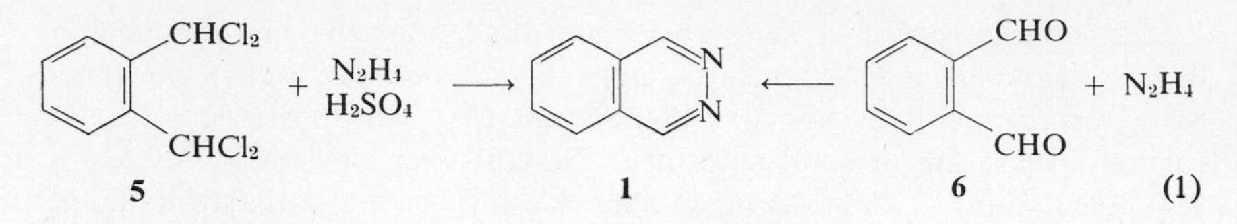

of *o*-phthaldehyde (**6**) with aqueous hydrazine sulfate (35, 36). Hirsch and Orphanos (1, 37) allowed alcoholic solutions of **6** to react with hydrazine hydrate at 0° to give **1** in high yield (96%) (Eq. 1).

2. *Dehalogenation*

Halogen atoms located at the 1 or the 1 and 4 positions of phthalazine are easily removed by chemical or catalytic reduction. This situation is similar to other nitrogen heterocycles. 1-Chlorophthalazine (**7**) on catalytic reduction with 5% palladium on charcoal (4) and 1,4-dichlorophthalazine (**8**) on chemical reduction (38) with red phosphorus and hydroiodic acid give **1** in 58% and 35% yields, respectively (Eq. 2).

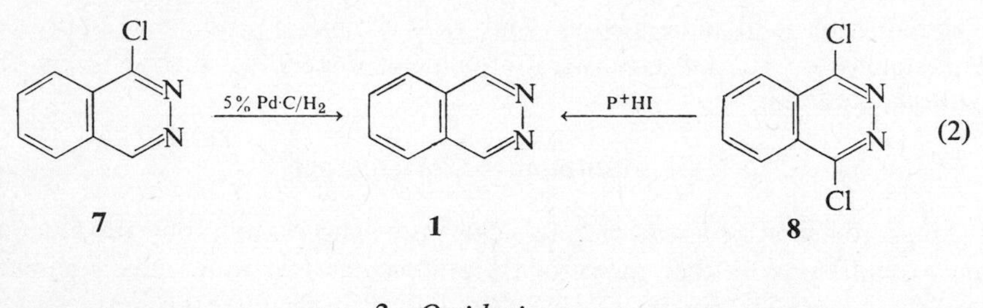

3. *Oxidation*

1-Hydrazinophthalazine (**9**) has been oxidized at room temperature by oxygen in ethanolic alkali (39) or by heating with copper sulfate (40) at *p*H 8 to give **1** in 56% and 60% yields, respectively. 1-*N*′-(*p*-toluenesulfonyl)-hydrazinophthalazine (**10**) on heating with aqueous alkali (38) was converted to **1** (Eq. 3).

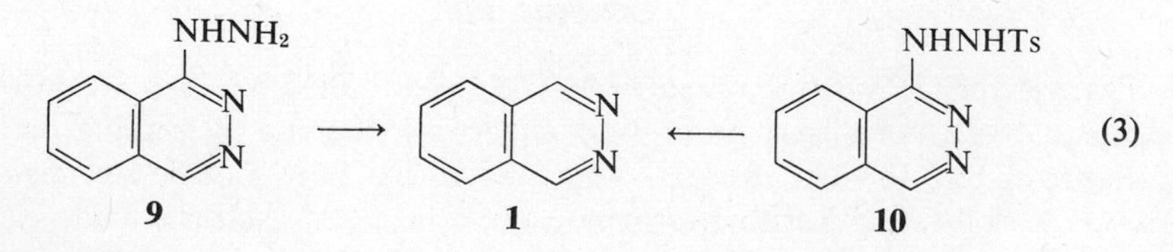

G. Reactions

1. *Salt Formation*

Phthalazine forms well-defined salts, functioning as a monoacid base, hydrochloride, mp 235–236° (dec.) (4); hydrobromide, mp 254–255° (with foaming) (41); hydriodide, mp 203° (3); picrate, mp 208–210° (2).

Phthalazine forms a stable complex with bromine, mp 122–123° (41), and the complex with iodine in chloroform solution has been observed spectrophotometrically (42). The phthalazine-bromine complex may be used for the bromination of organic substances by addition or substitution. An aqueous solution of phthalazine with an aqueous solution of $K_2[Pd(SCN)_4]$ gives $[Pd(Phthalazine)(SCN)_2]$, mp 193–194° (43).

Phthalazine reacts smoothly with various alkyl halides to give *N*-alkylphthalazinium halides which are discussed in Section 2K.

2. *Oxidation*

Phthalazine on treatment with monoperphthalic acid in ether furnished phthalazine *N*-oxide (**11**) (Eq. 4).

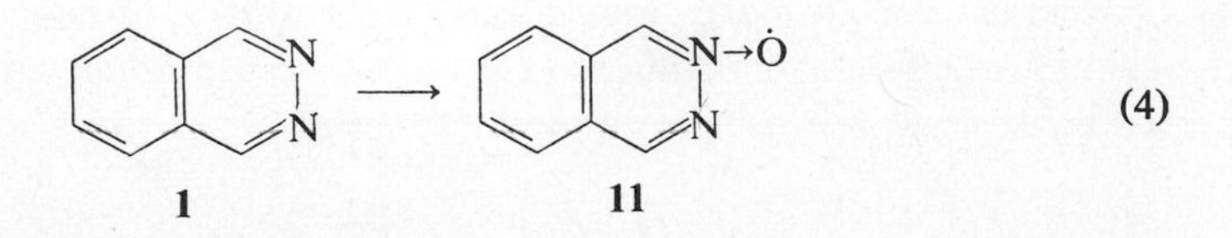

(4)

Phthalazine on oxidation with hot alkaline potassium permanganate gives the expected pyridazine-4,5-dicarboxylic acid **12** in 66% yield (44). This is the method of choice for the preparation of **12**. This reaction has been used occasionally in structure determination where an aromatic ring with electron-donating substituents is oxidized to **12** (45, 46) (Eq. 5).

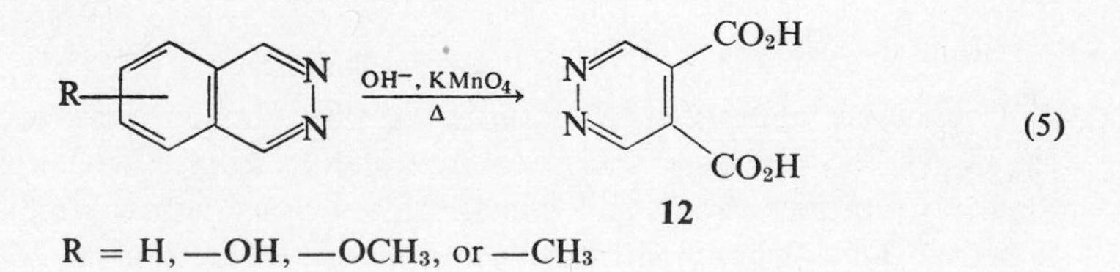

(5)

3. *Reduction*

Phthalazine undergoes chemical reduction of the heterocyclic ring to give 1,2-dihydrophthalazine (isolated as an intermediate) with lithium aluminum hydride (47) or to 1,2,3,4-tetrahydrophthalazine with sodium amalgam (2, 48). On treatment with zinc and hydrochloric acid, heterocyclic ring opening takes place giving *o*-aminomethylbenzylamine (2). Elder (48a) has carried out a detailed study on the catalytic hydrogenation of phthalazine at low pressure and temperature and at high pressure and temperature using various catalysts in neutral as well as acidic media. The following products are detected and isolated: 1,2-dihydrophthalazine, 1,2,3,4-tetrahydrophthalazine, α,α′-diamino-*o*-xylene, 2-methylbenzylamine, 1,3-dihydroisoindole, *o*-xylene, and *cis*- and *trans*-1,2-dimethylcyclohexane. A recent study (49) on the catalytic reduction of phthalazine has revealed a two-stage reduction process to *o*-aminomethylbenzylamine. Phthalazine on hydrogenation with Raney nickel or 20% palladium on carbon absorbs two moles of hydrogen giving presumably 1,2,3,4-tetrahydrophthalazine as an intermediate. In the second stage with Raney nickel, one mole of hydrogen is absorbed and presumably scission of the cyclic hydrazine bond takes place.

4. *Nitration*

Nitration of phthalazine with potassium nitrate in concentrated sulfuric acid affords 5-nitrophthalazine (**13**) as the main product and 5-nitro-1(2*H*)-phthalazinone (**14**) as a by product (50) (Eq. 6).

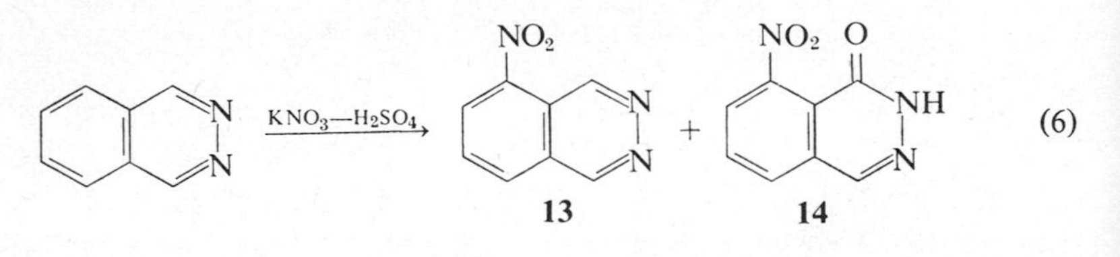

5. *Reactions with Organometallic Reagents*

Phthalazine undergoes 1,2-addition of phenylmagnesium bromide across the C=N bond accompanied by autooxidation to give 1-phenylphthalazine. Similarly, methyllithium and phenyllithium react with phthalazine to give 1-methyl- and 1-phenylphthalazine, respectively (see Section 2B).

6. *Addition of Dienophiles*

The addition reaction of maleic anhydride to phthalazine in 2:1 molar ratio leads to a pyrido[2,1-*a*]phthalazine derivative. This type of addition reaction is discussed in Section 2M. Phthalazine, 1-substituted phthalazine, and some 1,4-disubstituted phthalazines react with dimethylketene to give 1:2 adducts having oxazinophthalazine structures (see Section 2L).

7. *Reissert Compound Formation*

A phthalazine Reissert compound (**15**) has been isolated by reaction of phthalazine, benzoyl chloride, and potassium cyanide in a methylene chloride-water solvent system (51) (Eq. 7).

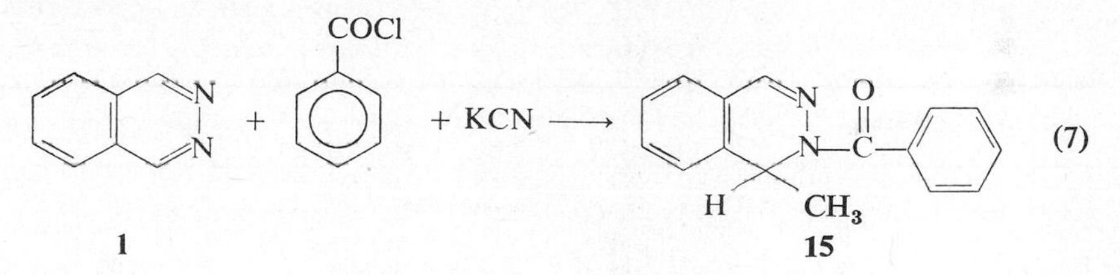

II. Phthalazines Unsubstituted in the Heterocyclic Ring

There are only a few compounds reported in this group where the heterocyclic ring is devoid of substituents.

The two main routes for the synthesis of this group of compounds are condensation of the properly substituted *o*-phthaldehyde with hydrazine and dehalogenation of the corresponding 1-halo- or 1,4-dihalophthalazines (Eq. 8).

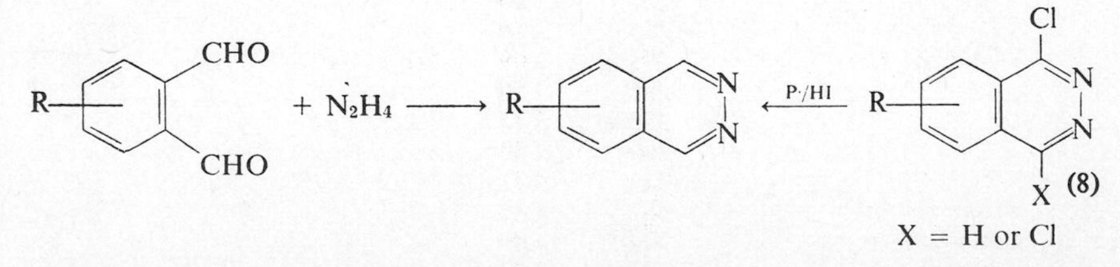

Direct nitration of phthalazine (Eq. 6) has been shown to introduce the nitro group in position 5 (50). Similarly, phenylation of phthalazine with *N*-nitrosoacetanilide gives 5-phenylphthalazine in poor yield (38).

III. Tables

TABLE 2A-1. Ultraviolet Absorption Spectrum of Phthalazine

Solvent	λ(nm)	log ε	Origin	References
Cyclohexane	387	1.20	1st $n \rightarrow \pi^*$	20
	379	1.38	1st $n \rightarrow \pi^*$	
	370(s)[a]	1.51	1st $n \rightarrow \pi^*$	
	364(s)	1.61	1st $n \rightarrow \pi^*$	
	357	1.73	1st $n \rightarrow \pi^*$	
	303	2.91	1st $\pi \rightarrow \pi^*$	
	296	2.90	1st $\pi \rightarrow \pi^*$	
	290	2.94	1st $\pi \rightarrow \pi^*$	
	277(s)	3.25	1st $\pi \rightarrow \pi^*$	
	267	3.52	2nd $\pi \rightarrow \pi^*$	
	259	3.61	2nd $\pi \rightarrow \pi^*$	
	252	3.56	2nd $\pi \rightarrow \pi^*$	
	212	4.77	3rd $\pi \rightarrow \pi^*$	
Methylcyclohexane	296.5	1.06	—	21
	290	1.11	—	
	267	3.59	—	
	259	3.67	—	
	252	3.63	—	
95% ethanol	305	3.04	1st $\pi \rightarrow \pi^*$	20
	297	3.06	1st $\pi \rightarrow \pi^*$	
	292	3.13	1st $\pi \rightarrow \pi^*$	
	269(s)	3.50	2nd $\pi \rightarrow \pi^*$	
	261	3.55	2nd $\pi \rightarrow \pi^*$	
	254(s)	3.51	2nd $\pi \rightarrow \pi^*$	
	218	4.76	3rd $\pi \rightarrow \pi^*$	
Methanol	304	3.15	—	22
	299	3.11	—	
	297	3.18	—	
	296	3.18	—	
	291	3.23	—	
	268(i)[b]	3.54	—	
	259	3.58	—	
	254(i)	3.52	—	
	216	4.38	—	
Methanol acidified with one drop of $HClO_4$	313	3.48	—	22
	305	3.41	—	
	271	3.40	—	
	226	3.20	—	
	205	3.89	—	
Water pH 7.0	305	3.09	1st $\pi \rightarrow \pi^*$	5, 20
	297	3.10	1st $\pi \rightarrow \pi^*$	
	292(s)	3.17	1st $\pi \rightarrow \pi^*$	
	260	3.53	2nd $\pi \rightarrow \pi^*$	
	219	4.71	3rd $\pi \rightarrow \pi^*$	
Water pH 0.0	312	3.46	—	5
Cationic form of phtahlazine	304(i)	3.40	—	
	271	3.40	—	
	230	4.62	—	

[a] (s) = shoulder.
[b] (i) = inflection.

TABLE 2A-2. Phthalazines Unsubstituted in the Heterocyclic Ring

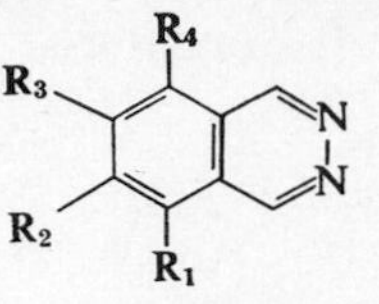

R_1	R_2	R_3	R_4	Preparation[a]	Yield (%)	MP[b]	Ref.
$—NO_2$	H	H	H	A	—	87° (ethanol)	50
—Cl	H	H	H	B	—	—	52
H	—Cl	H	H	C	29	White crystals, 132°	52
$—C_6H_5$	H	H	H		6	Colorless plates, 106° (ligroin)	38
H	$—C_6H_5$	H	H	B_1	18	Pale yellow needles, 137–139° (ligroin); picrate, 202°	53
H	—OH	H	H	B_2	80	~300° (dec.) (water)	54
H	$—OCH_3$	$—OCH_3$	H	C	—	181–183° (dec.) (benzene-pet. ether)	55
—OH	—OH	$—OCH_3$	H	C	—	Yellow needles, 228–230°	56
—OH	$—CH_3$	$—OCH_3$	H	C	72	Pale yellow prisms, 260–262° (dec.); (EtOH); HCl, 248–250° (dec.)	45, 57
$—OCH_3$	$—CH_3$	—OH	H	C	76	Yellow needles, 226–228° (dec.) (aq. ethanol)	58
—OH	$—CH_3$	—OH	H	D	33	HBr, colorless needles, 238–240° (dec.)	58
—OH	—OH	—OH	CH_3	C	76	Yellow rods, >300° (ethanol)	46
H	R_2-R_3 Methylenedioxy		H	C	50	Yellow crystalline powder, darkening at 175°, 255° (dec.) (xylene)	59
H	R_2-R_3 Ethylenedioxy		H	C	35	Yellow crystals, 260° (alcohol)	60
C_6H_5	C_6H_5	C_6H_5	C_6H_5	—	77	249° (ethanol)	58a

[a] A, phthalazine + KNO_3 + conc. H_2SO_4. B, 1,5-Dichloro or 1,8-dichlorophthalazine + P/HI. B_1, 1,4-Dichloro-6-phenylphthalazine + P/HI, reflux 15 min. B_2, 1-Chloro-7-methoxyphthalazine + P/HI, reflux 1 hr. C, The corresponding *o*-phthaladehyde + aqueous or alcoholic hydrazine kept at room temperature for a few hr or reflux 15 min to 3 hr. D, 5-Hydroxy-6-methyl-7-methoxy or 5-methoxy-6-methyl-7-hydroxyphthalazine + 50% HBr, reflux 3 hr.

[b] MP with physical description and recrystallization solvent in parentheses.

References

General reference, J. C. E. Simpson in *Condensed Pyridazine and Pyrazine Rings, The Chemistry of Heterocyclic Compounds*, Vol. 5, A. Weissberger, Ed., Interscience, N.Y., 1953, p. 69.

1. A. Hirsch and D. Orphanos, *J. Heterocyclic Chem.*, **2,** 206 (1965).
2. S. Gabriel and G. Pinkus, *Ber.*, **26,** 2210 (1893).
3. V. Paul, *Ber.*, **32,** 2014 (1899).
4. E. F. M. Stephenson, *Chem. Ind.* (*London*), **1957,** 174.
5. A. Albert, W. L. F. Armarego, and E. Spinner, *J. Chem. Soc.*, **1961,** 2689.
6. A. Albert, R. Goldacre, and J. Phillips, *J. Chem. Soc.*, **1948,** 2240.
7. C. Furlani, S. Bertola, and G. Morpurgo, *Ann. Chim.* (*Rome*), **50,** 858 (1960).
8. T. Nakajima and A. Pullman, *J. Chim. Phys.*, **55,** 793 (1958).
9. H. F. Hameka and A. M. Laquori, *Mol. Phys.*, **1,** 9 (1958).
10. J. Crossley and S. Walker, *Can. J. Chem.*, **46,** 2369 (1968).
11. H. C. Longuet-Higgins and C. A. Coulson, *J. Chem. Soc.*, **1949,** 971.
12. A. H. Gawer and B. P. Dailey, *J. Chem. Phys.*, **42,** 2568 (1965).
13. S. C. Wait and J. W. Wesley, *J. Mol. Spectrosc.*, **19,** 25 (1966).
14. L. Goodman and R. W. Harrell, *J. Chem. Phys.*, **30,** 1131 (1959).
15. D. A. Kearns and M. A. El-Bayoumi, *J. Chem. Phys.*, **38,** 1508 (1963).
16. N. Mataga, *Z. Phys. Chem.*, **18,** 285 (1958).
17. G. Favini, I. Vandoni, and M. Simonetta, *Theoret. Chim. Acta*, **3,** 45 (1965); **3,** 418 (1965).
18. S. F. Mason, *J. Chem. Soc.*, **1962,** 493.
19. R. M. Hochstrasser and C. Marzzacco, *J. Chem. Phys.*, **48,** 4079 (1968).
20. R. C. Hirt, F. T. King, and J. C. Cavagnol, *J. Chem. Phys.*, **25,** 574 (1956).
21. E. D. Amstutz, *J. Org. Chem.*, **17,** 1508 (1952).
22. R. M. Acheson and F. W. Foxton, *J. Chem. Soc.*, **1966,** 2218.
23. R. Muller and F. Dorr, *J. Electrochem.*, **63,** 1150 (1959).
24. P. J. Black and M. L. Heffernan, *Aust. J. Chem.*, **18,** 707 (1965).
25. K. Tori and T. Nakagawa, *J. Phys. Chem.*, **68,** 3163 (1964).
26. P. J. Black, R. D. Brown, and M. L. Heffernan, *Aust. J. Chem.*, **20,** 1305 (1967).
27. J. H. Bowie et al., *Aust. J. Chem.*, **20,** 2677 (1967).
28. E. W. Stone and A. H. Maki, *J. Chem. Phys.*, **39,** 1635 (1963).
29. G. Favini and A. Gomba, *Ric. Sci. Rend.*, Ser. A, **6,** 383 (1964).
30. J. F. Mucci, M. K. Orloff, and D. D. Fits, *J. Chem. Phys.*, **42,** 1841 (1965).
31. A. Hinchliffe, *Theor. Chim. Acta*, **5,** 208 (1966).
32. L. R. Snyder, *J. Chromatog.*, **17,** 73 (1965).
33. L. H. Klemm, C. E. Klopfenstein, and H. P. Kelly, *J. Chromatog.*, **23,** 428 (1966).
34. French Pat. 1,438,827 (1966); *Chem. Abstr.*, **66,** 95069 (1967).
35. A. Mustafa, A. H. Harhash, and A. A. S. Saleh, *J. Am. Chem. Soc.*, **82,** 2735 (1960).
36. R. F. Smith and E. D. Otremba, *J. Org. Chem.*, **22,** 879 (1962).
37. A. Hitzsch and D. Orphanos, U.S. Pat., 3,249,611 (1966); *Chem. Abstr.*, **65,** 726 (1966).
38. C. M. Atkinson and C. J. Sharpe, *J. Chem. Soc.*, **1959,** 3040.
39. A. Albert and G. Catterall, *J. Chem. Soc.*, (*c*), **1967,** 1533.
40. W. L. F. Armarego, *J. Appl. Chem.*, **11,** 70 (1961).

41. A. Hirsch and D. Orphanos, *Can. J. Chem.*, **46,** 1455 (1968); U.S. Pat. 3,528,979 (1970); *Chem. Abstr.*, **73,** 120653 (1970).
42. I. Ilmet and M. Krasij, *J. Phys. Chem.*, **70,** 3755 (1966).
43. P. Spacu and D. Camboli, *An. Univ. Bucur., Ser. Stiint. Nat.*, **14,** 101 (1965); *Chem. Abstr.*, **66,** 7964 (1967).
44. S. Gabriel, *Ber.*, **36,** 3373 (1903).
45. A. J. Birch and M. Kocar, *J. Chem. Soc.*, **1960,** 866.
46. H. Raistrick and P. Rudman, *Biochem. J.*, **63,** 395 (1956).
47. C. C. Leznoff, *Can. J. Chem.*, **46,** 1152 (1968).
48. I. Zugravescu et al., *An. Stiint Univ.* "*Al. I Cuza*" *Iasi, Sect. 1c*, **14,** 51 (1968); *Chem. Abstr.*, **70,** 77895 (1965).
48a. D. E. Elder, Ph.D. Thesis, Brigham Young University, Provo, Utah, 1969; *Diss. Abstr. Int., B*, **30,** 4040B (1970).
49. E. F. Elslager, D. F. Worth, N. F. Haley, and S. C. Perricone, *J. Heterocyclic Chem.*, **5,** 609 (1968).
50. S. Kanahara, *Yakugaku Zasshi*, **84,** 489 (1964); *Chem. Abstr.*, **61,** 8304 (1964).
51. F. D. Popp and J. M. Wefer, *Chem. Comm.*, **1967,** 59; F. D. Popp, J. M. Wefer, and C. W. Klinowski, *J. Heterocyclic Chem.*, **5,** 879 (1968).
52. G. Favini and M. Simonetta, *Gazz. Chim. Ital.*, **90,** 369 (1960).
53. C. M. Atkinson and C. J. Sharpe, *J. Chem. Soc.*, **1959,** 2858.
54. A. R. Osborn, K. Schofield, and L. N. Short, *J. Chem. Soc.*, **1956,** 4191.
55. A. V. El'tsov and A. A. Ginesina, *Zh. Org. Khim.*, **3,** 191 (1967); *Chem. Abstr.*, **66,** 94851 (1967).
56. J. A. Ballantine, C. H. Hassall, and G. Jones, *J. Chem. Soc.*, **1965,** 4672.
57. J. H. Birkinshaw, H. Raistrick, D. J. Ross, and C. E. Stickings, *Biochem. J.*, **50,** 610 (1952).
58. J. H. Birkinshaw, P. Chaplen, and R. Lahoz-Oliver, *Biochem. J.*, **67,** 155 (1957).
58a. W. Ried and H. Kohl, *Ann. Chim.*, **734,** 203 (1970).
59. F. Dallacker, K. W. Gombitza, and M. Lipp, *Ann.* **643,** 82 (1961).
60. F. Dallacker and J. Bloeman, *Monatsh. Chem.*, **92,** 640 (1961).
61. R. J. Pugmire et al., *J. Am. Chem. Soc.*, **91,** 6381 (1969).
62. H. Baba, I. Yamazaki, and T. Takemura, *Bull. Chem. Soc. Japan*, **42,** 276 (1969).
63. D. Van der Meer, *Rec. Trav. Chim.*, **88,** 136 (1969).
64. H. Lund and E. T. Jansen, *Acta Chem. Scand.*, **24,** 1867 (1970).

Part B. Phthalazines with Alkyl and/or Aryl Substituents in the 1- or the 1,4-Positions

I. 1-Alkyl- and 1-Arylphthalazines

A. Methods of Preparation

1. *Dehalogenation or Desulfurization*

Catalytic dehalogenation of 1-alkyl- or 1-aryl-4-chlorophthalazine with palladium on carbon (1–4) or Raney nickel (5, 6) gives 1-alkyl- or 1-arylphthalazine in good yield. 1-Methyl-4-chlorophthalazine on attempted dehalogenation with platinum catalyst resulted in cleavage of the N—N bond (2). 1-Phenyl-4-chlorophthalazine on chemical reduction with red phosphorus and hydriodic acid gave 1-phenylphthalazine (**1**) in 37% yield (10). 1-Phenyl-4-mercaptophthalazine on hydrogenation with Raney nickel (1) affords 1-phenylphthalazine (**1**) (83%) (Eq. 1).

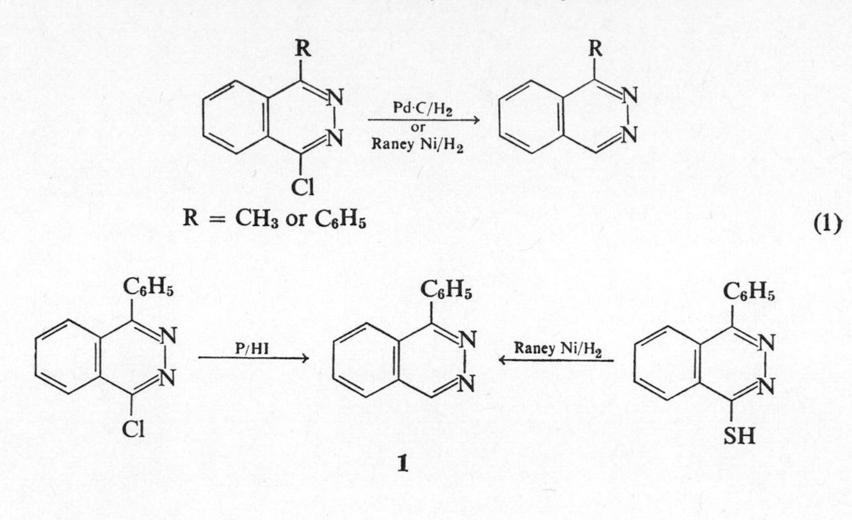

2. *From Phthalazine and Organometallic Reagents*

Mustafa and co-workers (7) first observed that phthalazine undergoes 1,2-addition of the Grignard reagent, phenylmagnesium bromide, across the C=N bond, accompanied by autooxidation to afford 1-phenylphthalazine (**1**) (44%) (Eq. 2).

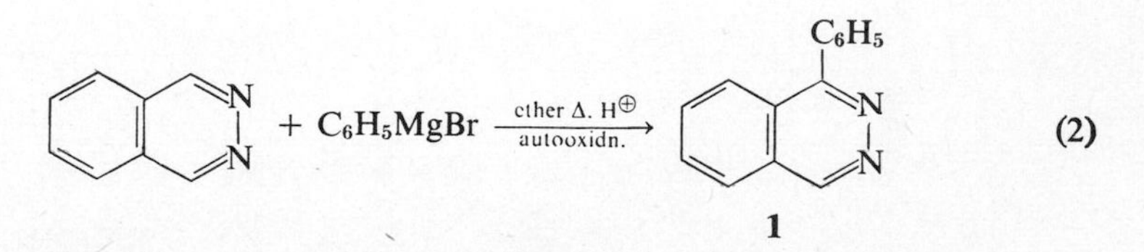

Marxer and co-workers (10) have aminoalkylated phthalazine with 3-dimethylaminopropylmagnesium chloride to give 1-(3-dimethylaminopropyl)-1,2-dihydrophthalazine, which is oxidized with potassium ferricyanide to 1-(3-dimethylaminopropyl)phthalazine (**2**) (55%) (see Section 2B-II-A-4-a). Marxer (11) has also prepared **2**, by reacting 1-chlorophthalazine with the same Grignard reagent, in 30% yield (Eq. 3).

Similarly, Hirsch and Orphanos (8) have reported the addition of organolithium compounds to phthalazine. Reaction of methyllithium with phthalazine in ether at 0°, followed by hydrolysis gave 1-methyl-1,2-dihydrophthalazine (**3**), which on autooxidation afforded a peroxyphthalazine compound (**4**) and 1-methylphthalazine (**5**) (8%). Later these workers (9) isolated the

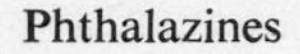

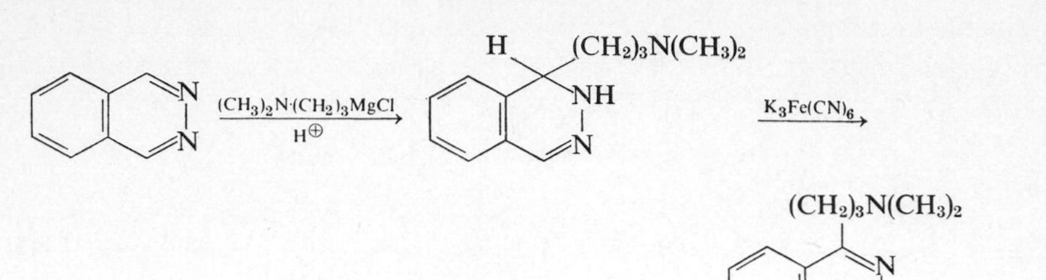

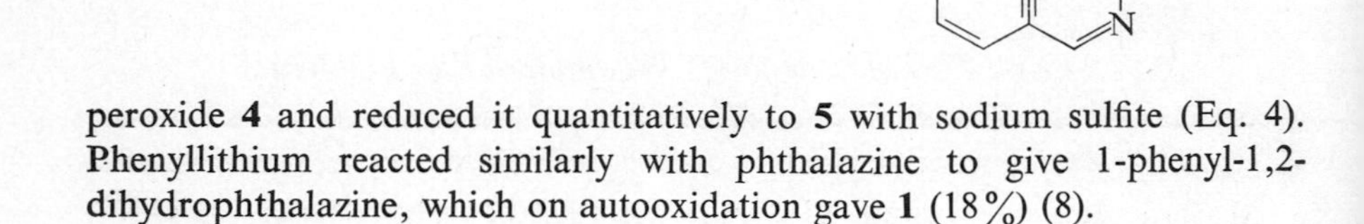

peroxide **4** and reduced it quantitatively to **5** with sodium sulfite (Eq. 4). Phenyllithium reacted similarly with phthalazine to give 1-phenyl-1,2-dihydrophthalazine, which on autooxidation gave **1** (18%) (8).

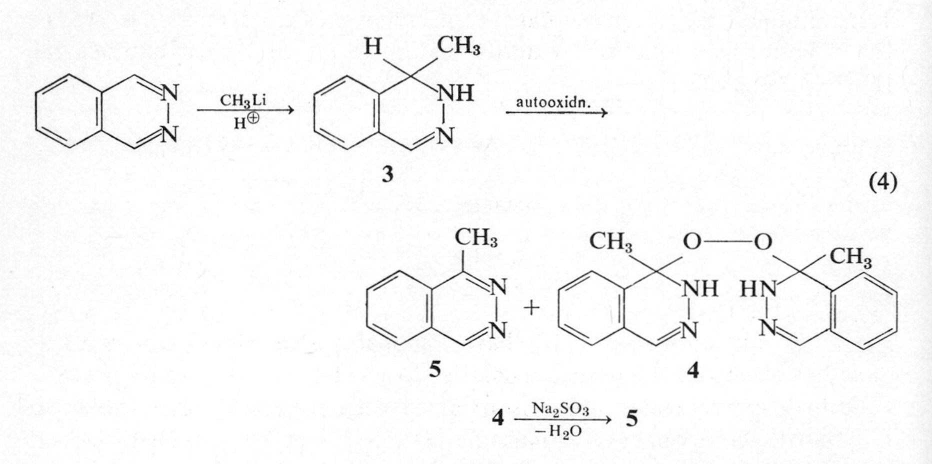

3. *From Phthalazine Reissert Compounds*

Popp and co-workers (12, 13) have reported that 1-cyano-1-methyl-2-benzoyl-1,2-dihydrophthalazine (**7**), obtained by the reaction of the phthalazine Reissert compound (**6**) with methyl iodide and sodium hydride in dimethylformamide, on alkaline hydrolysis gives 1-methylphthalazine (**5**) (76%). Similarly they have reacted **6** with benzaldehyde in presence of sodium hydride to give 1-(1-hydroxy-1-phenylmethyl)phthalazine (13) (Eq. 5).

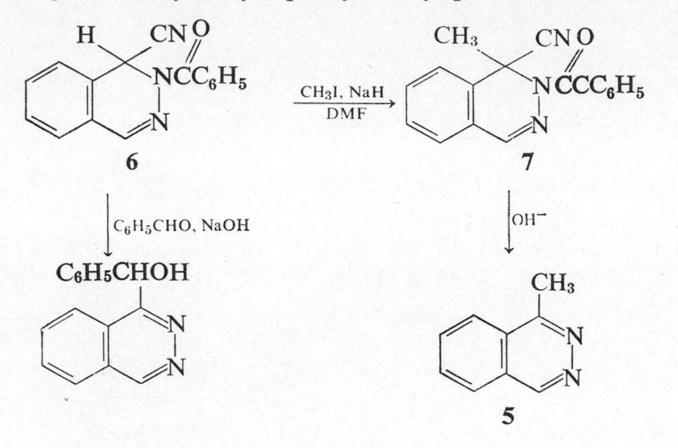

(5)

4. *From* 1-*Chloro- and* 1-(*Methylsulfonyl*)*phthalazine and Compounds with Active Methylene Groups*

Mizuno and co-workers (14) have extended the malonic ester synthesis to the reaction of 1-chlorophthalazine with active methylene compounds in presence of a strong base, such as sodium amide in benzene. The active methylene compounds used were benzyl cyanide, ethyl cyanoacetate, diethyl malonate, and malononitrile. Ethyl acetoacetate and diphenylmethane failed to react under the reported conditions. The yields of 4-substituted phthalazines (**8a–8d**) are above 65% except in the case of **8d** (32%). The compounds **8b–8d** on acidic hydrolysis followed by decarboxylation gave **5** in high yields (Eq. 6).

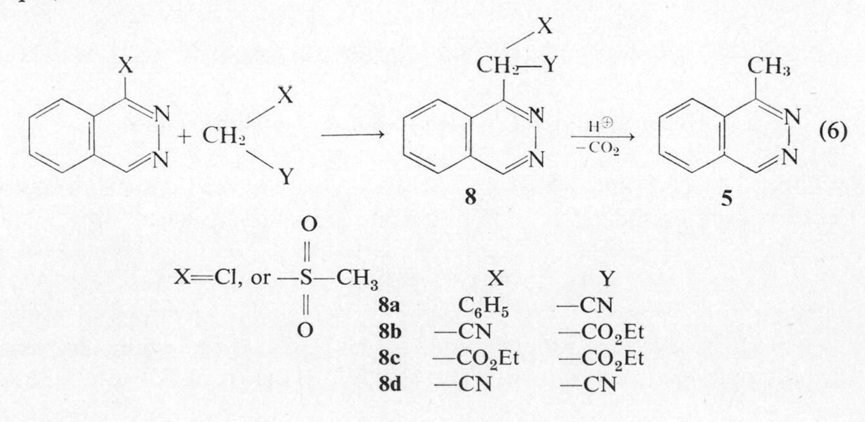

	X	Y
8a	C_6H_5	—CN
8b	—CN	—CO_2Et
8c	—CO_2Et	—CO_2Et
8d	—CN	—CN

Similarly, Oishi (43) has substituted the methylsulfonyl group from 1-(methylsulfonyl)phthalazine using the same carbanions from active methylene compounds as mentioned above (Eq. 6). In addition Oishi has also used successfully the carbonions derived from ethyl acetoacetate, acetone, and diethyl ketone to prepare ethyl α-acetyl-1-phthalazinylacetate, 1-(1-phthalazinyl)-2-propanone and 2-(1-phthalazinyl)-3-pentanone, respectively (see also Section 2B-II-A-2).

5. *Miscellaneous*

The action of dilute alkali on 1-phenyl-4-tosylhydrazinophthalazine hydrochloride gave 1-phenylphthalazine (**1**) in 80% yield (1) (Eq. 7).

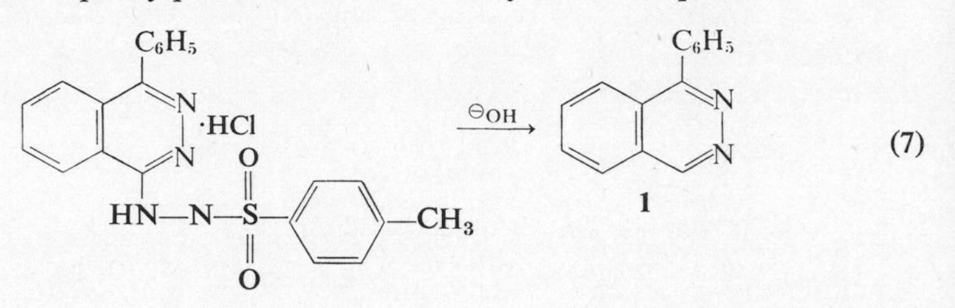

The preparation of 1-substituted phthalazines by Aggarwal's method (15), involving the cyclodehydration of acylhydrazones (Eq 8), has been critically reexamined (16). This method in fact does not lead to the formation of phthalazines. The products are azines resulting from the fragmentation of the starting materials. The cyclodehydration was studied with polyphosphoric acid and hydrochloric acid in amyl alcohol.

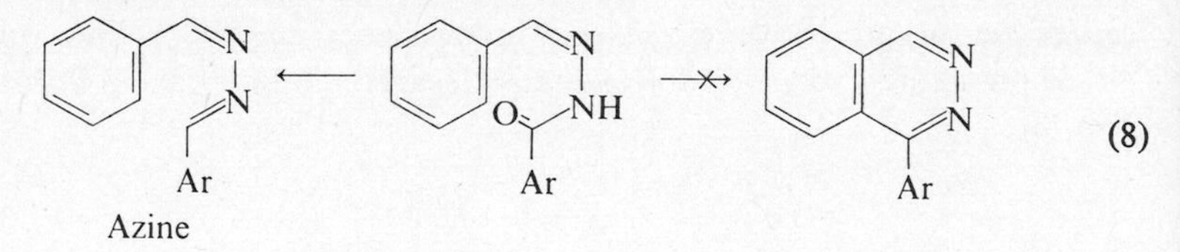

B. Properties of 1-Alkyl- and 1-Arylphthalazines

Since 1-methyl- and 1-phenylphthalazine are most studied, the properties of these two representative compounds are discussed here.

1. *Physical Properties*

1-Methylphthalazine sublimes at 125–135°/0.05 mm to a white, crystalline hygroscopic sublimate, mp 70.5–71.5° (2) [70–71° (13); 71–72° (8, 9); 72–74°

and 70–74° (14)]. 1-Methylphthalazine has a pK_a of 4.39 in water at 20° (17).

The UV spectrum of 1-methylphthalazine has been recorded in ethanol (2) and in water at 7.0 *p*H and 2.0 *p*H (cationic form) (17) as follows:

Ethanol: λ_{max} mμ (log ε); 263 (3.17), 217 (4.88).

H_2O *p*H 7.0: 305 (3.18); 292 (3.21); 270 (inflection) (3.57); 262 (3.59); 219 (4.70).

H_2O *p*H 2.0: 312 (3.46); 304 (inflection) (3.40); 271 (3.40); 230 (4.62).

1-Phenylphthalazine forms yellow crystals, mp 139–141° (1), [135–138° (10), 138–141° (1); 144° (7, 8)]. It is easily soluble in most organic solvents and can be recrystallized from benzene-ligroin, benzene-ether, or ether. It has a pK_a 3.51 in 1% ethanol (19).

The UV spectrum and polarographic reduction waves at different *p*H have been recorded (10); the UV spectrum in 95% ethanol (11) shows two maxima at 223 and 278 mμ with log ε values 4.8 and 4.1, respectively.

2. *Chemical Properties*

a. Salt Formation. 1-Methylphthalazine forms well-defined salts functioning as a monoacid base; hydrochloride, prisms, mp 204–205° (dec.) (2-propanol) (2); hydrobromide, prisms, mp 280–282° (dec.) (1-butanol) (2); perchlorate, mp 241–242° (dec.) (water) (2); picrate, mp 203–205° (dec.) (14) [204° (8), 206–207° (13)].

1-Phenylphthalazine forms a picrate, yellow needles, mp 180° (ethanol) (1, 7). It forms a stable complex with bromine, red powder, mp 105° (18).

Both 1-methyl- and 1-phenylphthalazine form quaternary salts with methyl iodide in benzene at room temperature. The quaternization takes place in the 3-position in both instances (Eq. 9) (see Section 2K).

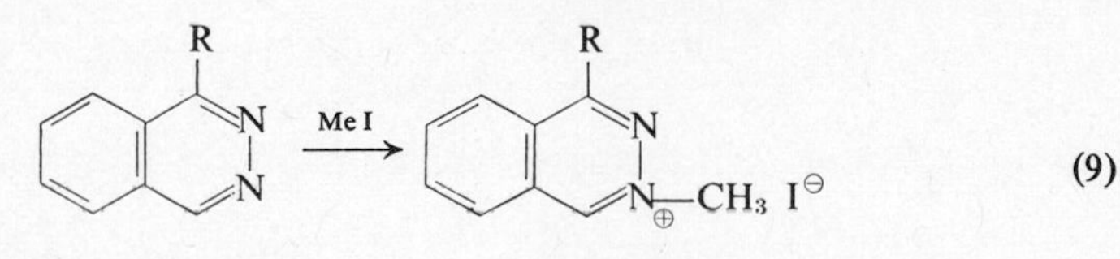

(9)

$R = CH_3, C_6H_5$

b. *N*-Oxidation. 1-Methyl and 1-phenylphthalazine both react with monoperphthalic acid to give the corresponding *N*-oxides. 1-Phenylphthalzine exclusively gives the 3-oxide, whereas 1-methylphthalazine gives *N*-oxides at both the 2- and the 3-position in approximately 1:2 ratio (see Section 2C-II).

c. The Reactivity of Methyl Group in 1-Methylphthalazine. 1-Methylphthalazine contains an active methyl group as expected from its structure. Castle and Takano (20) have studied the condensation of 1-methylphthalazine with aqueous formaldehyde. After heating at 63–65° for 4 hr it gives 1-(2-hydroxyethyl)phthalazine (**9**) and refluxing in pyridine for 16 hr 2-(1-phthalazinyl)-1,3-propanediol (**10**) was isolated. These authors have carried out several reactions on this side chain in order to prepare 1-(2-mercaptoethyl)phthalazine and its analogs.

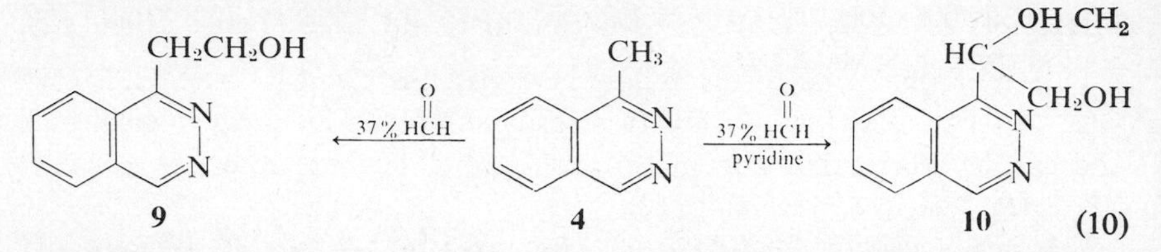

1-Methylphthalazine condenses with 2-furaldehyde and 5-nitro-2-furaldehyde. By heating on a steam bath, the ethylenic compounds, 1-(furylvinyl)phthalazine (**11**) and 1-(5-nitrofurylvinyl)phthalazine (**12**) were obtained (21) (see Eq. 11). When 1-methylphthalazine was heated on steam bath with 5-nitro-2-furaldehyde in presence of benzene for 30 min the intermediate 1-(2′-hydroxy-2′,5-nitrofurylethyl)phthalazine (**13**) was isolated and this in turn was dehydrated to **12**. Nitration of **11** with fuming nitric acid in acetic anhydride at −10° gave **12**. (Eq. 11).

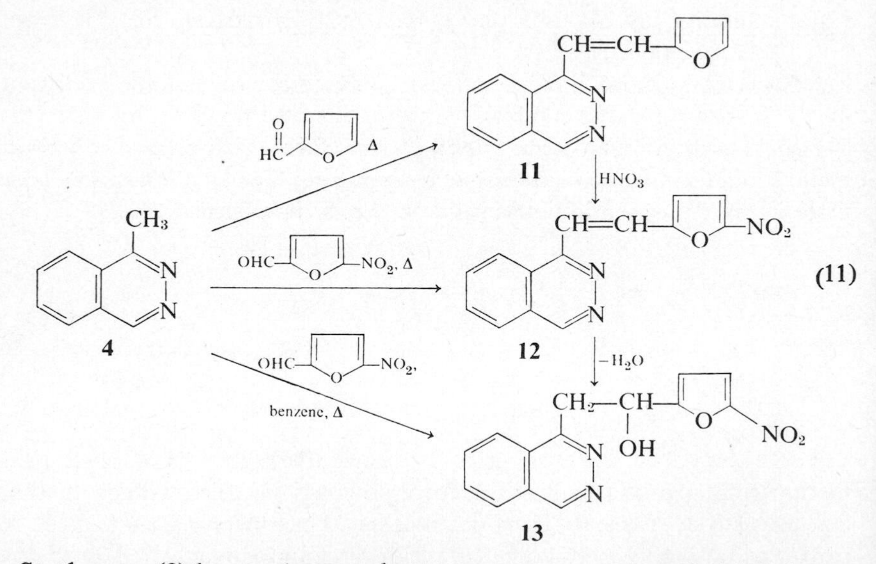

Stephenson (2) has made several attempts to convert 1-methylphthalazine into phthalazine-1-carboxaldehyde. The reaction with selenium dioxide led to 1,2,3-tri-1′-phthalazinylpropane: mercuric oxide and aqueous sodium hydroxide gave a mercury derivative, but attempts to hydrolyze it failed.

1-Methylphthalazine on reaction with pyridine and iodine afforded *N*-(1-phthalazinylmethyl)pyridinum iodide (**14a**), which was converted to perchlorate (**14b**). The treatment of **14b** with *N*,*N*-dimethyl-*p*-nitrosoaniline and alkali gave *N*,*N*-dimethyl-*N'*-(1'-phthalazinylmethylene)-*p*-phenylenediamine (**15**) (Eq. 12). It was not possible to isolate the free phthalazine-1-carboxaldehyde, but the anil **15** was directly converted into the semicarbazone (**16a**), thiosemicarbazone (**16b**), phenylhydrazone (**16c**), and 2,4-dinitro-phenylhydrazone (**16d**) (Eq. 12).

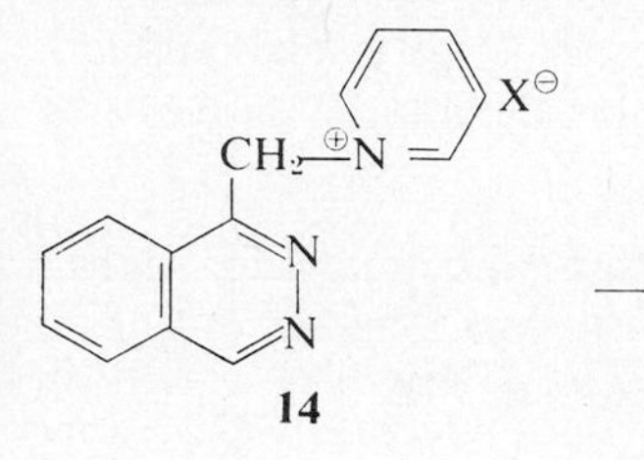

14

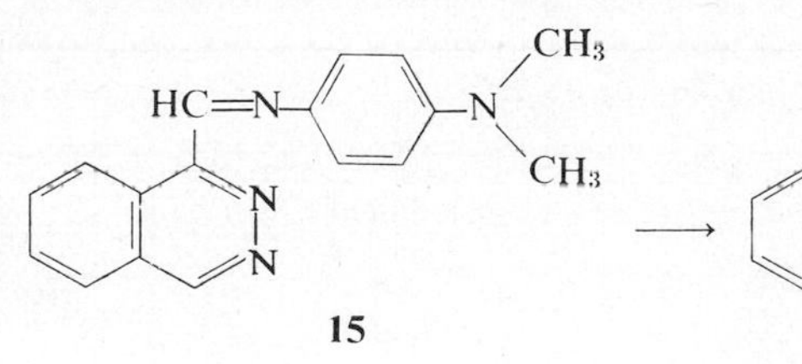

15

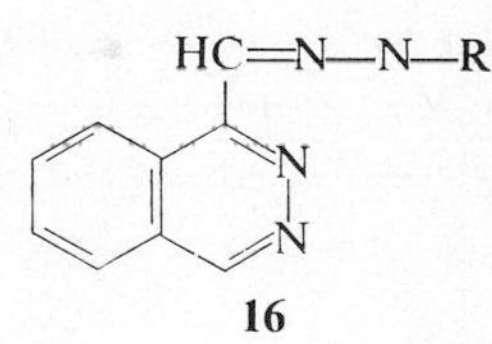

16

(12)

14a $X^{\ominus} = I^{\ominus}$

16a

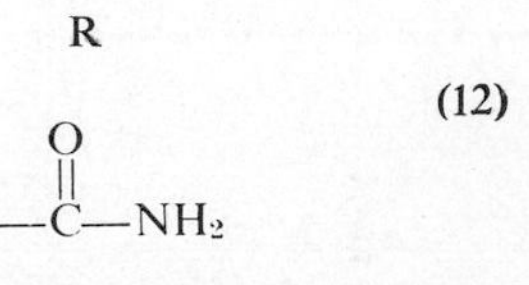

14b $X^{\ominus} = ClO_4^{\ominus}$

16b

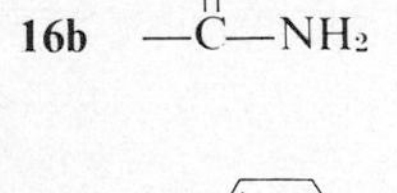

16c

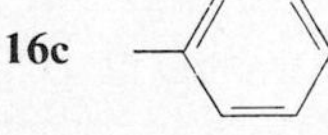

16d

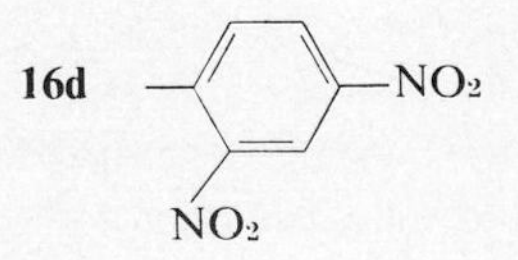

d. Reaction with Grignard Reagents. 1(3-Dimethylaminopropyl)-phthalazine (**5**) has been shown to undergo the addition of one mole of 3-dimethylaminopropylmagnesium chloride to give 1,2-dihydro-bis-1,4-(3-dimethylaminopropyl)phthalazine, which on oxidation gives the 1,4-disubstituted product, 1,4-bis(3-dimethylaminopropyl)phthalazine (10). Similarly, 1-phenyl- and 1-*p*-chlorophenylphthalazine undergo addition of the Grignard reagent (see Section 2B-II-A-4-a).

e. Addition of Dienophiles. Dimethyl acetylenedicarboxylate has been observed to add to 1-methylphthalazine in acetonitrile to give small amounts of pyrrolo[2,1-*a*]phthalazine and pyrido[2,1-*a*]phthalazine derivatives (40) (see Section 2M).

f. Nitration. 1-Methylphthalazine on nitration with potassium nitrate in concentrated sulfuric acid gives 1-methyl-5-nitrophthalazine (**17**) as a major product and some 8-nitro-1(2*H*)phthalazinone (**18**) (22) (Eq. 13).

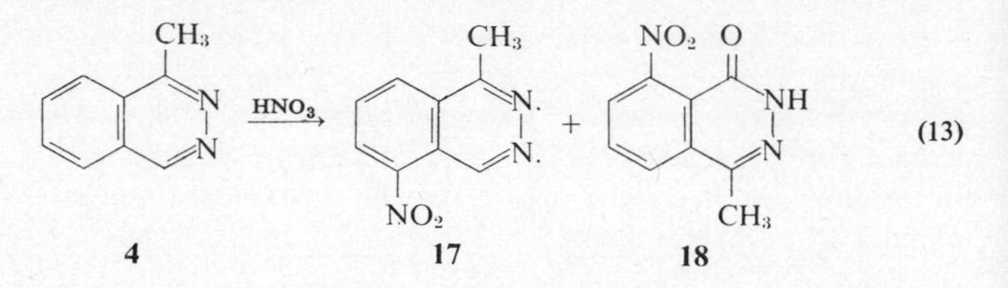

II. 1,4-Dialkyl-, 1,4-Diaryl-, and 1-Alkyl-4-arylphthalazines

A. Methods of Preparation

1. *Condensation of* 1,2-*Diacylbenzene with Hydrazine*

This is the most commonly used method for the preparation of this group of compounds because of its simplicity and high yields. The appropriate

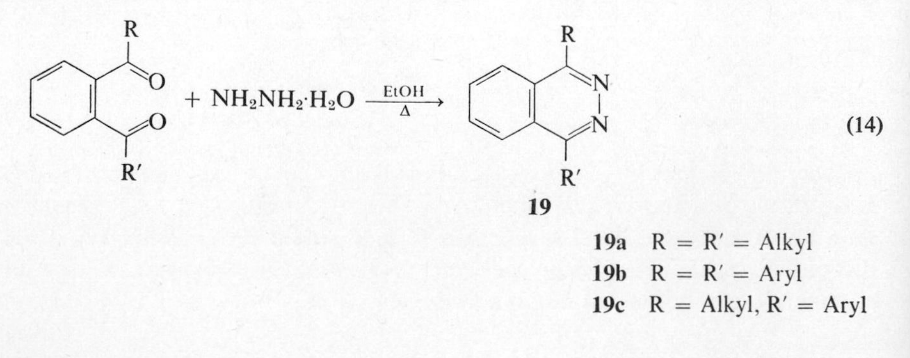

1,2-diacylbenzene is treated with hydrazine hydrate in ethanol or acetic acid and refluxed from a few minutes to a few hours to give 1,4-dialkyl- (**19a**) (23–26), 1,4-diaryl (**19b**) (1, 27–31), and 1-alkyl-4-arylphthalazines (**19c**) (32, 33) (Eq. 14).

Occasionally hydrazine hydrochloride along with sodium acetate is used to give equally good results. In one case (23) hydrazine has been replaced with semicarbazide or thiosemicarbazide to give the same compound.

2. *From 1-Methylsulfonyl-4-phenylphthalazine by Nucleophilic Substitution*

Hayashi and Oishi (34, 35) have studied the nucleophilic substitution reactions of 1-methylsulfonyl-4-phenylphthalazine (**20**) with the carbanions obtained from various ketones, for example, acetone, acetophenone, diethyl ketone, methyl ethyl ketone, cyclopentanone, and cyclohexanone, and from active methylene compounds such as malonodinitrile, benzylnitrile, diethyl malonate, ethyl acetoacetate, and ethyl cyanoacetate, to give 1-substituted alkyl-4-phenylphthalazines (Eq. 15). The carbanions are prepared in benzene in the presence of sodium amide and the reaction mixture is refluxed for about 2 hr. The methylsulfonyl group is smoothly displaced and the products are isolated in about 50% yields. This gives an excellent approach to the synthesis of these compounds, which are otherwise difficult to prepare.

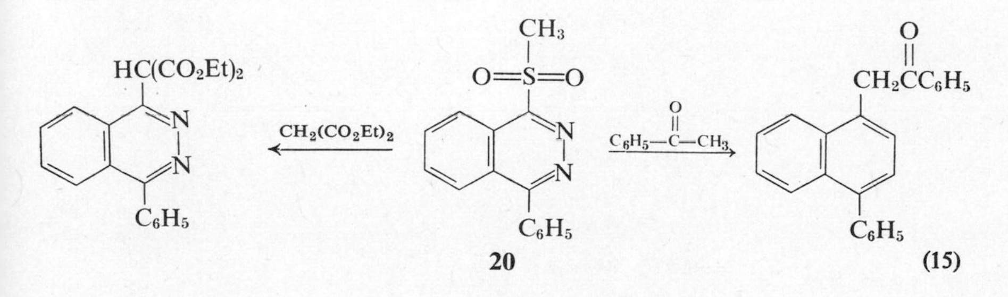

3. *Diels-Alder Condensation*

Sauer and Heinrich (36) reported a novel approach for the preparation of 1,4-diphenylphthalazine (**21**) involving a Diels-Alder reaction. 1,4-Diphenyl-*sym*-tetrazine reacted readily with benzyne, liberated from anthranilic acid diazonium betain, in dichloromethane to give **21** in 67% yield with the elimination of nitrogen from the adduct (Eq. 16).

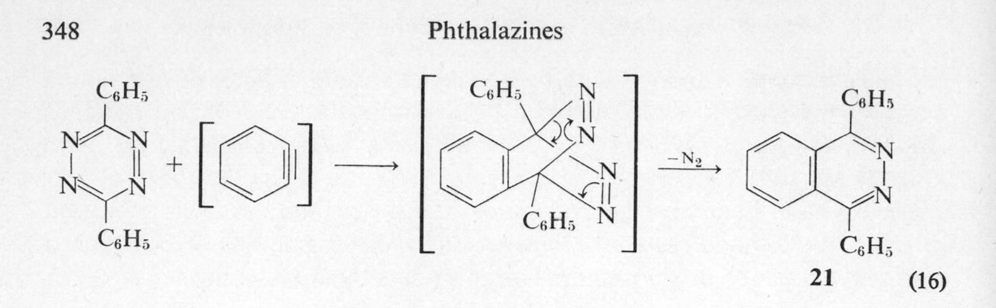

4. *Reaction of Grignard Reagents*

a. To 1-Alkyl- or 1-Arylphthalazine. Marxer and co-workers (10) have shown that 1-alkyl- or 1-arylphthalazine (**22**) undergoes a Grignard addition reaction with 3-dimethylaminopropylmagnesium chloride to furnish 3,4-dihydro derivatives (**23**) which on oxidation with potassium ferricyanide give 1,4-disubstituted phthalazines (**24**) (Eq. 17).

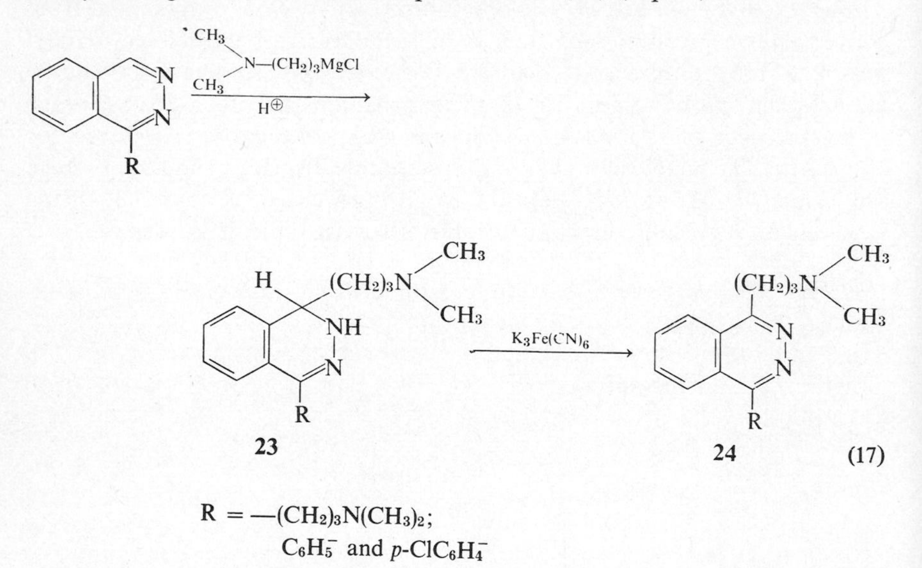

The inactivity of the second C=N in the dihydro derivatives (**23** and also when R=H) to undergo further addition of Grignard reagent has been explained (10) on the basis of polarizability. The polarization caused after the reaction of the NH group with the first mole of Grignard reagent (**24**) restricts the formation of two adjacent negative charges (see Eq. 18 as shown in **25**), that is, it causes an energetically unfavorable condition for the nucleophilic attack of the Grignard carbanion $[(CH_3)_2—N—CH_2—CH_2—CH_2^-]$ at the 4-position. At each stage after the reaction with one mole of

Grignard reagent it is necessary to oxidize the dihydro derivative for further reaction with the Grignard reagent. This can be seen by the fact that 1,4-disubstituted phthalazine can be further alkylated with one mole of Grignard reagent to give 1,2-dihydro-1,1,4-trisubstituted phthalazines (see Section 2B-II-B-2-b).

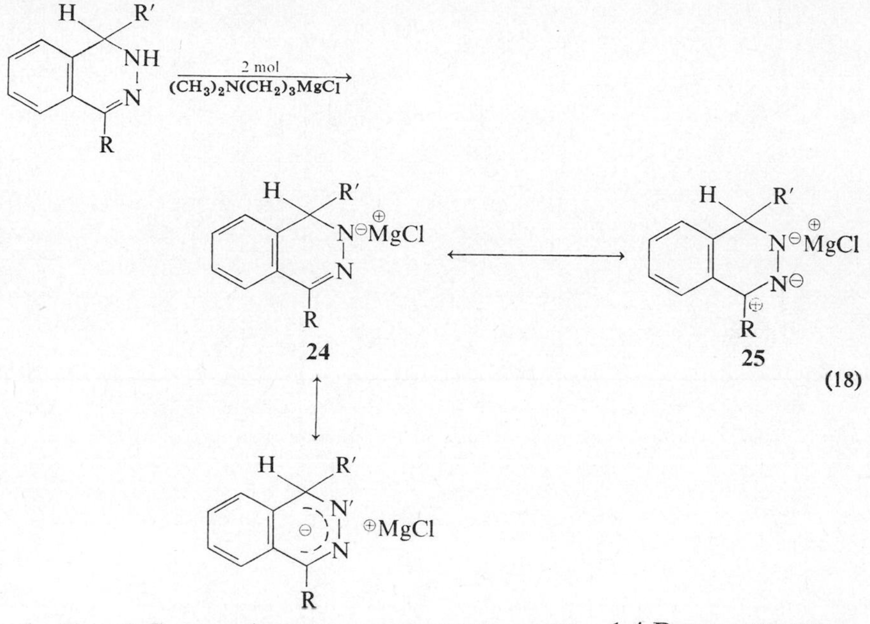

b. To 1-Chloro-4-phenylphthalazine and 1,4-Dichlorophthalazine. Mustafa and co-workers (7) have allowed 1-chloro-4-phenylphthalazine and 1,4-dichlorophthalazine to react with phenylmagnesium bromide to give 1,4-diphenylphthalazine (**21**) in 62 and 75% yields, respectively (Eq. 19).

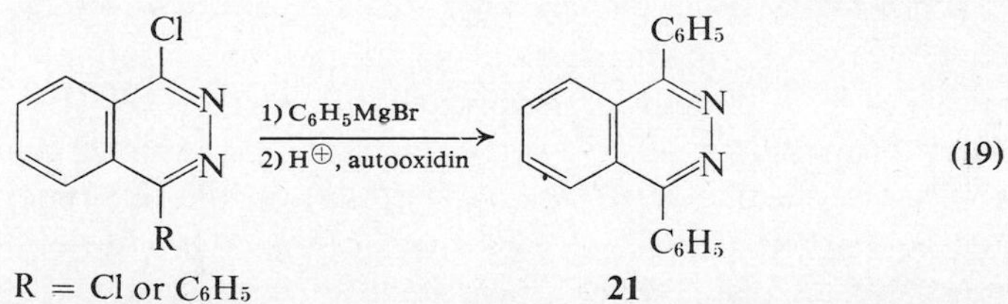

c. To 1-(2*H*)Phthalazinone and its Derivatives. Mustafa and co-workers (7) have added phenylmagnesium bromide to 1-(2*H*)phthalazinone (**26**) and several of its derivatives, 4-phenyl-1-(2*H*)phthalazinone (**27**), 4-phenyl-1-(2*H*)phthalazinethione (**27a**), 4-phenyl-2-acetyl-1-(2*H*)phthalazinone (**27b**) and 4-phenyl-2-*N*(piperidinomethyl)-1-(2*H*)phthalazinone (**27c**) (37) and isolated in each case 1,4-diphenylphthalazine in good yields (see

Eq. 20). In the case of 1-(2*H*)phthalazinone 2 moles of phenylmagnesium bromide add to the carbonyl group and to the C═N group; however, when the phenyl group is already present in the 4-position, addition of only 1 mole of phenylmagnesium bromide at the C═O group takes place due to the steric hindrance of the phenyl group (Eq. 20).

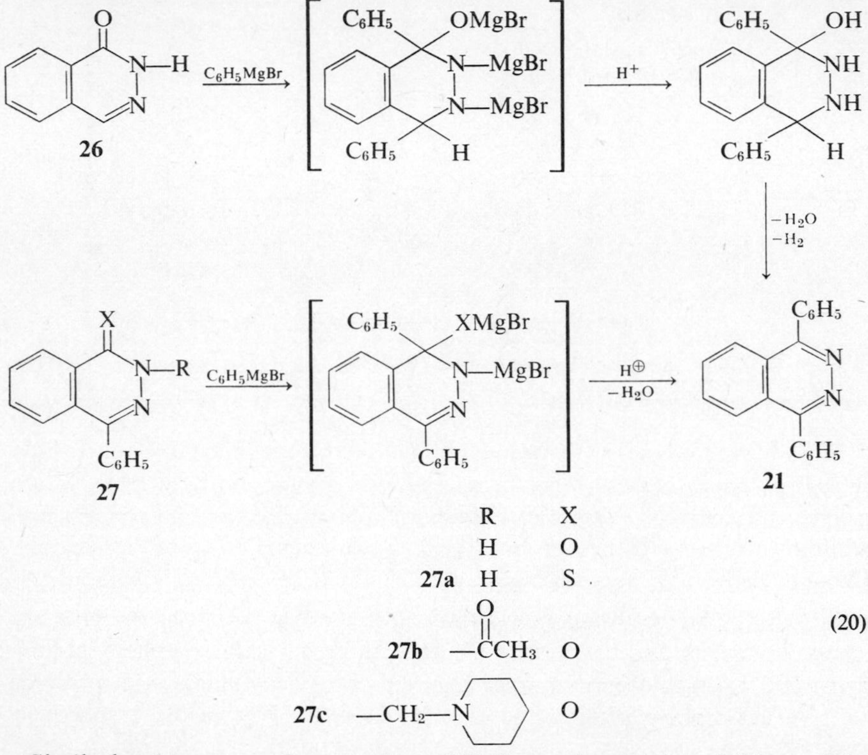

Similarly, Marxer and co-workers (10) have shown the addition of 3-dimethylaminoproplymagnesium chloride to 1-(2*H*)phthalazinone (**26**) gives 1,4-bis(3-dimethylaminopropyl)phthalazine (**28**) (Eq. 21). The mechanism proposed by Marxer and co-workers, using the arguments of polarizability (see Section 2B-II-A-4-a), differs from the one reported by Mustafa and co-workers (see Eq. 20) by the fact that the addition of the Grignard reagent takes place in successive steps. According to this mechanism, after the addition of 1 mole of Grignard reagent to the carbonyl group, aromatization proceeds to the intermediate 2. Then the addition of the second mole of the Grignard reagent to the C═N bond at the 4-position takes place, followed by hydrolysis and oxidation, to furnish **28**. However, the attempted preparation of **2** from **26** with 2 moles of Grignard reagent resulted in failure and the starting material was recovered, whereas the reaction with 3 moles of Grignard reagent led to the formation of the final product **28**.

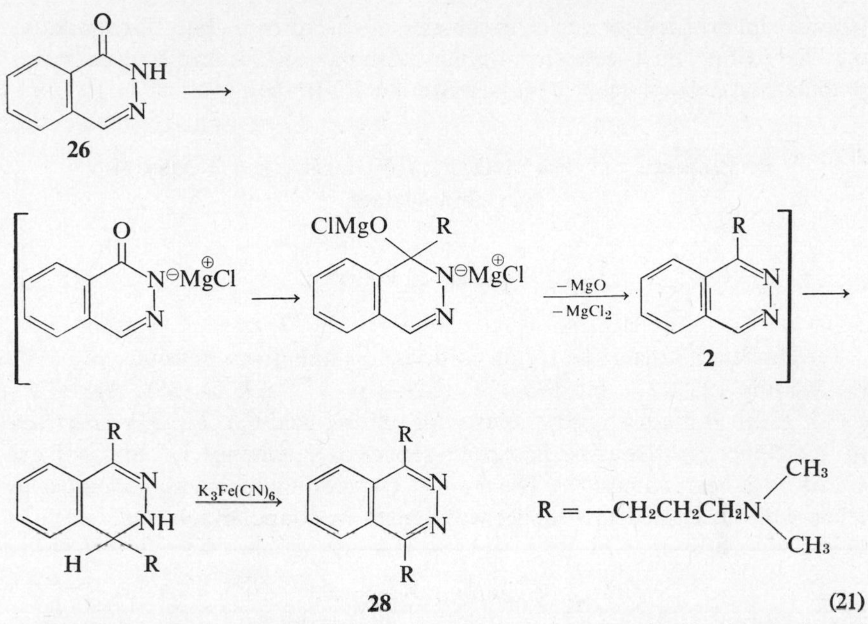

(21)

The addition of an organolithium compound to **26** resulted in 4-substituted 1-(2*H*)-phthalazones only, but in no case was 1,4-disubstituted phthalazine isolated (see Section 2D). Staunton and Topham (38) have carried out the addition of phenylmagnesium bromide to 2,4-diphenyl-1-(2*H*)phthalazinone (**29**) and converted the resulting quaternary bromide (**30**) into 1,4-diphenylphthalazine by refluxing with hydrazine hydrate (Eq. 22).

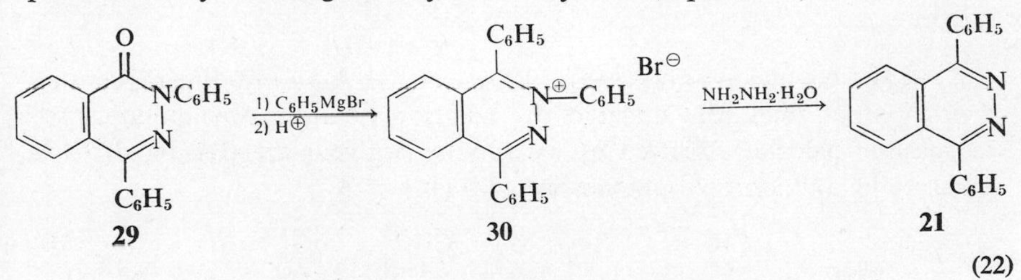

(22)

5. *From Phthalazine N-Oxides*

Hayashi and Oishi (1) have prepared 1,4-diphenylphthalazine (**21**) in 56% yield by cleaving the oxygen atom from 1,4-diphenylphthalazine 2-oxide

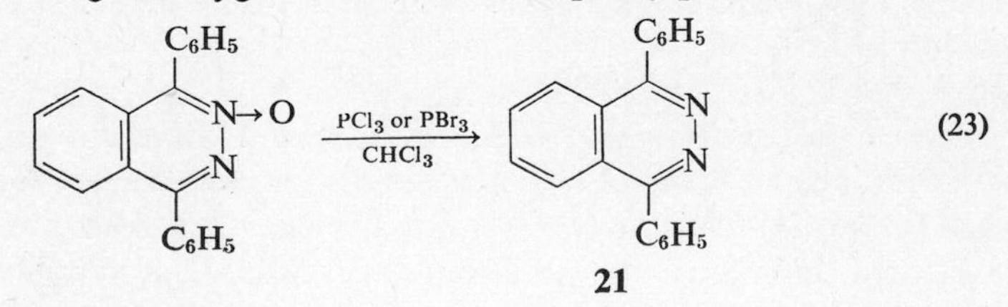

(23)

with phosphorus tribromide or trichloride in chloroform (Eq. 23). Similarly, they have obtained 1-isopropyl-4-phenylphthalazine (34) and 2-(4-phenyl-1-phthalazinyl)acetophenone (39) (see Section 2C-III-A).

B. Properties of 1,4-Dialkyl-, 1,4-Diaryl-, and 1-Alkyl-4-arylphthalazines

1. *Physical Properties*

1,4-Diphenylphthalazine forms colorless crystals from benzene, mp 194° (7, 37) [mp 192° (27), mp 193–194° (38); mp 197.5–198.5° (36); mp 197.5–199.5° (28)]. It gives a yellow color with sulfuric acid (7). The UV spectrum of 1,4-diphenylphthalazine has been reported in ethanol (1) and solvent effects have been noted (41). The $n \rightarrow \pi^*$ transition appears as a shoulder in ethanol and is shifted to a longer wavelength in nonpolar solvents.

2. *Chemical Properties*

a. Formation of Quaternary Salts and *N*-Oxides. 1,4-Diphenylphthalazine reacts with methyl iodide to give the quaternary salt, methiodide, mp 273° (42) [solvated with methanol, orange solid, mp 309–314° (dec.) (28)]. 1,4-Diphenylphthalazine and other 1,4-disubstituted phthalazines with alkyl or aryl groups form *N*-oxides with hydrogen peroxide (see Section 2C-II-C).

b. Reaction with Grignard Reagent. 1-Aryl-4-(3-dimethylamino-propyl)phthalazines (**31**) undergo the addition of 3-dimethylaminopropyl-magnesium chloride followed by hydrolysis to give 1-aryl-3,4-dihydro-4,4-bis(dimethylaminopropyl)phthalazine (**32**) (10).

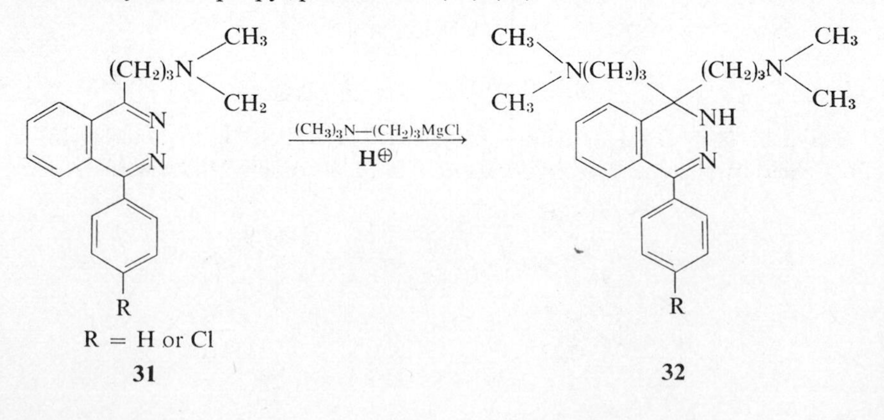

III. Tables

TABLE 2B-1. 1-Alkyl- or 1-Arylphthalazines

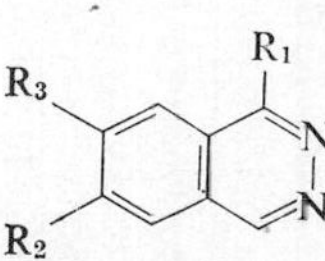

R_1	R_2	R_3	Preparation[a]	Yield (%)	MP[b]	Reference
CH_3—	H—	H	A		70.5–71.5°, 70–71°, 71–72°, 72–74°, 70–74°; HCl, prisms, 204–205°; $HClO_4$, 241–242° (dec.); picrate, 203–205°, 204°, 206–207°	2, 8, 9, 13, 14
C_6H_5	H—	H—	A		Yellow crystals, 139–141°, 135–138°, 138–141°, 144°; picrate, yellow needles, 180°	1, 7, 8, 10
p-Cl—C_6H_4—	H	H	B	60	Light green crystals, 155–156° (EtOH)	10
$(CH_3)_2N$—$(CH_2)_3$	H	H	C	30; 50	2 HCl, 231° (EtOH)	10, 11
C_6H_5—CH—OH (with a vertical bond line above the CH)	H	H—	D	100	172–175° (EtOH)	13
C_6H_5—	—OCH_3	H	E	100	111–112.5° (ligroin); picrate, 162–164°	5
C_6H_5—	H—	—OCH_3	E	—	151–152° (xylene)	5
C_6H_5—	—OCH_3	—OCH_3	E	—	127° (xylene); picrate, 193.5–194°	6
$C_6H_5CH_2$	—OCH_3	—OCH_3	F	52	Needles, 156°, 129–130°)(xylene); picrate, 180–182°, 193.5–194°	3, 6
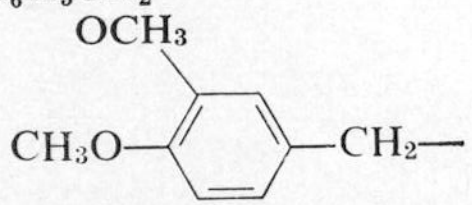 CH_3O—$C_6H_3(OCH_3)$—CH_2—	—OCH_3	—OCH_3	F	50	120–121°, 125–127° (ethyl acetate-pet. ether)	4

[a] A, see different methods of preparations in Section 2B-I-A. B, corresponding 4-chlorophthalazine + Red P + 57% HI, reflux 15 hr. C, 1-Chlorophthalazine + $(CH_3)_2N$—$(CH_2)_2MgCl$ in THF, followed by hydrolysis, and phthalazine + $(CH_3)_2N(CH_2)_3MgCl$ → hydrolysis and oxidation with $K_3Fe(CN)_6$. D, 1,2-Dihydro-1-cyano-2-benzoylphthalazine + benzaldehyde in dimethylformamide + 30% NaH in oil → stir at room temperature for 1.5 hr and hydrolyze. E, Corresponding 4-chlorophthalazine catalytic dehalogenation with Raney Ni/H_2 in EtOH with a little alkali. F, Same as E but dehalogenation with 5% Pd-C.

[b] Mp with physical appearance. Recrystallization solvent in parentheses.

TABLE 2B-2. 1-Substituted Alkylphthalazines; Condensation Products from 1-Methylphthalazine with Aldehydes

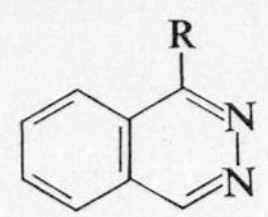

R	Preparation[a]	Yield (%)	MP[b]	Reference
$—CH_2CH_2OH$	A	39	Colorless needles, 90–91° (benzene)	20
$—CH(CH_2OH)_2$	B	47	Colorless pillars, 161–162° (ethanol)	20
$—CH_2CH(OH)CCl_3$	—	—	—	2
$—CH_2—CH(OH)—$(5-nitro-2-furyl)	C	50	165.5–166° (dec.) (acetonitrile)	21
$—CH{=}CH—$(5-nitro-2-furyl)	D	27	240–242° (acetone)	11
$—CH{=}CH—$(2-furyl)	E	58	103.5–104.5° (acetone)	11

[a] A, Heat with 37% Formalin at 63–65° for 4 hr or react with phenyllithium in benzene + gaseous formaldehyde. B, Reflux with 37% Formalin in pyridine for 16 hr. C, Heat with 5-nitro-2-furaldehyde in benzene on steam bath for 30 min. D, Same as C but heat on steam bath for 10 min in absence of solvent. E, Same as D but heat at 120–130° for 14 hr.
[b] Mp with physical description. Recrystallization solvent in parentheses.

TABLE 2B-3. 1-Substituted Alkylphthalazines; From Nucleophilic Substitution of 1-Chlorophthalazine with Active Methylene Compounds

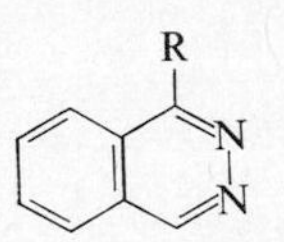

R	Preparation[a]	Yield (%)	Mp (Solvent of Recrystallization)	Reference
$—CH(CN)C_6H_5$	A	77	132–134° (benzene)	14
$—CH(CN)CO_2CH_2CH_3$	B	73	186–187° (ethanol)	14
$—CH(CO_2Et)_2$	B	68	70–71° (ligroin-benzene, 3:1)	14, 20
$—CH(CN)_2$	C	31	>300°	14

[a] A, Reflux 1 hr with benzyl cyanide in benzene with sodium amide. B, Same as A but reflux 20 hr with proper active methylene compound. C, Reflux 20 hr with malonodinitrile in benzene with sodium ethoxide.

TABLE 2B-4. 1-Substituted Alkylphthalazines. Reactions on the Side Chain

(Structure: phthalazine with substituent R at position 1)

R	Yield (%)	MP[a]	Reference
$-CH_2CO_2CH_2CH_3$	74	Pale yellow needles, 80–83° (ligroin-benzene)	20
$-CH_2CH_2SSO_3Na$	20	White powder, 261° (darkening) (ethanol)	20
$-CH_2CH_2-\overset{\overset{NH}{\|}}{S}NH_2$	60	2 HBr, colorless needles, 184–185° (dec.) (ethanol)	20
$-CH_2CH_2SH$	39	HCl, 205° (dec.)	20
$-CH_2CH_2SCH_2CH_2-$(1-phthalazinyl)	40	Cream colored needles, 138–139° (benzene)	20
$-CH_2CH_2S-$(2,4-dinitrophenyl: NO_2, NO_2)	—	—	20
$-CH(CH_2Cl)_2$	45	Pale yellow plates, 70.5° (benzene)	20
$-CH(CH_2S-\overset{\overset{O}{\|}}{C}-CH_3)_2$	—	Colorless pillars, 145–146° (dec.) (chloroform-acetone)	20
$-CH(CH_2SH)_2$	—	HCl, colorless pillars, 178° (dec.) (ethanol)	20
—(thietan-3-yl, S)	7	Colorless plates, 131–135° (benzene-ligroin)	20
$-CH(CH_2SSO_3Na)_2$	52	Hygroscopic white powder, 103° (swelling) darkening at 203° (ethanol)	20
$-CH(CH_2SH)-CH_2N(CH_3)_2$	32	Hygroscopic colorless needles, 116°; 153° (dec.) in a sealed tube (ethanol)	20
$-CH(CH_2SH)-CH_2-N\langle\ \rangle N-CH_3$ (piperazine)	—	Hygroscopic white precipitate	20

TABLE 2B-4 (*continued*)

R	Yield (%)	MP[a]	Reference
$-CH_2-\overset{\oplus}{N}C_5H_5\ I^{\ominus}$	86	Red powder, 220° (dec.)	2
$-CH_2-\overset{\oplus}{N}C_5H_5\ ClO_4^{\ominus}$	74	Bright orange crystals, 224–226° (dec.)	2
$-CH{=}N-C_6H_4-N(CH_3)_2$	44	Exists in two forms, yellow and red, 136–137° sublimed at 125–150°/0.01 mm	2
$-CH{=}N-NH-C(=O)NH_2$	—	Pale cream needles, 229–231° (dec.) (water)	2
$-CH{=}N-NH-C(=S)NH_2$	—	Cream colored, 234–235° (dec.)	2
$-CH{=}N-NH-C_6H_5$	—	Yellow, 205–205.5° (dec.) (1-butanol)	2
$-CH{=}N-NH-C_6H_3(NO_2)_2$ (2,4-dinitrophenyl)	86	Yellow prisms, 282–283° (ethyl benzoate)	2
$-CH_2-CH(\text{1-phthalazinyl})-CH_2(\text{1-phthalazinyl})$	—	Cream needles, 190–192° (sublimed at 160–170°/0.05 mm)	2
$-CH{\cdot}Hg_2O$	78	Yellow precipitate, >220° (dec.)	2

[a] MP with physical description. Recrystallization solvent in parentheses.

TABLE 2B-5. 1,4-Dialkylphthalazines

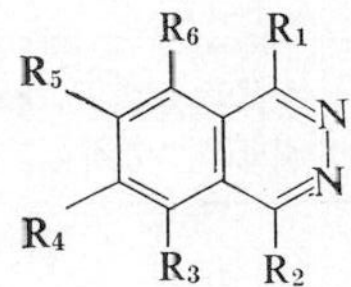

R_1	R_2	R_3	R_4	R_5	R_6	Preparation[a]	Yield (%)	MP	Reference
—CH_3	—CH_3	—H	—H	—H	—H	A, A-1	41	Light yellow needles, 108 (benzene-ligroin), 106–107° (water and sublimed)	23, 23a
—CH_2CH_3	—CH_2CH_3	—H	—H	—H	—H	B	82	Yellow crystals, 82–85° (ether)	24
—$(CH_2)_3N(CH_3)_2$	—$(CH_2)_3N(CH_3)_2$	—H	—H	—H	—H	C	—	3 HCl, 252–254° (very hygroscopic)	10
—$(CH_2)_3N(CH_3)_2$	—CH_2—C_6H_5	—H	—H	—H	—H	D	—	2 HCl, 229–234° (dec.)	10
—CH_3	—CH_3	—H	—CH_3	—CH_3	—H	E	—	Rectangular plates, 215° (benzene)	25
—CH_3	—CH_2CH_3	—OCH_3	—OCH_3	—H	—OCH_3	F	—	Needles, 120° (methanol)	26

[a] A, Corresponding *o*-diacylbenzene + semicarbazide hydrochloride + sodium acetate in water shaken at 50°. Same results with hydrazine hydrate and thiosemicarbazide. A-1, 1-chloro-4-methylphthalazine + diethyl malonate in ethanolic sodium ethoxide followed by decarboxylation in concentrated hydrochloric acid. B, Corresponding *o*-diacylbenzene refluxed 1 hr with 85% hydrazine hydrate in ethanol. C, By addition of 3-dimethylaminopropylmagnesium chloride to 1-(2*H*)phthalazinone followed by hydrolysis and oxidation with $K_3Fe(CN)_6$. D, Same as C but the starting material used was 1-benzylphthalazine. E, Corresponding *o*-diacylbenzene heated with hydrazine hydrate in acetic acid. F, Corresponding *o*-diacylbenzene + hydrazine hydrochloride + sodium acetate in aqueous ethanol heated for 10 min at 50°.

TABLE 2B-6. 1-Alkyl-4-Arylphthalazines

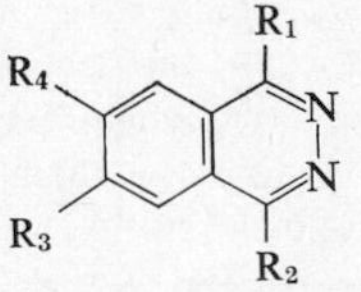

R_1	R_2	R_3	R_4	Preparation[a]	Yield (%)	MP[b]	Reference
$-CH_3$	$-C_6H_5$	$-H$	$-H$	A	77, 51	Colorless crystals, 125–126° (benzene-ligroin)	23a, 34
$-CH_2CH_3$	$-C_6H_5$	$-H$	$-H$	A-1	46	Colorless plates, 119–120° (ligroin)	34
$-(CH_2)_3N(CH_3)_2$	$-C_6H_5$	$-H$	$-H$	B	—	2 HCl, 243–248° (dec.)	10
$-(CH_2)_3N(CH_3)_2$	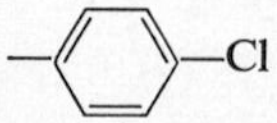 $-C_6H_4-Cl$ (p)	$-H$	$-H$	B	—	92–94° (heptane)	10
$-CH_2CH_3$	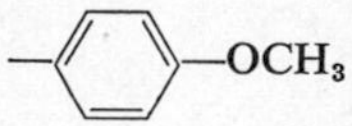 $-C_6H_4-OCH_3$ (p)	$-OCH_3$	$-H$	C	75	125.5–126°; 136° (dil. alcohol); HCl, rosettes, 93°; picrate, yellow needles, 179°	32, 33
$-CH_2CH_3$	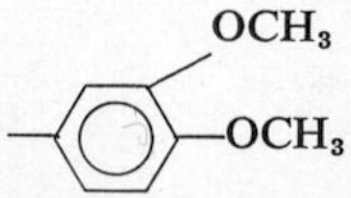 $-C_6H_3(OCH_3)_2$ (3,4)	$-OCH_3$	$-OCH_3$	D	89	Colorless needles, 200–201° (alcohol); picrate, yellow needles, 199–201° (dec.)	32

[a] A, 2-(4-Phenyl-1-phthalazinyl)acetophenone heated for 1 hr in aqueous ethanol containing alkali. A-1, Similar to A. B, 1-Arylphthalazine in tetrahydrofuran boiled for 7 hr with 3-dimethylaminopropylmagnesium chloride followed by hydrolysis and oxidation with potassium ferricyanide. C, Corresponding *o*-diacylbenzene was refluxed in ethanol for 5 min with hydrazine hydrate. D, Same as C but treatment in warm alcohol.

[b] MP with physical description. Recrystallization solvent in parentheses.

TABLE 2B-7. 1-Substituted Alkyl-4-phenylphthalazines. Obtained by Nucleophilic Substitution of 1-Methylsulfonyl-4-phenylphthalazine with Ketones (34).

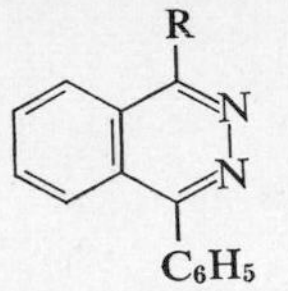

R	Preparation[a]	Yield (%)	Appearance, MP (Recrystallization Solvent)
$-CH_2C(=O)C_6H_5$	A	73	Yellow needles, 192–194° (benzene)
$-CH_2C(=O)CH_3$	B	42	Orange needles, 155–157° (benzene-ligroin)
$-CH(CH_3)-C(=O)-CH_3$	C	21	Colorless plates, 150–152° (benzene-ligroin)
$-CH_2C(=O)CH_2CH_3$	B	14	Yellow needles, 113–115° (ligroin)
$-CH(CH_3)C(=O)CH_2CH_3$	B	50	Colorless needles, 160–162° (benzene-ligroin)
$-CH_2C(=O)-CH(CH_3)CH_3$	B	46	Yellow scales, 164–165° (ligroin)
$-CH_2C(=O)CH_2CH_2CH_3$	B	18	Yellow plates, 96–98° (ligroin)
$-CH(CH_2CH_3)C(=O)CH_3$	C	trace	Colorless prisms, 108–110° (ligroin)
$-CH_2C(=O)CH_2CH(CH_3)CH_3$	B	64	Yellow scales, 137–138° (ligroin)
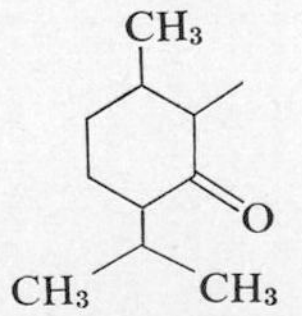	B	14	Colorless needles, 133–135° (benzene-ligroin)
$-(CH_2)_4CO_2H$	B-1	59	Colorless scales, 145–147° (ligroin)
$-(CH_2)_5CO_2H$	B-2	81	Colorless scales, 157–158° (benzene-ligroin)

[a] A, The corresponding ketone was refluxed with 1-methylsulfonyl-4-phenylphthalazine in benzene with sodium amide for 3 min. B, Same as A, reflux time 2 hr. B-1, same as B but the ketone used was cyclopentanone. B-2, Same as B-1. Ketone used was cyclohexanone. C, Same as A but stir in 50% aqueous sodium hydroxide at room temperature.

TABLE 2B-8. 1-Alkyl-4-phenylphthalazines. Obtained by Nucleophilic Substitution of 1-Methylsulfonyl-4-phenylphthalazine with Active Methylene Compounds (35).

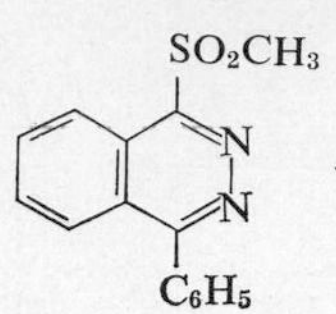

+ active methylene compd.

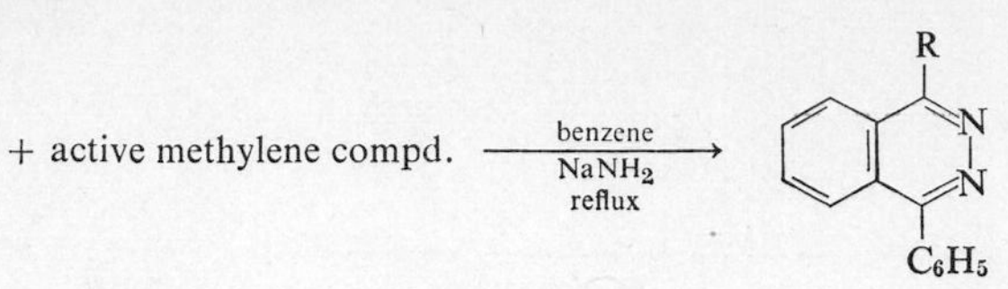

R	Reaction Conditions	Yield (%)	Appearance, MP (Recrystallization Solvent)
$-CH(CN)CO_2CH_2CH_3$	Reflux 4 hr	72	Pale yellow scales, 208° (benzene)
$-CH(COCH_3)(CO_2CH_2CH_3)$	Reflux 4 hr	14	Yellowish powder, 99–101° (chloroform-ligroin)
$-CH(CO_2CH_2CH_3)_2$	Reflux 2 hr	42	Pale yellow needles, 102–103° (isopropyl ether)
$-CH(CN)_2$	Reflux 1 hr	48	Yellow scales, 257° (dec.) (chloroform)
$-CH(CN)C_6H_5$	3 hr at room temp.	78	Pale yellow prisms, 216–218° (chloroform-benzene)
$-C{\equiv}C-C_6H_5$	Reflux 1 hr	26	Yellow needles, 180–181° (benzene-ligroin)
CHCN[a] C_6H_5	Reflux 10 hr	64	Orange powder, 215° (benzene)
CH_2—[b] N N C_6H_5	—	85	97–99° (isopropyl ether)

[a] The active methylene compound used was acetonitrile.

[b] Prepared by the hydrolysis followed by decarboxylation of the compound[a] by heating at 100° for 1 hr in concentrated hydrochloric acid.

TABLE 2B-9. 1,4-Diarylphthalazines

R_1	R_2	R_3	R_4	Preparation[a]	Yield (%)	MP	Reference
C_6H_5	C_6H_5	H	H	A	—	Colorless crystals 194° (benzene), 192°, 193–194°, 197.5–198.5°, 197.5–199.5°	7, 27, 28, 36, 37, 38
C_6H_5	$4\text{-}CH_3C_6H_4$	H	H	B	95	171–172°	30
$2\text{-}CH_3C_6H_4$	$2\text{-}CH_3C_6H_4$	H	H	C	76	Colorless crystals, 162–163°	29
C_6H_5	$2\text{-}ClC_6H_4$	H	H	B	95	131°	30
C_6H_5	$4\text{-}ClC_6H_4$	H	H	B	95	185–187°	30

TABLE 2B-9 (*continued*)

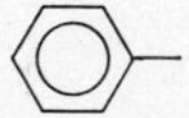	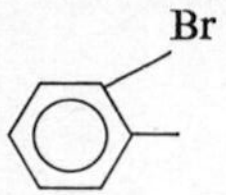	H	H	B	67	134–135°	30
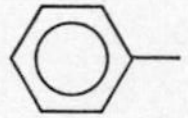	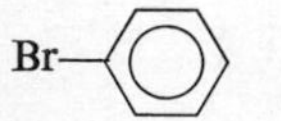	H	H	B	88	192–193°	30
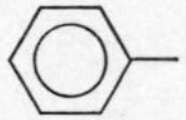	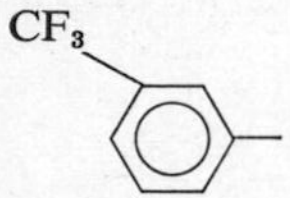	H	H	B	89	193–194°	30
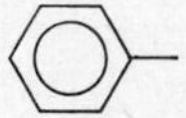	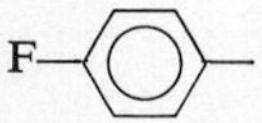	H	H	B	84	191–192°	30
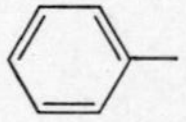	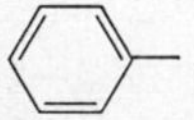	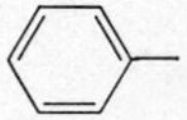	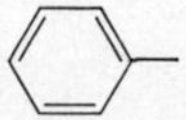	C_1, C_2	57, 95	338–341° (pyridine or toluene) 344° (acetic acid)	28, 31

[a] A, See different methods of preparation in Section 2B-II-A. B, The corresponding *o*-diacylbenzene was heated in acetic acid with 85% hydrazine hydrate. C, Same as B but refluxed for 1 hr in ethanol. C-1, Same as in C, reflux time 8 hr. C-2, Same as in C, reflux time 16 hr.

References

1. E. Hayashi and E. Oishi, *Yakuguku Zasshi*, **86,** 576 (1966); *Chem. Abstr.*, **65,** 15373 (1966).
2. E. F. M. Stephenson, *J. Chem. Soc.*, **1963,** 1913.
3. T. Ikeda and S. Kanahura, *Kanazawa Daigaku Yakugakubu Kenyu Nempo*, **9,** 6 (1959); *Chem. Abstr.*, **54,** 4607 (1960).
4. M. V. Sigal, Jr., P. Marchini, and B. L. Poet, U.S. Pat., 3,274,184, Sept., 1966; *Chem. Abstr.*, **65,** 18600 (1966).
5. T. Ikeda and S. Kanahara, *Kanazawa Daigaku Yakugakubu Kenkyu Nempo*, **7,** 6 (1957); *Chem. Abstr.*, **52,** 6361 (1958).
6. T. Ikeda, S. Kanahara, and T. Ujiie, *Kanazawa Daigaku Yakugakubu Kenkyu Nempo*, **8,** 1 (1958); *Chem. Abstr.*, **53,** 4287 (1959).
7. A. Mustafa, A. H. Harhash, and A. A. S. Saleh, *J. Am. Chem. Soc.*, **82,** 2735 (1960).
8. A. Hirsch and D. G. Orphanos, *J. Heterocyclic Chem.*, **3,** 38 (1966).
9. A. Hirsch and D. G. Orphanos, *Can. J. Chem.*, **44,** 2109 (1966).
10. A. Marxer, F. Hofer, and U. Salzman, *Helv. Chim. Acta*, **52,** 1376 (1969).
11. A. Marxer, *Helv. Chim. Acta.*, **49,** 572 (1966).
12. F. D. Popp and J. F. Wefer, *Chem. Comm.*, **1967,** 59.
13. F. D. Popp, J. M. Wefer, and C. W. Klinowski, *J. Heterocyclic Chem.*, **5,** 879 (1968).
14. Y. Mizuno, K. Adachi, and K. Ikeda, *Pharm. Bull.* (*Japan*), **2,** 225 (1954).
15. J. Aggarwal, N. Darbari, and J. Ray, *J. Chem. Soc.*, **1929,** 1941; J. Aggarwal, J. Khera, and J. Ray, *J. Chem. Soc.*, **1930,** 2354; C. V. Wilson, *J. Am. Chem. Soc.*, **70,** 1901 (1948); J. C. E. Simpson, in *Condensed Pyridazine and Pyrazine Rings*, *The Chemistry of Heterocyclic Compounds*, Vol. 5., A. Weissberger, Ed., Interscience, New York, 1953, p. 72; W. R. Vaughan, *Chem. Rev.*, **43,** 447 (1948).
16. H. J. Rodda and P. E. Rogash, *J. Chem. Soc.*, **1956,** 3927; N. K. Kochetkov and L. A. Vorotnikova, *J. Gen. Chem. USSR.*, **26,** 1297 (1956); T. Ikeda, S. Kanahara, and N. Nishikawa, *Ann. Rept. Fac. Pharm. Kanazawa Univ.*, **6,** 1 (1956); *Chem. Abstr.*, **51,** 3608 (1957).
17. A. Albert, W. L. F. Armarego, and E. Spinner, *J. Chem. Soc.*, **1961,** 2689.
18. A. Hirsch and D. Orphanos, *Can. J. Chem.*, **46,** 1455 (1968).
19. C. Furlani, S. Bertola, and G. Morpurgo, *Ann. Chim.* (*Rome*), **50,** 858 (1960).
20. R. N. Castle and S. Takano, *J. Heterocyclic Chem.*, **5,** 89 (1968).
21. A. Fujita et al., *Yakugaku Zasshi*, **86,** 427 (1966); *Chem. Abstr.*, **65,** 3870 (1966); H. Takumatsu et al., Japanese Pat. 28,337 (1964); *Chem. Abstr.*, **62,** 11826 (1965).
22. S. Kanahara, *Yakugaku Zasshi*, **84,** 483 (1964); *Chem. Abstr.*, **61,** 8304 (1964).
23. S. Goldschmidt and A. Zoebelien, *Chem. Ber.*, **94,** 169 (1961).

23a. M. A. Shah and G. A. Taylor, *J. Chem. Soc.* (*C*), **1970,** 1651.

24. C. R. Warner, E. J. Walsh, and R. F. Smith, *J. Chem. Soc.*, **1962,** 1232.
25. W. Reid, K. Wesselborg, and K. H. Schmidt, *Naturwiss.*, **46,** 142 (1959).
26. F. M. Dean, D. R. Randell, and G. Winfield, *J. Chem. Soc.*, **1959,** 1071.
27. G. M. Badger, R. S. Pearce, H. J. Rodda, and I. S. Walker, *J. Chem. Soc.*, **1954,** 3151.
28. T. H. Regan and J. B. Miller, *J. Org. Chem.*, **31,** 3053 (1966).
29. E. Clar and D. G. Stewart, *J. Chem. Soc.*, **1951,** 3215.
30. W. W. Zajac, Jr., and D. E. Pichler, *Can. J. Chem.*, **44,** 833 (1966).
31. H. Fletcher, Br. Pat. 1,028,920, May, 1966; *Chem. Abstr.*, **65,** 3799 (1966).
32. A. Müller, M. Lempert-Streter, and A. Karczag-Wilhelme, *J. Org. Chem.*, **19,** 1533 (1954).

33. J. M. Van der Zanden and G. De Vries, *Rec. Trav. Chim.*, **74,** 52 (1955).
34. E. Hayashi and E. Oishi, *Yakugaku Zasshi*, **87,** 807 (1967); *Chem. Abstr.*, **68,** 2871 (1968); E. Hayashi and E. Oishi, Japanese Pat. 7005,268 (1970); *Chem. Abstr.*, **72,** 121565 (1972).
35. E. Hayashi and E. Oishi, *Yakugaku Zasshi*, **88,** 83 (1968); *Chem. Abstr.*, **69,** 19104 (1968).
36. J. Sauer and G. Heinrichs, *Tetrahedron Lett.*, **1966,** 4979.
37. A. Mustafa et al., *Tetrahedron*, **20,** 531 (1964).
38. R. S. Staunton and A. Topham, *J. Chem. Soc.*, **1953,** 1889.
39. E. Hayashi and E. Oishi, *Yakugaku Zasshi*, **87,** 940 (1967); *Chem. Abstr.*, **68,** 39573 (1968).
40. R. M. Acheson and M. W. Foxton, *J. Chem. Soc.*, **1968,** 2218.
41. G. M. Badger and I. S. Walker, *J. Chem. Soc.*, **1956,** 122.
42. Bui-Khac-Diep and B. Cavin, *Comp. Rend.*, Sec. C., **262,** 1010 (1966).
43. E. Oishi, *Yakugaku Zasshi*, **89,** 959 (1969); *Chem. Abstr.*, **72,** 3450 (1970).

Part C. Phthalazine *N*-Oxides

The chemistry of phthalazine *N*-oxides has been investigated only in the last decade. This class of compounds is reported in only a few papers. Hayashi and co-workers in Japan have contributed much to our knowledge of the phthalazine *N*-oxides. They have investigated the reactions of some phthalazine *N*-oxides and the synthetic utility has been demonstrated. This group of compounds is prepared mainly by the action of monoperphthalic acid in ether or by hydrogen peroxide in acetic acid solution on the corresponding phthalazines.

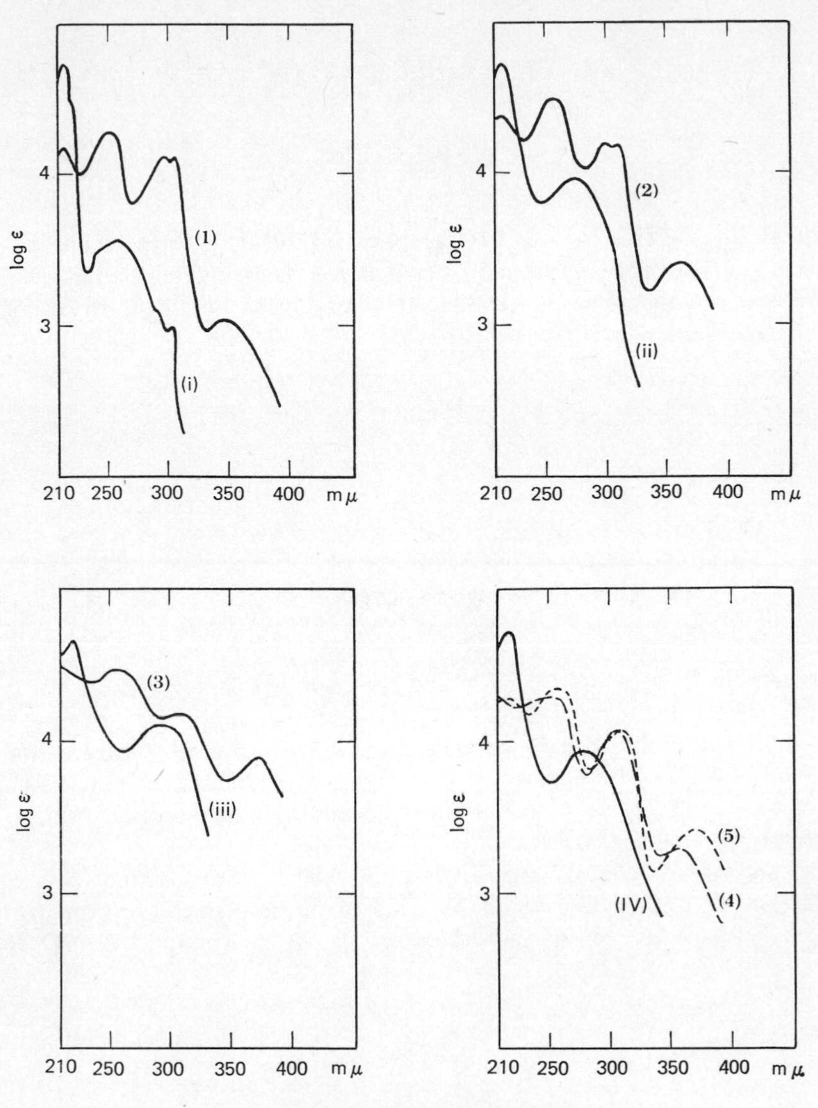

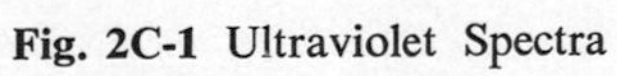

Fig. 2C-1 Ultraviolet Spectra

I. Phthalazine 2-Oxide

A. Preparation

Phthalazine 2-oxide (**1**) was first prepared in 1962 by Hayashi and co-workers (1). The benzene solution of phthalazine is treated with an etherial solution of monoperphthalic acid at 0° C and kept overnight to afford phthalazine 2-oxide in 73% yield (Eq. 1).

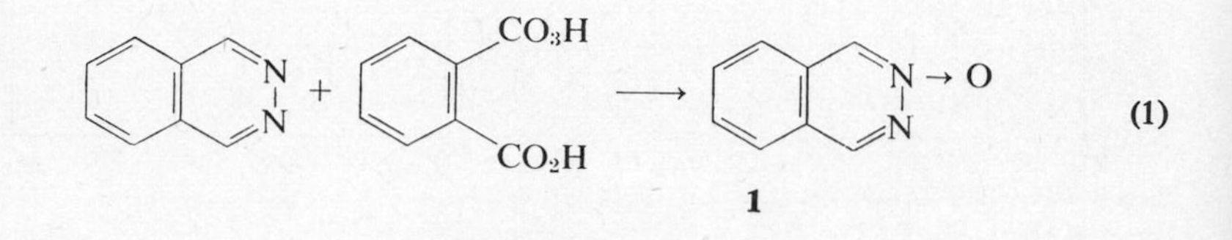

B. Properties

Phthalazine 2-oxide has mp 143° and the hydrated form melts at 94°. It forms a picrate melting at 152–155° (1).

Hayashi and Oishi (2) have compared the UV absorption spectra of the phthalazine *N*-oxides in ethanol with their parent phthalazines (Fig. 2C-1). All of these phthalazine *N*-oxides show an additional absorption peak at ~350 mμ with log ε = 3–4.

Tori and co-workers (3) have reported the pmr spectra of phthalazine 2-oxide (**1**) and 1-ethoxyphthalazine 3-oxide (**6**) in deuteriochloroform with tetramethylsilane as an internal reference, using a Varian A-60 NMR

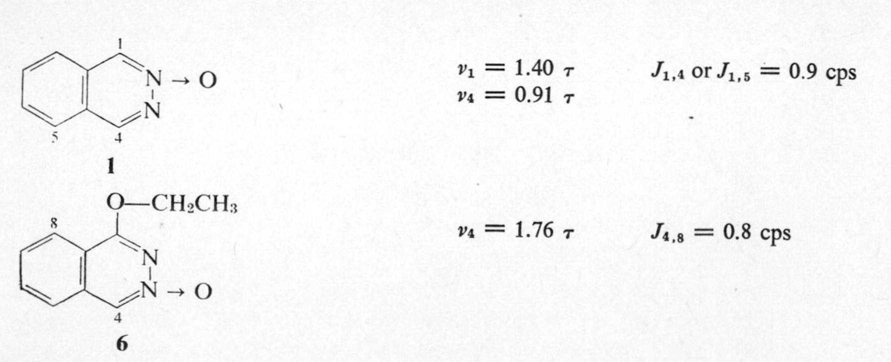

Fig. 2C-2

spectrometer. The assignment for the protons at the 1 and 4 positions in **1** and for proton at 4 position in **6** has been made (Fig. 2C-2).

II. Substituted Phthalazine *N*-Oxides

A. 1-Substituted Phthalazine 2-Oxides

These are difficult to prepare because of the steric hindrance caused by the substituents present at the 1-position. Hayashi and Oishi (4) have studied the *N*-oxidation of 1-methyl- and 1-benzylphthalazine with monoperphthalic acid in chloroform-ether solution. In both cases the *N*-oxidation occurs at the 2 and 3 positions; however, the product distribution is different. In the case of 1-methylphthalazine, *N*-oxidation at positions 2 and 3 occur in an approximately 1:2 ratio as compared to 1-benzylphthalazine where the ratio is 3:1 (Eq. 2).

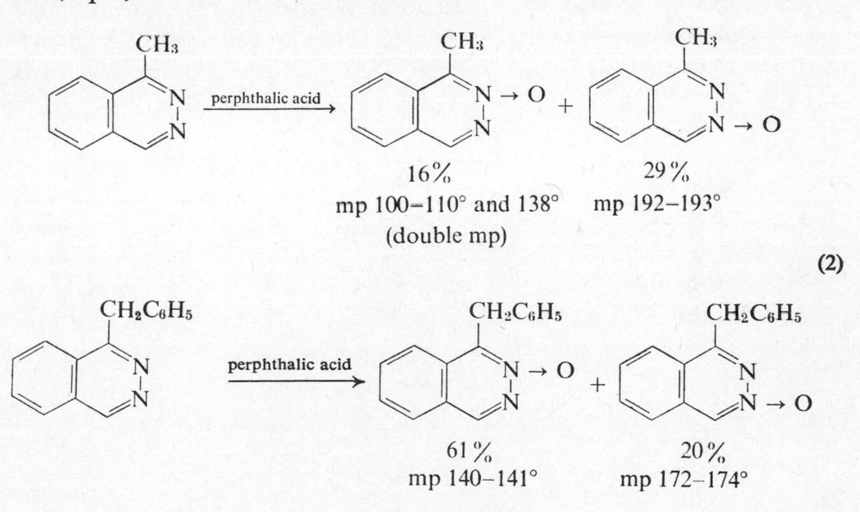

B. 1-Substituted Phthalazine 3-Oxides

The presence of an alkoxyl or a phenyl group at the 1 position in phthalazine has been shown, (1 2, 5), on *N*-oxidation, to give exclusively 1-substituted phthalazine 3-oxides in high yields. The steric effects caused by these groups play an important role in affording specifically the 3-oxides (Eq. 3).

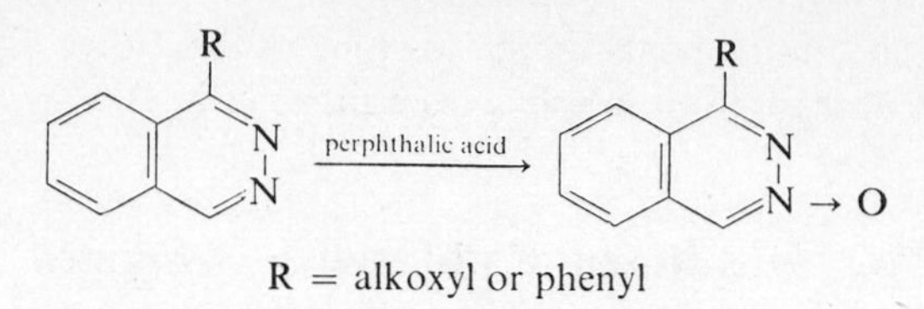

R = alkoxyl or phenyl

C. 1,4-Disubstituted Phthalazine Monoxides

In general it can be said that whenever there are two different groups present at the 1 and 4 positions of phthalazine the size of the groups plays an important role in the direction of *N*-oxidation as can be observed from the few cases reported in the literature.

1-Methyl-4-phenylphthalazine on treatment with perphthalic acid (6, 7) gives 1-methyl-4-phenylphthalazine 2-oxide (**7**) in 76% yield. The same product can be isolated by the addition of phenylmagnesium bromide to 1-methylphthalazine 2-oxide (4) (22%) or in a poor yield (2%) by the addition of methylmagnesium bromide (6, 7) to 1-phenylphthalazine 3-oxide (**2**) (Eq. 4).

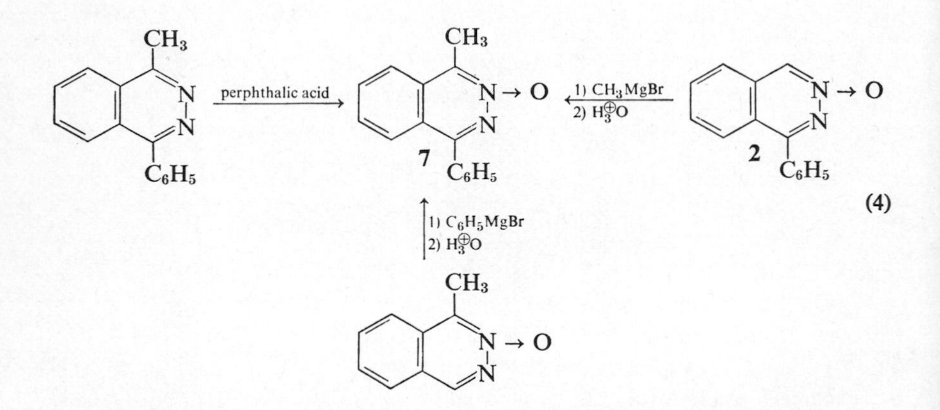

1,4-Diphenylphthalazine reacts with hydrogen peroxide in acetic acid to give the 2-oxide **3** (2). 1-Isopropyl-4-phenylphthalazine under identical conditions gives 1-isopropyl-4-phenylphthalazine 3-oxide (**5**) exclusively, in 77% yield (2). The preparation of 1-isopropyl-4-phenylphthalazine 2-oxide can be accomplished by the addition of isopropylmagnesium bromide to 1-phenylphthalazine 3-oxide (**1**) followed by hydrolysis (2) (Eq. 5).

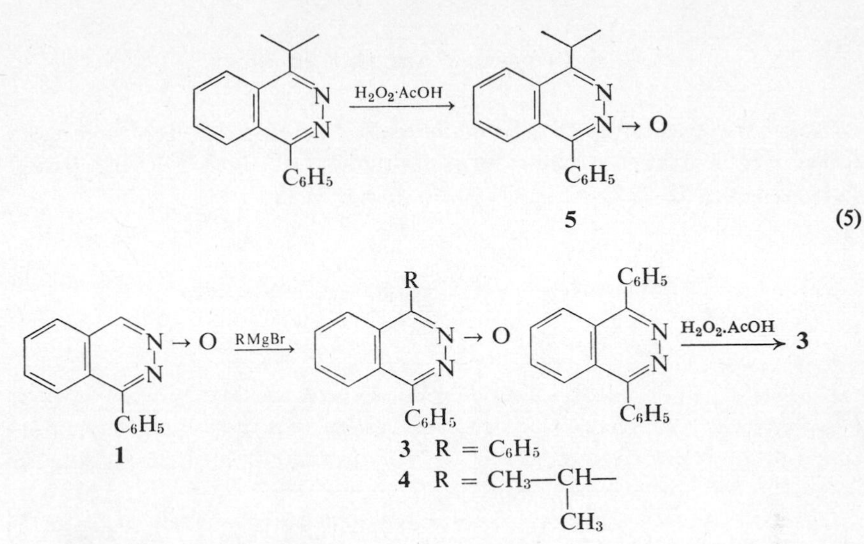

1-Chloro-4-methylphthalazine and 1-methoxy-4-methylphthalazine on treatment with hydrogen peroxide in acetic acid gave 1-chloro-4-methylphthalazine 3-oxide (**8**) and 1-methoxy-4-methylphthalazine 3-oxide (**9**) (5). Both **8** and **9** are shown to give 1-hydroxy-4-methylphthalazine 3-oxide (**10**) in high yields on alkaline hydrolysis (5) (Eq. 6).

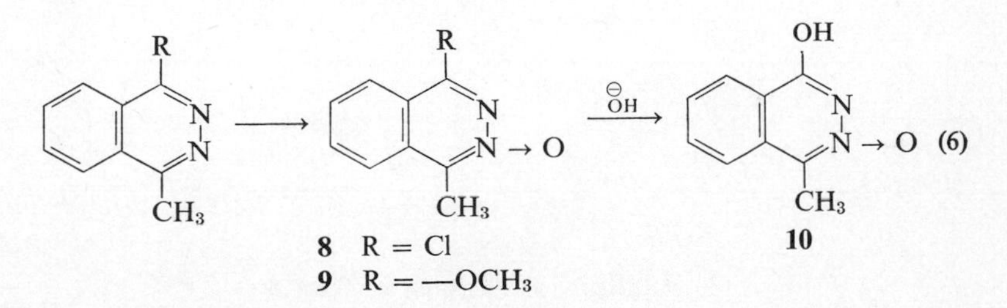

1,4-Dialkoxyphthalazines, due to the steric hindrance on both of the nitrogen atoms, are shown to be inert to the treatment of monoperphthalic acid in ether (1). In all cases attempted, the majority of the starting material is recovered unchanged and only part of it undergoes hydrolysis to 4-alkoxy-1(2*H*)phthalazinones.

III. Reactions of Phthalazine *N*-Oxides

Phthalazine *N*-oxides undergo reactions similar to the other aromatic heterocyclic *N*-oxides.

A. Removal of the Oxygen Atom

Phosphorus trichloride or tribromide easily cleaves the oxygen atom from phthalazine *N*-oxides in chloroform solution at 0° C to give the parent phthalazines in 52–72% yields (1, 2, 6, 7) (Eq. 7).

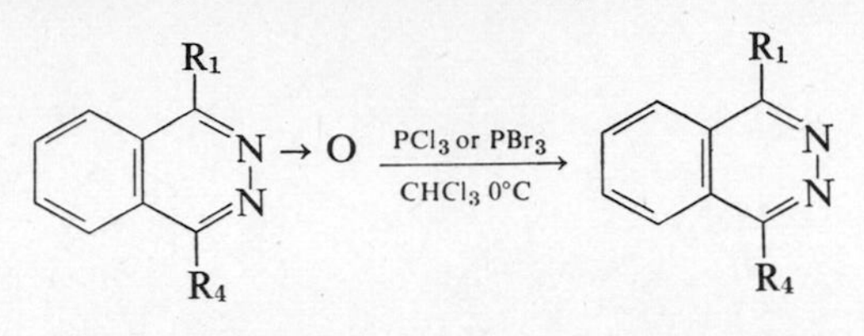

(7)

where:

R_1	R_4
H—	H—
H—	EtO—
$C_6H_5C(=O)CH_2$—	C_6H_5—
$CH_3CH(CH_3)$—	C_6H_5—
C_6H_5—	C_6H_5—
C_6H_5—	$CH_3CH(CH_3)$—

B. Addition of Grignard Reagents

Addition of Grignard reagents across the C=N bond at the C_4 position in 1-phenyl- or 1-alkylphthalazine 3-oxides (2,4) and 1-alkylphthalazine 2-oxide (4) gives 1,4-disubstituted phthalazine *N*-oxides (Eq. 8).

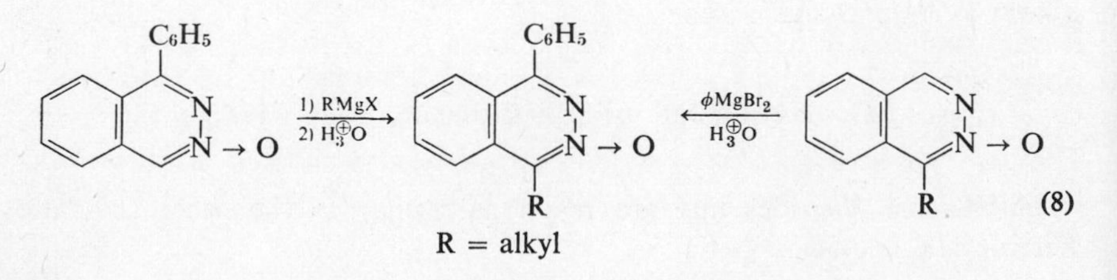

(8)

R = alkyl

C. Reaction with Acetic Anhydride

1-Hydroxyphthalazine 3-oxide on heating with acetic anhydride gives the expected 1,4-dihydroxyphthalazine (50%). Similarly 1-hydroxy-4-methylphthalazine 3-oxide gives 1-hydroxy-4-acetoxymethylphthalazine (8%) (5) (Eq. 9).

1-Phenylphthalazine 3-oxide (**2**) failed to give any product of known structure with acetic anhydride (2).

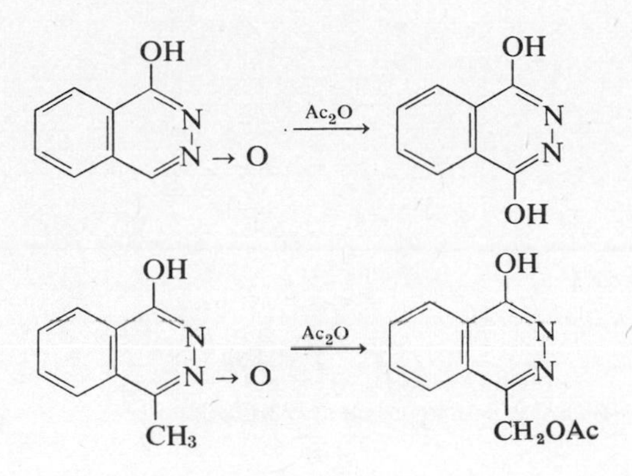

(9)

D. Miscellaneous Reactions

1-Phenylphthalazine 3-oxide (**2**) undergoes the Reissert reaction with benzoyl chloride in the presence of potassium cyanide to furnish 1-phenyl-4-cyanophthalazine (24%) mp 180–182° (2). Phosphorus oxychloride or sulfuryl chloride converts **2** into 1-phenyl-4-chlorophthalazine in 74 and 46% yields, respectively (2). Treatment of **2** with *p*-toluenesulfonyl chloride followed by treatment with sodium hydroxide gives 1-phenyl-4-hydroxyphthalazine (2). Reaction of **2** with phenyl isocyanate gives 1-phenyl-4-anilinophthalazine in poor yields (2) (Eq. 10).

The photolysis reaction of 1,4-diphenylphthalazine 2-oxide has been studied and the resulting products identified. The products, 1,3-diphenylisobenzofuran (~5%), 1,2-dibenzoylbenzene (40–60%), and 1,4-diphenylbenzene (0–15%) were isolated by preparative thin layer chromatography. The gas evolved during the reaction was identified as nitrogen. The author has also proposed a reaction sequence for the formation of these products (8).

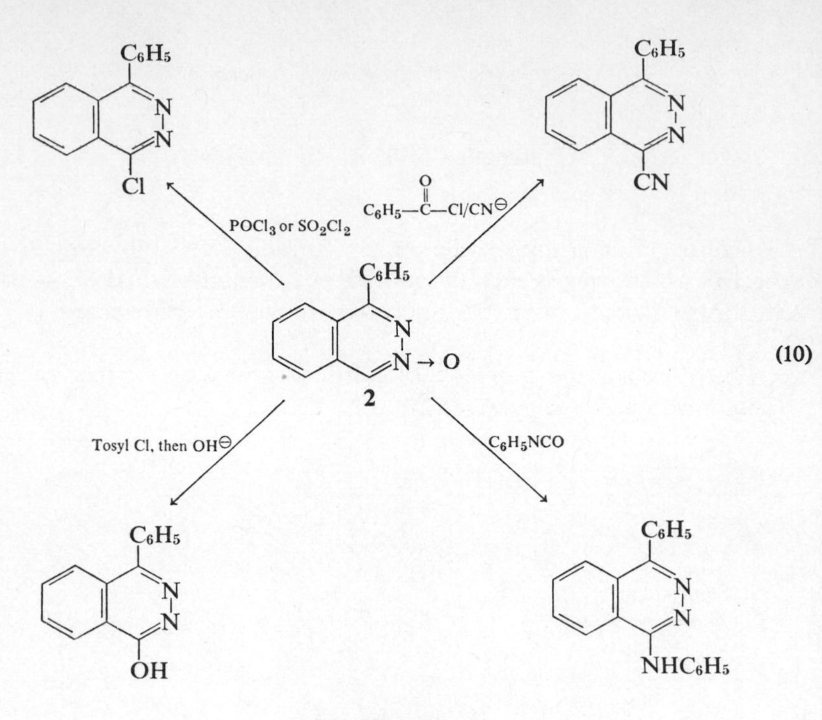

(10)

IV. Tables

TABLE 2C-1. 1-Alkoxy- or 1-Arylphthalazine 3-Oxides

R	Preparation[a]	Yield (%)	MP (Recrystallization Solvent)	Reference
CH_3O—	A, B	70, 39	140° (benzene-pet. ether); 147–148° (ethyl acetate); picrate 159°	1, 5
—OH	C	76	251–253°	5
CH_3CH_2O—	A, B	80	157°; picrate 149°	1
$CH_3CH_2CH_2O$—	A, B	61	141°	1
$CH_3CH(CH_3)O$—	A, B	65	151°	1
C_6H_5O—	A, B	77	187°	1
C_6H_5—	D	65	Pale yellow plates, 181–183° (chloroform-ligroin)	2

[a] A, Corresponding 1-substituted phthalazine + perphthalic acid in ether kept at 0° C for 24 hr. B, similar to A but used 34% hydrogen peroxide in acetic acid and kept at 80° C for 5 hr. C, alkaline hydrolysis of 1-methoxyphthalazine 3-oxide. D, same as in A but kept at 10° for 20 hr.

TABLE 2C-2. 1,4-Disubstituted Phthalazine *N*-Oxides

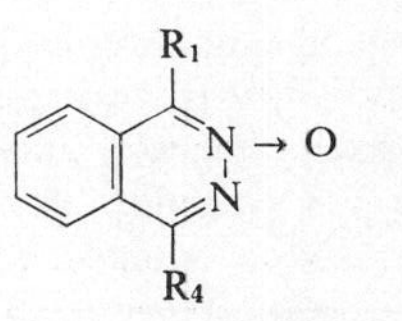

R_1	R_4	Preparation[a]	Yield (%)	MP[b]	Reference
CH_3O—	Cl—	—	—	179–182°	5
CH_3—	—OH	A	60–69	208–210° (dec.)	5
CH_3—	CH_3O—	B	44	Needles, 106–108° (ethyl acetate)	5
CH_3—	C_6H_5—	B_1, C, D, E	76, 22, 2, 72	Pale yellow needles 166–168° (benzene-pet. ether)	4, 6, 7
CH_3CH_2—	C_6H_5—	—	—	171–173°	4
$CH_3CH(CH_3)$—	C_6H_5—	D	26	Pale yellow needles, 169–170° (benzene-ligroin)	2, 4
$C_6H_5CH_2$—	C_6H_5—	—	—	167–168°	4
$C_6H_5C(=O)CH_2$—	C_6H_5—	F	13	Pale yellow, 184–185° (benzene)	6

TABLE 2C-2. (*continued*)

R_1	R_4	Preparation[a]	Yield (%)	MP[b]	Reference
C_6H_5—	CH_3—	C_1	—	181–183°	4
C_6H_5—	CH_3—CH(CH_3)—	B	77	Pale yellow needles 188–189° (benzene-ligroin)	2
C_6H_5—	$C_6H_5CH_2$—	—	—	201–212°	4
C_6H_5—	C_6H_5—	B, C	22, 20	Yellow needles, 188–190° (benzene-ligroin)	2, 4
—CH=CH—(5-nitro-2-furyl) (NO_2)	—$O(CH_2CH_2O)_3CH_3$	—	—	—	9

[a] A, alkaline hydrolysis of 1-methyl-4-methoxyphthalazine 3-oxide or 1-methyl-4-chlorophthalazine 3-oxide. B, corresponding 1,4-disubstituted phthalazine heated with 30% H_2O_2 in AcOH. B_1, same as B but kept with perphthalic acid in ether at 0° overnight. C, corresponding 1-substituted phthalazine 2-oxide treated with phenylmagnesium bromide in benzene followed by hydrolysis. C_1, same as C but used 1-substituted phthalazine 3-oxide. D, 1-phenylphthalazine 3-oxide treated with corresponding alkylmagnesium bromide in ether followed by hydrolysis. E, reflux 10 min 2-(2-oxo-4-phenyl-1-phthalazinyl)acetophenone in alcoholic sodium hydroxide. F, 1-phenylphthalazine 3-oxide treated with acetophenone in aqueous alkali at room temperature.

[b] MP with physical description. Recrystallization solvent in parentheses.

References

1. E. Hayashi et al., *Yakugaku Zasshi*, **82,** 584 (1962); *Chem. Abstr.*, **58,** 3427 (1963). E. Hayashi et al., Japanese Pat., 11,998, June, 1964; *Chem. Abstr.*, **62,** 2782 (1965).
2. E. Hayashi and E. Oishi, *Yakugaku Zasshi*, **86,** 576 (1966); *Chem. Abstr.*, **65,** 15373 (1966).
3. K. Tori, M. Ogata, and H. Kano, *Chem. Pharm. Bull.* (*Tokyo*), **11,** 681 (1963).
4. E. Hayashi et al., *Yakugaku Zasshi*, **88,** 1333 (1968); *Chem. Abstr.*, **70,** 47384 (1969).
5. S. Kanahara and M. Yamamoto, *Kanazawa Daigaku Kenkyu Nempo*, **12,** 1 (1962); *Chem. Abstr.*, **59,** 3917 (1963).
6. E. Hayashi and E. Oishi, *Yakugaku Zasshi*, **87,** 940 (1967); *Chem. Abstr.*, **68,** 39573 (1968).
7. E. Hayashi and E. Oishi, *Yakugaku Zasshi*, **87,** 807 (1967).
8. O. Buchardt, *Tetrahedron Lett.*, **1968,** 1911.
9. K. Eichenberger, P. Schmidt, and M. Wilhelm, German Pat., 2,002,024 (1970); *Chem. Abstr.*, **73,** 98979 (1970).

Part D. 1-(2*H*)Phthalazinone

1-(2*H*)Phthalazinone has been considered in the literature under different names, including 1-hydroxyphthalazine, 1-phthalazone, 1,2-dihydrophthalazone, and 1-(2*H*)phthalazinone. The name 1-(2*H*)phthalazinone more closely describes the structure of the molecule in the light of the known fact that it exists mainly in the lactam form. The compounds described and discussed in this section have the oxygen atom at the 1-position of phthalazine and having substituents in the 2-, the 4-, and the 2,4-positions. The substituents in the aromatic ring of phthalazine are not used as a criterion for the classification and they are reported with the classification based on the substituents in the heterocyclic ring. This section does not include 1-(2*H*)phthalazinones having halo, amino, thio, carboxyl, hydrazino, and hydroxy (alkoxy) groups attached at the 4-position. These compounds are reported in the sections concerning halophthalazine, aminophthalazine, thiophthalazine, phthalazine carboxylic acid, hydrazinophthalazine and 4-hydroxy-1(2*H*)phthalazinone, respectively.

I. 1-(2*H*)Phthalazinone and Its Derivatives

A. Methods of Preparation

1. 1-(2*H*)*Phthalazinone*

a. From Naphthalene. Naphthalene has been used very successfully as a starting material for the synthesis of 4-carboxyl-1-(2*H*)phthalazinone (**2**) (see Section 2J), which is easily decarboxylated in high yields (83–99%) to give 1-(2*H*)phthalazinone (**1**). The decarboxylation of **2** proceeds smoothly either by heating in the absence of solvent at 200–210° (1, 2) or in nitrobenzene (3), or in diphenyl ether (4) (Eq. 1).

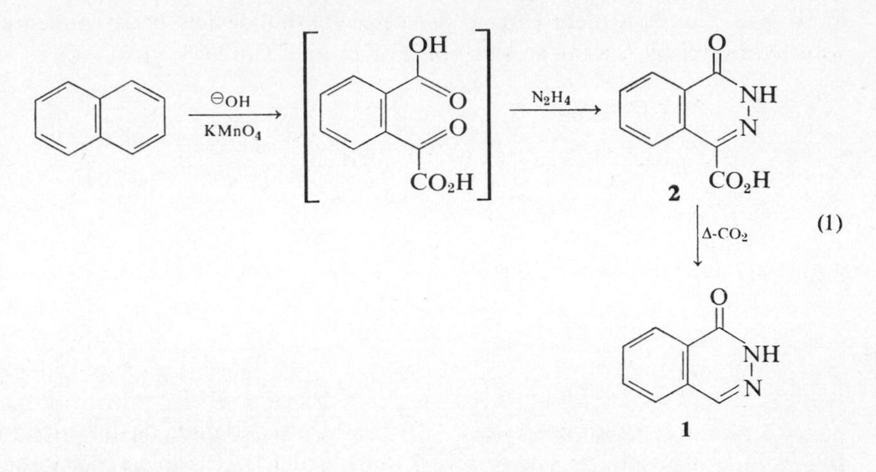

b. FROM PHTHALDEHYDIC ACID AND SIMILAR STARTING MATERIALS. Phthaldehydic acid, on condensation with hydrazine sulfate in hot aqueous solution containing sodium acetate, gives **1** in quantitative yield (5) (Eq. 2). Phthaldehydic acid having substituents, for example, chloro (6, 7), nitro (8, 9), groups can also be used to prepare the derivatives of **1** (Eq. 2).

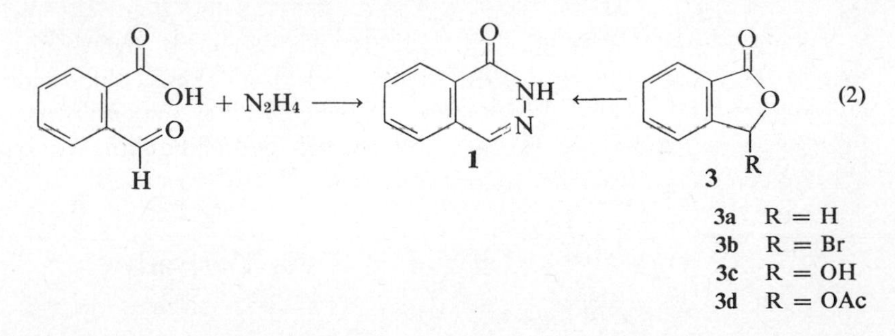

Phthalide (**3a**) has been converted to 3-hydroxyphthalide (**3c**) by bromination followed by hydrolysis, and it has been reacted with hydrazine to give **1** in 60% yield (10). By using this procedure 4,5,6- and 7-fluorophthalides have been converted to the corresponding fluoro-1-(2*H*)phthalazinones in ~ 60% yield (11). Similarly, 3-bromophthalide (**3b**) and 3-hydroxyphthalide (**3c**) have different substituents, such as 6-nitro (12), 6-chloro (6), 4 or 7-chloro (13), 7-carboxaldehyde (14), 7-carboxylic, and thus react with hydrazine to give the corresponding derivatives of **1**.

3-Hydroxyphthalimidine (**4**) (15) and phthalimidine (16) both have been used for the preparation of **1**. In the reaction of **4** the condensation proceeds at reflux temperature with 80% hydrazine hydrate. Phthalamidine requires vigorous conditions; it is necessary to heat in a bomb at 180–220° for 24 hr to give **1** in 60% yield (Eq. 3) 3-Acetoxyphthalide has been condensed with hydrazine by heating at 100° for 3 hr to give **1** in 96% yield (16a).

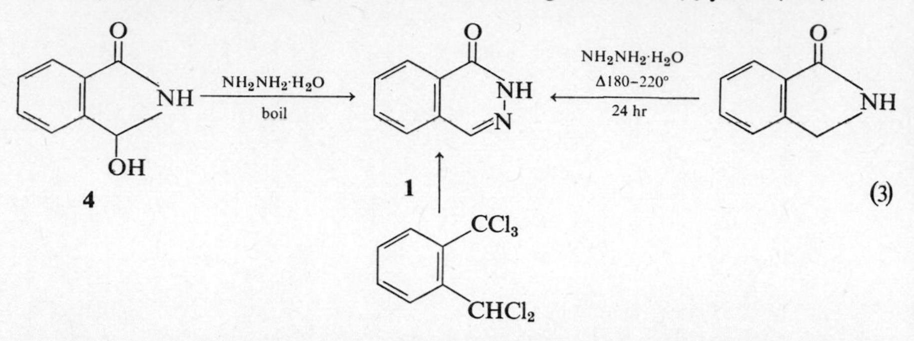

(3)

α,α,α,α′,α′-Pentachloro-*o*-xylene (17) has been condensed with hydrazine sulfate in >90% sulfuric acid to give **1** in 85% yield. Attempted preparation of the monohydrazone of benzocyclobutadienoquinone under various conditions resulted in the formation of **1** (63%) instead of the desired product (18).

2. 2-*Substituted*-1-(2*H*)*phthalazinones*

a. From Phthaldehydic Acid. Phthaldehydic acid condenses with alkylhydrazine (9, 30) or arylhydrazine (19) by refluxing for 2–3 hr in ethanol to give 2-substituted-1-(2*H*)-phthalazinones (**5**). Similarly, 3-nitrophthaldehydic acid condenses with alkylhydrazines to give **5** ($R{=}NO_2$) (9) (Eq. 4).

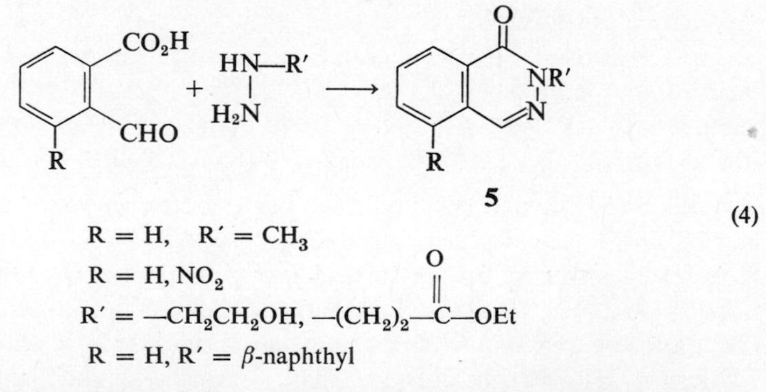

(4)

$R = H, \quad R' = CH_3$

$R = H, NO_2$

$R' = {-}CH_2CH_2OH, {-}(CH_2)_2{-}\overset{O}{\overset{\|}{C}}{-}OEt$

$R = H, R' = \beta$-naphthyl

Vaughan (20) has reviewed the reaction mechanism. In certain instances it has been suggested the formation of a true hydrazone as an intermediate in the reaction of phthalaldehydic acid with substituted hydrazines which may not always take place (Eq. 5). Arguments for the formation of a pseudohydrazone have been advanced. A proposed mechanism accounting for the cyclization of either form (pseudo or normal) of hydrazone into a 2-substituted phthalazinone has been suggested.

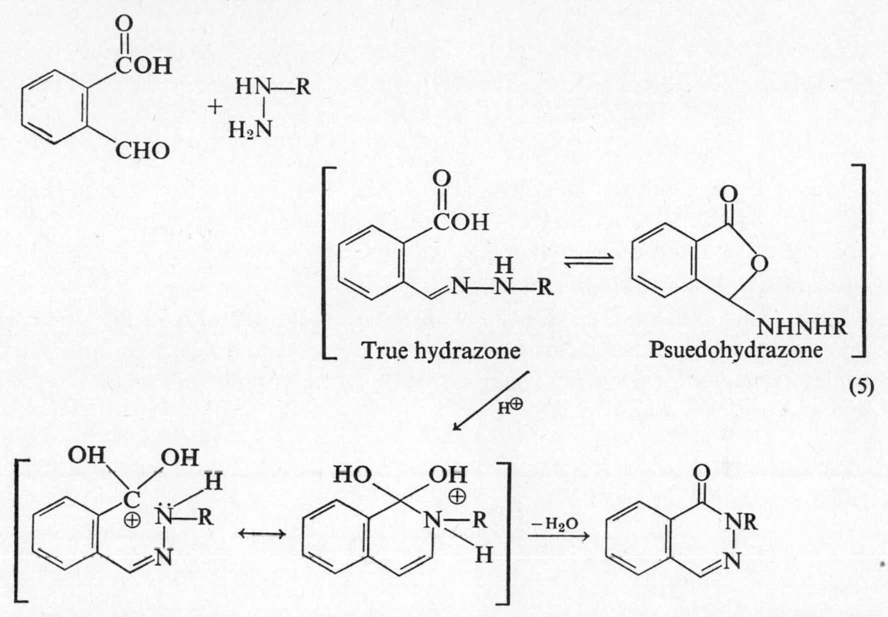

Since the review was written no work has been reported clarifying the formation of a true or pseudohydrazone. In a few cases the true hydrazones have been isolated and further cyclized to 2-substituted-1-(2*H*)phthalazinones. However, in all instances the hydrazine derivatives reported have electron-withdrawing groups which enhance the formation of the true hydrazone as suggested by Vaughan (20).

Cava and Stein (18) have isolated the tosylhydrazone of phthaldehydic acid (**6**) and it has been characterized by the infrared spectrum. The hydrazone **6** has been successfully cyclized to 2-tosyl-1-(2*H*)phthalazinone (**7**) on treatment with acetylchloride, and **7** on alkaline hydrolysis gave the starting hydrazone **6** (Eq. 6).

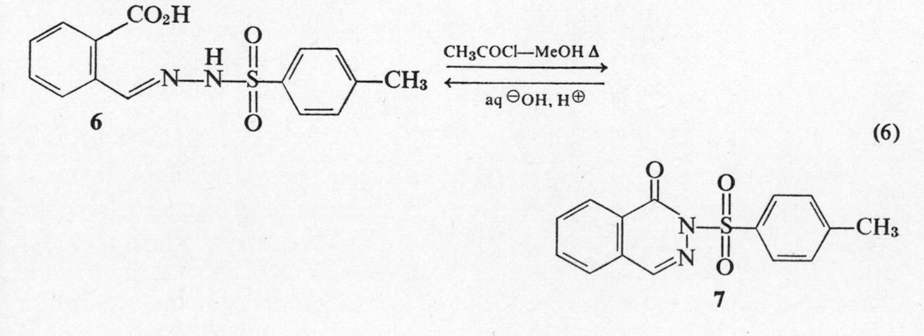

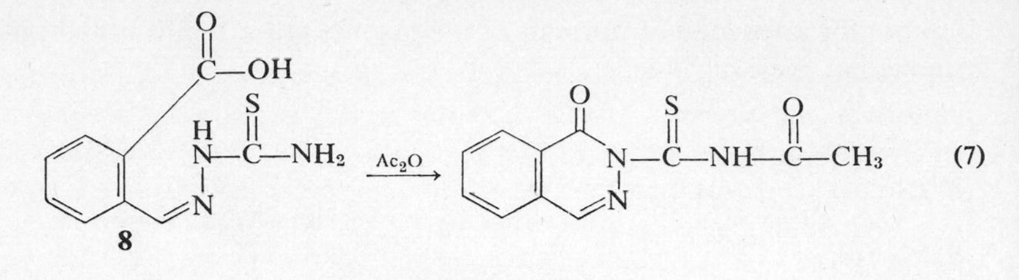

Musante and Fabbrini (21) have isolated the thiosemicarbazone of phthaldehydic acid (**8**). Characterization was accomplished by the formation of the amine salt. The thiosemicarbazone **8** on refluxing with acetic anhydride gave 2-substituted-1-(2*H*)phthalazinone (Eq. 7).

Ikeda and co-workers (22) have isolated 1-isonicotinoyl-2(2-carboxybenzylidene)hydrazine and observed that this compound decomposes into the cyclized product 1-(2*H*)phthalazinone (**1**) and isonicotinic acid when heated above its melting point.

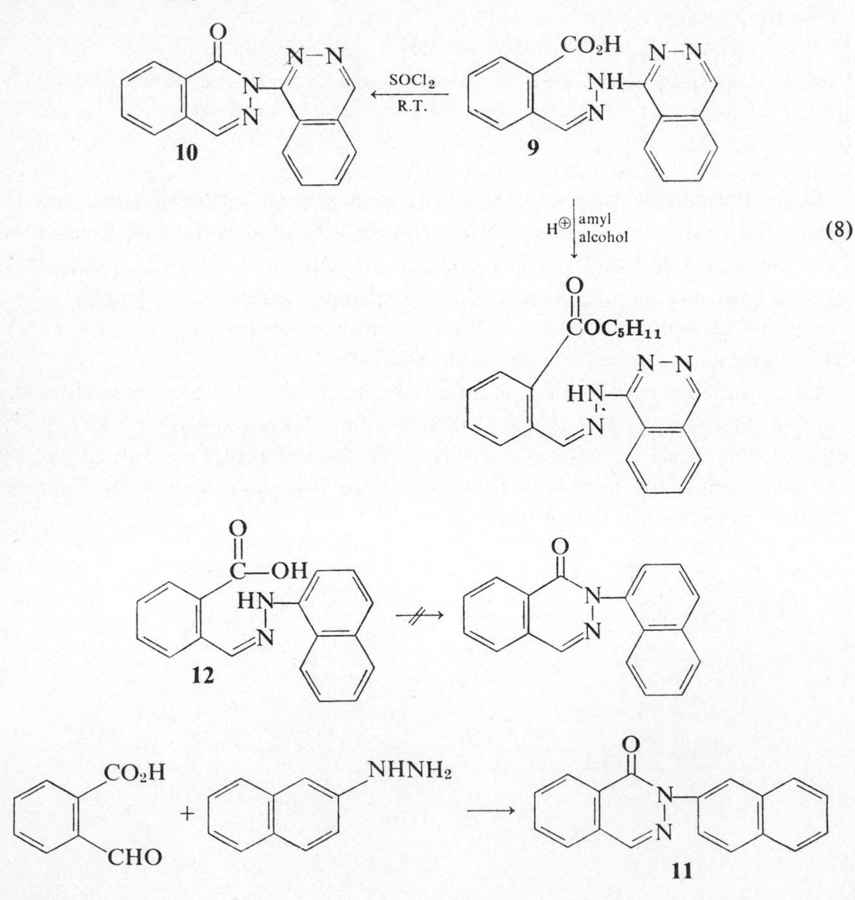

Rodda and Rogasch (19) have isolated phthaldehydic acid 1-phthalazinylhydrazone (**9**) and shown that there is steric hinderance in the formation of the cyclized product 2-phthalazinyl-1-(2*H*)phthalazinone (**10**), which might explain the stability and isolation of **9** (Eq. 8). The cyclization of **9** failed to give **10** on attempted thermal cyclodehydration by pyrolysis or in high boiling inert solvents. The reaction of **9** in amyl alcohol containing hydrochloric acid gave the amyl ester instead of the desired product **10**. However, treatment with thionyl chloride gave the cyclized product **10**. In order to test the steric influence the same authors (19) investigated the 1- and 2-naphthylhydrazones of phthalaldehydic acid and showed that the latter cyclized so readily that the reaction of 2-naphthylhydrazine with phthaldehydic acid gave directly 2-(2′-naphthyl)-1-(2*H*)phthalazinone (**11**), whereas all attempts to cyclodehydrate the 1-naphthylhydrazone of phthaldehydic acid (**12**) were unsuccessful.

b. BY ALKYLATION OF 1-(2*H*)PHTHALAZINONE. 1-(2*H*)Phthalazinone (**1**) has been successfully alkylated to give 2-substituted-1-(2*H*)phthalazinones. The reaction of **1** with diazomethane (23) or with methyl iodide (24, 25) in the presence of base proceeded smoothly to furnish 2-methyl-1-(2*H*)-phthalazinone (**13**) in 91 and 52% yields, respectively. Similarly, alkylation of 7-nitro (8) and 8-carboethoxy-1-(2*H*)phthalazinones (14) with dimethyl sulfate in the presence of base gave the corresponding 2-methyl derivatives. The reaction of *N*,*N*-dialkylaminoethyl chloride with **1** in the presence of a strong base gave 2-substituted-1-(2*H*)phthalazinones in high yields (24, 26, 27). Michael-type addition of **1** and its 4-methyl and 4-phenyl derivatives to acrylonitrile takes place in pyridine water system to afford the corresponding 2-(3-propionitrile)-1-(2*H*)phthalazinones in higher than 64% yields (29) (Eq. 9).

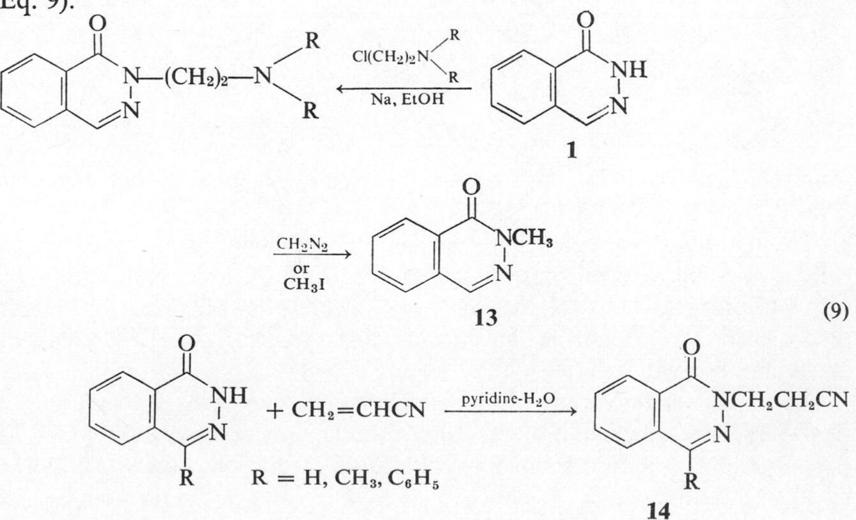

c. BY THE MANNICH REACTION. Mustafa and co-workers (29) have carried out the Mannich reaction on 1-(2*H*)phthalazinone to give 2-(*N*,*N*-dialkylaminomethyl-1-(2*H*)phthalazinones (**15**, R = H) in more than 63% yields. They were also able to condense 2-hydroxymethyl-1-(2*H*)phthalazinone, the expected intermediate in the Mannich reaction, with piperidine to give the desired product, which is also prepared by reacting 2-chloromethyl-1-(2*H*)phthalazinone with piperidine (Eq. 10).

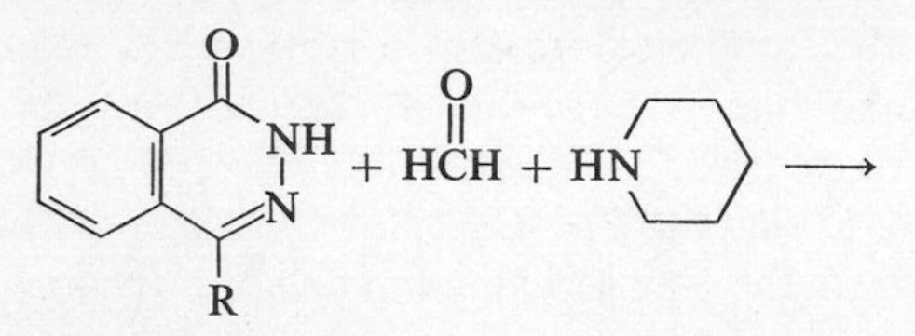

R = H, CH_3, C_6H_5

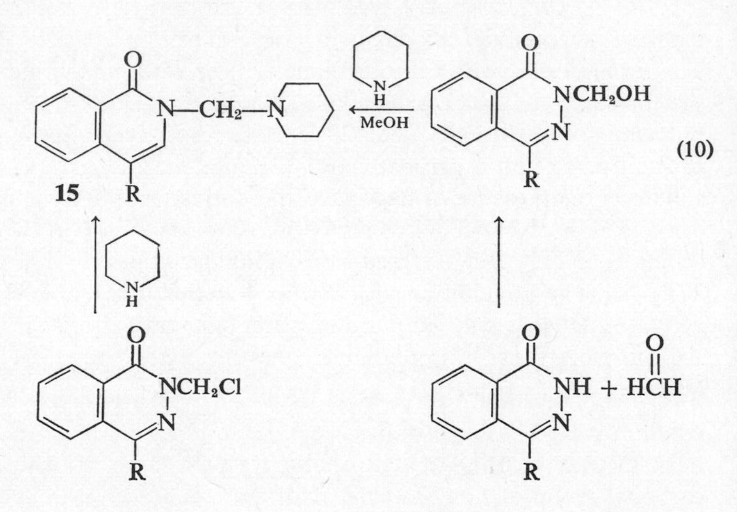

d. MISCELLANEOUS. 2-Methyl-5-nitro- and 2-methyl-8-nitrophthalazinium iodide on oxidation with potassium ferricyanide give the corresponding 2-methyl-1-(2*H*)phthalazinones (31).

The compound described by Gabriel and Eschenbach as the anhydride of 2,2′-dicarboxybenzaldazine has been identified by Griehl and Hecht (32) as 2-(3-phthalidyl)-1-(2*H*)phthalazinone, which is obtained by heating 2,2′-dicarboxybenzaldazine in acetic acid or by reacting 1-(2*H*)phthalazinone with phthaldehydic acid.

An unusual observation has been made in the condensation of 1 mole of 4,4′-dihydrazinodiphenylsulfone with 2 moles of opianic acid (33). The reaction in refluxing ethanol gave the dihydrazone of opianic acid, while refluxing in water it led to 4,4′-(7,8-dimethoxy-1-(2*H*)phthalazonyl)-diphenylsulfone.

3. 4-*Substituted*-1-(2*H*)*phthalazinones from 2-Acylbenzoic Acid and Similar Starting Materials*

The most commonly used method for the preparation of this group of compounds is to condense properly substituted 2-acylbenzoic acids with hydrazine in water, alcohol, or acetic acid by refluxing for about 2 hr. 3-Hydroxy-3-alkylphthalides as well as 3-alkylidinephthalides have also been condensed with hydrazine to give 4-substituted-1-(2*H*)phthalazinones (**16**) (Eq. 11).

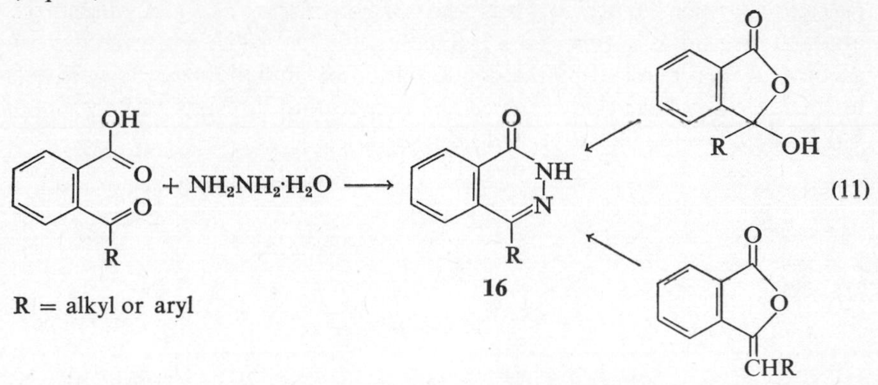

In several instances the intended reduction of the ketone carbonyl group of 2-acylbenzoic acids using Wolf-Kishner or its modified Huang-Minlon conditions have resulted in the formation of 4-substituted-1-(2*H*)phthalazinones instead of the desired products, 2-alkylbenzoic acids. The ketone carbonyl group of 2-acylbenzoic acids first condenses with hydrazine to give the hydrazone, which further reacts with the carboxyl carbonyl to give 4-substituted-1-(2*H*)phthalazinones. This is true with the unhindered ketones. Fuson and Hammann (34) have studied the ring closure involving highly hindered ketone carbonyl groups in methyl 2-duroylbenzoate (**17a**) and methyl 2(2′,4′,6′-trisopropylbenzoyl)benzoate (**17b**) and showed that the first important step was the formation of hydrazides (**18**), which are further cyclized to 4-duryl and 4-(2′,4′,6′-triisopropyl)-1-(2*H*)phthalazinone (**19a** and **19b**, respectively) (Eq. 12). Duryl phenyl ketone does not form a hydrazone nor does 2-duroylbenzoic acid condense with hydrazine, under normal conditions, to give **19a**. However, the pyrolysis of the hydrazine salt of *o*-durylbenzoic acid did give **19a**. Here again the hydrazide **18** must be formed

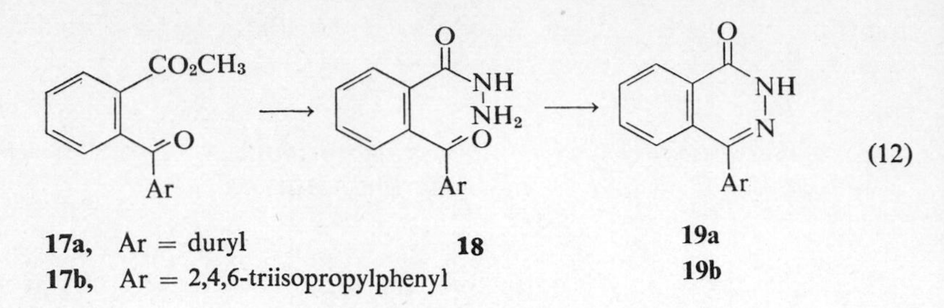

as an intermediate. The authors have emphasized the importance of the intramolecular cyclization and further studied the ease of the ring closure of the hydrazide **18**.

Foldeak and co-workers (123) have reported the synthesis of 4-methyl-1-(2*H*)phthalazinone in high yields by the decarboxylation of 1-(2*H*)phthalazinone-4-acetic acid in nitrobenzene by heating at 170°. 1-(2*H*)Phthalazinone-4-acetic acid was prepared by the condensation of phthalidineacetic acid with hydrazine. They have also reported the preparation of esters and amides of 1-(2*H*)phthalazinone-4-acetic acid (Eq. 12a).

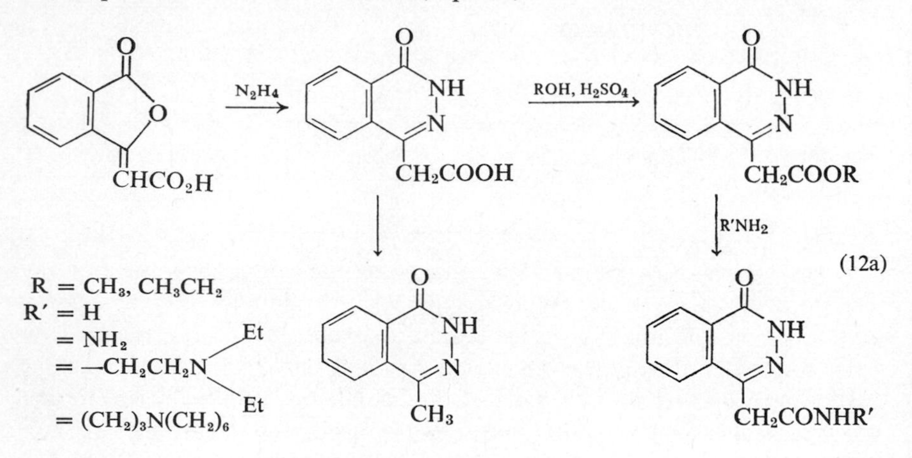

4. 2,4-*Disubstituted*-1-(2*H*)*phthalazinones*

Compounds in this group are prepared mainly by the condensation of 2-acylbenzoic acids with alkyl or arylhydrazines or by alkylating 4-substituted-1-(2*H*)phthalazinones with alkyl halides in the presence of alkali (Eq. 13). Recently (35) a condensation has been carried out by a different procedure using a Dean Stark trap to remove water formed during the reaction.

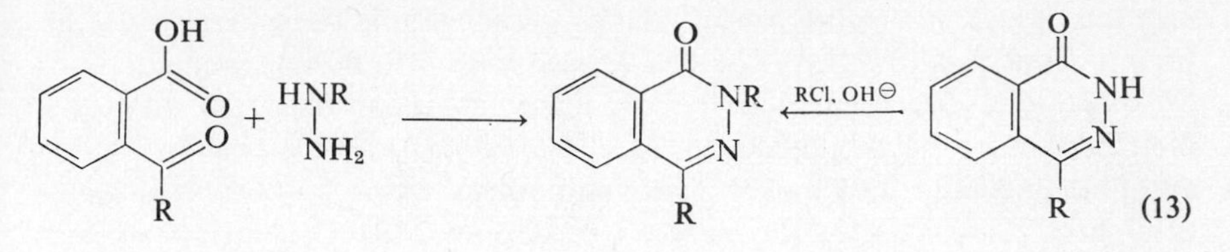

There is one case reported where a tertiary amide reacts like an ester (37): *N*,*N*-dialkyl-2-acetyl-4,6-dimethoxybenzamide has been condensed with phenylhydrazine and hydrazine to give corresponding 1-(2*H*)phthalazinones (Eq. 14). Knott (36) has observed an interesting transformation where the

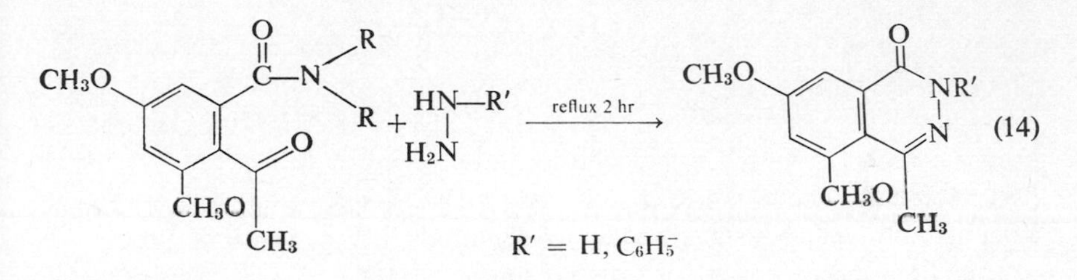

phenylhydrazone of isochromandione (**20**) rearranges to 2-phenyl-4-hydroxymethyl-1-(2*H*)phthalazinone with aqueous alkali (Eq. 15).

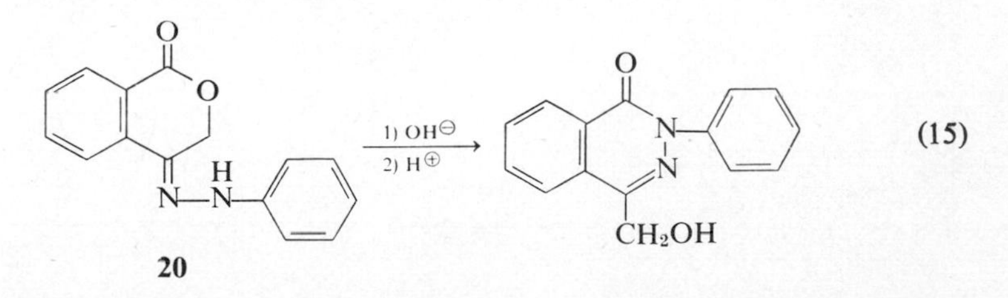

In almost all of the cases of the synthesis of 1-(2*H*)phthalazinone derivatives, the heterocyclic ring is synthesized starting from *o*-substituted benzene derivatives. However, Druey (38) has isolated a side product, 2,4-dimethyl-6,8-dihydroxy-1-(2*H*)phthalazinone (**21**), where the aromatic

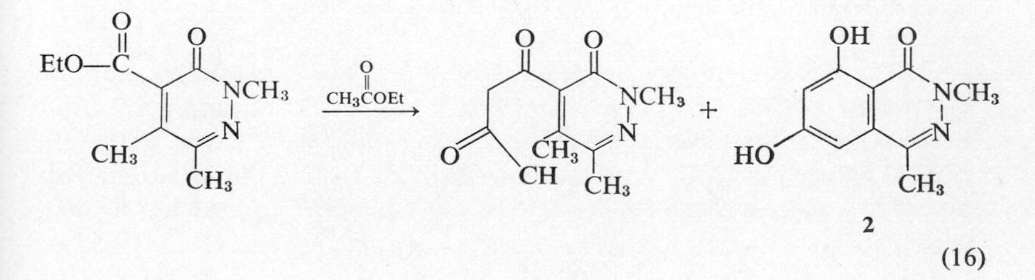

ring is synthesized starting from a 1-(2*H*)pyridazinone derivative using the Claisen condensation (Eq. 16). 2,4-Dimethyl-8-nitrophthalazinum iodide, on oxidation with potassium ferricyanide, has been shown to give 2,4-dimethyl-8-nitro-1-(2*H*)phthalazinone (31). Similarly, 2-methyl-4-(alkyl or aryl)phthalazinium iodide (39) gives the corresponding 2,4-disubstituted-1-(2*H*)phthalazinones (**22**) (Eq. 17) on oxidation. Mustafa and co-workers

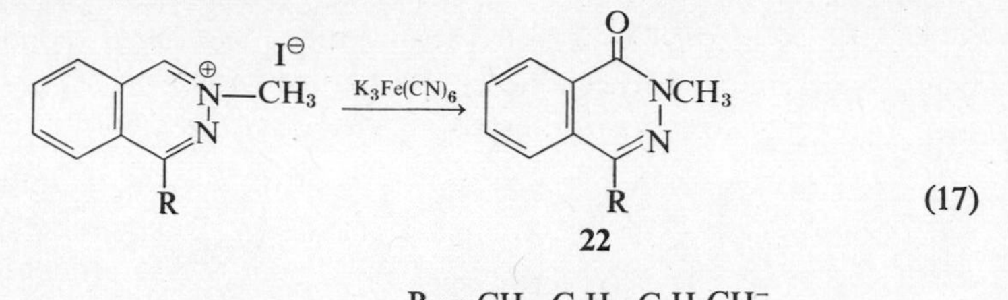

(17)

$R = CH_3, C_6H_5, C_6H_5CH_2$

(29) have carried out the Mannich reaction on 4-substituted-1-(2*H*)phthalazinones to give 2,4-disubstituted-1-(2*H*)phthalazinones (see Eq. 10).

The Rowe-rearrangement or Ψ-phthalazinone 1-(2*H*)phthalazinone rearrangement involving 3-aryl-Ψ-phthalazinones such as, 3-(4′-nitrophenyl)-4-methyl-Ψ-phthalazinone on heating under acidic conditions provides 2-(4′-nitrophenyl)-4-methyl-1-(2*H*)phthalazinone (see Section 2K).

B. Properties of 1-(2*H*)Phthalazinones

1. *Physical Properties and Structure*

1-(2*H*)Phthalazinone forms colorless needles when crystallized from alcohol, mp 182° (16) [mp 180° (21); 182–183° (2); 183–184° (5); 184° (1, 15); 184–185° (17); 186–187° (18)]. 1-(2*H*)Phthalazinone has a pK_a of 11.99 in water at 20° C (40). 1-(2*H*)Phthalazinones are devoid of basic properties and they do not form salts with acids. They are stable to hot

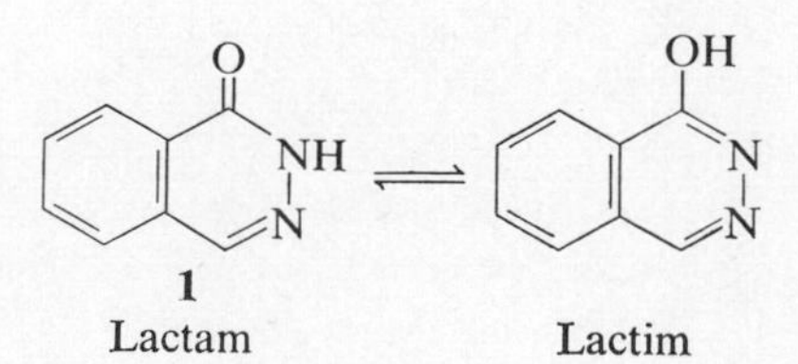

(18)

concentrated hydrochloric and sulfuric acid. 1-(2*H*)-Phthalazinone is soluble in alkali but substituents at the 2-position make 2-substituted-1-(2*H*)phthalazinones insoluble in alkali.

1-(2*H*)Phthalazinone (**1**) is capable of existing in two possible tautomeric structures (in the absence of substituents at the 2-position). It exists mainly in the lactam form, 1-(2*H*)phthalazinone, as shown by several workers using ir (41–43) and uv spectroscopy (43, 46). During the salt formation of **1** with metal ions it was revealed, using ir data (44), that the change from lactam to lactim form takes place. Regardless of the electropositive character of the metal ions, metal atoms were located on oxygen atom and not on nitrogen atom.

Mason (42) has studied the tautomerism of **1** along with several other *N*-heteroaromatic hydroxy compounds using ir spectra in solution as well as in the solid state; **1** absorbs in N—H and C═O stretching vibration regions and principally possesses the lactam structure. The ir spectrum of **1** in CCl_4 solution reveals absorption bands at NH, 3401 cm^{-1}; C═O, 1674 cm^{-1}; and in the solid state, NH—3292 (w), 3240 (m) cm^{-1}; C═O, 1658 (s) cm^{-1}.

Mori (43) has studied the ir spectrum of **1** in chloroform solution and in a nujol mull and the conclusions were similar. Recently Sohar (45) studied the ir spectrum of **1** and concluded that it exists as a dimer. The stretching vibration band of the associated NH group emerges between 3200 and 2700 cm^{-1}. The dimeric structure **23** was supported further by molecular weight determination. Cook (57), using ir data, concluded that the protonation of

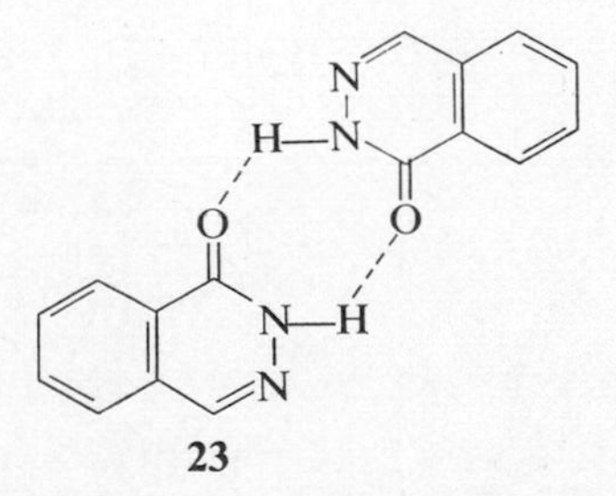

23

2-methyl-1-(2*H*)phthalazinone occurs at the N-3 position.

Albert and Barlin (46) have recorded the uv spectra of **1** as a neutral species in water at *p*H 7.0, as a cationic species at *p*H 4.65, and as an anionic species at *p*H 14.0 (Fig. 2D-1). They have also recorded the spectra of 2-methyl-1-(2*H*)phthalazinone (**24**) and 1-methoxyphthalazine as neutral and cationic species and showed that **1** and **24** have very similar spectra.

The uv spectrum of 1-(2*H*)phthalazinone (**1**) has been found to be very similar to 2-methyl-1-(2*H*)phthalazinone (**24**) (35) and also similar to that

1

1 Neutral species pH 7.0	208 (4.48), *240* (3.90), 276 (3.77),	222 (4.27) 249 (3.87), 298 (3.55),	*230* (4.18) 271 (3.77), 310 (3.46).
1 Cationic species pH-4.65	224 (4.46), 299 (3.67),	256 (3.46), 308 (3.68).	*288* (3.64)
1 Anionic species pH 14.0	223 (4.15), 301 (3.84).	246 (3.74),	*295* (3.83),
24			
24 Neutral species pH 7.0	*223* (4.30) 284 (3.81)	*241* (3.83) 299 (3.70)	250 (3.79), 301 (3.54)
24 Cationic species pH 6.6	217 (4.43) *303* (386)	227 (4.45) 312 (3.88)	*259* (3.52) *318* (3.85)
24 In 95% ethanol (35)	225 (4.17) 287 (3.84)	244 (3.77) 313 (4.18)	253 (3.77)

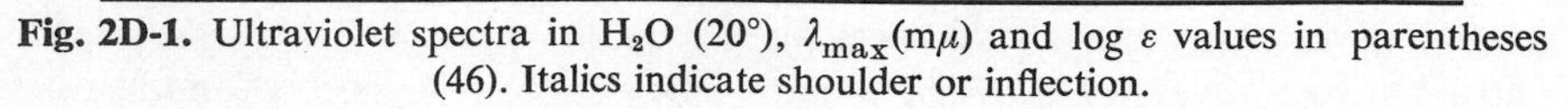

Fig. 2D-1. Ultraviolet spectra in H_2O (20°), λ_{max}(mμ) and log ε values in parentheses (46). Italics indicate shoulder or inflection.

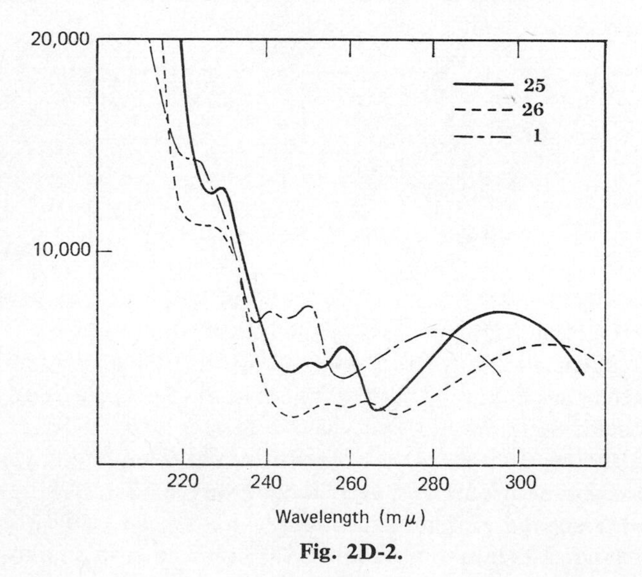

Fig. 2D-2.

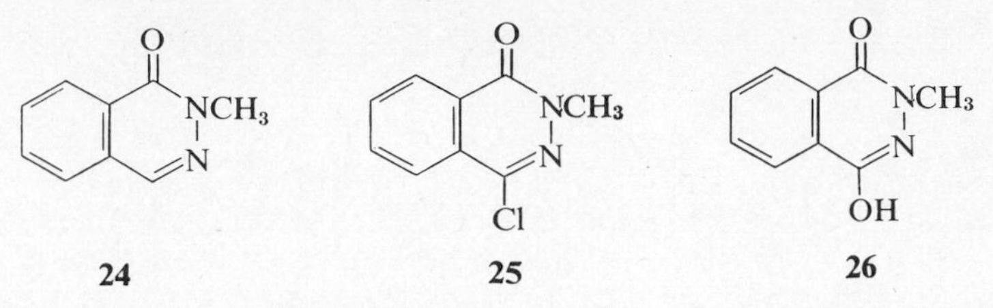

of 4-chloro-2-methyl-1-(2*H*)phthalazinone (**25**) and 4-hydroxy-2-methyl-1-(2*H*)phthalazinone (**26**) (43) (see Fig. 2D-2).

Parsons and Rodda (47) have studied the pmr spectrum of **1** and assigned the C-8 proton at 1.48 τ. Bowie and co-workers (58) have reported the mass spectra of **1** and 4-substituted-1-(2*H*)phthalazinones. Zyakun and co-workers (58a) have reported the mass spectra and the fragmentation schemes of 1-(2*H*)phthalazinone, 1-(2*H*)phthalazinethione, and their methylated products.

2. *Chemical Properties*

a. ALKYLATION AND ACYLATION. 1-(2*H*)Phthalazinone (**1**), as mentioned earlier (Eq. 9), on alkylation with alkyl halides gives 2-substituted-1-(2*H*)-phthalazinones. Reaction with diazomethane also gives 2-methyl-1-(2*H*)-phthalazinone (23). The ambident nature of the anion of 1-(2*H*)phthalazinone is revealed in the reaction of the silver salt and the potassium salt with α-acetobromoglucose to give *o*-alkylated (**27**) and *N*-alkylated (**28**) products (48) (R = tetraacetylglucosyl) (Eq. 18).

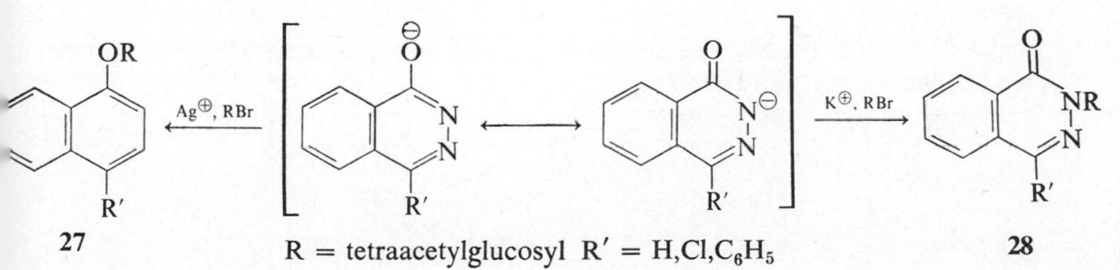

R = tetraacetylglucosyl R′ = H,Cl,C_6H_5 (18)

Parsons and Rodda (47) have studied the structure of the acetylated products of 1-(2*H*)phthalazinone, 4-methyl-1-(2*H*)phthalazinone, and 4-phenyl-1-(2*H*)phthalazinone using proton magnetic resonance and in each case the structure has been assigned to the 2-acetyl derivative, confirming the 1-(2*H*)phthalazinone structure.

The proton at the C_8 position is deshielded because of the anisotropic effects of the carbonyl group at C-1 as compared to the other benzenoid protons (2.2 τ). Since the acetylated products do not change the signal of the C_8 proton, this indicates acetylation has occurred at the nitrogen atom. The

NMR Data of the C-8 Proton and CH_3 Protons of Acetyl Groups

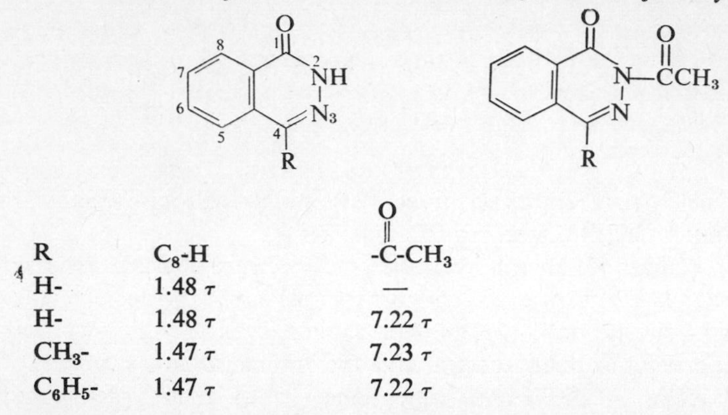

R	C_8-H	$\overset{O}{\parallel}$ -C-CH_3
H-	1.48 τ	—
H-	1.48 τ	7.22 τ
CH_3-	1.47 τ	7.23 τ
C_6H_5-	1.47 τ	7.22 τ

pmr spectrum of 2-methyl-1-(2*H*)phthalazinone also shows the C-8 proton at 1.53 τ.

Acetylation is carried out by refluxing with acetic anhydride. In similar acetylations of 4-substituted-1-(2*H*)phthalazinones the products have been assigned as 2-acetyl derivatives (28, 49, 50), except in one case where the authors, using ir data, concluded that *O*-acylation has occurred (51).

b. Reaction with Organometallic Reagents.* Hirsch and Orphanos (52) have studied the addition of organolithium compounds to 1-(2*H*)-phthalazinone. The addition took place across the C=N bond at C-4 position, followed by hydrolysis and oxidation, to give 4-substituted-1-(2*H*)phthalazinones (Eq. 19). However, in no case was addition to the carbonyl group at the C_1 position observed as was the case with the addition of Grignard reagents to **1**.

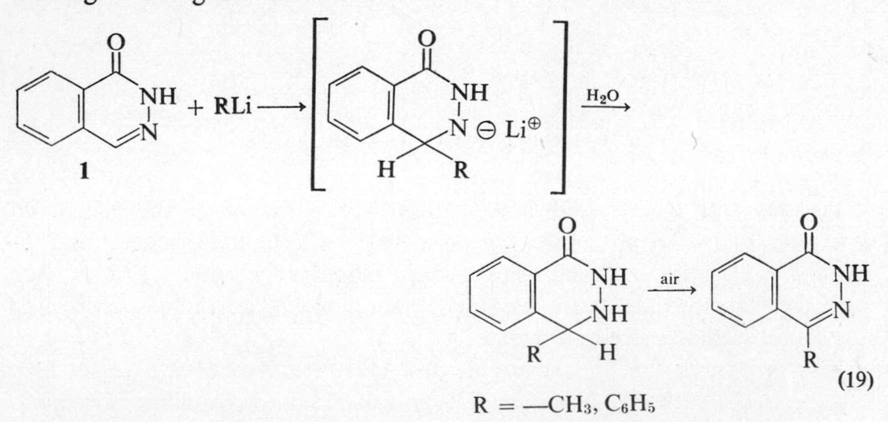

* For addition of Grignard reagents to 1-(2*H*)phthalazinones and derivatives, see Section 2B–II-A-4-c.

c. Reactions with Phosphorus Pentasulfide and Phosphorus Halides. 1-(2*H*)Phthalazinones and 4-substituted 1-(2*H*)phthalazinones are capable of tautomerization to the hydroxy form and react with phosphorus pentasulfide and phosphorus oxychloride to give the 1-thio and 1-chlorophthalazines, respectively (see Sections 2G and 2F) (Eq. 20).

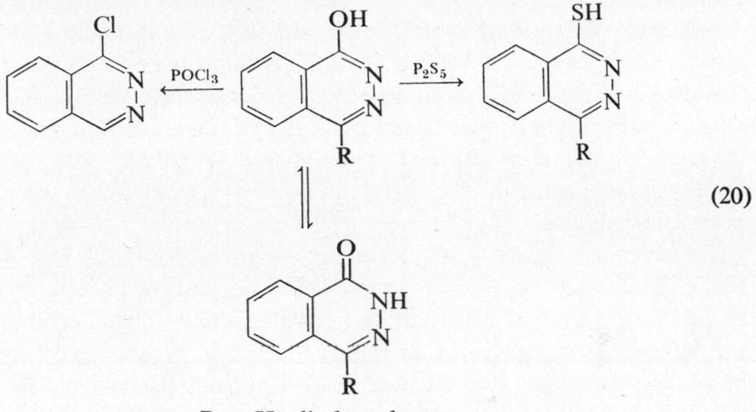

(20)

R = H, alkyl, aryl, etc.

d. Reduction. The reduction of 1-(2*H*)phthalazinones with strong reducing agents such as, tin or zinc and hydrochloric acid yields phthalimidines (16, 20) (Eq. 21).

O, NR, N, R′ ⟶ O, NR + NH_3, H, R′ (21)

R = H, C_6H_5; R′ = H, alkyl

Reduction of 1-(2*H*)phthalazinone with lithium aluminum hydride and catalytic reduction with platinum oxide has been shown to give 1,2-dihydrophthalazine and 3,4-dihydro-1-(2*H*)phthalazinone, respectively. In the former case the carbonyl group is reduced while in the latter case the C═N bond is reduced. Polarographic reduction of **1**, 2-phenyl-1-(2*H*)phthalazinone or 4-methyl-1-(2*H*)phthalazinone gives the corresponding 3,4-dihydro derivative (see Section 2L) in neutral and alkaline solutions. In mineral acid solutions the reduction products could be reduced further to give phthalimides.

Reduction of 4-methyl-8-nitro-1-(2*H*)phthalazinone with Raney Ni/H_2 at 3 atm pressure gives 4-methyl-8-amino-1-(2*H*)phthalazinone (53) where the nitro group is reduced without affecting the heterocyclic ring. Similarly,

4-nitromethyl-1-(2*H*)phthalazinone on reduction with phosphorus and hydriodic acid gives 4-aminomethyl-1-(2*H*)phthalazinone (50).

e. MISCELLANEOUS REACTIONS. 1-(2*H*)Phthalazinones and 4-substituted 1-(2*H*)phthalazinones undergo the Mannich reaction as discussed earlier (Eq. 10).

Nitration of 1-(2*H*)phthalazinones with 96% sulfuric acid and potassium nitrate gives 5-nitro- and 8-nitro-1-(2*H*)phthalazinones in poor yields (54).

Boltze and Dell (55) have attempted to condense 4-methyl-1-(2*H*)phthalazinone with benzaldehyde under a variety of conditions but in no case was the condensed product obtained, proving that the methyl group in 4-methyl-1-(2*H*)phthalazinone is not an active methyl group as in the case of 1-methylphthalazine.

The cyanoethylation of 1-(2*H*)phthalazinone, 4-methyl- and 4-phenyl-1-(2*H*)phthalazinone in pyridine-water solutions has been shown to take place at the N-2 position as discussed earlier (Eq. 9) (29). The addition product of acrylonitrile to 4-methyl-1-(2*H*)phthalazinone in dioxane with triton B has been formulated as 1-(β-cyanoethoxy)-4-methylphthalazine (**29**) (56) (see Eq. 22).

II. 1-Alkoxyphthalazines

A. Methods of Preparation

1. *Alkylation of* 1-(2*H*)*Phthalazinones*

1-(2*H*)Phthalazinones on alkylation (see Eq. 9) with methyl iodide, dimethyl sulfate, and diazomethane give 2-methyl-1-(2*H*)phthalazinones. The silver salts of 1-(2*H*)phthalazinones have been shown to react with α-acetobromoglucose to give 1-glucosyloxy derivatives (48) (Eq. 18). Zugravescu and co-workers (56) have reported the cyanoethylation of 4-methyl-1-(2*H*)phthalazinone to give the *O*-alkylated derivative (**29**) (Eq. 22).

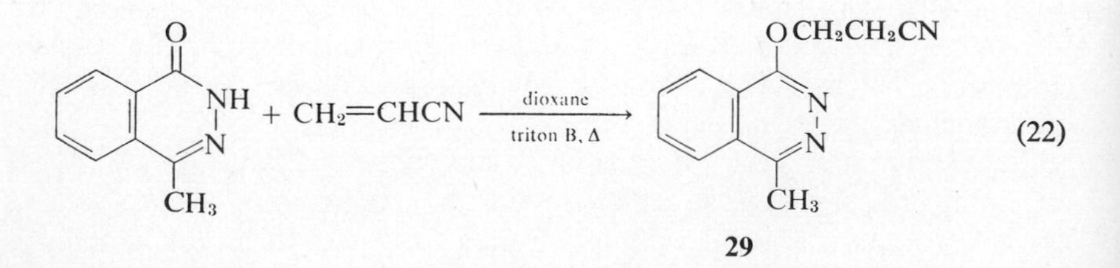

Aebi and Hofstetter (51) have obtained 1-acetoxy-4-isopropylphthalazine (**30**) by refluxing 4-isopropyl-1-(2*H*)phthalazinone with acetic anhydride and by reacting 1-chloro-4-isopropylphthalazine with sodium acetate (Eq. 23). The assignment has been made using ir data of **30**; $—O—\overset{O}{\overset{\|}{C}}CH_3$ (C=O, 1757 cm^{-1}); 1675 cm^{-1} (C=N, cyclic) and 1220 cm^{-1} ($—C—O—\overset{O}{\overset{\|}{C}}—$, C—O of acetyl group). However, similar acetylation of 4-(2-butyl)-1-(2*H*)phthalazinone has been shown to give 2-acetyl-4-(2-butyl)-1-(2*H*)phthalazinone where acetylation takes place at N-2 position (28) (see Section 2D-I-B-2-a).

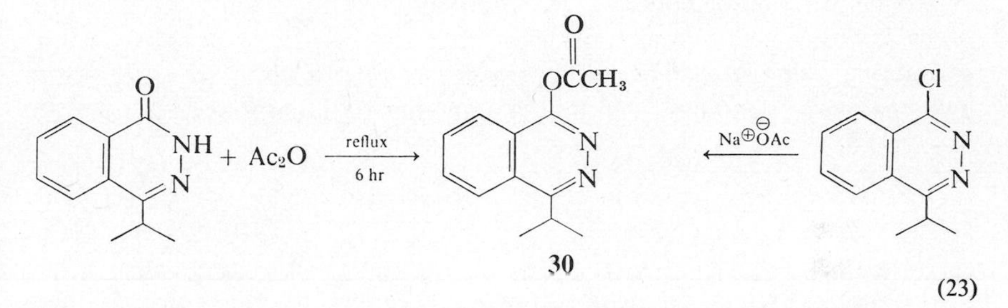

2. *By Nucleophilic Substitution*

a. 1-HALOPHTHALAZINES. 1-Alkoxyphthalazines are prepared primarily by the nucleophilic substitution of the chlorine atom of 1-chlorophthalazines with sodium alkoxides. A large number of 1-alkoxyphthalazines have been prepared by boiling the corresponding 1-chlorophthalazine with the sodium alkoxides in the appropriate alcohol for 20 min to 3 hr (28, 49, 59, 60) (Eq. 24). Chapman and Russell-Hill (61) have studied the kinetics of the reaction of 1-chlorophthalazine with ethoxide ion. 1-Chlorophthalazine (62) and 1-chloro-4-methylphthalazine (63) were allowed to react with phenol in

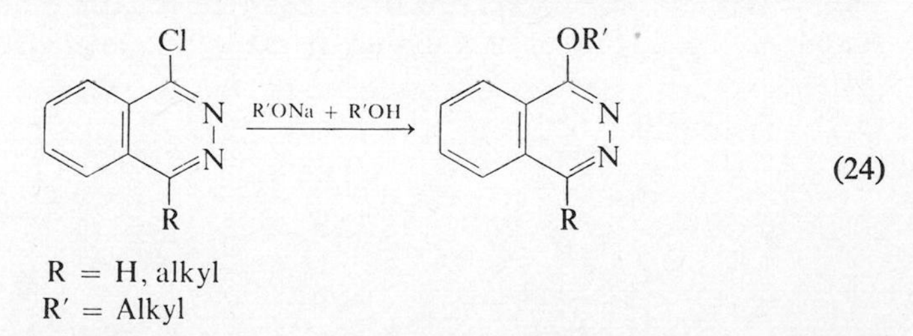

the presence of potassium carbonate by heating at 100° C for 1–2 hr to give the corresponding 1-phenoxyphthalazines (Eq. 25).

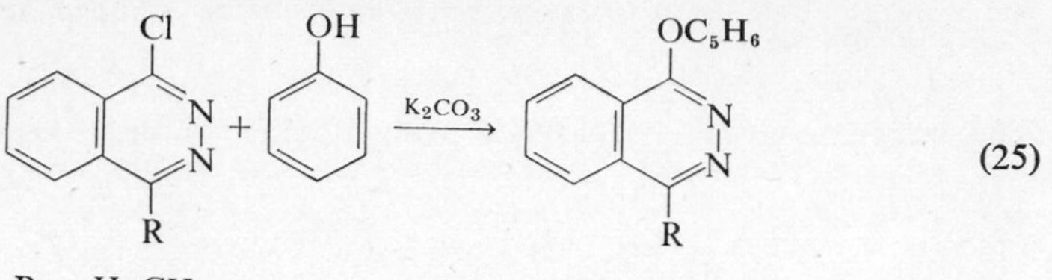

(25)

$R = H, CH_3$

N,*N*-Dialkylaminoalcohols have been allowed to react with 1-chlorophthalazines in presence of sodium without any solvent or in solvents such as benzene, toluene, xylene in the presence of either sodium or sodamide to give the corresponding *N*,*N*-dialkylaminoalkoxyphthalazines (27, 64–66) **31** (Eq. 26).

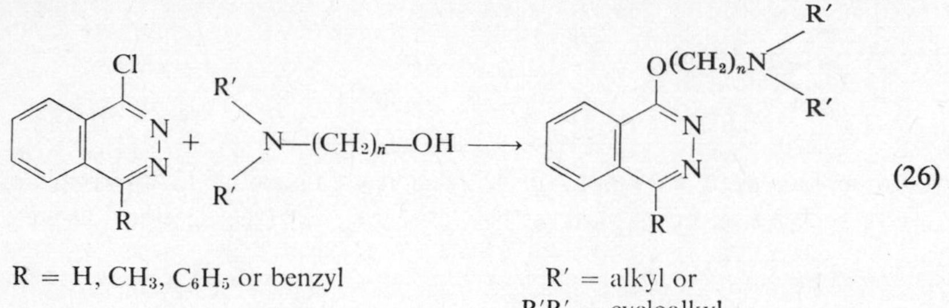

(26)

$R = H, CH_3, C_6H_5$ or benzyl

R′ = alkyl or
R′R′ = cycloalkyl
$n = 2$ or 3

b. 1-Methylsulfonylphthalazine. Barlin and Brown (67) have studied the kinetics of replacement of the methylsulfonyl group in 1-methylsulfonylphthalazine (**32**) with methoxide ion. It has been found that the methylsulfonyl group is very reactive compared to the corresponding chloro group because of the low energy of activation. These authors (67a) have also found that the rate of substitution with methoxide ion in the 1-position of phthalazine with different substituents proceeds as follows: $—SOCH_3 > —SO_2CH_3 > SCH_3 > Cl$. Oishi (68) has reported the nucleophilic substitution of the methylsulfonyl group in **32** with methoxide ion and other

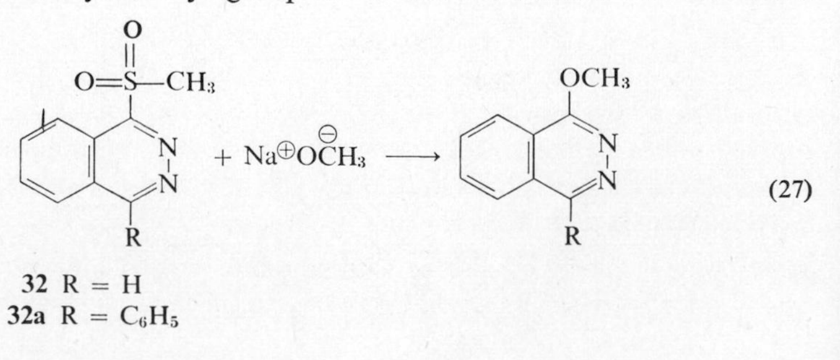

(27)

32 R = H
32a $R = C_6H_5$

nucleophiles. Hayashi and co-workers (69) have reacted 1-methylsulfonyl-4-phenylphthalazine (**32a**) with sodium methoxide to give 1-methoxy-4-phenylphthalazine (Eq. 27).

B. Properties of 1-Alkoxyphthalazines

1. *Physical Properties*

1-Methoxyphthalazine can be distilled under reduced pressure (130°/10 mm) (60) without decomposition, it has mp 60–61° (59); 60–62° (60). It has a pK_a of 3.77 in water at 20° (46).

Albert and Barlin (46) have recorded the uv spectrum of 1-methoxyphthalazine as a neutral species in water at pH 7.0 and as a cationic species at pH 1.18 (Fig. 2D-3).

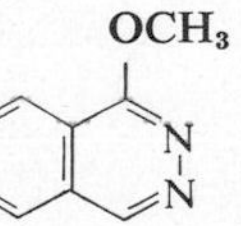

Neutral species *p*H 7.0	212 (4.69) 291 (3.41)	*216* (4.66) 304 (3.39)	265 (3.74)
Cationic species *p*H 1.18	224 (4.56) 311 (3.69)	283 (3.71)	*301* (3.62)

Fig. 2D-3. Ultraviolet spectra in water at 20°, λ_{max}(mμ) and log ε values in parentheses. Italics indicate shoulder or inflection.

2. *Chemical Properties*

a. SALT FORMATION. 1-Alkoxyphthalazines have marked basic properties. They form quaternary salts such as 1-alkoxy-3-alkylphthalazinium iodides with alkyl iodides (see Section 2K). 1-Methoxyphthalazine forms a picrate, mp 139–140° (methanol) (60); 161–162 (67). 1-Alkoxyphthalazines undergo *N*-oxidation with perphthalic acid to give exclusively 1-alkoxyphthalazine 3-oxides (see Section 2C).

b. TRANSETHERIFICATION. Adachi (59) has studied the transetherification of 1-methoxyphthalazine by refluxing 15 min with ethanol or 2-propanol in presence of sodium to give 1-ethoxy and 1-(2-propoxy)phthalazine, respectively (Eq. 28). The reaction proceeds with comparative ease and the yields are above 71%.

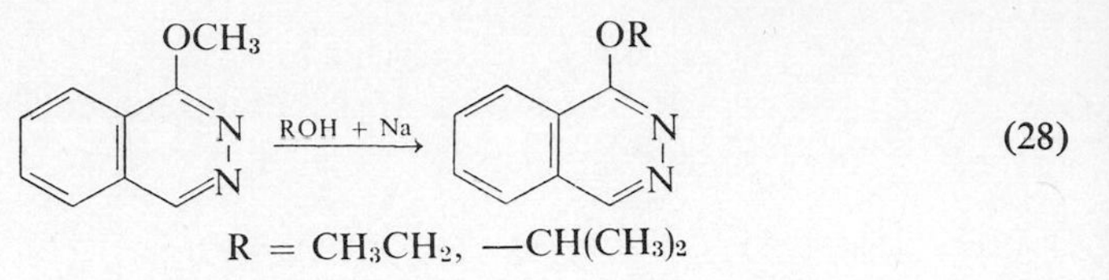

(28)

c. Amination. Nucleophilic substitution of the phenoxy group in 1-phenoxyphthalazine and some of its derivatives with ammonium acetate has been reported to give the corresponding 1-aminophthalazines in high yields (Eq. 29). The reaction is carried out by heating with ammonium acetate at 150–160° for 30 min (see Section 2H).

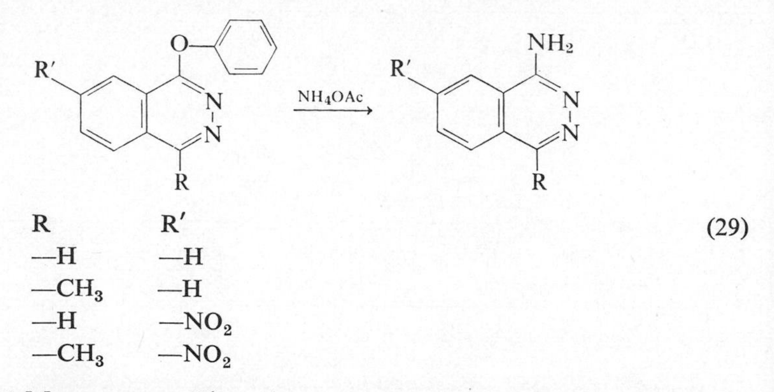

(29)

d. $O \rightarrow N$ Migration. There is one report of $O \rightarrow N$ migration of a substituent. Wagner and Heller (48) have shown the migration of the glucosyl group of 1-glucosyloxyphthalazine to the 2-position of phthalazine. The reaction has been carried out in toluene by refluxing for 5 hr in the presence of mercuric bromide (Eq. 30).

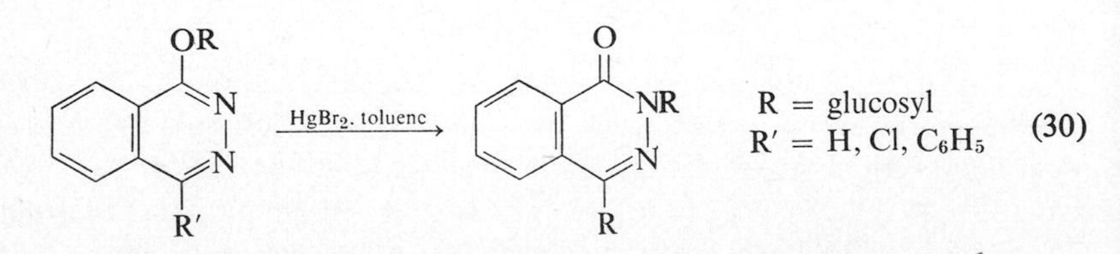

(30)

e. Nitration. Kanahara (31) has reported the nitration of 1-methoxyphthalazine with potassium nitrate and concentrated sulfuric acid. The major product isolated was 1-methoxy-5-nitrophthalazine with 5-nitro-1-(2*H*)phthalazinone as a byproduct (Eq. 31).

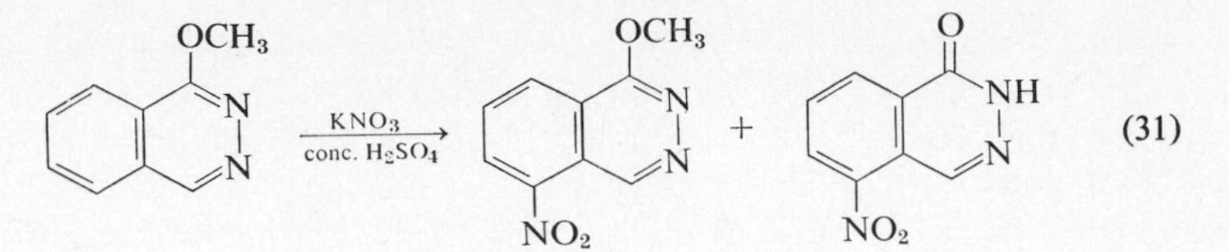

III. Tables

TABLE 2D-1. 1-(2*H*)Phthalazinones with Substituents in the Aromatic Ring

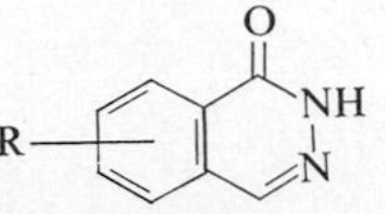

R	Preparation[a]	Yield (%)	MP[b] (Recrystallization Solvent)	Reference
—H	A		182°; 180°; 182–183°; 183–184°; 184°; 184–185°; 186–187°	1–5, 10, 15–18, 21, 22
5-F	B	60	245–246°	11
6-F	B	—	240–242°	11
7-F	B	—	191–193°	11
8-F	B	—	215–217°	11
5-Cl	C	—	263° (ethanol)	13
6-Cl	D	—	278–279° (acetic acid)	6
7-Cl	C	66	247.5–248° (ethanol)	6
8-Cl	C	23	198–199°	13
5-NO_2	D, E, F	79, —, 7	262–263°; 265–266°; 263–265°	9, 54
7-NO_2	C, D, E	50–70	237–239° (H_2O); 232–233°; 233°	8, 12, 54
8-NO_2	D, E, F	83, 92,	252–253°; 253–255°	9, 54
7-OH	G	—	310° (dec.)	70
8-CHO	H	56	235–237°; sublimes at 120°/0.05 mm	14
8-CO_2H	H	99	303.5–306°	14
8-CO_2CH_3	—	—	207–209°	14
8-$CO_2CH_2CH_3$	—	—	165–166°	14

[a] A, see different methods of preparation in this section. B, corresponding phthalide was brominated, hydrolyzed with alkali, and boiled in aqueous medium for 15 min with hydrazine sulfate. C, same as B but started with corresponding 3-bromophthalide. D, corresponding phthaldehydic acid condensed with hydrazine. E, decarboxylation of the corresponding 4-carboxyl-1-(2*H*)phthalazinone. F, nitration with KNO_3 + 96% H_2SO_4—heat at 90° for 5 hr. G, 1-chloro-7-methoxyphthalazine was hydrolyzed by refluxing with hydrobromic acid. H, corresponding 3-hydroxyphthalide was boiled with hydrazine in aqueous medium for 30 min.

[b] Recrystallization solvent given in parentheses.

TABLE 2D-2. 2-Substituted-1-(2*H*)phthalazinones

R	Preparation[a]	Yield (%)	MP[b]	Reference
$—CH_3$	A, B, A_1, A_2	52–91	Yellow columns, 112–114°, 112–114° (ethyl acetate); 110–111° (toluene); 113–115° (pet. ether)	23–25, 30, 39
$—C_6H_5$	C	—	105°	15, 71
$—CH_2OH$	D	83	125° (dec.)(benzene-pet. ether); 129°, 134.5–135°	1, 29, 72
$—CH_2Cl$	E	92	140°	1
$—CH_2—S—P(=S)—(OCH_3)_2$	F	87	87° (2-propanol)	1
$—CH_2—S—P(=S)—(OEt)_2$	F	93	58° (dil. 2-propanol)	1
$—CH_2—S—P(=O)—(OEt)_3$	F_1	72	—	1
$—CH_2—S—P(=S)(OEt)—CH_3$	F_2	86	91° (methanol)	1

$—CH_2—O—C(=O)$–(2,2-dimethyl-3-(2-methylpropenylidene)cyclopropyl)	G	—	79–81°	73
$—CH_2CH_2OH$	B	—	109–110°	9
$—CH_2CH_2Cl$	E_1	—	95°	9
$—CH_2CO_2Et$	B	—	79°	9
$—CH_2C(=O)NHNH_2$	H	—	245–246°	9
$—CH_2CH_2N(CH_3)_2$	I	91	BP 170–172°/2 mm; 135–137°/0.5 mm HCl, 218–219°; 220–221°; 224°	24, 26, 27
$—CH_2CH_3$	A_2	—	59°	70a
$—CH_2C_6H_5$	A_1, A_2	90	102–104°; 107°	39, 70a
$—CH_2CH_2N(CH_2CH_3)_2$	I	—	BP 156–159°/1 mm; HCl, 218°; 178°; HBr, 212–213°; picrate, 192–193°; gentisate, 145°	26, 27
$—CH(CH_3)CH_2N(CH_3)_2$	I	—	BP 145–150°/0.75 mm; HCl, 181–182°	26
$—CH_2CH_2N$(piperidino)	I	—	112°; HCl, 235–237°	26
$—CH_2CH_2CN$	J	64	112° (ether)	29
$—CH_2CH_2CO_2H$	K	76	Colorless needles, 144° (benzene)	29
$—CH_2N(CH_3)_2$	L	69	125° (benzene–pet. ether)	29

TABLE 2D-2. (*continued*)

R	Preparation[a]	Yield (%)	MP[b]	Reference
$-CH_2N(CH_2CH_3)C_6H_5$	L	65	108° (methanol)	29
$-CH_2-N(CH_2)_5$ [piperidino]	L	63	81° (pet. ether)	29
$-CH_2-N(CH_2CH_2)_2O$ [morpholino]	L	71	136° (benzene–pet. ether)	29
$-CH_2-N(CH_2CH_3)-CH_2-$N [1-oxophthalazin-2-yl]	L_1	63	139° (benzene–pet. ether)	29
$-C(=S)-NH-C(=O)CH_3$	M	—	250°	21
$-SO_2-C_6H_4-CH_3$ (p)	N	86	White plates 205–207° (DMSO)	18
[phthalazin-1-yl]	O	66	Yellow prisms, 274°, 279–280° (dil. acetic acid)	19, 74
[2-naphthyl]	B	—	168° (ethanol)	19

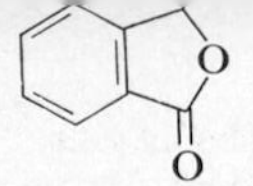	P	62	Colorless compound, 225–226° (acetic acid)	32
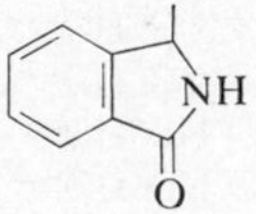	Q	—	266° (ethanol)	75
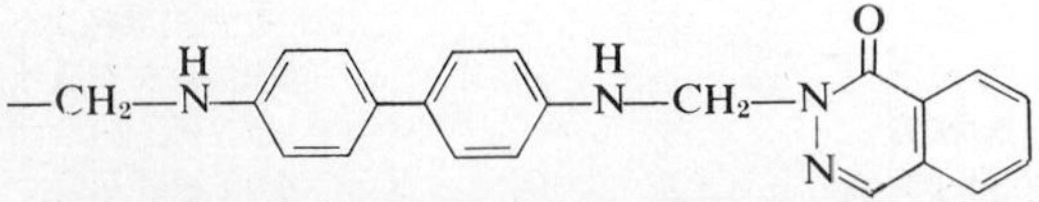	L_2	88	254–255°	76

[a] A, alkylation of 1-(2*H*)phthalazinone with diazomethane in ether or with methyl iodide in the presence of base. A_1, the corresponding 2-alkyl-4-ethoxyphthalazinium iodide pyrolyzed at 150–160° in stream of CO_2. A_2, decarboxylation of the corresponding 4-carboxylic acid derivative by heating at 240° for 30 min. B, phthaldehydic acid condensed with properly substituted hydrazine by refluxing in ethanol 2–3 hr. C, hydroxyphthalimidine boiled with phenylhydrazine in anisole or dehalogenation of 4-chloro-2-phenyl-1-(2*H*)phthalazinone with red phosphorus and hydriodic acid. D, 1-(2*H*)phthalazinone heated with aqueous formaldehyde for 20 min. E, 2-hydroxymethyl-1-(2*H*)phthalazinone heated with thionyl chloride in dichloromethane at 60° for 30 min. E_1, similar to E. Reflux in chloroform for 8 hr. F, 2-chloromethyl-1-(2*H*)phthalazinone heated with the corresponding $NH_4^{\oplus}S^{\ominus}\text{—}P(=S)\text{—}(OR)_2$ in acetone for 1 hr at 50–60°. F_1, same as F using $NH_4^{\oplus}S^{\ominus}\text{—}P(=O)\text{—}(OEt)_2$. F_2, same as F using $NH_4^{\oplus}S^{\ominus}\text{—}P(=S)(OEt)\text{—}CH_3$. G, 2-hydroxymethyl-1-(2*H*)phthalazinone refluxed with chrysanthemamic acid in toluene in the presence of *p*-toluenesulfonic acid. H, reflux the corresponding ethyl ester with hydrazine in ethanol for 4 hr. I, 1-(2*H*)phthalazinone refluxed with *N*,*N*-dialkylaminoalkanols in ethanol or benzene in the presence of a strong base. J, 1-(2*H*)phthalazinone refluxed with acrylonitrile in pyridine-water for 3 hr. K, alkaline hydrolysis of 2-(2-cyanoethyl)-1-(2*H*)phthalazinone. L, Mannich reaction of 1-(2*H*)phthalazinone. Heat with formaldehyde and proper secondary amine in methanol. L_1, same as L, using ethylamine. L_2, same as L, using 4,4′-diaminobiphenyl. M, thiosemicarbazone of phthaldehydic acid heated in acetic anhydride. N, tosylhydrazone of phthaldehydic acid heated with acetyl chloride in methanol on the steam bath. O, 1-phthalazinylhydrazone of phthaldehydic acid treated with thionyl chloride at room temperature, or hydrolyzed 1-chlorophthalazine with dilute hydrochloric acid. P, 2,2′-dicarboxybenzaldezine is refluxed for 1 hr in acetic acid; or fusion of 1-(2*H*)-phthalazinone with phthaldehydic acid at 230° for 15 min. Q, hydroxyphthalimidine refluxed in anisole with benzoic or isonicotinic acid hydrazide.

[b] Recrystallization solvent given in parentheses.

TABLE 2D-3. 2-Alkyl- or Aryl-1-(2*H*)phthalazinones with Substituents in the Benzenoid Ring

R	R′	Preparation[a]	Yield (%)	MP	Reference
$-CH_3$	5-NO_2	A	—	125–126° (ethanol)	31
$-CH_3$	7-NO_2	B	51	177–179° (ethanol)	8
$-CH_3$	8-NO_2	A	—	226–227°	31
$-CH_3$	8-$CO_2CH_2CH_3$	B	—	103.5–105°	14
$-CH_2CH_2OH$	5-NO_2	C	—	120–121°	9
$-CH_2CH_2OH$	7-NO_2	C	—	162–163°	9
$-CH_2CH_2OH$	8-NO_2	C	—	162–163°	9
$-CH_2CH_2Cl$	5-NO_2	D	—	125°	9
$-CH_2CH_2Cl$	7-NO_2	D	—	128.5°	9
$-CH_2CH_2Cl$	8-NO_2	D	—	134°	9
$-CH_2CO_2Et$	5-NO_2	C	—	109°	9
$-CH_2CO_2Et$	7-NO_2	C	—	158°	9
$-CH_2CONHNH_2$	5-NO_2	E	—	238–240°	9
$-CH_2CONHNH_2$	7-NO_2	E	—	250°	9

Phenyl (structure)	$8-CO_2H$	F	87	197–198° (benzene)	14, 77
Phenyl (structure)	$8-CO_2Et$	G	—	150–151°	14, 77
Phenyl (structure)	—S—(2-CO_2H-phenyl) (structure)	H	—	Straw colored needles, 199° (methanol)	78
(structure: tolyl sulfonyl phenyl bearing a phthalazinone N)	7,8-Dimethoxy	I	—	258–260°	33

[a] A, $K_3Fe(CN)_6$ oxidation of the corresponding methylphthalazinium iodide. B, alkylation of the corresponding 1-(2*H*)phthalazinone with the dimethyl sulfate in presence of alkali. C, condensation of properly substituted phthaldehydic acid with the corresponding hydrazine derivative by refluxing in ethanol for about 2 hr. D, reflux the corresponding 2-(2-hydroxyethyl)-1-(2*H*)phthalazinone with thionyl chloride in chloroform for 8 hr. E, reflux the corresponding ethyl ester with hydrazine hydrate in ethanol for 4 hr. F, reflux 3-hydroxy-7-carboxylphthalide with phenylhydrazine in acetic acid for 18 hr. G, esterification of 2-phenyl-8-carboxyl-1-(2*H*)phthalazinone by treatment with thionyl chloride followed by refluxing in chlorobenzene with ethanol for 2 hr. H, 2-dihydroxymethyl-2′,3-biphenyldicarboxylic acid dilactone condensed with phenylhydrazine. I, one mole of 4,4′-dihydrazinobiphenyl sulfone condensed with two moles of opianic acid in water.

TABLE 2D-4. 4-Substituted-1-(2*H*)phthalazinones

R	Preparation[a]	Yield (%)	MP	Reference
$—CH_3$	A	55–89	218–220°, 220°, 219–221° (ethanol)	28, 52, 62, 79, 80
	B(84)		223–224°, 221–222°, 219–220° (ethanol)	81–84
	C(85)		Colorless needles 222° (ethanol-water)	85, 123
	C_1, C_2			
$—CH_2CH_3$	A	—	168–170°	28
$—CH_2CH_2CH_3$	A, D	75	163–165°, 165°	28, 86
$—CH(CH_3)_2$	A, E	85–87	156°, 155–157° (ethanol); 158–159° (benzene-hexane)	28, 51, 87
$—CH(CH_3)CH_2CH_3$	A	—	136–138° (ethanol)	28
$—CH_2CH_2CH_2CH_3$	D	—	Prisms, 160–163° (ethanol)	88
$—CH(CH_3)CH_2CH_2CH_3$	A	—	140–141°	28
$—CH(CH_3)CH(CH_3)_2$	A	—	160–161°	28

$—CH(CH_3)CH_2CH_2CH_2CH_3$	A	—	95–96°	28
$—CH(CH_2CH_3)CH_2CH_2CH_3$	A	—	119–120°	28
$—CH(CH_2CH_3)CH(CH_3)_2$	A	—	182–183°	28
$—CH(CH_2CH_3)CH_2CH_2CH_2CH_3$	A	—	116–117°	28
$—CH(CH_3)CH_2CH_2CH_2CH_3$	A	—	87–89°	28
$—CH(CH_3)(CH_2)_7CH_3$	A	—	99–101°	28
$—CH_2CN$	A	—	—	56
$—CH_2—NO_2$	F	73	215° (formic acid)	25
$—C_6H_5$ (phenyl)	C_1, Q	11	231–233°, 235°	52, 109
$—C_6H_4—CH_3$ (p-tolyl)	A	—	Colorless plates, 258–259°	89

TABLE 2D-4. (*continued*)

R	Preparation[a]	Yield (%)	MP	Reference
4-fluorophenyl	A, A_1	100	267–268.5° (acetic acid); colorless plates, 264° (ethanol)	90, 91
4-chlorophenyl	A_1	100	Colorless prisms, 268° (xylene); 268–270° (toluene)	91, 92
4-methoxyphenyl (OCH_3)	A, P	—	241°	93, 108
4-($N(CH_3)_2$)phenyl	A	—	283–285°	93
2-thienyl	A_1	100	Colorless silky needles, 195° (alcohol)	91
5-methyl-2-thienyl (CH_3)	A_1	—	Fine colorless needles, 175° (toluene)	91
5-chloro-2-thienyl	A_1	100	Pale yellow needles 203° (alcohol)	91
5-bromo-2-thienyl	A_1	—	Fine yellow needles, 197° (methanol)	91
3,4-dichlorophenyl	A	—	277–278° (ethanol)	94
4-chloro-3-(SO_2NH_2)phenyl	A	80	323–325° (aqueous dimethylformamide)	94

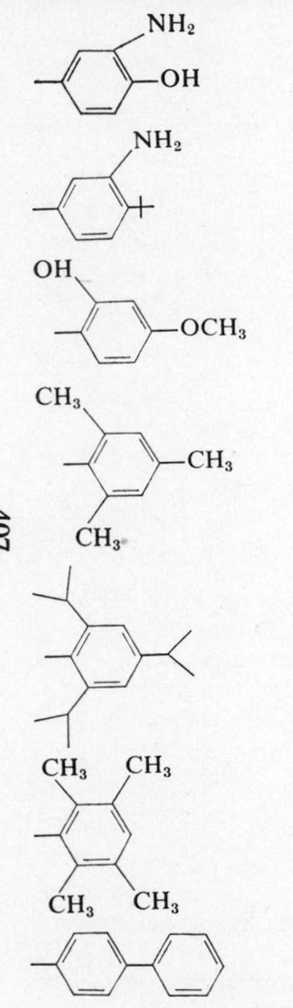

A	—	298–300° (darkening)	92
A	82	278–280° (toluene)	95
G	100	White crystals, 286–287° (ethanol)	96
H, I	41	Yellow crystals, 262–263° (ethanol)	34
H_1	80	240–242°	34
H_2, I	60–93	Bright yellow, ~320° (dec.), 335° (dec.)	34
A	100	White needles, 278°	97

TABLE 2D-4. (*continued*)

R	Preparation[a]	Yield (%)	MP	Reference
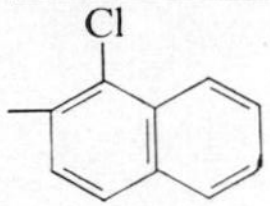 Cl	J	—	Colorless, 292°	98
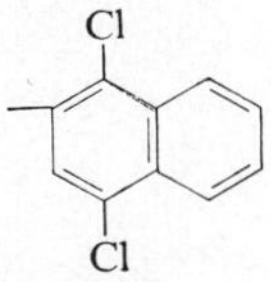 Cl, Cl	J	28	288° (acetic acid)	98
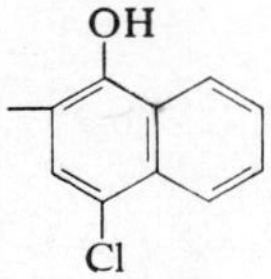 OH, Cl	A	—	280–281° (chlorobenzene)	99
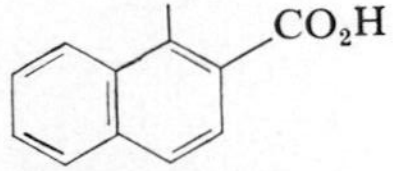 CO_2H	K	51	322° (acetone)	100
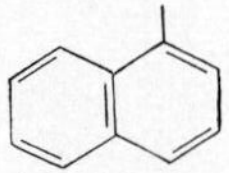	K_1	57	260°	100
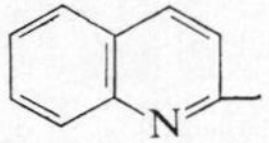 N	A	62	Needles, 286° (ethanol)	101
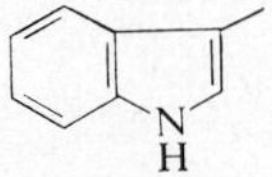 N H	L	86	315°	102

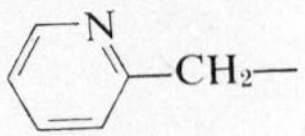 2-pyridyl-CH_2—	M	—	179–182°; HCl, 261–264°	103, 104
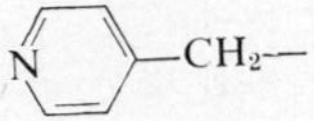 4-pyridyl-CH_2—	M	—	208–210°; HCl, 280–282°	103, 104
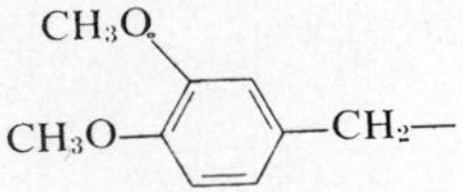 3,4-$(CH_3O)_2C_6H_3$-CH_2—	N	100	193–194° (ethanol)	105
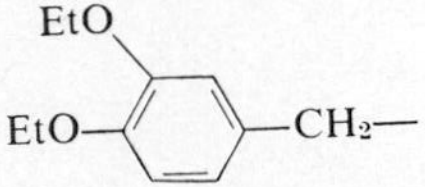 3,4-$(EtO)_2C_6H_3$-CH_2—	—	—	249–259° (dec.)	105
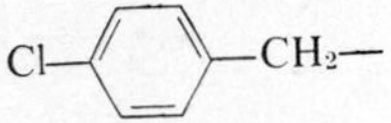 Cl-C_6H_4-CH_2—	D	—	218°	106
 CH_3O-C_6H_4-CH_2—	D	—	196°	106
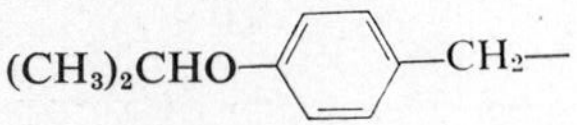 $(CH_3)_2CHO$-C_6H_4-CH_2—	D	—	208°	106
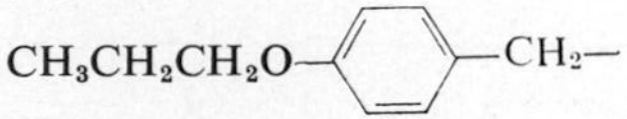 $CH_3CH_2CH_2O$-C_6H_4-CH_2—	D	—	190°	106
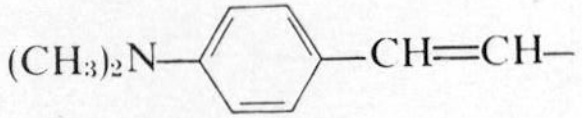 $(CH_3)_2N$-C_6H_4-CH=CH—	O	56	239–241° (dioxane)	107

TABLE 2D-4. (*continued*)

R	Preparation[a]	Yield (%)	MP	Reference
$C_6H_5CH_2$—	M_1	82	197–199° (acetic acid)	104a
2-Cl-C_6H_4—CH=CH—	O	62	237–238.5° (acetic acid)	107
4-Cl-C_6H_4—CH=CH—	O	94	233–234.5° (dioxane)	107
2-NO_2-C_6H_4—CH=CH—	O_1	65–78	316–317.5° (acetic acid)	107
4-O_2N-C_6H_4—CH=CH—	O_1	60–72	331–333° (acetic acid)	107
3-pyridyl—CH=CH—	A	66	157–158° (dec.)	55
—$CH_2OC(=O)CH_3$	Q	8	179–181°	49

$-CH_2NH_2$	R	—	HI, 266–267° (ethanol–ether) HCl, 308°	50
$-CH_2NHC(=O)CH_3$	—	—	226–227° (aqueous methanol)	50
$-CH_2NH-C(=O)-C_6H_5$	—	—	257–258°	50
$-CH_2N$ (phthalimido)	—	—	353° (acetic acid)	50
$-CH_2-NH-C(=O)-NH-CH_2-$ (4-phthalazin-1(2H)-one)	R	—	320–325° (water)	50
$-CH_2-NH-C(=O)-NH-C_6H_5$	R_1	—	300–301°	50
$-CH_2CO_2H$	S	75	185° (dec.)	123
$-CH_2CO_2CH_3$	T	91	187–191°	123
$-CH_2CO_2CH_2CH_3$	T	93	177–178.5°	123
$-CH_2CO_2(CH_2)_2Cl$	T_1	94	183–184.5°	123
$-CH_2CO_2CH_2CHCl(CH_3)$	T_1	84	122.5–124°	123
$-CH_2CONH_2$	U	99	162–164°	123
$-CH_2CONHNH_2$	U_1	96	250–252° (dec.)	123

TABLE 2D-4. (*continued*)

R	Preparation[a]	Yield (%)	MP	Reference
$—CH_2CONH(CH_2)_2N(CH_2CH_3)_2$	U_2	87	186°; HCl, 191–192°	123
$—CH_2CONH(CH_2)_3N(CH_2)_6$	U_2	74	202–203°; HCl, 201–202°	123

[a] A, the corresponding 2-acylbenzoic acid condensed with hydrazine hydrate in water or ethanol by heating for about 90 min on the steambath. A_1, similar to A, reflux in acetic acid for about 2 hr. B, dehydrohalogenation of 4-methyl-5,6,7,8-tetrahydro-6,7-dibromo-1-(2*H*)phthalazinone with methanolic hydroxide, reflux 30 min. C, condensation of

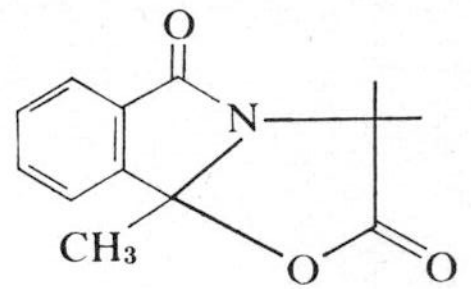

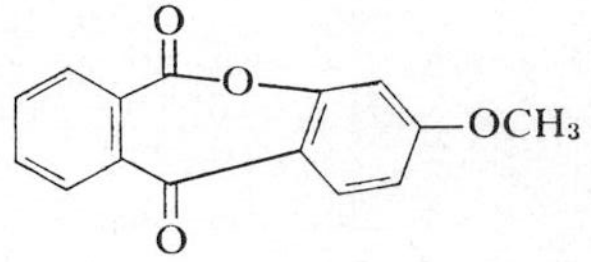

with hydrazine hydrate by refluxing for 6 hr in ethanol. C_1, 1-(2*H*)phthalazinone reacted with alkyl or aryllithium followed by hydrolysis and air oxidation. C_2, heat 1-(2*H*)phthalazinone-4-acetic acid in nitrobenzene at 170°. D, similar to A; condensed the corresponding 3-alkylidinephthalide. E, condensed 3-hydroxy-3-isopropylphthalide with hydrazine hydrate under Huang-Minlon reduction conditions. F, 2-nitro-1,3-indanedione refluxed with hydrazine hydrate in ethanol for 3 hr. G, condensed with 50% hydrazine hydrate at room temperature for 5 min. H, similar to A, reflux with 85% hydrazine hydrate for 8 hr. H_1, similar to A, reflux with 85% hydrazine hydrate for 24 hr. H_2, similar to A, reflux with 95% hydrazine hydrate for 2 hr; or reflux methyl 2-duroylbenzoate with 85% hydrazine hydrate for 2 days in ethanol; or reflux 2-duroylbenzoic acid hydrazide with 85% hydrazine hydrate in ethanol for 2 hr. I, same as A, heat strongly to fuse (without solvent). J, similar to A, reflux in nitrobenzene for 30–60 min. K, condense

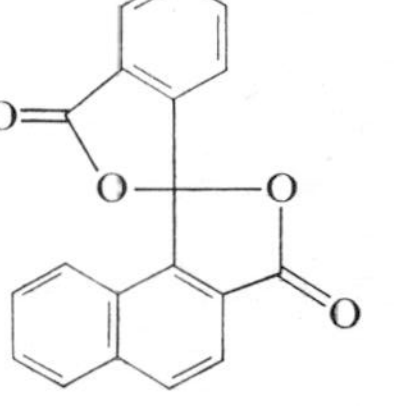

with hydrazine hydrate under Huang-Minlon reduction conditions. K_1, decarboxylation of the product obtained by method K with cupric oxide in quinoline. L, similar to A, heat with 72% hydrazine hydrate at 140°. M, condensed the corresponding 3-alkylidinephthalide with hydrazine hydrate by heating at 135° for 3 hr. M_1, 3-benzylidinephthalide reacted with hydrazine in acetic acid. N, condensed 3-(3′,4′-dimethoxybenzyl)phthalide with hydrazine in ethanol by refluxing for 4 hr. O, the corresponding 2-acylbenzoic acid condensed with acetylhydrazine by refluxing in dimethyl formamide for 7 hr. O_1, same as O but reflux with hydrazine in acetic acid for 7 hr. P, fused 2-(4′-methoxybenzoyl)benzoic acid with malonic acid dihydrazide, phenylsemicarbazide, phenylthiosemicarbazide, etc., or by alkaline hydrolysis of 2-(isonicotinoyl)-4-(4′-methoxy)-1-(2*H*)phthalazinone. Q, 1-alkyl- or 1-arylsulfonyl-4-phenylphthalazines hydrolyzed by heating with 2*N* sodium hydroxide solution at 100° for 4–5 hr. Q, 4-methyl-1-(2*H*)phthalazinone 3-oxide heated with acetic anhydride at 70° for 3 hr; also prepared in low yield by treatment of 4-hydroxymethyl-1-(2*H*)phthalazinone with acetic anhydride at 130° for 1 hr. R, 4-methylamino-1-(2*H*)phthalazinone hydrochloride refluxed in water with urea for 4 hr. R_1, same as R but refluxed with phenylurea. S, heat phthalidineacetic acid for 2 hr with hydrazine sulfate in aqueous potassium bicarbonate. T, 1-(2*H*)phthalazinone-4-acetic acid esterification with ROH-H_2SO_4. T_1, 1-(2*H*)phthalazinone-4-acetic acid esterification with corresponding chlorohydrin by refluxing in benzene with H_2SO_4. U, 1-(2*H*)phthalazinone-4-acetic acid stirred with aq. NH_4OH for 10 days. U_1, 1-(2*H*)phthalazinone-4-acetic acid refluxed with hydrazine hydrate in ethanol for 4 hr. U_2, 1-(2*H*)phthalazinone-4-acetic acid heated with corresponding *N*,*N*-dialkylaminoalkylamine at 140–170° for 4 hr.

TABLE 2D-5. 4-Alkyl- or Aryl-1-(2*H*)phthalazinones with Substituents in the Benzenoid Ring

R	R′	Preparation[a]	Yield (%)	MP	Reference
$—CH_3$	$5-NO_2$	A	—	273–275°, 282°	54, 110
$—CH_3$	$7-NO_2$	A	—	283°	110
$—CH_3$	$8-NO_2$	B	83	280–287°, 293°	53, 54
$—CH_3$	$8-NH_2$	C, C_1, C_2	57–70	276–278°, 277–280° (acetic acid)	53, 111
$—CH_3$	8-OH	D	46	228–230° (ethanol)	53
$—CH_3$	5,7-Dimethoxy	E	—	262° (propanol)	37
$—CH_3$	6,7-Dimethoxy	—	—	249–259° (dec.)	105
$—CH_3$	$5-NH_2$	J	—	288° (dec.) (ethanol)	110a
—C6H5	$5-NO_2$	A	—	303–305°	110
—C6H5	$6-OCH_3$	F	—	244–245° (acetone)	112
—C6H5	$7-OCH_3$	F	88	218–219°	112
—C6H5	$8-CO_2H$	G	—	257–259°	77

$—C_6H_5$	8-CO_2Et	H	—	198–202°	77
$—C_6H_5$	6,7-Dimethoxy	F_1	76	Needles, 257–260° (dec.) (ethanol)	113
$—C_6H_5$	5,6,7-Triphenyl	F_2	—	>290° (acetic acid)	114
$—C_6H_4$-o-CH_3	6,7-Dimethoxy	F_1	—	234–235 (acetic acid)	113
$—C_6H_4$-p-Br	5,6,7,8-Tetrachloro	G_2	—	Colorless plates, 305° (chlorobenzene)	91
$—CH_2—C_6H_5$	6-Cl	—	—	163°	106
$—CH_2—C_6H_5$	6,7-Dimethoxy	F_1	100	Needles, 182–184° (methanol)	113
$—CH_2—C_5H_4N$ (4-pyridyl)	7-NH_2	I	—	298°, HCl > 300°	104

TABLE 2D-5 (*continued*)

[a] A, properly substituted 2-acylbenzoic acid was treated with hydrazine hydrate in nitrobenzene. B, same as A but refluxed in water for 10 hr. C, 3-hydroxy-3-methyl-7-aminophthalide refluxed with hydrazine hydrate in water for 1 hr. C_1, catalytic reduction of 8-nitro-4-methyl-1-(2*H*)phthalazinone with Raney nickel in ethyl acetate. C_2, 2-acetyl-6-aminobenzoic acid condensed with hydrazine hydrate under Huang–Minlon conditions. D, same as C, using 3-hydroxy-3-methyl-7-hydroxyphthalide. E, *N*,*N*-dialkyl-2-acetyl-3,5-dimethoxybenzamide refluxed 2 hr with hydrazine. F, same as A but warmed with hydrazine hydrate in ethanol. F_1, same as A but refluxed with hydrazine hydrate in ethanol for 1–2 hr. F_2, same as A but refluxed with hydrazine hydrate in methanol for 3 hr. G, same as A but refluxed with hydrazine hydrate in acetic acid for 18 hr. G_1, same as A but refluxed with hydrazine hydrate in acetic acid for 2 hr. H, esterification of 4-phenyl-8-carboxy-1-(2*H*)phthalazinone by treatment with thionyl chloride followed by reflux in chlorobenzene with ethanol for 2 hr. I, refluxed

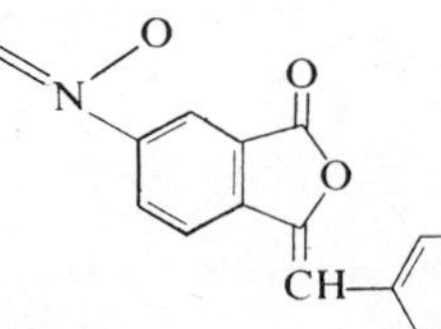

with excess hydrazine for 3 hr. J, reduction of the corresponding nitro compound by hydrazine in ethanol in presence of catalytic amount of Pd-C.

TABLE 2D-6. 2,4-Disubstituted-1-(2*H*)phthalazinones

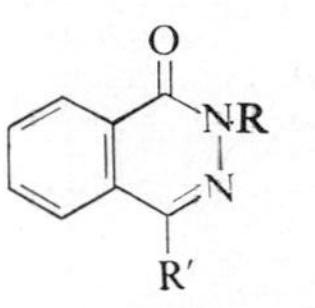

R	R′	Preparation[a]	Yield (%)	MP	Reference
$-CH_3$	$-CH_3$	A, B, C, C_1	80	110–111°, 112°	28, 39, 114a
$-CH_3$	$-CH_2CH_3$	A, B, C_1	44	84–95° 77–78°	28, 114a
$-CH_3$	$-CH_2(CH_2)_2CH_3$ with CH_2CH_3 branch	A or B	—	BP 150°/0.3 mm	28
$-CH_3$	$-C_6H_5$	C	—	—	39
$-CH_3$	$-C_6H_4Cl$ (4-Cl)	D	56–87	147–148° (pet. ether); 152–154° (CCl_4 + $CHCl_3$)	55, 115
$-CH_3$	$-C_6H_3Cl_2$ (3,4-Cl)	D	76	171–172° (ethyl acetate)	115

TABLE 2D-6 (*continued*)

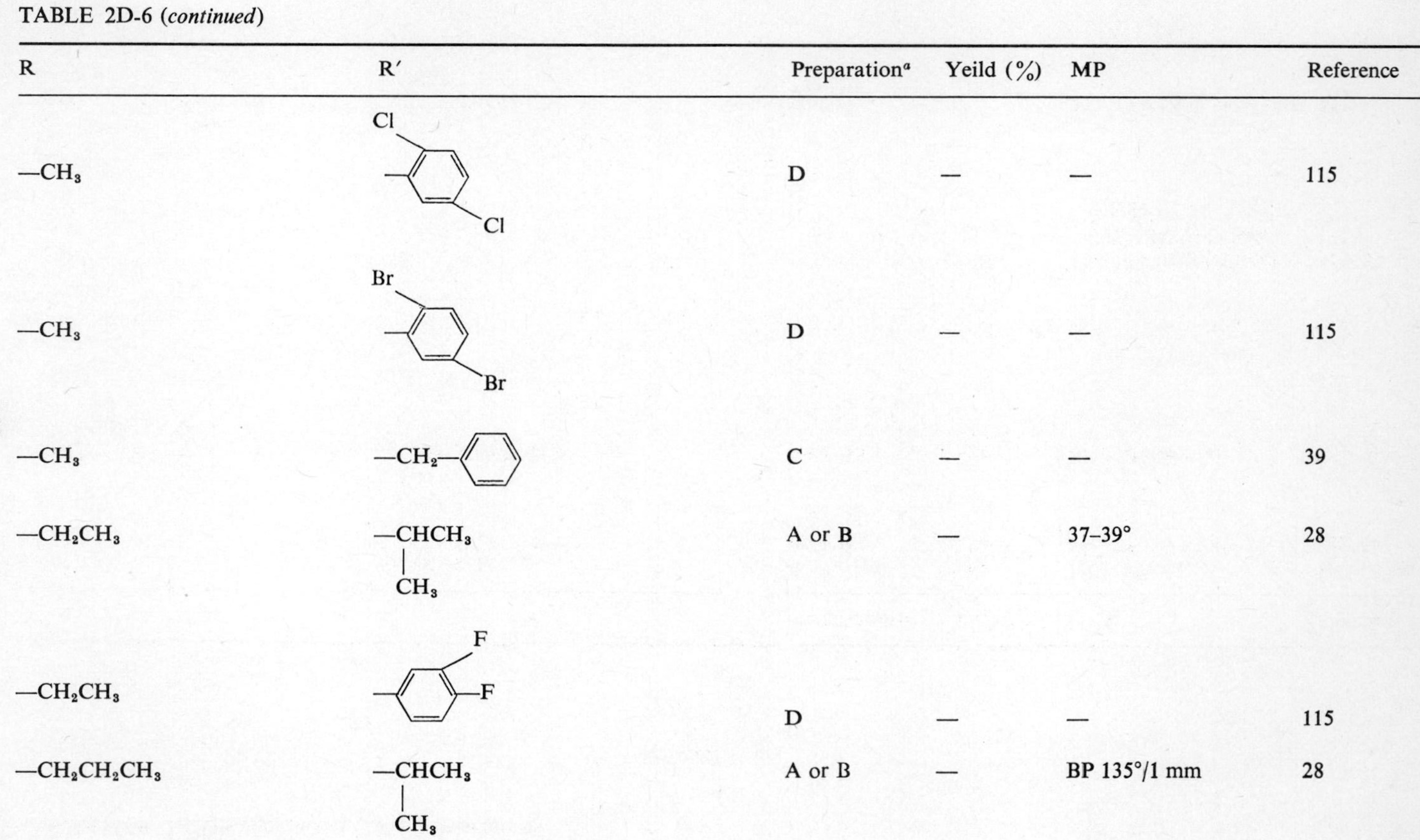

R	R′	Preparation[a]	Yeild (%)	MP	Reference
—CH_3	(ring with Cl, Cl)	D	—	—	115
—CH_3	(ring with Br, Br)	D	—	—	115
—CH_3	—CH_2—(ring)	C	—	—	39
—CH_2CH_3	—$CHCH_3$ (with CH_3 below)	A or B	—	37–39°	28
—CH_2CH_3	(ring with F, F)	D	—	—	115
—$CH_2CH_2CH_3$	—$CHCH_3$ (with CH_3 below)	A or B	—	BP 135°/1 mm	28

$-CH(CH_3)CH_3$	$-CH_3$	A or B	—	84–85°	28
$-CH(CH_3)CH_3$	$-CH_2CH_3$	A or B	—	55–56°	28
$-CH(CH_3)CH_3$	$-CH(CH_3)CH_2CH_3$	A, B	82–84	41–43° (pet. ether); sublimes at 70°/0.1 mm	28, 51
$-CH(CH_3)CH_3$	$-CH(CH_2CH_3)(CH_2)_2CH_3$	A or B	—	BP 123°/0.2 mm	28
$-CH(CH_3)CH_3$	2,5-diiodophenyl	D	—	—	115
-*n*-butyl	$-CH(CH_3)CH_3$	A or B	—	BP 135°/0.1 mm	28
-2-butyl	4-chlorophenyl	D	—	—	115
-*t*-butyl	2,3-dichlorophenyl	D	—	—	115

TABLE 2D-6 (*continued*)

R	R′	Preparation[a]	Yield (%)	MP	Reference
-*n*-pentyl	$-CH(CH_3)CH_3$	A or B	—	BP 125°/0.1 mm	28
-*n*-hexyl	$-CH(CH_3)CH_3$	A or B	—	BP 145°/0.2 mm	28
—Br	$-CH(Br)NO_2$	E	37	208° (dec.) (acetic acid)	25
—Br	$-CH_2NO_2$	F	—	168–170° (gl. acetic acid)	25
$-CH_2OH$	$-CH_3$	G	78	Colorless needles, 177° (dec.) (benzene)	29
$-CH_2OH$	$-C_6H_5$	G	80	Colorless needles, 182° (dec.) (benzene)	29
$-CH_2Cl$	$-CH_3$	H	86	Colorless needles, 173° (benzene–pet. ether)	29
$-CH_2Cl$	$-C_6H_5$	H	93	Colorless needles, 127° (benzene–pet. ether)	29
$-CH_2N(CH_3)_2$	$-CH_3$	I	72	172° (benzene)	29

$-CH_2N(CH_3)_2$	$-C_6H_5$	I	75	168° (ethanol)	29
$-CH_2-NC_5H_{10}$	$-CH_3$	I	83	106° (ethanol)	29
$-CH_2-NC_5H_{10}$	$-C_6H_5$	I	87	168° (ethanol)	29
$-CH_2-NC_4H_8O$	$-CH_3$	I	73	114° (pet. ether)	29
$-CH_2-NC_4H_8O$	$-C_6H_5$	I	80	177° (ethanol)	29
$-CH_2N(CH_3)-C_6H_5$	$-CH_3$	I	61	104° (ethanol)	29
$-CH_2N(CH_3)-C_6H_5$	$-C_6H_5$	I	68	140° (ethanol)	29
$-CH_2N(CH_2CH_3)-C_6H_5$	$-C_6H_5$	I	66	112° (ethanol)	29

TABLE 2D-6. (*continued*)

R	R′	Preparation[a]	Yield (%)	MP	Reference
$-CH_2N(CH_2CH_3)-CH_2-$ (N-2 of 5-methyl-1(2H)-phthalazinone; ring C=O at C-1, CH_3 at C-5)	$-CH_3$	I_1	61	102° (benzene–pet. ether)	29
$-CH_2N(CH_2CH_3)-CH_2-$ (N-2 of 5-methyl-1(2H)-phthalazinone; ring C=O at C-1, CH_3 at C-5)	$-C_6H_5$	I_1	65	166° (ethanol)	29
$-CH_2SP(=S)(OCH_3)_2$	$-CH_3$	J	93	113–115.5°	116
$-CH_2SP(=S)(OCH_2CH_3)_2$	$-CH_3$	J	—	79–82°	116
$-C(=O)NH_2$	$-CH_3$	K	—	224°	79
$-C(=O)CH_3$	$-CH(CH_3)CH_2CH_3$	L	—	BP 165–167°/0.5 mm	28

$-CCH_3$ (C=O)	$-CH_2N(CO)_2C_6H_4$	L_1	—	207°	50
$-C(=O)-C_5H_4N$ (4-pyridyl)	$-C_6H_4-OCH_3$ (p)	M	—	238°	108
$-CH_2CH_2OH$	$-C_6H_5$	N	—	158–160° (ethanol)	94
$-CH_2CH_2OH$	$-C_6H_4-Cl$ (p)	N	—	152.5–153.5° (ethanol)	94
$-CH_2CH_2Cl$	$-C_6H_5$	O_1	71	Colorless prisms, 125–126° (ethanol)	94
$-CH_2CH_2N(CH_3)_2$	$-C_6H_5$	B	—	101–103.5° (pet. ether)	94

TABLE 2D-6. (*continued*)

R	R′	Preparation[a]	Yield (%)	MP	Reference
$-CH_2CH_2N(CH_3)_2$	$-CH_2-C_6H_5$	B_1	91	107–108° (aq. acetone), BP 208–210°/0.1 mm; HCl, 175°, 178°	27, 106, 117
$-CH_2CH_2N(CH_3)_2$	$-CH_2-C_6H_4-Br$	—	—	HCl, 238°	27
$-CH_2CH_2N(CH_3)_2$	$-CH_2-C_6H_4-Cl$	B_1	—	BP 208–214°/0.1 mm; HCl, 142–143°	27, 106, 117
$-CH_2CH_2N(CH_3)_2$	$-CH_2-C_6H_4-OCH_3$	B_2	68	Bp 220–223°/0.2 mm; HCl, 202–203°, 248°	27, 106
$-CH_2CH_2N(CH_3)_2$	$-CH_2-C_6H_4-OCH_2CH_3$	—	—	HCl, 190°	27

$-CH_2CH_3N(CH_3)_2$	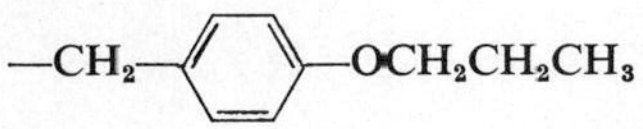 $-CH_2-C_6H_4-OCH_2CH_2CH_3$	B_2	—	HCl, 142°	106
$-CH_2CH_2N(CH_3)_2$	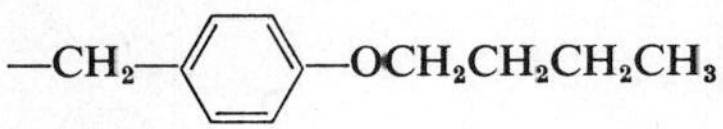 $-CH_2-C_6H_4-OCH_2CH_2CH_2CH_3$	—	—	HCl, 126°	27
$-CH_2C(=O)N(CH_2CH_3)_2$	$-CH_3$	B_3	55	160° (benzene)	118
$-CH_2C(=O)N(CH_2CH_3)_2$	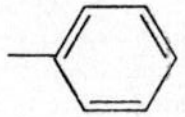 $-C_6H_5$	B_2	93	146–147° (aq. alcohol)	118
$-CH_2C(=O)N(CH_2CH_3)_2$	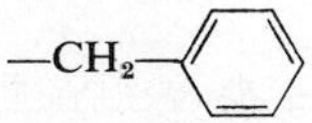 $-CH_2-C_6H_5$	B_2	84	158° (ethanol)	118

TABLE 2D-6. (*continued*)

R	R′	Preparation[a]	Yield (%)	MP	Reference
$-CH_2CH_2N(CH_2CH_3)_2$	$-CH_3$	—	—	CH_3I, 186°	27
$-CH_2CH_2N(CH_2CH_3)_2$	$-C_6H_5$	B_1	—	BP 225–230°/0.5 mm; HCl, 198°	106, 117
$-CH_2CH_2N(CH_2CH_2)_2O$ (morpholino)	$-CH_2-C_6H_4-OCHCH_3$	B_2	—	HCl, 100–101°	106
$CH_2CH_2CH_2OH$	$-C_6H_5$	P	—	138–139°	119
$-CH_2CH_2CH_2OH$	$-C_6H_4-Cl$	P	91	102–105° (methanol-H_2O); 104–106°	35, 119
$-CH_2CH_2CH_2OH$	$-C_6H_4-OCH_3$	P	—	117–118° (toluene-pentane)	119
$-CH_2CH_2CH_2Cl$	$-C_6H_5$	P_1	—	127–128° (CH_2Cl_2-ether)	119

$-CH_2CH_2CH_2Cl$	$-C_6H_4-Cl$	P_1	97	108–109° (CH_2Cl_2-ether); 112–113° (ether)	35, 119
$-CH_2CH_2CH_2Cl$	$-C_6H_4-OCH_3$	P_1	—	95–96.5° ($CHCl_3$-pentane)	119
$-CH_2CH_2CH_2N(CH_3)_2$	$-CH_2-C_6H_5$	B_1	—	BP 218°/0.5 mm; HCl, 173–174°	27, 117
$-CH_2CH_2CH_2-N(CH_2CH_2)_2N-CH_3$	$-C_6H_5$	B	—	117–119° (pet. ether)	94
$-CH_2CH_2CN$	$-CH_3$	Q	68	120° (pet. ether)	29
$-CH_2CH_2CN$	$-C_6H_5$	Q	73	142° (ethanol)	29
$-CH_2CH_2CO_2H$	$-CH_3$	Q_1	81	Colorless needles, 131° (benzene)	29
$-CHH_2C_2CO_2H$	$-C_6H_5$	Q_1	85	Colorless needles, 188° (benzene)	29
$-CH_2CH_2CH_2CH_2OH$	$-C_6H_4-Cl$	P	77	116–118° ($CHCl_3$-pentane)	35

TABLE 2D-6. (*continued*)

R	R	Preparation[a]	Yield (%)	MP	Reference
—$CH_2CH_2CH_2CH_2$—Cl	—C_6H_4—Cl (p-chlorophenyl)	P_1	97	148–151° (CH_2Cl_2-ether)	35
—(Tetra-*o*-acetyl-β-D-glucopyranosyl)	—C_6H_5 (phenyl)	R	20	155–158° (methanol)	48
(1-β-D-glucosyl)	C_6H_5 (phenyl)	R_1	90	148–151° (*n*-propanol)	48
—C_6H_5 (phenyl)	—CH_3	A	81	98–99°	81
—C_6H_5 (phenyl)	—$CH(CH_3)CH_3$	A	—	131–132°	28
—C_6H_5 (phenyl)	—CH_2OH	S	95	Cream colored needles, 171° (ethanol)	36
—C_6H_5 (phenyl)	2-thienyl	T	100	Pale yellow prisms, 190° (alcohol)	91
—C_6H_5 (phenyl)	5-methyl-2-thienyl (CH_3)	T	—	Pale yellow needles, 156° (toluene)	91
—C_6H_5 (phenyl)	5-chloro-2-thienyl (Cl)	T	100	Small yellow prisms, 156° (alcohol)	91

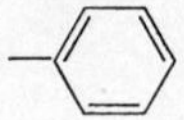 —	 S —Br	T	—	Brilliant yellow needles, 172° (alcohol)	91
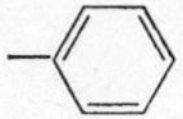 —	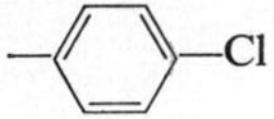 —Cl	T	—	Prisms, 169° (alcohol)	91
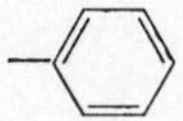 —	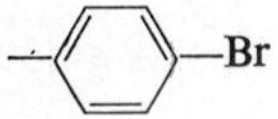 —Br	T	—	Needles, 152° (alcohol)	91
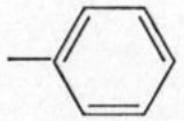 —	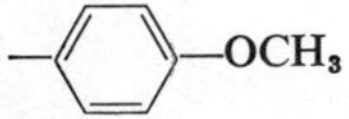 —OCH_3	U	81	127–128°	108
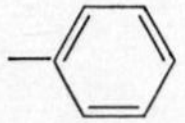 —	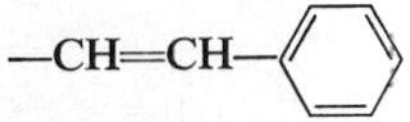 —CH=CH—	V	54	178–179° (acetic acid)	55, 120, 121
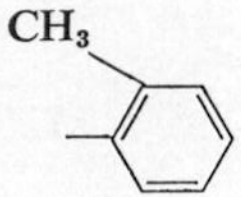 CH_3 —	—CH_3	V	85	103–104°	122
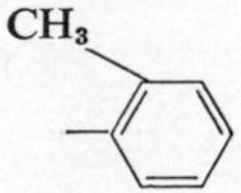 CH_3 —	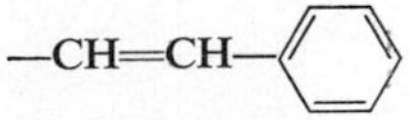 —CH=CH—	V	52-66	181–182° (2-propanol)	55, 121
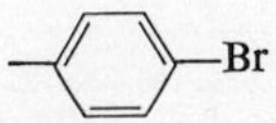 —Br	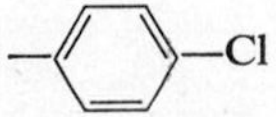 —Cl	T (acetic acid)	—	Needles, 218°	91

TABLE 2D-6. (*continued*)

R	R′	Preparation[a]	Yield (%)	MP	Reference
4-bromophenyl	4-methoxyphenyl ($—C_6H_4—OCH_3$)	U	80	179–180°	108
4-nitrophenyl	$—CH_3$	A	71	214–215° (ethanol-dioxane)	81
4-nitrophenyl	$—CH(CH_3)CH_3$	A	—	134–135°	28
4-nitrophenyl	4-methoxyphenyl	U	91	220°	108
2,4-dinitrophenyl	4-methoxyphenyl	M_1	43	95–105° (softening at 80°)	108
$—CH_2—C_6H_5$	phenyl	B	—	186–187° (DMF-EtOH)	94
$C_6H_5CH_2—$	$CH_3—$	C_1	80	113.5–114.5°	114a
2-pyridyl-$CH_2CH_2—$	$CH_3—$	C_1	92	140–141°	114a
$CH_3—$	$—CH_2CO_2H$	—	96	195°	114a

CH_3—	—$CH_2CO_2CH_3$	—	81	109–110°	114a
CH_3—	—CH_2CONH_2	—	90	248–249°	114a
CH_3—	—$CH_2CONHNH_2$	—	90	233.5°	114a
CH_3CH_2—	—CH_2CO_2H	—	99	178°	114a
CH_3CH_2—	—CH_2CO_2Et	—	72	71–72°	114a
CH_3CH_2—	—CH_2CONH_2	—	98	251–252°	114a
CH_3CH_2—	—$CH_2CONHNH_2$	—	83	200–201°	114a
C_6H_5—CH_2—	—CH_2CO_2H	—	99	147–148°	114a
C_6H_5—CH_2—	—CH_2CO_2Et	—	45	126–127°	114a
C_6H_5—CH_2—	—CH_2CONH_2	—	72	244.5–245.5°	114a
C_6H_5—CH_2—	—$CH_2CONHNH_2$	—	91	213.5–214.5°	114a
2-pyridyl—CH_2CH_2—	—CH_2CO_2H	—	83	177–178°	114a
2-pyridyl—CH_2CH_2—	—CH_2CO_2Et	—	82	108–109°	114a

TABLE 2D-6. (*continued*)

R	R′	Preparation[a]	Yield (%)	MP	Reference
2-pyridyl-CH_2CH_2—	—CH_2CONH_2	—	80	219°	114a
2-pyridyl-CH_2CH_2—	—$CH_2CONHNH_2$	—	90	184–185°	114a

[a] A, 2-acylbenzoic acid condensed with the substituted hydrazine by keeping at room temperature overnight then heating 110–120° for 5–10 min. B, 4-substituted-1-(2*H*)phthalazinone reacted with alkyl halide by heating at 120° in ethanol with sodium. B_1, same as B but reacted in aqueous sodium hydroxide by heating at 60° for 1 hr. B_2, same as B but refluxed in toluene with sodamide for 2 hr. B_3, same as B but boiled in xylene for 1–2 hr in presence of potassium. C, 4-substituted-2-methylphthalazinium iodide oxidized with $K_3Fe(CN)_6$ in the presence of potassium hydroxide, and alkylation with dimethyl sulfate of the corresponding 4-substituted-1-(2*H*)phthalazinone. C_1, the corresponding 2-substituted 1-(2*H*)phthalazinone-4-ylacetic acids decarboxylated by heating at 200° in nitrobenzene. D, same as A, reflux in benzene for 2–3 hr. E, bromination of 4-nitromethyl-1-(2*H*)phthalazinone in acetic acid. F, hydrolysis of 2-bromo-4-(1′-bromo-1′-nitromethyl)-1-(2*H*)phthalazinone by keeping in water at room temperature for 24 hr. G, the corresponding 1-(2*H*)phthalazinone condensed with aqueous formaldehyde in methanol. H, the corresponding 2-hydroxymethyl-1-(2*H*)phthalazinone refluxed with phosphorus oxychloride for 2 hr. I, Mannich reaction, the corresponding 1-(2*H*)phthalazinone treated with aqueous formaldehyde and proper secondary amine in methanol and heated on water bath. I_1, same reaction as I but using ethylamine. J, 2-chloromethyl-4-methyl-1-(2*H*)-phthalazinone refluxed with proper $Na^{\oplus}S^{\ominus}—P(=S)—(OR)_2$ in acetone for 3 hr. K, 2-acetylbenzoic acid condensed with $H_2NNH—C(=O)—NH_2$ in pyridine. L, reflux the corresponding 1-(2*H*)phthalazinone in acetic anhydride for 2 hr. L_1, same as L, reflux time 10 hr. M, 2-(*p*-methoxybenzoyl)benzoic acid fused with isonicotinic acid hydrazide at 150–170° for 10 min. M_1, similar to M but fused with 2,4-dinitrophenylhydrazine at 150–160°. N, same as A but heated for 30 min. O, 2-(2-hydroxyethyl)-1-(2*H*)phthalazinone refluxed in benzene with thionyl chloride for 2 hr. P, same as A but reflux in toluene using Dean-Stark trap. P_1, the corresponding 2-(ω-hydroxyalkyl)-1-(2*H*)phthalazinone was refluxed with thionyl chloride in chloroform for 20 hr. Q, the corresponding 1-(2*H*)phthalazinone refluxed with acrylonitrile in pyridine-water for 3 hr. Q_1, hydrolysis of the corresponding compound prepared by Q by refluxing with aqueous alkali for 1 hr. R, 4-phenyl-1-(2*H*)phthalazinone reacted with α-acetobromoglucose in dimethylformamide in presence of KOH at room temperature for 1 hr. R_1, the product obtained by R treated with sodium methoxide in methanol. S, phenylhydrazone of isochromanedione treated with alkali followed by acidification. T, similar to A but reflux in acetic acid for 2 hr. U, similar to A but reflux in ethanol + aqueous acetic acid. V, similar to A but heat 15 min in ethanol-acetic acid.

TABLE 2D-7. 2,4-Disubstituted-1-(2*H*)phthalazinones with Substituents in the Benzenoid Ring

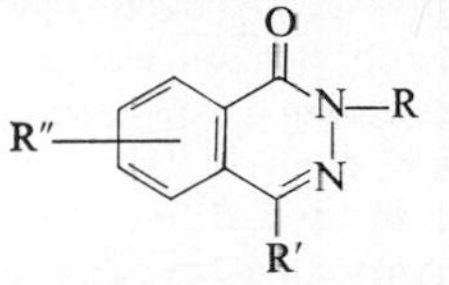

R	R′	R″	Preparation[a]	Yield (%)	MP	Reference
$-CH_3$	$-CH(CH_3)_2$	5,8-Dimethyl	A	—	60–61°	28
$-CH_2CH_3$	$-CH(CH_3)_2$	5,8-Dimethyl	A	—	54–55°	28
$-CH_2CH_2N(CH_3)_2$	$-CH_2-C_6H_5$	6-Cl	B	48	BP 230–240°/0.5 mm; HCl, 203°–204°	106
$-C_6H_5$	$-CH_3$	8-NO_2	C	60	146–148° (methanol)	53
$-C_6H_5$	$-CH_3$	8-NH_2	D	71	129–131° (ethanol)	53
$-C_6H_5$	$-CH_3$	8-OH	D	77	158–160° (ethanol)	53
$-C_6H_5$	$-CH_3$	5,7-Dimethoxy	E	—	184° (ethanol)	37

TABLE 2D-7. (*continued*)

R	R′	R″	Preparation[a]	Yield (%)	MP	Reference
$-C_6H_5$	$-C_6H_5$	6,7-Dimethoxy	C_1	—	198–198.5°	122
$-CH_2-C_6H_5$	$-CH_3$	8-NO_2	C_2	71	163–165° (benzene)	53
$-CH_2-C_6H_5$	$-CH_3$	8-NH_2	D	66	124–125° (ethanol)	53
$-CH_2-C_6H_5$	$-CH_3$	8-OH	D_1	68	170–172° (benzene)	53
CH_3-	CH_3-	5-NO_2	A_1	—	Crystals, 125–127° (methanol)	110a
CH_3-	CH_3-	5-NH_2	A_2	—	198–200° (ethanol)	110a
CH_3-	CH_3-	8-NH_2	A_2	—	154–156° (ethanol)	110a

[a] A, the corresponding 2-alkylbenzoic acid condensed with substituted hydrazine by keeping at room temperature overnight then heating 110–120° for 5–10 min or the corresponding 1-(2*H*)phthalazinone alkylated with the alkyl halide by heating at 120° in ethanol containing sodium. A_1, 2-acetyl-3-nitrobenzoic acid condensed with methylhydrazine in ethanol. A_2, reduction of the corresponding nitro derivative by hydrazine in ethanol in presence of catalytic amount of Pd—C. B, by alkylation of the corresponding 1-(2*H*)phthalazinone with 2-*N*,*N*-dimethylaminoethyl chloride by heating in toluene with sodamide. C, the corresponding 2-acylbenzoic acid condensed with substituted hydrazine by refluxing in acetic acid for 2 hr. C_1, same as C, reflux in ethanol for 2 hr. C_2, same as C, reflux in toluene for 2 hr. D, 3-methyl-3-hydroxy-7-amino or 7-hydroxyphthalide condensed with substituted hydrazine by refluxing in acetic acid for 2 hr. E, *N*,*N*-dialkyl-2-acetyl-3,5-dimethoxybenzamide refluxed with phenylhydrazine for 2 hr.

TABLE 2D-8. 1-Alkoxy or 1-Aryloxyphthalazines

R	Preparation[a]	Yield (%)	MP	Reference
$—CH_3$	A, B	65–72	60–61°, 60–62°, Bp 130°/10 mm; picrate, 139–140°, 161–162°	49, 59, 60, 67, 67a
$—CH_2CH_3$	A, C	49–93	23°, 55°, Bp 130°/10 mm; picrate 141°, 149°	59, 60
$—CH(CH_3)_2$	A, C	69–73	BP 157°/8 mm; picrate, 137°, 146°	59, 60
$—CH_2CH_2CH_2CH_3$	A	89	BP 164°/8 mm; picrate, 136°	60
$—CH_2CH_2N(CH_3)_2$	A_1	—	BP 150°/0.75 mm; picrate, 150–152°	64
$—CH_2CH_2N(CH_2CH_3)_2$	A_1	—	BP 150–155°/1 mm; 2HCl, 148°; picrate, 140–141°	64
$—CH_2CH_2N(n\text{-}C_4H_9)_2$	A_1	—	BP 175°/0.4 mm; 2HCl, 107–108°; picrate, 125°	64
$—(CH_2)_3N(CH_2CH_3)_2$	A	10–21	Citrate $—H_2O$, 129.5–130.5°; $2CH_3I$, 168.5–169.5°	65
$—(CH_2)_2O(CH_2)_2N(CH_2CH_3)_2$	A	16	2HCl, 118–120°; $2CH_3I$, 163–165°	65
—C6H5 (phenyl)	A_2	90	106–108° (methanol), 107°	62, 63

[a] A, reflux 1-chlorophthalazine with the corresponding alcohol for 1–3 hr with equimolar sodium. A_1, same as A but reflux in benzene for 3 hr. A_2, 1-chlorophthalazine was heated with phenol at 100° for 1–2 hr in the presence of potassium carbonate. B, 1-methylsulfonylphthalazine was refluxed for 5 min with sodium methoxide in methanol. C, 1-methoxyphthalazine was refluxed with the proper alcohol in the presence of sodium for 15 min (transetherification).

TABLE 2D-9. Substituted 1-Alkoxy- or 1-Aryloxyphthalazines

R	R′	R″	Preparation[a]	Yield (%)	MP	Reference
$—CH_3$	$—CH_3$	H	A	—	70–72°, 70–73° (ligroin) picrate, 199°	28, 49
$—CH_3$	$—CH_2CH_3$	H	A	—	45–46°	28
$—CH_3$	$—CH(CH_3)_2$	H	A	—	45–47°; HCl, 103–105°	28
$—CH_3$	$—C_6H_5$	H	A, B	60, 85	137–138° (pet. ether), 139°	69, 124
$—CH_3$	$—CH(CH_3)_2$	5,8-Dimethyl	A	—	129–130°	28
$—CH_3$	—H	6-F	—	—	159–160°	11
$—CH_2CH_3$	$—CH_3$	H	A	—	43–52°	28
$—CH_2CH_3$	$—CH_2CH_3$	H	A	—	45–47°	28
$—CH_2CH_3$	$—CH(CH_3)_2$	H	A		78–82°	28
$—CH_2CH_3$	$—C_6H_5$	H	A_1	—	91–93° (ligroin)	30
$—CH_2CH_3$	$—CH(CH_3)_2$	5,8-Dimethyl	A	—	132–133°	28
$—CH_2CH_2CH_3$	$—CH_2CH_3$	H	A	—	BP 140°/0.3 mm	28
$—CH_2CH_2CH_3$	$—CH(CH_3)_2$	H	A	—	37–38°	28
$—CH_2CH_2CH_3$	$—C_6H_5$	H	A_1	—	71–73°	30

$—CH(CH_3)_2$	$—CH_3$	H	A	—	BP 120°/0.2 mm	28
$—CH(CH_3)_2$	$—CH_2CH_3$	H	A	—	BP 130°/0.2 mm	51
					BP 138–140°/0.3 mm 78–80° (pet. ether); HCl, 100–101° (dec.)	
$—CH(CH_3)_2$	$—CH(CH_3)_2$	H	A	—	105–106°	28
$—CH(CH_3)_2$	$—C_6H_5$	H	A_1	—	102–104° (ethanol-water)	30
$—CH(CH_3)_2$	$—CH_2—C_6H_5$	H	A_1	—	91–93° (ligroin)	30
$—CH(CH_3)_2$	$—CH(CH_3)_2$	5,8-Dimethyl	A	—	114–116°	28
-n-C_4H_9	$—CH(CH_3)_2$	H	A	—	40–42°	28
-n-C_5H_{11}	$—CH(CH_3)_2$	H	A	—	BP 135°/0.1 mm	28
$—C_5H_9$ (cyclopentyl)	$—C_6H_5$	H	C	—	120–121°	30
-n-C_6H_{13}	$—CH(CH_3)_2$	H	A	—	BP 155°/0.1 mm	28
$—C_6H_{11}$ (cyclohexyl)	$—C_6H_5$	H	C	—	111–113° (ethanol-water)	30
-n-C_6H_{13}	$—CH(CH_3)_2$	5,8-Dimethyl	A	—	95–96°	28

TABLE 2D-9. (*continued*)

R	R′	R″	Preparation[a]	Yield (%)	MP	Reference
$—CH_2CH_2N(CH_3)_2$	$—CH_2—C_6H_5$	H	D	84	BP 201–208°/0.5 mm; HCl, 173°, 178°; CH_3I, 186°	27, 66
$—CH_2CH_2N(CH_3)_2$	$—CH_2—C_6H_4—Cl$	H	E	—	BP 218–220°/0.3 mm; HCl, 239–240°	66
$—CH_2CH_2N(CH_3)_2$	$—CH_2—C_6H_4—OCH_3$	H	E	63	BP 219–226°/0.15 mm	66
$—CH_2CH_2N(CH_2CH_3)_2$	$—CH_3$	H	F	—	BP 170–173°/0.7 mm; 2HCl, 175–176°	64
$—CH_2CH_2N(CH_2CH_3)_2$	$—C_6H_5$	H	E	75	BP 229–233°/0.5 mm; HCl, 202°	66
$—CH_2CH_2N(CH_2CH_3)_2$	$—CH_2—C_6H_5$	H	E	83	BP 226–233°/0.5 mm; HCl, 131–132°, 132°; benzyl chloride, 169–170°	27, 66

—CH_2CH_2—N(piperidino)	—CH_2—C_6H_4—Cl (p)	H	D	—	HCl, 165–166°	66
—$CH_2CH_2CH_2N(CH_3)_2$	—CH_2—C_6H_5	H	A_2	70	BP 220–225°/0.4 mm; HCl, 170–171°, 171°; CH_3I, 211°	27, 66
—$CH_2CH_2CH_2N(CH_3)_2$	—CH_2—C_6H_4—OCH_3 (p)	H	A_2	—	BP 224–230°/0.6 mm	66
—$C(=O)CH_3$	—$CH(CH_3)_2$	H	G	90	77–79° (pet. ether)	51
—CH_2CH_2CN	—CH_3	H	H	—	White crystals, 129–130° (methanol)	56
—$CH_2CH_2CO_2H$	—CH_3	H	I	—	145–146° (water)	56
—$CH_2CH_2CONHNH_2$	—CH_3	H	J	—	White crystals, 107–108° (water)	56
—CH_2CH_2—$C(=NOH)$—NH_2	—CH_3	H	K	—	White crystals, 145° (benzene)	56
-(Tetraacetyl-1-β-D-glucosyl)	—C_6H_5	H	L	14	Needles, 189–190° (1-propanol)	48

TABLE 2D-9. (*continued*)

R	R′	R″	Preparation[a]	Yield (%)	MP	Reference
-(1-β-D-glucosyl)	$-C_6H_5$	H	M	70	Needles, 145–155° (dec.) (methanol)	48
$-C_6H_5$	$—CH_3$	H	N	82	Needles, 137–139° (ethanol-water)	62
$CH_3CH_2—$	$-C_6H_4-Cl$ (p)	H	A_1	—	156–157° (ethyl acetate-ligroin)	30
$(CH_3)_2CH—$	$-C_6H_4-Cl$ (p)	H	A_1	—	122–124° (ligroin)	30
$—O(CH_2CH_2O)_3CH_3$	$CH_3—$	H	—	—	—	125
$—O(CH_2CH_2O)_3CH_3$	O_2N-(2,5-furylene)-CH=CH—	H	—	—	—	125

[a] A, the corresponding 1-chlorophthalazine refluxed in the proper alcohol with equimolar sodium for 20 min. A_1, same as A, reflux 2–3 hr. A_2, same as A, heat to 150°. B, 4-phenyl-1-methylsulfonylphthalazine refluxed with sodium methoxide in methanol for 15 min. C, same as A, heat at 70–80° for 3 hr in dimethylformamide with sodium hydride. D, same as A, heat in xylene at 140° with sodamide. E, same as A, reflux few hours in toluene with sodium. F, same as A, reflux 3 hr in benzene with sodium. G, reflux 4-isopropyl-1-(2*H*)phthalazinone with acetic anhydride for 6 hr or heat 1-chloro-4-isopropylphthalazine with sodium acetate at 200° for 10 min. H, heat 4-methyl-1-(2*H*)phthalazinone and acrylonitrile in dioxane with TRITON B for 3 hr. I, alkaline hydrolysis of the product prepared by method H. J, the corresponding ethyl ester was treated with hydrazine. K, 1-(β-cyanoethoxy)-4-methylphthalazine was stirred at room temperature with hydroxylamine hydrochloride in ethanol with sodium ethoxide. L, silver salt of 1-hydroxy-4-phenylphthalazine was refluxed in toluene for 5 min with α-acetobromoglucose. M, the product prepared by method L was treated with sodium methoxide in methanol. N, 1-chloro-4-methylphthalazine was heated with phenol in presence of potassium carbonate at 100° for 1–2 hr.

References

1. W. Lorenz, French Pat., 1,335,759 (1963); *Chem Abstr.*, **60,** 558 (1964).
2. K. Adachi, *J. Pharm. Soc. Japan*, **75,** 1423 (1955); *Chem. Abstr.*, **50,** 10,105 (1956).
3. V. B. Brasyunas and A. S. Podzhyunas, *Med. Prom. SSSR*, **12,** 47 (1958); *Chem. Abstr.*, **53,** 13160 (1959).
4. T. P. Sycheva et al., *Med. Prom. SSSR.*, **14,** 13 (1960); *Chem. Abstr.*, **54,** 22,669 (1960).
5. S. Biniecki and B. Gutkowska; *Acta Polon. Pharm.*, **11,** 27 (1955); *Chem. Abstr.*, **50,** 12,062 (1956).
6. S. Biniecki and L. Rylski, *Ann. Pharm. Fr.*, **16,** 21 (1958); *Chem. Abstr.*, **52,** 18446 (1958).
7. G. Favini and M. Simoneta, *Gazz. Chim. Ital.*, **90,** 369 (1960); *Chem. Abstr.*, **55,** 11426 (1961).
8. C. M. Atkinson, C. W. Brown, and J. C. E. Simpson, *J. Chem. Soc.*, **1956,** 1081.
9. V. Kolesnikov and E. Bisagni, *Chim. Ther.*, **2,** 250 (1967); *Chem. Abstr.*, **69,** 19105 (1968).
10. S. Biniecki and M. Moll, *Acta Polon. Pharm.*, **25,** 105 (1968); *Chem. Abstr.*, **69,** 77198 (1968); Polish Pat. 60,625 (1970); *Chem. Abstr.*, **74,** 76435 (1971); A. W. Sogn, German Pat., 1,958,805 (1970); *Chem. Abstr.*, **73,** 66598 (1970).
11. B. Cavalleri and E. Bellasio, *Farmaco, Ed. Sci.*, **24,** 833 (1969); *Chem. Abstr.*, **71,** 101,798 (1969).
12. S. S. Berg and E. W. Parnel!, *J. Chem. Soc.*, **1961,** 5275.
13. S. Biniecki, M. Moll, and L. Rylski, *Ann. Pharm. Franc.*, **16,** 421 (1958); *Chem. Abstr.*, **53,** 7187 (1959).
14. Neth. Pat. Appl., 6,604,484 (1966); *Chem. Abstr.*, **66,** 65,519 (1967). Neth. Pat. Appl., 6,604,482 (1966); *Chem. Abstr.*, **66,** 65,517 (1967).
15. A. Dunet and A. Willemart, *Bull. Soc. Chim. France*, **1948,** 1081.
16. W. Triebs, H. M. Barchet, G. Bach, and W. Korchhof, *Ann. Chem.*, **574,** 54 (1951).

16a. P. Raoul, German Pat. 2,045,111 (1971); *Chem. Abstr.*, **74,** 141839 (1971).

17. French Pat. 1,438,827 (1966); *Chem. Abstr.*, **66,** 95069 (1967).
18. M. P. Cava and R. P. Stein, *J. Org. Chem.*, **31,** 1866 (1966).
19. H. J. Rodda and P. E. Rogasch, *Aust. J. Chem.*, **19,** 1291 (1966).
20. W. R. Vaughan, *Chem. Rev.*, **43,** 447 (1948).
21. C. Musante and L. Fabrini, *Il Farmico (Pavia) Ed. Sci.*, **8,** 264 (1953); *Chem. Abstr.*, **48,** 4536 (1954).
22. T. Ikeda, M. Ishikawa, S. Hirai, and M. Yoshioka, *Ann. Repts. Shionogi Res. Labs.*, **1955,** 27; *Chem. Abstr.*, **50,** 15,530 (1956).
23. R. Gompper, *Chem. Ber.*, **93,** 187 (1960).
24. K. Fujii and S. Sato, *Ann. Rept. G. Tanabe Co., Ltd.*, **1,** 3 (1956); *Chem. Abstr.*, **51,** 6650 (1957).
25. G. Vanags and M. A. Matskanova, *J. Gen. Chem. USSR*, **26,** 1963 (1956); *Chem. Abstr.*, **51,** 14746 (1957).
26. W. Zerweik, W. Kunze, and R. E. Nitz, German Pat. 1,005,078 (1957); *Chem. Abstr.*, **53,** 18,073 (1959); British Pat. 808,638 (1959); *Chem. Abstr.*, **53,** 14,129 (1959).
27. D. Lenke, *Arzneimittel-Forsch.*, **7,** 678 (1957).
28. Belgian Pat. 628,255 (1963); *Chem. Abstr.*, **60,** 14516 (1964).
29. A. Mustafa et al., *Tetrahedron*, **20,** 531 (1964).
30. H. M. Hollava and R. P. Partyka, *J. Med. Chem.*, **12,** 555 (1969).

31. S. Kanahara, *Yakugaku Zasshi*, **84,** 489 (1964); *Chem. Abstr.*, **61**, 8304 (1964).
32. W. Griehl and J. Hecht, *Chem. Ber.*, **91,** 1816 (1958), and references therein.
33. T. V. Gortinskaya, V. G. Samolovova, and M. N. Schukina, *Zh. Obshch. Khim.*, **27,** 1960 (1957); *Chem. Abstr.*, **52,** 5384 (1958).
34. R. C. Fuson and W. C. Hammann, *J. Am. Chem. Soc.*, **74,** 1626 (1952).
35. P. Aeberli and W. J. Houlihan, *J. Org. Chem.*, **34,** 2715 (1969).
36. E. B. Knott, *J. Chem. Soc.*, **1963,** 402.
37. C. Broquet and J. P. Genet, *C.R. Acad. Sci., Paris Ser. C.*, **265,** 117 (1967).
38. J. Druey, *Angew. Chem.*, **70,** 5 (1958).
39. T. Ikeda, S. Kanahara, and K. Aoki, *Yakuguku Zusshi*, **88,** 521 (1968); *Chem. Abstr.*, **69,** 86934 (1968).
40. A. Albert and J. N. Phillips, *J. Chem. Soc.*, **1956,** 1294.
41. Y. N. Sheinker and Y. I. Pomerantscu, *Zh. Fiz. Khim.*, **33,** 1819 (1959); *Chem. Abstr.*, **54,** 12156 (1960).
42. S. F. Mason, *J. Chem. Soc.*, **1957,** 4874.
43. K. Mori, *Yakugaku Zasshi*, **82,** 1161 (1962); *Chem. Abstr.*, **58,** 4555 (1963).
44. Y. N. Sheinker and Y. I. Pomerantscu, *Zh. Fiz. Khim.*, **30,** 79 (1956); *Chem. Abstr.*, **50,** 14780 (1956).
45. P. Sohar, *Acta Chim. Acad. Sci. Hung.*, **40,** 317 (1964); *Chem. Abstr.*, **62,** 14457 (1965).
46. A. Albert and G. B. Barlin, *J. Chem. Soc.*, **1962,** 3129.
47. P. G. Parsons and H. J. Rodda, *Aust. J. Chem.*, **17,** 491 (1964).
48. G. Wagner and D. Heller, *Arch. Pharm.*, **299,** 768 (1966).
49. S. Kanahara and Y. Yamamoto, *Kanazawa Daigaku Yakugakubu Kenkyu Nempo*, **12,** 1 (1962); *Chem. Abstr.*, **59,** 3917 (1963).
50. M. Makanova and G. Vanags, *Zh. Obshch. Khim.*, **28,** 2798 (1958); *Chem. Abstr.*, **53,** 9229 (1959).
51. A. Aebi and E. Hofstetter, *Pharm. Acta Helv.*, **40,** 241 (1965).
52. A. Hirsch and D. G. Orphanos, *J. Heterocyclic Chem.*, **3,** 38 (1966).
53. J. Finkelstein and J. A. Romano, *J. Med. Chem.*, **11,** 398 (1968).
54. S. Kanahara, *Yakugaku Zasshi*, **84,** 483 (1964); *Chem. Abstr.*, **61,** 8304 (1964).
55. K. Boltze and H. D. Dell, *Arch. Pharm.*, **299,** 702 (1966).
56. I. Zugravescu, M. Petrovanu, and E. Rucinschi, *Rev. Chim., Acad. Rep. Pop. Roum.*, **7,** 1405 (1962); *Chem. Abstr.*, **61,** 7009 (1965).
57. D. Cook, *Can. J. Chem.*, **42,** 2292 (1964).
58. J. H. Bowie et al., *Aust. J. Chem.*, **17,** 491 (1964).
58a. A. M. Zyakun et al., *Izv. Akad. Nauk SSSR, Ser. Khim.*, **1970,** 2208; *Chem. Abstr.*, **74,** 140376 (1971).
59. K. Adachi, *J. Pharm. Soc. Japan*, **75,** 1426 (1955); *Chem. Abstr.*, **50,** 10105 (1956).
60. E. Hayashi et al., *Yakugaku Zasshi*, **82,** 584 (1962); *Chem. Abstr.*, **58,** 3425 (1963).
61. N. B. Chapman and D. Q. Russell-Hill, *J. Chem. Soc.*, **1956,** 1563.
62. I. Satoda, N. Yoshida, and K. Mori, *Yakugaku Zasshi*, **77,** 703 (1957); *Chem. Abstr.*, **51,** 17,927 (1957).
63. M. Hartman and J. Druey, U.S. Pat. 2,484,029 (1949); *Chem. Abstr.*, **44,** 4046 (1950).
64. British Pat., 810,108 (1959); *Chem Abstr.*, **53,** 16172 (1959).
65. S. O. Wintrop, S. Sybulski, R. Gaudry, and G. A. Grant, *Can. J. Chem.*, **34,** 1557 (1956).
66. H. J. Engelbrecht, D. Lenke, and H. Miller, German (East) Pat. 17,319 (1959); *Chem. Abstr.*, **55,** 590 (1961); German Pat. 1,046,626 (1958); *Chem. Abstr.*, **55,** 7447 (1961).

67. G. B. Barlin and W. V. Brown, *J. Chem. Soc.*, **B, 1967,** 736.
67a. G. B. Barlin and W. V. Brown, *J. Chem. Soc. B*, **1968,** 1435.
68. E. Oishi, *Yakugaku Zasshi*, **89,** 859 (1969); *Chem. Abstr.*, **72,** 3450 (1970).
69. E. Hayashi et al., *Yakugaku Zasshi*, **87**, 687 (1967); *Chem. Abstr.*, **67**, 90,750 (1967).
70. A. R. Osborn, K. Schofield, and L. N. Short, *J. Chem. Soc.*, **1956,** 4191.
70a. A. N. Kost, S. Foldeak, and K. Grabliauskas, *Khim. Farm. Zh.*, **1,** 43 (1967); *Chem. Abstr.* **67,** 100090 (1967).
71. T. Ohta, *J. Pharm. Soc. Japan*, **63,** 239 (1943); *Chem. Abstr.*, **45** 5163 (1951).
72. H. Sakamoto, H. Tsuchiya, M. Nakagawa, and T. Mizutani, Japanese Pat. 21,254 (1964); *Chem. Abstr.*, **62,** 9148 (1965).
73. K. Ueda et al., Japanese Pat. 1791 (1966); *Chem. Abstr.*, **64,** 12741 (1966).
74. G. M. Badger, I. J. McCarthy, and H. J. Rodda, *Chem. Ind.* (*London*), **1954,** 964.
75. P. Lechat, *Compt. Rend.*, **246,** 2771 (1958).
76. L. A. Walker, U.S. Pat. 3,225,045 (1965); *Chem. Abstr.*, **64,** 12,690 (1966).
77. K. J. Doebel and J. E. Francis, S. African Pat. 67 05,858 (1968); from *Chem. Abstr.*, **70,** 106,567 (1969); S. African Pat. 67 05,856 (1968); *Chem. Abstr.*, **70,** 115,190 (1969); French Pat. 1,538,290 (1968); *Chem. Abstr.*, **71,** 61,407 (1969); French Pat. 1,550,404 (1968); *Chem. Abstr.*, **71,** 112,964 (1969).
78. T. Pozzo-Balbi, *Gazz. Chim. Ital.*, **81,** 125 (1951); *Chem. Abstr.*, **45,** 9022 (1951).
79. B. B. Elsmer, H. E. Strauss, and E. J. Forbes, *J. Chem. Soc.*, **1957,** 578.
80. G. Berti and E. Mancini, *Gazz. Chim. Ital.*, **88,** 714 (1958); *Chem. Abstr.*, **53,** 19965 (1958).
81. W. Reid and K. U. Bonnighausen, *Ann.*, **639,** 56 (1961).
82. R. Huisgen and E. Rauenbausch, *Ann.*, **641,** 51 (1961).
83. Y. S. Shabarov, N. I. Vasilev, N. K. Mamaeva, and R. Y. Levina, *Zh. Obshch. Khim.*, **33,** 1206 (1963); *Chem. Abstr.*, **59,** 11490 (1963).
84. S. Dixon and L. F. Wiggins, *J. Chem. Soc.*, **1954,** 594.
85. J. Honzl, *Chem. Listy*, **49,** 1671 (1955); *Chem. Abstr.*, **50,** 5622 (1956); *Coll. Czech. Chem. Comm.*, **21,** 725 (1956); *Chem. Abstr.*, **50,** 16,740 (1956).
86. M. Yasue, M. Itaya, and Y. Takari, *Bull. Nagoya City Univ., Pharm. School, No. 2.*, 53 (1954); *Chem. Abstr.*, **50,** 12063 (1956).
87. R. L. Lestinger and W. J. Vullo, *J. Org. Chem.*, **25,** 1844 (1960).
88. H. Mitsuhashi, U. Nagai, T. Muramatsu, and H. Tashiro, *Chem. Pharm. Bull.* (*Tokyo*), **8,** 243 (1960); *Chem. Abstr.*, **55,** 9333 (1961).
89. H. G. Rule, N. Campbell, A. G. McGregor, and A. A. Woodham, *J. Chem. Soc.*, **1950,** 1816.
90. K. Suzuki, E. K. Weisburger, and J. H. Weisburger, *J. Org. Chem.*, **26,** 2239 (1961).
91. N. G. Buu-Hoi, N. Hoan, and N. D. Xuong, *Rec. Trav. Chim.*, **69,** 1083 (1950).
92. W. Bradley and H. E. Nursten, *J. Chem. Soc.*, **1951,** 2177.
93. J. Druey and B. H. Riniger, *Helv. Chim. Acta*, **34,** 195 (1951).
94. J. M. Loynes, H. F. Ridley, and R. C. W. Spickett, *J. Med. Chem.*, **8,** 691 (1965).
95. W. Bradley and H. C. Nursten, *J. Chem. Soc.*, **1951,** 2170.
96. M. Lamchen, *J. Chem. Soc.*, **1962,** 4695.
97. E. D. Bergmann and S. Pinchos, *J. Org. Chem.*, **15,** 1023 (1950).
98. C. Marschalk and C. Stumm, *Bull. Soc. Chim. France*, **1948,** 418.
99. W. Logemann, F. Lauria, and E. Fachinelli, *Il. Farmaco* (*Pavia*) *Ed. Sci.*, **11,** 274 (1956); *Chem. Abstr.*, **50,** 13,940 (1956).
100. H. E. Schroeder, F. B. Stilmar, and F. S. Palmer, *J. Am. Chem. Soc.*, **78,** 446 (1956).
101. E. Ochiai and S. Suzuki, *Pharm. Bull.* (*Tokyo*), **5,** 405 (1957).

102. A. N. Kost, V. N. Eraksima, and E. V. Vinogradova, *Zh. Org. Khim.*, **1**, 129 (1965); *Chem. Abstr.*, **62**, 14611 (1965).
103. J. Druey and A. Marker, U.S. Pat. 2,960,504 (1960); *Chem. Abstr.*, **55**, 11,446 (1961).
104. J. Druey and A. Marker, German Pat. 1,061,788 (1959); *Chem. Abstr.*, **55**, 13,458. (1961); Swiss Pat. 361,812 (1962); *Chem. Abstr.*, **59**, 8761 (1963).
104a. F. G. Baddar, M. F. El-Newaihy, and M. R. Salem, *J. Chem. Soc. (C)*, **1971**, 716.
105. M. V. Sigal, Jr., P. Marchini, and B. L. Poet, U.S. Pat. 3,274,185 (1966); *Chem. Abstr.*, **65**, 18,600 (1966).
106. H. J. Engelbrecht, D. Lenke, and H. Miller, German (East) Pat. 17,075 (1959); *Chem. Abstr.*, **55**, 2700 (1961); German Pat. 1,046,625 (1958); *Chem. Abstr.*, **55**, 7446 (1961).
107. J. Majer and V. Rehak, *Sb. Ved. Praci, Vysoka Skola Chem. Technol. Pardubice*, **1**, 123 (1965); *Chem. Abstr.*, **65**, 2254 (1966).
108. C. Runti and S. Galimberti, *Ann. Chim. (Rome)*, **47**, 250 (1957); *Chem. Abstr.*, **51**, 12924 (1957).
109. N. P. Bednyagina and S. V. Sokolov, *Nauk. Dokl. Vyssh. Shk. Khim. Khim. Tekhnol.*, No. 2, 338 (1959); *Chem. Abstr.*, **53**, 21971 (1959).
110. J. Tirouflet, *Bull. Soc. Sci. Bretagne Spec.*, No. 26, 69 (1951); *Chem. Abstr.*, **47**, 8694 (1953).
110a. M. Ghelardoni and V. Pastellini, *Ann. Chim. (Rome)*, **60**, 775 (1970).
111. B. R. Baker et al., *J. Org. Chem.*, **17**, 164 (1952).
112. T. Ikeda and S. Kanahara, *Kanazawa Daigaku Yakugakubu Kenkya Nempo*, **7**, 6 (1957); *Chem. Abstr.*, **52**, 6361 (1958).
113. T. Ikeda, S. Kanahara, and T. Ujiie, *Kanazawa Daigaku Yakugakubu Kenkya Nempo*, **8**, 1 (1958); *Chem. Abstr.*, **53**, 4287 (1959); *Kanazawa Daigaku Yakugakubu Kenkya Nempo*, **9**, 6 (1959); *Chem. Abstr.*, **54**, 4607 (1960).
114. C. F. H. Allen and J. A. VanAllen, *J. Org. Chem.*, **20**, 328 (1955).
114a. S. Foldeak et al., *Khim. Farm. Zh.*, **4**, 22 (1970); *Chem. Abstr.*, **73**, 77173 (1970).
115. H. R. Sullivan, U.S. Pat. 3,222,365 (1965); *Chem. Abstr.*, **64**, 5109 (1966).
116. H. Sakamoto, H. Tsuchiya, M. Nakagawa, and T. Fujimoto, Japanese Pat. 5707 (1965); *Chem. Abstr.*, **63**, 1799 (1965).
117. French Pat. 1,512,879 (1968); *Chem. Abstr.*, **70**, 106,545 (1969); H. J. Engelbrecht, German (East) Pat. 55,955 (1967); *Chem. Abstr.*, **67**, 90,826 (1967).
118. H. J. Engelbrecht and K. Lenke, German (East) Pat. 19,629 (1960); *Chem. Abstr.*, **55**, 22,346 (1961).
119. W. J. Houlihan, French Pat. 1,530,074 (1968); *Chem. Abstr.*, **71**, 61,434 (1969).
120. G. A. Hanson and P. Tarte, *Bull. Soc. Chim. Belges*, **66**, 619 (1957).
121. K. H. Boltze et al., *Arzneimittel-Forsch.*, **13**, 688 (1963).
122. W. Gensler, E. M. Healy, I. Onshuus, and A. L. Bluhm, *J. Am. Chem. Soc.*, **78**, 1713 (1956).
123. S. Foldeak et al., *Khim. Farm. Zh.*, **3**, 5 (1969); *Chem. Abstr.*, **72**, 100626 (1970).
124. M. A. Shah and G. A. Taylor, *J. Chem. Soc. (C)*, **1970**, 1651.
125. K. Eichenberger, P. Schmidt, and M. Wilhelm, German Pat. 2,002,024 (1970); *Chem. Abstr.*, **73**, 98979 (1970).

Part E. 4-Hydroxy-1-(2*H*)phthalazinone

4-Hydroxy-1-(2*H*)phthalazinone has been referred in the literature under various names including phthalichydrazide, phthalhydrazide, phthalaz-1,4-dione, 2,3-dihydrophthalazine-1,4-dione, and 1,4-dihydroxyphthalazine.

Elvidge and Redman (1) have commented on the nomenclature and structural representation of this compound. Due to its tautomeric nature, it has been structurally represented as 1,4-dihydroxy-, 1,4-diketo-, and 4-hydroxy-1-keto forms. Recent studies have clarified this situation and this compound can be best represented as 4-hydroxy-1-(2*H*)phthalazinone, the name used throughout the present work.

I. 4-Hydroxy-1-(2*H*)phthalazinone and Its Derivatives

A. Methods of Preparation

1. 4-*Hydroxy*-1-(2*H*)*phthalazinone from Phthalic Anhydride and Similar Starting Materials*

Phthalic anhydride condenses smoothly with an equimolar amount of hydrazine hydrate by refluxing in ethanol (2) for 1 hr or by heating in acetic acid (3) to give 4-hydroxy-1-(2*H*)phthalazinone (**1**) in 85 and 52% yields, respectively. Similarly, diethyl phthalate reacts with hydrazine hydrate by heating at 100° for 30 min (3a) to give **1** (Eq. 1).

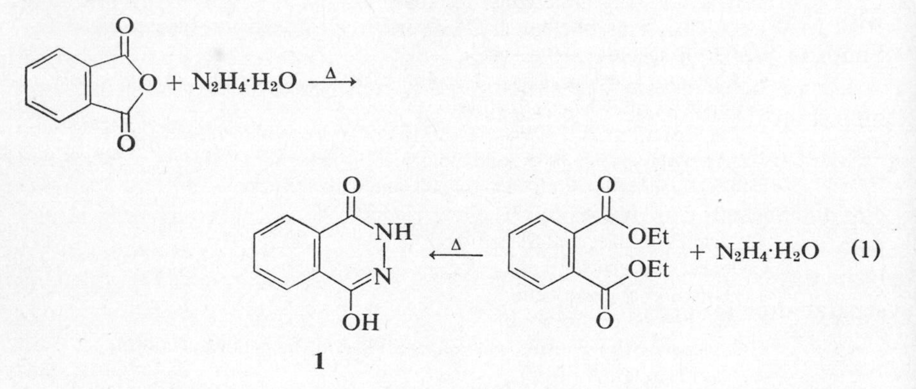

A new approach to the synthesis of **1** has been reported where phthalic acid is condensed with 3,3-dimethyl-1,2-diazacyclopropane in aqueous solution to give **1** in 97% yield (4) (Eq. 2).

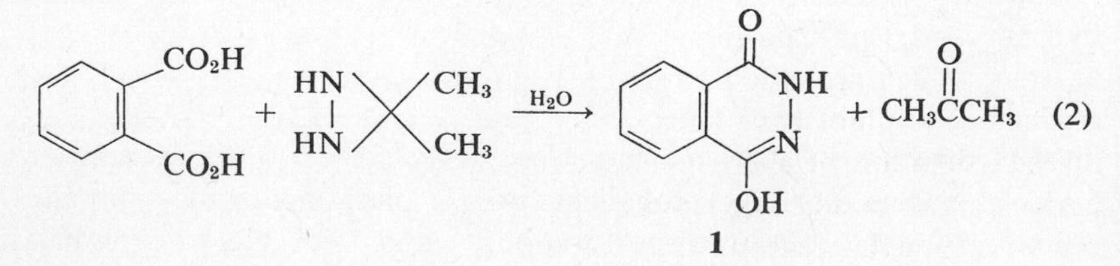

Addition of excess hydrazine hydrate to ethyl 2-chloroindan-1,3-dione-2-carboxylate immediately furnishes **1** in 62% yield (5). Phthalimide derivatives, such as 2-aminophthalimide (**2**) and 2-benzoylphthalaimide (**3**) are transformed into **1** in high yields. Heating **2** with 2*N* sodium hydroxide or 2*N* hydrochloric acid rearranges it to **1** in quantitative yield (6) or heating **3** in ethanol with hydrazine hydrate for 4 hr furnishes **1** in 80% yield (7) (Eq. 3).

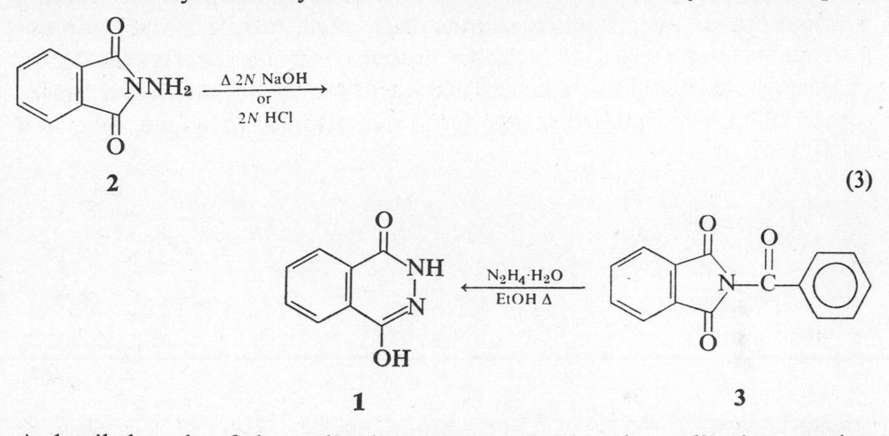

A detailed study of the earlier literature concerning the cyclization reaction with hydrazine, involving different intermediates, has been summarized by Simpson (8).

The phthaloyl group has been used frequently as a protective group for amines in the synthesis of α-amino acids and in the synthesis of peptides. The final step for the removal of the phthaloyl group has been accomplished by the reaction with hydrazine under very mild conditions. Boiling with an ethanolic solution of hydrazine hydrate for 2 hr or standing at room temperature for 1–2 days in aqueous or alcoholic solution of hydrazine hydrolyzes the phthaloyl group as 4-hydroxy-1-(2*H*)phthalazinone (**1**), which is separated on acidification (9).

2. 5-*Amino*-4-*hydroxy*-1-(2*H*)*phthalazinone* (*Luminol*) *and Its Analogs*

The most exhaustively studied compounds in this group are 5-amino-4-hydroxy-1-(2*H*)phthalazinone (**4**), commonly known as luminol, and its analogs, which are studied for their chemiluminescence properties. Several analogs of luminol have been synthesized to study either the enhancement or the inhibition of chemiluminescence properties. Alkaline oxidation of **4** under certain conditions is accompanied by a characteristic blue luminescence and for this reason, this compound is commonly known by the trivial name luminol. The mechanism of chemiluminescence and a discussion of the phenomenon is not included in present work* but all the compounds synthesized for this purpose are recorded in the proper table.

An excellent method for the preparation of **4** has been reported in organic synthesis by Redemann and Redemann (10). The disodium salt of 3-nitrophthalic acid has been condensed with hydrazine sulfate and the resulting 5-nitro-4-hydroxy-1-(2*H*)phthalazinone has been reduced with sodium hydrosulfite to give **4** (Eq. 4). A similar procedure involves catalytic reduction of 3-nitrophthalic acid to 3-aminophthalic acid with Raney nickel and further condensation with hydrazine, without isolating the intermediate, to give **4** (11).

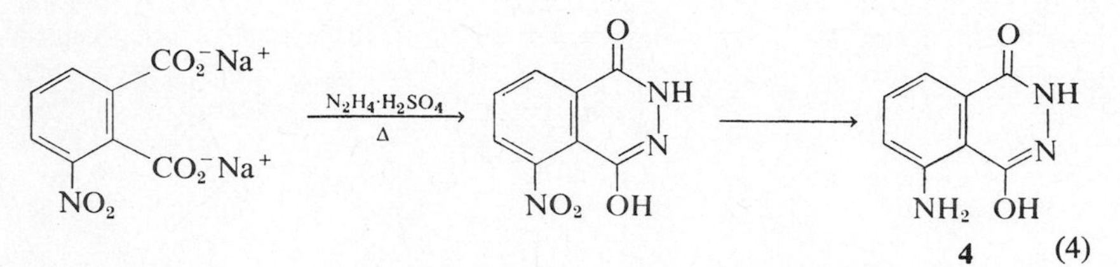

The analogs of **4** having alkoxy groups in the aromatic ring such as 8-methoxy- (12), 7,8-dimethoxy- (13), 6,7,8-trimethoxy- (13), and 6,7-methylenedioxy-5-amino-4-hydroxy-1-(2*H*)phthalazinone (14), have been synthesized in high yields by the condensation of the corresponding phthalimide (12), phthalic anhydride (13), and dimethyl phthalate (14) by refluxing with hydrazine hydrate in alcoholic or aqueous solutions. Similarly, 5-dimethylamino (12), 6-dimethylamino (12), 6-diethylamino (12, 15), and 6-cycloalkylamino-4-hydroxy-1-(2*H*)phthalazinone have been synthesized from the corresponding phthalimide (12) and dialkyl phthalate (15).

3. 4-*Hydroxy*-1-(2*H*)*phthalazinone with Groups Other than Amino in the Benzenoid Ring*

The compounds belonging to this group are synthesized mainly from the corresponding phthalic anhydride or dialkyl phthalate in the usual manner. 4-Phenylphthalic acid smoothly condensed with hydrazine hydrate by refluxing in 50% acetic acid to give 6-phenyl-4-hydroxy-1-(2*H*)phthalazinone in 72% yield (16). However, 3-phenylphthalic acid and 3-phenylphthalic anhydride failed to give the corresponding derivative of **4** (16). 4-Hydroxy (17), 4-alkoxy (17), 4-alkylthio (18), and dimethyl 4-chlorophthalate (17)

* See "ChemiLuminescence of Organic Compounds," *Monograph on Organic Chemistry*, Vol. 11, K. D. Gundermann, Springerverlag, Berlin, 1968, p. 63.

reacted with hydrazine hydrate in methanol to yield the desired products (Eq. 5).

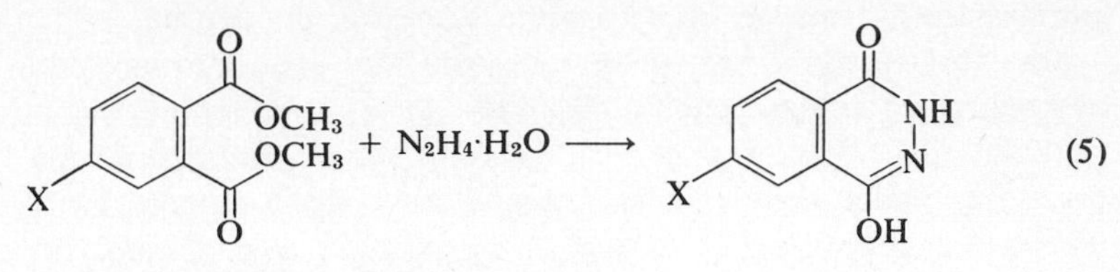

X = Cl, OH, alkoxy, alkylthio

6-Methyl (19), 6-trifluoromethyl (20), 5- and 6-fluoro (21), and 3,4,5,6-tetrachlorophthalic anhydride (22) have been condensed with hydrazine hydrate in acetic acid, by refluxing for about 2 hr, to yield the desired products (Eq. 6).

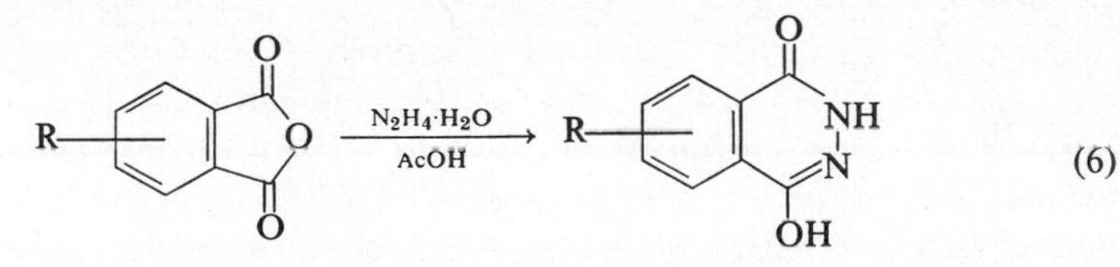

R = 6-CH_3, 6-CF_3, 5- or 6-F, and 3,4,5,6-tetrachloro

In general it can be said that the substituted phthalic anhydrides and phthalimides lead to the desired derivatives of **1**. In some cases the dialkyl phthalates do not give the expected cyclization for unknown reasons; for example, dimethyl 3-amino-4,5-methylenedioxyphthalate on refluxing in ethanol with hydrazine hydrate gave the expected product (14) but dimethyl 4,5-ethylenedioxyphthalate failed to give 6,7-ethylenedioxy-4-hydroxy-1-(2*H*)phthalazinone (**5**). When refluxed in 2-propanol with hydrazine hydrate instead the 4,5-ethylenedioxyphthalic dihydrazide was isolated (23).

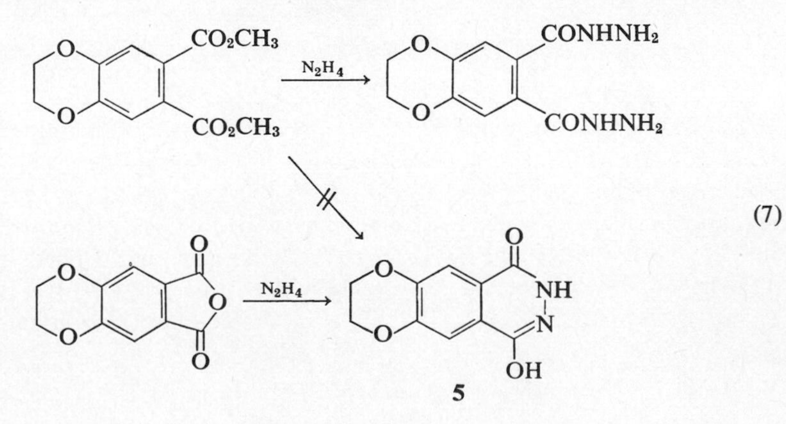

However **5** was obtained by reacting 4,5-ethylenedioxyphthalic anhydride in 2-propanol with hydrazine hydrate (23) (Eq. 7). Diethyl 5-chloro-6-sulfonamidophthalate, on the other hand, smoothly condensed with hydrazine hydrate in benzene to give 6-chloro-7-sulfonamido-4-hydroxy-1-(2*H*)-phthalazinone in high yields (95%) (24).

4. *4-Hydroxy-1-(2H)phthalazinones with Substituents at the 2- or 4-Position*

There are very few derivatives known having substituents only on the oxygen atom at the 4-position of 4-hydroxy-1-(2*H*)phthalazinone.

a. CONDENSATION OF PHTHALIC ANHYDRIDE WITH ALKYL- OR ARYLHYDRAZINE. Methylhydrazine condenses with phthalic anhydride in acetic acid (19, 25) or in the absence of a solvent (1) to give 2-methyl-4-hydroxy-1-(2*H*)phthalazinone (**6**) in high yields. Methylation of 4-hydroxy-1-(2*H*)-phthalazinone (**1**) with dimethyl sulfate in aqueous base furnished **6** (Eq. 8). The alkylation of **1** with diazomethane (26) and alkylation of the silver salt of **1** with methyl iodide gave 4-methoxy-1-(2*H*)phthalazinone (1).

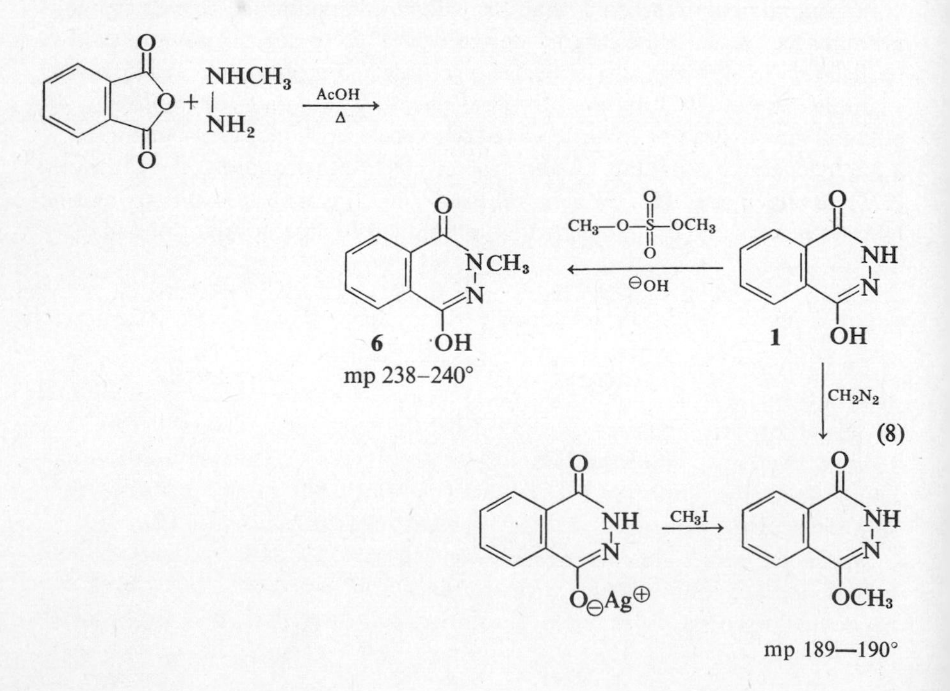

Phenylhydrazine, on condensation with phthalic anhydride by refluxing in ethanol, gave a small amount of the desired product 2-phenyl-4-hydroxy-1-(2*H*)phthalazinone (11%) (**7**), but the main product isolated was 2-anilinophthalimide (79%) (**8**) (27, 28) (Eq. 9). The treatment of **8** with sodium ethoxide in ethanol (27) or heating **8** with phenylhydrazine (28) gave excellent yields of the desired product **7**. Similarly, 2-(2′-aminopyridyl)phthalimide (**8a**) has been transformed into 2-(2′-pyridyl)-4-hydroxy-1-(2*H*)phthalazinone in 86% yield on treatment with sodium ethoxide (29).

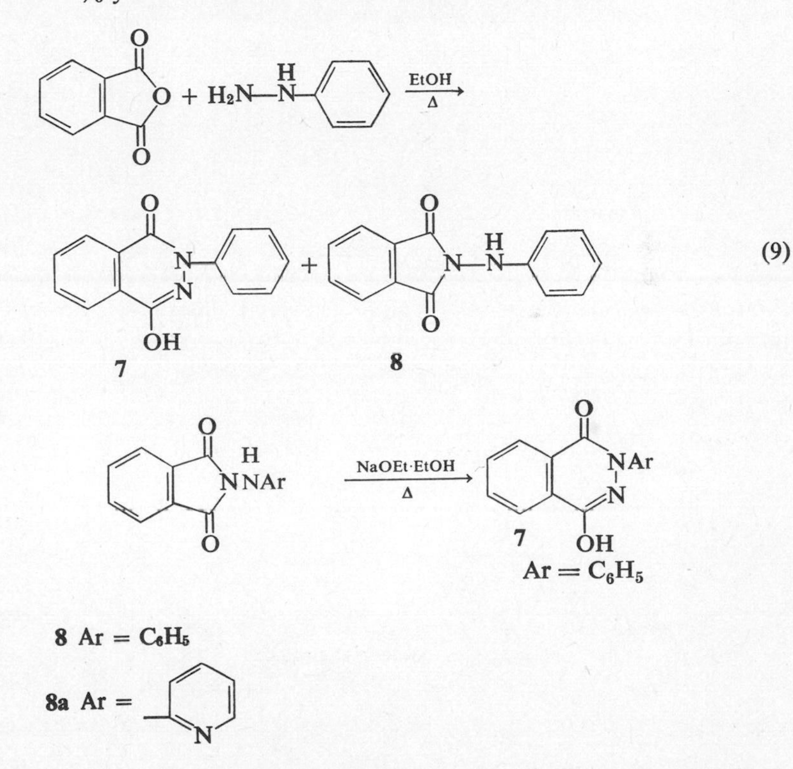

This type of isomerization has been observed with concentrated sulfuric acid in the case of 2-(3′-nitroanilino) and 2-(3′,4′-dinitroanilino)phthalimide to give the corresponding 2-aryl-4-hydroxy-1-(2*H*)phthalazinone (30, 31). It is of interest to note that similar treatment of 2-(2′-nitroanilino)phthalimide gave the sulfonated product 4-(*N*-Phthalyl)amino-3-nitrobenzenesulfonic acid (**9**) instead of the desired rearranged product (31) (Eq. 10).

Arylhydrazides, such as benzoic acid hydrazide and isonicotinic acid hydrazide, on condensation with phthalic anhydride or 2-carbomethoxybenzoylchloride in dioxane or pyridine, exclusively led to 2-aroylaminophthalimide (6) (Eq. 11). The treatment of 2-benzoylphthalimide with

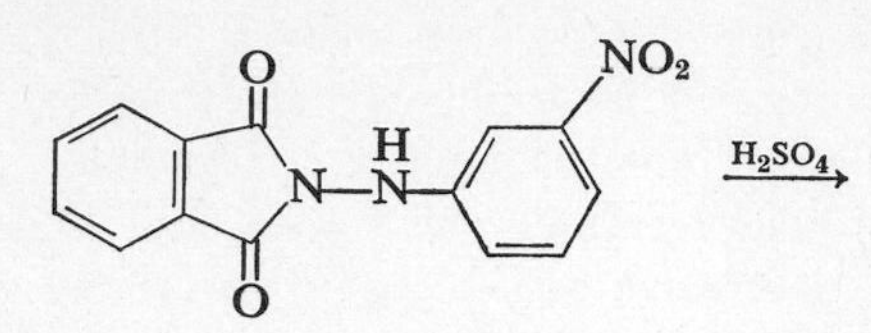

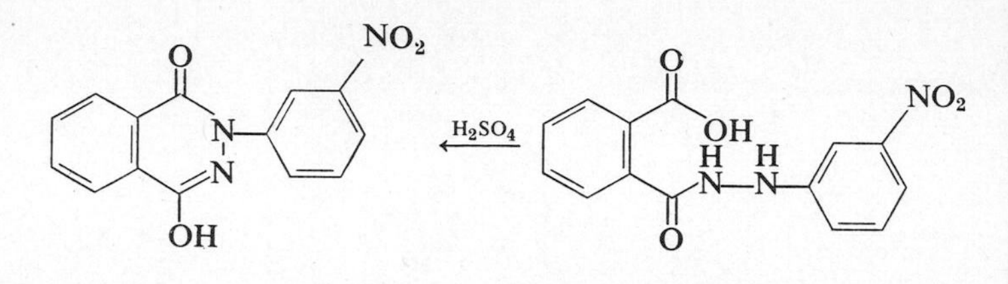

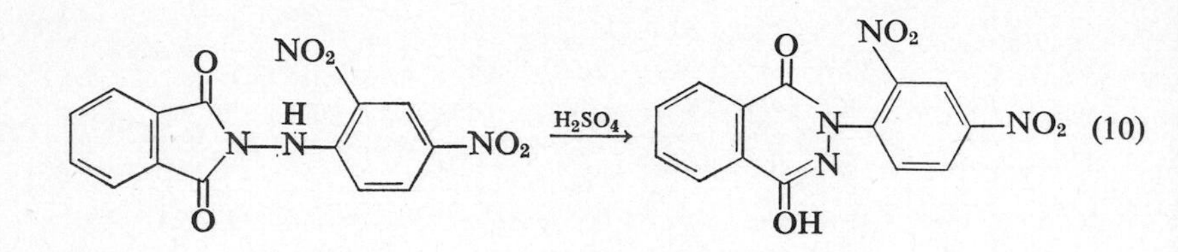

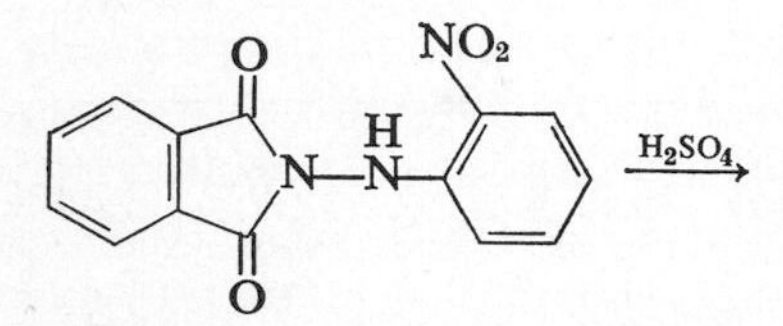

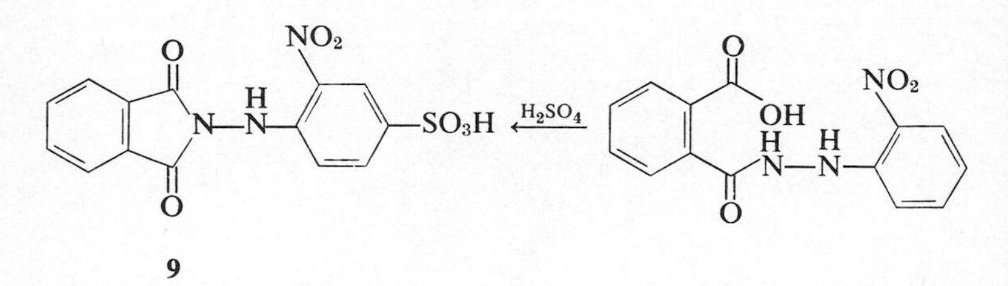

phenylhydrazine in ethanol gave 2-phenyl-4-hydroxy-1-(2*H*)phthalazinone (7) in very poor yield (5%) (7); the main product isolated was 2-anilinophthalimide (50%).

These results of the condensation of phthalic anhydride with methylhydrazine, arylhydrazine, and aroylhydrazine indicate that the basicity of the substituted nitrogen of hydrazine plays an important role in the formation of 2-substituted-4-hydroxy-1-(2*H*)phthalazinones.

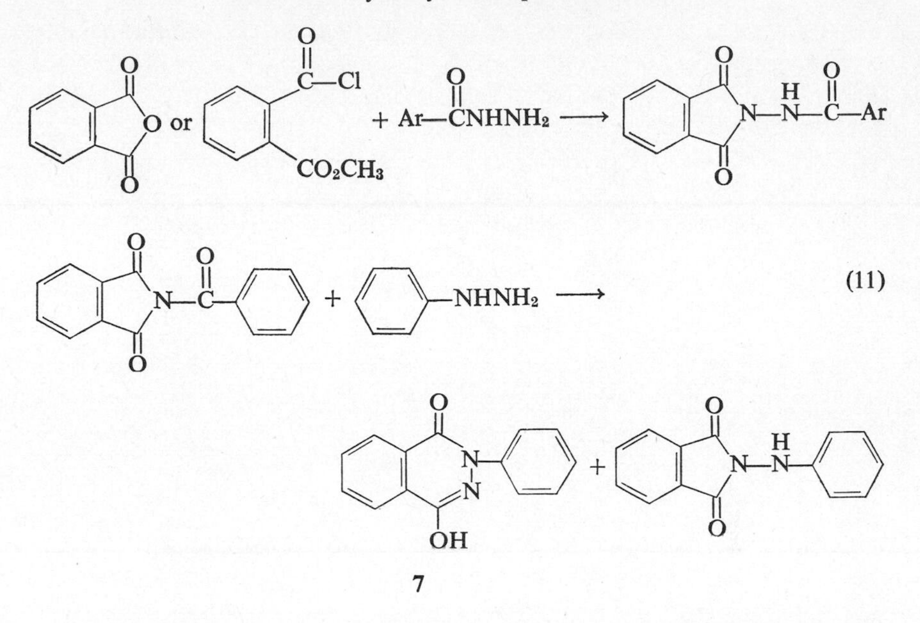

b. Alkylation or Acylation of 4-Hydroxy-1-(2*H*)phthalazinone. Alkylation of 4-hydroxy-1-(2*H*)phthalazinone with an equimolar amount of primary halide in the presence of a base leads to the alkylation at nitrogen giving 2-alkyl-4-hydroxy-1-(2*H*)phthalazinones (**10**), whereas excess of the halide affords 2-alkyl-4-alkoxy-1-(2*H*)phthalazinone (**11**) (32, 33). Buu-Hoi and co-workers (33) have shown that **11** could be easily dealkylated to give **10** simply by heating with pyridine hydrochloride (Eq. 12).

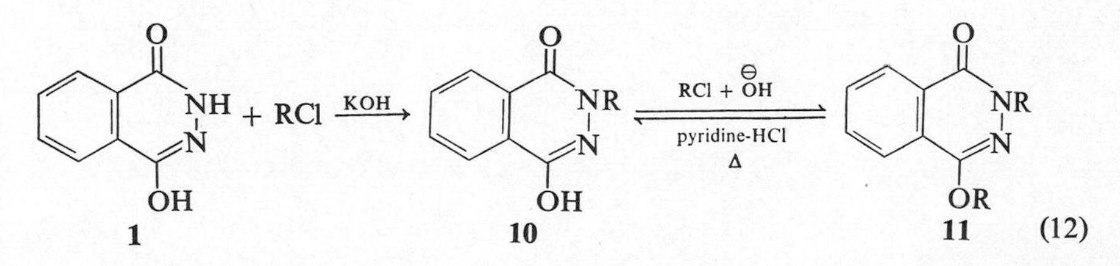

Similarly, reactive aromatic halides, such as 4-nitrochlorobenzene and 2,4-dinitrochlorobenzene have been reacted with alkali salts of **1** and shown to give products similar to **10** and **11** (R = 4-nitrophenyl, or 2,4-dinitrophenyl) (34, 35).

Rosseels and co-workers (36) have studied the effect of pH of the reaction medium on alkylation of **1** with chloroacetic acid. Reaction of **1** with equimolar sodium chloroacetate on refluxing in aqueous alcoholic solution at pH $\leqslant$ 8 gave a 2-*N*-alkylated product (**12**), whereas reaction with ethyl α-chloroacetate at pH > 8 afforded a 4-*O*-alkylated product (**13**) (Eq. 13). The reaction of **1** with 2 moles of sodium chloroacetate at pH > 8 gave the expected 2,4-disubstituted product **14**.

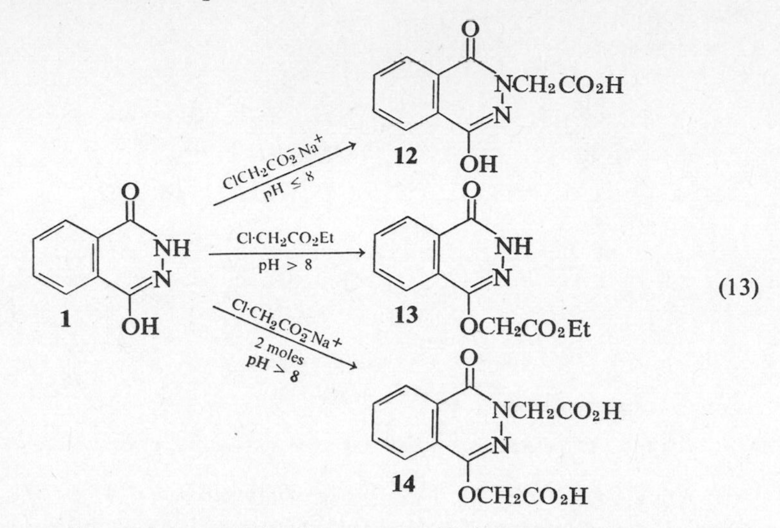

(13)

Le Berre and co-workers (37, 38) have studied the acylation of **1** with α,β-unsaturated acid chlorides and have shown that heating in nitrobenzene at 160° for 3 min furnishes the 4-*O*-acylated products (38), whereas heating at 160° for a longer time in nitrobenzene leads to a 2-*N*-acylated product (37). Galoyan and co-workers (22) have studied the reaction of **1** with alkyl and arylsulfonyl chloride by refluxing in 10% sodium hydroxide. The products have been characterized as 1-(2*H*)phthalazinone-4-alkyl or 4-arylsulfonates (**15**) (Eq. 14).

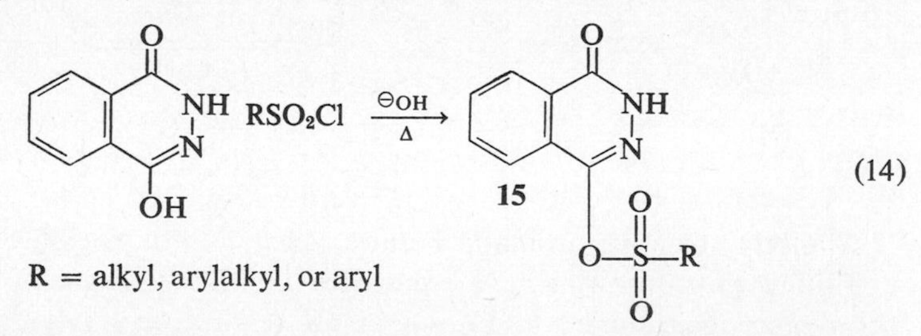

(14)

Le Berre and co-workers (39, 40) prepared the 2-*N*-acetyl, 4-*O*-acetyl, and 2,4-*N*,*O*-diacetyl derivatives of **1** and clarified the confusion about the acetylated products of **1** which existed in the earlier literature. The structures of these products have been assigned using pmr spectroscopy (see Section 2E-I-B-2-b). Acetylation of **1** with acetyl chloride in pyridine gave 2-acetyl-4-hydroxy-1-(2*H*)phthalazinone (**16**) and acetylation of the potassium salt of **1** with acetyl chloride in chloroform yielded 4-acetoxy-1-(2*H*)phthalazinone (**17**) (Eq. 15). Refluxing **1** with excess acetic anhydride furnished 2-acetyl-4-acetoxy-1-(2*H*)phthalazinone (**18**) in 90% yield. Further, it has been shown that **18** can be selectively hydrolyzed to give **17** in high yields by refluxing in ethanol.

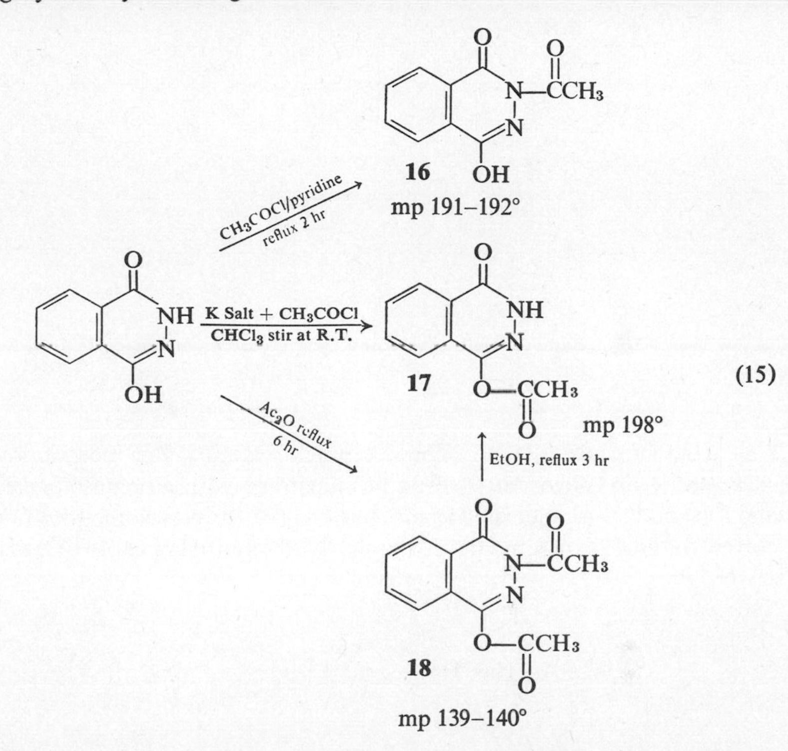

(15)

c. Michael-Type Addition of 4-Hydroxy-1-(2*H*)phthalazinone to Activated Double Bonds. 4-Hydroxyl-1-(2*H*)phthalazinone has been shown to add to activated double bonds for example of 2- or 4 -vinylpyridine (41, 42), methyl vinyl ketone (43), acrolein and sodium acrylate (37, 38, 44) by refluxing in ethanol or acetic acid to give the corresponding 2-substituted 4-hydroxy-1-(2*H*)phthalazinones (Eq. 16).

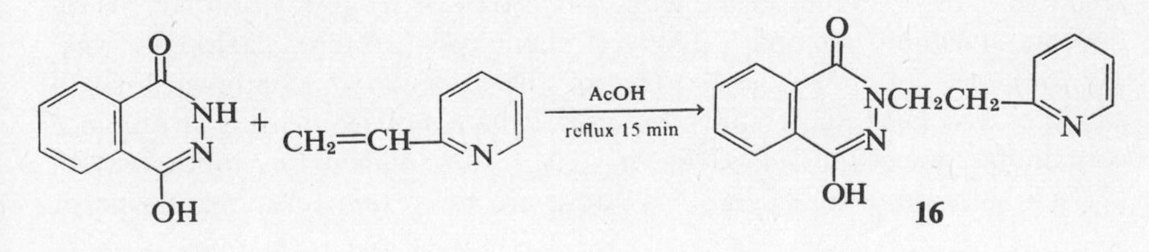

Addition of acrolein to **1** has been carried out by heating the reactants together in absence of any solvent (44) to give 2-(2′-propionaldehyde)-4-hydroxy-1-(2*H*)phthalazinone (**19**). The product **19** has been shown to exist in equilibrium with the cyclic form **19a** (Eq. 17).

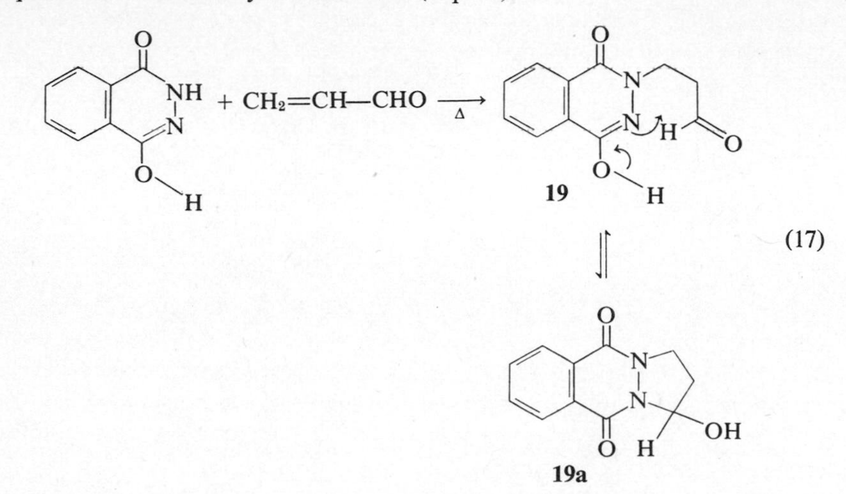

d. Mannich Reaction. The Mannich reaction has been carried out successfully on 4-hydroxy-1-(2*H*)phthalazinone with aqueous formaldehyde and a secondary amine, such as morpholine (45) or piperidine (46) to give the corresponding 2-substituted 4-hydroxyl-1-(2*H*)phthalazinones (Eq. 18).

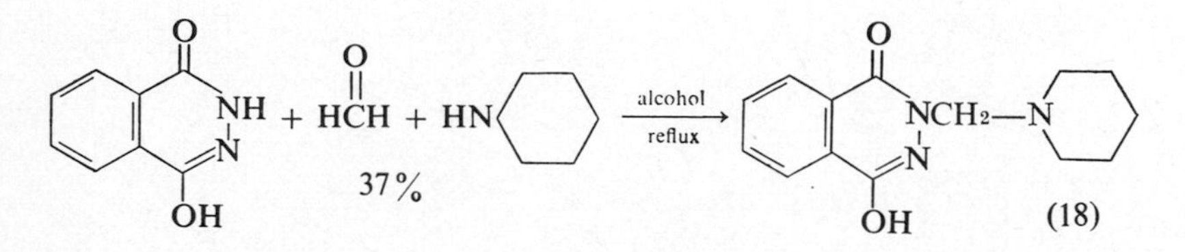

5. 4-*Hydroxy*-1-(2*H*)*phthalazinones with Substituents at the* 2- *and* 4-*Positions*

a. Alkylation of 4-Hydroxy-1-(2*H*)phthalazinone. Primary and secondary alkyl halides react with 4-hydroxy-1-(2*H*)phthalazinone (**1**) in alkaline medium to give 2-alkyl-4-alkyloxy-1-(2*H*)phthalazinones (**11**) (32, 33) (see Eq. 12). The tosylate of the secondary alcohol, cyclohexyl tosylate, has been successfully replaced when refluxed with **1** in ethanol containing potassium ethoxide to give 2-cyclohexyl-4-cyclohexyloxy-1-(2*H*)phthalazinone (32) (Eq. 19). There are few examples of this type of replacement, but this seems to be the method of choice where tosylates of secondary alcohols can be allowed to react with **1** as opposed to certain secondary alkyl halides to prepare this type of compound.

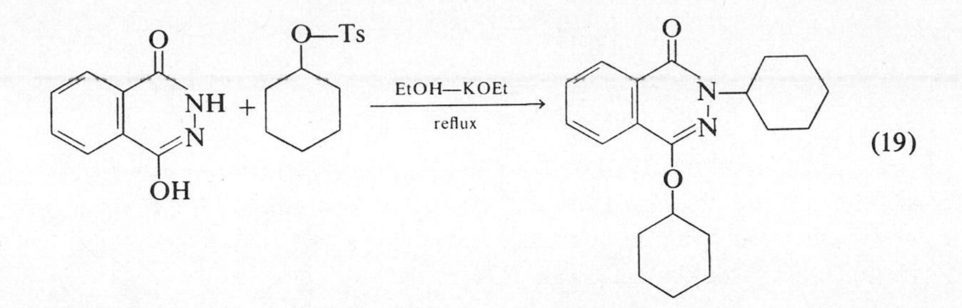

Methylation (1) of 2-methyl-4-hydroxy-1-(2*H*)phthalazinone (**6**) with dimethyl sulfate in aqueous alkali leads to a mixture of products containing 2-methyl-4-methoxy-1-(2*H*)phthalazinone (**20**) and 2,3-dimethylphthalazine-1,4-dione (**21**). On the other hand, methylation (1) of 4-methoxy-1-(2*H*)phthalazinone with dimethyl sulfate under similar conditions exclusively gives the desired product **20** (Eq. 20).

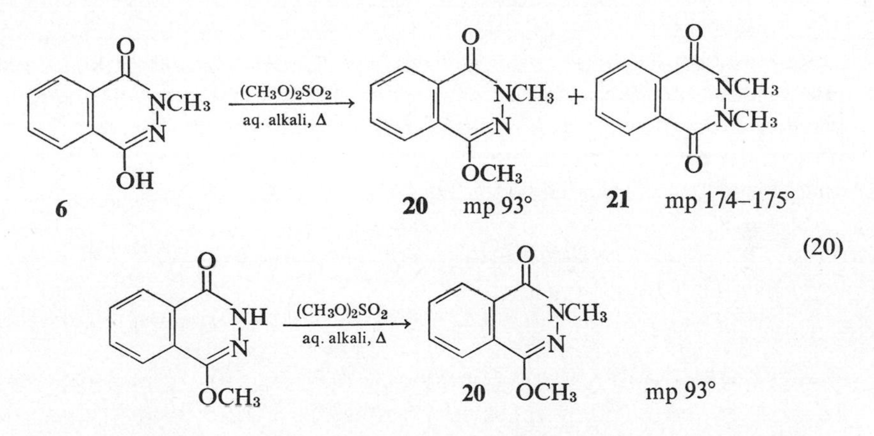

Reactive aromatic halides such as 4-nitrochlorobenzene and 2,4-dinitrochlorobenzene react with the disodium salt of **1** to give the corresponding 2,4-disubstituted derivatives of **1** (34, 35). Acetylation of **1** with acetic anhydride has been shown conclusively to give the 2,4-disubstituted product **18** as discussed earlier (Eq. 15).

b. 2-SUBSTITUTED-4-HYDROXY-1-(2*H*)PHTHALAZINONES. 2-Alkyl- or 2-aryl-4-hydroxy-1-(2*H*)phthalazinone react with alkyl or acyl halides to yield the corresponding 2,4-disubstituted products (29, 30, 47, 48) (Eq. 21).

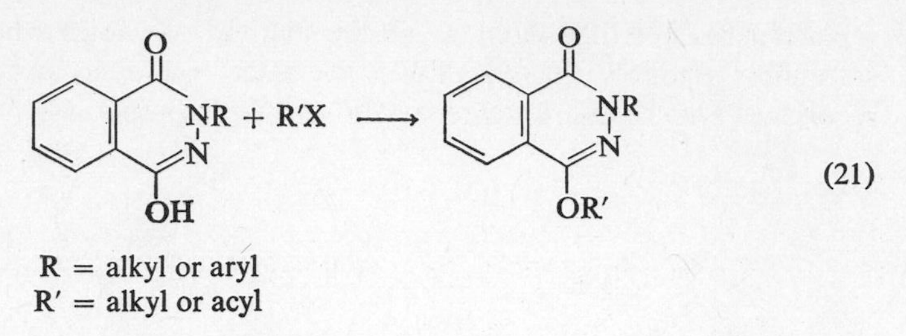

c. 2-SUBSTITUTED-4-CHLORO-1-(2*H*)PHTHALAZINONES. Nucleophilic substitution of the chlorine atom in 2-phenyl-4-chloro-1-(2*H*)phthalazinone (**22**) has been accomplished with alkoxides to give 2-phenyl-4-alkoxy-1-(2*H*)-phthalazinone (27, 49, 50) (Eq. 22).

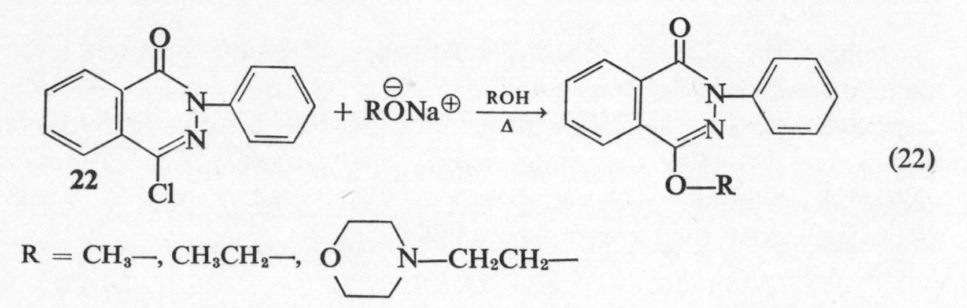

d. 4-ALKOXY-1-(2*H*)PHTHALAZINONES. 4-Methoxy-1-(2*H*)phthalazinone reacts with dimethyl sulfate as shown earlier to give the 2,4-disubstituted product **20** (1) (see Eq. 20). Since the preparation of 4-alkoxy-1-(2*H*)-phthalazinones is rather difficult, there is only one report of 2,4-disubstituted compounds being prepared using this procedure (51).

6. 2,3-*Disubstituted Phthalazine*-1,4-*diones*

a. CONDENSATION OF PHTHALIC ANHYDRIDE OR PHTHALOYL CHLORIDE WITH 1,2-DISUBSTITUTED HYDRAZINES. Phthalic anhydride and phthaloyl chloride with nitro or hydroxy groups in the 3- or 4-positions have been allowed to react with 1,2-diphenylhydrazine (52, 53) in pyridine or benzene containing an organic base to give the corresponding 2,3-diphenylphthalazine-1,4-diones (**23**) (Eq. 23).

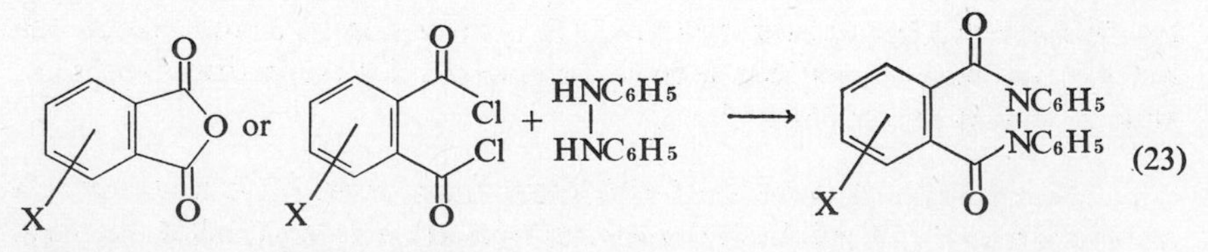

X, 5-NO_2; 5-OH; 6-NO_2; 6-OH

Similarly, phthaloyl chloride was allowed to react with diethylpyrazoline-2,4-dione in nitrobenzene and with 3-hydroxy-1-(2*H*)pyridazinone in dioxane to give the corresponding condensed ring systems, 2,2-diethylpyrazolo-[1,2-*b*]phthalazine-1,3,5,10-tetrone (**24**) and pyridazino[1,2-*b*]phthalazine-1,4,6,11-tetrone (**25**) (40), respectively (Eq. 24).

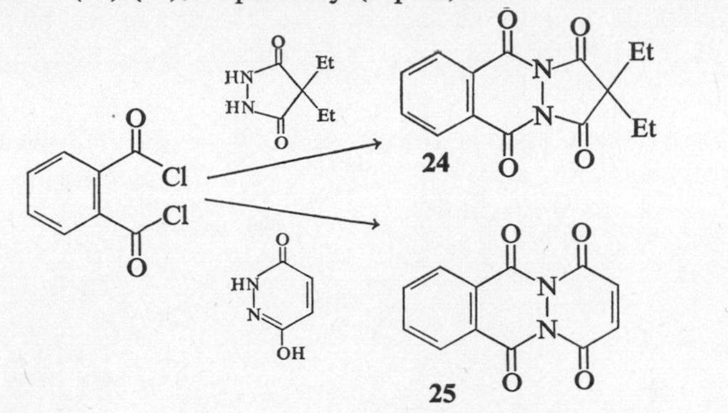

(24)

b. Acylation of 4-Hydroxy-1-(2*H*)phthalazinone with Malonyl Dichloride and Alkylation with Dihalides. Godin and Le Berre (40) have allowed **1** to react with malonyl dichloride, ethylmalonyl dichloride, and diethylmalonyl dichloride in nitrobenzene to give the corresponding pyrazolo[1,2-*b*]phthalazine-1,3,5,10-tetrones (**24**) (Eq. 25).

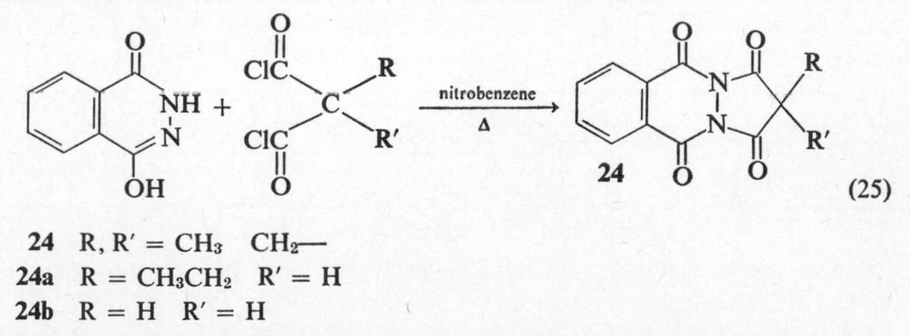

(25)

There are examples where **1** has been allowed to react with a dihalide to give a cyclic 2,3-disubstituted phthalazinedione derivative (**25**) (54) (Eq. 26).

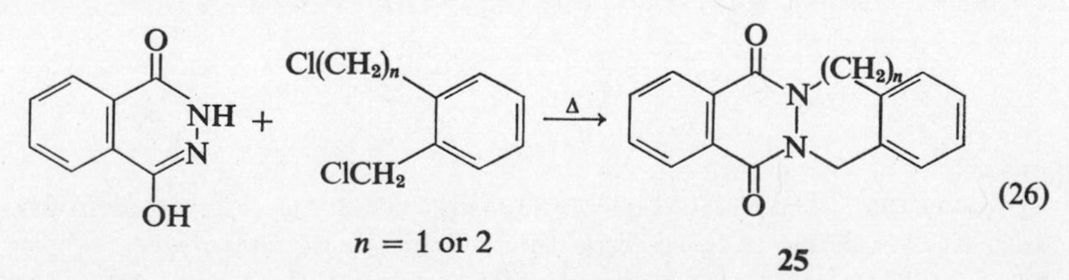

(26)

c. The Reaction of 4-Hydroxy-1-(2*H*)phthalazinone with α,β-Unsaturated Aldehydes, Acids, and Acid Chlorides. Le Berre and co-workers (37, 38) have shown that acrylic acid and substituted acrylic acids undergo Michael-type addition reactions with **1** to give 2-substituted-4-hydroxy-1-(2*H*)phthalazinones (**26**), which on treatment with thionyl chloride acylate the N-3 position of **26** to give the corresponding pyrazolo-[1,2-*b*]phthalazine-1,5,10-trione (**27**) (Eq. 27). They have also reacted **1** with acrylyl chloride in nitrobenzene to give **27** directly or by heating the intermediate 2-acylated-4-hydroxy-1-(2*H*)phthalazinone (**28**) (Eq. 27).

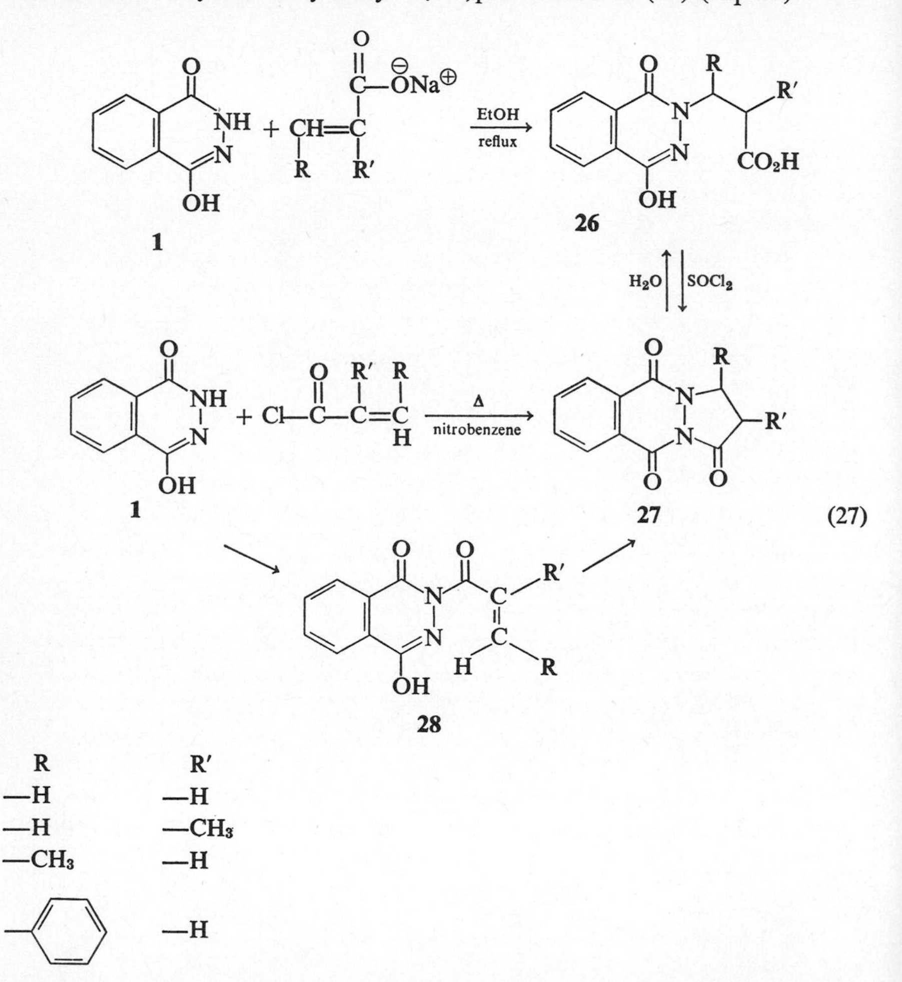

Godin and Le Berre (44) have similarly reacted **1** with acrolein and substituted acroleins to give first the Michael-type addition products (**19**, R = R′ = H) which exist in equilibrium with the cyclic isomer **19a** as shown earlier (see Eq. 17). These products have been dehydrated using sulfuric acid and hydrogenated with palladium to give the corresponding pyrazolo[1,2-*b*]phthalazine-5,10-diones (**29**) (Eq. 28).

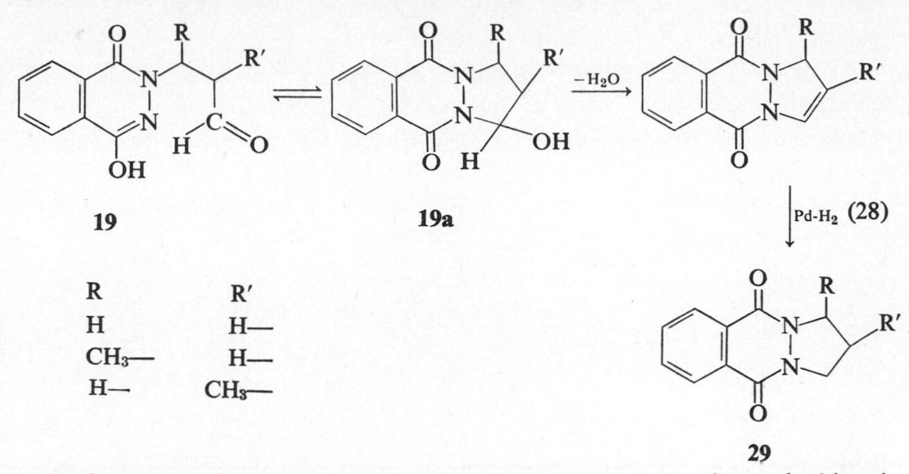

Nakamura and Kamiya (44a), using nmr spectroscopy, have elucidated the stereospecificity in ring chain tautomerism of **19** (R = CH_3, C_6H_5; R' = H). They have concluded that **19a** exists predominantly in an envelope conformation in which the methyl or phenyl group and the hydroxyl groups are *cis* and quasi axial.

d. The Reaction of Phthalazine-1,4-dione with Dienes. Clement (55) has successfully oxidized 4-hydroxy-1-(2*H*)phthalazinone with lead tetracetate at 0° C in acetonitrile to an unstable green intermediate which lost color and deposited a white polymer. The intermediate was identified as a diazaquinone, phthalazine-1,4-dione (**30**). Because of its extraordinary ability to act as a Diels-Alder dienophile, it was trapped with butadiene to give 1,4-dihydropyridazino[1,2-*b*]phthalazine-6,11-dione (Eq. 29).

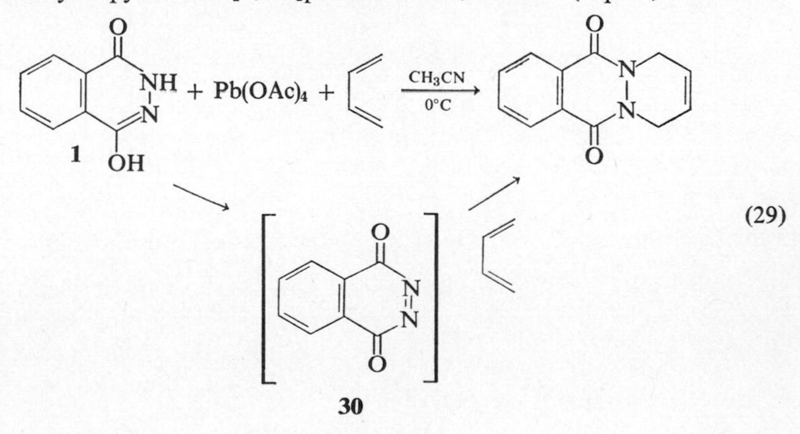

Clement (56) then reacted phthalazine-1,4-dione with anthracene, but this failed to react with thiophene, naphthalene, hexamethylbenzene, anisole, and bicyclo[2.2.1]hept-2-ene, thus showing the limitations of **30** as a dienophile.

Kealy (57) has prepared phthalazine-1,4-dione by reacting the sodium salt of **1** with *t*-butylhypochlorite at −50 to −65° C in acetone. He successfully isolated phthalazine-1,4-dione as a green, crystalline solid and was able to

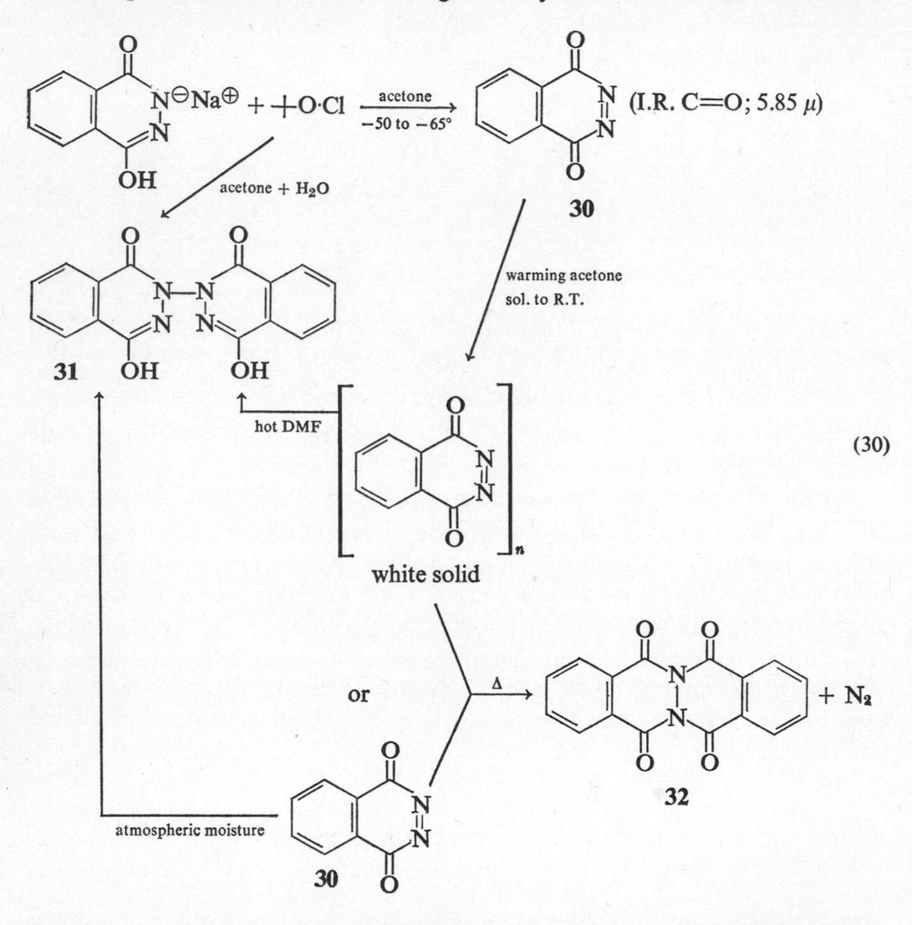

(30)

study its chemistry (Eq. 30). On warming the acetone solution of phthalazine-1,4-dione polymerized with evolution of a gas. Thermal decomposition of the polymeric form of phthalazine-1,4-dione or phthalazine-1,4-dione itself gave phthalazino[2,3-*b*]phthalazine-5,6,11,12-tetrone (**32**) (Eq. 30). Atmospheric moisture converted **30** into 2-(4′-hydroxy-1′-(2*H*)phthalazinoyl-2′)-4-hydroxy-1-(2*H*)phthalazinone (**31**).

At a later date, phthalazine-1,4-dione was reacted with 2,3-dimethylbutadiene (57, 58), cycloocta-1,3-diene (59), cyclopentadiene (60), and 5,5-diethoxycyclopentadiene (61). Popper and co-workers (62) have carried out the Diels-Alder addition of **30** to steroidal 14,16-dien-20-ones and reported the formation of both the 14α,17α and 14β,17β adducts in 9 and 47% yields, respectively. Hassall and co-workers (63) have prepared the adduct of **30** with 1,3-butadiene-4-carboxylic acid and transformed it into piperazine-3-carboxylic acid in order to compare it with the new amino acids derived from the monomycins (Eq. 31).

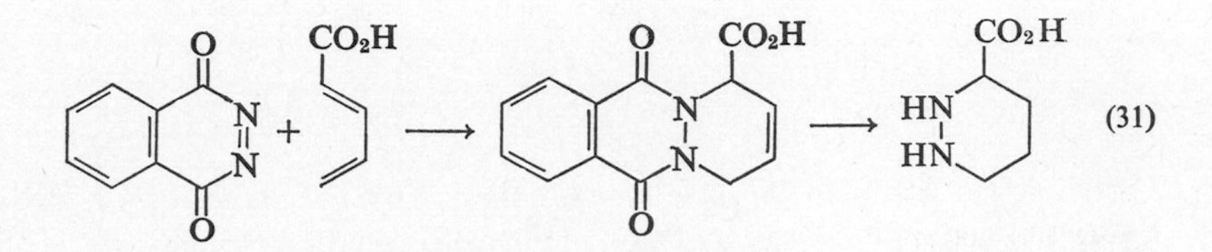

Chapman and Dominiani (59) have shown the addition of phthalazine-1,4-dione to an olefin to obtain in one step a derivative of diazacyclobutane (**33**). The reaction has been reported with indene (Eq. 32) and phenanthrene.

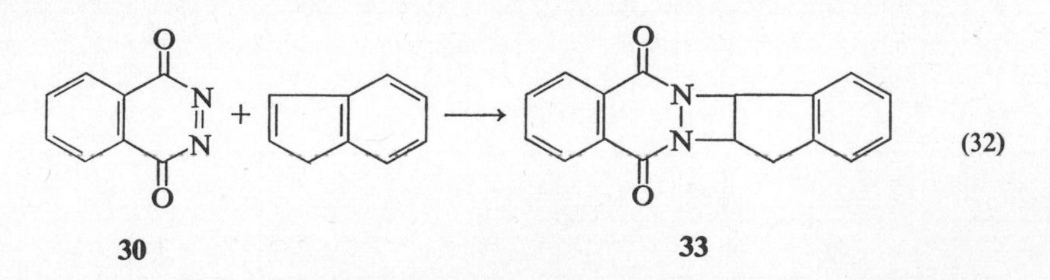

Recently Takase and Motoyama (63a) have prepared **30** by oxidation of 4-hydroxy-1-(2*H*)phthalazinone (**1**) using nickel peroxide and have reacted with dienes.

B. Properties of 4-Hydroxy-1-(2*H*)phthalazinone

1. *Physical Properties and Structure*

4-Hydroxy-1-(2*H*)phthalazinone is a high-melting solid forming white crystals from ethanol, mp 330–333° (3): [sublimes > 300° (3a); 333–334° (1) (DMF); 336° (2); 342–345° (5); 343–346° (7)]. It has a pK_a of 5.95 (aqueous ethanol, at 24° C) (2) and reacts as a monobasic acid. It forms metallic salts as well as stable salts with ammonia and hydrazine. Because of this property, Vaughan (64) had suggested that at least in aqueous solution it exists as a lactam-lactim structure. It has very little solubility in organic solvents except in alkaline solutions.

4-Hydroxy-1-(2*H*)phthalazinone is capable of existing in three possible tautomeric structures (**1**, **1a**, and **1b**).

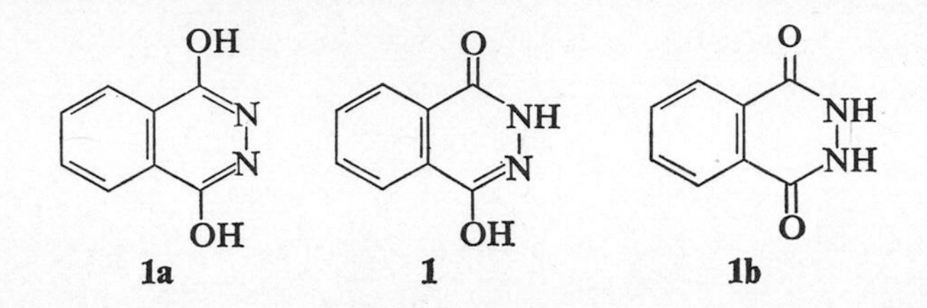

Elvidge and Redman (1) have studied the ir and uv spectra of 4-hydroxy-1-(2*H*)phthalazinone and compared them with 2-methyl-4-methoxy-1-(2*H*)-phthalazinone (**20**), 2,3-dimethylphthalazine-1,4-dione (**21**), and 1,4-dimethoxyphthalazine (**34**). These dimethyl derivatives **20**, **21**, and **34**

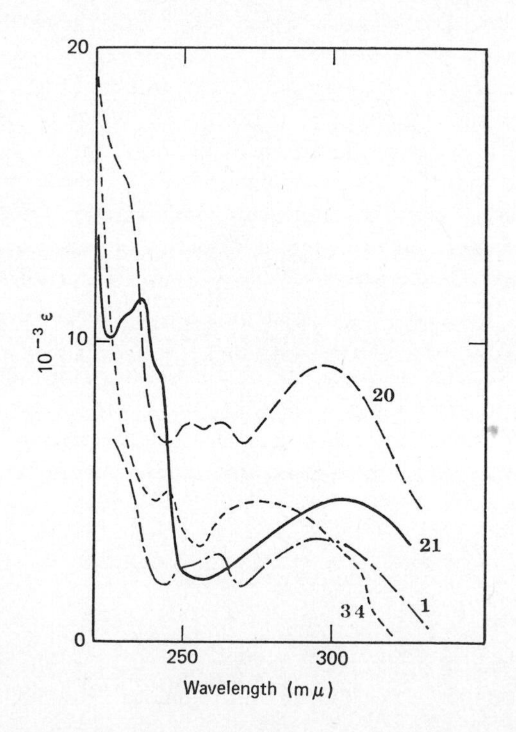

Fig. 2E-1. Ultraviolet absorption curves of **1**, **20**, **21**, and **34**.

represent (as fixed structures) the three tautomeric forms possible for **1**. The uv absorption curve of **1** (see Fig. 2E-1) has a shape very similar to that of **20**, which indicates it exists in solution in the lactam-lactim form rather than **1a** or **1b**.

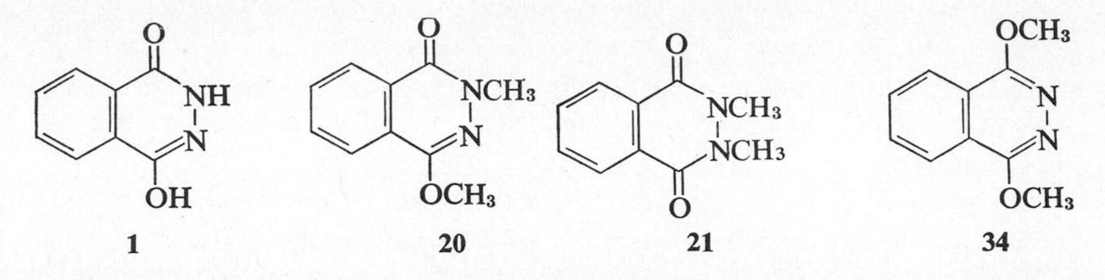

The ultraviolet absorption spectrum of **1** shows λ_{max} (mμ) (in propylene carbonate) (1), 263 (ε, 2930) and 298 (ε, 3600); (in ethanol) (40) 268 (3020) and 296 (5370). The ir spectra of these compounds in potassium bromide disks have been reported (1). The ir spectrum of **1** (KBr disk) shows max cm^{-1} NH, 3130 (m), C═O, 1655 (s), along with the other peaks.

Sohar (65) has suggested from the ir data that 4-hydroxy-1-(2*H*)phthalazinone tautomerizes between **1** and **1b** but not **1a** and that **1** predominates. This indicates that one of the amide groups can tautomerize, both cannot tautomerize at the same time, even though such tautomerization might lead to a stable aromatic structure. This is due to the pyramidal orientation of the nitrogen bonds in the nontautomerized amide group, which permits only one C═N—N group to fit without distortion of the bond angles into the nonplanar molecule and that the C═N—N group is not coplanar with the aromatic ring. Tautomerism of both groups would involve a change in C═N—N angles, and since structure **1a** has a planar configuration, this would require more energy than would be compensated by the formation of an aromatic structure **1a**.

Sheinker and co-workers (66) have studied this problem using ir spectroscopy and suggested structure **1** based on the following arguments. Hydrogen bonding is retained in weakly polar solvents, whereas in dioxane, hydrogen bonding is broken. Formation of metal salts resulted in replacement of hydrogen atom in the hydroxy group but no enolization of amide group took place.

2. *Chemical Properties*

a. Oxidation and Reduction. 4-Hydroxy-1-(2*H*)phthalazinone **(1)** has been found to be inert to oxidation and reduction. It has been oxidized by lead tetraacetate or *t*-butyl hypochlorate to give phthalazine-1,4-dione

(**30**) and because of its extraordinary reactivity as a Diels-Alder dienophile the reaction has been thoroughly investigated with various dienes (see Section 2E-I-A-6-d).

Recently Omote and co-workers (66a) have irradiated **1**, 2-phenyl-4-hydroxy-1(2*H*)phthalazinone and luminol in dimethylformamide in the presence of oxygen, and they observed for the first time a ring contraction leading to phthalimide, 2-phenylphthalimide, and 3-aminophthalimide, respectively. The authors have proposed a tentative mechanism for the photoreaction.

Ainsworth (67) has studied the hydrogenolysis of the nitrogen-nitrogen bond with Raney nickel in various heterocyclic compounds containing nitrogen-nitrogen bonds. It was found that **1** did not undergo hydrogenolysis, being recovered unchanged. The inertness of **1** makes it possible to perform several reductions on substituents either in the aromatic ring or in the side chain at the *N*-2 position; these include catalytic reduction of nitro groups with Raney nickel (52, 68), hydrogenation of the olefinic double bonds with palladium on carbon (44, 55), and platinum oxide reduction of a ketone to an alcohol (43). Reduction of 2-methyl-4-hydroxy-1-(2*H*)phthalazinone with lithium aluminum hydride has been shown to reduce the amide carbonyl to give 1-hydroxy-3-methyl-3,4-dihydrophthalazine. Similar reduction of 2,3-dimethylphthalazine-1,4-dione reduces both the amide functions to give 2,3-dimethyl-1,2,3,4-tetrahydrophthalazine. The polarographic reduction of 2-phenyl-4-hydroxy-1-(2*H*)phthalazinone and 2,3-dimethylphthalazine-1,4-dione afford 1-hydroxy-3-phenyl-3,4-dihydrophthalazine and 2,3-dimethyl-1,2,3,4-tetrahydrophthalazin-1-one, respectively (see Section 2L).

b. Alkylation and Acylation. Alkylation and acylation have been discussed under methods of preparation (see Section 2E-I-A-4, 5, 6). Methylated products of **1** such as 2-*N*-methyl (**6**), 4-*O*-methyl, 2,4-*N*,*O*-dimethyl (**20**), and 2,3-*N*,*N*-dimethyl (**21**) are well characterized and identified (see Eqs. 8 and 20).

It has been possible to acetylate **1** selectively to give 2-acetyl (**16**), 4-acetyl (**17**), or 2,4-diacetyl (**18**) derivative of **1** (see Eq. 15). In general, monoacetylation takes place easily either on oxygen or nitrogen as desired. Further acetylation to prepare diacetyl derivative of **1** is difficult and the resulting product is *O*,*N*-diacetylated as opposed to *O*,*O*-diacetyl or *N*, *N*-diacetyl derivative of **1**. The pmr spectra of the acetylated products are reported by Parsons and Rodda (69) and Goden and Le Berre (40). The pmr spectra have been found to be extremely useful for the assignment of *N*-acetyl or *O*-acetyl products. The assignment of the singlet from the methyl groups in deuteriochloroform is shown in tau units (40, 69), shown at top of facing page.

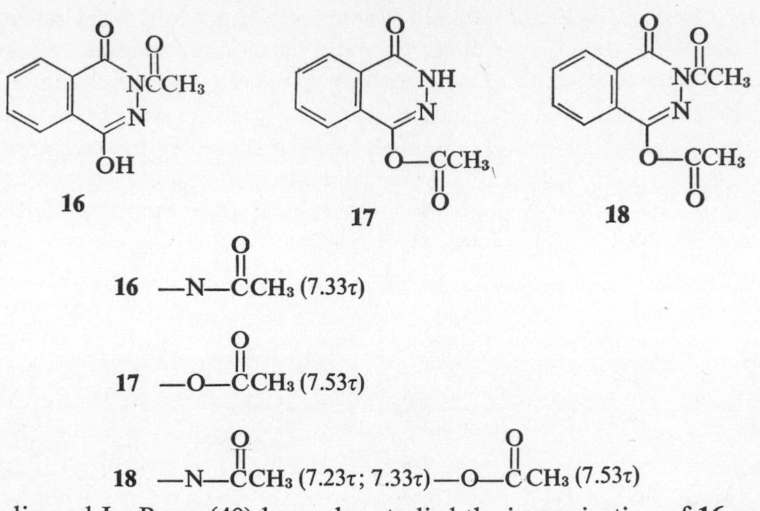

Godin and Le Berre (40) have also studied the isomerization of **16** and **17** using pmr spectroscopy. It was found that in pyridine the equilibrium (**16** ⇋ **17**) was reached in 20 min at 90–100° C. On the other hand, the isomerization was very slow in chloroform; the equilibrium is reached in 24 hr at 90° C.

Omote and co-workers (70, 71) have studied the acetylation of 5-amino-4-hydroxyl-1-(2*H*)phthalazinone (luminol, **4**) in detail. Monoacetylation with acetyl chloride and pyridine in presence of dimethylformamide gives 5-acetamido-4-hydroxy-1-(2*H*)phthalazinone as expected. The amino group in the aromatic ring behaves normally and undergoes various reactions as an aromatic amine, for example, diazotization (72, 73), reaction with isocyanate to give urea (74), and amidine formation with dimethylformamide (75).

c. Reactions with Phosphorus Pentasulfide and Phosphorus Halides. 4-Hydroxy-1-(2*H*)phthalazinone (**1**) reacts with phosphorus pentasulfide and phosphorus oxychloride or phosphorus pentachloride to give the 1,4-dithio- and 1,4-dichlorophthalazines, respectively (see Sections 2F and 2G) (Eq. 33).

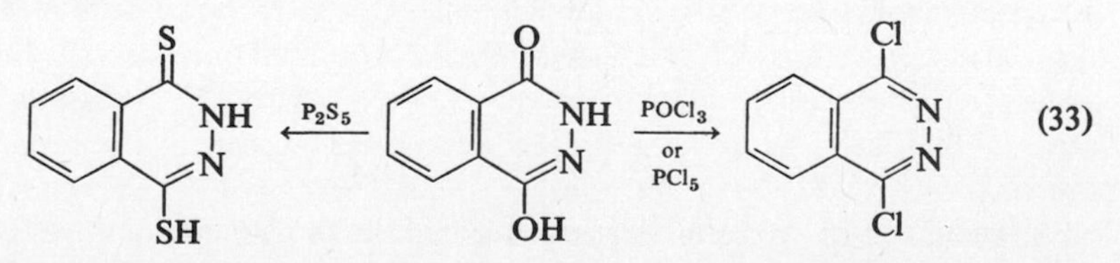

d. Isomerization. The conversion of *N*-aminophthalimide to the isomeric **1** and *N*-(arylamino)phthalimide to the isomeric 2-aryl-4-hydroxy-1-(2*H*)phthalazinone by the treatment with acid or alkali has been discussed earlier (see Eqs. 3, 9, 10, and 11). There are examples of 2-aryl-4-hydroxyphthalazinones being converted to *N*-(arylamino)phthalimide on heating with alkali where aryl groups in 2-position are electron-withdrawing in nature (29, 34) (Eq. 34).

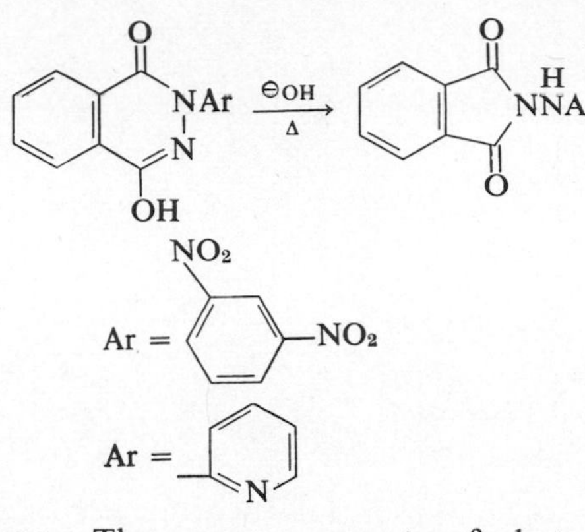

(34)

e. MISCELLANEOUS. There are no reports of electrophilic substitution reactions on 4-hydroxy-1-(2*H*)phthalazinone; in general it can be said that the compound is very inert. There are reports where 2-phenyl-4-hydroxy-1-(2*H*)phthalazinone on nitration gave 2-(4′-nitrophenyl)-4-hydroxyl-1-(2*H*)-phthalazinone (34, 76), and 2-phenyl-4-methoxy-1-(2*H*)phthalazinone has been shown to give 2-(2′,4′-dinitrophenyl-4-methoxy-1-(2*H*)phthalazinone (49). In both cases, nitration takes place on the phenyl ring substituted at the 2-position and not on the phthalazine nucleus.

4-Hydroxy-1-(2*H*)phthalazinone undergoes Michael-type addition reactions with activated double bonds, and it has also been shown to undergo a Mannich reaction as discussed earlier (see Section 2E-I-A-4-c, d).

Addition of a Grignard reagent to 2-methyl- and 2-phenyl-4-hydroxy-1-(2*H*)phthalazinone has been studied using alkyl and phenylmagnesium bromides to give 1-hydroxy-1,2-dihydrophthalazines (**35**) (77) (Eq. 35) (see Section 2L).

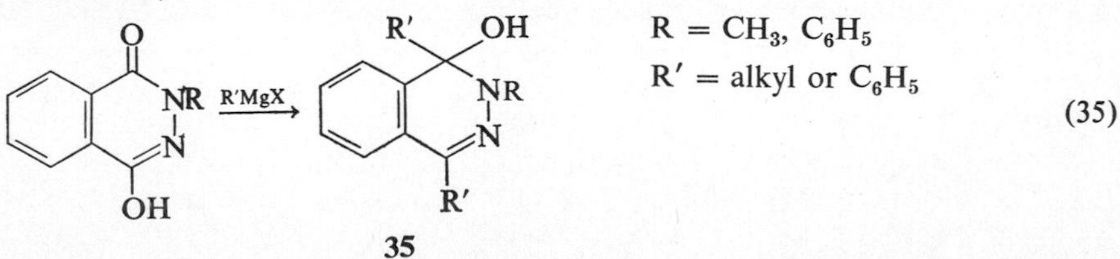

(35)

Higuchi and co-workers (78) have studied the molecular complex formation of **1** and its methyl derivatives with phenol, catechol, resorcinol, and hydroquinone. The phase diagrams were obtained and the relative stability of the molecular complexes calculated in terms of apparent equilibrium constant.

McIsaac (79) has studied the metabolic fate of radioactive **1**. He has found that 30–34% of **1** is excreted unchanged, 50% is excreted as *o*-glucosonide, and 3–5% undergoes *N*-methylation. The compound is found to be relatively nontoxic with a minimum lethal dose being >320 mg/kg.

II. 1,4-Dialkoxyphthalazines

There are only a few compounds reported in this group. The preparation and properties of this group of compounds are similar to those of 1-alkoxyphthalazines (Section 2D-II).

A. Preparation

1,4-Dichlorophthalazine is heated with sodium alkoxide in the corresponding alcohol to give 1,4-dialkoxyphthalazines (**34**, **36**) (Eq. 36) (80, 81). The yields of this reaction (80) are usually higher (**34**, R = CH_3: 81%, mp 92°, 93° (1); **36**, R = CH_3CH_2—, 71%, mp 120°).

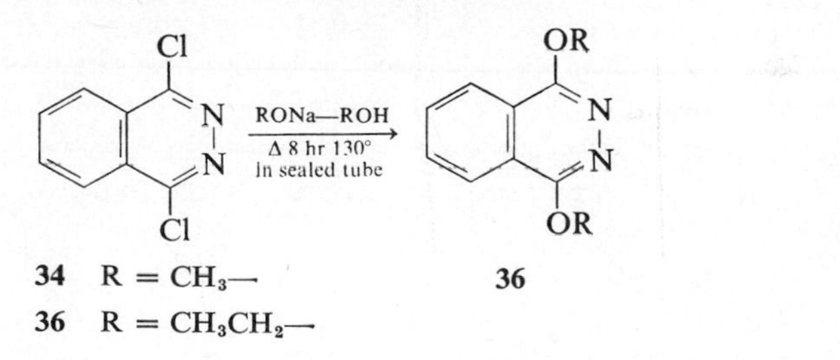

(36)

Phenol, upon reaction with 1,4-dichlorophthalazine in the presence of sodium amide (1) or potassium carbonate (82), or 1,4-dichlorophthalazine on refluxing in toluene with sodium phenoxide (81), gives 1,4-diphenoxyphthalazine (**37**) (Eq. 37).

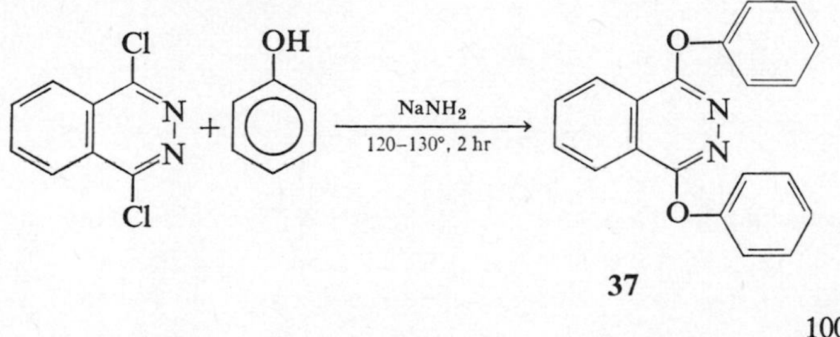

100%, mp
222° (DMF—H_2O)(1)
mp 216°
(MeOH-acetone)(82)

Recently Chambers and co-workers (83) have studied the nucleophilic substitution of 1,4,5,6,7,8-hexafluorophthalazine (**38**) with sodium methoxide in methanol at different reaction temperatures. All six fluorine atoms were replaced progressively. The fluorine atoms at positions 1 and 4 were most reactive, and those at 5 and 8 were the least reactive (Eq. 38).

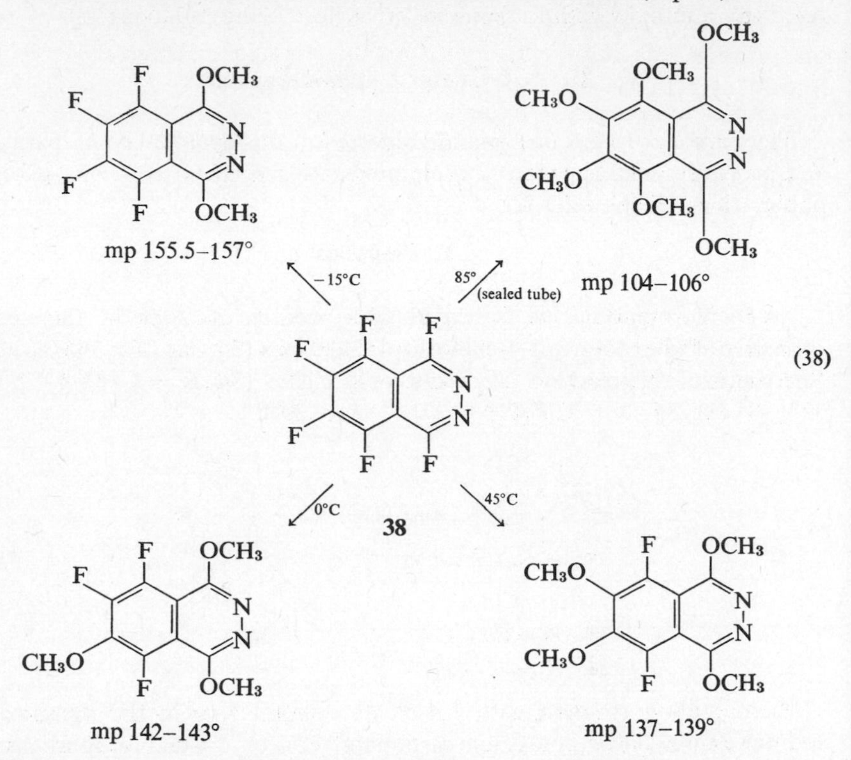

Vigevani and co-workers (108) have allowed a mixture of 6-trifluoromethyl- and 7-trifluoromethyl-1-chloro-4-methoxyphthalazine to react with an equimolar amount of sodium methoxide in methanol to give 1,4-dimethoxy-6-trifluoromethylphthalazine, mp 176°, in 83% yield.

B. Properties

1,4-Dimethoxyphthalazine (**34**) can be distilled at 125°/10 mm having mp 92° (80) [mp 93° (1)].

Elvidge and Redman (1) have reported the ir spectrum of **34** and uv spectrum in ethanol. The uv absorption curve (see Fig. 2E-1) shows maxima at 245 and 280 mμ with ε values 5320 and 4750, respectively.

Bowie and co-workers (84) have reported the mass spectra of 1,4-diethoxyphthalazine.

Hayashi and co-workers (80) have shown that 1,4-dialkoxyphthalazines are inert to *N*-oxidation and partial hydrolysis takes place to 4-alkoxy-1-(2*H*)phthalazinone.

Nucleophilic displacement of the ethoxy group in 1,4-diethoxyphthalazine (**36**) with hydroxylamine gives 1-(hydroxylamino)-4-ethoxyphthalazine (**39**) (Eq. 39). However, Elvidge and Redman have made an unusual observation on reaction of morpholine with 1,4-dimethoxyphthalazine (**34**) where instead of 1,4-dimorpholinophthalazine (**40**), 4-hydroxy-1-(2*H*)phthalazinone (1) and *N*-methylmorpholine were isolated (Eq. 39). The 1,4-dimethoxyphthalazine (**34**) had methylated the amine, itself being transformed into **1**. These authors have also shown that 4-methoxy-1-(2*H*)phthalazinone on heating with aniline gave **1** and *N*-methylaniline.

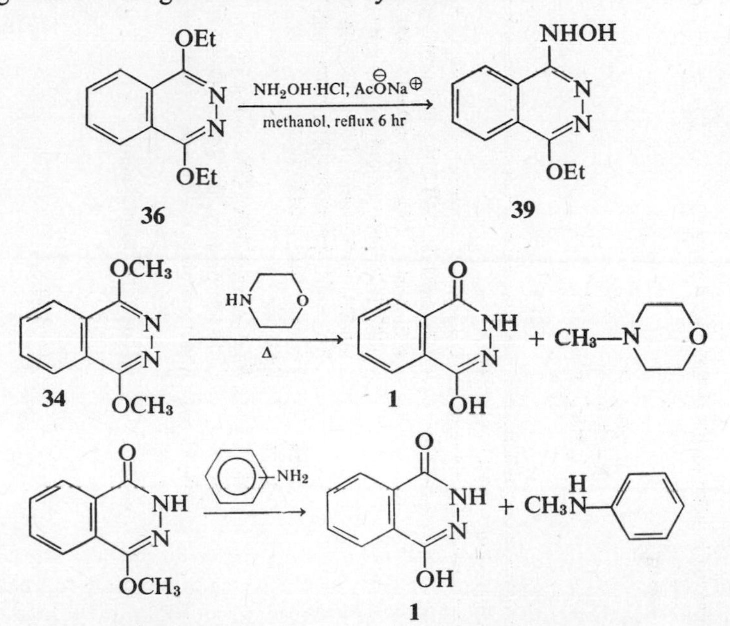

Nucelophilic substitution on 1,4-diphenoxyphthalazine (**37**) as opposed to 1,4-dialkoxyphthalazines appears to proceed smoothly to give the desired products; for example, **37** on heating with morpholine gave 1,4-dimorpholinophthalazine (**40**) (Eq. 40).

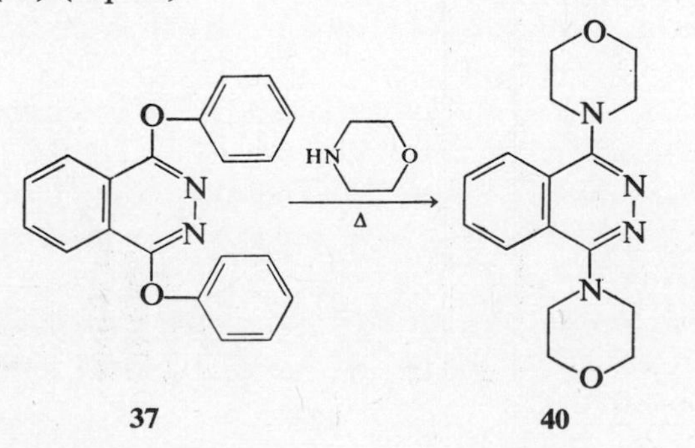

(40)

III. Tables

TABLE 2E-1. 4-Hydroxy-1-(2*H*)phthalazinones with Substituents in Benzenoid Ring

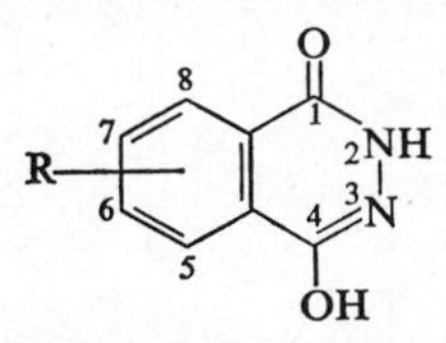

R	Preparation[a]	Yield (%)	MP	Reference
H	A	52–100	White crystals, 330–333° (ethanol); sublimes >300°; thin prisms, 333–334° (DMF); 336°; 342–345°; 343–346°; triethylbenzylammonium salt, 162–164.5° (dec.); methylsulfonic acid, 170–171°; phenylsulfonic acid, 175–176°; 2-chloroethylsulfonic acid, 134–136°	1–7, 19, 84a
5-NO_2	B	75–78	315–316° (dec.); 316° (dec.)	10, 85
5-NH_2	C	70–75	329–332°; 332°	10, 85
5-$N(CH_3)_2$	D	64	Pale yellow prisms, 208° (toluene)	12
6-$N(CH_3)_2$	D	86	Colorless prisms, 297° (dec.) (DMF-H_2O)	12
6-$N(CH_2CH_3)$	D, E	68, 55	Colorless prisms, 278° (dec.) (DMF-H_2O)	12, 15
6-N (pyrrolidino ring)	E	68	AcOH, 311° (acetic acid)	15
6-N (piperidino ring)	E	75	285° (dec.)	15

5-N(H)—C(=O)CH_3	F	—	Pale yellow needles 315° (dec.)	70, 71
$5\text{-N(H)—C(=O)CH}_2\text{CH}_3$	F	—	Pale yellow powder, 278° (dec.)	70
$5\text{-N(H)—C(=O)(CH}_2)_8\text{CH=CH}_2$	F	—	Light yellow powder, 275°	70
$5\text{-N(H)—C(=O)CH}_2\text{Cl}$	F_1	82	AcOH, 263° (dec.) (gl. acetic acid)	86
$5\text{-N(H)—C(=O)CH}_2\text{NH}_2$	G	—	244–245°	86
$5\text{-N(H)—C(=O)CH}_2\text{N(H)—C(=O)CH}_3$	—	—	White needles, 270° (dec.) (alcohol)	86
$5\text{-N(H)—C(=O)NHNH}_2$	G_1	—	White microcrystals, blackening at 310°	86
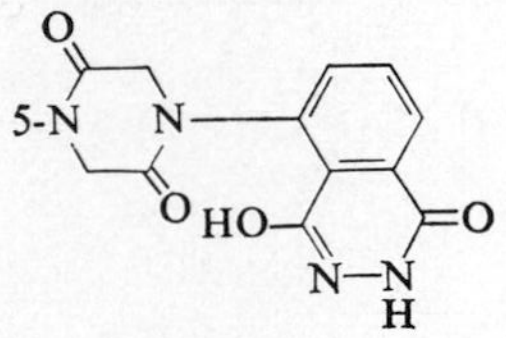 5-N(2,6-dioxopiperazine)N— linked to 5-position of 1-hydroxy-4-oxo-3H-phthalazine (HO, N—N(H), =O)	G_2	90	White crystals, >325° (gl. acetic acid)	86

TABLE 2E-1 (*continued*)

R	Preparation[a]	Yield (%)	MP	Reference
5-N(H)—C(=O)—N(H)—C_6H_5	H	60	334° (dec.)	74
5-N(N)—C(=S)—N(H)—C_6H_5	H	56	246° (dec.)	74
5-N(H)—C(=O)—N(H)—(1-naphthyl)	H	—	284° (dec.)	74
5-N=C—N—$(CH_3)_2$	1	—	HCl, colorless needles, 271–272° (dec.) (water)	75
5-N=N—(1,4-naphthylene)—N(H)—$(CH_2)_2N(Et)_2$	J	27	Greenish black crystals, 211–213°	72
5-N=N—C(C_6H_5)=N—N(H)C_6H_5	J_1	—	—	73
5-Amino-8-methoxy	D_1	74	242° (dec.)	12
5-Amino-7,8-dimethoxy	K	81	Darkens at 240–260°, melts at 290–292° (with dec.) (DMF-H_2O)	13

5-Amino-6,7,8-trimethoxy	K	68	174–175°	13
5-Amino-6,7-methylenedioxy	M	87	308° (dec.) (acetic acid)	14
5,7-Diamino	D	55	Colorless prisms, 300° (dec.)	12
5,6-Ethylenediamine	L	—	2-AcOH, colorless plates, sintered at 280° (acetic acid)	18
5,6-*N*,*N*′-Dimethylenediamine	L	96	Sintered at 298°, 310° (dec.)	18
5-F	K_1	—	350° (dec.)	21
6-F	K_1	—	306–308°	21
6-CF_3	K_1	85	298–300° (dec.) (methanol)	20
6-Cl	M_1	—	339°	17
6-OH	M_1	—	366–370°	17
6-OCH_3	M_1	52	293°	17
6-OCH_2CH_3	M_1	—	266°	17
6-$OCH_2CH_2CH_3$	M_1	—	243°	17
6-$OCH(CH_3)_2$	M_1	—	236°	17
6-$O(CH_2)_3CH_3$	M_1	—	226°	17
6-$OCH_2CH(CH_3)_2$	M_1	—	221°	17
6-$SCH_2(CH_2)_{10}CH_3$	M_2	85	186–188.5° (acetic acid)	18
6,7-Dimethoxy	K_2	75	333–335° (dec.) (acetic acid)	87
5,6,7-Trimethoxy	K_3	72	Colorless crystals, 235° (methanol-water)	88
5,6-Methylenedioxy	K_4	—	Colorless crystals, 277.5° (dec.) (acetic acid)	89
6,7-Methylenedioxy	K_4	—	>350° (dec.) (acetic acid)	89
6,7-Ethylenedioxy	K_5	82–92	Needles, 352–354° (dec.) (gl. acetic acid)	23
6-CH_3	K_1	—	>350° (sublimed)	19
6-CH_2—CH(O—)(—O) (1,3-dioxolan-2-yl)	N	80	211° (dec.) (DMF-benzene)	90

TABLE 2E-1 (*continued*)

R	Preparation[a]	Yield (%)	MP	Reference
6-CH_2—CH(=O)	O	65–75	300° (dec.)	90
6-CH=CHN(n·C_5H_{11})$_2$	P	38–55	Yellow crystals, 150° (dec.)	90
6-CH=CHN[CH(CH_3)$_2$]$_2$	P	58	192° (dec.)	90
6-CH=CHN(CH(CH_3)CH_2CH_3)$_2$	P	41	195° (dec.)	90
6-CH=CH—N(—cyclohexyl)$_2$	P	60–70	207° (dec.)	90
6-phenyl	Q	72		16
6-Chloro-7-sulfonamido	E_1	95	325–326° (dec.)	24
6-C(=O)- (4-hydroxy-1(2H)-phthalazinone-6-yl)	K_4	—	350°–360° (dec.)	91

6-CH_2—CH_2— (structure: 4-hydroxy-1-(2H)phthalazinone ring; O, NH, N, OH)	M_3	96	White, >300	19
5,6,7,8-Tetrachloro	R	—	—	22

[a] A, preparations of **1** are reported in Section 2E-I-A-1-a. B, disodium salt of 3-nitrophthalic acid + $N_2H_4 \cdot H_2SO_4$ in water → evaporate to dryness (10) or heat 3-nitrophthalic acid with hydrazine in water + triethyleneglycol → heat to remove water (85). C, reduction of the corresponding nitro compound with sodium hydrosulfite dihydrate in water. D, the corresponding phthalimide is refluxed with $N_2H_4 \cdot H_2O$ for 1 hr. D_1, same as D, heat for 2 hr at 140°. E, the corresponding diethyl phthalate is refluxed with $N_2H_4 \cdot H_2O$ for 6–8 hr. E_1, same as E, heat in benzene. F, the corresponding amino compound + acyl chloride + pyridine kept in DMF overnight at room temperature. F_1, same as F but refluxed in acetic acid for 30 min. G, 5-chloroacetamido-4-hydroxy-1-(2*H*)phthalazinone treated with ammonia for 8 days. G_1, 5-chloroacetamido-4-hydroxy-1-(2*H*)phthalazinone treated with $NH_2NH_2 \cdot H_2O$ for 3 days. G_2, 5-chloroacetamido-4-hydroxy-1-(2*H*)phthalazinone refluxed for 4 hr in aqueous potassium carbonate. H, 5-amino-4-hydroxy-1-(2*H*)phthalazinone + isocyanate or isothiocyanate + pyridine, heat at 60° for 4 hr in dimethylformamide. J, 5-amino-4-hydroxy-1-(2*H*)phthalazinone is diazotized with HNO_2 at 0° C and condensed with 1-(N,N-diethylethylamino)naphthalene. J_1, same as J but coupled with ϕ—C═N—N—ϕ. K, the corresponding phthalic anhydride refluxed with $N_2H_4 \cdot H_2O$ in water for 3–5 hr. K_1, reflux in acetic acid 1–2 hr, otherwise same as K. K_2, reflux in ethanol for 12 hr, otherwise same as K. K_3, reflux in dioxane 3 hr, otherwise same as K. K_4, heat with $N_2H_4 \cdot H_2O$, otherwise same as K. L, the corresponding *N*-methylphthalimide heated with 95% hydrazine for 1–2 hr on water bath. M, the corresponding dimethyl phthalate + $N_2H_4 \cdot H_2O$, reflux in ethanol for 1 hr. M_1, same as M, heat in methanol 12 hr. M_2, heat in $N_2H_4 \cdot H_2O$ for 24 hr on water bath, otherwise same as M. M_3, heat with N_2H_4 in a scaled tube on steam bath, otherwise same as M. N, the corresponding dimethyl phthalate + 80% $N_2H_4 \cdot H_2O$ reflux in methanol for 5 hr. O, hydrolysis of the product obtained by method N with acid + dimethylsulfoxide. P, the aldehyde obtained by method O was refluxed for 4 hr with the proper amine in DMSO-H_2O-benzene system. Q, 4-phenylphthalic acid + $N_2H_4 \cdot H_2O$ → 50% acetic acid, reflux for 5 hr. R, by the reaction of 3,4,5,6-tetrachlorophthalic anhydride + $N_2H_4 \cdot H_2SO_4$ in acetic acid + sodium acetate.

TABLE 2E-2. 2-Substituted-4-hydroxy-1-(2*H*)phthalazinones

R	Preparation[a]	Yield (%)	MP	Reference
$—CH_3$	A, A_1, A_2	~90	Plates, 238–240° (aq. methanol); white crystals, 238.5–239.5° (aq. ethanol); leaves, 239–240° (aq. acetic acid)	1, 19, 25, 92, 93
$—CH_2CH_2CH_3$	B	—	Colorless needles, 156° (ethanol)	33
$—CH(CH_3)_2$	B	—	Colorless prisms, 183° (ethanol); 120–121°	33, 92
$—CH_2CH(CH_3)_2$	B	—	Colorless needles, 150° (ethanol)	33
$—CH_2CH_2CH(CH_3)_2$	B	—	Colorless needles, 152° (ethanol)	33
$—CH_2(CH_2)_4CH_3$	B	—	Colorless needles, 164° (ethanol)	33
$—CH_2(CH_2)_{10}CH_3$	B	—	96° (ligroin)	33
$—CH_2N$ (morpholino)	C	59	Needles, 184° (ethanol)	45
$—CH_2N$ (piperidino)	C	—	169°	46
$—CH_2N$ (piperazine) $N—CH_2—$ (4-hydroxy-1-oxophthalazin-2-yl)	C_1	—	235–236°	46

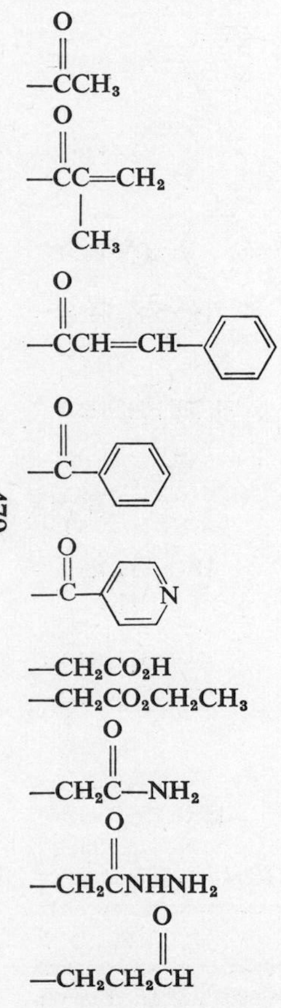

$-C(=O)CH_3$	D	34	Colorless needles, 191–192° (benzene); 191°	39, 40
$-C(=O)C(CH_3)=CH_2$	D_1	—	210° (ethyl acetate)	37
$-C(=O)CH=CH-C_6H_5$	D_1	—	216° (toluene)	37
$-C(=O)-C_6H_5$	D	82	231–232°	6
$-C(=O)-C_5H_4N$ (4-pyridyl)	D_2	—	226–227°	6
$-CH_2CO_2H$	E	70	235° (aq. ethanol)	2, 36
$-CH_2CO_2CH_2CH_3$	—	—	151°	36, 94
$-CH_2C(=O)-NH_2$	E_1	77	295° (aq. ethanol)	36, 94
$-CH_2C(=O)NHNH_2$	E_2	81	318° (aq. ethanol)	36, 94
$-CH_2CH_2CH(=O)$	F	67	Colorless prisms, 213° (ethanol)	44

TABLE 2E-2 (*continued*)

R	Preparation[a]	Yield (%)	MP	Reference
$—CH_2CH_2CH{=}NOH$	—	—	160° (ethanol)	44
$—CH_2CH_2CN$	F_1	57	181–182° (ethanol)	42
$—CH_2CH_2CO_2H$	G	99	206° (water)	37
$—CH_2CH_2CO_2CH_2CH_3$	G_1	90	Colorless crystals, 173° (ethanol)	38
$—CH(CH_3)CH_2CO_2H$	G	—	209° (water)	37
$—CHCH(CH_3)CO_2H$	G	—	148° (water)	37
$—CH(C_6H_5)CH_2CO_2H$	G	—	207° (methanol)	37
$—CH_2CH_2CH(OH)CH_3$	H	36, 97	154–155° (ethanol)	43
$—CH_2CH_2C(=O)CH_3$	I	82	153–154° (ethanol), 152.5–154°	43, 44
$—(CH_2)_5O—C_6H_4—NH_2$ (para)	J	43	169–170°	68
$—CH_2—C_6H_5$	K	—	BP 290–292°/10 mm, colorless needles, 203° (xylene)	33
$—CH_2CH_2—C_6H_5$	K	—	Colorless needles, 174–175° (xylene)	47

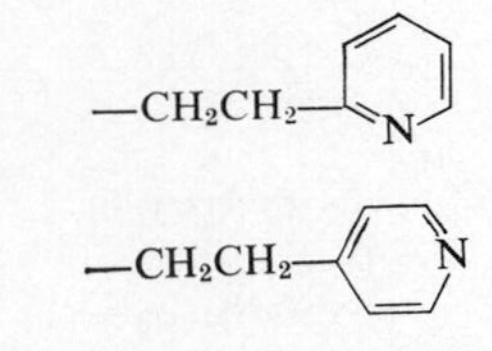	K	—	Colorless needles, 196° (xylene)	47
	L	78	Colorless needles, 156–157° (acetone); HBr, 223–224°	41, 42
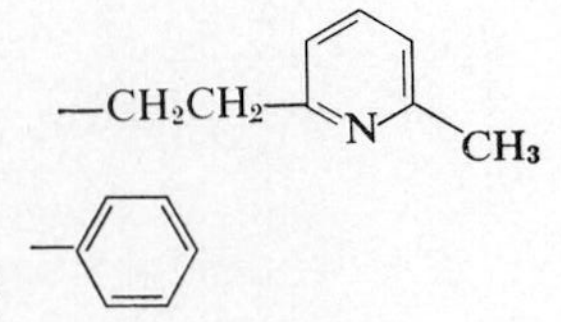	L	93	Colorless needles, 216–217° (ethanol); 146° (aq. ethanol); HBr, 250°	41, 42
	L	62	173–174°	42
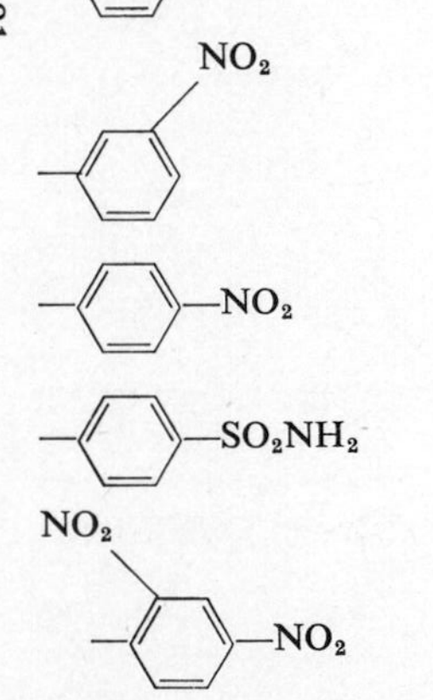	M	90–92	Prisms, 210°; 210–211°; piperzinium salt, 106°	7, 27, 28, 46
	M_1, N_1	40	280° (acetic acid)	30
	N, N_1	—	297–298°; pale yellow needles, 299°	31, 76
	O	—	273°	95
	K, M_2	—	282–283°	31, 34

TABLE 2E-2 (*continued*)

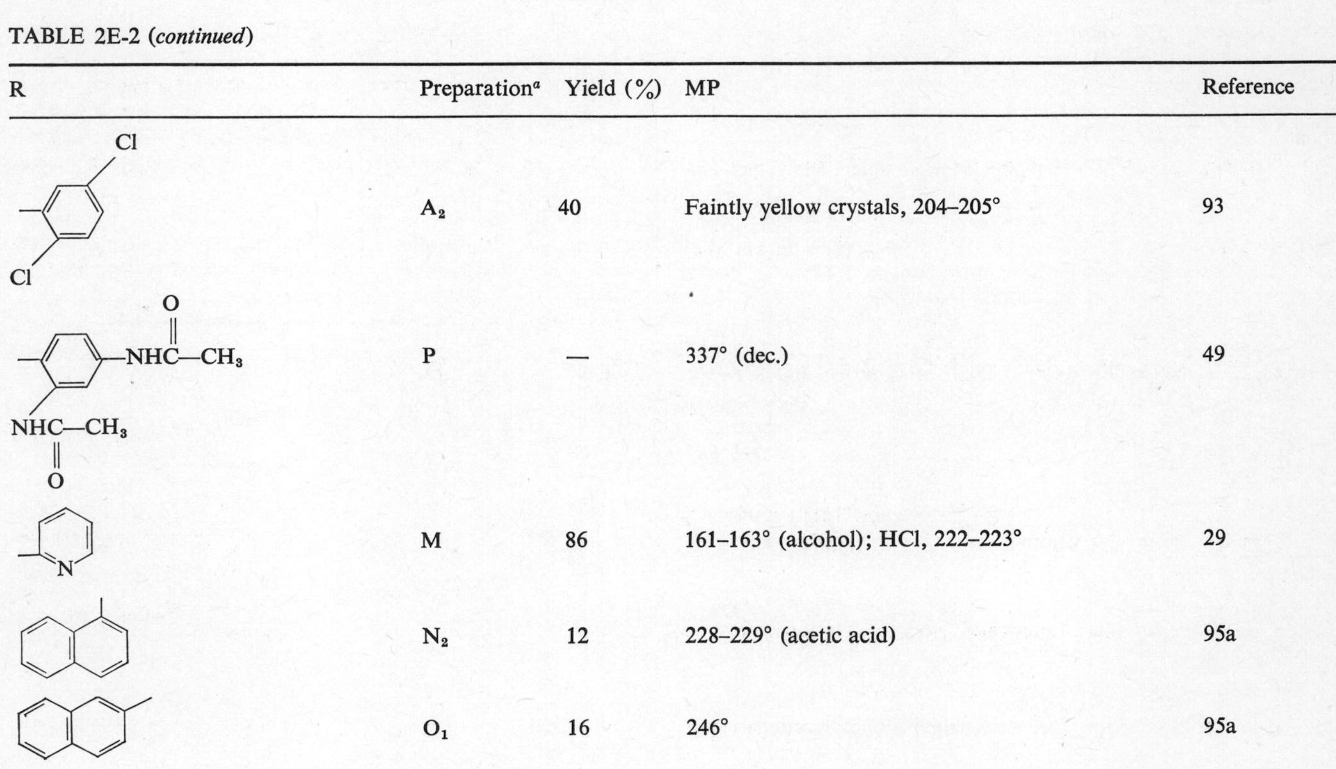

R	Preparation[a]	Yield (%)	MP	Reference
	A_2	40	Faintly yellow crystals, 204–205°	93
	P	—	337° (dec.)	49
	M	86	161–163° (alcohol); HCl, 222–223°	29
	N_2	12	228–229° (acetic acid)	95a
	O_1	16	246°	95a

9-anthryl (structure)	Q	61	Pale yellow microcrystalline powder >300° (1-propanol)	56
4-hydroxy-1-(2H)phthalazinon-2-yl (structure)	R	85	215–220° (dimethylsulfoxide)	57
—$Pb(CH_2CH_3)_3$	S	60	—	96
$CH_2CH_2SO_3H$	—	63	Needles, 279° (acetic acid)	84a
$CH_2CH_2SO_3C_6H_5$	—	53	Needles, 168° (ethanol)	84a
$CH_2CH_2SO_2$—N(morpholino)	—	74	Needles, 196° (ethanol)	84a
$CH_2CH_2SO_2N(Et)_2$	—	66	148° (ethyl acetate)	84a

[a] A, phthalic anhydride + CH_3NHNH_2. Reflux 2 hr in acetic acid. A_1, 4-hydroxy-1-(2*H*)phthalazinone treated with dimethyl sulfate in aq. base. A_2, 2-phthalimidoacetic acid + corresponding hydrazine. Reflux 12–20 hr in ethanol with tributylamine. B, the corresponding 2-alkyl-4-alkoxy-1-(2*H*)phthalazinone, heat with pyridine hydrochloride. C, Mannich reaction: 4-hydroxy-1-(2*H*)phthalazinone + 37% aq. HCH(=O) + amine. Warm on water bath or reflux in ethanol. C_1, same as C. Reflux in ethanol + dimethylformamide. D, 4-hydroxy-1-(2*H*)-phthalazinone + corresponding acyl chloride. Reflux 2 hr in pyridine. D_1, same as D, heat for < 1 hr in nitrobenzene at 160°. D_2, potassium salt of 4-hydroxy-1-(2*H*)phthalazinone + isonicotinoyl chloride. Heat at 210° for 4 hr. E, 4-hydroxy-1-(2*H*)phthalazinone + $ClCH_2CO_2^-Na^+$ in aq. ^-OH at pH 8, reflux 4 hr. E_1, the corresponding ethyl ester stirred at room temperature with 20% NH_4OH for 24 hr. E_2, same as E_1, reflux with $N_2H_4 \cdot H_2O$ in ethanol for 3 hr. F, 4-hydroxy-1-(2*H*)phthalazinone + $CH_2{=}CHCHO$. Heat 7 hr. F_1, reflux 15 min with $CH_2{=}CH{-}CN$ in Triton B, otherwise same as F. G, 4-hydroxy-1-(2*H*)phthalazinone + sodium acrylate or its derivative. Reflux in ethanol

TABLE 2E-2 (*continued*)

for 48–72 hr. G_1, heat with ethyl acrylate in ethanol + sodium, at 100° for 24 hr, otherwise same as G. H, phthalic anhydride + $H_2N-NH-CH_2-CH(OH)-CH_3$, stir 4 hr in acetic acid, or the ketone, prepared by method I, is reduced with PtO_2/H_2 in ethanol. I, 4-hydroxy-1-(2*H*)phthalazinone + $CH_2=CH-C(=O)-CH_3$. Reflux 15 hr in ethanol, or the alcohol prepared by method H is oxidized with CrO_3-acetone-H_2SO_4 by stirring for 10 min. J, reduction of the corresponding nitro compound with Raney Ni/H_2 in ethanol. K, 4-hydroxy-1-(2*H*)phthalazinone, treat with corresponding halide in ethanolic potassium hydroxide. L, 4-hydroxy-1-(2*H*)phthalazinone + corresponding vinylpyridine. Heat at 100° for 30 min in 75% aq. acetic acid. M, *N*-arylphthalimide treated with sodium ethoxide in ethanol or heated with phenylhydrazine. M_1, *N*-(corresponding-nitroaniline)phthalimide + conc. H_2SO_4 kept overnight at room temperature. M_2, same as M_1 but heat 130–160°. N, 2-phenyl-4-hydroxy-1-(2*H*)phthalazinone + fuming HNO_3 (1.45*d*) at 2–3° C. N_1, the corresponding phthalic acid monohydrazide, heat with conc. H_2SO_4 at 100–160°. N_2, same as N_1 kept at room temperature for 2 hr in acetic acid + sulfuric acid. O, phthalic anhydride + *p*-benzenesulfonamidophenylhydrazine hydrochloride. Reflux for 3.5 hr in water. O_1, phthalic anhydride + 2-naphthylhydrazine. Reflux 2 hr in acetic acid with sodium acetate. P, 4-methoxy-2-(2′,4′-diaminophenyl)-1-(2*H*)phthalazinone + NH_2NH_2 + Ac_2O (attempted substitution of methoxy group resulted in hydrolysis). Q, the Diels-Alder adduct of anthracene with phthalazine-1,4-dione. Reflux for 1 hr in aqueous acid. R, sodium salt of 4-hydroxy-1-(2*H*)phthalazinone treated with *t*-butyl hypochlorite in aqueous acetone. S, 4-hydroxy-1-(2*H*)phthalazinone + $ClPb(Et)_3$ in aqueous base.

TABLE 2E-3. 2-Substituted-4-hydroxy-1-(2*H*)phthalazinones with Substituents in Benzenoid Ring

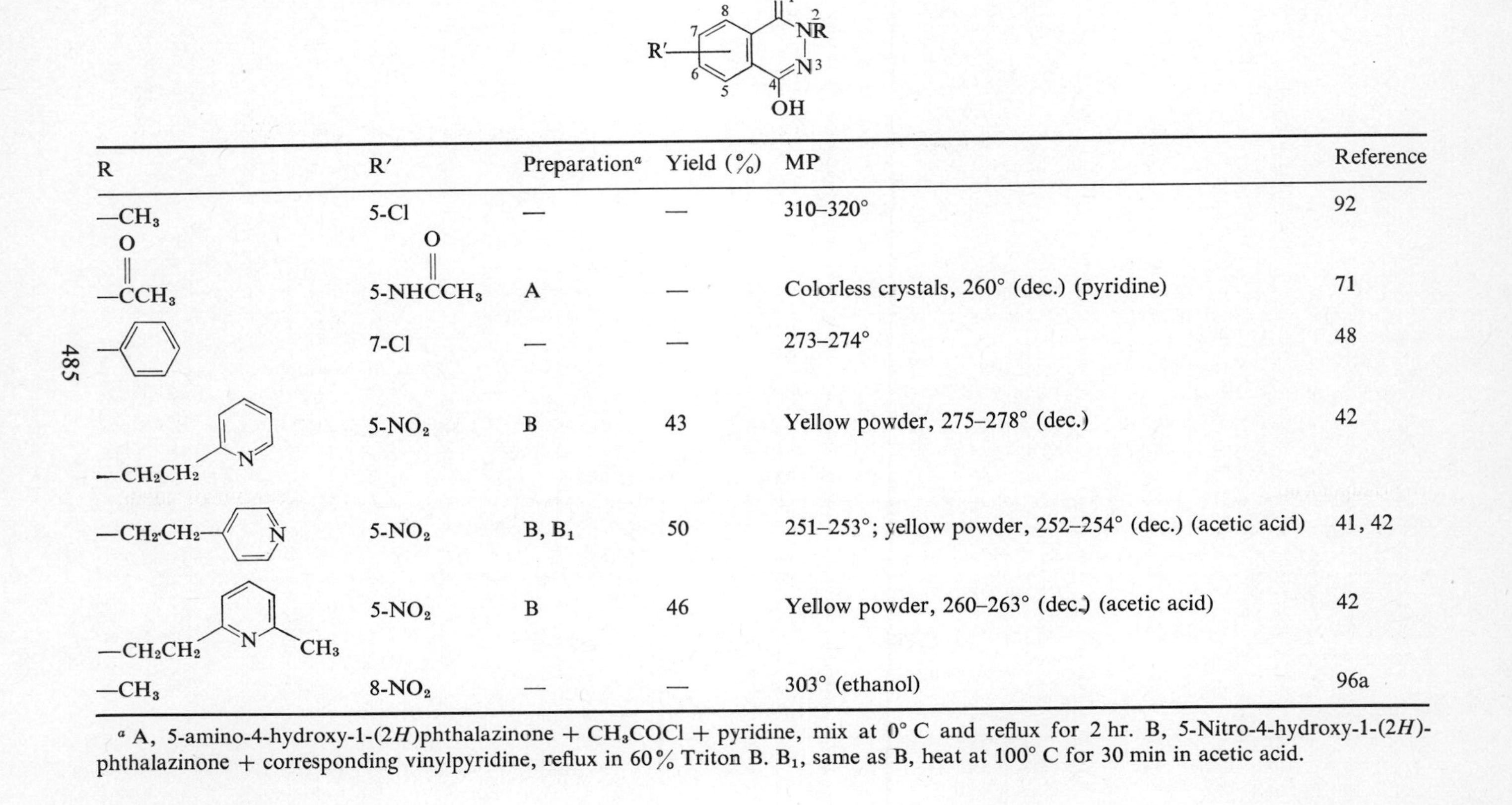

R	R′	Preparation[a]	Yield (%)	MP	Reference
$-CH_3$	5-Cl	—	—	310–320°	92
$-\overset{O}{\overset{\parallel}{C}}CH_3$	$5\text{-}NH\overset{O}{\overset{\parallel}{C}}CH_3$	A	—	Colorless crystals, 260° (dec.) (pyridine)	71
	7-Cl	—	—	273–274°	48
$-CH_2CH_2-$ N	$5\text{-}NO_2$	B	43	Yellow powder, 275–278° (dec.)	42
$-CH_2CH_2-$ N	$5\text{-}NO_2$	B, B_1	50	251–253°; yellow powder, 252–254° (dec.) (acetic acid)	41, 42
$-CH_2CH_2-$ N CH_3	$5\text{-}NO_2$	B	46	Yellow powder, 260–263° (dec.) (acetic acid)	42
$-CH_3$	$8\text{-}NO_2$	—	—	303° (ethanol)	96a

[a] A, 5-amino-4-hydroxy-1-(2*H*)phthalazinone + CH_3COCl + pyridine, mix at 0° C and reflux for 2 hr. B, 5-Nitro-4-hydroxy-1-(2*H*)-phthalazinone + corresponding vinylpyridine, reflux in 60% Triton B. B_1, same as B, heat at 100° C for 30 min in acetic acid.

TABLE 2E-4. 4-*O*-Substituted-1-(2*H*)phthalazinones

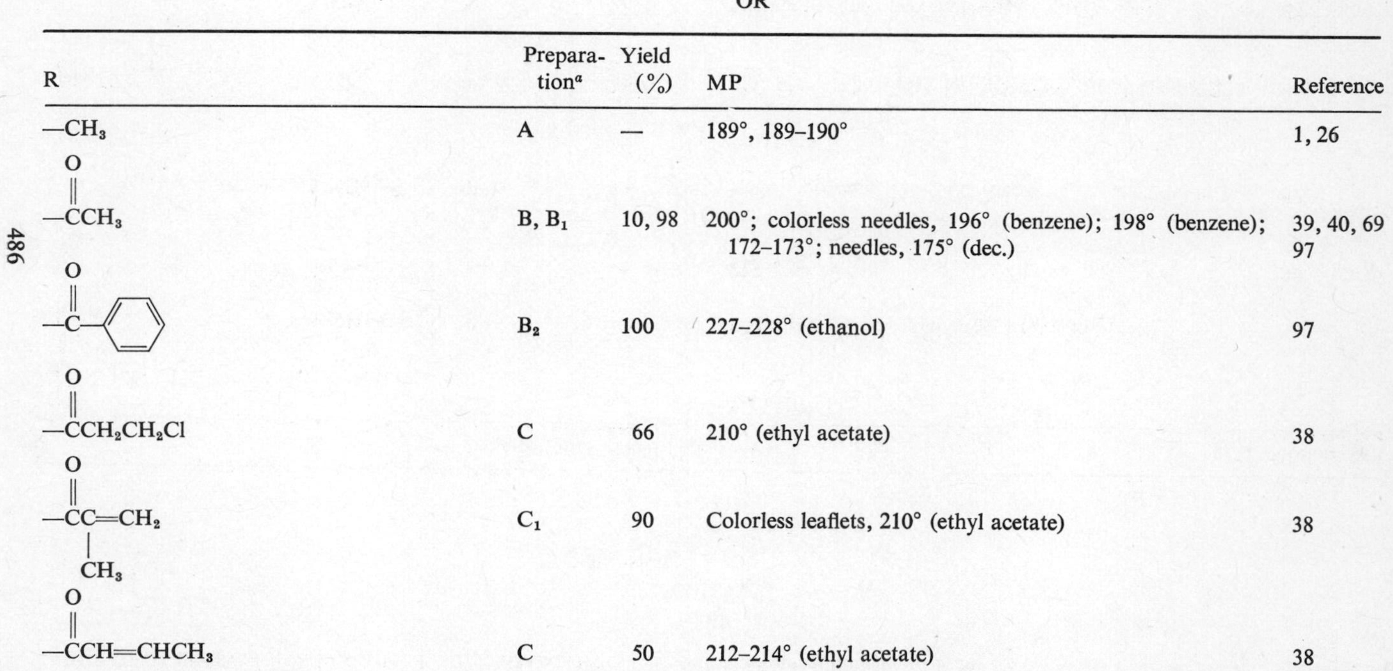

R	Preparation[a]	Yield (%)	MP	Reference
—CH_3	A	—	189°, 189–190°	1, 26
—C(=O)CH_3	B, B_1	10, 98	200°; colorless needles, 196° (benzene); 198° (benzene); 172–173°; needles, 175° (dec.)	39, 40, 69 97
—C(=O)— 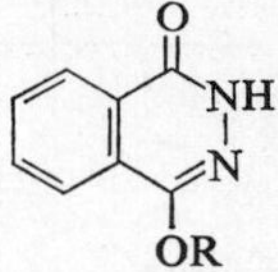	B_2	100	227–228° (ethanol)	97
—C(=O)CH_2CH_2Cl	C	66	210° (ethyl acetate)	38
—C(=O)C(CH_3)=CH_2	C_1	90	Colorless leaflets, 210° (ethyl acetate)	38
—C(=O)CH=CHCH_3	C	50	212–214° (ethyl acetate)	38

$-CH_2CO_2CH_2CH_3$	D	50	190° (ethanol)	2, 36
$-CH_2CO_2H$	—	—	281° (ethanol)	2, 36
$-CH_2C(=O)NH_2$	E	82	298° (aq. ethanol)	94
$-CH_2C(=O)NHNH_2$	E_1	80	235° (ethanol)	94
$-P(=S)-(OCH_2CH_3)_2$	F	—	99–100°	98
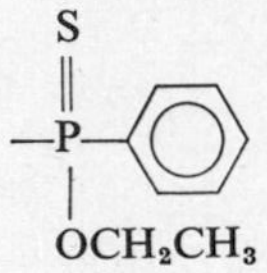 $-P(=S)(C_6H_5)OCH_2CH_3$	F_1	—	135.5–136.5°	99
$-S(=O)_2CH_2(CH_2)_2CH_3$	G	64	65°	22
$-S(=O)_2-CH_2CH_2CH(CH_3)_2$	G	54	145°	22

TABLE 2E-4 (*continued*)

R	Preparation[a]	Yield (%)	MP	Reference
$-SO_2CH_2C_6H_5$	G	70	156°	22
$-SO_2C_6H_5$	G	76	88–90° (ethanol); needles, 197° (ethanol)	22, 84a
$-SO_2C_6H_4Cl$ (p)	G	82	175°	22
$-SO_2C_6H_4CH_3$ (p)	G	62	96–98°	22
$-SO_2CH_3$	—	72	Needles, 197° (ethanol)	84a
$-SO_2CH_2CH_2Cl$	—	83	Needles, 250° (ethanol)	84a
$-SO_2CH_2CH_2-N^{\oplus}C_5H_5\ Cl^{\ominus}$	—	40	Crystals, 228–230°	84a

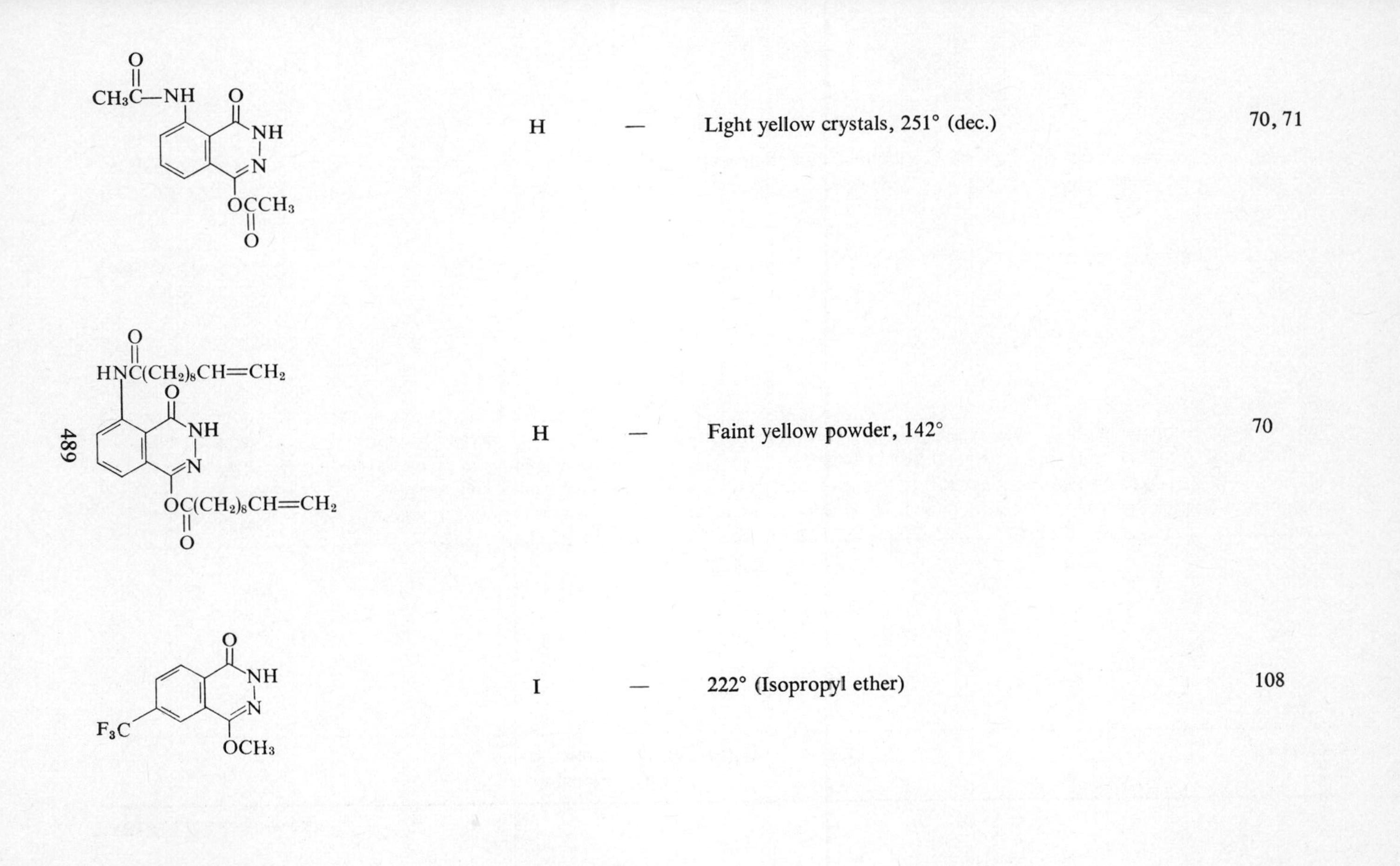

	H	—	Light yellow crystals, 251° (dec.)	70, 71
	H	—	Faint yellow powder, 142°	70
	I	—	222° (Isopropyl ether)	108

TABLE 2E–4 (*conitnued*)

R	Preparation	Yield (%)	MP	Reference
7-F_3C-4-OCH_3-1-(2*H*)-phthalazinone (structure: F_3C, O, NH, N, OCH_3)	I	—	226° (Isopropyl ether)	108

[a] A, methylation of 4-hydroxy-1-(2*H*)phthalazinone with diazomethane in acetone or methylation of its silver salt with methyl iodide. B, acetylation of 4-hydroxy-1-(2*H*)phthalazinone with acetyl chloride in chloroform, stir 5 hr. B_1, reflux 2-acetyl-4-acetoxy-1-(2*H*)phthalazinone in ethanol for 3 hr. B_2, same as B, stir in pyridine overnight at room temperature. C, 4-hydroxy-1-(2*H*)phthalazinone + corresponding acroyl chloride—heat at 160° for 3 min. C_1, same as C, heat for 20 min. D, 4-hydroxy-1-(2*H*)phthalazinone + ethyl-α-chloroacetate + NaOH. Reflux in aqueous ethanol for 8 hr. E, the corresponding ethyl ester was stirred with 20% NH_4OH at room temperature for 24 hr. E_1, same as E, reflux in ethanol for 3 hr. F, 4-hydroxy-1-(2*H*)phthalazinone + $Cl—P(=S)—(OEt)_2$, heat 60–65° for 2 hr in presence of base. F_1, same as F, reflux with $Cl—P(=S)(C_6H_5)—OEt$ in methyl ethyl ketone with potassium carbonate for 30 hr. G, 4-hydroxy-1-(2*H*)phthalazinone + corresponding sulfonyl chloride, reflux in 10% aq. sodium hydroxide solution for 12 hr. H, 5-amino-4-hydroxy-1-(2*H*)phthalazinone + acyl chloride, stir at room temperature overnight in dimethylformamide with pyridine. I, corresponding 1-chloro-4-methoxyphthalazine was refluxed with glacial acetic acid for 2 hr.

TABLE 2E-5. 2,4-*N*,*O*-Disubstituted-4-hydroxy-1-(2*H*)phthalazinones

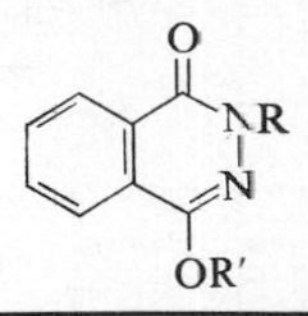

R	R′	Preparation[a]	Yield (%)	MP	Reference
$—CH_3$	$—CH_3$	A, A_1	—	93°, 94–95°, 97–98°	1, 107
$—CH_3$	$—SCCl_3$	A_2	—	128–129°	100
$—CH_2CH_3$	$—SCCl_3$	A_2	—	105–106°	100
$—CH_2CH_2CH_3$	$—CH_2CH_2CH_3$	B	—	BP 229–231°/20 mm	33
$—CH(CH_3)_2$	$—CH(CH_3)_2$	B	—	BP 222–224°/16 mm	33
$—CH_2(CH_2)_2CH_3$	$—CH_2(CH_2)_2CH_3$	B	—	BP 232–234°/20 mm	33
$—CH_2CH(CH_3)_2$	$—CH_2CH(CH_3)_2$	B	—	BP 228–230°/15 mm	33
$—CH_2CH_2CH(CH_3)_2$	$—CHCH_2CH(CH_3)_2$	B	90	65°; BP 228–235°/15 mm	33
$—CH_2(CH_2)_4CH_3$	$—CH_2(CH_2)_4CH_3$	B	—	BP 240°/15 mm	33
—(cyclohexyl)	—(cyclohexyl)	B_1	—	—	32
-n-$C_{12}H_{25}$	n-$C_{12}H_{25}$	B	—	BP 270–275°/0.5 mm	47
$—CH_2CH{=}CH_2$	$—CH_2CH{=}CH_2$	B	—	BP 200–202°/12 mm	47
$—CH_2CO_2H$	$—CH_2CO_2H$	B_2	83	281° (acetic acid)	2, 36
$—C(=O)CH_3$	$—C(=O)CH_3$	C	90	Colorless needles, 139–140° (benzene) (133°)	39, 40, 97
$—CH_2CH_2OC(=O)CH_3$	$—C(=O)CH_3$	C_1	71	126–127° (ethanol)	48

TABLE 2E-5 (*continued*)

R	R′	Preparation[a]	Yield (%)	MP	Reference
$—CH_2CH_2N(Et)_2$	$—CH_3$	D	—	BP 165–167°/1.5 mm	51
$—CH_2CH_2N(Et)_2$	$—CH_2CH_2N(Et)_2$	D_1	—	BP 198–201°/1 mm; 2 HCl, 231°; picrate, 159–160°	51
$—CH_2—C_6H_5$	$—CH_2CH_3$	A_1, E	—	74–76°	107
$—CH_2—C_6H_5$	-*iso*-Amyl	E	—	73%; BP 245–260°/1.5 mm	33
$—CH_2—C_6H_5$	-*n*-Dodecyl	E	—	BP 323–325°/11 mm; 325°/10 mm	33, 47
$—CH_2—C_6H_5$	$—CH_2—C_6H_5$	E	—	95–96°	33
$—CH_2CH_2—C_5H_4N$ (4-pyridyl)	$—CH_2CH_2OH$	—	—	110–113°	48
$—C_6H_5$	$—CH_3$	F	96	116–117°	49
$—C_6H_5$	$—CH_2CH_3$	F	77	Needles, 104° (pet. ether)	27
$—C_6H_5$	-*n*-Propyl	E	—	Colorless needles, 75° (methanol); BP 255–258°/18 mm	47

—C6H5 (phenyl)	*iso*-Propyl	E	—	Colorless needles, 112° (alcohol); BP 265–275°/18 mm	47
—C6H5 (phenyl)	-Allyl	E	—	Pale yellow needles, 141° (alcohol)	47
—C6H5 (phenyl)	-*iso*-Butyl	E	—	Long yellow needles, 99° (methanol); BP 258–260°/12 mm	47
—C6H5 (phenyl)	-*iso*-Amyl	E	71	Needles, 58–59° (methanol); 56–58°; BP 244–245°/5 mm; BP 265–270°/12 mm	47, 101
—C6H5 (phenyl)	-*n*-Hexyl	E	—	BP 230–235°/0.5 mm	47
—C6H5 (phenyl)	-*n*-Octyl	E	—	BP 242–247°/0.5 mm	47
—C6H5 (phenyl)	-*n*-Dodecyl	E	—	BP 255–260°/0.5 mm	47
—C6H5 (phenyl)	-*n*-Hexadecyl	E	—	BP 285–295°/0.5 mm	47
—C6H5 (phenyl)	-*n*-Octadecyl	E	—	55° (ligroin)	47
—C6H5 (phenyl)	-*n*-9-Octadecenyl	E	—	BP 298–300°/0.5 mm	47

TABLE 2E-5 (*continued*)

R	R′	Preparation[a]	Yield (%)	MP	Reference
$-C_6H_5$	-Chaulmoogryl	E	—	BP 300–320°/1 mm	47
$-C_6H_5$	$-C(=O)CH_2Cl$	G	68	127–129° (ligroin)	48
$-C_6H_5$	$-C(=O)CH_2-N$(piperidino)	G_1	28	170–172° (aq. ethanol)	48
$-C_6H_5$	$-C(=O)CH_2-N$(morpholino)	—	—	205–207°	48
$-C_6H_5$	$-CH_2CH_2OH$	E	20	165–167° (ethanol)	48
$-C_6H_5$	$-CH_2CH(OH)CH_2OH$	—	—	Needles, 82°	47
$-C_6H_5$	$-CH_2CH_2-N$(morpholino)	F_1	83	HCl, 185–18.65°	50
$-C_6H_5$	$-C(=O)C_6H_5$	—	—	144–145°	48

$-C_6H_5$	$-C(=O)-C_6H_4-NO_2$	—	—	202–203°	48
$-C_6H_5$	$-C(=O)-C_6H_4-O(CH_2)_4CH_3$	—	—	75–77°	48
$-C_6H_5$	$-C(=O)-C_6H_2(OCH_3)_3$	—	—	217–218°	48
$-C_6H_5$	$-C(=O)CH_2-C_6H_5$	—	—	123–125°	48
$-C_6H_5$	$-C(=O)CH_2-C_6H_4-NO_2$	—	—	179–181°	48
$-C_6H_5$	$-S(=O)_2-C_6H_4-CH_3$	—	—	120–122° (benzene-ligroin)	48, 102
$-C_6H_5$	$-CH_2-C_6H_5$	E	—	Needles, 208° (ethanol)	47

TABLE 2E-5 (*continued*)

R	R′	Preparation[a]	Yield (%)	MP	Reference
$—C_6H_5$	$—CH_2—C_6H_4—OCH_3$ (p)	E	—	Yellow needles, 215° (ethanol); BP ~280°/0.5 mm	47
$—C_6H_5$	$—CH_2—C_6H_3(CH_3)_2$ (2,5)	E	—	Yellow needles, 206–207° (ethanol); BP 255–258°/0.5 mm	47
$—C_6H_5$	$—CH_2CH_2—C_6H_5$	E	—	Yellow prisms, 124° (ethanol); BP 304–305°/12 mm	47
2-pyridyl	$—CH_3$	H	27	185–186°	29
2-pyridyl	$—C(=O)CH_3$	I	44	134–136°	29
$—C_6H_4NO_2$ (m)	$—C(=O)CH_3$	I	77	165° (alcohol)	30

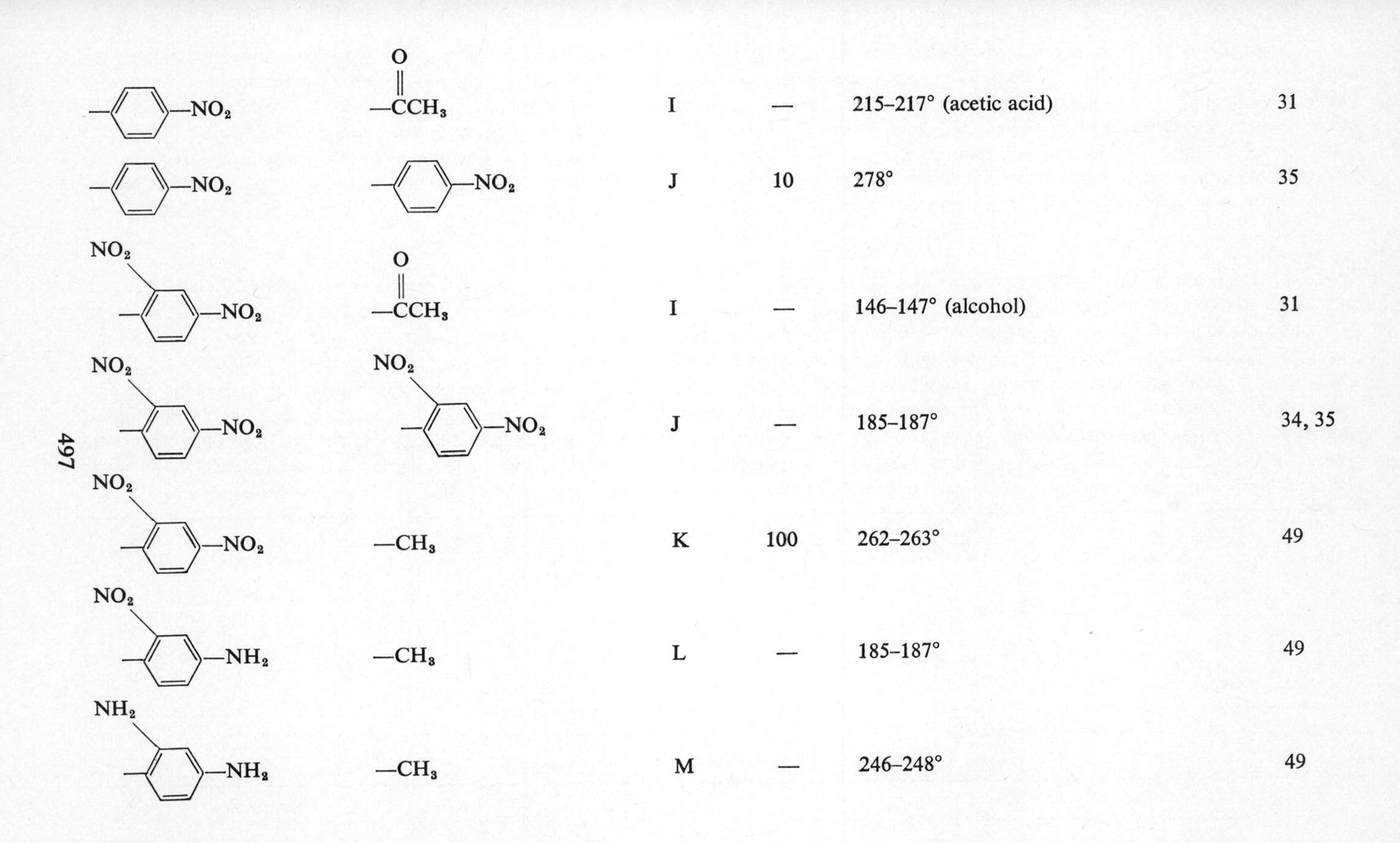

$-C_6H_4-NO_2$ (p)	$-C(=O)CH_3$	I	—	215–217° (acetic acid)	31
$-C_6H_4-NO_2$ (p)	$-C_6H_4-NO_2$ (p)	J	10	278°	35
$-C_6H_3(NO_2)_2$ (2,4)	$-C(=O)CH_3$	I	—	146–147° (alcohol)	31
$-C_6H_3(NO_2)_2$ (2,4)	$-C_6H_3(NO_2)_2$ (2,4)	J	—	185–187°	34, 35
$-C_6H_3(NO_2)_2$ (2,4)	$-CH_3$	K	100	262–263°	49
$-C_6H_3(NO_2)(NH_2)$ (2-NO_2, 4-NH_2)	$-CH_3$	L	—	185–187°	49
$-C_6H_3(NH_2)_2$ (2,4)	$-CH_3$	M	—	246–248°	49

TABLE 2E–5 (*continued*)

R	R′	Preparation[a]	Yield (%)	MP	Reference
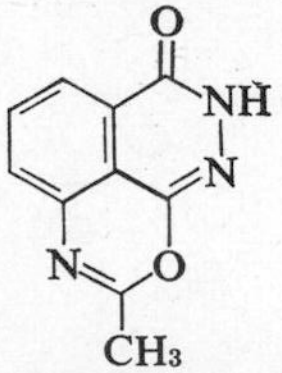 O, NH, N, N, O, CH_3	—	N	—	Faint yellow needles, 195°	71

[a] A, reflux 2-methyl-4-hydroxy- or 4-methoxy-1-(2*H*)phthalazinone with dimethyl sulfate in aqueous potassium hydroxide for 2–4 hr. A_1, 1-alkoxy-3-alkylphthalazinium iodide treated with $K_3Fe(CN)_6$. A_2, stir 2-alkyl-4-hydroxy-1-(2*H*)phthalazinone with trichloromethanesulfenyl chloride in chloroform with alkali at room temperature. B, 4-hydroxy-1-(2*H*)phthalazinone + alkyl halide, reflux 40 hr in ethanolic potassium hydroxide. B_1, same as B, reflux with cyclohexyl tosylate in ethanol containing potassium ethoxide. B_2, same as B. Reflux with chloroacetic acid for 12 hr in ethanol containing alkali, pH > 8. C, 4-hydroxy-1-(2*H*)phthalazinone refluxed with acetic anhydride for 6 hr. C_1, 2-(2′-hydroxyethyl)-4-hydroxy-1-(2*H*)phthalazinone treated as in C. D, 4-alkoxy-1-(2*H*)phthalazinone + Cl—CH_2—CH_2—$N(Et)_2 \cdot HCl$. Stir 3 hr at 70° in ethanol containing sodium. D_1, same as D, stir 4 hr at 55°. E, the corresponding 2-substituted-4-hydroxy-1-(2*H*)phthalazinone + proper alkyl halide. Reflux in ethanolic potassium hydroxide for ~40 hr. F, 2-phenyl-4-chloro-1-(2*H*)-phthalazinone refluxed for ~1 hr in corresponding alcohol containing sodium alkoxide. F_1, same as F. Heat in xylene with sodium. G, 2-phenyl-4-hydroxy-1-(2*H*)phthalazinone treated with Cl—C(=O)CH_2Cl. G_1, product obtained by method G, treated with piperidine. H, 2-(2′-pyridyl-1,4-hydroxy-1-(2*H*)phthalazinone + CH_3I. Reflux in methanolic potassium hydroxide for 2 hr. I, 2-substituted-4-hydroxy-1-(2*H*)phthalazinone refluxed with acetic anhydride 5–30 min. J, the disodio salt of 4-hydroxy-1-(2*H*)phthalazinone + 4-nitrochlorobenzene, heat at 270°–300° for 2 hr. J_1, same as J, heat with 2,4-dinitrochlorobenzene at 210–220° for 4 hr. K, 2-phenyl-4-methoxy-1-(2*H*)phthalazinone treated with HNO_3 at 10–20° C; or 2-(2′,4′-dinitrophenyl)-4-hydroxy-1-(2*H*)phthalazinone treated with dimethyl sulfate in aqueous sodium carbonate. L, 2-(2′,4′-dinitrophenyl)-4-methoxy-1-(2*H*)phthalazinone treated with $N_2H_4 \cdot H_2O$ in methanol. M, reduction of 2-(2′,4′-dinitro- or 2′-nitroso-4-amino)-4-methoxy-1-(2*H*)phthalazinone with hydrogen sulfide. N, 5-amino-4-hydroxy-1-(2*H*)phthalazinone refluxed with excess of acetic anhydride.

ABLE 2E-6. 2,3-Disubstituted Phthalazine-1,4-dione

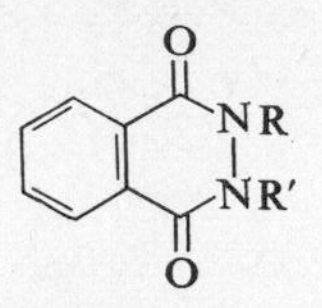

	R′	Preparation[a]	Yield	MP	Reference
$-CH_3$	$-CH_3$	—	—	Needles, 174–175°	1
$-CH_3$	$-\overset{\text{O}}{\overset{\parallel}{\text{C}}}\text{CH}_3$	A	—	White, 139.5–140.5°	93
$-\overset{\text{S}}{\overset{\parallel}{\text{P}}}(\text{OCH}_3)_2$	$-\overset{\text{O}}{\overset{\parallel}{\text{C}}}\text{CH}_3$	—	—	112–113° (ether)	103
$-\overset{\text{S}}{\overset{\parallel}{\text{P}}}(\text{OCH}_2\text{CH}_3)_2$	$-\overset{\text{O}}{\overset{\parallel}{\text{C}}}\text{CH}_3$	—	—	96–98° (ether)	103
$-\overset{\text{S}}{\overset{\parallel}{\text{P}}}(\text{OCH}_3)_2$	$-\overset{\text{S}}{\overset{\parallel}{\text{P}}}(\text{OCH}_3)_2$	B	—	112–114° (ether)	103
$-\overset{\text{S}}{\overset{\parallel}{\text{P}}}(\text{OCH}_2\text{CH}_3)_2$	$-\overset{\text{S}}{\overset{\parallel}{\text{P}}}(\text{OCH}_2\text{CH}_3)_2$	B	—	106–107° (cyclohexane-ether)	103

[a] A, 2-methyl-4-hydroxy-1-(2*H*)phthalazinone, heat with acetic anhydride. B, phthaloyl chloride +

$$\begin{array}{l} \text{(N}-\overset{\text{S}}{\overset{\parallel}{\text{P}}}-(\text{OR})_2 \\ \;\; | \\ \text{(N}-\underset{\text{S}}{\underset{\parallel}{\text{P}}}-(\text{OR})_2 \end{array}$$

. Heat in benzene for 5 hr at 50° in presence of triethylamine.

TABLE 2E-7. 2,3-Diphenylphthalazine-1,4-dione with Substituents in the Benzenoid Ring

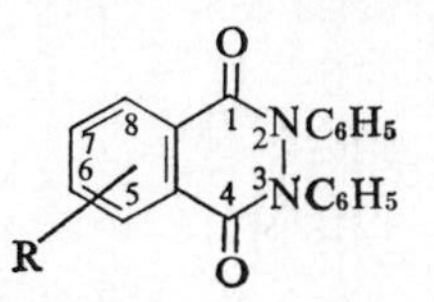

R	Preparation[a]	Yield (%)	MP	Reference
5-OH	A	—	188–189° (methanol)	52, 53
6-OH	A	—	268–270° (methanol)	52, 53
5-NO_2	B	81	250–254° (acetic acid)	52, 53
6-NO_2	A_1, B_1	69	189–191° (methanol)	52, 53
5-NH_2	C	—	190–192°	52, 53
6-NH_2	C	—	238–240°	52, 53
5-NH—$(CH_2)_2N(Et)_2$	D	—	HCl, 251–253°	52, 53
5-NH—C(=O)—CH_2Cl	D_1	100	184–187°	53, 104
5-NH—C(=O)—$CH_2N(Et)_2$	D_2	76	156–157°	53, 104
5-O—C(=O)—CH_2Br	E	68	185–189°	53, 104
5-O—C(=O)—CH=CH_2	E_1	70	193–194°	53

$5\text{-}O{-}C(=O){-}(CH_2)_{16}CH_3$	E_2	57	80–82° (ether)	53, 104
$5\text{-}O{-}C(=O){-}(CH_2)_7CH{=}CH(CH_2)_7CH_3$	E_2	42	69–70° (cyclohexane); 71–73° (aq. methanol)	53, 104
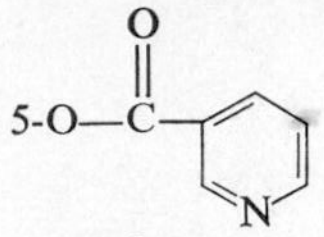 $5\text{-}O{-}C(=O){-}$(3-pyridyl)	E_3	83	187–190°; CH_3I, 245–250° (dec.); CH_3OSO_3H, 225–228°; CH_3CH_2Br, 240° (dec.)	53, 104
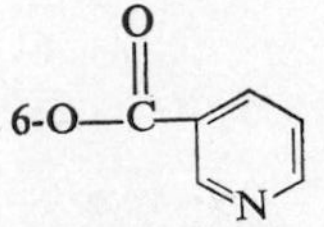 $6\text{-}O{-}C(=O){-}$(3-pyridyl)	E_3	60	169–171°	53, 104
$6\text{-}OCH_2CH_2N(Et)_2$	F	—	120–121°	52, 53
$5\text{-}O{-}C(=O){-}CH_2{-}N^{\oplus}(CH_3)_3\ Cl^{\ominus}$	G	76	2.5–3 H_2O, 163–165° (dec.) (2-propanol)	53, 104
$5\text{-}O{-}C(=O){-}CH_2{-}N^{\oplus}H(Et)_2\ Cl^{\ominus}$	G_1	—	116–118° (dec.) (abs. alcohol); free base, 129–131° (benzene-pet. ether); CH_3CH_2Br, 168° (2-propanol)	53, 104
$5\text{-}O{-}C(=O){-}CH_2{-}N^{\oplus}(Et)_3\ Br^{\ominus}$	E_4	—	166–168° (acetone)	53, 104

TABLE 2E-7 (*continued*)

R	Preparation[a]	Yield (%)	MP	Reference
$5\text{-O—C(=O)—O—(CH}_2)_2\overset{\oplus}{N}(CH_3)_3\ I^{\ominus}$	H	66	—	53, 104
$5\text{-O—C(=O)—O—(CH}_2)_2\overset{\oplus}{N}(Et)_3\ I^{\ominus}$	H	74	; nitrate, H_2O, 150° (dec.)	53, 104

[a] A, the corresponding phthalic anhydride with 1,2-diphenylhydrazine in refluxing pyridine for 7–10 hr. A_1, same as A, heat at 100° for 3 hr. B, the corresponding phthaloyl chloride + 1,2-diphenylhydrazine + *N,N*-dimethylaniline; keep in benzene for 24 hr. B_1, same as B, heat at 100° for 2 hr. C, the corresponding nitro compound reduced with Raney Ni/H_2 in methanol. D, 5-amino-2,3-diphenylphthalazine-1,4-dione + Cl—$(CH_2)_2$ $N(Et)_2$, heat 3 hr at 150°. D_1, same as D; treat with chloroacetyl chloride. D_2, the product obtained by method D_1 is refluxed with diethylamine in acetone containing potassium carbonate for 8 hr. E, 5-hydroxy-2,3-diphenylphthalazine-1,4-dione + BrC(=O)—CH_2Br. Reflux in toluene for 1 hr. E_1, the sodium salt of 5-or (6) hydroxy-2,3-diphenylphthalazine-1,4-dione was heated at 60–65° with acryloyl chloride in toluene for 3 hr. E_2, same as E_1; the sodium salt was refluxed in dichloromethane for 2 hr with the acid chloride. E_3, same as E_1; the sodium salt was refluxed in toluene with acid chloride for 1.5–3 hr. E_4, the compound prepared by method E was heated for 20 hr at 100° with triethylamine in a sealed tube. F, 5-hydroxy-2,3-diphenylphthalazine-1,4-dione + $Cl(CH_2)_2N(Et)_2$. Reflux 8 hr in toluene with sodium ethoxide. G, 5-hydroxy-2,3-diphenylphthalazine-1,4-dione + Cl—C(=O)—CH_2—$\overset{\oplus}{N}(CH_3)_3\ Cl^{\ominus}$; stir in pyridine. G_1, same as G but using Cl—C(=O)—CH_2—$\overset{\oplus}{N}$(H)$(Et)_2Cl^{\ominus}$.

H, sodium salt of 5-hydroxy-2,3-diphenylphthalazine-1,4-dione + Cl—C(=O)—O—$(CH_2)_2\overset{\oplus}{N}(R)_3\ Cl^{\ominus}$; stir in dichloromethane 3 days at room temperature; treat with aqueous potassium iodide.

TABLE 2E-8. Phthalazine-1,4-dione in a Fused Ring System at N_2,N_3 Positions

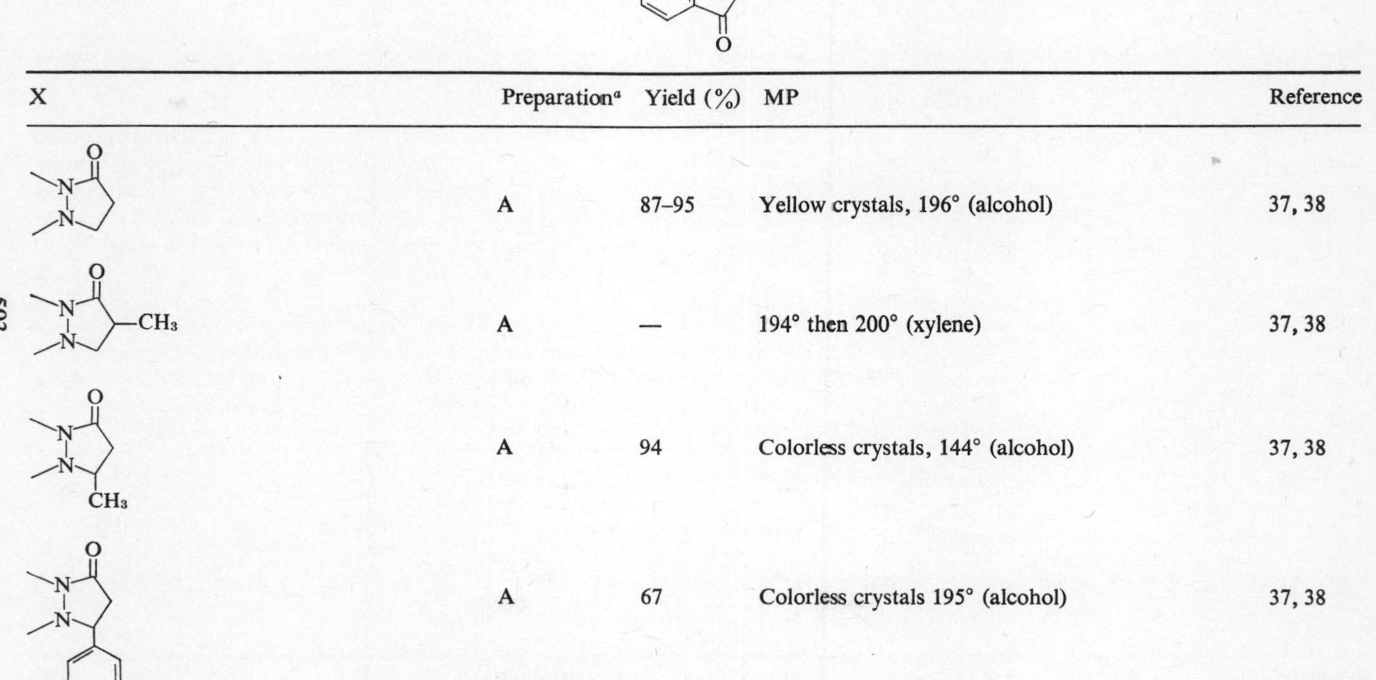

X	Preparation[a]	Yield (%)	MP	Reference
	A	87–95	Yellow crystals, 196° (alcohol)	37, 38
	A	—	194° then 200° (xylene)	37, 38
	A	94	Colorless crystals, 144° (alcohol)	37, 38
	A	67	Colorless crystals 195° (alcohol)	37, 38

TABLE 2E-8 (*continued*)

X	Preparation[a]	Yield (%)	MP	Reference
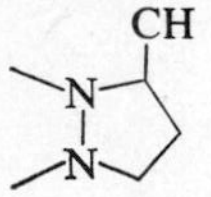 CH, N, N	B	67	Colorless prisms, 213° (ethanol)	44
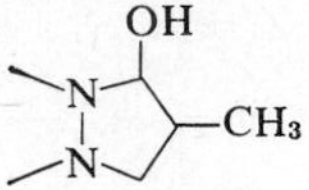 OH, CH_3, N, N	B	78	Colorless needles, 178°	44
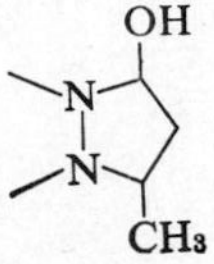 OH, CH_3, N, N	B	84	152° (ethanol)	44
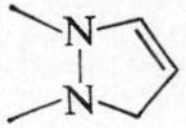 N, N	C	44	230° (ethyl acetate)	44
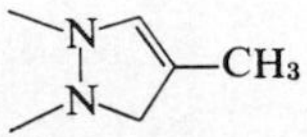 CH_3, N, N	C	85	270° (ethanol)	44
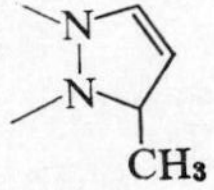 CH_3, N, N	C	83	128–130° (ethyl acetate)	44
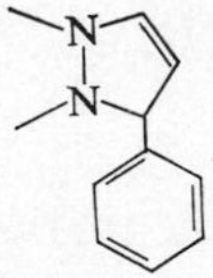 N, N	B_1	25	228° (ethanol)	44

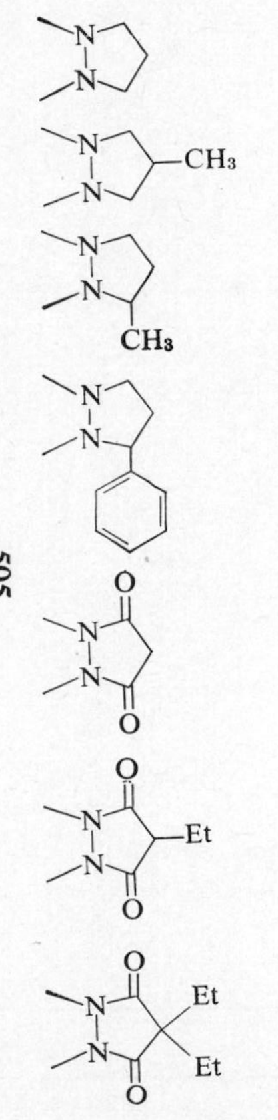

	D	99	206° (ethyl acetate)	44
	D	93	190° (ethyl acetate)	44
	D	87	138° (ethyl acetate)	44
	D	72	154° (ethyl acetate)	44
	E	82	Yellow compound, 330° (dec.)	40
	E_1	84	Yellow needles, 188° (ethanol)	40
	E_2	78–90	174° (ethanol)	40

TABLE 2E-8 (*continued*)

X	Preparation[a]	Yield (%)	MP	Reference
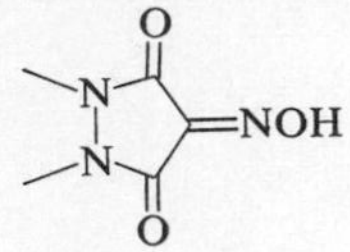 (O, O, =NOH)	F	83	300° (dec.)	40
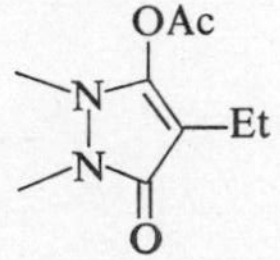 (OAc, —Et, O)	G	97	Yellow needles, 180° (ethanol)	40
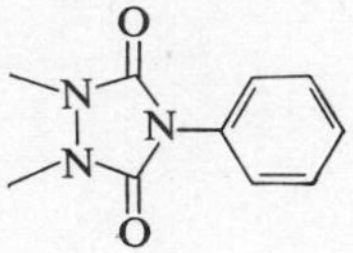 (O, O, N—Ph)	H	25	Colorless crystals, 316° (DMF)	105
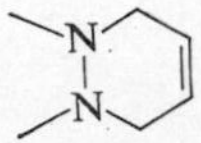	I, I_1, I_2	77–90	Fine white needles, 272–275° (dec.) (acetic acid + water); 263–268° (dec.) (acetone); 272–275°	55, 57, 58
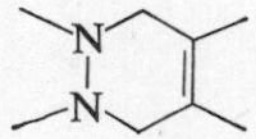	I_4	50	200° (acetone); dimorphic, colorless prisms, 198° and yellow needles, 198°	57, 58
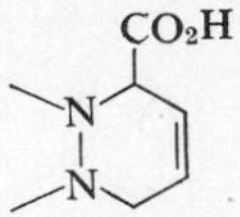 (CO_2H)	—	—	—	63

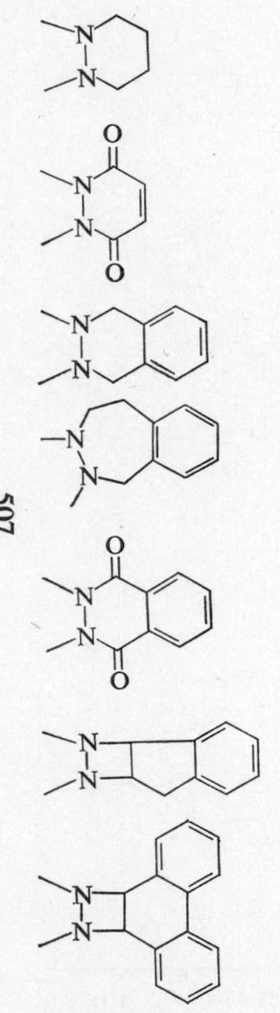

	J	56–82	160° (water); two crystalline forms, white swords, 145.5–146.5° and white needles, 156–156.5°, mixed mp 156–156.5°	55, 106
	K	83	Yellow plates, 244° (acetic acid)	40
	L	27	195°	54
	L	25	Colorless needles, 181–182° (ethanol)	54
	M	33–66	>300° (dec.) (DMF); 430°	40, 57
	N	58	White needles, 256–258° (dec.) (chloroform-hexane)	59
	N	5	White needles, 308, 310° (toluene)	59

TABLE 2E-8 (*continued*)

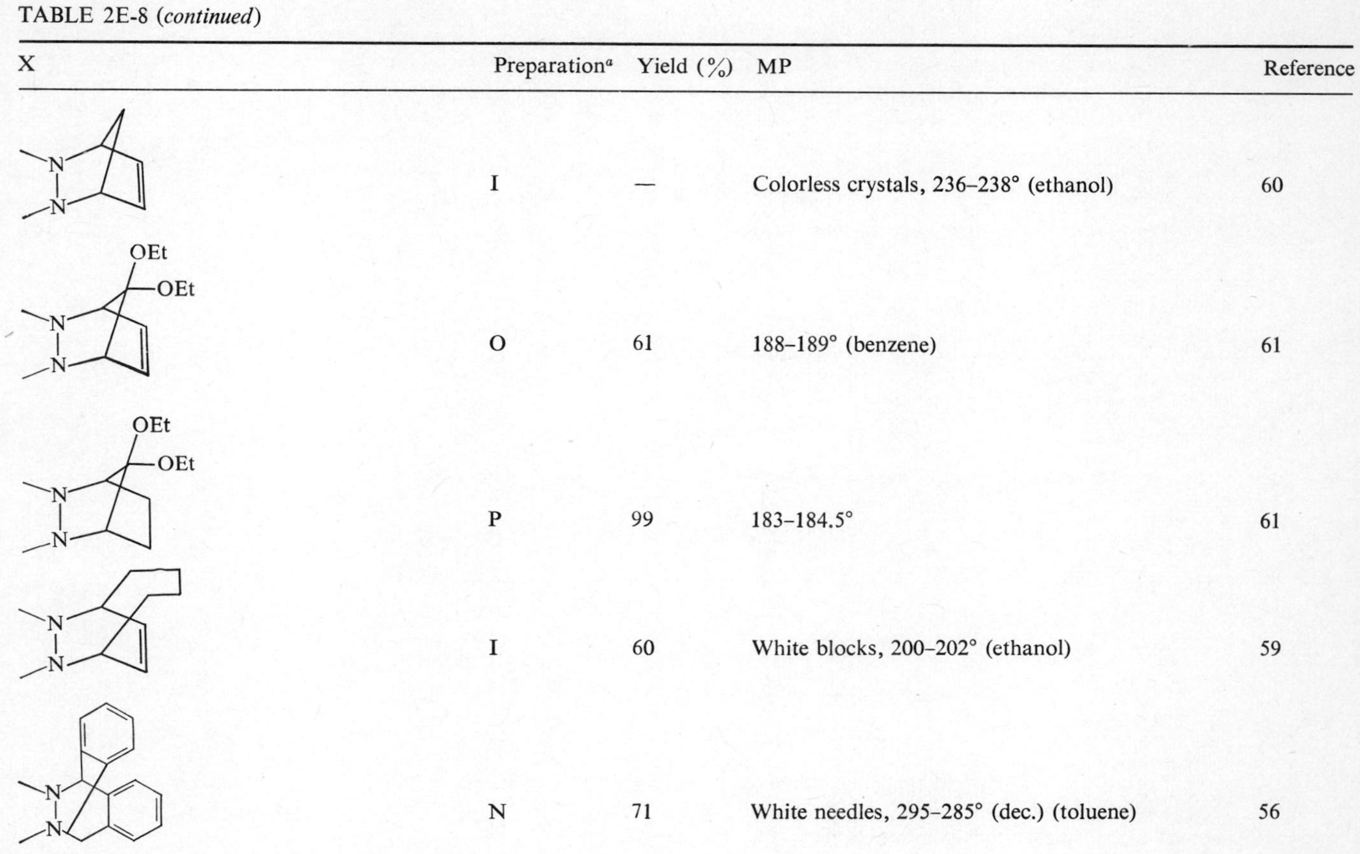

X	Preparation[a]	Yield (%)	MP	Reference
	I	—	Colorless crystals, 236–238° (ethanol)	60
	O	61	188–189° (benzene)	61
	P	99	183–184.5°	61
	I	60	White blocks, 200–202° (ethanol)	59
	N	71	White needles, 295–285° (dec.) (toluene)	56

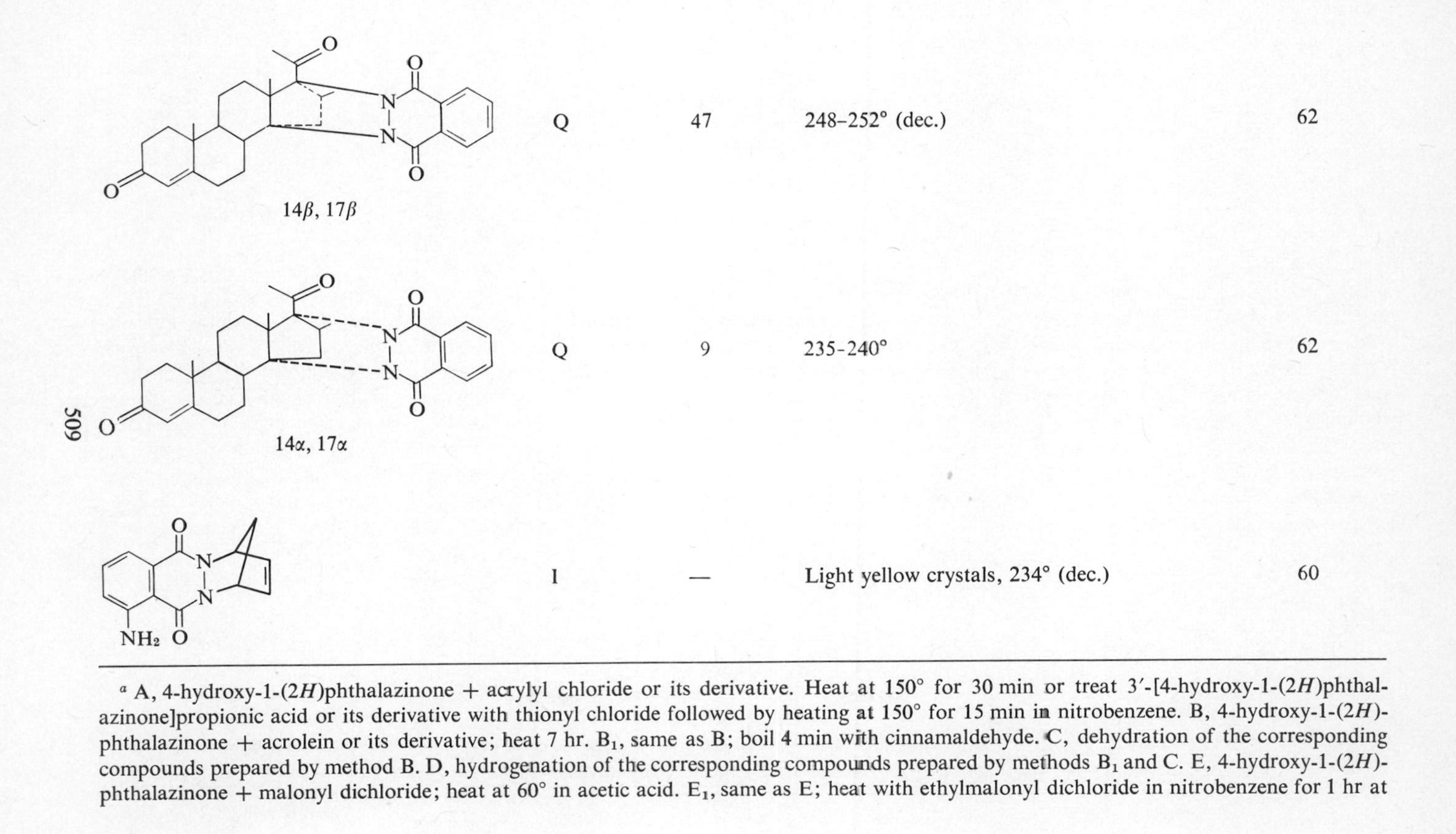

Compound	Method[a]	Yield (%)	Properties	Ref.
14β, 17β	Q	47	248–252° (dec.)	62
14α, 17α	Q	9	235–240°	62
	I	—	Light yellow crystals, 234° (dec.)	60

[a] A, 4-hydroxy-1-(2*H*)phthalazinone + acrylyl chloride or its derivative. Heat at 150° for 30 min or treat 3′-[4-hydroxy-1-(2*H*)phthalazinone]propionic acid or its derivative with thionyl chloride followed by heating at 150° for 15 min in nitrobenzene. B, 4-hydroxy-1-(2*H*)-phthalazinone + acrolein or its derivative; heat 7 hr. B_1, same as B; boil 4 min with cinnamaldehyde. C, dehydration of the corresponding compounds prepared by method B. D, hydrogenation of the corresponding compounds prepared by methods B_1 and C. E, 4-hydroxy-1-(2*H*)-phthalazinone + malonyl dichloride; heat at 60° in acetic acid. E_1, same as E; heat with ethylmalonyl dichloride in nitrobenzene for 1 hr at

TABLE 2E–8 (*continued*)

120° E_2, same as E; heat with diethylmalonyl chloride in nitrobenzene for 1 hr at 140–160° or treat phthaloyl chloride with 4,4-diethylpyrazolidine 3,5-dione. F, treat 2*H*-pyrazolo[1,2-*b*]phthalazine-1,3,5,10-tetrone with aqueous sodium nitrite at 0° C followed by acid treatment. G, 2-ethyl 2*H*-pyrazolo[1,2-*b*]phthalazine-1,3,5,10-tetrone treated with acetic anhydride. H, phthalic anhydride + (HN–C(=O)–N(C_6H_5)–C(=O)–NH) , heat at 180–200° for 2 hr. I, 4-hydroxy-1-(2*H*)phthalazinone + $Pb(OAc)_4$ + corresponding diene in acetonitrile at ice bath temperature. I_1, sodium salt of 4-hydroxy-1-(2*H*)phthalazinone + *t*-butyl hypochlorite + butadiene in acetone at −77 to −70° C. I_2, phthalic anhydride condensed with 1,2,3,6-tetrahydropyridazine by refluxing in aqueous hydrochloric acid for 11 hr. I_4, similar to I_1, using 2,3-dimethylbutadiene. J, reduction of 1,4-dihydropyridazino[1,2-*b*]phthalazine-6,11-dione with Pd-C/H_2 in acetic acid or phthalic anhydride condensed with hexahydropyridazine by heating in ethanol. K, phthaloyl chloride heated with 3-hydroxy-6-pyridazone in dioxane at 120° for 1 hr. L, 4-hydroxy-1-(2*H*)phthalazinone + corresponding dihalide; heat ~200–215° C. M, by exothermic decomposition of phthalazine-1,4-dione or by heating its polymeric form at 133°; or heating phthaloyl chloride with hexahydropyridazine-3,6-dione in dioxane at 110–120° for 30 min. N, 4-hydroxy-1-(2*H*)-phthalazinone + $Pb(OAc)_4$ + corresponding aromatic hydrocarbon; stir in dichloromethane at ice bath temperature for 3 hr and at room temperature for 3 hr. O, similar to I_1; stir with 1,1-diethoxycyclopentadiene at −70° till the color disappeared. P, reduction of the product obtained by method O with Pt/H_2 in ethyl acetate. Q, 4-hydroxy-1-(2*H*)phthalazinone + $Pb(OAc)_4$ + steroidal 14,16-dien-20-one; stir in dichloromethane at 0–5°; two adducts are isolated.

References

1. J. A. Elvidge and A. P. Redman, *J. Chem. Soc.*, **1960,** 1710.
2. G. Rosseels, *Bull. Soc. Chim. Belges*, **74,** 91 (1965).
3. F. C. Brown, C. K. Bradsher, B. F. Moser, and S. Forrester, *J. Org. Chem.*, **24,** 1056 (1959).

3a. G. Cavallini, E. Massarani, F. Mazzucchi, and F. Revina, *Farm. Sci. e. Tec.* (*Pavia*), **7,** 397 (1952); *Chem. Abstr.*, **47,** 8015 (1953).

4. Netherlands Pat. appl. 6,502,841 (1965); *Chem. Abstr.*, **64,** 5012 (1966).
5. R. Silaraja, A. Arens, and G. Vanags, *Latv. PSR Zinat. Akad. Vestis Khim. Sec.*, **1964,** 707; *Chem. Abstr.*, **62,** 14563 (1965).
6. J. Nishe and S. Kamimoto, *Nippon Kagaku Zasshi*, **79,** 1403 (1958); *Chem. Abstr.*, **54,** 5665 (1960).
7. J. Bosjnak, R. I. Mamuzic, and M. L. Mihailovic, *Glas. Khim. Drus. Beogr.*, **27,** 313 (1963); *Chem. Abstr.*, **60,** 5361 (1964).
8. J. C. E. Simpson, in *Condensed Pyridazine and Pyrazine Rings*, *The Chemistry of Heterocyclic Compounds*, Vol. 5, A. Weissberger, Ed., Interscience, New York, 1953, p. 140.
9. B. A. Boissonas, in *Advances in Organic Chemistry*, *Methods and Results*, Vol. 3, R. A. Raphael, E. C. Taylor, and H. Wynberg, Eds., Interscience, New York-London, 1963, p. 180.
10. C. T. Redemann and C. E. Redemann, in *Organic Synthesis*, Collective Vol. III, E. C. Horning Ed., John Wiley & Sons, N.Y. 1955 pp. 69 and 656.
11. E. P. Krysin, L. I. Mizrakh, and L. D. Sabkaya, USSR Pat. 130,903 (1960); *Chem. Abstr.*, **55,** 6504 (1961).
12. K. D. Gundermann and M. Drawert, *Chem. Ber.*, **95,** 2018 (1962).
13. E. H. White and M. M. Bursey, *J. Org. Chem.*, **31,** 1912 (1966).
14. F. Dallacker, L. Doyen, and G. Schmets, *Ann. Chem.*, **694,** 117 (1966).
15. K. D. Gundermann, W. Horstmann, and G. Bergmann, *Ann. Chem.*, **684,** 127 (1965).
16. C. M. Atkinson and C. J. Sharpe, *J. Chem. Soc.*, **1959,** 2858.
17. J. Nishie, *Nippon Kagaku Zasshi*, **87,** 1239 (1966); *Chem. Abstr.*, **66,** 115662 (1967).
18. E. H. White and K. Matsuo, *J. Org. Chem.*, **32,** 1921 (1967).
19. E. H. White, D. F. Roswell, and O. G. Zafiriou, *J. Org. Chem.*, **34,** 2462 (1969).
20. B. Cavalleri, E. Bellasio, and A. Sareli, *J. Med. Chem.*, **13,** 148 (1970).
21. B. Cavalleri and E. Bellasio, *Farm. Ed. Sci.*, **24,** 833 (1969); *Chem. Abstr.*, **71,** 101798 (1969).
22. G. A. Galoyan, S. G. Agbalyan, and G. T. Esayan, *Arm. Khim. Zh.*, **22,** 334 (1969); *Chem. Abstr.*, **71,** 49882 (1969).
23. F. Dallacker and J. Bloemen, *Monatsh. Chem.*, **92,** 640 (1961); M. Lipp, F. Dallacker, and R. Schaffranek, *Chem. Ber.*, **91,** 2274 (1958).
24. M. Ohara, K. Yamamoto, and T. Fujisuwa, Japanese Pat. 19,742 (1963); *Chem. Abstr.*, **60,** 4161 (1964).
25. I. Satoda, N. Yoshida, and K. Mori, *Yakugaku Zasshi*, **72,** 703 (1957); *Chem. Abstr.*, **51,** 17927 (1957).
26. F. Arndt, L. Loewe, and L. Ergener, *Rev. fac. Sci.*, *Univ. Istanbul*, **13A,** 103 (1948); *Chem. Abstr.*, **43,** 579 (1949).
27. D. J. Drain and D. E. Seymour, *J. Chem. Soc.*, **1955,** 852.
28. T. Ohta, *J. Pharm. Soc. Japan*, **62,** 452 (1942); *Chem. Abstr.*, **45,** 5163 (1951).

29. T. Kubaba and R. S. Ludwiczak, *Roczniki Chem.*, **38,** 367 (1964).
30. S. Baloniak and R. Stella, *Roczniki Chem.*, **38,** 373 (1964).
31. S. Baloniak and R. S. Ludwiczak, *Roczniki Chem.*, **36,** 411 (1962).
32. Belgium Pat. 490,069 (1949); *Chem. Abstr.*, **51,** 1305 (1957).
33. N. P. Buu-Hoi, H. L. Bihan, and F. Binon, *Rec. Trav. Chim.*, **70,** 1099 (1951).
34. S. Baloniak and E. Domagalina, *Roszniki Chem.*, **33,** 725 (1959).
35. S. Baloniak, *Roszniki Chem.*, **37,** 1655 (1963).
36. G. Rosseels, G. Thuillier, and P. Rumpf, *Compt. Rend.*, **255,** 1453 (1962).
37. A. Le Berre, M. Dormoy, and J. Godin, *Compt. Rend.*, **261,** 1872 (1965).
38. M. Dormoy, J. Godin, and A. Le Berre, *Bull. Soc. Chim. France*, **1968,** 4222.
39. A. Le Berre, J. Godin, and R. Garreau, *Compt. Rend. Acad. Sci. Paris*, Ser. C, **265,** 570 (1967); *Chem. Abstr.*, **68,** 68948 (1968).
40. J. Godin and A. Le Berre, *Bull. Soc. Chim. France*, **1968,** 4210.
41. L. Bauer, A. Shoeb, and V. C. Agwada, *J. Org. Chem.*, **27,** 3153 (1962).
42. E. Profft and R. Kader, *Arch. Pharm.*, **297,** 673 (1963).
43. H. Feuer, G. B. Silverman, H. P. Angstadt, and A. R. Fauke, *J. Org. Chem.*, **27,** 2081 (1962).
44. J. Godin and A. Le Berre, *Bull. Soc. Chim. France*, **1968,** 4222.
44a. A. Nakamura and S. Kamiya, *Yakugaku Zasshi*, **90,** 1069 (1970); *Chem. Abstr.*, **73,** 130537 (1970).
45. H. Hellman and I. Loschmann, *Chem. Ber.*, **89,** 594 (1956); German Pat. 1,028,127 (1958); *Chem. Abstr.*, **54,** 18564 (1960).
46. E. Domagalina, I. Kyrpiel, and N. Koktysz, *Roczniki Chem.*, **43,** 775 (1969); *Chem. Abstr.*, **71,** 61315 (1969).
47. N. P. Buu-Hoi et al., *Compt. Rend.*, **228,** 2037 (1949).
48. G. Hugh, W. H. Hunter, J. King, and B. J. Millard, British Pat. 1,100,911 (1968); *Chem. Abstr.*, **69,** 19181 (1968).
49. E. Domagalina and S. Baloniak, *Roczniki Chem.*, **36,** 253 (1962).
50. L. Rylski et al., *Acta Polon. Pharm.*, **22,** 111 (1965); *Chem. Abstr.*, **63,** 8350 (1965).
51. W. Zerweck, W. Kunze, and R. E. Nitz, German Pat. 1,005,072 (1957); *Chem. Abstr.*, **53,** 18073 (1959); British Pat. 808,636 (1959); *Chem. Abstr.*, **53,** 14129 (1959).
52. H. Ruschig et al., German Pat. 956,044 (1957); *Chem. Abstr.*, **53,** 4319 (1959).
53. H. Ruschig et al., U.S. Pat. 2,874,156 (1959); *Chem. Abstr.*, **53,** 12316 (1959).
54. E. Schmitz and R. Ohme, *Chem. Ber.*, **95,** 2012 (1965); R. Ohme and E. Schmitz, *Z. Anal. Chem.*, **220,** 105 (1966).
55. R. A. Clement, *J. Org. Chem.*, **25,** 1724 (1960).
56. R. A. Clement, *J. Org. Chem.*, **27,** 1115 (1962).
57. T. J. Kealy, *J. Am. Chem. Soc.*, **84,** 966 (1962).
58. T. J. Kealy, U.S. Pat. 3,062,820 (1962); *Chem. Abstr.*, **58,** 9101 (1963).
59. O. L. Chapman and S. J. Dominiani, *J. Org. Chem.*, **31,** 3862 (1966).
60. Y. Omote, T. Miyake, and N. Sugiyama, *Bull. Chem. Soc. Japan*, **40,** 2446 (1967).
61. N. P. Marullo and J. A. Alford, *J. Org. Chem.*, **33,** 2368 (1968).
62. T. L. Popper, F. E. Carlson, H. M. Marigliano, and M. D. Yudis, *Chem. Comm.*, **1968,** 1434.
63. C. H. Hassall, R. B. Morton, Y. Ogihara, and W. A. Thomas, *Chem. Comm.*, **1969,** 1079.
63a. S. Takase and T. Motoyama, *Bull. Chem. Soc., Japan*, **43,** 3926 (1970).
64. W. R. Vaughan, *Chem. Rev.*, **43,** 447 (1948).
65. P. Sohar, *Acta Chim. Acad. Sci. Hung.*, **40,** 317 (1964); *Chem. Abstr.*, **62,** 14457 (1965); **63,** 1344 (1965).

66. Y. N. Sheinker and Y. I. Pomerantseu, *Zh. Fiz. Khim.*, **30,** 79 (1956); Y. N. Sheinker, T. V. Gortinskaya, and T. P. Sycheva, **31,** 599 (1957) [*J. Chem. Phys.*, **55,** 217 (1958)]; Y. N. Sheinker and Y. N. Pomerantseu, *Zh. Fiz. Khim.*, **33,** 1819 (1959); *Chem. Abstr.*, **50,** 14780 (1956); **52,** 877 (1958); **54,** 12156 (1960).
66a. Y. Omote, H. Yamamoto, and N. Sugiyama, *J. Chem. Soc. D*, **1970,** 914.
67. C. Ainsworth, *J. Am. Chem Soc.*, **78,** 1636 (1956).
68. J. N. Ashley, R. E. Collins, M. Davis, and N. E. Sirett, *J. Chem. Soc.*, **1959,** 3880.
69. P. G. Parsons and H. J. Rodda, *Aust. J. Chem.*, **17,** 491 (1964).
70. Y. Omote, T. Miyake, S. Ohmori, and N. Sugiyama, *Bull. Chem. Soc. Japan*, **39,** 932 (1966).
71. Y. Omote, T. Miyake, S. Ohmori, and N. Sugiyama, *Bull. Chem. Soc. Japan*, **40,** 899 (1967).
72. E. F. Elslager et al., *J. Med. Chem.*, **6,** 646 (1963); U.S. Pat. 3,139,421 (1964); *Chem. Abstr.*, **61,** 6972 (1964).
73. V. M. Ostrovskaya, T. A. Maryashkina, and Y. S. Ryabokobylko, *Zh. Org. Khim.*, **5,** 1307 (1969); *Chem. Abstr.*, **71,** 101799 (1969).
74. Y. Omote, T. Miyake, and N. Sugiyama, *Nippon Kagaku Zasshi*, **87,** 621 (1966); *Chem. Abstr.*, **65,** 15373 (1966).
75. Y. Omote, H. Yamamoto, S. Tomioka, and N. Sugiyama, *Bull. Chem. Soc. Japan*, **42,** 2090 (1969).
76. T. Ohta, *J. Pharm. Soc., Japan*, **64,** 50 (1944); *Chem. Abstr.*, **46,** 118 (1952).
77. B. K. Diep and B. Cauvin, *Compt. Rend. Ser. C.*, **262,** 1010 (1966).
78. T. Higuchi, B. J. Sciarrone, and A. F. Haddad, *J. Med. Pharm. Chem.*, **3,** 195 (1961).
79. W. M. McIsaac, *Biochem. Pharmacol.*, **13,** 1113 (1963).
80. E. Hayashi et al., *Yakugaku Zasshi*, **82,** 584 (1962); *Chem. Abstr.*, **58,** 3425 (1963).
81. D. G. Parsons, A. F. Turner, and B. G. Murray, Brit. Pat. 1,133,406 (1968); *Chem. Abstr.*, **70,** 57,872 (1969); Fr. Pat. 1,532,163 (1968); *Chem. Abstr.*, **71,** 112,963 (1969).
82. I. Satoda, F. Kusada, and K. Mori, *Yakugaku Zasshi*, **82,** 233 (1962); *Chem. Abstr.*, **58,** 3427 (1963).
83. R. D. Chambers, J. A. A. MacBride, W. K. R. Musgrave, and I. S. Reilly, *Tetrahedron Lett.*, **1970,** 55.
84. J. H. Bowie et al., *Aust. J. Chem.*, **20,** 2677 (1967).
84a. A. LeBerre, B. Dumaitre, and J. Petit, *Bull. Soc. Chim., France*, **1970,** 4376.
85. L. F. Fieser, in *Organic Experiments*, D. C. Heath & Co., Boston, 1964, p. 240.
86. B. E. Cross and H. D. K. Drew, *J. Chem. Soc.*, **1949,** 638.
87. M. V. Sigal, Jr., P. Manchini, and B. L. Poet, U.S. Pat. 3,274,185 (1966); *Chem. Abstr.*, **65,** 18600 (1966).
88. F. Dallacker, E. Meunier, J. Limpens, and M. Lipp, *Monatsh. Chem.*, **91,** 1077 (1960).
89. F. Dallacker, *Ann. Chem.*, **633,** 14 (1960).
90. K. D. Gundermann and D. Schedlitzki, *Chem. Ber.*, **102,** 3241 (1969).
91. R. A. Dine-Hart and W. W. Wright, *Chem. Ind. (London)*, **1967,** 1565.
92. S. A. Agripot, French Pat. 1,460,552 (1966); *Chem. Abstr.*, **67,** 31774 (1967).
93. A. F. Rosenthal, *J. Org. Chem.*, **22,** 89 (1957).
94. G. Rosseels, *Bull. Soc. Chim. Belges*, **74,** 101 (1965).
95. R. Behnisch et al., *Med. Chem. Abhandl Med. Chem. Forschungs.*, Farbwerke Hoecht A G, 7, 296 (1963); *Chem. Abstr.*, **61,** 5628 (1964).
95a. S. Baloniak, *Roczniki. Chem.*, **38,** 1295 (1964).
96. W. B. Ligett, R. D. Clusson, and C. N. Wolf, U.S. Pat. 2,595,798 (1952); *Chem. Abstr.*, **46,** 7701 (1952).

96a. M. Ghelardoni and V. Pestellini, *Ann. Chim.* (*Rome*), **60,** 775 (1970).
97. K. Kormendy, P. Sohar, and J. Volford, *Acta Chim. Acad Sci., Hung.*, **39,** 109 (1963); *Chem. Abstr.*, **60,** 1744 (1964).
98. S. DuBreuil, U.S. Pat. 2,759,938 (1956); *Chem. Abstr.*, **51,** 2885 (1957).
99. C. Harukawa and K. Konishi, Japanese Pat. 17,677 (1963); *Chem. Abstr.*, **60,** 6870 (1964).
100. British Pat. 769,181 (1957); *Chem. Abstr.*, **51,** 14835 (1957).
101. T. P. Sycheva and M. N. Shuchukina, *Zh. Obshch. Khim*, **30,** 608 (1968); *Chem. Abstr.*, **54,** 24, 783 (1960).
102. H. Cairns, W. H. Williams, J. King, and B. J. Millard, British Pat. 1,100,911 (1968); *Chem. Abstr.*, **69,** 19181.
103. T. Tolkmith, U.S. Pat. 2,967,180 (1961); *Chem. Abstr.*, **55,** 14489 (1961).
104. H. Ruschig, R. Fugmann, and E. Linder, German Pat. 1,029,379, addition to German Pat. 956,044 (1958); *Chem. Abstr.*, **54,** 19726 (1960).
105. C. Zinner and W. Deucker, *Arch. Pharm.*, **296,** 13 (1963).
106. M. Rink, S. Mehta, and K. Brabowski, *Arch. Pharm.*, **292,** 225 (1959).
107. T. Ikeda, S. Kanahara, and A. Aoki, *Yakugaku Zasshi*, **88,** 521 (1968); *Chem. Abstr.*, **69,** 86934 (1968).
108. A. Vigevani, B. Cavalleri, and G. G. Gallo, *J. Heterocyclic Chem.*, **7,** 677 (1970).

Part F. Halophthalazines

I. 1-Halo- and 1,4-Dihalophthalazines

The most extensively studied halophthalazines are the 1-chloro- and 1,4-dichlorophthalazines. The chlorine atoms in these compounds undergo nucleophilic substitution by various anions quite readily, characteristic of heterocyclic compounds containing the —N=C—Cl group. Some of the corresponding bromo-, iodo-, and fluorophthalazines have been prepared recently. A few phthalazines with halogen atoms in the benzenoid ring have been reported, but only those having halogen atoms in the heterocyclic ring are discussed here.

A. Preparation of 1-Chlorophthalazines

1-(2*H*)Phthalazinone has been converted to 1-chlorophthalazine (**1**) by refluxing with phosphorus oxychloride for 2–30 min (1–3) or by heating with phosphorus oxychloride on a steam bath until a clear solution is obtained (4) (Eq. 1). High yields of **1** (78–96%) are obtained. 1-Chloro-7-methoxy- (5, 6), 1-chloro-7-nitro- (1, 7), 1,5-dichloro- (8, 9), 1,6-dichloro- (10), 1,7-dichloro- (10), and 1,8-dichlorophthalazines (8, 9) have been similarly prepared (Eq. 1).

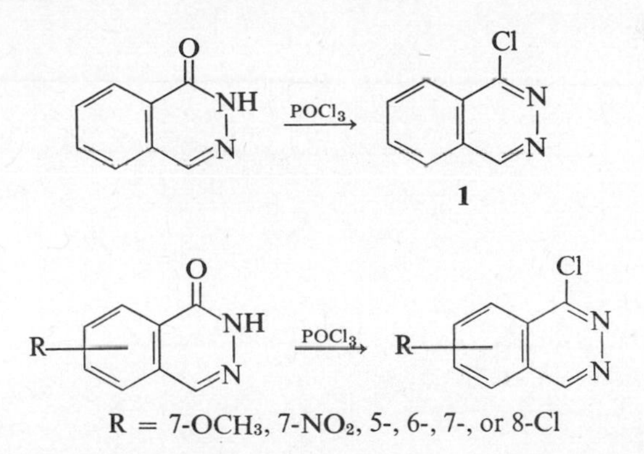

R = 7-OCH_3, 7-NO_2, 5-, 6-, 7-, or 8-Cl

Some of the spectroscopic properties of 1,5-, 1,6-, 1,7-, and 1,8-dichlorophthalazine are recorded but physical properties are not reported (8–10).

1-Chloro-4-alkyl- (11–15), 1-chloro-4-aryl- (16–20), and 1-chloro-4-aralkylphthalazines (21–23) are prepared from the corresponding 4-substituted 1-(2*H*)phthalazinones by refluxing with phosphorus oxychloride for 20 min to 1 hr (Eq. 2).

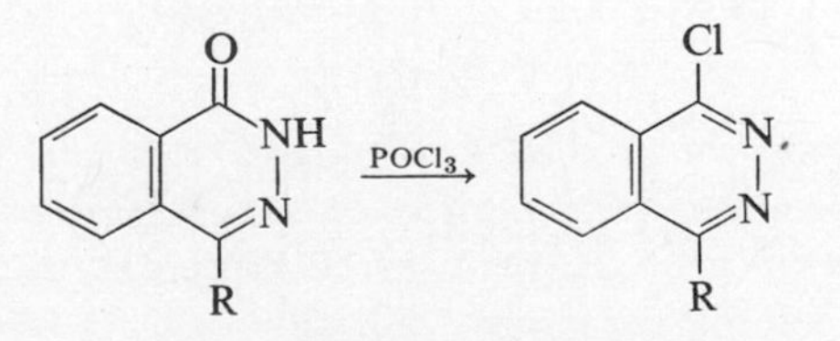

(2)

R = alkyl, aryl, or aralkyl

1-Phenylphthalazine 3-oxide, on treatment with phosphorus oxychloride or sulfuryl chloride, gives the expected 1-chloro-4-phenylphthalazine in 74 and 46% yields, respectively (24) (Eq. 3).

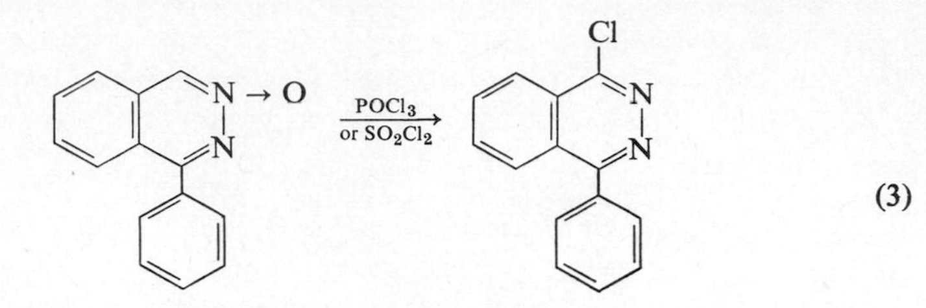

2-Methyl- (11, 25), 2-phenyl- (26), and 2-(2′-pyridyl)-4-chloro-1-(2*H*)-phthalazinone (27) are prepared in high yields by refluxing the corresponding 2-substituted 4-hydroxy-1-(2*H*)phthalazinone (**2**) with phosphorus oxychloride for 2–4 hr (Eq. 4).

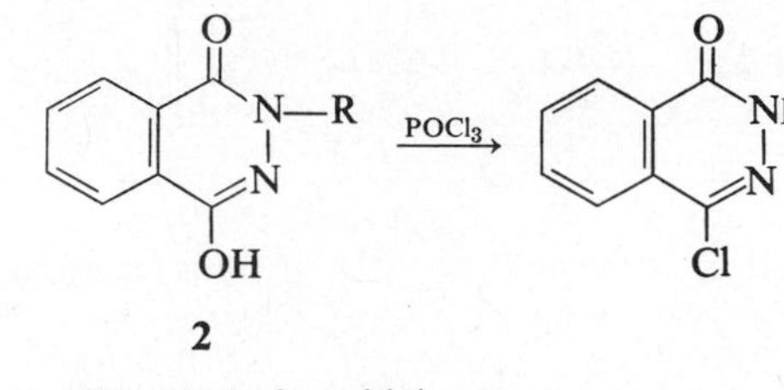

R = CH_3, C_6H_5, 2-pyridyl

B. Preparation of 1,4-Dichlorophthalazines

1,4-Dichlorophthalazine (**3**) has been prepared by heating 4-hydroxy-1-(2*H*)phthalazinone with phosphorus oxychloride (28) or phosphorus pentachloride (29–31). Hirsch and Orphanos (29, 30) have carried out the reaction with phosphorus pentachloride by stirring the reaction mixture in a pressure bottle at 140–150° for 4 hr to give **3** in 96% yield (Eq. 5). Treatment of

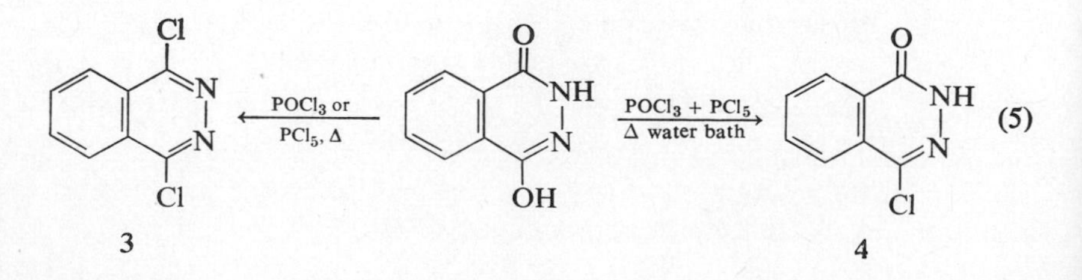

4-hydroxy-1-(2*H*)phthalazinone with a mixture of phosphorus oxychloride and pentachloride on a hot water bath leads mainly to 4-chloro-1-(2*H*)-phthalazinone (**4**) (32).

6-Phenyl- (33) and 6-trifluoromethyl-1,4-dichlorophthalazine (34) have been prepared in 43 and 84% yields, respectively, from the corresponding 6-substituted 4-hydroxy-1-(2*H*)phthalazinones by heating with phosphorus pentachloride (Eq. 6). 6-Fluoro-4-hydroxy-1-(2*H*)phthalazinone, on treatment with phosphorus oxychloride, affords 1,4-dichloro-6-fluorophthalazine and a single monochlorinated product, **5a** or **5b**, the structure of which has not been assigned (35) (Eq. 6). 6-Sulfonyl (36), 6-carboxy- (36), and 6-chlorocarbonyl-1,4-dichlorophthalazine (37–39) have also been reported.

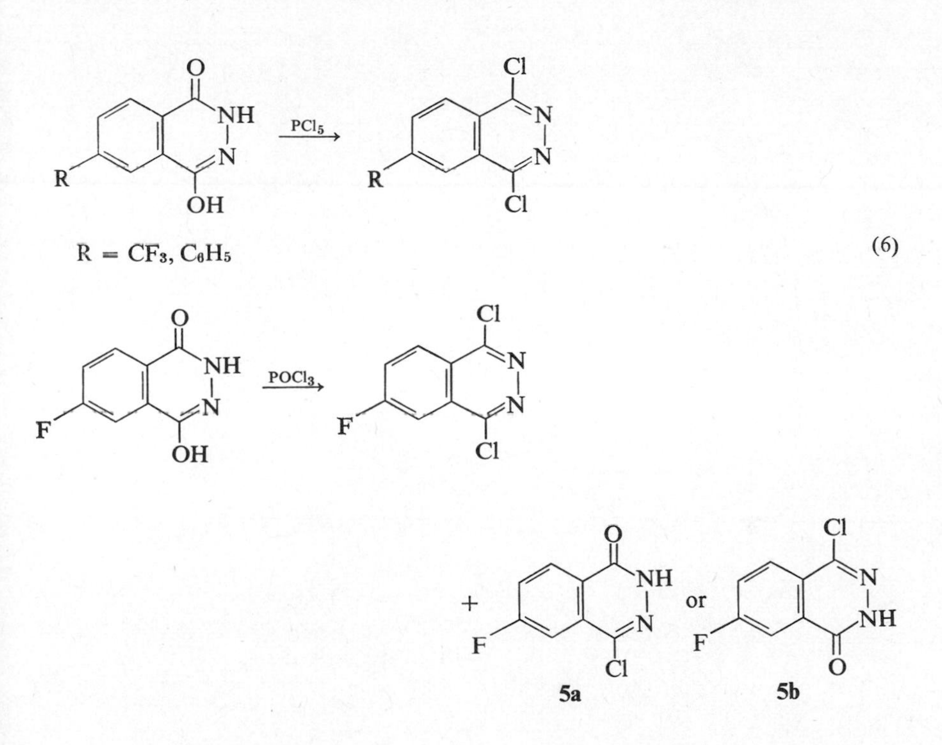

C. Preparation of Bromo-, Iodo-, and Fluorophthalazines

Hirsch and Orphanos (29) have carried out the reaction of 1-(2*H*)phthalazinone with phosphorus oxybromide to give 1-bromophthalazine and have shown that the reaction of 4-hydroxy-1-(2*H*)phthalazinone with phosphorus pentabromide gives 1,4-dibromophthalazine (Eq. 7). The same authors (30) have also reported the reaction of 1-chloro- and 1,4-dichlorophthalazine with sodium iodide in acetone containing 50% aqueous hydriodic acid. In these cases, there are obtained 1-iodo- and 1,4-diiodophthalazine in 88 and 76% yields, respectively (Eq. 7).

Recently Chambers and co-workers (40) have prepared 1,4,5,6,7,8-hexafluorophthalazine (**6**) and have carried out an excellent study of its

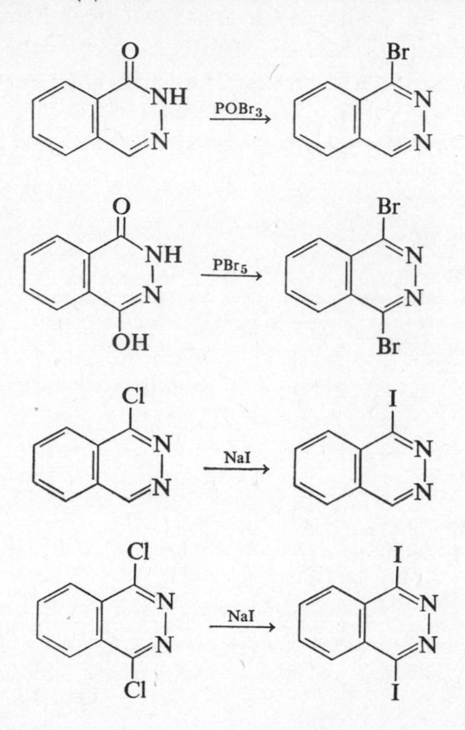

(7)

nucleophilic substitution by sodium methoxide (see Section 2F-II-B-3). 1,4-Dichlorophthalazine (**3**) and chlorine, in the presence of aluminum chloride at 200°, gives 1,4,5,6,7,8-hexachlorophthalazine (**7**), which upon interaction with potassium fluoride at 290° gives hexafluorophthalazine (**6**) in 60% yield (Eq. 8).

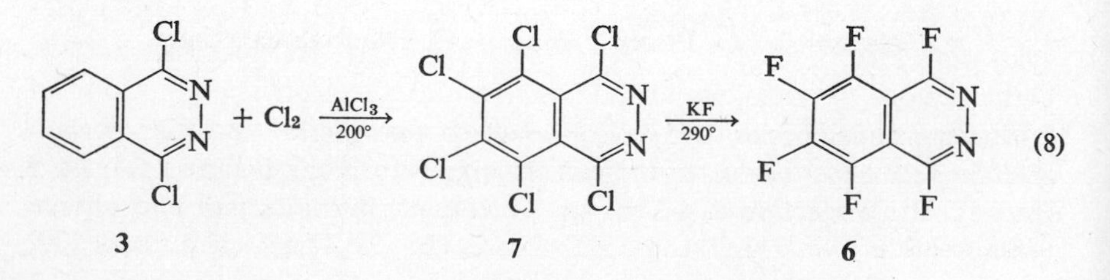

II. Properties of 1-Halo- and 1,4-Dihalophthalazines

A. Physical Properties

1-Chlorophthalazine (**1**) should be used freshly prepared because of its instability (32, 41). The melting points of **1** reported in the literature vary from 109 to 121° C. On standing at room temperature it is transformed into higher melting substances. It forms a picrate having mp 135° (42). Favini and Simonetta (9) have reported the UV spectrum of **1** in isooctane and methanol together with the spectra of 5-chloro- and 6-chlorophthalazine. The UV spectrum of **1** in isooctane shows λ_{max} at 307.5 (log ε 2.97) and 265 mμ (log ε, 3.70); in methanol 307 (log ε, 3.02) and 267.5 mμ (log ε, 3.65).

1,4-Dichlorophthalazine (**3**) forms long, colorless needles when recrystallized from tetrahydrofuran, mp 164° (29, 30) [160–162° (28), 162–163° (31)]. The ir spectra of **3** and of 1-bromo-, 1-iodo-, 1,4-dibromo-, and 1,4-diiodophthalazine have been reported by Hirsch and Orphanos (29, 30). Bowie and co-workers (43) have reported the mass spectra of 6-bromophthalazine and of 4-substituted 1-halophthalazines. 1,4-Dichloro- and 1,4-dibromophthalazine form stable complexes with bromine in chloroform-carbon tetrachloride, yellow powder melting at 155° and orange powder melting at 195°, respectively (44).

B. Chemical Properties

1. *Hydrolysis*

1-Chlorophthalazine (**1**) is surprisingly resistant to 10% aqueous sodium hydroxide, but on vigorous boiling it is hydrolyzed to 1-(2*H*)phthalazinone (**8**) (45) (Eq. 9). The hydrolysis of **1** with water or with dilute hydrochloric acid proceeds abnormally to give principally 2-(1-phthalazinyl)-1-(2*H*)-phthalazinone (**9**) along with a small amount of **8** (1, 45). The formation of **9** has been explained on the basis that 1-chlorophthalazine first couples with itself to give a quaternary salt which undergoes further hydrolysis to give **9** (Eq. 4). The end product, **9**, has been identified by independent synthesis (46).

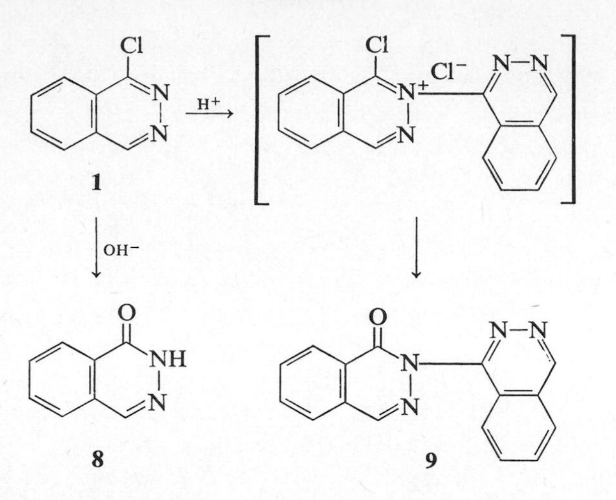

1,4-Dichlorophthalazine (**3**) is stable in water and aqueous alkaline solution but is rapidly hydrolyzed to 4-chloro-1-(2*H*)phthalazinone (**4**) by warm 2-*N* hydrochloric acid (47). 1,4,5,6,7,8-Hexafluoro- and hexachlorophthalazine are hydrolyzed by atmospheric moisture or by dilute sulfuric acid to the corresponding 4,5,6,7,8-pentahalo-1-(2*H*)phthalazinones (40).

2. *Reduction and Dehalogenation*

1-Chlorophthalazine and 1-chloro-4-alkylphthalazines, on reduction with zinc and hydrochloric acid, give dihydroisoindoles with loss of ammonia. 1-Chloro-4-methylphthalazine, on treatment with red phosphorus and hydriodic acid, gives the corresponding methylisoindole (32), while 1,4-dichlorophthalazine and 1-chloro-4-phenylphthalazine give phthalazine and 4-phenylphthalazine, respectively (see Sections 2A-I-A-1 and 2B-I-F-2).

Catalytic dehalogenation of 1-alkyl- or 1-aryl-4-chlorophthalazine or 1-chlorophthalazine over 5% palladium on charcoal, or over Raney nickel, gives the corresponding phthalazines in high yields (see Sections 2A-I-A-1 and 2B-I-F-2).

3. *Nucleophilic Substitutions*

1-Chlorophthalazine and 4-substituted 1-chlorophthalazines, on treatment with sodium alkoxides in the corresponding alcohols, give 1-alkoxy- and 4-substituted 1-alkoxyphthalazines (see Section 2D-II-A-2). Similarly,

1,4-dichlorophthalazine reacts with excess sodium alkoxides to give 1,4-dialkoxyphthalazines (see Section 2E-II-A). 1,4-Dichlorophthalazine can be selectively converted to 1-chloro-4-alkoxyphthalazine (**10**) by reaction with an equivalent amount of sodium alkoxide (31, 48) (Eq. 10). A few examples are known wherein 4-chloro-1-(2*H*)phthalazinone (**4**) has been successfully *O*-alkylated to give 1-chloro-4-*O*-alkylated products (**10**). The reaction of **4** with trichloromethanesulfinyl chloride (49) and that of the silver salt of **4** with α-acetobromoglucose (50) give the corresponding **10** (R = —$SCCl_3$, tetracetyl-1-β-D-glucosyl).

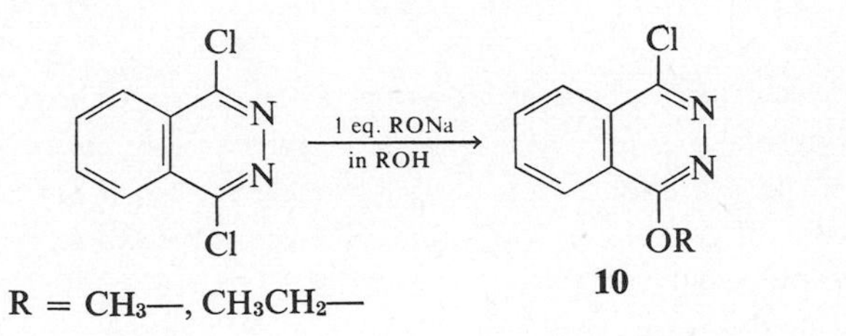

(10)

Nucleophilic substitution of halogen atoms in the heterocyclic ring occurs much more readily than expected. Chambers and co-workers (40) have studied the reaction of sodium methoxide with 1,4,5,6,7,8-hexafluorophthalazine (**6**) using different molar ratios and different temperatures. Methanolic sodium methoxide progressively replaced all six fluorine atoms. The lability of the fluorine atoms varied according to predictions of π-electron density distribution in the phthalazine ring. Positions 1 and 4 were found to be most reactive and 5 and 8 were least reactive. Chambers and co-workers (40) also isolated and identified the mono-, di-, tri-, tetra-, and hexamethoxyphthalazines produced in this study (see also Section 2E-II-A). 1-Chlorophthalazine reacts with alkyl- or arylmercaptans to give 1-alkyl- or 1-arylthiophthalazines. 1,4-Dichlorophthalazine, on reaction with potassium or sodium sulfhydride, gives 4-chloro-1-(2*H*)phthalazinethione (50) and 4-mercapto-1-(2*H*)phthalazinethione (28) (see Section 2G).

Nucleophilic substitution of the chlorine atom in 1-chlorophthalazine with ammonia, alkylamines, arylamines, hydroxylamine, and hydrazine has been accomplished. Reaction of 1,4-dichlorophthalazine with ammonia or an alkylamine normally leads to monosubstituted 1-chloro-4-amino- or 1-chloro-4-alkylaminophthalazines, whereas reactions with weaker amines such as the aromatic amines leads to disubstituted products, 1,4-bis(diarylamino)phthalazines (see Sections 2H and 2I).

Cavaleri and Bellasio (35) have examined the reactions of 1-(4)-methoxy-4-(1)-chloro-5-fluorophthalazine (**11**) and of 1-(4)-methoxy-4-(1)-chloro-6-fluorophthalazine (**12**) with hydrazine hydrate in refluxing ethanol. In both cases the fluorine atom located at the 5- or 6-position is displaced. The chlorine atom and methoxy group located at the 1- and 4-positions are not displaced (Eq. 11).

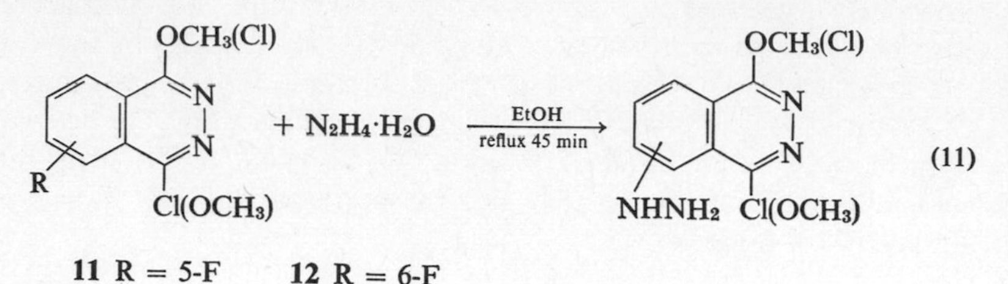

11 R = 5-F **12** R = 6-F

In an extension of the malonic ester synthesis, the chlorine atom in 1-chlorophthalazine has been displaced by various active methylene compounds in the presence of a strong base, such as sodium amide in benzene (see Section 2B-I-A-4). 1-Iodo- and 1,4-diiodophthalazine will react with copper cyanide in pyridine to give 1-cyano- and 1,4-dicyanophthalazine in 18 and 35% yields, respectively (30) (Eq. 12).

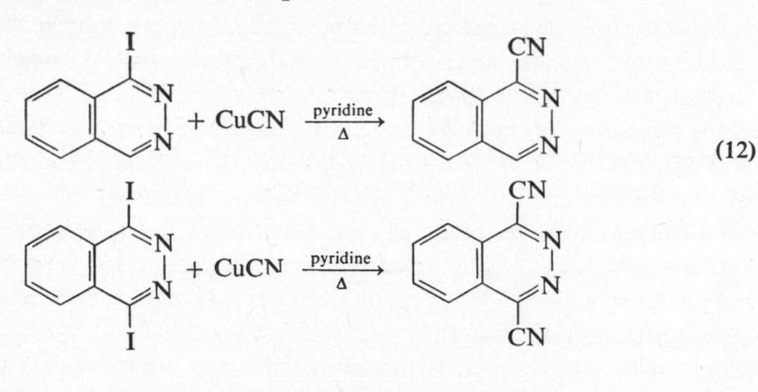

4. *Miscellaneous*

1,4-Dichlorophthalazine and 1-chloro-4-phenylphthalazine have been shown to undergo addition of phenylmagnesium bromide to give 1,4-diphenylphthalazine in high yields (see Section 2B-II-A-4-b). Similarly, 1-chlorophthalazine undergoes addition of Grignard reagents (see Section 2A-I-A-2).

Chapman and Russell-Hill (51) have studied the kinetics of the reactions of 1-chlorophalazine with ethoxide ion and piperidine. These authors have compared the reactivity of ethoxide ions with 1-chlorophthalazine, 4-chlorocinnoline, and 4-chloroquinazoline and found that 4-chloroquinazoline is more reactive than 1-chlorophthalazine, which has approximately the same reactivity as 4-chlorocinnoline. Badger and co-workers (45) have observed that 4-chloroquinazoline reacts with ammonia at room temperature, whereas 1-chlorophthalazine and 4-chlorocinnoline do not react with ammonia at room temperature. 1,4-Dichlorophthalazine is used in an improved process for the production of vat dyestuffs of anthraquinone where one of the chlorine atoms of the phthalazine is replaced by aminodibenzanthrone (52) and 1,4-diamino-2-acetylanthraquinone (53). 1,4-Dichlorophthalazine-6-carboxylic acid and 1,4-dichlorophthalazine-6-sulfonic acid react with sulfamide and *p*-toluidine, respectively, and the resulting products are used for decreasing the dye affinity for polyamide fibers (36). 1,4-Dichlorophthalazine-6-carbonyl chloride is used in the preparation of dyes where the halogen atom of the carbonyl chloride group is replaced (37, 54).

III. Tables

TABLE 2F-1. 1-Chlorophthalazine with Substituents in the Benzenoid Ring

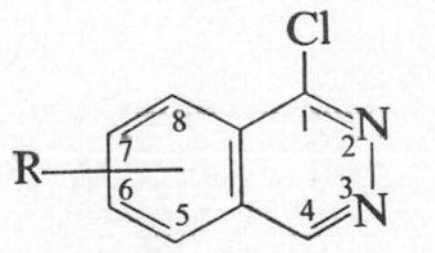

R	Preparation[a]	Yield (%)	MP	Reference
H	A	78–96	109–111°; 110–111° (ligroin); 113°; 114.5–115.5°; 117–118° (ligroin); yellow needles, 120–121°; picrate, 135°	1–4, 9, 41, 42
7-NO_2	B	53–78	155–157° (dec.); buff solid, 156–159° (dec.); 155–157°	1, 7, 42a
7-OCH_3	C	76	136°; 142° (dec.)	5, 6
5-Cl	D	—	—	8, 9
6-Cl	D	—	—	10
7-Cl	D	—	—	10
8-Cl	D	—	—	8, 9
5-F	A	—	156–157°	35
6-F	A	—	138–140°	35
7-F	A	—	165–168°	35
8-F	A	—	169–170°	35

[a] A, 1-(2*H*)phthalazinone is refluxed with $POCl_3$ for 2–30 min or heated on a steam bath until a clear solution is obtained. B, 7-nitro-1-(2*H*)phthalazinone is heated with $POCl_3$ on a steam bath for 1 hr or heated in the presence of diethylamine for 20 min. C, 7-methoxy-1-(2*H*)phthalazinone is refluxed with $POCl_3$ for 30 min. D, the corresponding chloro-1-(2*H*)phthalazinone is refluxed with $POCl_3$ for 15 min.

TABLE 2F-2. 1-Chlorophthalazine with Substituents in the 4-Position

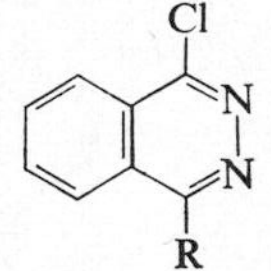

R	Preparation[a]	Yield (%)	MP	Reference
CH_3—	A, B	81	Needles 130°; 133°; 124°	11, 12, 42a
—CH_2CH_3	A	—	90–92°	12
—$CH(CH_3)_2$	A	—	81–83° (ethanol)	12, 15
—CH_2CN	B	—	—	13
—$(CH_2)_4OC_6H_5$	C	61	Pale brown, 109–110° (ethanol)	14
C_6H_5—	—	—	159–161°; 160°	16, 20
p-ClC_6H_4—	—	—	190–191°	23
p-$CH_3OC_6H_4$—	B	70	147°	16–18
p-$(CH_3)_2NC_6H_4$—	B	—	195–197°	16, 17
$2,4,6$-$(CH_3)_3C_6H_2$—	B	64	Colorless crystals, 192–193° (ether)	19
$C_6H_5CH_2$—	D	—	152°	20, 42a

p-$CH_3OC_6H_4CH_2$—	D	—	196°	20
3,4-$(CH_3O)_2C_6H_3CH_2$—	E	81	130–132°	22
3,4-$(C_2H_5O)_2C_6H_3CH_2$—	E	—	102°	22
—CH_2—(2-pyridyl)	F	—	—	21
—CH_2—(4-pyridyl)	F	—	170° (ethanol-water)	21
$C_6H_5CH{=}CH$—	—	—	Colorless crystals, 120–123° (DMF)	55
—OCH_3	G	83	107–108°; 108–109°	31, 48
—OCH_2CH_3	G	—	78°	48
—$OCH_2CH{=}CH_2$	—	—	87–89°	56
—O(tetracetyl-1-β-D-gluosyl)	H	14	Needles, 189–190° (1-propanol)	50
—$OSCCl_3$	I	—	Colorless crystals, 154–156° (benzene)	49

[a] A, the corresponding 4-substituted 1-(2*H*)phthalazinone is refluxed with $POCl_3$ for 20 min. B, same as A, heat on water bath for 30 min. C, same as A, heat at 100° for 3.5 hr. D, same as A, heat (time not specified). E, same as A, reflux for 1 hr. F, Same as A, heat on water bath for 2.5 hr. G, 1,4-dichlorophthalazine is refluxed with equimolar RONa in ROH for 30 min. H, silver salt of 1-chloro-4-hydroxyphthalazine + α-acetobromoglucose; reflux in toluene for 5 hr. I, 4-chloro-1-(2*H*)phthalazinone + trichloromethanesulfinyl chloride + 1*N* NaOH; stir in chloroform at room temperature for several hours.

TABLE 2F-3. 4-Substituted 1-Chlorophthalazines with Substituents in the Benzenoid Ring

R	R_1	R_2	Preparation[a]	Yield (%)	MP	Reference
CH_3	H	NO_2	A	73	Pink solid, 160–161° (shrinks at 140°)	7
C_6H_5	OCH_3	H	B	—	144–145° (dil. alcohol); 143–144°	57, 42a
C_6H_5	H	OCH_3	B	93	143–144° (dil. alcohol)	57
—CH_2—(4-pyridyl)	H	—NHC(=O)CH_3	C	—	—	21, 58
C_6H_5	OCH_3	OCH_3	D	—	Light yellow needles, 178–179° (dec.) (alcohol)	59
$C_6H_5CH_2$	OCH_3	OCH_3	D	—	154–155° (methanol); light yellow needles 160–162° (dil. alcohol)	59, 60
3,4-$(CH_3O)_2C_6H_3CH_2$	OCH_3	OCH_3	D	95	Light yellow needles, 175–176° (benzene-ethanol)	61
OCH_3	Cl	Cl	E	70	112° (cyclohexane)	62, 42a

Miscellaneous Compounds and Compounds of Unknown Structure

Structure	Preparation[a]	Yield	MP	Reference
Cl; NO_2; $OSCCl_3$ (phthalazine)	F	—	203–204°	49

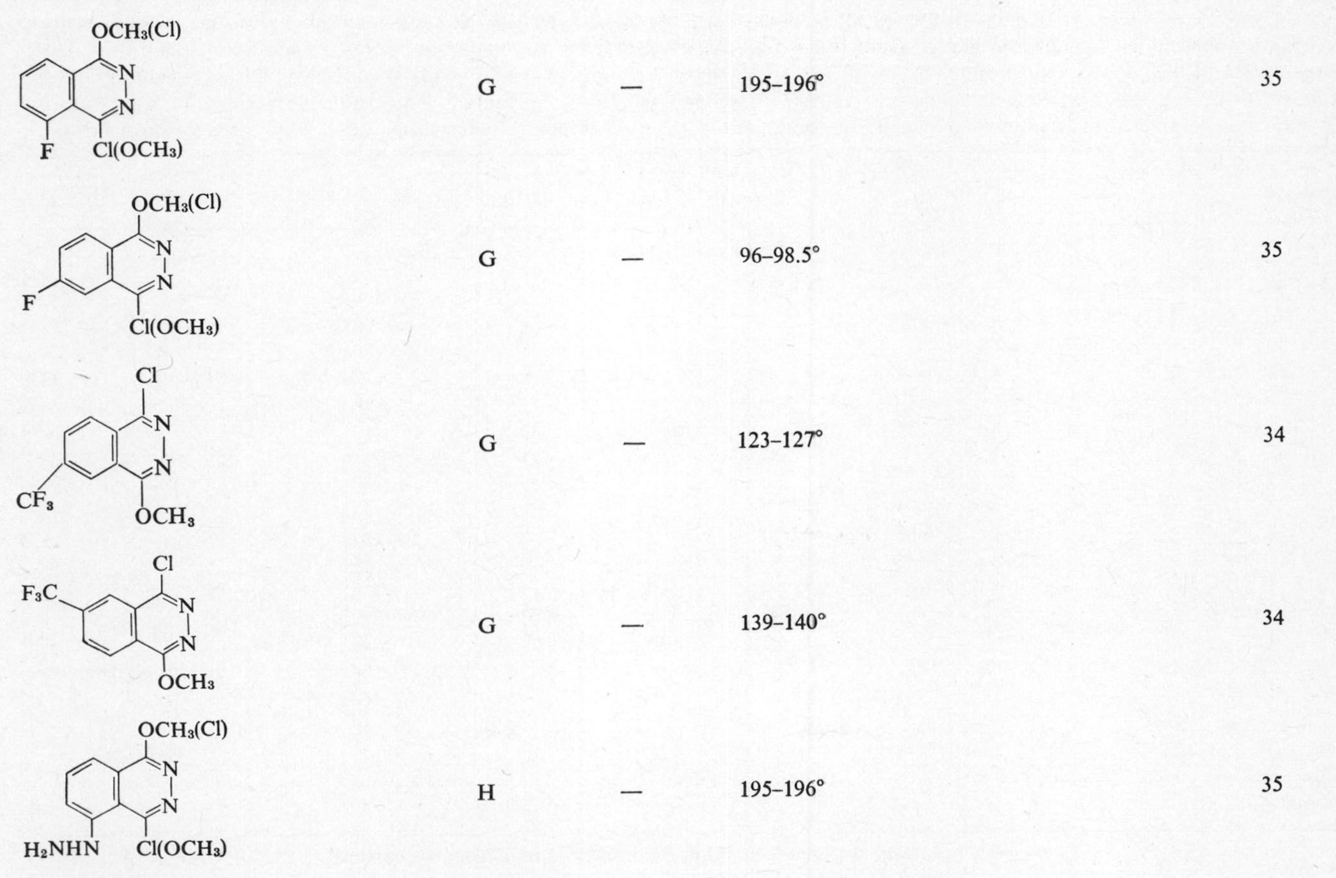

Structure				
1-OCH₃(Cl)-4-Cl(OCH₃)-5-F-phthalazine	G	—	195–196°	35
1-OCH₃(Cl)-4-Cl(OCH₃)-6-F-phthalazine	G	—	96–98.5°	35
1-Cl-4-OCH₃-6-CF₃-phthalazine	G	—	123–127°	34
1-Cl-4-OCH₃-7-CF₃-phthalazine	G	—	139–140°	34
1-OCH₃(Cl)-4-Cl(OCH₃)-5-H₂NHN-phthalazine	H	—	195–196°	35

TABLE 2F-3 (*continued*)

Miscellaneous Compounds and Compounds of Unknown Structure (*continued*)

Structure	Preparation[a]	Yield	MP	Reference
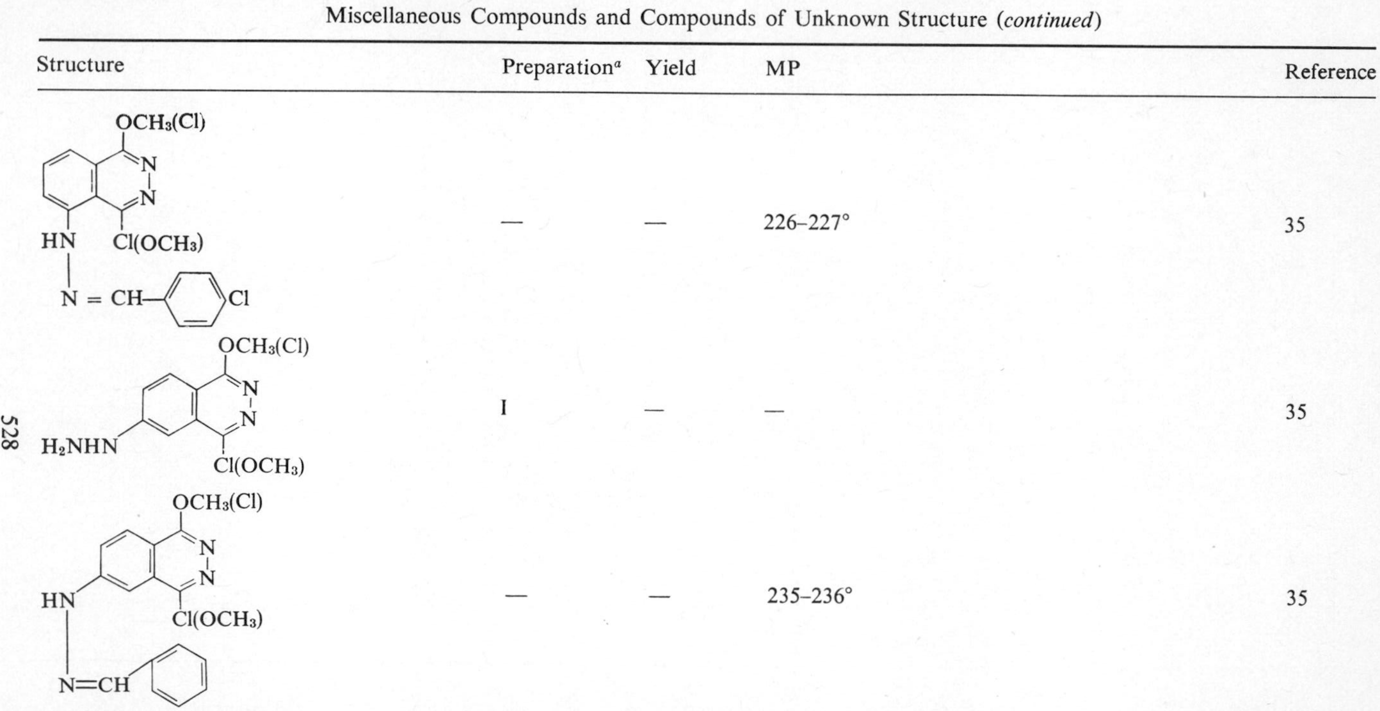	—	—	226–227°	35
	I	—	—	35
	—	—	235–236°	35

[a] A, the properly substituted 1-(2*H*)phthalazinone refluxed with $POCl_3$ for 15 min. B, same as A, boil for 30 min. C, same as A, heat 3 hr on water bath. D, same as A, heat a few minutes. E, 1,4,6,7-tetrachlorophthalazine + equimolar CH_3ONa, reflux 3 hr in methanol. F, 4-chloro-5-nitro-1-(2*H*)phthalazinone + trichloromethanesulfinyl chloride + 1*N* NaOH, stir in chloroform at room temperature for several hours. G, properly substituted 1,4-dichlorophthalazine + equimolar CH_3ONa, reflux about 45 min in methanol. H, 1-methoxy(chloro)-4-chloro(methoxy)-5-fluorophthalazine is refluxed with 98% $NH_2NH_2 \cdot H_2O$ in ethanol for 45 min. I, same as H, using 1-methoxy(chloro)-4-chloro(methoxy)-6-fluorophthalazine.

TABLE 2F-4. 4-Chloro-1-(2*H*)phthalazinones

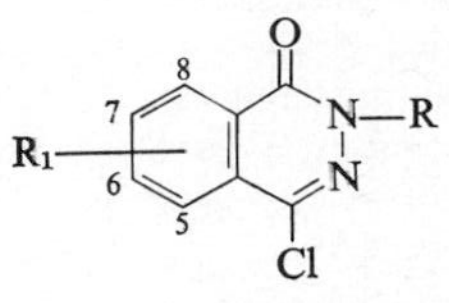

R	R_1	Preparation[a]	Yield (%)	MP	Reference
H	H	A	—	274°	32, 47
H	5-NO_2	B	78	Pale yellow needles, 249–250° (dec.) DMF	63
H	6 or 7-F	B_1	—	225–230°	35
H	4,5,6,7-Tetrachloro	A	—	—	40
H	6,7-Ethylenedioxy	B_2	90	Yellow crystals, 336° (dec.)	64
CH_3	H	B_3	92	128–129.5°; 129° (ethanol)	11, 25
CH_2Cl	H	C	100	White crystals, 145.5–146.5°	65
—$CH_2SP(=S)(OEt)_2$	H	D	100	78.5°–79° (hexane)	65
—$C(=O)OCH(CH_3)_2$	H	—	—	92–95°	66
—CH_2—N(piperidino)	H	E	—	103°	67
—CH_2—N(morpholino)	H	E	—	91°	67

TABLE 2F-4 (*continued*)

R	R_1	Preparation[a]	Yield (%)	MP	Reference
$-CH_2-N(CH_2CH_2)_2NH$ (piperazinylmethyl)	H	E_1	—	222° (dec.)	67
$-CH_2-N(CH_2CH_2)_2N-CH_2-$ (4-chloro-1-(2H)phthalazinon-2-yl)	H	E_1	—	257–258°	67
C_6H_5	H	B_4	82	Needles, 136° (aq. ethanol); colorless long prisms, 130° (ethanol)	26, 68
2-pyridyl	H	B	70	176–178° (aq. ethanol)	27
p-$NO_2C_6H_4$—	H	F	—	243° (dec.) (chloroform-methanol)	69
p-$NH_2C_6H_4$—	H	—	—	213–214°; acetamide, 248°	69

[a] A, acidic hydrolysis of properly substituted 1,4-dichlorophthalazine. B, the corresponding 4-hydroxy-1-(2*H*)phthalazinone + $POCl_3$, reflux 4 hr. B_1, the same as B, obtained as a byproduct. B_2, the same as B, heat 3 hr on water bath. B_3, the same as B, heat at 100° for 30 min. B_4, the same as B, or heat phenylhydrazide of phthalic acid with $POCl_3$. C, 4-chloro-2-hydroxymethyl-1-(2*H*)phthalazinone + $SOCl_2$, reflux 1 hr. D, 4-chloro-2-chloromethyl-1-(2*H*)phthalazinone + $KS-P(=S)-(OEt)_2$ in acetone, heat at 58° for 5 hr. E, Mannich reaction: 4-chloro-1-(2*H*)phthalazinone + 37% aqueous formaldehyde + amine, reflux in ethanol. E_1, same as E, reflux in ethanol + DMF. F, 2-phenyl-4-chloro-1-(2*H*)phthalazinone + HNO_3 + H_2SO_4.

TABLE 2F-5. 1,4-Dichlorophthalazine with Substituents in the Benzenoid Ring

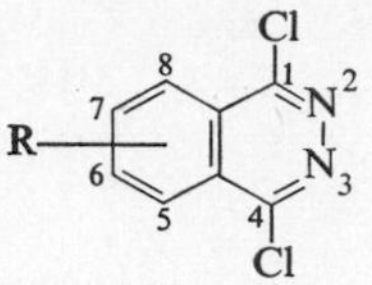

R	Preparation[a]	Yield (%)	MP	Reference
H	A	88–96	160–162°; 162–163°; colorless long needles 164° (tetrahydrofuran)	28–31
6-C_6H_5	B	43	Buff-colored compound 150–153°	33
6-CO_2H	—	—	—	36
6-SO_3H	—	—	—	36
6-$\overset{O}{\overset{\parallel}{C}}Cl$	C	94	130–132° (pet. ether); 133°	37–39
6-F	A	—	148–153°	35
6-CF_3	D	84	129–130° (isopropyl ether); sublimes at 70°/0.2 mm	34
6,7-Dichloro	E	62	140° (benzene)	62, 42a
5,6,7,8-Tetrachloro	F	—	194–195.5°; 120°	40, 42a
5-Cl	—	—	170–175°	42a

[a] A, see different methods of preparation in this section. B, 6-phenyl-4-hydroxy-1-(2*H*)phthalazinone + PCl_5, heat at 150–160° for 4 hr. C, 4-hydroxy-1-(2*H*)phthalazinone-6-carboxylic acid, heat with $POCl_3$ + PCl_5 at 110–120° for 4 hr or heat with PCl_3 + Cl_2 at 117°, then pass in SO_2. D, 6-trifluoromethyl-4-hydroxy-1-(2*H*)phthalazinone + PCl_5, heat at 170–175° for 6 hr. E, 6,7-dichloro-4-hydroxy-1-(2*H*)phthalazinone + PCl_5, heat at 155–160° for 4 hr. F, 1,4-dichlorophthalazine + Cl_2 + $AlCl_3$, heat at 200°.

TABLE 2F-6. Phthalazines with Bromo-, Iodo-, and Fluoro-substituents in the Heterocyclic Ring

Structure	Preparation[a]	Yield (∞)	MP	Reference
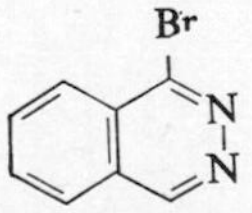	A	88	175° (with foaming)	29
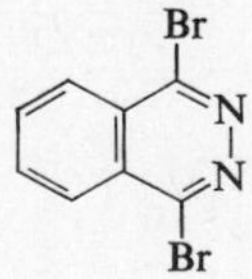	B	78–97	Colorless long needles, 160–161°; pale yellow long needles, 160° (tetrahydrofuran)	29

TABLE 2F-6 (*continued*)

Structure	Preparation[a]	Yield (∞)	MP	Reference
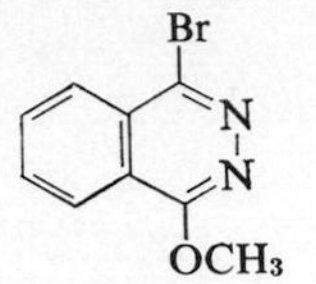 Br, N, N, OCH_3	—	—	108–112°	56
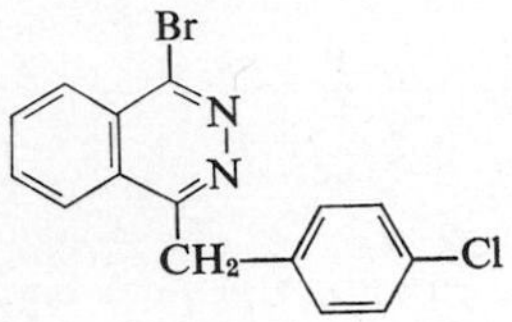 Br, N, N, CH_2, Cl	C	—	218°	20
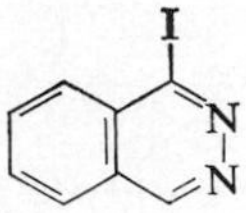 I, N, N	D	88	78° (dec.)	30
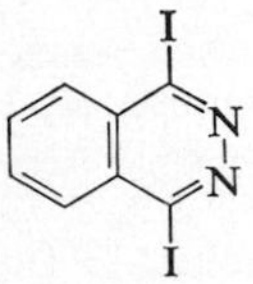 I, N, N, I	D	76	White prisms, 161° (dichloromethane)	30
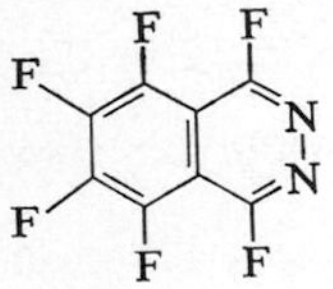 F, F, F, F, F, F, N, N	E	60	Colorless plates, 91–93° (vacuum sublimed)	40
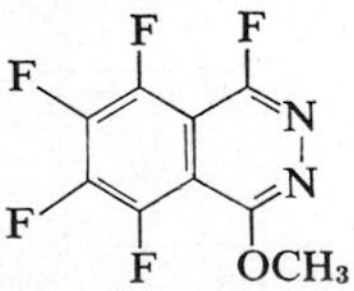 F, F, F, F, F, N, N, OCH_3	F	—	67.5–69.5[c]	40
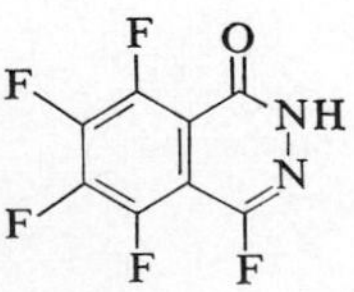 F, O, F, NH, N, F, F, F	G	—	244–246°	40

TABLE 2F-6 (*continued*)

Structure	Preparation[a]	Yield (%)	MP	Reference
4,5,6,7,8-pentafluoro-2-methyl-1-(2H)phthalazinone (F, F, F, F, F; O; NCH_3; N)	H	—	150–152°	40

[a] A, 1-(2*H*)phthalazinone + $POBr_3$ in ether-tetrahydrofuran, heat at 55–60° for 3 hr with stirring. B, 4-hydroxy-1-(2*H*)phthalazinone + PBr_5, heat in a pressure bottle with stirring at 140° for 4 hr or heat in carbon tetrabromide at 130° for 6 hr. C, 4-(*p*-chlorobenzyl)-1-(2*H*)phthalazinone + PBr_5, heat. D, the corresponding chlorophthalazine + NaI + acetone + 50% aqueous HI, stir at room temperature for 18 hr. E, 1,4,5,6,7,8-hexachlorophthalazine + KF, heat at 290°. F, 1,4,5,6,7,8-hexafluorophthalazine + equimolar CH_3ONa in methanol at −15° C. G, 1,4,5,6,7,8-hexafluorophthalazine, hydrolysis with aqueous acid. H, 4,5,6,7,8-pentafluoro-1-(2*H*)phthalazinone + diazomethane in ether.

References

1. C. M. Atkinson, C. W. Brown, and J. C. E. Simpson, *J. Chem. Soc.*, **1956,** 1081.
2. E. Hayashi et al., *Yakugaku Zasshi*, **82,** 584 (1962); *Chem. Abstr.*, **58,** 3425 (1963).
3. S. Biniecki and B. Gutkowaska, *Acta Polon. Pharm.*, **11,** 27 (1955); *Chem. Abstr.*, **50,** 12062 (1956).
4. V. B. Brasyunas and A. S. Podzhyunas, *Med. Prom. SSSR*, **13,** 38 (1959); *Chem. Abstr.*, **53,** 16144 (1959).
5. A. R Osborn, K. Schofield, and L. N. Short, *J. Chem. Soc.*, **1956,** 4191.
6. Swiss Pat. 266,287 (1950); *Chem. Abstr.*, **45,** 7605 (1951).
7. S. S. Berg and E. W. Parnell, *J. Chem. Soc.*, **1961,** 5275.
8. S. Biniecki, M. Moll, and L. Rylski, *Ann. Pharm. Franc.*, **16,** 421 (1958); *Chem. Abstr.*, **53,** 7187 (1959).
9. G. Favini and M. Simonetta, *Gazz. Chim. Ital.*, **90,** 369 (1960).
10. S. Biniecki and L. Rylski, *Ann. Pharm. Franc.*, **16,** 21 (1958); *Chem. Abstr.*, **52,** 18446 (1958).
11. I. Satoda, N. Yoshida, and K. Mori, *Yakugaku Zasshi*, **77,** 703 (1957); *Chem. Abstr.*, **51,** 17927 (1957).
12. Belgian Pat. 628,255 (1963); *Chem. Abstr.*, **60,** 14516 (1964).
13. I. Zugravescu, M. Petrovanu, and E. Rucinschi, *An. Stiint. Univ., A. I. Cuza, Iasi Sect. 17*, 169 (1961); *Rev. Chim., Acad. Rep. Pop. Roum.*, **2,** 1405 (1962); *Chem. Abstr.*, **59,** 6399 (1963); **61,** 7009 (1964).
14. D. G. Parsons and A. F. Turner, *J. Chem. Soc. (C)*, **1966,** 2016.
15. A. Aebi and E. Hofstetter, *Pharm. Acta. Helv.*, **40,** 241 (1965); *Chem. Abstr.*, **63,** 7004 (1965).
16. British Pat. 629,177 (1949); *Chem. Abstr.*, **44,** 4516 (1950).
17. British Pat. 732,581 (1955); *Chem. Abstr.*, **50,** 8747 (1956).
18. J. Druey and B. H. Riniger, *Helv. Chim. Acta.*, **34,** 195 (1951).
19. R. C. Fuson and W. C. Hammann, *J. Am. Chem. Soc.*, **74,** 1626 (1952).
20. H. J. Engelbrecht, D. Lenke, and H. Muller, German (East), Pat. 17319 (1959); *Chem. Abstr.*, **55,** 590 (1961).

21. J. Druey and A. Marxer, U.S. Pat. 2,960,504 (1960); German Pat. 1,061,788 (1959); *Chem. Abstr.*, **55,** 11446, 13458 (1961).
22. M. V. Sigal Jr., P. Marchini, and B. L. Poet, U.S. Pat. 3,274,185 (1966); *Chem. Abstr.*, **65,** 18,600 (1966).
23. H. M. Holva, Jr. and R. A. Partyka, *J. Med. Chem.*, **12,** 555 (1969).
24. E. Hayashi and E. Oishi, *Yakugaku Zasshi*, **86,** 576 (1966); *Chem. Abstr.*, **65,** 15373 (1966).
25. H. Morishita et al., Japanese Pat. 2981 (1959); *Chem. Abstr.*, **54,** 13150 (1960).
26. D. J. Drain and D. E. Seymour, *J. Chem. Soc.*, **1955,** 852.
27. T. Kubaba and R. S. Luwiczak, *Roczniki Chem.*, **38,** 367 (1964).
28. H. L. Yale, *J. Am. Chem. Soc.*, **75,** 675 (1953).
29. A. Hirsch and D. Orphanos, *Can. J. Chem.*, **43,** 2708 (1965).
30. A. Hirsch and D. Orphanos, *Can. J. Chem.*, **43,** 1551 (1965).
31. S. Biniecki and J. Izdebski, *Acta Polon. Pharm.*, **15,** 421 (1957); *Chem. Abstr.*, **52,** 15540 (1958).
32. J. C. E. Simpson, in *Condensed Pyridazine and Pyrazine Rings*, *The Chemistry of Heterocyclic Compounds*, Vol. 5, A. Weisberger, Ed., Interscience, New York, 1953, p. 178.
33. C. M. Atkinson and C. J. Sharpe, *J. Chem. Soc.*, **1959,** 2858.
34. B. Cavalleri, E. Bellasio, and A. Sardi, *J. Med. Chem.*, **18,** 148 (1970), B. Vigevani, B. Cavalleri, and G. G. Gallo, *J. Heterocyclic Chem.*, **7,** 677 (1970).
35. B. Cavalleri and E. Bellasio, *Farm. Ed. Sci.*, **24,** 833 (1969); *Chem. Abstr.*, **71,** 101,798 (1969).
36. French Pat. 1,458,332 (1966); *Chem. Abstr.*, **67,** 22,822 (1967).
37. J. Cole, Jr. and W. H. Gumprecht, French Pat. 1,359,644 (1964); U.S. Pat. 3,184,282 (1965); *Chem. Abstr.*, **62,** 13278 (1965); **63,** 8526 (1965).
38. R. Roe, A. Richard, M. Rouget, and G. Thirst, French Pat. 1,357,726 (1964); *Chem. Abstr.*, **61,** 4371 (1964).
39. L. A. Rothman, French Pat. 1,384,789 (1965); *Chem. Abstr.*, **63,** 10101 (1965).
40. R. D. Chambers, J. A. A. MacBride, W. K. R. Musgrave, and I. S. Reilly, *Tetrahedron Lett.*, **1970,** 55.
41. S. O. Wintrop, S. Sybulski, R. Gaudry, and G. A. Grant, *Can. J. Chem.*, **34,** 1557 (1956).
42. K. Adachi, *J. Pharm. Soc. Japan*, **75,** 1423 (1955); *Chem. Abstr.*, **50,** 10105 (1956).
42a. Japanese Pat. 7,106,800 (1971); from *Derwent AGDOC—Complete Spec.*, 14803s (1971).
43. J. H. Bowie et al., *Aust. J. Chem.*, **20,** 2697 (1967).
44. A. Hirsch and D. Orphanos, *Can. J. Chem.*, **46,** 1455 (1968).
45. G. M. Badger, J. C. McCarthy, and M. J. Rodda, *Chem. Ind.*, **1954,** 964.
46. H. J. Rodda and P. E. Rogasch, *Aust. J. Chem.*, **19,** 1291 (1961).
47. R. D. Haworth and S. Robinson, *J. Chem. Soc.*, **1948,** 777.
48. J. Druey, U.S. Pat. 2,484,785 (1949); British Pat. 654,821 (1951); *Chem. Abstr.*, **44,** 2573 (1950); **46,** 9622 (1952).
49. A. Margot and H. Gysin, U.S. Pat. 2,835,626 (1958); *Chem. Abstr.*, **52,** 20,213 (1958).
50. G. Wagner and D. Heller, *Arch. Pharm.*, **299,** 768 (1966).
51. N. B. Chapmann and D. Q. Russell-Hill, *J. Chem. Soc.*, **1956,** 1563; *Chem. Ind.*, **1954,** 1298.
52. Belgian Pat. 635,078 (1964); *Chem. Abstr.*, **61,** 16204 (1964).
53. British Pat. 719,282 (1954); *Chem. Abstr.*, **49,** 6616 (1955).

54. M. Jirou, C. Brouarel, and P. Bouvet., French Pat. 1,336,679 (1963); French Addn. **89,** 178 (1967) (Addn. to Fr. 1,394,020); French Pat. 1,459,111 (1966); *Chem. Abstr.*, **60,** 4281 (1964); **68,** 14,092 (1968); **67,** 22,889 (1967). E. Siegel and K. Susse, Belgian Pat. 614,375 (1962); *Chem. Abstr.*, **59,** 6553 (1963).
55. K. H. Boltze and H. D. Dell, *Arch. Pharm.*, **299,** 702 (1966).
56. British Pat. 822,069 (1959); *Chem. Abstr.*, **55,** 2005 (1961).
57. T. Ikeda and S. Kanahara, *Kanazawa Yakugakubu Kenkyu Nempo*, **7,** 6 (1957); *Chem. Abstr.*, **52,** 6361 (1958).
58. J. Druey and A. Marxer, Swiss Pat. 361,812 (1962); *Chem. Abstr.*, **59,** 8761 (1963).
59. T. Ikeda, S. Kanahara, and T. Ujiie, *Kanazawa Daigaku Yakugakubu Kenkyu Nempo*, **8,** 1 (1958); *Chem. Abstr.*, **53,** 4287 (1959).
60. T. Ikeda and S. Kanahara, *Kanazawa Daigaku Yakugakubu Kenkyu Nempo*, **9,** 6 (1959); *Chem. Abstr.*, **54,** 4607 (1960).
61. T. Ikeda and S. Kanahara, *Kanazawa Daigaku Yakugakubu Kenkyu Nempo*, **10,** 15 (1960); *Chem. Abstr.*, **55,** 11,426 (1961).
62. S. Kumada, N. Watanabe, K. Yamamoto, and H. Zenno, *Yakugaku Kenkyo*, **30,** 635 (1958); *Chem. Abstr.*, **53,** 20,554 (1959). M. Ohara, K. Yamamoto, and A. Sugihara, Japanese Pat. 10,027 (1960); *Chem. Abstr.*, **55,** 8439 (1961).
63. C. Volker, J. Marth, and H. Bayer, *Chem. Ber.*, **100,** 875 (1967).
64. F. Dallacker and J. Bloemen, *Monatsh. Chem.*, **92,** 640 (1961).
65. S. DuBruil, U.S. Pat. 2,938,902 (1960); *Chem. Abstr.*, **54,** 21,146 (1960).
66. S. A. Agripat, French Pat. 1,460,552 (1966); *Chem. Abstr.*, **67,** 31,774 (1967).
67. E. Domagalina, I. Kurpiel, and N. Koktysz, *Roczniki Chem.*, **43,** 775 (1969); *Chem. Abstr.*, **71,** 61,315 (1969).
68. T. Ohta, *J. Pharm. Soc., Japan*, **63,** 239 (1943); *Chem. Abstr.*, **45,** 5163 (1951).
69. E. Domagalina and S. Baloniak, *Roczniki Chem.*, **36,** 253 (1962); *Chem. Abstr.*, **57,** 16609 (1962).

Part G. 1-(2*H*)Phthalazinethione and 4-Mercapto-1-(2*H*)phthalazinethione

The study of these compounds is relatively new; generally their chemistry has been reported only since 1950. They are named here according to the dominant tautomeric structure in which they exist. For example, an equilibrium between the structures 1-(2*H*)phthalazinethione and 1-phthalazinethiol may operate, but Albert and Barlin (1) have shown that this compound exists principally as a thione (═S) rather than a thiol (—SH). Because of this, the compound is referred to as 1-(2*H*)phthalazinethione. No such study of the tautomeric forms of phthalazine having thione (or thiol) groups in both the 1- and 4-positions has been reported, but it is likely that this compound exists principally as 4-mercapto-1-(2*H*)phthalazinethione, and is named as such here, since its oxygen analog is known to exist as 4-hydroxy-1-(2*H*)phthalazinone (see Section 2E-I-B). Those 1-(2*H*)phthalazinethiones which have a carboxylic acid group in the 4-position are discussed in Section 2J, while phthalazines having alkylsulfinyl and alkyl- or arylsulfonyl groups in the 1-position are discussed in this section.

I. Methods of Preparation

A. Preparation of 1-(2*H*)Phthalazinethione

1-(2*H*)Phthalazinethione (**1**) can be prepared in high yields (75%) by refluxing 1-(2*H*)phthalazinone with phosphorus pentasulfide in dry pyridine for 1.5 hr (1), in toluene for 6 hr (2), or in xylene for 2 hr (3). Another approach to the synthesis of **1** has been reported (4) in which 1-(2*H*)-phthalazinethione-4-carboxylic acid (**2**) is decarboxylated to give **1** in 80% yield (Eq. 1). The preparation of **2** has been achieved by reacting ethyl 1-(2*H*)phthalazinone-4-carboxylate with phosphorus pentasulfide in xylene followed by alkaline hydrolysis (4).

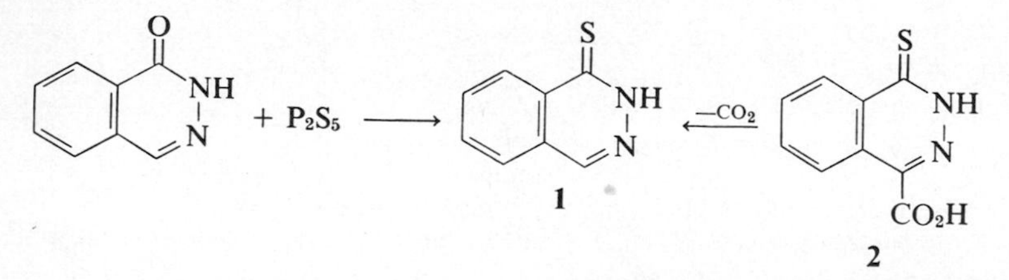

The synthesis of 4-methyl- and 4-phenyl-1-(2*H*)phthalazinethione (**2a**; R = CH_3, C_6H_5) has been carried out by refluxing 4-methyl- and 4-phenyl-1-(2*H*)phthalazinone with phosphorus pentasulfide in benzene for 6 hr (2). 4-Phenyl-1-(2*H*)phthalazinethione has also been prepared in 83% yield (5) from 1-chloro-4-phenylphthalazine by refluxing with thiourea in methanol for 10 min then decomposing the intermediate *S*-thiouronium salt (**3**) in alkaline solution (Eq. 2).

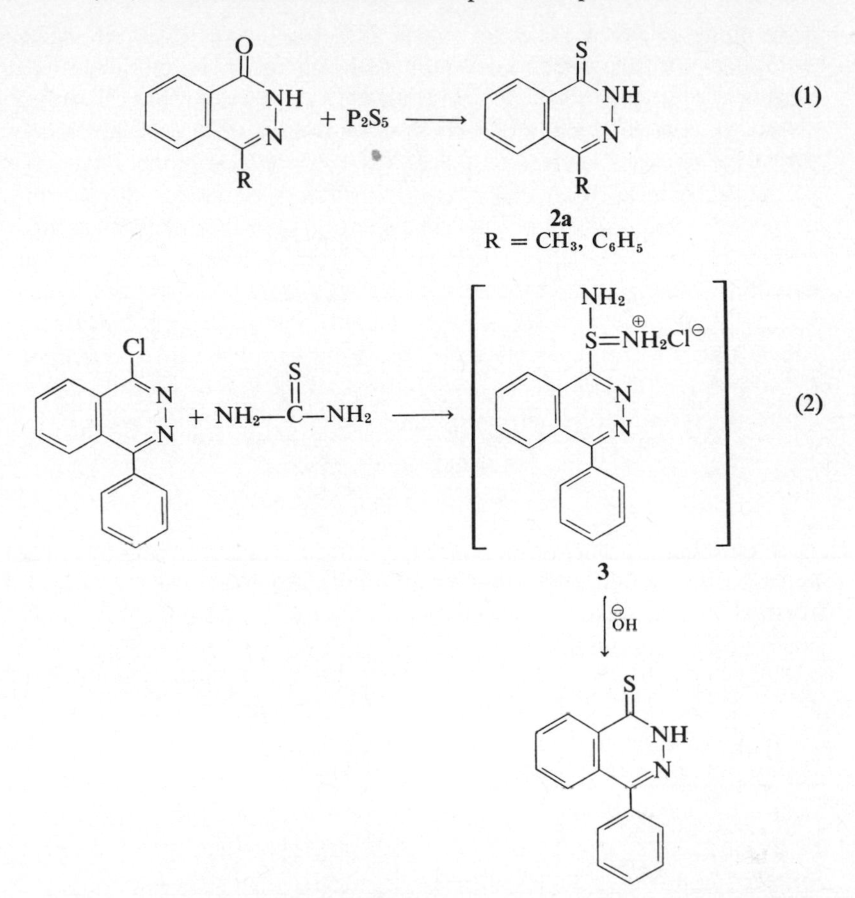

4-Chloro-1-(2*H*)phthalazinethione has been obtained by refluxing 1,4-dichlorophthalazine for 2 hr with potassium sulfhydride in methanol (6). Similarly, 4-hydrazino-1-(2*H*)phthalazinethione is synthesized from 1-chloro-4-hydrazinophthalazine and potassium sulfhydride (7). 4-Hydrazino-1-(2*H*)-phthalazinethione is also prepared from 4-mercapto-1-(2*H*)phthalazinethione and an equimolar amount of hydrazine hydrate (8). 7-Nitro-1-(2*H*)phthalazinethione has been prepared by the reaction of 7-nitro-1-(2*H*)phthalazinone with phosphorus pentasulfide (9).

B. Preparation of 4-Mercapto-1-(2*H*)phthalazinethione

Radulescu and Georgescu (10) first prepared 4-mercapto-1-(2*H*)phthalazinethione (**4**) by heating 4-hydroxyl-1-(2*H*)phthalazinone with phosphorus pentasulfide at 200° C. However, Yale (11) found that this method gave a very impure product, which was difficult to purify. Yale obtained **4** in 83% yield by refluxing 1,4-dichlorophthalazine with sodium hydrosulfide in ethanol for 4 hr after the initial reaction had subsided (11) (Eq. 3).

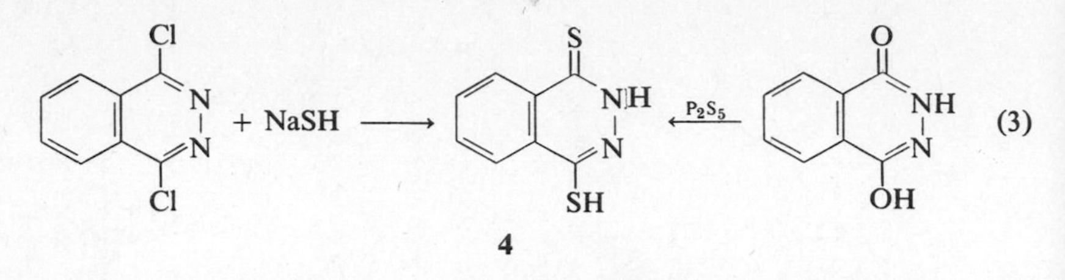

C. Preparation of 2- and 2,4-Disubstituted 1-(2*H*)phthalazinethiones

This group of compounds is usually prepared by the reaction of phosphorus pentasulfide on the corresponding 1-(2*H*)phthalazinones. Thus 2-methyl- and 2-phenyl-1-(2*H*)phthalazinethione are obtained by refluxing 2-methyl-1-(2*H*)phthalazinone with phosphorus pentasulfide in pyridine (1) or xylene (8) and by refluxing 2-phenyl-1-(2*H*)phthalazinone with phosphorus pentasulfide in benzene (2). Similarly, 4-methyl- (2), 4-phenyl- (2), and 4-alkoxy-2-phenyl-1-(2*H*)phthalazinethiones (12) are obtained from the corresponding 1-(2*H*)phthalazinones (Eq. 4).

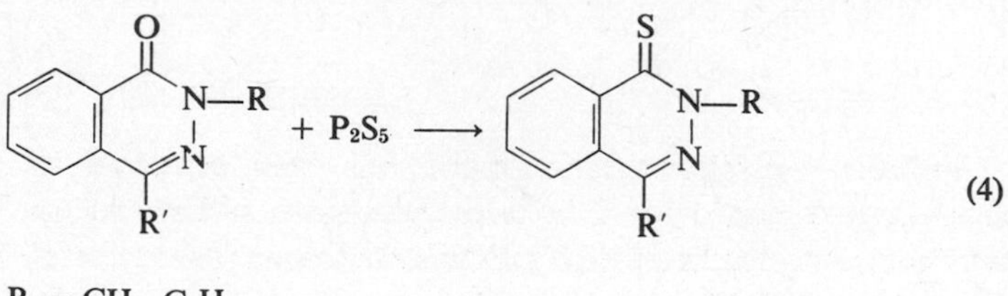

R = CH_3, C_6H_5
R′ = H, CH_3, C_6H_5, alkoxy

Treatment of 2-phenyl-4-hydroxy-1-(2*H*)phthalazinone (12) with phosphorus pentasulfide in xylene gave the expected 2-phenyl-4-mercapto-1-(2*H*)phthalazinethione in poor yield (17%) and a by-product, bis(1,2-dihydro-2-phenyl-1-thio-4-phthalazinyl)sulfide (**5**) in 22% yield (Eq. 5). Alkylation of 2-phenyl-4-mercapto-1-(2*H*)phthalazinethione with ethyl chloride, benzyl chloride, and 2,4-dinitrochlorobenzene in ethanolic sodium ethoxide gives 2-phenyl-4-alkyl(or aryl)thio-1-(2*H*)phthalazinethiones (12) (Eq. 5).

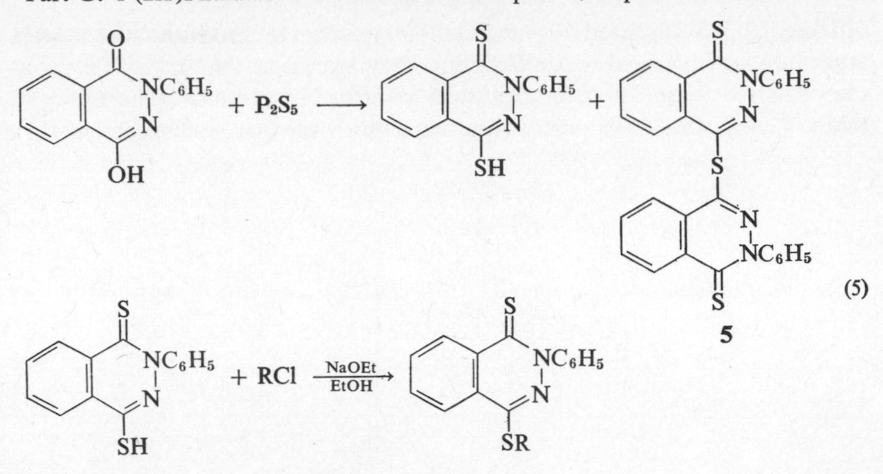

R = Ethyl, benzyl, 2,4-dinitrophenyl

2-Phenyl-4-mercapto- and 2-phenyl-4-benzylthio-1-(2*H*)phthalazinone are synthesized from 2-phenyl-4-chloro-1-(2*H*)phthalazinone by reaction of this with potassium sulfhydride and benzyl mercaptan, respectively (12) (Eq. 6). Alkylation of 4-ethylthio-1-(2*H*)phthalazinone with *N*,*N*-diethylaminoethyl chloride in benzene-sodium ethoxide furnished 2-(*N*,*N*-diethylaminoethyl)-4-ethylthio-1-(2*H*)phthalazinone (**6**) (13) (Eq. 6).

Wagner and Heller (6, 14) have reacted the alkali metal salts of 1-(2*H*)-phthalazinethione, 4-chloro-1-(2*H*)phthalazinethione, and 4-phenyl-1-(2*H*)-phthalazinethione with α-acetobromoglucose and obtained in each case a

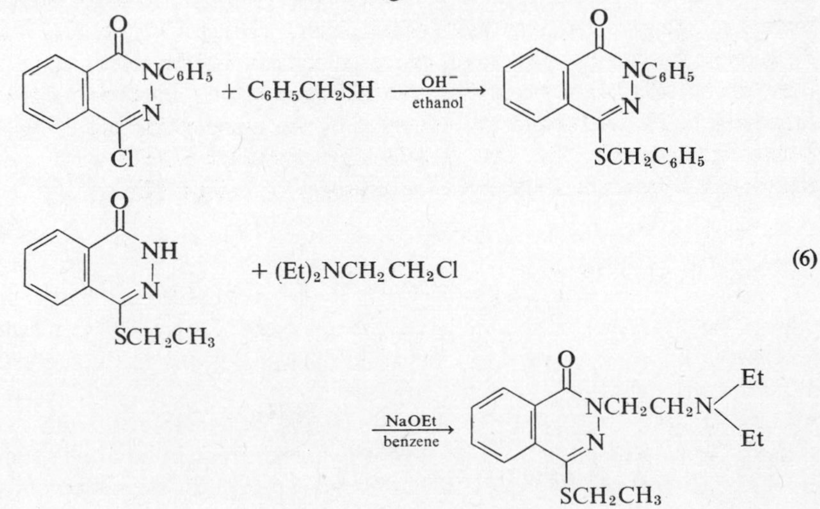

mixture of 1-*S*-alkylated (**7**) and 2-*N*-alkylated (**8**) products which were separated by fractional crystallization. They have also shown S → N transglucosylation when **7** was refluxed with mercuric bromide in toluene to give **8** (Eq. 7), and they have proposed a mechanism for this rearrangement (15).

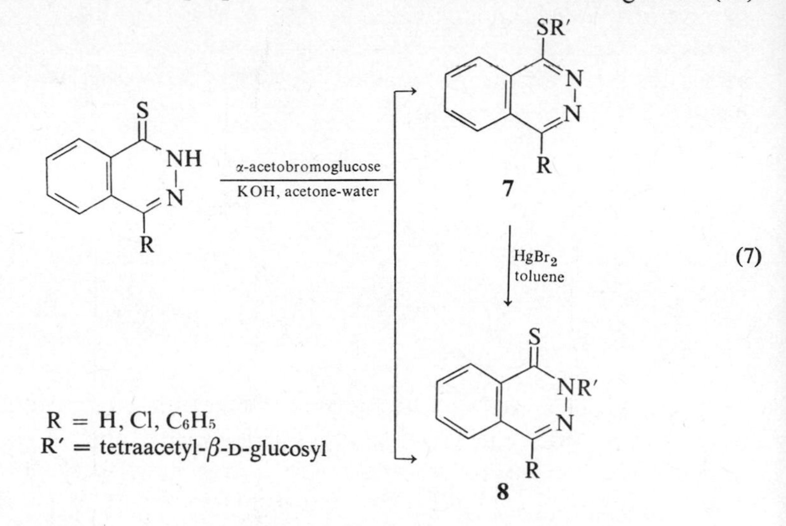

D. Preparation of Alkylthio- and Arylthiophthalazines

Alkylation of 1-(2*H*)phthalazinthione with methyl iodide (1) or with dimethyl sulfate (16) in aqueous sodium hydroxide gives 1-methylthiophthalazine (**9**) in 65% yield. Similarly, 4-mercapto-1-(2*H*)phthalazinthione and methyl iodide give 1,4-dimethylthiophthalazine (**10**) in 87% yield (17) (Eq. 8). In both cases alkylation takes place principally on the sulfur atom, and little or no alkylation occurs on the nitrogen atom. Another method of preparing **9**, **10**, and similar thioethers is by the nucleophilic substitution of 1-chlorophthalazine (16) and 1,4-dichlorophthalazine (17) with methyl mercaptan in benzene solution in the presence of sodium (Eq. 8).

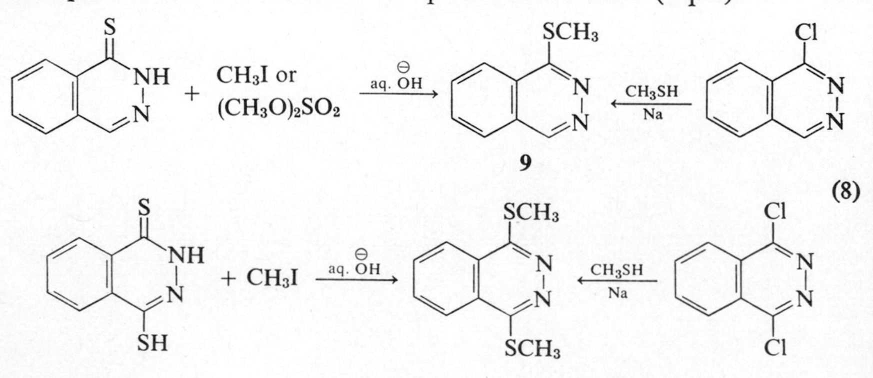

Several heterocyclic mercaptans, for example, 4- and 4,5-disubstituted 2-mercaptothiazoles, 2-mercaptobenzoxazole, 2-mercaptobenzthiazole, 2-mercaptobenzimidazole, and 2-mercapto-4,6-dimethylpyridine, have been allowed to react with 1-chlorophthalazine in ethanolic potassium hydroxide to give the corresponding 1-arylthiophthalazines in 45–88% yields (18). Similarly, *N*,*N*-dialkyldithiocarbamic acids will react with 1-chlorophthalazine to give the corresponding *N*,*N*-dialkyl-(1-phthalazinyl)dithiocarbamates (**11**) (Eq. 9) (18, 19).

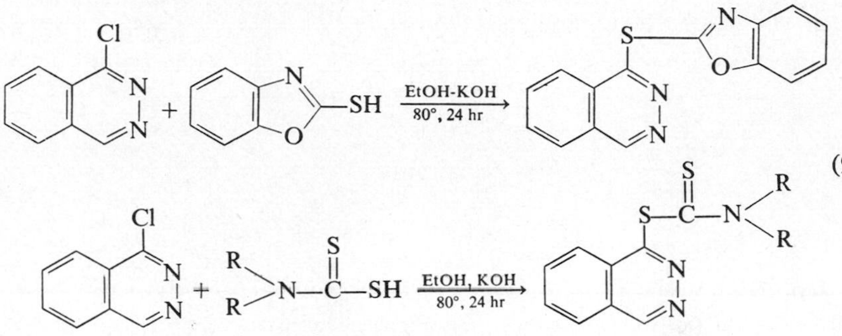

(9)

11

R = CH_3, CH_3CH_2, $(CH_3)_2CH$, cycloalkyl, etc.

4-Substituted 1-alkylthio- or arylthiophthalazines have been synthesized by alkylation of the corresponding 1-(2*H*)phthalazinethiones or by nucleophilic substitution of the corresponding 1-chlorophthalazines. 4-Phenyl-1-(2*H*)phthalazinethione will react with methyl iodide, ethyl bromide, and

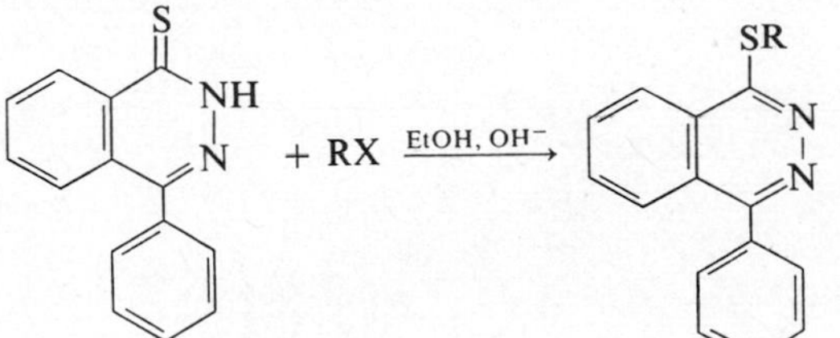

R = CH_3, CH_3CH_2, n-C_4H_9 (10)

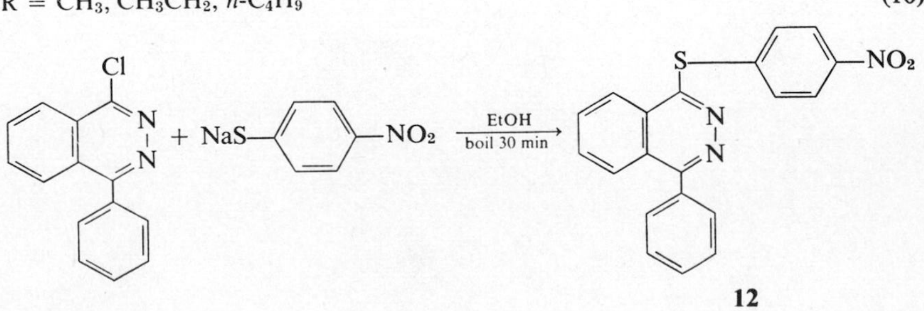

12

n-butyl bromide to give the corresponding 1-alkylthio-4-phenylphthalazines (20). 1-Chloro-4-phenylphthalazine reacts with the sodium salt of 4-nitrothiophenol to give 1-(4-nitrophenylthio)-4-phenylphthalazine (**12**) in 97% yield (21) (Eq. 10).

E. Preparation of 1-Methylsulfinyl- and 1-Alkyl- or Arylsulfonylphthalazines

Barlin and Brown (22) have synthesized 1-methylsulfinylphthalazine (**13**), the only alkylsulfinylphthalazine known, by oxidation of 1-methylthiophthalazine (**9**) with 3-chloroperbenzoic acid in chloroform solution at 0° C

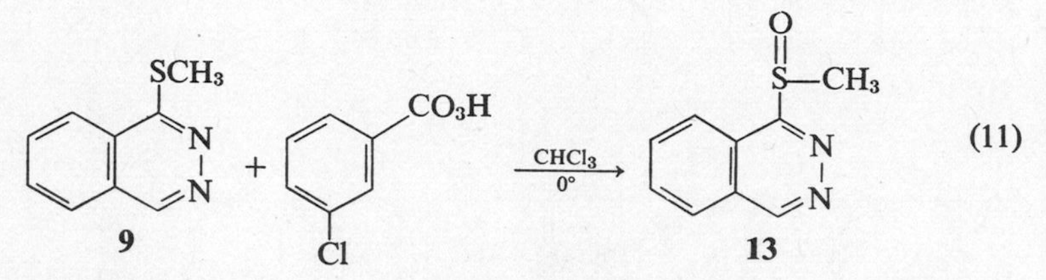

in 41% yield (Eq. 11). The main interest in preparing **13** and 1-alkylsulfonylphthalazines is due to the high reactivity of these compounds with various anions (see Section 2G-II-B).

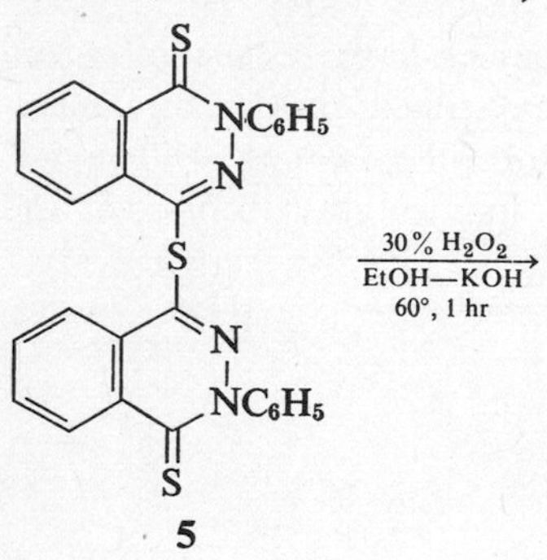

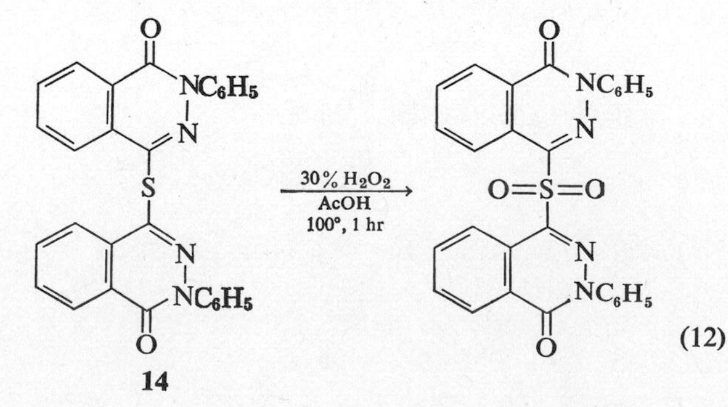

Drain and Seymour (12) first studied the oxidation of bis(1,2-dihydro-2-phenyl-1-thio-4-phthalazinyl)sulfide (**5**), which on treatment with hydrogen peroxide in alkaline medium gave bis(1,2-dihydro-2-phenyl-1-oxo-4-phthalazinyl)sulfide (**14**). Further oxidation of **14** with hydrogen peroxide in acetic acid furnished bis(1,2-dihydro-2-phenyl-4-phthalazinyl)sulfone (Eq. 12).

Bednyagina and co-workers (20, 21) successfully oxidized 1-alkylthio- and 1-(4′-nitrophenyl)thio-4-phenylphthalazine with 7% potassium permanganate in glacial acetic acid to give the corresponding sulfones (Eq. 13). Hayashi and co-workers (23) prepared 1-methylsulfonyl-4-phenylphthalazine using this procedure, and they also observed that oxidation of 1-methylthio-4-phenylphthalazine with hydrogen peroxide or perphthalic acid does not give the desired sulfone. Instead, the alkylthio group is replaced by a hydroxy group (Eq. 13).

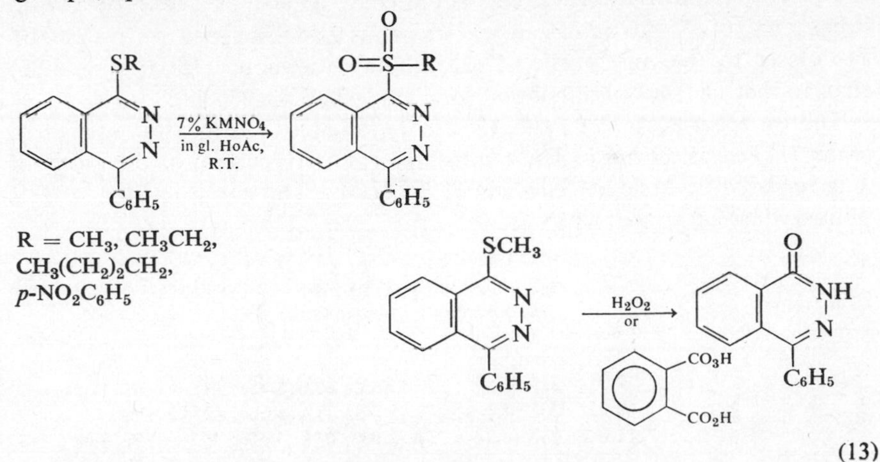

(13)

1-Methylsulfonylphthalazine has been prepared by the oxidation of 1-methylthiophthalazine with potassium permanganate (24).

II. Properties

A. Physical Properties

1-(2*H*)Phthalazinethione (**1**) forms yellow crystals when crystallized from ethanol, mp 169–170° (1) [170–171° (3); 171° (4); 174° (2)]. 4-Mercapto-1-(2*H*)phthalazinethione (**4**) forms yellow needles, mp 262–264° (toluene) (11); 262–265° (10). Both **1** and **4** are soluble in alkali. Phthalazinethiones in general are alkali soluble in the absence of substituents at the 2-position which would prevent tautomerism to the thiol form. They are yellow compounds and usually have high melting points.

Albert and Barlin (1) have concluded from a study of ionization constants and uv spectra of **1** that the tautomeric equilibrium favors the thione rather than the thiol form, that is, 1-(2*H*)phthalazinethione (**1**) as opposed to 1-mercaptophthalazine.

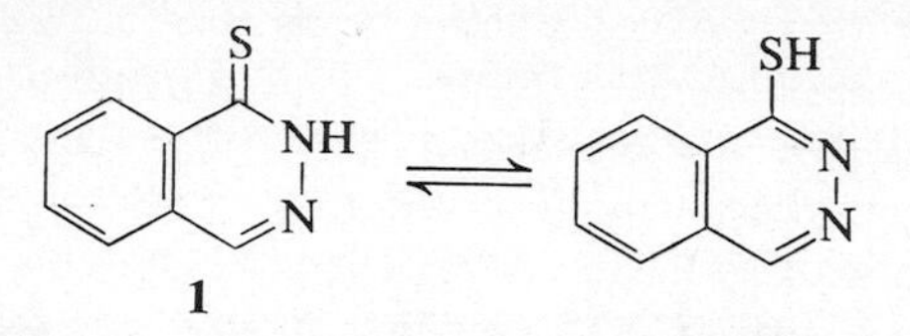

1-(2*H*)Phthalazinethione (**1**) may act as either an acid or a base. As a base, it has a pK_a of −3.43, and as an acid its pK_a is 9.98. The basic pK_a value of **1** lies closer to that of 2-methyl-1-(2*H*)phthalazinethione (**15**) (pK_a −3.98) than to that of 1-methylthiophthalazine (**9**) (pK_a 3.48) (3, 48).

The uv spectra of **1**, **9**, and **15** are reported as the neutral species in water (1) and as cations in acidic solution. The uv spectrum of the anion of **1** is reported in basic solution (1) (Fig. 2G-1). The uv spectrum of **1** is almost identical to that of **15**.

Compound	pH	λ_{max} (mμ) with log ε values in parentheses; italics indicate shoulders or inflections (1)
1	7.0	217 (4.57); *281* (3.83); 289 (3.95); 347 (3.99).
	−6.05	231 (4.40); 273 (3.76); 317 (3.66); *327* (3.63).
	12.0	216 (4.74); *236* (3.93); 273 (3.57); *282* (3.55); 334 (3.89)
15	7.0	219 (4.47); *280* (3.71); 289 (3.82); 339 (3.83).
	−6.0	232 (4.50); 274 (4.00); 317 (3.65); 327 (3.64).
9	7.0	216 (4.64); 263 (3.73); 296 (3.81).
	1.0	225 (4.45); 286 (3.87); 321 (3.81).

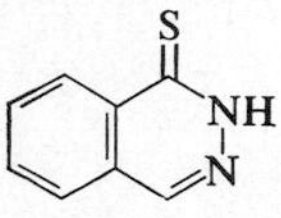

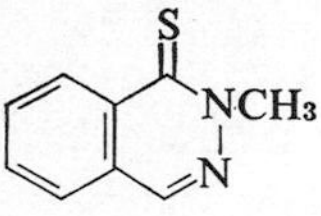

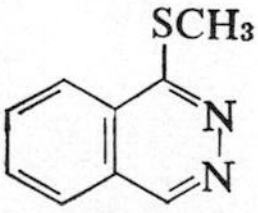

Fig. 2G-1. Ultraviolet spectra of **1**, **9**, and **15**.

1-Methylthiophthalazine (**9**) forms light yellow crystals, mp 75–77° (ligroin) (1); 74–75° (ether) (16). It forms a picrate, mp 202–203° (26). 1,4-Dimethylthiophthalazine (**10**) forms colorless scales, mp 163°(ethanol) (17), and this is the only 1,4-dialkylthiophthalazine reported in the literature. 1-Methylsulfinylphthalazine (**13**) is a colorless compound having mp 105° (cyclohexane) (22). 1-Methylsulfonylphthalazine (**16**) is a hygroscopic compound with mp 156° (benzene-pet. ether) (24). It reacts with atmospheric moisture and hydrolyzes to 1-(2*H*)phthalazinone, although it is stable in solution. The instability of **13** and **16** in strongly acidic solutions has prevented pK_a determinations (22, 24). Barlin and Brown (22, 24) have reported the uv spectra of **13** and **16** in water. The spectrum of 1-methylsulfinylphthalazine (**13**) shows λ_{max} 225 and 279 mμ with log ε values of 4.56 and 3.58, respectively. The spectrum of 1-methylsulfonylphthalazine (**16**) shows λ_{max} 228 and 281 mμ with log ε values of 4.58 and 3.49, respectively. The same authors (22, 24) have also reported the pmr spectra of 1-methylthiophthalazine (**9**), and of **13**, and **16** in deuteriochloroform. Singlet absorptions for the methyl protons are found at 7.12, 6.71, and 6.28 τ, respectively.

B. Chemical Properties

Alkylation of 1-(2*H*)phthalazinethione (1), 4-mercapto-1-(2*H*)phthalazinethione (17), and 4-phenyl-1-(2*H*)phthalazinethione (20) with methyl iodide has been reported to give exclusively *S*-alkylated products (see Eqs. 8 and 10). Similar results are obtained on alkylation of 4-phenyl-1-(2*H*)phthalazinethione (see Eq. 10). However, alkylation of 1-(2*H*)phthalazinethione with dimethyl sulfate (16), or 7-nitro-1-(2*H*)phthalazinethione with methyl iodide (9) and reaction of 1-(2*H*)phthalazinethione or 4-chloro- and 4-phenyl-1-(2*H*)phthalazinethione with α-acetobromoglucose (see Eq. 7) lead to two products, 1-alkylthiophthalazines and 2-alkyl-1-(2*H*)phthalazinethione, where the *S*-alkylated product is the major isomer. This can be attributed to the higher nucleophilic character of the sulfur atom as compared to the oxygen atom in the corresponding oxygen analogs.

Nucleophilic substitution on 1-(2*H*)phthalazinethione, 4-mercapto-1-(2*H*)phthalazinethiones, and corresponding thioethers with ammonia, alkylamines, hydrazine, and so on, gives the corresponding amino- and hydrazinophthalazines (see Sections 2H and 2I).

The desulfurization of 4-phenyl-1-(2*H*)phthalazinethione with Raney nickel affords 1-phenylphthalazine in 83% yield (see Section 2B-I-A-1).

The addition of phenylmagnesium bromide to 4-phenyl-1-(2*H*)phthalazinethione affords 1,4-diphenylphthalazine (see Section 2B-II-A-4-c).

The oxidation of 1-(2*H*)phthalazinethione with bromine in acetic acid gives 1,1′-diphthalazinyl disulfide (7). The oxidation of alkylthio- and arylthiophthalazines to the corresponding sulfones with 7% potassium permanganate in glacial acetic acid has been discussed earlier (see Section 2G-I-E).

Bednyagina and co-workers (20, 21) studied the stability of 4-phenyl-1-alkylthiophthalazines by boiling them in dilute alkaline and acidic solutions and observed that these compounds were stable under these conditions, whereas 4-phenyl-1-alkylsulfonylphthalazines are hydrolyzed by boiling 2*N* aqueous sodium hydroxide to give 4-phenyl-1-(2*H*)phthalazinone. These authors have also found that the stability of these sulfones increased with the length of the alkyl group (Eq. 14).

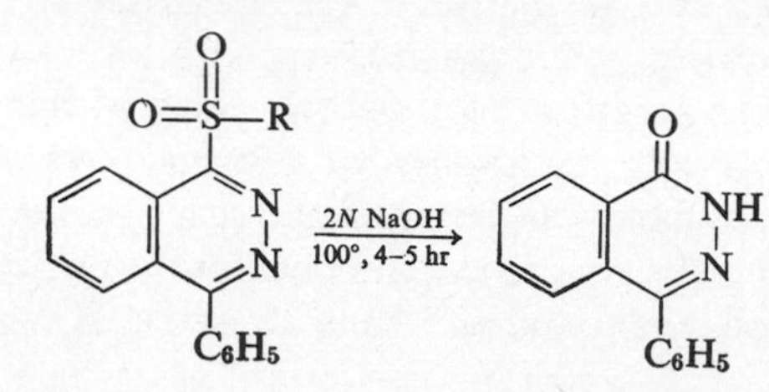

(14)

Stability of sulfones: R = *n*-butyl > ethyl > methyl > *p*-nitrophenyl.

Recently two groups of workers, Barlin and Brown in Australia and Hayashi and Oishi in Japan, have studied the nucleophilic displacement of the methylsulfonyl group in 1-methylsulfonylphthalazine. Barlin and Brown (25) reported the substitution of 1-methylsulfonylphthalazine with *n*-propylamine and with the hydroxide ion. They also studied (24) the kinetics of the reaction of 1-methylsulfonylphthalazine with methoxide ion and found the methylsulfonylphthalazine to be much more reactive than 1-chlorophthalazine because of a low energy of activation. They have compared the reactivity of the methylsulfonyl group in other benzodiazines and found that 2-methylsulfonylquinoxaline was more reactive than 4-methylsulfonylcinnoline, which in turn was more reactive than 1-methylsulfonylphthalazine. Barlin and Brown (26) have studied the displacement of the methylsulfinyl group with butylamine and hydroxide ion. In the reaction of 1-methylsulfinylphthalazine with 1*N* sodium hydroxide at 50° for 15 min, they isolated the expected product, 1-(2*H*)phthalazinone (41%), and an unexpected product, 1-methylthiophthalazine (**9**) (26%). The mechanism of reduction in this reaction—the formation of 1-methylthiophthalazine (**9**) from 1-methylsulfinylphthalazine—is not clear, but it is believed that the oxidation of methanesulfinate anion, produced by the nucleophilic displacement, causes simultaneous reduction of 1-methylsulfinylphthalazine. The same authors (22) have compared the reactivity of 1-methylsulfinyl-, 1-methylsulfonyl-, and 1-methylthiophthalazines with the methoxide ion at 30° C and found that 1-methylsulfinylphthalazine is ~6.5 times more reactive than 1-methylsulfonylphthalazine, which is more reactive than 1-methylthiophthalazine. Hayashi and co-workers (23) have substituted the methylsulfonyl group in

4-phenyl-1-methylsulfonylphthalazine with methoxide ion and aniline to give the corresponding 1-methoxy- and 1-anilino-4-phenylphthalazines.

Oishi (27) has studied the nucleophilic substitution of 1-methylsulfonylphthalazine and 1-chlorophthalazine with methoxide ion, hydroxylamine, hydrazine, and *n*-butylamine. Hayashi and Oishi (28) studied the displacement of the methylsulfonyl group in 1-methylsulfonyl-4-phenylphthalazine with various active methylene compounds and with carbanions derived from ketones to give the expected 1-substituted 4-phenylphthalazines (see Section 2B-I-A-4). However, Oishi (27) found that the reaction of 1-methylsulfonylphthalazine itself with ketone carbanions unexpectedly led to 1-substituted 1,2-dihydro-2-(1′-phthalazinyl)-4-sulfonylphthalazines (**17**) (Eq. 15). It was concluded that in reactions with ketone carbanions, where 4-phenyl-1-sulfonylphthalazine gives the expected 1-substituted products, the phenyl group at the 4-position affects the direction of reaction. The unexpected products, **17**, were characterized by substitution of the methylsulfonyl group with cyanide ion to give 1-substituted 1,2-dihydro-2-phthalazinyl-4-cyanophthalazines (**18**) (Eq. 15).

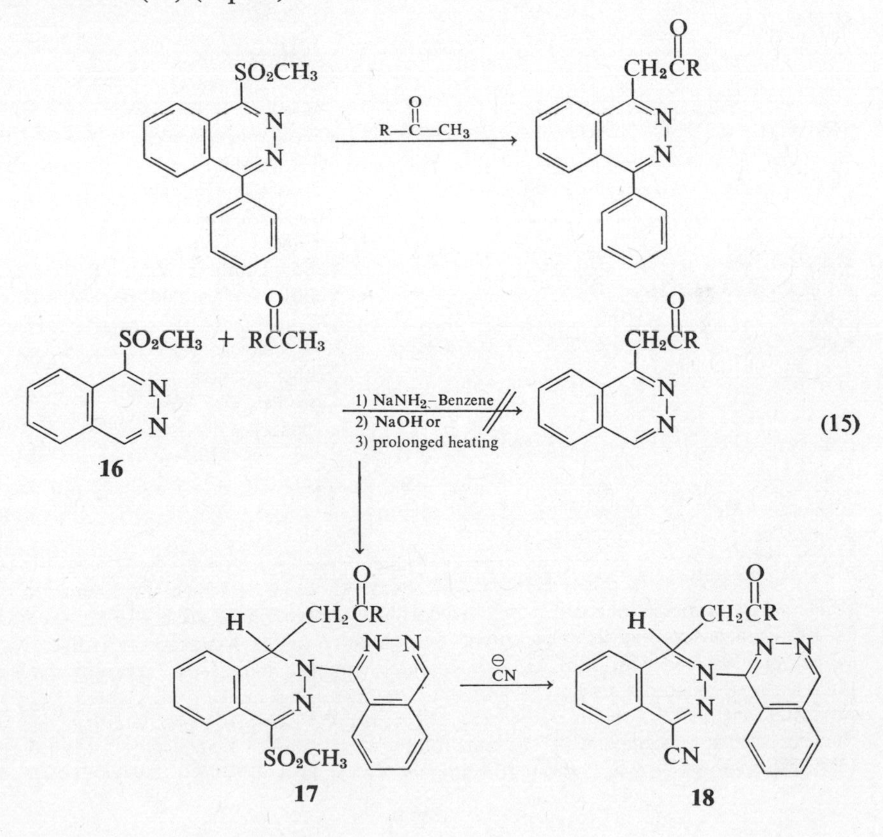

III. Tables

TABLE 2G-1. 1-(2*H*)Phthalazinethione with Substituents in the 4-Position

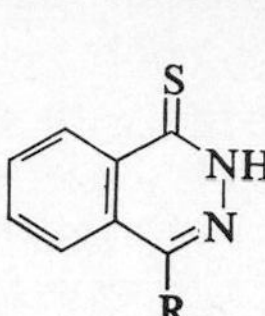

R	Preparation[a]	Yield (%)	MP	Reference
H	A	74–80	Yellow crystals, 169–170° (ethanol); yellow needles, 170–171° (ethanol); 171°; 174° (benzene-pet. ether)	1–4
SH	A	83	Yellow needles, 262–264° (toluene); 262–265°	10, 11
OH	—	—	—	29
Cl	B	50–60	Yellow crystals, 222–224° (acetic acid)	6
$NHNH_2$	C	32–85	Yellow microneedles, 253° (dec.) (propylene glycol); 255–260° (dec.) (DMF)	7, 8
$C_6H_5CH_2NH$—	D	—	Yellow columns, 205°	8
p-$CH_3OC_6H_4CH_2NH$—	D	—	Yellow needles, 205°	8
CH_3	E	81	238°	2
C_6H_5	E, F	78–83	202° (benzene-pet. ether); yellow needles, 206–208° (benzene)	2, 5
O_2N S NH N	G	80	171°	4

[a] A, see discussion in this section for different methods. B, 1,4-dichlorophthalazine + KSH, reflux in methanol for 2 hr. C, 1-chloro-4-hydrazinophthalazine + KSH, heat at 90–100° in a pressure vessel using ethanol solvent for 6 hr, or 4-mercapto-1-(2*H*)phthalazinethione warmed with $N_2H_4{\cdot}H_2O$ on a water bath for 4 hr. D, 4-mercapto-1-(2*H*)-phthalazinone + proper amine warmed on water bath for 4 hr. E, 4-substituted 1-(2*H*)-phthalazinone + P_2S_5, reflux in benzene for 6 hr. F, 1-chloro-4-phenylphthalazine + thiourea, reflux in methanol for 10 min followed by alkaline hydrolysis. G, 7-nitro-1-(2*H*)phthalazinone + P_2S_5, heat at 100° in pyridine for 1.5 hr.

TABLE 2G-2. Phthalazine-1-thioethers and Other 1-*S*-Substituted Derivatives

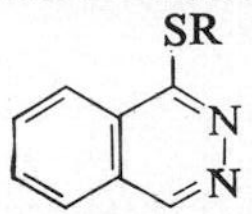

SR

N
N

R	Preparation[a]	Yield (%)	MP	Reference
CH_3	A, B	65	Light yellow crystals, 76–77° (ligroin); 74–75° (ether); picrate, 202–203°	1, 16, 26
$(CH_3)_2NCH_2CH_2$—	C	93	Colorless needles, 68–69°; HCl, 216–217°	16
Tetraacetyl-1-β-D-glucosyl-	D	37	Needles, 204–205°	6, 14
1-β-D-glucosyl-	D_1	80	Yellow amorphous powder, 110–120° (unsharp)	6, 14
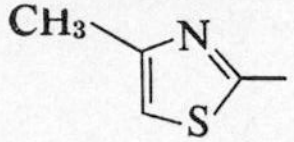 CH_3 N S	E	70	124–125° (ethanol)	18
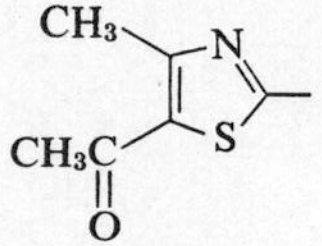 CH_3 N S CH_3C O	E	91	206–207° (DMF)	18
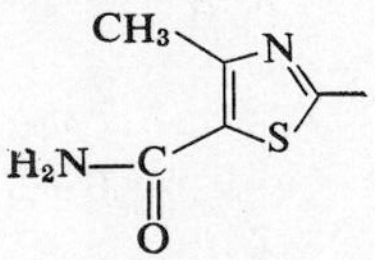 CH_3 N S H_2N—C O	E	73	228–229°	18
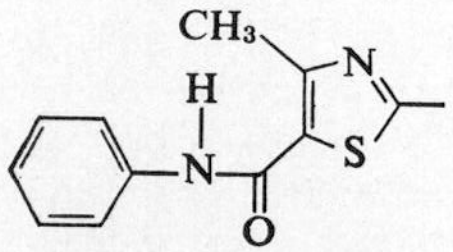 CH_3 N S H N O	E	70	224–226° (DMF)	18

TABLE 2G-2 (*continued*)

R	Preparation[a]	Yield (%)	MP	Reference
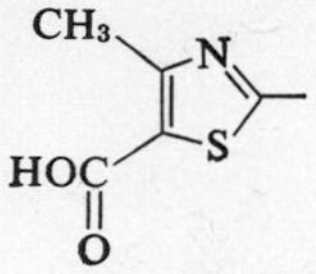 CH_3, N, S, HOC, O	F	76	174–175° (ethyl acetate)	18
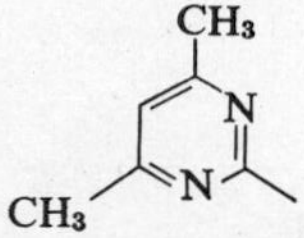 CH_3, N, N, CH_3	E	19	189–190° (ethanol)	18
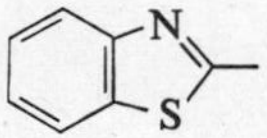 N, S	E	88	172–173° (ethanol)	18
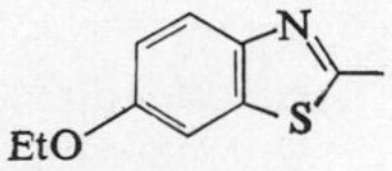 EtO, N, S	E	62	141–143° (ethanol-acetone)	18
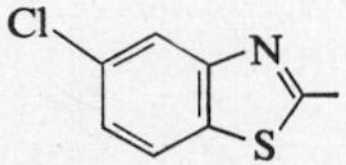 Cl, N, S	E	70	204–205° (benzene)	18
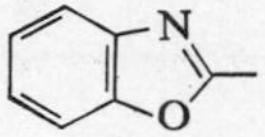 N, O	E	68	162–163° (ethanol)	18
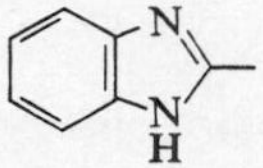 N, N, H	E	45	191–192° (ethanol)	18

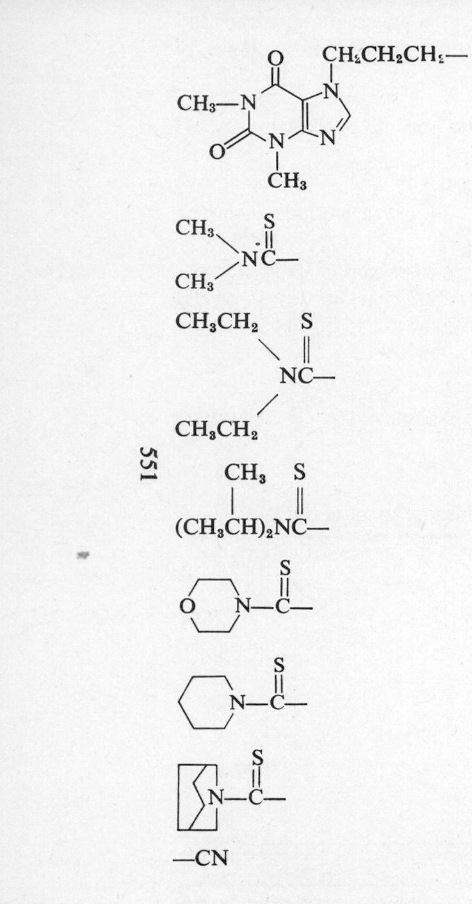

Structure	Method	Yield	mp	Ref.
$CH_2CH_2CH_2$— (theophylline derivative)	A_1	—	210°	30
$(CH_3)_2NC(=S)$—	G	88	150–152° (ethanol)	18
$(CH_3CH_2)_2NC(=S)$—	G	65	109–110° (ethanol)	18
$(CH_3CH)_2NC(=S)$—	G	39	191–192° (ethanol-acetone)	18
Morpholino-C(=S)—	G	83	183–184° (DMF)	18
Piperidino-C(=S)—	G	99	153–154° (ethanol)	18
N—C(=S)— (bicyclic)	G	97	150–151°	19
—CN	H	—	BP 180–205°/0.008 mm; mp 130°	31

TABLE 2G-2 (*continued*)

R	Preparation[a]	Yield (%)	MP	Reference
$-P(=S)(OEt)_2$	I	78	Orange crystals, 85° (ether)	32
1-phthalazinyl-S—	J	—	Yellow crystals, 220–221° (carbitol-ethanol)	7
7-NO_2-1-SCH_3-phthalazine	K	—	Yellow needles 180–181° (ethanol)	9

[a] A, 1-(2*H*)phthalazinethione + CH_3I, stir in 1*N* aqueous sodium hydroxide for 1 hr, or stir with dimethyl sulfate in aqueous sodium hydroxide. A_1, 1-(2*H*)phthalazinethione treated with [7-($CH_2CH_2CH_2Br$)-1-CH_3-3-CH-xanthine structure: CH_3—N, O, $CH_2CH_2CH_2Br$, N, CH] in ethanolic sodium hydroxide. B, 1-chlorophthalazine + CH_3SNa. C, 1-chlorophthalazine + $(CH_3)_2N—CH_2CH_2—Cl\cdot HCl$ + NaOEt, reflux in ethanol for 90 min. D, 1-(2*H*)phthalazinethione + tetracetyl-α-D-glucosyl bromide shaken with KOH in aqueous acetone for 1 hr. D_1, the compound obtained by method D hydrolyzed with $NaOCH_3$ in methanol. E, 1-chlorophthalazine + corresponding mercapto compound heated with KOH in ethanol at 78–80° for 24 hr. F, 1-chlorophthalazine + [4-CH_3-5-EtOC(=O)-thiazol-2-yl]—SH stirred with KOH in ethanol at 25–30° for 24 hr, then hydrolysis of the ester group with concentrated hydrochloric acid (pH 4). G, 1-chlorophthalazine + corresponding *N*,*N*-dialkyldithiocarbamic acid, heat with KOH in ethanol at 75–80° for 4 hr. H, 1-chlorophthalazine + NaSCN. I, 1-chlorophthalazine + $HS—P(=S)—(OEt)_2$, stir in aqueous acid at 26° C. J, 1-(2*H*)-phthalazinethione boiled with bromine in acetic acid. K, 7-nitro-1-(2*H*)phthalazinethione + CH_3I in ethanolic sodium ethoxide.

TABLE 2G-3. Phthalazine-1-thioethers with Substituents in the 4-Position

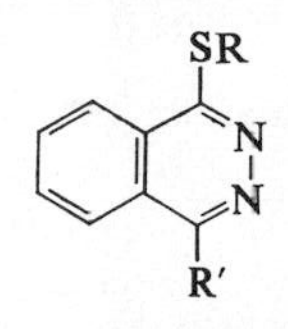

R	R′	Preparation[a]	Yield (%)	MP	Reference
CH_3	C_6H_5	A	60	143–144° (ethanol)	20
CH_3	[pyridin-4-yl]—CH_2—	B	—	—	33
CH_3	CH_3S	C	—	Colorless scales, 163° (ethanol)	17
CH_3CH_2	C_6H_5	A_1	—	95–96°	20
CH_3CH_2	Cl	—	—	—	13
CH_3CH_2	OH	D	—	166–167°	13
n-Butyl	C_6H_5	A_1	—	72–73°	20
p-$NO_2C_6H_4$	C_6H_5	E	97	Coarse yellow needles, 158–160° (alcohol-benzene)	21
Tetracetyl-1-*β*-D-glucosyl	Cl	F	41	Colorless, 189–190° (ethanol)	6, 14
Tetracetyl-1-*β*-D-glucosyl	C_6H_5	F	14	Colorless needles, 152–153° (ethanol)	6, 14
1-*β*-D-Glucosyl	Cl	F_1	44	Needles, 187–190° (methanol)	6, 14
1-*β*-D-Glucosyl	C_6H_5	F_1	—	Colorless syrup	6, 14

[a] A, 4-phenyl-1-(2*H*)phthalazinone + CH_3I, stir in alcoholic NaOH. A_1, same as A, stir with corresponding bromide in presence of KI. B, 1-chloro-4-(4′-pyridylmethyl)phthalazine + $NaSCH_3$, heat in a sealed tube with methanol at 135–140°. C, 4-mercapto-1-(2*H*)phthalazinthione + CH_3I in aqueous sodium hydroxide, or 1,4-dichlorophthalazine + $NaSCH_3$ in benzene. D, 1-chloro-4-ethylthiophthalazine hydrolyzed with 5*N* HCl. E, 1-chloro-4-phenylphthalazine + *p*-$NO_2C_6H_4SNa$, boil in alcohol for 30 min. F, 4-substituted 1-(2*H*)phthalazinethione + tetracetyl-*α*-D-glucosyl bromide shaken with KOH in aqueous acetone for 3–4 hr. F_1, products obtained by method F hydrolyzed with $NaOCH_3$ in methanol.

TABLE 2G-4. 1-(2*H*)Phthalazinethiones with Substituents at the 2- and 4-Positions

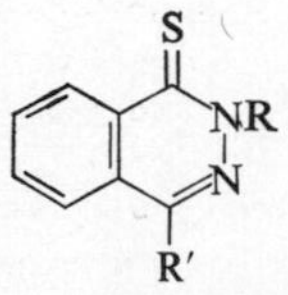

R	R′	Preparation[a]	Yield (%)	MP	Reference
CH_3	H	A, A_1	82–91	Yellow crystals, 128–129°; yellow needles, 126–127° (ethanol)	1, 8
$(Et)_2N—CH_2CH_2$	H	A_1	63	BP 185–188°/2 mm; HCl, 209–210°	8
C_6H_5	H	A_2	81	138° (benzene-pet. ether)	2
C_6H_5	CH_3	A_2	73	150° (benzene-pet. ether)	2
C_6H_5	C_6H_5	A_2	67	177° (benzene-pet. ether)	2
C_6H_5	OH	B	72	Yellow needles, 227–230° (ethanol)	12
C_6H_5	CH_3CH_2O	A_1	89	Orange yellow rods, 153° (ethanol)	12
C_6H_5	$CH_3CH_2CH_2O$	A_1	—	Yellow needles, 128° (ethanol)	12
C_6H_5	$CH_3(CH_2)_2CH_2O$	A_1	—	Yellow plates, 96° (ethanol)	12
Tetracetyl-1-β-D-glucosyl	H	C	19	Yellow needles, 165–167° (ethanol)	6, 14
Tetracetyl-1-β-D-glucosyl	Cl	C, C_1	10–20	Yellow needles, 165–167°; 164–166°	6, 14
Tetracetyl-1-β-D-glucosyl	C_6H_5	C, C_1, C_2	31–80	Yellow needles, 200–202° (ethanol); 199–200°; 202°	6, 14

TABLE 2G-4 (*continued*)

1-β-D-Glucosyl	H	D	68	Yellow needles, 110–120° (unsharp, ethanol)	6, 14
1-β-D-Glucosyl	Cl	D	55	Yellow needles, 130–140° (unsharp, ethanol)	6, 14
1-β-D-Glucosyl	C_6H_5	D	82	Yellow needles, 135–145° (ethanol)	6, 14
7-NO_2-1-thioxo-2-NCH_3 phthalazine (structure)	—	E	63–70	Orange prisms 161–162° (dec.) (benzene-pet. ether); yellow prisms, 160–162° (dec.)	9

[a] A, 2-methyl-1-(2*H*)phthalazinone + P_2S_5, reflux in pyridine for 1 hr. A_1, 2-methyl-1-(2*H*)phthalazinone or the corresponding 4-alkoxy-2-phenyl-1-(2*H*)phthalazinone + P_2S_5, reflux in xylene for 1 hr. A_2, 4-substituted 2-phenyl-1-(2*H*)phthalazinone + P_2S_5, reflux in benzene for 6 hr. B, 2-phenyl-4-ethoxy-1-(2*H*)phthalazinethione was refluxed in acetic acid with HBr for 1 hr. C, the corresponding 1-(2*H*)phthalazinethione + tetracetyl-α-D-glucosyl bromide was shaken in aqueous acetone with KOH 1–4 hr (obtained as by products). C_1, the corresponding 1-*S*-alkylated phthalazine was refluxed in toluene with $HgBr_2$ for 20 min (*trans*-glucosylation). C_2, the corresponding 2-substituted 4-phenyl-1-(2*H*)phthalazinone + P_2S_5 was refluxed in pyridine for 1–5 hr. D, the products obtained by method C were hydrolyzed by treatment with $NaOCH_3$ in methanol. E, 7-nitro-1-(2*H*)phthalazinethione + $(CH_3O)_2SO_2$, stir 1 hr in aqueous sodium hydroxide, or 7-nitro-1-(2*H*)phthalazinone + P_2S_5 in pyridine was stirred with heating on a steam bath for 1–5 hr.

TABLE 2G-5. 4-Alkyl- or Arylthio-2-phenyl-1-(2*H*)phthalazinethiones

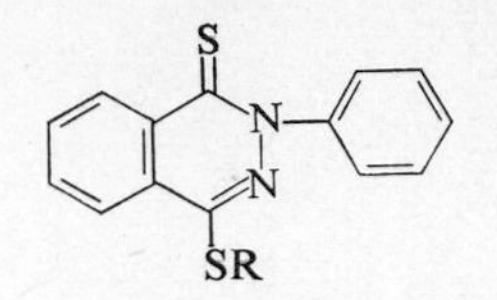

R	Preparation[a]	Yield (%)	MP	Reference
H	A	17	Yellow needles, 167–168° (ethanol)	12
CH_3CH_2	B	—	Yellow needles, 151° (ethanol)	12
$C_6H_5CH_2$	B	89	Yellow needles, 135° (ethanol)	12
2,4-$(NO_2)_2C_6H_3$	B	—	Orange prisms, 206° (acetic acid)	12
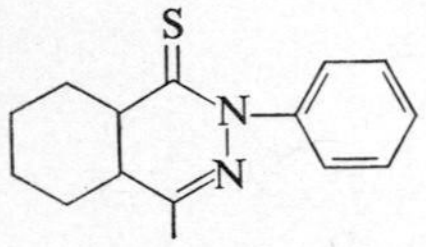	A_1	22	Orange needles, 256.5° (dec.) (acetic acid)	12

[a] A, 4-hydroxy-2-phenyl-1-(2*H*)phthalazinone + P_2S_5, reflux in xylene for 6 hr. A_1, obtained as by product from method A. B, 4-mercapto-2-phenyl-1-(2*H*)phthalazinethione + corresponding alkyl or aryl chloride, reflux in ethanol with NaOEt for 1 hr.

TABLE 2G-6. 2-Substituted 4-Alkylthio-1-(2*H*)phthalazinones

R	R′	Preparation[a]	Yield (%)	MP	Reference
H	CH_3CH_2	A	—	166–167°	13
H	C_6H_5	B	48	Yellow needles, 135° (alcohol)	12
$(Et)_2NCH_2CH_2$	CH_3CH_2	C	—	BP 175–185° (0.8 mm); HCl, 169–170°	13
C_6H_5	$C_6H_5CH_2$	D	82	Needles, 117–118° (ethanol)	12
C_6H_5	2-phenyl-1-(2*H*)phthalazinon-4-yl (structure)	E	90	Yellow needles, 198–200° (alcohol)	12

[a] A, 1-chloro-4-ethylthiophthalazine treated with 5*N* HCl. B, 4-chloro-2-phenyl-1-(2*H*)phthalazinone + KSH, reflux 2 hr in ethylene glycol. C, 4-ethylthio-1-(2*H*)phthalazinone + *N*,*N*-diethylaminoethyl chloride stirred in benzene at 70° for 3 hr with sodium ethoxide. D, 4-mercapto-2-phenyl-1-(2*H*)phthalazinethione + benzyl chloride, reflux in ethanol for 1 hr with NaOEt, or 4-benzylthio-2-phenyl-1-(2*H*)phthalazinethione treated with 30% H_2O_2 at 60° in alcoholic KOH. E, bis-(1,2-dihydro-2-phenyl-1-thio-4-phthalazinyl) sulfide + 30% H_2O_2, heat at 60° for 1 hr in ethanolic KOH.

TABLE 2G-7. 4-Substituted 1-Alkylsulfonylphthalazines

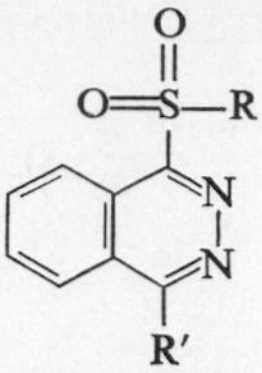

R	R′	Preparation[a]	Yield (%)	MP	Reference
CH_3	H	A	32	156° (benzene–pet. ether)	24, 27
CH_3	C_6H_5	B	85–98	205–207° (alcohol); 210–212° (methanol)	20, 23
CH_3CH_2	C_6H_5	B	—	178–180°	20
$CH_3(CH_2)_2CH_2$	C_6H_5	B	—	125–128°	20
p-$O_2N—C_6H_4$	C_6H_5	B	50	Colorless crystals, 147–150° (ethanol)	21
Miscellaneous					
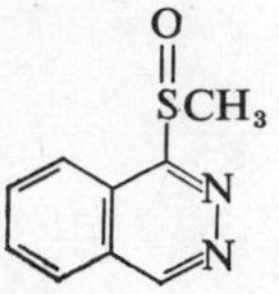 1-(methylsulfinyl)phthalazine ($S(O)CH_3$)		C	41	Colorless product 105° (cyclohexane)	22
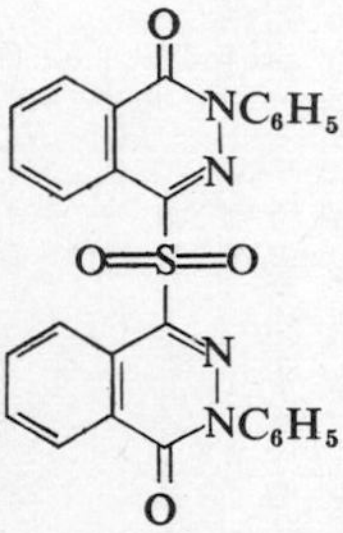 bis(1,2-dihydro-1-oxo-2-phenyl-4-phthalazinyl) sulfone (NC_6H_5, O=S=O)		D	93	Feathery needles 265° (acetic acid)	12

[a] A, 1-methylthiophthalazine + 4% $KMnO_4$, stir in acetic acid at room temperature. B, the corresponding 1-alkylthio-4-phenylphthalazine + 7% $KMnO_4$, stir in acetic acid as in A. C, 1-methylthiophthalazine + 3-chloroperbenzoic acid, stir in chloroform at 0° C, then at room temperature overnight. D, di-(1,2-dihydro-2-phenyl-1-oxo-4-phthalazinyl) sulfide + 30% H_2O_2, heat at 100° C for 1 hr in acetic acid.

References

1. A. Albert and G. B. Barlin, *J. Chem. Soc.*, **1962,** 3129.
2. A. Mustafa, A. H. Harhash, and A. A. S. Saleh, *J. Am. Chem. Soc.*, **82,** 2735 (1960).
3. Japanese Pat. 5088 (1954); *Chem. Abstr.*, **50,** 6522 (1956).
4. V. Brasiunas and K. Grabliauskas, *Khim. Farm., Zh.*, **1,** 40 (1967); *Chem. Abstr.*, **67,** 100089 (1967).
5. E. Hayashi and E. Oishi, *Yakugaku Zasshi*, **86,** 576 (1966); *Chem. Abstr.*, **65,** 15373 (1966).
6. G. Wagner and D. Heller, *Arch. Pharm.*, **299,** 768 (1966).
7. British Pat. 735,899 (1955); *Chem. Abstr.*, **50,** 10801 (1956).
8. K. Fujii and S. Sato, *Ann. Rept., G. Tanabe Co., Ltd.*, **1,** 1 (1956); *Chem. Abstr.*, **51,** 6650 (1957).
9. S. S. Berg and E. W. Parnell, *J. Chem. Soc.*, **1961,** 5275.
10. D. Radulescu and V. Georgescu, *Bull. Soc. Chim.*, **37,** 381 (1925); *Chem. Abstr.*, **20,** 184 (1926).
11. H. L. Yale, *J. Am. Chem. Soc.*, **75,** 675 (1953).
12. D. J. Drain and D. E. Seymour, *J. Chem. Soc.*, **1955,** 852.
13. British Pat. 808,636 (1959); *Chem. Abstr.*, **53,** 14129 (1959).
14. G. Wagner and D. Heller, *Z. Chem.*, **4,** 28 (1964); *Chem. Abstr.*, **60,** 14597 (1964).
15. G. Wagner and D. Heller, *Z. Chem.* **4,** 71 (1964); *Chem. Abstr.*, **60,** 14579 (1964).
16. K. Fujii and S. Sato, *Ann. Rept., G. Tanabe Co., Ltd.*, **1,** 3 (1956); *Chem. Abstr.*, **51,** 6650 (1957).
17. K. Asano and S. Asai, *Yakugaku Zasshi*, **78,** 450 (1958); *Chem. Abstr.*, **52,** 18428 (1958).
18. J. J. D'Amico, U.S. Pat. 3,379,700 (1968); *Chem. Abstr.*, **69,** 11257 (1968).
19. J. J. D'Amico, French Pat. 1,463,732 (1966); *Chem. Abstr.*, **67,** 54045 (1967).
20. N. P. Bednyagina and S. V. Sokolov, *Nauk. Dokl. Vyssh. Shkoly, Khim. Khim. Technol.*, **1959,** 338; *Chem. Abstr.*, **53,** 21971 (1959).
21. N. B. Bednyagina and I. Y. Postovski, *J. Gen. Chem., USSR*, **26,** 2549 (1956); *Chem. Abstr.*, **52,** 5418 (1958); *Zh. Obsh. Khim.*, **26,** 2279 (1956); *Chem. Abstr.*, **51,** 5086 (1957).
22. G. B. Barlin and W. V. Brown, *J. Chem. Soc. (B)*, **1968,** 1435.
23. E. Hayashi, T. Higashino, E. Oishi, and M. Sano, *Yakugaku Zasshi*, **87,** 687 (1967); *Chem. Abstr.*, **67,** 90750 (1967).
24. G. B. Barlin and W. V. Brown, *J. Chem. Soc. (B)*, **1967,** 736.
25. G. B. Barlin and W. V. Brown, *J. Chem. Soc. (C)*, **1967,** 2473.
26. G. B. Barlin and W. V. Brown, *J. Chem. Soc. (C)*, **1969,** 921.
27. E. Oishi, *Yakugaku Zasshi*, **89,** 959 (1969); *Chem. Abstr.*, **72,** 3450 (1970).
28. E. Hayashi and E. Oishi, *Yakugaku Zasshi*, **88,** 83 (1968); *Chem. Abstr.*, **69,** 19104 (1968); *Yakugaku Zasshi*, **87,** 807 (1967); *Chem. Abstr.*, **68,** 2871 (1968).
29. J. D. Kendall and D. J. Fry, U.S. Pat. 2,573,027 (1951); *Chem. Abstr.*, **46,** 844 (1952).
30. M. Eckstein and J. Sulko, *Diss. Pharm.*, **17,** 7 (1965); *Chem. Abstr.*, **63,** 8363 (1965).
31. S. Biniecki and B. Gutkowski, *Acta Polon. Pharm.*, **11,** 27 (1955); *Chem. Abstr.*, **50,** 12062 (1956).
32. Belgian Pat. 635,443 (1964); *Chem. Abstr.*, **61,** 16096 (1964).
33. J. Druey and A. Marxer, German Pat. 1,061,788 (1959); *Chem. Abstr.*, **55,** 13458 (1961).

Part H. Aminophthalazines

I. 1-Amino- and 1,4-Diaminophthalazines

Phthalazines with amino or substituted amino groups in the 1- and 1,4-positions are discussed in this section. 1-Aminophthalazines having chloro, hydroxy, and alkoxy groups in the 4-position and 1-hydroxyaminophthalazines are also included in this section. Hydrazinophthalazines and 4-amino-1-(2*H*)phthalazinethiones are discussed in Sections 2I and 2G, respectively.

A. Preparation of 1-Aminophthalazines

1-Chlorophthalazine reacts with methanolic ammonia to give 1-aminophthalazine (**1**) in 64% yield (1). 1-Phenoxyphthalazine, on fusion with ammonium acetate at 150–160° for 30 min, provides **1** in 72–92% yields (2, 3) (Eq. 1).

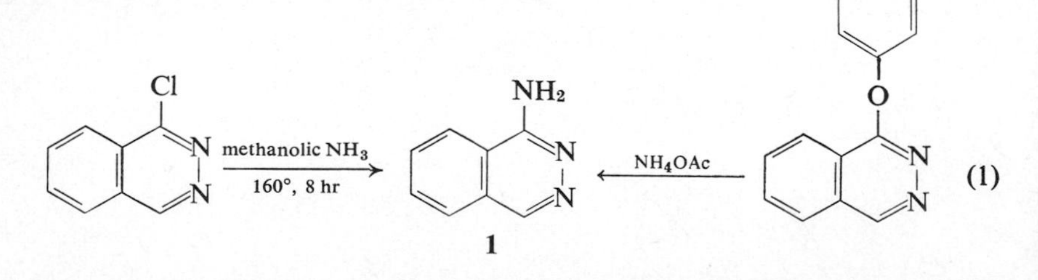

Similar nucleophilic substitution in 7-nitro-1-(2*H*)phthalazinethione and 1-methylthio-7-nitrophthalazine with ammonium acetate gives 1-amino-7-nitrophthalazine (**2**) (4). 1-Phenoxy-7-nitrophthalazine, on fusion with ammonium acetate, also gives **2** in 83% yield (2) (Eq. 2).

1,4-Dichlorophthalazine reacts with ammonium hydroxide in a sealed tube at 120° to give 1-amino-4-chlorophthalazine (**3**) in 79% yield (5). The

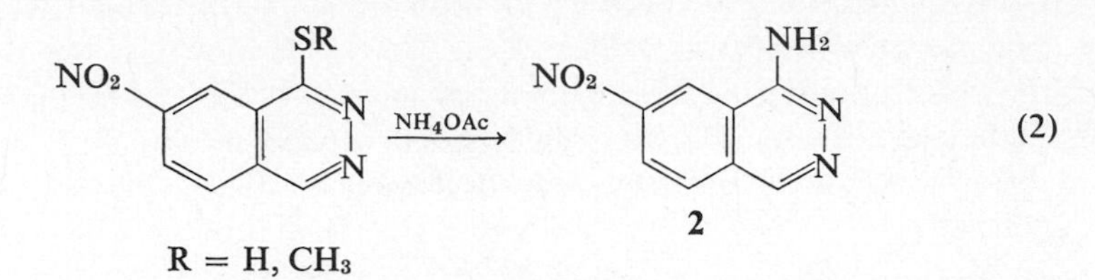

reaction of 1,4-dichlorophthalazine with stronger amines usually stops after the replacement of one of the chlorine atoms. Heating **3** with sodium methoxide furnishes 1-amino-4-methoxyphthalazine in excellent yields (92%) (5) (Eq. 3). 1-Chloro-4-methylphthalazine, on treatment with ammonia, gives 1-amino-4-methylphthalazine (6, 7), which is also obtained by the fusion of 1-phenoxy-4-methylphthalazine with ammonium acetate (3).

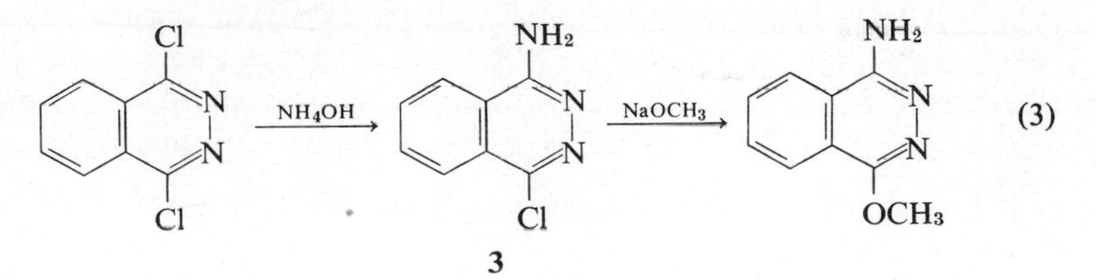

Recently Eberle and Houlihan (7a) prepared 1-amino-4-arylphthalazines in high yields from the reaction of 1-aryl-3-ethoxyisoindolenine with excess of hydrazine (Eq. 3a).

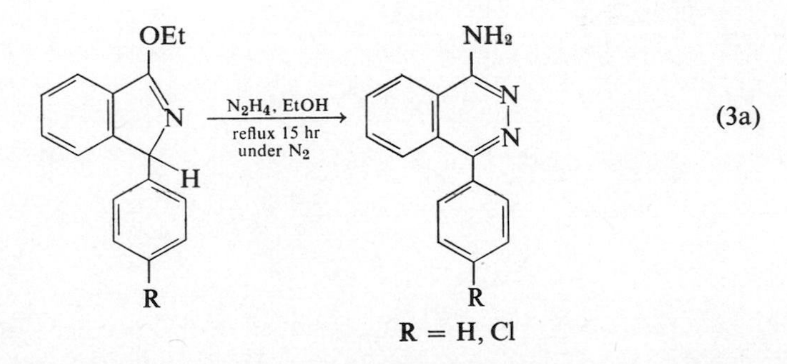

B. Preparation of 1-Alkyl- or Arylaminophthalazines

1-Chlorophthalazine reacts with primary or secondary alkylamines at about 100° C in the absence of solvent or in ethanol or benzene. Thus the nucleophilic substitution in 1-chlorophthalazine with *N*,*N*-dimethylaminoethylamine (8), 2-morpholinoethylamine (9), 2-phenylethylamine (10), the potassium salt of 4-aminobenzenesulfonamide (3), aziridine (11), piperidine (9), morpholine (9), *N*-ethyl-*N*,*N*-diethylethylamine (12), and so on, gives the corresponding 1-alkylaminophthalazines (**4**) (Eq. 4). Similarly, 1-(2*H*)-phthalazinethione (13) has been reacted with benzylamine, 2-(4′-methoxylphenyl)ethylamine and 2-hydroxyethylamine to furnish the corresponding **4** (Eq. 4).

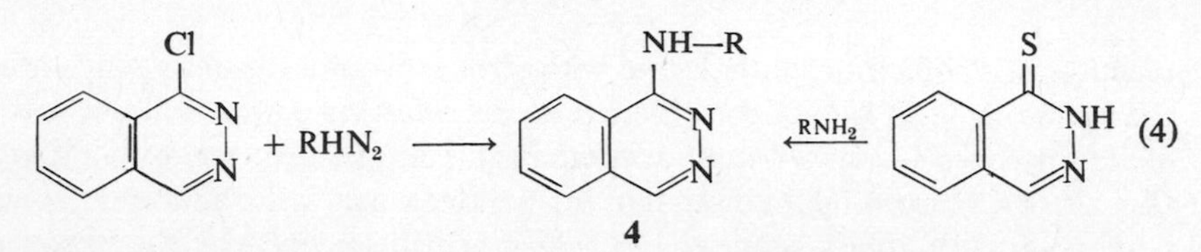

Recently Barlin and Brown (14, 15) studied the nucleophilic substitution in 1-methylsulfinylphthalazine and 1-methylsulfonylphthalazine with *n*-butylamine and *n*-propylamine to give the corresponding 1-alkylaminophthalazines (**4**). Similarly, Oishi (16) reported the reaction of 1-methylsulfonylphthalazine and 1-chlorophthalazine with *n*-butylamine and aniline (Eq. 5).

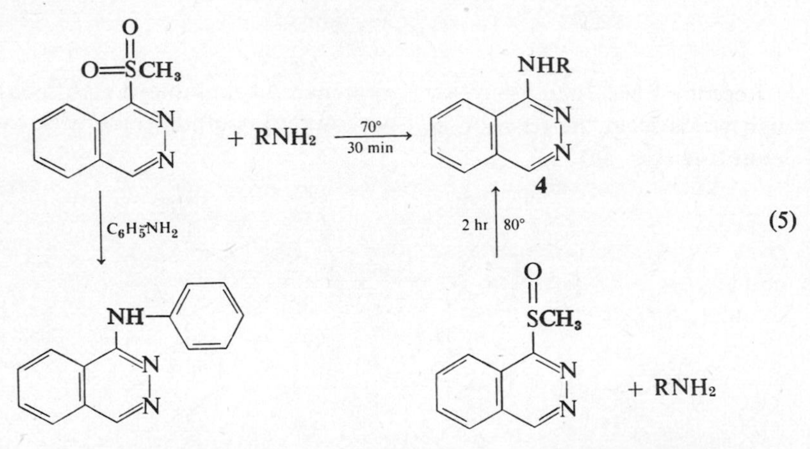

1-Chloro-4-methylphthalazine reacts with 2-hydroxyethylamine (17), 2,3-dimethylaniline, and 3-trifluoromethylaniline (18) to give the corresponding 1-alkylamino- and 1-arylamino-4-methylphthalazines in 60–82% yields. Similarly, 1-chloro-4-*n*-butylphthalazine reacts with aniline and *p*-toluidine to give the corresponding 1-arylaminophthalazines (19). Holava and Partyka (20) have shown that 1-chloro-4-phenylphthalazine will react with a number of primary and secondary alkylamines under a variety of conditions to give 1-alkylamino-4-phenylphthalazines (Eq. 6). Hayashi and co-workers (21) have studied the nucleophilic substitution in 1-methylsulfonyl-4-phenylphthalazine with aniline by heating at 100° for 2 hr to give 1-anilino-4-phenylphthalazine in 48% yield (Eq. 6).

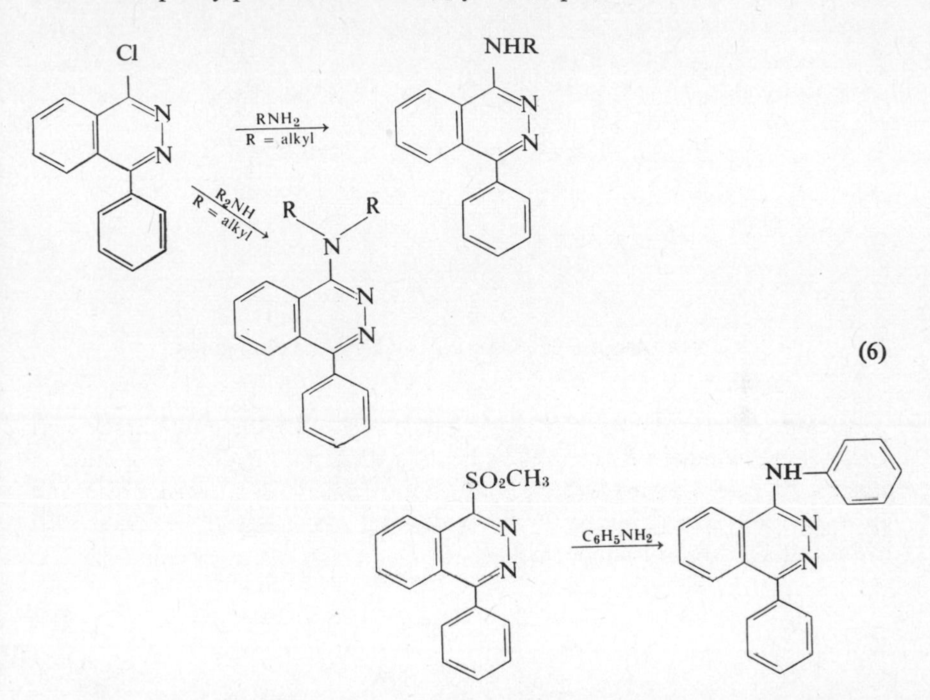

(6)

1,4-Dichlorophthalazine, with excess of methylamine (22), the sodium salt of glycine (22), or the potassium salt of *p*-aminobenzenesulfonamide (5, 23), gives 1-alkylamino-4-chlorophthalazines (**5**). Reaction of 1,4-dichlorophthalazine with piperidine (7) or 2-phenylethylamine (10) gives the corresponding **5** (Eq. 7).

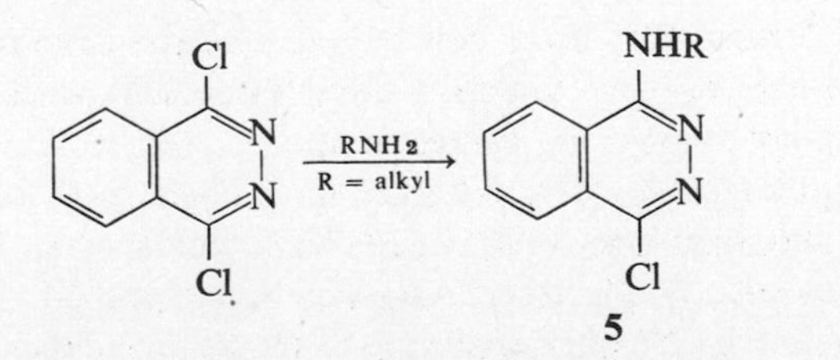

(7)

Haworth and Robinson (24) have shown that in the reaction of arylamines such as aniline and 4-chloroaniline with 1,4-dichlorophthalazine it is important to use the reactants in equimolar quantities in order to obtain 1-chloro-4-arylaminophthalazines (Eq. 8); use of excess of arylamines leads to 1,4-diarylaminophthalazines (see also Section 2H-I-E).

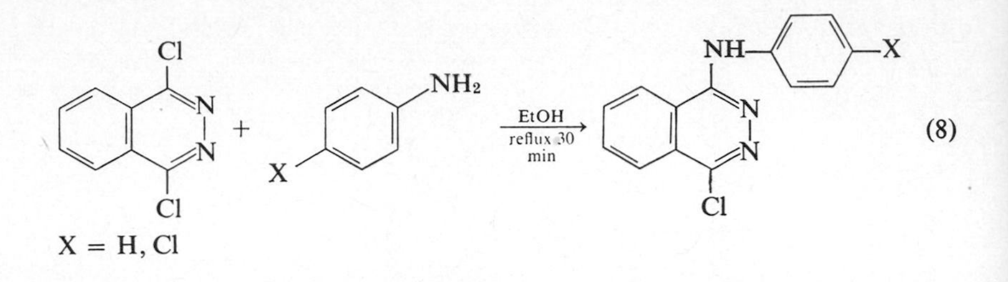

C. Preparation of 4-Amino-1-(2*H*)phthalazinones

Koehler and co-workers (25) have condensed monothiophthalimide (**6a**) and 3-iminoisoindolin-3-one (**6b**) by heating them with hydrazine hydrate in ethanol to give 4-amino-1-(2*H*)phthalazinone (**7**) in 88% yield. Filho and daCosta (26) have obtained **7** by condensing ethyl 2-cyanobenzoate with hydrazine hydrate in ethanol (Eq. 9).

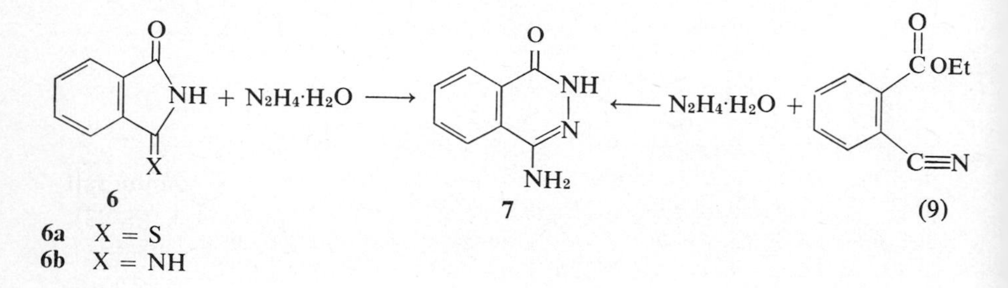

Flitsch and Peters (22) have condensed *N*-[3-oxo-isoindolinyliden-(1)]-glycine ethyl ester (**8a**) and α-alanin ethyl ester (**8b**) with 52% hydrazine hydrate in ethanol to give the corresponding 3,6-dioxo-2,3,5,6-tetrahydro-imidazo[2,1-*a*]phthalazines (**9**). Alkaline hydrolysis of **9** furnished 4-alkyl-amino-1-(2*H*)phthalazinones (**10**) in 76–78% yields (Eq. 10). These same authors (22) have also synthesized **7** and 4-methylamino-1-(2*H*)phthalazinone (**11a**) and 4-anilino-1-(2*H*)phthalazinone (**11b**) by the acidic hydrolysis of the corresponding 1-chloro-4-aminophthalazines (Eq. 10).

In the Ing-Manske reaction, used in the synthesis of amines and amino acids, and *N*-substituted phthalimide is condensed with hydrazine and the intermediate is subjected to acid hydrolysis to give the desired amine hydrochloride and 4-hydroxy-1-(2*H*)phthalazinone. The intermediate was assumed to be a compound similar to **10** or **11** (27, 28) (Eq. 11). However, Flitsch and Peters (22) attempted to hydrolyze **10** and **11** under the conditions of the Ing-Manske reaction, heating them with both dilute and

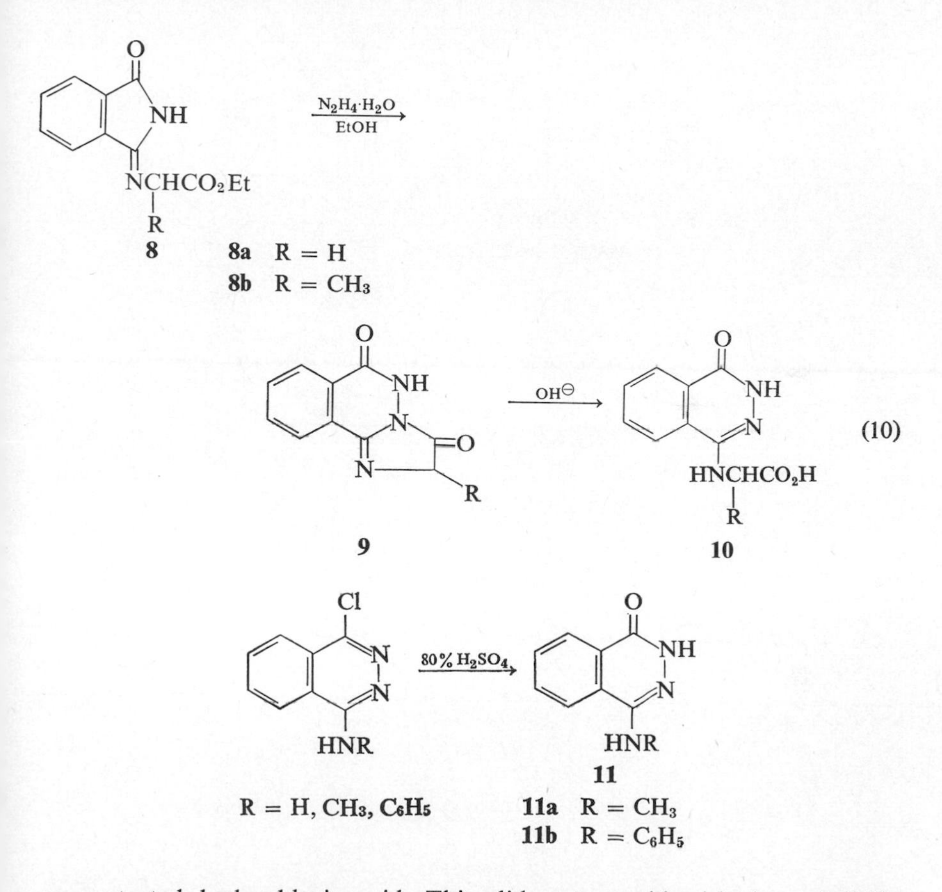

concentrated hydrochloric acid. This did not provide 4-hydroxy-1-(2*H*)-phthalazinone. Flitsch and Peters concluded therefore that compounds similar to **10** and **11** cannot be intermediates in this reaction. On the other hand, Bagal and associates (29) have claimed the isolation of 4-(2-bromo-2-nitropropylamino)-1-(2*H*)phthalazinone, the suggested intermediate in the Ing-Manske reaction, and subsequently hydrolyzed it to 4-hydroxy-1-(2*H*)-phthalazinone and 2-bromo-2-nitropropylamine hydrochloride. Klamerth (30) isolated an unexpected product, 4-(2-hydroxyethylamino)-1-(2*H*)-phthalazinone (**12**), from the reaction of *N*-2-pyrrolidinoethylphthalimide

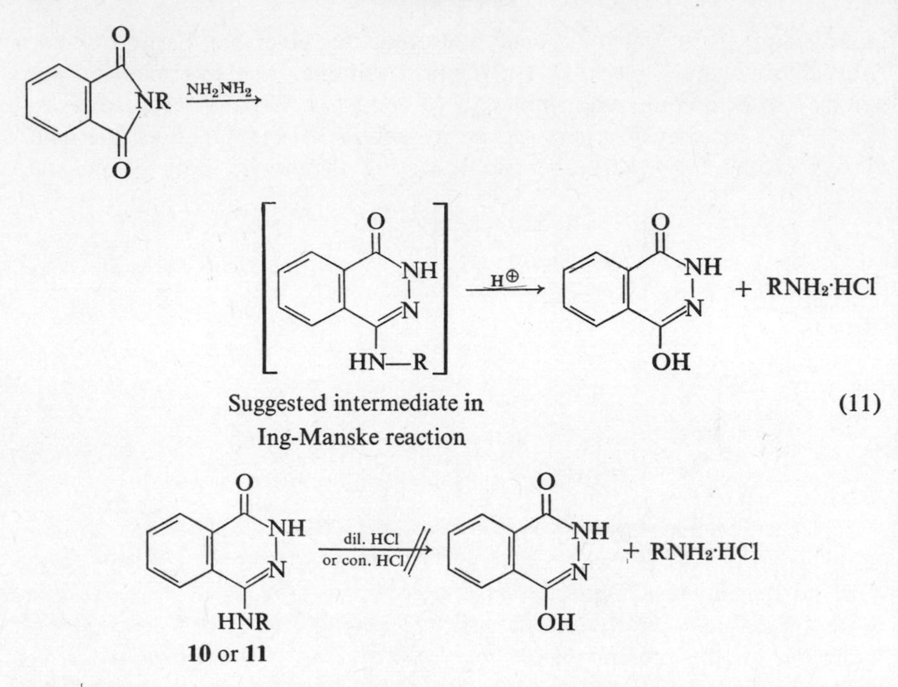

with hydrazine. The phthalazinone **12** formed a hydrochloride, but Klamerth was unable to hydrolyze it to 4-hydroxy-1-(2*H*)phthalazinone and 2-hydroxyethylamine hydrochloride, the expected products of the Ing-Manske reaction (Eq. 12).

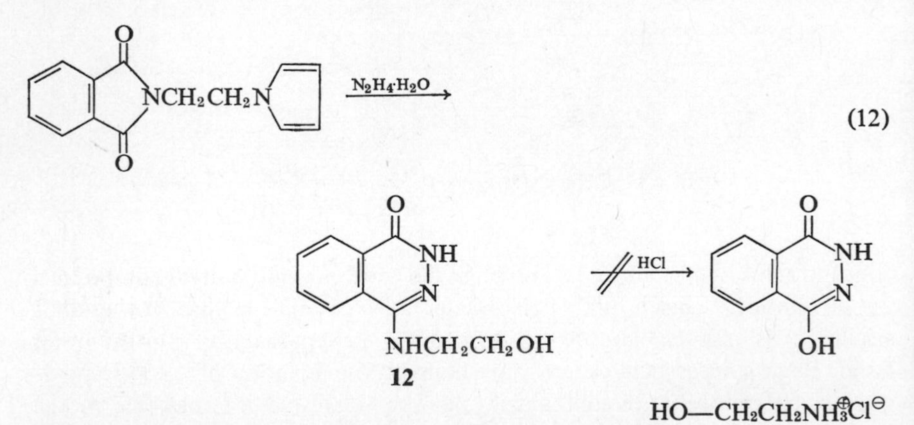

The intermediate in the Ing-Manske reaction may be similar to **13**, which Connors and Ross (31) isolated, characterized, and hydrolyzed in acidic media to give 4-hydroxy-1-(2*H*)phthalazinone and the desired amine hydrochloride (Eq. 13).

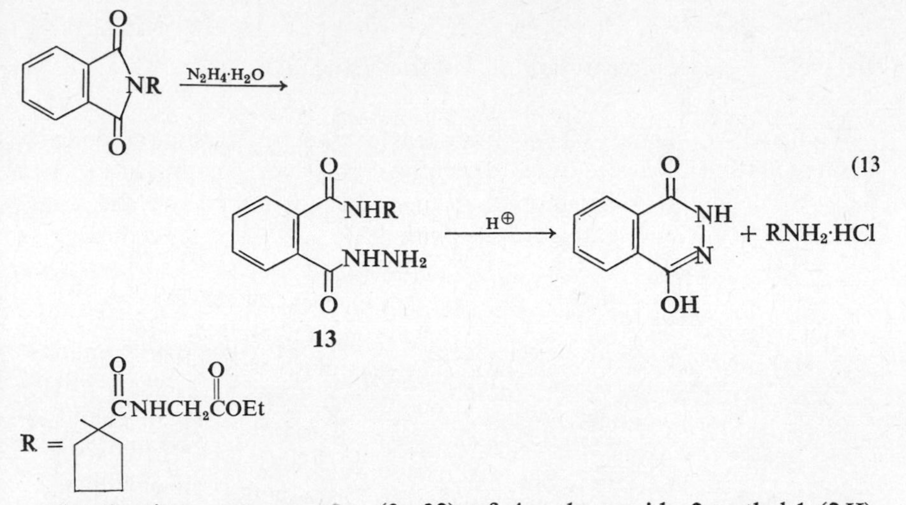

The Curtius rearrangement (3, 32) of 4-carboxyazido-2-methyl-1-(2*H*)-phthalazinone (**14**) in boiling ethanol followed by acid hydrolysis gives 4-amino-2-methyl-1-(2*H*)phthalazinone (**15**) in 66% yield (Eq. 14). 2-Methyl-4-chloro-1-(2*H*)phthalazinone (3) on heating with ammonium hydroxide in the presence of copper bronze also gives **15**. Similarly, 4-chloro-2-phenyl-1-(2*H*)phthalazinone (33) has been reacted with morpholine, 2-morpholinoethylamine, and 2-phenylethylamine to give the corresponding 4-amino-2-phenyl-1-(2*H*)phthalazinones (**16**) (Eq. 14).

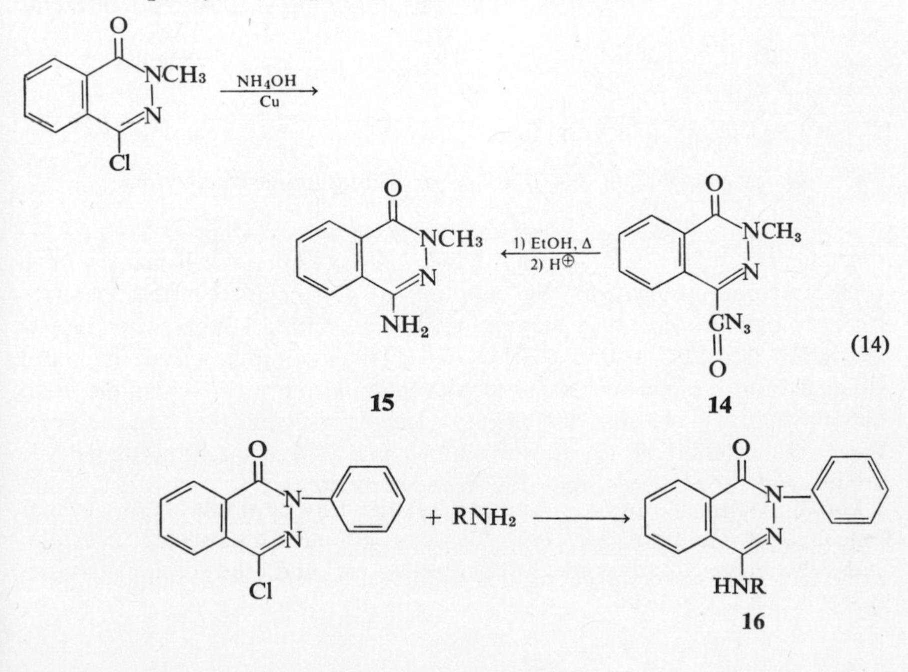

D. Preparation of 1,4-Diaminophthalazines

Wolf and Vollmann (34) have condensed 1-amino-3-iminoisoindolenine nitrate (**17**) and its 5-substituted derivatives with hydrazine hydrate to give the corresponding 1,4-diaminophthalazines (Eq. 15). DiStefano and Castle (35) have condensed phthalonitrile with 95% hydrazine by refluxing in

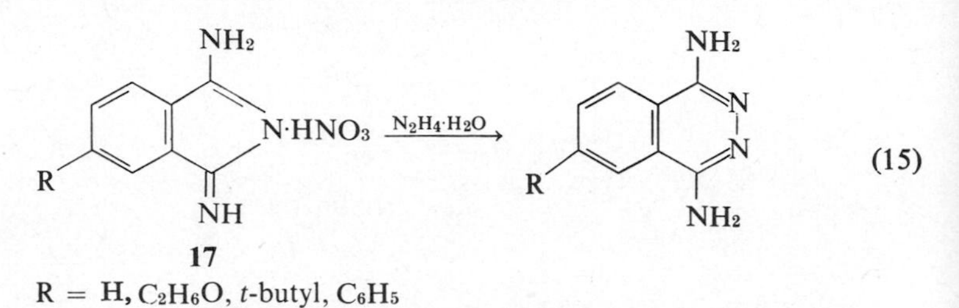

R = H, C_2H_6O, *t*-butyl, C_6H_5

methanol to give 1,4-diaminophthalazine (**18**) in 40% yield. Satoda and co-workers (5) have prepared **18** by the fusion of 1,4-diphenoxyphthalazine with ammonium acetate at 180° for 1 hr (Eq. 16).

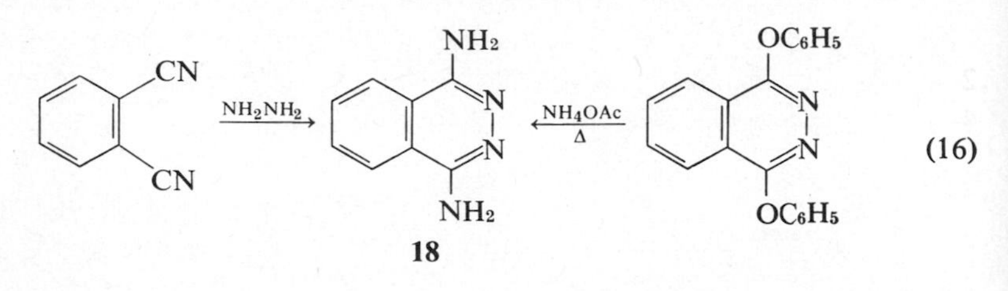

E. Preparation of 1,4-Dialkyl- or Diarylaminophthalazines

A small number of 1,4-dialkyl- and 1,4-diarylaminophthalazines have been prepared. In general, the reaction of 1,4-dichlorophthalazine with strongly basic amines such as ammonia or alkylamines stops after mono-alkylation (see Eq. 7). In exception, 1,4-diethylaminophthalazine has been isolated from the reaction of 1,4-dichlorophthalazine with ethylamine in an autoclave at 105–115° for 16 hr using xylene as a solvent (36). On the other hand, the reaction of 1,4-dichlorophthalazine (24) with the weakly basic amine, aniline, proceeds smoothly in refluxing ethanol to give after 30 min 1,4-dianilinophthalazine (**19**) in 58% yield (Eq. 17). Haworth and Robinson (24) have studied this reaction in detail, obtaining 1-anilino-4-chlorophthalazine (see Eq. 8) and subjecting this to further reaction with aniline under a variety of conditions. They found that 1-anilino-4-chlorophthalazine does not react with aniline in refluxing ethanol, even after 7 hr of refluxing, but it does react with refluxing aniline in the absence of any other solvent or with aniline hydrochloride for 7 hr in refluxing ethanol. The product is 1,4-dianilinophthalazine (**19**). Aniline hydrochloride appears to have a catalytic effect upon the reaction (Eq. 17). Kautsky and Kaiser (37) have isolated **19** as its hydrochloride in 77% yield by refluxing 1,4-dichlorophthalazine with aniline in ether containing sodium carbonate.

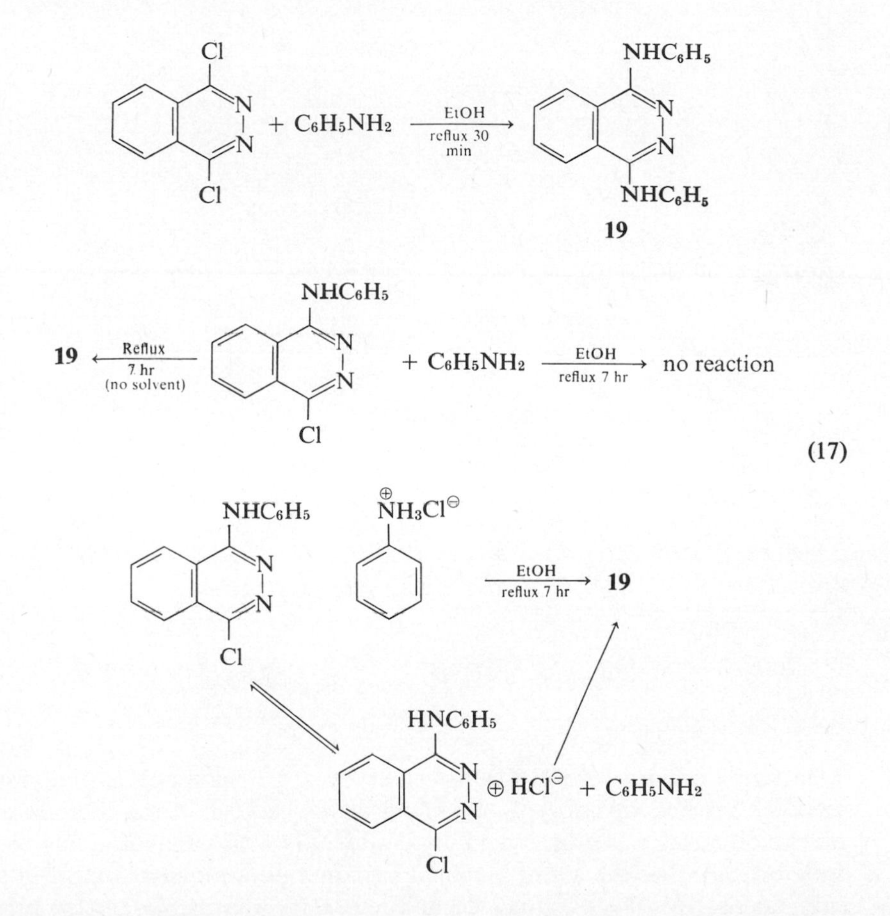

Haworth and Robinson (24) have prepared 1-anilino-4-(*N*,*N*-dialkylaminoalkylamino)phthalazines by heating 1-anilino-4-chlorophthalazine with *N*,*N*-dialkylaminoalkylamines at about 150° C. It is interesting that 1,4-dichloropyridazine behaves in the same way as 1,4-dichlorophthalazine towards alkyl- and arylamines (38).

1,4-Diphenoxyphthalazine reacts smoothly with morpholine to give 1,4-dimorpholinophthalazine in 87% yield (39). 4-Mercapto-1-(2*H*)-phthalazinethione similarly reacts with 2-hydroxyethylamine to give 1,4-bis(2-hydroxyethylamino)phthalazine (13) (Eq. 18).

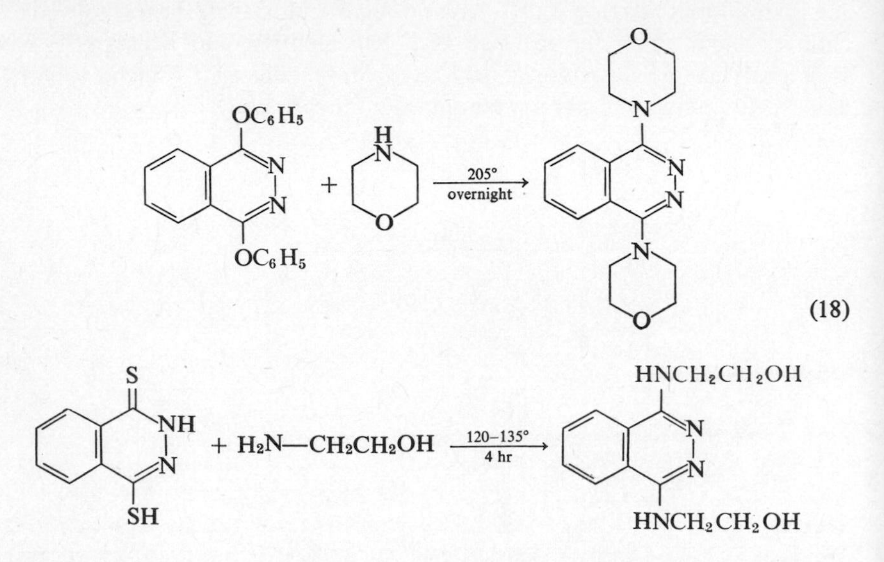

Inspection of the literature indicates that the best preparative route to 1,4-diarylaminophthalazines is by the reaction of 1,4-dichlorophthalazine with aromatic amines, whereas 1,4-dialkylaminophthalazines might be prepared most successfully by the reaction of 4-mercapto-1-(2*H*)phthalazinethione or 1,4-diphenoxyphthalazine with alkylamines.

F. Preparation of 1-Hydroxyaminophthalazines

Parsons and Turner (40–46) have synthesized and studied 1-hydroxyaminophthalazine and several of its derivatives because of their important pharmacological properties, including antipyretic, antiinflammatory, hypotensive, and bronchodilatory effects and respiratory stimulation. These compounds have been synthesized and tested for pharmacological activity only in the last few years. 1-Chlorophthalazine or 1-chlorophthalazines having alkyl, aryl, aralkyl, and arylamino groups in the 4-position react with hydroxylamine (40, 41) at room temperature in methanol to give the corresponding 1-hydroxyaminophthalazines (**20**) (Eq. 19). Grabliauskas (47) has prepared compounds similar to **20**. Oishi (16) has studied the nucleophilic substitution reaction of 1-methylsulfonylphthalazine and 1-chlorophthalazine with hydroxylamine and obtained **20** (R = H) in 55 and 57% yields, respectively (Eq. 19).

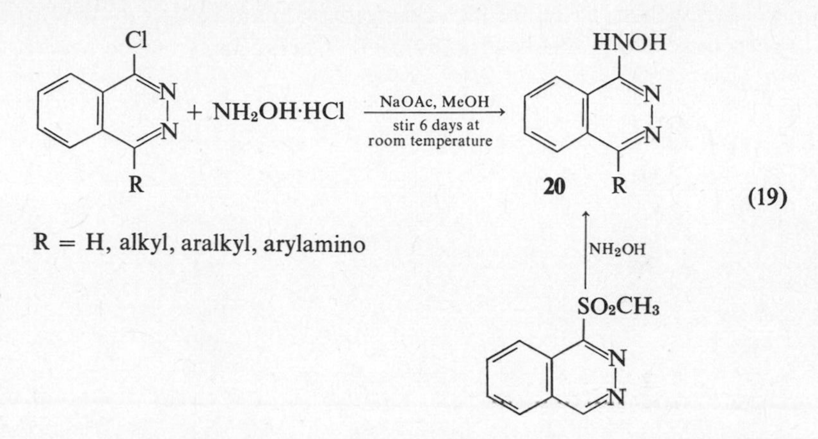

1,4-Diethoxyphthalazine and 1,4-dimorpholinophthalazine (45, 46) react with hydroxylamine hydrochloride in refluxing methanol in the presence of sodium acetate to give 1-hydroxyamino-4-ethoxyphthalazine (**21**) and 1-hydroxyamino-4-morpholinophthalazine (**22**) in 97 and 85% yields,

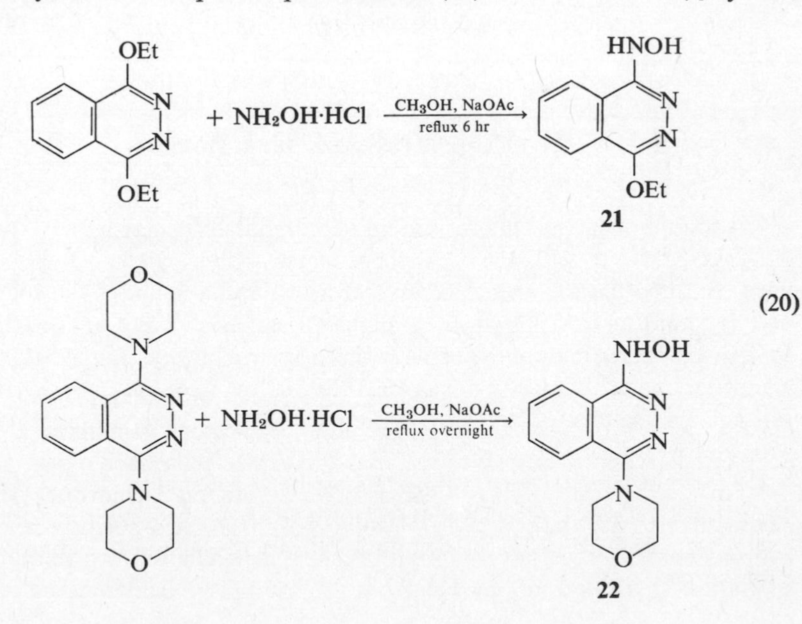

respectively (Eq. 20). Similarly, *O*-methylhydroxylamine reacts with 1,4-dialkoxyphthalazines to give compounds similar to **21** (45, 46).

1-Hydroxyamino-4-methylphthalazine (42) reacts with ethyl chlorocarbonate in dimethylformamide to give *O*-acylated derivative **23** (R = CH_3). Further heating of **23** in dimethylformamide at 140–145° for 30 min gives oxadiazolophthalazines (**24**) (43, 44) (Eq. 21).

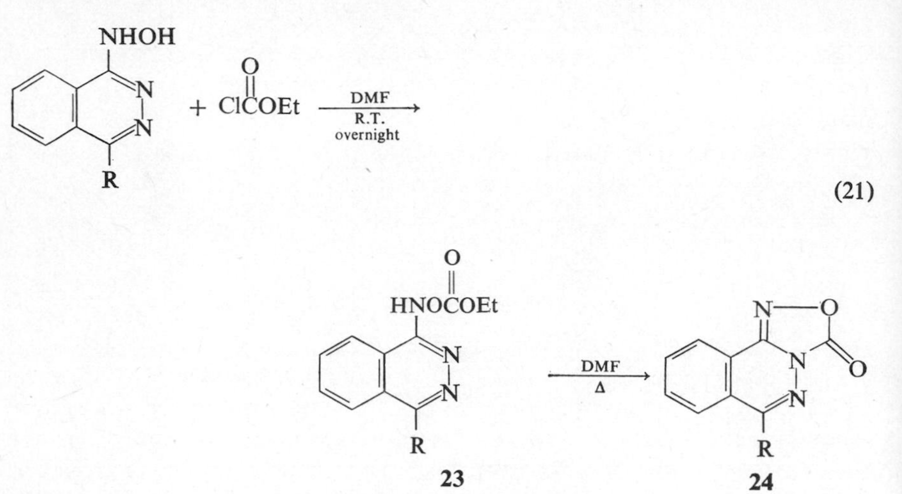

II. Properties and Reactions

1-Aminophthalazine (**1**) forms cream colored needles having mp 210–211° (1) [212–213° (2), 212–214° (3)] from water. Albert and co-workers (48) have recorded the pK_a values of several heterocyclic amines. The pK_a value of **1** is found to be 6.60, which is slightly less basic than 4-aminocinnoline (pK_a = 6.84). 1-Aminophthalazine forms a hydrochloride, mp 205–206° (1); a nitrate, mp 220° (dec.); a picrate, mp 301° (1); and a *p*-nitrobenzenesulfonic acid salt, mp 226° (1). 1-Aminophthalazine forms a quaternary salt with methyl iodide, mp 251–252° (2).

1,4-Diaminophthalazine (**18**) forms white needles mp 254° (35) from water [256° (ethanol) (34); 243° (ethanol) (5)]. It forms a hydrochloride salt having mp 230–231° (23). Aminophthalazines in general are high-melting compounds.

No studies of the pmr, ir, and uv spectral parameters of the 1-amino- and 1,4-diaminophthalazines have been reported in the literature. Bowie and co-workers (49) have reported the mass spectra of 1-amino- and 1,4-diaminophthalazines. Thomson and co-workers (55) have carried out an X-ray structural analysis of the copper chloride derivative of 1,4-bis(2′-pyridyl)aminophthalazine and found that the compound exists in a tautomeric form where both the hydrogen atoms are located on exocyclic nitrogen atoms.

No studies have been reported concerning alkylation or diazotization of 1-amino or 1,4-diaminophthalazines. Berg and Parnell (4) allowed 1,7-diamino-4-methylphthalazine to react with 2-amino-4-chloro-6-methylpyrimidine and obtained the corresponding 7-pyrimidinylaminophthalazine (**25**) in 91% yield (Eq. 22).

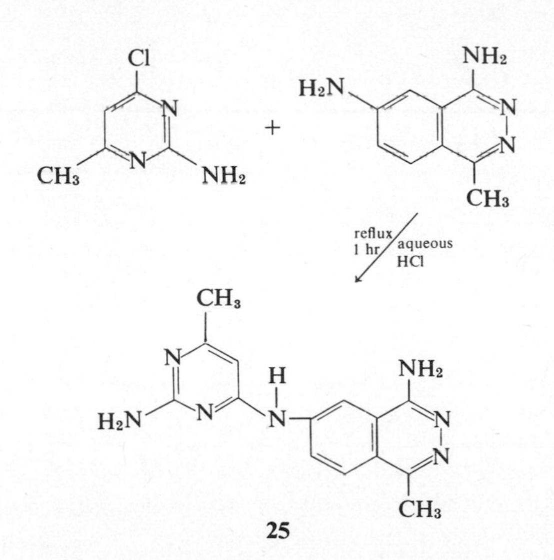

(22)

Acylation of 1-aminophthalazine with acetic anhydride (1, 2), aroyl chlorides (1, 50), *p*-nitrobenzenesulfonyl chloride (1), and *p*-acetamidobenzenesulfonyl chloride (1, 3, 51) proceeds smoothly to give the corresponding carboxamides and sulfonamides. Similarly, 1,4-diaminophthalazine gives 1,4-diacetamidophthalazine (34).

Nucleophilic substitution of 1,4-dimorpholinophthalazine with hydroxylamine gives 1-hydroxyamino-4-morpholinophthalazine (45, 46) (see Eq. 20). Zugravescu and co-workers (6) have obtained 1-guanidino-4-methylphthalazine (**26**) from the reaction of 1-amino-4-methylphthalazine with

S-methylpseudothiourea (Eq. 23). Flitsch and Peters (22) have obtained 4-anilino-1-(2*H*)phthalazinone in 74% yield by reacting 4-amino-1-(2*H*) phthalazinone with a mixture of aniline and aniline hydrochloride (Eq. 23).

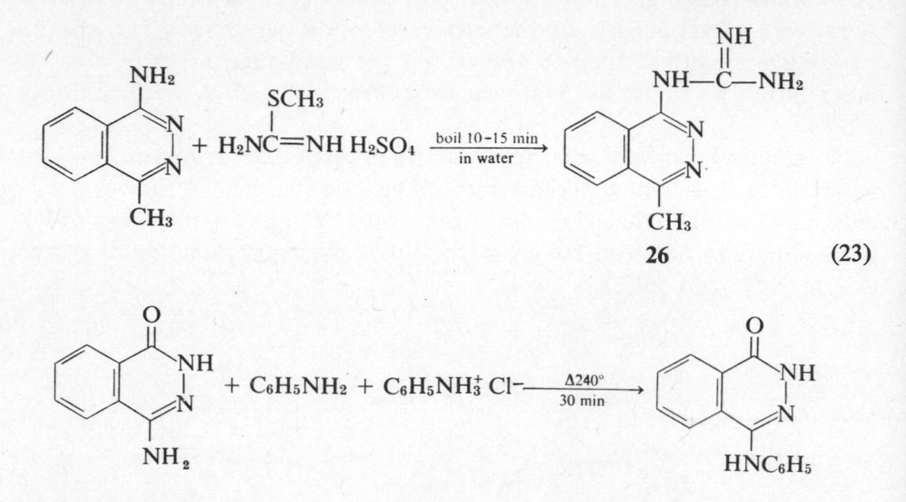

DiStefano and Castle (35) attempted to prepare the nitramine derivative of 1,4-diaminophthalazine in concentrated nitric acid-sulfuric acid (1:3) and in 100% nitric acid at 0–5° C. A nitramine derivative was not obtained. Instead, the products isolated were 4-hydroxy-1-(2*H*)phthalazinone and phthalic anhydride, respectively. Klamerth (30), attempting to hydrolyze 4-(2-hydroxyethylamino)-1-(2*H*)phthalazinone (**12**) by heating it with hydrobromic acid, isolated phthalic anhydride. Ing-Manske reaction products were not found (see Eq. 12).

Kautsky and Kaiser (37) have shown that the interaction of 1,4-dianilinophthalazine hydrochloride in 5*N* sodium hydroxide with iodine at less than −10° C yields 1,4-dehydrodianilinophthalazine (**27**) as black needles, mp 95–97° (Eq. 24).

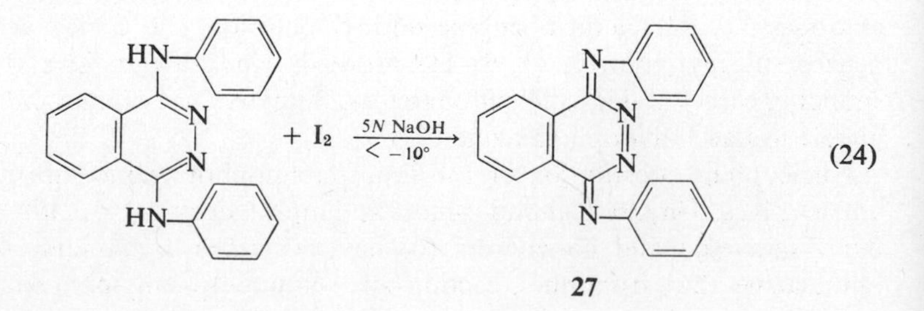

III. Tables

TABLE 2H-1. 1-Aminophthalazines with or without Substitutents in the 4- and 7-Positions

R	R′	Preparation[a]	Yield (%)	MP	Reference
H	H	A	64–92	Cream colored needles, 210–211° (water); 212–213°; 212–214°; HCl, 205–206°; HNO_3, 220 (dec.); picrate, 301°; *p*-nitrobenzene-sulfonic acid salt, 226°; methiodide, 251–252°	1–3
H	NO_2	B, C	30–83	Orange needles, 309–310° (dec.) (ethanol); 313–315° (dec.); 303–304° (dec.); methiodide, 268–270°	2, 4
H	NH_2	D	77	Methiodide, pale brown plates, 297–299° (dec.) (aqueous KI solution)	4
H	2-amino-6-methylpyrimidin-4-yl-NH— (H_2N, CH_3, NH—)	E	100	Bis-methiodide, 281–283°; bis-methochloride, 327°	4
CH_3	H	C_1, F	81	Needles, 212° (water); 218–220°; 178–180°	3, 6, 7
CH_3	NO_2	C	73	Yellow needles, 282° (dec.) (ethanol); methochloride, 310–312° (dec.); ethiodide, 283–285° (dec.)	4

TABLE 2H-1 (*continued*)

R	R′	Preparation[a]	Yield (%)	MP	Reference
CH_3	NH_2	D_1	82	Brown needles, 274–275° (dec.) (water); ethiodide, 293–295°	4
CH_3	(4-amino-6-methylpyrimidin-2-yl)amino: CH_3, N, N, H_2N, NH—	E_1	91	Gray needles, 315–316° (dec.); bis-methochloride, 322–324°; bis-ethiodide, 305°	4
CH_2— (4-pyridyl, N)	H	G	—	—	52
Cl	H	H	79	202°	5
CH_3O	H	I	92	156° (ethanol–hexane)	5
C_6H_5	H	A	90	93–94°	7a
p-ClC_6H_4	H	A	83	100–101°	7a

[a] A, see different methods of preparation in text of this section. B, 1-methylthio-7-nitrophthalazine + NH_4OAc, heat at 130–135° for 45 min; or 7-nitro-1-(2*H*)phthalazinethione + $Hg(OAc)_2$ + NH_4OAc, heat at 140° for 30 min. C, the corresponding 1-phenoxyphthalazine + NH_4OAc, fuse at 180–190° for 15 min. C_1, same as C, fuse at 150–160° for 30 min. D, 1-amino-7-nitrophthalazine methiodide reduced with Pt/H_2 in methanol at 40° C. D_1, 1-amino-4-methyl-7-nitrophthalazine reduced with Pt/H_2 in acetic acid. E, 1,7-diaminophthalazine methiodide + 2-amino-4-chloro-6-methylpyrimidine methiodide, reflux in aqueous hydrochloric acid for 1 hr. E_1, 1,7-diamino-4-methylphthalazine + 2-amino-4-chloro-6-methylpyrimidine, reflux in aqueous hydrochloric acid for 1 hr. F, 1-chloro-4-methylphthalazine + NH_4OH (25%), heat in sealed tube at 120–130° for 15 hr; or heat with ammonia at 170–180° for a few hours. G, 1-chloro-4-(4′-pyridylmethyl)phthalazine + ethanolic ammonia, heat at 175° for 5 hr in pressure vessel. H, 1,4-dichlorophthalazine + NH_4OH, heat in sealed tube at 120° for 8 hr. I, 1-amino-4-chlorophthalazine + $NaOCH_3$—CH_3OH, heat in sealed tube at 120° for 20 hr.

TABLE 2H-2. 1-Alkylamino- and 1-Arylaminophthalazines

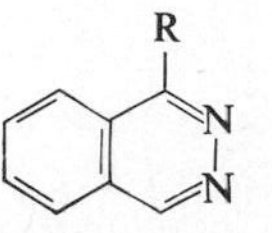

R	Preparation[a]	Yield (%)	MP	Reference
$-NHCH_2CH_2OH$	A	—	Colorless grains	13
$-NHCH_2CHSSO_3Na$	B	35	White powder, 72° (swells), 210° (dec.) (ethanol-ether)	11
$-NHCH_2CH_2CH_3$	C	—	Picrate, 138–139° (water)	15
$-NHCH_2CH_2CH_2CH_3$	C_1	82	White needles, 99–100° (benzene–ligroin); picrate, 134–135°	14, 16
$-NHCH_2C_6H_5$	A	—	Colorless needles, 162–163°	13
$-NHCH_2CH_2C_6H_5$	D	93	Yellow, 151–152°; HCl, 93–94°; hydrate, 91–95°	10
$-NHCH_2CH_2C_6H_4OCH_3$-*p*	A	—	Colorless needles, 179°	13
$-NHCH_2CH_2(CH_3)_2$	D_1	27	2 HCl; 242–243°; bis-methiodide, 235–237°	8
$-NHCH_2CH_2-N$ O	D_2	—	169.5–170.5°	9
$-NHC_6H_5$	C_2	30–68	Pale yellow needles, 184–186° (benzene-ligroin)	16
—N	D_3	7	Colorless needles, 115° (ether)	11
—N	D_2	—	97–98°	9
—N O	D_2	—	113–114°	9

TABLE 2H-2. (*continued*)

R	Preparation[a]	Yield (%)	MP	Reference
$-N(CH_2CH_3)(CH_2CH_2N(CH_3)_2)$	D_4	16	BP 150–155°/0.1 mm dipicrate, 193–195° (dec.)	12
$-N(CH_2C_6H_5)(CH_2CH_2N(CH_3)_2)$	D_4	15	BP 172–176°/0.03 mm dipicrate, 209–211° (dec.)	12
$-N(CH_2C_6H_4OCH_3)(CH_2CH_2N(CH_3)_2)$	D_4	29	BP 200–205°/0.03 mm dipicrate, 162–164° (dec.)	12
$-NHC(=O)CH_3$	E	—	Colorless needles, 183–184° (ethanol); 185°; methiodide 253–254°	1, 2
$-NHC(=O)C_6H_5$	—	—	Yellow prisms, 146° (aqueous pyridine)	1

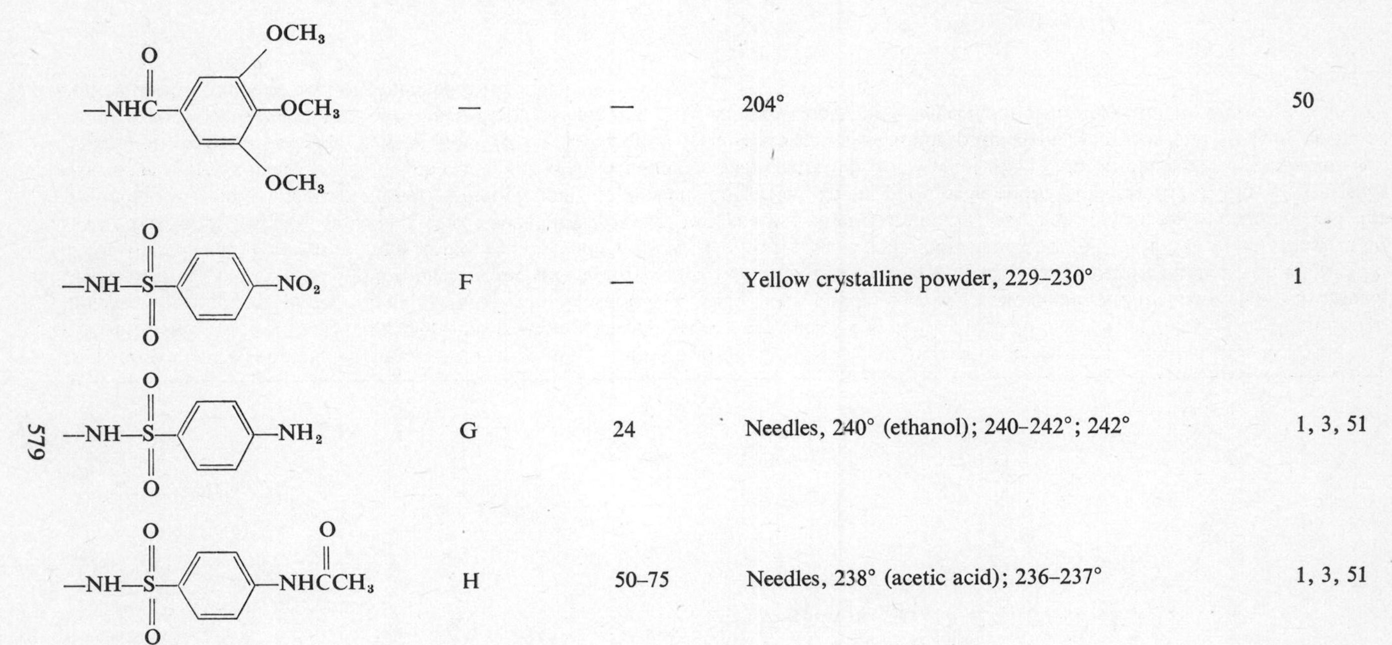

	—	—	204°	50
	F	—	Yellow crystalline powder, 229–230°	1
	G	24	Needles, 240° (ethanol); 240–242°; 242°	1, 3, 51
	H	50–75	Needles, 238° (acetic acid); 236–237°	1, 3, 51

TABLE 2H-2. (*continued*)

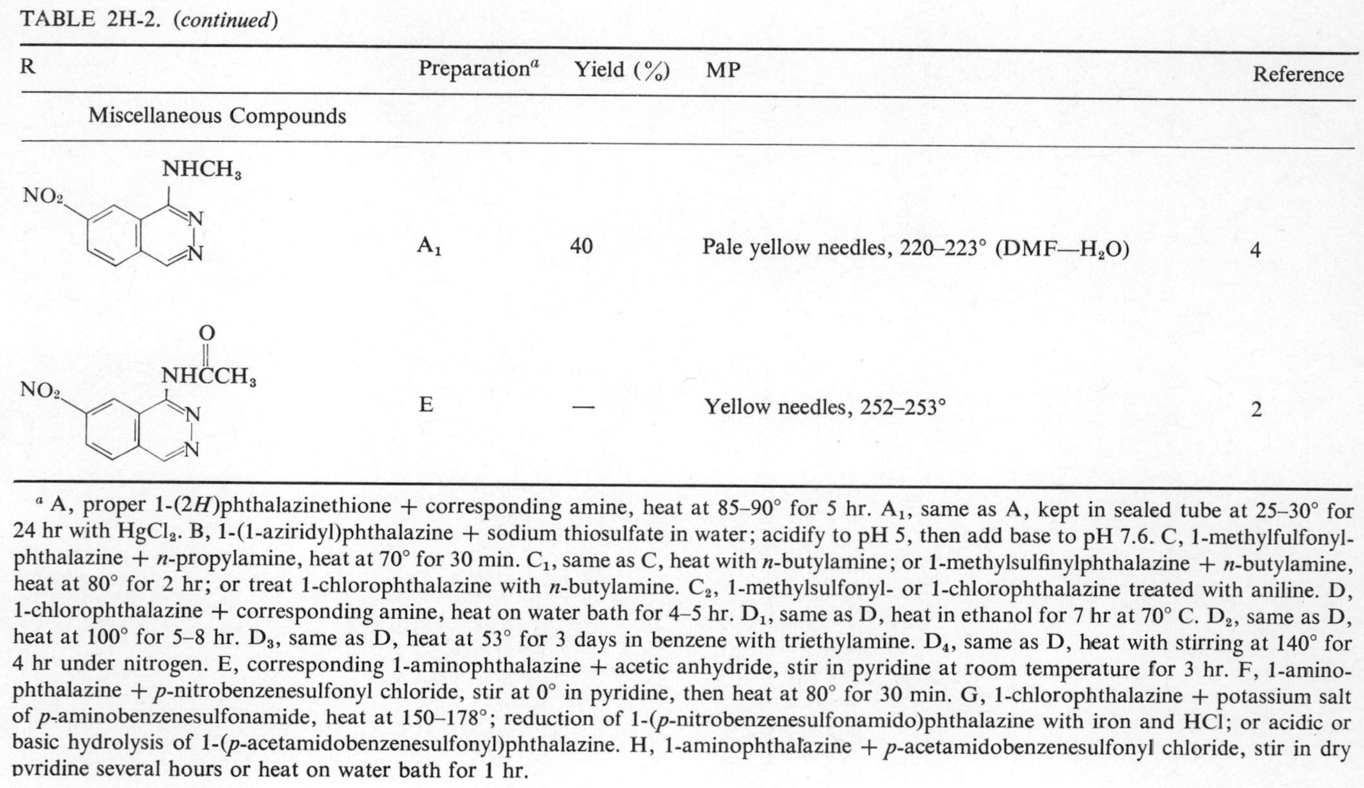

R	Preparation[a]	Yield (%)	MP	Reference
Miscellaneous Compounds				
1-NHCH$_3$, 7-NO$_2$-phthalazine	A_1	40	Pale yellow needles, 220–223° (DMF—H_2O)	4
1-NHCOCH$_3$, 7-NO$_2$-phthalazine	E	—	Yellow needles, 252–253°	2

[a] A, proper 1-(2*H*)phthalazinethione + corresponding amine, heat at 85–90° for 5 hr. A_1, same as A, kept in sealed tube at 25–30° for 24 hr with $HgCl_2$. B, 1-(1-aziridyl)phthalazine + sodium thiosulfate in water; acidify to pH 5, then add base to pH 7.6. C, 1-methylfulfonylphthalazine + *n*-propylamine, heat at 70° for 30 min. C_1, same as C, heat with *n*-butylamine; or 1-methylsulfinylphthalazine + *n*-butylamine, heat at 80° for 2 hr; or treat 1-chlorophthalazine with *n*-butylamine. C_2, 1-methylsulfonyl- or 1-chlorophthalazine treated with aniline. D, 1-chlorophthalazine + corresponding amine, heat on water bath for 4–5 hr. D_1, same as D, heat in ethanol for 7 hr at 70° C. D_2, same as D, heat at 100° for 5–8 hr. D_3, same as D, heat at 53° for 3 days in benzene with triethylamine. D_4, same as D, heat with stirring at 140° for 4 hr under nitrogen. E, corresponding 1-aminophthalazine + acetic anhydride, stir in pyridine at room temperature for 3 hr. F, 1-aminophthalazine + *p*-nitrobenzenesulfonyl chloride, stir at 0° in pyridine, then heat at 80° for 30 min. G, 1-chlorophthalazine + potassium salt of *p*-aminobenzenesulfonamide, heat at 150–178°; reduction of 1-(*p*-nitrobenzenesulfonamido)phthalazine with iron and HCl; or acidic or basic hydrolysis of 1-(*p*-acetamidobenzenesulfonyl)phthalazine. H, 1-aminophthalazine + *p*-acetamidobenzenesulfonyl chloride, stir in dry pyridine several hours or heat on water bath for 1 hr.

TABLE 2H-3. 4-Substituted 1-Alkylamino- and 1-Arylaminophthalazines

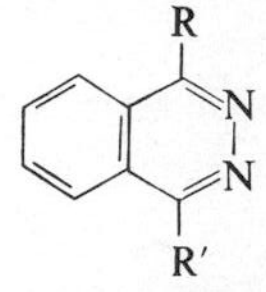

R	R′	Preparation[a]	Yield (%)	MP	Reference
$—NHCH_3$	Cl	A	71	223–224°	22
$—NHCH_3$	C_6H_5	B	—	195–197° (ethyl acetate)	20
$—N(CH_3)_2$	C_6H_5	B	—	111–113° (ligroin)	20
$—NHCH_2CH_3$	C_6H_5	B	—	150–151° (acetonitrile–H_2O)	20
$—NHCH_2CH_2CH_3$	C_6H_5	B	—	103–104° (2-propanol–H_2O)	20
$—NHCH(CH_3)_2$	C_6H_5	B	—	191–192° (ethyl acetate–ligroin)	20
$—NHCH(CH_3)_2$	*p*-ClC_6H_4	B	—	199–202° (ethyl acetate–ligroin)	20
$—NHCH(CH_3)_2$	$C_6H_5CH_2—$	B	—	229–231° (ethyl acetate–ligroin)	20
—NH—◁ (cyclopropyl)	C_6H_5	C	—	188–190° (acetonitrile)	20
—NH—◁ (cyclopropyl)	*p*-ClC_6H_4	B	—	189–191° (ethyl acetate)	20
—NH—◁ (cyclopropyl)	$C_6H_5CH_2—$	B_1	—	149–150 (ethyl acetate–ligroin)	20, 20a
$—NHCH_2CH(CH_3)_2$	C_6H_5	D	—	143–144° (acetonitrile)	20
$—NHCH_2CH(CH_3)_2$	*p*-$CH_3OC_6H_4$	—	—	203°	7
$—N(CH_2CH_2CH_2CH_3)_2$	C_6H_5	C_1	—	HCl, 173–175° (ethyl acetate)	20
$—NHCH_2C_6H_5$	C_6H_5	C_2	—	219–220° (acetonitrile)	20

TABLE 2H-3. (*continued*)

R	R′	Preparation[a]	Yield (%)	MP	Reference
$—NHCH_2CH_2C_6H_5$	Cl	A_1	49	192–194° (methanol); HCl, 243–244°	10
$—NHCH_2CH_2C_6H_5$	C_6H_5	C_1	—	147–149° (acetonitrile)	20
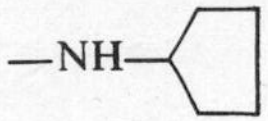 —NH—(cyclopentyl ring)	C_6H_5	D	—	189–191° (2-propanol–ligroin)	20
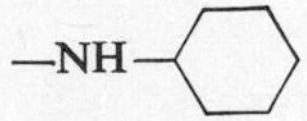 —NH—(cyclohexyl ring)	C_6H_5	E	—	HCl, 303–305° (ethanol)	20
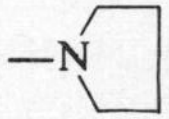 —N(pyrrolidine ring)	C_6H_5	C_1	—	1/4 H_2O, 106–108° (2-propanol-H_2O)	20
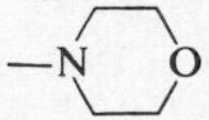 —N(morpholine ring, O)	C_6H_5	C_1	—	192–194° (acetonitrile)	20
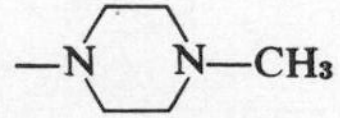 —N(piperazine ring)N—CH_3	C_6H_5	C_1	72	155–158° (acetonitrile)	20, 20a
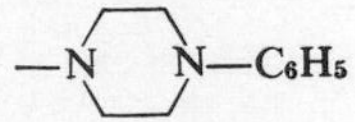 —N(piperazine ring)N—C_6H_5	C_6H_5	C_1	—	217–219° (ethanol)	20
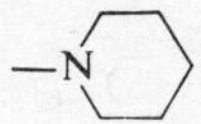 —N(piperidine ring)	Cl	—	—	132°	7
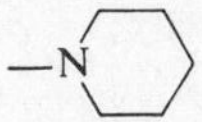 —N(piperidine ring)	C_6H_5	F	55	152–153° (acetonitrile)	20, 20a

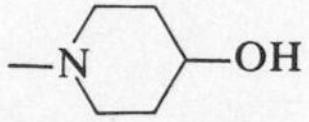 4-hydroxypiperidino ($—N\langle\rangle—OH$)	C_6H_5	F	—	159–160° (2-propanol–water)	20
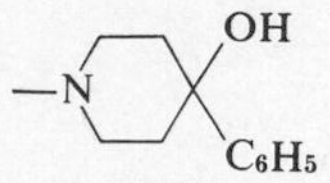 4-hydroxy-4-phenylpiperidino ($—N$, 4-OH, 4-C_6H_5)	C_6H_5	E	—	208–210° (acetonitrile)	20
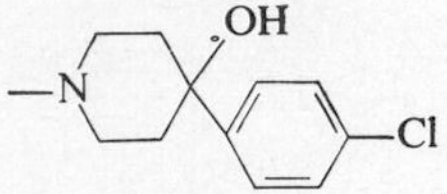 4-hydroxy-4-(p-chlorophenyl)piperidino ($—N$, 4-OH, 4-C_6H_4Cl)	C_6H_5	E	—	232–234° (ethanol)	20
$—NHCH_2CH_2OH$	CH_3	G	82	White needles, 195–196° (1-butanol); picrate, 158–160°	17
$—NHCH_2CH_2OH$	C_6H_5	C_1	—	161–163° (acetonitrile)	20
$—NHCH_2CH_2Cl$	CH_3	H	27	Cream colored microcrystals, 300° (1-butanol)	17
$—NHCH_2CO_2H$	Cl	I	78	130–140° (dec.) (ethanol); sodium salt, 160–170°	22
$—NHCH_2CH_2N(CH_3)_2$	C_6H_5	C_1	—	126–129° (ethyl acetate–ligroin)	20
$—NHCH_2CH_2N(CH_3)_2$	p-ClC_6H_4	D_1	—	2 HCl, 265–267° (ethanol–ether)	20
$—NHCH_2CH_2CH_2N(CH_3)_2$	C_6H_5	D_1	—	2 HCl · H_2O, 257–259° (2-propanol)	20
$—NHCH_2CH_2CH_2N(CH_3)_2$	p-ClC_6H_4	D_1	—	2 HCl, 269–270° (ethanol–ether)	20
$—NHCH_2CH_2CH_2N(CH_3)_2$	$—CH_2C_6H_5$	D_1	—	2 HCl, 234–236° (ethanol–ether)	20
$—NHC_6H_5$	Cl	A_2	79	Colorless needles, 200°; HCl, 270°	24
$—NHC_6H_5$	$CH_2CH_2CH_2CH_3$	J	48	Pale yellow needles, 161–162° (ethanol)	19
$—NHC_6H_5$	C_6H_5	K	48	230.5–231° (methanol)	21, 53
$—NHC_6H_4Cl$-p	Cl	A_2	—	Colorless needles, 241° (acetone)	24
$—NHC_6H_4CH_3$-p	$CH_2CH_2CH_2CH_3$	J	87	Felted needles, 152–154° (ethanol)	19

TABLE 2H-3. (*continued*)

R	R′	Preparation[a]	Yield (%)	MP	Reference
$—NHC_6H_4CF_3$-*m*	CH_3	J_1	60	217–218° (1-propanol)	18
$2,3-(CH_3)_2C_6H_3NH—$	CH_3	J_1	68	Colorless crystals, 234–235° (ethanol)	18
$—NH—C(=NH)—NH_2$	CH_3	L	—	H_2SO_4, white 280° (water)	6
$—NH—SO_2—C_6H_4—NH_2$	Cl	M	51–60	Yellow crystals, 216° (dec.) (ethanol); 234–235°	5, 23
$—NH—SO_2—C_6H_4—NH_2$	CH_3	N	80	Needles 262° (acetone–ethanol)	3, 51
$—NH—SO_2—C_6H_4—NH_2$	OCH_3	O	—	238° (methanol)	5
$—NH—SO_2—C_6H_4—NH—C(=O)CH_3$	Cl	—	—	259–260°	23

$—NH—SO_2—C_6H_4—NH—C(=O)CH_3$ (p-)	CH_3	P	47	253°	3, 51
$—N(CH_2CH_2)_2N—CH_2CH_2OH$	C_6H_5	—	79	—	20a
$—N(CH_2CH_2)_2N—CH_2CH_2OH$	p-ClC_6H_4	—	64	—	20a
$—NH(CH_2)_3—N(CH_2CH_2)_2O$	C_6H_5	—	78	—	20a
$—N(CH_3)_2$	CH_3	—	60	43° (pet. ether) BP 116°/0.1 mm	20b

[a] A, 1,4-dichlorophthalazine + excess methylamine, heat 24 hr. A_1, 1,4-dichlorophthalazine + 2-phenylethylamine, heat at 120° for 1 hr. A_2, 1,4-dichlorophthalazine + equimolar aniline, reflux 30 min in ethanol. B, 4-substituted 1-chlorophthalazine + equimolar CH_3NH_2 and Na_2CO_3, heat in a sealed tube for 4 hr on a steam bath. B_1, same as B, heating time 16 hr. C, 1-chloro-4-phenylphthalazine + excess amine, heat at 60° for 3–4 hr. C_1, same as C, heat at 130–160° for 3–4 hr. C_2, same as C, heat at 200° for 3–4 hr. D, 4-substituted 1-chlorophthalazine + equimolar amine and sodium carbonate, heat at 130° in dimethyl sulfoxide for 2–3 hr. D_1, same as D, heat at 160° in dimethyl sulfoxide for 2–3 hr. E, 1-chloro-4-phenylphthalazine + equimolar amine and sodium carbonate, reflux in dimethyl sulfoxide for 3 hr. F, 1-chloro-4-phenylphthalazine + equimolar amine and sodium carbonate, reflux in methyl isobutyl ketone for 18 hr. G, 1-chloro-4-methylphthalazine + 2-hydroxyethylamine, heat 36 hr in 1-butanol on a water bath. H, 1-(2-hydroxyethylamino)phthalazine + $SOCl_2$, reflux in chloroform for 1 hr. I, 1,4-dichlorophthalazine + glycine, reflux 5 hr in 2*N* aqueous sodium hydroxide. J, 4-substituted 1-chlorophthalazine + corresponding amine, reflux in ethanol for 90 min. J_1, same as J, heat at 150° for 2 hr without solvent. K, 1-methylsulfonyl-4-phenylphthalazine + aniline, heat at 100° for 2 hr; or react 1-chloro-4-phenylphthalazine + aniline. L, 1-amino-4-methylphthalazine + *S*-methylpseudothiourea sulfate, boil in water for 10–15 min. M, 1,4-dichlorophthalazine + potassium salt of *p*-aminobenzenesulfonamide, heat at 160–180° for 1 hr; or heat sodium salt of *p*-amino-benzenesulfonamide in acetamide at 110–115° for 4 hr. N, 1-chloro-4-methylphthalazine + potassium salt of *p*-aminobenzenesulfonamide, heat at 150–170°; or alkaline hydrolysis of 1-(4′-acetamidobenzenesulfonamido)-4-methylphthalazine with 10% sodium hydroxide. O, 1-(4′-aminobenzenesulfonamido)-4-chlorophthalazine + sodium methoxide, heat at 120° for 13 hr. P, 1-amino-4-chlorophthalazine + *p*-acetamidobenzenesulfonyl chloride, stir in pyridine.

TABLE 2H-4. 4-Amino-1-(2*H*)phthalazinone and Derivatives

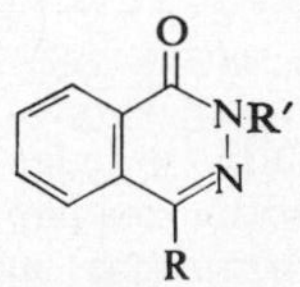

R	R′	Preparation[a]	Yield (%)	MP	Reference
NH_2	H	A, C	57–88	260–262°; 265–266°; 274° (ethanol)	22, 25, 26
NH_2	CH_3	B	100	158–160°; 160° (water)	3, 32
$NHCH_3$	H	C	39	238°	22
$—NHCH_2CH_2OH$	H	D	—	Pale violet needles 250–252° (methanol); HCl, 190–200°	30
$—NHCH_2CO_2H$	H	E	78	200° (dec.)	22
$—NHCH(CH_3)CO_2H$	H	E	76	270° (dec.)	22
$—NH—C(Br)(NO_2)—CH_3$	H	—	—	—	29
$—NHCH_2CH(OH)CH_2OC_6H_5$	H	D_1	100	White rosettes, 177° (ethanol-methanol)	28

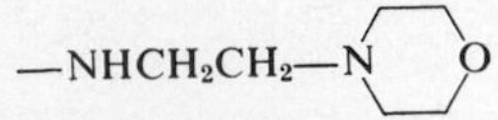 $-NHCH_2CH_2-N(CH_2CH_2)_2O$ (morpholino)	C_6H_5	F	66	109.5–111° (ethanol-benzene)	33
$-NHCH_2CH_2C_6H_5$	C_6H_5	F	84	136.5–138° (ethanol)	33
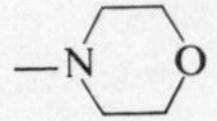 $-N(CH_2CH_2)_2O$ (morpholino)	C_6H_5	F	—	155–156° (ethanol)	33
$-NHC_6H_5$	H	C, C_1	4–74	256–257° (ethanol); 258°	22
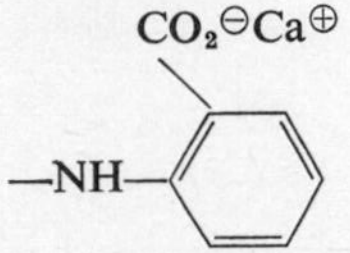 $-NH-C_6H_4-CO_2^{\ominus}Ca^{\oplus}$ (ortho)	H	G	66	Yellow needles, 450–460° (dec.) (water)	54
$-NH-C(=O)-OCH_2CH_3$	CH_3	B_1	66	158° (ethanol)	3, 32
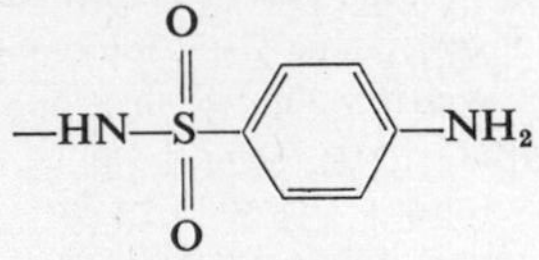 $-HN-SO_2-C_6H_4-NH_2$ (para)	CH_3	H	—	218–221° (acetone–alcohol); 220°	3, 51
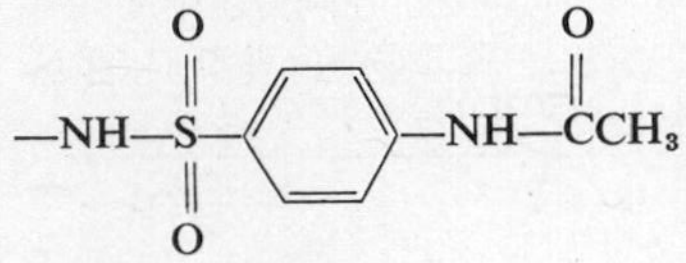 $-NH-SO_2-C_6H_4-NH-C(=O)CH_3$ (para)	CH_3	H_1	47–62	256°	3, 51

TABLE 2H-4. (*continued*)

R	R′	Preparation[a]	Yield (%)	MP	Reference
—HN-(2-pyridyl)	H	I	47	304° (DMF)	55

[a] A, ethyl 2-cyanobenzoate + $N_2H_4{\cdot}H_2O$ in ethanol, heat 6 hr on a water bath; monothiophthalimide or 3-iminoisoindolin-3-one + $N_2H_4{\cdot}H_2O$ in ethanol, heat. B, 4-carboethoxyamino-2-methyl-1-(2*H*)phthalazinone, heat with 20% hydrochloric acid. B_1, 4-carboxyazido-2-methyl-1-(2*H*)phthalazinone, boil in ethanol for 5 hr. C, the corresponding 1-amino-4-chlorophthalazine is heated with 80% sulfuric acid for 30 min. C_1, 4-amino-1-(2*H*)phthalazinone + aniline + aniline hydrochloride, heat at 240° for 30 min. D, 2-pyrrolidinoethylphthalimide + $N_2H_4{\cdot}H_2O$ in ethanol, heat at 160–180°. D_1, the corresponding *N*-substituted phthalimide + 22% $N_2H_4{\cdot}H_2O$ in methanol, reflux 40 min. E, the corresponding 3,6-dioxo-2,3,5,6-tetrahydroimidazo[2,1-*a*]phthalazine is hydrolyzed with aqueous sodium hydroxide. F, 4-chloro-2-phenyl-1-(2*H*)phthalazinone + corresponding amine, heat at 170–180° for 26 hr. G, 5-chloro-8*H*-phthalazino[1,2-*b*]quinazoline-8-one is heated with 20% alkali for 3 hr. H, 4-(*p*-acetamidobenzenesulfonamido)-2-methyl-1-(2*H*)phthalazinone + 5% sodium hydroxide for 1 hr. H_1, 4-amino-2-methyl-1-(2*H*)phthalazinone + *p*-acetamidobenzenesulfonyl chloride, stir in dioxane. I, 3-oxo-1-(2′-pyridyl)iminoisoindoline in methanol + $NH_2NH_2{\cdot}H_2O$ (64%) at 0° for 24 hr; or 1,4-di(2′-pyridyl)aminophthalazine was hydrolyzed by refluxing in concentrated hydrochloric acid for 6 hr.

TABLE 2H-5. 1,4-Diaminophthalazine and Derivatives

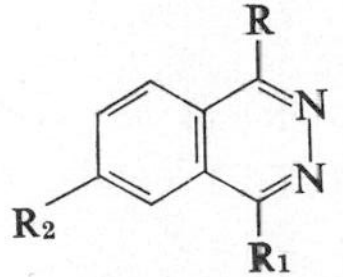

R	R_1	R_2	Preparation[a]	Yield (%)	MP	Reference
NH_2	NH_2	H	A	40	White needles 254° (water); 256°; 243°; HCl, 230–231°	5, 23, 34, 35
NH_2	NH_2	$C(CH_3)_3$	B	—	250°	34
NH_2	NH_2	C_6H_5	B	—	238°	34
NH_2	NH_2	OCH_2CH_3	B	—	231°	34
NH_2	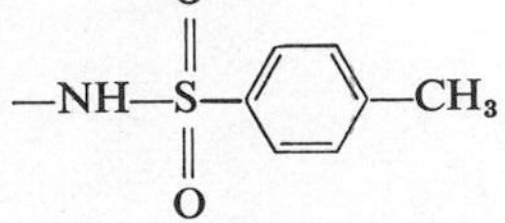 $-NH-SO_2-C_6H_4-CH_3$	H	C	55	230–231°	23
NH_2	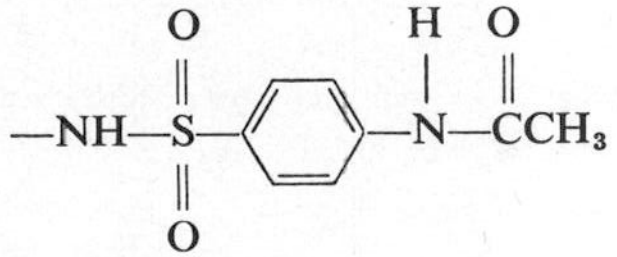 $-NH-SO_2-C_6H_4-NH-C(=O)CH_3$	H	C	61	243–245°	23
$NHC(=O)CH_3$	$NHC(=O)CH_3$	H	—	—	264°	34
$NHCH_2OH$	$NHCH_2OH$	H	—	—	64–67°	34
$NHCH_2CH_3$	$NHCH_2CH_3$	H	D	—	BP 134°/0.03 mm	36
$NHCH_2CH_2OH$	$NHCH_2CH_2OH$	H	E	—	Colorless needles, 192°	13

TABLE 2H-5. (*continued*)

R	R_1	R_2	Preparation[a]	Yield (%)	MP	Reference
$—NH(CH_2)_2N(CH_2CH_3)_2$	$C_6H_5NH—$	H	F	30	Needles, 148–149° (ligroin)	24
$—NH(CH_2)_3N(CH_2CH_3)_2$	$C_6H_5NH—$	H	F_1	50	Colorless platelets, 174° (ligroin–acetone)	24
$—NH(CH_2)_2N(CH_2CH_3)_2$	*p*-$ClC_6H_4NH—$	H	F_2	40	Needles, 145° (ligroin)	24
$—NH(CH_2)_3N(CH_2CH_3)_2$	*p*-$ClC_6H_4NH—$	H	F_3	—	Oil, bp 215–230°/ 0.002 mm; dipicrate, 203–204°	24
—N(morpholino) O	—N(morpholino) O	H	G	87	Needles, 204° (DMF); 205–207° (DMF)	39, 45, 46
$C_6H_5NH—$	$C_6H_5NH—$	H	H	58–77	Yellow plates, 223° (ethanol); HCl, 288° (alcohol)	24, 37
—NH(2-pyridyl) N	—NH(2-pyridyl) N	H	I	96	210° (acetone); $2HNO_3$, 187° (dec.)	55

[a] A, 1,4-diphenoxyphthalazine + NH_4OAc, fuse at 180° for 1 hr; or phthalonitrile + 95% NH_2NH_2, reflux 2 hr in methanol; or 1-amino-3-iminoisoindolenine nitrate + $N_2H_4 \cdot H_2O$, heat at 95° in aqueous solution. B, the corresponding 1-amino-3-imino-5-substituted isoindolenine is treated with hydrazine hydrate. C, 1,4-diaminophthalazine + the proper benzenesulfonyl chloride, heat at 75–80° for 2 hr in pyridine. D, 1,4-dichlorophthalazine + ethylamine, heat at 105–115° in xylene in an autoclave. E, 4-mercapto-1-(2*H*)phthalazinethione + 2-hydroxyethylamine, heat at 120–135° for 4 hr. F, 1-chloro-4-arylaminophthalazine + corresponding *N*,*N*-dialkylalkyldiamine, heat at 150–160° for 3 hr. F_1, same as F, heat at 150–160° for 7 hr. F_2, same as F, heat at 150° for 3 hr or at 100° for 24 hr. F_3, same as F, heat at 140° for 3 hr. G, 1,4-diphenoxyphthalazine + morpholine, reflux overnight. H, see different methods of preparation within this section. I, 1,3-di(2′-pyridyl)iminoisoindoline in methanol + $N_2H_4 \cdot H_2O$ (64%), let stand overnight at room temperature; or 1,4-diphenoxyphthalazine + 2-aminopyridine, reflux for 13 hr.

TABLE 2H-6. 4-Substituted 1-Hydroxyaminophthalazines

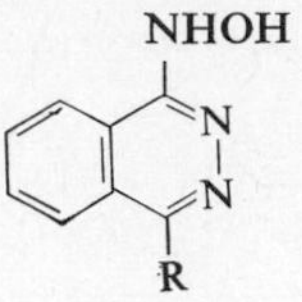

R	Preparation[a]	Yield (%)	MP	Reference
H	A, B	43–55	Yellow prisms, 183–185° (dec.) 188–190° (methanol)	16, 40, 41
CH_3	A	63	Yellow 226–228° (dioxane)	40, 41
CH_2CH_3	A	—	204–207°	40, 41
$CH_2CH_2CH_3$	A	—	220–224°	40, 41
$CH(CH_3)_2$	A	—	193–195°	40, 41
$CH_2CH_2CH_2CH_3$	A	—	180–183°; 189–191°	40, 41, 46
$CH_2CH(CH_3)_2$	A	—	205–215°	46
CH_2OCH_3	A	—	268–270°	46
$CH_2CH_2N(CH_3)_2$	A	—	161–162°	45, 46
$CH_2CH_2N(CH_2CH_3)_2$	A	—	162–163°	45, 46
CH_2CH_2—N (piperidine ring) 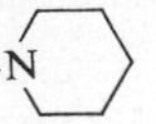	A	—	HCl, 260–263°	45, 46
CH_2CH_2—N O (morpholine ring) 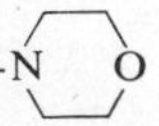	A	—	HCl·H_2O, 260–261°	45, 46
CH_2CO_2H	C	—	218–222°	40, 41
$CH_2CO_2CH_3$	—	—	115–117°	40, 41
CH_2CONH_2	—	—	188–191°	40, 41
$CH_2CONHNH_2$	D	99	254–257° (dec.)	40, 41
OCH_3	E	—	128–130°	45, 46
OCH_2CH_3	E	97	201–204° (dec.)	45, 46
OCH_2CH_2OH	E	—	212–222°	45, 46
$OCH_2CH_2OCH_3$	E	—	182–185°	45, 46
$OCH_2CH_2N(CH_3)_2$	E	—	230–233°	45, 46
$OCH_2CH_2N(CH_2CH_3)_2$	E	—	90–91°	45, 46
OCH_2CH_2—N (piperidine ring) 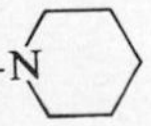	E	—	119°	45, 46
OCH_2CH_2—N O (morpholine ring) 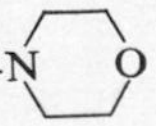	E	—	151°	45, 46
C_6H_5O—	E	—	196–200°	45, 46

TABLE 2H-6. (*continued*)

R	Preparation[a]	Yield (%)	MP	Reference
—N(morpholino, ring with O)	F	85	210–230° (DMF)	45, 46
—N(piperazine ring)N—CH_3	F	—	215–217°	45, 46
C_6H_5NH—	A	—	195–197°; 208–210°; HCl, 195–197° (?)	40, 41
p-ClC_6H_4NH—	A	—	190–194°; HCl, 190–194° (?)	40, 41
C_6H_5	A	—	251–253°	40, 41
m-$NH_2C_6H_4$	A	—	237–239°	40, 41
$C_6H_5CH_2$—	A	—	235–236°	40, 41
p-$CH_3OC_6H_4CH_2$—	A	—	211–213°	40, 41
p-$NO_2C_6H_4CH_2$—	A	—	235–236°	40, 41
—CH_2(2-pyridyl, ring with N)	A	—	205–207°	40, 41
—CH_2—(4-pyridyl, ring with N)	A	—	215–218°	40, 41

[a] A, the corresponding 1-chlorophthalazine + $NH_2OH{\cdot}HCl$ + NaOAc, stir 6 days in methanol; or reflux in methanol for 2 hr with hydroxylamine and aqueous potassium carbonate. B, treat 1-methylsulfonylphthalazine with hydroxylamine hydrochloride in presence of sodium hydroxide. C, methyl 1-hydroxyaminophthalazin-4-ylacetate hydrolyzed with 2*N* hydrochloric acid. D, methyl 1-hydroxyaminophthalazin-4-ylacetate + $N_2H_4{\cdot}H_2O$ in dioxane, heat at 80–85° for 3 hr. E, proper 1,4-dialkoxyphthalazine + $NH_2OH{\cdot}HCl$, reflux in methanol for 6 hr in presence of sodium acetate. F, the corresponding 1,4-diaminophthalazine + $NH_2OH{\cdot}HCl$, reflux in methanol overnight in presence of sodium acetate.

TABLE 2H-7. *O*- and *N*-Substituted 1-Hydroxyaminophthalazines

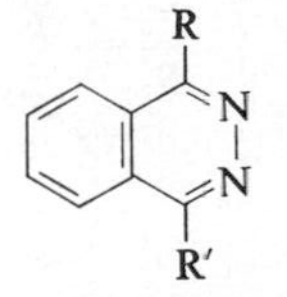

R	R′	Preparation[a]	Yield (%)	MP	Reference
$NHOCH_3$	CH_3	A	—	190–210° (dec.)	40, 41
$NHOCH_3$	$CH_2CH_2CH_3$	A	—	94–95°	46
$NHOCH_3$	OCH_2CH_3	B	—	70–72°	45, 46
$NHOCH_3$	OCH_2CH_2OH	B	—	99–101°	45, 46
$NHOCH_3$	$OCH_2CH_2N(CH_3)_2$	B	—	93–94°	45, 46
$NHOC(=O)CH_3$	CH_3	C	81	179–182° (alcohol)	40, 41
$NHOC(=O)CH_3$	$CH_2C(=O)OCH_3$	D	—	115–117°	40, 41
$NHOC(=O)OCH_2CH_3$	CH_3	E	47	220–225° (dec.) (ethanol)	42, 43
$-N(OH)CH_3$	CH_3	—	—	188–191°	40, 41

[a] A, the corresponding 1-chlorophthalazine + $NH_2OCH_3 \cdot HCl$ + NaOAc, reflux in methanol for 6 hr. B, the corresponding 1,4-dialkoxyphthalazine + $NH_2OCH_3 \cdot HCl$ + NaOAc, reflux overnight in methanol. C, 1-hydroxyamino-4-methylphthalazine + acetic anhydride, stir at room temperature. D, methyl 1-hydroxyaminophthalazin-4-ylacetate + acetic anhydride, stir at room temperature. E, 1-hydroxyamino-4-methylphthalazine + ethyl chloroformate kept in dimethylformamide overnight.

TABLE 2H-8. 4-*H*-3-Keto-1,2,4-oxadiazolo[3,4-*a*]phthalazines[a]

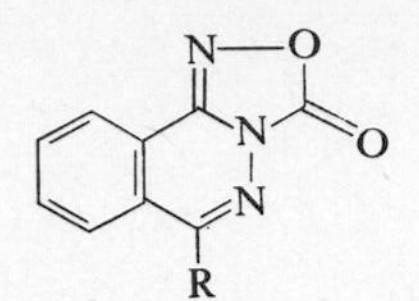

R	Yield (%)	MP	Reference
H	—	228–230°	43, 44
CH_3	25	222–224° (chloroform)	43, 44
$CH_2CH_2CH_2CH_3$	73	128–129°	43, 44
$C_6H_5CH_2$—	61	169–171°	43, 44
p-$CH_3OC_6H_4CH_2$	67	190–192°	43, 44
C_6H_5	42	229–231°	43, 44

[a] See methods of preparation discussed in text of this section.

References

1. H. J. Rodda, *J. Chem. Soc.*, **1956,** 3509.
2. C. M. Atkinson, C. W. Brown, and J. C. E. Simpson, *J. Chem. Soc.*, **1956,** 1081.
3. I. Satoda, N. Yoshida, and K. Mori, *Yakugaku Zasshi*, **77,** 703 (1957); *Chem. Abstr.*, **51,** 17927 (1957).
4. S. S. Berg and E. W. Parnell, *J. Chem. Soc.*, **1961,** 5275.
5. I. Satoda, F. Kusuda, and K. Mori, *Yakugaku Zasshi*, **82,** 233 (1962); *Chem. Abstr.*, **58,** 3427 (1963).
6. I. Zugravescu, M. Petrovanu, E. Rucinschi, and M. Caprosu, *Rev. Roum. Chim.*, **10,** 641 (1965); *Chem. Abstr.*, **64,** 732 (1966).
7. British Pat. 732,581 (1955); *Chem. Abstr.*, **50,** 8747 (1956).
7*a*. M. Eberle and W. J. Houlihan, *Tetrahedron Lett.*, **1970,** 3167.
8. S. O. Wintrop, S. Sybulski, R. Gaudry, and G. A. Grant, *Can. J. Chem.*, **34,** 1557 (1956).
9. L. E. Rylski, K. Gajewski, J. Traczynska, and T. Orlowska, *Farm. Polska.*, **21,** 758 (1965); *Chem. Abstr.*, **65,** 3865 (1966).
10. S. Biniecki et al., *Acta Polon. Pharm.*, **18,** 261 (1961); *Chem. Abstr.*, **57,** 16, 613 (1962).
11. R. N. Castle and S. Takano, *J. Heterocyclic Chem.*, **5,** 89 (1968).
12. N. B. Chapman, K. Clarke, and K. Wilson, *J. Chem. Soc.*, **1963,** 2256.
13. K. Fujii and S. Sato, *Ann. Repts., G. Tanabe Co., Ltd.*, **1,** 1 (1956); *Chem. Abstr.*, **55,** 6650 (1957).
14. G. B. Barlin and W. V. Brown, *J. Chem. Soc. (C)*, **1969,** 921.
15. G. B. Barlin and W. V. Brown, *J. Chem. Soc. (C)*, **1967,** 2473.
16. E. Oishi, *Yakugaku Zasshi*, **89,** 959 (1969); *Chem. Abstr.*, **72,** 3450 (1970).
17. J. N. Singh and A. B. Lal, *J. Indian Chem. Soc.*, **43,** 308 (1966).
18. K. H. Boltze and H. D. Dell, *Arch. Pharm.*, **299,** 702 (1966).

19. A. F. Turner and D. G. Parsons, U.S. Pat. 3,240,781 (1966); *Chem. Abstr.*, **64,** 17, 616 (1966).
20. H. M. Holava, Jr. and R. A. Partyka, *J. Med. Chem.*, **12,** 555 (1969).
20a. R. E. Rodway and R. G. Simmonds, German Pat. 2,021,195; *Chem. Abstr.*, **74,** 22865 (1971).
20b. M. A. Shah and G. A. Taylor, *J. Chem. Soc.* (*C*), **1970,** 1651.
21. E. Hayashi, T. Higashino, E. Oishi, and M. Sano, *Yakugaku Zasshi*, **87,** 687 (1967); *Chem. Abstr.*, **67,** 90750 (1967).
22. W. Flitsch and H. Peters, *Angew. Chem. Int. Ed. Eng.*, **6,** 173 (1967); *Chem. Ber.*, **102,** 1304 (1967).
23. K. Belniak, E. Domagalina, and H. Hopkala, *Roczniki. Chem.*, **41,** 831 (1967).
24. R. D. Haworth and S. Robinson, *J. Chem. Soc.*, **1948,** 777.
25. W. Koehler, M. Bubner, and G. Ulbricht, *Chem. Ber.*, **100,** 1073 (1967).
26. G. M. Filho and J. G. daCosta, *An. esc. Super quim. Univ. Recife*, **1,** 41 (1959); *Chem. Abstr.*, **55,** 3601 (1961).
27. G. Spielberger in *Methoden der Organischen Chemie*, Vol. XI-1, G. Thieme, Stuttgart, 1957, p. 80.
28. H. J. Roth, *Arch. Pharm.*, **292,** 194 (1954).
29. L. I. Bagal et al., *Zh. Org. Khim.*, **5,** 93 (1969); from *Index Chem.*, **33** (285), 111, 729 (1969).
30. O. Klamerth, *Chem. Ber.*, **84,** 254 (1951).
31. T. A. Connors and W. C. J. Ross, *J. Chem. Soc.*, **1960,** 2119.
32. H. Morishita et al., Japanese Pat. 6,280 (1959); *Chem. Abstr.*, **54,** 15,411 (1960).
33. L. Rylski et al., *Acta Polon. Pharm.*, **22,** 111 (1965); *Chem. Abstr.*, **63,** 8350 (1965).
34. W. Wolf and H. Vollmann; German Pat. 941,845 (1956); *Chem. Abstr.*, **53,** 6251 (1959).
35. L. DiStefano and R. N. Castle, *J. Heterocyclic Chem.*, **5,** 111 (1968).
36. British Pat. 822,069 (1959); *Chem. Abstr.*, **55,** 2005 (1961).
37. H. Kautsky and K. H. Kaiser, *Z. Naturforsch.*, **56,** 353 (1950); *Chem. Abstr.*, **45,** 2780 (1951).
38. M. Tisler and B. Stanovnik, in *Advances in Heterocyclic Chemistry*, Vol. 9, A. R. Katritzky and A. J. Boulton, Eds., Academic Press, New York, 1968, p. 270.
39. J. A. Elvidge and A. P. Redman, *J. Chem. Soc.*, **1960,** 1710.
40. Neth. Appl., 6,609,718 (1967); *Chem. Abstr.*, **68,** 49,632 (1968).
41. D. G. Parsons and A. F. Turner, British Pat. 1,094,044 (1967); *Chem. Abstr.*, **69,** 19180 (1968).
42. D. G. Parsons and A. F. Turner, British Pat. 1,094,045 (1967); *Chem. Abstr.*, **69,** 10456 (1968).
43. D. G. Parsons and A. F. Turner, British Pat. 1,094,046 (1967); *Chem. Abstr.*, **69,** 10457 (1968).
44. A. F. Turner and D. G. Parsons, U.S. Pat. 3,359,268 (1967); *Chem. Abstr.*, **69,** 27439 (1968).
45. D. G. Parsons, A. F. Turner, and B. G. Murray, British Pat. 1,133,406 (1968); *Chem. Abstr.*, **70,** 57872 (1969).
46. French Pat. 1,532,163 (1968); *Chem. Abstr.*, **71,** 112963 (1969).
47. K. Grabliauskas, USSR Pat. 218,190 (1968); *Chem. Abstr.*, **69,** 87005 (1968).
48. A. Albert, R. Goldacre, and J. Phillips, *J. Chem. Soc.*, **1948,** 2240.
49. J. H. Bowie et al., *Aust. J. Chem.*, **20,** 2677 (1967).
50. S. Kumada, N. Watanabe, K. Yamamoto, and H. Zenno, *Yakugaku Kenkyu*, **30,** 635 (1958); *Chem. Abstr.*, **53,** 20554 (1959).

51. H. Morishita et al., Japanese Pat. 2981 (1959); *Chem. Abstr.*, **54,** 13150 (1960).
52. J. Druey and E. Marxer, German Pat. 1,061,788 (1959); *Chem. Abstr.*, **55,** 13,458 (1961).
53. E. Hayashi and E. Oishi, *Yakugaku Zasshi*, **86,** 576 (1966); *Chem. Abstr.*, **65,** 15373 (1966).
54. C. D. Volcker, J. Marth, and H. Beyer, *Chem. Ber.*, **100,** 875 (1967).
55. L. K. Thomson et al., *Can. J. Chem.*, **47,** 414 (1969).

Part I. Hydrazinophthalazines

I. 1-Hydrazino- and 1,4-Dihydrazinophthalazines

Hydrazinophthalazines have been studied intensively only since 1950. Research in this field was prompted by the discovery that 1-hydrazinophthalazine (hydralazine, Apresoline®) and 1,4-dihydrazinophthalazine (dihydralazine, Nepresol®), are excellent hypotensive and antihypertensive agents. Numerous derivatives have been synthesized and biologically tested but compounds with improved properties have not been found. Druey and associates in Switzerland have carried out excellent research in this field and have written review articles (1, 2) concerning structure-activity relationships, biochemistry, and pharmacology.

A. Preparation of 1-Hydrazinophthalazines

1. *Nucleophilic Substitutions*

1-Chlorophthalazine reacts with hydrazine hydrate in ethanol on a hot water bath for 2 hr to give 1-hydrazinophthalazine (**1**) in 63–78% yields (3–8). In one case the formation of a by-product, 1,2-diphthalazinylhydrazine, has been reported (5). 1-Phenoxyphthalazine, on reaction with hydrazine hydrate, affords **1** (3). 1-(2*H*)Phthalazinethione, when heated with hydrazine hydrate on a steam bath for 5 hr (9, 10) or refluxed in ethanol (11) for 3 hr, furnished **1** in 65–80% yields. Similarly, diphthalazin-1-yl disulfide reacts with hydrazine hydrate to give **1** (11). 1-Methylsulfonylphthalazine also reacts with hydrazine hydrate to give **1** in 80% yield (8) (Eq. 1).

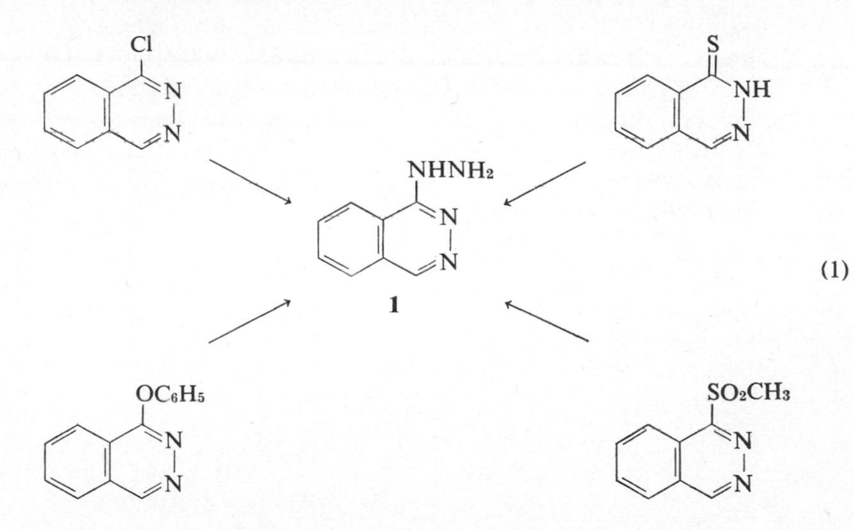

(1)

A few 1-hydrazinophthalazines having substituents in the benzenoid ring have been reported. These are usually prepared from the appropriately substituted 1-chlorophthalazine. Thus 5-chloro- (12), 6-chloro- (13, 14), 7-chloro- (13, 14), and 8-chloro-1-hydrazinophthalazines (12) are obtained by refluxing the corresponding dichlorophthalazines with hydrazine hydrate in ethanol for about 30 min. Similarly, 5-, 6-, and 7-fluoro-1-hydrazinophthalazine are obtained from the corresponding 1-chlorophthalazines (15). However, the reaction of 1-chloro-8-fluorophthalazine with hydrazine resulted in the formation of 1,8-dihydrazinophthalazine (**2**) (15) (Eq. 2). 1-Chloro-7-methoxylphthalazine (5a), on reaction with hydrazine hydrate, gives

1-hydrazino-7-methoxyphthalazine, which on heating with pyridine hydrochloride is demethylated to 1-hydrazino-7-hydroxyphthalazine (5).

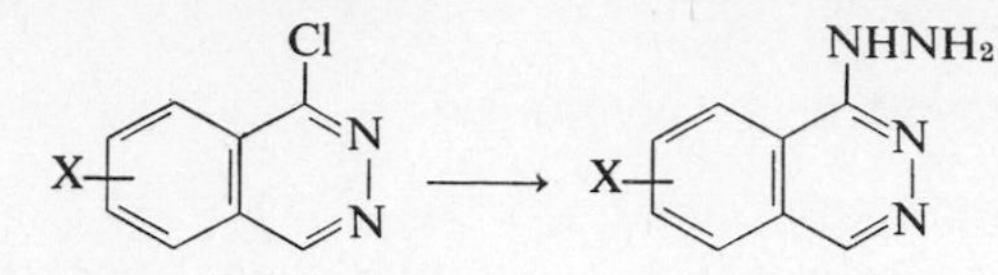

X = 5-, 6-, 7-, or 8-Cl;
5-, 6-, or 7-F;
7-OCH_3

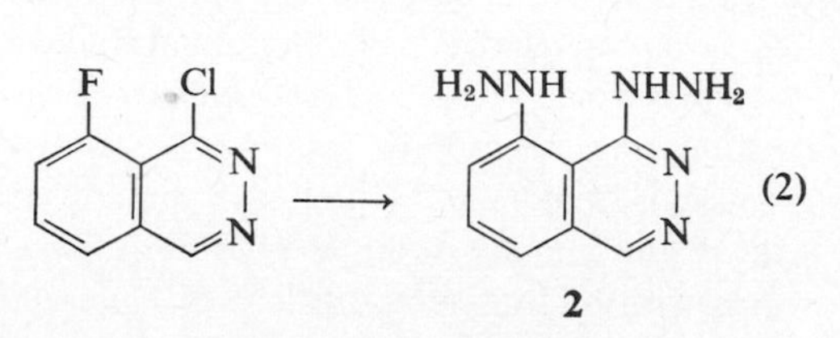

Several 4-substituted 1-hydrazinophthalazines have been obtained in an analogous way from the corresponding 4-substituted 1-chlorophthalazines. 1-Chloro-4-alkylphthalazines react with hydrazine hydrate in ethanol to give 1-hydrazino-4-alkylphthalazines (3–5). α-(1-Chloro-4-phthalazinyl)acetonitrile reacts with hydrazine hydrate in 50% dioxane to give α-(1-hydrazino-4-phthalazinyl)acetonitrile (16, 17). 1-Chloro-4-aralkylphthalazines, on heating with hydrazine hydrate in ethanol or methanol, give 1-hydrazino-4-aralkylphthalazines (3–5, 18–20). 1-Hydrazino-4-arylphthalazines have been obtained in similar fashion (3–5, 21) (Eq. 3). 1,4-Dichlorophthalazine reacts with hydrazine hydrate in ethanol to give 1-hydrazino-4-chlorophthalazine (5, 22, 23).

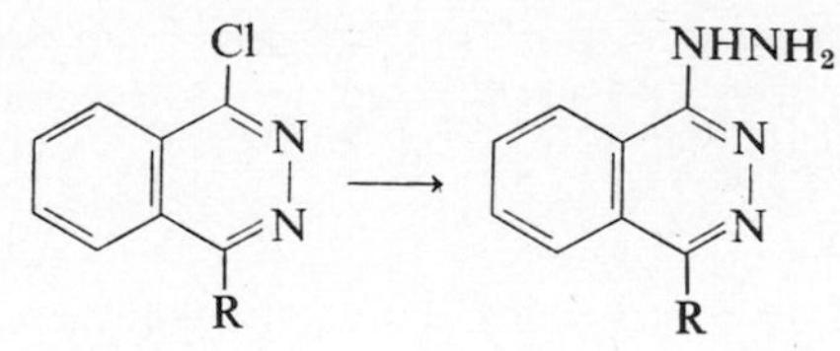

(3)

R = Cl, alkyl, aryl, aralkyl

4-Methyl-1-(2*H*)phthalazinethione, on refluxing in ethanol with hydrazine hydrate for 45 min, gives 1-hydrazino-4-methylphthalazine (**3**) in 83% yield (11). Similarly, 1-methylthio-4-(4′-pyridylmethyl)phthalazine yields the corresponding 1-hydrazinophthalazine (**4**) (19) (Eq. 4). 1-Amino-4-methyl- (21) and 1-amino-4-(4′-pyridylmethyl)phthalazine (19) are converted by hydrazine hydrate in refluxing ethanol to **3** and **4**, respectively (Eq. 4). A similar nucleophilic substitution of the alkylamino group in 1-isobutylamino-4-(*p*-methoxyphenyl)phthalazine by hydrazine hydrate in refluxing propanol gives 81% of 1-hydrazino-4-(*p*-methoxyphenyl)phthalazine (21) (Eq. 4).

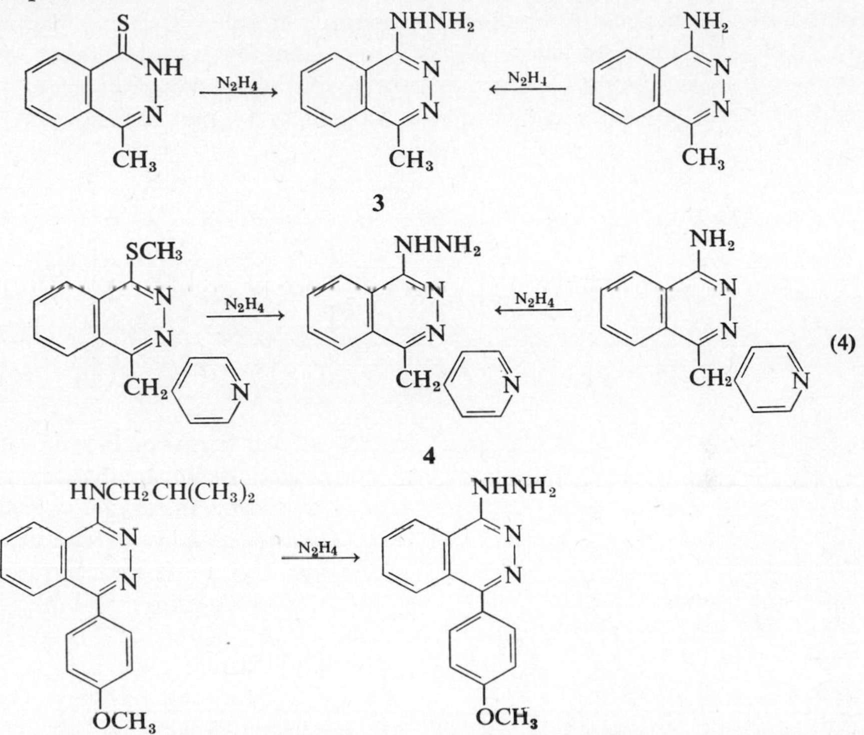

1-Hydrazino-4-(4′-pyridylmethyl)-7-acetamidophthalazine (18, 19, 24) and 1-hydrazino-4-(3′,4′-dimethoxybenzyl)-6,7-dimethoxyphthalazine (20) are obtained by refluxing the corresponding 1-chlorophthalazines with hydrazine hydrate in alcohol.

2. *Ring Closure of* 2-*Substituted Benzonitriles*

2-(Diacetoxymethyl)benzonitrile (**5**) condenses with hydrazine hydrate in refluxing ethanol to give 1-hydrazinophthalazine (**1**) in one step (25) (Eq. 5).

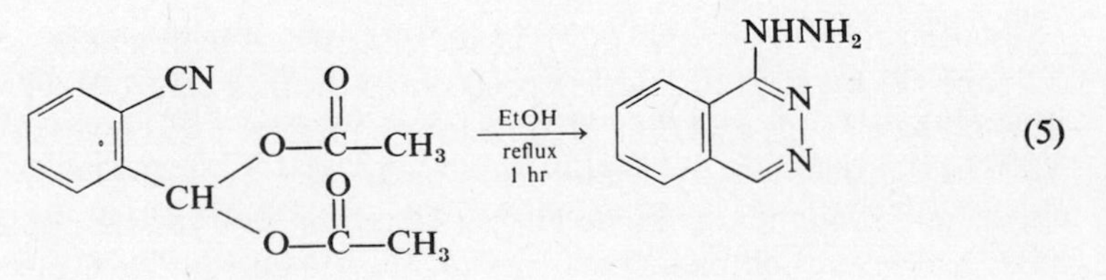

Elvidge and Jones (26) have reported an interesting cyclization of 2-cyano-*cis*-cinnamamide with 3 moles of hydrazine to give α-(1-hydrazino-4-phthalazinyl)acetic acid hydrazide (**7**). The reaction occurs with elimination of 2 moles of ammonia and loss of 2 hydrogen atoms. Elvidge and Jones (26) suggest that β-addition of hydrazine to the double bond takes place initially, and the resulting benzylhydrazine dehydrogenates to the hydrazone (**6**),

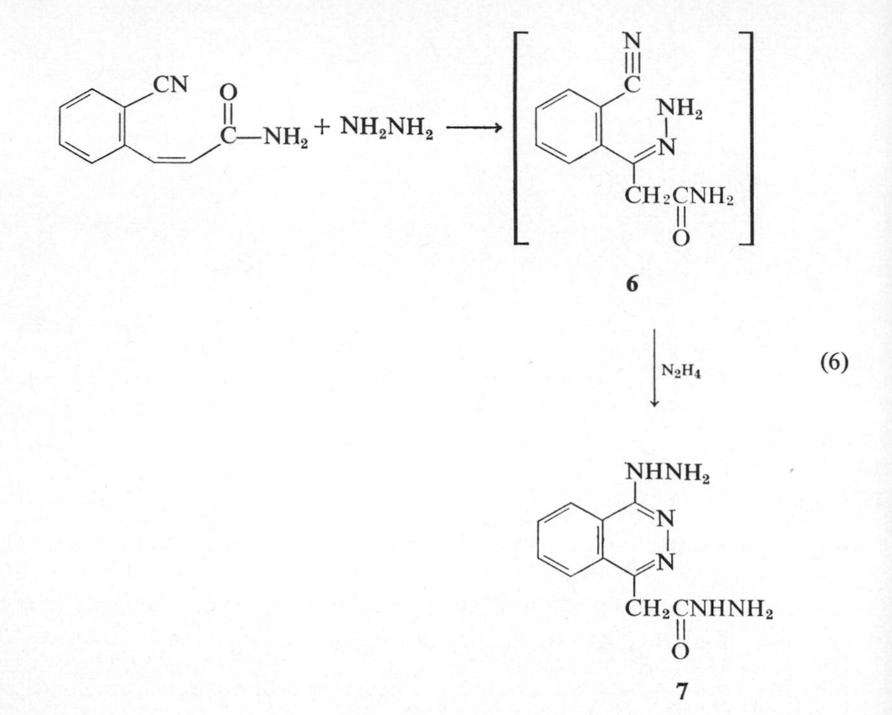

hydrazine acting as hydrogen acceptor. Cyclization then occurs, followed by the obvious reactions at the functional groups to give **7** (Eq. 6).

B. Preparation of 4-Hydrazino-1-(2*H*)phthalazinones

Koehler and associates (27) have obtained 4-hydrazino-1-(2*H*)phthalazinone (**8**) by several different procedures. Both 4-chloro- and 4-amino-1-(2*H*)phthalazinone give **8** in 90% yield when refluxed with hydrazine hydrate for 5 hr. Koehler and associates also prepared **8** by condensing 3-thiophthalimide (**9**), 3-iminophthalimide (**10**), *N*-alkyl- or *N*-aryl-3-iminophthalimides (**11**), and several phthalimide-azines (**12**) with hydrazine hydrate in 80–85%

yields (Eq. 7). Reaction of 1,4-dichlorophthalazine with hydrazine hydrate gives **8** as a by product (28).

A small number of 2-alkyl-4-hydrazino-1-(2*H*)phthalazinones have been reported in the literature (1). 2-Phenyl-4-hydrazino-1-(2*H*)phthalazinone has been prepared in 81% yield by refluxing 4-chloro-2-phenyl-1-(2*H*)-phthalazinone with hydrazine hydrate in ethylene glycol for 3 hr (29). By a

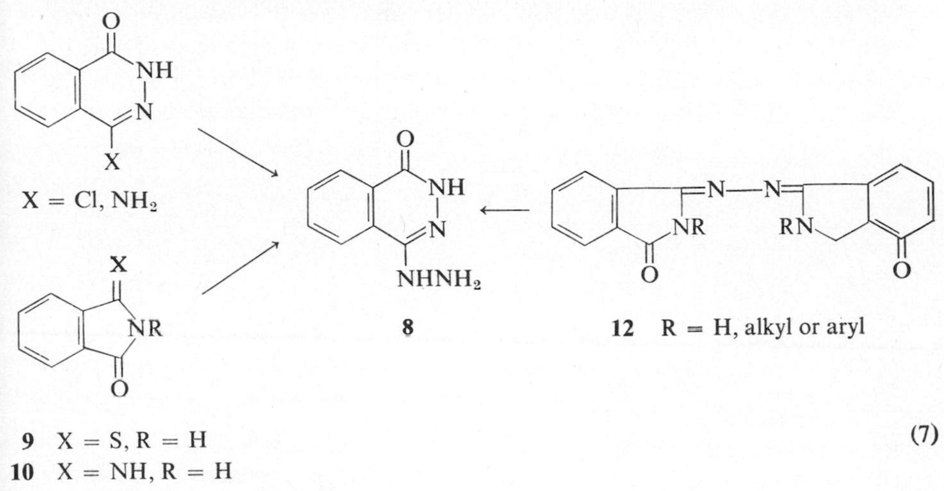

9 X = S, R = H (7)
10 X = NH, R = H
11 X = NH, R = alkyl or aryl

similar procedure 2-(2′-pyridyl)-4-hydrazino-1-(2*H*)phthalazinone has been obtained in 41% yield (30) by refluxing 2-(2′-pyridyl)-4-chloro-1-(2*H*)-phthalazinone with 80% hydrazine hydrate for 3 hr. The nucleophilic substitution of various groups in the 4-position of 1-(2*H*)phthalazinones by hydrazine generally requires stringent conditions, the reaction being carried out in absence of solvent or in high-boiling alcohols. Such conditions may cause additional reactions, as in the case of the reaction of 2-(2,4-diaminophenyl)-4-methoxy-1-(2*H*)phthalazinone with hydrazine hydrate in hot ethylene or propylene glycol. The product from this reaction is 2-(2,4-diaminophenyl)-4-hydroxy-1-(2*H*)phthalazinone, rather than the expected 4-hydrazinophthalazinone (31). Reduction of the nitro groups in 2-(2,4-dinitrophenyl)-4-methoxy-1-(2*H*)phthalazinone by hydrazine also has been reported (31).

C. Preparation of Alkyl- and Arylhydrazinophthalazines

1-Chlorophthalazine reacts with methylhydrazine when refluxed in methanol for 1 hr to give 1-methyl-1-(1′-phthalazinyl)hydrazine (**13**) in 70% yield (32). Similarly, methylhydrazine (3–5) and benzylhydrazine (5) react with

1-chloro-4-methylphthalazine to give 1-methyl- and 1-benzyl-(1-4′-methyl-1′-phthalazinyl)hydrazines (**13a** and **13b**), respectively (Eq. 8). Alkylhydrazines react at the secondary nitrogen atom of hydrazine because the basicity of this atom is greater than that of the primary nitrogen atom. The reaction of *p*-toluenesulfonylhydrazide with 1-chlorophthalazine (33) and 1-chloro-4-phenylphthalazine (34) gives 1,2-disubstituted hydrazino derivates (**14**) as expected (Eq. 8). 1,2-Dimethylhydrazine and *N*-aminomorpholine react with 1-chloro-4-methylphthalazine to give the expected products, **15** and **16**, respectively (5), while *N*-aminomorpholine and *N*-aminopiperidine react with 1-chlorophthalazine in the same fashion to give analogous products (35).

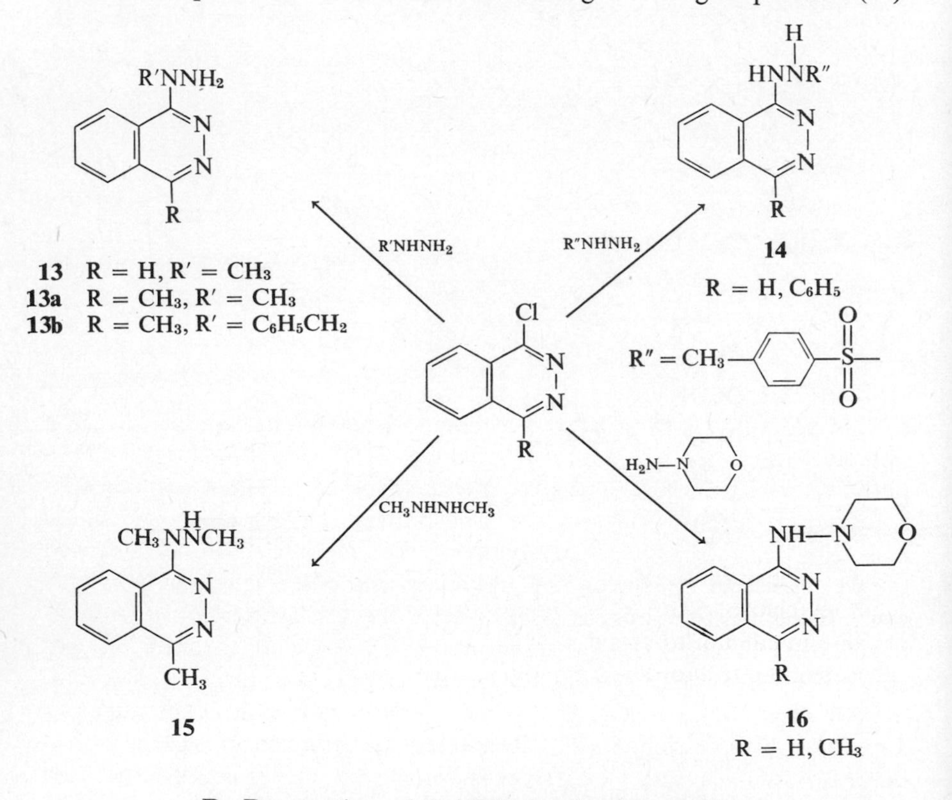

D. Preparation of 1,4-Dihydrazinophthalazines

1. *Nucleophilic Substitutions*

Druey and Ringier (5) first obtained 1,4-dihydrazinophthalazine (**17**) by refluxing 1,4-dichlorophthalazine with hydrazine hydrate for 2 hr in ethanol. Similarly, **17** has been prepared in 64% yield using anhydrous hydrazine in methanol (36). Druey (22, 23) has also shown that 1-chloro-4-hydrazinophthalazine will react with hydrazine hydrate at 150° to give **17**. 1-Chloro-4-alkoxyphthalazines and 1,4-diphenoxyphthalazine also react with hydrazine hydrate to give **17** (22, 23). Similarly, when 1,4-dichlorophthalazine is treated with equimolar sodium methoxide and the intermediate 1-chloro-4-methoxyphthalazine (without isolation) is allowed to react with hydrazine hydrate, **17** is obtained in 60–80% yields (37). 1-Chloro-4-piperidinophthalazine (21) reacts with hydrazine hydrate in refluxing *n*-propanol to give **17** (Eq. 9). The reaction of 4-hydrazino-1-(2*H*)phthalazinethione (11) and 4-mercapto-1-(2*H*)phthalazinethione (10, 11, 38) with hydrazine hydrate in refluxing ethanol provides **17** in 65–85% yields (Eq. 9).

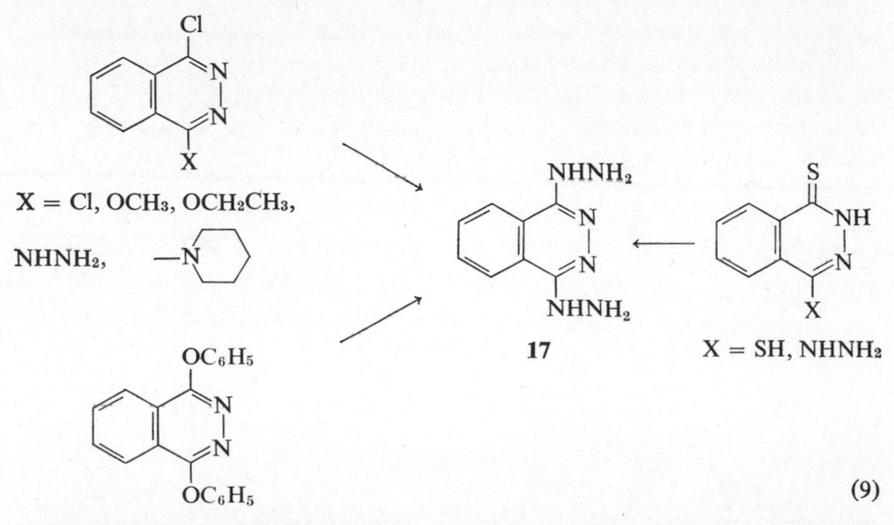

(9)

1,6,7-Trichloro-4-methoxyphthalazine (28, 39) and 1-chloro(methoxy)-4-methoxy(chloro)-6-trifluoromethylphthalazine (40) react with hydrazine hydrate in ethanol to give 6,7-dichloro-1,4-dihydrazinophthalazine and 1,4-dihydrazino-6-trifluoromethylphthalazine, respectively.

2. *Condensation*

The condensation of phthalonitrile with hydrazine hydrate in refluxing carbon disulfide (41) or in dioxane-acetic acid (42, 43) on a steam bath for 3 hr gives **17** in one step in 58–89% yields (Eq. 10). The mechanism of this reaction has not been investigated but it may be similar to the formation of 5,8-dihydrazinopyrido[2,3-*d*]pyridazine from pyridine-2,3-dinitrile, a reaction which has been studied by Paul and Rodda (44). These authors have shown that 5,8-diaminopyrido[2,3-*d*]pyridazine is an isolable intermediate in this reaction and that it further reacts with hydrazine to give 5,8-dihydrazinopyrido[2,3-*d*]pyridazine. The condensation of phthalonitrile with hydrazine in refluxing methanol has been reported to give 1,4-diaminophthalazine (see Section 2H-I-D).

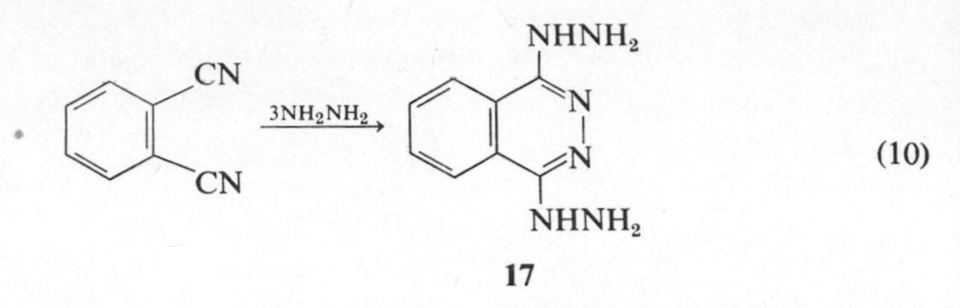

(10)

Zerweck and Kunze (45) have obtained **17** by the condensation of 1-amino-3-iminoisoindolenine nitrate with hydrazine hydrate by heating at 95° for 4 hr. Similarly, Kunze (46) has condensed 1,1-bis(alkylthio)-3-aminoisoindolenine with hydrazine hydrate in dioxane-acetic acid to prepare **17** in 79–87% yields (Eq. 11).

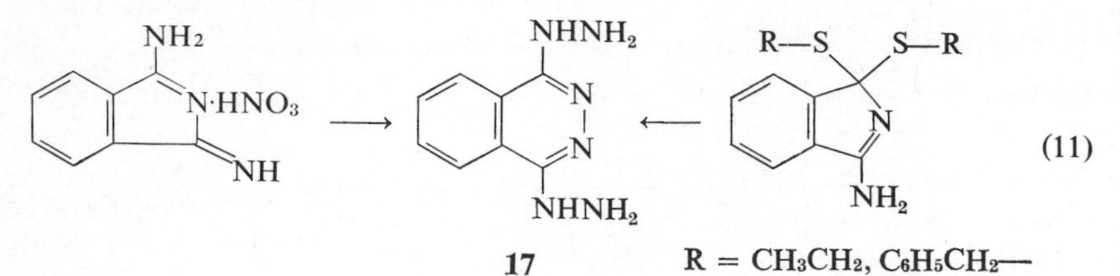

(11)

II. Properties

A. Physical Properties

Both 1-hydrazinophthalazine (**1**) and 1,4-dihydrazinophthalazine (**17**) are unstable (2), although they form stable salts with strong acids. 1-Hydrazinophthalazine crystallizes as yellow needles from methanol, mp 172–173° (3–5) [172° (11), 165–172° (6), 173–175° (8)]. It forms a monohydrochloride, colorless needles having mp 273° (dec.) (3–5, 9, 10) [271–272° (dec.) (6), 265° (11)].

1,4-Dihydrazinophthalazine (**17**) forms orange needles from water, mp 180° (22, 23) [180° (dec.) (5), 182° (36), 189° (dec.) (21), 189–191° (41), 190° (dec.) (43), 191–193° (dec.) (42)]. It forms a dihydrochloride, colorless needles, mp 257° (dec.) (10, 38) [255° (dec.) (22, 23, 36)], and a sulfate, yellow crystals, mp 263° (dec.) (11) [233° (dec.) (5), 238–239° (37)]. It also forms ditheobromine and ditheophylline salts having mp 327–334° (dec.) and 244–247° (dec.), respectively (47. 48). Similarly, 1-hydrazinophthalazine forms a monotheobromine salt (48). The UV absorption spectra of 1-hydrazinophthalazine hydrochloride in water shows λ_{max} at 240 (ε, 11, 100), 260 (ε, 10, 650), and 303 mμ (ε, 5, 150), while 1,4-dihydrazinophthalazine sulfate in water shows λ_{max} at 321 mμ (ε, 5, 910) (2).

B. Chemical Reactions

1. *Aldehydes and Ketones*

Hydrazinophthalazines react smoothly with aldehydes and ketones in alcohol solvents to form hydrazones. Druey and associates (1, 3, 4) have condensed 1-hydrazinophthalazine (**1**) with aldehydes and ketones. Sycheva and Shchukina (49) have prepared hydrazones of **1** with benzaldehyde derivatives, while Biniecki and co-workers (14, 50) have used benzaldehyde and acetophenone derivatives. Menziani (51) has prepared several sugar derivatives of **1**.

Koehler and colleagues (27) have formed hydrazones of 4-hydrazino-1-(2*H*)phthalazinone (**8**) with aldehydes, ketones, phthalimide, and *N*-substituted phthalimides. Similarly, 1,4-dihydrazinophthalazine (**17**) gives the corresponding dihydrazones with acetone (1), pyruvic acid (1), acetophenone (42), benzaldehyde (42), and so on. Recently, Prescott and associates (52) have prepared several dihydrazones of **17** with commercially available aliphatic and aromatic aldehydes. Ruggieri (53) has prepared sulfomethylhydrazino derivatives of **1** and **17** by treating them with formaldehyde in the presence of sodium hydrogen sulfite.

2. *Organic Acid Derivatives*

1,4-Dihydrazinophthalazine (**17**), on acetylation with acetic anhydride at 90° and 130°, gives diacetyl and triacetyl derivatives, respectively (42). Biniecki and co-workers (14, 54) have obtained *N*-carbethoxy derivatives of 1-hydrazinophthalazine (**1**) and 1,4-dihydrazinophthalazine (**17**) by reacting **1** and **17** with ethyl chlorocarbonate. Ruggieri (55) claims to have diacylated **17** with 3-phenylpropionyl chloride. On the other hand, Druey and Ringier (5) have shown conclusively that the reaction of 1-hydrazinophthalazine with acids, acid chlorides, anhydrides, esters, and so on, gives *s*-triazolo[3,4-*a*]-phthalazines (**18**) in high yield (Eq. 12).

Reynolds and co-workers (43) have refluxed 1,4-dihydrazinophthalazine with formic acid and obtained 6-(2-formylhydrazino)-*s*-triazolo[3,4-*a*]-phthalazine (**19**). In this case, cyclization of only one of the formylhydrazino substituents occurred (Eq. 13). On the other hand, Potts and Lovelette (32) allowed **17** to react with triethyl orthoformate by refluxing the mixture for

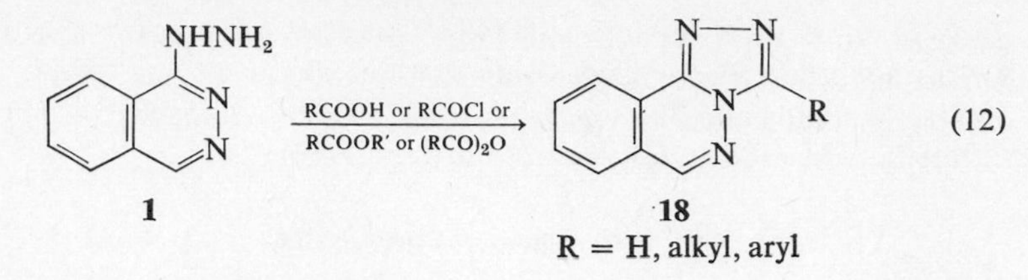

17 hr to give bis-*s*-triazolo[3,4-*a*:4,3-*c*]phthalazine (**20**) in 80% yield. Here, cyclization of both substituents occurred (Eq. 13). Similarly, reaction of **17** with cyanogen bromide gives 3,6-diamino-bis-*s*-triazolo[3,4-*a*:4,3-*c*]phthalazine (**21**). Reaction of **17** with triethyl orthoacetate yields ethyl acetate (3-methyl-*s*-triazolo[3,4-*a*]phthalazin-6-yl)hydrazone (**22**) where only one of the substituent groups has cyclized. The authors (32) postulate that the failure of one of the substituent groups in **22** to cyclize is due to steric effects (see Eqs. 13 and 18). Potts and Lovelette (32) have carried out excellent work on the synthesis of the *s*-triazolo[3,4-*a*]phthalazine ring system and have studied several of its reactions, including oxidation, reduction, hydrolysis, alkylation, and electrophilic and nucleophilic substitutions.

The same authors (32) have studied the reaction of 1-hydrazinophthalazine with dichloroacetic acid at room temperature and by heating on a steam bath

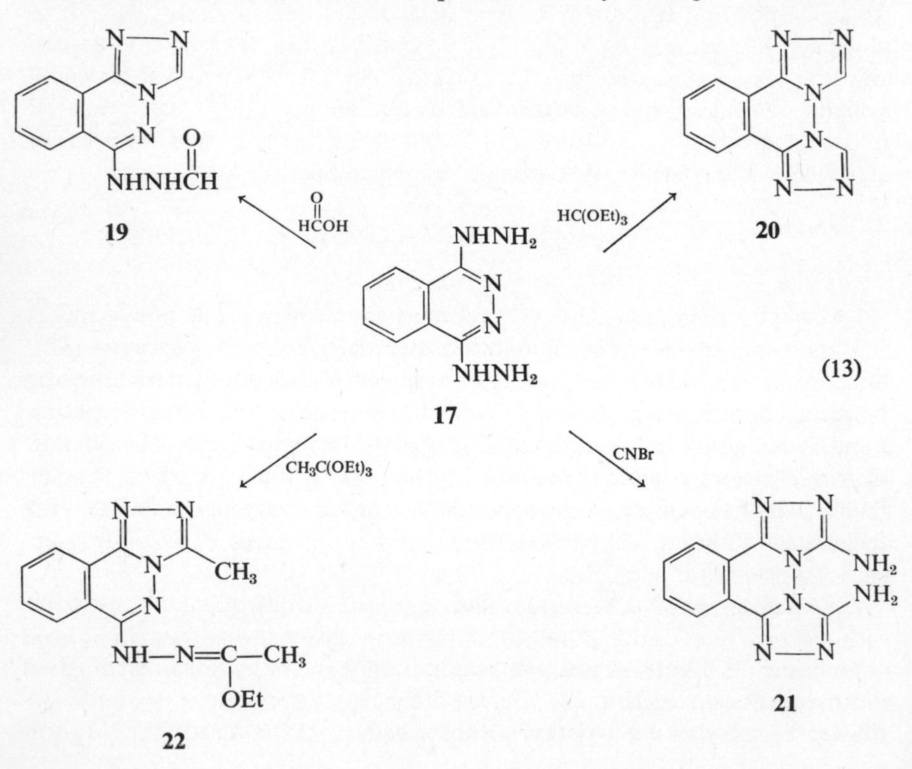

for 1 hr. In the former case, acylation takes place to give 1-dichloroacetamido-2-(1′-phthalazinyl)hydrazine (**23**), whereas in the latter case, 3-dichloromethyl-*s*-triazolo[3,4-*a*]phthalazine (**24**) is isolated (Eq. 14).

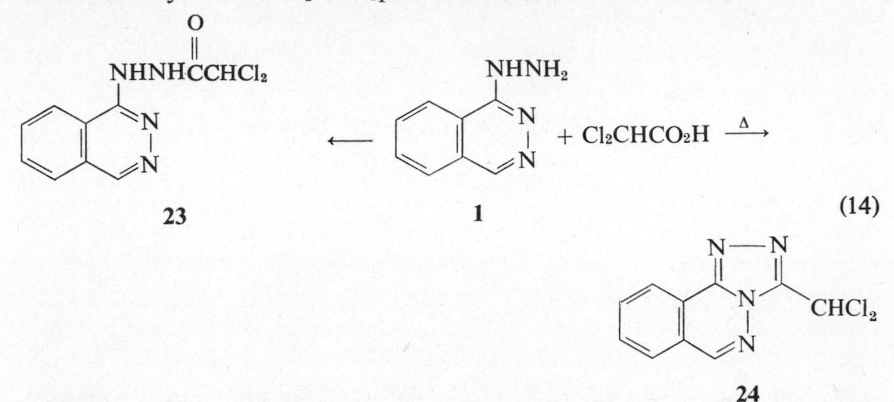

Kunze (56) has allowed 1-amino-3-iminophthalimide to react with formic acid and hydrazine with or without dimethylformamide by heating at 90°. In both cases he isolated 6-hydrazino-*s*-triazolo[3,4-*a*]phthalazine (**25**) (Eq. 15).

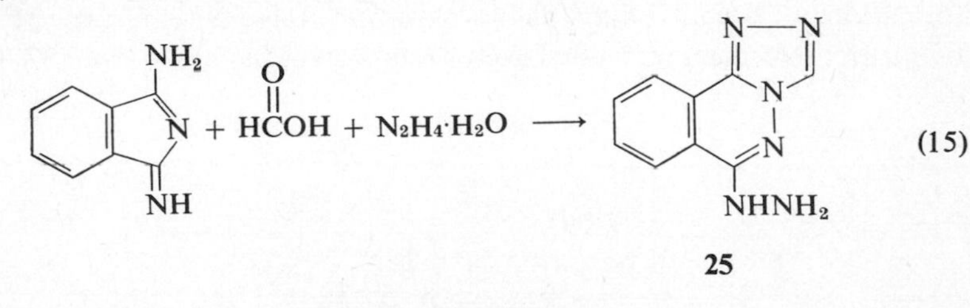

3. *Nitrous Acid*

1-Hydrazinophthalazine (5) and 4-hydrazino-1-(2*H*)phthalazinone (27), on treatment with nitrous acid, give tetrazolo[4,5-*a*]phthalazine (**26**) and 6-hydroxytetrazolo[4,5-*a*]phthalazine (**27**), respectively (Eq. 16). Similarly, 1,4-dihydrazinophthalazine (43) furnished 6-azidotetrazolo[4,5-a]phthalazine (**28**). Here again, cyclization of only one substituent occurred. The reaction of 1,4-dichlorophthalazine with sodium azide also leads to the formation of **28** (Eq. 16).

4. *Diacetonitrile and Ethyl Acetoacetate*

Ruggiere (57) carried out the reaction of 1-hydrazinophthalazine with diacetonitrile in dry ether and alkylated the resulting product to give 1-(3′-methyl-5′-dialkylamino-1-pyrazolye)phthalazine (**29**). Similarly, 1,4-di-

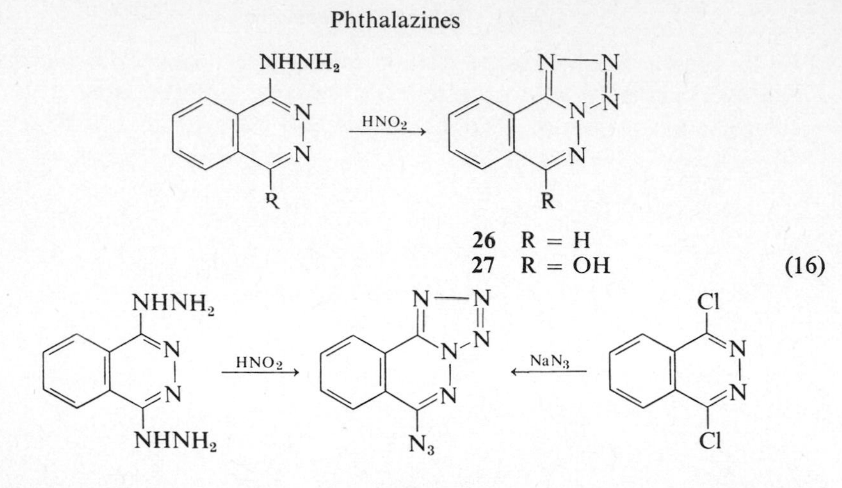

26 R = H
27 R = OH (16)

28

hydrazinophthalazine gives 1,4-bis(3′-methyl-5′-dialkylamino-1-pyrazolyl)-phthalazine (**30**) (Eq. 17). In the same way, reaction of 1,4-dihydrazinophthalazine with ethyl acetoacetate gives 1,4-bis(3-methyl-4*H*-5-pyrazolone-1-yl)phthalazine (**31**) (42) (the details of this reaction are not reported), but heating in propanol (58) for 1 hr at 60° and 3 hr at 95° leads to the formation of 3-methyl-6-hydrazino-*s*-triazolo[3,4-*a*]phthalazine (**32**), which on further treatment with ethyl acetoacetate gives the hydrazone (**33**) (Eq.18).

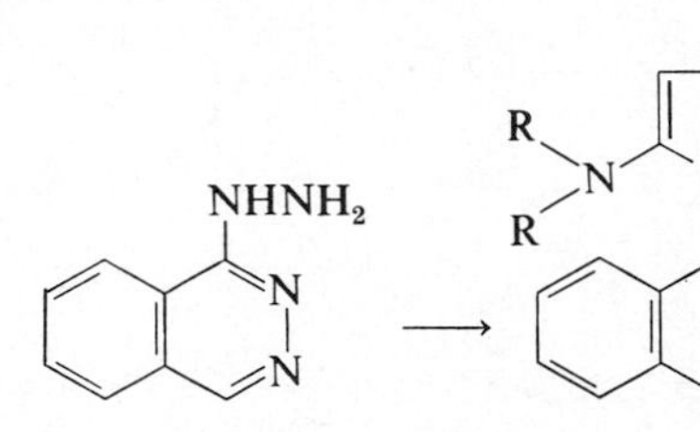

29 R = H, CH_3, CH_3CH_2

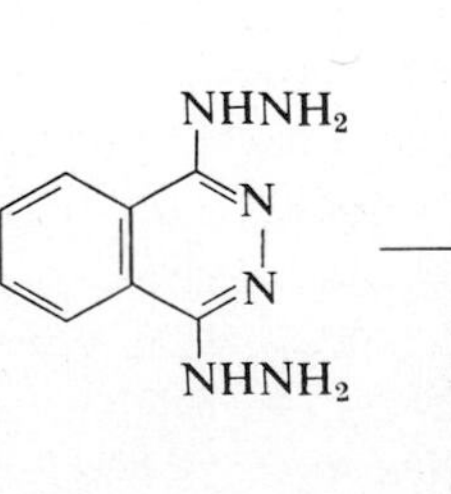

(17)

30 R = H, CH_3, CH_3CH_2

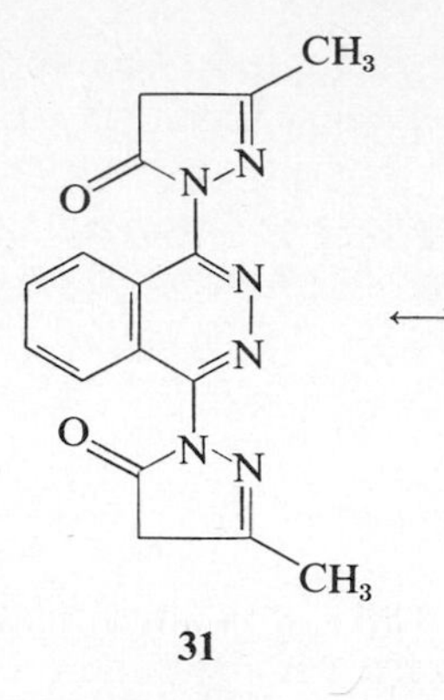

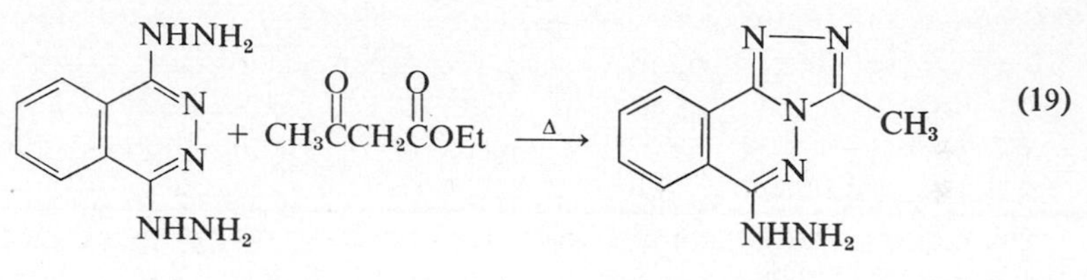

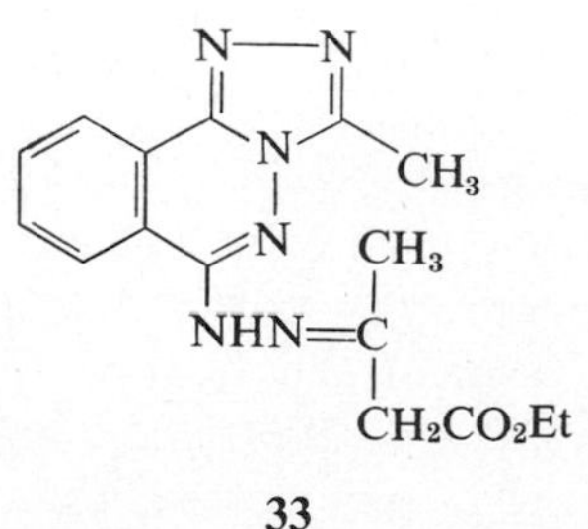

5. *Miscellaneous*

Oxidation of 1-hydrazinophthalazine with oxygen in ethanolic alkali or with copper sulfate at *p*H 8 furnishes phthalazine. Similarly, 1-*N*′-(*p*-toluenesulfonyl)hydrazinophthalazine, on heating with aqueous alkali, gives phthalazine (see Section 2A-I-F-3).

The reaction of 1-hydrazinophthalazine with potassium cyanate gives 4-(1′-phthalazinyl)semicarbazone (5), while 1-methyl-4-hydrazinophthalazine and allyl isothiocyanate yield 1-allyl-4-(1′-phthalazinyl)thiosemicarbazide (5).

Treatment of 1-hydrazinophthalazine with carbon disulfide, cyanogen bromide, and urea yields 3-thio-, 3-amino-, and 3-hydroxy-*s*-triazolo-[3,4-*a*]phthalazine, respectively (32).

The pyruvic acid hydrazone of 1-hydrazinophthalazine, on heating above its melting point, gives 3-methyl-4-keto-*as*-triazino[3,4-*a*]phthalazine, mp 250–252° (5).

6. *Metal Complexes*

1-Hydrazinophthalazine (**1**) gives a blue complex with ferrous ion (59). The reaction of this complex with oxygen has been studied and a detailed reaction mechanism proposed (60). The reaction of **1** with oxygen in presence of Fe^{2+} gives a dimeric product with a tetrazine structure which forms a deep red coordination compound with Fe^{2+} (61). The structure and magnetic properties of the Co^{2+} and Ni^{2+} complexes of 1,4-dihydrazinophthalazine (**17**) has been studied recently (62).

Metallic thiocyanate complexes of 1,4-dihydrazinophthalazine (**17**) of the type [metal **17** $(SCN)_2 \cdot H_2O$] have been reported. The various metals used include copper, cadmium, cobalt, nickel, palladium, zinc, mercury, and lead. Their preparation, color, and analytical applications are discussed (63). The metal sulfate complexes of **17** of the type [metal **17** SO_4] are prepared with Cu^{2+}, Cd^{2+}, Zn^{2+}, Fe^{2+}, Co^{2+}, Ni^{2+} and Pd^{2+} (64). Similarly, metallic benzoate complexes of **17** of the type [metal **17** $(C_6H_5CO_2)_2] \cdot nH_2O$ with divalent metals are reported (65).

1-Hydrazinophthalazine (**1**) has been used as an organic reagent for the determination of cobalt (66) and **1** and **17** both are used for the determination of iron and vanadium (66, 67).

Several qualitative tests for **1** have been developed. For example, with sodium nitroprusside it gives a red color, and with cinnamaldehyde and *p*-nitrobenzaldehyde it gives characteristic colored crystals (68).

Since both **1** and **17** are used as medicinal agents, several titrimetric procedures have been developed for their determination: determination of **17** with silver nitrate (69), determination of **1** and **17** with potassium bromate in the presence of potassium bromide and hydrochloric acid, or with hydriodic acid and potassium iodide followed by back titration with sodium thiosulfate (70), potentiometric titration of **1** in dimethylformamide with sodium methoxide (71), and potentiometric titration of **17** with potassium bromate (72). The conductometric titration of **1** has also been reported (73).

1-Hydrazinophthalazine (**1**) has been oxidized with potassium iodate in the presence of $0.2N$ H_2SO_4 and the evolved nitrogen measured (74), or **1** and **17** have been oxidized by potassium permanganate in the presence of sulfuric acid and the evolved nitrogen is again measured (75). Coulometric microdetermination of **17** has been carried out by electrolytically deposited halogens (76).

On heating **1** with zinc and sulfuric acid in a Kjeldahl flask, it is converted to dihydroisoindole, thereby releasing 3 moles of ammonia, which is determined by titration (77).

Several colorometric methods have been developed where **1**, **17** and 1-hydrazino-4-methylphthalazine from highly colored complexes with ninhydrin (78) *p*-dimethylaminobenzaldehyde (79), ferric chloride (80), and ammonium molybdate (69). Characteristic coloration has been measured at different wavelengths.

Paper and thin-layer chromatographic methods for the determination of 1-carbethoxy-2-(1′-phthalazinyl)hydrazine have been developed (81), and a specific orange-yellow spot on paper chromatograms as a color test with mercurous nitrate is reported (82).

III. Tables

TABLE 2I-1. 1-Hydrazinophthalazine with Substituents in the Benzenoid Ring

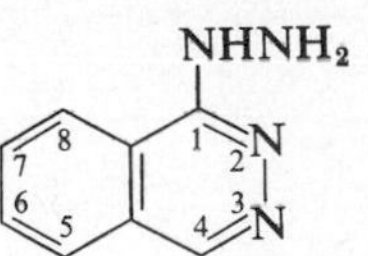

R	Preparation[a]	Yield (%)	MP	Reference
H	A	53–78	Yellow needles, 172–173°; 172°; 173–175°; 165–172°; 2HCl, 273° (dec.); 271–272° (dec.); 265°	3–11
5-Cl	B	80	HCl, 232–233° (2NHCl)	12
6-Cl	B	59	HCl, 248°	13, 14
7-Cl	B	74	HCl, 248°	13, 14
8-Cl	B	72	HCl, 197° (solidified and remelted at 270°)	12
5-F	B	—	HCl, 232–234°	15
6-F	B	—	HCl, 272–273°	15
7-F	B	—	HCl, 301–303°	15
8-$NHNH_2$	B_1	—	HCl, 238–240°	15
7-OCH_3	C	—	160° (dec.); HCl, 242.5° (dec.)	5a
7-OH	D	53	HCl, yellow crystalline powder 307° (dec.) (water)	5

[a] A, see different methods of preparation within this section. B, properly substituted 1-chlorophthalazine, reflux with hydrazine hydrate in ethanol for 30 min. B_1, same as B, 1-chloro-8-fluorophthalazine was treated with hydrazine hydrate. C, 1-chloro-7-methoxyphthalazine + hydrazine hydrate, heat 2 hr on a water bath. D, 1-hydrazino-7-methoxyphthalazine · HCl + pyridine hydrochloride, heat 2 hr at 150°.

TABLE 2I-2. 1-Hydrazinophthalazines with Substituents in the 4-Position

R	Preparation[a]	Yield (%)	MP	Reference
Cl	A	—	Blackening >200°; HCl, 150°; ~220°	5, 22, 23
CH_3	B, C	83–92	310° (dec.); HCl, 285° (dec.).	3, 4, 11, 21
CH_2CH_3	C	—	HCl, 205°	5
$CH_2CH_2CH_3$	C	—	185°; HCl, 209–211°	5
$CH_2CH(CH_3)_2$	C	—	HCl, 190–193°	5
C_6H_5	C	—	Yellow needles, 135° (benzene); HCl, 290–291°	3–5
p-$CH_3OC_6H_4$	C, C_1	81–93	Yellow needles, 174–175° (benzene); HCl, 212–213°; 155–160°	3–5, 21
p-$(CH_3)_2NC_6H_4$	C	—	203° (chlorobenzene); HCl, 168–172°	3–5
$C_6H_5CH_2$	C	—	145–146° (benzene–pet. ether); HCl, 148° (dec.)	3–5
—CH_2—(2-pyridyl)	D	—	2HCl, 230–232° (dec.)	18, 19
—CH_2—(4-pyridyl)	D, E	—	287–289° (liquifies at 88–133°, followed by resolidification at 183°); 2HCl, $2H_2O$, 279–281°	18, 19
—CH_2—(3,4-di-OCH_3-phenyl)	F	—	150–152°	20

$—CH_2CN$	G	—	197° (ethanol)	16, 17
$—CH_2C(=NOH)NH_2$	G_1	—	202°	16, 17
$—CH_2CO_2H$	G_2	—	245°	16
$—CH_2CO_2Et$	G_3	—	243°	16
$—CH_2CONHNH_2$	H	33	265°; H_2O, 280°; H_2SO_4, 220° (foaming)	1, 26
Miscellaneous				
1-Hydrazino-4-(4-pyridylmethyl)-7-($CH_3C(=O)$—NH)phthalazine [structure: $NHNH_2$, CH_2-pyridyl]	I	—	Yellow crystals, 143–145° (with foaming); 2HCl, hygroscopic, 264–267°	18, 19, 24
1-Hydrazino-4-(4-pyridylmethyl)-7-(H_2N)phthalazine [structure: $NHNH_2$, CH_2-pyridyl]	I_1	—	3HCl, 285°	18, 19, 24
1-Hydrazino-6,7-di(CH_3O)-4-(3,4-di(OCH_3)benzyl)phthalazine [structure: $NHNH_2$, CH_2]	J	34%	178–179° (methanol)	20

TABLE 2I-2. (*continued*)

[a] A, 1,4-dichlorophthalazine + $N_2H_4 \cdot H_2O$, boil in ethanol. B, 1-amino-4-methylphthalazine + $N_2H_4 \cdot H_2O$, reflux 2 hr in dioxane + 50% acetic acid, or reflux in ethanol for 20 hr, or 4-methyl-1-(2*H*)phthalazinethione + $N_2H_4 \cdot H_2O$, reflux 45 min in ethanol. C, corresponding 4-substituted 1-chlorophthalazine + $N_2H_4 \cdot H_2O$, heat 1–2 hr on a water bath in ethanol. C_1, 1-isobutylamino-4-(*p*-methoxyphenyl)phthalazine + $N_2H_4 \cdot H_2O$, boil 3 hr in propanol. D, same as C, reflux 1 hr in methanol. E, 1-methylthio-4-(pyridyl-4′-methyl)phthalazine + $N_2H_4 \cdot H_2O$, reflux 4 hr in ethanol, or reflux 1-amino-4-(pyridyl-4′-methyl)phthalazine with hydrazine hydrate in ethanol. F, same as C, reflux 10 min. G, α-(4′-chloro-1′-phthalazinyl)acetonitrile + $N_2H_4 \cdot H_2O$, boil 2 hr in dioxane. G_1, α-(4′-hydrazino-1′-phthalazinyl)acetonitrile + NH_2OH, heat 24 hr in ethanol. G_2, α-(4′-hydrazino-1′-phthalazinyl)acetonitrile + $NH_2OH \cdot HCl$ and sodium ethoxide. G_3, α-(4′-hydrazino-1′-phthalazinyl)acetonitrile + $NH_2OH \cdot HCl$ in ethanol, or treat with gaseous hydrochloric acid in ethanol. H, *o*-cyano-*cis*-cinnamamide + $N_2H_4 \cdot H_2O$, reflux 90 min in methanol. I, 1-chloro-4-(pyridyl-4′-methyl)-7-acetylaminophthalazine + $N_2H_4 \cdot H_2O$ in methanol, reflux 2 hr. I_1, acidic hydrolysis of the compound prepared by method I by refluxing in concentrated hydrochloric acid for 90 min. J, the corresponding 1-chloro-4,6,7-trisubstituted phthalazine + $N_2H_4 \cdot H_2O$, reflux in ethanol for 10 min.

TABLE 2I-3. Alkyl- or Acylmonohydrazinophthalazines

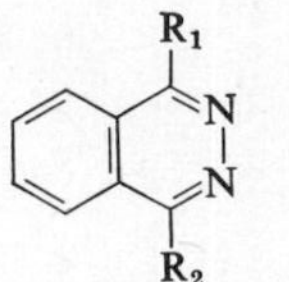

R_1	R_2	Preparation[a]	Yield (%)	MP	Reference
$-N(NH_2)CH_3$	H	A	70	Yellow needles, 119–120° (benzene)	32
$-N(NH_2)C_6H_5$	H	—	—	170°; HCl, 190° (dec.)	1
$-NHNHCH(CH_3)_2$	H	—	—	HCl, 198–200°	1
—NH—N(piperidine ring)	H	B	—	148.5–149.5°	35
—NH—N(morpholine ring, O)	H	B_1	—	175–176°	35
$-NHNHC(=O)CHCl_2$	H	B_2	95	166–168°	32
—NHNH—(phthalazinyl ring as drawn)	H	—	—	Orange colored (by-product), 272–275°; 2HCl, 235° (dec.)	5

TABLE 2I-3. (*continued*)

R_1	R_2	Preparation[a]	Yield (%)	MP	Reference
$-NHNHC(=O)NH_2$	H	C	72	Yellow crystals, 278–280°	5
$-NHNHC(=O)NHC_6H_5$	H	—	—	275°	1
$-NHNHC(=S)SCH_3$	H	—	—	200° (dec.)	1
$-NHNHC(=O)OCH_2CH_3$	H	D	52–75	HCl, 212° (dec.) (5% HCl)	14, 54, 83
$-NHNHC(=S)NHNHCH_2CH{=}CH_2$	H	—	—	HCl, 165° (dec.)	1
$-NHNH-S(=O)_2-C_6H_4-CH_3$ (p)	H	E	78	HCl, 220° (dec.) (glacial acetic acid)	33
$-NHNHCH_2SO_3Na$	H	F	61	Brown-yellow crystals, 105–108° (dec.) (hygroscopic)	53
$-NCH_2SO_3Na$ $HNCH_2SO_3Na$	H	F_1	—	Brown-yellow, 99–100° (dec.) (hygroscopic)	53

$-NHNHCH_2SO_3H$	H	F_2	—	Yellow-brown, 135°	53
$-N(NH_2)CH_3$	CH_3	A	—	145° (benzene); HCl, 236–237°	3–5
$-N(NH_2)CH_2C_6H_5$	CH_3	—	—	145–148°; HCl, 215°	5
$-N(CH_3)NHCH_3$	CH_3	—	—	79–80°; HCl, 202–203°	5
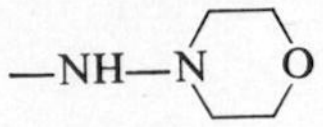 —NH—N(morpholino)	CH_3	G	55	Yellow crystals, 123°; HCl, 270°	5
$-NHNHC(=S)NHCH_2CH{=}CH_2$	CH_3	H	91	HCl, 165° (dec.)	5
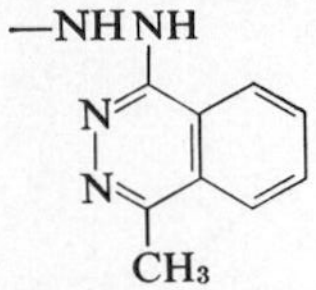 —NHNH—(4-methyl-1-phthalazinyl)	CH_3	—	—	290° (dec.); HCl, 280° (dec.)	1
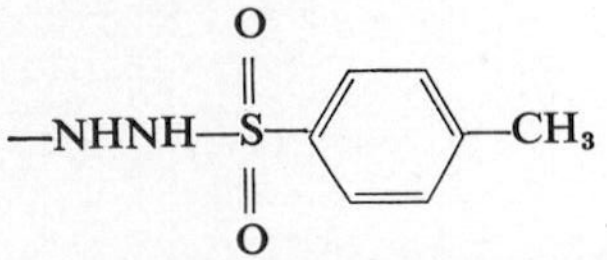 $-NHNH-SO_2-C_6H_4-CH_3$ (p)	CH_3	E_1	80	HCl, 174–175° (dec.) (methanol)	34

[a] A, corresponding 1-chlorophthalazine + methylhydrazine, reflux in methanol for 1 hr. B, 1-chlorophthalazine + *N*-aminopiperidine, heat at 110° for 16 hr. B_1, same as B, heat with *N*-aminomorpholine at 110° for 8 hr. B_2, 1-hydrazinophthalazine + dichloroacetic acid, stir at room temperature for 1 hr. C, 1-hydrazinophthalazine hydrochloride + KCNO, heat on water bath. D, 1-hydrazinophthalazine + ethyl chlorocarbonate, stir in ethanol at −10° for 2 hr then at room temperature for 2 hr, or heat to boil and keep at room temperature for 30 min. E, corresponding 1-chlorophthalazine + *p*-toluenesulfonylhydrazide, reflux 2 hr in chloroform. E_1, same as E, kept in chloroform for 4 days at room temperature. F, 1-hydrazinophthalazine + $NaHSO_3$ + aqueous formaldehyde, warm for 30 min. F_1, The by-product obtained from method F. G, 1-chloro-4-methylphthalazine + *N*-aminomorpholine, heat at 110° for 16 hr. H, 1-hydrazino-4-methylphthalazine + alkyl isothiocyanate, stir in methanol with warming.

TABLE 2I-4. 4-Hydrazino-1-(2*H*)phthalazinones

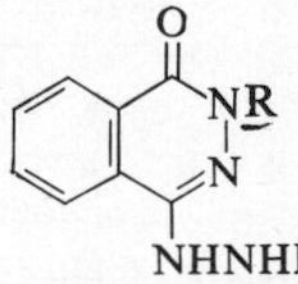

R	R′	Preparation[a]	Yield (%)	MP	Reference
H	H	A	80–90	269° (dec.); 214°, HCl, 267° (dec.)	27, 28
CH_3	H	—	—	230°; HCl, 240–242°	1
$CH_2CH{=}CH_2$	H	—	—	64–65°	1
$CH_2CH_2N(Et)_2$	H	—	—	144–145°	1
$C_6H_5CH_2$	H	—	—	200–202°; HCl, 220° (dec.)	1
C_6H_5	H	B	81	Pale yellow prismatic needles, 190°; HCl, 254–255°	29, 84
p-ClC_6H_4	H	—	—	202–204°	1
p-$NH_2C_6H_4$	H	C	—	196°; diacetamide, 298–300° (dec.)	31
C_6H_5	*p*-$CH_3C_6H_4SO_2$—	D	80	204°	84
C_6H_5	*p*-CH_3—C(=O)NH$C_6H_4SO_2$—	E	84	202–203° (aqueous ethanol)	84
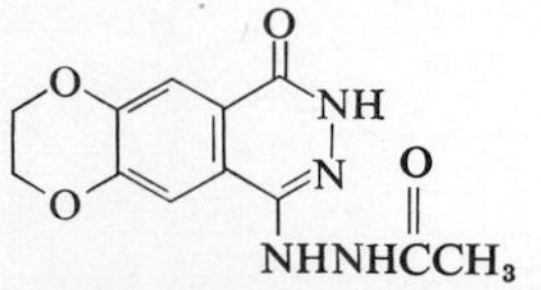 O, O, O, NH, N, O, NHNHC(=O)CH_3		F	70	Colorless crystals, 285° (dec.) (acetic anhydride)	85

[a] A, see different methods of preparation within this section. B, 4-chloro-2-phenyl-1-(2*H*)phthalazinone + hydrazine hydrate, reflux 3 hr in ethylene glycol. C, 4-chloro-2-(*p*-aminophenyl)-1-(2*H*)phthalazinone + hydrazine hydrate, heat in propylene glycol. D, 4-hydrazino-2-phenyl-1-(2*H*)phthalazinone + *p*-toluenesulfonyl chloride, reflux in ethanol for 7 hr. E, same as D, reflux with *p*-acetamidobenzenesulfonyl chloride in ethanol. F, 4-chloro-6,7-ethylenedioxy-1-(2*H*)phthalazinone + hydrazine hydrate, heat in water bath for 2 hr; product crystallized from acetic anhydride.

TABLE 2I-5. 1,4-Dihydrazinophthalazines

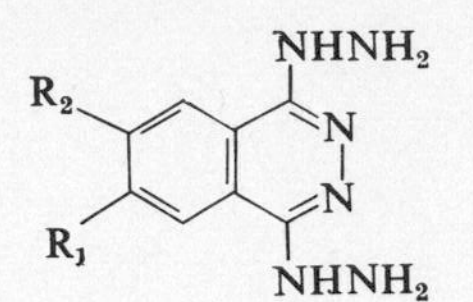

R_1	R_2	Preparation[a]	Yield (%)	MP	Reference
H	H	A	60–90	Orange needles, 180° (water); 180° (dec.); 182°; 189° (dec.); 190° (dec.); 191–193° (dec.); 189–191°, 2HCl, 257° (dec.), 255° (dec.), H_2SO_4, 263° (dec.); 233° (dec.); 238–239°.	5, 10, 11, 21–23, 36–38, 41–43
CF_3	H	B	39	155–156°; 2HCl, 212° (dec.)	40
Cl	Cl	C	58	H_2O, light yellow needles, 222–223° (dec.), HCl, 208–215° (dec.); 222° (dec.).	28, 39

[a] A, see different methods of preparation within this section. B, 1-chloro(methoxy)-4-methoxy(chloro)-6-trifluoromethylphthalazine + $N_2H_4 \cdot H_2O$, reflux for 2 hr in absolute alcohol. C, 1-methoxy-4,6,7-trichlorophthalazine or 1,4,6,7-tetrachlorophthalazine + $N_2H_4 \cdot H_2O$, reflux for 2 hr in ethanol.

TABLE 2I-6. *N*-Acyl and *N*-Sulfomethyl Derivatives of 1,4-Dihydrazinophthalazine

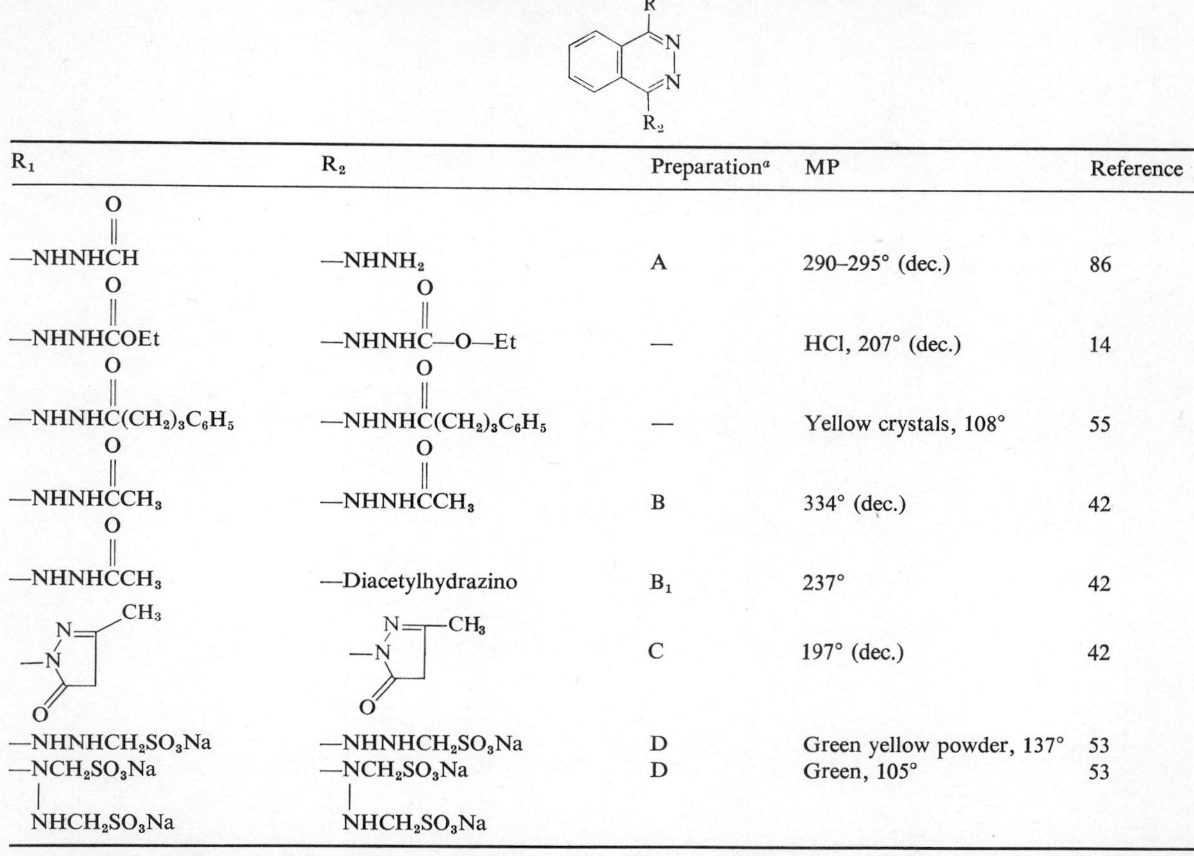

R_1	R_2	Preparation[a]	MP	Reference
—NHNHC(=O)H	$—NHNH_2$	A	290–295° (dec.)	86
—NHNHC(=O)OEt	—NHNHC(=O)—O—Et	—	HCl, 207° (dec.)	14
$—NHNHC(=O)(CH_2)_3C_6H_5$	$—NHNHC(=O)(CH_2)_3C_6H_5$	—	Yellow crystals, 108°	55
$—NHNHC(=O)CH_3$	$—NHNHC(=O)CH_3$	B	334° (dec.)	42
$—NHNHC(=O)CH_3$	—Diacetylhydrazino	B_1	237°	42
3-methyl-5-oxo-2-pyrazolin-1-yl	3-methyl-5-oxo-2-pyrazolin-1-yl	C	197° (dec.)	42
$—NHNHCH_2SO_3Na$	$—NHNHCH_2SO_3Na$	D	Green yellow powder, 137°	53
$—N(CH_2SO_3Na)NHCH_2SO_3Na$	$—N(CH_2SO_3Na)NHCH_2SO_3Na$	D	Green, 105°	53

[a] A, 1,4-dihydrazinophthalazine + 85% formic acid, heat at 40° for 3 hr and 80° for 3 hr. B, 1,4-dihydrazinophthalazine + acetic anhydride, heat at 90°. B_1, same as B, heat at 130°. C, 1,4-dihydrazinophthalazine + ethyl acetoacetate. D, 1,4-dihydrazinophthalazine + $NaHSO_3$ + aqueous formaldehyde warmed in water.

TABLE 2I-7. Hydrazones of 1-Hydrazinophthalazine

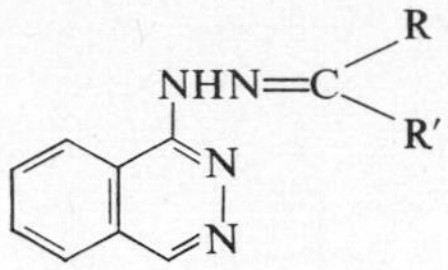

R
NHN=C
R'
N
N

Aldehyde or Ketone Used	MP	Reference
CH_2O	116–117° (aqueous methanol)	3, 5
CH_3CHO	147°	1
C_6H_5CHO	177–178°	1
o-HOC_6H_4CHO	211–212°; 211.5–212.5°	14, 50
o-$(COOH)C_6H_4CHO$	208° (aqueous carbitol)	87
p-HOC_6H_4CHO	195–196° (dec.); 197°	14, 50
p-$(CH_3)_2NC_6H_4CHO$	181–183°	49
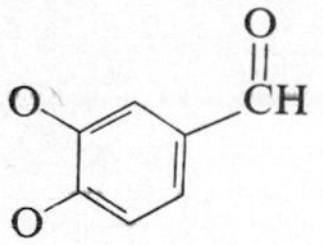 O ‖ O CH O	236–238°; 236–238° (dec.)	14, 50
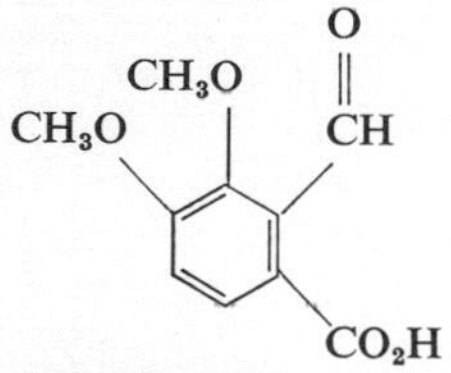 O CH_3O ‖ CH_3O CH CO_2H	236–239°; diethylamine salt, 244–245°	49
$C_6H_5CH_2CHO$	165.5°–166.5°	50
D-Glucose	150° (dec.)	51
D-Galactose	131° (dec.)	51
D-Mannose	179°	51
L-Rhamnose	138°	51
D-Xylose	186°	51
D-Ribose	138°	51
D-Arabinose	143° (dec.)	51
D-Fructose	118° (dec.)	51
D-Glucuronolactone	188° (dec.)	51
$CH_3\overset{\overset{O}{\|}}{C}COOH$	Yellow crystals, 180° (dec.)	3, 5
$CH_3\overset{\overset{O}{\|}}{C}CH_3$	114° (acetone); bp 119°/12 mm	3, 5
$CH_3\overset{\overset{O}{\|}}{C}—\overset{\overset{O}{\|}}{C}CH_3$	195–197°	5

TABLE 2I-7. (*continued*)

Aldehyde or Ketone Used	Mp	Reference
$CH_3C(=O)CH_2CH_3$	76–77°	1
$CH_3C(=O)(CH_2)_4CH_3$	Yellow crystals, 85–86° (ethyl acetate)	5
Cyclopentanone	111°–112° (cyclohexane)	5
Cyclohexanone	87–88°	1
$C_6H_5C(=O)CH_3$	143–144°; 144–145°	5, 14
$C_6H_5C(=O)CH_2CH_3$	92–93°	5
$C_6H_5C(=O)(CH_2)_4CH_3$	84°	1
o-$HOC_6H_4C(=O)CH_3$	203–205°; 203.5–205.5°	14, 50
p-$HOC_6H_4C(=O)CH_3$	229–231°	14, 50
o-$(CH_3COO)C_6H_4C(=O)CH_3$	182.5–183°	50
o-$NH_2C_6H_4C(=O)CH_3$	268–270°	14
p-$NH_2C_6H_4C(=O)CH_3$	183.5–184.5°; 183.5–185.5°	14, 50
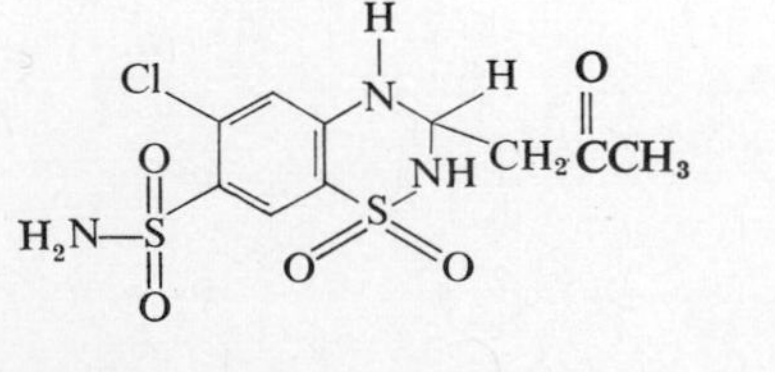	172–174°	88

TABLE 2I-7. (*continued*)

Aldehyde or Ketone Used	MP	Reference
Cl, H, H, O, N, CH_2CCH_3, H O, NCH_3, $CH_3N—S$, S, O, O, O	159–161°	88
F_3C, H, H, O, N, CH_2CCH_3, O, N—CH_3, $H_2N—S$, S, O, O, O	180–182°	88
Miscellaneous Compounds		
$C_6H_5N—N=CH_2$, N, N	120°	1
$NHN=CCH_2CN$, N, CH_3, N, CH_2CN	220° (dec.)	17
NH_2, $HNN=CCH_2C=NOH$, N, CH_3, N, $CH_2C=NOH$, NH_2	225°	17
$NHN=CHC_6H_5$, N, N, $CH_2CNHN=CHC_6H_5$, O	260°	26

TABLE 2I-7. (*continued*)

Miscellaneous Compounds	MP	Reference
CH_3 NHN=C CH_3 N N Cl	117–118°	1
OCH_3 NHN=CH– –OCH_3 N N OCH_3 CH_2– –OCH_3 OCH_3	159–160° (ethanol)	20
HO CH_3 NH—N=CH– N N N CH_2OH CH_2– –OCH_3 OCH_3	HCl, 157–159° (dec.)	20
NHN=CCH_3 CH_3O N CO_2H N CH_3O CH_2– –OCH_3 OCH_3	210–215°	20
NHN=CH– –Cl F N N	220–223°	15

TABLE 2I-8. Hydrazones of 4-Hydrazino-1-(2*H*)phthalazinones

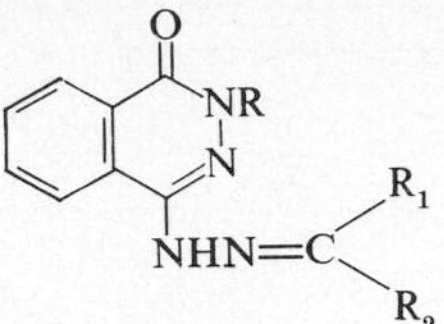

R	Aldehyde, Ketone, or Imide Used	MP	Reference
H	C_6H_5CHO	272°	27
H	$C_6H_5CH{=}CHCHO$	248.5° (dec.)	27
H	*p*-$CH_3OC_6H_4CHO$	213°	27
H	O, O, CHO	248° (dec.)	27
H	$CH_3\overset{O}{\overset{\|}{C}}CH_3$	259° (dec.)	27
H	$C_6H_5\overset{O}{\overset{\|}{C}}CH_3$	217°	27
H	O, NH, O	417° (dec.) (DMF—H_2O)	27
H	O, NCH_3, O	350° (dec.) (DMF—H_2O)	27
H	O, NEt, O	350° (dec.) (DMF—H_2O)	27
H	O, $NCH_2C_6H_5$, O	216.5° (dec.) (DMF—H_2O)	27

TABLE 2I-8. (*continued*)

R	Aldehyde, Ketone, or Imide Used	MP	Reference
C_6H_5	p-$NO_2C_6H_4CHO$	Yellow needles, 237–238° (ethanol)	29
C_6H_5	p-$(CH_3CONH)C_6H_4CHO$	Pale yellow, 246–247° (ethanol–water)	29

TABLE 2I-9. Dihydrazones of 1,4-Dihydrazinophthalazine

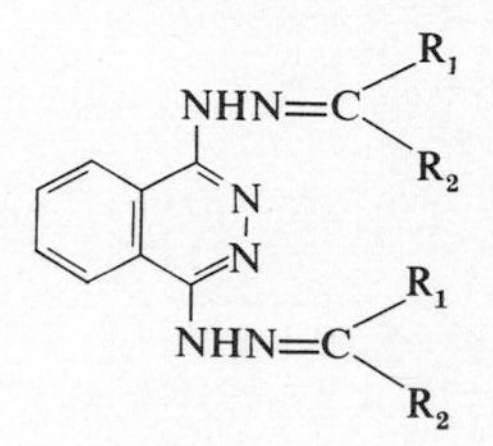

Ketone or Aldehyde Used	MP	Reference
$CH_3\overset{O}{\overset{\|}{C}}CH_3$	96–97°	1
$CH_3\overset{O}{\overset{\|}{C}}COOH$	227–228° (dec.)	1
$C_6H_5\overset{O}{\overset{\|}{C}}CH_3$	156°	42
$CH_3CH{=}C(CH_3)CHO$	186–188°	52
$CH_3CH(OCH_3)CH_2CHO$	199–202°	52
$CH_3(CH_2)_7CHO$	253°	52
$CH_3(CH_2)_{12}CHO$	247–250°	52
$CH_3(CH_2)_{16}CHO$	165–170°	52
(cyclohexane ring)—CHO	164–166°	52
(dihydropyran ring, O) with CHO and CH_3	>300°	52

TABLE 2I-9. (*continued*)

Ketone or Aldehyde Used	MP	Reference
C_6H_5CHO	197°	42
$C_6H_5CH{=}CHCHO$	185–188°	52
$C_6H_5CH_2CH_2CHO$	105°	52
$C_6H_5CH{=}C(CH_3)CHO$	238–240°	52
$C_6H_5CH{=}C((CH_2)_4CH_3)CHO$	248°	52
$C_6H_5CH{=}C((CH_2)_5CH_3)CHO$	158–160°	52
p-$(CH_3)_2NC_6H_4CH{=}CHCHO$	209°	52
o-ClC_6H_4CHO	199–200°	52
m-ClC_6H_4CHO	203–205°	52
p-ClC_6H_4CHO	285–287°, 284–287°	1, 52
p-BrC_6H_4CHO	264–266°	52
m-IC_6H_4CHO	245–248°	52
p-IC_6H_4CHO	248–250°	52
o-$CH_3C_6H_4CHO$	225–227°	52
p-$C_6H_5C_6H_4CHO$	144–147°	52
o-$NO_2C_6H_4CHO$	205–208°; 240–242°	42, 52
m-$NO_2C_6H_4CHO$	255–258°	52
p-$NO_2C_6H_4CHO$	280–285°	52
o-HOC_6H_4CHO	186–190°	52
m-HOC_6H_4CHO	225–228°	52
o-$CH_3OC_6H_4CHO$	242°	52
m-$CH_3OC_6H_4CHO$	239°	52
o-$EtOC_6H_4CHO$	195–197°	52
m-$(C_6H_5CH_2O)C_6H_4CHO$	100–103°	52
p-$(CH_3CONH)C_6H_4CHO$	234–238°	52
p-$(CH_3)_2NC_6H_4CHO$	202°; 226–269° (dec.)	1, 52
p-$(Et)_2NC_6H_4CHO$	200°	52
p-$(ClCH_2CH_2)_2NC_6H_4CHO$	215–216°	52
o-$(COOH)C_6H_4CHO$	200°	52
p-$(COOH)C_6H_4CHO$	290–293°	52
2,4-$Cl_2C_6H_3CHO$	262°	52
2,6-$Cl_2C_6H_3CHO$	198–201°	52
3,4-$Cl_2C_6H_3CHO$	277–280°	52
3,4-$NO_2(Cl)C_6H_3CHO$	248–251°	52
2,6-$NO_2(Cl)C_6H_3CHO$	192–195°	52
2,5-$NO_2(HO)C_6H_3CHO$	>300°	52
2,6-$(NO_2)_2C_6H_3CHO$	175–178°	52
2,5-$HO(Cl)C_6H_3CHO$	253–255°	52
2,5-$HO(Br)C_6H_3CHO$	253°	52
2,5-$HO(NO_2)C_6H_3CHO$	268–270°	52

TABLE 2I-9. (*continued*)

Ketone or Aldehyde Used	MP	Reference
2,4-$(HO)_2C_6H_3CHO$	271–274°	52
3,4-$(HO)_2C_6H_3CHO$	243–245°	52
3,4-$HO(CH_3O)C_6H_3CHO$	208–211°	52
2,3-$(CH_3O)_2C_6H_3CHO$	172–175°	52
2,4-$(CH_3O)_2C_6H_3CHO$	245–247°	52
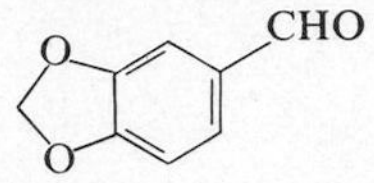 CHO; O; O	242–245°	52
3,4-$(EtO)_2C_6H_3CHO$	222–225°	52
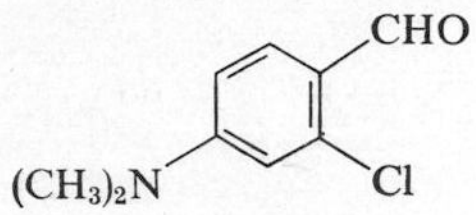 CHO; $(CH_3)_2N$; Cl	213–215°	52
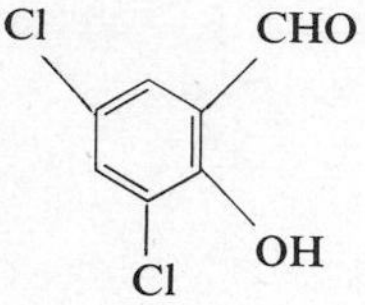 Cl; CHO; OH; Cl	>300°	52
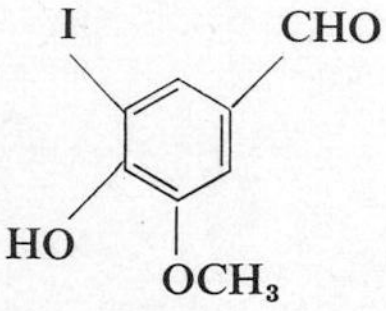 I; CHO; HO; OCH_3	240–241°	52
C_6F_5CHO	235°	52
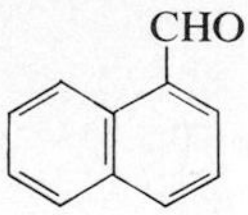 CHO	225–230°	52
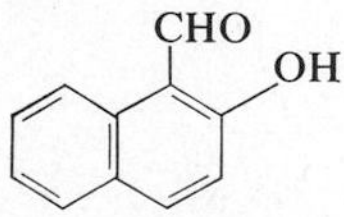 CHO; OH	268–270°	52
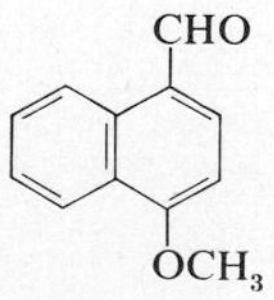 CHO; OCH_3	231–233°	52

TABLE 2I-9. (*continued*)

Ketone or Aldehyde Used	MP	Reference
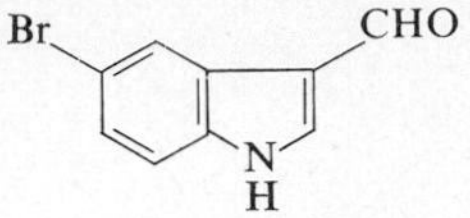 Br, CHO, N, H (indole structure)	299°	52
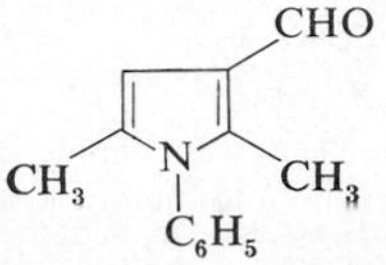 CHO, CH_3, N, CH_3, C_6H_5 (pyrrole structure)	>300°	52
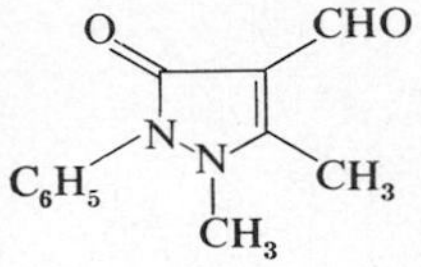 O, CHO, N, N, C_6H_5, CH_3, CH_3 (pyrazolone structure)	241–243°	52

TABLE 2I-10. 1-(3′-Methyl-5′-amino-1-pyrazolyl)phthalazine (I) and 1,4-Di(3′-methyl-5′-amino-1-pyrazolyl)phthalazine (II)

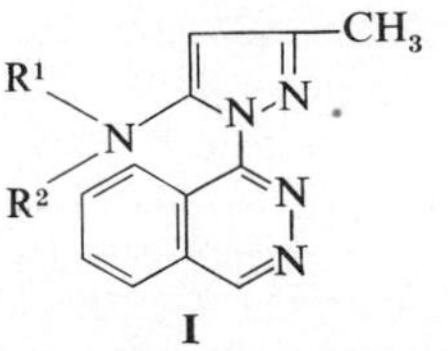

I: R^1, R^2, N, CH_3, N, N, N, N

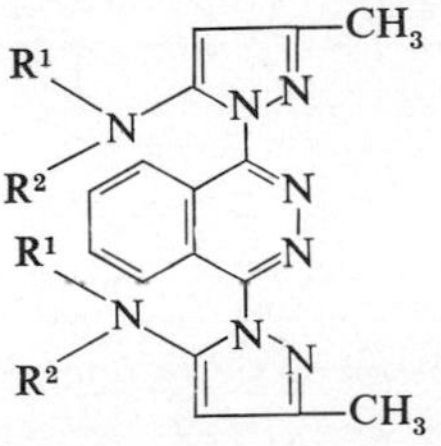

II: R^1, R^2, N, CH_3, N, N, N, N, R^1, R^2, N, N, N, CH_3

R^1	R^2	I, MP	II, MP	Reference
H	H	158–159°	187–188°	57
CH_3	CH_3	156–157°	192–193°	57
CH_2CH_3	CH_2CH_3	142–143°	178–180°	57
H	$-C(=O)CH_3$	248°	215–216°	57
H	$-C(=O)CH_2Cl$	>300°	>300°	57
H	$-C(=O)CHCl_2$	155°	196°	57
H	$-C(=O)CH_2N(CH_3)_2$	>300°	199–200°	57

TABLE 2I-10. (*continued*)

R^2	R^1	I, MP	II, MP	Reference
H	$—\overset{O}{\overset{\parallel}{C}}C_6H_5$	278–280°	274°	57
H	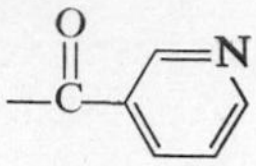 —C(=O)–(3-pyridyl)	157–158°	201–203°	57
H	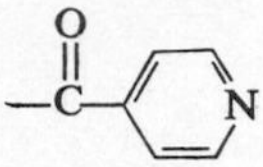 —C(=O)–(4-pyridyl)	162–163°	205–206°	57

TABLE 2I-11. Tetrazolo[4,5-*a*]phthalazines

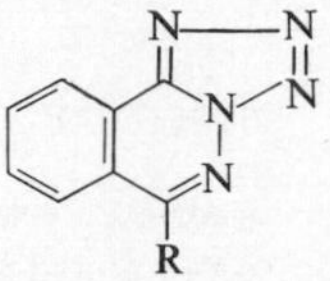

R	Preparation[a]	Yield (%)	MP	Reference
H	A	90	209–210° (50% alcohol)	5
OH	B	91	287° (dec.) (ethanol)	27
N_3	C	—	152°	43

[a] A, 1-hydrazinophthalazine in acetic acid treated with sodium nitrite in water followed by warming. B, 4-hydrazino-1-(2*H*)phthalazinone in hydrochloric acid at 0° was treated with sodium nitrite in water. C, 1,4-dihydrazinophthalazine in acidic solution treated with aqueous sodium nitrite at 10° C, or 1,4-dichlorophthalazine, reflux in alcohol with sodium azide.

TABLE 2I-12. *s*-Triazolo[3,4-*a*]phthalazines

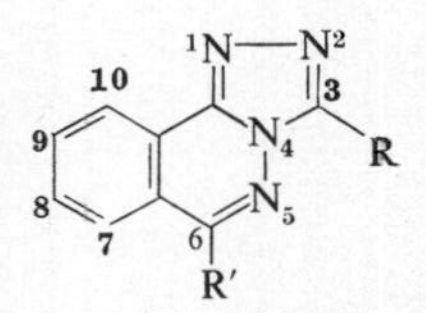

R	R′	Preparation[a]	Yield (%)	MP	Reference
H	H	A	70–82	Small needles, 190–191°; 192°; HBr, 289–290°; CH_3I, 246–248°	3, 5, 32
CH_3	H	B	82–92	170–171°; 171–172° (2-propanol); CH_3I, 284° (dec.)	5, 32
CH_2Cl	H	—	—	188–189°	5
$CHCl_2$	H	C	75	Yellow needles, 224° (methanol); 213–215°	5, 32
$CH_2(CH_2)_2CH_3$	H	D	31	101–102° (cyclohexane); HCl, 197–198°	5
$CH_2CH(CH_3)_2$	H	—	—	91–92°	5
$CH_2(CH_2)_4CH_3$	H	—	—	71–73°	5
C_6H_5	H	—	—	208–209°; 212° (50% acetic acid)	3, 5
3-pyridyl (structure)	H	D	—	215–216°	14
4-pyridyl (structure)	H	D	—	253–254°	14

TABLE 2I-12 (*continued*)

R	R′	Preparation[a]	Yield (%)	MP	Reference
Cl	H	E	70	Colorless needles, 223–224° (water); 205° (acetone)	5, 32
Br	H	F	80	206–208° (water)	32
OH	H	G	80	Yellow needles, 280° (ethanol)	5, 32
$-OC(=O)CH_3$	H	H	10	Colorless irregular prisms, 269° (water)	32
NH_2	H	I	30	Yellow irregular prisms, 291° (water)	32
$NHNH_2$	H	J	90	Yellow needles, 224–225° (methanol)	32
NHNHCHO	H	K	85	Colorless needles, 277–278°	32
NHN=CHOEt	H	L	60	Buff colored irregular prisms, 205–206° (benzene)	32
SH	H	M	85–98	Yellow needles, 305° (ethanol); yellow plates 305°	32
SCH_3	H	N	80	Pale yellow irregular prisms, 192–193°; CH_3I-; dimethylsulfate, 194–195°	32
CHO	H	O	60	242–243° (benzene)	32
H	CH_3	P	70–75	Cream plates, 185° (benzene); 184–186°	32
H	$NHNH_2$	Q	56–85	319–320°; >298°	32, 43, 56
H	NHNHCHO	Q_1	70	White crystals, 300° (dec.) (H_2O)	43
H	N_3	R	73	195°	43

H	p-$ClC_6H_4CH{=}NNH$—	—	—	275°	56
C_6H_5	CH_3	—	—	210°	5
SH	CH_3	S	38	Colorless, 300° (dec.)	5
$NHCH_2CH{=}CH_2$	CH_3	S_1	38	Yellow, 139–140° (ethyl acetate)	5
CH_3	$NHNH_2$	T	—	Red colored, 303° (dec.) (DMF); 305–307° (dec.)	58
CH_3	$—NHN{=}C(OEt)CH_3$	U	60	Tan needles, 195–196° (benzene)	32
CH_3	$—NHN{=}C(CH_3)CH_2CO_2CH_3$	V	—	198° (dec.)	58
CH_3	$—NHN{=}C(CH_3)CH_2CO_2Et$	V_1	—	Colorless, 207–208° (dec.)	58
CH_3	$—NHN{=}CHC_6H_5$	—	—	280–281°	58
Miscellaneous					
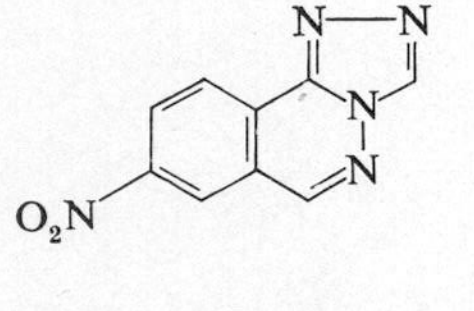		W	51	Yellow irregular prisms, 285° (acetone)	32
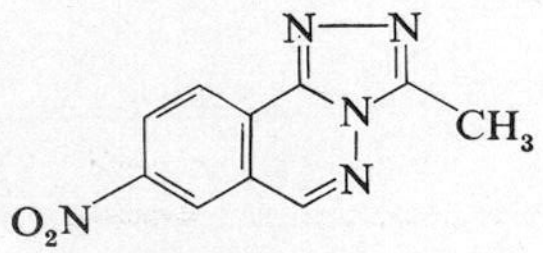		W	43	Yellow irregular prisms, 280–282° (acetone)	32

TABLE 2I-12 (*continued*)

R	R′	Preparation[a]	Yield (%)	MP	Reference
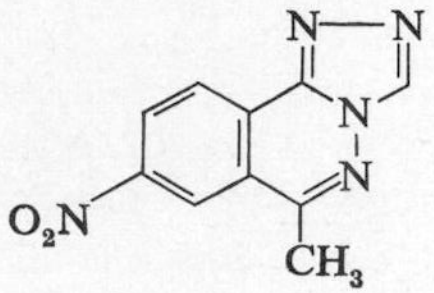		W	50	Yellow irregular prisms, 315–316° (acetone)	32
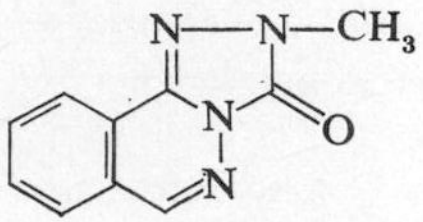		X	44	263° (methanol)	32
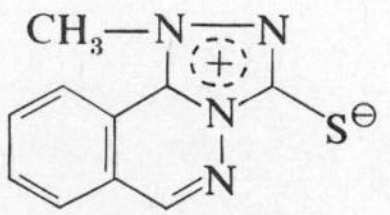		Y	85	Brown plates, 265° (H_2O); dimethyl sulfate, 183°	32
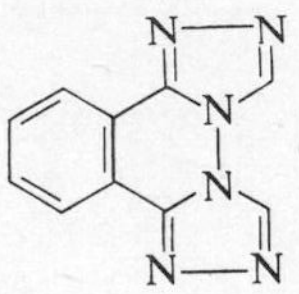		Z	80	Pale green prisms, 342° (methanol-DMF)	32

	Z_1	80	Pale green prisms, 345° (dec.)	32

[a] A, 1-hydrazinophthalazine + 85% formic acid, boil 1 hr; or 3-chloro-*s*-triazolo[3,4-*a*]phthalazine, hydrogenate with 10% Pd/C in ethanol; or *s*-triazolo[3,4-*a*]phthalazine-3-carboxaldehyde, treated with silver oxide. B, 1-hydrazinophthalazine + acetyl chloride kept at 0° for 1 hr in pyridine; or heat with ethyl acetate for 4 hr at 100°; or heat with acetic anhydride on a water bath for 30 min; or heat with methyl benzyl ketone at 150–160° for 6 hr in a pressure vessel. C, 1-hydrazinophthalazine + $CHCl_2CO_2H$, heat on a steam bath for 1 hr. D, 1-hydrazinophthalazine + corresponding acid chloride, heat with pyridine on a water bath. E, 3-hydroxy-*s*-triazolo[3,4-*a*]phthalazine + $POCl_3$ + PCl_5, heat at 190–200° for 6 hr; or 3-thio-*s*-triazolo[3,4-*a*]phthalazine treated with chlorine at 0° for 3 hr in chloroform-water. F, *s*-Triazolo[3,4-*a*]phthalazine + *N*-bromosuccinimide, reflux for 3 hr in carbon tetrachloride; or reflux the triazolophthalazine in acetic acid with bromine. G, 1-hydrazinophthalazine + urea, heat at 180° for 1 hr; or 1-hydrazinophthalazine + ethyl chlorocarbonate, heat in pyridine followed by treatment with 1*N* sodium hydroxide. H, attempted nitration of *s*-triazole[3,4-*a*]phthalazine with copper nitrate in acetic anhydride. I, 1-hydrazinophthalazine + CNBr, reflux 2 hr in methanol. J, 3-bromo-*s*-triazolo[3,4-*a*]phthalazine + 85% $N_2H_4 \cdot H_2O$, reflux 3 hr. K, 3-hydrazino-*s*-triazolo[3,4-*a*]phthalazine + HCO_2H, reflux 2 hr. L, same as K, reflux with triethyl orthoformate for 17 hr, M, 1-hydrazinophthalazine + CS_2, reflux 48 hr in chloroform, or reflux 3 hr in ethanol containing potassium hydroxide. N, 3-thio-*s*-triazolo[3,4-*a*]phthalazine + CH_3I, stir at room temperature for 3 hr in presence of sodium hydroxide. O, 3-dichloromethyl-*s*-triazolo-[3,4-*a*]phthalazine + morpholine, heat 5 hr on steam bath followed by acidic hydrolysis. P, 1-hydrazino-4-methylphthalazine, reflux 3 hr with formic acid or reflux 90 min with triethyl orthoformate. Q, 1-Amino-3-imino- or 1-ethoxy-3-iminoisoindolenine + 80% $N_2H_4 \cdot H_2O$ + HCO_2H, heat at 90° for 3 hr with or without DMF. Reflux *bis*-*s*-triazolo[3,4-*a*:4,3-*c*]phthalazine in concentrated hydrochloric acid for 17 hr. Q_1, 1,4-dihydrazinophthalazine + HCO_2H, reflux for 3 hours. R, 6-hydrazino-*s*-triazolo[3,4-*a*]phthalazine in acidic medium treated with aqueous sodium nitrite at 10° C. S, 1-allyl-4-(4′-methyl-1′-phthalazinyl)thiosemicarbazide boiled in acetic acid. S_1, second product obtained using method S. T, 1,4-dihydrazinophthalazine + ethyl acetoacetate, heat at 60° for 1 hr and at 95° for 3 hr in 1-propanol. U, 1,4-dihydrazinophthalazine + triethyl orthoacetate, reflux 48 hr. V, 6-hydrazino-3-methyl-*s*-triazolo[3,4-*a*]phthalazine + ethyl acetoacetate (excess), heat at 60° for 1 hr and 95° for 3 hr in 1-propanol. V_1, same as V, used excess methyl acetoacetate. W, nitration of the corresponding *s*-triazolo[3,4-*a*]phthalazine, stir at 0° C in H_2SO_4-HNO_3 for 45 min. X, 2-methyl-*s*-triazolo[3,4-*a*]phthalazinium iodide or 3-methylthio-2-methyl-*s*-triazolo[3,4-*a*]phthalazinium iodide oxidized in aqueous solution with potassium ferricyanide; or methylation of 3-hydroxy-*s*-triazolo[3,4-*a*]phthalazine with diazomethane at room temperature, or with methyl iodide in refluxing acetone for 48 hr. Y, 1-methyl-1-(1′-phthalazinyl)hydrazine + CS_2, reflux in water for 72 hr. Z, 1,4-dihydrazinophthalazine + triethyl orthoformate, reflux 17 hr. Z_1, 1,4-dihydrazinophthalazine + cyanogen bromide, reflux in methanol for 36 hr.

References

1. J. Druey and A. Marxer, *J. Med. Pharm. Chem.*, **1,** 1 (1959).
2. J. Druey and J. Tripod, *Med. Chem.* (*New York*), **7,** 223 (1967).
3. British Pat. 629,177 (1949); *Chem. Abstr.*, **44,** 4516 (1950).
4. M. Hartman and J. Druey, U.S. Pat. 2,484,029 (1949); *Chem. Abstr.*, **44,** 4046 (1950).
5. J. Druey and J. Ringier, *Helv. Chim. Acta*, **34,** 195 (1951).

5a. Swiss Pat. 266,287 (1950); *Chem. Abstr.*, **45,** 7605 (1951).

6. S. Biniecki and B. Gutkowska, *Acta Polon. Pharm.*, **11,** 27 (1955); *Chem. Abstr.*, **50,** 12062 (1956).
7. T. P. Sycheva et al., *Med. Prom. SSSR*, **14,** 13 (1960); *Chem. Abstr.*, **54,** 22669 (1960).
8. E. Oishi, *Yakugaku Zasshi*, **89,** 959 (1969); *Chem. Abstr.*, **72,** 3450 (1970).
9. T. Fujii and H. Sato, Japanese Pat. 5526 (1954); *Chem. Abstr.*, **49,** 15982 (1955).
10. K. Fujii and H. Sato, *Ann. Rept., G. Tanabe Co., Ltd.*, **1,** 1 (1956); *Chem. Abstr.*, **51,** 6650 (1957).
11. British Pat. 735,899 (1955); *Chem. Abstr.*, **50,** 10801 (1956).
12. S. Biniecki, M. Moll, and L. Rylski, *Ann. Pharm. Fr.*, **16,** 421 (1958); *Chem. Abstr.*, **53,** 7187 (1959).
13. S. Biniecki and L. Rylski, *Ann. Pharm. Fr.*, **16,** 21 (1958); *Chem. Abstr.*, **52,** 18446 (1958).
14. S. Biniecki et al., *Bull. Acad. Polon. Sci., Ser. Sci., Chim. Geol. Geog.*, **6,** 227 (1958); *Chem. Abstr.*, **52,** 18424 (1958).
15. B. Cavalleri and E. Bellasio, *Farm. Ed. Sci.*, **24,** 833 (1969); *Chem. Abstr.*, **71,** 101,798 (1969).
16. I. Zugravescu, M. Petrovanu, and E. Rucinschi, *An. Stiint. Univ., "A. I. Cuza", Iasi Sect.*, **17,** 189 (1961); *Chem. Abstr.*, **59,** 6399 (1963).
17. I. Zugravescu, M. Petrovanu, and E. Rucinschi, *Rev. Chim. Acad. Rep. Pop. Roum.*, **7,** 1405 (1962); *Chem. Abstr.*, **61,** 7009 (1964).
18. J. Druey and A. Marxer, U.S. Pat. 2,960,504 (1960); *Chem. Abstr.*, **55,** 11446 (1961).
19. J. Druey and A. Marxer, German Pat. 1,061,788 (1959); *Chem. Abstr.*, **55,** 13458 (1961).
20. M. V. Sigal, Jr., P. Marchini, and B. L. Poet, U.S. Pat. 3,274,185 (1966); *Chem. Abstr.*, **65,** 18600 (1966).
21. British Pat. 732,581 (1955); *Chem. Abstr.*, **50,** 8747 (1956).
22. J. Druey, U.S. Pat. 2,484,785 (1949); *Chem. Abstr.*, **44,** 2573 (1950).
23. British Pat. 654,821 (1959); *Chem. Abstr.*, **46,** 9622 (1952).
24. J. Druey and A. Marxer, Swiss Pat. 361,812 (1962); *Chem. Abstr.*, **59,** 8761 (1963).
25. British Pat. 719,183 (1954); *Chem. Abstr.*, **49,** 15982 (1955).
26. J. A. Elvidge and D. E. Jones, *J. Chem. Soc.* (*C*), **1968,** 1297.
27. W. Koehler, M. Bubner, and G. Ulbricht, *Chem. Ber.*, **100,** 1073 (1967).
28. S. Kumada, N. Watanabe, K. Yamamoto, and H. Zenno, *Yakugaku Kenkyo*, **30,** 635 (1958); *Chem. Abstr.*, **53,** 20554 (1959).
29. D. J. Drain and D. E. Seymour, *J. Chem. Soc.*, **1955,** 852.
30. T. Kubaba and R. S. Ludwiczak, *Roczniki Chem.*, **38,** 367 (1964).
31. E. Domagalina and S. Baloniak, *Roczniki Chem.*, **36,** 253 (1962).
32. K. T. Potts and C. Lovelette, *J. Org. Chem.*, **34,** 3221 (1969).
33. C. M. Atkinson and C. J. Sharpe, *J. Chem. Soc.*, **1959,** 3040.

34. E. Hayashi and E. Oishi, *Yakugaku Zasshi*, **86,** 576 (1966); *Chem. Abstr.*, **65,** 15373 (1966).
35. L. E. Rylski, F. Gajewski, J. Tarczynska, and T. Orlowska, *Farm. Polska*, **21,** 758 (1965); *Chem. Abstr.*, **65,** 3865 (1966).
36. S. Biniecki and J. Izdebski, *Acta Polon. Pharm.*, **15,** 421 (1957); *Chem. Abstr.*, **52,** 15540 (1958).
37. H. Starke, German (East) Pat. 15116 (1958); *Chem. Abstr.*, **54,** 3464 (1960).
38. T. Fujii and H. Sato, Japanese Pat. 6725 (1954); *Chem. Abstr.*, **50,** 1933 (1956).
39. M. Ohara, K. Yamamoto, and A. Sugihari, Japanese Pat. 10,027 (1960); *Chem. Abstr.*, **55,** 6439 (1961).
40. R. Cavalleri, E. Bellasio, and A. Sardi, *J. Med. Chem.*, **13,** 148 (1970).
41. British Pat. 743,204 (1956); *Chem. Abstr.*, **51,** 502 (1957).
42. British Pat. 707,337 (1954); *Chem. Abstr.*, **49,** 7606 (1955).
43. G. A. Reynolds, J. A. Van Allan, and J. F. Tinker, *J. Org. Chem.*, **24,** 1205 (1959).
44. D. B. Paul and H. V. Rodda, *Aust. J. Chem.*, **21,** 1291 (1968).
45. W. Zerweck and W. Kunze, German Pat. 952,810 (1956); *Chem. Abstr.*, **53,** 4313 (1959).
46. W. Kunze, German Pat. 951,995 (1956); *Chem. Abstr.*, **53,** 4313 (1959).
47. W. Persch, U.S. Pat. 2,742,467 (1956); German Pat. 948,976 (1956); *Chem. Abstr.*, **51,** 1305 (1957); **53,** 2263 (1959).
48. British Pat. 753,783 (1956); *Chem. Abstr.*, **51,** 7442 (1957).
49. T. P. Sycheva and M. N. Shchukina, *Zh. Obshch. Khim.*, **30,** 608 (1960); *Chem. Abstr.*, **54,** 24783 (1960).
50. E. Kesler and E. Biniecki, *Acta Polon. Pharm.*, **16,** 93 (1959); *Chem. Abstr.*, **53,** 18046 (1959).
51. E. Menziani, *Bol. Sci. Fac. Chim. Ind.* (*Bologna*) **12,** 162 (1954); E. Menziani, *Med. Sper.*, **25,** 277 (1954); *Chem. Abstr.*, **50,** 355, 828 (1956).
52. B. Prescott, G. Lones, and G. Caldes, *Antimicrob. Ag. Chemother.*, **1968,** 262; *Chem. Abstr.*, **72,** 109,766 (1970).
53. R. Ruggieri, *Giorn. Med. Militare*, **107,** 239 (1957); *Chem. Abstr.*, **52,** 2021 (1958).
54. S. Biniecki et al., Belgian Pat. 647,722 (1964); *Chem. Abstr.*, **64,** 2102 (1966).
55. R. Ruggieri, *Giorn. Med. Militare*, **107,** 4 pp (1957) (Reprint); *Chem. Abstr.*, **53,** 21792 (1959).
56. W. Kunze, German Pat. 951,993 (1956); *Chem. Abstr.*, **53,** 5298 (1959).
57. R. Ruggieri, *Boll. Chim. Farm.*, **103,** 196 (1964); *Chem. Abstr.*, **61,** 4344 (1964).
58. W. Zerweck and W. Kunze, German Pat. 951,992 (1956); *Chem. Abstr.*, **53,** 4314 (1959).
59. S. Fallab, *Helv. Chim. Acta*, **45,** 1957 (1962).
60. D. Walz and S. Fallab, *Helv. Chim. Acta*, **44,** 13 (1961).
61. D. Walz and S. Fallab, *Helv. Chim. Acta*, **43,** 540 (1960).
62. J. E. Andrew, P. W. Bell, and A. B. Balke, *Chem. Commun.*, **1969,** 143.
63. I. Grecu and E. Curea, *Rev. Chim.* (*Bucharest*), **16,** 348 (1965); *Farmacia* (*Bucharest*), **15,** 193 (1967); *Chem. Abstr.*, **63,** 15560 (1965); **67,** 87303 (1967).
64. I. Grecu and E. Curea, *Rev. Roum. Chim.*, **13,** 781 (1968).
65. I. Grecu and E. Curea, *Farm. Ed. Prat.*, **23,** 603 (1968); *Chem. Abstr.*, **70,** 3172 (1969).
66. V. S. Kublanovski and E. A. Mazurenko, *Tr. Odessk Gidrometeorol. Inst.*, **1961,** 39; *Chem. Abstr.*, **60,** 12651 (1964).
67. R. Ruggieri, *Anal. Chim. Acta*, **16,** 242 (1952); *Chem. Abstr.*, **51,** 11914 (1957).
68. V. G. Belikov et al., *Aptechn. Delo.*, **12,** 60 (1963); *Chem. Abstr.*, **61,** 9357 (1964).
69. I. Grecu and E. Curea, *Rev. Chim* (*Bucharest*), **16,** 389 (1965); *Chem. Abstr.*, **64,** 11031 (1966).

70. G. Favichi, *Boll. Chim. Farm.*, **96,** 431 (1957); *Chem. Abstr.*, **52,** 4109 (1958).
71. Y. M. Pardman and K. I. Evstratova, *Aptehn. Delo.*, **12,** 45 (1963); *Chem. Abstr.*, **61,** 9360 (1964).
72. F. Janick, B. Budensinsky, and J. Krobl, *Ceskoslov. Farm.*, **9,** 304 (1960); *Chem. Abstr.*, **55,** 9790 (1961).
73. B. Artamanov, T. Y. Komenkova, and S. L. Maifotis, *Med. Prom. SSSR.*, **19,** 57 (1965); *Chem. Abstr.*, **63,** 17800 (1965).
74. H. Mekennis, Jr. and A. S. Yard, U.S. Dept. Com., Office Tech. Service *P. B. Report*, **143,** 914 (1957); *Chem. Abstr.*, **55,** 15375 (1961).
75. A. Viala, *Trav. Chim. Pharm. Montpellier*, **18,** 96 (1958); *Chem. Abstr.*, **55,** 8534 (1959).
76. Z. E. Kalinowska, *Chem. Anal.* (*Warsaw*), **9,** 831 (1964); *Chem. Abstr.*, **62,** 4620 (1965).
77. A. K. Ruzhentseva, I. S. Tubina, and L. S. Bragina, *Med. Prom. SSSR.*, **14,** 34 (1960); *Chem. Abstr.*, **55,** 9789 (1961).
78. H. M. Perry, Jr., *J. Lab. Clin. Med.*, **41,** 566 (1956); *Chem. Abstr.*, **47,** 10601 (1953).
79. B. Wesley-Hadzija and F. Abuffy, *Croat. Chem. Acta*, **30,** 15 (1958); *Chem. Abstr.*, **54,** 2661 (1960).
80. W. Kitzing, *Pharmazie*, **16,** 401 (1961).
81. T. Cieszynski, *Diss. Pharm. Pharmacol*, **19,** 563 (1967); *Chem. Abstr.*, **68,** 43221 (1968).
82. T. Cieszynski, *Acta Polon. Pharm.*, **25,** 551 (1968); *Chem. Abstr.*, **70,** 60887 (1969).
83. S. Biniecki et al., Polish Pat. 50,987 (1966); *Chem. Abstr.*, **66,** 115722 (1967).
84. E. Belniak, E. Domagalina, and H. Hopkala, *Roczniki. Chem.*, **41,** 831 (1967).
85. F. Dallacker and J. Bloemen, *Monatsh. Chem.*, **92,** 640 (1961).
86. W. Kunze, German Pat. 958,561 (1957); *Chem. Abstr.*, **53,** 8178 (1959).
87. H. J. Rodda and P. E. Rogasch, *Aust. J. Chem.*, **19,** 1291 (1966).
88. J. E. Robertson and J. H. Biel, U.S. Pat. 3,243,343 (1966); *Chem. Abstr.*, **64,** 15902 (1966).

Part J. Phthalazinecarboxylic Acids and Related Compounds

Phthalazinecarboxylic acids having carboxyl groups attached either at the 1- or 1,4-positions of the phthalazine nucleus are discussed in this section. In addition, ester, amide, hydrazide, and nitrile derivatives are discussed, as are 1-(2*H*)phthalazinone- and 1-(2*H*)phthalazinethione-4-carboxylic acids. Phthalazine-1-carboxyaldehyde has not been isolated but several of its derivatives such as phenylhydrazone and semicarbazone have been prepared through the formation of the anil, *N*,*N*-dimethyl-*N*′-1′-phthalazinylmethylene-*p*-phenylenediamine (see Section 2B-I-B-2-c). There are no reports of phthalazine having an acetyl group at the 1-position.

I. Phthalazine-1-carboxylic Acid and 1,4-Dicarboxylic Acid

Popp and co-workers (1) have reported the only synthesis of phthalazine-1-carboxylic acid (**1**) where the phthalazine Reissert compound (see Eq. 1) has been hydrolyzed with hydrobromic acid in acetic acid to give the hydrobromide of **1**, in 89% yield.

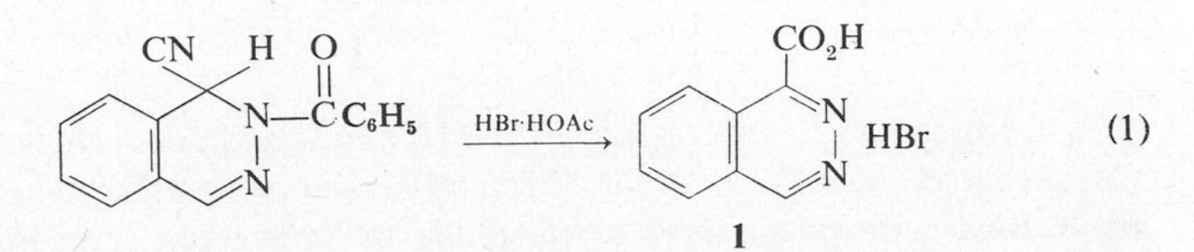

Phthalazine-1,4-dicarboxylic acid has not been reported, but Sauer and Heinrichs (2) have synthesized diethyl phthalazine-1,4-dicarboxylate (**2**) by a novel route involving the Diels-Alder reaction. Diethyl tetrazine-1,4-dicarboxylate in dichloromethane is allowed to react with benzyne, liberated from anthranilic acid diazonium betain, to give **2** in 73% yield (Eq. 2).

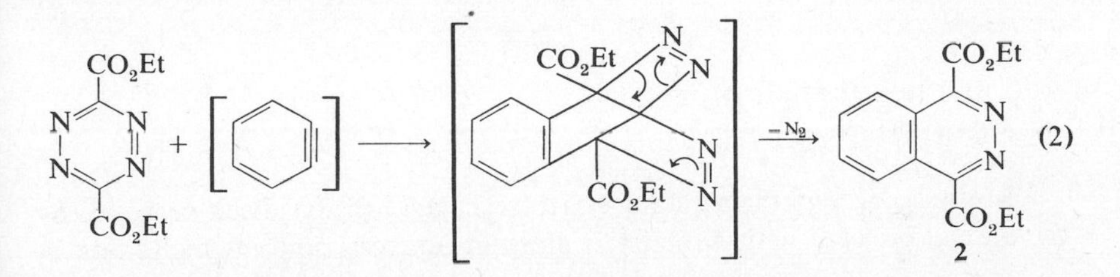

II. 1-(2*H*)Phthalazinone- and 1-(2*H*)Phthalazinethione-4-carboxylic Acids and Derivatives

1-(2*H*)Phthalazinone-4-carboxylic acid (**3**) and some of its 2-substituted derivatives have been prepared very conveniently by oxidation of naphthalene with potassium permanganate in presence of alkali to give phthalonic acid, which is condensed with hydrazine (3–7), an alkylhydrazine (8–10), or an arylhydrazine (10, 11) to give **3** and its derivatives (Eq. 3). Similarly, tetralin has been oxidized with alkaline potassium permanganate and condensed with hydrazine to give **3** in 96% yield (11a).

1-(2*H*)Phthalazinone-4-carboxylic acid (**3**) has been prepared in 73% yield by the oxidation of 4-nitromethyl-1-(2*H*)phthalazinone with dilute sulfuric acid on a steam bath (12). Alkaline permanganate oxidation of 3-methyl-5,6-dihydro-6-ketopyrrolo[2,1-*a*]phthalazine at room temperature also gives **3** (13).

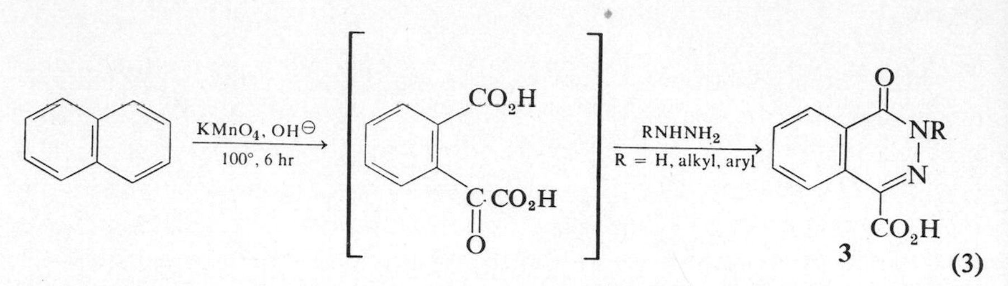

The methyl and ethyl esters of 1-(2*H*)phthalazinone-4-carboxylic acid (**3**) (14) and 2-alkyl- or aryl-1-(2*H*)phthalazinone-4-carboxylic acid (15), on heating with phosphorus pentasulfide in xylene for 1 hr followed by alkaline hydrolysis, furnish 1-(2*H*)phthalazinethione-4-carboxylic acid (**4**) and its 2-substituted derivatives (Eq. 4).

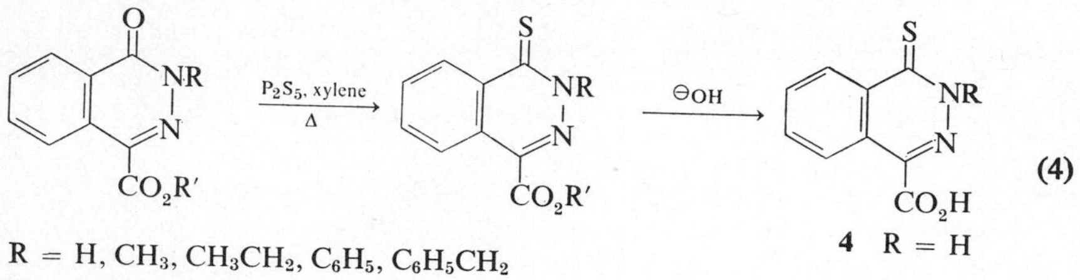

2-Methyl-1-(2*H*)phthalazinone-4-carboxylic acid (**5**) has been prepared by the alkylation of **3** with dimethyl sulfate in aqueous sodium hydroxide in

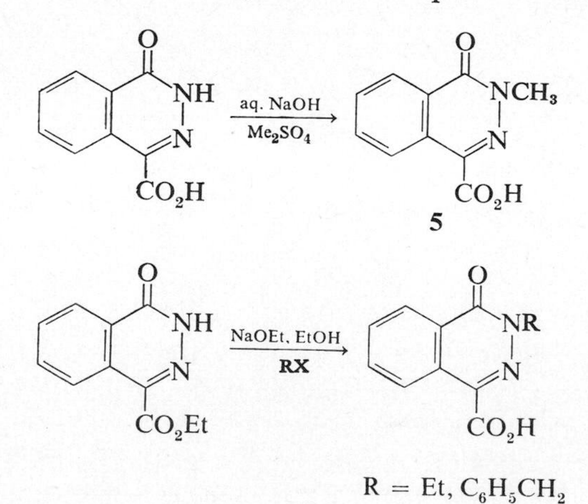

62% yield (10). Similarly, alkylation of ethyl 1-(2*H*)phthalazinone-4-carboxylate with diethyl sulfate or with benzyl chloride in ethanolic sodium ethoxide gives (with simultaneous hydrolysis of the ester group) 2-ethyl- and 2-benzyl-1-(2*H*)phthalazinone-4-carboxylic acids in 91 and 85% yield, respectively (Eq. 5) (10).

1-(2*H*)Phthalazinethione-4-carboxamide (15,16) and ethyl 1-(2*H*)phthalazinethione-4-carboxylate (15), on alkylation with methyl iodide or ethyl bromide in acetone in presence of alkali, provide the *S*-alkylated products **6** (Eq. 6). The reaction of ethyl 1-(2*H*)phthalazinethione-4-carboxylate with methyl iodide or benzyl chloride in ethanolic sodium hydroxide also yields *S*-alkylated products (**7**) in which the ester group has been simultaneously hydrolyzed (15) (Eq. 6).

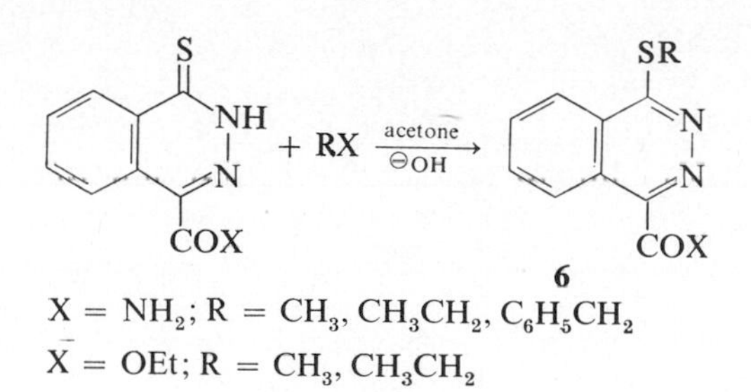

(6)

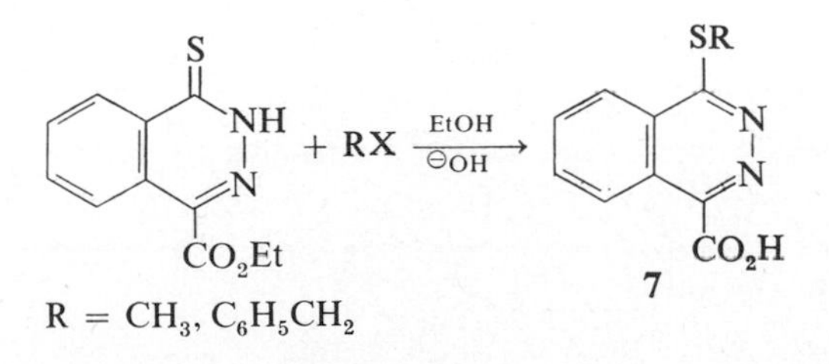

The decarboxylation of 1-(2*H*)phthalazinone-4-carboxylic acid (**3**) and 1-(2*H*)phthalazinethione-4-carboxylic acid (**4**) in an inert solvent proceeds smoothly to give 1-(2*H*)phthalazinone (3, 6, 7, 12) and 1-(2*H*)phthalazinethione (14), respectively, in high yields. Similarly, 2-alkyl-1-(2*H*)phthalazinone-4-carboxylic acids, on heating at 240° for 30 min, give 2-alkyl-1-(2*H*)phthalazinone in high yields (10) (Eq. 7).

1-(2*H*)Phthalazinethione-4-carboxylic acid, on refluxing with methanol in presence of acid, provided methyl 1-(2*H*)phthalazinethione-4-carboxylate (14). Similarly, ethyl 1-(2*H*)phthalazinone-4-carboxylate (12) and ethyl 2-methyl-1-(2*H*)phthalazinone-4-carboxylate (7, 8) are obtained from the corresponding phthalazinone-4-carboxylic acids and ethanol in presence of

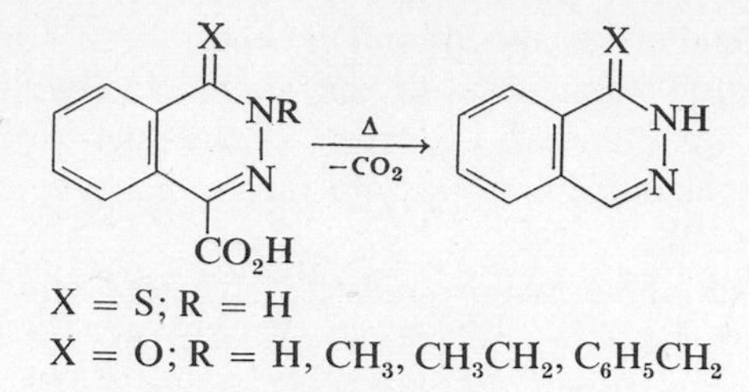

(7)

$X = S; R = H$

$X = O; R = H, CH_3, CH_3CH_2, C_6H_5CH_2$

acid. 1-(2*H*)Phthalazine-4-carboxylic acid and 2-alkyl- and 2-aryl-1-(2*H*)-phthalazinone-4-carboxylic acids, on treatment with thionyl chloride, with (9, 17) or without (17) a catalytic amount of dimethylformamide, give the corresponding acid chlorides. The reaction of 1-(2*H*)phthalazinethione-4-carboxylic acid with thionyl chloride failed to give the acid chloride (14).

Grabliuskas and co-workers (17, 18) have allowed 2-methyl-, 2-phenyl-, and 2-benzyl-1-(2*H*)phthalazinone-4-carbonyl chlorides to react with several *N*,*N*-dialkylamino alcohols to furnish the corresponding esters. Ethyl 1-(2*H*)phthalazinone-4-carboxylate, on treatment with 25% aqueous ammonium hydroxide (5) or ethanolic ammonia (19) at room temperature, furnished 1-(2*H*)phthalazinone-4-carboxamide in 80–84% yields. Similarly, ethyl 1-(2*H*)phthalazinethione-4-carboxylate and ethyl 2-methyl-1-(2*H*)-phthalazinone-4-carboxylate, on treatment with ammonia, gave 1-(2*H*)-phthalazinethione-4-carboxamide (14) and 2-methyl-1-(2*H*)phthalazinone-4-carboxamide (7, 8), respectively. Morishita and co-workers (20) have allowed ethyl 1-(2*H*)phthalazinone-4-carboxylate to react with several akylamines, including methylamine, dimethylamine, alkylamine, and isopropylamine, by heating in a sealed tube for 8 hr at 100° C to give the corresponding amides. 2-Alkyl- and 2-aryl-1-(2*H*)phthalazinone-4-carbonyl chlorides also react with ammonia (10) and *N*,*N*-dialkylalkylamines (17, 21) to furnish carboxamides and various amide derivatives.

1-(2*H*)Phthalazinone-4-carboxyhydrazide (5) and 1-(2*H*)phthalazinethione-4-carboxyhydrazide (14) are obtained in 75 and 95% yields, respectively, by heating ethyl 1-(2*H*)phthalazinone-4-carboxylate and ethyl 1-(2*H*)phthalazinethione-4-carboxylate with hydrazine hydrate. These hydrazides condense with various aldehydes to furnish the hydrazones (4, 14).

Treatment of 2-methyl-1-(2*H*)phthalazinone-4-carboxyhydrazide with nitrous acid at 0° furnished 2-methyl-1-(2*H*)phthalazinone-4-carboxyazide (8, 22) in 63% yield. The azide rearranged smoothly to ethyl *N*-(2-methyl-1-oxo-1,2-dihydro-4-phthalazinyl)carbamate on boiling in ethanol.

Nitration of 1-(2*H*)phthalazinone-4-carboxylic acid with potassium nitrate in sulfuric acid gave two products, 5-nitro- and 7-nitro-1-(2*H*)phthalazinone-4-carboxylic acid. The former, on further nitration, yielded 4,5-dinitro-1-(2*H*)phthalazinone (23).

III. 1-Cyano- and 1,4-Dicyanophthalazines

Hirsch and Orphanos (24) have shown that 1-iodo- and 1,4-diiodophthalazine will react with cuprous cyanide in pyridine to give 1-cyanophthalazine (**8**) and 1,4-dicyanophthalazine (**9**) in 17 and 35% yields, respectively (Eq. 8). The hydrolysis of **8** with 10% sodium hydroxide and of **9** in dioxane with 20% sodium hydroxide provided 1-(2*H*)phthalazinone and 4-cyano-1-(2*H*)-phthalazinone (**10**), respectively. Further hydrolysis of **10** with methanolic

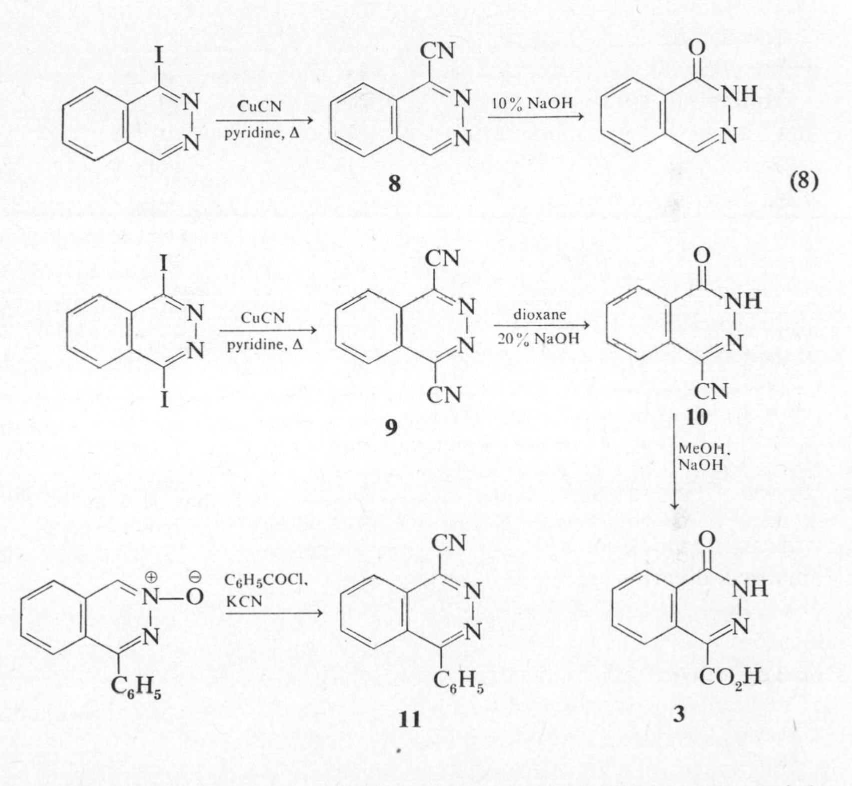

sodium hydroxide gave **3** (Eq. 8). Hayashi and Oishi (25) have treated 4-phenylphthalazine 2-oxide with benzoyl chloride and potassium cyanide to obtain 1-cyano-4-phenylphthalazine (**11**) (Eq. 8).

Oishi (24a) has shown that 1-methylsulfonyl-4-phenylphthalazine will react with potassium cyanide in dimethyl sulfoxide to give 1-cyano-4-phenylphthalazine (**11**) and 1-carboxyamido-4-phenylphthalazine in 25 and 24% yields, respectively.

IV. Tables

TABLE 2J-1. Phthalazinecarboxylic Acids and Nitriles

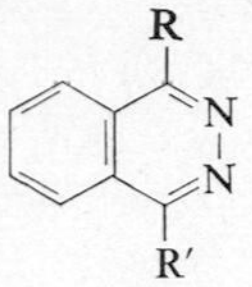

R	R′	Preparation[a]	Yield (%)	MP	Reference
CO_2H	H	A	89	HBr, 198–200°	1
$CO_2CH_2CH_3$	$CO_2CH_2CH_3$	B	73	176–177°	2
CN	H	C	17	White needles, 156–157° (ether)	24
CN	CN	C	35	Grey yellowish shiny crystals, 204–205° (ether)	24
CN	C_6H_5	D	24	180–182° (benzene)	24a, 25
$CONH_2$	C_6H_5	—	24	203°	24a

[a] A, 1-cyano-2-benzoyl-1,2-dihydrophthalazine in glacial acetic acid + aqueous hydrobromic acid, reflux 15 min after exothermic reaction subsides. B, diethyl tetrazine-1,4-dicarboxylate + benzyne, generated from anthranilic acid diazonium betain, in dichloromethane. C, proper iodophthalazine + cuprous cyanide, heat in pyridine at 110–120° for 10 min. D, 4-phenylphthalazine 2-oxide + benzoyl chloride + potassium cyanide.

TABLE 2J-2. 1-(2*H*)Phthalazinone-4-carboxylic Acids and Esters

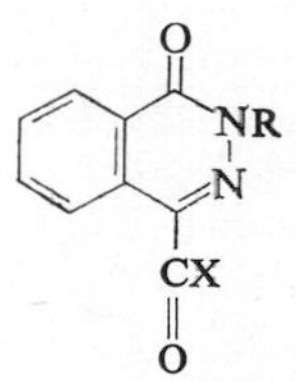

R	X	Preparation[a]	Yield (%)	MP	Reference
H	OH	A	~65, 96	229–230° (ethanol); 232° (dec.); 234° (foaming); >250° (dec.); 230–232 (dec.) ammonium salt, 257–260°; piperidine salt, 237–240° (dec.)	3–6, 11a, 12, 24
H	OCH_3	B	—	208–209°	5
H	OCH_2CH_3	B	—	169°; 170–171°	5, 12
CH_3	OH	A, C	60–65	236° (dec.) (acetic acid); 237° (dec.)	8–10
CH_3	OCH_3	D	94	129°	9, 10
CH_3	OCH_2CH_3	B	94	85° (aqueous ethanol)	8, 9
CH_3	$O(CH_2)_2N(CH_2CH_3)_2$	D	76	220–221° (abs. ethanol)	17
CH_3	$OCH(CH_3)CH_2N(CH_3)_2$	D	66	218–219° (propanol)	17
CH_3	$O(CH_2)_3N(CH_3)_2$	D	66	205–206° C (propanol)	17
CH_3	$O(CH_2)_2$—N(piperidine ring)	D	71	219–220° (dec.) (butanol)	17
CH_2CH_3	OH	A, C_1	91	220–221° (dec.)	10
CH_2CH_3	OCH_3	D	95	103–104°	10
C_6H_5	OH	A	—	Orange, 219° (ethanol-water); 208° (dec.)	4, 11, 26

TABLE 2J-2 (*continued*)

R	X	Preparation[a]	Yield (%)	MP	Reference
C_6H_5	OCH_3	D	93	129°	10
C_6H_5	$O(CH_2)_2N(CH_2CH_3)_2$	D	75	221–222° C (abs. ethanol)	17
C_6H_5	$OCH(CH_3)CH_2N(CH_3)_2$	D	66	184–185° (abs. ethanol-ether)	17
C_6H_5	$O(CH_2)_3N(CH_3)_2$	D	85	197–198° (abs. ethanol)	17
C_6H_5	$O(CH_2)_2$—N(piperidino)	D	61	240° (dec.) (propanol)	17
m-$CH_3C_6H_4$	OH	A	73	203° (dec.)	10
m-$CH_3C_6H_4$	$O(CH_2)_2N(CH_2CH_3)_2$	D	64	200–201° (abs. ethanol)	17
$C_6H_5CH_2$	OH	A, C_1	34–85	200–201° (dec.)	10
$C_6H_5CH_2$	$O(CH_2)_2N(CH_2CH_3)_2$	D	64	212–213° (propanol)	17
$C_6H_5CH_2$	$OCH(CH_3)CH_2N(CH_3)_2$	D	65	234° (dec.) (isobutyl alcohol)	17
$C_6H_5CH_2$	$O(CH_2)_3N(CH_3)_2$	D	85	149° (isobutyl alcohol)	17
$C_6H_5CH_2$	$O(CH_2)_2$—N(piperidino)	D	47	200–201° (dec.) (propanol)	17

[a] A, phthalonic acid obtained from alkaline oxidation of naphthalene is heated with proper hydrazine in aqueous medium for 1–3 hr on steam bath. B, proper 1-(2*H*)phthalazinone-4-carboxylic acid boiled with corresponding alcohol in presence of sulfuric acid. C, 1-(2*H*)phthalazinone-4-carboxylic acid in aqueous sodium hydroxide stirred with dimethyl sulfate at room temperature for 1 hr. C_1, ethyl 1-(2*H*)phthalazinone-4-carboxylate heated with corresponding halide in ethanolic sodium ethoxide. D, 2-alkyl- or 2-aryl-1-(2*H*)phthalazinone-4-carboxylic acid treated with thionyl chloride in chloroform containing a catalytic amount of dimethylformamide, followed by heating the acid chloride with proper alcohol. D_1, same as D, but reflux for 1–2 hr without dimethylformamide. The acid chloride was taken up in benzene and refluxed with corresponding *N*,*N*-dialkylaminoalcohol for 1 hr.

TABLE 2J-3. 1-(2*H*)Phthalazinone-4-carboxamide and Hydrazide Derivatives

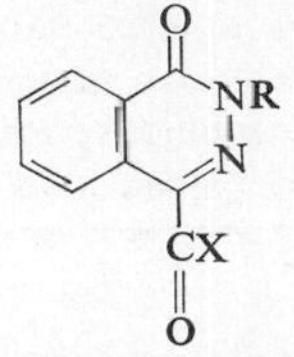

R	X	Preparation[a]	Yield (%)	MP	Reference
H	NH_2	A	80–84	318° (dec.) (water or ethanol); 312–313° (dec.)	4, 5, 19
H	$NHCH_3$	B	100	Needles, 281° (ethanol)	4, 20
H	$N(CH_3)_2$	B	—	Plates, 221–223° (dec.)	20
H	$NHCH(CH_3)_2$	B	—	273–274°	20
H	$NHCH_2CH{=}CH_2$	B	—	250–252°; 250–253°	4, 20
H	$NHCH_2CH_2OH$	B	—	Needles, 229–231°	20
H	$NHCH_2CO_2H$	—	—	253.5°–254.5°	27
H	$N(CH_2CH_2OH)_2$	B	—	181–183°	20
H	$NHCH_2CH_2CH_2OH$	—	—	224.5–225.5°	27
H	$NHCH_2CH(OH)CH_2OH$	B	—	210–212°	4, 20
H	$NHCH(CH_2OH)_2$	B	—	222–223°	4
H	$NC(OH)_3$ with CH_3	B	—	222–223°	20
H	$NH(CH_2)_2N(CH_2CH_3)_2$	C	—	208–5°–209° (benzene) HCl, 245–246° (dec.)	17
CH_3	NH_2	A, C	92	Needles, 274° (methanol)	8–10
CH_3	$NH(CH_2)_2N(CH_2CH_3)_2$	C_1	52	57.5–58.5° (pet. ether) HCl, 225°	17
CH_2CH_3	NH_2	C	85	209°	10
C_6H_5	NH_2	C	95	214°; 214–215°	4, 10

TABLE 2J-3 (*continued*)

R	X	Preparation[a]	Yield (%)	MP	Reference
C_6H_5	$NH(CH_2)_2N(CH_2CH_3)_2$	C_1	88	91.5°; HCl, 199–200°	17
m-$CH_3C_6H_4$	NH_2	C	92	223°	10
$C_6H_5CH_2$	NH_2	C	93	179–180°	10
$C_6H_5CH_2$	$NH(CH_2)_2N(CH_2CH_3)_2$	C_1	71	101° (pet. ether), HCl, 168°	17
H	$NHNH_2$	D	75	235–236° (dec.) (ethanol-water); 241° (dec.)	4, 5
H	NHN=CH— (furan ring, O)	E	—	304–305° (alcohol)	5
H	NHN=CH— (benzene ring) —OH, OCH_3	E	—	280° (alcohol)	5
H	NHN=CH— (benzene ring) —OCH_3, OCH_3, HOC(=O)	E	—	262–263° (alcohol)	5
CH_3	$NHNH_2$	D	84	Long needles, 160° (ethanol)	8, 28
C_6H_5	$NHNH_2$	—	—	192–193°	4
CH_3	N_3	F	63	—	8, 22

[a] A, proper ethyl 1-(2*H*)phthalazinone-4-carboxylate + ethanol or methanol saturated with ammonia, kept at room temperature in sealed tube for 8 hr. B, ethyl 1-(2*H*)phthalazinone-4-carboxylate + corresponding amine, heat in a sealed tube at 100° for 8 hr. C, proper 1-(2*H*)phthalazinone-4-carboxylic acid + $SOCl_2$, reflux in chloroform with catalytic amount of dimethylformamide for 1–2 hr, followed by treatment with corresponding amine. C_1, same as C, but dimethylformamide was not used. D, proper ethyl 1-(2*H*)phthalazinone-4-carboxylate + hydrazine hydrate, reflux for 2 hr. E, 1-(2*H*)phthalazinone-4-carboxyhydrazide + proper aldehyde, boil 10 min in aqueous media. F, 2-methyl-1-(2*H*)phthalazinone-4-carboxyhydrazide treated with nitrous acid at 0° in aqueous media.

TABLE 2J-4. 1-(2*H*)Phthalazinethione-4-carboxylic Acid and Derivatives

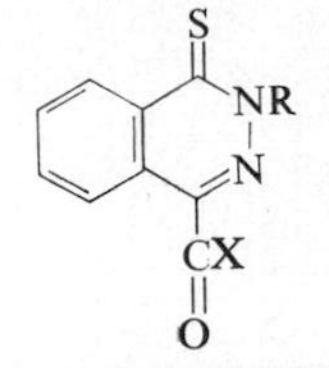

R	X	Preparation[a]	Yield (%)	MP	Reference
H	OH	A	92	212° (dec.)	14
H	OCH_3	B	73	215°	14
H	OCH_2CH_3	C	100	176°	14
CH_3	OH	A	79	209–210° (ethanol)	15
CH_3	OCH_3	C	26	129–130°	15
CH_2CH_3	OH	A	81	148–149° (dec.) (ethanol)	15
CH_2CH_3	OCH_3	C	28	119–120°	15
C_6H_5	OH	A	82	215–216° (dec.) (ethanol)	15
C_6H_5	OCH_3	C	53	189–190°	15
$C_6H_5CH_2$	OH	A	85	156–157° (aqueous methanol)	15
$C_6H_5CH_2$	OCH_3	C	39	132–133° (methanol)	15
H	NH_2	D	92	287–288° (dec.)	14
H	$NHNH_2$	—	97	261° (dec.)	14
H	$NHN{=}C(CH_3)_2$	—	—	241°	14
H	NHN=CH—(4-OH-3-OCH_3-phenyl)	—	—	263°	14

[a] A, alkaline hydrolysis of the corresponding ester. B, 1-(2*H*)phthalazinethione-4-carboxylic acid in refluxing methanolic hydrochloric acid. C, corresponding ester of 1-(2*H*)phthalazinone-4-carboxylic acid, heat 1 hr with phosphorus pentasulfide in xylene. D, ethyl 1-(2*H*)-phthalazinone-4-carboxylate treated with 25% ammonia at 20° C.

TABLE 2J-5. 1-Alkylthiophthalazine-4-carboxylic Acid Derivatives

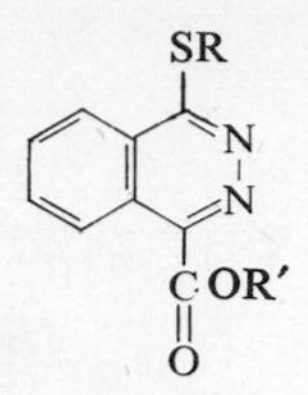

R	R′	Preparation[a]	Yield (%)	MP	Reference
CH_3	OH	A	100	106–107°	15
CH_3	OCH_2CH_3	B	54	127–128° (cyclohexane)	15
CH_2CH_3	OCH_2CH_3	B	65	103–104° (cyclohexane)	15
$C_6H_5CH_2$	OH	A	86	114° (dec.) (methanol)	15
CH_3	NH_2	B_1	—	210–211° (propanol)	15
CH_2CH_3	NH_2	B_1	64	221° (propanol)	15
$C_6H_5CH_2$	NH_2	B_1	88	170° (propanol)	15

[a] A, ethyl 1-(2*H*)phthalazinethione-4-carboxylate treated in ethanolic alkali with proper alkyl halide (simultaneous hydrolysis of ester). B, ethyl 1-(2*H*)phthalazinethione-4-carboxylate treated in acetone containing 1*N* sodium hydroxide with proper halide at room temperature. B_1, same as B, but used 1-(2*H*)phthalazinethione-4-carboxamide.

TABLE 2J-6. Miscellaneous Compounds

Reaction	Product, Yield	MP	Reference
(structure: OH, C=O, C=O, C–NH–, CO_2H) + $N_2H_4 \cdot H_2O$; steam bath, 15 min → (structure: O, NH, N, HN–C(=O)–, CO_2H)	54%	333° (acetic acid)	26
(structure: $CH_2CC_6H_5$ (C=O), N, N, C_6H_5); monoperphthalic acid, $CHCl_3$, < −10°, overnight → (structure: O O, C–CC_6H_5, N, N, C_6H_5)	67%	150–151° (benzene-pet. ether)	29

TABLE 2J-6 (*continued*)

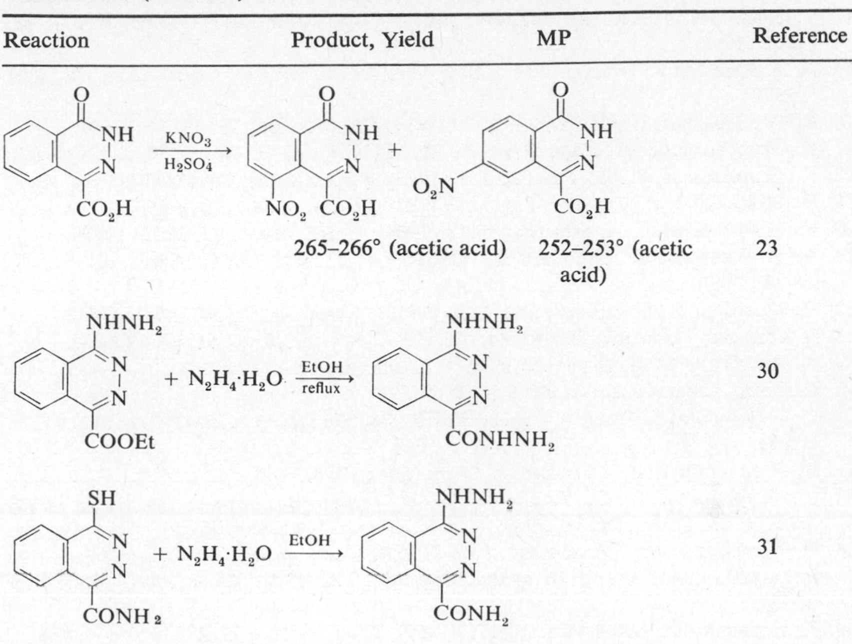

Reaction	Product, Yield	MP	Reference
$\xrightarrow[H_2SO_4]{KNO_3}$		265–266° (acetic acid) 252–253° (acetic acid)	23
$+ N_2H_4 \cdot H_2O \xrightarrow[reflux]{EtOH}$			30
$+ N_2H_4 \cdot H_2O \xrightarrow{EtOH}$			31

References

1. F. D. Popp, J. M. Wefer, and C. W. Klinowski, *J. Heterocyclic Chem.*, **5,** 879 (1968).
2. J. Sauer and G. Heinrichs, *Tetrahedron Lett.*, **1966,** 4979.
3. K. Adachi, *J. Pharm. Soc. Japan* **75,** 1423 (1955); *Chem. Abstr.*, **50,** 10105 (1956).
4. I. Satoda N. Yoshida and K. Mori, *Yakugaku Zasshi*, **28,** 613 (1956); *Chem. Abstr.*, **51,** 16483 (1957).
5. V. B. Brasyunas and K. Grabliauskas, *Kaunas Med. Inst. Darbai*, **6,** 277 (1958); *Chem. Abstr.*, **55,** 4517 (1961).
6. T. P. Sychiva et al., *Med. Prom. SSSR.*, **14,** 13 (1960); *Chem. Abstr.*, **54,** 22669 (1960).
7. W. Lorenz, French Pat. 1,335,759 (1963); *Chem. Abstr.*, **60,** 558 (1964).
8. I. Satoda, N. Yoshida, and K. Mori, *Yakugaku Zasshi*, **77,** 703, (1957); *Chem. Abstr.*, **51,** 17927 (1957).
9. H. Morishita et al., Japanese Pat. 3324 (1959); *Chem. Abstr.*, **54,** 595 (1960).
10. A. N. Kost, S. Foldeak, and K. Grabliauskas, *Khim. Farm. Zh.*, **1,** 43 (1967); *Chem. Abstr.*, **67,** 100090 (1967).
11. T. C. Bruice and F. M. Ridhards, *J. Org. Chem.*, **23,** 145 (1958).

11a. H. Kitamikado, H. Hattori, and M. Kuwayama, Japanese Pat. 10,348 (1970); *Chem. Abstr.*, **73,** 45530 (1970).

12. G. A. Vanag and M. A. Matskanova, *Zh. Obshc. Khim.*, **26,** 1749 (1956); *J. Gen. Chem., USSR*, **26,** 1963 (1956); *Chem. Abstr.*, **51,** 1944, 14746 (1957).
13. V. Spiro, *Ric. Sci. Rend., Sez A.*, **8,** 197 (1965); *Chem. Abstr.*, **63,** 13188 (1965).

14. V. Brasiunas and K. Grabliauskas, *Khim. Farm. Zh.* **1**, 40 (1967); K. Grabliauskas and V. Brasiunas, USSR Pat. 202,959 (1967); *Chem. Abstr.*, **67**, 100089 (1967); **69**, 59266 (1968).
15. A. N. Kost and K. Grabliauskas, *Khim. Farm Zh.*, **2**, 3 (1968); *Chem. Abstr.*, **70**, 4001 (1969).
16. K. Grabliauskas, USSR Pat. 210,172 (1969); *Chem. Abstr.*, **71**, 13134 (1969).
17. K. Grabliauskas et al., *Khim Farm. Zh.*, **2**, 35 (1968); *Chem. Abstr.*, **70**, 28884 (1969).
18. K. Grabliauskas, USSR, Pat. 210,173 (1968); *Chem. Abstr.*, **70**, 47477 (1969).
19. H. Morishita et al., Japanese Pat. 7783 (1957); *Chem. Abstr.*, **52**, 13807 (1958).
20. H. Morishita et al., Japanese Pat. 5121 (1959); *Chem. Abstr.*, **53**, 22024 (1959).
21. A. N. Kost and K. Grabliauskas, USSR Pat. 223,098 (1968); *Chem. Abstr.*, **70**, 28934 (1969).
22. H. Morishita et al., Japanese Pat. 6280 (1959); *Chem. Abstr.*, **54**, 15411 (1960).
23. S. Kanahara, *Yakugaku Zasshi*, **84**, 483 (1964); *Chem. Abstr.*, **61**, 8304 (1964).
24. A. Hirsch and D. G. Orphanos, *Can. J. Chem.*, **43**, 1551 (1965).
24a. E. Oishi, *Yakugaku Zasshi*, **89**, 959 (1969).
25. E. Hayashi and E. Oishi, *Yakugaku Zasshi*, **86**, 576 (1966); *Chem. Abstr.*, **65**, 15374 (1966).
26. R. A. Staunton and A. Tophan, *J. Chem. Soc.*, **1953**, 1889.
27. K. Grabliauskas, V. G. Vinokurov, N. D. Konevskaya, and A. N. Kost, *Vestn. Mosk. Univ., Khim.*, **24**, 58 (1969); *Chem. Abstr.*, **71**, 81289 (1969).
28. H. Morishita et al., Japanese Pat. 5679 (1959); *Chem. Abstr.*, **54**, 1570 (1960).
29. E. Hayashi and E. Oishi, *Yakugaku Zasshi*, **87**, 940 (1967); *Chem. Abstr.*, **68**, 39573 (1968).
30. K. Grabliauskas, USSR Pat. 210,171 (1969); *Chem. Abstr.*, **71**, 13133 (1969).
31. K. Grabliauskas and V. Brasiunas, USSR Pat. 243,609 (1969); *Chem. Abstr.*, **72**, 132,764 (1970).

Part K. Pseudophthalazinones and Phthalazine Quaternary Compounds

I. Pseudophthalazinones

Pseudophthalazinones (Ψ-phthalazinones) are dipolar compounds as shown in **1**. There has been very little work done in this field since 1950.

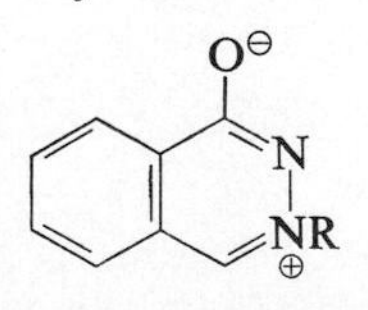

1 R = alkyl, aryl

Most of the earlier work has been reviewed by Simpson (1). The compounds prepared thereafter are reported here. Vaughan (2) first suggested the name pseudophthalazone for this group of compounds, but in this work the name Ψ-phthalazinone (3) is used in order to distinguish it easily from 1-(2*H*)-phthalazinone and to have consistency in the nomenclature.

Originally Rowe and co-workers (1) studied this group of compounds, preparing them as follows. 1-Hydroxy-3-aryl-3,4-dihydrophthalazine-4-acetic acids (**2**), on heating with aqueous sulfuric acid at 140°, provide 3-aryl-Ψ-phthalazinones (**3**) with the elimination of acetic acid. Treatment of **2** in cold acidic dichromate solution furnishes 3-aryl-4-methyl-Ψ-phthalazinones (**4**) with loss of formic acid (Eq. 1).

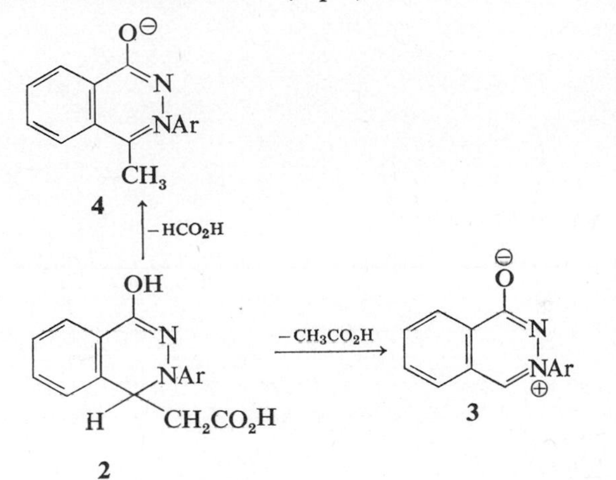

Lund (3) has reported a new procedure for the preparation of aryl-Ψ-phthalazinones starting from *N*-arylaminophthalimides (**5**). The electrolytic reduction of **5** at a mercury cathode or sodium borohydride reduction gives *N*-arylamino-3-hydroxyphthalimidines (**6**), which rearrange smoothly to **3** by heating at 120° or by boiling the aqueous solution for a few hours. Rearrangement occurs with loss of a water molecule (Eq. 2).

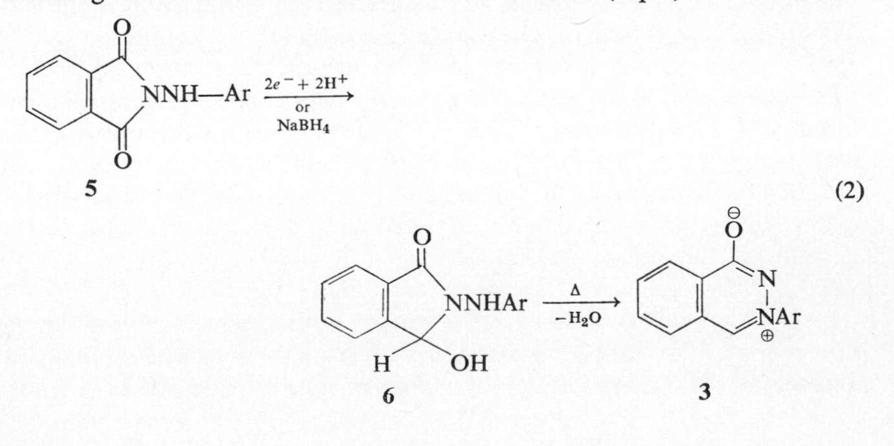

Lund (3) has also helped to establish the structure of the 3-aryl-Ψ-phthalazinones by showing absence of carbonyl absorption in the ir spectra of these compounds between 1,610 and 2,000 cm^{-1} as well as the absence of —NH or —OH absorption in the 3,100–3,600 cm^{-1} region. Further two-electron reduction of 3-phenyl-Ψ-phthalazinone (**7**) at controlled potential gives 3-phenyl-3,4-dihydro-1-(2*H*)phthalazinone (**8**), which is also formed by a four-electron reduction of 4-hydroxy-2-phenyl-1-(2*H*)phthalazinone (**9**) (Eq. 3). Two-electron reduction of 2-phenyl-1-(2*H*)phthalazinone gives 2-phenyl-3,4-dihydro-1-(2*H*)phthalazinone (**10**).

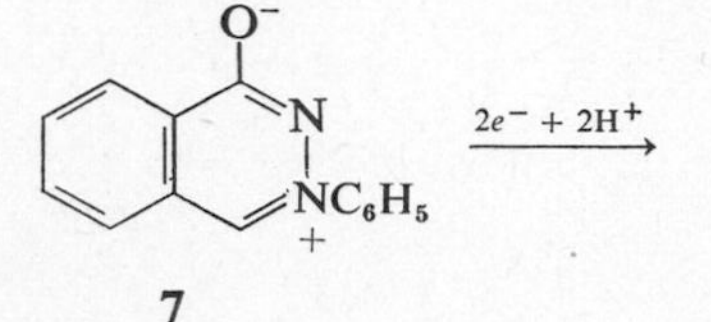

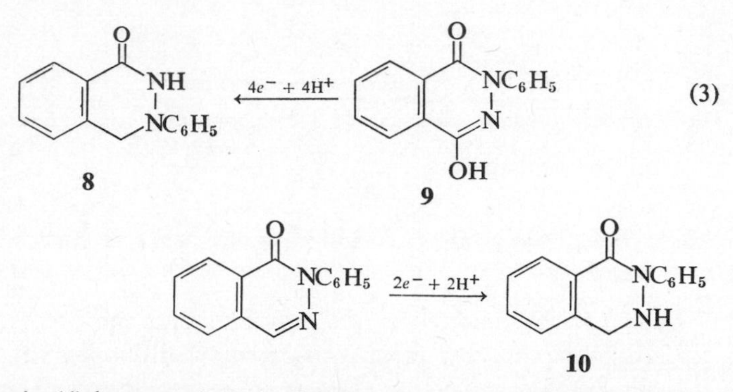

Diep and Cauvin (4) have studied the lithium aluminum hydride reduction of **9** and found that the main product is **8** with a small amount of **7**.

Ikeda and co-workers (5) have reported the first synthesis of 3-alkyl-Ψ-phthalazinones in their studies of quaternization of 1-alkoxyphthalazines. Thus 1-methoxy-3-methylphthalazinium iodide (**11**) on heating at 120° *in vacuo*, or on reaction with silver oxide in the cold, furnished 3-methyl-Ψ-phthalazinone (**12**), whereas heating **11** in a stream of carbon dioxide gave 2-methyl-1-(2*H*)phthalazinone (Eq. 4). 1-(2*H*)Phthalazinone, on treatment with methyl iodide and silver oxide, also provide **12**. Alkylation of **12** with dimethyl sulfate gave the expected 1-methoxy-3-methylphthalazinium methyl sulfate **13**, which is also prepared from 1-methoxyphthalazine and dimethyl sulfate in benzene (Eq. 4).

The UV spectrum of **12** shows absorptions at 268 (s) (log ε, 3.65), 280 (min.) (log ε, 3.03), and 330 mμ (max) (log ε, 3.94). Ikeda and co-workers (5) have also reported the uv spectra of 3-benzyl-Ψ-phthalazinone and 3,4-dimethyl-Ψ-phthalazinone, both of which are similar to that of **12**.

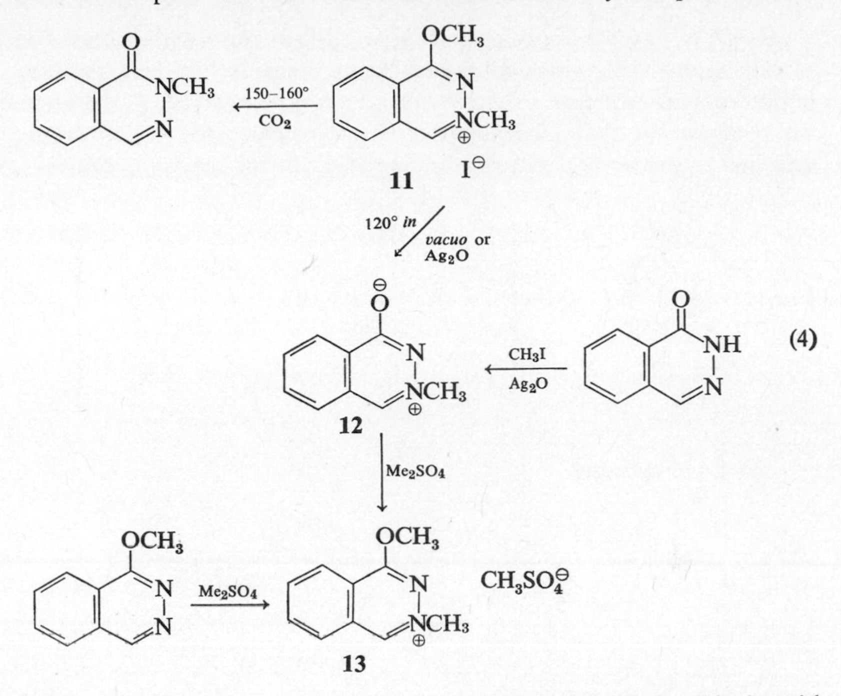

3-Aryl-Ψ-phthalazinones, on heating at 180° with 1.2*N* hydrochloric acid in a sealed tube for 6 hr, rearrange to give 2-aryl-1-(2*H*)phthalazinones (Rowe rearrangement) (Eq. 5). Vaughan (2) has summarized the effects of substituents on the aryl group in this rearrangement. The presence of a nitro group at the 2- and 4-positions facilitates this rearrangement but

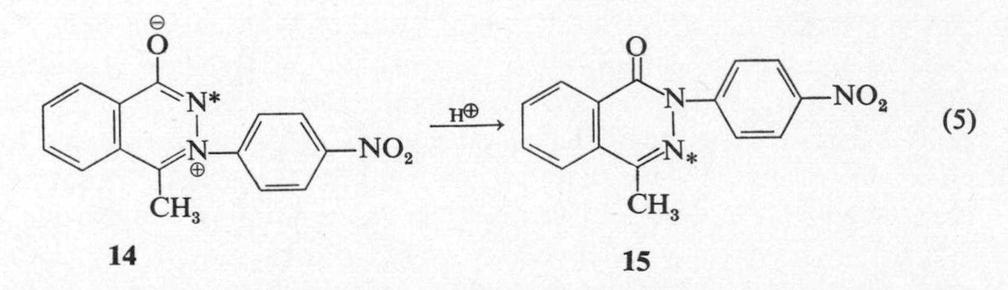

a 3-nitro group does not; halogen atoms in the 2- and/or 4-positions and a methyl group in the 2- and 4-positions appear to retard the reaction.

A mechanism for this reaction, originally proposed by Rowe (1, 2), involved migration of the aryl group. However, Vaughan (2) proposed an alternative mechanism (Eq. 6), which he substantiated by the use of labeled nitrogen as shown in Eq. 5. Vaughan and co-workers (6) studied the rearrangement of 3-(4′-nitrophenyl)-4-methyl-Ψ-phthalazinone (**14**) containing an excess of N^{15} in the 2-position to give 2-(4′-nitrophenyl)-4-methyl-1-(2*H*)-phthalazinone (**15**), which contained the excess of N^{15} in the 3-position. This indicates that migration of the aryl group is not involved. Vaughan and co-workers proposed the mechanism of Eq. 6 in which the aryl group remains attached to the same nitrogen atom throughout the reaction (Eq. 6).

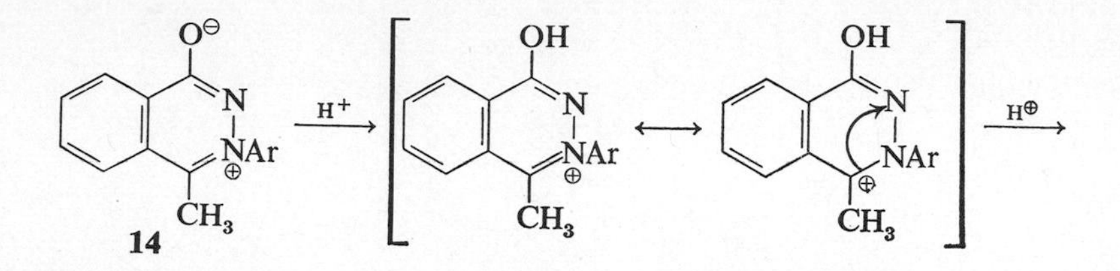

Ar = *p*-nitrophenyl

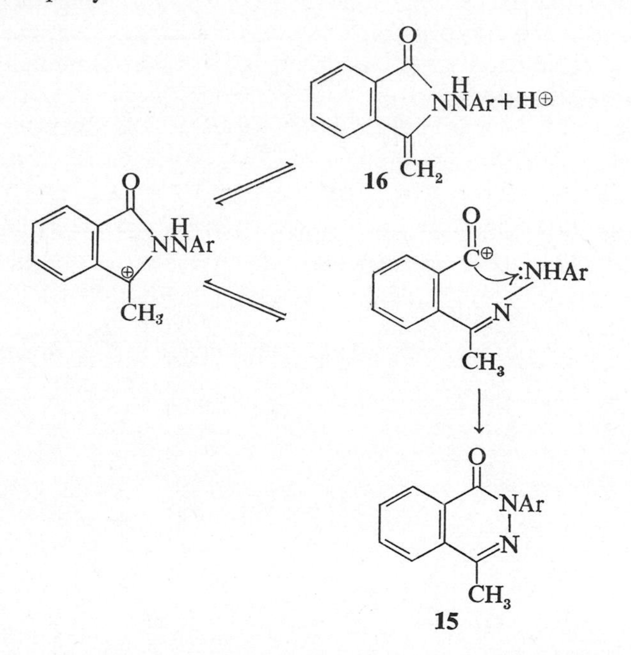

(6)

Rowe had isolated compounds similar to **16**, which on heating with sulfuric acid gave **15**. But he discarded the mechanism outlined in Eq. 6 because the rearrangement also proceeded with 3-aryl-Ψ-phthalazinones in which the 4-methyl group is replaced by a hydrogen atom. In this case, an intermediate like **16** is impossible. However, Vaughan argued that this objection is valid only if one considers that an intermediate like **16** must be stable, capable of finite existence, or actually isolable.

II. Phthalazine Quaternary Compounds

The most common method of preparing phthalazine quaternary compounds is to heat phthalazine or one of its derivatives with an alkyl halide in the absence of solvent or in acetonitrile, benzene, toluene, and so on. Thus phthalazine has been quaternized with methyl iodide (7), ethyl iodide (7), benzyl chloride (7), and several substituted benzyl chlorides (8) (Eq. 7).

N, N + RX ⟶ $NR^{\oplus}$, N $X^{\oplus}$ (7)

Druey and Danieker (9–11) have shown that phthalazine will react with butyl iodide, decyl bromide, dodecyl bromide, and other long-chain alkyl halides in refluxing acetonitrile. Subsequently adjusting the *p*H to ~12, they isolated and characterized the pseudo bases, 1-hydroxy-2-alkyl-1,2-dihydrophthalazines (**17**). Treatment of **17** with ethanolic hydrochloric acid provided the corresponding *N*-alkylphthalazinium chlorides (**18**). The same authors also condensed phthalaldehyde with long-chain alkylhydrazine sulfates in refluxing aqueous alcohol followed by *p*H adjustment to ~12 to give **17** (Eq. 8). In this way it is possible to replace bromide, iodide, and sulfate anions with a chloride anion in the quaternary salts. Similarly, *N*,*N*-ethylene-, tetra-

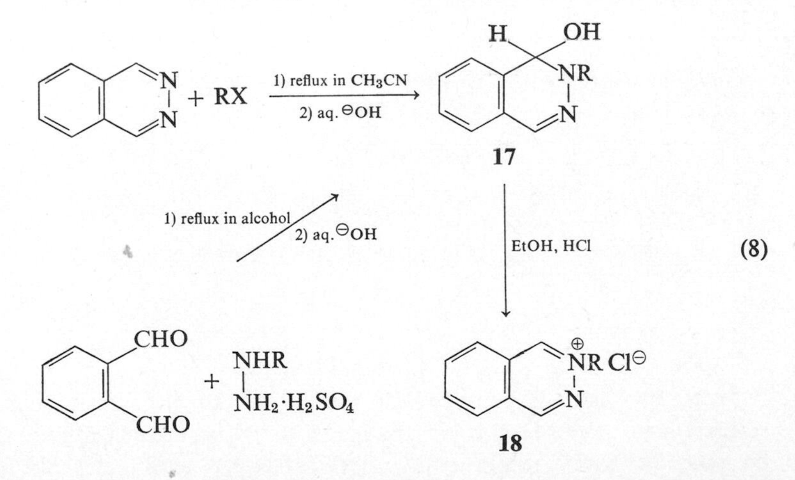

methylene-, and hexamethylene-bis-phthalazinium halides are prepared (11). Treatment of *N*,*N*′-tetramethylene-bis-phthalazinium chloride with alkali gives the pseudo base 2,2′-tetramethylene-bis-(1-hydroxy-1,2-dihydrophthalazine), but *N*,*N*′-ethylene-bis-phthalazinium halides (**19**) on treatment with alkali give anhydro-2,2′-ethylene-bis-(1-hydroxy-1,2-dihydrophthalazine) (**20**), which on treatment with ethanolic hydrochloric acid reverts to **19** (X = Cl) (Eq. 9).

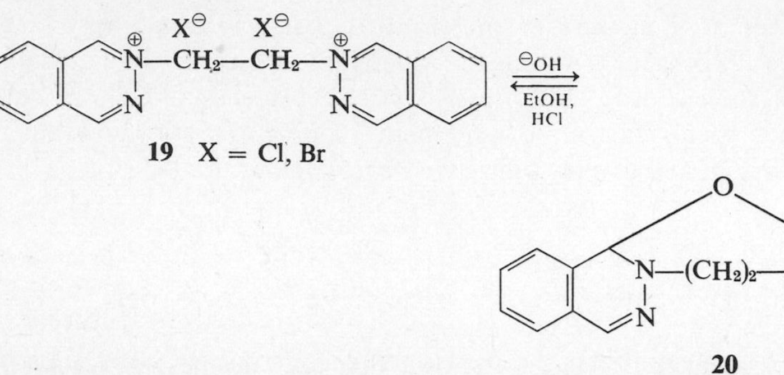

Ikeda and co-workers (5) have allowed 1-alkoxy-, 1-phenyl-, 1-benzyl-, and 1-methylphthalazine to react with methyl iodide in benzene and showed that in these cases quaternization takes place at the 3-position. Structural proof was obtained by oxidation of 1-substituted 3-methylphthalazinium iodides (**21**)

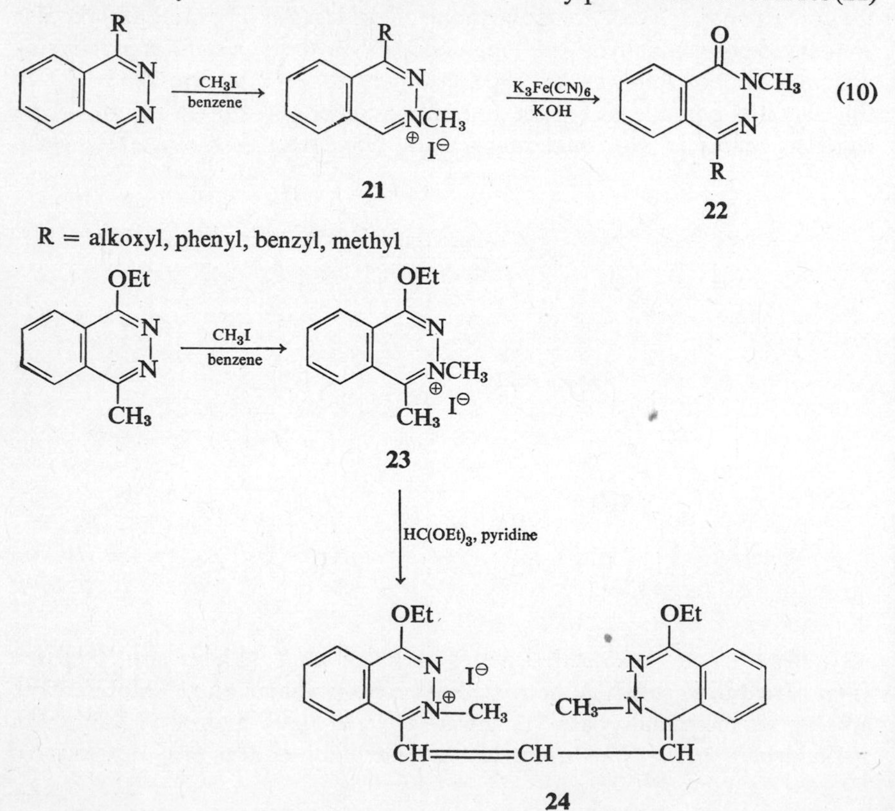

to the known 4-substituted 2-methyl-1-(2*H*)phthalazinones (**22**) (Eq. 10) and by heating **21** above the decomposition temperature or by reaction of **21** with silver oxide (see Eq. 4). The same authors (5) quaternized 1-methyl-4-ethoxyphthalazine with methyl iodide and obtained 1-ethoxy-3,4-dimethylphthalazinium iodide (**23**). The position of quaternization in **23** is confirmed by the formation of the trimethylcyanine dye (**24**), mp 225–226°, on reaction of **23** with triethyl orthoformate in pyridine (Eq. 10).

Smith and Otremba (7) have shown that 2-methyl- and 2-ethylphthalazinium iodide, on reduction with aqueous sodium borohydride, give 2-methyl- and 2-ethyl-1,2-dihydrophthalazines. Similar reduction of 2-benzylphthalazinium chloride furnishes 1,2-dihydrophthalazine with simultaneous debenzylation. 2-Methyl-1,2-dihydrophthalazine, on treatment with methyl iodide, gives 1,2-dihydro-2,2-dimethylphthalazinium iodide (**25**). Structure proof for **25** is provided by the reaction of **25** with alkali to give *o*-(*N*,*N*-dimethylaminomethyl)benzonitrile (**26**) by beta-elimination (Eq. 11).

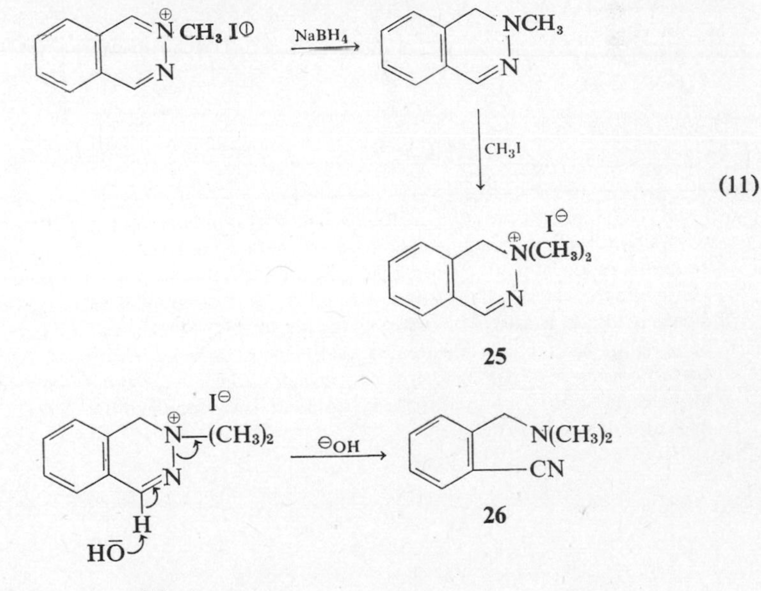

Quaternization of certain aminophthalazines has been reported, but the structures of the products are not known. These compounds are reported as derivatives of the aminophthalazines (see Section 2H).

Kanahara (16) has allowed 5-nitrophthalazine and 1-methyl-5-nitrophthalazine to react with methyl iodide. In each case, both 2-methyl and 3-methyl quaternary salts were obtained. The isomers were separated and their structures established by oxidation with potassium ferricyanide to give the corresponding 2-methyl-1-(2*H*)phthalazinones.

III. Tables

TABLE 2K-1. Ψ-Phthalazinones

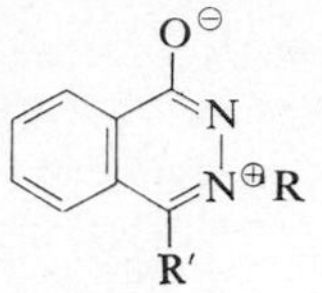

R	R′	Preparation[a]	MP	Reference
CH_3	H	A	236°	5
CH_3	CH_3	A_1	H_2O, 216–217°	5
$C_6H_5CH_2$	H	A_2	218–220°	5
C_6H_5	H	B	208°; 210°	3, 4
o-$CH_3C_6H_4$	H	B	201°	3
m-$CH_3C_6H_4$	H	B	184°	3
p-$CH_3C_6H_4$	H	B	215°	3
p-$NO_2C_6H_4$	CH_3	—	251–252°	6
p-$NH_2C_6H_4$	CH_3	C	277° (dec.)	6
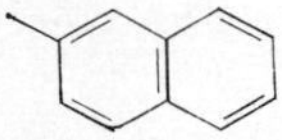	H	B	245°	3

[a] A, aqueous solution of 1-methoxy-3-methylphthalazinium iodide stirred with silver oxide at room temperature; or heat 1-methoxy-3-methylphthalazinium iodide at 120° *in vacuo*; or 1-(2*H*)phthalazinone is treated with methyl iodide followed by silver oxide. A_1, 1,2-dimethyl-4-ethoxyphthalazinium iodide heated with silver oxide as in method A. A_2, 1-ethoxy-3-benzylphthalazinium iodide treated with silver oxide as in method A. B, corresponding *N*-arylamino-3-hydroxyphthalimidines, heat at 120° or reflux in water solution. C, 3-(*p*-nitrophenyl)-4-methyl-Ψ-phthalazinone is reduced with aqueous sodium sulfide.

TABLE 2K-2. Phthalazine Quaternary Salts

(Phthalazinium structure: $\overset{\oplus}{N}R$, $X^{\ominus}$)

R	$X^{\ominus}$	Preparation[a]	Yield (%)	MP	Reference
CH_3	I	—	86	240–243° (dec.)	7
CH_3	Picrate	A	75	Orange crystals, 199–200° (dec.) (ethanol)	7
CH_2CH_3	I	—	95	Yellow crystals, 225–228° (dec.) (ethanol)	7
CH_2CH_3	Picrate	A	93	Yellow crystals, 167–169° (dec.) (ethanol)	7
CH_2CH_2OH	Cl	B	—	232–234° (ethanol)	12–14
CH_2CH_2Cl	Cl	B_1	—	Hygroscopic crystals, 178–179° (dec.) (ethanol-ether)	12–14
CH_2CH_2Br	Br	B_2	—	155° (dec.)	12–14
$CH_2(CH_2)_2CH_3$	Cl	C	74	64°	9–11
$CH_2(CH_2)_8CH_3$	Cl	C	53	Colorless needles, 93° (acetone)	9, 10
$CH_2(CH_2)_9CO_2H$	Cl	C_1	—	100° (acetone)	9
$CH_2(CH_2)_9CO_2H$	Br	D	78	166.5°–167° (alcohol)	9
$CH_2(CH_2)_9CO_2CH_3$	Br	D_1	68	94–95°	9
$CH_2(CH_2)_{10}CH_3$	Br	D_2	77	78° (acetone)	9
$CH_2(CH_2)_{14}CH_3$	Cl	C	—	Colorless crystals, 70° (acetone)	9, 10
$CH_2(CH_2)_{16}CH_3$	Cl	C	83	Colorless crystals, 97° (acetone)	9, 10
$CH_2(CH_2)_7CH{=}CH(CH_2)_7CH_3$	Cl	C	—	76° (acetone)	9, 10
$CH_2(CH_2)_{20}CH_3$	I	D_3	97	105° (acetone)	9
$C_6H_5CH_2$	Cl	E	89	White crystals, hygroscopic, 171–175° (ethanol-ether)	7
$CH_2(CH_2)_{14}CH_3$					
$C_6H_5CH_2$	Picrate	A	100	Yellow crystals, 183–184° (methanol)	7
$C_6H_5CH_2$	CO_2H–$CO_2^{\ominus}$	—	—	170°	8

TABLE 2K-2 (*continued*)

R	$X^{\ominus}$	Preparation[a]	Yield (%)	MP	Reference
o-$NO_2C_6H_4CH_2$	Cl	D_4	—	192°	8
m-$NO_2C_6H_4CH_2$	Cl	D_4	—	232–234°	8
p-$NO_2C_6H_4CH_2$	Cl	D_4	—	170–171°	8
m-$FC_6H_4CH_2$	Cl	D_4	—	171–173°	8
p-$ClC_6H_4CH_2$	Cl	D_4	—	121–123°	8
2,4-$Cl_2C_6H_3CH_2$	Cl	D_4	—	208–210° (2-propanol)	8
3,4-$Cl_2C_6H_3CH_2$	Cl	D_4	—	181–182°	8
$C_6Cl_5CH_2$	Cl	D_4	—	235–237°	8
$C_6Cl_5CH_2$	HSO_4	—	—	250°	8
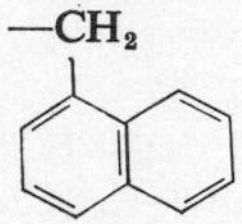 —CH_2	Cl	D_4	—	110°	8
$C_6H_5CH_2CH_2$	Br	D_4	—	198–199°	8
—$CH_2C(=O)C_6H_5$	Br	E_1	—	—	15
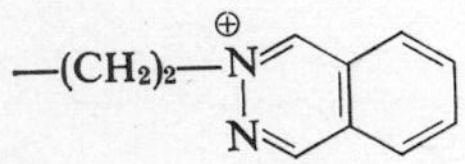 —$(CH_2)_2$—$N^{\oplus}$ N	2Cl	F	74	Hygroscopic yellow prisms, 221° (alcohol)	11–14
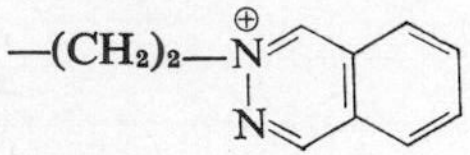 —$(CH_2)_2$—$N^{\oplus}$ N	2Br	G	55	Hygroscopic yellow crystals, 277°–278° (alcohol)	11–14

TABLE 2K-2 (*continued*)

R	$X^{\ominus}$	Preparation[a]	Yield (%)	MP	Reference
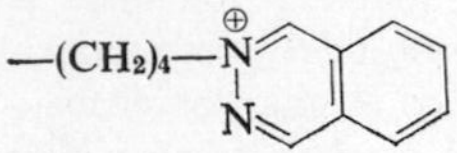 $-(CH_2)_4-N^{\oplus}$ (phthalazinium)	2Cl	F	82	Hygroscopic, 285–286° (alcohol)	11–14
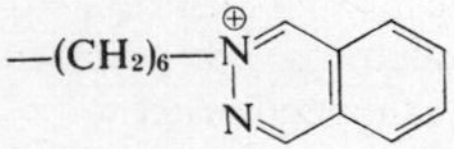 $-(CH_2)_6-N^{\oplus}$ (phthalazinium)	2Cl	F	62	Hygroscopic colorless crystals, 248°	11–14
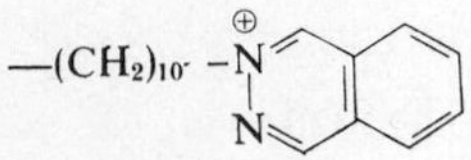 $-(CH_2)_{10}-N^{\oplus}$ (phthalazinium)	2Br	G	—	182–183°	12, 14
Miscellaneous Compounds					
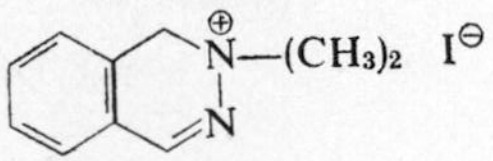 (1,2-dihydrophthalazine) $N^{\oplus}-(CH_3)_2$ $I^{\ominus}$		E_2	83	White crystals, 175–176° (methanol)	7
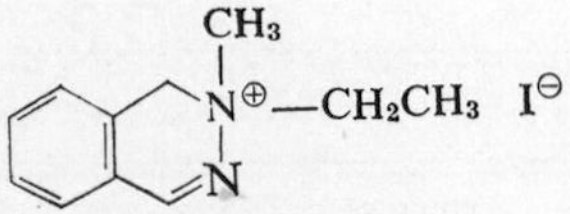 (1,2-dihydrophthalazine) $N^{\oplus}(CH_3)-CH_2CH_3$ $I^{\ominus}$		E_2	—	White crystals, 155–157° (ethanol)	7

TABLE 2K-2 (*continued*)

R	$X^{\ominus}$	Preparation[a]	Yield (%)	MP	Reference
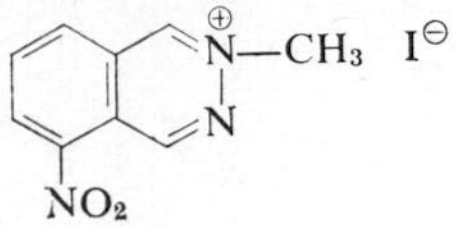 5-Nitro-2-methylphthalazinium ($\overset{\oplus}{N}$—CH_3 $I^{\ominus}$, NO_2)		H	—	Yellow needles, 195–202.5° (dec.)	16
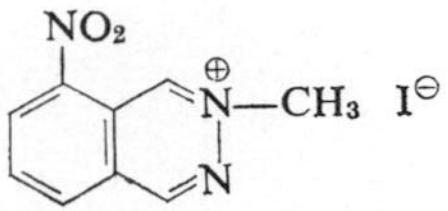 8-Nitro-2-methylphthalazinium ($\overset{\oplus}{N}$—CH_3 $I^{\ominus}$, NO_2)		H	—	Reddish orange needles, 170–177° (dec.)	16
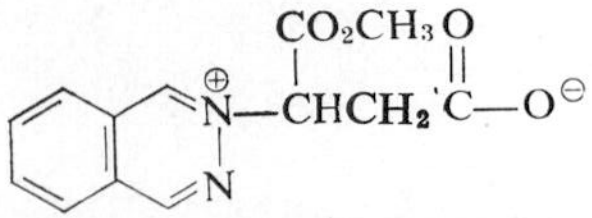 Phthalazinium $\overset{\oplus}{N}$—$CH(CO_2CH_3)CH_2C(=O)$—$O^{\ominus}$		I	—	82°	17

[a] A, corresponding alkylphthalazinium halide is warmed in a saturated ethanolic solution of picric acid (~20 ml/g of halide). B, phthalazine + 2-chloroethanol, reflux in toluene for 10 hr. B_1, 2-hydroxyethylphthalazinium chloride + thionyl chloride, reflux 2 hr. B_2, same as B_1, heat with 48% hydrobromic acid in a sealed tube at 180° for 20 hr. C, the corresponding 1-hydroxy-2-alkyl-1,2-dihydrophthalazine is treated with ethanolic hydrochloric acid. C_1, 1-hydroxy-2-(ω-carbomethoxy-*n*-decyl)-1,2-dihydrophthalazine treated with methanolic hydrochloric acid. D, phthalazine + ω-bromodecanoic acid, reflux 18 hr in acetonitrile. D_1, phthalazine + methyl ω-bromodecanoate, reflux 48 hr in acetonitrile. D_2, phthalazine + dodecyl bromide, reflux 48 hr in acetonitrile. D_3, phthalazine + docosanyl iodide, reflux 3 days in acetonitrile. D_4, phthalazine + corresponding aralkyl chloride, reflux in acetonitrile for 1 hr. E, phthalazine + benzyl chloride, reflux 3 hr in methanol. E_1, phthalazine + bromoacetophenone kept in methanol at room temperature for 24 hr. E_2, corresponding 2-alkyl-1,2-dihydrophthalazine + methyl iodide, reflux 3 hr in ethanol. F, phthalaldehyde + proper $NH_2NH(CH_2)_nNHNH_2 \cdot 2HCl$, reflux in aqueous alcohol for 5 hr. G, phthalazine + $Br(CH_2)_nBr$, reflux in acetonitrile for 2 hr. H, 5-nitrophthalazine and methyl iodide in refluxing methanol. Quaternization takes place at both nitrogen atoms. I, the ylide obtained by mixing phthalazine with maleic anhydride was refluxed with methanol.

TABLE 2K-3. 1-Substituted and 1,4-Disubstituted Phthalazine Quaternary Salts

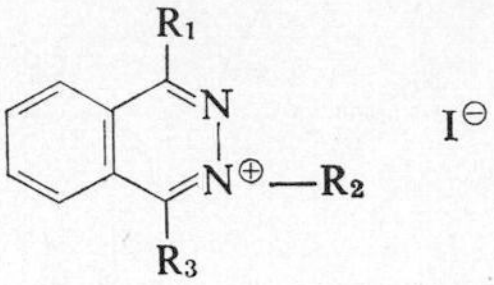

R_1, N, $N^{\oplus}$—R_2, R_3, $I^{\ominus}$

R_1	R_2	R_3	Preparation[a]	MP	Reference
CH_3	CH_3	H	—	142–143°	1
CH_3	$C_6Cl_5CH_2$	H	A	244–245°	8
OCH_3	CH_3	H	B	158–160°	5
OCH_3	CH_2CH_3	H	B	122°	5
OCH_3	$CH_2CH_2CH_3$	H	B	130–133°	5
OCH_2CH_3	CH_3	H	B	147°	5
OCH_2CH_3	$C_6H_5CH_2$	H	B	142–144°	5
$OCH_2CH_2CH_3$	CH_3	H	B	150–151°	5
C_6H_5	CH_3	H	B	245–246°	5
$C_6H_5CH_2$	CH_3	H	B	212–214°	5
OCH_3	CH_3	C_6H_5	C	340°	4
OCH_2CH_3	CH_3	CH_3	B	204°	5
C_6H_5	C_6H_5	CH_3	D	273°	4

Miscellaneous Compounds

Compound	Preparation[a]	MP	Reference
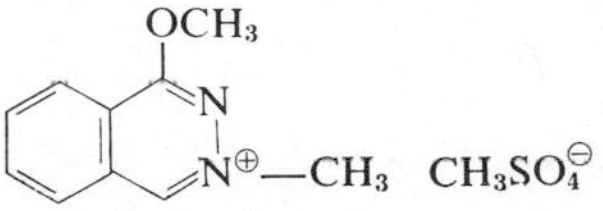 OCH_3, N, $N^{\oplus}$—CH_3 $CH_3SO_4^{\ominus}$	E	146–148°	5
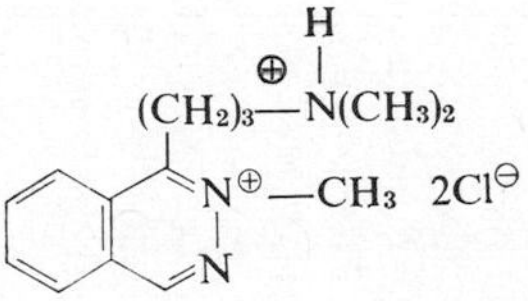 H, $\oplus$, $(CH_2)_3$—$N(CH_3)_2$, $N^{\oplus}$—CH_3 $2Cl^{\ominus}$, N	F	195–198° (ethanol)	18
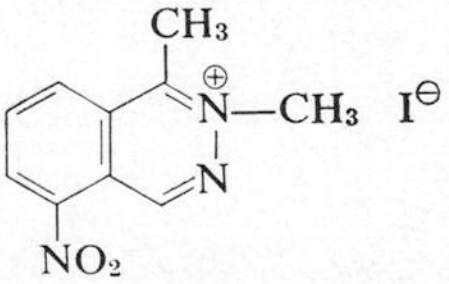 CH_3, $\oplus$, N—CH_3 $I^{\ominus}$, N, NO_2	G	208–213° (dec.) (ethanol)	16
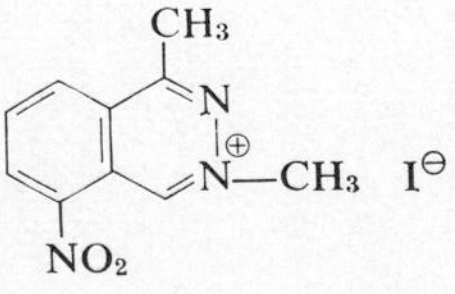 CH_3, N, $\oplus$, N—CH_3 $I^{\ominus}$, NO_2	G	167–170° (dec.) (ethanol)	16

TABLE 2K-3 (*continued*)

R_1	R_2	R_3	Preparation[a]	MP	Reference
C_6H_5 C_6H_5 $\overset{\oplus}{N}$—CH_3 $I^{\ominus}$ N C_6H_5 C_6H_5			H	Orange solid, 309–314° (dec.)	19
OCH_3 N $^{\oplus}CH_2CH_3$ $Br^{\ominus}$ N $CH(CH_3)_2$			I	123–124° (ethanol-pet. ether)	20

[a] A, 1-methylphthalazine + pentachlorobenzyl chloride, reflux in acetonitrile for 1 hr. B, properly substituted phthalazine + alkyl iodide, let stand overnight at room temperature in benzene. C, Grignard addition of methylmagnesium iodide to 2-phenyl-4-hydroxy-1-(2*H*)phthalazinone in tetrahydrofuran. D, 1,4-diphenylphthalazine + methyl iodide. E, 3-methyl-Ψ-phthalazinone + dimethyl sulfate, heat in methanol at 60° for 4 hr. F, 1-hydroxy-1-(3′-*N*,*N*-dimethylaminopropyl)-2-methyl-1,2-dihydrophthalazine treated with ethanolic hydrochloric acid. G, 1-methyl-5-nitrophthalazine + methyl iodide, reflux in methanol. Quaternization takes place at both nitrogen atoms. H, 1,4,5,8-tetraphenylphthalazine + methyl iodide, let stand overnight in chloroform. I, 1-methoxy-4-isopropylphthalazine + ethyl bromide, let stand for 3 days in ethanol-acetone.

References

1. J. C. E. Simpson, in *Condensed Pyridazine and Pyrazine Rings*, *The Chemistry of Heterocyclic Compounds*, Vol. 5, A. Weissberger, Ed., Interscience, New York, 1953, p. 119.
2. W. R. Vaughan, *Chem. Rev.*, **43,** 447 (1948).
3. H. Lund, *Tetrahedron Lett.*, **1965,** 3973; *Coll. Czech. Chem. Commun.*, **30,** 4237 (1965).
4. B. K. Diep and B. Cauvin, *Compt. Rend.*, *Ser. C.*, **262,** 1010 (1960).
5. T. Ikeda, S. Kanahara, and A. Aoki, *Yakugaku Zasshi*, **88,** 521 (1968); *Chem. Abstr.*, **69,** 86934 (1968).
6. W. R. Vaughan, D. I. McCane, and G. J. Sloan; *J. Am. Chem. Soc.*, **73,** 2298 (1951).
7. R. C. Smith and E. D. Otremba, *J. Org. Chem.*, **27,** 879 (1962).
8. Belgian Pat. 648,344 (1964); *Chem. Abstr.*, **68,** 39638 (1968).
9. J. Druey and H. U. Daeniker, U.S. Pat. 2,945,037 (1960); *Chem. Abstr.*, **55,** 4546 (1961).
10. J. Druey and H. U. Daeniker, Swiss Pat. 353,369 (1957); *Chem. Abstr.*, **59,** 5177 (1963).
11. H. U. Daeniker and J. Druey, *Helv. Chim. Acta*, **40,** 918 (1957).
12. J. Druey and H. U. Daeniker, U.S. Pat. 2,945,036 (1960); *Chem. Abstr.*, **55,** 4550 (1961).
13. J. Druey and H. U. Daeniker, German Pat. 1,056,614 (1959); *Chem. Abstr.*, **55,** 13456 (1961).

14. J. Druey and H. U. Daeniker, Swiss Pat. 358,430 (1962); *Chem. Abstr.*, **59,** 11529 (1963).
15. M. Petrovanu, A. Sauciuc, I. Gabe, and I. Zugravescu, *Rev. Roum. Chim.*, **14,** 1153 (1969); *Chem. Abstr.*, **72,** 43591 (1970).
16. S. Kanahara, *Yakugaku Zasshi*, **84,** 483 (1964); *Chem. Abstr.*, **61,** 8304 (1964).
17. I. Zugravescu, M. Petrovanu, A. Cavaculacu, and A. Sauciuc, *Rev. Roum. Chim.*, **12,** 109 (1967); *Chem. Abstr.*, **68,** 49537 (1968).
18. A. Marxer, *Helv. Chim. Acta*, **49,** 572 (1966).
19. T. H. Regan and J. B. Miller, *J. Org. Chem.*, **31,** 3053 (1966).
20. Belgian Pat. 628,255 (1963); *Chem. Abstr.*, **60,** 14516 (1964).

Part L. Reduced Phthalazines

The subsections of this review are placed according to the position of reduction in the phthalazine nucleus, 1,2-dihydro-, 1,2,3,4-tetrahydro-, 5,6,7,8-tetrahydro-, 4a,5,8,8a-tetrahydro-, and 4a,5,6,7,8,8a-hexahydrophthalazines. The reduced phthalazines may have hydrogen atoms or other substituents at these positions. For example, 1-hydroxy-1,2,4-trimethyl-1,2-dihydrophthalazine (**1**) is classified as a 1,2-dihydrophthalazine even though there are no hydrogen atoms at positions 1 and 2. Compounds such as 1,2,3,4-tetrahydro-4-oxophthalazine (**2**) are classified as 1,2-dihydrophthalazines.

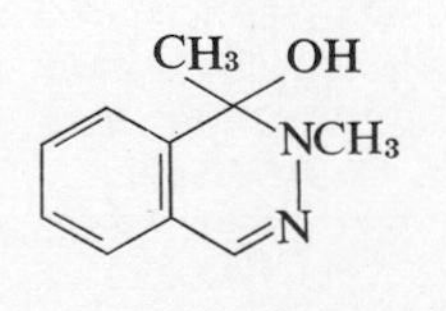

1

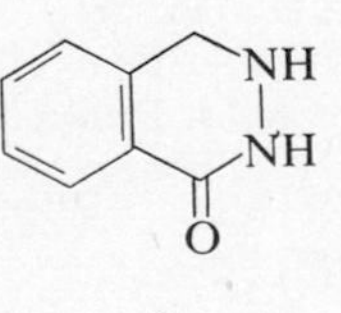

2

I. 1,2-Dihydrophthalazines

A. Preparation

Shabarov and co-workers (1) first obtained 1,2-dihydrophthalazine (**3**) by the reduction of 1-(2*H*)phthalazinone with lithium aluminum hydride. In an attempt to prepare benzocyclobutadiene, Leznoff (2) isolated *N*-(1,2-dihydrophthalazinyl)-*p*-tolylsulfone (**4**) from the reaction of the ditosylhydrazone of phthalaldehyde with sodium hydride (Eq. 1). In order to confirm the structure of **4**, the author reduced phthalazine with lithium aluminum hydride to give **3** as an intermediate; this was reacted further with *p*-toluenesulfonyl chloride to give **4**. Smith and Otremba (3) have reduced

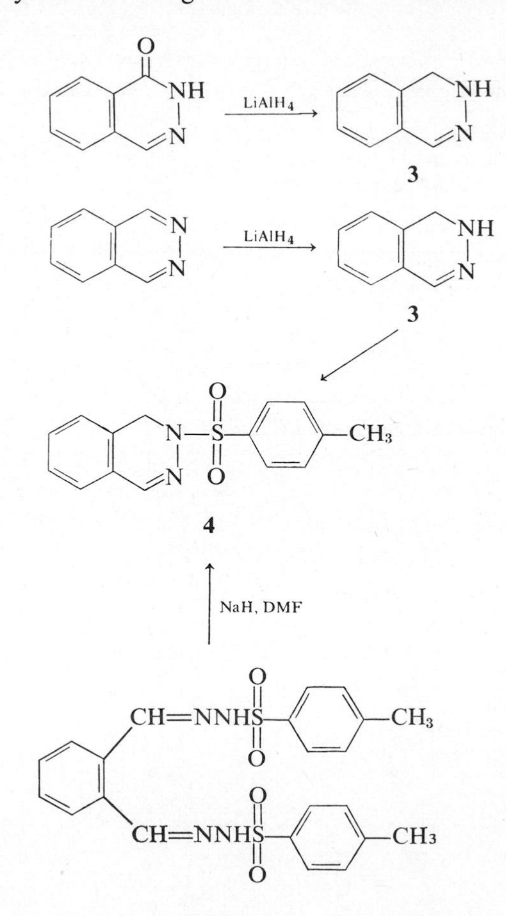

(1)

phthalazine quaternary compounds with sodium borohydride to give 2-alkyl-1,2-dihydrophthalazines (**5**). The reduction of 2-benzylphthalazinium chloride (**6**) with sodium borohydride gave **3** with simultaneous debenzylation (Eq. 2).

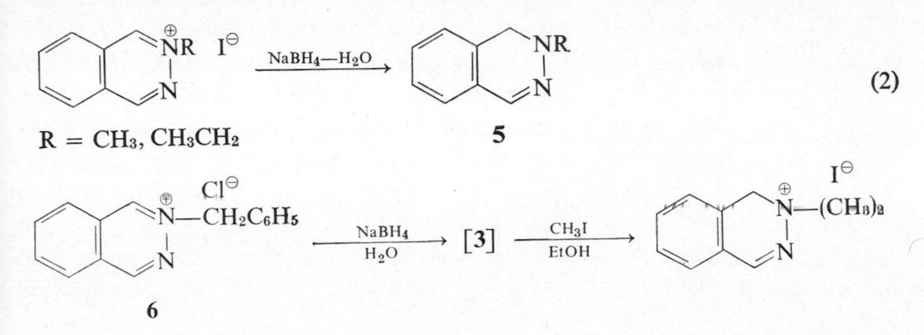

Lund (4, 5) has studied the polarographic reduction of 4-hydroxy-2-phenyl-1-(2*H*)phthalazinone (**7**) to give 2-phenyl-1,2,3,4-tetrahydro-4-oxophthalazine (**8**), which is also obtained by a two-electron reduction of 3-phenyl-Ψ-phthalazinone at controlled potential (Eq. 3). He has also shown

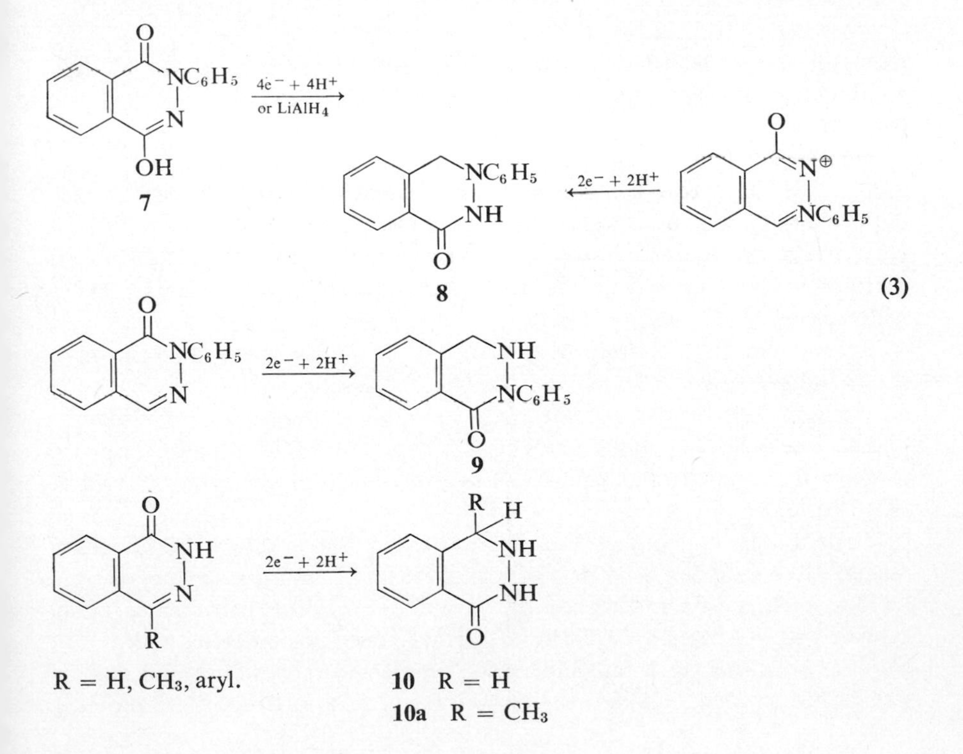

that two-electron reductions of 2-phenyl-1-(2*H*)phthalazinone and 4-methyl-1-(2*H*)phthalazinone provide 3-phenyl-1,2,3,4-tetrahydro-4-oxophthalazine (**9**) and 1-methyl-1,2,3,4-tetrahydro-4-oxophthalazine (**10a**), respectively. Lithium aluminum hydride reduction of **7** also provided **8** (6).

The lithium aluminum hydride reductions (1) of 1-(2*H*)phthalazinone and 4-phenyl-1-(2*H*)phthalazinone have been shown to reduce the carbonyl group only to give **3** and 4-phenyl-1,2-dihydrophthalazine. However, similar

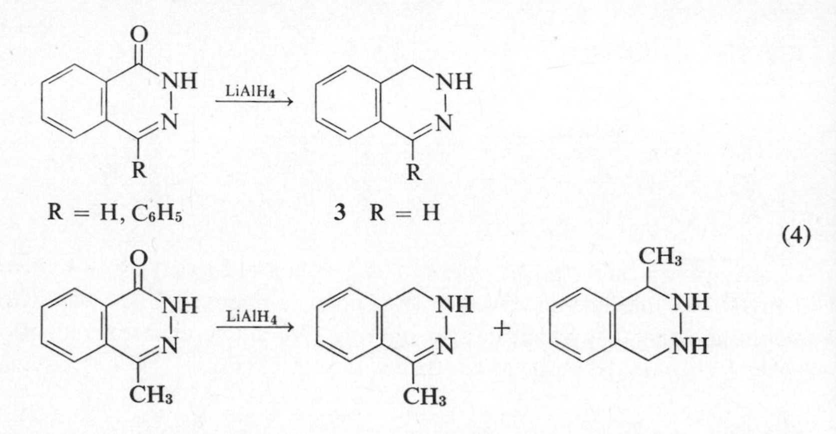

reduction of 4-methyl-1-(2*H*)phthalazinone (40) afforded 57% of 4-methyl-1,2-dihydrophthalazine and 41% of 1-methyl-1,2,3,4-tetrahydrophthalazine (Eq. 4).

Bellasio and Testa (7) have hydrogenated 1-(2*H*)phthalazinone with platinum oxide in acetic acid to give 2,3-diacetyl-1,2,3,4-tetrahydro-4-oxophthalazine (**11**) in poor yield (5%), which on acidic hydrolysis provided 1,2,3,4-tetrahydro-4-oxophthalazine (**10**) (Eq. 5). The condensation of 2-bromomethylbenzoyl chloride with *N*,*N*′-diacetylhydrazine gave **11** in better yields (7, 8).

Marxer (9) has reported the addition of *N*,*N*-dimethylaminopropylmagnesium chloride to 2-methyl-1-(2*H*)phthalazinone, followed by hydrolysis, to give 1-hydroxy-1-(*N*,*N*-dimethylaminopropyl)-2-methyl-1,2-dihydrophthalazine (**12**). Diep and Cauvin (6) have studied the addition of Grignard reagents to 2-methyl- and 2-phenyl-4-hydroxy-1-(2*H*)phthalazinones to give (**13**) (Eq. 6).

Mustafa and co-workers (10) have studied the addition of Grignard reagents to 2,4-diphenyl-1-(2*H*)phthalazinone and found that arylmagnesium halides such as phenylmagnesium bromide gave the expected product 1-hydroxy-1,2,4-triphenyl-1,2-dihydrophthalazine (**14**), whereas addition of alkyl- or arylalkylmagnesium halides such as methylmagnesium iodide gave an unexpected product, 1,1-dimethyl-2,4-diphenyl-1,2-dihydrophthalazine (**15**).

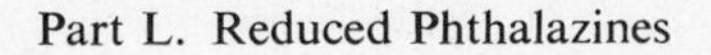

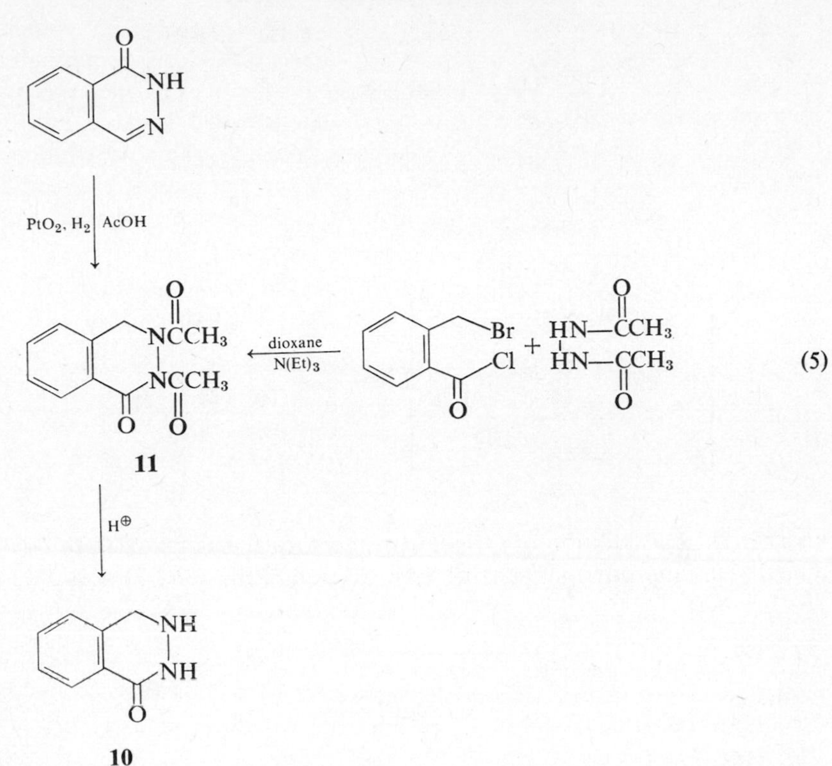

The authors have proposed the intermediate **16** for the formation of **15** (Eq. 7), where ring opening and recyclization takes place.

Marxer and co-workers (11) have studied the addition of Grignard reagents of phthalazine and its alkyl derivatives to give the corresponding 1,2-dihydrophthalazines in high yields. The reaction of phthalazine with *N*,*N*-dimethylaminopropylmagnesium chloride in tetrahydrofuran afforded 1-(*N*,*N*-dimethylaminopropyl)-1,2-dihydrophthalazine (Eq. 8), which does not

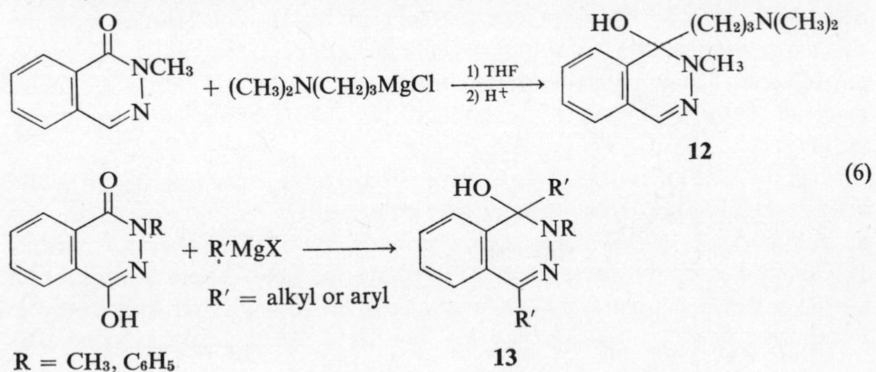

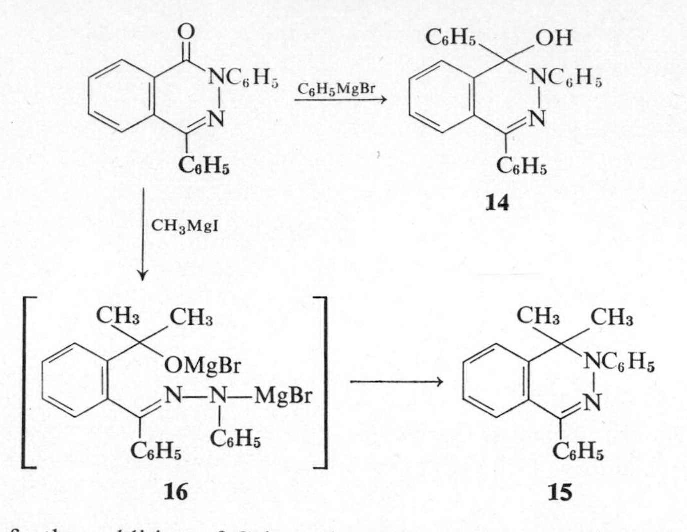

undergo further addition of Grignard reagent until it is oxidized to 1-(*N*,*N*-dimethylaminopropyl)phthalazine (see Sections 2B-I-A-2, 2B-I-B-2-d, and 2B-II-A-4-a).

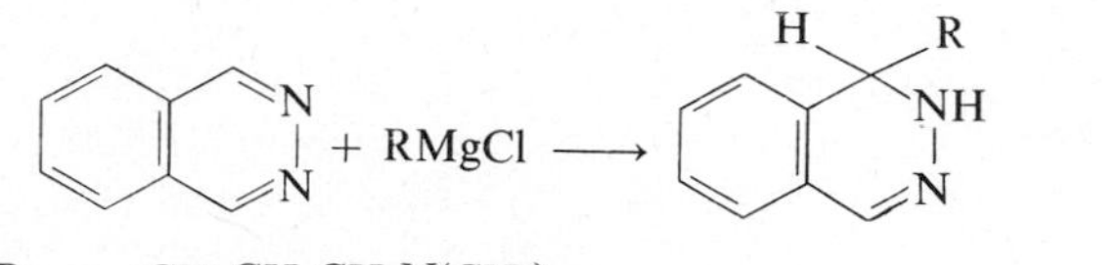

R = $—CH_2CH_2CH_2N(CH_3)_2$

Hirsch and Orphanos (12) have reported the addition of organolithium compounds such as vinyllithium to phthalazine and 1-(2*H*)phthalazinone. The reaction with phthalazine gave 1-vinyl-1,2-dihydrophthalazine (**17**), but the reaction with 1-(2*H*)phthalazinone gave 1-vinyl-1,2,3,4-tetrahydro-4-oxophthalazine (**18**) (Eq. 9) where addition of vinyllithium took place at

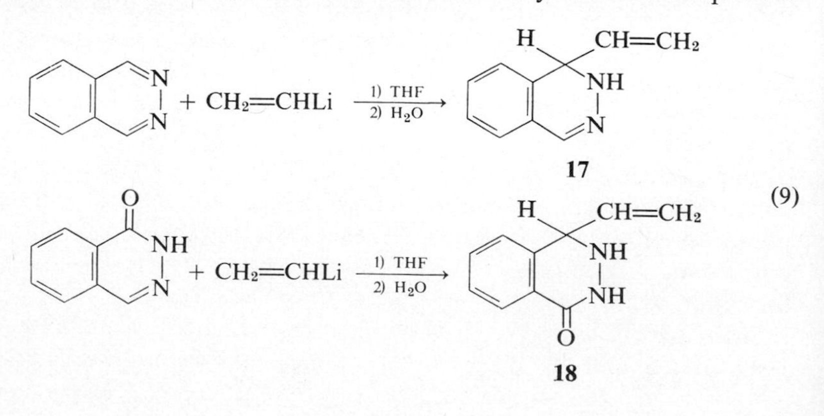

—C=N— instead of at the carbonyl group which was the case in addition of Grignard reagents.

Popp and co-workers (13) have studied the reaction of acid chlorides with phthalazine and potassium cyanide to prepare Reissert compounds in high yields. For example, benzoyl chloride afforded 1-cyano-2-benzoyl-1,2-dihydrophthalazine (**19**) (Eq. 10).

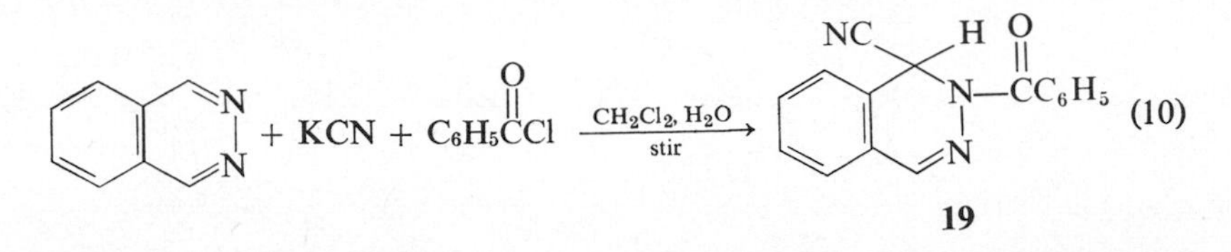

Druey and Daeniker (14) have shown that the reaction of phthalaldehyde with alkylhydrazine hydrobromides, followed by treatment with alkali, gives 1-hydroxy-2-alkyl-1,2-dihydrophthalazines (**20**). They have also shown the formation of **20** by treatment of phthalazine quaternary salts with aqueous base (Eq. 11).

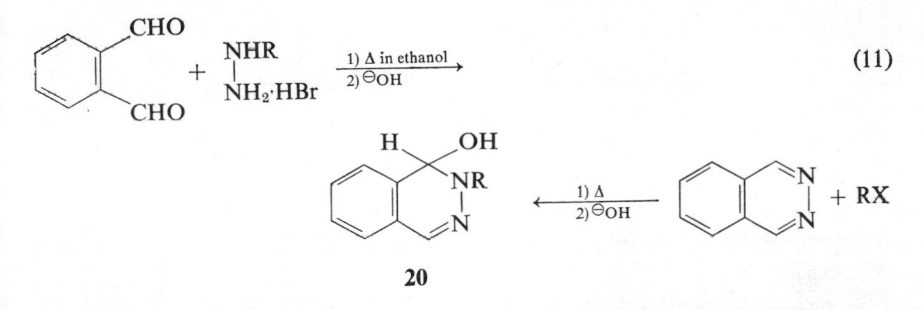

Oishi (15), in an attempt to substitute the methylsulfonyl group in 1-methylsulfonylphthalazine with certain ketone carbanions, observed the formation of unexpected derivatives of 1,2-dihydrophthalazines. Similar reactions with 1-methylsulfonyl-4-phenylphthalazine gave the expected nucleophilic substitution of the methylsulfonyl group (see Section 2B-I-A-4). The reaction of 1-methylsulfonylphthalazine with acetophenone in presence of aqueous base gave 1-phenacyl-2-(1′-phthalazinyl)-4-methylsulfonyl-1,2-dihydrophthalazine (**20a**). The structure of **20a** was established by spectroscopy and nucleophilic substitution of the methylsulfonyl group with potassium cyanide to give 1-phenacyl-2-(1′-phthalazinyl)-4-cyano-1,2-dihydrophthalazine (Eq. 12). Oishi has also proposed a mechanism for the formation of **20a** as shown in Eq. 12.

Shah and Taylor (15a) have studied the reaction of phthalazine and 1-substituted and 1,4-disubstituted phthalazines with dimethylketene in

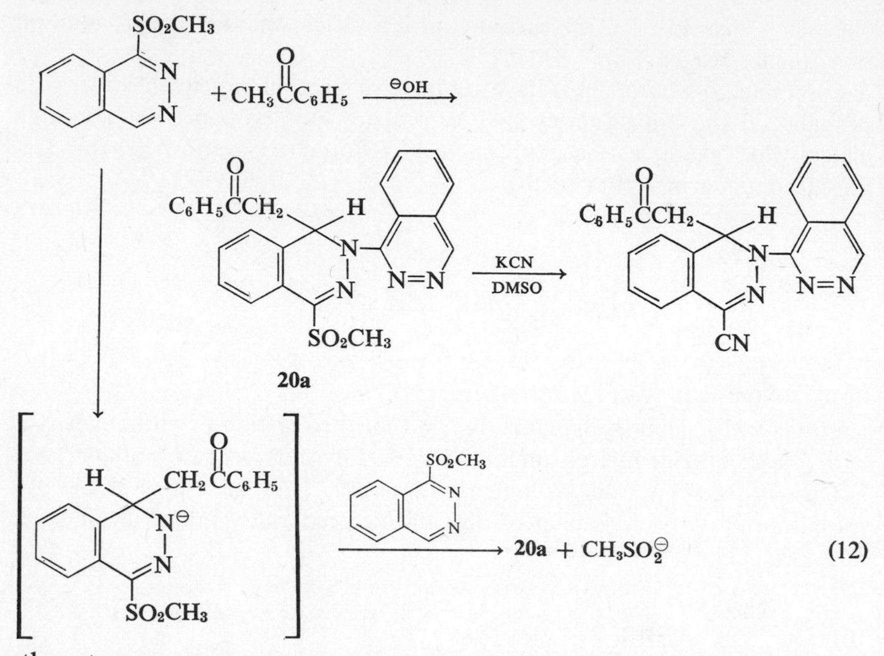

ether at room temperature to give quantitative yields of 1:2 adducts having oxazinophthalazine structures. Hydrolysis of these adducts furnished the corresponding amino acids (Eq. 12a).

Potts and Lovelette (16) have reduced *s*-triazolo[3,4-*a*]phthalazines with excess of lithium aluminum hydride in tetrahydrofuran to give 5,6-dihydro-*s*-triazolo[3,4-*a*]phthalazines in high yields (Eq. 13).

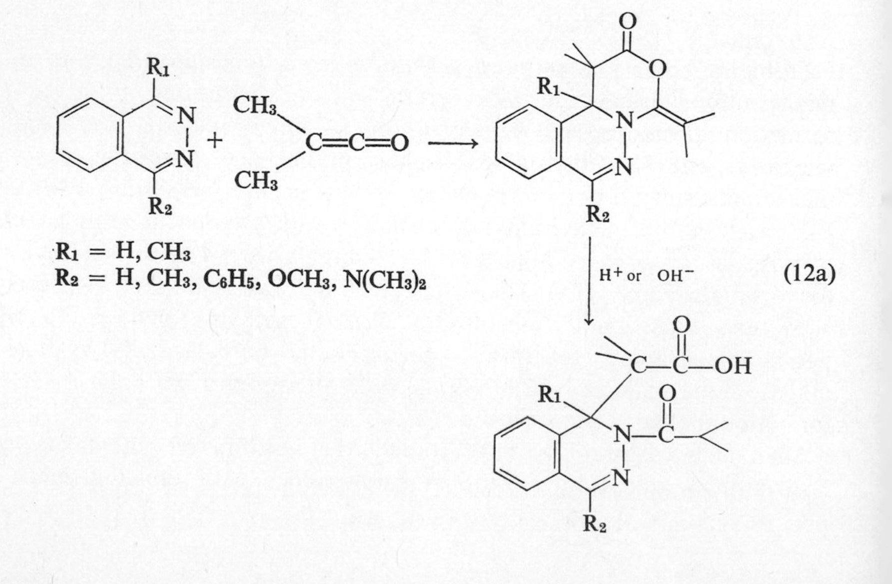

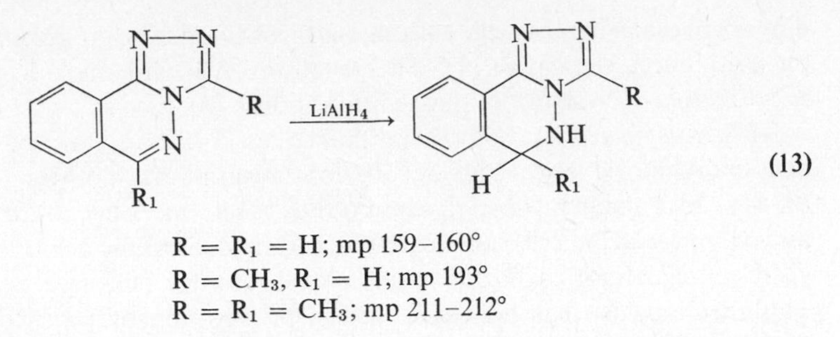

$R = R_1 = H$; mp 159–160°
$R = CH_3, R_1 = H$; mp 193°
$R = R_1 = CH_3$; mp 211–212°

Lund (17) has proposed the structure 1,1′-di(4-hydroxyphenyl)-1,2,3,4-tetrahydro-4-oxophthalazine (**21**) for the product obtained from the reaction of phenolphthalin with hydrazine (Eq. 14).

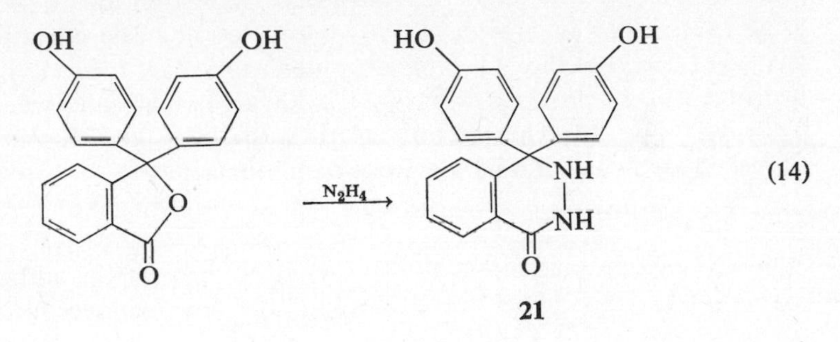

B. Reactions

In general, 1,2-dihydrophthalazines are unstable compounds. 2-Methyl-1,2-dihydrophthalazine easily oxidizes in air to give 2-methyl-1-(2*H*)phthalazinone (3). 1-Alkyl-1,2-dihydrophthalazines are also sensitive to air oxidation, and potassium ferricyanide oxidation converts them to 1-alkylphthalazines (11, 12) (Eq. 15) (see also Section 2B-I-A-2).

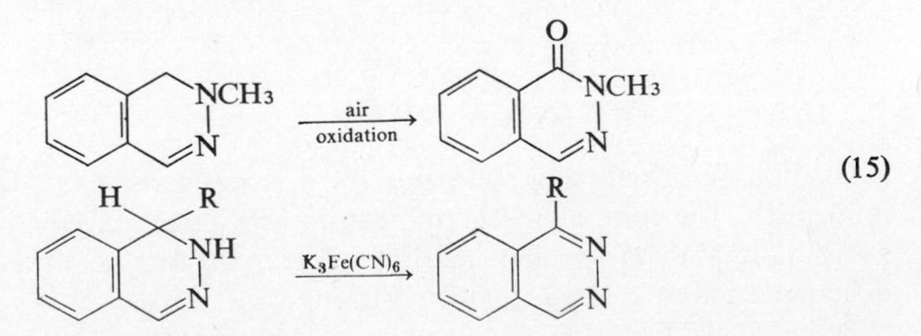

1,2-Dihydrophthalazine (**3**) reacts with *p*-toluenesulfonyl chloride (2) to give *N*-(1,2-dihydrophthalazinyl)-*p*-toluenesulfone (**4**) (see Eq. 1). The dihydrophthalazine **3** reacts with phenyl isothiocyanate to give the expected phenylthiourea derivative (1). The reaction of **3** with methyl iodide yields 2,2-dimethyl-1,2-dihydrophthalazinium iodide (3).

1-Hydroxy-2-alkyl-1,2-dihydrophthalazines (14), on treatment with ethanolic hydrochloric acid, provide alkylphthalazinium chlorides (see Section 2K-II). Phthalazine Reissert compounds such as 1-cyano-2-benzoyl-1,2-dihydrophthalazine (**19**), on treatment with hydrobromic acid in acetic acid, yield phthalazine-1-carboxylic acid hydrobromides (13) (see Section 2J-I). This reaction also has been used in the preparation of 1-alkylphthalazines (see Section 2B-I-A-3).

Bellasio and Testa (7, 8, 18–20) have studied the reactions of 1,2,3,4-tetrahydro-4-oxophthalazine (**10**) in detail. They showed that the reaction of **10** with acetyl chloride (7) gave 2-acetyl-1,2,3,4-tetrahydro-4-oxophthalazine (**22**), which on treatment with boron tetrafluoride etherate (18, 19) yielded 2-acetyl-4-ethoxy-1,2-dihydrophthalazine (**23**) (Eq. 16). Alkaline hydrolysis of **23** provided 4-ethoxy-1,2-dihydrophthalazine, whereas acidic hydrolysis gave **10**. The reaction of **10** with boron tetrafluoride etherate furnished 2-ethyl-1,2,3,4-tetrahydro-4-oxophthalazine.

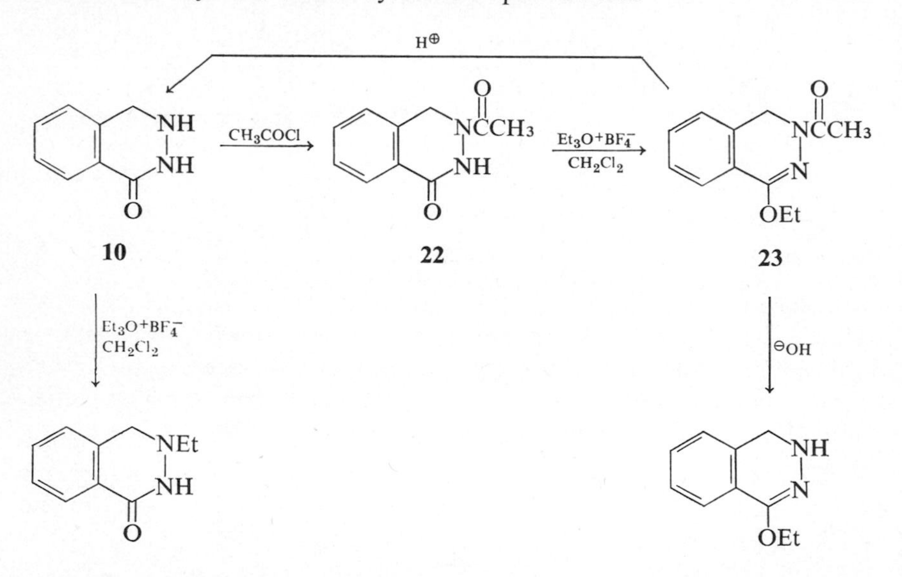

Heating **10** in refluxing hydrazine for 6 hr converts it to *N*-aminophthalimide (20). The compound **23**, on heating with hydrazine for 1 hr, provided 1-hydrazinophthalazine and 1,2-diphthalazinylhydrazine with simultaneous dehydrogenation at the 1- and 2-positions.

Bellasio and Testa (8, 20) have also studied the condensation reactions of **10** with 3-bromopropionyl chlorides to give the pyrazolo[1,2-*b*]phthalazine ring system (**24**), which is also prepared by the condensation of 2-bromomethylbenzoyl chloride with pyrazolidone (Eq. 17). Similarly, the reaction of **10** with 4-bromobutyryl chloride and 2-bromomethylbenzoyl chloride yields derivatives of pyridazino[1,2-*b*]phthalazine (**25**) and phthalazino[2,3-*b*]-phthalazines (**26**), respectively.

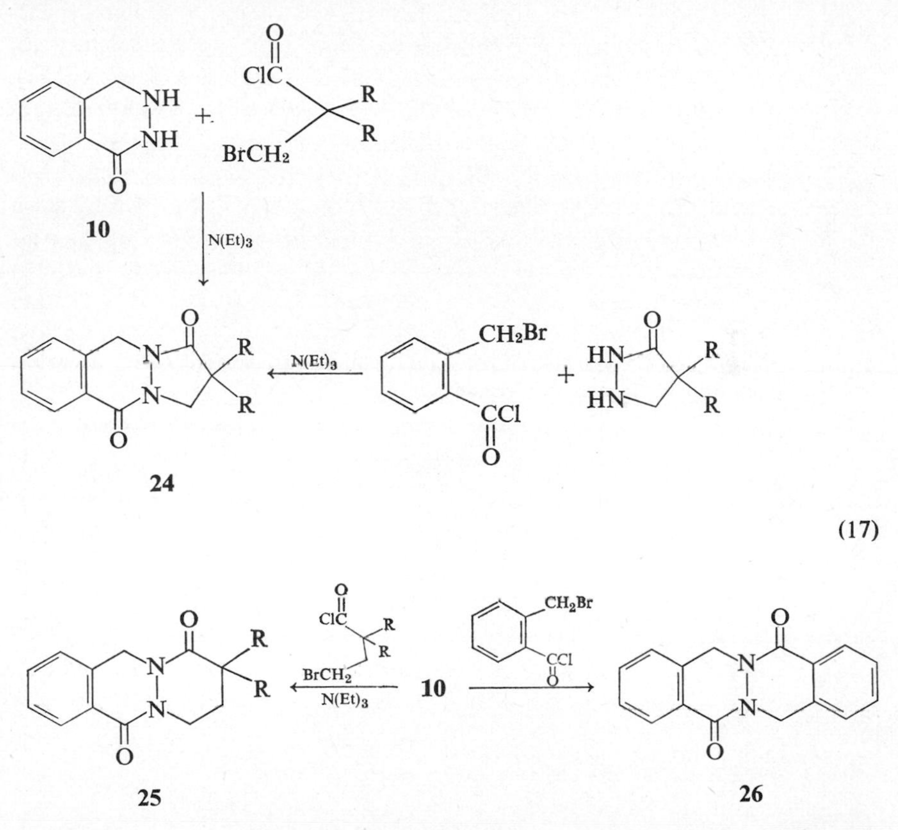

Singh (21) has studied the photochemical rearrangement of 1,1-dibenzyl-1,2-dihydro-2,4-diphenylphthalazine to give the dark red azomethine imine **27** (Eq. 18). The structure of **27** was supported by UV and pmr spectra and by 1,3-dipolar additions with various dipolarophiles.

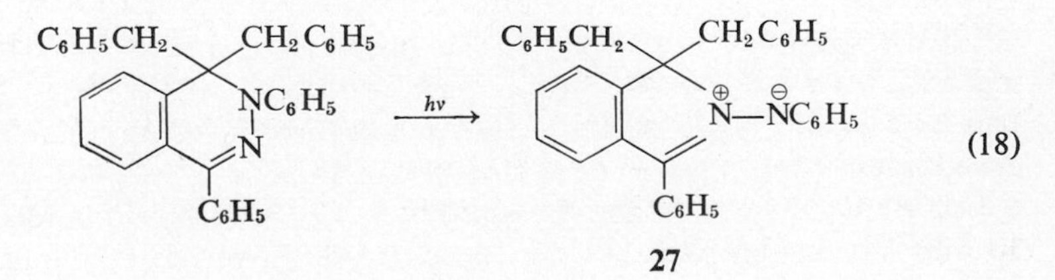

II. 1,2,3,4-Tetrahydrophthalazines

A. Preparation

Reduction of phthalazine with sodium amalgam gives 1,2,3,4-tetrahydrophthalazine (**28**), isolated as the hydrochloride salt (22). Carpino (23) has developed a synthesis of **28** where α,α′-dibromo-*o*-xylene is allowed to react with the dipotassium salt of *t*-butyl hydrazodiformate to give di-*t*-butyl-1,2,3,4-tetrahydrophthalazine-2,3-dicarboxylate (**29**), which on acidic hydrolysis provided **28** in high yields (Eq. 19). Ohme and Schmitz (24) have employed a similar approach in which α,α′-dichloro-*o*-xylene is allowed to

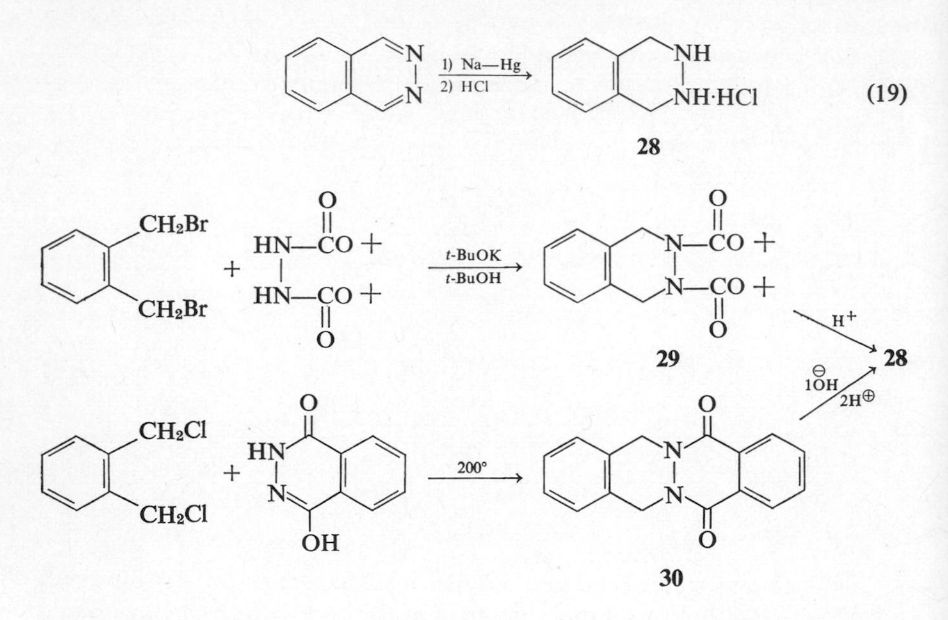

react with 4-hydroxy-1-(2*H*)phthalazinone to give 7*H*,12*H*-phthalazino-[2,3-*b*]phthalazine-5,14-dione (**30**), which on alkaline hydrolysis provided **28** (Eq. 19).

2,3-Dimethylphthalazine-1,4-dione, on reduction with lithium aluminum hydride, furnished 2,3-dimethyl-1,2,3,4-tetrahydrophthalazine (25). Wittig and co-workers (26) have shown that α,α′-dibromo-*o*-xylene will react with *N*,*N*′-diphenylhydrazine to give 2,3-diphenyl-1,2,3,4-tetrahydrophthalazine, which is also obtained by the reduction of 2,3-diphenylphthalazine-1,4-dione with lithium alanate (Eq. 20).

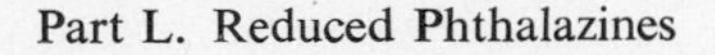

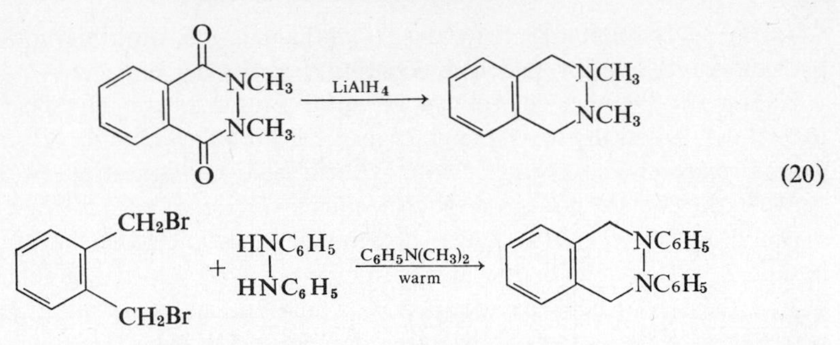

(20)

Aeberli and Houlihan (27) have similarly reduced 2-methyl-1-(2*H*)-phthalazinone and **30** with lithium aluminum hydride to give 2-methyl-1,2,3,4-tetrahydrophthalazine and 5,7,12,14-tetrahydrophthalazino[2,3-*b*]-phthalazine (**31**) in 92 and 46% yields, respectively (Eq. 21). Furthermore, the reduction of 2-(3-hydroxypropyl)-4-*p*-chlorophenyl-1-(2*H*)phthalazinone (27, 28) provided 1-*p*-chlorophenyl-3-(3-hydroxypropyl)-1,2,3,4-tetrahydro-phthalazine (**32**), which on treatment with thionyl chloride furnished 5-*p*-chlorophenyl-2,3,5,10-tetrahydro-1*H*-pyrazolo[1,2-*b*]phthalazine (**33**). 2-(3-

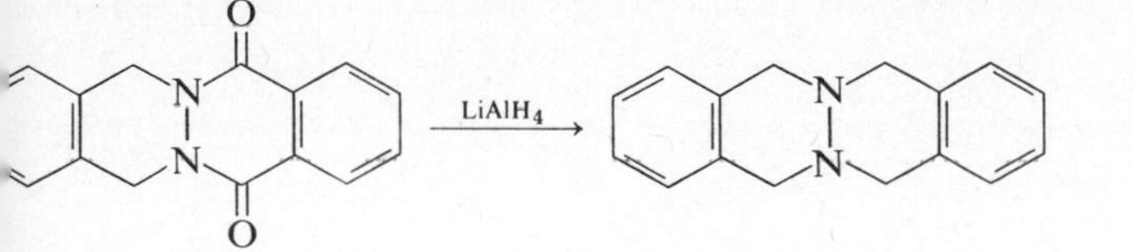

30 **31**

(21)

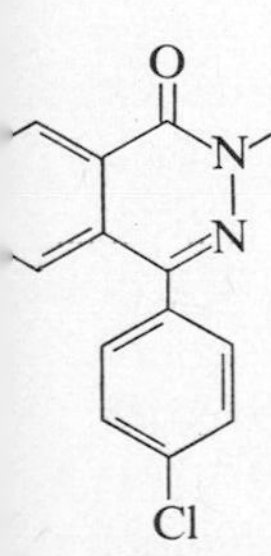

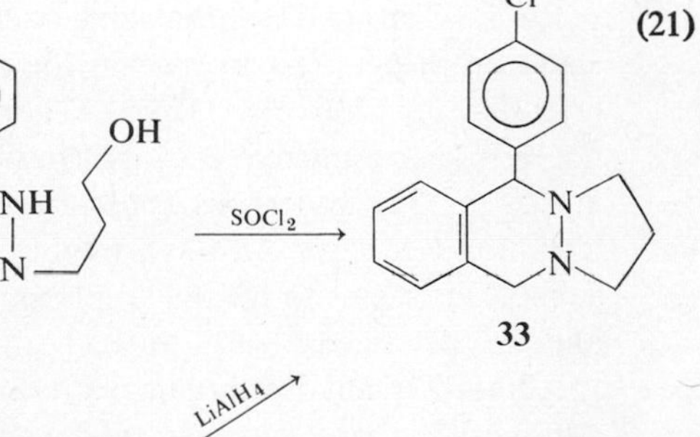

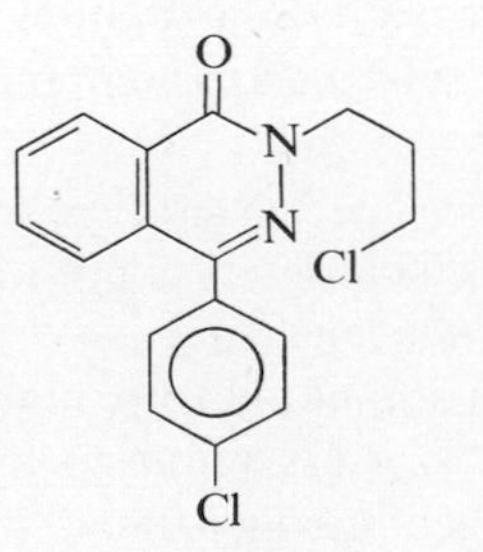

Chloropropyl)-4-*p*-chlorophenyl-1-(2*H*)phthalazinone, on lithium aluminum hydride reduction, also provided **33** (Eq. 21).

Nakamura (28a) has determined by variable temperature nmr spectroscopy that 2,3,5,10-tetrahydro-1-*H*-pyrazolo[1,2-*b*]phthalazine and its 3-methyl analog exist as conformers with trans-fused configuration, converting between the two antipodes via synchronous inversion of both nitrogen atoms.

Bould and Farr (28b) have shown that 5,6-dimethylenebicyclo[2.2.0]-hexene-2 will react with the dienophile, 2-phenyl-2,4,5-triazoline-1,3-dione to give the Dewar benzene adduct **33a**, which on brief warming in methanol rearranged to the triazolophthalazine derivative **33b** (Eq. 21a).

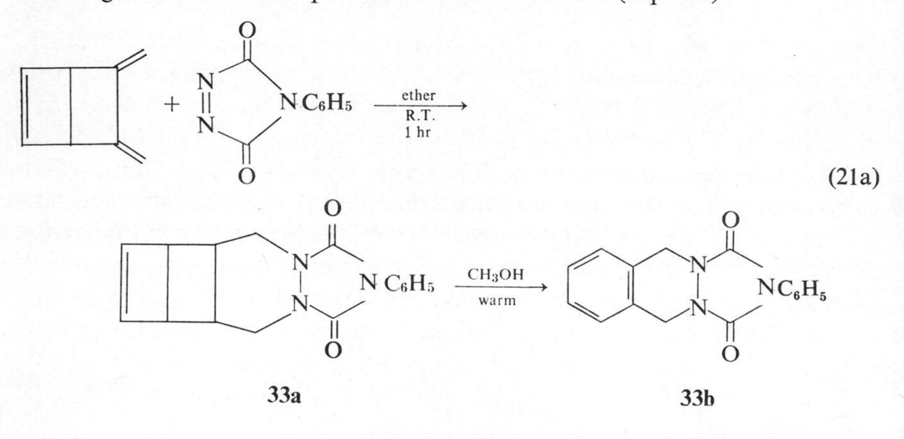

B. Reactions

1,2,3,4-Tetrahydrophthalazine (**28**) is unstable and decomposes on standing in air. The hydrochloride salt is stable and forms feathery white crystals, mp 236–238° (23). Aeberli and Houlihan (27) have observed that 1-(*p*-chlorophenyl)-3-(4-hydroxybutyl)-1,2,3,4-tetrahydrophthalazine, an analog of **32**, undergoes spontaneous dehydrogenation on standing for a few hours to give 2-(4-hydroxybutyl)-4-*p*-chlorophenyl-1,2-dihydrophthalazine. This seems to be an exception, since **32** is stable enough to be isolated and characterized.

1,2,3,4-Tetrahydrophthalazine, on oxidation with mercuric oxide, gave *o*-quinodimethane, but even at 0° the azo compound, 1,4-dihydrophthalazine, was not isolated (23).

Treatment of 2,3-diphenyl-1,2,3,4-tetrahydrophthalazine with stannous chloride and hydrochloric acid causes cleavage of the nitrogen-nitrogen bond to give α,α′-diphenylamino-*o*-xylene (26).

Hatt and Stephenson (29) attempted to prepare **31** by reduction of **30** and found that it was resistant to most reducing agents, whereas clemmenson reduction severed the nitrogen-nitrogen bond to give **34**, which on further reduction furnished *N*-(2-aminomethylbenzyl)phthalimidine (Eq. 22). Subsequently, **30** was successfully reduced to **31** with lithium aluminum hydride (see Eq. 21).

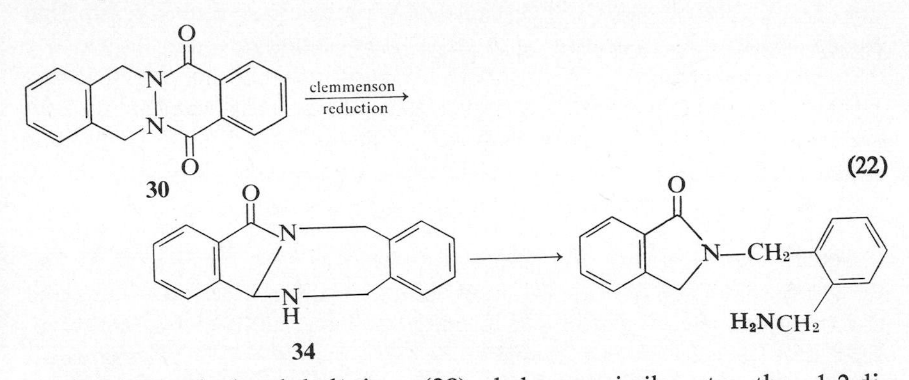

1,2,3,4-Tetrahydrophthalazine (**28**) behaves similar to the 1,2-disubstituted hydrazines. It reacts with acid chlorides, anhydrides, and alkyl halides as expected. The 2,3-dibenzoyl derivative (23) and the monobenzoyl derivative (30) of **28** have been prepared. α,α′-Dibromo-*o*-xylene reacts with **28** at 140–150° (29) or in refluxing ethanol containing potassium acetate (22) to give **31**. Condensation of 1,2,3,4-tetrahydrophthalazine (**28**) with succinic anhydride and phthalic anhydride in presence of potassium acetate at 160–170° for 5 hr gave 2,3,6,11-tetrahydropyridazine[1,2-*b*]phthalazine-1,4-dione (**35**) and its analog, **30**, respectively (22) (Eq. 23).

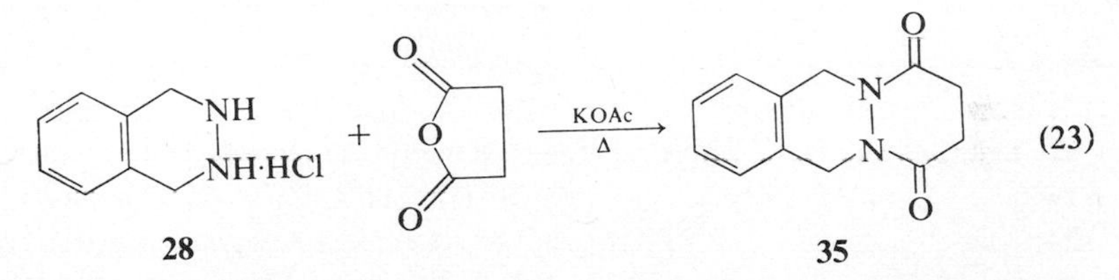

Panneman (30) has carried out several reactions on 1,2,3,4-tetrahydrophthalazine; for example, reactions with *S*-methylthiourea or cyanamide to give 1,2,3,4-tetrahydrophthalazine-2-carboxamidine and with ammonium isothiocyanate to give 1,2,3,4-tetrahydrophthalazine-2-carbothiamide (Eq. 24). Ohme and Schmitz (24, 31) showed that 1,2,3,4-tetrahydrophthalazine

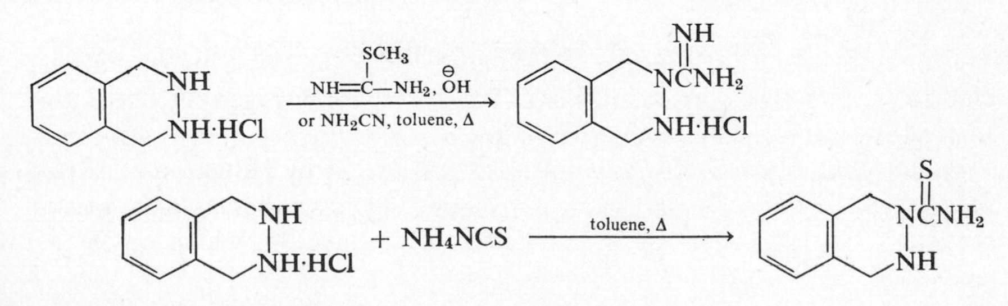

reacts quantitatively with formaldehyde, even at low concentrations, at pH 2–4. Presence of other aldehydes does not interfere in this reaction. The product with formaldehyde is identified as bis-(1,2,3,4-tetrahydrophthalazino[2,3-*a*:2′,3′]perhydro-1,2,4,5-tetrazine) (**36**) (Eq. 25). Compound **36** forms white prisms from water and melts at 265° with decomposition.

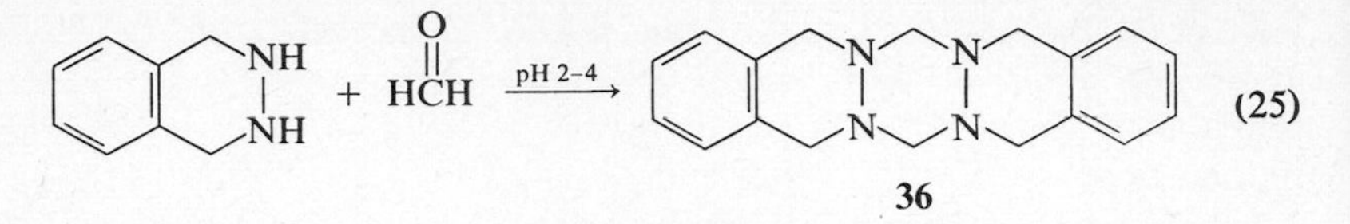

Aeberli and Houlihan (27) prepared quaternary salts of **31** and **33** and then subjected the salts to a Hoffman elimination to prepare diazonine and diazocine ring systems. Thus 5-*p*-chlorophenyl-11-methyl-2,3,5,10-tetrahydro-1*H*-pyrazolo-[1,2-*b*]phthalazinium iodide and 6-methyl-5,7,12-14-tetrahydrophthalazino[2,3-*b*]phthalazinium bromide, on heating with sodium methoxide in methanol, gave 1-(*p*-chlorophenyl)-6-methyl-4,5,6,7-tetrahydro-3*H*-2,6-benzodiazonine (**37**) and 6-methyl-5,6,7,14-tetrahydrodibenzo[*c*,*h*][1,6]-diazecine (**38**), respectively (Eq. 26).

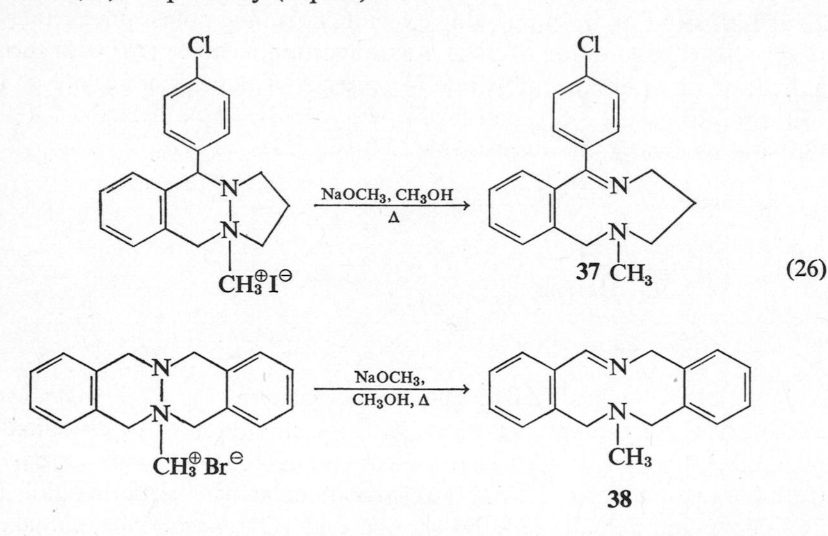

III. 5,6,7,8-Tetrahydrophthalazines

Horning and Amstutz (32) have prepared 5,6,7,8-tetrahydrophthalazine (**39**) and some of its derivatives. They also studied the UV spectra of these compounds. 3,4,5,6-Tetrahydrophthalic anhydride smoothly condensed with hydrazine hydrate in benzene (32) or with aqueous hydrazine sulfate (33) to give 4-hydroxy-5,6,7,8-tetrahydro-1-(2*H*)phthalazinone (**40**) in high yields. The reaction of **40** with phosphorus oxychloride provided 1,4-dichloro-5,6,7,8-tetrahydrophthalazine (**41**) in 84% yield (32, 33). Dehalogenation of **41** with phosphorus and hydriodic acid provided **39** in 39% yield, together with a lesser amount of 1-iodo-5,6,7,8-tetrahydrophthalazine. Satoda and co-workers (33) have allowed **41** to react with ethanolic ammonia at 160° for 56 hr in sealed tube to give 1-amino-4-chloro-5,6,7,8-tetrahydrophthalazine and at 160–170° for 88 hr to give 1,4-diamino-5,6,7,8-tetrahydrophthalazine. Steck and co-workers (34) replaced both chlorine atoms of 1,4-dichloro-5-methyl-5,6,7,8-tetrahydrophthalazine with *N*,*N*-dimethylaminoethanol by refluxing the dichlorophthalazine in xylene in the sodium to

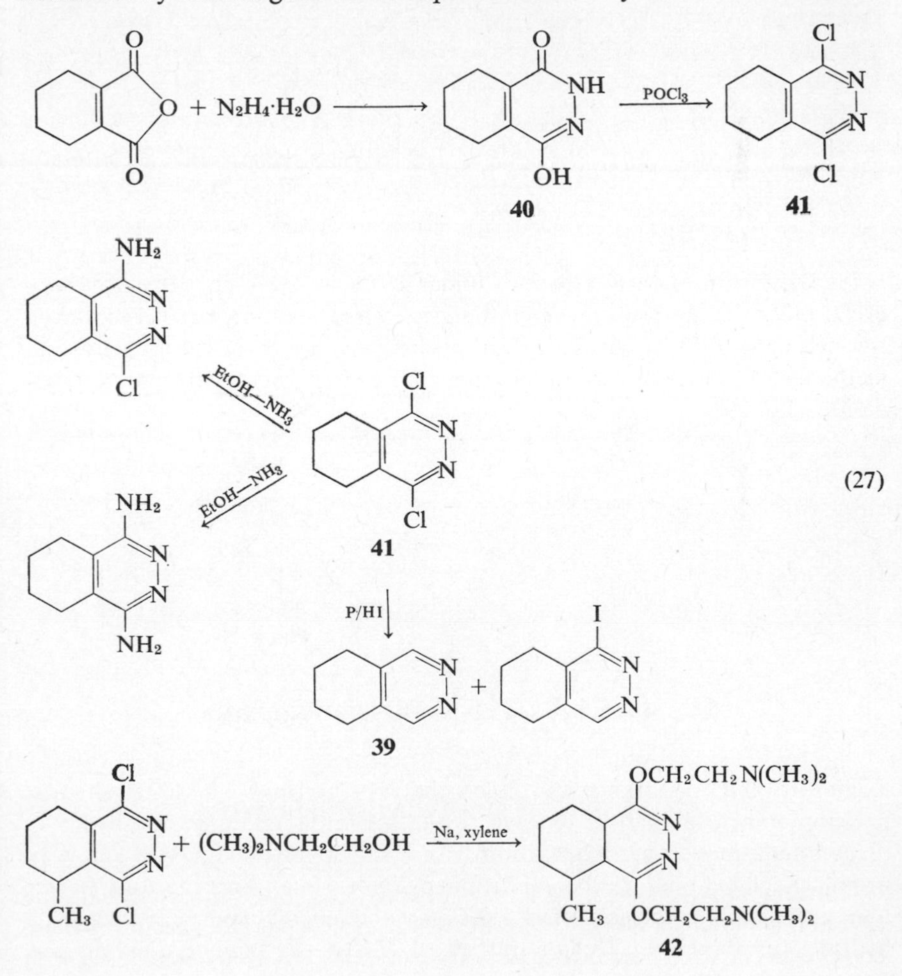

(27)

give **42** (Eq. 27). Mori (35) observed that 3,4,5,6-tetrahydrophthalic anhydride gave 2-phenyl-5,6,7,8-tetrahydro-1-(2*H*)phthalazinone with phenylhydrazine in 20% sulfuric acid, but the reaction failed in acetic acid. He has reported ir data on some of the tetrahydrophthalazines.

Dehydrogenation of 4*a*,5,6,7,8,8*a*-hexahydro-1-(2*H*)phthalazinone (36) and *cis*- or *trans*-2-(*N*-methyl-4′-piperidyl)-4-phenyl-4*a*,5,6,7,8,8*a*-hexa-, hydro-1-(2*H*)phthalazinones (37), on heating with bromine in acetic acid, furnished 5,6,7,8-tetrahydro-1-(2*H*)phthalazinone and 2-(*N*-methyl-4′-piperidyl)-4-phenyl-5,6,7,8-tetrahydro-1-(2*H*)phthalazinone, respectively (Eq. 28)

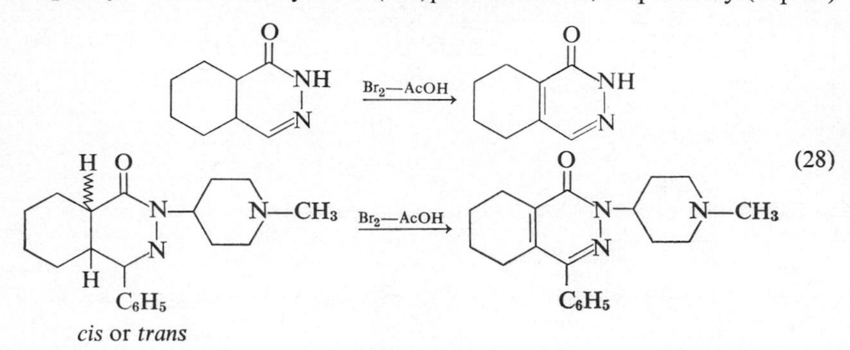

(28)

1,4-Disubstituted *sym*-tetrazines undergo Diels-Alder condensation with cyclic olefins to give hexahydrophthalazines (see Eq. 34), but 1,4-diphenyl-*sym*-tetrazine (38) condensed with norbornene to give 1,4-diphenyl-5,8-methano-5,6,7,8-tetrahydrophthalazine (**43**) in 94% yield, through autooxidation (Eq. 29).

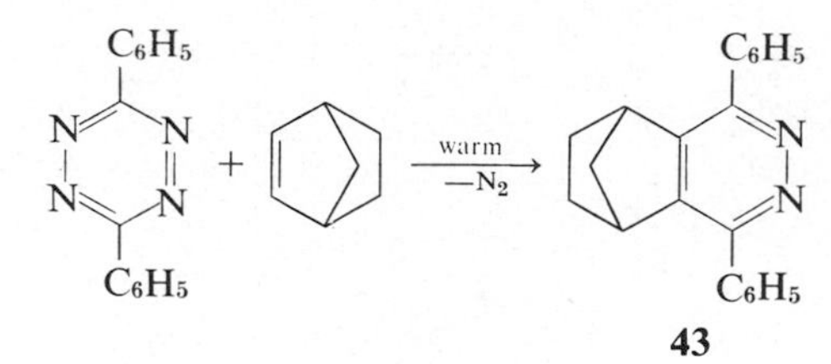

(29)

IV. 4*a*,5,8,8*a*-Tetrahydrophthalazines

Hufford and co-workers (39) noted that *cis*- or *trans*-1,2,3,6-tetrahydrophthalic anhydride failed to give 2-*p*-nitrophenyl-4*a*,5,8,8*a*-tetrahydro-1-(2*H*)phthalazinone on condensation with *p*-nitrophenylhydrazine. The only product isolated was the bis-*p*-nitrophenylhydrazone. Andreev and Usova (36) successfully condensed *cis*-1,2,3,6-phthalaldehydic acid with hydrazine hydrate to obtain 4*a*,5,8,8*a*-tetrahydro-1-(2*H*)phthalazinone. Dixon and

Wiggins (41) allowed 2-acetyl-4-cyclohexenecarboxylic acid and its ethyl ester to react with hydrazine hydrate to give 4-methyl-4*a*,5,8,8*a*-tetrahydro-1-(2*H*)phthalazinone (**44**) in ~90% yield. Bromination of **44** provided 4-methyl-6,7-dibromo-4*a*,5,6,7,8,8*a*-hexahydro-1-(2*H*)phthalazinone (**45**) (Eq. 30). Arbuzov and co-workers (42) observed that *trans*-2-benzoyl-4-cyclohexenecarboxylic acid, on attempted Huang-Minlon reduction, gave 90%

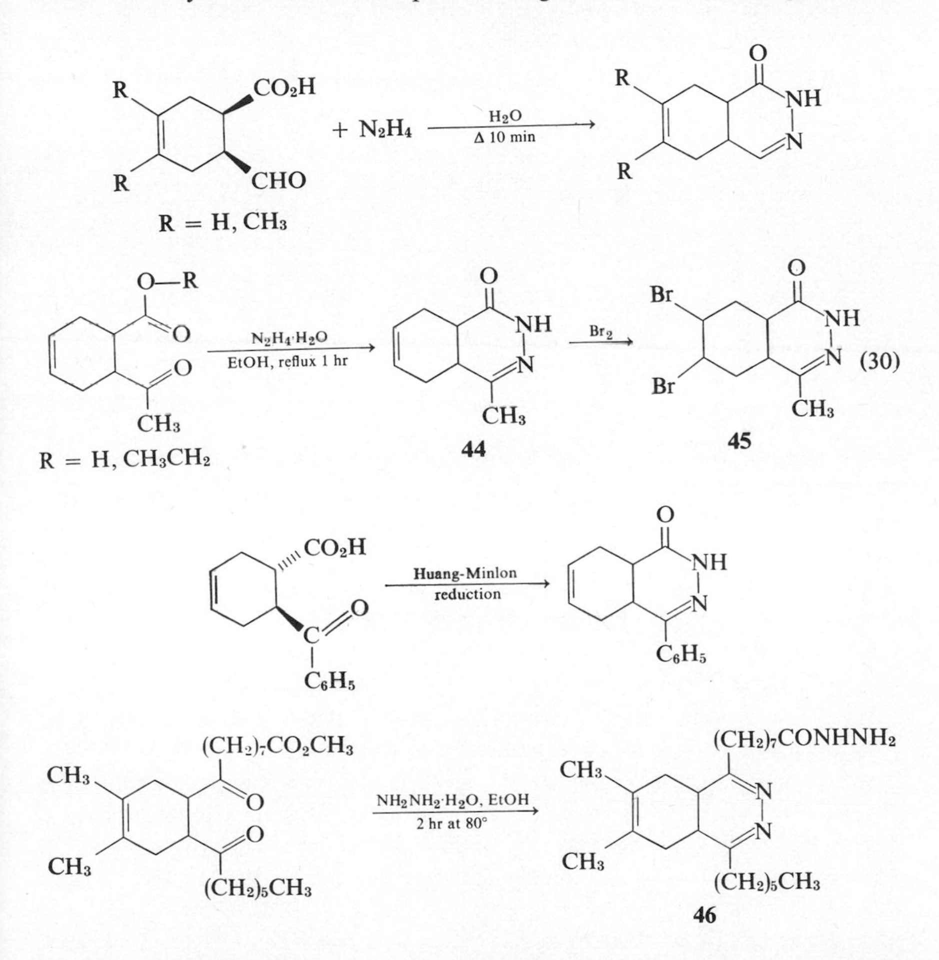

4-phenyl-4*a*,5,8,8*a*-tetrahydro-1-(2*H*)phthalazinone. Whitfield (43) prepared 4-hexyl-4*a*,5,8,8*a*-tetrahydro-6,7-dimethylphthalazine-1-octanoic acid hydrazide (**46**) by condensation of the corresponding 1,2-diacetyl-4-cyclohexene with hydrazine hydrate (Eq. 30).

Fujimoto and Okabe (44, 45) condensed endo-*cis*-3,6-ethano-1,2,3,6-tetrahydrophthalic anhydride with hydrazine hydrate and its derivatives to give the corresponding tetrahydrophthalazines (Eq. 31). Druey and co-workers (46) have reported the Diels-Alder condensation of cyclopentadiene

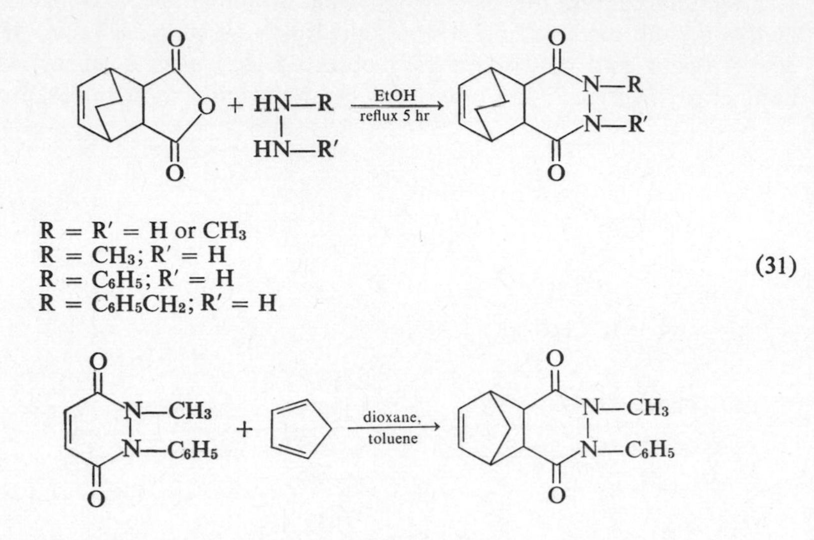

with 1-methyl-2-phenylpyridazine-3,6-dione to prepare similar tetrahydrophthalazines (Eq. 31).

V. 4*a*,5,6,7,8,8*a*-Hexahydro- and Other Hexahydrophthalazines

Lora-Tamayo and co-workers (47) observed that *cis*-hexahydrophthalic acid, on condensation with hydrazine, failed to give 4-hydroxy-4*a*,5,6,7,8,8*a*-hexahydro-1-(2*H*)phthalazinone but gave hexahydro-*N*-aminophthalimide. Andreev and Usova (40) successfully condensed *cis*-hexahydrophthaldehydic acid with hydrazine hydrate to obtain 4*a*,5,6,7,8,8*a*-hexahydro-1-(2*H*)-phthalazinone (**47**) (Eq. 32). Jucker and Suess (37) have synthesized *cis*- and *trans*-4-phenyl-2-(*N*-methyl-4-piperidyl)-4*a*,5,6,7,8,8*a*-hexahydro-1-(2*H*)-phthalazinones (**48*a***; **48*b***) by reacting *cis*- and *trans*-2-benzoylcyclohexanecarboxylic acid with *N*-methyl-4-hydrazinopiperidine by refluxing in ethanol for 24 hr. Both compounds **48*a*** and **48*b***, on treatment with bromine in acetic acid, provided a single product, 4-phenyl-2-(*N*-methyl-4-piperidyl)-5,6,7,8-tetrahydro-1-(2*H*)phthalazinone (Eq. 32).

Zinner and Duecker (49) have condensed hexahydrophthalic anhydride with 1,2-dihydro-4-phenyl-1,2,4-triazoline-3,5-dione to prepare **49** by

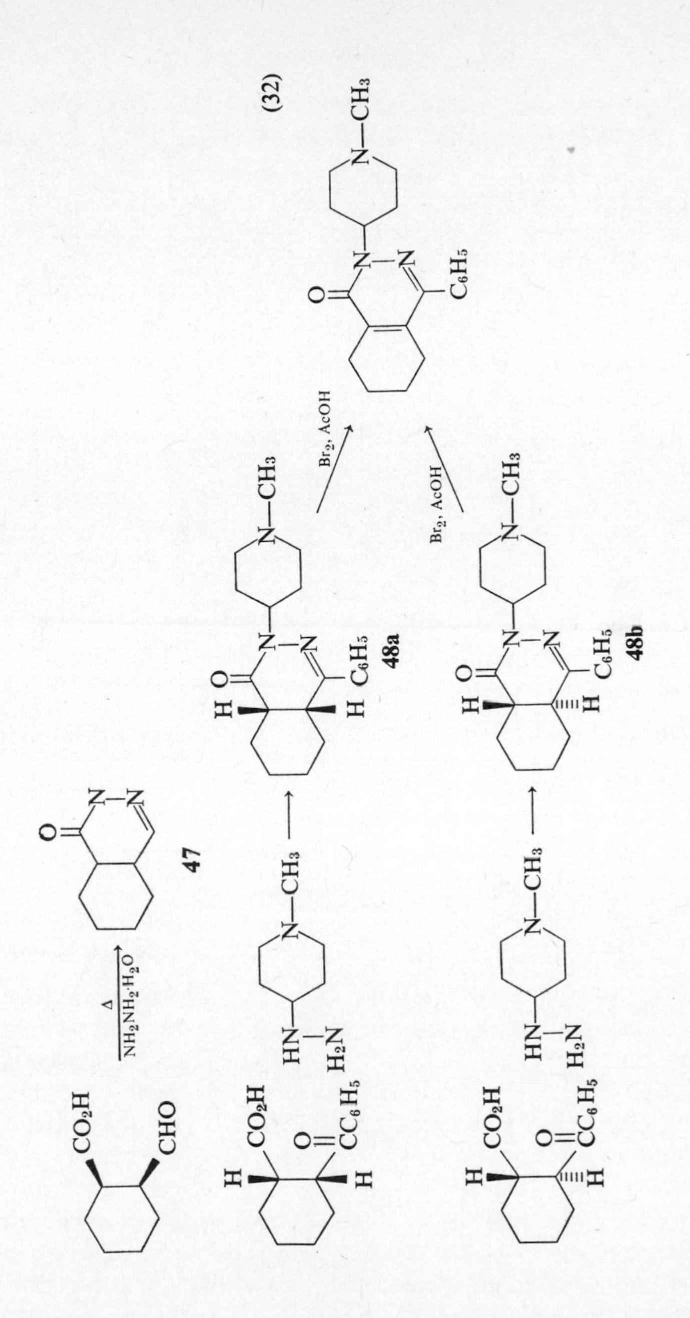

CO2H
CHO
NH2NH2·H2O
Δ
47
(32)
Br2, AcOH
48a
48b
N—CH3
C6H5
H2N
HN
CC6H5

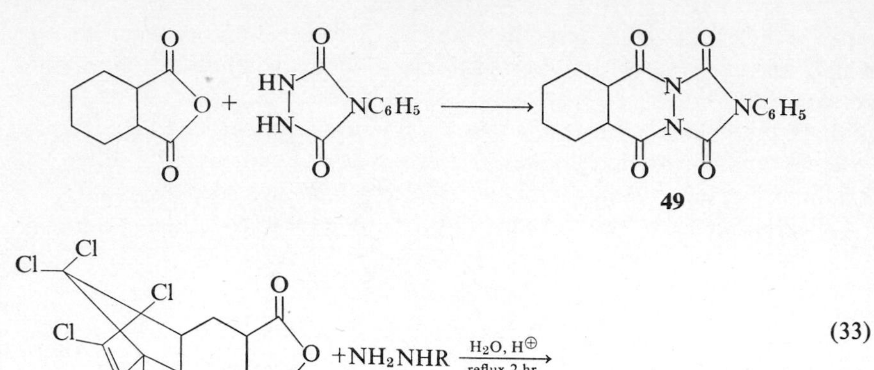

(33)

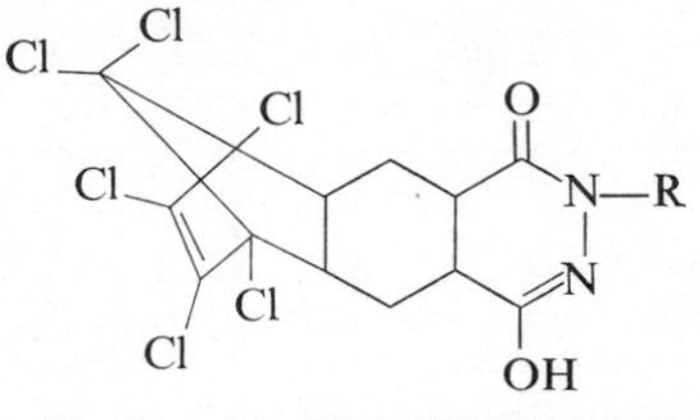

50 R = H, CH_3, CH_3CH_2, C_6H_5

heating at 180–200° for 2 hr (Eq. 33). Similarly, 1,2,3,4,9,9-hexachloro-1,4,4*a*,5,6,7,8,8*a*-octahydro-1,4-methanonaphthalene-6,7-dicarboxylic anhydride (50) reacts with hydrazine and its derivatives to give **50** by refluxing in aqueous media with acid for ~2 hr (Eq. 33).

Carboni and Lindsey (51, 52) first reported the Diels-Alder reaction of cyclohexene with 3,6-di(1,2,2,2-tetrafluoroethyl)-*sym*-tetrazine to

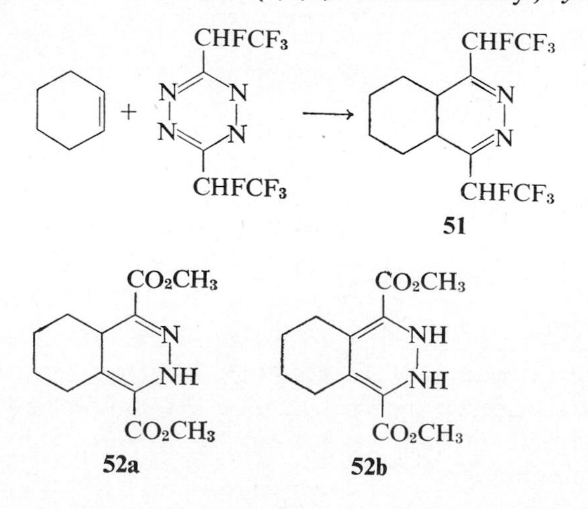

(34)

give 1,4-di(1,2,2,2-tetrafluoroethyl)-4*a*,5,6,7,8,8*a*-hexahydrophthalazine (**51**), which may coexist with other tautomeric forms. Similarly, Avram and co-workers (53) have synthesized 1,4-di(carbomethoxy)hexahydrohydrophthalazine, which exists in the tautomeric forms **52a** and **52b** (Eq. 34).

Sauer and Heinrich (38) have shown that norbornene will react with 3,6-di(carbomethoxy)-*sym*-tetrazine. The product has been represented as 1,4-dicarbomethoxy-5,8-methano-3,5,6,7,8,8*a*-hexahydrophthalazine.

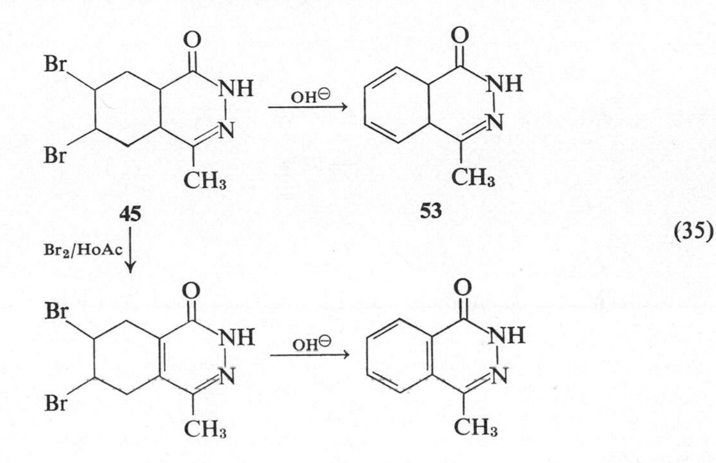

Dixon and Wiggins (41) have dehydrobrominated **45** with alcoholic potassium hydroxide to give 4-methyl-4*a*,8*a*-dihydro-1-(2*H*)phthalazinone (**53**), mp 223–224°, in 82% yield. Treatment of **45** with bromine in acetic acid gave 6,7-dibromo-4-methyl-5,6,7,8-tetrahydro-1-(2*H*)phthalazinone, which on dehydrobromination provided 4-methyl-1-(2*H*)phthalazinone (Eq. 35).

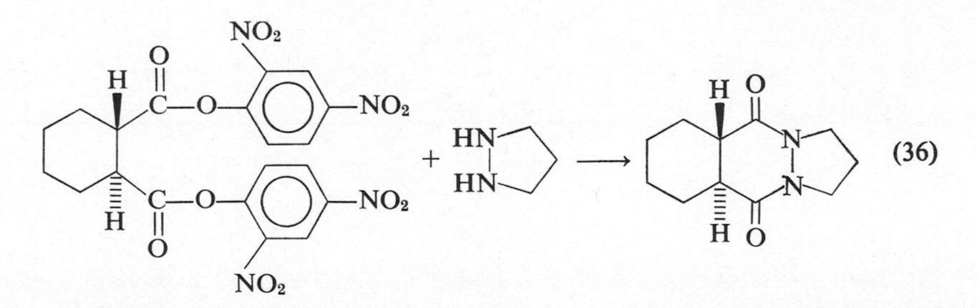

Overberger and co-workers (54) have studied the optical rotatory properties of some derivatives of hexahydrophthalazines. Synthesis of these compounds was achieved by reaction of di-2,4-dinitrophenyl *trans*-hexahydrophthalate with pyrazolidine in chloroform with triethylamine at 0° to avoid racemization (Eq. 36).

VI. Miscellaneous Reduced Phthalazines

Hexahydrophthalaldehyde (39) has been condensed with *p*-nitrophenylhydrazine in presence of methanol to give 1-methoxy-2-(*p*-nitrophenyl)-1,2,4*a*,5,6,7,8,8*a*-octahydrophthalazine (**54**) (Eq. 37). 1,2,4,5-Tetrabenzoyl-3,6-dihydrobenzene (55) on condensation with 1 mole of hydrazine and 2

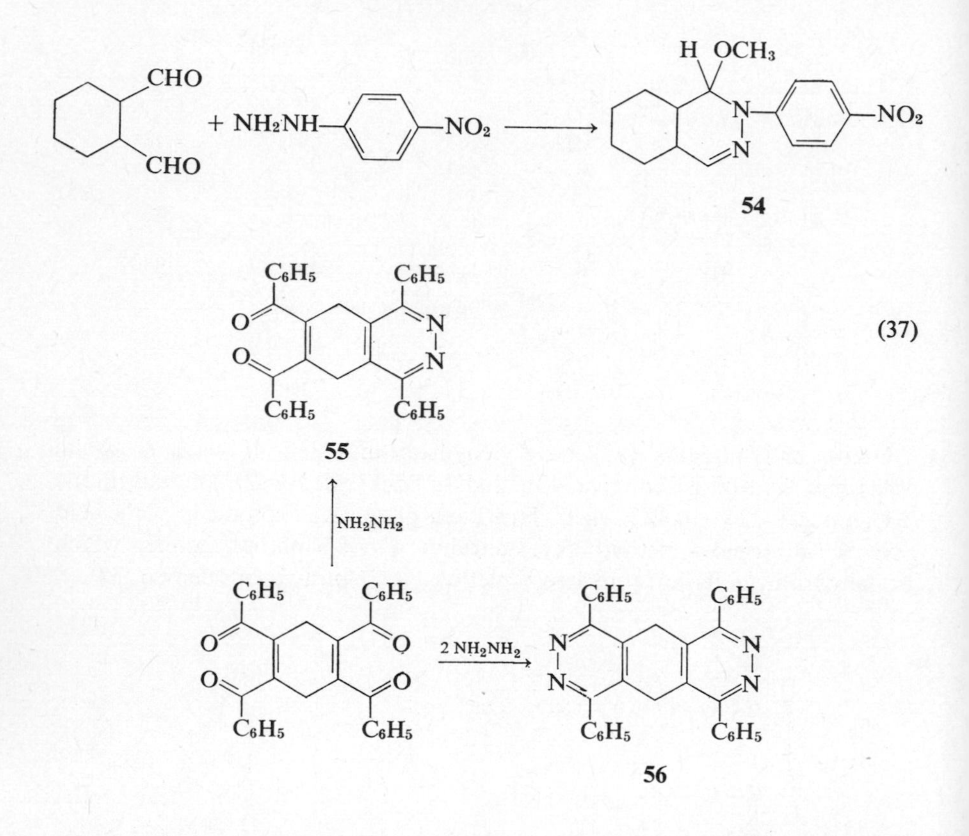

moles of hydrazine provided 6,7-dibenzoyl-5,8-dihydro-1,4-diphenylphthalazine (**55**) and 5,10-dihydro-1,4,6,9-tetraphenylpyridazino[4,5-*g*]phthalazine (**56**), respectively (Eq. 37).

Price and co-workers (56) have carried out the Diels-Alder condensation of 1,2-dimethylenecyclohexane with diethyl azodicarboxylate to give 2,3-dicarbethoxy-1,2,3,4,5,6,7,8-octahydrophthalazine, which on bromination provided 4*a*,8*a*-dibromo-2,3-dicarbethoxydecahydrophthalazine (57) (Eq. 38). These authors (56) have also carried out conformational studies of these compounds using proton magnetic resonance.

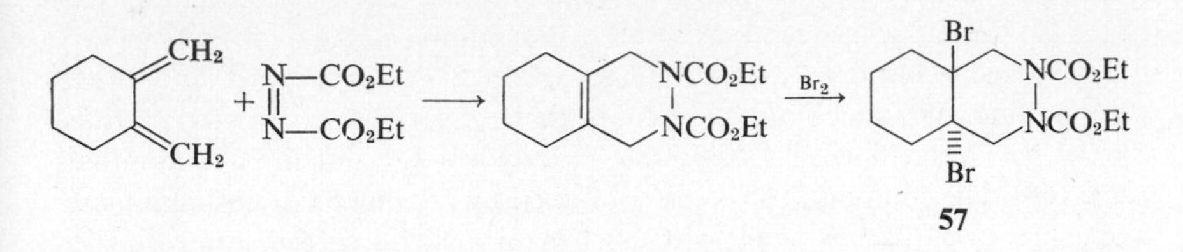

Kuderna and co-workers (57) have reported the Diels-Alder condensation of hexachlorocyclopentadiene with 2,3-diazabicyclo[2,2,1]hept-5-ene to give

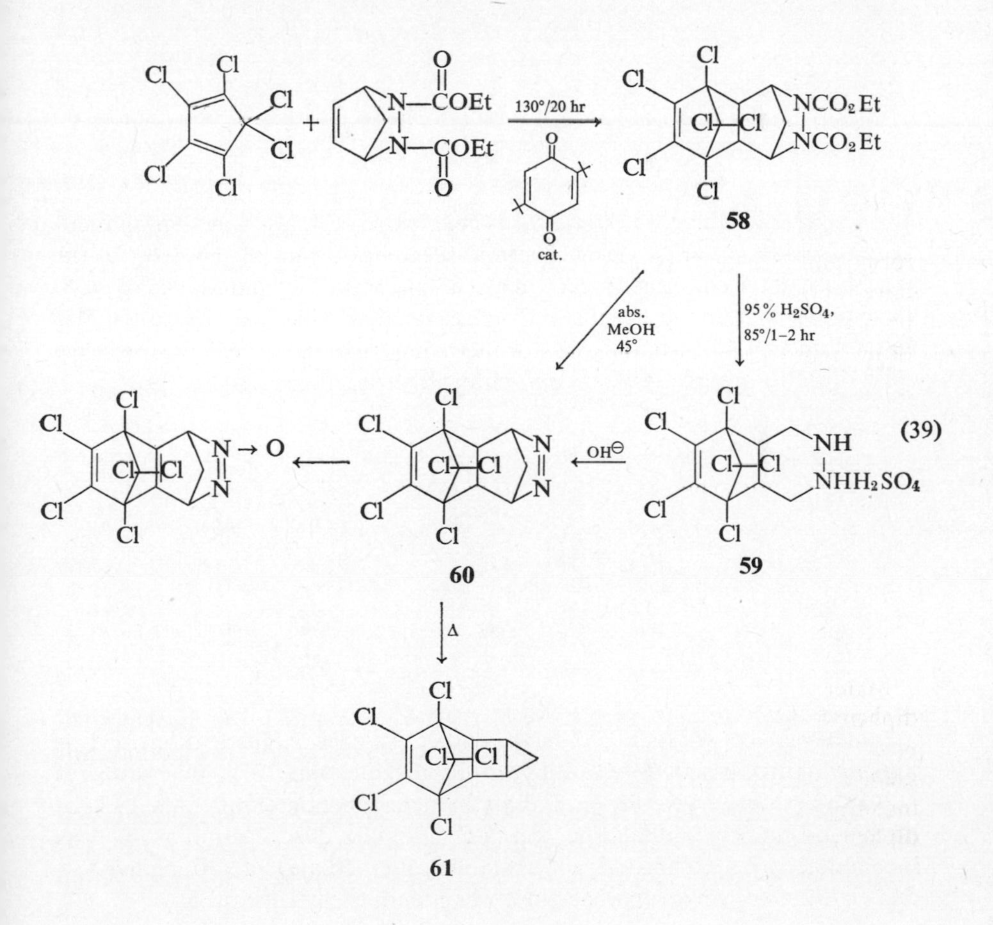

diethyl 5,6,7,8,9,9-hexachloro-1,2,3,4,4*a*,5,8,8*a*-octahydro-1,4,5,8-dimethano-2,3-phthalazinedicarboxylate (**58**) in 95% yield (Eq. 39). Similarly, they have used other dienes such as cyclopentadiene and tetrachlorocyclopentadienone dimethyl acetal to prepare bridged phthalazines. These authors (57) have also studied some reactions of **58**; for example, acidic hydrolysis gave 5,6,7,8,9,9-hexachloro-1,2,3,4,4*a*,5,8,8*a*-octahydro-1,4,5,8-dimethanophthalazine sulfate (**59**), which on treatment with 10% sodium hydroxide provided 5,6,7,8,9,9-hexachloro-1,4,4*a*,5,8,8*a*-hexahydro-1,4,5,8-dimethanophthalazine (**60**). The compound **60** is also obtained by heating **58** in absolute methanol. The *N*-oxidation of **60** with perbenzoic acid in chloroform furnished

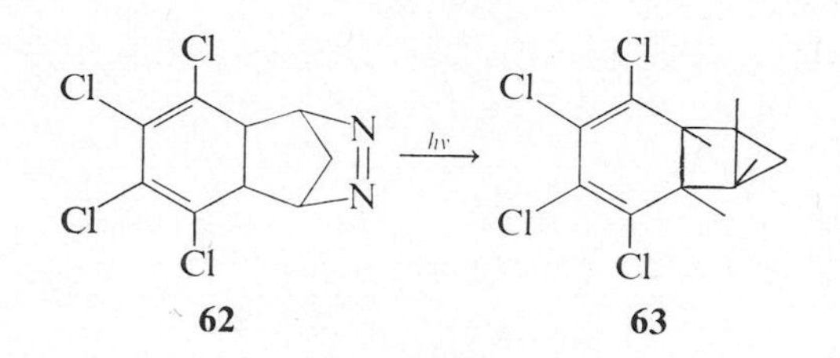

5,6,7,8,9,9- hexachloro- 1,4,4*a*,5,8,8*a*- hexahydro- 1,4,5,8- dimethanophthalazine 2-oxide in 91% yield. Thermal decomposition of **60** led to the formation of the chlorohydrocarbon **61** and nitrogen. Lay and Mackenzie (58) have reported a similar thermal retrogressive Diels-Alder reaction of 5,6,7,8-tetrachloro-1,4,4*a*,8*a*-tetrahydro-1,4-methanophthalazine (**62**) in presence of uv light to provide *syn*- and *anti*-chlorohydrocarbons, **63** (Eq. 39).

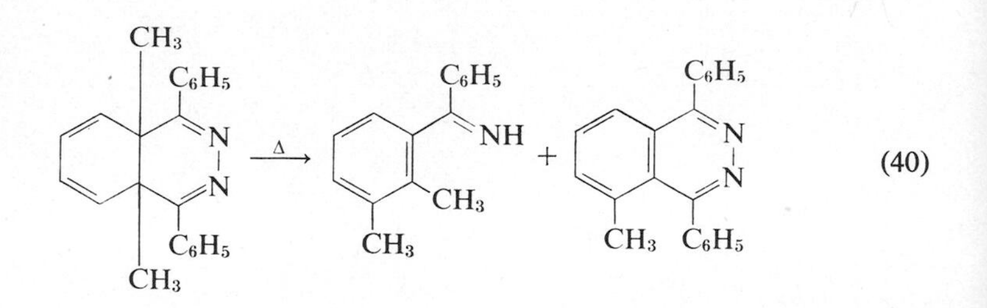

Maier and co-workers (59) have studied valence isomerization of 1,4-diphenyl-4*a*,8*a*-dimethyl-4*a*,8*a*-dihydrophthalazine, mp 121°, by heating it at 100° for 5 hr to give 83% 2,3-dimethylbenzophenone imine and 14% 5-methyl-1,4-diphenylphthalazine, mp 206° (Eq. 40). The reaction at 60° was incomplete after 10 hr and an unstable intermediate, 4*a*,5-dimethyl-1,4-diphenyl-2,4*a*-dihyhydrophthalazine, was identified by spectral data.

VII. Tables

TABLE 2L-1. 1,2-Dihydrophthalazines

Structure	Preparation[a]	Yield (%)	MP	Reference
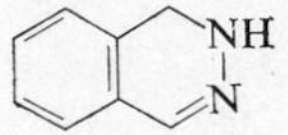 NH, N	A	63	47–48°, phenylthiourea, 212–213°; 2 *N-p*-tolylsulfone, 172–174°	1, 2, 3
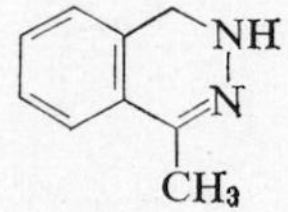 NH, N, CH_3	A	57	34–35°; phenylthiourea, 235–236°	4a
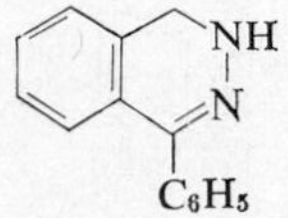 NH, N, C_6H_5	A	73	67–68°; phenylthiourea; 174.5–175°	1
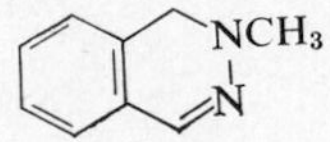 NCH_3, N	B	75	Yellow oil, bp 129–130°/17 mm, HCl 133–135°; picrate, 93–95°; methiodide, 175–176°	3
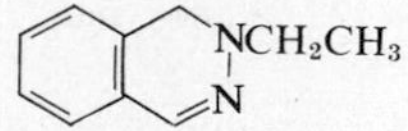 NCH_2CH_3, N	B	78	HCl, 142–144°; methiodide, 155–157°	3
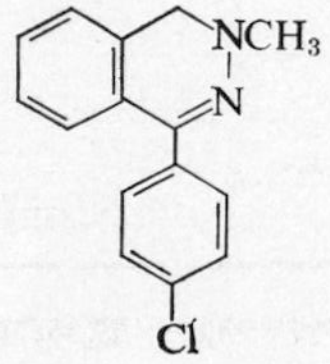 NCH_3, N, Cl	A	81	137–138° (CH_2Cl_2-pentane); methiodide, 163–166°	27

TABLE 2L-1 (*continued*)

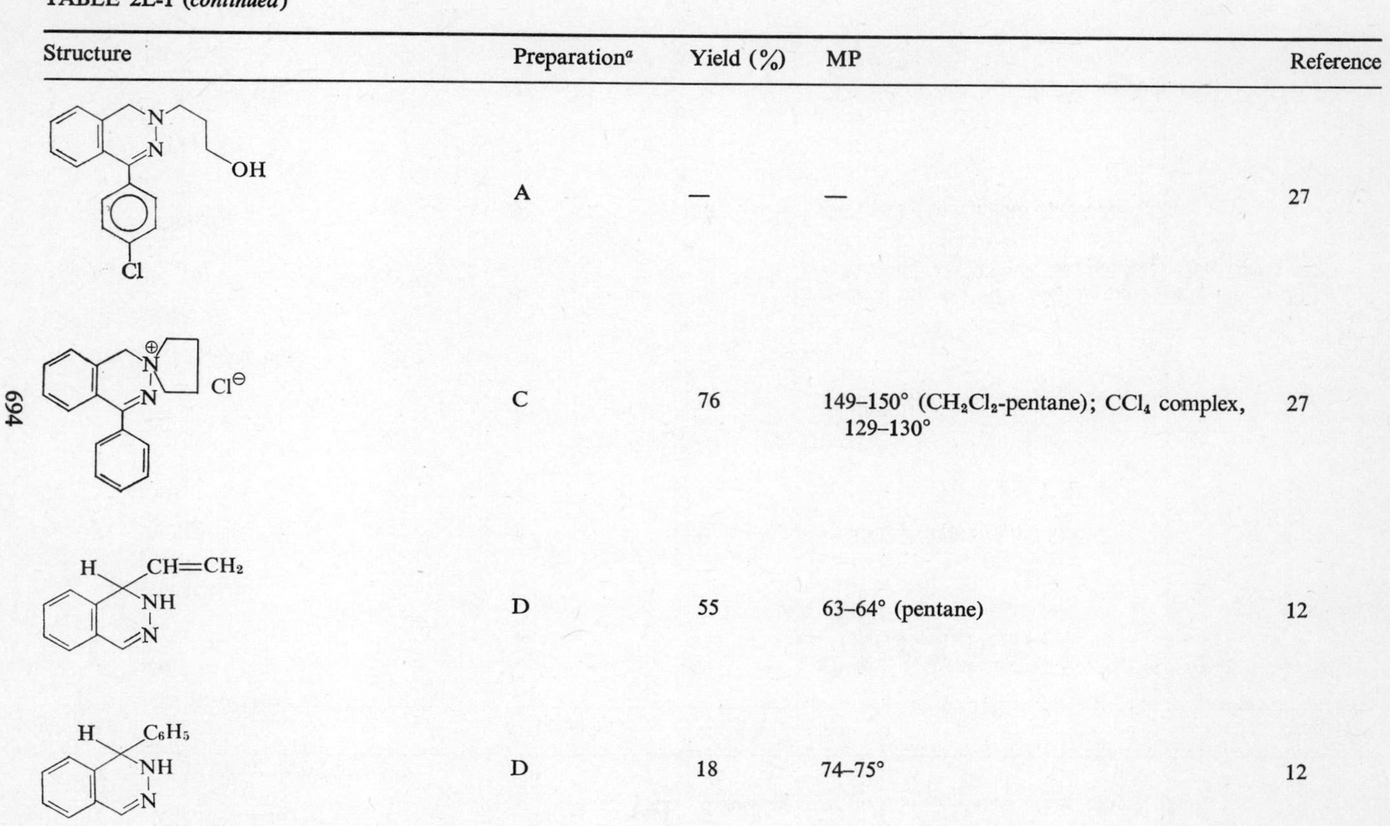

Structure	Preparation[a]	Yield (%)	MP	Reference
	A	—	—	27
	C	76	149–150° (CH_2Cl_2-pentane); CCl_4 complex, 129–130°	27
	D	55	63–64° (pentane)	12
	D	18	74–75°	12

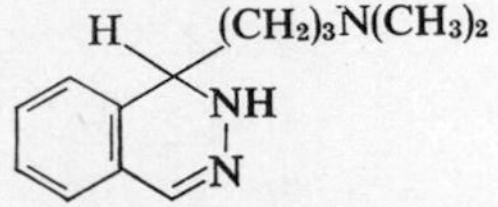 H, $(CH_2)_3N(CH_3)_2$, NH, N	E	91	Hygroscopic, 54–57° (pet. ether); HCl, 140–150°	11
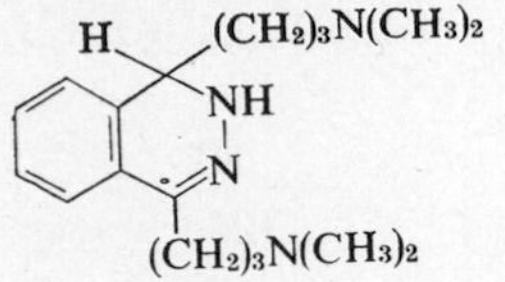 H, $(CH_2)_3N(CH_3)_2$, NH, N, $(CH_2)_3N(CH_3)_2$	E_1	84	3 HCl, H_2O, 169–172° (dec.)	11
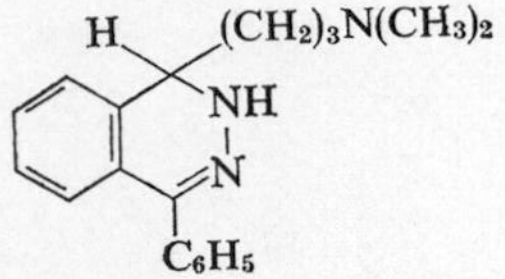 H, $(CH_2)_3N(CH_3)_2$, NH, N, C_6H_5	E_2	90	HCl, 173–176° (dec.)	11
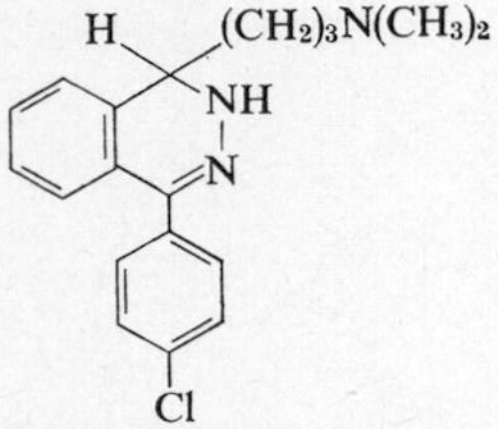 H, $(CH_2)_3N(CH_3)_2$, NH, N, Cl	E_2	—	HCl, 227–230° (dec.)	11
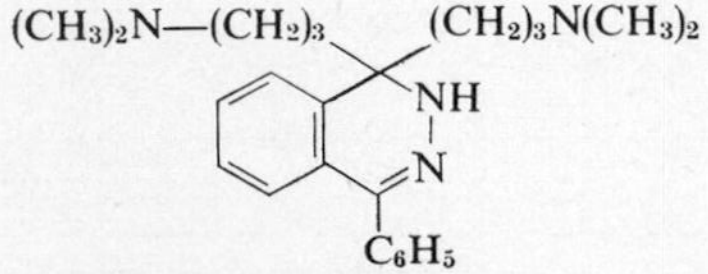 $(CH_3)_2N—(CH_2)_3$, $(CH_2)_3N(CH_3)_2$, NH, N, C_6H_5	E_3	30	97–98.5° (heptane); 2 HCl, 231–232°	11

TABLE 2L-1 (*continued*)

Structure	Preparation[a]	Yield (%)	MP	Reference
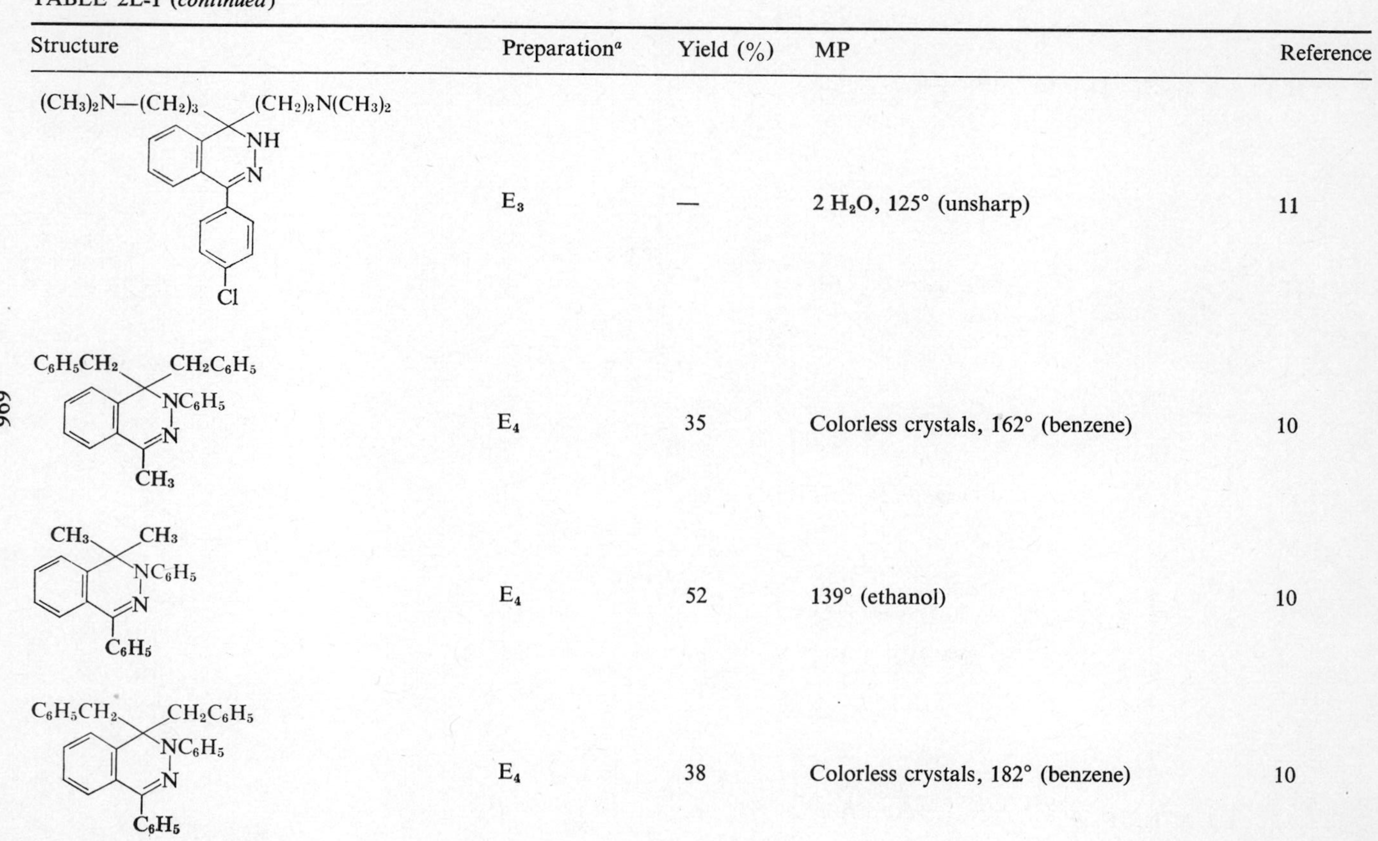	E_3	—	2 H_2O, 125° (unsharp)	11
	E_4	35	Colorless crystals, 162° (benzene)	10
	E_4	52	139° (ethanol)	10
	E_4	38	Colorless crystals, 182° (benzene)	10

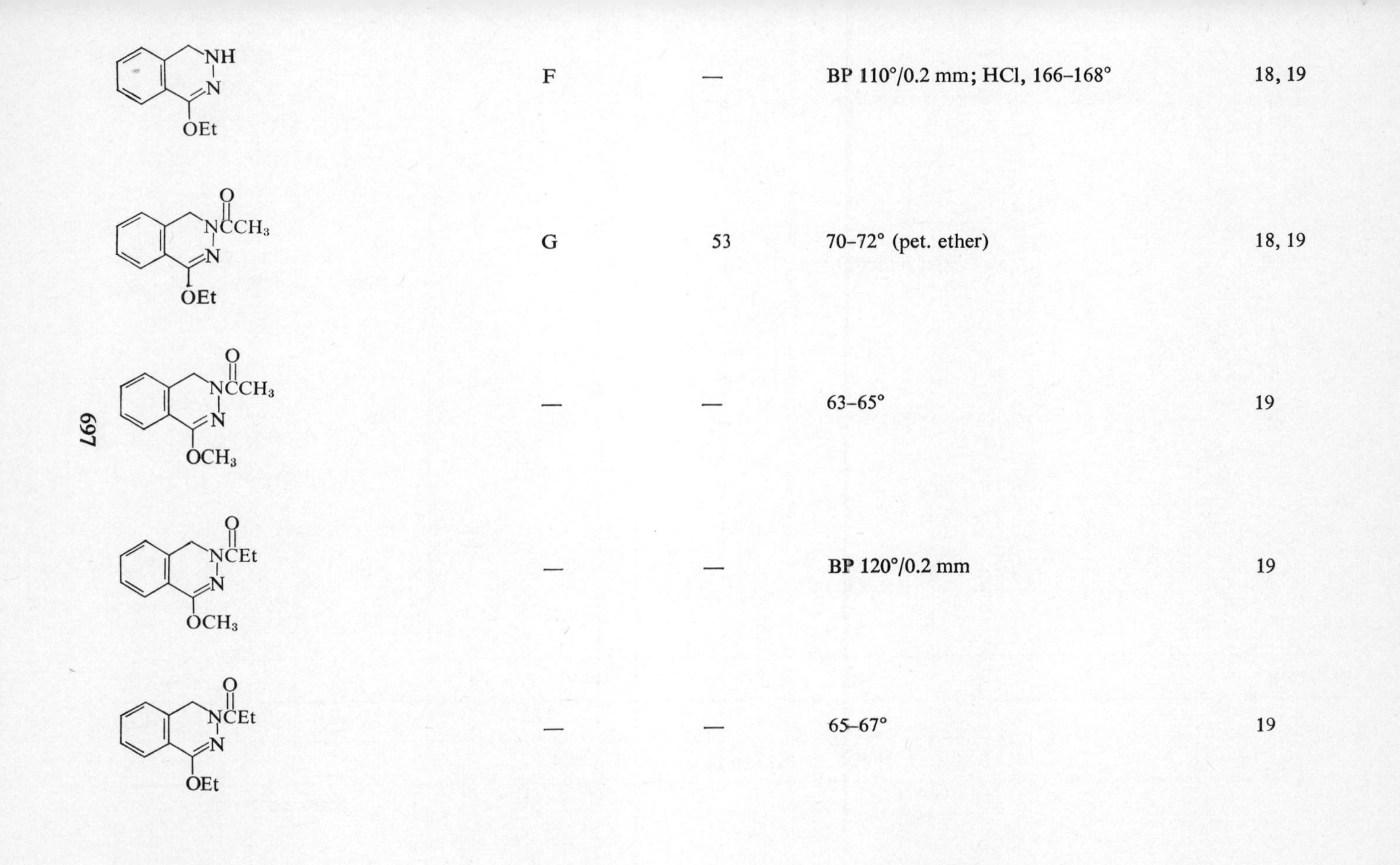

	F	—	BP 110°/0.2 mm; HCl, 166–168°	18, 19
	G	53	70–72° (pet. ether)	18, 19
	—	—	63–65°	19
	—	—	BP 120°/0.2 mm	19
	—	—	65–67°	19

TABLE 2L-1 (*continued*)

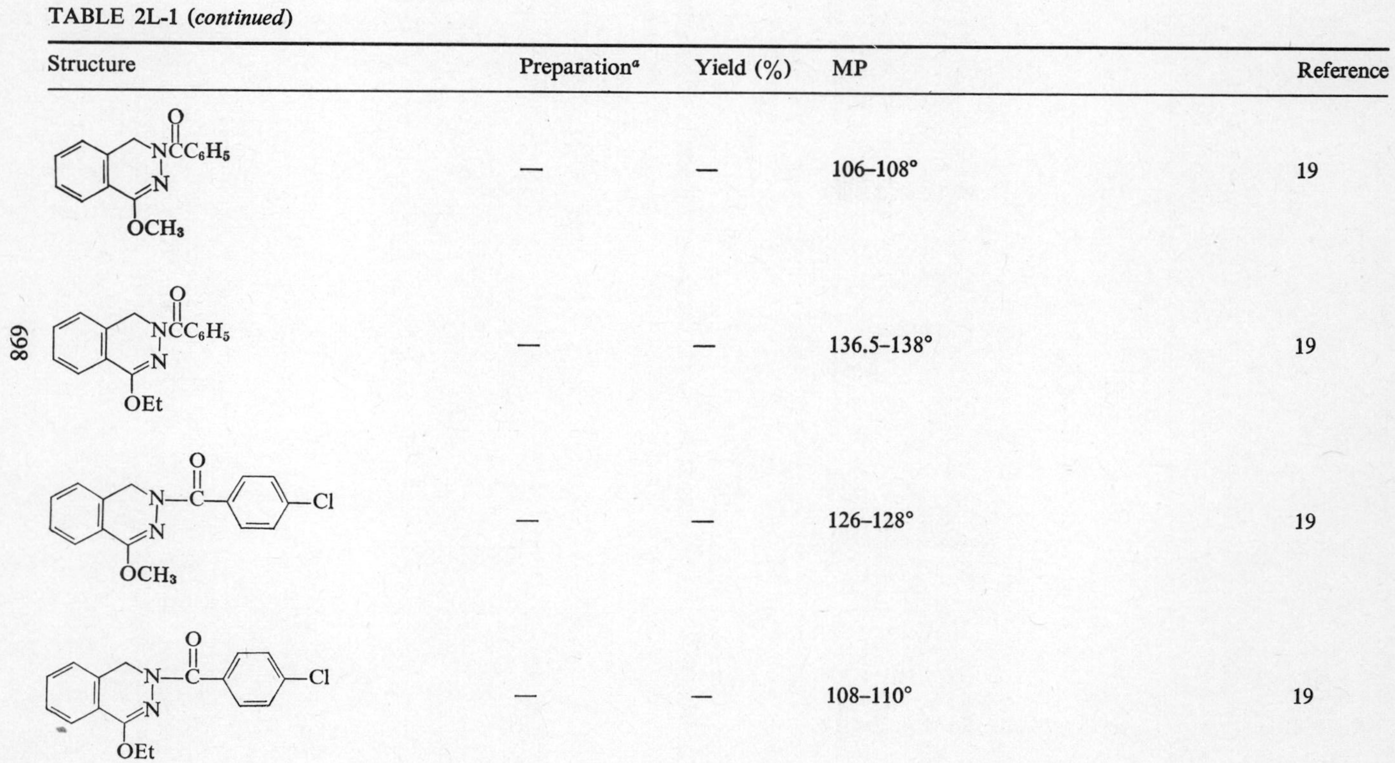

Structure	Preparation[a]	Yield (%)	MP	Reference
	—	—	106–108°	19
	—	—	136.5–138°	19
	—	—	126–128°	19
	—	—	108–110°	19

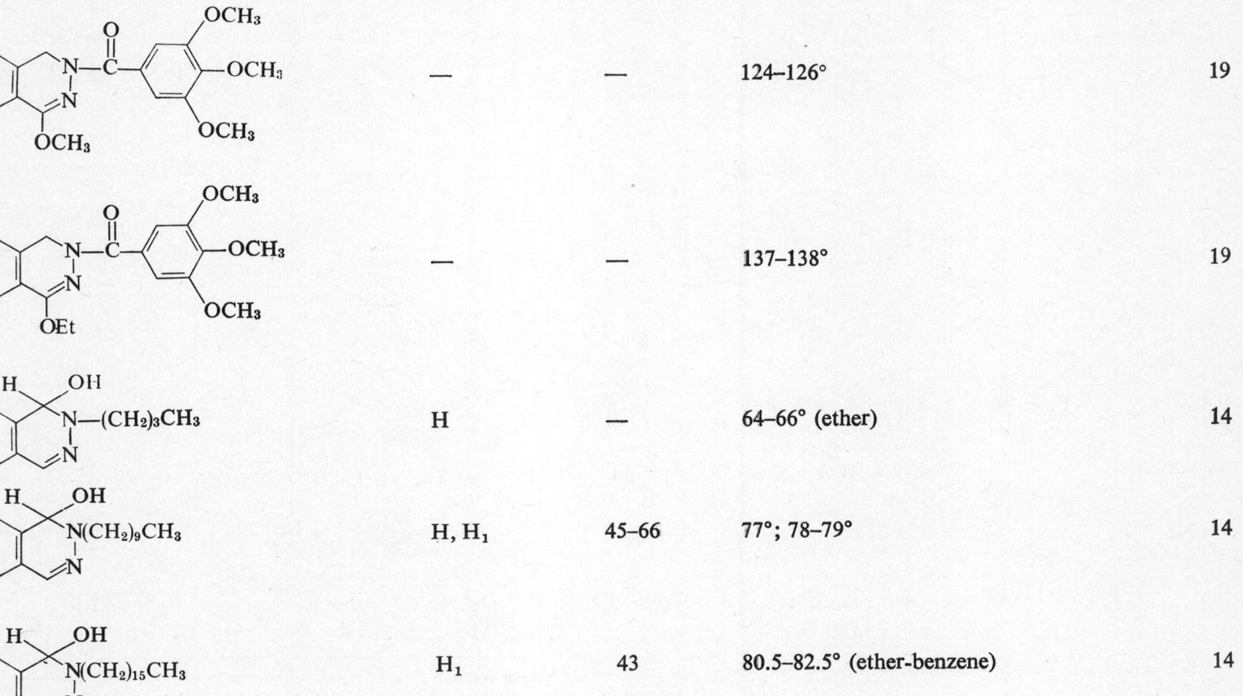

	—	—	124–126°	19
	—	—	137–138°	19
	H	—	64–66° (ether)	14
	H, H_1	45–66	77°; 78–79°	14
	H_1	43	80.5–82.5° (ether-benzene)	14

TABLE 2L-1 (*continued*)

Structure	Preparation[a]	Yield (%)	MP	Reference
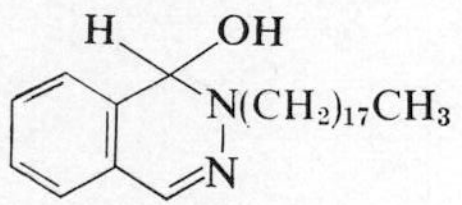 1-H, 1-OH, 2-$N(CH_2)_{17}CH_3$ phthalazine	H, H_1	37–100	80° (benzene-ether); 82°	14
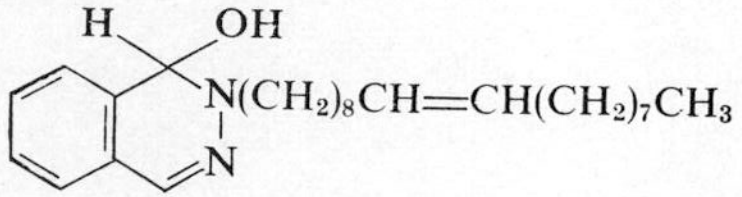 1-H, 1-OH, 2-$N(CH_2)_8CH{=}CH(CH_2)_7CH_3$ phthalazine	H_1	93	78–80° (ether-benzene)	14
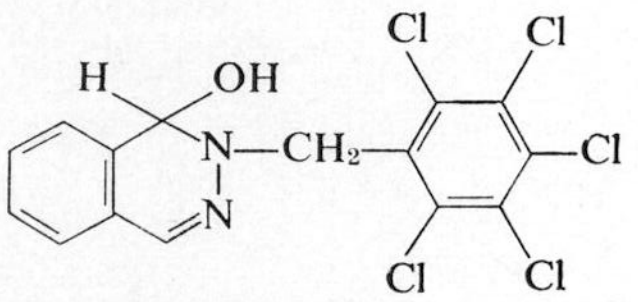 1-H, 1-OH, 2-N—CH_2—(pentachlorophenyl) phthalazine (Cl, Cl, Cl, Cl, Cl)	—	—	213–215°	60
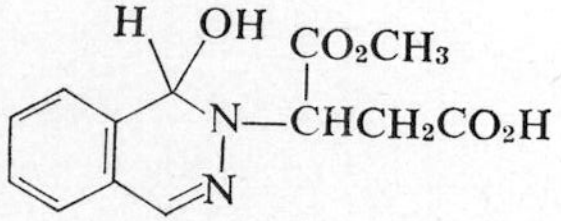 1-H, 1-OH, 2-N—$CH(CO_2CH_3)CH_2CO_2H$ phthalazine	—	—	90–92°	61
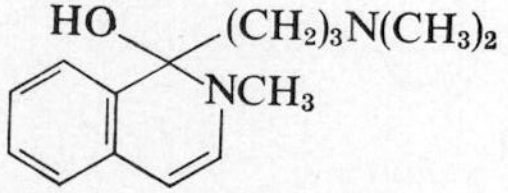 HO, $(CH_2)_3N(CH_3)_2$, NCH_3 isoquinoline	A	—		9

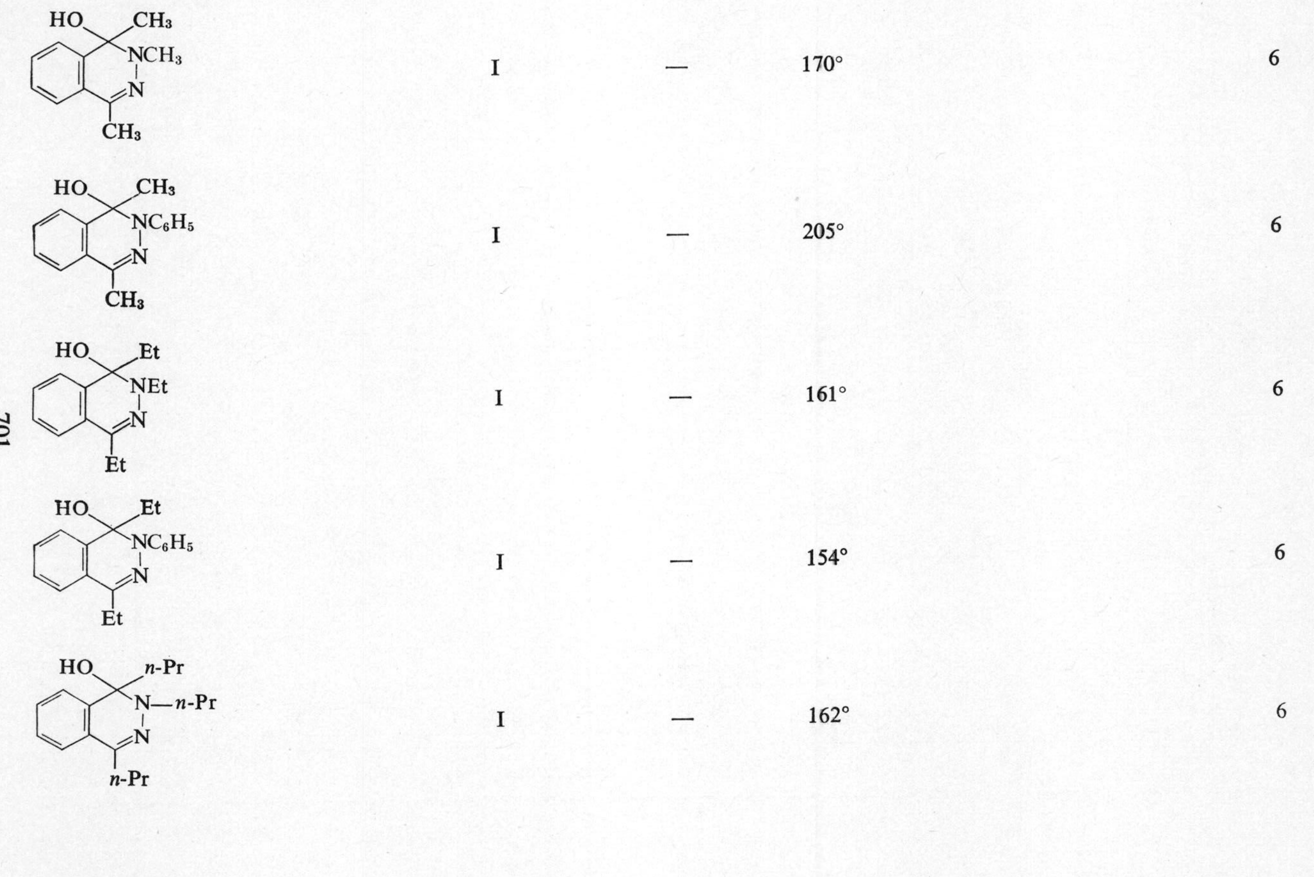

Structure				
1-hydroxy-1,2,4-trimethyl (HO, CH_3; NCH_3; CH_3)	I	—	170°	6
HO, CH_3; NC_6H_5; CH_3	I	—	205°	6
HO, Et; NEt; Et	I	—	161°	6
HO, Et; NC_6H_5; Et	I	—	154°	6
HO, n-Pr; N—n-Pr; n-Pr	I	—	162°	6

TABLE 2L-1 (*continued*)

Structure	Preparation[a]	Yield (%)	MP	Reference
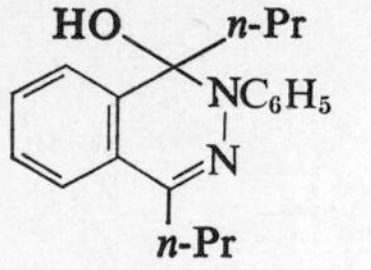 HO, *n*-Pr; NC_6H_5; N; *n*-Pr	I	—	189°	6
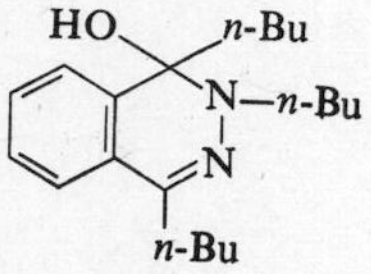 HO, *n*-Bu; N—*n*-Bu; N; *n*-Bu	I	—	130°	6
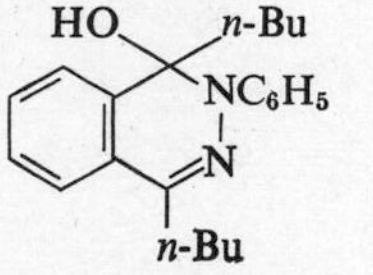 HO, *n*-Bu; NC_6H_5; N; *n*-Bu	I	—	177°	6
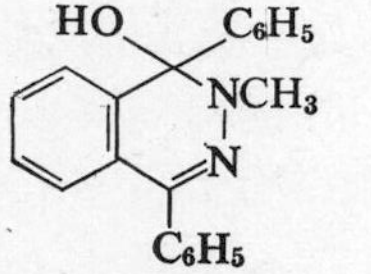 HO, C_6H_5; NCH_3; N; C_6H_5	I	—	129°	6
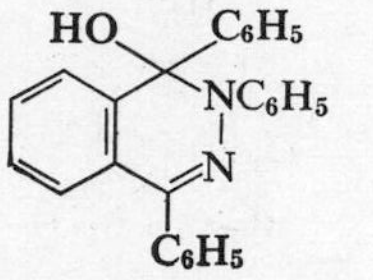 HO, C_6H_5; NC_6H_5; N; C_6H_5	I, E_4	55	Colorless crystals, 182° (ethanol); 189°	6, 10

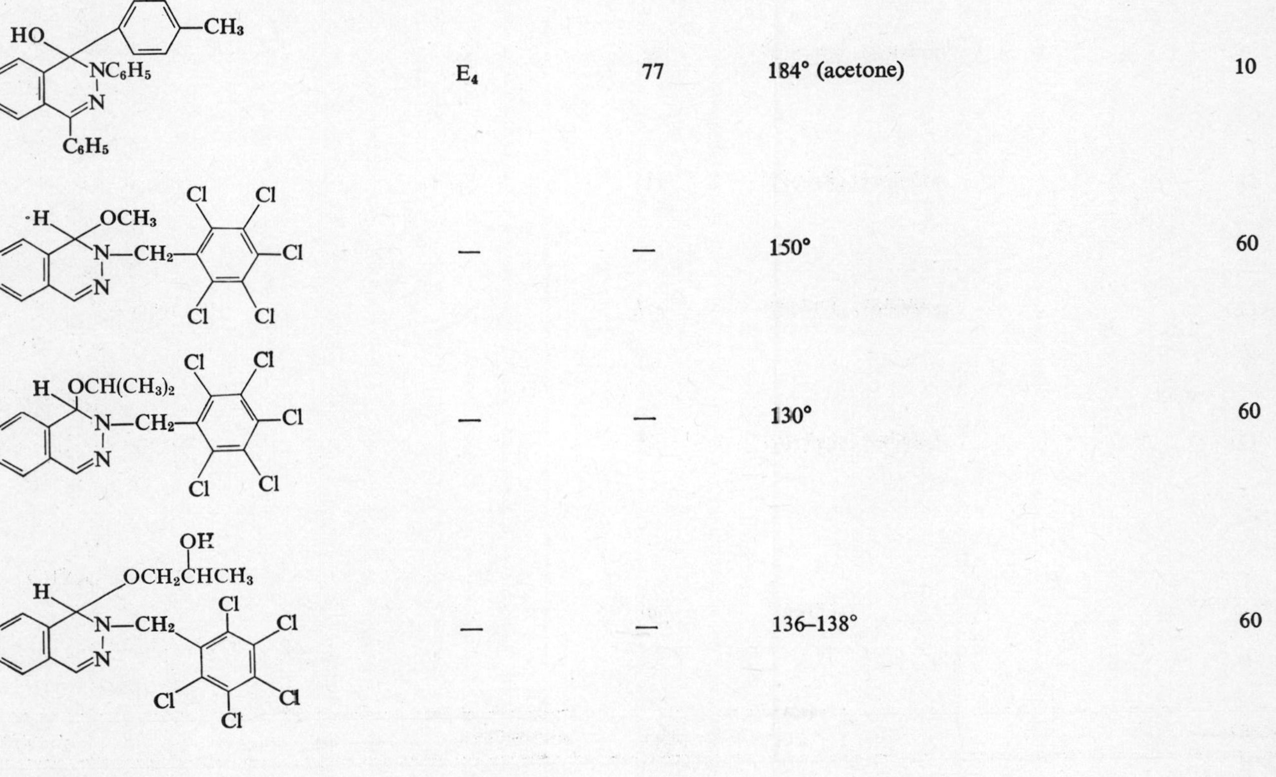

	E_4	77	184° (acetone)	10
	—	—	150°	60
	—	—	130°	60
	—	—	136–138°	60

TABLE 2L-1 (*continued*)

Structure	Preparation[a]	Yield (%)	MP	Reference
1-H, 1-OCH$_3$, 2-[C(COOCH$_3$)=CH(COOCH$_3$)]-1,2-dihydrophthalazine (structure: H, OCH$_3$, N—C(=C(H)COCH$_3$... CH$_3$OC(=O)), COOCH$_3$)	J	39	136–137°	62, 63
1-H, 1-CN, 2-(COC$_6$H$_5$)-1,2-dihydrophthalazine	K	55	163–164° (ethanol)	13
1-H, 1-CN, 2-(CON(C$_6$H$_5$)$_2$)-1,2-dihydrophthalazine	K	72	206–208° (ethanol)	13
1-H, 1-CN, 2-(CON(Et)$_2$)-1,2-dihydrophthalazine	K	11	136–138° (ethanol)	13
1-H, 1-CN, 2-(SO$_2$C$_6$H$_5$)-1,2-dihydrophthalazine	K	99	151–154° (ethanol)	13

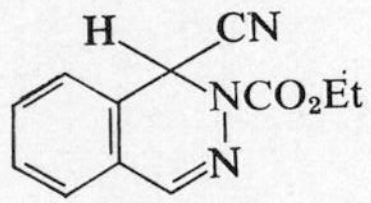 H, CN, NCO_2Et, N	K	35	137–138° (ethanol-water)	13
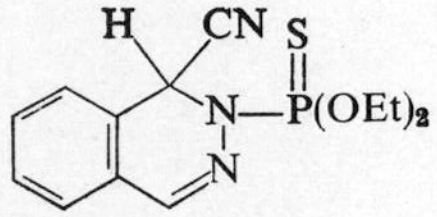 H, CN, S, N—$P(OEt)_2$, N	K	74	83–86° (ethanol)	13
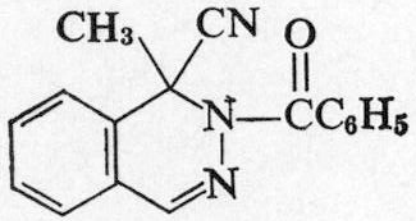 CH_3, CN, O, N—CC_6H_5, N	K_1	100	143–145° (ethanol)	13
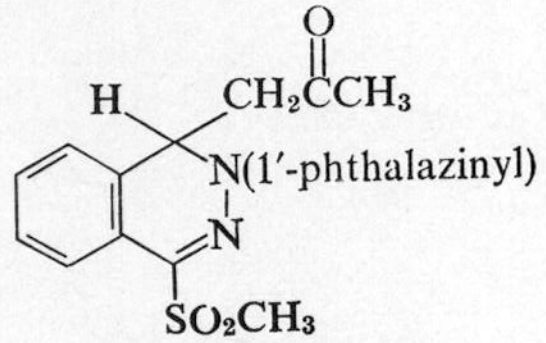 H, O, CH_2CCH_3, N(1′-phthalazinyl), N, SO_2CH_3	L	49	183–185°	15
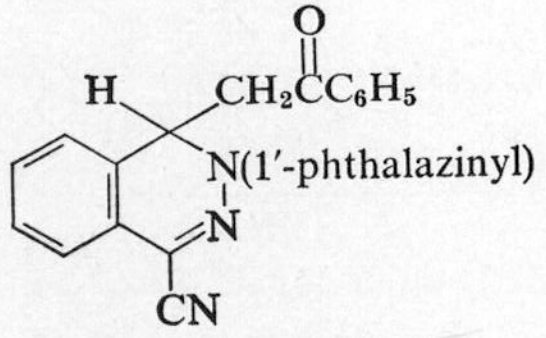 H, O, $CH_2CC_6H_5$, N(1′-phthalazinyl), N, CN	L_1	87	163–164°	15

TABLE 2L-1 (*continued*)

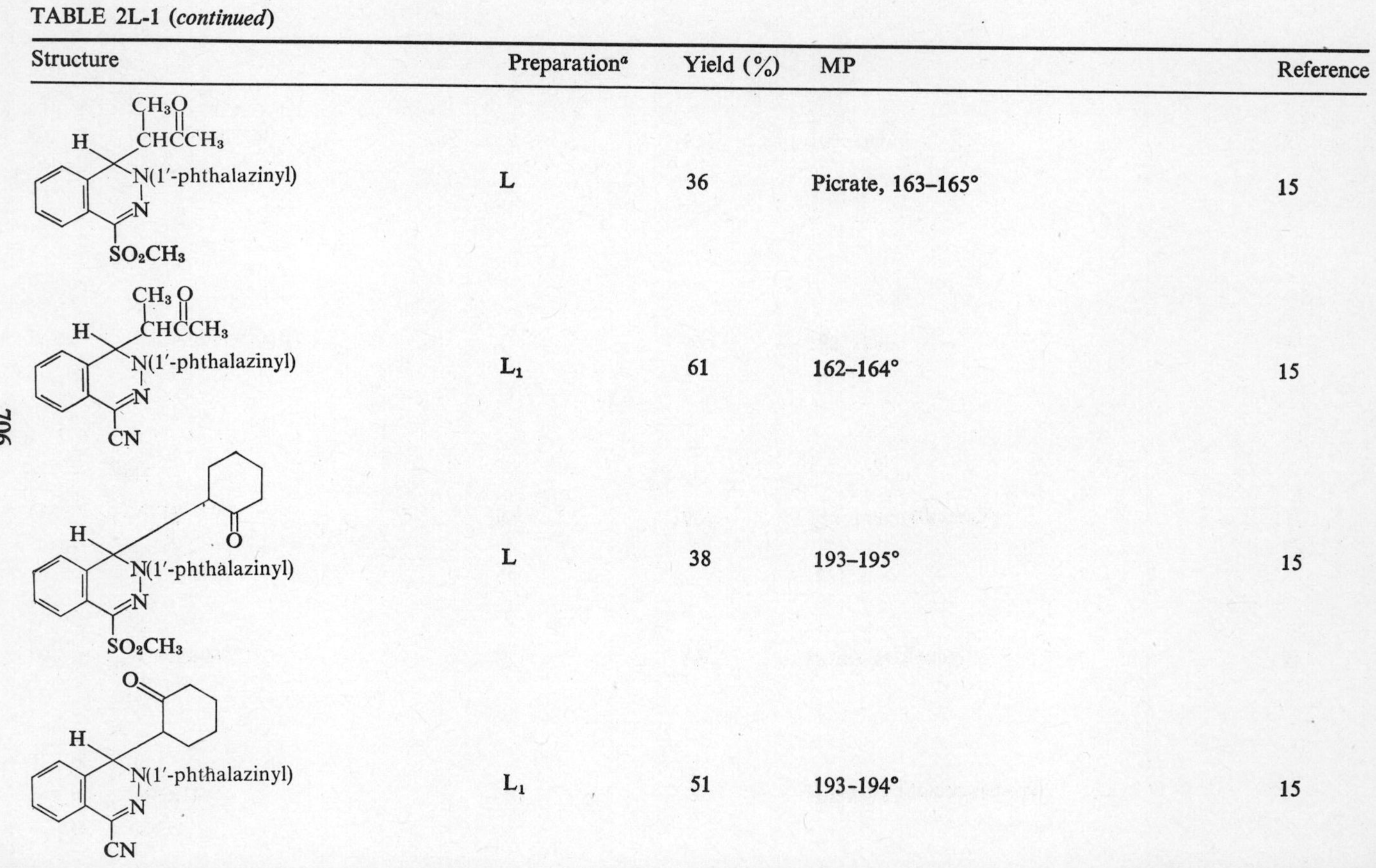

Structure	Preparation[a]	Yield (%)	MP	Reference
H, CH(CH3)C(=O)CH3, N(1'-phthalazinyl), SO2CH3	L	36	Picrate, 163–165°	15
H, CH(CH3)C(=O)CH3, N(1'-phthalazinyl), CN	L_1	61	162–164°	15
H, 2-oxocyclohexyl, N(1'-phthalazinyl), SO2CH3	L	38	193–195°	15
H, 2-oxocyclohexyl, N(1'-phthalazinyl), CN	L_1	51	193-194°	15

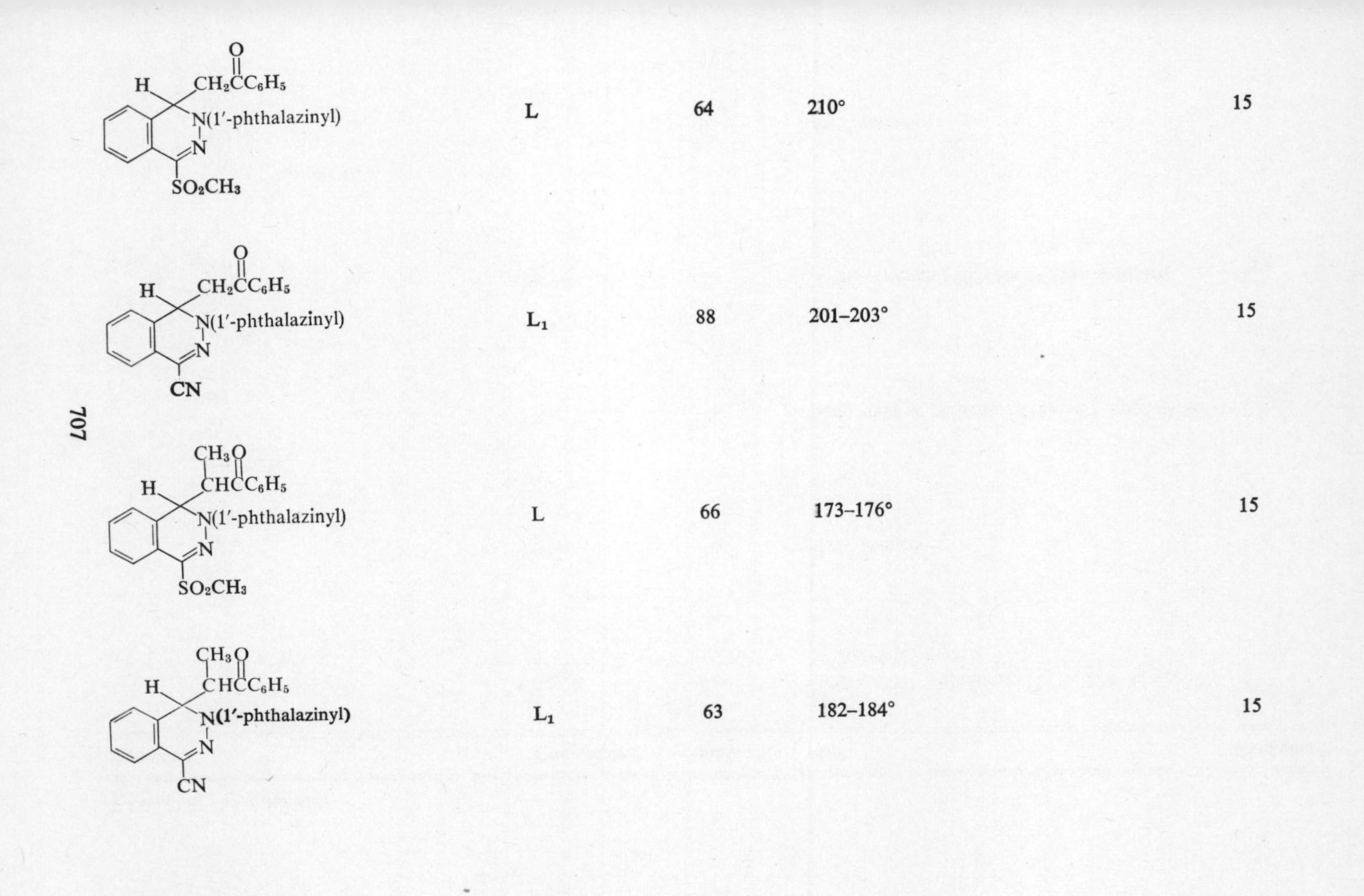

L	64	210°	15
L_1	88	201–203°	15
L	66	173–176°	15
L_1	63	182–184°	15

TABLE 2L-1 (*continued*)

Structure	Preparation[a]	Yield (%)	MP	Reference
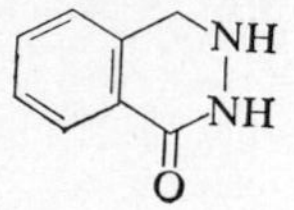 NH, NH, O	M, M_2	—	169°; 170° (ethanol) HCl, 222–224°; HI, 200–202°	5, 7, 8
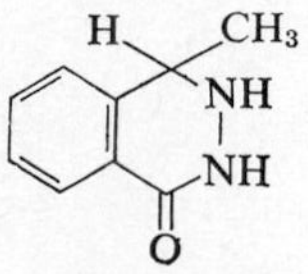 H, CH_3, NH, NH, O	M	75	130° (benzene)	5
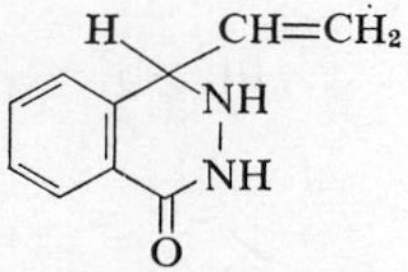 H, $CH{=}CH_2$, NH, NH, O	N	8	Pale yellow prisms, 138–140° (pet. ether)	12
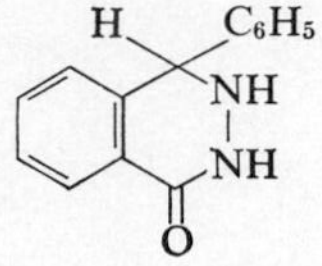 H, C_6H_5, NH, NH, O	N	12	White crystals, 165–166° (ether-pentane)	12
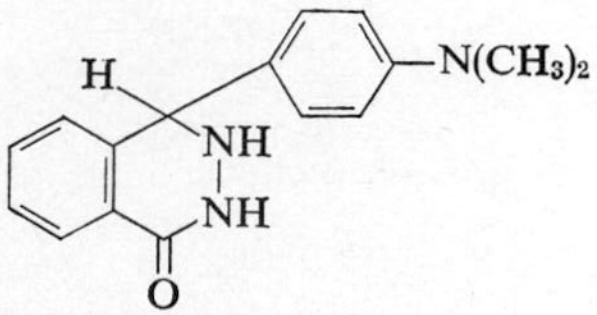 H, $N(CH_3)_2$, NH, NH, O	M	95	205° (ethanol)	5

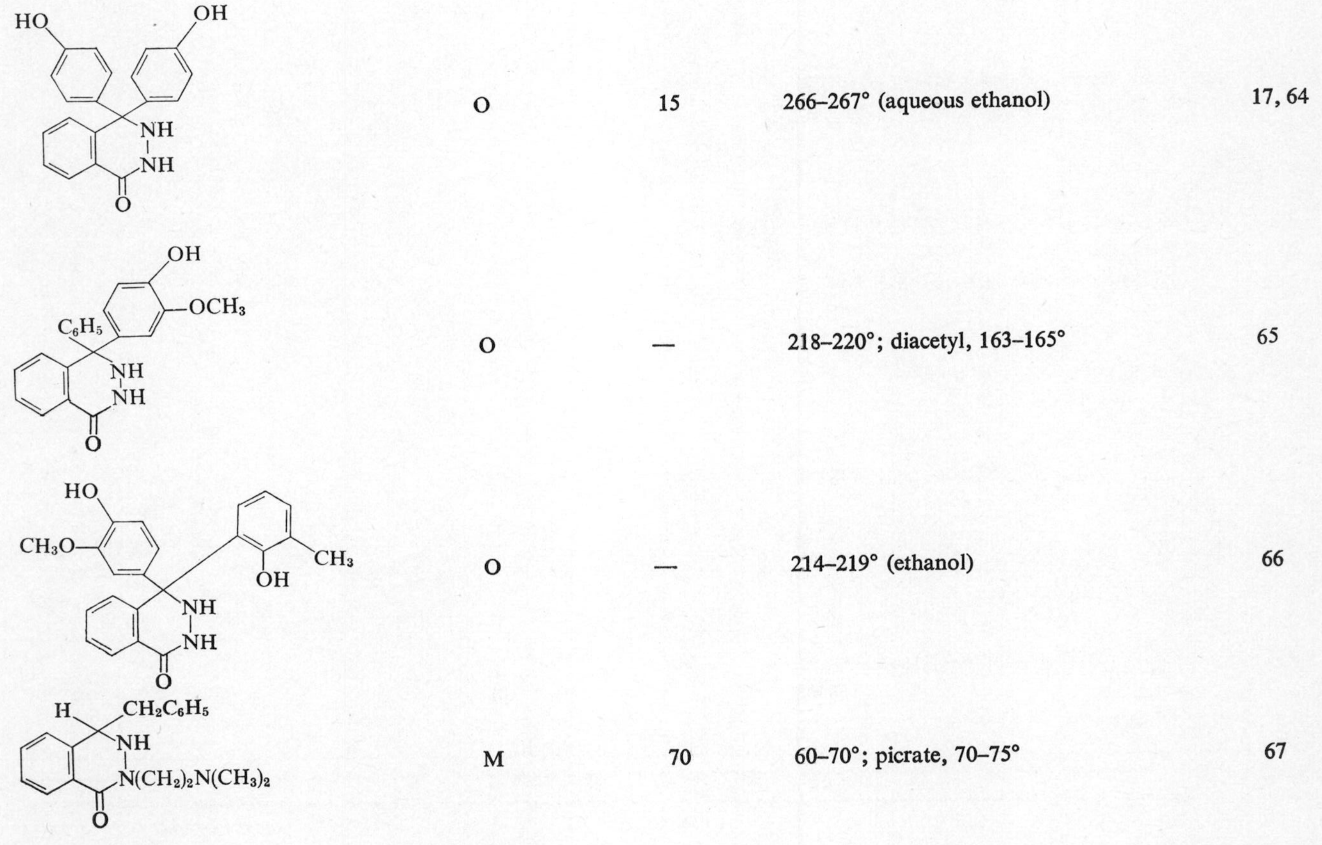

Compound		Yield (%)	m.p.	Ref.
	O	15	266–267° (aqueous ethanol)	17, 64
	O	—	218–220°; diacetyl, 163–165°	65
	O	—	214–219° (ethanol)	66
	M	70	60–70°; picrate, 70–75°	67

TABLE 2L-1 (*continued*)

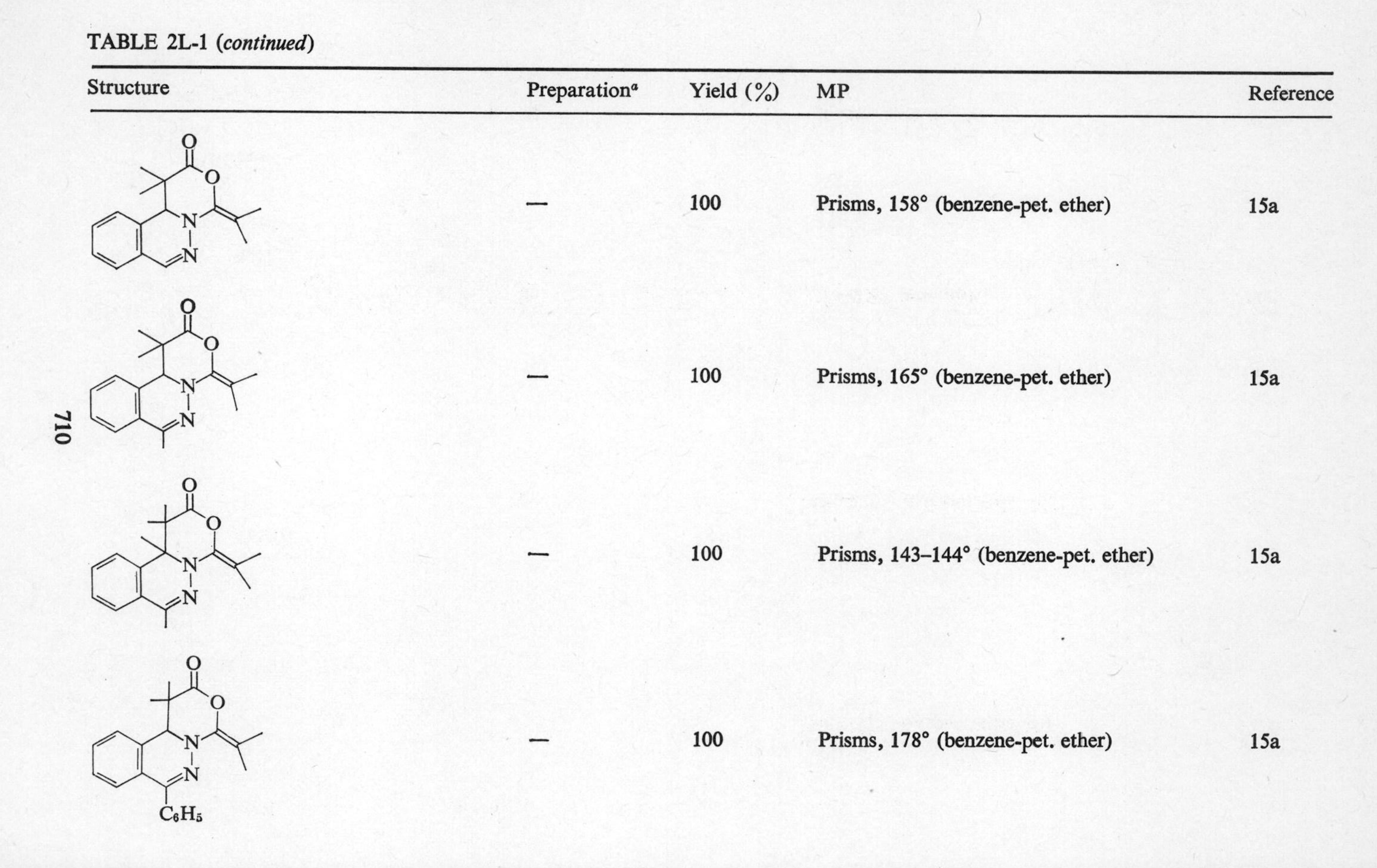

Structure	Preparation[a]	Yield (%)	MP	Reference
	—	100	Prisms, 158° (benzene-pet. ether)	15a
	—	100	Prisms, 165° (benzene-pet. ether)	15a
	—	100	Prisms, 143–144° (benzene-pet. ether)	15a
C_6H_5	—	100	Prisms, 178° (benzene-pet. ether)	15a

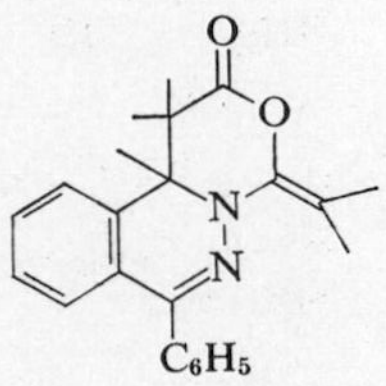 Structure (C_6H_5)	—	100	Prisms, 178° (benzene-pet. ether)	15a
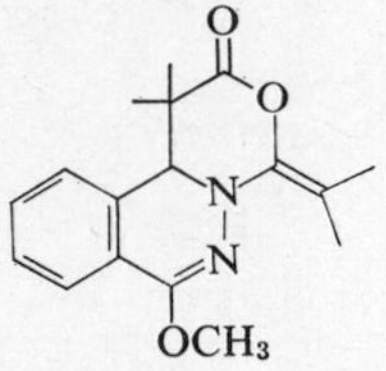 Structure (OCH_3)	—	84	Prisms, 125° (benzene-pet. ether)	15a
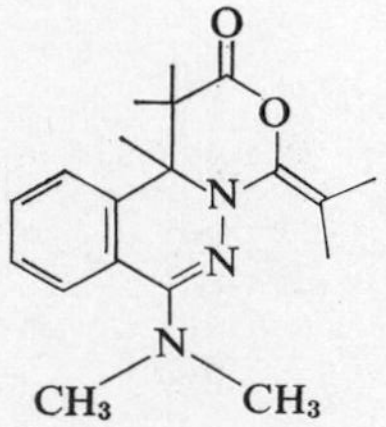 Structure ($N(CH_3)_2$)	—	76	Prisms, 140°–142° (benzene-pet. ether)	15a
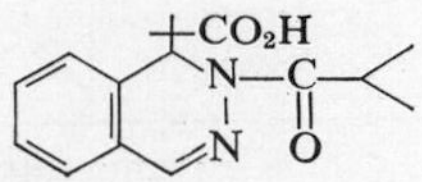 Structure (CO_2H)	—	66	Needles, 117° (benzene-pet. ether): methyl ester, 70°	15a

TABLE 2L-1 (*continued*)

Structure	Preparation[a]	Yield (%)	MP	Reference
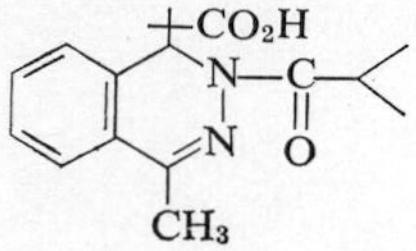 CO_2H; N—C(=O); N; CH_3	—	75	Needles, 143–144° (benzene-pet. ether)	15a
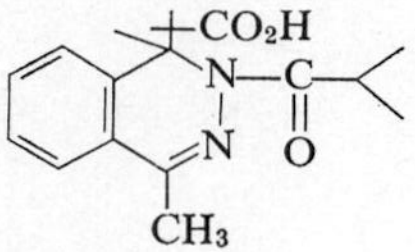 CO_2H; N—C(=O); N; CH_3	—	71	Needles, 158–159° (benzene-pet. ether)	15a
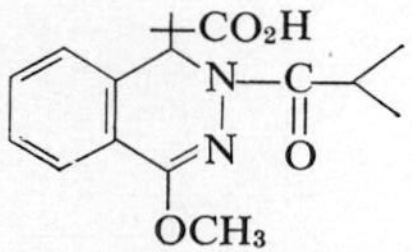 CO_2H; N—C(=O); N; OCH_3	—	66	Needles, 154–155° (benzene-pet. ether)	15a
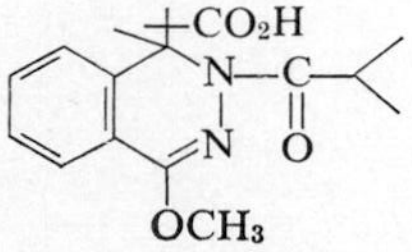 CO_2H; N—C(=O); N; OCH_3	—	73	Needles, 145° (pet. ether)	15a
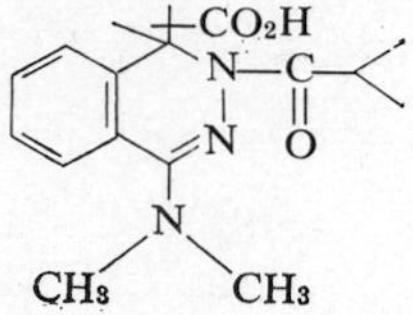 CO_2H; N—C(=O); N; N; CH_3, CH_3	—	71	Needles, 198° (EtOH-pet. ether)	15a

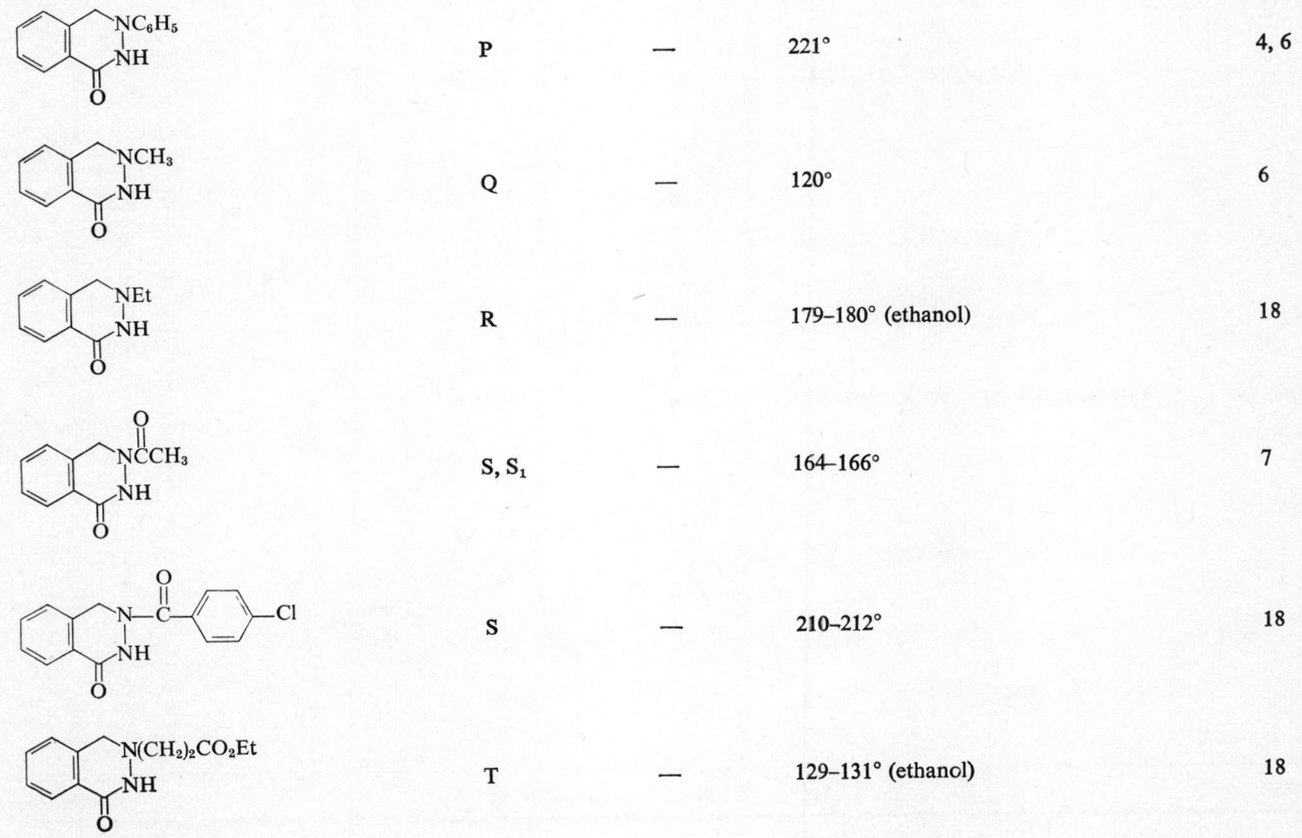

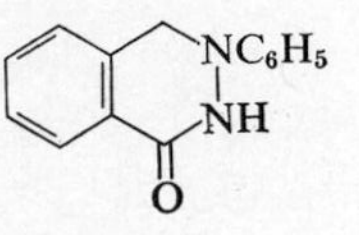	P	—	221°	4, 6
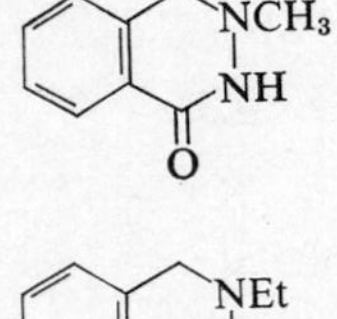	Q	—	120°	6
	R	—	179–180° (ethanol)	18
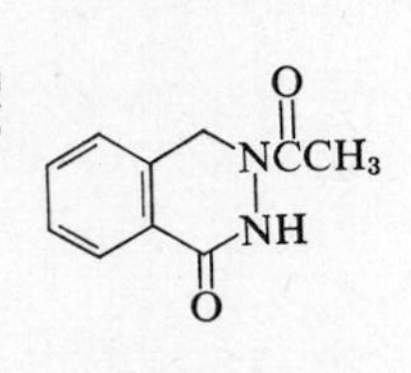	S, S_1	—	164–166°	7
	S	—	210–212°	18
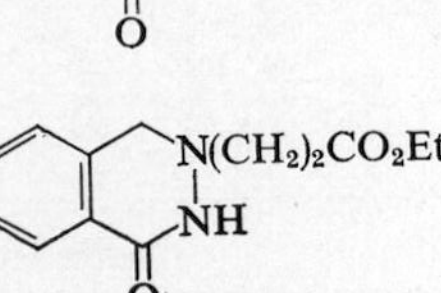	T	—	129–131° (ethanol)	18

TABLE 2L-1 (*continued*)

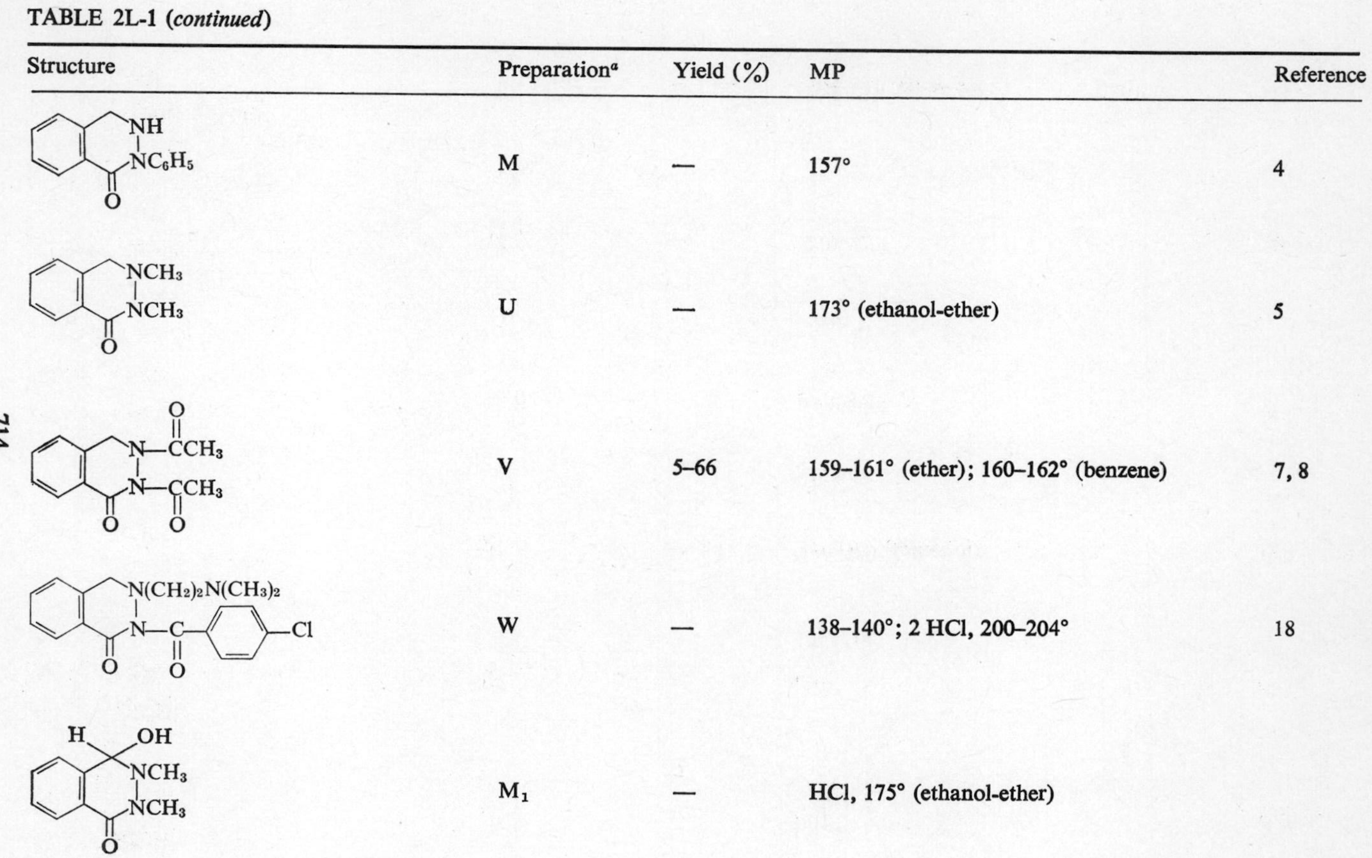

Structure	Preparation[a]	Yield (%)	MP	Reference
	M	—	157°	4
	U	—	173° (ethanol-ether)	5
	V	5–66	159–161° (ether); 160–162° (benzene)	7, 8
	W	—	138–140°; 2 HCl, 200–204°	18
	M_1	—	HCl, 175° (ethanol-ether)	

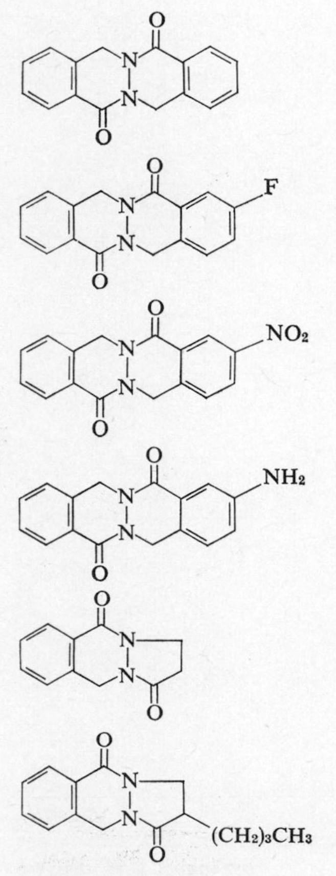

Compound				
	X	75	226–227° (ethanol), 218–220° (dioxane)	8, 20
	X	60	228–229° (ethanol)	8, 20
	X	55	233–234°, 239–241° (ethanol)	8, 20
	—	—	275–277°	8
	—	40	114–116° (ethyl acetate)	20
	—	47	108–110° (ether)	20

TABLE 2L-1 (*continued*)

Structure	Preparation[a]	Yield (%)	MP	Reference
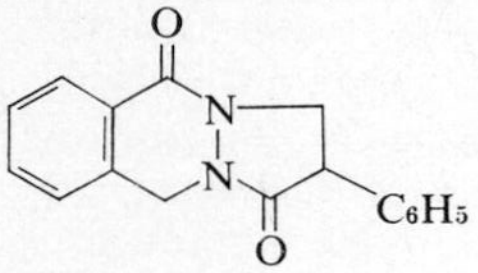 C_6H_5	—	60	160–161° (ethanol)	20
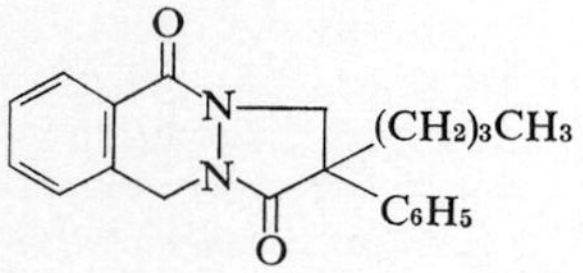 $(CH_2)_3CH_3$, C_6H_5	—	35	140–141° (ethanol)	20
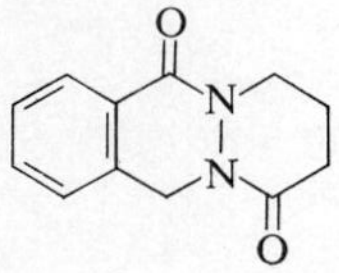	—	51	103–104° (ether)	20
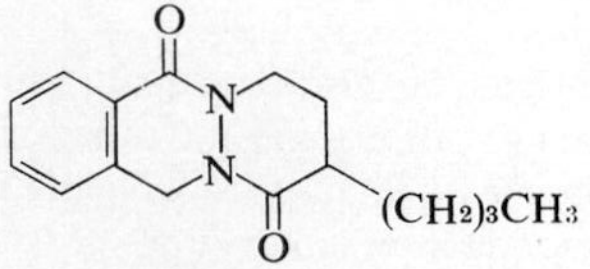 $(CH_2)_3CH_3$	—	50	61–63° (ether-pet. ether), bp 170°/0.2 mm	20
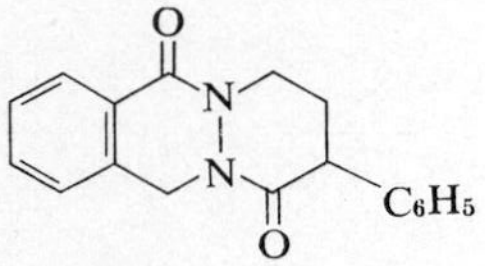 C_6H_5	—	60	148–150° (isopropyl ether)	20

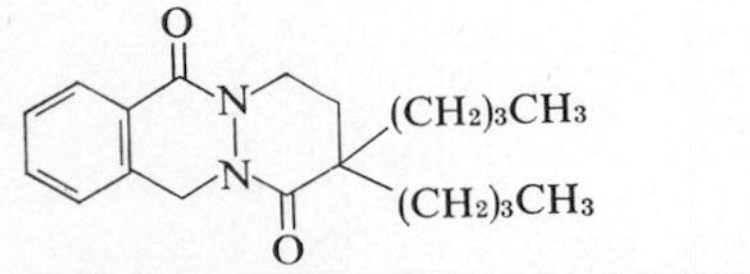	—	30	60° (ether-pet. ether), bp 165°/0.1 mm	20

[a] A, lithium aluminum hydride reduction of properly substituted 1-(2*H*)phthalazinone in ether or tetrahydrofuran. B, sodium borohydride reduction of 2-alkylphthalazinium halides in aqueous medium. C, 2-(4-hydroxybutyl)-4-(*p*-chlorophenyl)-1-(2*H*)phthalazinone + $SOCl_2$, reflux in chloroform for 18 hr. D, phthalazine + organolithium compound in ether at 0° followed by hydrolysis. E, phthalazine + 3-*N*,*N*-dimethylaminopropylmagnesium chloride in tetrahydrofuran, heat 5 hr followed by hydrolysis. E_1, 1-(3-*N*,*N*-dimethylaminopropyl)phthalazine + Grignard reagent, same as E; or 1-(2*H*)phthalazinone + Grignard reagent, similar to E. E_2, 1-arylphthalazine + Grignard reagent, similar to E. E_3, 1-aryl-4-(3-*N*,*N*-dimethylaminopropyl)phthalazine + Grignard reagent, similar to E; or 4-aryl-1-(2*H*)phthalazinone + Grignard reagent, similar to E. E_4, corresponding 2,4-disubstituted 1-(2*H*)phthalazinone + proper Grignard reagent. F, alkaline hydrolysis of 2-acetyl-4-ethoxy-1,2-dihydrophthalazine. G, 2-acetyl-1,2,3,4-tetrahydro-4-oxophthalazine + boron trifluoride etherate stirred in dichloromethane at room temperature for 4 hr. H, phthalazine + corresponding alkyl halide, reflux in acetonitrile followed by treatment with base. H_1, phthalaldehyde + corresponding alkylhydrazine sulfate, reflux in aqueous ethanol followed by treatment with base. I, 2-substituted 4-hydroxy-1-(2*H*)phthalazinone + corresponding Grignard reagent. J, phthalazine + dimethyl acetylenedicarboxylate kept in methanol at 0° for 2 weeks. K, phthalazine + potassium cyanide + proper chloro compound stirred several hours at room temperature in CH_2Cl_2—H_2O. K_1, 1-cyano-2-benzoyl-1,2-dihydrophthalazine + sodium hydride in dimethylformamide followed by treatment with methyl iodide. L, 1-sulfonylphthalazine + corresponding ketone in aqueous sodium hydroxide. L_1, 1-substituted 2-(1′-phthalazinyl)-4-methylsulfonyl-1,2-dihydrophthalazine treated with potassium cyanide in dimethyl sulfoxide. M, two-electron polarographic reduction of corresponding 1-(2*H*)-phthalazinone. M_1, similar reduction of 2,3-dimethylphthalazine-1,4-dione at 8.3 pH. M_2, 2,3-diacetyl-1,2,3,4-tetrahydro-4-oxophthalazine refluxed 5 min with 10% hydrochloric acid. N, 1-(2*H*)phthalazinone + corresponding organolithium compound in tetrahydrofuran. O, the corresponding phthalide is treated with hydrazine hydrate. P, lithium aluminum hydride reduction of 4-hydroxy-2-phenyl-1-(2*H*)phthalazinone in tetrahydrofuran; or its four-electron polarographic reduction; or two-electron polarographic reduction of 3-phenylpseudophthalazinone. Q, lithium aluminum hydride reduction of 4-hydroxy-2-methyl-1-(2*H*)phthalazinone in tetrahydrofuran. R, 1,2,3,4-tetrahydro-4-oxophthalazine + boron trifluoride etherate stirred in dichloromethane at room temperature for 4 hr. S, 1,2,3,4-tetrahydro-4-oxophthalazine treated with corresponding acid chloride. S_1, byproduct from PtO_2/H_2 reduction of 1-(2*H*)phthalazinone in acetic acid. T, 1,2,3,4-tetrahydro-4-oxophthalazine treated with ethyl acrylate in presence of sodium ethoxide. U, four-electron polarographic reduction of 2,3-dimethylphthalazine-1,4-dione at pH 5. V, 2-bromomethylbenzoyl chloride + *N*,*N*′-diacetylhydrazine, heat in dioxane at 80° for 2 hr in presence of triethylamine; or PtO_2/H_2, reduction of 1-(2*H*)phthalazinone in acetic acid. W, 2-(*p*-chlorobenzyl)-1,2,3,4-tetrahydro-4-oxophthalazine + 2-*N*,*N*-dimethylaminochloroethylamine, reflux in ethanol with sodium ethoxide for 7 hr. X, 1,2,3,4-tetrahydro-4-oxophthalazine treated with 5-substituted 2-bromomethylbenzoyl chloride in presence of triethylamine.

TABLE 2L-2. 1,2,3,4-Tetrahydrophthalazines

Structure	Preparation[a]	Yield (%)	MP	Reference
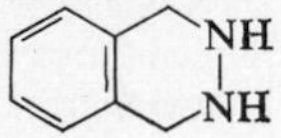 NH, NH	A	85–98	HCl, 233–236°; 236–238°; monobenzoyl, 90–91°; dibenzoyl, 210–212°	22–24, 30
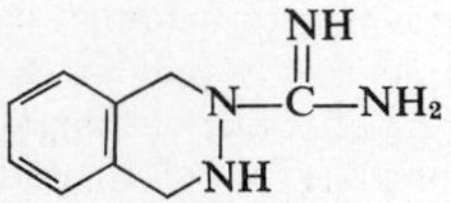 N—C(=NH)—NH_2, NH	B	—	HCl, H_2O, 206–208°; acetate, 237–238°	30
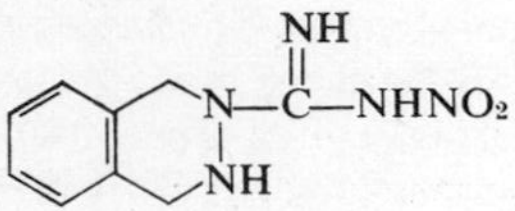 N—C(=NH)—$NHNO_2$, NH	C	—	170–171° (ethanol)	30
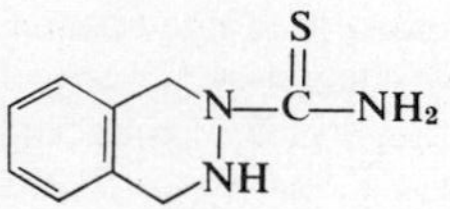 N—C(=S)—NH_2, NH	D	83	191–192° (aqueous butanol)	30
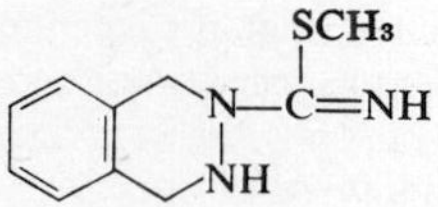 N—C(SCH_3)=NH, NH	E	—	HI, 177–178°; HCl, 214–215°	30
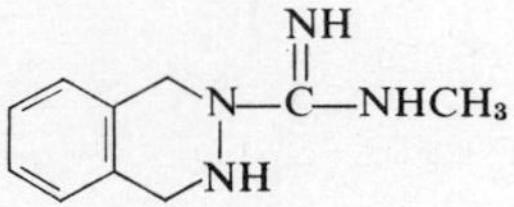 N—C(=NH)—$NHCH_3$, NH	F	—	232–233° (ethanol-ether)	30

Structure				
NH ‖ N—C—N$(CH_3)_2$ NH	F_1	—	HCl, 220°	30
S ‖ N—C—$NHCH_3$ NH	G	80	226–227° (aqueous butanol)	30
SCH_3 | N—C=NCH_3 NH	H	—	140–143°	30
N—CH_3 ‖ N—C—$NHCH_3$ NH	—	—	HCl, H_2O, 145–147° (ethanol-ether) H_2SO_4 108–109°	30
NH NH CH_3	I	41	BP 131–134°/10 mm, phenylthiourea, 191–192°	40
NH ‖ N—C—NH_2 NH CH_3	—	—	225°	30

TABLE 2L-2 (*continued*)

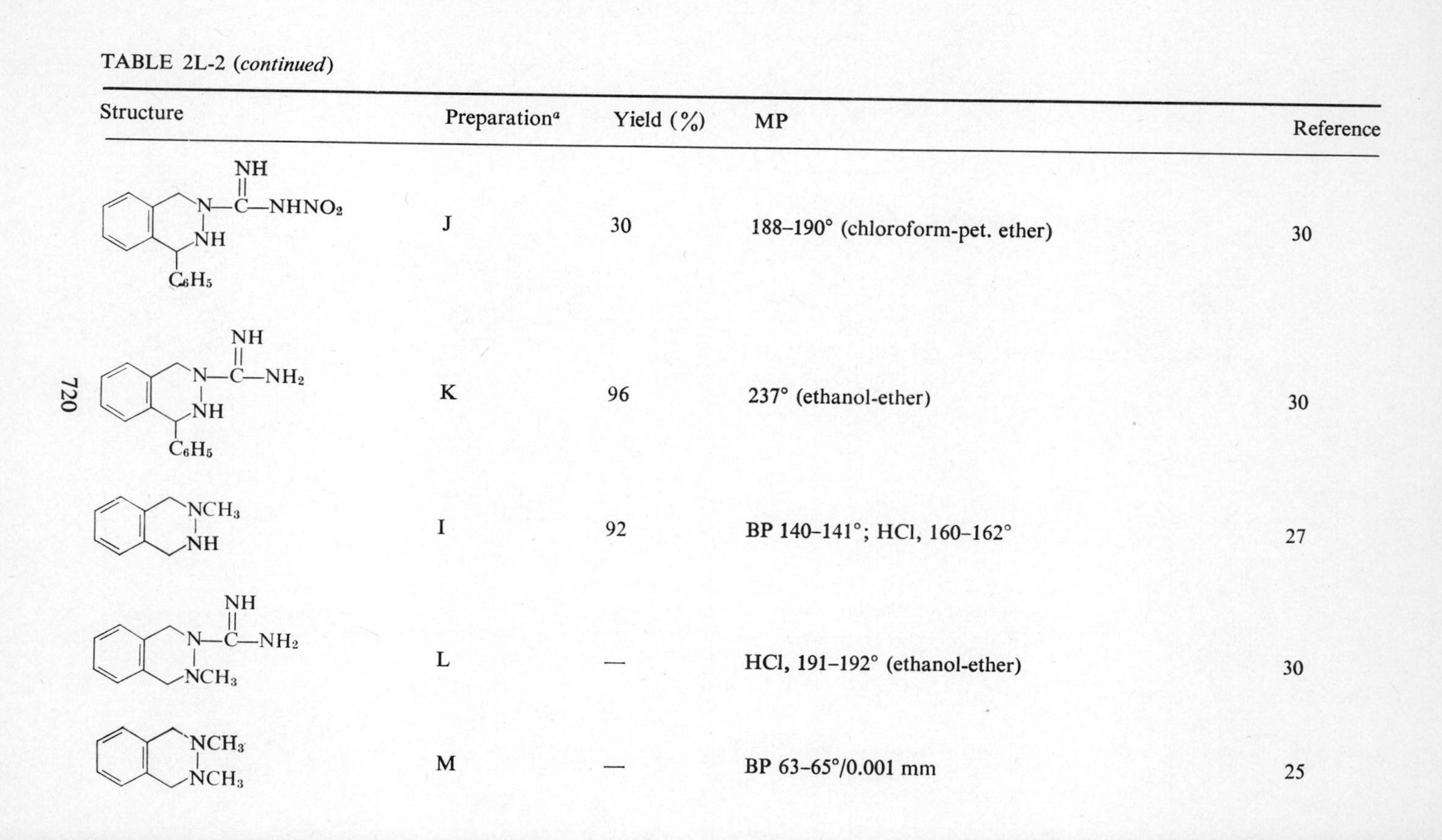

Structure	Preparation[a]	Yield (%)	MP	Reference
	J	30	188–190° (chloroform-pet. ether)	30
	K	96	237° (ethanol-ether)	30
	I	92	BP 140–141°; HCl, 160–162°	27
	L	—	HCl, 191–192° (ethanol-ether)	30
	M	—	BP 63–65°/0.001 mm	25

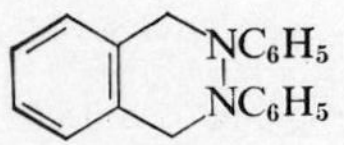 NC_6H_5 / NC_6H_5	N	40–93	92–92.5° (methanol)	26
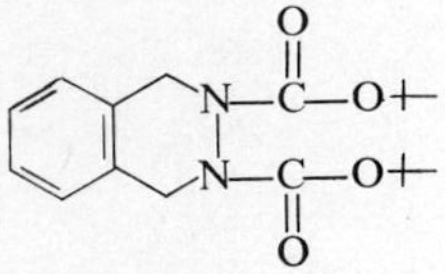 $N{-}C(=O){-}O{-}t\text{-}Bu$ / $N{-}C(=O){-}O{-}t\text{-}Bu$	O	63	White crystals, 118–120° (ligroin)	23
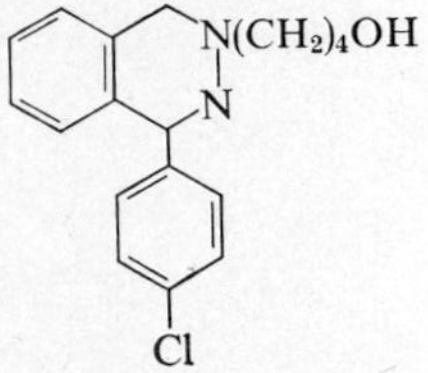 $N(CH_2)_4OH$, N, Cl	P	95	Oil	27, 28
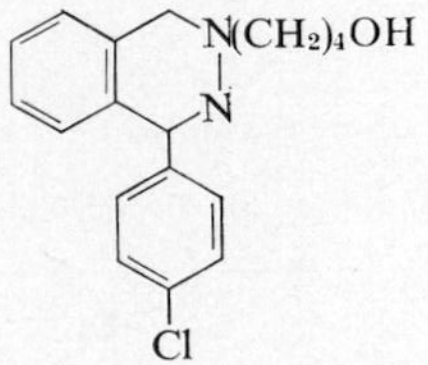 $N(CH_2)_4OH$, N, Cl	P	—	—	27
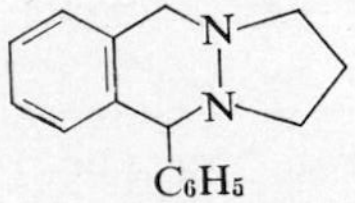 N, N, C_6H_5	—	—	Oil, HCl, 197–199°	28

TABLE 2L-2 (*continued*)

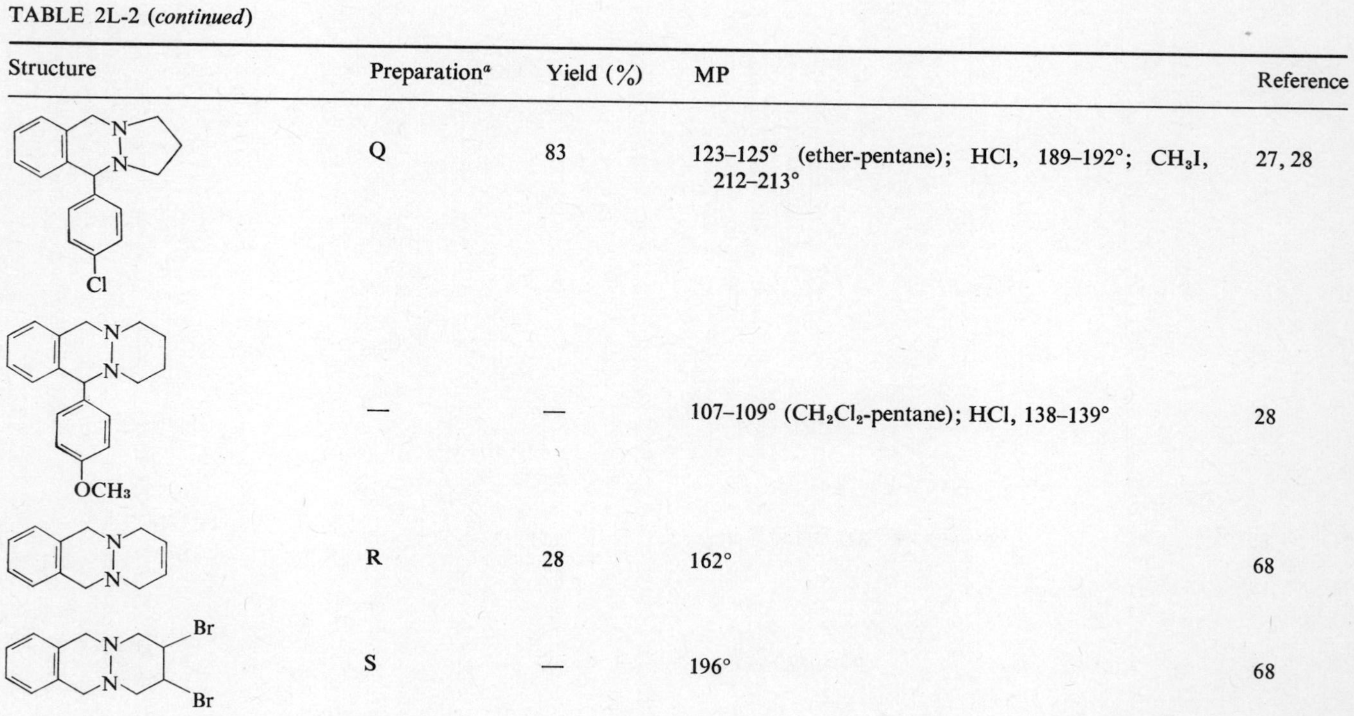

Structure	Preparation[a]	Yield (%)	MP	Reference
	Q	83	123–125° (ether-pentane); HCl, 189–192°; CH_3I, 212–213°	27, 28
	—	—	107–109° (CH_2Cl_2-pentane); HCl, 138–139°	28
	R	28	162°	68
	S	—	196°	68

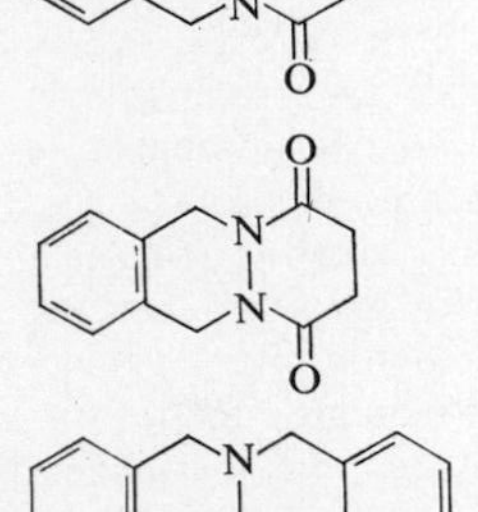

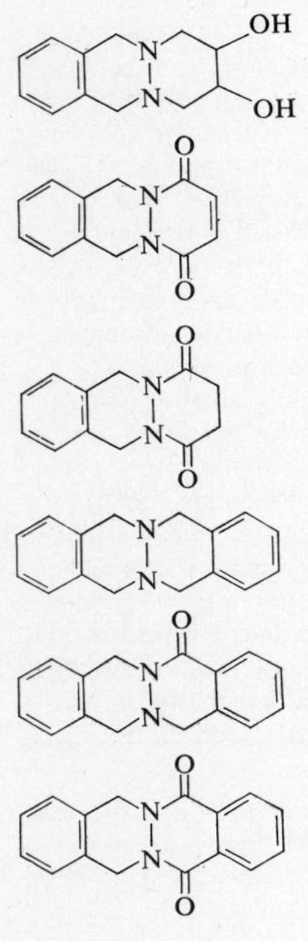

Compound		Yield (%)	m.p.	Refs.
	T	40	279° (ethanol)	68
	U	—	140°	22
	V	—	202–204°	22
	W	40–46	127–129°; 132–133°, HCl, 248–250°; HBr, 253–256°, CH_3I, 221–223°	22, 27, 29
	X	16	Colorless crystals, 198–199° (dec.) (benzene)	29
	Y	27	190–192°; 195°	22, 24

TABLE 2L-2 (*continued*)

Structure	Preparation[a]	Yield (%)	MP	Reference
[structure: 2-phenyl-1H-[1,2,4]triazolo[1,2-b]phthalazine-1,3(2H)-dione ring with N—C_6H_5]	—	—	185–187° (methanol)	28b

[a] A, See discussion within text of this section. B, 1,2,3,4-tetrahydrophthalazine hydrochloride treated with *S*-methylisothiourea in 2*N* sodium hydroxide; reflux with cyanamide in toluene for 6 hr. C, 1,2,3,4-tetrahydrophthalazine hydrochloride treated with 2-methyl-1-nitroisothiourea in 1*N* sodium hydroxide. D, 1,2,3,4-tetrahydrophthalazine refluxed with NH_4NCS in toluene for 24 hr. E, product obtained by method D + methyl iodide, reflux 2 hr in ethanol. F, product obtained by method D treated with methylamine in ethanol. F_1, same as F, using dimethylamine. G, 1,2,3,4-tetrahydrophthalazine hydrochloride treated with methyl isothiocyanate in ethanol containing sodium ethoxide. H, product obtained by method G + methyl iodide refluxed in ethanol for 3 hr. I, properly substituted 1-(2*H*)phthalazinone reduced with lithium aluminum hydride. J, 4-phenyl-1,2,3,4-tetrahydrophthalazine hydrochloride treated with 2-methyl-1-nitroisothiourea as in method C. K, product obtained by method J was reduced with 10% Pd-C/H_2. L, 2-methyl-1,2,3,4-tetrahydrophthalazine hydrochloride refluxed with cyanamide in toluene. M, 2,3-dimethylphthalazine-1,4-dione reduced with lithium aluminum hydride. N, 2,3-diphenylphthalazine-1,4-dione reduced with lithium alanate; or α,α′-dibromo-*o*-xylene + *N,N′*-diphenylhydrazine, warm with *N,N*-dimethylaniline. O, α,α′-dibromo-*o*-xylene + *t*-butyl hydrazodiformate, stir at room temperature in *t*-butanol containing potassium *t*-butoxide. P, 2-substituted 4-(*p*-chlorophenyl)-1-(2*H*)phthalazinone reduced with lithium aluminum hydride in ether. Q, 4-(*p*-chlorophenyl)-2-(3-chloropropyl)-1-(2*H*)phthalazinone reduced with lithium aluminum hydride; or 4-(*p*-chlorophenyl)-2-(3-hydroxypropyl)-1,2,3,4-tetrahydrophthalazine + $SOCl_2$, reflux 18 hr in chloroform. R, α,α′-dibromo-*o*-xylene + 1,2,3,6-tetrahydropyridazine, reflux 4 hr in ethanol with sodium acetate. S, bromination of product obtained by method R with bromine in chloroform in presence of light and benzoyl peroxide. T, α,α′-dibromo-*o*-xylene + 4,5-dihydroxyhexahydropyridazine, reflux 10 hr in ethanol with potassium carbonate. U, same as S, used 3,6-dihydroxypyridazine. V, 1,2,3,4-tetrahydrophthalazine hydrochloride heated with succinic anhydride at 160° for 5 hr in presence of potassium acetate. W, lithium aluminum hydride reduction of product obtained by method X; or 1,2,3,4-tetrahydrophthalazine hydrochloride + α,α′-dibromo-*o*-xylene heated at 140–150° for 1 hr. X, 1,2,3,4-tetrahydrophthalazine hydrochloride + 2-bromomethylbenzoyl bromide, heat at 200–210° for 30 min. Y, 1,2,3,4-tetrahydrophthalazine hydrochloride + phthalaic anhydride, heat at 170° for 5 hr in presence of potassium acetate; or α,α′-dichloro-*o*-xylene + 4-hydroxy-1-(2*H*)phthalazinone, heat at 200° for 16 hr.

TABLE 2L-3. 5,6,7,8-Tetrahydrophthalazines

Structure	Preparation[a]	Yield (%)	MP	Reference
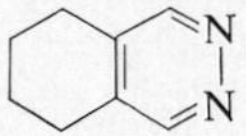 N, N	A	39	88.5–89°	32
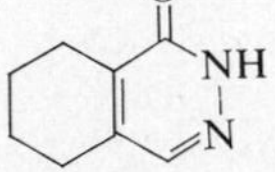 O, NH, N	B	—	144–145°; 144–146° (dil. methanol) HBr, 200–203°	33, 36
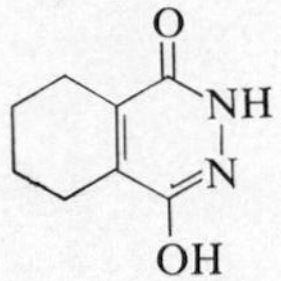 O, NH, N, OH	C	76–100	295–298° (ethanol); 298°	32, 33
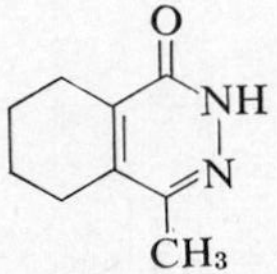 O, NH, N, CH_3	D	33	260–262° (methanol)	33
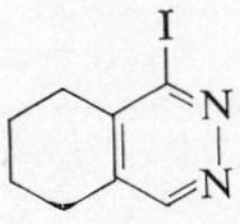 I, N, N	A_1	11	174–174.5° (dec.)	32
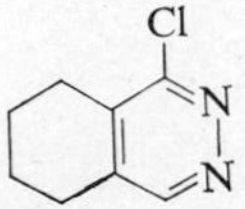 Cl, N, N	E	—	118–119°	36

TABLE 2L-3 (*continued*)

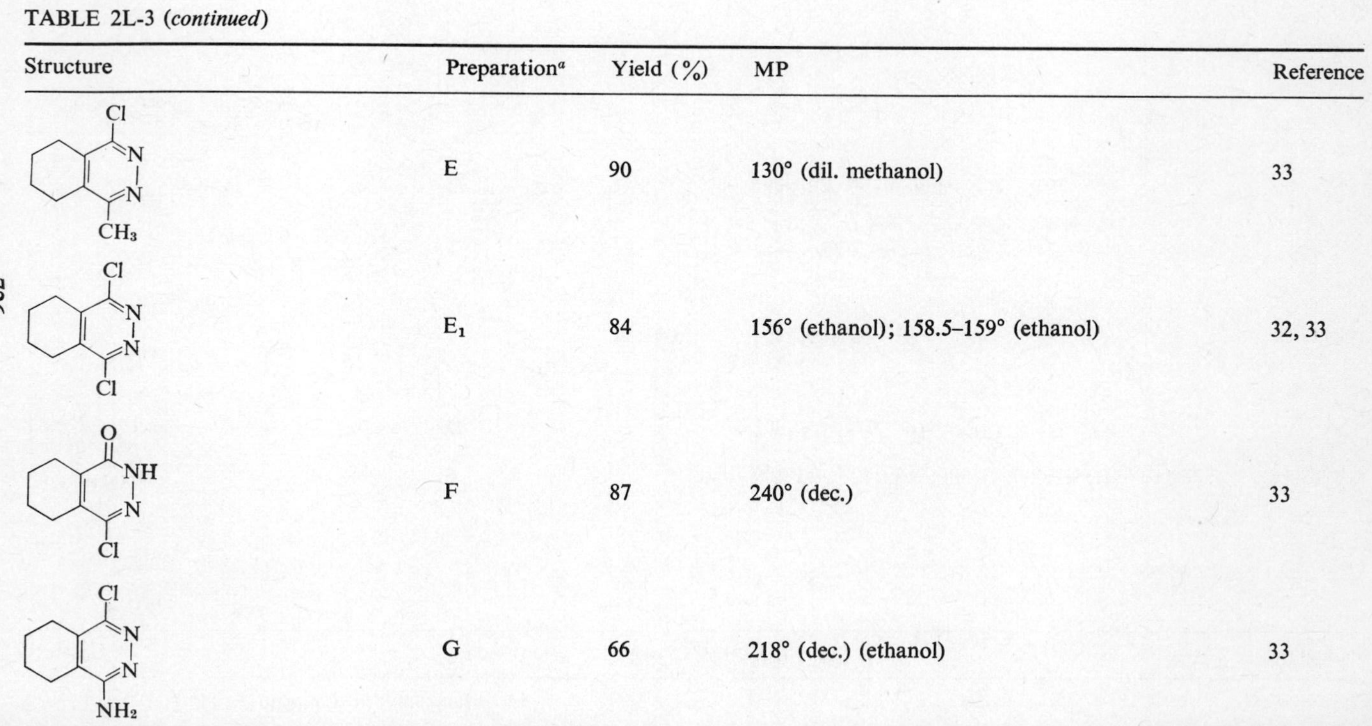

Structure	Preparation[a]	Yield (%)	MP	Reference
	E	90	130° (dil. methanol)	33
	E_1	84	156° (ethanol); 158.5–159° (ethanol)	32, 33
	F	87	240° (dec.)	33
	G	66	218° (dec.) (ethanol)	33

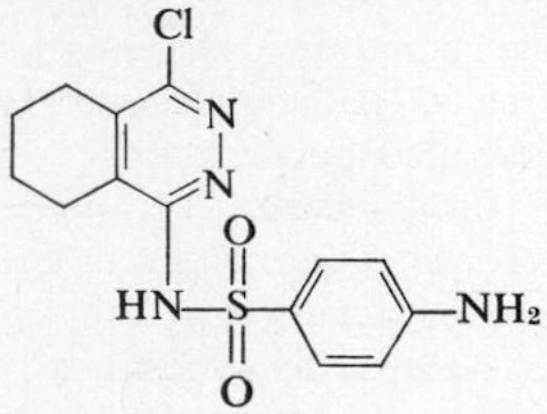 Cl, N, N, HN–S(O)(O)–, –NH_2	H	—	Yellow needles, 226° (dec.)	33
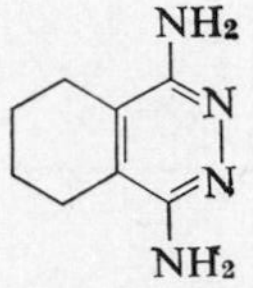 NH_2, N, N, NH_2	G_1	73	265° (dec.) (dil, ethanol, HCl, 160–190°)	33
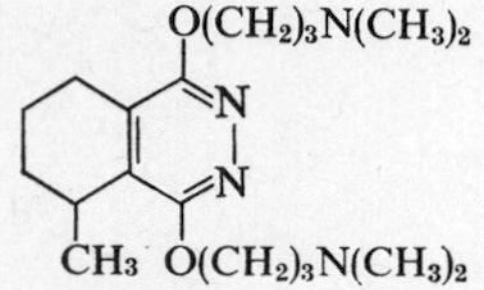 $O(CH_2)_3N(CH_3)_2$, N, N, CH_3, $O(CH_2)_3N(CH_3)_2$	I	73	Pale yellow oil, 114–116°/5 × 10^{-4} mm; triphosphate, 155–156.5°; 2-*p*-nitrobenzyl bromide, 191–192°	34
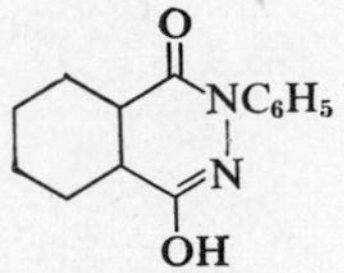 O, NC_6H_5, N, OH	J	93	187° (methanol)	35

TABLE 2L-3 (*continued*)

Structure	Preparation[a]	Yield (%)	MP	Reference
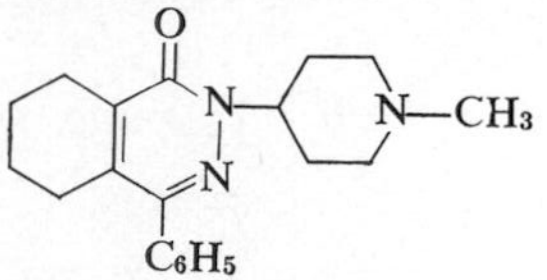 [structure: C6H5, N-CH3]	K	62–69	179–180° (acetone); HBr, 220–222°	37
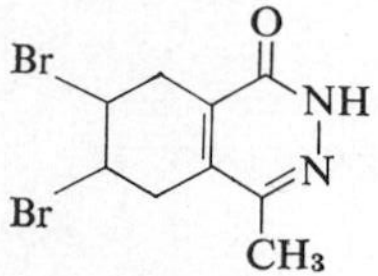 [structure: Br, Br, NH, CH3]	L	30	Colorless prisms, 204° (aqueous ethanol)	41
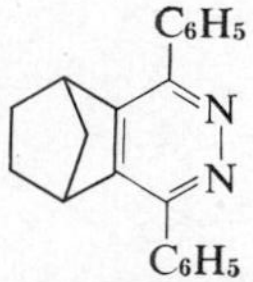 [structure: C6H5, C6H5]	M	94	Colorless needles 232–233°	38

[a] A, 1,4-dichloro-5,6,7,8-tetrahydrophthalazine refluxed 14 hr with phosphorus and hydriodic acid. A_1, by-product from method A. B, 4*a*,5,6,7,8,8*a*-hexahydro-1-(2*H*)phthalazinone treated with bromine in acetic acid; or reductive dehalogenation of 4-chloro-5,6,7,8-tetrahydro-1-(2*H*)phthalazinone with 10% Pd-C/H_2. C, 3,4,5,6-tetrahydrophthalic anhydride + hydrazine sulfate, reflux 4–8 hr in aqueous medium. D, condense 3,4,5,6-tetrahydrophthalic anhydride with malonic acid by heating in pyridine for 3 hr, then treat with hydrazine in ethanol. E, corresponding hydroxy compound heated with phosphorus oxychloride for 0.5–1.5 hr. E_1, same as E, heat at 80–85° for 2 hr. F, 1,4-dichloro-5,6,7,8-tetrahydrophthalazine, reflux 6 hr in acetic acid. G, 1,4-dichloro-5,6,7,8-tetrahydrophthalazine + 16% ammonia in ethanol, heat in sealed tube at 160° for 16 hr. G_1, same as G, heat at 160–170° for 17 hr. H, 1-chloro-4-amino-5,6,7,8-tetrahydrophthalazine + *p*-*N*-acetylaminobenzenesulfonyl chloride in pyridine followed by alkaline hydrolysis of product. I, corresponding dichloro compound + 2-*N*,*N*-dimethylaminoethanol + sodium, reflux in xylene. J, 3,4,5,6-tetrahydrophthalic anhydride + phenylhydrazine in 20% sulfuric acid. K, corresponding *cis*- or *trans*-4*a*,5,6,7,8,8*a*-hexahydrophthalazine, heat with bromine in acetic acid. L, corresponding 6,7-dibromo-4*a*,5,6,7,8,8*a*-hexahydrophthalazine, treat with bromine in hot glacial acetic acid. M, norbornene + 1,4-diphenyl-*sym*-tetrazine, warm.

TABLE 2L-4. 4*a*,5,8,8*a*-Tetrahydrophthalazines

Structure	Preparation[a]	Yield (%)	MP	Reference
$(CH_2)_7CONHNH_2$; CH_3; CH_3; N; N; $(CH_2)_5CH_3$	A	—	122° (benzene-pet. ether)	43
O; NH; N	B	—	218–219°	36
O; CH_3; CH_3; NH; N	B	—	114–115°	36
O; NH; N; CH_3	C	88–95	Long colorless needles, 225–226° (alcohol)	41
O; NH; N; C_6H_5	D	90	231–232°	42

TABLE 2L-4 (*continued*)

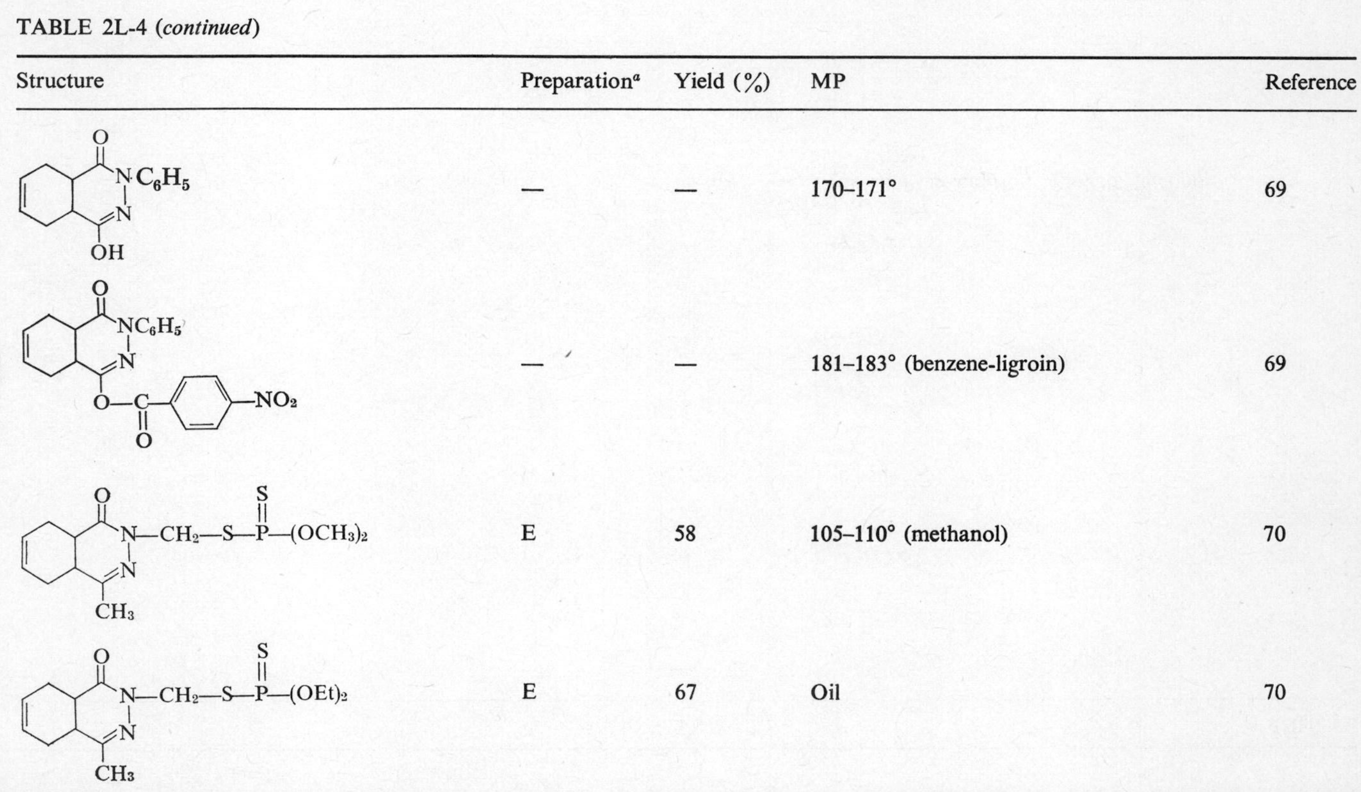

Structure	Preparation[a]	Yield (%)	MP	Reference
	—	—	170–171°	69
	—	—	181–183° (benzene-ligroin)	69
	E	58	105–110° (methanol)	70
	E	67	Oil	70

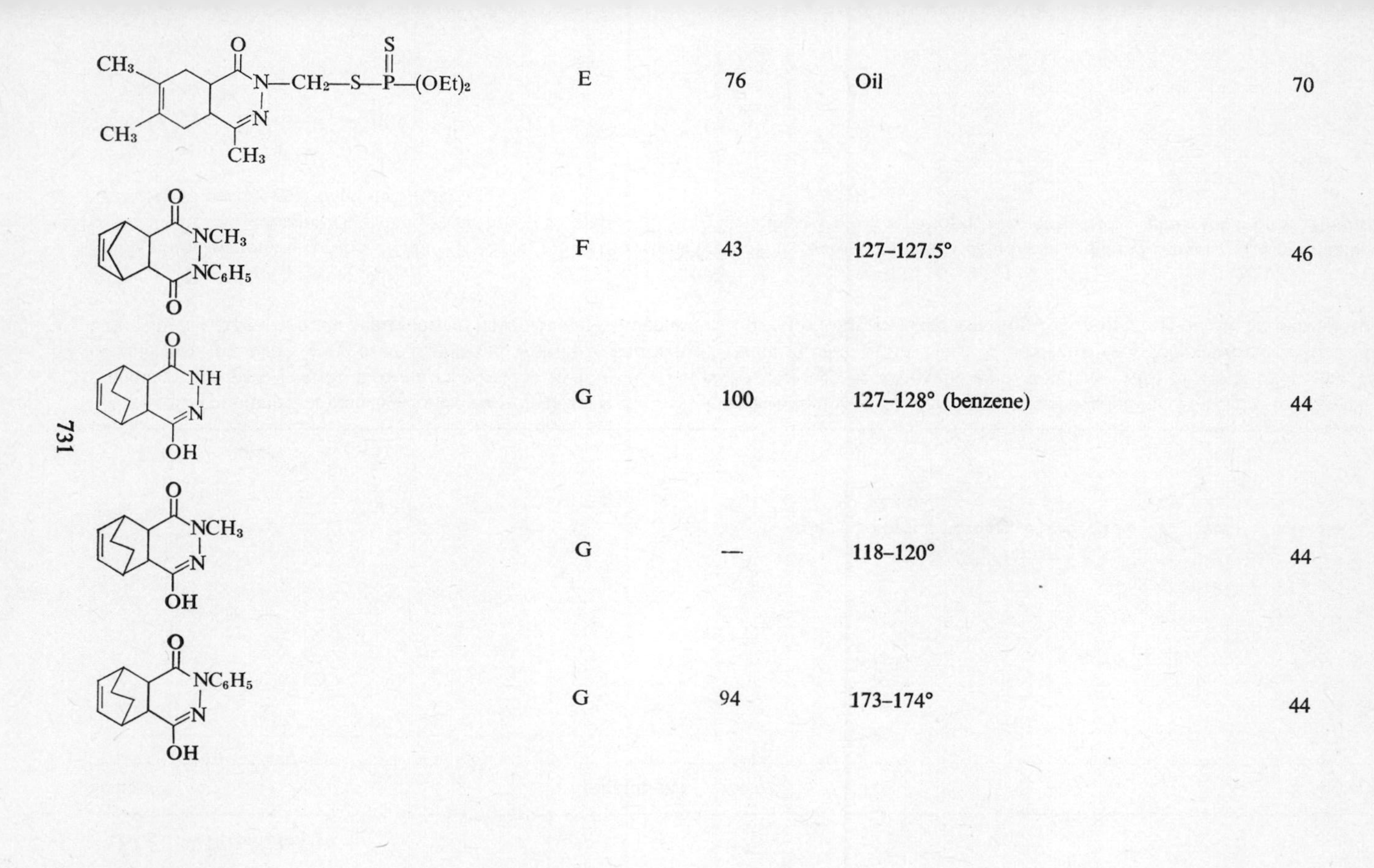

	E	76	Oil	70
	F	43	127–127.5°	46
	G	100	127–128° (benzene)	44
	G	—	118–120°	44
	G	94	173–174°	44

TABLE 2L-4 (*continued*)

Structure	Preparation[a]	Yield (%)	MP	Reference
Bicyclo[2.2.2]octene-fused pyridazinone: C=O, $NCH_2C_6H_5$, =N, OH	G	—	111–113°	44
Bicyclo[2.2.2]octene-fused pyridazinedione: C=O, NCH_3, NCH_3, C=O	G	89	Colorless needles, 84–86°; 114–115° (benzene)	44, 45

[a] A, corresponding 1,2-diacylcyclohexene + hydrazine hydrate, heat in ethanol at 80° for 2 hr. B, corresponding *cis*-1,2,3,6-tetrahydrophthaldehydic acid, heat 10 min with hydrazine hydrate. C, 2-acetyl-4-cyclohexenecarboxylic acid, condense with aqueous hydrazine by heating for 1 hr; or its ethyl ester is refluxed with 50% hydrazine hydrate in ethanol for 1 hr. D, 2-benzoyl-4-cyclohexenecarboxylic acid, treat with hydrazine hydrate in presence of base (Huang-Minlon reduction conditions). E, corresponding 2-chloromethyl-4*a*,5,8,8*a*-tetrahydro-1-(2*H*)phthalazinone + proper $(RO)_2$—P(=S)—SNa, reflux in acetone for 4 hr. F, cyclopentadiene treated with 1-methyl-2-phenyl-3,6-pyridazinedione in dioxane-toluene using methylene blue as catalyst. G, 1,2,3,6-tetrahydro-3,6-ethanophthalic anhydride, reflux for 5 hr in ethanol with corresponding hydrazine derivative.

TABLE 2L-5. Hexahydrophthalazines

Structure	Preparation[a]	Yield (%)	MP	Reference
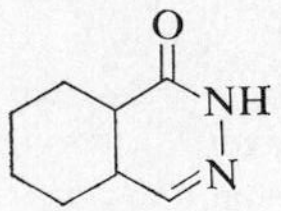 O, NH, N	A	—	BP 150–151°/9 mm	36
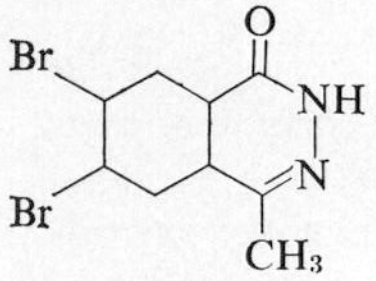 Br, Br, O, NH, N, CH_3	B	59	Colorless plates, 194° (dec.) (alcohol)	41
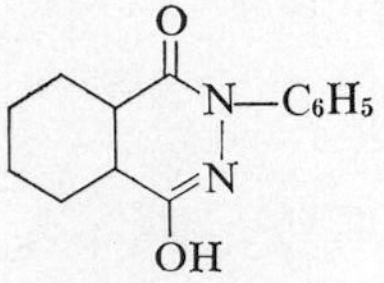 O, N—C_6H_5, N, OH	—	—	214–215°	69
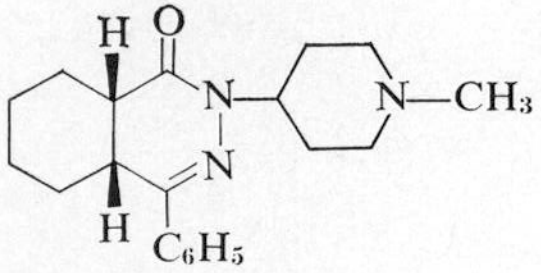 H, O, N, N—CH_3, N, H, C_6H_5	C	74	146° (ethanol)	37
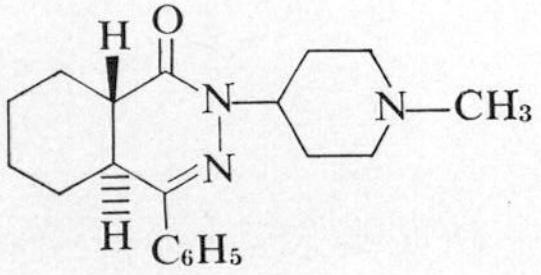 H, O, N, N—CH_3, N, H, C_6H_5	C_1	78	155° (ethanol)	37

TABLE 2L-5 (*continued*)

Structure	Preparation[a]	Yield (%)	MP	Reference
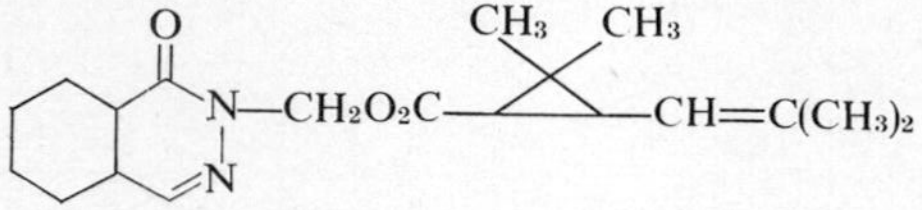 O, N—CH_2O_2C—, CH_3, CH_3, —CH=C$(CH_3)_2$, N	D	81	$n_D^{23°}$ 1.5391	48
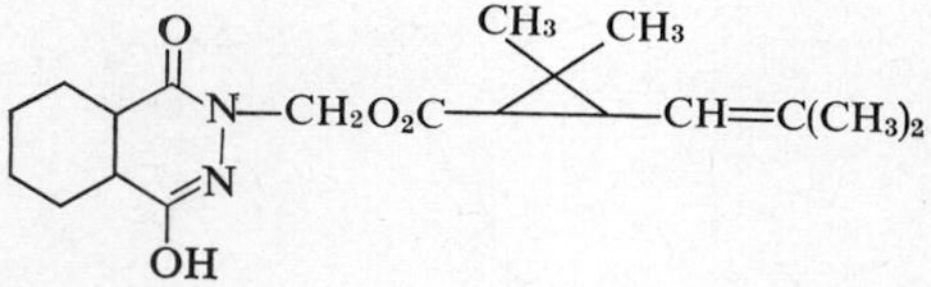 O, N—CH_2O_2C—, CH_3, CH_3, —CH=C$(CH_3)_2$, N, OH	D	77	$n_D^{25°}$ 1.5358	48
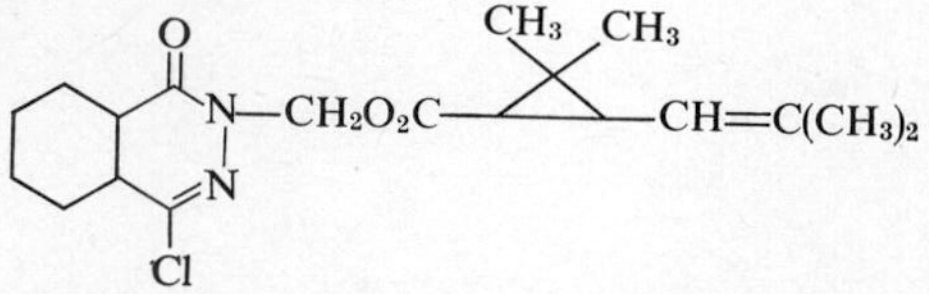 O, N—CH_2O_2C—, CH_3, CH_3, —CH=C$(CH_3)_2$, N, Cl	D	79	$n_D^{18°}$ 1.5413	48
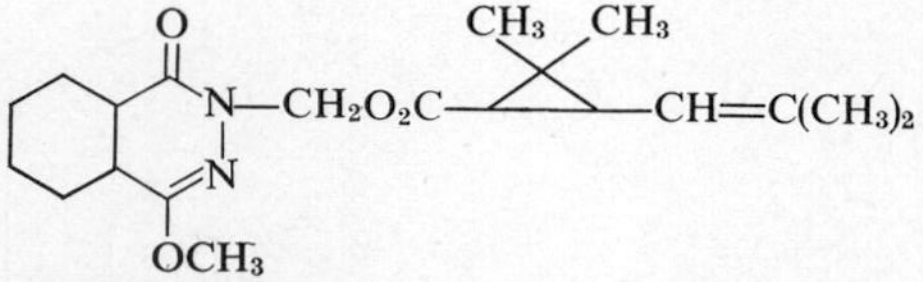 O, N—CH_2O_2C—, CH_3, CH_3, —CH=C$(CH_3)_2$, N, OCH_3	D	79	$n_D^{20.5°}$ 1.5315	48
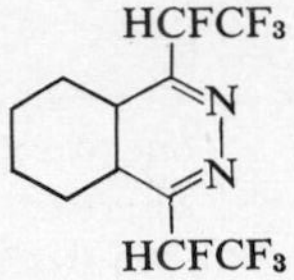 $HCFCF_3$, N, N, $HCFCF_3$	E	—	—	51, 52

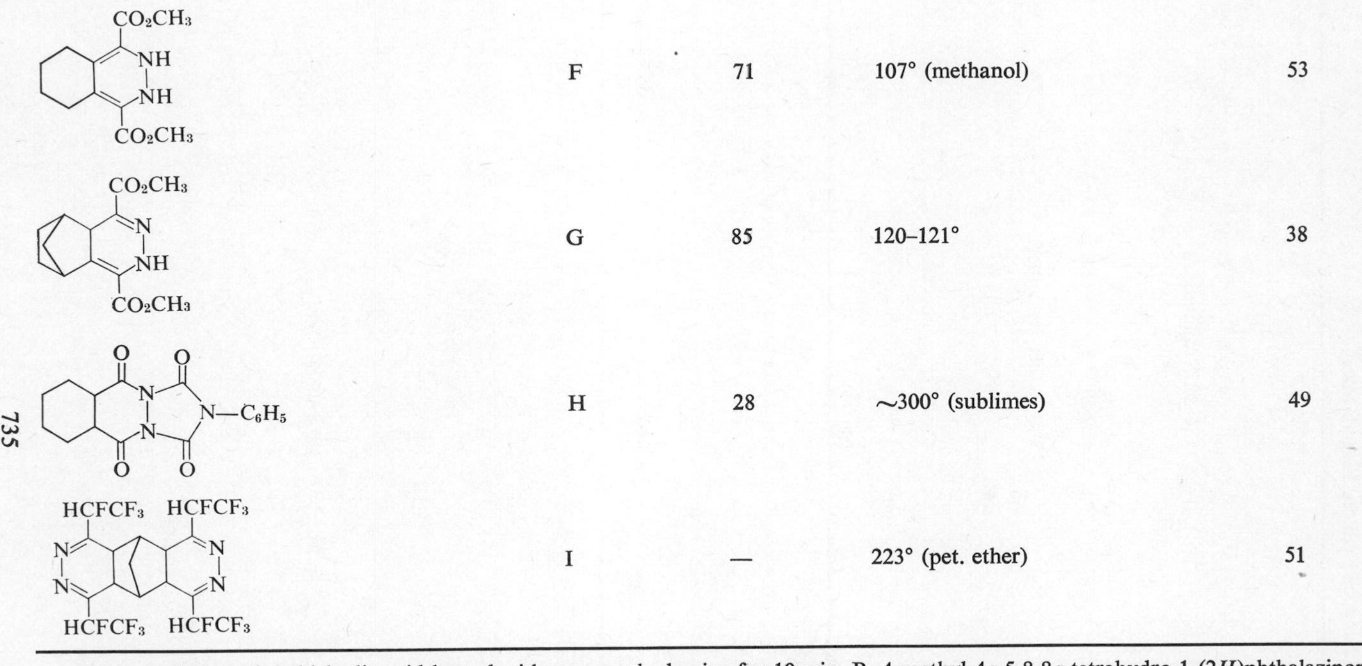

Structure	Method[a]	Yield (%)	M.p.	Ref.
	F	71	107° (methanol)	53
	G	85	120–121°	38
	H	28	~300° (sublimes)	49
	I	—	223° (pet. ether)	51

[a] A, hexahydro-*cis*-phthaldehydic acid heated with aqueous hydrazine for 10 min. B, 4-methyl-4*a*,5,8,8*a*-tetrahydro-1-(2*H*)phthalazinone treated with bromine in glacial acetic acid. C, *cis*-2-benzoylcyclohexanecarboxylic acid + *N*-methyl-4-hydrazinopiperidine, reflux in ethanol for 24 hr. C_1, same as C, using *trans*-2-benzoylcyclohexanecarboxylic acid. D, corresponding 2-hydroxymethyl-1-(2*H*)phthalazinone treated with chrysanthemumcarboxyl chloride in toluene with pyridine, let stand overnight at room temperature. E, cyclohexene + 1,4-di(1,2,2,2-tetrafluoroethyl)-*sym*-tetrazine, reflux 3 hr. F, cyclohexene treated with 1,4-dicarbomethoxy-*sym*-tetrazine in CH_2Cl_2. G, norbornene treated with 1,4-dicarbomethoxy-*sym*-tetrazine at 20°. H, phthalic anhydride + 1,2-dihydro-4-phenyl-1,2,4-triazoline-3,5-dione, heat at 180–200° for 2 hr. I, bicycloheptadiene mixed with 2 moles of 1,4-di(1,2,2,2-tetraflurorethyl)-*sym*-tetrazine.

TABLE 2L-6. Miscellaneous Reduced Phthalazines

Structure	MP	Reference
H, OCH_3, N, N, NO_2	126–127° (methanol)	39
Br, NCO_2Et, NCO_2Et, Br	100° (ether-pet. ether)	56
C_6H_5, C, O, O, C, C_6H_5, C_6H_5, N, N, C_6H_5	164°	55
C_6H_5, C_6H_5, N, N, N, N, C_6H_5, C_6H_5	240°	55
Cl, Cl, NH, NH, Cl, Cl	200–203° (dec.)	57
N—CO_2Et, N – CO_2Et	BP 160°/0.1 mm	57
Cl, Cl, OCH_3, OCH_3, Cl, Cl, N—CO_2Et, N—CO_2Et	125–126°	57

TABLE 2L-6 (*continued*)

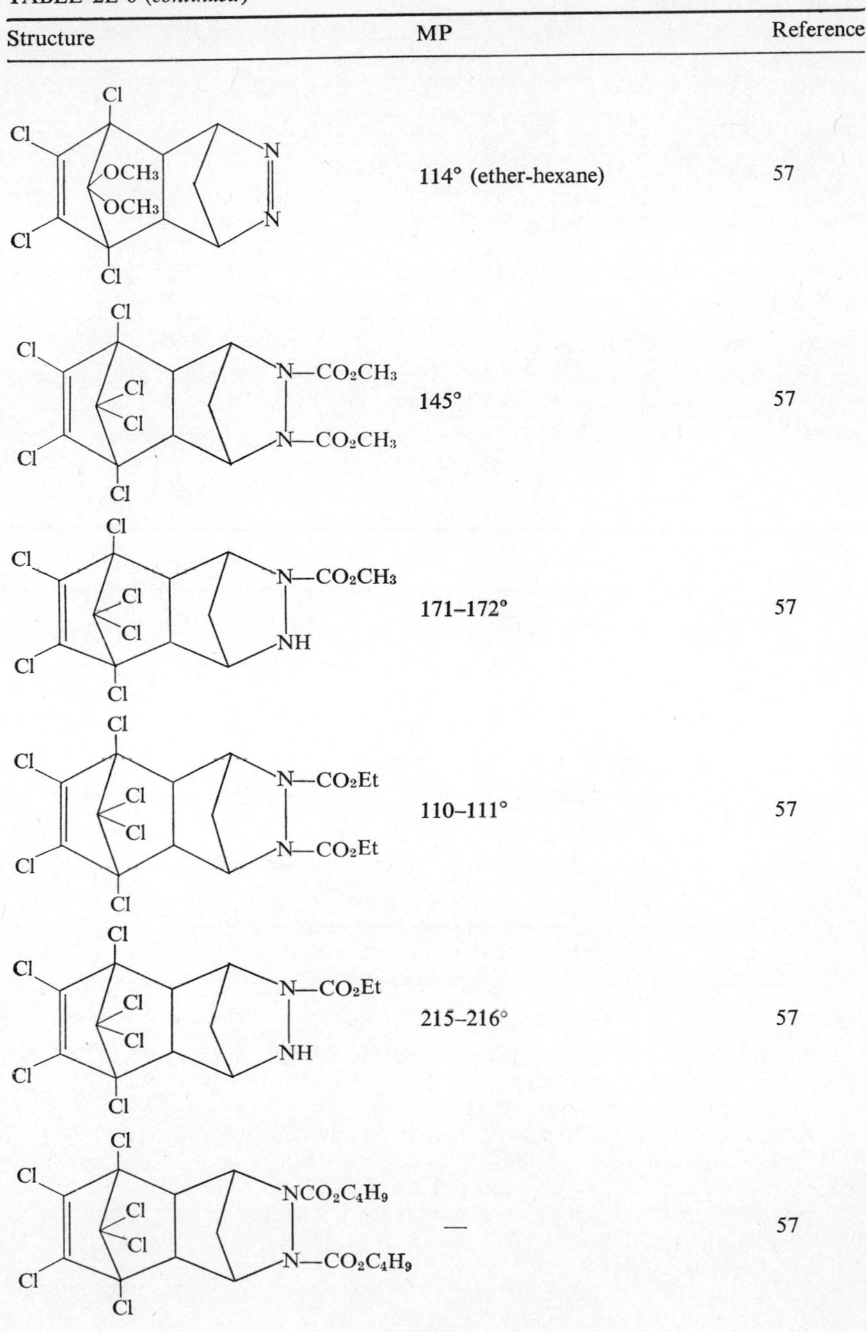

Structure	MP	Reference
	114° (ether-hexane)	57
	145°	57
	171–172°	57
	110–111°	57
	215–216°	57
	—	57

TABLE 2L-6 (*continued*)

Structure	MP	Reference
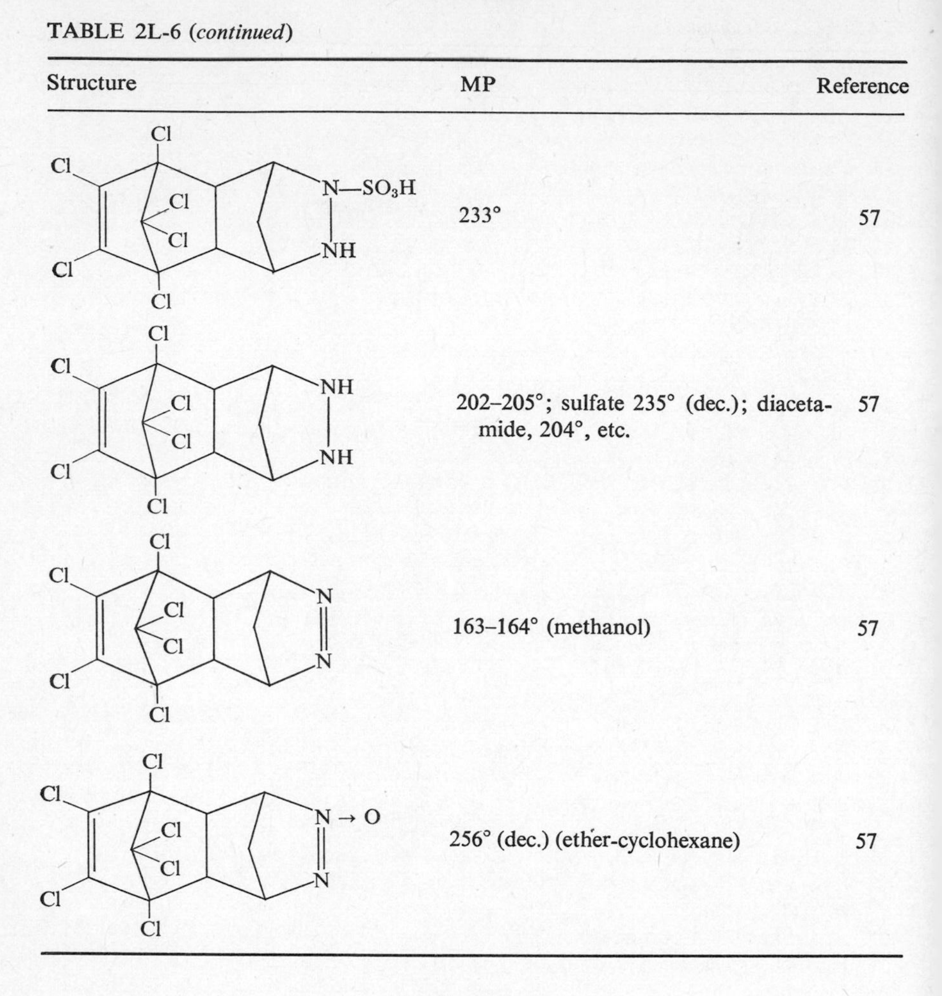	233°	57
	202–205°; sulfate 235° (dec.); diacetamide, 204°, etc.	57
	163–164° (methanol)	57
	256° (dec.) (ether-cyclohexane)	57

References

1. Y. S. Shabarov, N. I. Vasilev, and R. Y. Levina, *Zh. Obshch. Khim.*, **31,** 2478 (1961); *Chem. Abstr.*, **56,** 7312 (1962).
2. C. C. Leznoff, *Can. J. Chem.*, **46,** 1152 (1968).
3. R. E. Smith and E. D. Otremba, *J. Org. Chem.*, **27,** 879 (1962).
4. H. Lund, *Tetrahedron Lett.*, **1965,** 3973.
5. H. Lund, *Coll. Czech. Chem. Commun.*, **30,** 4237 (1965).
6. B. K. Diep and B. Cauvin, *Compt. Rend. Ser. C.*, **262,** 1010 (1966).
7. E. Bellasio and E. Testa, *Ann. Chim.* (*Rome*), **59,** 443(1969); *Chem. Abstr.*, **71,** 91411 (1969).
8. E. Bellasio et al., S. African Pat. 6800,195 (1968); *Chem. Abstr.*, **70,** 77993 (1969).
9. A. Marxer, *Helv. Chim. Acta*, **49,** 572 (1966).

10. A. Mustafa, A. H. Harhash, and A. A. S. Saleh, *J. Am. Chem. Soc.*, **82,** 2735 (1960).
11. A. Marxer, F. Hofer, and U. Salzmann, *Helv. Chim. Acta*, **52,** 1376 (1969).
12. A. Hirsch and D. G. Orphanos, *J. Heterocyc. Chem.*, **3,** 38 (1966).
13. F. D. Popp, J. M. Wefer, C. W. Klinowski, *J. Heterocyclic Chem.*, **5,** 879 (1968).
14. J. Druey and H. H. Daeniker, U.S. Pat. 2,945,037 (1960); Swiss Pat. 353,369 (1957); *Chem. Abstr.*, **55,** 4546 (1961); **59,** 5177 (1963).
15. E. Oishi, *Yakugaku Zasshi*, **89,** 959 (1969); *Chem. Abstr.*, **72,** 3450 (1970).
15a. M. A. Shah and G. A. Taylor, *J. Chem. Soc. (C)*, **1970,** 1651.
16. K. T. Potts and C. Lovelette, *J. Org. Chem.*, **34,** 3221 (1969).
17. H. Lund, *Acta Chem. Scand.*, **8,** 1307 (1954).
18. E. Bellasio and E. Testa, *Ann. Chim. (Rome)*, **59,** 451 (1969); *Chem. Abstr.*, **71,** 124361 (1969).
19. E. Bellasio, G. Maffii, and E. Testa, German Pat. 1,929,499 (1969); *Chem. Abstr.*, **72,** 55843 (1970).
20. E. Bellasio and E. Testa, *Farm. Ed. Sci.*, **25,** 305 (1970); *Chem. Abstr.*, **73,** 25836 (1970); E. Bellasio, German Pat. 2,010,040 (1970); *Chem. Abstr.*, **73,** 98962 (1970).
21. B. Singh, *J. Am. Chem. Soc.*, **91,** 3670 (1969).
22. I. Zugravescu et al., *An. Stiint. Univ. "Al. I. Cuza" Iasi, Sect. Ic*, **14,** 51 (1968); *Chem. Abstr.*, **70,** 77895 (1969).
23. L. A. Carpino, *J. Am. Chem. Soc.*, **85,** 2144 (1963).
24. R. Ohme and E. Schmitz, *Z. Anal. Chim.*, **220,** 105 (1966).
25. B. Junge and H. A. Staab, *Tetrahedron Lett.*, **1967,** 709.
26. G. Wittig, W. Joss, and P. Rathfelder, *Ann. Chem.*, **610,** 180 (1957).
27. A. Aeberli and W. J. Houlihan, *J. Org. Chem.*, **34,** 2715 (1969).
28. W. J. Houlian, French Pat. 1,530,074 (1968); *Chem. Abstr.*, **71,** 61434 (1969).
28a. A. Nakamura, *Chem. Pharm. Bull.*, **18,** 1426 (1970).
28b. N. L. Bould and F. Farr, *J. Am. Chem. Soc.*, **92,** 6695 (1970).
29. H. H. Hatt and E. F. M. Stephenson, *J. Chem. Soc.*, **1952,** 199.
30. H. J. Pannemann, German Pat. 1,937,219 (1970); *Chem. Abstr.*, **72,** 100,731 (1970).
31. E. Schmitz and R. Ohme, *Monatsber, deut. Akad. Wiss. Berlin*, **1,** 366 (1959); *Chem. Abstr.*, **54,** 14263 (1960).
32. R. H. Horning and E. D. Amstutz, *J. Org. Chem.*, **20,** 707, 1069 (1955).
33. I. Satoda, F. Kusuda, and K. Mori, *Yakugaku Zasshi*, **82,** 233 (1962); *Chem. Abstr.*, **58,** 3427 (1963).
34. E. A. Steck, R. P. Brundage, and L. T. Fletcher, *J. Am. Chem. Soc.*, **76,** 4454 (1954).
35. K. Mori, *Yakugaku Zasshi*, **82,** 1161 (1962); *Chem. Abstr.*, **58,** 4555 (1963).
36. V. Andreev and A. V. Usova, *Izv. Akad. Nauk, SSSR Ser. Khim.*, **1966,** 1410; *Chem. Abstr.*, **66,** 54756 (1967).
37. E. Jucker and R. Suess, *Helv. Chim. Acta*, **42,** 2506 (1959).
38. S. Sauer and G. Heinrichs, *Tetrahedron Lett.*, **1966,** 4979.
39. D. L. Hufford, D. S. Tarbell, and T. R. Koszalka, *J. Am. Chem. Soc.*, **74,** 3014 (1952).
40. Y. S. Shabarov et al., *Zh. Obshch. Khim.*, **33,** 1206 (1961); *Chem. Abstr.*, **59,** 11490 (1963).
41. S. Dixon and L. F. Wiggins, *J. Chem. Soc.*, **1954,** 594.
42. Y. A. Arbuzov et al., *Izv. Akad. Nauk SSSR, Ser. Khim.*, **1964,** 310; *Chem. Abstr.*, **60,** 11959 (1964).
43. G. F. Whitfield, *J. Chem. Soc. (C)*, **1968,** 1781.
44. M. Fujimoto and K. Okabe, Japanese Pat., 11,386 (1964); *Chem. Abstr.*, **61,** 13325 (1964).
45. M. Ishikawa, M. Fujimoto, K. Okabe, and M. Sasaski, *Chem. Pharm. Bull.*, **16,** 227 (1968).

46. J. Druey, K. Meier, and A. Staehelin, *Helv. Chim. Acta*, **45,** 1485 (1962).
47. M. Lora-Tamayo, J. L. Soto, and E. D. Toro, *An. Quim.*, **65,** 1125 (1969); *Current Abstr. Chem.*, **37,** (335), 138348 (1970).
48. K. Ueda et al., Japanese Pat., 17,913 (1967); *Chem. Abstr.*, **69,** 19332 (1968).
49. C. Zinner and W. Duecker, *Arch. Pharm.*, **296,** 13 (1963).
50. U.S. Pat. 3,364,217 (1968); *Chem. Abstr.*, **68,** 49631 (1968).
51. R. A. Carboni and R. V. Lindsey, Jr., *J. Am. Chem. Soc.*, **81,** 4342 (1959).
52. R. A. Carboni, U.S. Pat. 3,022,305 (1962); *Chem. Abstr.*, **58,** 9102 (1963).
53. M. Avram, I. G. Dinulescu, E. Marcia, and C. D. Nenitzescu, *Chem. Ber.*, **95,** 2248 (1962).
54. C. G. Overberger, G. Montaudo, J. Sebenda, and R. A. Veneski, *J. Am. Chem. Soc.*, **91,** 1256 (1969).
55. F. B. Gaudeman, *Ann. Chim. (Paris)*, **3,** 52 (1958); *Chem. Abstr.*, **52,** 20038 (1958).
56. B. Price, I. O. Sutherland, and F. G. Williamson, *Tetrahedron*, **22,** 3477 (1966).
57. J. G. Kuderna, J. W. Sims, J. F. Wikstrom, and S. B. Soloway, *J. Am. Chem. Soc.*, **81,** 382 (1959).
58. W. P. Lay and K. MacKenzie, *Chem. Commun.*, **1970,** 398.
59. G. Meier, I. Fuss, and M. Schneider, *Tetrahedron Lett.*, **1970,** 1057.
60. Belgian Pat. 646,344 (1964); *Chem. Abstr.*, **68,** 39638 (1968).
61. I. Zugravescu, M. Petrovanu, A. Caraculacu, and A. Sauciuc, *Rev. Roum Chim.*, **12,** 109 (1967); *Chem. Abstr.*, **68,** 49537 (1968).
62. R. M. Acheson and M. W. Foxton, *J. Chem. Soc.*, **1966,** 2218.
63. M. Petrovanu, A. Sauciuc, I. Gabe, and I. Zugravescu, *Rev. Roum. Chim.*, **13,** 513 (1968); *Chem. Abstr.*, **70,** 57757 (1969).
64. S. Kubota and T. Akita, *Yakugaku Zasshi*, **81,** 521 (1961); *Chem. Abstr.*, **55,** 19926 (1961).
65. J. Gronowska, *Roczniki Chem.*, **39,** 245 (1965).
66. J. Gronowska, *Roczniki Chem.*, **39,** 375 (1965).
67. P. Pflegel and G. Wagner, *Pharmazie*, **22,** 147 (1967).
68. E. Carp, M. Pahomi, and I. Zugravescu, *An. Stinnt. Univ. "Al. I. Cuza", Tashi, Sect. Ic. Chim.*, **12,** 171 (1966); *Chem. Abstr.*, **68,** 49533 (1968).
69. H. Cairns, W. H. Williams, J. King, and B. J. Milliard, British Pat. 1,100,911 (1968); *Chem. Abstr.*, **69,** 19181 (1968).
70. H. Sakamoto et al., Japanese Pat. 5633 (1964); *Chem. Abstr.*, **63,** 1800 (1965).

Part M. Condensed Phthalazines

Several condensed phthalazines have been reported as phthalazine derivatives in previous sections of this chapter: oxazolo[3,4-*a*]phthalazines in section 2H; tetrazolo[4,5-*a*]-*s*-triazolo[3,4-*a*]-, bis-triazolo[3,4-*a*:4,3-*c*]-phthalazines in Section 2I; phthalazine-1,4-dione in fused ring systems at

N_2,N_3-position in Section 2E. In the present section, condensed phthalazines and their derivatives that have not been reported earlier are discussed.

I. Benzo[*f*]phthalazine and 5,6-Dihydrobenzo[*f*]-phthalazine

The chemistry of benzo[*f*]phthalazine (**1**) is very similar to phthalazine. Recently it has been synthesized by Kobe and co-workers (1) by catalytic dehalogenation of 1,4-dichlorobenzo[*f*]phthalazine (**2**) over palladium on carbon. Naphthalene-1,2-dicarboxylic anhydride, on condensation with hydrazine hydrate in acetic acid, provided 1,4-dihydroxybenzo[*f*]phthalazine (**3**), which is also obtained by dehydrogenation of 5,6-dihydro-1,4-dihydroxybenzo[*f*]phthalazine with bromine in acetic acid. Refluxing **3** with phosphorus ocychloride in the presence of *N*,*N*-dimethylaniline furnished **2** in quantitative yield (Eq. 1). Nitration of **1** in concentrated nitric acid-sulfuric

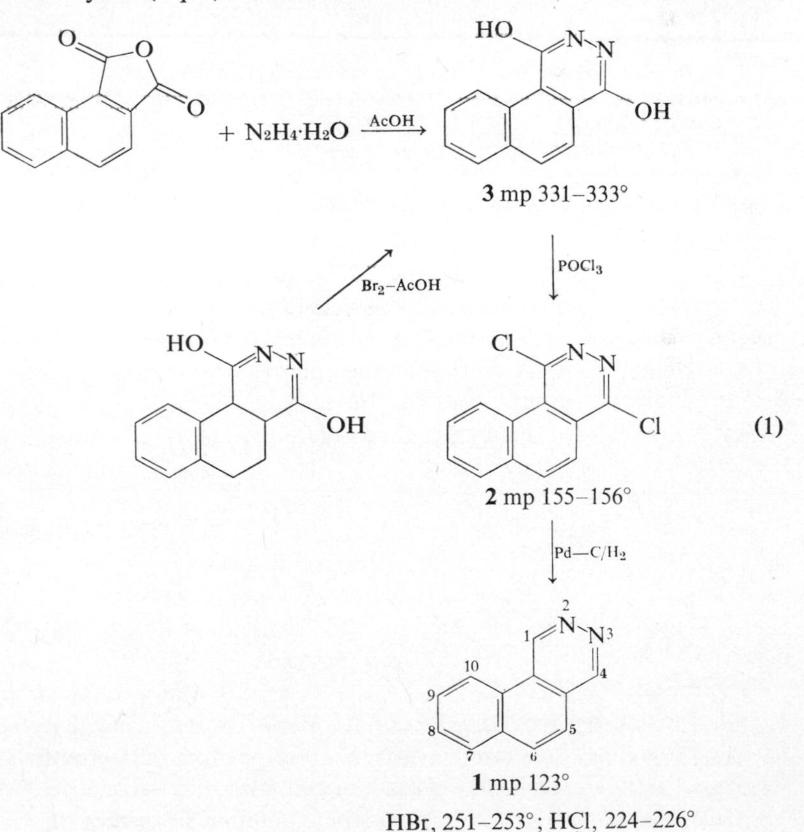

acid provided 9-nitrobenzo[*f*]phthalazine, mp 210°, in 68% yield. Gunderman and co-workers (2) have prepared several 9-substituted 1,4-dihydroxybenzo[*f*]phthalazines from 9-substituted naphthalene-1,2-dicarboxylic anhydrides or dimethyl esters by condensing these with hydrazine hydrate (Eq. 2).

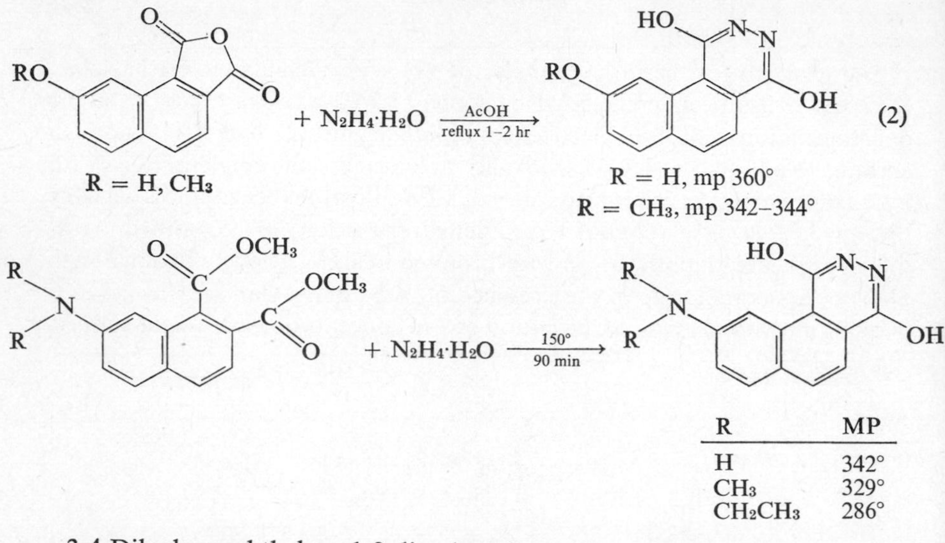

3,4-Dihydronaphthalene-1,2-dicarboxylic acid condenses with hydrazine by heating in alcohol to give 5,6-dihydro-1,4-dihydroxybenzo[*f*]phthalazine (3), which reacts with phosphorus oxychloride to give 5,6-dihydro-1,4-dichlorobenzo[*f*]phthalazine (**4**). The reaction of **4** with excess of alkoxides provided 1,4-dialkoxy-5,6-dihydrobenzo[*f*]phthalazines (3) (Eq. 3).

Stanovik and Tišler (4, 5) studied the nucleophilic substitution reactions of **4** with various nucleophiles and concluded that it behaves in the same way as 4-alkyl-5-aryl-3,6-dichloropyridazines where the alkyl substituents exert a moderate deactivating effect and the aryl group exerts a moderate activating effect. Substitution of only one of the chlorine atoms in **4** occurs preferentially so that only one isomer, 1-substituted 4-chloro-5,6-dihydrobenzo[*f*]-phthalazines, is obtained (Eq. 4). Catalytic dehalogenation of **4** over palladium on carbon provided 5,6-dihydrobenzo[*f*]phthalazine (**5**). The nitration (1) of **4** and **5** in a sulfuric acid-nitric acid mixture at −10° provided the corresponding 9-nitro derivatives in high yields (Eq. 4).

In addition, Stanovik and Tišler (3, 4) have synthesized an azasteroid, 4,5-dihydrobenzo[*f*]tetrazolo[5,1-*a*]phthalazine (**6**), by allowing 4-chloro-5,6-dihydrobenzo[*f*]phthalazine to react with sodium azide, or by treating 4-hydrazino-5,6-dihydrobenzo[*f*]phthalazine with nitrous acid (Eq. 5). They have also synthesized a number of other 5,6-dihydrobenzo[*f*]phthalazines fused with various nitrogen heterocycles (Eq. 5).

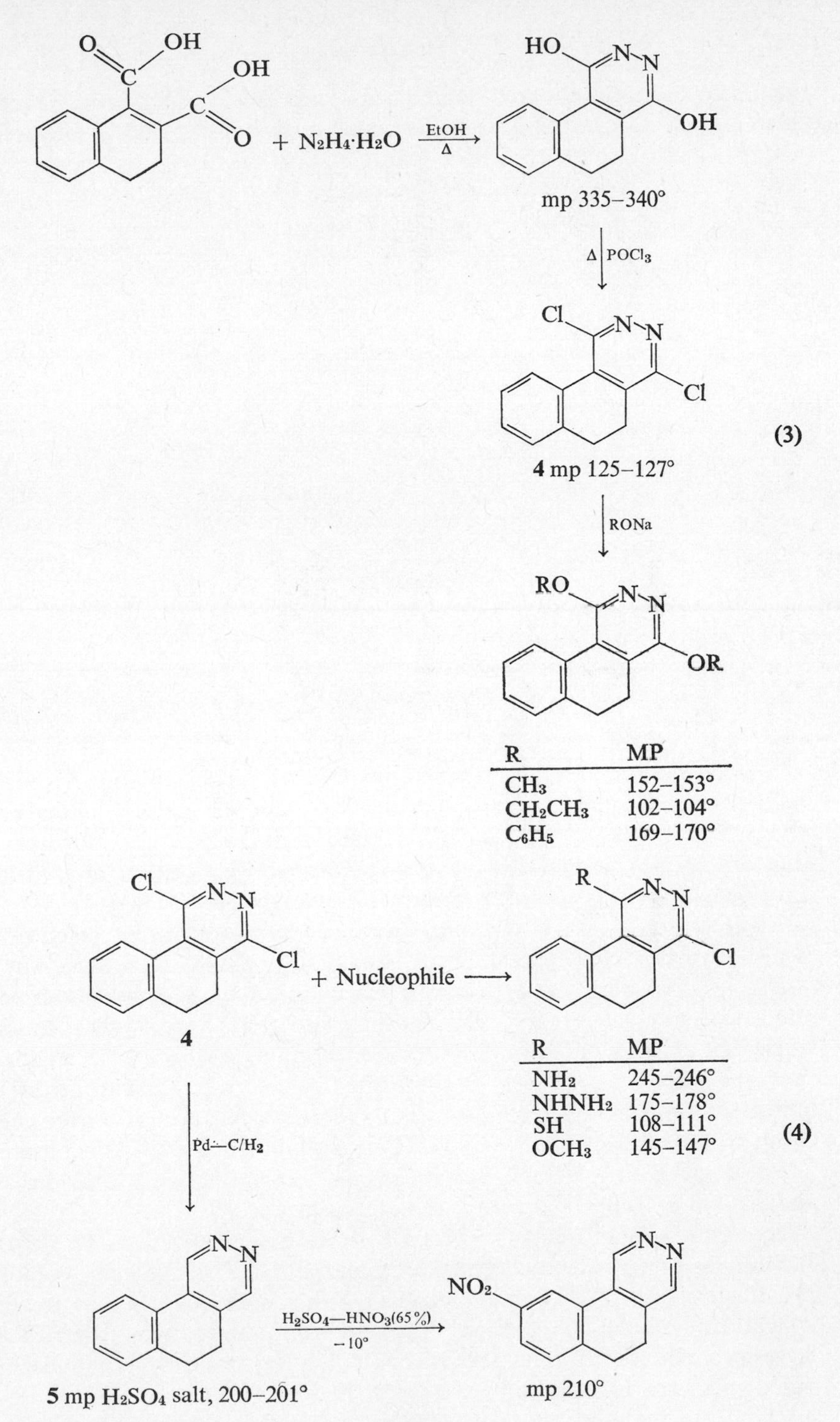

R	MP
CH_3	152–153°
CH_2CH_3	102–104°
C_6H_5	169–170°

R	MP
NH_2	245–246°
$NHNH_2$	175–178°
SH	108–111°
OCH_3	145–147°

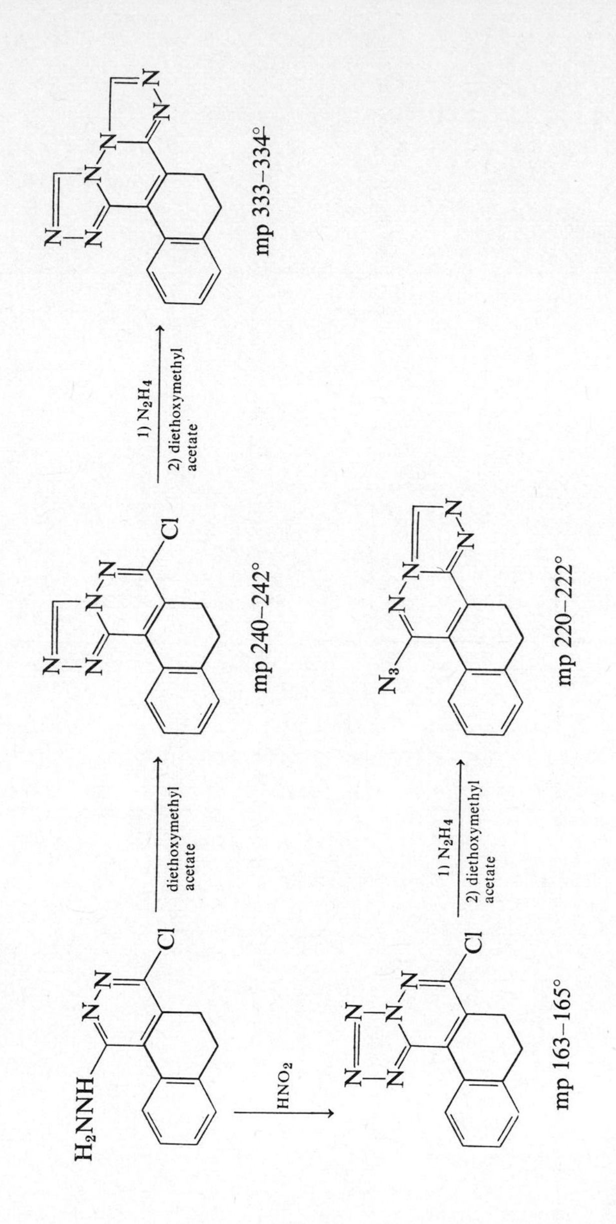
diethoxymethyl acetate
1) N_2H_4
2) diethoxymethyl acetate
HNO$_2$
1) N_2H_4
2) diethoxymethyl acetate
mp 240–242°
mp 333–334°
mp 163–165°
mp 220–222°

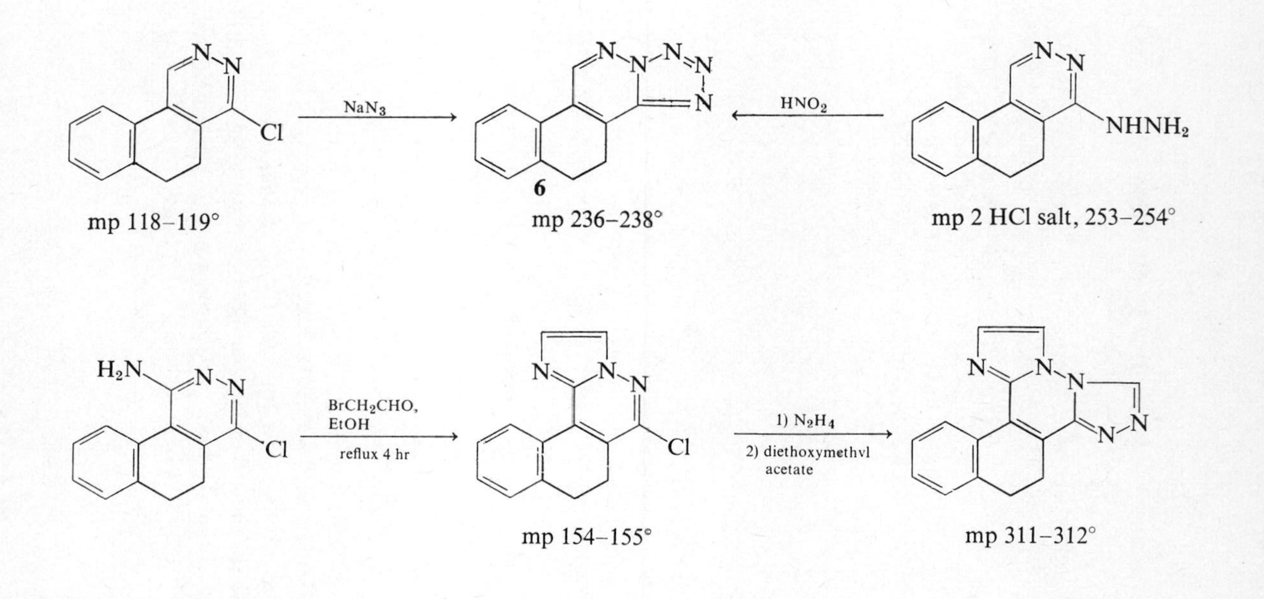
NaN3
Cl
mp 118–119°
6
mp 236–238°
HNO2
NHNH2
mp 2 HCl salt, 253–254°
H2N
BrCH2CHO,
EtOH
reflux 4 hr
Cl
mp 154–155°
1) N2H4
2) diethoxymethyl acetate
mp 311–312°

II. Benzo[*g*]phthalazine

The chemiluminescent properties of 1,4-dihydroxybenzo[*g*]phthalazine (**7**) were reported by Drew and Garwood (6) in 1939. Recently, White and co-workers (7) have oxidized **7** with chlorine to give a diazaquinone, benzo-[*g*]phthalazine-1,4-dione (**8**) (Eq. 6), which also undergoes a chemiluminescent reaction. The authors have proposed **8** as an intermediate in the chemiluminescent reaction of **7**. The diazaquinone **8** forms Diels-Alder adducts with

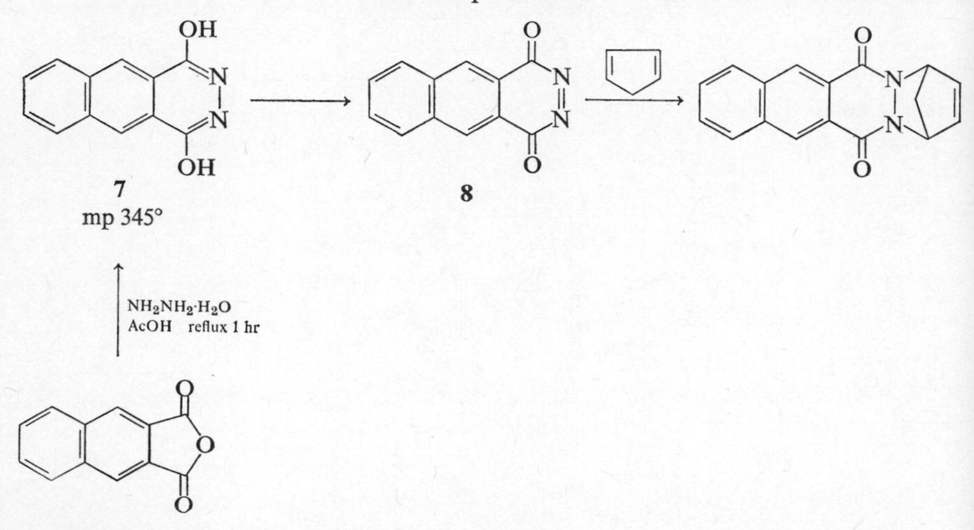

butadiene and cyclopentadiene in the same way that phthalazine-1,4-dione forms such adducts (see Section 2E-I-A-6-d). The dione **8** is stable for months at −20° but for only a few fours at room temperature in a dilute diglyme solution. Phthalazine-1,4-dione also is known to be very unstable.

Regan and Miller (8) have condensed some 1,3-disubstituted 1,3-dimethoxy-1,3-dihydronaphtho[2,3-*c*]furans, after acidic hydrolysis, with hydrazine

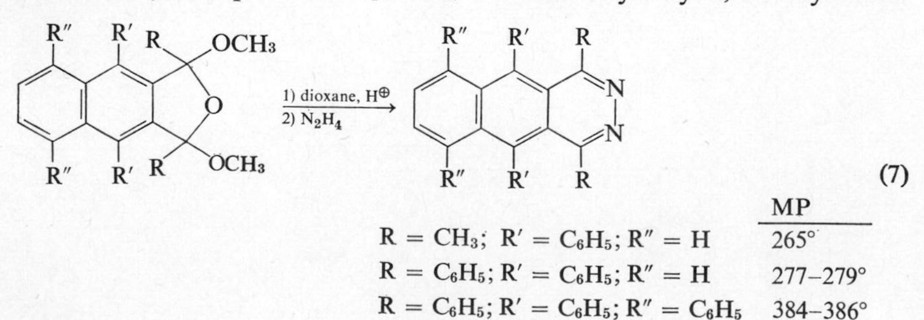

in refluxing ethanol to give benzo[g]phthalazine derivatives in 40–70% yields (Eq. 7).

Badder and co-workers (9) have condensed some naphthalene-2,3-dicarboxylic anhydrides with hydrazine hydrate by heating in refluxing 1-butanol for 5 hr. These authors observed that 1-phenylnaphthalene-2,3-dicarboxylic anhydride gave 55% 1,4-dihydroxy-5-phenylbenzo[g]phthalazine and 40% *N*-amino-1-phenylnaphthalene-2,3-dicarboxamide (Eq. 8). A similar reaction of 1-(*p*-methoxyphenyl)-7-methoxynaphthalene-2,3-dicarboxylic anhydride gave only *N*-amino derivative. On the other hand, 1-(*o*-chlorophenyl)-5-chloro- and 1-(*p*-chlorophenyl)-7-chloronaphthalene-2,3-dicarboxylic anhydride gave exclusively the corresponding 1,4-dihydroxybenzo[g]phthalazines in ~90% yields. It is also interesting to note that all these anhydrides, on condensation with hydrazine in glacial acetic acid, gave

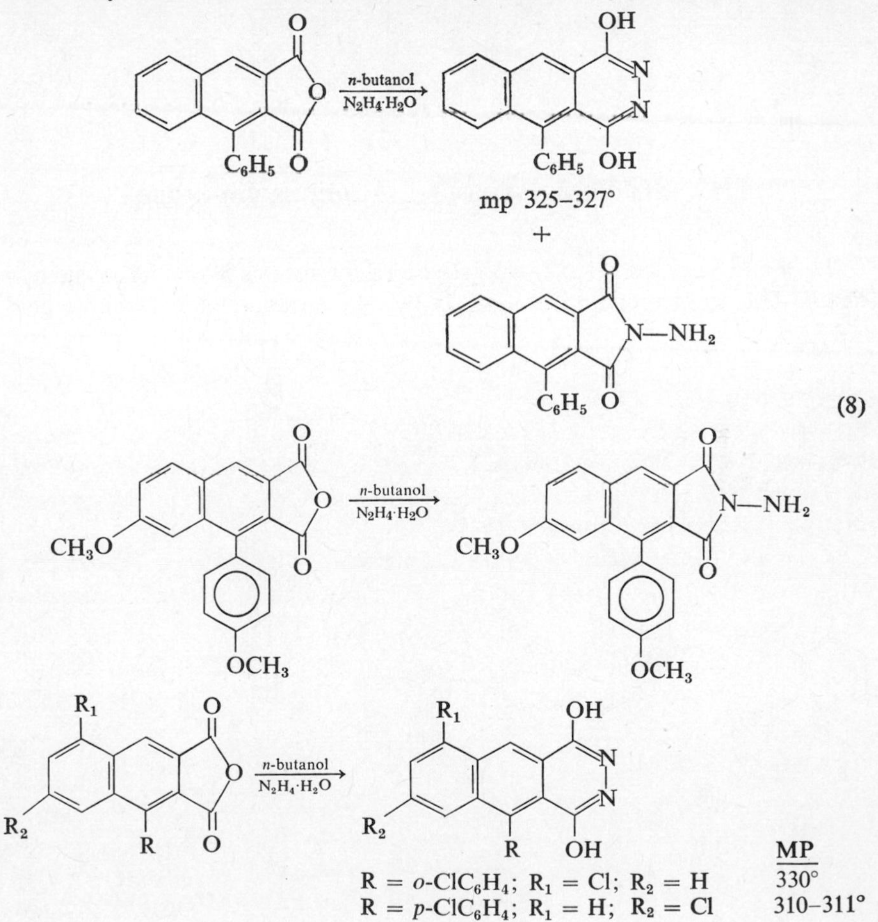

only the corresponding *N*-acetamido imides. 1,4-Dihydroxy-7-dimethylaminobenzo[*g*]phthalazine, mp 311°, has been obtained (2) by heating 5-dimethylaminonaphthalene-2,3-dicarboxylic acid dimethyl ester with hydrazine hydrate at 150° for 90 min.

Lomme and Lepage (10) have reported the Diels-Alder condensation reactions of 1,4,5,7-tetramethylfuro[3,4-*d*]pyridazine with benzoquinone and maleic anhydride to give condensed phthalazine derivatives. The epoxide obtained on condensation with benzoquinone was dehydrated with sulfuric acid to give 1,4,5,10-tetramethylbenzo[*g*]phthalazine-6,9-dione (**9**) in poor yield (Eq. 9).

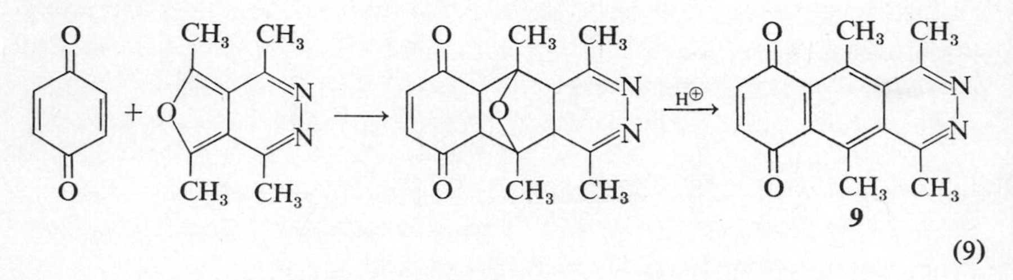

(9)

III. 8*H*-Phthalazino[1,2-*b*]quinazolin-8-one

Beyer and Voelcker (11, 12) have studied the synthesis of this ring system in detail. The main preparative route is by condensation of anthranilic acid

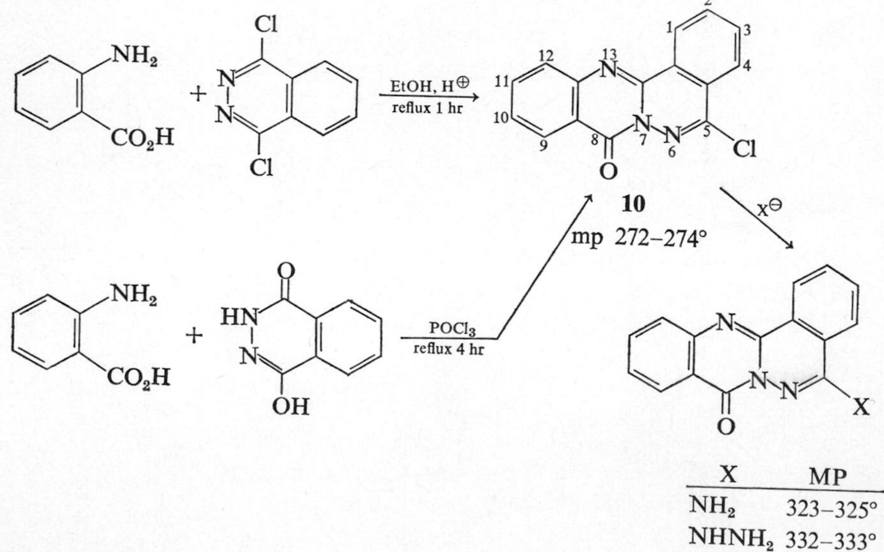

X	MP
NH_2	323–325°
$NHNH_2$	332–333°
OCH_3	219–220°

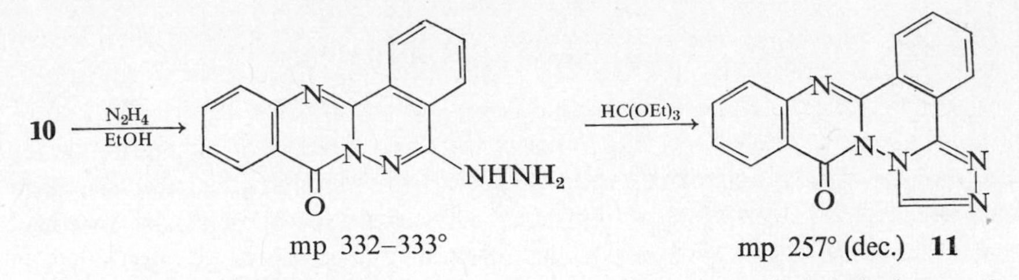

with 1,4-dichlorophthalazine or with 4-hydroxy-1-(2*H*)phthalazinone in presence of phosphorus oxychloride to give 5-chloro-8*H*-phthalazino[1,2-*b*]-quinazoline-8-one (**10**). Compound **10** undergoes nucleophilic substitutions (12) with ammonia, hydrazine, alkoxides, and so on. 5-Hydrazino-8*H*-phthalazino[1,2-*b*]quinazoline-8-one condenses with triethyl orthoformate to give a triazolo derivative, **11** (Eq. 10).

Anthranilic acid similarly reacts with 1-chlorophthalazine (12) and 1-chloro-4-methylphthalazine (12, 12a) to give 8*H*-phthalazino[1,2-*b*]-quinazolin-8-one, mp 260–261°, and 5-methyl-8*H*-phthalazino[1,2-*b*]-quinazolin-8-one, mp 266–268°, respectively. Fused dihydroisoindolo-benoxazinones (**12**) react with hydrazine hydrate in refluxing ethanol to give 5-substituted 8*H*-phthalazino[1,2-*b*]quinazolin-8-ones (13, 14) (Eq. 11). Lamchen (14) has proposed a mechanism for this reaction and has prepared several derivatives having aryl groups in the 5-position.

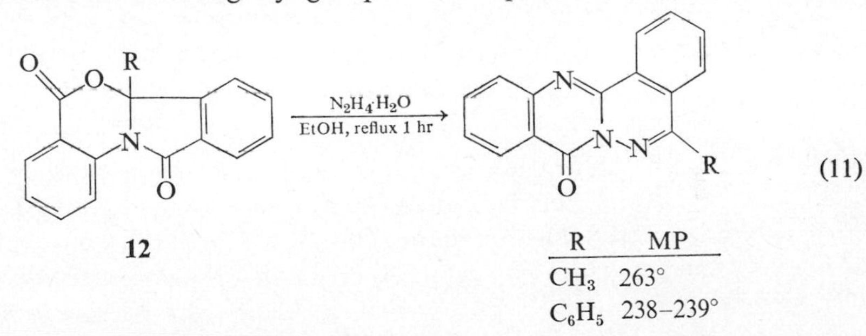

R	MP
CH_3	263°
C_6H_5	238–239°

(11)

Kirchner and Zalay (15) have condensed anthraniloyl hydrazide with phthaldehydic acid by heating in refluxing ethanol for 3 hr to give 5,6,7,8,13,13*a*-hexahydrophthalazino[1,2-*b*]quinazoline-5,8-dione (**13**) (Eq. 12).

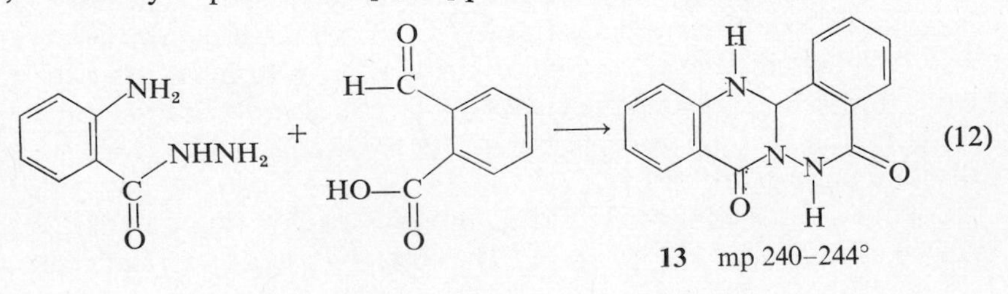

(12)

IV. 1,2,8,9-Tetraazaphenalenes

Doebel and Frances (16–22) have synthesized 1,2,8,9-tetraazaphenalene and a number of its derivatives and found them to be blood-pressure reducing agents. 1,2,8,9-Tetraazaphenalene (**14**) (16–18) has been obtained by the condensation of 3-hydroxy-7-formylphthalide with 2,6-bis(α,α-dibromo-methyl) benzoic acid and by condensation of 1-(2*H*)phthalazinone-8-carboxaldehyde with hydrazine hydrate in refluxing aqueous solvents. Yields range from 55 to 99% (Eq. 13). The synthesis of **14** has also been achieved by Raney nickel desulfurization of 3-mercapto-1,2,8,9-tetraazaphenalene (17, 18), by dehalogenation of 3-chloro-1,2,8,9-tetraazaphenalene (18), and by treatment of 3-hydrazino-1,2,8,9-tetraazaphenalene with copper sulfate in the presence of alkali (17, 18).

The quaternization of **14** proceeds smoothly in the usual manner (16) to give a methiodide, mp 279–281°; ethiodide, mp 260–262°; and isopropyl

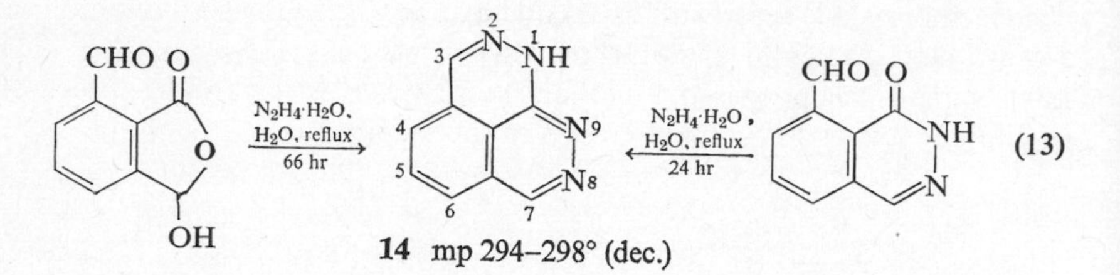

14 mp 294–298° (dec.)

iodide, mp 256–257°. Several acidic addition salts of **14** are also described (16). Alkylation of **14** with benzyl bromide in presence of sodium methoxide provides 9-benzyl-1,2,8,9-tetraazaphenalene (17). Nuclear halogenation of **14** has been studied with *N*-bromosuccinimide and *N*-chlorosuccinimide in 50% sulfuric acid to give the corresponding 4-halo-1,2,8,9-tetraazaphenalenes in low yields (17, 19) (Eq. 14). The bromination of 9-methyl-1,2,8,9-tetraazaphenalene with bromine in acetic acid provided 3-bromo-9-methyl-1,2,8,9-tetraazaphenalene (18) similarly the 9-phenyl analog is obtained (19).

4-Substituted 2,6-bis(α,α-dibromomethyl)benzoic acids condense with hydrazine hydrate to provide 5-substituted 1,2,8,9-tetraazaphenalenes (**15**) (16, 17). Similarly, 3-hydroxy-7-benzoylphthalide, with hydrazine hydrate, gave 7-phenyl-1,2,8,9-tetraazaphenalene, which is also obtained by catalytic dehalogenation of 3-chloro-7-phenyl-1,2,8,9-tetraazaphenalene (16, 17) over palladium on carbon (Eq. 15).

9-Methyl-1,2,8,9-tetraazaphenalene, mp 145–147° (149°–151°) [HI salt, 247–248°; CH_3I salt, 247–248° (17, 18)] has been prepared by several methods including desulfurization of 3-mercapto-9-methyl-1,2,8,9-tetraazaphenalene over Raney nickel, chemical reduction of 7-chloro-9-methyl-1,2,8,9-tetraazaphenalene with phosphorus and hydriodic acid, catalytic reduction of 3-chloro-9-methyl-1,2,8,9-tetraazaphenalene over palladium on carbon, and treatment of the methiodide salt of **14** with silver nitrate in alkaline medium (16). Similarly, 9-phenyl-1,2,8,9-tetraazaphenalene, mp 168.5–170°, has

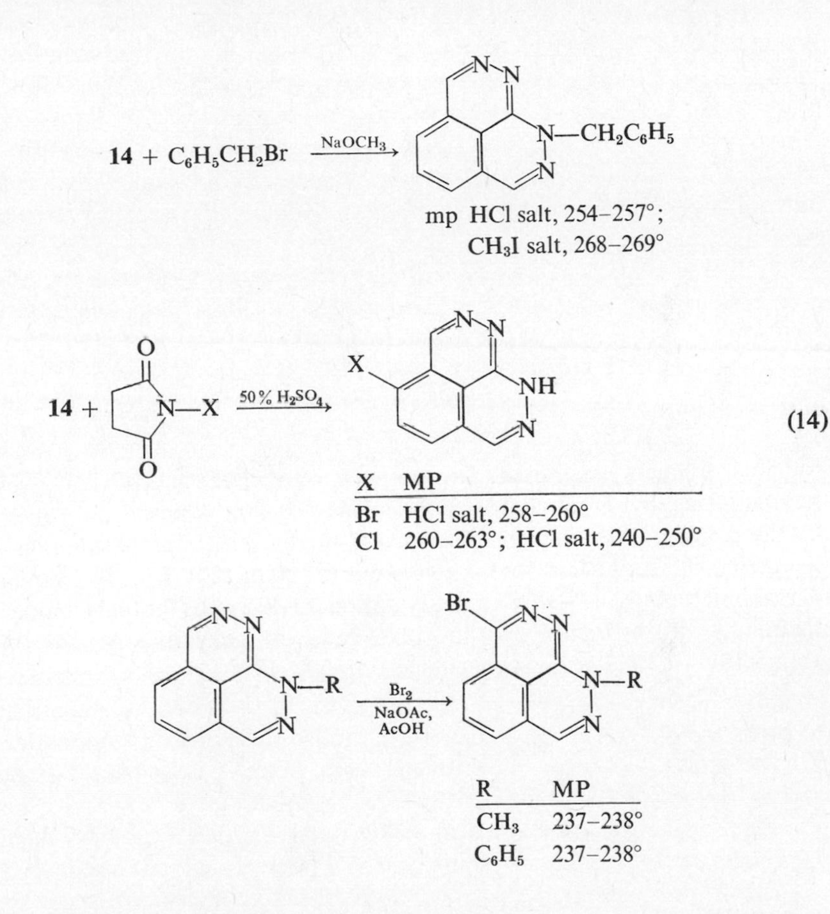

been obtained by chemical and catalytic dehalogenation of 3-chloro-9-phenyl-1,2,8,9-tetraazaphenalene (17, 18).

The important intermediate 3-hydroxy-1,2,8,9-tetraazaphenalene (**16**) has been synthesized by condensing 3-(dibromomethyl)phthalic anhydride, 2-carbomethoxy-6-(dibromomethyl)benzoic acid, 8-carbethoxy-1-(2*H*)phthalazinone, and properly substituted phthalide derivatives with hydrazine hydrate in aqueous ethanol (20) (Eq. 16). Alkylation of **16** with methyl

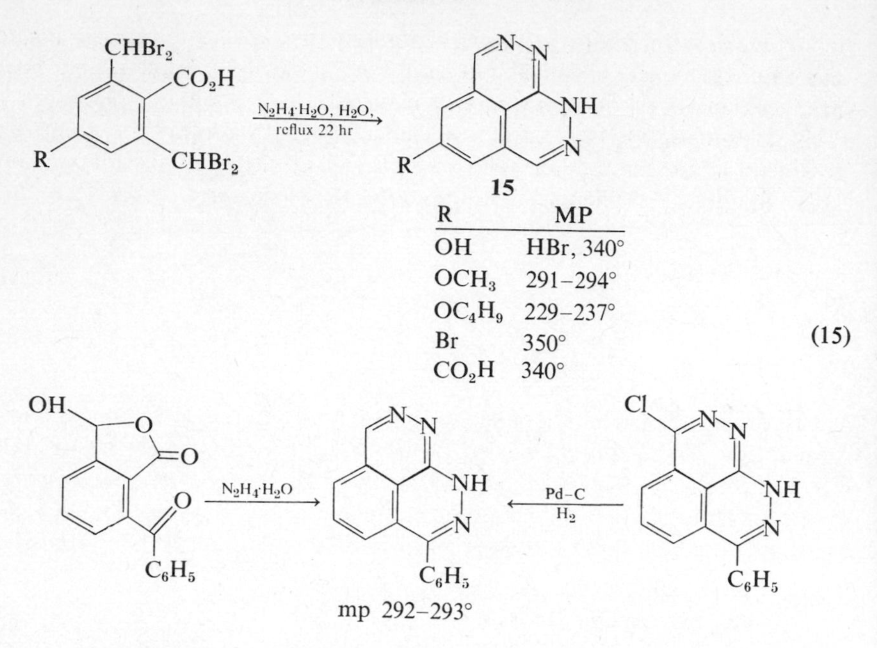

R	MP
OH	HBr, 340°
OCH_3	291–294°
OC_4H_9	229–237°
Br	350°
CO_2H	340°

(15)

iodide in dimethyl sulfoxide in presence of sodium methoxide gives 3-hydroxy-9-methyl-1,2,8,9-tetraazaphenalene (18–20), which is also obtained by the condensation of 8-carbomethoxy-2-methyl-1-(2*H*)phthalazinone with hydrazine by heating in refluxing cellosolve for 26 hr (20) (Eq. 16). Similarly, 8-carbomethoxy-2-(2′-*N*,*N*-dimethylaminoethyl)-1-(2*H*)phthalazinone, on treatment with hydrazine hydrate, provided 3-hydroxy-9-(2′-*N*,*N*-dimethylaminoethyl)-1,2,8,9-tetraazaphenalene, mp 197–199° (21).

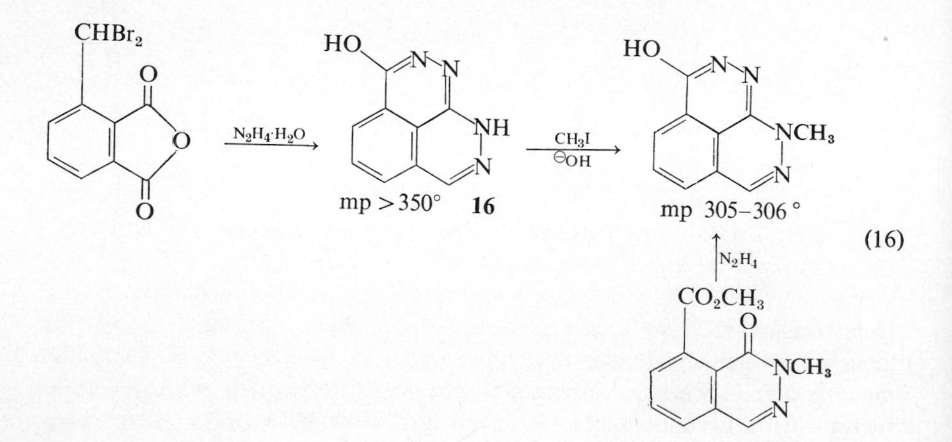

(16)

3-Hydroxy-1,2,8,9-tetraazaphenalene (**16**) and its 9-substituted derivatives smoothly react with phosphorus pentasulfide by heating in refluxing pyridine for 2–3 hr to give the corresponding 3-mercapto-1,2,8,9-tetraazaphenalene **17** and its derivatives (18–20) in high yields (Eq. 17).

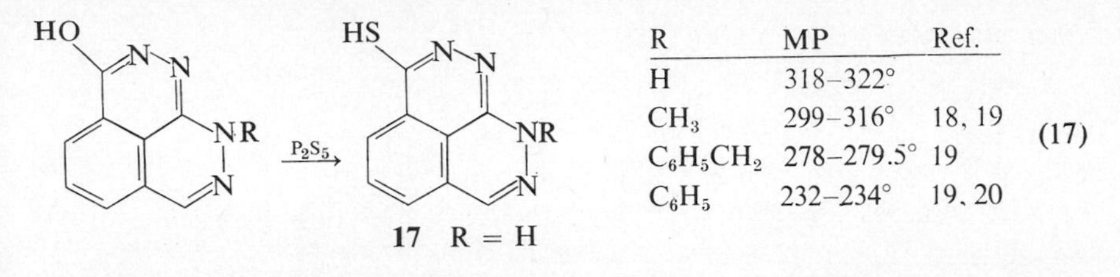

R	MP	Ref.
H	318–322°	
CH_3	299–316°	18, 19
$C_6H_5CH_2$	278–279.5°	19
C_6H_5	232–234°	19, 20

3-Hydroxy-1,2,8,9-tetraazaphenalene (**16**), on refluxing with phosphorus oxychloride for 20 hr, furnished 3-chloro-1,2,8,9-tetraazaphenalene (**18**) (18, 19). Similarly, 9-substituted derivatives of **16**, on refluxing with a phosphorus oxychloride and phosphorus pentachloride mixture for 2 hr, gave the corresponding 3-chloro-1,2,8,9-tetraazaphenalenes (17–19) (Eq. 18). Nucleo-

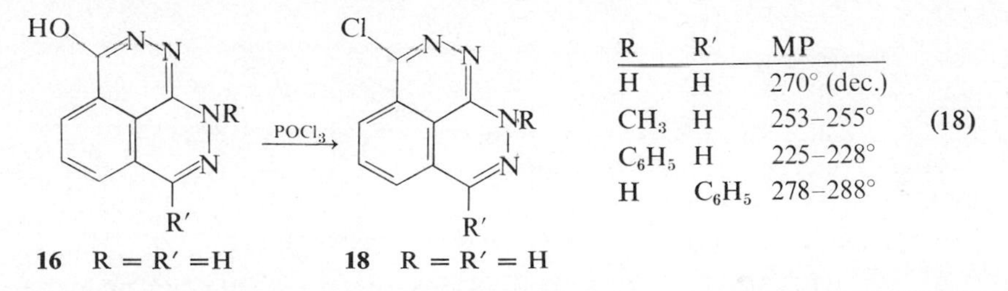

R	R′	MP
H	H	270° (dec.)
CH_3	H	253–255°
C_6H_5	H	225–228°
H	C_6H_5	278–288°

philic substitution of the chlorine atom in **18** with various primary and secondary amines is easily achieved; for example, refluxing with morpholine gave 3-morpholino-1,2,8,9-tetraazaphenalene (19) (Eq. 19).

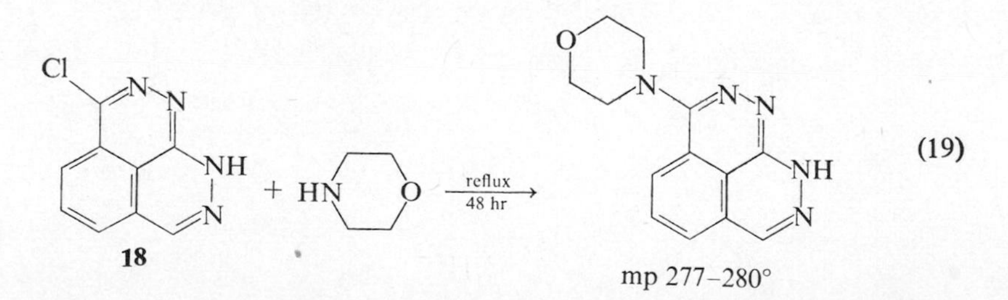

3-Hydrazino-1,2,8,9-tetraazaphenalene has been similarly prepared in high yield by substitution of the chloro or mercapto group in 3-chloro- or 3-mercapto-1,2,8,9-tetraazaphenalene (18, 19, 22). Various aromatic aldehydes have been condensed with **18** by heating them with **18** in refluxing

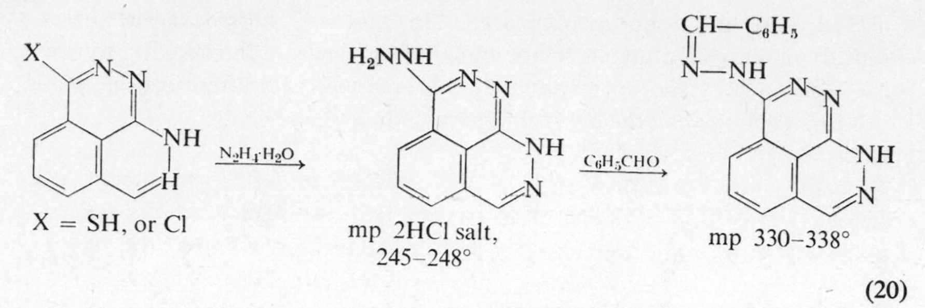

(20)

aqueous ethanol for 30 min to give the corresponding hydrazones (22) (Eq. 20).

V. Miscellaneous Condensed Phthalazines

Acheson and Foxton (23) studied the addition of dimethyl acetylenedicarboxylate to 1-methylphthalazine in refluxing acetonitrile to give tetramethyl 7-methyl-11-*bH*-pyrido[2,1-*a*]phthalazine-1,2,3,4-tetracarboxylate (**18**) and

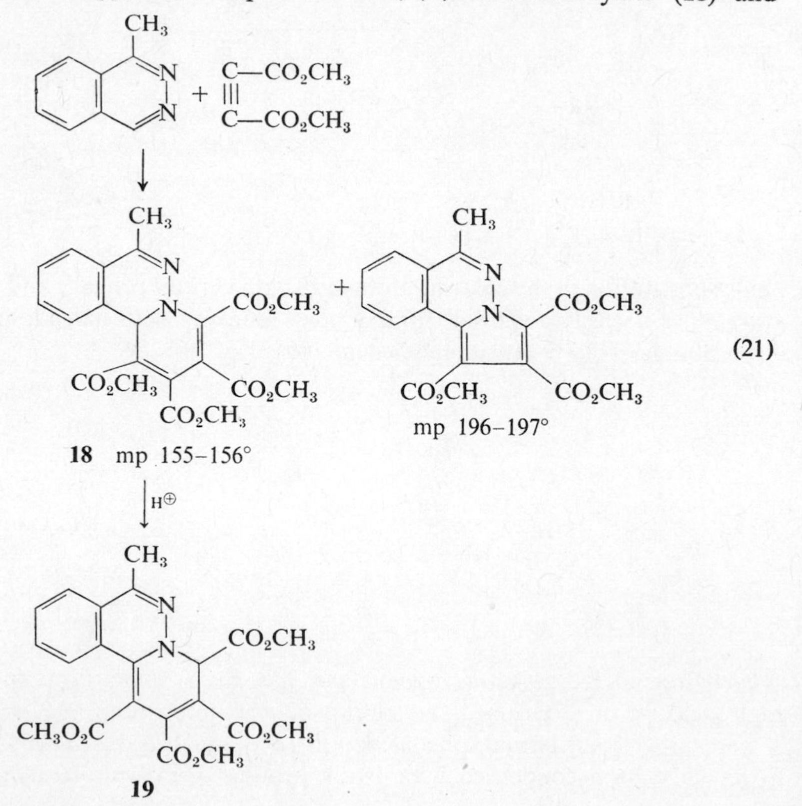

(21)

a small amount of trimethyl 6-methylpyrrolo[2,1-*a*]phthalazine-1,2,3-tricarboxylate. The compound **18**, on treatment with acid at room temperature, undergoes isomerization of the double bonds in the dihydropyridine ring to give tetramethyl 7-methyl-4*H*-pyrido[2,1-*a*]phthalazine-1,2,3,4-tetracarboxylate (**19**) (Eq. 21).

Petrovanu and co-workers (24) studied the addition of dimethyl acetylenedicarboxylate to phthalazine and found that in equimolar amounts the reaction proceeded at room temperature in ether to provide the ylide **20**, whereas

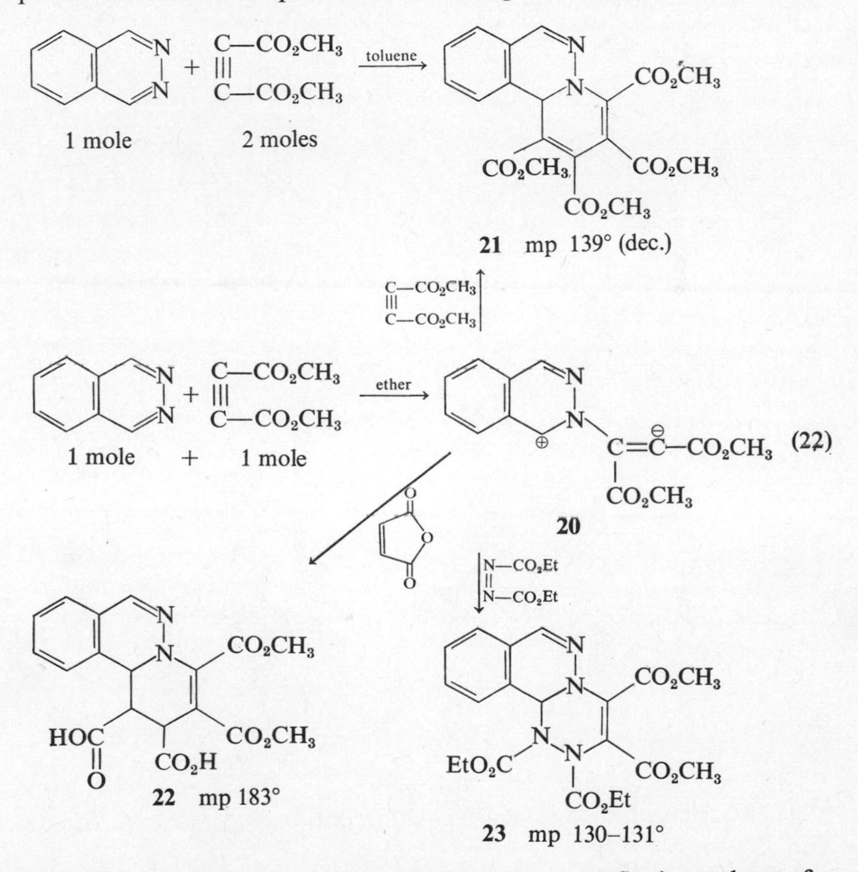

reaction with 2 moles of acetylenedicarboxylate in refluxing toluene for 10–12 hr furnished tetramethyl 11*bH*-pyrido[2,1-*a*]phthalazine-1,2,3,4-tetracarboxylate (**21**). The ylide **20** was further treated with dimethyl acetylenedicarboxylate to give **21**. Similarly, the ylide reacts with maleic anhydride and diethyl azodicarboxylate to give **22** and 1,2-diethyl-3,4-dimethyl 11*bH*-1,2-dihydro-*as*-triazino[3,4-*a*]phthalazine-1,2,3,4-tetracarboxylate (**23**), respectively (Eq. 22). The same group of workers (25) have similarly carried out the reactions of phthalazine with maleic anhydride and *N*-phenylmaleimide (Eq. 23) and studied the chemistry of the products. In addition, these authors (26) have prepared phthalazinium phenacylide (**24**) by the treatment of 2-phenacylphthalazinium bromide with alkali. They also carried out 1,3-cycloaddition reactions of **24** with maleic anhydride, dimethyl acetylenedicarboxylate, diethyl azodicarboxylate and phenyl isocyanate. The

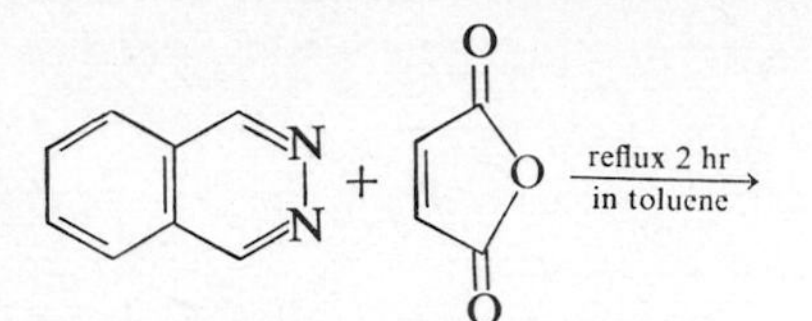

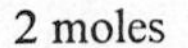

1 mole 2 moles

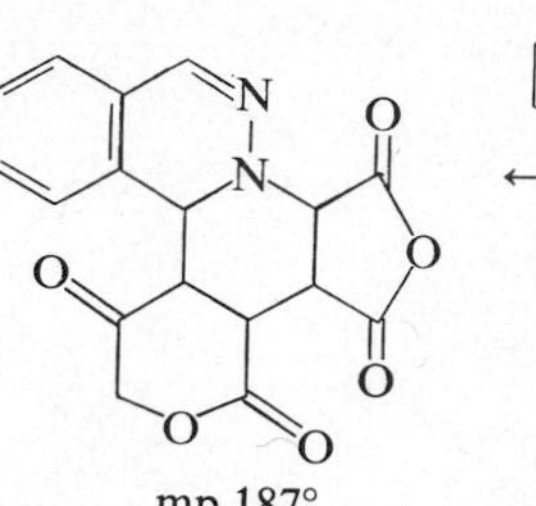

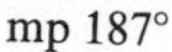

mp 187°

mp 80–85° (swells), 135° (dec.)

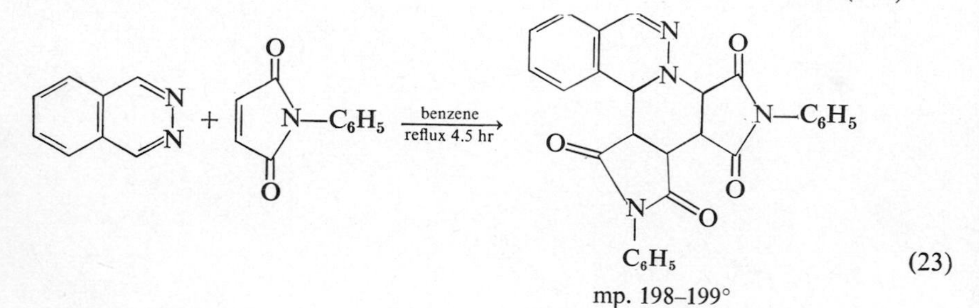

mp. 198–199° (23)

ylide **24** has been shown to undergo dimerization and reaction with ammonia (Eq. 24).

Spiro (27) has reported a novel synthesis of 3-methylpyrrolo[2,1-*a*]-phthalazin-6-one (**25**) by fusing 2-methyl-5-phenyl-1-ureidopyrrole-4-carboxylic acid. The structure of **25** is confirmed by oxidation to 1-(2*H*)phthalazinone-4-carboxylic acid by treatment with alkaline potassium permanganate in the cold and to phthalic acid in refluxing alkaline potassium permanganate. The alkylation of **25** with dimethyl sulfate provided 3,5-dimethylpyrrolo-[2,1-*a*]phthalazin-6-one (Eq. 25).

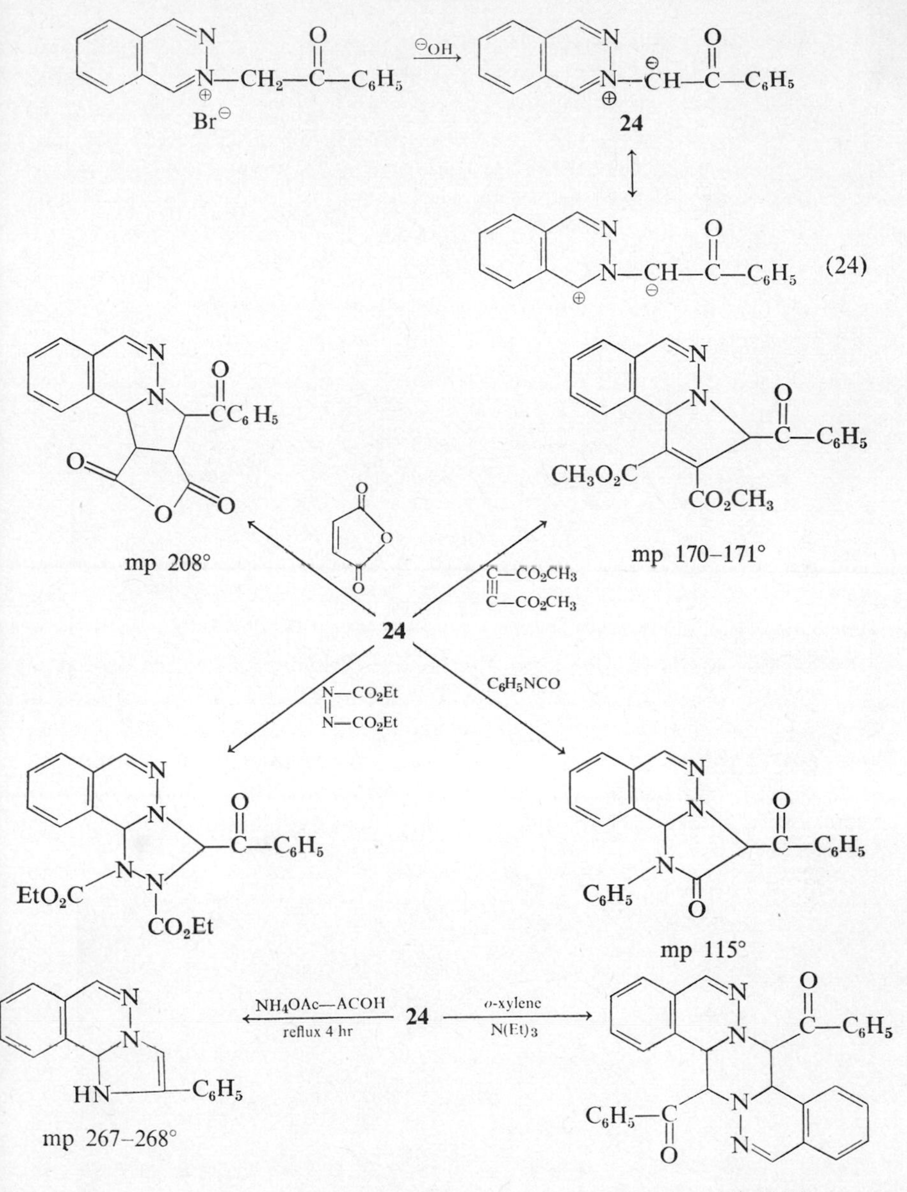

⊖OH
Br⊖
24
(24)
mp 208°
mp 170–171°
24
C₆H₅NCO
mp 115°
NH₄OAc—ACOH
reflux 4 hr
24
o-xylene
N(Et)₃
mp 267–268°
mp 256–258°

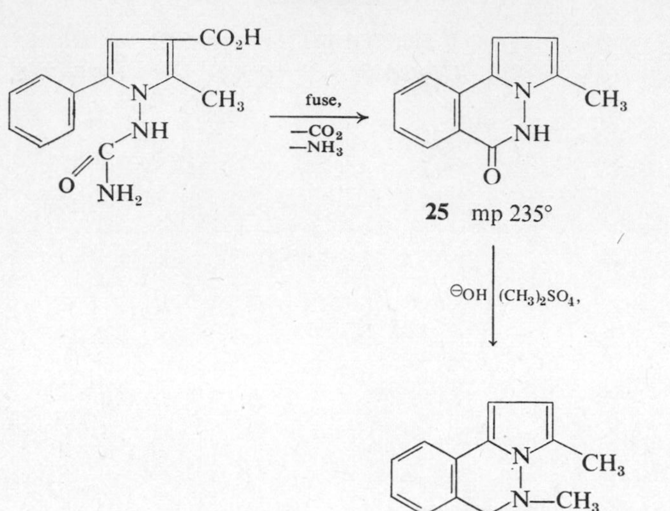

(25)

Potts and Lovelette (28) have shown that 3-hydrazino-*s*-triazolo[3,4-*a*]-phthalazine will react with cyanogen bromide in refluxing methanol for 8 hr to give 3-amino-*s*-triazolo[4,3-*b*]-*s*-triazolo[3,4-*a*]phthalazine (**26**), which undergoes phthalazine *N—N* bond fission on treatment with barium hydroxide (Eq. 26). Similarly, *s*-triazolo[3,4-*a*]phthalazine gave 3-*o*-cyanophenyl-*s*-triazole on boiling in alcohol with a catalytic amount of potassium hydroxide.

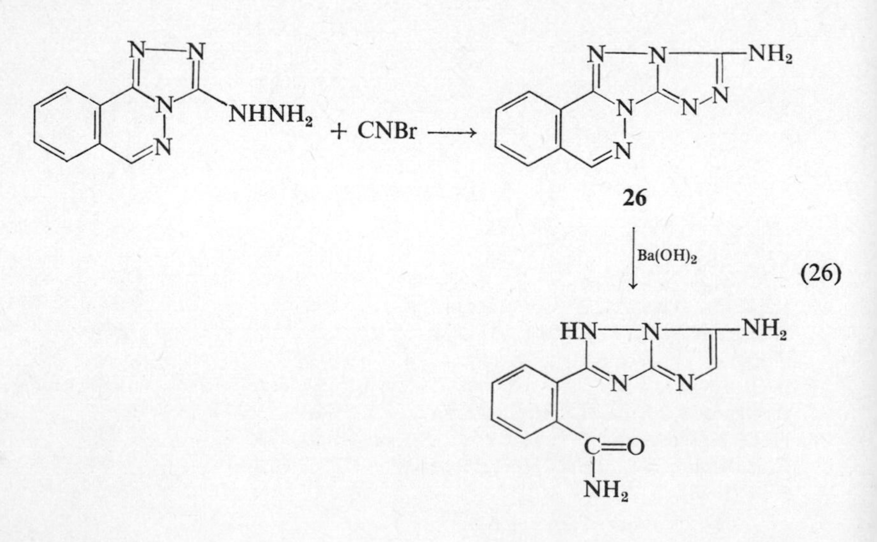

(26)

Some of the condensed phthalazines prepared by the reaction of the *o*-carbonyl compounds with hydrazine or its derivatives are shown in Eq. 27.

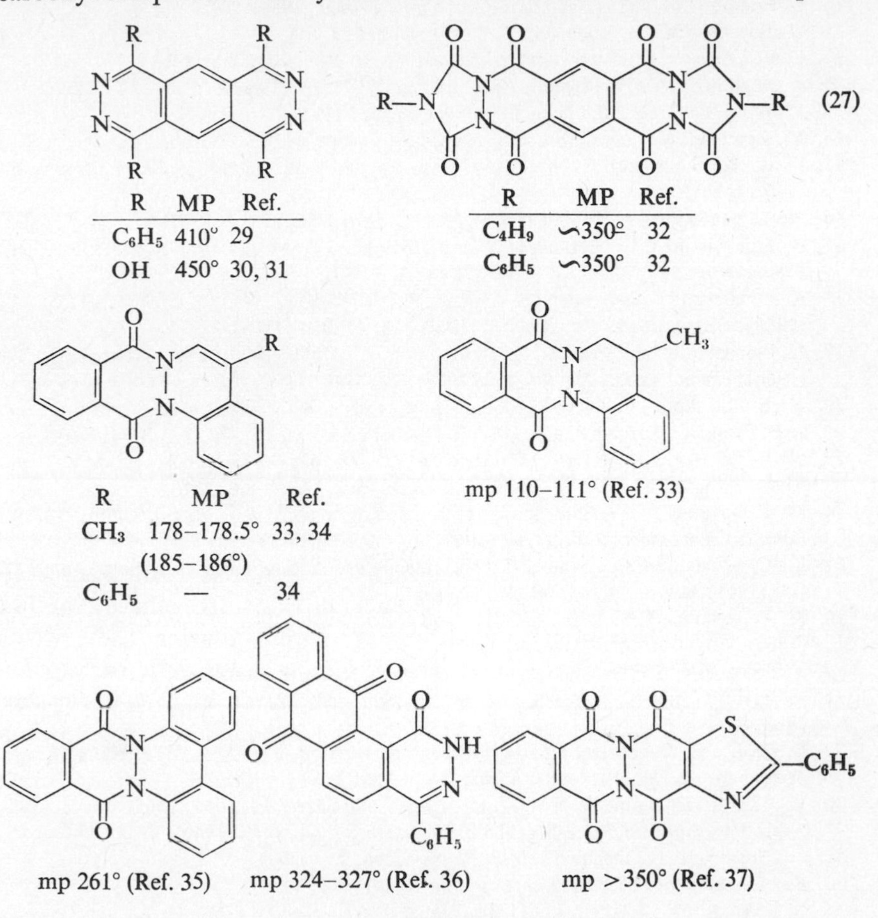

(27)

R	MP	Ref.
C_6H_5	410°	29
OH	450°	30, 31

R	MP	Ref.
C_4H_9	∽350°	32
C_6H_5	∽350°	32

R	MP	Ref.
CH_3	178–178.5° (185–186°)	33, 34
C_6H_5	—	34

mp 110–111° (Ref. 33)

mp 261° (Ref. 35) mp 324–327° (Ref. 36) mp >350° (Ref. 37)

References

1. J. Kobe, A. Krbavacic, B. Stanovic, and M. Tisler, *Croat Chem. Acta*, **41,** 245 (1969).
2. K. D. Gundermann, W. Horstmann, and G. Bergmann, *Ann. Chem.*, **684,** 127 (1965).
3. J. Dreiseitel and A. Koewa, *Diss. Pharm.*, **11,** 157 (1959); *Chem. Abstr.*, **54,** 4593 (1960).
4. B. Stanovik and M. Tisler, *Chimia*, **22,** 141 (1968); *Chem. Abstr.*, **69,** 19371 (1968).
5. B. Stanovik, M. Tisler, and P. Skufca, *J. Org. Chem.*, **33,** 2910 (1968).
6. H. D. K. Drew and R. F. Garwood, *J. Chem. Soc.*, **1939,** 836.
7. E. H. White, E. G. Nash, D. R. Roberts, and O. C. Zafiriou, *J. Am. Chem. Soc.*, **90,** 5932 (1968).

8. T. H. Regan and J. B. Miller, *J. Org. Chem.*, **31,** 3053 (1966).
9. F. G. Baddar, M. F. El-Newaihy, and M. R. Salem, *J. Chem. Soc.* (*C*), **1969,** 838.
10. L. Lomme and Y. Lepage, *Bull. Soc. Chim. France*, **1969,** 4183.
11. H. Beyer and C. E. Voelcker, *Z. Chem.*, **1,** 224 (1961).
12. C. E. Voelcker, J. Marth, and H. Beyer, *Chem. Ber.*, **100,** 875 (1967).
12a. K. H. Boltze and H. D. Dell, *Arch. Pharm.*, **299,** 702 (1966).
13. J. Honzl, *Coll. Czech. Chem. Commun.*, **21,** 725 (1956).
14. M. Lamchen, *J. Chem. Soc.* (*C*), **1966,** 537, and references therein.
15. F. K. Kirchner and A. W. Zalay, U.S. Pat. 3,375,250 (1968); *Chem. Abstr.*, **69,** 52170 (1968).
16. Neth. Pat. Appl., 6,604,484 (1966) *Chem. Abstr.*, **66,** 65519 (1967).
17. K. Doebel and J. E. Francis, U.S. Pat. 3,429,882 (1969); U.S. Pat. 3,479,355 (1969); *Chem. Abstr.*, **70,** 115177 (1969); **72,** 43736 (1970).
18. K. Doebel and J. E. Francis, S. African. Pat. 6,705,856 (1968); French Pat. 1,550,404 (1968); *Chem. Abstr.*, **70,** 115,190 (1969); **71,** 112964 (1969).
19. K. Doebel and J. E. Francis, S. African Pat. 6,705,858 (1968); French Pat. 1,538,290 (1968); *Chem. Abstr.*, **70,** 106,567 (1969); **71,** 61407 (1969).
20. Neth. Pat. Appl., 6,604,482 (1966); *Chem. Abstr.*, **66,** 65517 (1967).
21. J. E. Francis, German Pat. 1,940,827 (1970); *Chem. Abstr.*, **72,** 111519 (1970).
22. Neth. Pat. Appl., 6,604,483 (1966); *Chem. Abstr.*, **66,** 65518 (1967).
23. R. M. Acheson and M. W. Foxton, *J. Chem. Soc.* (*C*), **1966,** 2218.
24. M. Petrovanu, A. Sauciuc, I. Gabe, and I. Zugravescu, *Rev. Roum. Chim.*, **13,** 513 (1968); *Chem. Abstr.*, **70,** 57757 (1969).
25. I. Zugravescu, M. Petrovanu, A. Caraculacu, and A. Sauciuc, *Rev. Roum. Chim.*, **12,** 109 (1967); *Chem. Abstr.*, **68,** 49537 (1968).
26. M. Petrovanu, A. Sauciuc, I. Gabe, and I. Zugravescu, *Rev. Roum. Chem.*, **14,** 1153 (1969); *Chem. Abstr.*, **72,** 43591 (1970).
27. V. Spiro, *Ric. Sci. Rend., Sez. A.*, **8,** 197 (1965); *Chem. Abstr.*, **63,** 13188 (1965).
28. K. T. Potts and C. Lovelette, *J. Org. Chem.*, **34,** 3221 (1969); *Chem. Commun.*, **1968,** 845.
29. E. Profft, G. Drechsler, and H. Oberender, *Wiss. Z. Tech. Hochsch. Chim. Leuna-Merseberg*, **2,** 259 (1959–60); *Chem. Abstr.*, **55,** 2565 (1961).
30. R. A. Dine-Hart and W. W. Wright, *Chem. Ind.* (*London*), **1967,** 1565.
31. Neth. Pat. Appl., 6,502,841 (1965); *Chem. Abstr.*, **64,** 5012 (1966).
32. C. Zinner and W. Deucker, *Arch. Pharm.*, **296,** 13 (1963).
33. R. N. Castle and M. Onda, *J. Org. Chem.*, **26,** 4465 (1961).
34. L. S. Bresford, G. Allen, and J. M. Bruce, *J. Chem. Soc.*, **1963,** 2867.
35. A. Etienne and R. Piat, *Bull. Soc. Chim. France*, **1962,** 292.
36. H. G. Rule, N. Campbell, A. G. McGregor, and A. A. Woodham, *J. Chem. Soc.*, **1950,** 1816.
37. M. Robba and Y. L. Guen, *Bull. Soc. Chim. France*, **1970,** 4317.

CHAPTER III

Azolo- and Azinopyridazines and Some Oxa and Thia Analogs

M. TIŠLER AND B. STANOVNIK

Department of Chemistry, University of Ljubljana, Ljubljana, Yugoslavia

Part A. Introduction

In view of the rapid expansion of the chemistry of heterocyclic compounds, it is worthwhile to review bicyclic systems with fused pyridazine rings. A pyridazine ring may be fused on the different atoms of an azole, furan, thiophene, or azine and in this way a number of bicyclic azaheterocycles can be designed. A number of systems with these structural characteristics remain unknown and, on the other hand, many of the known systems have been thoroughly investigated and require attention.

All fully aromatic systems of this kind are best considered as resonance hybrids and a number of uncharged and charged resonance structures can be written for every system. Some conclusions about which structures make important contributions can be reached by an examination of the π-electron densities at various positions in the ring. For each system the results of MO (molecular orbital) calculations of total and frontier π-electron densities are presented and give, in the absence of more reliable information, useful approximations in predicting the reactivity of the azaheterocycles. Certainly, these approximations are valid for isolated molecules and for more accurate predictions other factors such as the attacking species and the environment have to be considered.

The rings are oriented and the positions are numbered according to the IUPAC Rules and Revised Ring Index (RRI). Every system, if registered in the Ring Index (second edition), is accompanied with the Revised Ring Index number. Other names that have been used in the literature for several systems are included in parentheses. In several cases the orientation of the rings and the numbering may be different from those in original communications.

All systems in which the pyridazine part resembles maleic hydrazide and where lactam-lactim tautomerism is possible are, for simplicity, assigned the dioxo structure. It is probable that one oxo group is present in the enolized form, but no studies concerning these structures have been performed so far.

Part B. Pyrrolopyridazines

I. Pyrrolo[1,2-*b*]pyridazines

Ring Index Suppl. I, 7989
(5-Azaindolizine)

The most highly investigated synthetic approach to form this bicyclic system is the reaction between pyridazines and esters of acetylenedicarboxylic acid.

Letsinger and Lasco (1) first reported that pyridazine reacts readily with methyl acetylenedicarboxylate in ether solution, although in early experiments they were not able to isolate products, other than those with high molecular weights. If the same reaction was conducted in methanol at 0° for several days a crystalline adduct (**1**, R = H) could be isolated in low yield (26%). A somewhat better yield was obtained in the case of the 3-methylpyridazine adduct (37%) (**1**, R = Me). The structure of the bicyclic product

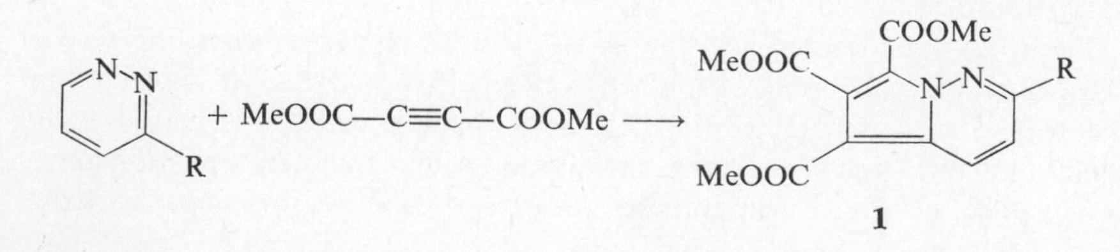

was proven by an independent synthesis by condensing 3,6-dimethylpyridazine with ethyl bromopyruvate. From the reaction mixture an acid and its ester were isolated and the acid was found to be identical with **13** (R = Me) obtained by decarboxylation of **12** (R = Me).

Experiments of Letsinger and Lasco were reexamined by Acheson and Foxton (2), who employed acetonitrile as the solvent. Evidence was presented that the major product from the reaction between pyridazine and acetylenedicarboxylic ester was a pyrido[1,2-*b*]pyridazine derivative, whereas the pyrrolo[1,2-*b*]pyridazine (**1**, R = H) was obtained only in small quantity (2%). The pyrrolo[1,2-*b*]pyridazine structure was confirmed from nmr data (2).

In an analogous way the bicycle **3** (R = Me) was formed in a reaction between *N*-(carbomethoxymethyl)pyridazinium bromide (**2**, R = Me) and dimethyl acetylenedicarboxylate in boiling methylene chloride and in the presence of triethylamine (3). That the reaction occurred through a 1,3-dipolar addition was established when *N*-(carbethoxymethyl)pyridazinium bromide (**2**, R = Et) was used as the starting compound. The product was the expected 7-carbethoxy analog (**3**, R = Et), although the less probable alternative **4** was also mentioned as a possibility (3). In a similar way *N*-

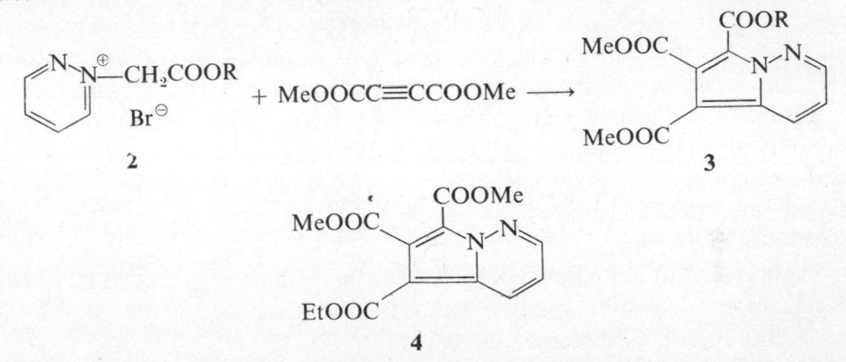

(carbomethoxymethyl)-3,6-dimethylpyridazinium bromide reacted under mild reaction conditions with dimethyl acetylenedicarboxylate to form the bicycle of the type **3**, a methyl group being lost during this transformation (3).

Another possible way of building pyrrolo[1,2-*b*]pyridazines was reported recently (4, 25, 26). 1-Aminopyrrole reacts with 1,3-diketones to give the bicyclic compounds of the type **5**.

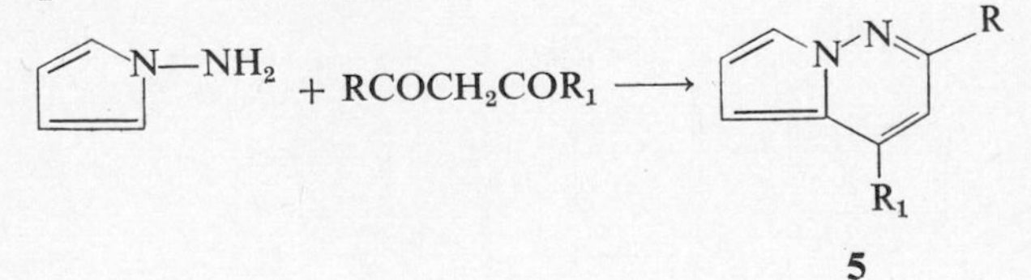

In the case of asymmetric 1,3-diketones two isomeric products are expected and, in fact, with benzoylacetaldehyde both isomers (**5**, R = H, $R_1 = C_6H_5$, and R = C_6H_5, R_1 = H) were isolated in the ratio of 2:1. The structure of one of the isomers was verified by an independent synthesis from 1-aminopyrrole and **6**, via **7**, which upon oxidation with potassium ferric cyanide yielded the aromatized bicycle **8**. On the other hand, from the reaction with

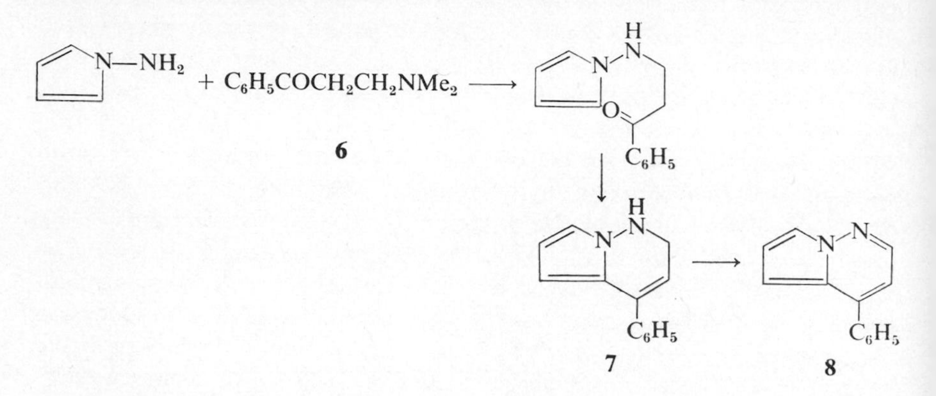

benzoylacetone only one isomer could be isolated. The structure was assigned on the basis of differences in UV spectra of 2- and 4-substituted pyrrolo-[1,2-*b*]pyridazines and the results were consistent also with nmr spectra.

Recently, this system was synthesized by 1,3-dipolar cycloaddition of 3-substituted pyridazinium methylides with dimethyl acetylenedicarboxylate or cyanoacetylene (27). When 3,6-dialkoxypyridazinium methylides were used as starting material, one of two alkoxyl groups was expelled during the formation of adducts.

There is also an old report about a doubtful formation of a pyrrolo[1,2-*b*]-pyridazinone derivative (**9**) (5), but later the same author envisaged the possibility of other structures (**10**, **11**) (6).

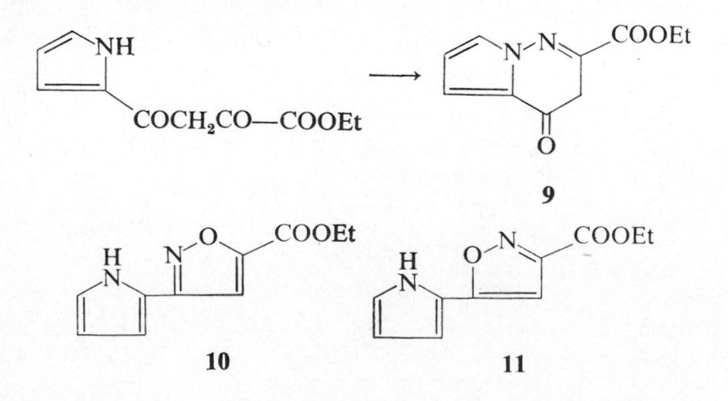

The results of MO calculations of total and frontier π-electron densities for the parent compound (26) is presented in Table 3B-1. On this basis, it is expected that pyrrolo[1,2-*b*]pyridazine represents a 10 π-electron aromatic system with the positions 7 or 5 most susceptible for electrophilic attack.

Pyrrolo[1,2-*b*]pyridazine has been prepared and in the nmr spectrum the protons attached to the pyrrole ring appear as an ABX system (in CS_2: $\tau = 3.69$ (H_5), $\tau = 3.33$ (H_6), $\tau = 2.48$ (H_7); $J_{5,6} = 4$ cps, $J_{6,7} = 2.5$ cps, and $J_{5,7} = 1$ cps). Similarly, the protons attached to the pyridazine ring appear as an ABX system (in CS_2: $\tau = 2.21$ (H_2), $\tau = 3.68$ (H_3), $\tau = 2.49$ (H_4); $J_{2,3} = 4.5$ cps, $J_{2,4} = 2$ cps, and $J_{3,4} = 9$ cps).

Several reactions on this system have been performed. Adducts of the type **1** were saponified with alkali at room temperature to the tricarboxylic acid **12**, which when heated alone or with hydrochloric acid is decarboxylated to the monocarboxylic acid **13** (1).

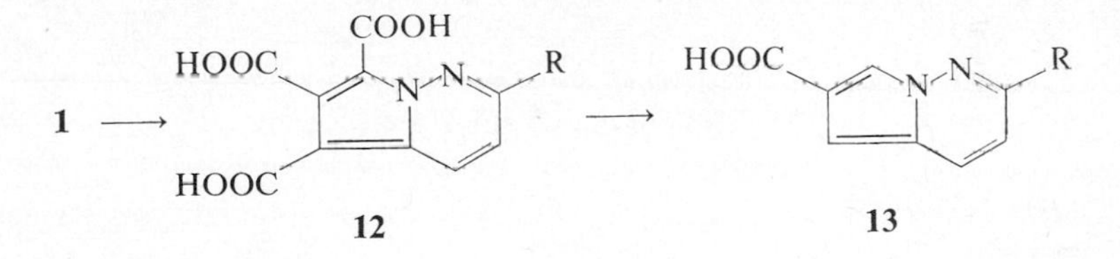

It is reported that neither **1** nor the methyl ester of **13** reacted with methyl iodide or phenacyl bromide under conditions that are common in the case of simple azines (1). It is also claimed that reaction with methyl acetylenedicarboxylate was not amenable, although other authors (4, 25, 26) were able to isolate addition products of the type **14** when applying this addition reaction to compounds of the type **5**. As minor products, derivatives of 5-azacycl[3.2.2]azine were isolated. The addition takes place at position 7 and

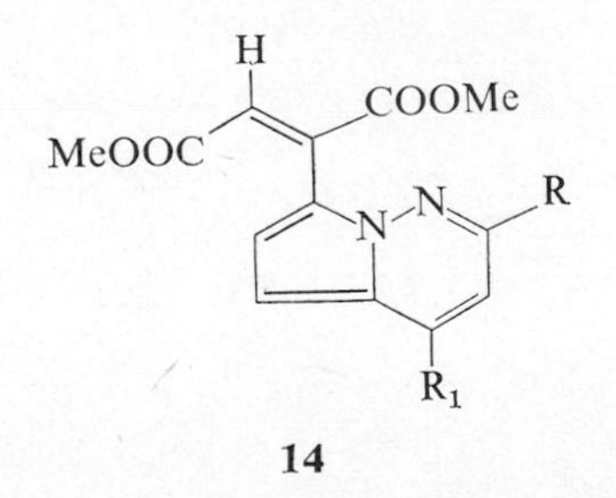

such a conclusion has been reached on the basis of nmr spectra correlations. Tetracyanoethylene is claimed to give similar products (**15**), however, with the abstraction of HCN.

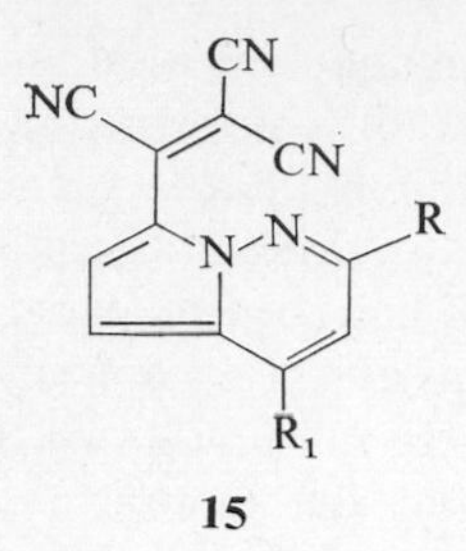

15

Protonation of pyrrolo[1,2-*b*]pyridazines takes place chiefly at position 7 and to a small extent at position 5 (25). A similar substitution at position 7 has been observed for several electrophilic substitutions, such as nitrosation, coupling, acylation with trifluoroacetic acid, and addition of tetracyano-ethylene (26). Nitration afforded a 5,7-dinitro product, whereas bromination under different reaction conditions can lead to di-, tri-, or tetrabromo derivatives (26).

II. 1*H*-Pyrrolo[2,3-*d*]pyridazines

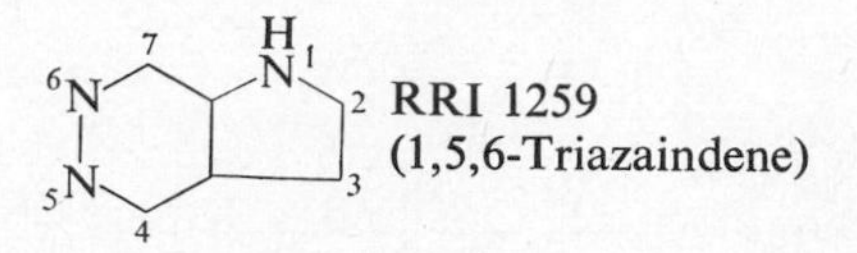

In principle, the same synthetic approaches that were considered for the isomeric pyrrolo[3,4-*d*]pyridazine system were applied for the formation of 1*H*-pyrrolo[2,3-*d*]pyridazines.

Pyrrolo[2,3-*d*]pyridazines (**16**) were obtained from 2-formyl-3-acylpyrroles and hydrazine (8, 9) and their 4- (**17**) or 7-oxo analogs (**18**) from 2-

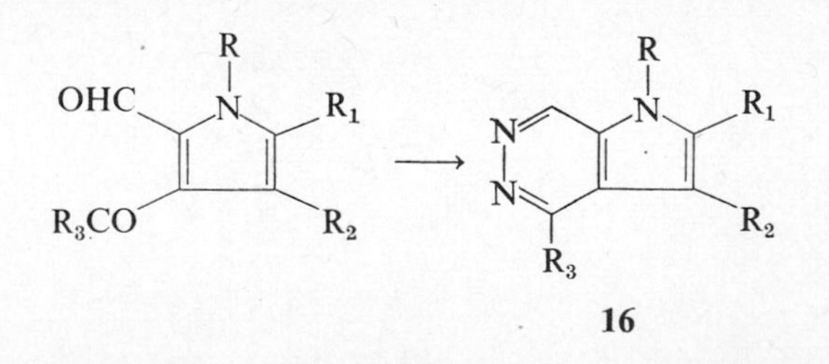

16

formyl-3-carbethoxy- (10, 11, 24) and 2-acetyl-3-carbethoxypyrroles (12) of 2-carboxy (or carbalkoxy)-3-acetylpyrroles (13), respectively. To accomplish the reaction, the appropriate pyrroles are heated with hydrazine or phenyl-hydrazine in a solution of ethanol or acetic acid for several hours.

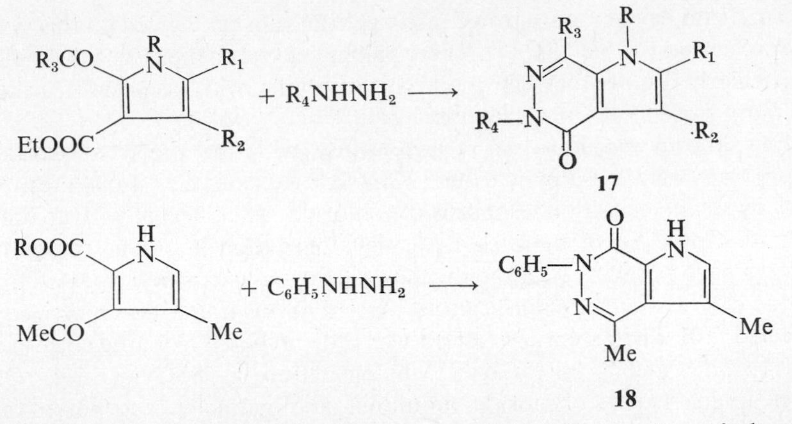

Similarly, 2,3-dicarbethoxypyrroles, when heated for longer periods with hydrazine, afforded 4,7-dioxopyrrolo[2,3-*d*]pyridazines (**19**, or tautomeric form) (14, 15). Shorter reaction periods cause the formation of the corresponding hydrazides only (15, 24).

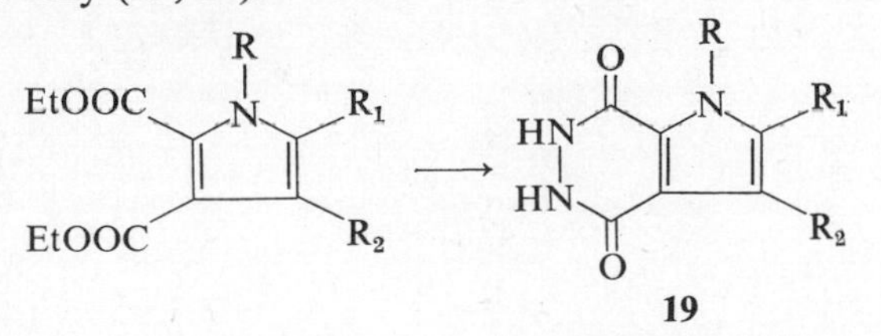

There are only few investigations about the reactivity of the pyrrolo[2,3-*d*]-pyridazine system. There are no reports about studies on the structure of 4,7-dioxo derivatives which could exist in several tautomeric forms (**20a**–**20d**). Taking into account the established hydroxy-oxo form of maleic hydrazide, it appears most likely that form **20b** or **20c** should predominate.

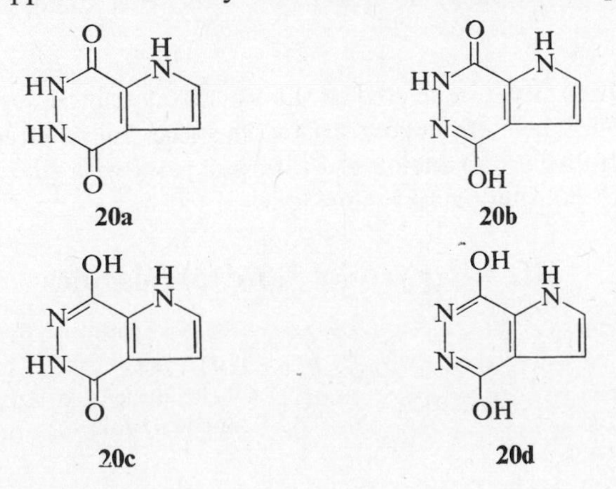

Total and frontier π-electron density calculations performed on this system are collected in Table 3B-2 (7). These values suggest that position 3 should be particularly reactive for electrophilic substitutions, whereas position 4 should be more susceptible for nucleophilic attack than position 7.

Few known reactions are in accordance with the predicted reactivity. 4-Chloropyrrolo[2,3-*d*]pyridazines (**22**), obtained from their 4-oxo precursors (**21**) by treatment with phosphorus oxychloride, when heated with an amine, thiourea, or P_2S_5, or upon catalytic dehalogenation at room temperature, always gave products with the displaced halogen at position 4 (10, 23, 24). In the case of another chlorine atom present at position 3, this remained unaffected (10). The parent compound was thus prepared from the corresponding 4-chloro analog by catalytic dehalogenation in 38% yield (24). Other nucleophiles such as phenoxide, ammonia, and hydrazine are reported to fail to react with **22**.

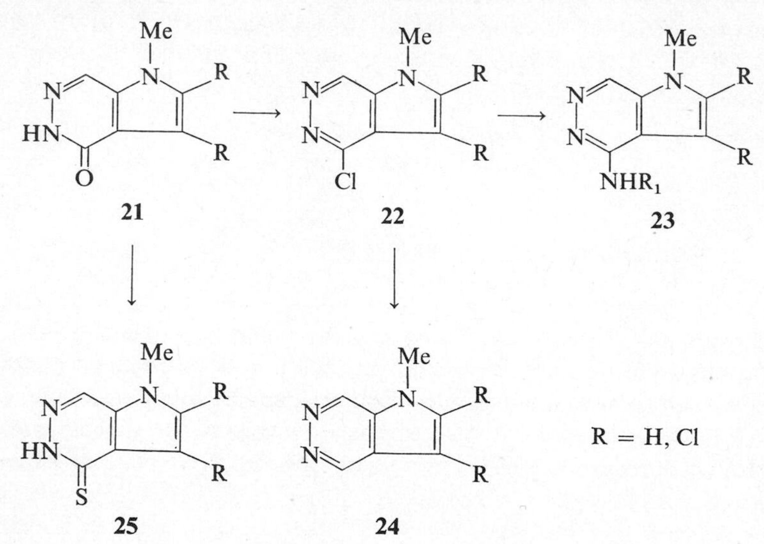

The oxo group could be thiated in the usual way using phosphorus pentasulfide method to give **25**. In contrast to the successful reactions mentioned, the 1-unsubstituted 4-oxo analog of **21** did not react with phosphorus pentasulfide or with phosphorus oxychloride.

III. 6*H*-Pyrrolo[3,4-*d*]pyridazines

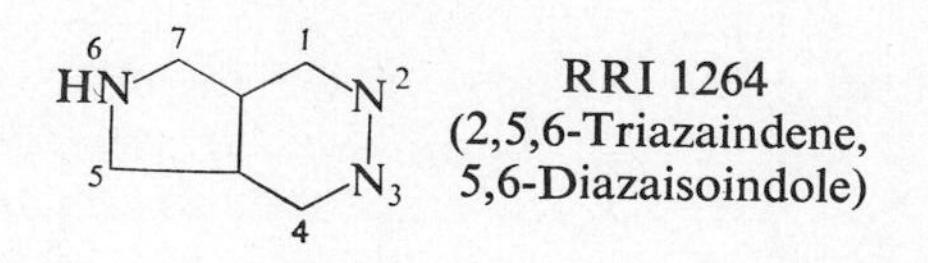

The most widely used synthesis of this bicyclic system consists of treating pyrroles with appropriate functional groups capable of ring closure to form the pyridazine ring at positions 3 and 4 with hydrazine.

3,4-Diformyl- or 3,4-diacylpyrroles when refluxed with a solution of hydrazine hydrate in alcohol gave the corresponding pyrrolo[3,4-*d*]pyridazines (**26**) as reported by Rips and Buu-Hoi (16, 17) and Kreher and Vogt

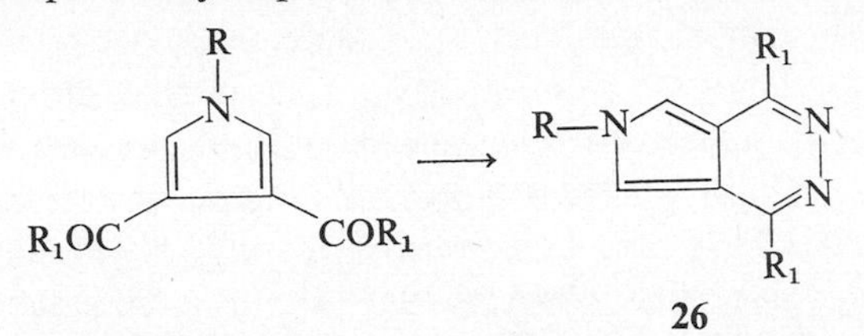

26

(28). In a like manner, 3,4-dicarbethoxypyrroles yielded the corresponding pyrrolo[3,4-*d*]pyridazine-1,4(2*H*,3*H*)diones (**27**) (or enolic form) (14, 18, 19).

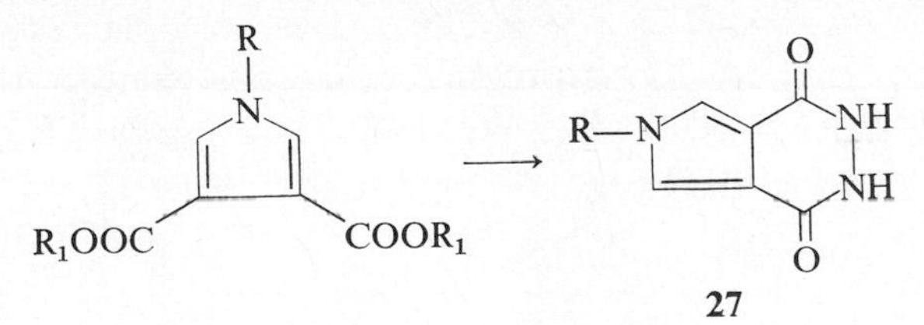

27

Jones (14), who studied the reactions between many heterocyclic 1,2-dicarboxylic esters and hydrazine in detail, observed that unsubstituted dialkyl pyrrole-3,4-dicarboxylates when treated with hydrazine afforded only dihydrazides. However, if in addition to both carbalkoxy groups methyl or phenyl substituents were attached on the ring, the bicyclic compounds were formed. It was also possible to convert the dihydrazides to the corresponding pyrrolo[3,4-*d*]pyridazines either by heating the dihydrazide with excess of hydrazine or by treatment of the dihydrazide with dilute hydrochloric acid.

It is claimed that the same bicyclic system (**28**) is formed in a reaction

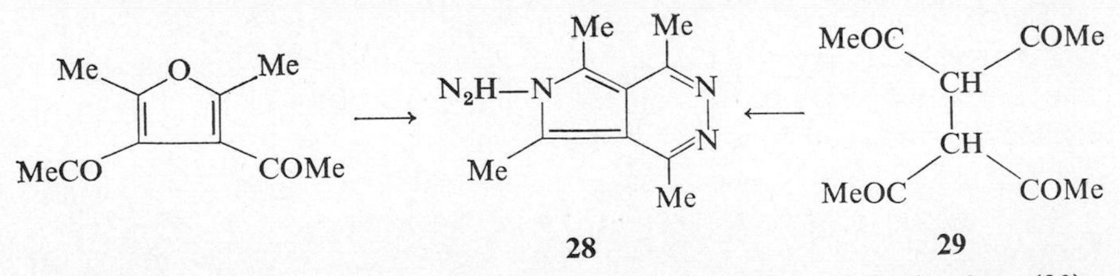

28 **29**

between 3,4-diacetyl-2,5-dimethylfuran and 2 moles of hydrazine (20). Since the corresponding furopyridazine could be obtained with 1 mole of hydrazine, it is most probable that the additional quantity of hydrazine caused hydrazinolysis of the furan ring. The other possible structure (**30**) could not

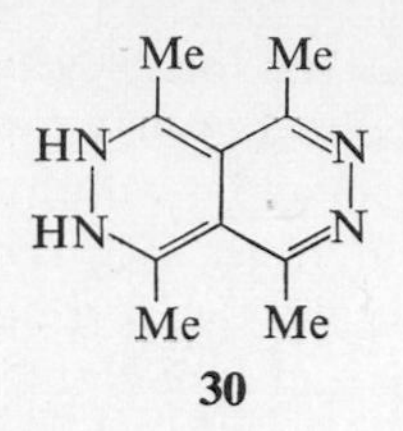

30

be rejected on the basis of uv or ir evidence, but deamination with nitrous acid greatly favors the pyrrolopyridazine structure **28** (20). However, the reaction conducted only toward the exclusive formation of compounds of the type **28**—the reaction using *N*,*N*-dimethylhydrazine instead of hydrazine—failed. On the other hand, the same pyrrolo[3,4-*d*]pyridazine (**28**) could be obtianed from 3,4-diacetylhexane-2,5-dione (**29**) after treatment with hydrazine in hot acetic acid. Recently, it was shown (29) that the pyrrolopyridazine **28** is formed along with the fully aromatic analog of **30** when tetraacetyl ethylene is heated with a methanolic solution of hydrazine hydrate.

In a similar treatment of compound **29** with benzhydrazide two products were obtained, **31** and **32** (20). Furthermore, **31** when treated with hydrazine gave a 40% yield of **32** and a 60% yield of dibenzoylhydrazine.

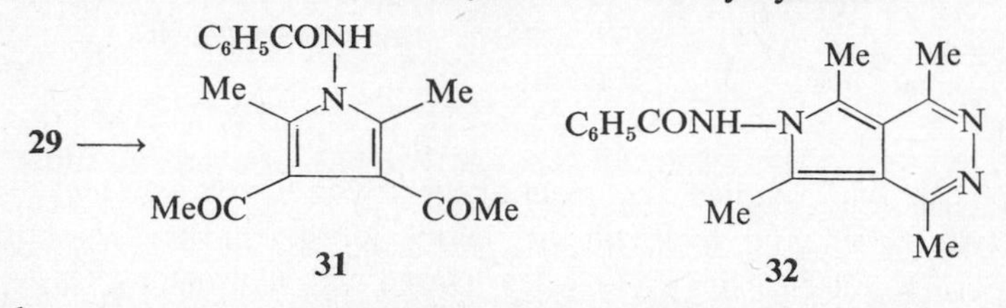

Another type of ring closure is reported by Bilton and Linstead (21). The diamide **33** at its melting point or when heated at 240°/10 mm was converted into the cyclic imide **34**. The five-membered ring of the bicycle **34** is readily opened with cold aqueous ammonia to give back the diamide **33**.

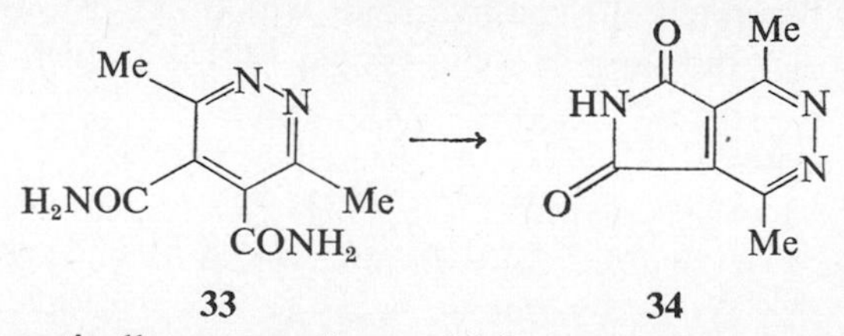

33 **34**

There are practically no reactions which would disclose the reactivity of pyrrolo[3,4-*d*]pyridazines. So far, all known representatives of this system are substituted and the parent compound remains still unknown.

Total and frontier π-electron densities for the parent highly symmetric system have been calculated by the simple HMO method (7) and are tabulated in Table 3B-3.

Studies of chemiluminiscence of some pyrrolo[3,4-*d*]pyridazines are also reported (22).

IV. 7*H*-Pyrrolo[2,3-*c*]pyridazines

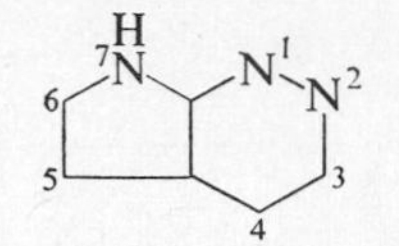

The only representative of this system is reported by Lund and Gruhn (23) to result from the degradation of 8-chloro-6,7-dihydro-3-methyldipyridazino-[2,3-*a*:4,3-*d*]pyrrole (**35**, R = Cl). Compound **35** is formed as a byproduct in the preparation of 3-chloro-6-methylpyridazine from 6-methyl-3(2*H*)-pyridazinone by means of phosphorus oxychloride. Catalytic halogenation of **35** (R = Cl) in the presence of palladized carbon and ammonia afforded the dehalogenated product (**35**, R = H) together with a derivative of 7*H*-pyrrolo[2,3-*c*]pyridazine (**36**) (23). A possible explanation of this trans-

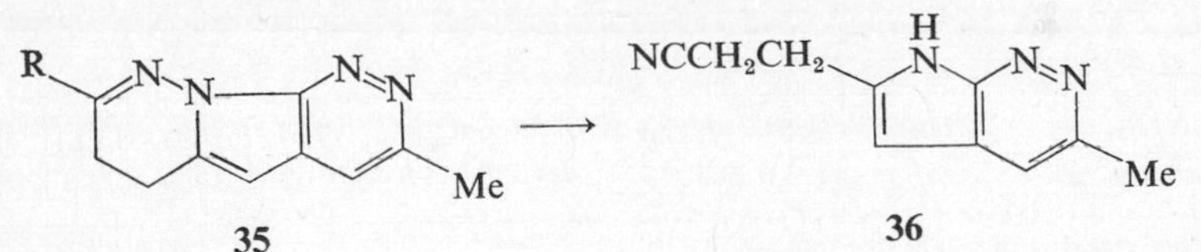

formation is that the N—N bond of compound **35** is cleaved and subsequently hydrogen chloride is eliminated.

The calculated (7) total and frontier π-electron densities for this system are listed in Table 3B-4. On the basis of these values one would expect an electrophilic attack at position 5 and this has indeed been observed. 3-Methyl-6-(β-carbethoxyethyl)-7*H*-pyrrolo[2,3-*c*]pyridazine (**37**, R = H) was readily brominated at position 5 to give **37** (R = Br). Assignment of the position of

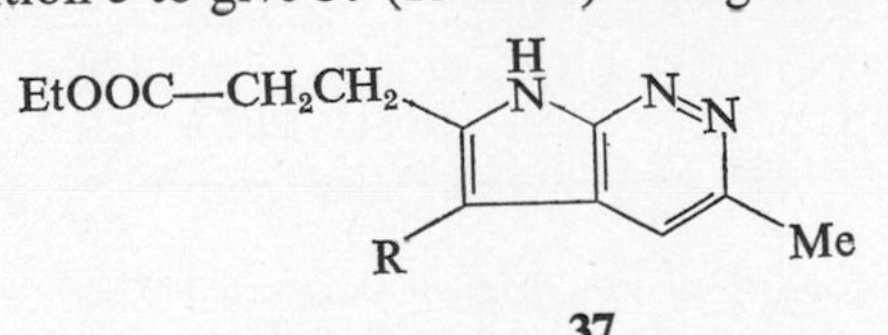

the entering halogen could be made on the basis of nmr spectra correlations. NMR spectra and the polarographic behavior of pyrrolo[2,3-*c*]pyridazines and dipyridazinopyrroles are very similar and uv spectra resemble those of indoles (23).

With compound **37** (R = H) a long-range coupling constant (J = 2–3 cps) between the hydrogen atom at N_7 and that at C_5 was observed. This is similar to the coupling constant between *N*-bonded hydrogen and the hydrogen at C_3 in indole (23).

V. Tables

TABLE 3B-1. Total and Frontier π-Electron Densities of Pyrrolo[1,2-*b*]pyridazine

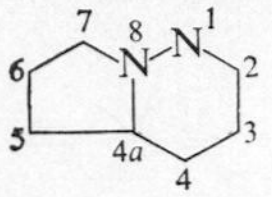

Position	1	2	3	4	4*a*	5	6	7	8
Total	1.1757	0.9570	1.0155	0.9425	1.0260	1.1292	1.0625	1.0994	1.5922
Frontier	0.1734	0.0094	0.1483	0.0675	0.0757	0.2340	0.0049	0.2638	0.0230

TABLE 3B-2. Total and Frontier π-Electron Densities of 1*H*-Pyrrolo[2,3-*d*]pyridazine

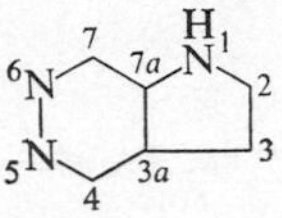

Position	1	2	3	3*a*	4	5	6	7	7*a*
Total	1.6499	0.9914	1.1220	1.0173	0.9403	1.1730	1.1691	0.9547	0.9824
Frontier	0.0695	0.2311	0.3073	0.0081	0.1203	0.0143	0.1439	0.0321	0.0734

TABLE 3B-3. Total and Frontier π-Electron Densities of 6*H*-Pyrrolo[3,4-*d*]pyridazine

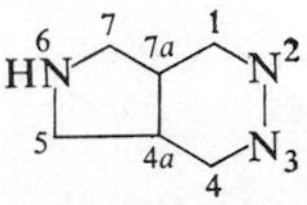

Position	1	2	3	4	4*a*	5	6	7	7*a*
Total	0.9146	1.1921	1.1921	0.9146	1.0496	1.0646	1.5580	1.0646	1.0496
Frontier	0.0612	0.0899	0.0899	0.0612	0.0430	0.3059	0.0000	0.3059	0.0430

TABLE 3B-4. Total and Frontier π-Electron Densities of 7*H*-Pyrrolo[2,3-*c*]pyridazine

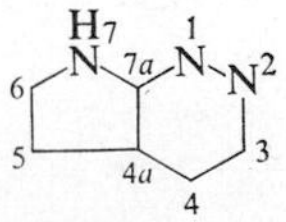

Position	1	2	3	4	4*a*	5	6	7	7*a*
Total	1.1759	1.1433	0.9954	0.9482	1.0112	1.1318	0.9832	1.6663	0.9441
Frontier	0.0680	0.0231	0.0137	0.0243	0.0953	0.0550	0.1099	0.4063	0.2044

TABLE 3B-5. Pyrrolo[1,2-*b*]pyridazines

R	R_1	R_2	R_3	R_4	R_5	MP (°C) or BP (°C/mm)	References
None						93/12	4, 25
						90/18	
		Phenyl				49	4, 25
Methyl		Methyl				69/0.1	
						72°/0.5	4, 25, 26
Methyl		Phenyl				51, 52°	4, 25, 26
Phenyl						81	4, 25
Phenyl		Phenyl				107	4, 25
				COOH		243	1
				COOMe		93–94	1
Methyl				COOH		225.5–226	1
Methyl				COOMe		84–85	1
Methyl				COOEt		67.5–68	1
			COOH	COOH	COOH	223	1
			COOMe	COOMe	COOMe	163–164	2, 3
						160–161	1, 27
			COOMe	COOMe	COOEt	133.5–134.5	3
			CN		CN	204–205	27
Methoxy			CN		CN	233–235	27
Ethoxy			CN		CN	166–168	27
			COOMe	COOMe	CN	141–143	27
Methoxy			COOMe	COOMe	CN	155–156	27
Ethoxy			COOMe	COOMe	CN	170–171	27
Phenoxy			COOMe	COOMe	CN	155–157	27
	Methyl		COOMe	COOMe	CN	164–167	27
Methyl		Methyl	Bromo		Bromo	104–106	26

TABLE 3B-5. (*continued*)

R	R_1	R_2	R_3	R_4	R_5	MP (°C) or BP (°C/mm)	References
Methyl		Phenyl	Bromo		Bromo	133–135°	26
Methyl	Bromo	Methyl	Bromo		Bromo	183	26
Methyl	Bromo	Phenyl	Bromo		Bromo	115	26
Methyl	Bromo	Methyl	Bromo	Bromo	Bromo	140	26
Methyl		Methyl			Nitroso	160	26
Methyl		Methyl			CF_3CO	149–152	26
Methyl		Methyl			*p*-$NO_2C_6H_4$—N=N—	180	26
Methyl		Methyl			C_6H_5—N=N—	178–180°	26
Methyl		Methyl	Nitro		Nitro	137	26
Methyl		*p*-NO_2—C_6H_4—	Nitro		Nitro	92	26
Methyl			COOMe	COOMe	COOMe	164–165 166–167	1, 3 2
Methyl		Methyl			—C(COOMe)=CH(COOMe)	96	4, 25, 26
Methyl		Phenyl			—C(COOMe)=CH(COOMe)	97	4, 25
Phenyl		Phenyl			—C(COOMe)=CH(COOMe)	133	4, 25
					—C(COOMe)=CH(COOMe)	79	25

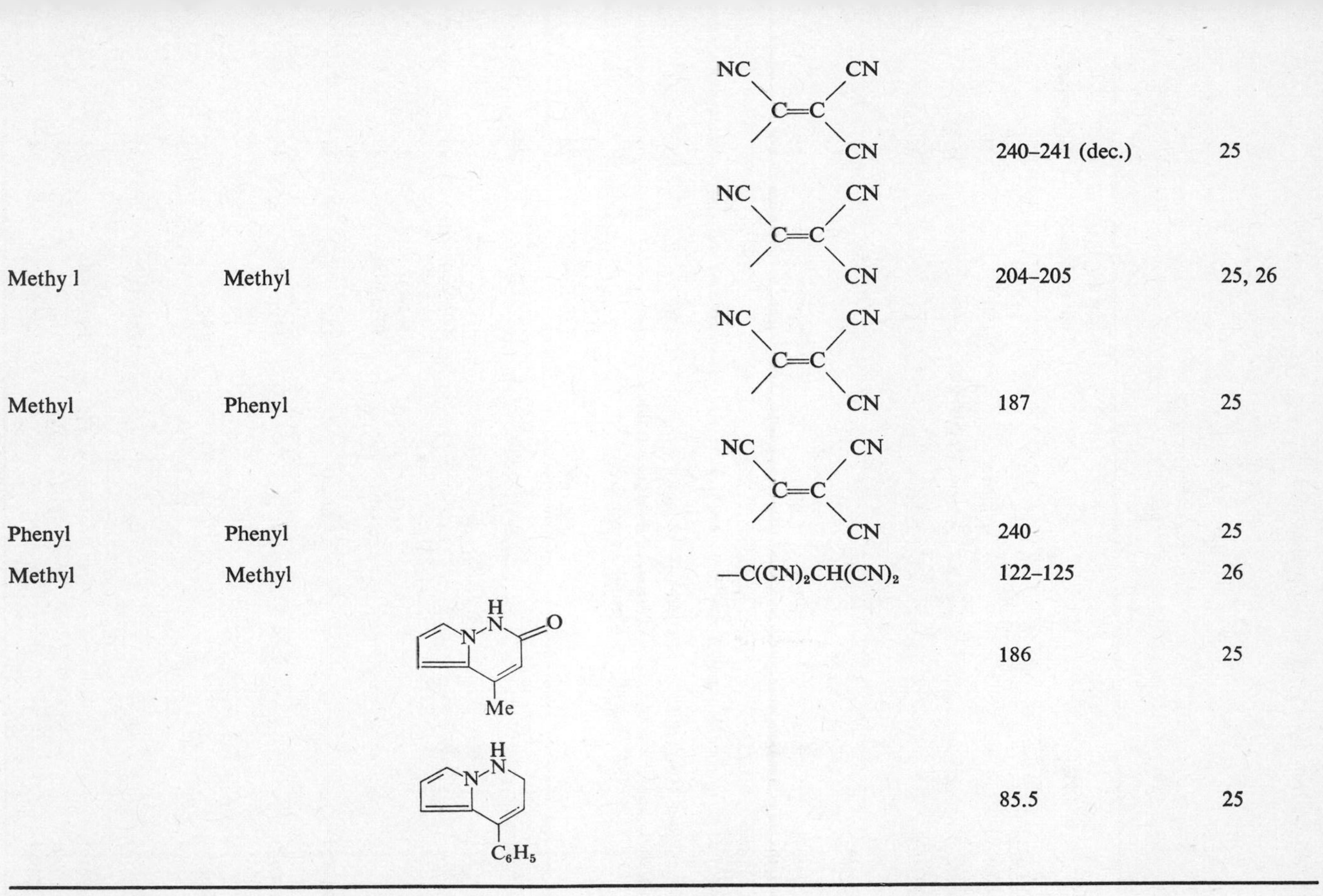

			$-C(CN)=C(CN)_2$	240–241 (dec.)	25
Methy l	Methyl		$-C(CN)=C(CN)_2$	204–205	25, 26
Methyl	Phenyl		$-C(CN)=C(CN)_2$	187	25
Phenyl	Phenyl		$-C(CN)=C(CN)_2$	240	25
Methyl	Methyl		$-C(CN)_2CH(CN)_2$	122–125	26
				186	25
				85.5	25

TABLE 3B-6. 1*H*-Pyrrolo[2,3-*d*]pyridazines

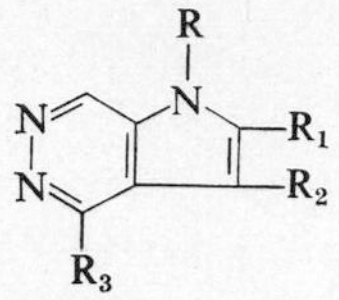

R	R_1	R_2	R_3	MP (°C) or BP (°C/mm)	References
Methyl				111	10
Methyl			Chloro	170	10
Methyl			$C_6H_5CH_2NH$—	218	10
Methyl			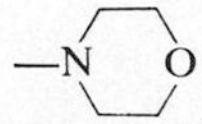 —N O	113	10
Methyl			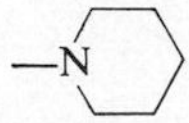 —N	91	10
Methyl	Chloro	Chloro	Chloro	218	10
Methyl	Chloro	Chloro	$C_6H_5CH_2NH$—	107	10
Methyl	Chloro	Chloro	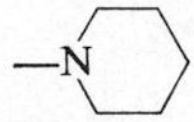 —N	160	10
Methyl	Chloro	Chloro	Phenyl	176	8
Methyl	Chloro	Chloro	*p*-MeO—C_6H_4	142	8
	$CONHNH_2$	Methyl	Methyl	—	9
				169–172	24
			Chloro	145 (dec.)	24
			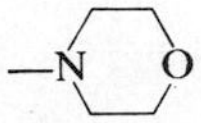 —N O	273—283	24
			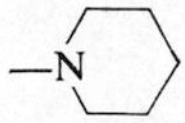 —N	229–234	24
			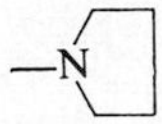 —N	307–309	24
Benzyl			Chloro	122	24
Benzyl			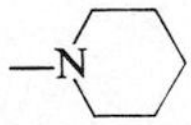 —N	119.5	24
Benzyl			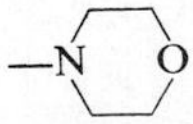 —N O	140	24
Benzyl			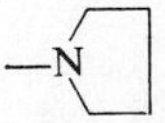 —N	160–163	24
Benzyl			—NH—C_4H_9—*n*	143–144	24
Benzyl			—$NHCH_2Ph$	166	24
Benzyl				118.5	24

TABLE 3B-7. 1*H*-Pyrrolo[2,3-*d*]pyridazin-4(5*H*)ones

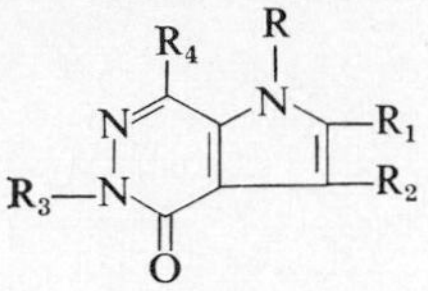

R	R_1	R_2	R_3	R_4	MP (°C) or BP (°C/mm)	References
None					347	10, 24
	Chloro	Chloro			>350	10
	COOEt	Methyl			318 (dec.)	11
	Methyl			Methyl	>300	12
	Methyl		Phenyl	Methyl	324	12
Methyl					239	10, 24
Methyl	Chloro	Chloro			318	10
Benzyl					233	24
$Me_2NCH_2CH_2—$					142–143	24
$Me_2N(CH_2)_3—$					MeI 255	24
Methyl			$—CH_2N$  (piperidine ring)		144–145	24
Benzyl			$—CH_2N$  (piperidine ring)		187	24
$Me_2NCH_2CH_2—$			$—CH_2N$  (piperidine ring)		135–136	24
Methyl			$Me_2NCH_2CH_2—$		97	24
Benzyl			$Me_2NCH_2CH_2—$		121	24

TABLE 3B-8. 1*H*-pyrrolo[2,3-*d*]pyridazine-4(5*H*)thiones

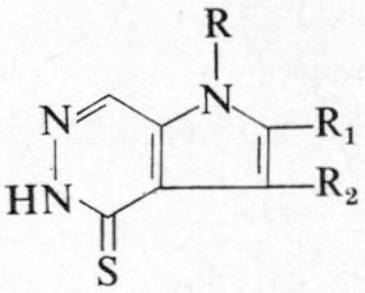

R	R_1	R_2	MP (°C) or BP (°C/mm)	References
None			277	24
Methyl			282	10
Methyl	Chloro	Chloro	360[a]	10
Benzyl			218	24

[a] Chars without melting at 360° C.

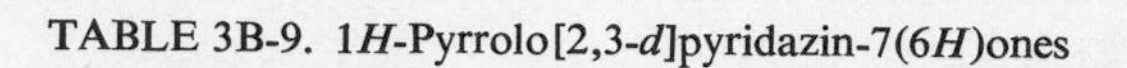
TABLE 3B-9. 1*H*-Pyrrolo[2,3-*d*]pyridazin-7(6*H*)ones

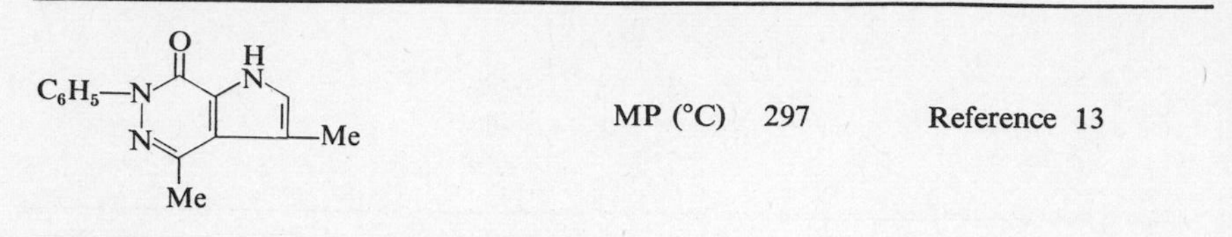

MP (°C) 297 Reference 13

TABLE 3B-10. 1*H*-Pyrrolo[2,3-*d*]pyridazine-4,7(5*H*,6*H*)-diones

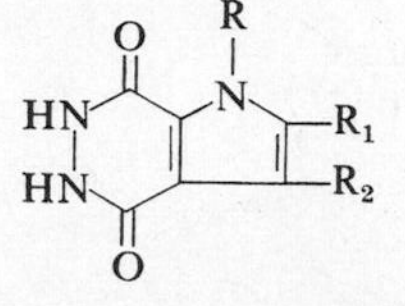

R	R_1	R_2	MP (°C) or BP (°C/mm)	References
	Methyl		355 (dec.)	14
		Methyl	>360	15
Phenyl	Methyl		355–357	14

TABLE 3B-11. 6*H*-Pyrrolo[3,4-*d*]pyridazines

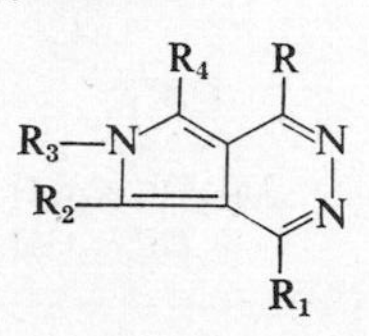

R	R_1	R_2	R_3	R_4	MP (°C) or BP (°C/mm)	References
		Methyl	Methyl	Methyl	184–190	28
		Methyl	Phenyl	Methyl	288	17
Methyl	Methyl	Methyl		Methyl	>300 (dec.)	20
Methyl	Methyl	Methyl	Amino	Methyl	294–295	20
					Picrate, 189.5–191	20
Methyl	Methyl	Methyl	C_6H_5CONH	Methyl	303.5–305	20
Methyl	Methyl	Methyl	Phenyl	Methyl	318	16
Ethyl	Ethyl	Methyl	Phenyl	Methyl	190	16
Phenyl	Phenyl	Methyl	Phenyl	Methyl	294	16
p-$MeOC_6H_4$	*p*-$MeOC_6H_4$	Methyl	Phenyl	Methyl	295	16
Methyl	Methyl	Methyl	Phenyl	Phenyl	239	16
Phenyl	Phenyl	Methyl	Phenyl	Phenyl	277	16
p-$MeOC_6H_4$	*p*-$MeOC_6H_4$	Methyl	Phenyl	Phenyl	301	17

TABLE 3B-12. 6*H*-Pyrrolo[3,4-*d*]pyridazine-1,4(2*H*,3*H*)diones

R	R_1	R_2	MP (°C) or BP (°C/mm)	References
None			>310	14
	Methyl		339–340	14
Methyl	*p*-$AcNHC_6H_4$-	Methyl	>300	18
Phenyl	Amino	Phenyl	>320	22
Methyl		Methyl	359 (dec.)	19
Phenyl		Phenyl	324	19

TABLE 3B-13. 6*H*-Pyrrolo[3,4-*d*]pyridazine-5,7-diones

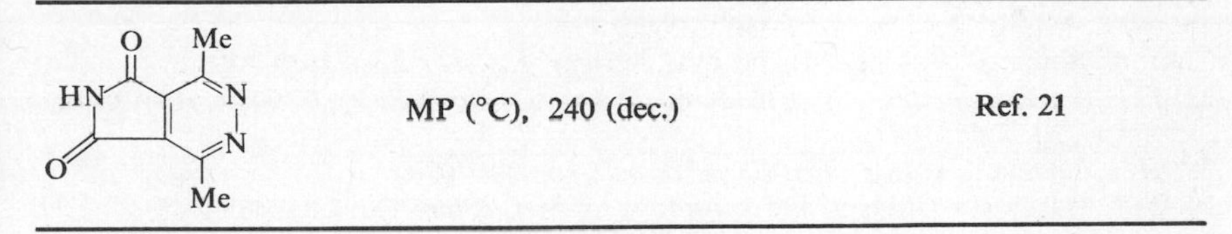

MP (°C), 240 (dec.) Ref. 21

TABLE 3B-14. 7*H*-Pyrrolo[2,3-*c*]pyridazines

R	R_1	MP (°C) or BP (°C/mm)	References
	CN	186	23
	COOEt	157	23
Bromo	COOEt	202	23

References

1. R. L. Letsinger and R. Lasco, *J. Org. Chem.*, **21,** 764 (1956).
2. R. M. Acheson and M. W. Foxton, *J. Chem. Soc.* (*C*), **1966,** 2218.
3. D. G. Farnum, R. J. Alaimo, and J. M. Dunston, *J. Org. Chem.*, **32,** 1130 (1967).
4. W. Flitsch and U. Kraemer, *Tetrahedron Lett.*, **1968,** 1479.
5. A. Angeli, *Ber.*, **23,** 1793 (1890).

6. A. Angeli, *Ber.*, **23,** 2154 (1890).
7. Unpublished data from this laboratory. Parameters for LCAO Calculations were taken from A. Streitwieser, *Molecular Orbital Theory for Organic Chemists*, Wiley, New York, 1961, p. 135.
8. E. Bisagni, J. P. Marquet, J. André-Louisfert, A. Cheutin, and F. Feinte, *Bull. Soc. Chim. France*, **1967,** 2796.
9. H. Fischer and E. Adler, *Z. physiol. Chem. Hoppe-Seyler's*, **210,** 139 (1932); *Chem. Abstr.*, **26,** 5961 (1932).
10. J. P. Marquet, J. André-Louisfert, and E. Bisagni, *Compt. rend.*, **C265,** 1271 (1967).
11. H. Fischer, A. Kirstahler, and B. v. Zychlinski, *Ann.*, **500,** 1, (1932); *Chem. Abstr.*, **27,** 987 (1933).
12. H. Fischer, H. Beyer, and E. Zaucker, *Ann.*, **486,** 55 (1931).
13. H. Fischer, E. Sturm, and H. Friedrich, *Ann.*, **461,** 244 (1928).
14. R. G. Jones, *J. Am. Chem. Soc.*, **78,** 159 (1956).
15. H. Fischer and O. Wiedemann, *Z. physiol. Chem. Hoppe-Seyler's*, **155,** 52 (1926).
16. R. Rips and N. P. Buu-Hoi, *J. Org. Chem.*, **24,** 551 (1959).
17. R. Rips and N. P. Buu-Hoi, *J. Org. Chem.*, **24,** 372 (1959).
18. C. Bülow and W. Dick, *Ber.*, **57B,** 1281 (1924).
19. R. Seka and H. Preissecker, *Monatsh. Chem.*, **57,** 71 (1931); *Chem. Abstr.*, **25,** 1826 (1931).
20. W. L. Mosby, *J. Chem. Soc.*, **1957,** 3997.
21. J. A. Bilton, R. P. Linstead, and J. M. Wright, *J. Chem. Soc.*, **1937,** 922.
22. E. S. Vasserman and G. P. Miklukhin, *J. Gen. Chem.* (*USSR*), **9,** 606 (1939); *Chem. Abstr.*, **33,** 7666 (1939).
23. H. Lund and S. Gruhn, *Acta Chem. Scand.*, **20,** 2637 (1966).
24. J. P. Marquet, E. Bisagni, and J. André-Louisfert, *Chim. Ther.*, **3,** 348 (1968).
25. W. Flitsch and U. Kraemer, *Ann.*, **735,** 35 (1970).
26. M. Zupan, B. Stanovnik, and M. Tišler, *J. Heterocyclic Chem.*, **8,** 1 (1971).
27. T. Sasaki, K. Kanematsu, Y. Yakimoto, and S. Ochiai, *J. Org. Chem.*, **36,** 813 (1971).
28. R. Kreher and G. Vogt, *Angew. Chem.*, **82,** 958 (1970).
29. G. Adembri, F. DeSio, R. Nesi, and M. Scotton, *J. Chem. Soc.* (*C*), **1970,** 1536.

Part C. Pyrazolopyridazines

I. 1*H*-Pyrazolo[1,2-*a*]pyridazines

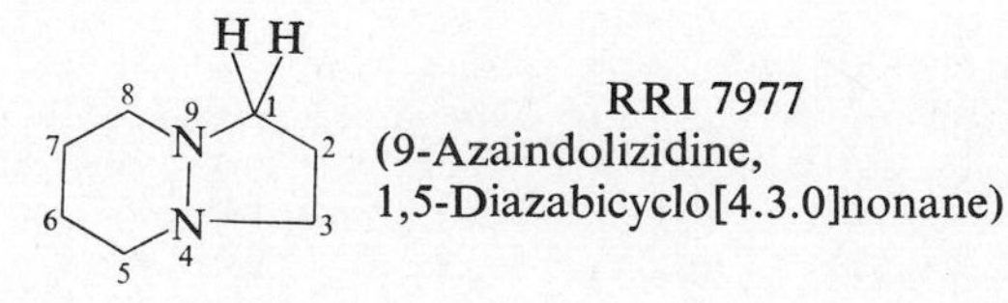

RRI 7977
(9-Azaindolizidine,
1,5-Diazabicyclo[4.3.0]nonane)

All known compounds with the pyrazolo[1,2-*a*]pyridazine skeleton are partially or fully hydrogenated or contain oxo groups.

The most widely used method of synthesis uses as starting material reduced pyridazines with an —NH—NH— structural element which serves for the elaboration of the condensed five-membered ring. For the ring closure various open-chain compounds such as derivatives of dicarboxylic, β-keto-, and α,β-unsaturated acids, 1,3-dihaloalkanes or epichlorohydrins, and α-acyl lactones were used.

According to Büchi, Vetsch, and Fabiani (1), malonylchlorides condense with 1,2,3,6-tetrahydropyridazine in the presence of triethylamine in cold benzene solution to form compounds of the type **1**. In a similar way from

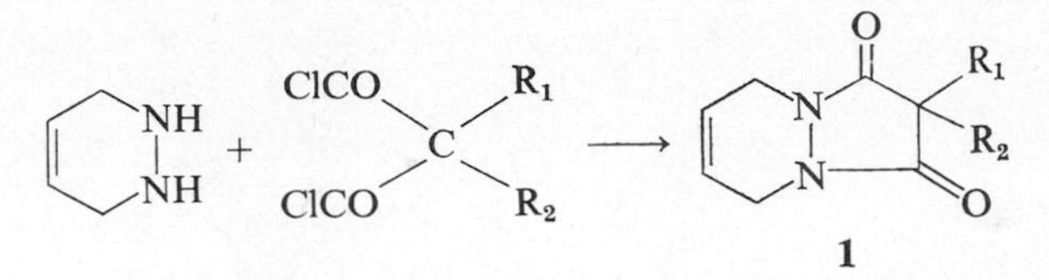

maleic hydrazide and hexahydro-3,6-pyridazinedione products of the structural types **2** and **3** could be obtained (2, 3, 29). The reaction is best carried

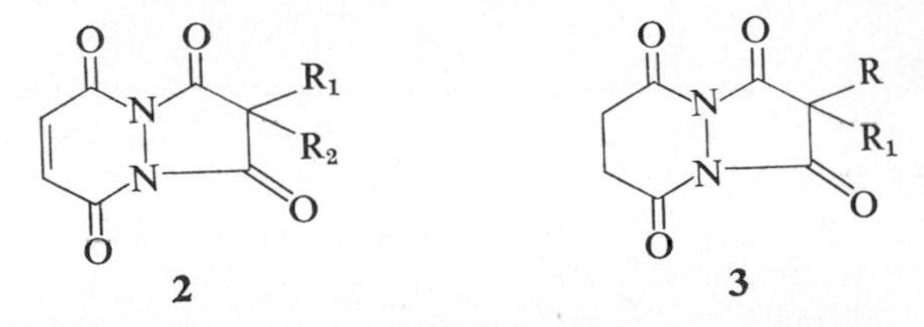

out using stoichiometric amounts of reactants heating the reaction mixture in nitrobenzene until hydrogen chloride is evolved (60–150°) (2).

Molnar and Wagner-Jauregg (4) have prepared similar bicyclic compounds from reduced pyridazines and malonylchlorides, but the position of the double bond in the pyridazine part of the molecule has not been established.

Carpino and co-workers (5, 6) used for cyclization methyl α-phenylbenzoylacetate and obtained from tetrahydropyridazine **4** which was identical with the product **13** from the diene synthesis. Similarly, α,β-unsaturated aldehydes react with maleic hydrazide or its dihydro analog to form the related 1-hydroxy pyrazolo[1,2-*a*]pyridazine-5,8-diones (30).

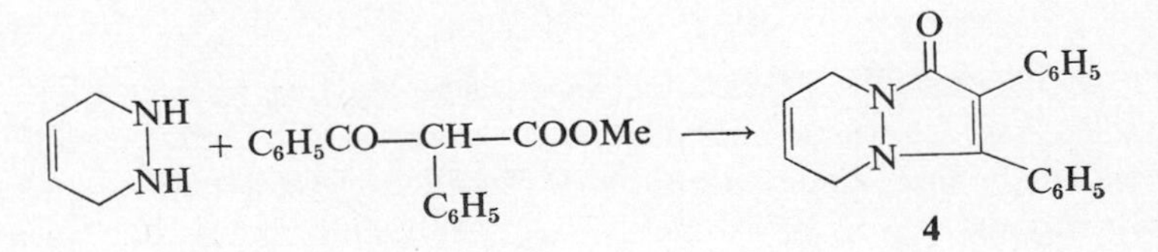

Le Berre et al. (7, 28) studied the reaction between maleic hydrazide and α,β-unsaturated acid chlorides in nitrobenzene at about 160°. Cyclic products (**5**) were obtained with acrylic, methacrylic, crotonic, or cinnamic acid chlorides. On the other hand, **5** also could be obtained from **6** (prepared by the addition of acrylic acid to maleic hydrazide) after treatment with thionyl chloride.

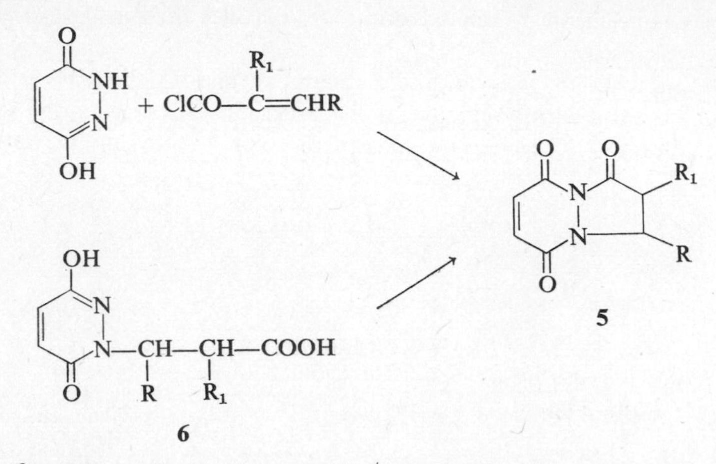

The formation of perhydropyrazolo[1,2-*a*]pyridazine is reported to result from treatment of hexahydropyridazine with 1,3-dibromopropane in the presence of sodium hydroxide solution (8). Similarly, the use of epichlorohydrin is reported to give the corresponding 2-hydroxy derivative (9), but the product was characterized only by its nitrogen analysis. Ring closure of *N*-(hydroxypropyl)pyridazines has been accomplished with thionyl chloride (35).

Wamhoff and Korte (10) condensed hexahydropyridazine with α-acyl lactones, which under the influence of basic catalysts such as sodium methoxide undergo ring opening. The pyrazolo[1,2-*a*]pyridazine structure of the products (**7**) has been confirmed from nmr and ir spectroscopical data.

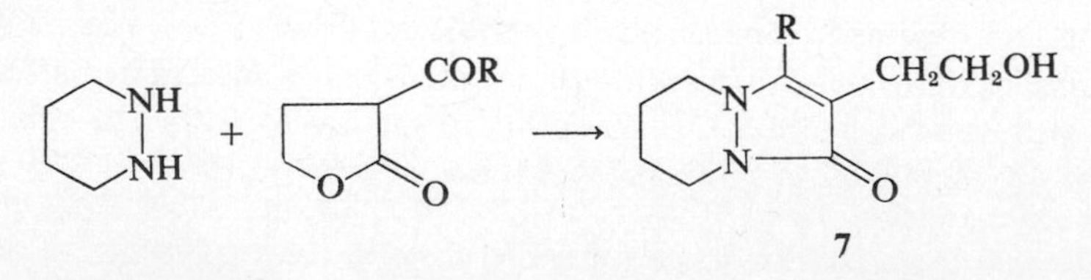

The other possibility of forming the bicycle—starting with an appropriate pyrazole—has also been elaborated in few cases. Büchi, Vetsch, and Fabiani (1) treated the monopotassium salts of 3,5-dioxopyrazolidines (**8**) with 1,4-dibromobutane in hot *N*,*N*-dimethylformamide and obtained **9**, sometimes accompanied with 1,4-bis(dioxopyrazolidinyl)butane derivatives. In a

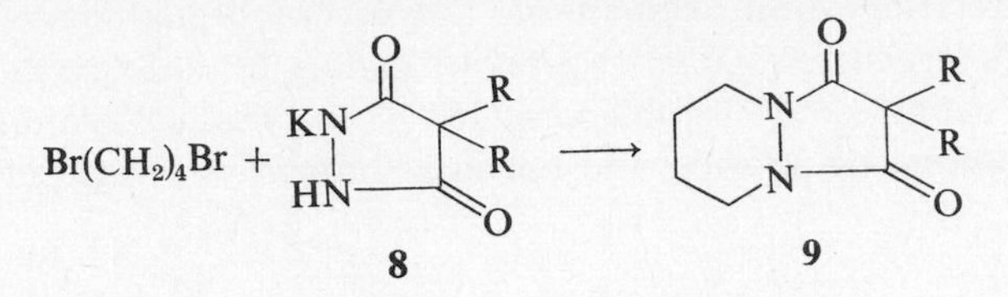

similar reaction with pyrrazolinones, in addition to the expected bicyclic products, pyrazolo[3,2-*b*]oxazepines also were formed (36).

When instead of 1,4-dihaloalkanes, succinylchlorides have been utilized (2, 3), compounds of the type **10** were obtained.

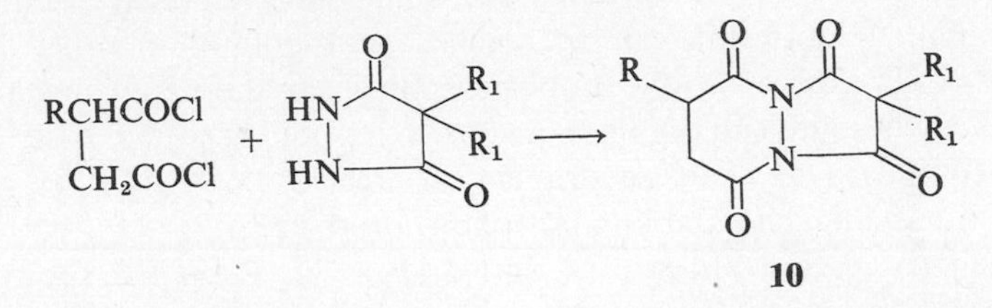

In a like manner, pyrazolidine when heated with dimethyl succinate (160°, 5 hr) afforded 5,8-dioxoperhydropyrazolo[1,2-*a*]pyridazine (**15**) in moderate yield (11).

Pyrazol-3-ones (**12**), although unstable to isolation, when formed *in situ* act as potent dienophiles and Diels-Alder adducts with dienes are readily formed. The existence of *cis*-azodienophiles has been demonstrated from treatment of 3,4-diphenyl-4-chloro-2-pyrazolin-5-one (**11**) with triethylamine and the intermediate (**12**) trapped with butadiene to form the addition product **13** (5, 6).

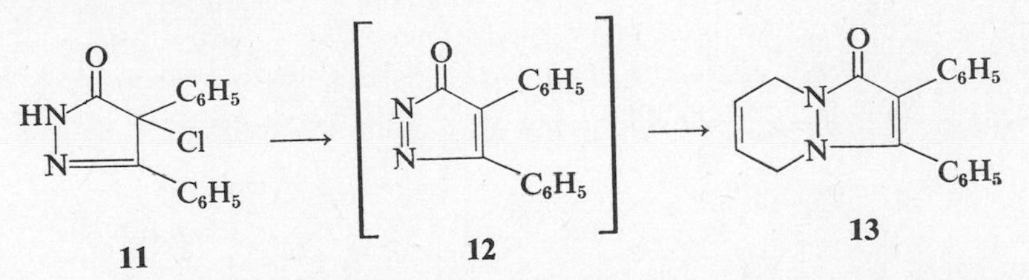

Gillis and Weinkam (12) generated pyrazol-3-ones through oxidation of the appropriate pyrazolin-3-ones with lead tetraacetate at −10° in methylene chloride and these intermediates were also trapped in the presence of dienes. Among these, 1,4-diphenyl- and 2,3-dimethylbutadiene yielded pyrazolo-[1,2-*a*]pyridazin-1-ones. Structures of these bicyclic compounds are in accord with the nmr chemical shifts. An extension of this reaction has been reported recently (31, 37, 38).

Pyrazolo[1,2-*a*]pyridazinium cation (**14**), which would represent an interesting 10 π-electron cationic system, has thus far not been prepared and only a substituted derivative (39) or its dibenzo analog, benzo[*c*]pyrazolo[1,2-*a*]-cinnolinium cation, could be obtained (13). Theoretical calculations provide some data concerning physical and chemical properties of the parent cation (32).

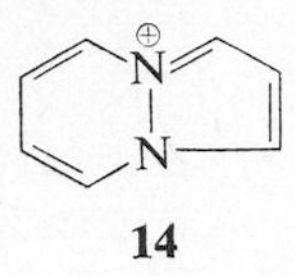

14

There are transformations from which some conclusions about the structure-reactivity relationship can be reached. Compounds of types **2** and **3** ($R = R_1 = H$) possess acidic properties, they are thermally unstable and are very weakly soluble in organic solvents. With boiling water, **2** ($R = R_1 = H$) is decomposed to maleic hydrazide, whereas **3** ($R = R_1 = H$) yields a mixture of succinic acid and an amorphous product (2).

Monoalkylated derivatives of **2** and **3** ($R = H$, $R_1 = Et$) show similar properties upon hydrolytic treatment. Whereas the monoethyl derivative of **3** ($R = H$, $R_1 = Et$) is more stable than the unsubstituted analog and forms a monoacetyl derivative, the unsaturated analog (**2**, $R = H$, $R_1 = Et$) is hygroscopic, thermally unstable, and does not form an acetyl derivative. At position 2 dialkylated derivatives of **2** and **3** are neutral and hydrolyze in boiling water. Compound **2** ($R = R_1 = Et$) afforded quantitatively maleic hydrazide, while **3** ($R = R_1 = Et$) upon cleavage of the six-membered ring yielded diethylmalonic acid hydrazide (2).

5,8-Dioxoperhydropyrazolo[1,2-*a*]pyridazine (**15**) has been reduced with lithium aluminum hydride to perhydropyrazolo[1,2-*a*]pyridazine (**16**). Further hydrogenation in the presence of Raney nickel caused the cleavage of the *N—N* bond and the cyclic diamine (**17**) was produced and identified as the bis-tosylate (11). Reduction of 1,3-dioxo derivatives with lithium alumi-

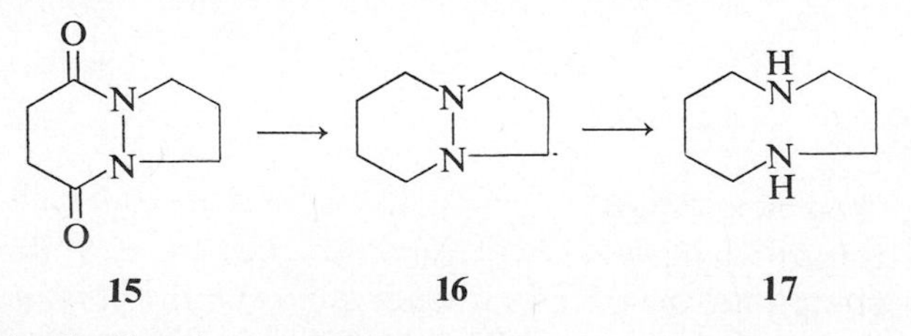

15 **16** **17**

num hydride also has been performed (31).

At position 2 unsubstituted compounds of types **2** and **3** are claimed to give with aqueous sodium nitrite solution the corresponding hydroxyimino derivatives **18** and **19**, respectively (3).

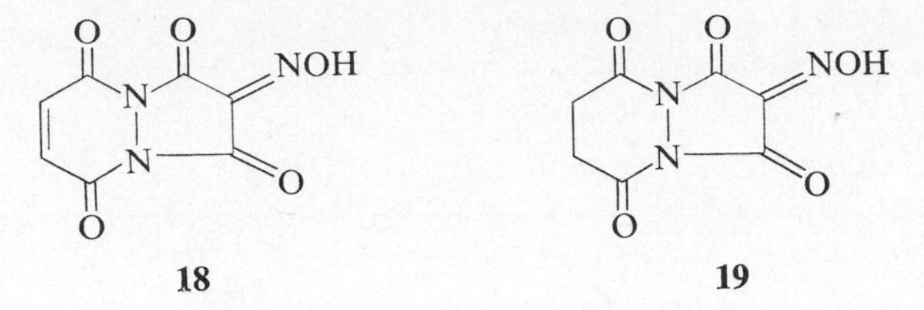

Pyrazolo[1,2-*a*]pyridazine derivatives were tested for their antipyretic properties (1) and some are claimed to be useful pesticides (3).

II. Pyrazolo[3,4-*d*]pyridazines

Based upon the position of the extra hydrogen atom, several isomeric systems are possible. Representatives of the following are known:

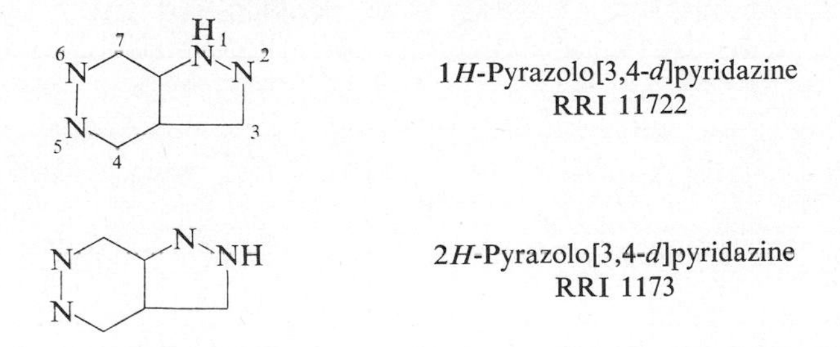

The most important and unambiguous synthetic approach to obtaining pyrazolo[3,4-*d*]pyridazines utilizes pyrazoles or pyrazolones with appropriate *ortho* functional groups, capable of forming the pyridazine ring, as starting material.

Ito (14) described the synthesis of **21** from the diformylpyrazolone **20**

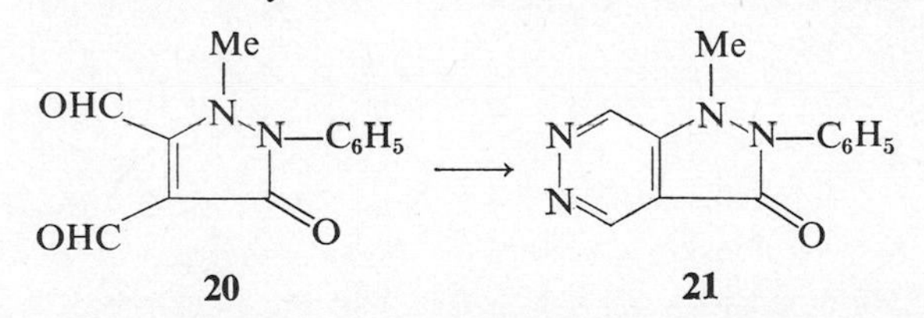

and 60% hydrazine hydrate (3 hr), whereas pyrazoles with *ortho* acyl and carbalkoxy groups yielded the corresponding 1*H*-pyrazolo[3,4-*d*]pyridazines (**22**) (15). In a like manner derivatives of 2*H*-pyrazolo[3,4-*d*]pyridazine (**24**) were synthesized (16, 17).

However, Rossi et al. (17) reported that in a particular case heating with a solution of hydrazine in alcohol (0.5 hr) caused the elimination of the

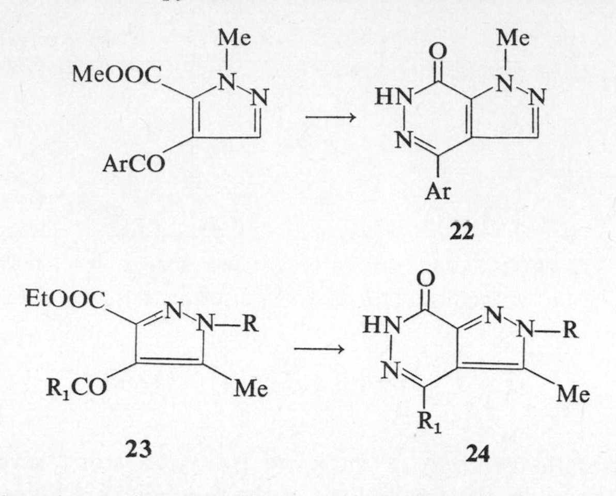

2,4-dinitrophenyl substituent (**23**, R = 2,4-di-NO_2-C_6H_3-) and the 2-unsubstituted product was isolated (**24,** R = H) together with 2,4- dinitrophenylhydrazine.

Moreover, it has been reported (17) that diethyl 1-(2′,4′-dinitrophenyl)-5-methyl-3,4-pyrazoledicarboxylate did not form the corresponding pyrazolopyridazine even in the presence of excess hydrazine, but according to Jones (18) diethyl 1-phenyl-4,5-pyrazoledicarboxylate afforded the bicyclic product (**25**, or enolic form) after heating with hydrazine in fairly good yield. Similarly,

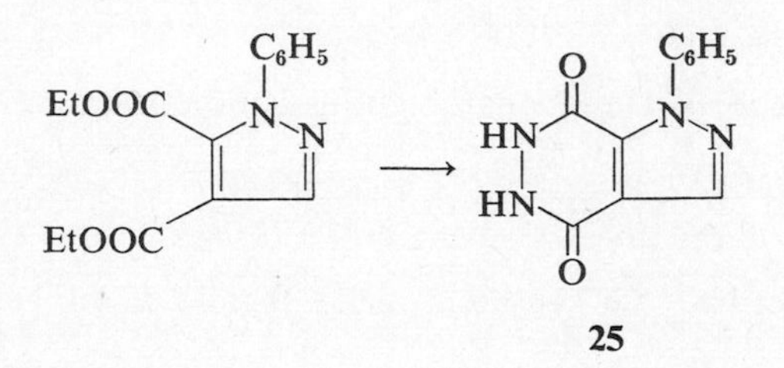

dimethyl 1-phenyl-3,4-pyrazoledicarboxylate has been transformed into the 2-phenyl analog of **25** (33).

The use of 3,4-diacylpyrazoles or esters of pyrazole-3,4-dicarboxylic acids is reported for the synthesis of simple bicyclic compounds (40) or for the synthesis of *C*-nucleosides (41) and here also the ester-amide could be transformed into the corresponding pyrazolo[3,4-*d*]pyridazine-3-glucoside (41). The 1-unsubstituted analogs were prepared similarly from pyrazoledicarboxylic acid dihydrazides after heating with excess hydrazine hydrate (18) or in dilute hydrochloric acid (34, 42).

A 1*H*-pyrazolo[3,4-*d*]pyridazine derivative (**27**) has been reported to result from treatment of 4,7-bis(methylthio)-3-phenylisoxazolo[4,5-*d*]pyridazine

(**26**) with 95% hydrazine in 2-propanol in an attempt to obtain the corresponding 4,7-dihydrazino derivative from **26** (19). Similarly, dimethyl 3-phenyl-4,5-

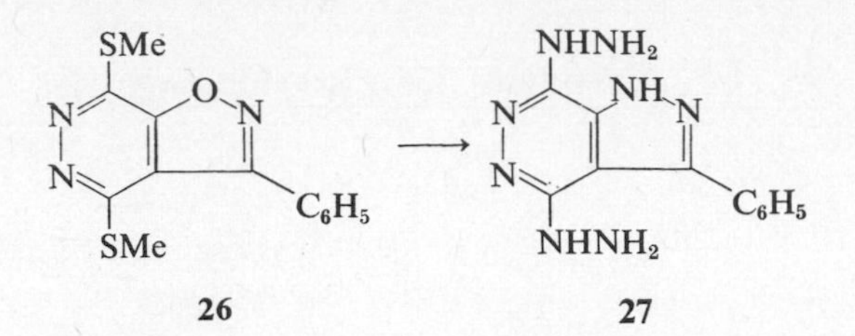

isoxazoledicarboxylate (**28**) when heated with 2 moles of 85% hydrazine hydrate (45 min) afforded a mixture of the expected isoxazolo[4,5-*d*]pyridazine (**29**, 50%) and pyrazolo[4,5-*d*]pyridazine (**30**, 40%) (19). Both **27** and **30** resulted from hydrazinolysis of the isoxazolo ring.

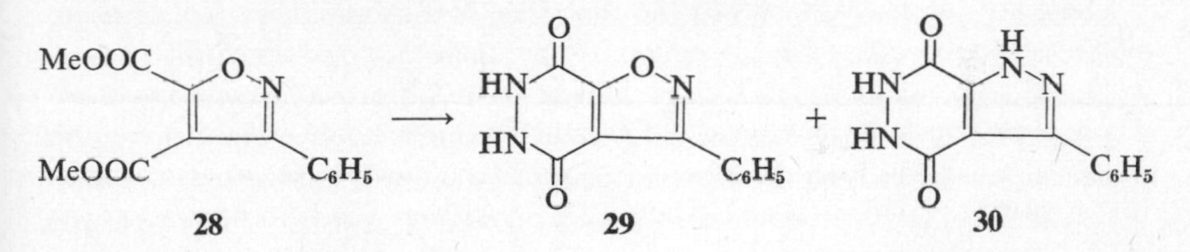

A synthetic approach from pyridazines as starting material has been also observed. Thus 2-methyl-6-phenylpyridazin-3(2*H*)one reacted in a 1,3-dipolar addition reaction with 2-diazopropane to afford a pyrazolo[3,4-*d*]-pyridazinone derivative, which, however, was found to be unstable and underwent several reactions (thermally, photochemically) in which pyridazinones, cyclopropanopyridazines, or diazepinones were formed (43).

There are two reports in the early literature describing the formation of two pyrazolo[3,4-*d*]pyridazine derivatives by some obscure reactions. The products to which structures **31** (20) and **32** (21) have been assigned should have their structures reexamined.

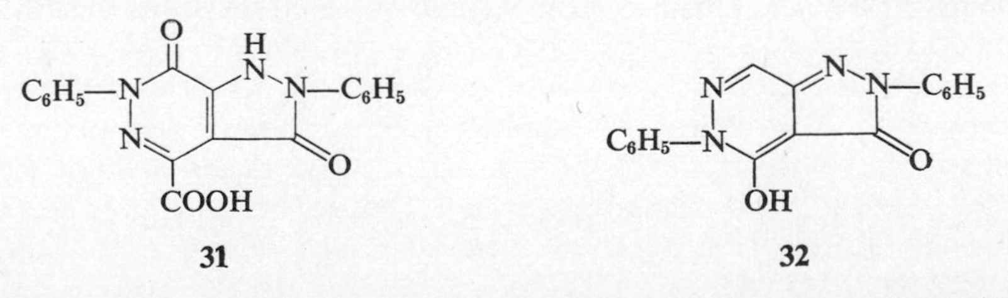

The parent system has not been synthesized to date and reactions which would describe the reactivity or tautomerism of the synthetic products are lacking.

The calculated total and frontier π-electron densities (22) for 1*H*-pyrazolo-[3,4-*d*]pyridazine are listed in Table 3C-1.

III. Pyrazolo[3,4-*c*]pyridazines

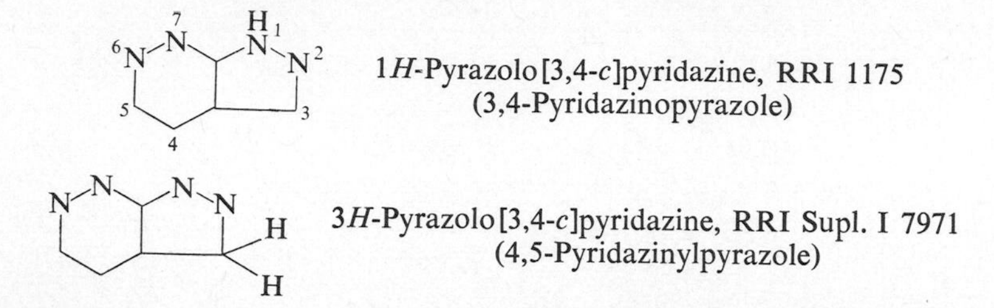

There are not many reports on the chemistry of this bicyclic system. Among synthetic approaches the formation of pyrazolo[3,4-*c*]pyridazines can be achieved starting with pyridazines having *ortho* chloro and cyano groups. These pyridazines (**33**) when treated with hydrazine are converted through the intermediate cyano-hydrazino derivatives into the bicycles **34**. An example of this is given by Dornow and Abele (23). A similar reaction sequence was

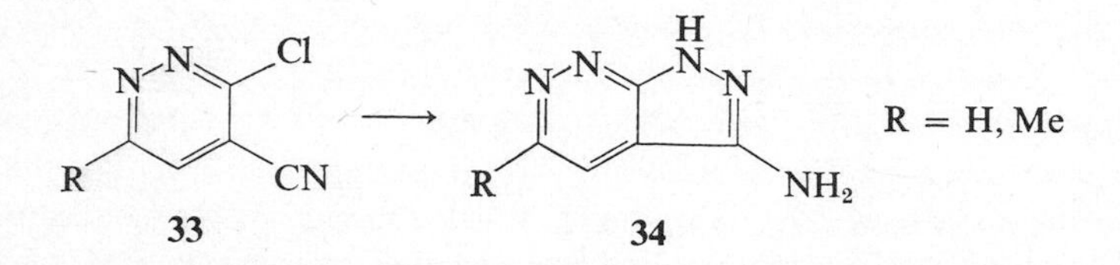

mentioned starting with 3,4-dimethyl-5-cyano-6-chloropyridazine, but no experimental details were given (24, 25).

From pyridazines with *ortho* chloro and carbethoxy substituents the corresponding 3-hydroxy (or tautomeric form) derivatives of 1*H*-pyrazolo[3,4-*c*]-pyridazine (**35**) were obtained (23). Similarly, 3-chloro-4-hydroxymethyl-6-

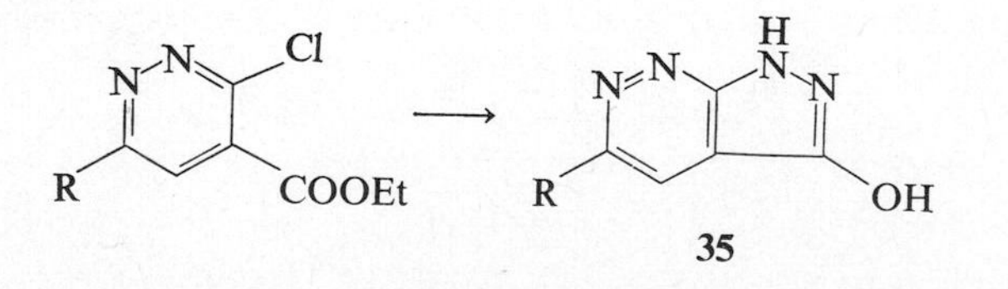

methylpyridazine, when heated with hydrazine for 6 hr, afforded the partially reduced bicycle **36**, which in a subsequent reaction with nitrous acid was transformed to the aromatized compound **37** (23).

The parent compound, 1*H*-pyrazolo[3,4-*c*]pyridazine, was obtained by Dornow and Abele (23) in 50% yield by diazotizing **34** (R = H) with isopropylnitrite and subsequent heating in ethanol.

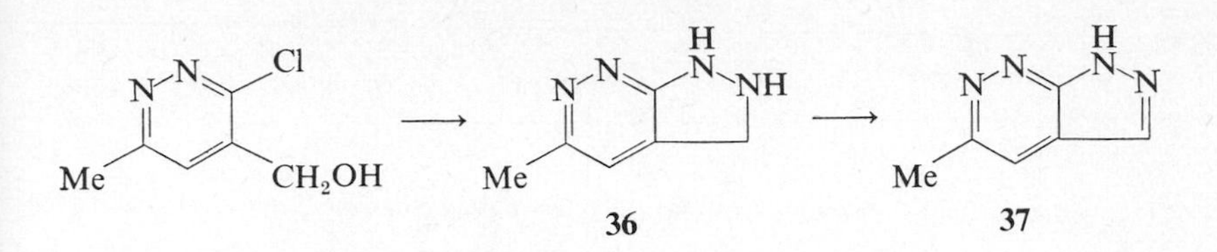

Two papers (26, 27) describe the formation of the bicyclic system from two isomeric 4-acetyl-5-aminopyrazol-3-ones (**38**, R = Me, $R_1 = C_6H_5$, or $R = C_6H_5$, R_1 = Me). These, when treated with a solution of sodium nitrite in the presence of acetic acid and hydrochloric acid, are claimed to be transformed into **39** (R = Me, $R_1 = C_6H_5$ or $R = C_6H_5$, R_1 = Me). Only the scant analytical data for these products suggest the proposed structures.

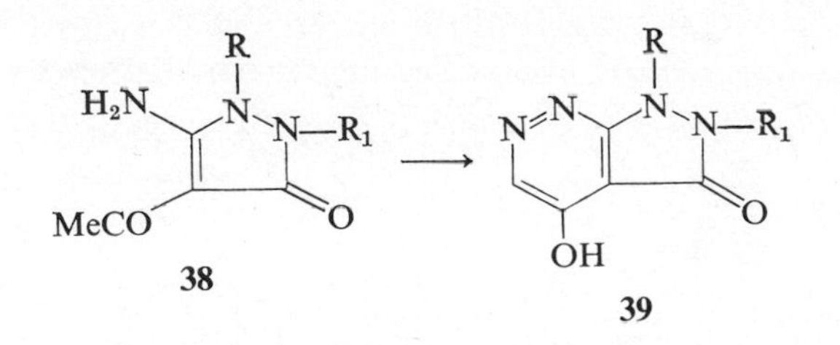

Except the few reactions mentioned above, there are no investigations that would allow a complete picture of the reactivity of pyrazolo[3,4-*c*]pyridazines. Total and frontier π-electron density calculations using the HMO method (22) for 1*H*-pyrazolo[3,4-*c*]pyridazine are listed in Table 3C-2.

IV. Tables

TABLE 3C-1. Total and Frontier Electron Densities of 1*H*-Pyrazolo[3,4-*d*]pyridazine

Position	1	2	3	3*a*	4	5	6	7	7*a*
Total	1.6659	1.1690	1.0168	1.0154	0.9228	1.1718	1.1409	0.9554	0.9420
Frontier	0.1361	0.1661	0.2113	0.0035	0.1603	0.0599	0.1352	0.1156	0.0121

TABLE 3C-2. Total and Frontier Electron Densities of 1*H*-Pyrazolo[3,4-*c*]pyridazine

Position	1	2	3	3*a*	4	5	6	7	7*a*
Total	1.6846	1.1638	1.0189	1.0109	0.9194	0.9935	1.1127	1.1819	0.9144
Frontier	0.1427	0.1634	0.1997	0.0075	0.1881	0.0703	0.0601	0.1613	0.0069

TABLE 3C-3. Pyrazolo[1,2-*a*]pyridazines

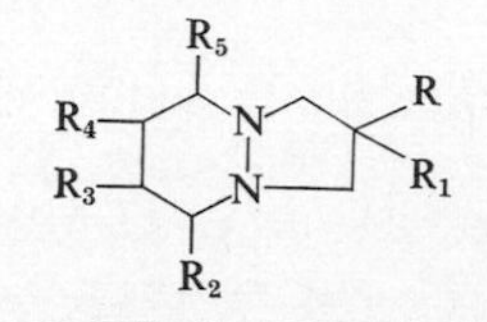

R	R_1	R_2	R_3	R_4	R_5	MP (°C) or BP (°C/mm)	References
None						66/13	8
						MeI, 262–263, 71–73/25	11
						Monopicrate, 200—202 (dec.)	11
Hydroxy						105 (dec.)	9
$OCOCH_3$						81	9
Methyl	Methyl					60/12.5	31
						MeI, 237	31
Methyl	Methyl		Methyl			68/10	31
						MeI, 191	31
Methyl	Methyl	Methyl			Methyl	86–87/14	31
						Monopicrate, 178—179°	31
Methyl	Methyl		Methyl	Methyl		86–87/14	31
						Monopicrate, 168	31
Ethyl	Ethyl					94–96°/10	31
						HCl, 172.5–173	31
Ethyl	Ethyl		Methyl			100–102/9	31
						HCl, 112–113°	31
Ethyl	Ethyl	Methyl			Methyl	114/11	31
						Monopicrate, 128–129	31
Ethyl	Ethyl		Methyl	Methyl		135.5–136.5/17	31
						HCl, 151–151.5	31
						p-MeO—C_6H_4, 150–152/2.0	35
						HCl, 135–138	35
						MeI, 194–197	35

TABLE 3C-4. Pyrazolo[1,2-*a*]pyridazines

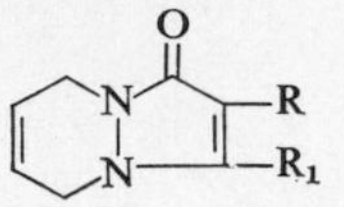

R	R_1	MP (°C) or BP (°C/mm)	References
CH_2CH_2OH	Methyl	86–88	10
CH_2CH_2OH	Cyclopropyl	176–178	10
CH_2CH_2OH	Phenyl	155–156	10
CH_2CH_2OH	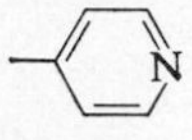 N	152	10
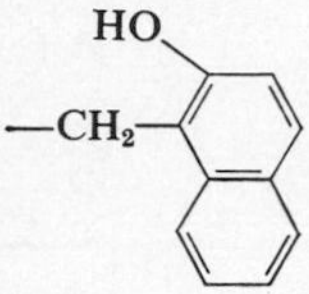 HO —CH_2—	Phenyl	249–250	10
	Methyl	147–148/0.5	36
		Picrate, 213–214	
			36
Methyl	Methyl	155–158/0.5	36
Ethyl	Methyl	147–148/0.5	36
—CH_2—CH=CH_2	Methyl	145–147/0.5	36
Benzyl	Methyl	120–123/0.5	36
Phenyl	Methyl	105–106/0.5	36
		Picrate, 161–162	36
	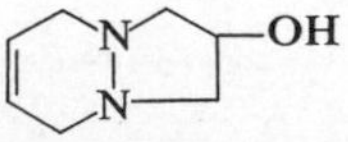 N N —OH	78	9

TABLE 3C-5. Pyrazolo[1,2-*a*]pyridazines

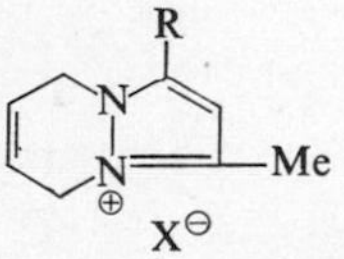

R	X	MP (°C)	References
Chloro	Chloro	227–229	39
SeEt	BF_4	81–82	39
Ethoxy	BF_4	101–102	39
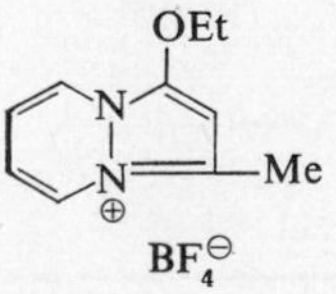 OEt, N, N⊕, —Me, $BF_4^{\ominus}$		129–131	39

TABLE 3C-6

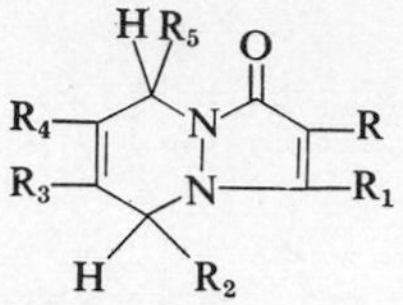

R	R_1	R_2	R_3	R_4	R_5	MP (°C) or BP (°C/mm)	References
	Methyl		Methyl	Methyl		90–92	12
	Methyl	Phenyl			Phenyl	176–177	12
Phenyl	Phenyl					187–189 (dec.)	5, 6
	Methyl					51–53	39
					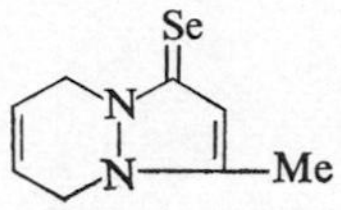	170–172°	39
					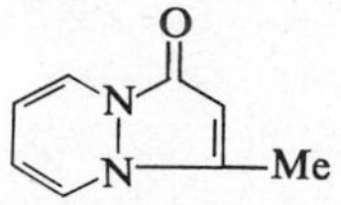	115–117	39

TABLE 3C-7

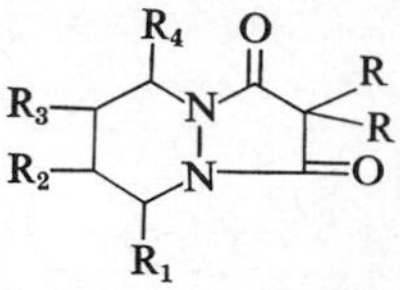

R	R_1	R_2	R_3	R_4	MP (°C) or BP (°C/mm)	References
Methyl					126–127	1
					120–122	31
Ethyl					101–102/0.03	1
					99/2	31
n-Propyl					52–53	1
					108/0.005	1
n-Butyl					124–129/0.02	1
n-Amyl					56–56.6	1
					130/0.005	1
Methyl		Methyl			85.5–86	31
Ethyl		Methyl			111/2	31
Methyl	Methyl			Methyl	122.5/4	31
Methyl		Methyl	Methyl		95–96	31
Ethyl	Methyl			Methyl	126/2	31
Ethyl		Methyl	Methyl		120/2	31

TABLE 3C-8. Pyrazolo[1,2-*a*]pyridazine

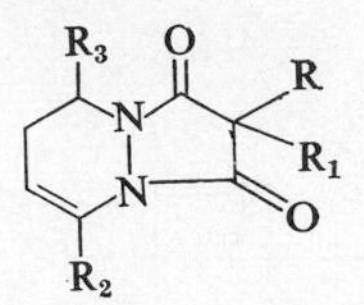

R	R_1	R_2	R_3	MP (°C) or BP (°C/mm)	References
	n-Butyl	Phenyl		116–117	4
	n-Butyl	Phenyl	Phenyl	113–114	4
	n-Butyl	COOEt	Methyl	140–155/0.01	4
Ethyl	Ethyl	Phenyl		79	4
Ethyl	Ethyl	COOMe	Methyl	113–115/0.01	4
Ethyl	Ethyl	COOEt	Methyl	140/0.01	4
Ethyl	Ethyl	$CONHNH_2$	Methyl	108–111	4
Ethyl	Ethyl	$CONHN{=}CMe_2$	Methyl	177–178	4

TABLE 3C-9

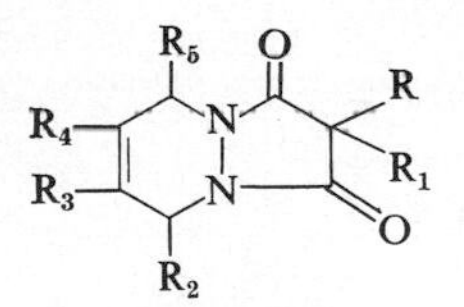

R	R_1	R_2	R_3	R_4	R_5	MP (°C) or BP (°C/mm)	References
Ethyl						74–76	1
n-Butyl						120/0.005	1
Methyl	Methyl					133–135	31
Ethyl	Ethyl					67.5–68.5	1
						65–66	31
						116/2	31
Methyl	Methyl		Methyl			80	31
						125/2.5	31
Methyl	Methyl	Methyl			Methyl	41–44	31
						111/2.5	31
Methyl	Methyl		Methyl	Methyl		133.5–134	31
Ethyl	Ethyl		Methyl			54–55	31
						58.5–60	37
						53–55	38
Ethyl	Ethyl	Methyl			Methyl	121.5/2	31
Ethyl	Ethyl		Methyl	Methyl		105.5–106.5	31
						107–108	37
Ethyl	Ethyl	Phenyl			Phenyl	147–148	37
						154–156	38

TABLE 3C-10

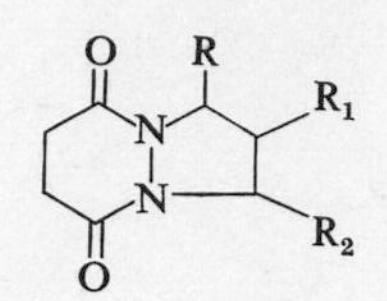

R	R_1	R_2	MP (°C) or BP (°C/mm)	References
None			113–114	11
Hydroxy			104–106	30
Hydroxy	Methyl		134	30
Hydroxy		Methyl	112	30

TABLE 3C-11

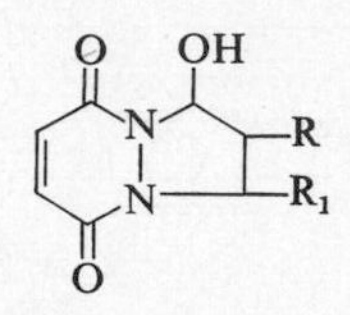

R	R_1	MP (°C) or BP (°C/mm)	References
None		180	30
		Oxime, 195	30
		Semicarbazone, 235 (dec.)	30
Methyl		166	30
	Methyl	162	30
	Phenyl	178	30

TABLE 3C-12

R	R_1	MP (°C) or BP (°C/mm)	References
None		197, 202	7, 28
	Methyl	164	7, 28
	Phenyl	230	7, 28
Methyl		158	7, 28

TABLE 3C-13. Pyrazolo[1,2-*a*]pyridazines

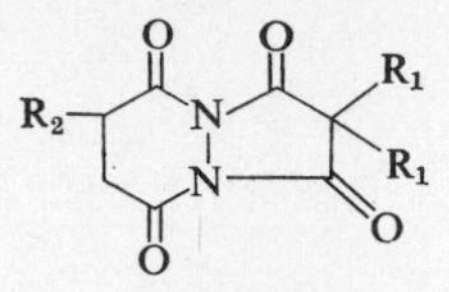

R	R_1	R_2	MP (°C) or BP (°C/mm)	References
None			290 (dec.)	2, 3, 29
	Ethyl		202	2, 3, 29
Ethyl	Ethyl		252	2, 3, 29
Ethyl	Ethyl	Methyl	166	2, 3, 29
Ethyl	Ethyl	Phenyl	185	2, 3, 29
=NOH (R, R_1)			275	3, 29

TABLE 3C-14

R	R_1	MP (°C) or BP (°C/mm)	References
None		360 (dec.)	2, 3, 29
	Ethyl	135 (dec.)	2, 3, 29
Ethyl	Ethyl	174	2, 3, 29
=NOH (R, R_1)		282	3, 29

TABLE 3C-15. Pyrazolo[3,4-*d*]pyridazines

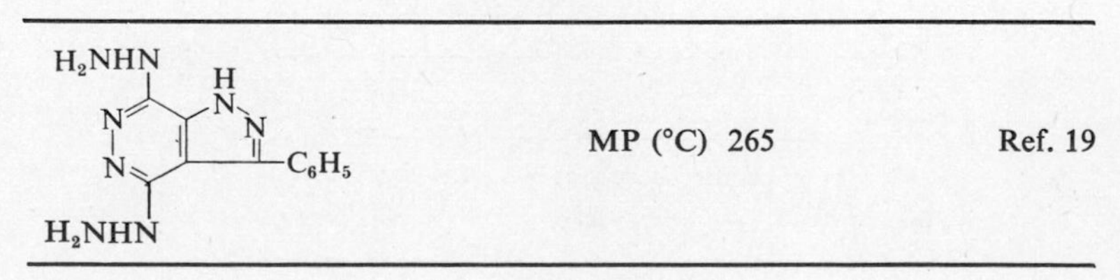

MP (°C) 265 — Ref. 19

TABLE 3C-16

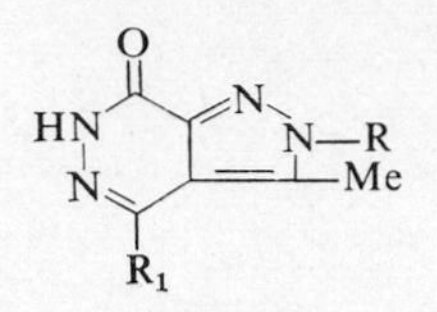

R	R_1	MP (°C) or BP (°C/mm)	References
	Methyl	220	17
	Phenyl	240	17
—NO_2, CN	Phenyl	274–276	16

TABLE 3C-17

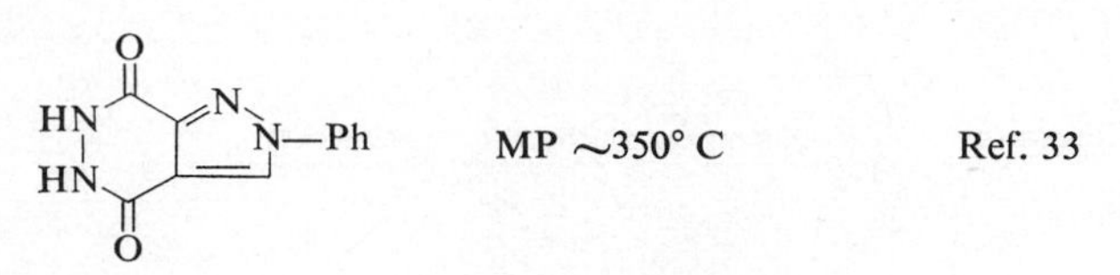

MP ~350° C Ref. 33

TABLE 3C-18

R	MP (°C)	References
H	>300	34
$PhCH_2$—	258	34

TABLE 3C-19

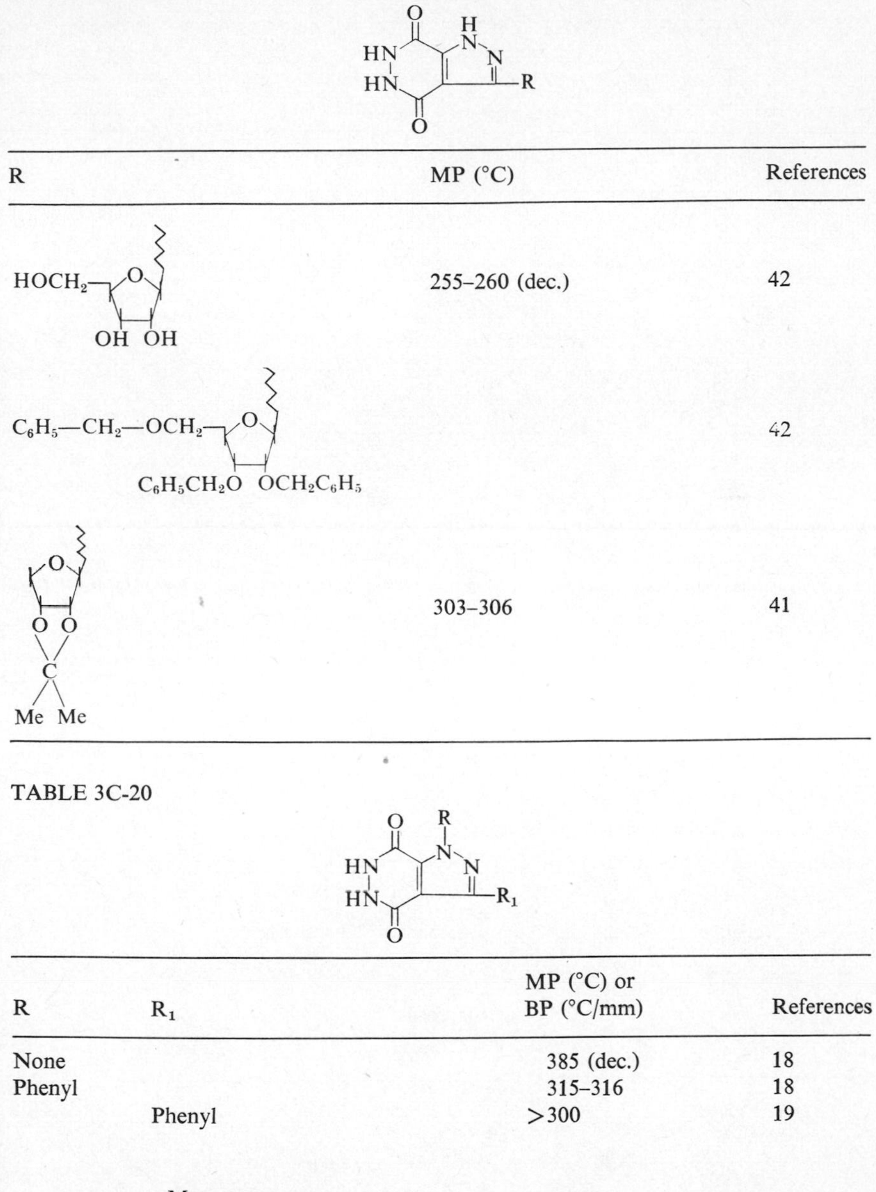

R	MP (°C)	References
HOCH$_2$– (furanose ring) OH OH	255–260 (dec.)	42
C_6H_5—CH_2—OCH_2– (furanose ring) $C_6H_5CH_2O$ $OCH_2C_6H_5$		42
(furanose ring) O O C Me Me	303–306	41

TABLE 3C-20

R	R_1	MP (°C) or BP (°C/mm)	References
None		385 (dec.)	18
Phenyl		315–316	18
	Phenyl	>300	19
(structure: Me, N, N—C_6H_5, O)		186	14

TABLE 3C-20 (*continued*)

R	R_1	MP (°C) or BP (°C/mm)	
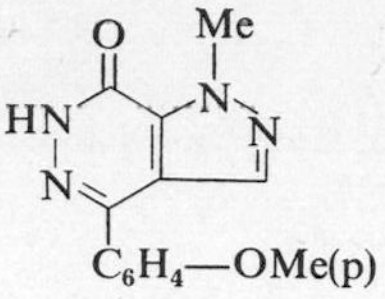 O, Me, HN, N, N, N, N, C_6H_4—OMe(p)		274–275 (dec.)	15
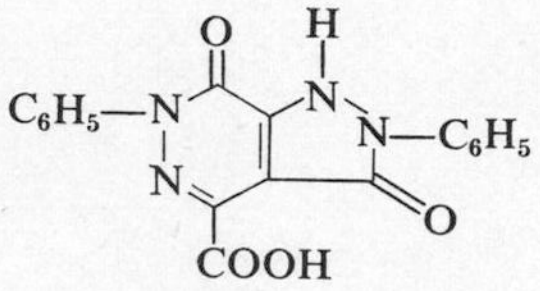 O, H, C_6H_5—N, N, N, N—C_6H_5, O, COOH		245	20
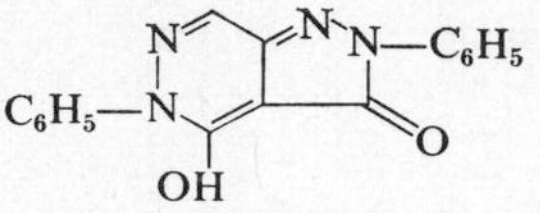 N, N, N—C_6H_5, C_6H_5—N, O, OH		283 (dec.)	21

TABLE 3C-21. Pyrazolo[3,4-*c*]pyridazines

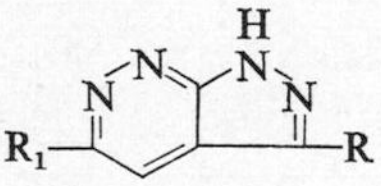

H, N, N, N, N, R_1, R

R	R_1	MP (°C) or BP (°C/mm)	References
None		196	23
		HBr 165 (dec.)	
	Methyl	227	23
Amino		236	23
		HCl 290–300 (dec.)	
Amino	Methyl	258 (dec.)	23
		HCl 260–280 (dec.)	

TABLE 3C-22

R	MP (°C)	References
H	280–300 (dec.)	23
Methyl	190 (dec.)	23

TABLE 3C-23

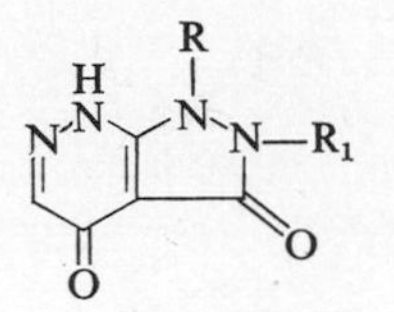

R	R_1	MP (°C)	References
Phenyl	Methyl	140–141	27
Methyl	Phenyl	250 (dec.)	26[a]
		193	23

[a] Structure not proved.

References

1. J. Büchi, W. Vetsch, and P. Fabiani, *Helv. Chim. Acta*, **45,** 37 (1962).
2. A. Le Berre and J. Godin, *Compt. rend.*, **260,** 5296 (1965).
3. French Pat. 1,441,519 (1966); *Chem. Abstr.*, **66,** 10950 (1967).
4. I. Molnar and T. Wagner-Jauregg, *Helv. Chim. Acta*, **39,** 155 (1964).
5. L. A. Carpino, P. H. Terry, and S. D. Thatte, *Tetrahedron Lett.*, 1964, 3329.
6. L. A. Carpino, P. H. Terry, and S. D. Thatte, *J. Org. Chem.*, **31,** 2867 (1966).
7. A. Le Berre, M. Dormoy, and J. Godin, *Compt. rend.*, **261,** 1872 (1965).
8. M. Rink, S. Mehta, and K. Grabowski, *Arch. Pharm.*, **292,** 225 (1959).
9. I. Zugravescu and E. Carp, *Anal. Stiint Univ. "Al. I. Cuza", Iasi, Sect I, Chem.*, **11c,** 59 (1965); *Chem. Abstr.*, **63,** 14855 (1965).
10. H. Wamhoff and F. Korte, *Chem. Ber.*, **101,** 778 (1968).
11. H. Stetter and H. Spangenberger, *Chem. Ber.*, **91,** 1982 (1958).
12. B. T. Gillis and R. Weinkam, *J. Org. Chem.*, **32,** 3321 (1967).

13. D. G. Farnum, R. J. Alaino, and J. M. Dunston, *J. Org. Chem.*, **32**, 1130 (1967).
14. I. Ito, *J. Pharm. Soc. Japan*, **76**, 822 (1956); *Chem. Abstr.*, **51**, 1149 (1957).
15. V. Zikan, M. Semonsky, E. S. Svatek, and V. Jelinek, *Coll. Czech. Chem. Commun.*, **32**, 3587 (1967); *Chem. Abstr.*, **62**, 100061 (1967).
16. G. Bianchetti, D. Pocar, and P. Dalla Croce, *Gazz. chim. ital.*, **94**, 340, (1964).
17. S. Rossi, S. Maiorana, and G. Bianchetti, *Gazz. chim. ital.*, **94**, 210, (1964).
18. R. G. Jones, *J. Am. Chem. Soc.*, **78**, 179 (1956).
19. L. Erichomovitch and F. L. Chubb, *Can. J. Chem.*, **44**, 2095 (1966).
20. S. Ruhemann and K. J. P. Orton, *Ber.*, **27**, 3449 (1894).
21. D. Vorländer, W. Zeh, and H. Enderlein, *Ber.*, **60B**, 849 (1927).
22. Unpublished data from this laboratory. Parameters for LCAO Calculations were taken from *Molecular Orbital Theory for Organic Chemists*, A. Streitwieser, Wiley, New York, 1961, p. 135.
23. A. Dornow and W. Abele, *Chem. Ber.*, **97**, 3349 (1964).
24. J. Druey, *Angew. Chem.*, **70**, 5 (1958).
25. P. Schmidt, K. Eichenberger, and M. Wilhelm, *Angew. Chem.*, **73**, 15 (1961).
26. H. Stenzl, A. Staub, C. Simon, and W. Baumann, *Helv. Chim. Acta*, **33**, 1183 (1950).
27. B. Janik, A. Kocwa, and I. Zagata, *Diss. Pharm.*, **10**, 143 (1958); *Chem. Abstr.*, **52**, 20134 (1958).
28. M. Dormoy, J. Godin, and A. Le Berre, *Bull. Soc. Chim. France*, **1968**, 4222.
29. J. Godin and A. Le Berre, *Bull. Soc. Chim. France*, **1968**, 4210.
30. J. Godin and A. Le Berre, *Bull. Soc. Chim. France*, **1968**, 4229.
31. H. Stetter and P. Woernle, *Ann.*, **724**, 150 (1969).
32. V. Galasso and G. De Alti, *Tetrahedron*, **5**, 2259 (1969).
33. G. Coispeau, J. Elguero, and R. Jacquier, *Bull. Soc. Chim. France*, **1969**, 2061.
34. M. Sprinzl, J. Farkaš, and F. Šorm, *Tetrahedron Lett.*, **1969**, 289.
35. P. Aeberli and W. J. Houlihan, *J. Org. Chem.*, **34**, 2720 (1969).
36. Z. Aboul-Ella, M. N. Aboul-Enein, A. I. Eid, and A. Ibrahim, *J. Chem. UAR*, **11**, 289 (1968); *Chem. Abstr.*, **71**, 3319 (1969).
37. B. T. Gillis and R. A. Izydore, *J. Org. Chem.*, **34**, 3181 (1969).
38. S. Takase and T. Motoyama, *Bull. Chem. Soc. Japan*, **43**, 3926 (1970).
39. D. G. Farnum, A. T. Au, and K. Rasheed, *J. Heterocyclic Chem.*, **8**, 25 (1971).
40. J. Bastide and J. Lematre, *Compt. rend. C*, **267**, 1620 (1968).
41. E. M. Acton, K. J. Ryan, and L. Goodman, *J. Chem. Soc. (D), Chem. Commun.*, **1970**, 313.
42. M. Bobek, J. Farkas, and F. Šorm, *Tetrahedron Lett.*, **1970**, 4611.
43. M. Franck-Neumann and G. Leclerc, *Tetrahedron Lett.*, **1969**, 1063.

Part D. Imidazopyridazines

I. Imidazo[1,2-*b*]pyridazines

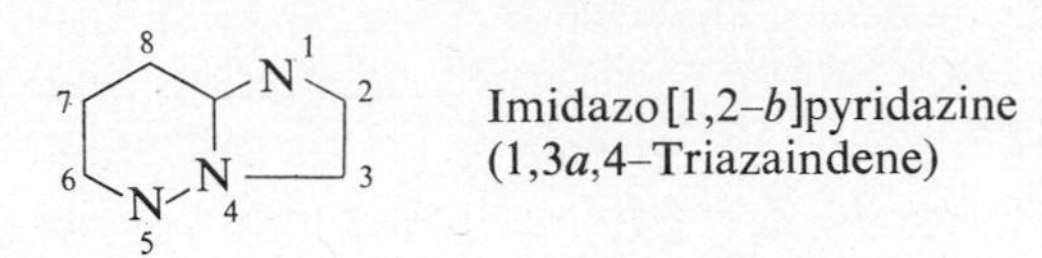

It was not until 1964 that the first synthesis of compounds belonging to this system was reported by Yoneda et al. (1). Since then the chemistry of imidazo-[1,2-*b*]pyridazines has quickly developed because of theoretical and practical interest.

The bicyclic ring system is built up exclusively from appropriate 3-amino-pyridazines which have been condensed with α-halocarbonyl compounds such as α-haloaldehydes or their acetals, α-haloketones, esters of β-halo-α-keto acids or α-halo-β-keto acids, α-haloesters and 1,2-dibromoethane. The components are usually heated together in a solution of alcohol or dialkoxyethane for several hours.

The reaction was shown to involve a two-step reaction and the inter-mediate 3-amino-1-phenacylpyridazin-1-ium bromide has been isolated. When refluxed in an aqueous solution (4 hr) this was transformed into 2-phenylimidazo[1,2-*b*]pyridazine hydrobromide (**2**, $R = C_6H_5$) (1). It is obvious that the proposed intermediate is not capable of ring closure, which proceeds only with the isomeric pyridazin-2-ium salt with the quaternized ring nitrogen close to the amino group. The initial step thus involves the formation of a mixture of N_1 and N_2-quaternized compounds and only the N_2-quaternized isomer (**1**) is then capable of cyclization to **2**. An indication for the formation of a mixture of quaternized compounds is the yield of the final product, which is usually not higher than 60%.

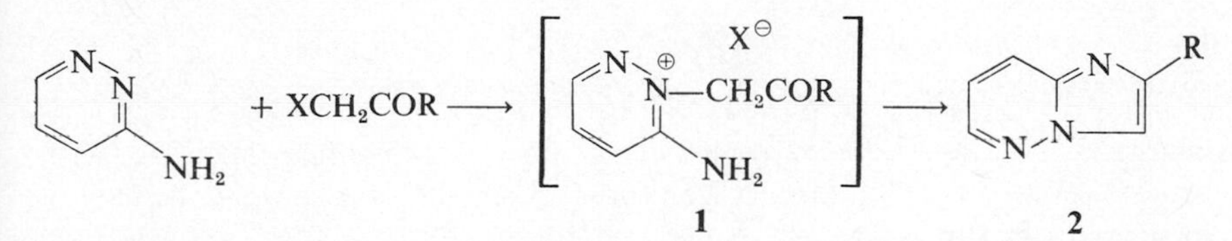

Another possibility, the initial condensation between the amino group and the halogen function, is ruled out since structural examinations on some imidazo[1,2-*b*]pyridazines have clearly eliminated the formation of isomeric products (2).

In all cases the choice of substituents on the pyridazine ring of imidazo-[1,2-*b*]pyridazines depends on the starting substituted 3-aminopyridazine, whereas substituents in the imidazole part can be varied with the choice of the appropriate α-halocarbonyl compound. In this manner, with chloro- or bromoacetaldehyde at positions 2 and 3, unsubstituted imidazo[1,2-*b*]-pyridazines have been prepared (3–6, 50). α-Haloketones or esters of β-halo-α-keto acids yielded the corresponding 2-substituted imidazo[1,2-*b*]-pyridazines (1, 5–10), and with esters of α-halo-β-keto acids the corresponding 2-substituted 3-carbalkoxy derivatives are formed (6, 7). Recently, it was shown that the bicyclic system can be formed also by employing 1,2-dibromoethane or ethyl chloroacetate (51).

The parent compound, imidazo[1,2-*b*]pyridazine (**6**) was prepared in two different ways. Armarego (11) used 3-methylthiopyridazine as starting compound and displaced the methylthio group with ethanolamine. The 3-(hydroxyethylamino)pyridazine (**3**), upon treatment with thionyl chloride, was cyclized via **4** into the partially reduced imidazo[1,2-*b*]pyridazinium chloride (**5**). When oxidized with potassium ferricyanide, this afforded **6** in 5% yield.

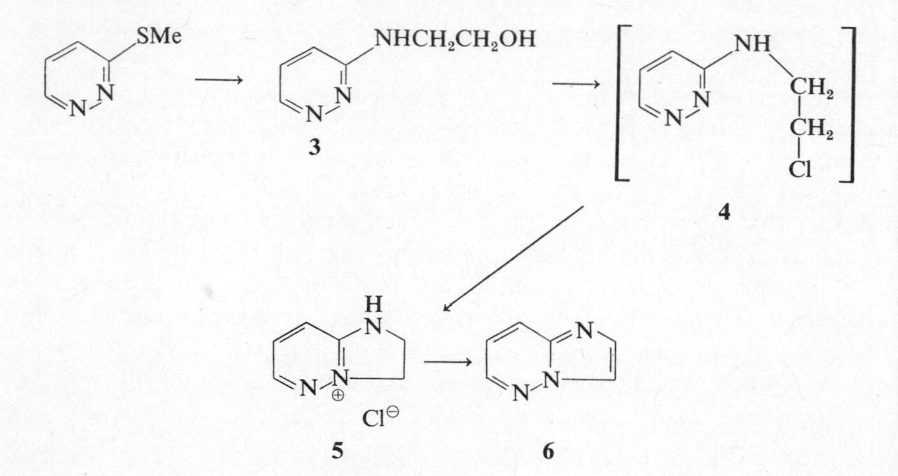

The other synthetic approach, described by Stanovnik and Tišler (2, 4) employed 3-amino-6-chloropyridazine as the starting material. Thus with bromoacetaldehyde compound **7** has been formed and when dehalogenated in the presence of hydrogen, palladized charcoal and triethylamine yielded the parent bicycle (**6**) in 75% yield.

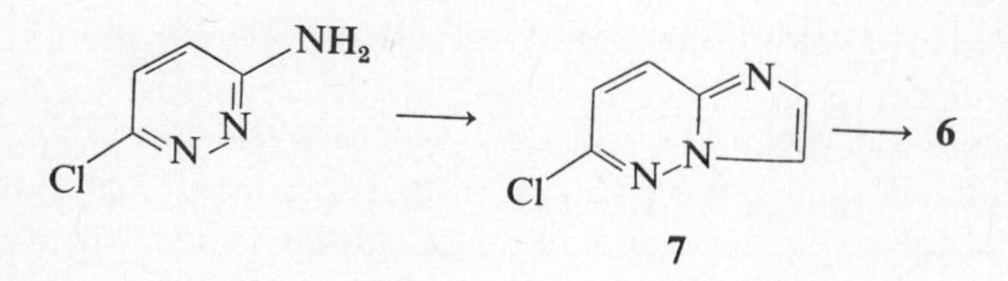

Recently, it was shown (60) that this system can be prepared from tetrazolo[1,5-*b*]pyridazines involving valence isomerization. Thus when 6-dimethoxyethylaminotetrazolo[1,5-*b*]pyridazine was treated with polyphosphoric acid the condensed imidazo ring was formed with simultaneous ring opening of the tetrazolo ring and formation of an azido group to give 6-azidoimidazo[1,2-*b*]pyridazine.

Imidazo[1,2-*b*]pyridazine represents a stable, aromatic 10 π-electron system, capable of several important transformations. It is a weak base [pK_a 4.4 (2), 4.57 (11)] and forms stable hydrochloride or hydrobromide salts, quaternary salts with methyl iodide (2, 52), and a perchlorate salt (11). It is a weaker base than imidazo[1,2-*a*]pyridine (pK_a 6.72) (12) by about 2.2 pK units. This is compatible with the insertion of an additional nitrogen atom, which lowers the basic strength. Protonation and quaternization occur at N_1 as concluded from ionization constant and uv and nmr spectral correlations. The possibility of covalent hydration has been excluded (11).

The uv spectrum of the neutral species of imidazo[1,2-*b*]pyridazine is very similar to that of imidazo[1,2-*a*]pyridine (indolizine). In the nmr spectrum the protons attached on the imidazole part appear as an AB pattern [in $CDCl_3$, $\tau = 2.21$ (H_2), 2.01 (H_3); $J_{2,3} = 1.0$ cps] and those attached to the pyridazine part as an ABX system [in $CDCl_3$, $\tau = 1.70$ (H_6), 3.00 (H_7), 2.05 (H_8); $J_{6,7} = 4.5$, $J_{7,8} = 10.0$, $J_{6,8} = 2.0$, and $J_{3,8} = 0.8$)] (2). NMR spectra correlations have been useful in determining the structure of substituted imidazo[1,2-*b*]pyridazines (2, 5, 6). Mass spectra of some imidazo-[1,2-*b*]pyridazines have been recorded and fragmentation patterns outlined (13).

Total and frontier π-electron densities, calculated by the simple HMO method (52) are listed in Table 3D-1 and predict the position 3 to be the most susceptible for electrophilic substitutions, whereas positions 6 and 8 should be involved in nucleophilic substitutions. In fact, all investigated electrophilic substitutions have been established to take place at position 3.

Electrophilic substitutions on imidazo[1,2-*b*]pyridazines give exclusively 3-substituted products, and in no case have disubstitution products been isolated. Studies included bromination with bromine in acetic acid or with *N*-bromosuccinimide, nitration, and sulfonation (2, 4). From some bromination experiments with bromine in acetic acid stable complexes with bromine could be isolated, a property common to several other azaheterocycles.

Electrophilic substitutions on 2-phenylimidazo[1,2-*b*]pyridazines are of particular interest since in addition to the substitution at position 3, with excess nitrating agent the *p*-position of the phenyl group can be substituted (15). Similarly, bromination of benzylidene derivatives of 6-hydrazinoimidazo[1,2-*b*]pyridazines gives dibromo derivatives, the second halogen being introduced into the side chain methine group (4, 15). A proof for such substitutions was presented in the following sequence of reactions. Bromination of **8** or **9** afforded the same dibromo compound **10**. Cyclization of **8** to the polyazaheterocycle **11** by means of lead tetraacetate and subsequent bromination afforded a product **12**, identical in all respects with that obtained from treatment of **10** with sodium acetate in acetic acid (4).

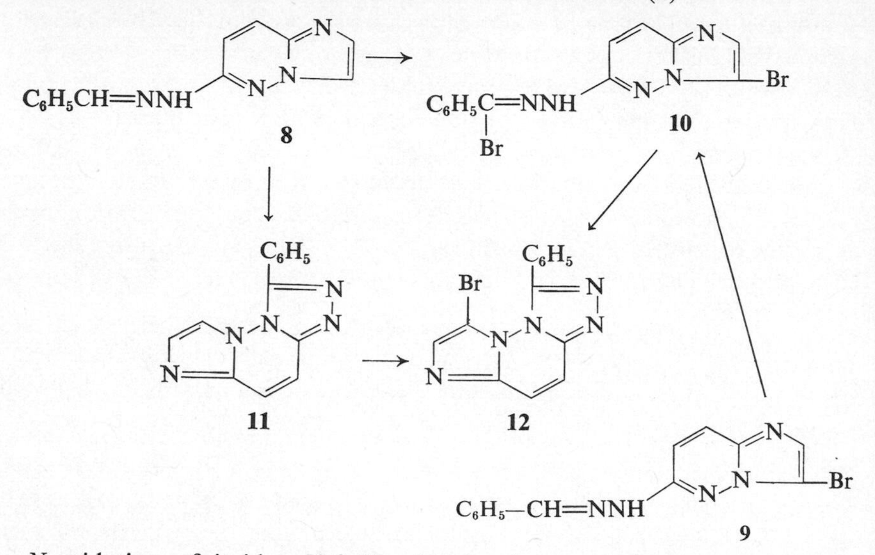

N-oxidation of imidazo[1,2-*b*]pyridazine has recently been successfully accomplished by Pollak, Stanovnik, and Tišler (61). With concentrated hydrogen peroxide in polyphosphoric acid the parent compound afforded its 5-oxide in moderate yield. A competitive reaction is the degradation of the bicycle and the corresponding pyridazines or pyridazine *N*-oxides were formed. Moreover, 6-aminoimidazo[1,2-*b*]pyridazine was transformed by this procedure into its 6-nitro analog.

Nucleophilic substitutions have been limited to the displacement of the halogen at position 6. Thus replacements with hydrazine (2–4, 7, 15), sodium thiophenolate (4), sodium methoxide (1, 7, 16), and other alkoxides (16), amines (1, 8, 17), or sodium azide (3, 4, 18) have been realized. Hydrolysis of **13** with a solution of potassium hydroxide in ethanol (160–170°, 4 hr) afforded the corresponding oxo derivative (**14**, R = H), which could be

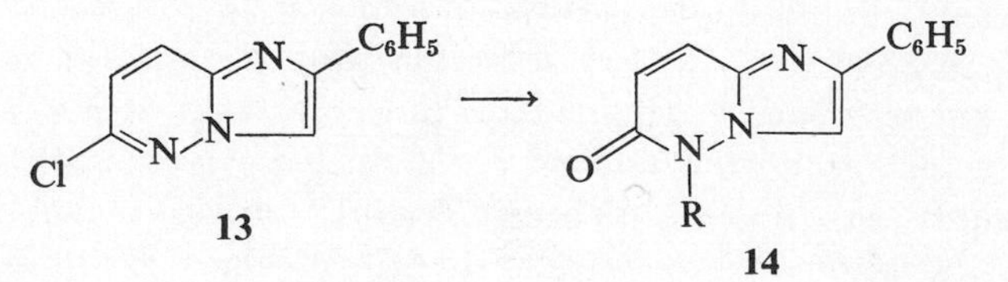

methylated with methyl iodide to the product assigned the structure of N_5-methyl derivative **14** (R = Me) (1).

3-Bromo-6-chloroimidazo[1,2-*b*]pyridazine displayed a different displaceability of halogen atoms in the reaction with hydrazine. The 6-chloro group is more readily displaced than is the 3-bromo group (4, 15).

Attempts to replace a halogen substituent at position 6 with an amino group in a direct reaction with ammonia failed, and 6-aminoimidazo[1,2-*b*]pyridazine (**16**) could be synthesized from its 6-chloro analog via the 6-azido derivative (**15**) by hydrogen sulfide reduction (4).

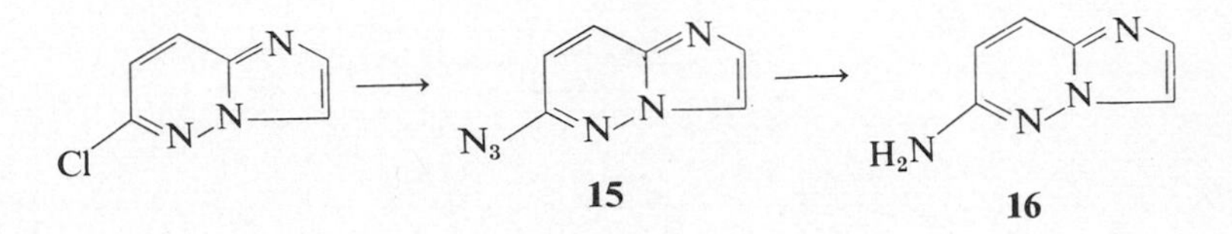

The 2-methyl group of 2-methylimidazo[1,2-*b*]pyridazines displays a reactivity similar to those of activated *ortho* or *para* to ring nitrogen methyl groups in azines. In this manner, 2-methyl-6-methoxyimidazo[1,2-*b*]pyridazine was condensed with chloral (58 hr at reflux temperature) to form the alcohol **17** (6).

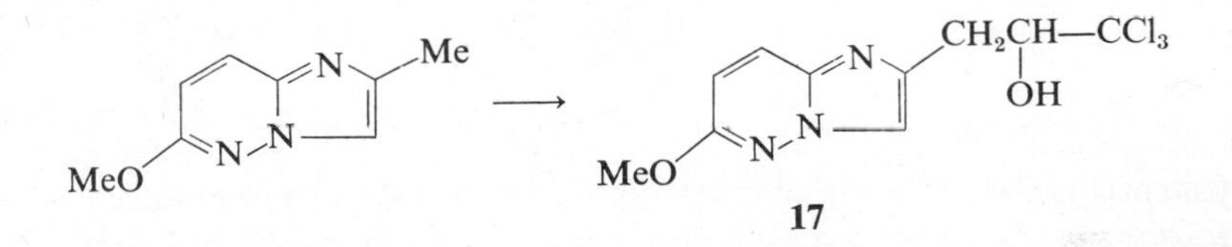

It is also possible to oxidize a 2-methyl group to the carboxyl group by means of potassium permanganate or chromium trioxide. This transformation, together with subsequent decarboxylation, proves the structure determinations in the outlined reaction sequence **18**–**22**. Compound **20** proved to be identical with the product obtained from the reaction between 3-amino-6-chloropyridazine and bromoacetone. Moreover, **22** was identical with the reaction product from the same pyridazine and bromoacetaldehyde (7).

Another widely used transformation is hydrogenolysis of the chlorine atom at position 6, which proceeded smoothly in the presence of palladized charcoal as catalyst (1, 2, 5, 15, 19). The aromatic ring is not affected and this reaction was particularly helpful for the synthesis of several methyl-substituted imidazo[1,2-*b*]pyridazines.

6-Methoxyimidazo[1,2-*b*]pyridazine and its 2-methyl analog have been found to undergo Mannich reaction (in some cases under forcing reaction

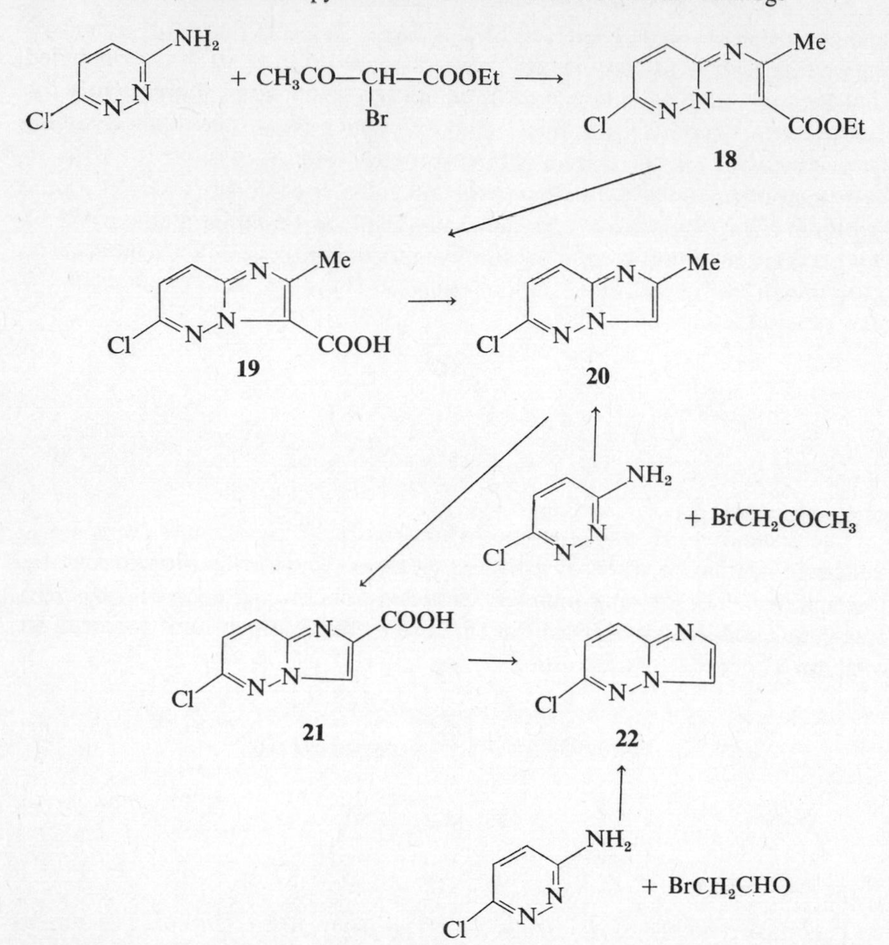

conditions), and 3-dialkylaminomethyl derivatives (**23**) were obtained in various yields (8–88%) (6, 50).

Imidazo[1,2-*b*]pyridazine has been also submitted to homolytic methylation and as source of methyl radicals, thermal homolysis of diacetyl peroxide in glacial acetic acid solution at 70° has been employed (5). The crude reaction product was separated into components by gas chromatography

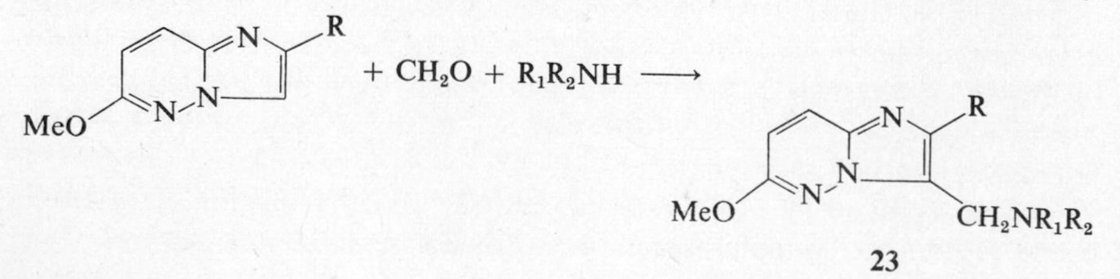

and the products were identified as 8-methyl-, 7-methyl, and 7,8-dimethylimidazo[1,2-*b*]pyridazine (in the ratio of about 1:2:5). It was concluded that the methyl radical attacks preferentially position 8 and subsequently the 7,8-dimethyl derivative is formed. A preferential *para*-substitution in homolytic reactions is known from phenylation studies on pyridazine (20), in contrast to the 2-substitution on pyridines (21).

Imidazo[1,2-*b*]pyridazines with appropriate functional groups, such as hydrazino, alkylidene, or benzylidene hydrazino or thiosemicarbazido group, attached at position 6 served for the preparation of derivatives of the new polyazaheterocyclic system **24** (3, 4, 7, 53).

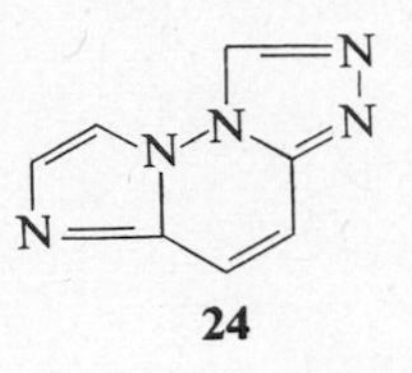

24

Some imidazo[1,2-*b*]pyridazines are reported to possess biological activity. Antipyretic and hypothermal activity (9), anticonvulsant activity (9); analgesic, sedative, and antispasmodic activity (10, 16, 19) are reported as well, as is their use as inhibitors for the central nervous system (7).

II. Imidazo[4,5-*d*]pyridazines

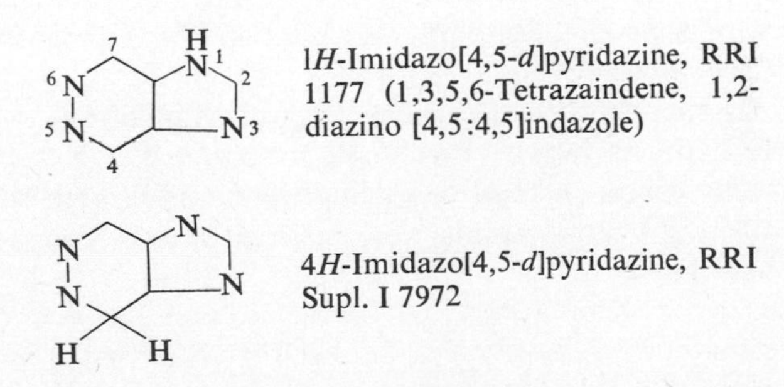

1*H*-Imidazo[4,5-*d*]pyridazine, RRI 1177 (1,3,5,6-Tetrazaindene, 1,2-diazino [4,5:4,5]indazole)
4*H*-Imidazo[4,5-*d*]pyridazine, RRI Supl. I 7972

As a consequence of the marked similarity of purine and imidazo[4,5-*d*]-pyridazine ring systems the chemistry of the latter has been thoroughly investigated.

As with imidazo[4,5-*c*]pyridazines, the mobile hydrogen in the imidazole ring is the cause of tautomerism. Since the imidazo[4,5-*d*]pyridazine molecule is symmetric, one would expect two isomeric products only when the mobile hydrogen is substituted and fixed structures occur, as in **25** and **26**.

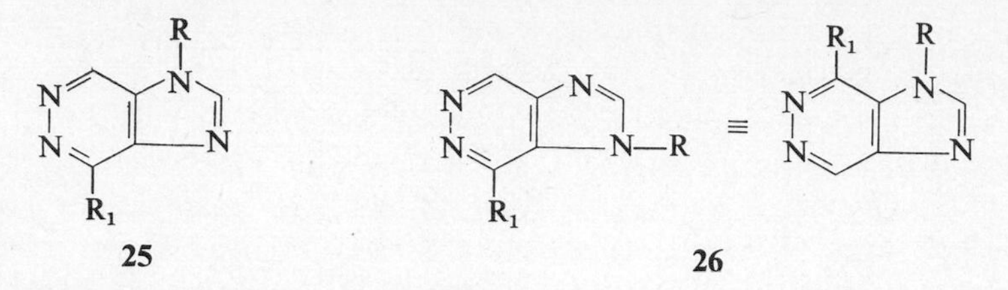

Another possibility of tautomerism arises with the introduction of two oxo groups at positions 4 and 7. By analogy with the structure of maleic hydrazide it should be expected that compounds would exist in the hydroxy-oxo form. One could thus expect that 1-substituted derivatives exist either as **27** or as **28**.

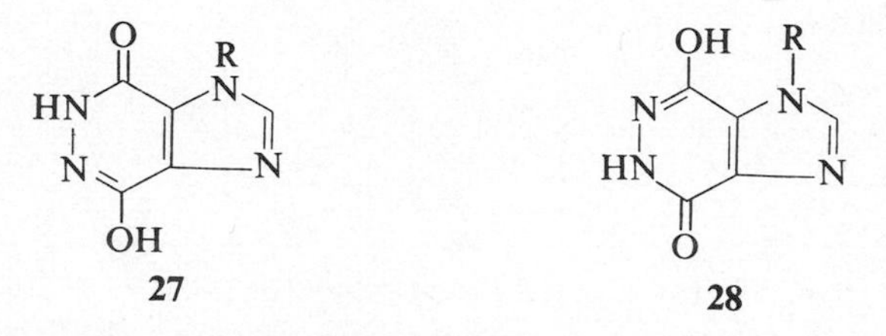

There are no studies concerning the possibility of such tautomerism and for the sake of simplicity all such compounds are referred to here as having the dioxo form.

There are two principal synthetic approaches for the formation of imidazo[4,5-*d*]pyridazines. In the first, imidazoles with functional groups capable of ring closure of the pyridazine ring, such as carbonyl or modified carboxyl groups, are used as starting compounds. In another approach pyridazines with *ortho*-amino groups were used and here the imidazole part was formed through ring closure.

There is only one report on the use of a 4,5-diacyl imidazole, which upon treatment with hydrazine was converted into the bicycle **29** (22, 23). Although the same starting compound and procedure were used in both communications, two different melting points were reported for the products, which should be identical.

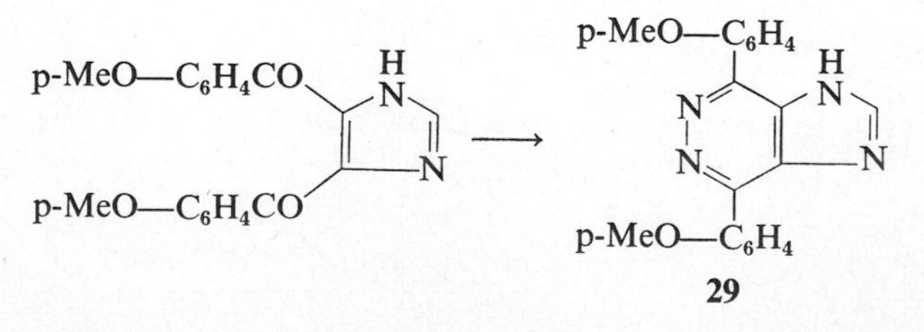

In many syntheses imidazole-4,5-dicarboxylic acid derivatives were utilized. Imidazole-4,5-dicarboxylic acid does not form a cyclic hydrazide (24), but the corresponding methyl or ethyl esters were readily transformed into imidazo[4,5-*d*]pyridazines. Jones (25) reported that a quantitative yield of imidazole-4,5-dicarboxylic acid dihydrazide was obtained after treatment of the corresponding dimethyl ester with hydrazine at room temperature. However, when applying higher temperatures, for example, in boiling 2-propanol (26), the cyclic hydrazide (**31**, R = H) was easily obtained. Similarly, the dihydrazide when heated with excess of hydrazine hydrate or with dilute hydrochloric acid could be cyclized to **31** (R = H) (25). Reaction

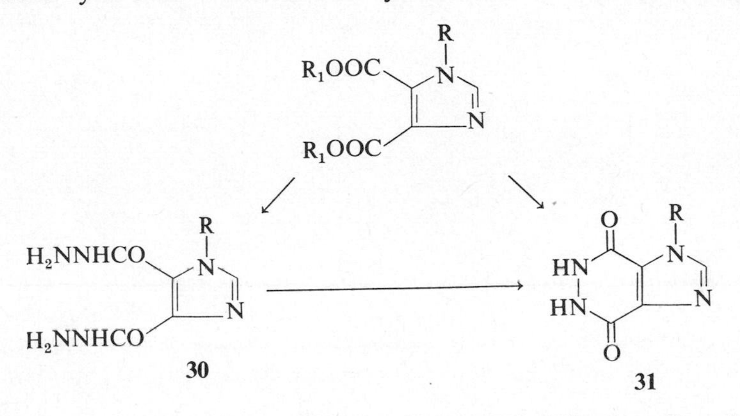

between the diester of imidazoline-2(3*H*)-thione-4,5-dicarboxylic acid and hydrazine gave only the cyclic product (25), whereas from diethyl 1-methylimidazole-4,5-dicarboxylate either the dihydrazide (**30**, R = Me) or the corresponding cyclic product (**31**, R = Me) were obtained (27).

There are several other reports on the application of the synthesis of imidazo[4,5-*d*]pyridazine-4,7(5*H*,6*H*)diones from diesters (28, 29), ester-hydrazides (54) or dihydrazides of imidazole-4,5-dicarboxylic acids (30–34).

Dimethyl imidazole-4,5-dicarboxylate afforded with methylhydrazine only the corresponding bis-methylhydrazide (**32**). The structure of **32** has been reinvestigated (27) and found to be correct as proposed by Castle and Seese (30). The attempted cyclization of **32** to the corresponding imidazo[4,5-*d*]-pyridazinedione by refluxing with 10% hydrochloric acid failed and the imidazole-4,5-dicarboxylic acid was obtained (30). In a similar manner, the

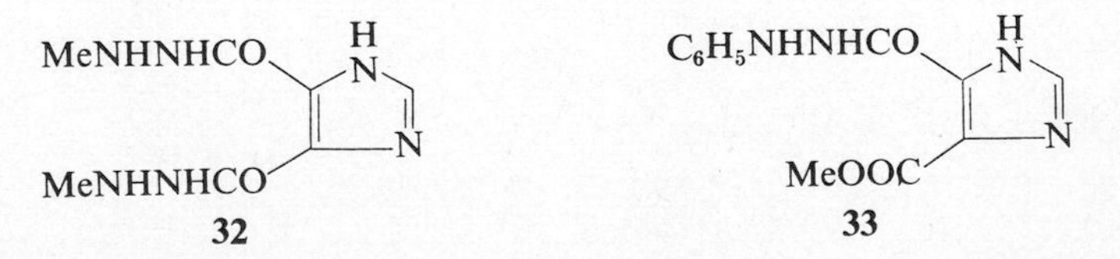

monophenylhydrazide **33** could not be cyclized (30), although there is a recent contrary statement (54).

In several cases for the formation of the bicyclic system appropriate pyridazines have been used. 4,5-Diaminopyridazines when refluxed with ethyl orthoformate and acetic anhydride or formamide or formic acid readily formed the corresponding imidazo[4,5-*d*]pyridazines (35, 36, 54, 62). If instead ethyl orthoformate-98% formic acid was used, 3,6-dimethoxy-4,5-diaminopyridazine was reported to afford (4 hr, reflux temperature) a mixture consisting of the cyclic product (**34**, 30%), 4-amino-5-formamido-3,6-dimethoxypyridazine (**35**, 3.2%), and 6-methyl derivative of **34** (**36**, 2.3%) (36). Similar behavior was observed with 3-methoxy-4,5-diamino-

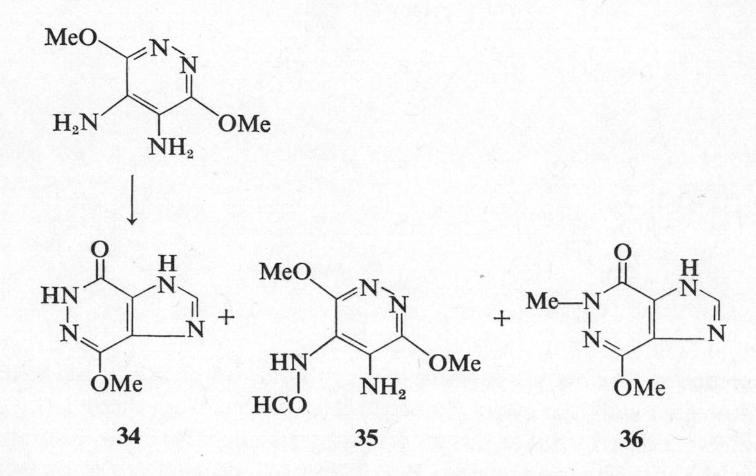

pyridazine (62). Evidently, hydrolytic cleavage of one methoxy group took place during this reaction. With 4,5-diamino-3,6-dihydroxypyridazine the expected bicyclic product is reported to result in good yield (56).

Ethyl orthoacetate in the presence of acetic anhydride (120–130°, 1 hr) yielded with the same pyridazine derivative a mixture of 3,6-dimethoxy-4,5-bis(1′-ethoxyethylidene)iminopyridazine (**37**, 54%) and a small quantity of a compound to which the structure of the bicycle **38** has been assigned (36).

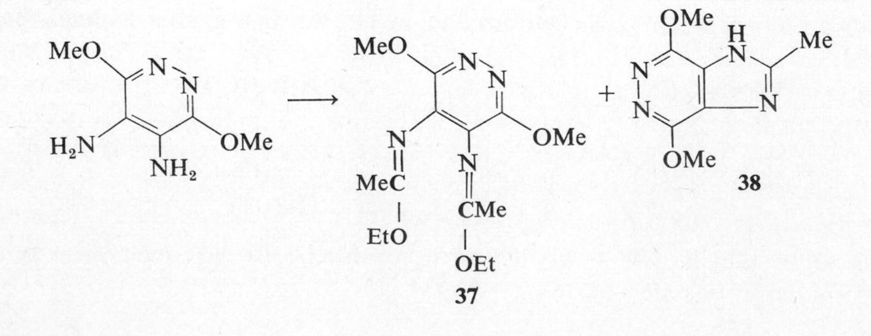

There is also a report about the possibility of imidazo[4,5-*d*]pyridazine ring formation from 4,5-diamino-3(2*H*)pyridazinones and carboxylic acids, but no experimental details are given (37).

Diaminopyridazines, when heated under reflux with carbon disulfide in pyridine and in the presence of sodium hydroxide, yielded the corresponding thio derivatives (**39**) (35, 36, 62). Instead of carbon disulfide, arylisothiocyanate or thiourea was used (57).

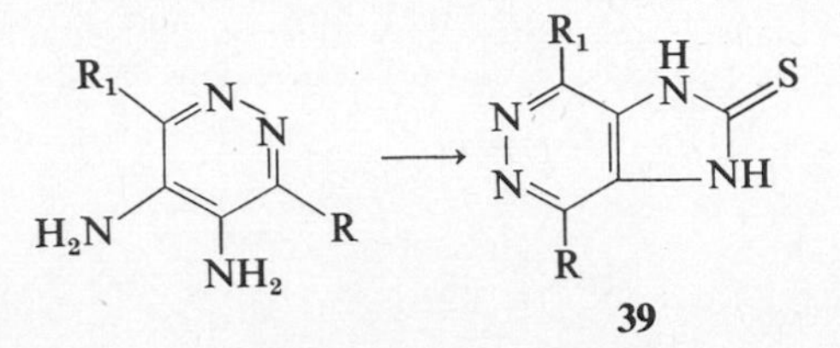

The parent compound, imidazo[4,5-*d*]pyridazine, was synthesized recently by desulfurization of the corresponding 4-thione (55). There are many reactions which have been performed on substituted imidazo[4,5-*d*]pyridazines and which permit some conclusions about the reactivity of this system to be drawn. Calculations of total and frontier π-electron densities are presented in Table 3D-2 (14). Included also are calculations performed by Alonso and Sebastian (38), including bond orders. The SCF LCAO-MO with CNDO II approximation method was used recently in the calculation of electron densities of imidazo[4,5-*d*]pyridazine (55).

The lowest electron density is at position 2, as displayed by imidazo[4,5-*c*]-pyridazines, and thus this position is apparently most susceptible for nucleophilic attack. There is, however, a slight difference in electron densities at position 2 in both systems and the higher value for imidazo[4,5-*d*]pyridazines probably accounts for the better stability of the imidazole part under the influence of nucleophiles. There are no cases of imidazole ring opening reported in the imidazo[4,5-*d*]pyridazines under the influence of bases.

The most versatile derivatives of imidazo[4,5-*d*]pyridazines for further transformations are chloro derivatives, which are obtained by the standard procedure of treating an oxo heterocycle with phosphorus oxychloride. 4,7-Dichloroimidazo[4,5-*d*]pyridazine was reported first to be not obtainable (28) or available only in low yield (30) by treatment of the corresponding dione with phosphorus oxychloride in *N*,*N*-dimethylaniline. Sometimes a very small amount of 7-chloroimidazo[4,5-*d*]pyridazine-4(5*H*)one accompanied the final product. In a later communication, Castle et al. (33) reported that good yields of the 4,7-dichloro compound (72%) are obtainable if the starting imidazo[4,5-*d*]pyridazine-4,7(5*H*,6*H*)dione has been prepared by cyclization of imidazole-4,5-dicarboxylic acid bishydrazide with anhydrous hydrazine. Poor or no yields were recorded if the dione was prepared from the bishydrazide by heating it in 10% aqueous hydrochloric acid solution.

Apparently there are no such difficulties in the preparation of the chloro analogs if the mobile hydrogen is substituted by another group. In this manner, a smooth conversion of the 1-substituted diones **40** (R = Me, C_6H_5, $C_6H_5CH_2$) into the 4,7-dichloro counterparts **41** is possible (28, 39).

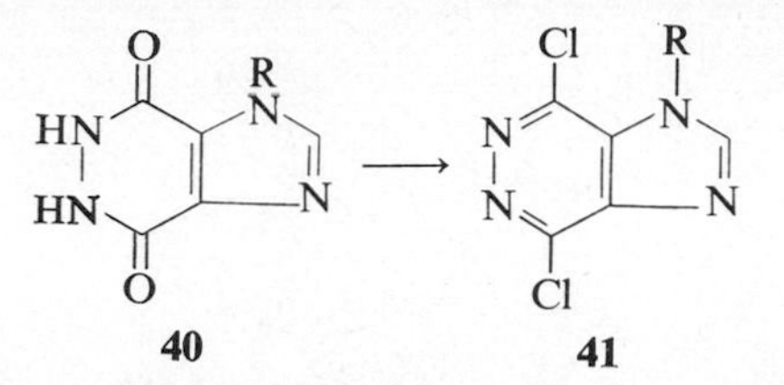

A good yield of the dichloro compound is reported for the 2-phenyl analog (34), whereas introduction of three chlorine atoms by the same method—the synthesis of 2,4,7-trichloroimidazo[4,5-*d*]pyridazine—required prolonged heating and yields were 34–45% (31, 40).

Chloroimidazo[4,5-*d*]pyridazines are useful for the preparation of amino, substituted amino, oxo, thio, hydrazino, or alkoxy derivatives and for the replacement with hydrogen.

Selective displacement of chlorine atoms of 4,7-dichloroimidazo[4,5-*d*]-pyridazines enables the synthesis of the corresponding amino or substituted amino derivatives. Carbon (28) treated 1-benzyl-4,7-dichloroimidazo[4,5-*d*]-pyridazine with ammonia or different amines in a solution of ethanol at 100–110° and obtained the corresponding amino derivatives (**42**) accompanied in most cases with small quantities of the 4-ethoxy derivative (**43**) as a result of competitive nucleophilic displacement.

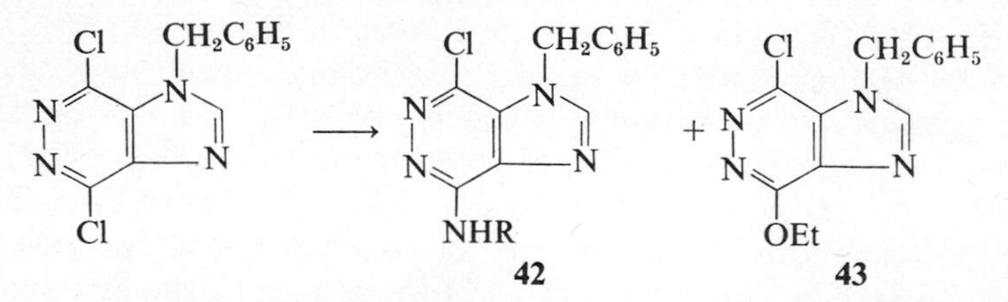

In the case of ammonolysis of 1-benzyl- (39) or 1-methyl-4,7-dichloro-imidazo[4,5-*d*]pyridazines (28) temperature of 150° was applied to obtain the 4-amino derivatives in moderate yield.

Following the observation of Robins (41) that blocking of the mobile hydrogen in the imidazole part of purines caused a more facile nucleophilic substitution, Patel, Rich, and Castle (33) treated 1-(tetrahydro-2′-pyranyl)-4,7-dichloroimidazo[4,5-*d*]pyridazine (**44**) with dialkylaminoalkylamines at

50–60° for 24–50 hr. The reaction between 3-dimethylaminopropylamine and **44** afforded a mixture of **45** and 4,7-dichloroimidazo[4,5-*d*]pyridazine (**46**). The formation of the byproduct **46** could not be completely eliminated even under modified reaction conditions. The protective tetrahydropyranyl group could be removed with hydrochloric acid at room temperature to give **47**.

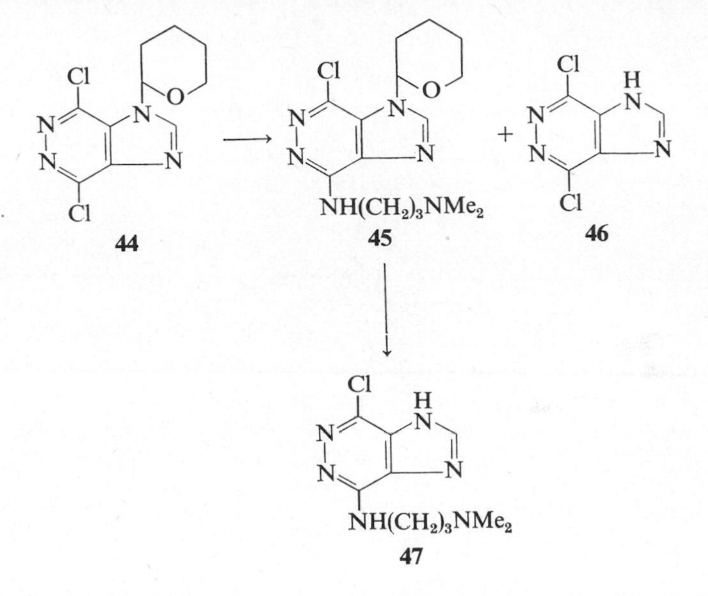

In no case of the mentioned selective aminolysis has there been a rigorous proof of the structure of the resulting products if there was a substituent at N_1 other than hydrogen. The assignment of the structures has been made on the basis of steric considerations and on the basis of the stabilization of the transition state (28, 33). However, this problem becomes irrelevant if the 1-protective group is removed.

The replacement of both chlorine atoms in the case of 1-unsubstituted 4,7-dichlorimidazo[4,5-*d*]pyridazines with amines requires rather vigorous reaction conditions (110–170°, 12–22 hr) (33) and 2-phenyl-4,7-dichloro-[4,5-*d*]pyridazine is reported to fail to aminate with ammonia under pressure (32). Moreover, 4(7)-chloro-7(4)-ethoxyimidazo[4,5-*d*]pyridazine (**48**) when allowed to react with 1–2 moles of dimethylaminopropylamine under varying reaction conditions afforded as the principal products always the 4,7-bis-(dimethylaminopropylamino) (**50**) and 4(7)-dimethylamino-propylamino-7(4)-ethoxy derivative (**49**). Only with the amine in excess could the bis-substituted amino derivative be obtained (33).

Carbon (28) has not been able to obtain 1-benzyl-4,7-bis(diethylamino)-imidazo[4,5-*d*]pyridazine by treatment of the corresponding dichloro

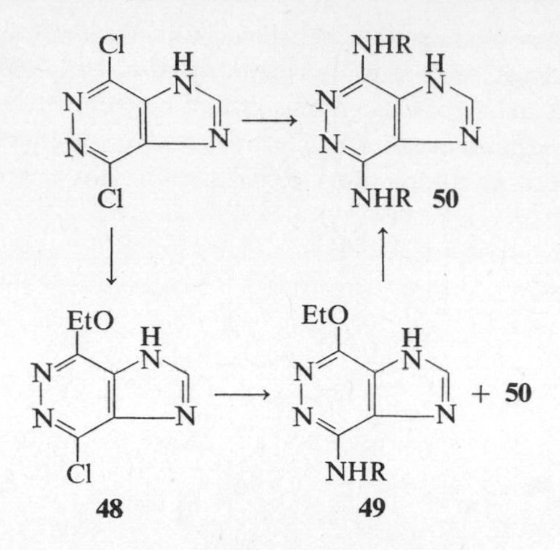

derivative with diethylamine. However, an easy displacement was possible with benzylamine as the nucleophile and 1-benzyl-4,7-bis-(benzylamino)-imidazo[4,5-*d*]pyridazine was obtained in good yield.

Amination studies on 2,4,7-trichloroimidazo[4,5-*d*]pyridazine revealed that two chlorine atoms are readily displaced and that only with morpholine were all three chlorine atoms replaced (31). The first two replaceable chlorine atoms have been found to reside at positions 4 and 7. These and the previously discussed results on aminolysis of chloroimidazo[4,5-*d*]pyridazines allow a conclusion on the ease of displaceability of the halogens at different positions to be made. The order of reactivity is consequently $4 > 7 > 2$ and this is also in agreement with the calculated frontier electron densities.

Another possible method of preparing 4,7-diamino or disubstituted amino derivatives of imidazo[4,5-*d*]pyridazines has been employed in several cases, that is, the displacement of an alkylthio group (29, 30, 32, 34, 55). Prolonged heating at temperatures of about 200° were generally necessary. Even under such conditions, 4,7-bis-(methylthio)imidazo[4,5-*d*]pyridazine, with 6–30 molar excess of some amines, afforded only monoaminated products (29). Furthermore, with a molar excess of di-*n*-propylamine, di-*n*-butylamine, or diethylamine and under similar reaction conditions, the sole isolated product was the imidazo[4,5-*d*]pyridazinone derivative **51** (29).

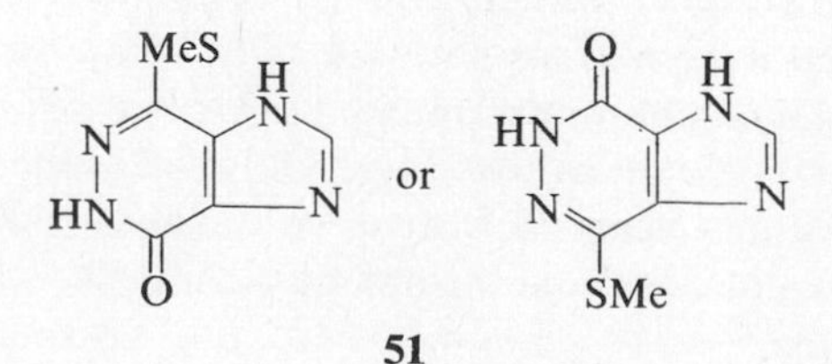

Hydrazinoimidazo[4,5-*d*]pyridazines can be prepared in a similar fashion. 1-Benzyl-4-hydrazino-7-chloroimidazo[4,5-*d*]pyridazine was prepared by hydrazinolysis of the 4,7-dichloro analog (42) and the structure has been assigned on the basis of steric considerations. 4,7-Dihydrazino derivatives, however, have been prepared easily by hydrazinolysis of the corresponding 4-methylthio (55) or 4,7-bis(methylthio) derivatives (32, 34).

Further examples of nucleophilic displacement on haloimidazo[4,5-*d*]-pyridazines include the introduction of a "hydroxy" or alkoxy group. 4,7-Dichloroimidazo[4,5-*d*]pyridazine when treated with 2 moles of sodium ethoxide afforded the monoethoxy compound in good yield (33). 1-Benzyl-(**52**, R = $C_6H_5CH_2$, R_1 = H) (28) and 2-phenylimidazo[4,5-*d*]pyridazin-

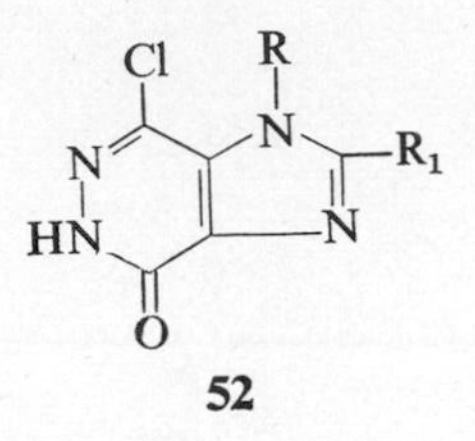

52

4(5*H*)one (**52**, R = H, R_1 = C_6H_5) (32) have been obtained from the corresponding 4,7-dichloro derivatives with either boiling 10% sodium hydroxide or glacial acetic acid.

Imidazo[4,5-*d*]pyridazinethiones were also prepared in good yields by either chlorine displacement with phosphorus pentasulfide in boiling pyridine solution (32, 62) or a solution of sodium hydrogen sulfide in ethanol (28) or

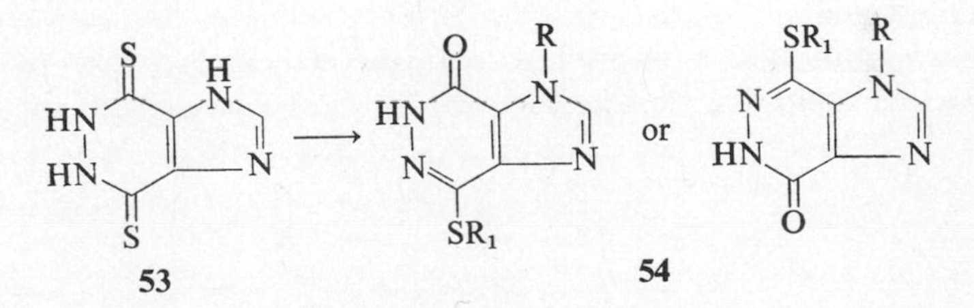

53 **54**

by thiation of the corresponding 4,7-dioxo analogs with phosphorus penta-sulfide in boiling pyridine (30, 32, 34).

Common transformations of the thioxo group have been performed, such as mono- or dialkylations to form *S*-alkylthio derivatives (29, 30, 32, 34, 35, 55) depending on the amount of the alkylating agent employed. However, it should be mentioned that when **53** reacted with 2 moles of substituted benzyl iodides in 2.5*N* potassium hydroxide solution compounds of the type **54** have been obtained (29).

A methylthio group could not be replaced with chlorine, but instead oxidation to a methylsulfonyl group was observed to take place along with some other transformations, all depending on reaction conditions (43). In this manner, when 4-methylthio- (**55**) or 4,7-bis-(methylthio)imidazo[4,5-*d*]-pyridazine (**56**) was allowed to react with chlorine gas at 55° in 50% aqueous solution of methanol, **57** was obtained. An increase in the reaction time favored the formation of **58**. In anhydrous methanol the corresponding methoxy analog **59** was formed. The formation of **58** is interpreted in terms

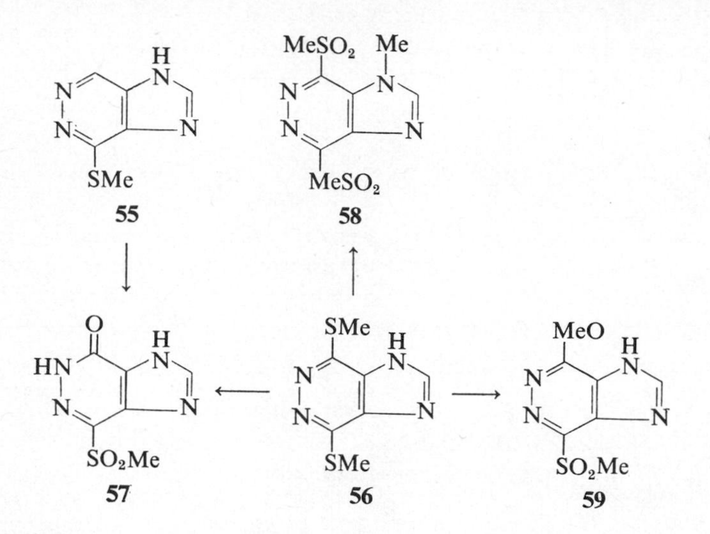

of the possibility of methyl carbonium ion formation and a possible mechanism is presented.

A thioxo group can be replaced by hydrogen with Raney nickel in boiling ethanol as shown for the preparation of **60** (30) or the parent compound (55).

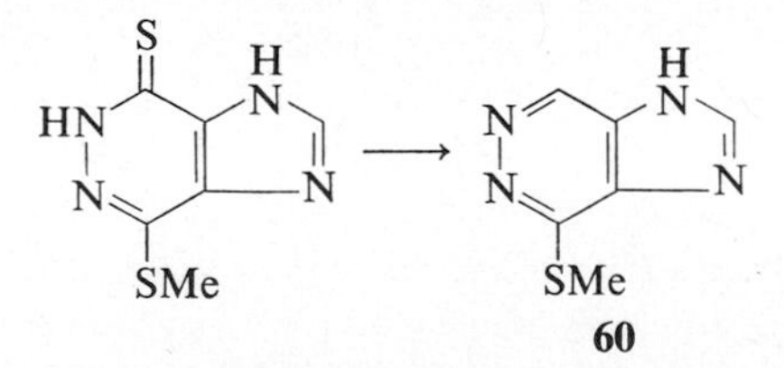

More frequently hydrogenolysis of chloroimidazo[4,5-*d*]pyridazines, pyridazinones, or pyridazinethiones has been used. Carbon (28, 39) used for such purposes treatment with sodium in liquid ammonia and under these conditions the 1-benzyl group was simultaneously cleaved and compounds of the type **61** were formed. However, this reaction failed with 4-dimethylamino or 4-diethylamino derivatives and 1-benzyl-4-7-dichloroimidazo[4,5-*d*]-pyridazine. This method was also used for the removal of the chlorine atom at position 2 (31), but otherwise the common catalytic dechlorination with

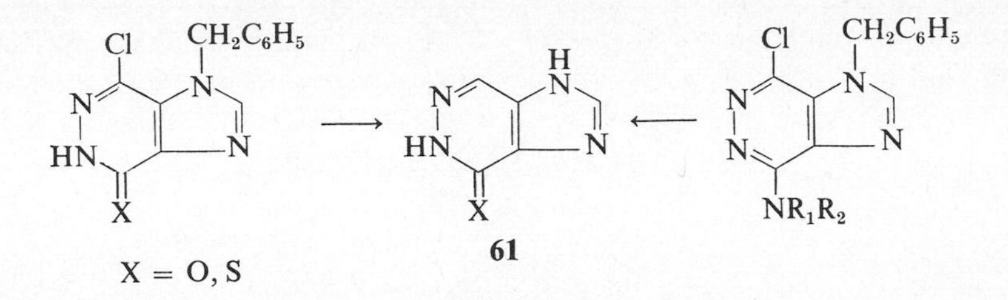

hydrogen in the presence of palladized charcoal was successfully used (28, 32, 39).

In view of the similarity of the imidazo[4,5-*d*]pyridazine ring system with that of purine, 7-amino- (**62**) and 4-amino-1-β-D-ribofuranosylimidazo-[4,5-*d*]pyridazine (**63**) have been synthesized as possible adenosine antimetabolites (42). 4(7)-Benzoylaminoimidazo[4,5-*d*]pyridazine was transformed according to the procedure of Baker (44) into the chloromercury derivative and thereafter converted with 2,3,5-tri-*O*-benzyoyl-D-ribofuranosyl chloride into a mixture of benzoylated nucleosides. After catalytic debenzoylation the nucleosides were isolated as picrates, with the aid of an ion exchange resin the free nucleosides were obtained and fractional crystallization was used to obtain pure isomers **62** and **63**, respectively.

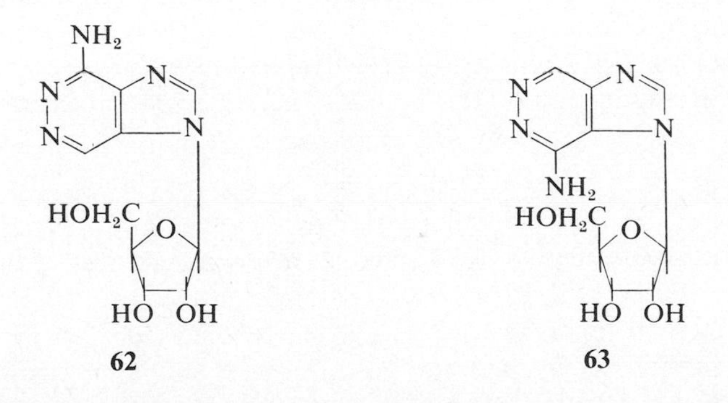

Structures to both isomers have been assigned on the basis of uv data and the conceivable ribosidation at the pyridazine part of the bicycle was considered to be unlikely.

Several other imidazo[4,5-*d*]pyridazines have been tested for their biological activity. Some compounds were claimed to be useful as central nervous system depressants (39) and others have been screened for antitumor activity (29, 35) or diuretic and other effects (32, 34, 58). 4-Aminoimidazo-[4,5-*d*]pyridazine is reported to cause feedback inhibition of purine biosynthesis at relatively low concentrations (45).

Uses of imidazo[4,5-*d*]pyridazines in color photography are also reported (46, 59).

III. Imidazo[4,5-*c*]pyridazines

7*H*-Imidazo[4,5-*c*]pyridazine

5*H*-Imidazo[4,5-*c*]pyridazine

This ring system was reported for the first time in 1964 by Kuraishi and Castle (47), and the chemistry of imidazo[4,5-*c*]pyridazines was studied in the following years by Castle and co-workers.

Since the system is capable of tautomerism in the imidazole part the imino-hydrogen atom could reside on either nitrogen atom (**64** or **65**, R = H).

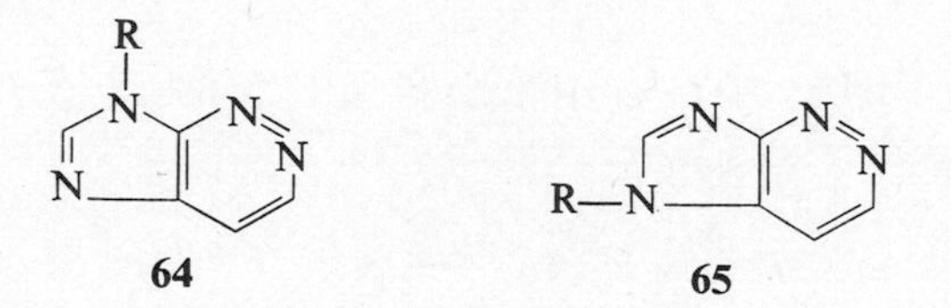

64 **65**

However, isomers of this kind have not been isolated, nor is precise location of the imino-hydrogen possible. Therefore, all representatives of this bicyclic system are referred as to 7*H*-imidazo[4,5-*c*]pyridazines.

Clarity in structure arises only when the mobile hydrogen in the imino group is substituted by another group, for example, a methyl group. Representatives of both isomers, 5*H*- and 7*H*-imidazo[4,5-*c*]pyridazine (**64** and **65**), are known.

Thus far, only one synthetic approach for building up this bicyclic system has been developed. The imidazole part of the bicycle is build up from appropriate pyridazines with *ortho* amino or modified amino groups. Ring closure has been accomplished with the aid of ethyl orthoformate or formic acid, whereas cyanogen bromide or carbon disulfide in pyridine yielded the corresponding 6-amino or thio derivatives.

In this way, diaminopyridazines when heated under reflux with ethyl orthoformate yielded the corresponding imidazo[4,5-*c*]pyridazines (**66**) (47, 48). Boiling formic acid was also used for cyclization purposes, although the yield was somewhat less satisfactory (49).

Cyclization with ethyl orthoformate proceeded satisfactorily when one amino group was replaced with a benzylidene hydrazino substituent, and the

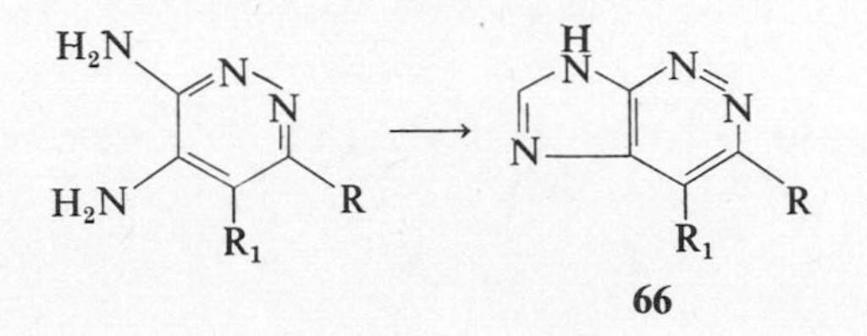

reaction has been applied to the preparation of **67** (48). 6-Aminoimidazo-[4,5-*c*]pyridazine (**68**) was obtained from 3,4-diaminopyridazine after treatment with cyanogen bromide (49).

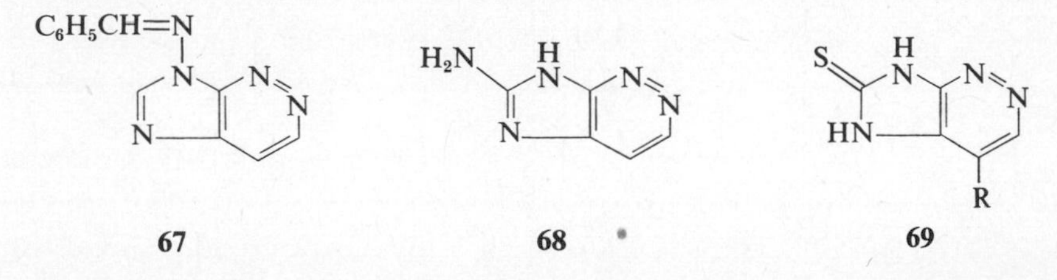

Similar treatment of the same pyridazine or its 5-chloro analog with carbon disulfide in pyridine and potassium hydroxide afforded after 1.5–2.5 hr of heating under reflux the corresponding 6-thioxo derivative (**69**, R = H) or its 4-chloro analog (**69**, R = Cl) (47, 49).

From 3,4,5-triaminopyridazine and following the same procedure with carbon disulfide as reagent only one product has been obtained (**70**), although the other isomer or a mixture of both could be expected. Structure proof of the product followed from dethiation with Raney nickel to the known 4-aminoimidazo[4,5-*c*]pyridazine **71** (49).

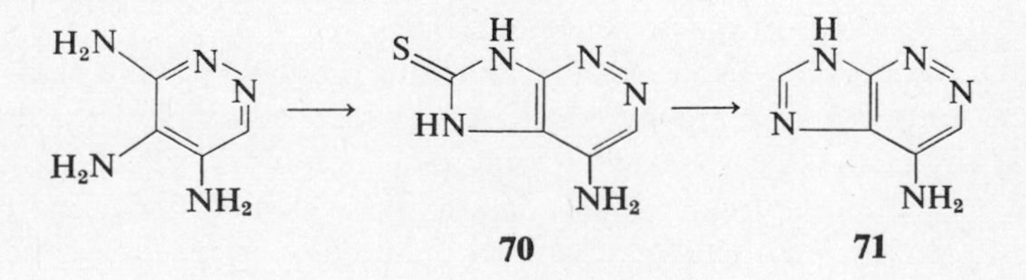

By starting with *ortho* amino and methylamino substituents on the pyridazine ring, derivatives of 5*H*- or 7*H*-imidazo[4,5-*c*]pyridazine with a fixed structure as a result of the impossibility of tautomerism could be prepared. In this manner 3-amino-4-methylaminopyridazines afforded the corresponding derivatives of the 5*H*-isomer (**72** and **73**), and 3-methylamino-4-aminopyridazines afforded the corresponding derivatives of the 7*H*-isomer (**74** and **75**) (49).

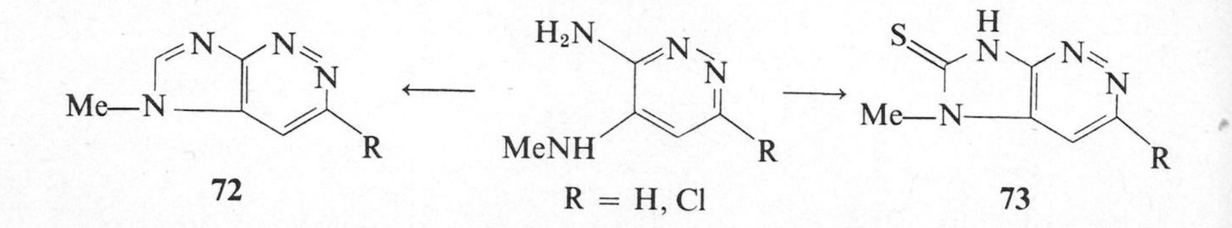

The parent compound was prepared in three different ways. Direct cyclization of 3,4-diaminopyridazine with formic acid and catalytic dechlorination of 3-chloroimidazo[4,5-*c*]pyridazine yielded imidazo[4,5-*c*]pyridazine in the same yield (68%), whereas dechlorination with palladized charcoal

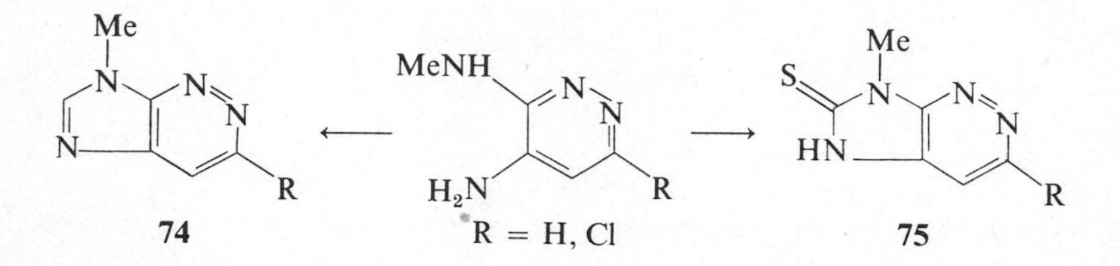

of the isomeric 4-chloroimidazo[4,5-*c*]pyridazine was less successful (36%) (49).

Although the parent compound has not been submitted to different reactions in order to study its reactivity, some data are available for its derivatives. HMO calculations (14) of total and frontier π-electron densities of 5*H*-imidazo[4,5-*c*]pyridazine are listed in Table 3D-3.

Experimental observations on chemical reactivity of this bicyclic system generally fit these theoretical calculations. The most susceptible position for nucleophilic attack is expected to be position 6, and this is evident from several ring openings in the presence of a base.

Normal nucleophilic displacements should be expected to take place on 3- and 4-chloroimidazo[4,5-*c*]pyridazines. A very low reactivity of the halogen has been observed in the reaction with ammonia or amines, whereas against a solution of sodium hydrogen sulfide in ethanol a greater susceptibility was observed (33, 47). With ammonia as nucleophile forcing reaction conditions were necessary (210–220°, 25 hr) (47); similar reactions with dialkylamino-alkylamines also require rather vigorous treatment. An interesting case is the replacement of the halogen in 3-chloro and 4-chloroimidazo[4,5-*c*]pyridazine with sodium hydrogen sulfide. The 4-chloro isomer afforded with a solution of sodium hydrogen sulfide in ethanol (140°, 1 hr) the corresponding thioxo compound (**76**) in moderate yield (55%). On the other hand, the 3-chloro

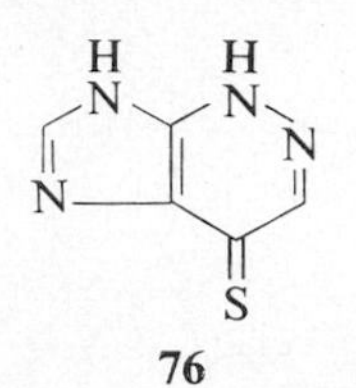

isomer yielded under similar reaction conditions (140°, 8 hr) a mixture of the expected thioxo compound (**77**) and 5,6-diamino-3(2*H*)pyridazinethione (**78**) (33).

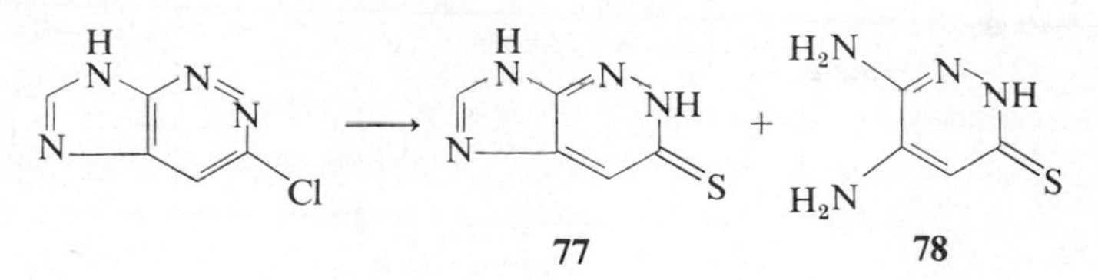

The formation of the pyridazine derivative **78** as a byproduct can be interpreted in terms of a greater susceptibility toward nucleophilic attack of the position 6 than position 3 (or 4) and on account of the prolonged action of the reagent. An easier displacement of the 4-chlorine than 3-chlorine atom in the discussed reactions is certainly due to stabilization of the transition state, which has a better resonance stabilization for the intermediate leading to 4- than to 3-substitution products.

The great susceptibility of the position 6 for nucleophilic attack is reflected also in imidazole ring openings which have been demonstrated in other cases. In an attempt to synthesize **76** from the corresponding 4-chloro analog by the phosphorus pentasulfide-pyridine method (8 hr under reflux) the replacement of a chlorine atom and ring opening resulted and 5,6-diamino-4(1*H*)pyridazinethione (**79**) was formed in low yield (49).

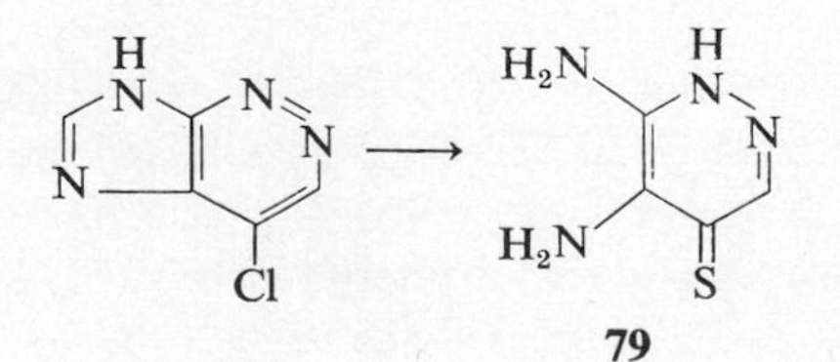

Similarly, under the influence of alkali ring opening took place in the attempted dechlorination of 3-chloro-5-methyl- or 3-chloro-7-methylimidazo-[4,5-*c*]pyridazine in the presence of palladized charcoal as catalyst (49). Both products, **80** and **81**, were obtained in moderate yield. On the other

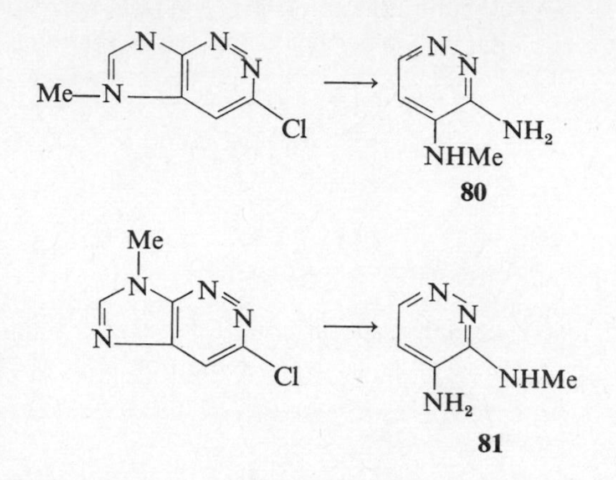

hand, an attempt to obtain the 7-amino derivative from **82** by a hydrolytic cleavage with 10% hydrogen chloride in absolute ethanol (30 min, reflux) resulted in the cleavage of the imidazole ring to produce **83** (48).

In a recent report, however, it was observed that after heating **82** under reflux with 1*N* hydrochloric acid or with acetic acid, 8-amino-6-chloro-*s*-triazolo[4,3-*b*]pyridazine was formed and accompanied with compound **83** as byproduct (63). Apparently recyclization with the elimination of the phenyl group took place. A reasonable explanation for this behavior could be in the diminished stability of the bicycle due to the impeded resonance stabilization due to protonation.

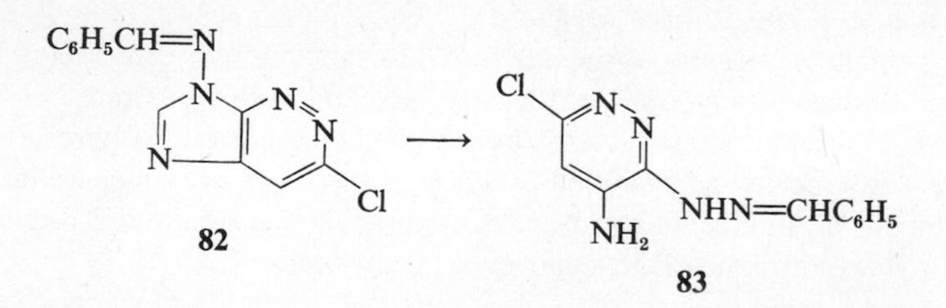

The mobile hydrogen in 4-chloro-7*H*(or 5*H*)imidazo[4,5-*c*]pyridazine has been blocked in the reaction with 2,3-dihydro-4*H*-pyran. It is feasible either that substitution at N_5 or N_7 took place or that a mixture of both isomers was formed. However, only one isomer was formed and isolated and on steric grounds and uv spectra correlations the N_7-tetrahydropyranyl structure **84** has been assigned to this isomer (33). During the replacement

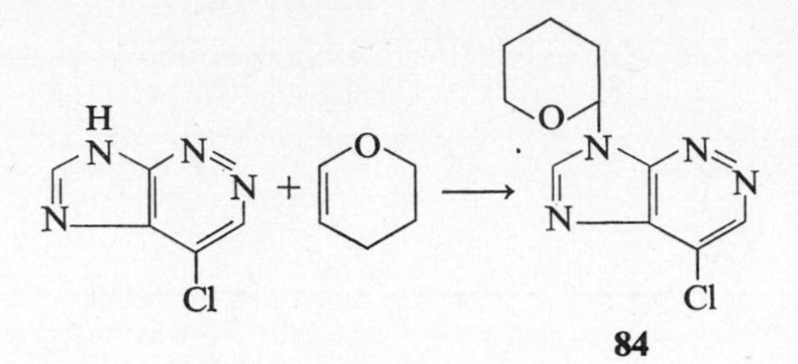

reaction of the 4-chlorine atom in **84** with amines, followed by acid treatment, the protective tetrahydropyranyl group was split off.

Catalytic dehalogenations of 3- or 4-chloroimidazo[4,5-*c*]pyridazines proceeded normally in the presence of hydrogen and palladized charcoal (49). It should be mentioned that under these conditions and in the presence of sodium hydroxide a thioxo group remained unaltered, as exemplified by the transformation of **85** to **86** (49).

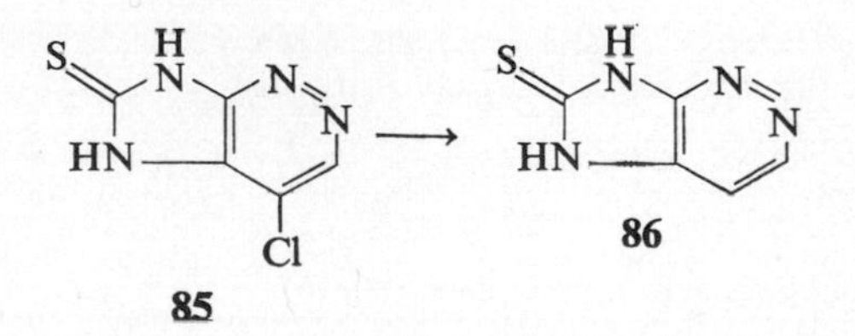

With Raney nickel, however, it is possible to remove only the sulfur from the thioxo group as shown by the preparation of 4-chloro- (47) and 4-aminoimidazo[4,5-*c*]-pyridazine (**87**) from the corresponding 6-thioxo derivatives (49).

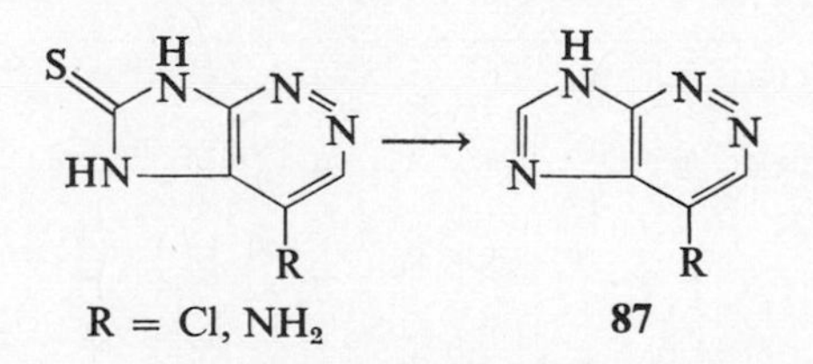

All 6-thioxo derivatives could be converted into the corresponding 6-methylthio derivatives by the usual methylation procedure with methyl iodide (33, 47, 49).

The interest in this ring system originates from its similarity to the purine skeleton and several adenine analogs have been prepared.

IV. Tables

TABLE 3D-1. Total and Frontier π-Electron Densities of Imidazo[1,2-*b*]pyridazine

Position	1	2	3	4	5	6	7	8	8*a*
Total	1.3214	1.0477	1.0789	1.5646	1.1443	0.9559	0.9767	0.9547	0.9558
Frontier	0.2118	0.0392	0.2991	0.0074	0.1677	0.0079	0.1373	0.0831	0.0464

TABLE 3D-2. Total and Frontier π-Electron Densities of 1*H*-Imidazo[4,5-*d*]pyridazine

Position	1	2	3	3*a*	4	5	6	7	7*a*
Total	1.6307	0.8526	1.3170	1.0038	0.9442	1.1636	1.1618	0.9539	0.9724
Frontier	0.0298	0.1776	0.2517	0.0628	0.1519	0.0014	0.1965	0.0336	0.0946

Total π-Electron Densities and Bond Orders (38)

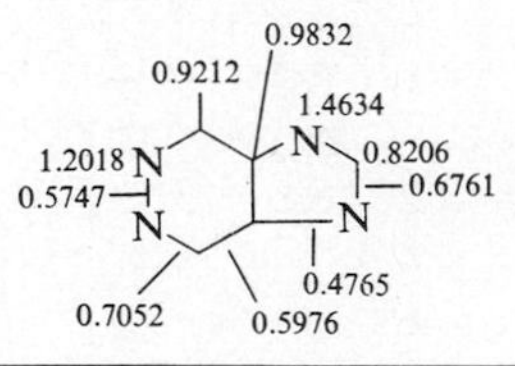

TABLE 3D-3. Total and Frontier π-Electron Densities of 5*H*-Imidazo[4,5-*c*]pyridazine

Position	1	2	3	4	4*a*	5	6	7	7*a*
Total	0.1673	1.1395	0.9841	0.9597	0.9691	1.6410	0.8359	1.3334	0.9699
Frontier	0.2336	0.0001	0.1702	0.0882	0.0273	0.0673	0.1146	0.2714	0.0273

TABLE 3D-4. Imidazo[1,2-*b*]pyridazines

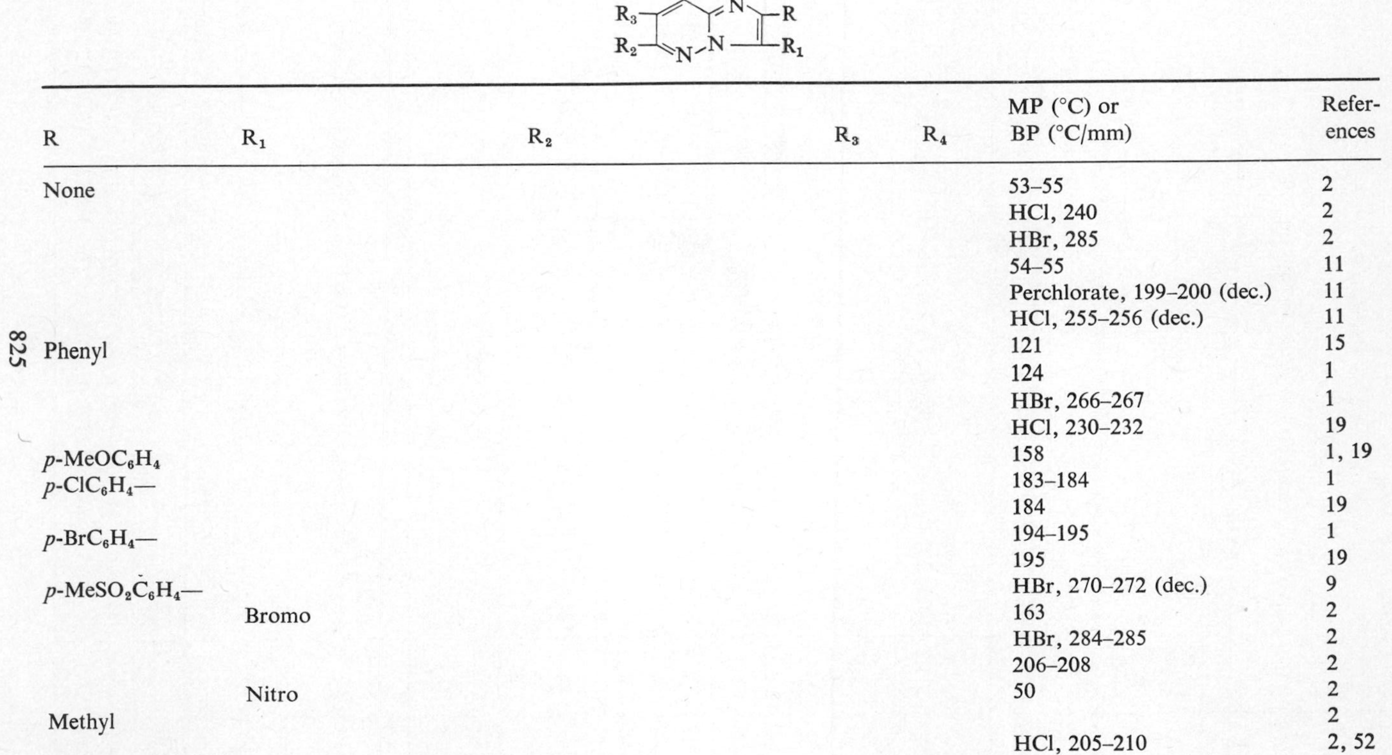

R	R_1	R_2	R_3	R_4	MP (°C) or BP (°C/mm)	References
None					53–55	2
					HCl, 240	2
					HBr, 285	2
					54–55	11
					Perchlorate, 199–200 (dec.)	11
					HCl, 255–256 (dec.)	11
Phenyl					121	15
					124	1
					HBr, 266–267	1
					HCl, 230–232	19
p-$MeOC_6H_4$					158	1, 19
p-ClC_6H_4—					183–184	1
					184	19
p-BrC_6H_4—					194–195	1
					195	19
p-$MeSO_2C_6H_4$—					HBr, 270–272 (dec.)	9
	Bromo				163	2
					HBr, 284–285	2
					206–208	2
	Nitro				50	2
Methyl						2
					HCl, 205–210	2, 52

TABLE 3D-4 (*continued*)

R	R_1	R_2	R_3	R_4	MP (°C) or BP (°C/mm)	References
		Chloro			115	3, 4
					HCl, 240	3, 4
					HBr, >320	3, 4
		Hydrazino			225	3, 4
					2 HCl, 261–262	3, 4
		NHNHCOH			218	3, 4
		NHN=CHOEt			137–138	3, 4
		Azido			108	3, 4, 60
		Amino			196	3, 4
		$NHNHCOC_6H_5$			258	4
		NHNHCSSH			—	4
		NHNHCSSMe			170	4
		$C_6H_5NHCSNHNH$—			179	4
		p-$EtOC_6H_4NHCSNHNH$—			120	4
		n-BuNHCSNHNH—			190	4
		$NH_2CSNHNH$—*x*2 HCl			—	4
		—NHN=CHMe			172	4
		—NHN=CMe_2			162	4
		—NHN=CHC_6H_5			224	4
		p-$NO_2C_6H_4CH$=NNH—			307–309	4
		p-BrC_6H_4CH=NNH—			270–271	4
		p-OHC_6H_4CH=NNH—			310–311	4
		Phenylthio-			100	4
		—NH—N=C(CH_3)—COOEt			189–190	53

		—NH—N=C(—CH_3)—COOH			178–179	53
		Nitro			145–146	61
			Methyl		132	5
				Methyl	43–44	5
		Methyl			125	5
		Methoxy			106–108	6, 50
			Methyl	Methyl	124	5
					HCl, 245	52
Methyl			Methyl		78	5
Methyl	Methyl				58–59	5
Methyl				Methyl	132	5
Methyl		Methyl			96	5
Methyl		Methoxy			87–88	6, 50
COOEt		Methoxy			HBr, 154–155	6
COONa		Methoxy			>320	6
—$CH_2CH(OH)$—CCl_3		Methoxy			HCl, 228–230	6
Phenyl		Chloro			199–201	8
					199	7
					200	10
Phenyl		—N(piperidino)			168.5–170.5	8
Phenyl		—$NHCH_2CH_2NEt_2$			116.5–118.5	8
Phenyl		—$NH(CH_2)_3NMe_2$			169–170	8
	Bromo	C_6H_5—C(Br)=N—NH—			154	4

TABLE 3D-4 (*continued*)

R	R_1	R_2	R_3	R_4	MP (°C) or BP (°C/mm)	References
	Bromo	Chloro			156	4
					HCl, 233–235	4
	Bromo	Hydrazino			236	4
	Bromo	$NHN{=}CHC_6H_5$			241	4
Methyl		Chloro			127–128	7
					HBr, 238–240	7
					HCl, 238	52
Methyl		Hydrazino			187–189	7
					2 HCl, 295	7
					2 HBr, 296–297	7
Methyl		$C_6H_5CH{=}NHN—$			235–236	7
Methyl		$MeCH{=}NHN—$			218–220	7
Methyl		$Me_2C{=}N—NH—$			198–200	7
Methyl		$p\text{-}OHC_6H_4CH{=}NNH—$			287–289	7
Methyl		$p\text{-}NO_2C_6H_4CH{=}NNH—$			288–289	7
Phenyl		Hydrazino			208	7
					2 HCl, 255–256	7
					2 HBr, 277–279	7
Phenyl		$—NHN{=}CHC_6H_5$			208–210	7
Phenyl		$MeCH{=}NNH—$			206	7
Phenyl		$Me_2C{=}NNH$			244	7
Phenyl		$p\text{-}OHC_6H_4CH{=}NNH—$			295	7
Phenyl		$p\text{-}NO_2C_6H_4CH{=}NNH—$			230	7
Phenyl		$HOOCCH{=}NNH—$			210–211	7
Phenyl		$HOOC—C(Me){=}NNH—$			186–187	7

Phenyl		—NHNHCOOEt			HCl, 189	7
Phenyl		$C_6H_{11}NHCSNHNH$—			226	7
Phenyl		o-$MeC_6H_4NHCSNHNH$—			222	7
Phenyl		m-$MeC_6H_4NHCSNHNH$—			210	7
Phenyl		o-$MeOC_6H_4NHCSNHNH$—			215–217	7
Phenyl		p-$EtOC_6H_4NHCSNHNH$—			228–230	7
Methyl		HOOC—CH=NNH—			218	7
Methyl		$C_6H_5NHCSNHNH$—			230	7
Methyl		o-$MeC_6H_4NHCSNHNH$—			143	7
Methyl		p-$MeC_6H_4NHCSNHNH$—			138	7
Methyl		p-$MeOC_6H_4NHCSNHNH$—			208	7
Methyl		p-$EtOC_6H_4NHCSNHNH$—			187	7
COOH		Chloro			260	7
	Nitro	Chloro			210	2
	Nitro	Hydrazino			188–189	2
	SO_3H	Chloro			328–330	2
		Chloro	Methyl		108–109	5
					HCl, 218	52
		Chloro		Methyl	38–39	5
					Oil	6
	$—CH_2N$ (ring)	Methoxy			2 HCl x0.5 H_2O 202–205 (dec.)	6
					191–193 (dec.)	50
	$—CH_2N(n\text{-}Bu)_2$	Methoxy			2 HCl, 155–157	50
	$—CH_2N(CH_2CH_2OH)_2$	Methoxy			2 HCl, 209–211 (dec.)	50
	$—CH_2NMe_2$	Methoxy			94–96	6, 50
	$—CH_2N$ (ring) O	Methoxy			—	6
					2 HCl·H_2O, 213–215 (dec.)	6, 50

TABLE 3D-4 (*continued*)

R	R_1	R_2	R_3	R_4	MP (°C) or BP (°C/mm)	References
	$—CH_2N$ (piperazinyl) $N—Me$	Methoxy			—	6
					3 $HCl \cdot H_2O$, 204–206 (dec.)	6, 50
	$—CH_2N(Me)CH_2C_6H_4$	Methoxy			—	6
					2 $HCl \cdot 0.5$ H_2O, 166–169 (dec.)	6
p-$MeSO_2C_6H_4$		Chloro			251–253 (dec.)	9
p-$MeSO_2C_6H_4$		Methoxy			216–219	9
Phenyl		Chloro			190	1
p-$MeOC_6H_4$		Chloro			214	1, 10
p-ClC_6H_4		Chloro			219	1, 10
p-BrC_6H_4		Chloro			222	1, 10
Phenyl		Methoxy			137.5	1, 16
Phenyl		$Me_2NCH_2CH_2O—$			107	1
p-$MeOC_6H_4$		Methoxy			182	1, 16
p-ClC_6H_4		Methoxy			179	1, 16
p-BrC_6H_4		Methoxy			178	1, 16
Phenyl		Ethoxy			132	1, 16
p-$MeOC_6H_4$		Ethoxy			131	1, 16
p-ClC_6H_4		Ethoxy			162	1, 16
p-BrC_6H_4		Ethoxy			171	1
					174	16
Phenyl		n-Propoxy			107	1, 16
p-$MeOC_6H_4$		n-Propoxy			98	1, 16

p-ClC_6H_4	n-Propoxy	138–139	1, 16
p-BrC_6H_4	n-Propoxy	168	1, 16
Phenyl	Iso-Propoxy	98	1, 16
p-$MeOC_6H_4$	Iso-Propoxy	110	1
		165	16
p-ClC_6H_4	Iso-Propoxy	138–139	1
p-BrC_6H_4	Iso-Propoxy	148–149	1, 16
p-$MeOC_6H_4$	$Me_2NCH_2CH_2O$—	103–104	1
p-ClC_6H_4	$Me_2NCH_2CH_2O$—	158	1
p-BrC_6H_4	$Me_2NCH_2CH_2O$—	172	1
Phenyl	$Et_2NCH_2CH_2O$—	102	1
p-$MeOC_6H_4$	$Et_2NCH_2CH_2O$—	75	1
p-ClC_6H_4	$Et_2NCH_2CH_2O$—	129	1
p-BrC_6H_4	$Et_2NCH_2CH_2O$—	132.5	1
Phenyl	Me_2N—	195	1, 17
p-$MeOC_6H_4$	Me_2N—	204	1, 17
p-ClC_6H_4	Me_2N—	203	1, 17
p-BrC_6H_4	Me_2N—	197	1, 17
Phenyl	—N O (morpholino ring)	183–184	1, 17
p-$MeOC_6H_4$	—N O (morpholino ring)	198–199	1, 17
p-ClC_6H_4	—N O (morpholino ring)	227	1, 17
p-BrC_6H_4	—N O (morpholino ring)	234	1
		243	17

TABLE 3D-4 (*continued*)

R	R_1	R_2	R_3	R_4	MP (°C) or BP (°C/mm)	References
Phenyl		$-N(CH_2)_5$			169	1, 17
p-$MeOC_6H_4$		$-N(CH_2)_5$			153	1, 17
p-ClC_6H_4		$-N(CH_2)_5$			222	1, 17
p-BrC_6H_4		$-N(CH_2)_5$			221	1, 17
Phenyl		$-N(CH_2)_4$			187–188	1, 17
p-$MeOC_6H_4$		$-N(CH_2)_4$			190	1, 17
p-ClC_6H_4		$-N(CH_2)_4$			210	1, 17
p-BrC_6H_4		$-N(CH_2)_4$			211	1, 17
Phenyl		Bromo			213	10
Phenyl		Azido			200–204	18
Phenyl	Bromo				HCl, 250–252	15
p-BrC_6H_4		Chloro			228–230	15
p-$NO_2C_6H_4$	Nitro				239–240	15

Methyl	Methyl	Methyl			56	5
Methyl			Methyl	Methyl	123	5
					HCl, 267–268	52
Methyl	Methyl			Methyl	145	5
Methyl	Methyl		Methyl		129	5
Methyl	Methyl	Chloro			140–141	5
Methyl		Chloro		Methyl	105–106	5
Methyl		Chloro	Methyl		190	5
		Chloro	Methyl	Methyl	85–86	5
					HCl >310	52
Methyl	COOEt	Chloro			99–100	7
					HCl, 165	7
					HBr, 220	7
Methyl	COOH	Chloro			255	7
Methyl	COOH	Methoxy			237	7
Methyl	$CONHNH_2$	Hydrazino			304–307	7
					3 HCl >320	
Methyl	$CONHNHCSNHC_6H_5$	$NHNHCSNHC_6H_5$			280–282	7
Methyl	$CONHNHCSNHC_6H_{11}$	$NHNHCSNHC_6H_{11}$			276–277	7
Methyl	$CONHNH_2$	$NHN{=}CHC_6H_4{-}OH{-}p$			301–304	7
Methyl	$CONHN{=}CHC_6H_5$	$NHN{=}CHC_6H_5$			270–271	7
Methyl	$CONHN{=}CMe_2$	$NHN{=}CMe_2$			290–292	7
Methyl	p-$BrC_6H_4CH{=}NHNCO$	p-$BrC_6H_4CH{=}NNH$			320–321	7
Methyl	p-$NO_2C_6H_4CH{=}NHNCO$	p-$NO_2C_6H_4CH{=}NNH$			330	7
Methyl	Bromo	Chloro			154	2
Methyl	Nitro	Chloro			196	2
Methyl	Nitro	Hydrazino			252–255	2
Methyl	Nitro	p-$OHC_6H_4CH{=}NNH$			>340	2
Methyl	Nitro	$C_6H_{11}NHCSNHNH$			238–240	2
Methyl	COOEt	Methoxy			110–112	6
Methyl	COONa	Methoxy			315 (dec.)	6

TABLE 3D-4 (*continued*)

R	R_1	R_2	R_3	R_4	MP (°C) or BP (°C/mm)	References
Methyl	CH_2NMe_2	Methoxy			68–70 2 HCl, 202–206 (dec.)	6, 50 6, 50
Methyl	$CH_2N(CH_2CH_2OH)_2$	Methoxy			130–131 129–130	6 50
Methyl	CH_2N (ring)	Methoxy			— 2 HCl·0.5 H_2O, 211–212 (dec.) 209–210.5	6 6 50
Methyl	$CH_2N(n\text{-}C_4H_9)_2$	Methoxy			— 2 HCl·0.5 H_2O, 168–170 (dec.)	6 6
Methyl	CH_2N (ring) O	Methoxy			77–79, 76–78	6, 50
Methyl	$CH_2N(Me)—CH_2C_6H_5$	M thoxy			— 2 HCl·0.5 H_2O, 206–207 (dec.) 204–205 (dec.) 136–140	6 6 50 50
Methyl	CH_2N (ring) N—C_6H_5	Methoxy			3 HCl, 213–214 (dec.) 211–212°	6 50

Methyl	CH_2—N(piperazine ring)N—Me	Methoxy			3 HCl·H_2O, 207–208 (dec.)	50
Methyl	$CH_2CH_2N(n\text{-}Bu)_2$	Methoxy			2 HCl·0.5 H_2O, 167.5–168.5	50
Phenyl	Nitro	Chloro			280	15
p-$NO_2C_6H_4$	Nitro	Chloro			235–237	15
Phenyl	Bromo	Chloro			190	15
Phenyl	Bromo	Hydrazino			204–205	15
					2 HCl >32 O	
Phenyl	Bromo	$NHN{=}CHC_6H_5$			236–238	15
Phenyl	Bromo	$NHNHCSNHC_6H_5$			193–194	15
Phenyl	Bromo	C_6H_5—C(Br)=NNH—			220–221	15
p-BrC_6H_4	Bromo	Chloro			200–201	15
Methyl	$CH_2\overset{\oplus}{N}Me_3I^{\ominus}$	Methoxy			260–267 (dec.)	6
Methyl	Methyl		Methyl	Methyl	127–128	5
Methyl		Chloro	Methyl	Methyl	112	5
					HCl, 280–282	52
Methyl	Methyl	Chloro		Methyl	125–127	5
Methyl	Methyl	Chloro	Methyl		142	5
Methyl	Methyl	Chloro	Methyl	Methyl	80	5

TABLE 3D-5

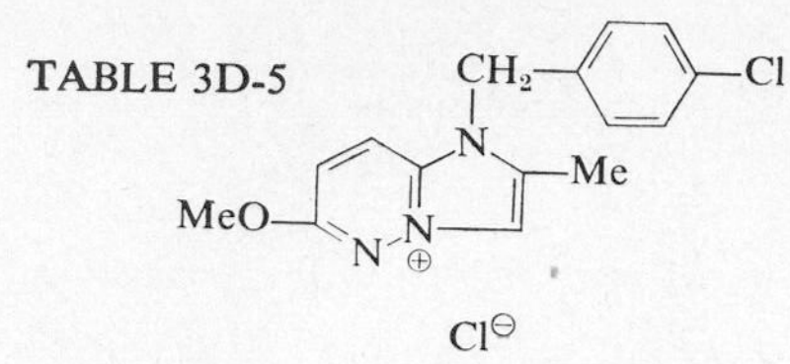

MP 194–195° C Ref. 6

TABLE 3D-6

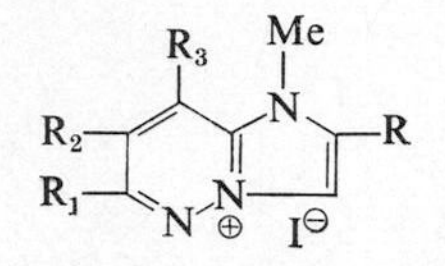

R	R_1	R_2	R_3	MP (°C) or BP (°C/mm)	References
None				285–286	2
Methyl				285–286	52
	Methyl			207	52
	Chloro			276–277	52
Methyl	Chloro			255–257	52
	Chloro	Methyl		268–270	52
	Chloro		Methyl	261–263	52
		Methyl	Methyl	275	52
Methyl		Methyl	Methyl	>330	52
	Chloro	Methyl	Methyl	280	52
Methyl	Chloro	Methyl	Methyl	292–294	52

TABLE 3D-7. Imidazo[1,2-*b*]pyridazin-6(5*H*)ones

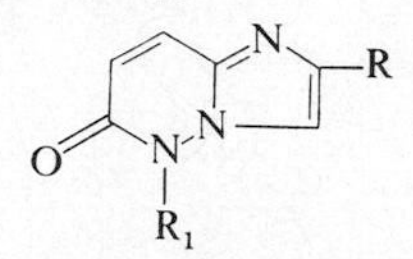

R	R_1	MP (°C) or BP (°C/mm)	References
Phenyl		300	1
p-$MeOC_6H_4$		287	1
p-BrC_6H_4		300	1
p-ClC_6H_4		300	1
Phenyl	Methyl	120	1
p-$MeOC_6H_4$	Methyl	140–141	1
p-ClC_6H_4	Methyl	246–247	1
p-BrC_6H_4	Methyl	261–262	1
Methyl		— HBr, 204 (dec.)	6

TABLE 3D-8

Structure	MP (°C)	References
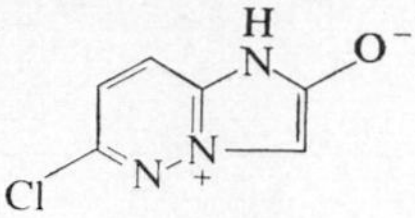	230–231	51
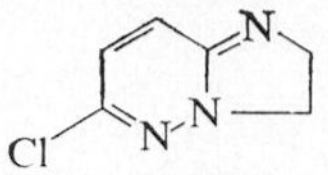	113–115	51
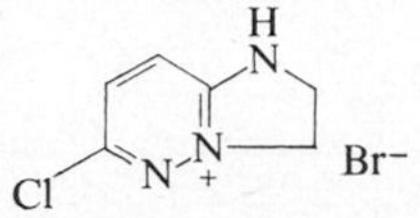	242–245	51
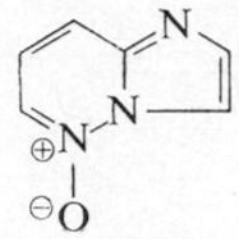	175–176	61

TABLE 3D-9. 1*H*-Imidazo[4,5-*d*]pyridazines

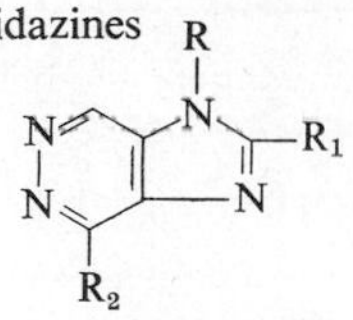

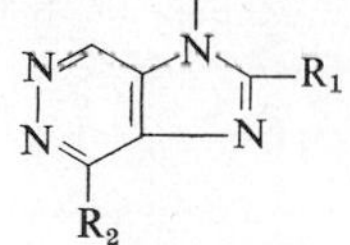

R	R_1	R_2	MP (°C) or BP (°C/mm)	References
None			0.5 H_2O, 308–309	62
			279–280 (dec.)	55
		Benzylthio	199–200	55
		$NH(CH_2)_2NMe_2$	2 HCl, 270–270.5 (dec.)	55
		$NH(CH_2)_3NMe_2$	2 HCl, 277–279	55
		$NH(CH_2)_3NEt_2$	2 $HCl{\cdot}H_2O$, 129–131	55
		Methoxy	283–285	62
		$NHCOC_6H_5$	298–299	42
		Hydrazino	HCl, >350	42, 55
		Methylthio	230–232, 235–236	30, 55
			229–231	62
		Amino	262–263 (dec.)	28
			HCl, 334–337	28, 29
			315 (dec.)	30
		NHMe	298–300 (dec.)	28, 39
		NHEt	280–281 (dec.)	28, 39
		NHPr-*n*	215–216	28
		N(*n*-Pr)$_2$	146.5–147	28
Benzyl			206–207	28
	Phenyl		330–332	32

TABLE 3D-10. 1*H*-Imidazo[4,5-*d*]pyridazines

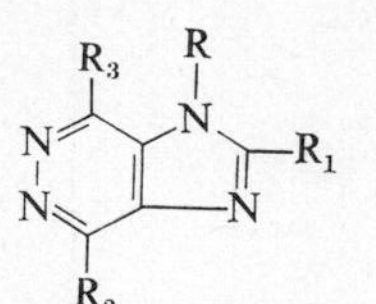

R	R_1	R_2	R_3	MP (°C) BP (°C/mm)	References
		Methoxy	Methyl	—	35[a], 62
		Chloro	Chloro	240–242	33, 56
				240.5–241.5	30
		Methoxy	Methoxy	238–239	36
		Methylthio	Methylthio	243–245	30
		Amino	Amino	340–342	30
				H_2O, 314–316	30
		SPr-*n*	SPr-*n*	126–128	29
		SPr-*i*	SPr-*i*	127–129	29
		$SCH_2C_6H_5$	$SCH_2C_6H_5$	125	29
		o-$NO_2C_6H_4CH_2S$—	*o*-$NO_2C_6H_4CH_2S$—	97–98	29
		3,4-di$ClC_6H_3CH_2S$—	3,4-di$ClC_6H_3CH_2S$—	186–187.5	29
		p-$ClC_6H_4CH_2S$—	*p*-$ClC_6H_4CH_2S$—	147–149	29
		p-$IC_6H_4CH_2S$—	*p*-$IC_6H_4CH_2S$—	315–317	29
		2,4-di$NO_2C_6H_3S$—	2,4-di$NO_2C_6H_3S$—	323–324	29
		p-$BrC_6H_4COCH_2S$—	*p*-$BrC_6H_4COCH_2S$—	216–217	29
		p-$IC_6H_4COCH_2S$—	*p*-$IC_6H_4COCH_2S$—	212–214	29
		MeNH	MeNH	314–315	29
		EtNH	EtNH	247–249	29
		n-BuNH	*n*-BuNH	110	29
		Me_2N	Me_2N	209–211	29

[a] No definite m.p., shrinking at about 210° C.

	Ethoxy	Methyl	>360	62
	$HOCH_2CH_2NH$—	$HOCH_2CH_2NH$—	251–253	29
	$Me_2N(CH_2)_3NH$	$Me_2N(CH_2)_3NH$	174–176	29
			175–176	33
	$Et_2N(CH_2)_3NH$	$Et_2N(CH_2)_3NH$	140–142	29
			143–144	33
	Chloro	Ethoxy	275	33
	Chloro	$Me_2N(CH_2)_3NH$	2 HCl, 249–251	33
	Chloro	$Et_2N(CH_2)_3CH(Me)NH$	2 HCl, 168–170	33
	Methylthio	—$N(Pr\text{-}i)_2$	305–307	29
	Methylthio	—$N(3\text{-}Me\text{-}Bu)_2$	319–321	29
	Hydrazino	Hydrazino	>360	29
	p-$MeOC_6H_4$—	*p*-$MeOC_6H_4$—	146–147	22
			262–263	23
	$C_6H_5CH_2NH$	$C_6H_5CH_2NH$	—	28
			HCl, 201–203	28
			HCl, 200–202	31
	Methoxy	$MeSO_2$	312–313	43
Methyl	Amino		295–296 (dec.)	28, 39
Methyl		Amino	318–319 (dec.)	28
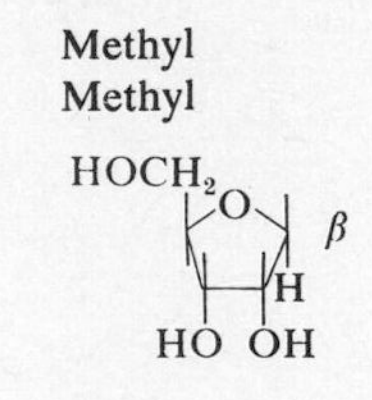	Amino		229–230 (dec.)	42

TABLE 3D-10 (*continued*)

R	R_1	R_2	R_3	MP (°C) BP (°C/mm)	References
$HOCH_2$ furanose (β) ring with H, HO, OH			Amino	218–220 (dec.)	42
Methyl		Amino	Chloro	273–275 (dec.)	28, 39
Methyl		Chloro	Chloro	236–237.5	28
Ethyl		Ethylthio	Ethylthio	122–124	30
				HI, 256–259	30
Methyl		$MeSO_2$	$MeSO_2$	236	43
Benzyl		$C_6H_5CH_2NH$	$C_6H_5CH_2NH$	202–203	28
Phenyl		Chloro	Chloro	231–231.5	28
Benzyl		Ethoxy	Chloro	190–192	28
Benzyl		Chloro	Chloro	161–161.5	28
				160–161	39
Benzyl		Amino	Chloro	271–273 (dec.)	28
				269–270	39
Benzyl		NHMe	Chloro	185–187	28, 39
Benzyl		NHEt	Chloro	215–217	28, 39
Benzyl		NH—Pr-*n*	Chloro	169–171	28, 39
Benzyl		NMe_2	Chloro	181–183	28, 39
Benzyl		$N(n\text{-}Pr)_2$	Chloro	160–161	28, 39
Benzyl		NEt_2	Chloro	139–140	28, 39
Benzyl		$N(CH_2CH_2OH)_2$	Chloro	158–159	28
Benzyl		Hydrazino	Chloro	191.5–192	42

[tetrahydropyran-2-yl structure]		Chloro	Chloro	129–130	33
[tetrahydropyran-2-yl structure]		$Me_2N(CH_2)_3NH$	Chloro	164–165	33
	Phenyl	Methylthio	Methylthio	195–197	32, 34
	Phenyl	Ethylthio	Ethylthio	182–184	32, 34
	Phenyl	$SCH_2C_6H_5$	$SCH_2C_6H_5$	163–165	32
	Phenyl	Amino	Amino	283–285	32, 34
				Hydrate, 183–200	32, 34
	Phenyl	NHMe	NHMe	310–313	32, 34
	Phenyl	Hydrazino	Hydrazino		
				HCl, 260–261	32, 34
	Phenyl	$NHN{=}CHC_6H_5$	$NHN{=}CHC_6H_5$	318–321	32
	Phenyl	Chloro	Chloro	282–284	32, 34
	Methyl	Methoxy	Methoxy	212	36
	Chloro	Chloro	Chloro	230–232	31
	—N [morpholino ring]	—N [morpholino ring]	—N [morpholino ring]	284–285 (dec.)	31
	Chloro	—N [piperidino ring]	—N [piperidino ring]	171.5–173	31
	Chloro	$—NHCH_2—$[2-furyl ring]	$—NHCH_2—$[2-furyl ring]	107–109	31
	Chloro	$C_6H_5CH_2NH—$	$C_6H_5CH_2NH—$	—	31
				HCl, 180–182	31
	Chloro	Hydrazino	Hydrazino	>300	31
	Methylthio	Methoxy	Methyl	182–184	35
	$C_6H_5CH_2S$	Methoxy	Methyl	178–181	35

TABLE 3D-11. 1*H*-Imidazo[4,5-*d*]pyridazin-4(5*H*)ones

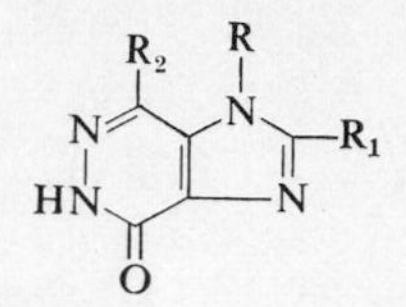

R	R_1	R_2	MP (°C) or BP (°C/mm)	References
		$MeSO_2$	342–343	43
Benzyl		Chloro	232–233	28
None			>300	28, 55
			>340	62
		Chloro	358–360	30
			>350	62
		Methylthio	322–323	29
	Phenyl		347–349	32
	Phenyl	Chloro	358–360	32, 34
		Methyl	>360	62
(structure: Ph—N pyridazinone fused imidazole-2-thione, H N, S, NH, O)			305–307	57

TABLE 3D-12. 1*H*-Imidazo[4,5-*d*]pyridazin-7(6*H*)ones

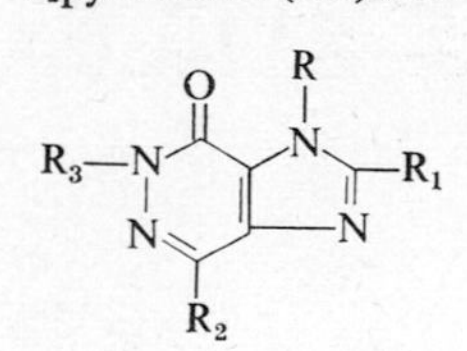

R	R_1	R_2	R_3	MP (°C) or BP (°C/mm)	References
o-$ClC_6H_4CH_2$		o-$ClC_6H_4SCH_2$		268–270	29
p-$ClC_6H_4CH_2$—		p-$ClC_6H_4SCH_2$		269–271	29
2,4-di$ClC_6H_3CH_2$		2,4-di$ClC_6H_3CH_2S$		275–277	29
		Methoxy		>360	36
		Methoxy	Methyl	262–264	36
	Chloro	Chloro		>300	31
		Methyl	Methyl	>300	62

TABLE 3D-13. 1*H*-Imidazo[4,5-*d*]pyridazine-4,7(5*H*,6*H*)diones

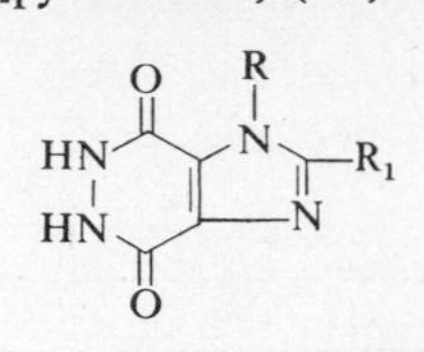

R	R_1	MP (°C) or BP (°C/mm)	References
None		>260	26, 56
		>400	25, 30, 33
	Phenyl	390	32, 34
Methyl		354–356	25
Phenyl		315–316	25
Benzyl		318–320	28
Benzyl	Methylthio	312–315	29
Benzyl	Ethylthio	258–260	29
Benzyl	*n*-Propylthio	255–257	29
Benzyl	*n*-Butylthio	250	29
Benzyl	*sec*-Butylthio	283	29
Benzyl	*t*-Amylthio	256–258	29
Benzyl	Cyclopentylthio	262–265	29
Benzyl	*p*-$ClC_6H_4CH_2S$—	265–268	29
Benzyl	H_2NCOCH_2S—	261–263	29
Benzyl	$HOOCCH_2CH_2S$—	210 (dec.)	29
Benzyl	$HOOCCH_2S$—	277 (dec.)	29
Benzyl	3-MeBuS—	240–241	29
Benzyl	*n*-Amylthio	254–256	29
Benzyl	*i*-Butylthio	254–256	29
Benzyl	*i*-Propylthio	260–262	29
Benzyl	$C_6H_5CH_2CH_2S$—	250	29
Benzyl	$C_6H_5CH_2S$	261	29
Benzyl	$C_6H_{11}S$—	262–262.5	29
Benzyl	*o*-$ClC_6H_4CH_2S$—	263–264	29
Benzyl	*o*-$BrC_6H_4CH_2S$—	260–261	29
Benzyl	*o*-$IC_6H_4CH_2S$—	262–263	29
Benzyl	*o*-$FC_6H_4CH_2S$—	236–238	29
Benzyl	*m*-$FC_6H_4CH_2S$—	240–242	29
Benzyl	*p*-$FC_6H_4CH_2S$—	258	29
Benzyl	*p*-$BrC_6H_4CH_2S$—	292–293	29
Benzyl	2,4-di$ClC_6H_3CH_2S$—	269	29
Benzyl	3,4-di$ClC_6H_3CH_2S$—	299	29
Benzyl	2,6-di$ClC_6H_3CH_2S$—	282–283	29

TABLE 3D-14. 1*H*-Imidazo[4,5-*d*]pyridazine-4(5*H*)-thiones

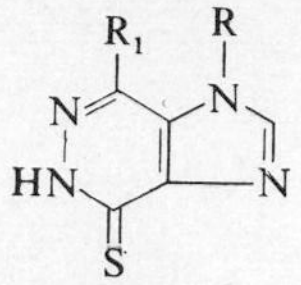

R	R_1	MP (°C) or BP (°C/mm)	References
None		>315 (dec.), >300, 310	28, 55, 62
Methyl	Chloro	>300	28
Benzyl	Chloro	207–209 (dec.)	28

TABLE 3D-15. 1*H*-Imidazo[4,5-*d*]pyridazine-7(6*H*)thiones

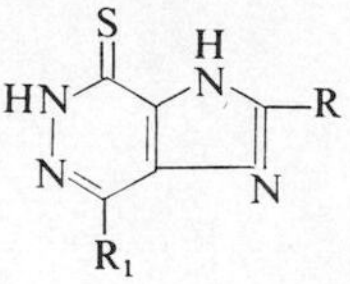

R	R_1	MP (°C) or BP (°C/mm)	References
	Methylthio	331–333	30
	Ethylthio	279–282	30
	n-Propylthio	257–258	29
	i-Propylthio	254–255.5	29
	n-Butylthio	244–246	29
	Allylthio	244.5–245	29
	3-Me-BuS	239–241	29
	β-Hydroxyethylthio	360	29
	$HOOCCH_2S$—	248 (dec.)	29
	$HOOCCH_2CH_2S$—	256 (dec.)	29
	H_2NCOCH_2S—	360	29
	Benzylthio	283–285	29
	o-$ClC_6H_4CH_2S$—	244.5–246	29
	o-$FC_6H_4CH_2S$—	239–241	29
	m-$FC_6H_4CH_2S$—	270–272	29
	p-$FC_6H_4CH_2S$—	276–277	29
	p-$IC_6H_4CH_2S$—	288–289	29
	p-$NO_2C_6H_4CH_2S$—	298–299	29
	2,4-di$ClC_6H_3CH_2S$—	269–271	29
	3,4-di$ClC_6H_3CH_2S$—	301–303	29
	p-$ClC_6H_4COCH_2S$—	263–264	29
	p-$BrC_6H_4COCH_2S$—	261–262	29
Phenyl	Methylthio	303–305	32, 34
Phenyl	Ethylthio	306–307	32, 34
Phenyl	*p*-$ClC_6H_4CH_2S$—	235–238	32

TABLE 3D-16. 1*H*-Imidazo[4,5-*d*]pyridazine-4,7(5*H*,6*H*)dithiones

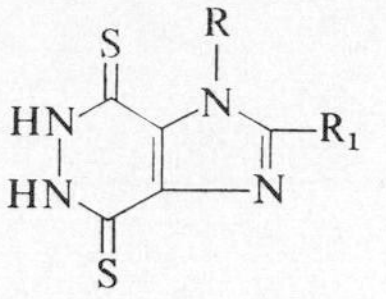

R	R_1	MP (°C) or BP (°C/mm)	References
None		309–311	30
	Phenyl	230–232	32, 34
Methyl		249–251	28

TABLE 3D-17. 1*H*-Imidazo[4,5-*d*]pyridazine-2(3*H*)thiones

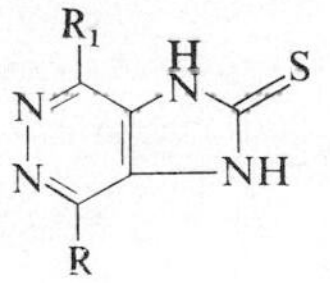

R	R_1	MP (°C) or BP (°C/mm)	References
Methoxy	Methyl	—	35, 62[a]
Methoxy	Methoxy	273–275 (dec.)	36
Ethoxy	Methyl	268–272	62
Ethoxy	Ethoxy	241–244	62
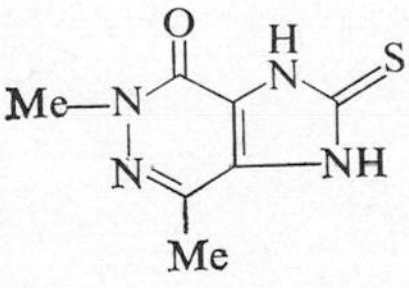		>300	62

[a] No definite mp, shrinking at about 250° C.

TABLE 3D-18. 2-Oxo- or 2-Thioxo-1*H*-imidazo[4,5-*d*]pyridazine-4,7(5*H*,6*H*)diones

R	X	MP (°C) or BP (°C/mm)	References
H	O	>300	31
H	S	>400	25
Methyl	S	>330	25
Phenyl	S	>367 (dec.)	25
Benzyl	S	336 (dec.)	29

TABLE 3D-19

Structure	MP (°C) or BP (°C/mm)	References
	315 (dec.)	29
	253–255	29

R	R_1	R_2	R_3	MP (°C) or BP (°C/mm)	References
None				225–226	49
		Amino		173–174	49
		Methylthio		212–213	49
			Methyl	170–176	49
	$Et_2N(CH_2)_3NH—$			—	33
				2 HCl, 257–259	33
Chloro				254–255 (dec.)	48
Methylthio				276–277	33
	Chloro			>360	47
				Picrate, 139–140 (dec.)	47
	Amino			335–338 (dec.)	47
				H_2O, 290–295 (dec.)	47
				336–338 (dec.)	49
	Methylthio			240–242	33
	$Me_2N(CH_2)_3NH—$			221–223	33
	$Et_2N(CH_2)_3CH(Me)NH—$			—	33
				2 $HCl \cdot H_2O$, 102–105	33
Chloro		Methylthio		262 (dec.)	49
	Chloro	Methylthio		>360 (dec.)	47
Chloro			Methyl	193–194	49
Chloro			$—N{=}CHC_6H_5$	184	48
	Chloro			90	33
Chloro		Methylthio	Methyl	175–176	49

TABLE 3D-21. 5*H*-Imidazo[4,5-*c*]pyridazines

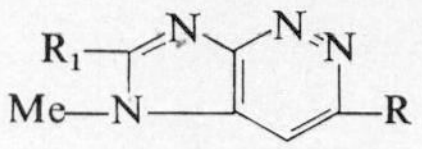

R	R_1	MP (°C) or BP (°C/mm)	References
None		—	49
		HCl, 252–253	49
Chloro		182–183	49
Chloro	Methylthio	217–218	49

TABLE 3D-22. 7*H*-Imidazo[4,5-*c*]pyridazine-6(5*H*)thiones

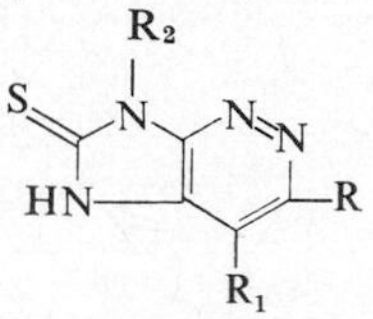

R	R_1	R_2	MP (°C) or BP (°C/mm)	References
None			360	49
		Methyl	319 (dec.)	49
Chloro			245–246	49
	Chloro		>360 (dec.)	47
	Amino		—	49
			x1.5 HCl, 310–311 (dec.)	49
Chloro		Methyl	295 (dec.)	49

TABLE 3D-23. 5*H*-Imidazo[4,5-*c*]pyridazine-6(7*H*)thiones

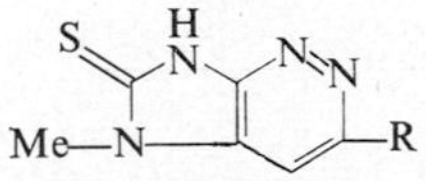

R	MP (°C) or BP (°C/mm)	References
None	308 (dec.)	49
Chloro	245–246	49
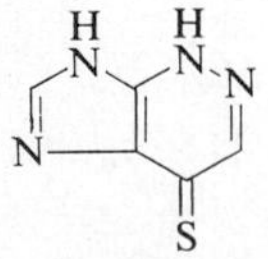	273–275 (dec.)	33
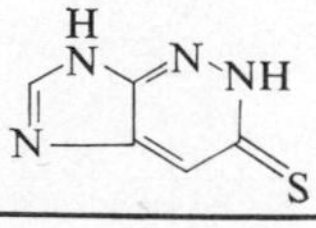	210 (dec.)	33

References

1. F. Yoneda, T. Otaka, and Y. Nitta, *Chem. Pharm. Bull.* (*Tokyo*), **12,** 1351 (1964); *Chem. Abstr.*, **62,** 5273 (1965).
2. J. Kobe, B. Stanovnik, and M. Tišler, *Tetrahedron*, **24,** 239 (1968).
3. B. Stanovnik and M. Tišler, *Tetrahedron Lett.*, **1966,** 2403.
4. B. Stanovnik and M. Tišler, *Tetrahedron*, **23,** 387 (1967).
5. A. Pollak, B. Stanovnik, and M. Tišler, *Tetrahedron*, **24,** 2623 (1968).
6. J. G. Lombardino, *J. Heterocyc. Chem.*, **5,** 35 (1968).
7. B. Stanovnik and M. Tišler, *Tetrahedron*, **23,** 2739 (1967).
8. L. M. Werbel and M. L. Zamora, *J. Heterocyclic Chem.*, **2,** 287 (1965).
9. L. Almirante, L. Polo, A. Mugnaini, E. Provinciali, P. Rugarli, A. Gamba, A. Olivi, and W. Murmann, *J. Med. Chem.*, **9,** 29 (1966).
10. Japanese Pat. 22,264 (1965); *Chem. Abstr.*, **64,** 3566 (1966).
11. W. L. F. Armarego, *J. Chem. Soc.*, **1965,** 2778.
12. J. G. Lombardino, *J. Org. Chem.*, **30,** 2403 (1965).
13. M. Tišler and W. W. Paudler, *J. Heterocyclic Chem.*, **5,** 695 (1968).
14. Unpublished data from this laboratory. Parameters for LCAO Calculations were taken from A. Streitwieser, *Molecular Orbital Theory for Organic Chemists*, Wiley, New York, 1961, p. 135.
15. B. Stanovnik and M. Tišler, *Croat. Chem. Acta*, **40,** 1 (1968).
16. Japanese Pat. 22,265 (1965); *Chem. Abstr.*, **64,** 3566 (1966).
17. Japanese Pat. 22,267 (1965); *Chem. Abstr.*, **64,** 3567 (1966).
18. A. Kovačič, B. Stanovnik, and M. Tišler, *J. Heterocyclic Chem.*, **5,** 351 (1968).
19. Japanese Pat. 22,263 (1965); *Chem. Abstr.*, **64,** 3566 (1966).
20. C. M. Atkinson and C. J. Sharpe, *J. Chem. Soc.*, **1959,** 3040.
21. S. Goldschmidt and M. Minsinger, *Chem. Ber.*, **87,** 956 (1954).
22. I. Hagedorn and H. Tönjes, *Pharmazie*, **12,** 567 (1957); *Chem. Abstr.*, **52,** 6363 (1958).
23. I. Hagedorn, U. Eholzer, and A. Lüttringhaus, *Chem. Ber.*, **93,** 1584 (1960).
24. R. Seka and H. Preissecker, *Monatsh. Chem.*, **57,** 71 (1931).
25. R. G. Jones, *J. Am. Chem. Soc.*, **78,** 159 (1956).
26. T. S. Gardner, F. A. Smith, E. Wenis, and J. Lee, *J. Org. Chem.*, **21,** 530 (1956).
27. J. Nematollahi, W. Guess, and J. Autian, *J. Med. Chem.*, **9,** 660 (1966).
28. J. A. Carbon, *J. Am. Chem. Soc.*, **80,** 6083 (1958).
29. G. A. Gehardt, D. L. Aldous, and R. N. Castle, *J. Heterocyclic Chem.*, **2,** 247(1965).
30. R. N. Castle and W. S. Seese, *J. Org. Chem.*, **23,** 1534 (1958).
31. P. H. Laursen and B. E. Christensen, *J. Org. Chem.*, **27,** 2500 (1962).
32. M. Malm and R. N. Castle, *J. Heterocyclic Chem.*, **1,** 182 (1964).
33. N. R. Patel, W. M. Rich, and R. N. Castle, *J. Heterocyclic Chem.*, **5,** 13 (1968).
34. U.S. Pat. 3,244,715 (1966); *Chem. Abstr.*, **64,** 19632 (1966).
35. D. L. Aldous and R. N. Castle, *Arzneimittel-Forsch.*, **13,** 878 (1963).
36. T. Itai and S. Suzuki, *Chem. Pharm. Bull.* (*Tokyo*), **8,** 999 (1960); *Chem. Abstr.*, **55,** 24750 (1961).
37. K. Dury, *Angew. Chem.*, **77,** 282 (1965).
38. J. I. F. Alonso and R. D. Sebastian, *Anal. real soc. españ. fis. y. quim.* (*Madrid*), **56B,** 759 (1960); *Chem. Abstr.*, **55,** 2267 (1961).
39. U.S. Pat. 2,918,469 (1959); *Chem. Abstr.*, **54,** 6767 (1960).
40. P. H. Laursen, *Diss. Abstr.*, **22,** 1015 (1961); *Chem. Abstr.*, **56,** 4763 (1962).

41. E. Y. Sutcliffe and R. K. Robins, *J. Org. Chem.*, **28,** 1662 (1963).
42. J. A. Carbon, *J. Org. Chem.*, **25,** 579 (1960).
43. D. L. Aldous and R. N. Castle, *J. Heterocyclic Chem.*, **2,** 321 (1965).
44. B. R. Baker, K. Hewson, H. J. Thomas, and J. A. Johnson, *J. Org. Chem.*, **22,** 954 (1957).
45. L. L. Bennett and D. Smithers, *Biochem. Pharmacol.*, **13,** 1331 (1964); *Chem. Abstr.*, **61,** 11195 (1964).
46. Belgian Pat. 570,978 (1958); *Chem. Abstr.*, **54,** 137 (1960).
47. T. Kuraishi and R. N. Castle, *J. Heterocyclic Chem.*, **1,** 42 (1964).
48. T. Kuraishi and R. N. Castle, *J. Heterocyclic Chem.*, **3,** 218 (1966).
49. H. Murakami and R. N. Castle, *J. Heterocyclic Chem.* **4,** 555 (1967).
50. British Pat. 1,135,893 (1968); *Chem. Abstr.*, **70,** 57870 (1969).
51. S. Ostroveršnik, B. Stanovnik, and M. Tišler, *Croat. Chem. Acta*, **41,** 135 (1969).
52. M. Japelj, B. Stanovnik, and M. Tišler, *J. Heterocyclic Chem.*, **6,** 559 (1969).
53. B. Stanovnik and M. Tišler, *J. Heterocyclic Chem.*, **6,** 413 (1969).
54. J. Nematollahi and J. R. Nulu, *J. Med. Chem.*, **12,** 43 (1969).
55. S. F. Martin and R. N. Castle, *J. Heterocyclic Chem.*, **6,** 93 (1969).
56. I. Sekikawa, *J. Heterocyclic Chem.*, **6,** 129 (1969).
57. N. B. Galstukhova, G. S. Predvoditeleva, I. M. Berzina, T. V. Kartseva, T. N. Zykova, M. N. Shchukina, and G. N. Pershin, *Khim. Farm. Zh.*, **3,** 7 (1969).
58. H. Fillion, A. Boucherle, G. Carraz, J. C. Donzeau, and S. Lebreton, *Chim. Ther.*, **3,** 309 (1968).
59. Y. Kuwabara, S. Irie, S. Sugita, M. Yanai, S. Takeda, and H. Sadaki, *Nippon Shashin Gakkai Kaishi*, **31,** 74 (1968); *Chem. Abstr.*, **70,** 15997 (1969).
60. B. Stanovnik, M. Tišler, M. Ceglar, and V. Bah, *J. Org. Chem.*, **35,** 1138 (1970).
61. A. Pollak, B. Stanovnik, and M. Tišler, *J. Org. Chem.*, **35,** 2478 (1970).
62. M. Yanai, T. Kinoshita, S. Takeda, M. Mori, H. Sadaki, and H. Watanabe, *Chem. Pharm. Bull.*, **18,** 1685 (1970).
63. M. Yanai, T. Kuraishi, T. Kinoshita, and M. Nishimura, *J. Heterocyclic Chem.*, **7,** 465 (1970).

Part E. Triazolopyridazines

I. *s*-Triazolo[4,3-*b*]pyridazines

s-Triazolo[4,3-*b*]pyridazine, RRI 1085 (2,3,7-Triazaindolizine, 2,3-Triazolo-7.0-pyridazine, 1,2,3*a*,4-Tetrazaindene, 2,3-Triazo–7.0-pyridazine, *s*-Triazolo [*b*]pyridazine; 1,2,4-Triazolo[*b*]pyridazine)

Compounds of this ring system were obtained first in 1909 by Bülow, but after his classical work there was practically no interest in this system until recently. Most of the work on *s*-triazolo[4,3-*b*]pyridazines has been done since 1950.

There are two approaches to the formation of this bicycle, the classical one from aminotriazoles and the newer one from hydrazinopyridazines.

4-Amino-1,2,4-triazole has been condensed with various β-ketoaldehydes in the form of their acetals, with 1,3-diketones or with β-keto esters. Condensation between 4-amino-1,2,4-triazole and 4,4-dimethoxy-2-butanone in xylene at the boiling point has been reported by Allen et al. (1, 2) to yield 6-methyl-*s*-triazolo[4,3-*b*]pyridazine in 16% yield. For the reaction product two structures can be written, **1** or **2**.

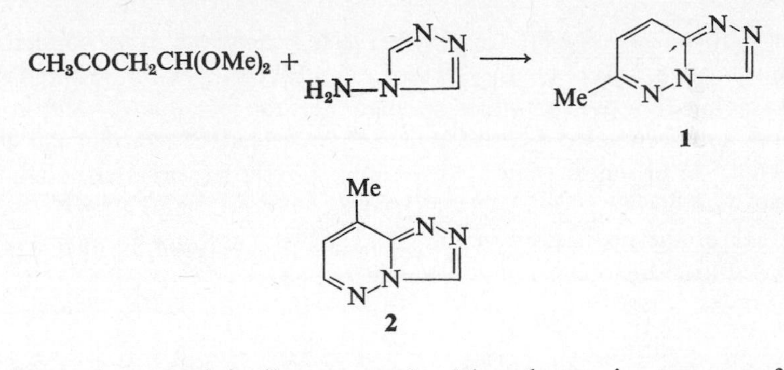

Structure **1** was originally proposed, although no rigorous proof was presented. Later, Libermann and Jacquier (3) were able to demonstrate that the product has indeed the originally proposed structure since an identical product (**1**) could be obtained either by cyclizing 3-hydrazino-6-methylpyridazine with formic acid or by converting compound **3** via **4** into **1** and by repeating the reaction of Allen.

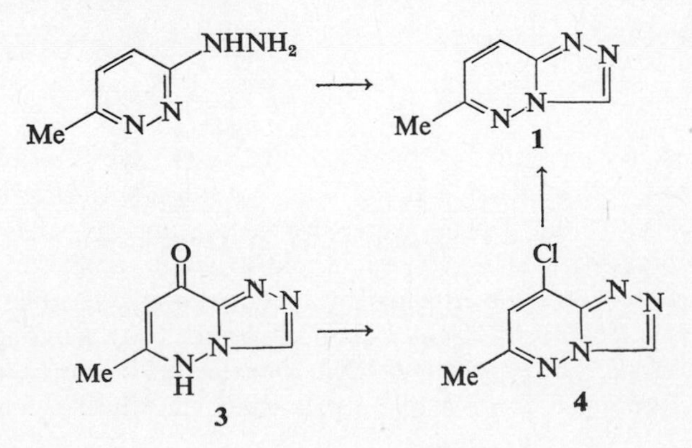

Reaction between 4-amino-1,2,4-triazole and 1,3-diketones leads to 6,8-disubstituted *s*-triazolo[4,3-*b*]pyridazines (4–6). With symmetrically substituted diketones there are no structural problems relative to the final products, whereas with the unsymmetrical diketones two isomers are feasible. Bülow and Weber (4), on the basis of analogy and without rigorous proof, assigned structure **5** to the product obtained from benzoylacetone as the keto component and 4-amino-1,2,4-triazole.

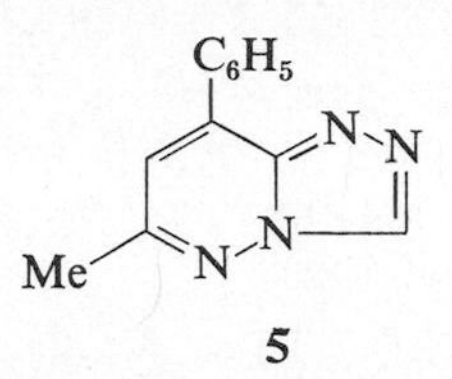

There was also some controversy regarding the structure of condensation products between 4-amino-1,2,4-triazole and β-keto esters. Bülow and Weber (7) formulated the product which resulted from the reaction of 4-amino-1,2,4-triazole and acetoacetic ester as a 6-methyl derivative (**6a**) and many other condensation products were assigned analogous structures, although in no case was a rigorous proof of their structures given. The alternative possible structure of the product from the reaction with acetoacetic ester (**7a**) was proposed in 1958 by Kost and Gents (8).

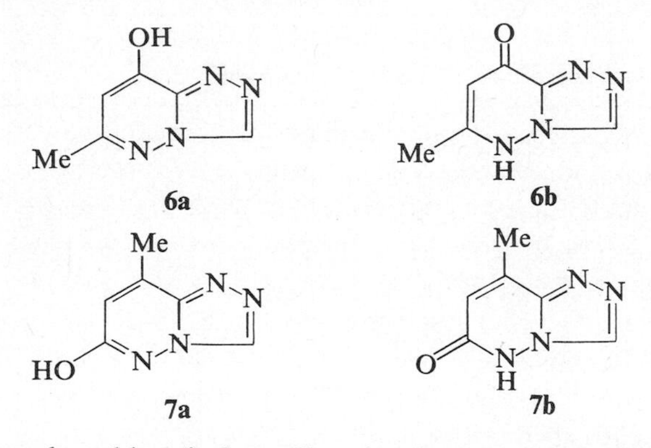

This structural problem is formally related to the synthesis of quinolines according to Conrad-Limpach and Knorr, and it has been shown by Hauser and Reynolds (9) that depending on reaction conditions, anils or anilides could be obtained from β-keto esters and aromatic amines. In 1962 this structural problem was investigated by Libermann and Jacquier (3) and by Linholter and Rosenørn (10). An identical product (**1**) was obtained from condensation of 3-hydrazino-6-methylpyridazine with formic acid (3) or when transforming Bülow's reaction product **3** via **4** into **1**. This indicates that in the first step of the cyclization reaction between 4-amino-1,2,4-triazole and acetoacetic ester anils are formed and that Bülow's structure assignment of **6** is correct. Authentic **7**, prepared from 3-chloro-4-hydrazino-5-methylpyridazine and formic acid via **8**, is distinctly different from **6** as judged from the comparison of ir spectra (10). Similar conclusions about the correct structure were also reached from uv spectra correlations (11).

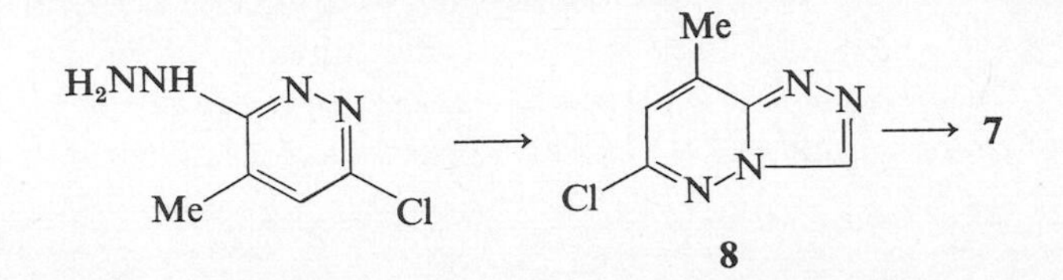

On the basis of ir evidence Linholter and Rosenørn (10) concluded that in the solid state **6** exists predominantly as **6a**, whereas for **7** the lactam form **7b** is predominant and in both cases indication of strong hydrogen bonding is given.

Several β-keto esters in addition to acetoacetic ester, have been employed for the preparation of *s*-triazolo[4,3-*b*]pyridazines from 4-amino-1,2,4-triazole (3, 7, 8, 12–18, 83). Reactants are generally heated in the absence of a solvent at high temperatures and the yield is claimed to be dependent on the rate of heating.

In 1955 Takahayashi presented the first example of an *s*-triazolo[4,3-*b*]-pyridazine synthesis from pyridazine precursors (19). Since then largely because of the ready accessibility of 3-hydrazinopyridazines, this synthetic approach has been further developed and used in many cases. A suitable 3-hydrazinopyridazine can be transformed into the corresponding *s*-triazolo-[4,3-*b*]pyridazine by heating with either formic acid (3, 5, 10, 19–23, 74) or its methyl ester (19) or triethyl orthoformate (19). Depending on reaction conditions, the reaction can proceed toward the intermediate formyl-hydrazinopyridazine, which is isolated and subsequently cyclized, or the bicycle can be obtained in a direct one-step reaction sequence. NMR spectra of some *s*-triazolo[4,3-*b*]pyridazines have been recorded and are consistent with the proposed structures (24, 74).

3-Substituted *s*-triazolo[4,3-*b*]pyridazines are obtainable when acids other than formic are used for cyclization. Acetic anhydride or ethyl acetate has been thus employed for the synthesis of 3-methyl derivatives (19, 21, 25, 74) and with benzoyl chloride in pyridine (22, 23) or acid chlorides of nicotinic or isonicotinic acid (26, 27) the corresponding 3-benzoyl or pyridyl derivatives have been prepared.

Takahayashi has observed (28) that 6-chloro-3-formylhydrazino and 6-chloro-3-acetylhydrazinopyridazine do not cyclize in the presence of hot 5*N* or 90% sulfuric acid, whereas 3-benzoylhydrazino-6-chloropyridazine could be thermally cyclized when heated at about 160° (24).

There is one report of cyanogen chloride being employed to effect cyclization. In the presence of sodium acetate and acetic acid 3-amino derivatives are thus obtained (29). Similarly, with carbon disulfide, 3-mercapto derivatives were obtained (29).

It is also possible to cyclize alkylidene or arylidene derivatives of 3-hydrazinopyridazines to **9** either with hot acetic anhydride or benzoyl chloride (23) or by dehydrogenative cyclizations using bromine in acetic acid or lead tetraacetate for these purposes (24). In the last case relatively mild reaction

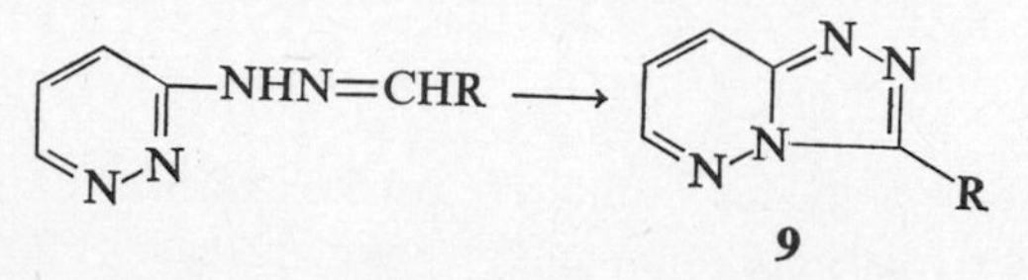

conditions are employed and 3-substituted *s*-triazolo[4,3-*b*]pyridazines are obtained in reasonably good yields.

There is one report about the triazolo ring formation from thiosemicarbazidopyridazines in boiling acetic acid; the structure of the product is indicated by **10** (30).

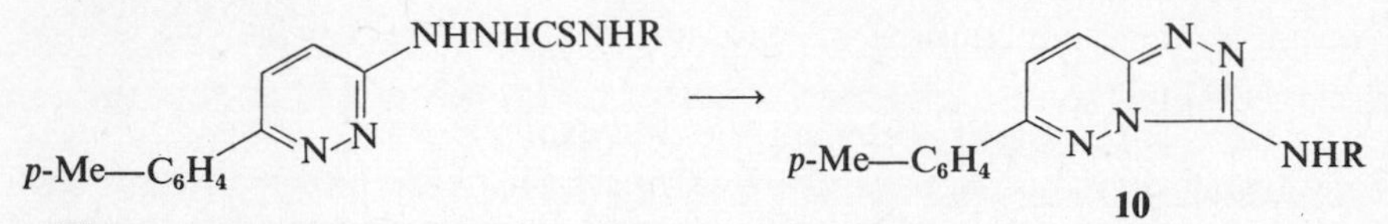

The two products (**10**, R = C_6H_5 or allyl) have been analyzed only for their nitrogen content; however, since similar cyclizations when applied to some tri- and polycyclic aza analogs (31, 32) invariably proceeded with the elimination of the amine and not hydrogen sulfide, a rigorous proof of the structure **10** should be needed. Moreover, although Basu and Rose (29) reported the failure of cyclization of **11** to **12**, the reaction in fact does proceed when **11** is heated in ethylene glycol under reflux for 10 min (33).

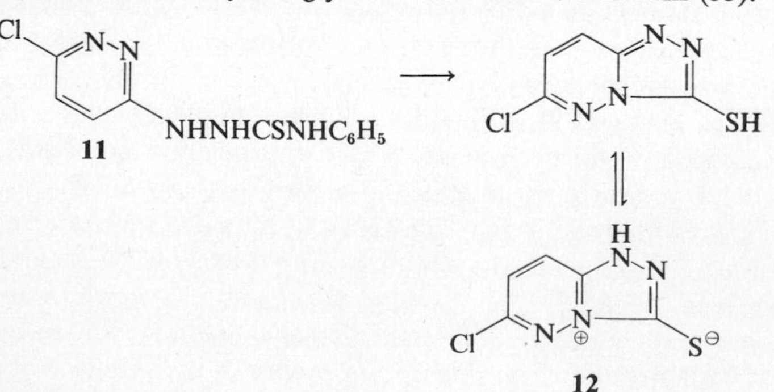

An interesting transformation concerning *s*-triazolo[4,3-*b*]pyridazines has been observed recently (34). When 6-hydrazinotetrazolo[1,5-*b*]pyridazine (**13**) was heated with diethoxymethyl acetate and the triazole ring was formed in this reaction, the tetrazole part of the molecule was simultaneously opened and 6-azido-*s*-triazolo[4,3-*b*]pyridazine (**14**) was obtained. Spectroscopic data, the synthesis of the 6-azido compound from the 6-chloro derivative, and the reduction of **14** with hydrogen sulfide into the 6-amino analog are in

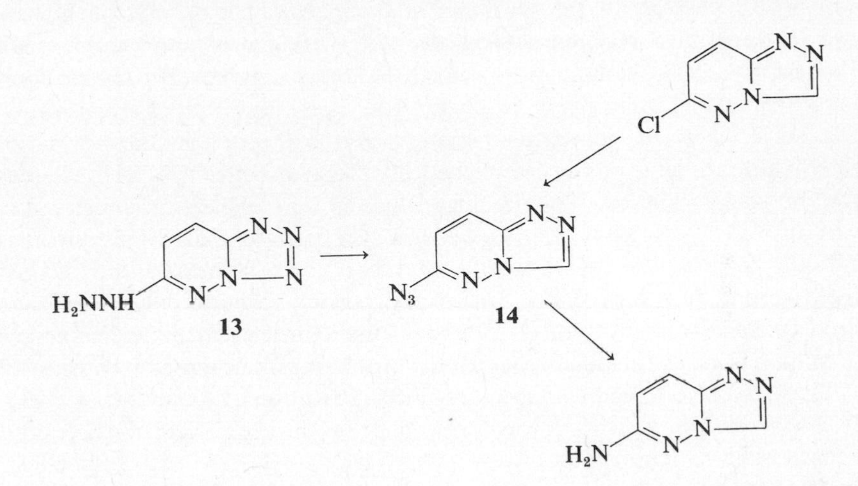

accordance with the foregoing structural changes. Similarly, 6-hydrazinotetrazolo[1,5-*b*]pyridazine, when treated with cyanogen bromide, is transformed into 3-amino-6-azido-*s*-triazolo[4,3-*b*]pyridazine. Oxidative cyclization of 6-benzylidenehydrazino analogs represents an analogous reaction (84).

The parent compound, *s*-triazolo[4,3-*b*]pyridazine, was obtained by catalytic dehalogenation of the 6-chloro analog (28), but no details about its behavior and reactivity are available. The results of MO calculations (76) of total and frontier π-electron densities are presented in Table 3E-1. These results suggest low or no reactivity of this system for electrophilic substitutions and the greatest susceptibility for nucleophilic attack would be expected preferentially at position 6 and then at position 8.

Most of the available experimental data concern nucleophilic displacements of chlorine atoms at positions 6 and 8. In this manner, the 6-chlorine atom has been displaced by an alkoxy (19, 20, 29), hydroxy (10, 28), amino or substituted amino (24, 28, 29, 36) azido (34, 36), hydrazino (10, 21, 24, 28, 29), or mercapto and alkyl or arylthio group (19, 20, 24, 28, 81). Likewise, reactions are described in which the 8-chlorine atom is replaced by an alkoxy (37), amino or substituted amino (12, 37, 38), hydrazino (3, 37), or mercapto group (13, 14, 39).

As one can judge from the experimental data describing some of these displacements, the 6-chloro derivatives are somewhat more reactive than the 8-chloro analogs, although data from preparative work are poor guides for such correlations. Data from kinetic measurements are completely lacking.

For example, 8-amino-6-methyl-*s*-triazolo[4,3-*b*]pyridazine was prepared from the corresponding 8-chloro analog with a solution of ammonia in methanol (130°, 8 hr) (37) or by applying the phenol melt method (165–170°, 6 hr) (12), whereas for the synthesis of 6-amino-3-phenyl-*s*-triazolo[4,3-*b*]-pyridazine milder reaction conditions (75°, 2 hr) were satisfactory (24). For the synthesis of 6-amino-*s*-triazolo[4,3-*b*]pyridazine (28) higher temperatures (150°, 3 hr) were applied. The preparation of 6-hydrazino-*s*-triazolo[4,3-*b*]pyridazine or its 7-methyl analog through hydrazinolysis of the 6-chloro analog was accomplished in several minutes (10, 28), whereas for the transformation of the 8-chloro analog 1 hr of heating under reflux was applied (3). 6-Alkoxy derivatives are also easily prepared (19, 20). It is difficult to compare the reactivity of the 6- or 8-chlorine atom against hydrogensulfide ion or thiols, since very varying reaction conditions have been applied. More data and particularly kinetic measurements are necessary to make a clear distinction in the reactivity of halogens at these two positions.

There is a recent report on electrophilic substitution of 8-amino-*s*-triazolo-[4,3-*b*]pyridazine which afforded with bromine in acetic acid the 7-bromo derivative (85). Here apparently the amino group exerts an activating effect. In these terms coupling of aryl diazonium salts to 8-hydroxy analogs (86, 83) is understandable.

Several additional reactions common to other heterocycles were performed. Hydrogenolysis of a chlorine atom attached at position 6, 7, or 8 has been performed in the presence of palladized charcoal (22, 23, 28, 29, 74) or with zinc (39). A hydrogen atom was also introduced by replacement of a hydrazino group with copper sulfate solution (28). Beside the displacement reaction of a chlorine atom by a mercapto group, direct thiation of the corresponding

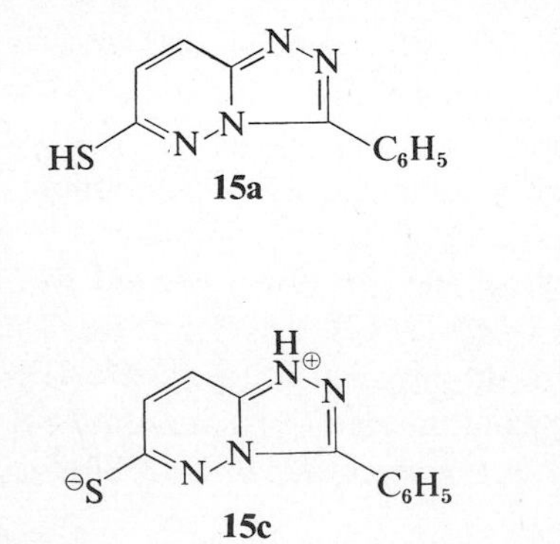

15a

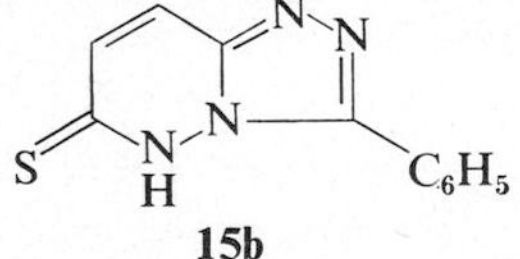

15b

15c

oxo-bicycle with phosphorus pentasulfide has been performed (3). The structure of 3-phenyl-6-mercapto-*s*-triazolo[4,3-*b*]pyridazine, as indicated from the ir spectrum, seems to be more in accord with the mercapto **15a** than thioxo **15b** or zwitterionic **15c** form.

The 3-unsubstituted analog of **15** has been found capable of addition to quinones (40) and the sulfide **16** could be oxidized to the corresponding quinone **17** by means of lead tetraacetate.

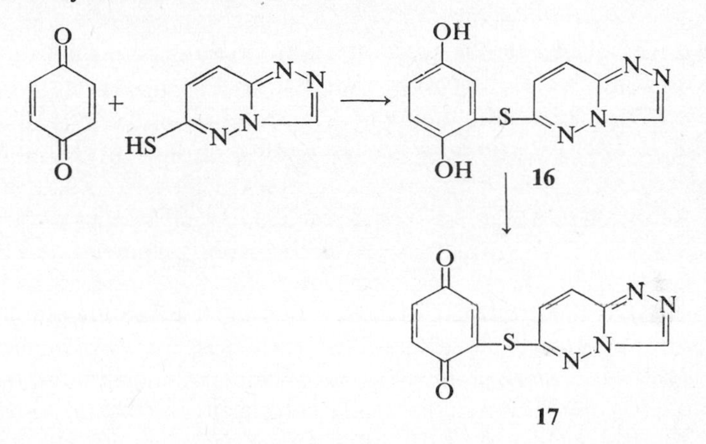

3-Methyl-*s*-triazolo[4,3-*b*]pyridazine reacted with heteroaromatic aldehydes in the presence of acetic anhydride to give the corresponding heteroarylidene derivatives (41), a reaction which is known to occur with many azines with a methyl group, α or γ to ring nitrogen.

Compound **6**, when oxidized with an alkaline solution of potassium permanganate, gives 1,2,4-triazole-3-carboxylic acid and hydrogenation of 6-methyl- and 6,8-dimethyl-*s*-triazolo[4,3-*b*]pyridazine (169 atm, at 180°) and in the presence of Raney nickel catalyst is reported to give the corresponding methylpyrrolidones **18** (6), most probably via an intermediate py-

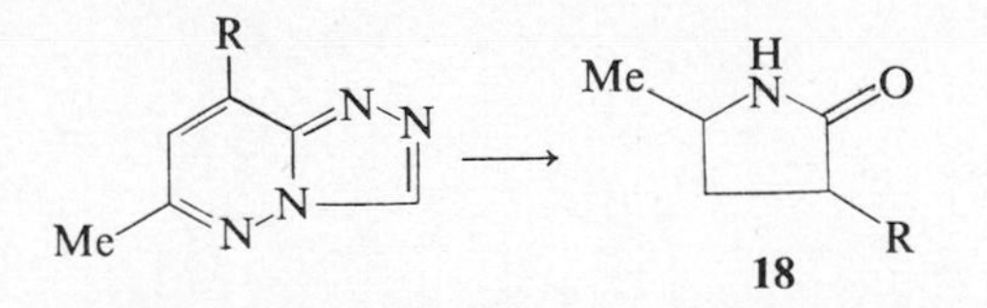

ridazine derivative. 3-Aminopyridazines can be also obtained from similar experiments (190–200°, 150–200 atm) (6, 75, 83).

s-Triazolo[4,3-*b*]pyridazines were submitted also to homolytic methylations. On the grounds of varying reaction conditions and analyses of mixtures of the resulting methyl derivatives the positional selectivity of $8 > 6 > 7$ was established (74).

s-Triazolo[4,3-*b*]pyridazines are capable of salt formation and the site of protonation and quaternization has been studied. It has been found on the basis of nmr spectral correlations that protonation takes place invariably at N_1 and the same holds for quaternization, with some exceptions (76). This is in accordance with the calculated frontier electron densities. Inductive and steric effects may influence the site of quaternization since 3- and 8-methyl derivatives form quaternary salts also at N_2 (25, 76). Other 8-methyl- or 8-phenyl-substituted *s*-triazolo[4,3-*b*]pyridazines are reported to be quaternized at N_2, the evidence for the quaternization site being obtained from chemical transformations (5). Depending on the reagent used and upon reaction conditions the quaternized compound (**19**) could be transformed either to *β*-formylhydrazinopyridazines (**20**) or to 3-aminopyridazines (**21**) (5, 77).

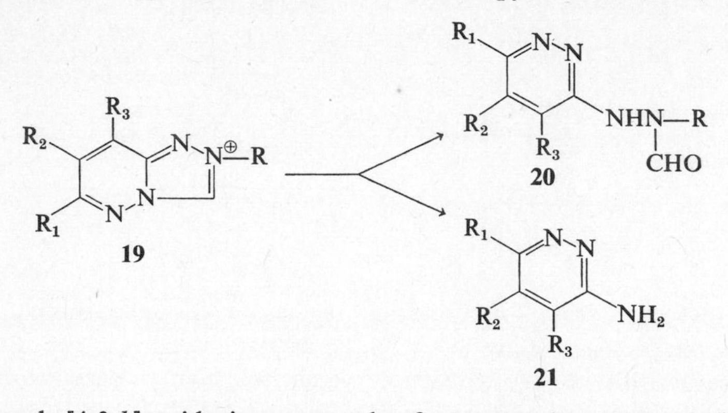

s-Triazolo[4,3-*b*]pyridazines were the first example among azoloazines where an *N*-oxide could be isolated and characterized. *s*-Triazolo[4,3-*b*]-pyridazine 5-oxides were obtained either from 3-hydrazinopyridazine 1-oxide, which was condensed with aldehydes, and the derivatives **22** were then cyclized by means of lead tetraacetate to the bicyclic *N*-oxide **23**. Another possible way is the direct cyclization of 3-hydrazinopyridazine

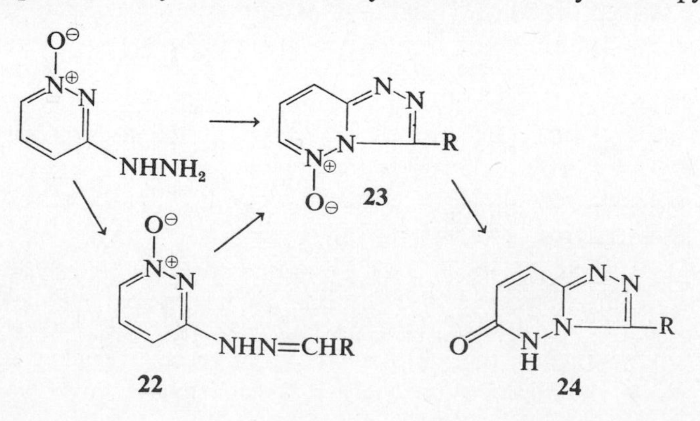

1-oxide with cyanogen bromide to yield the 3-amino derivative **23** (R = NH_2) (42). A direct *N*-oxidation procedure has been published recently (87).

The structure of the bicyclic *N*-oxide is consistent with the recorded ir and nmr spectra. *s*-Triazolo[4,3-*b*]pyridazine 5-oxides are relatively stable when stored in the dark, but in solution upon irradiation with uv light or after exposure to sunlight they are easily rearranged into the 6-oxo analogs **24**. Mass spectra of several *s*-triazolo[4,3-*b*]pyridazines are reported (89) and two main fragmentation patterns were observed to take place under electron impact.

s-Triazolo[4,3-*b*]pyridazines have been used for the preparation of cyanine dyes (25) and are claimed to be useful in photography when added to photographic emulsions (13–16, 43–47). They were also investigated for diuretic (37) and antimicrobial (88) activity. There is a report on potentiometric titrations with *s*-triazolo[4,3-*b*]pyridazines (48).

II. *s*-Triazolo[1,2-*a*]pyridazines

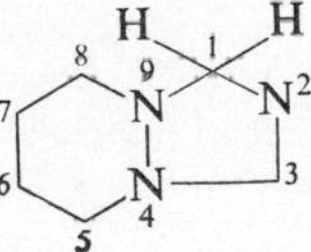

1*H*-s-Triazolo[1,2-*a*]pyridazine, RRI 1088 (2,8,9-Triazaindan,1, 3,5-triazabicyclo[3.4.0] nonane, 1,6,8-Triaza[4.3.0]bicyclononane)

Representatives of this bicycle are known only in the reduced form or as cyclic amides. Generally they are prepared by a Diels-Alder reaction (49) or by cyclizing appropriate pyridazine derivatives.

In addition to *trans*-azo dienophiles such as ethyl azodicarboxylate, cyclic azo-dienophiles which are compelled to possess a *cis* configuration were introduced recently as dienophiles in the Diels-Alder reaction. In 1962 Cookson, Gilani, and Stevens (50) discovered that 4-phenyl-1,2,4-triazoline-3,5-dione reacted *in situ* with butadiene in acetone solution at −50° to form the adduct **25**. The enhanced reactivity of this dienophile is certainly due to

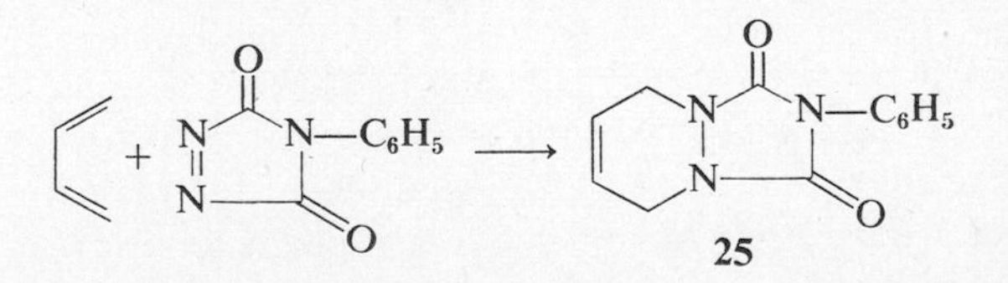

the *cis* configuration of the N—N double bond and other dienes also reacted very rapidly. The structure of the adduct **25** has been confirmed by chemical transformations (hydrogenation) and from nmr spectroscopic evidence (51).

In several other articles the extension of this reaction has been reported with the use of different dienes or different 4-substituted 1,2,4-triazoline-3,5-diones (52–54, 78, 91–94). 2,5-Dimethyl-2,4-hexadiene did not form a cyclic addition product and only the most unreactive dienes are reported to elude the Diels-Alder reaction (52). Kinetic studies of this addition revealed that 4-phenyl-1,2,4-triazoline-3,5-dione is in part a better dienophile than tetracyano ethylene (55).

s-Triazolo[1,2-*a*]pyridazines can also be obtained from reduced pyridazines when these contain groups such as carbethoxy, carbamyl, thiocarbamyl, or their derivatives attached at both ring nitrogens. Cyclization proceeds thermally, for example, 1,2-dicarbamyl-1,2,3,6-tetrahydropyridazine when heated at 275–285° for 30 min is transformed into **26** (56, 57).

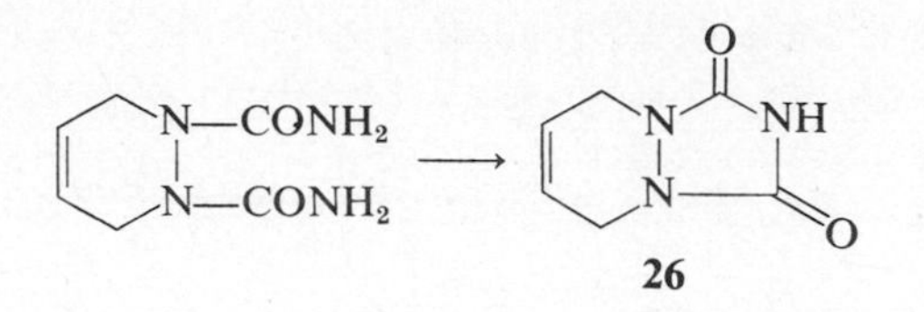

Pyridazines of the type **27** are reported to cyclize upon heating at 150–180° to **28** (58). It is also possible to obtain **28** in a one-step reaction from **29** and

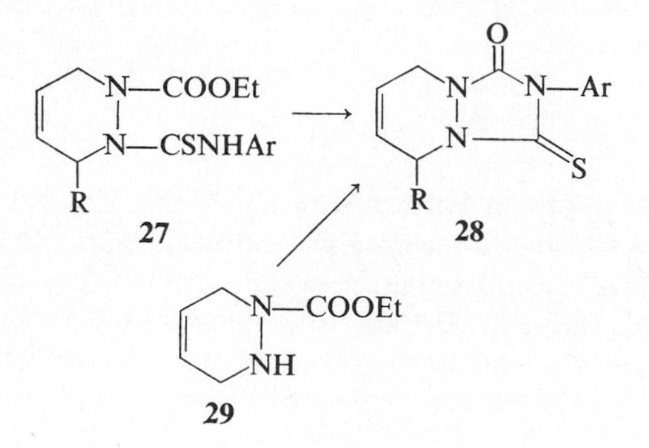

aromatic isothiocyanates (59). Verifications of these bicyclic structures are lacking and among transformations reported are reductions with lithium aluminum hydride. Depending on the quantity of the reducing agent used, oxo or thioxo groups could be reduced to methylene groups (59). The double bond of **26** allows addition reactions, and in this way mercurated derivatives were prepared (54, 57, 60).

There are some other reports concerning the synthesis of compounds related to this bicyclic system. Zinner and Deucker (61) reported that the dipotassium salt of 4-phenyl- or 4-butyl urazole reacted with 1,4-dibromobutane to form the bicycle **30**. Similar reactions with γ-bromobutyric acid chloride or succinic acid dichloride afforded the corresponding bicycle **31** or **32**, respectively (61). There are no rigorous proofs to substantiate the proposed structures.

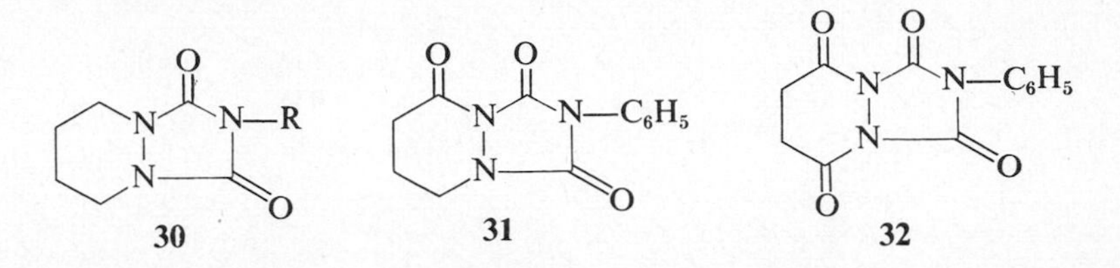

There are also some other reports in which for the reaction products structures of *s*-triazolo[1,2-*a*]pyridazine have been proposed. Compound **33**, obtained from addition of azobisformamidine dinitrate to butadiene, is transformed in a solution of methanol, preferentially upon heating, into **34** (62).

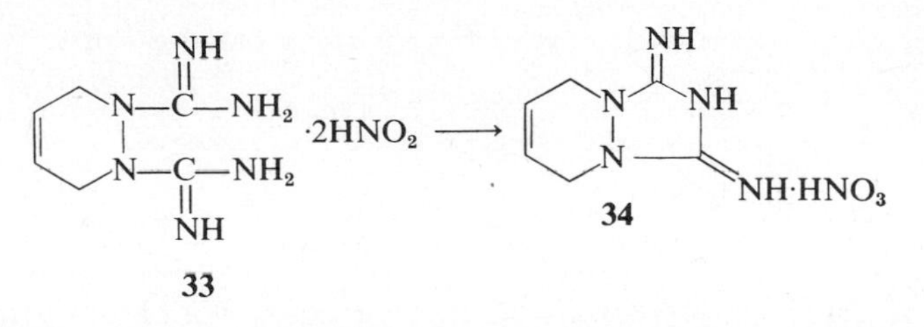

Similarly, **35** (probable structure), obtained from azobisformamidine dinitrate upon chlorination with hypochlorous acid and upon treatment with nitric acid, is claimed to react with 1,3-dimethylbutadiene to form an adduct for which the structure **36** was proposed (63).

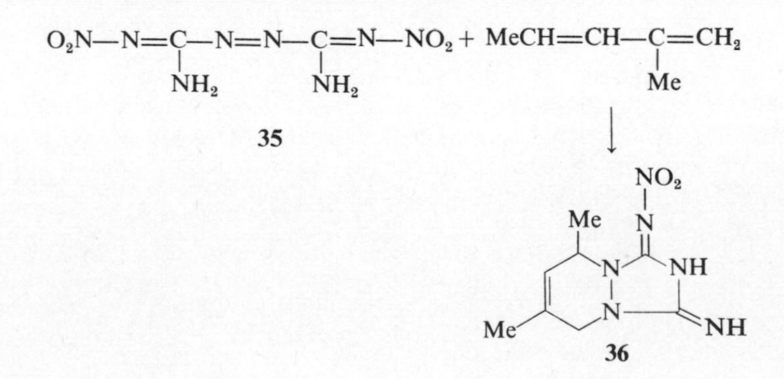

III. *v*-Triazolo[4,5-*d*]pyridazines

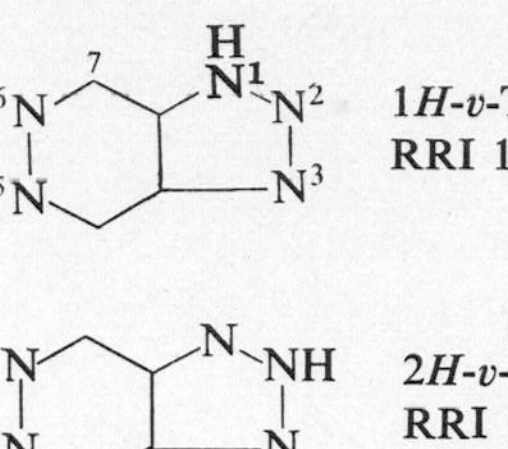

1*H*-*v*-Triazolo[4,5-*d*]pyridazine, RRI 1083

2*H*-*v*-Triazolo[4,5-*d*]pyridazine, RRI 1084

It is most probable that the first compound belonging to this heterocyclic system was synthesized in 1929 by Bertho and Hölder (64). These authors treated dimethyl 1-*p*-xylyl-1,2,3-triazolo-4,5-dicarboxylate with hydrazine hydrate and obtained the open chain dihydrazide as well as a "secondary" hydrazide to which formula **37** was assigned.

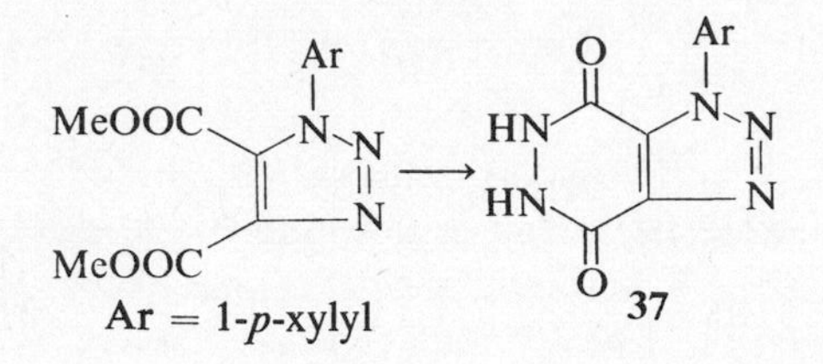

There are only a few reports regarding the chemistry of this heterocyclic system. Seka and Preissecker (65) in 1931 described the synthesis of a derivative of 2*H*-*v*-triazolo[4,5-*d*]pyridazine (**38**) from 2-phenyl-1,2,3-triazole-4,5-dicarboxylic acid dihydrazide which was heated *in vacuo* for 2 hr.

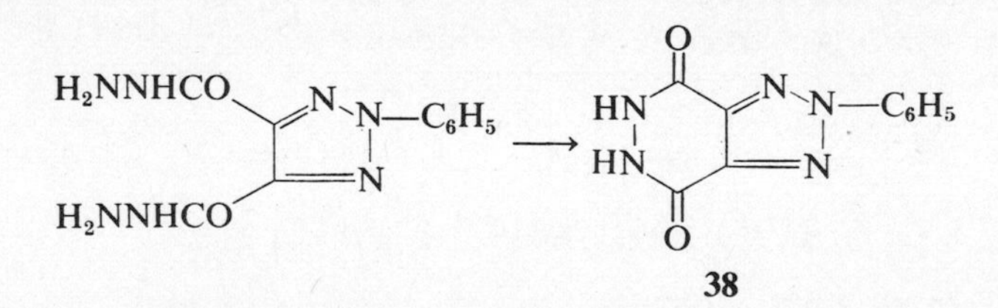

A similar synthetic approach was used by Erichomovitch and Chubb (66) for the preparation of derivatives of 1*H*-*v*-triazolo[4,5-*d*]pyridazines. 1-Phenyl-1,2,3-triazole-4,5-dicarboxylic acid dihydrazide could be converted into the bicycle **39** either by heating alone (260°, 6 hr) or in the presence of hydrazine or 2*N* hydrochloric acid; these methods were used by Jones (67) for the synthesis of related bicyclic systems.

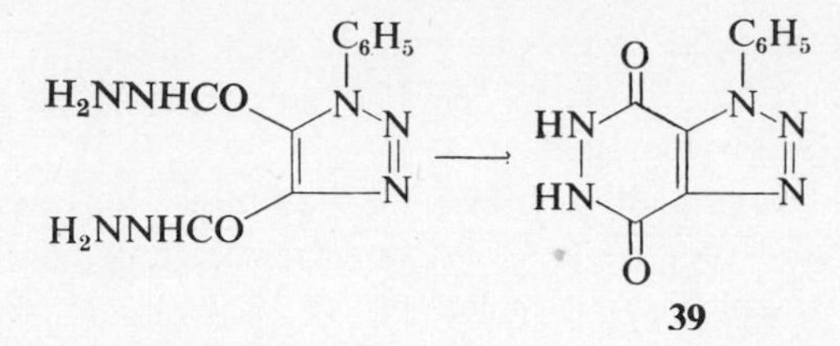

39

The possibility of conversion of appropriate pyridazines into *v*-triazolo-[4,5-*d*]pyridazines was demonstrated first by Itai and Suzuki (68). The authors treated a solution of 3,6-dimethoxy-4,5-diaminopyridazine in acetic acid dropwise with a solution of sodium nitrite and after the mixture was heated for 15 min at 100°, 4,7-dimethoxy-*v*-triazolo[4,5-*d*]pyridazine (**40**) was formed. This synthetic principle was then extended for the preparation of analogous compounds (69, 70, 79, 95).

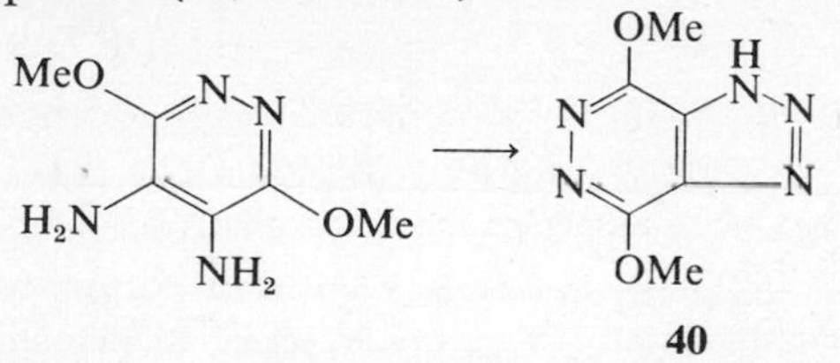

40

There are only few reactions connected with this bicyclic system. Compound **39** has been converted by means of phosphorus oxychloride into 1-phenyl-4,7-dichloro-*v*-triazolo[4,5-*d*]pyridazine (50%), accompanied by some (9%) monochloro derivative. The dichloro compound served for the preparation of 4,7-bis-thioxo, 4,7-bis-methylthio, or 4,7-bis-alkoxy derivatives (66). Aminolysis and hydrazinolysis of 4,7-bis-alkoxy derivatives gave chiefly monosubstituted derivatives (95). In cases where substituents at positions 4 and 7 are different and thus two isomeric structures are possible, structure determinations are lacking.

MO calculations (35) of total and frontier π-electron densities for 1*H*-*v*-triazolo[4,5-*d*]pyridazine are presented in Table 3E-2. The effect of triazolo-[4,5-*d*]pyridazines as addenda in photographic emulsions has been studied (80).

IV. *v*-Triazolo[4,5-*c*]pyridazines

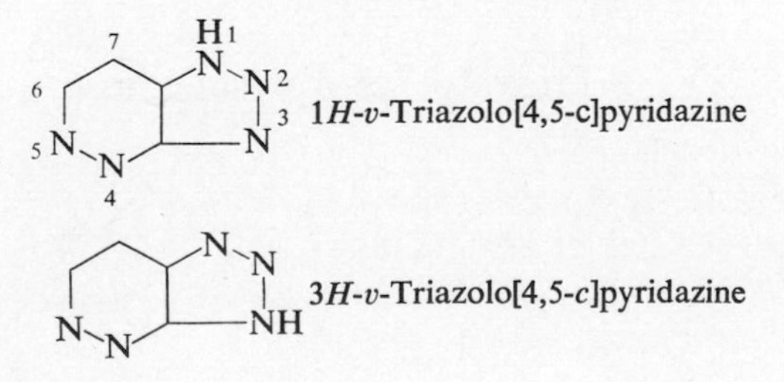

The parent compound and its derivatives were obtained for the first time in 1964 by Gerhardt and Castle (71). They are represented as derivatives of 3*H*-*v*-triazolo[4,5-*c*]pyridazine.

5-Chloro-3,4-diamino- or 6-chloro-3,4-diaminopyridazine served as starting compounds and they were easily transformed into the corresponding 7- and 6-chloro-*v*-triazolo[4,5-*c*]pyridazines **41** ($R = H$, $R_1 = Cl$ or $R = Cl$, $R_1 = H$) after they were treated below 10° with a solution of sodium nitrite in the presence of sulfuric acid.

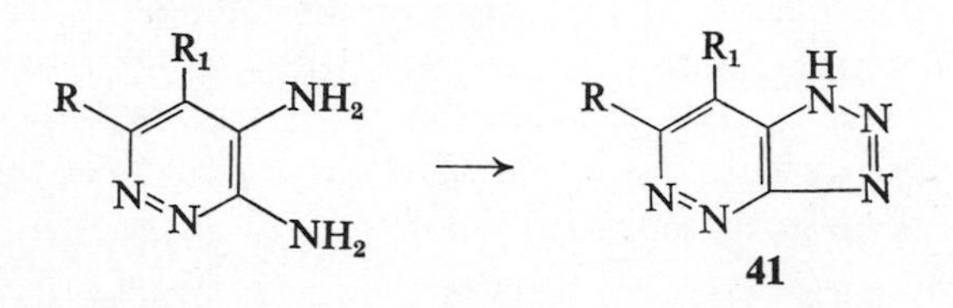

Catalytic dehalogenation in the presence of hydrogen and palladized charcoal yielded the parent compound **41** ($R = R_1 = H$). The structure for this ring system has been assigned from uv absorption spectra correlations. As one can judge from the reaction conditions employed for the displacement of the 7-chlorine atom, these nucleophilic displacements proceed very easily. In this manner the 7-"hydroxy" (with acetic acid, 10 min under reflux), 7-"mercapto" (with thiourea in warm methanol, 10 min) derivatives were obtained (71). More vigorous reaction conditions (115–150°, 3–12 hr) were employed for aminolyses with different amines (72). 7-Amino-*v*-triazolo[4,5-*c*]pyridazine was prepared by heating the 7-methylthio analog with ammonia (120–130°, 8 hr) (71).

Unfortunately, there are no available data for similar displacements on 6-chloro-*v*-triazolo[4,5-*c*]pyridazine which would allow some reactivity correlations. Nevertheless, the 6-chloro compound is stated to be much less reactive. It should be also mentioned that in view of the mobile hydrogen in the imidazole part of the bicycle, tautomerism is expected. This should result in the possibility of obtaining substituted derivatives at N_1 or N_3.

The results of MO calculations (35) of total and frontier π-electron densities are given in Table 3E-3.

V. *v*-Triazolo[1,5-*b*]pyridazines

v-Triazolo[1,5-*b*]pyridazine
(Pyridazo[3,2-*c*]1,2,3]triazole)

The only representative of this system was obtained recently by Evans, Johns, and Markham (73). 3,6-Diphenyl-*v*-triazolo[1,5-*b*]pyridazine (**43**) was obtained in 45% yield together with the hydrazone (**44**) when the oxime **42** was treated with excess hydrazine.

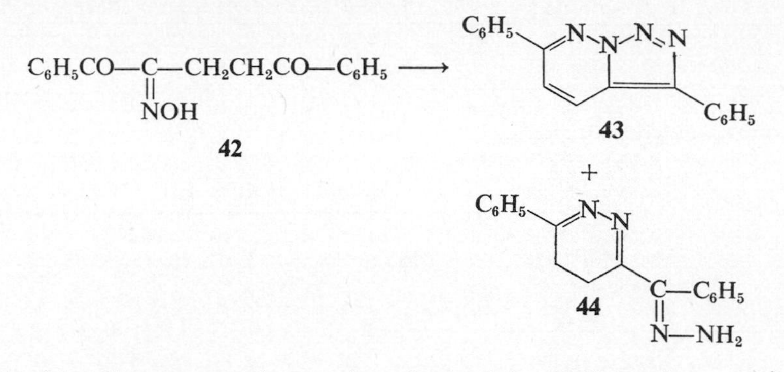

The bicyclic compound was also obtained in the foregoing reaction if only one equivalent of hydrazine was used and if acetic acid was used instead of hydrochloric acid in the cyclization step. The reaction mixture then consisted of **43** (15%), **45** (3%), **46** (20%), and **47** (20%).

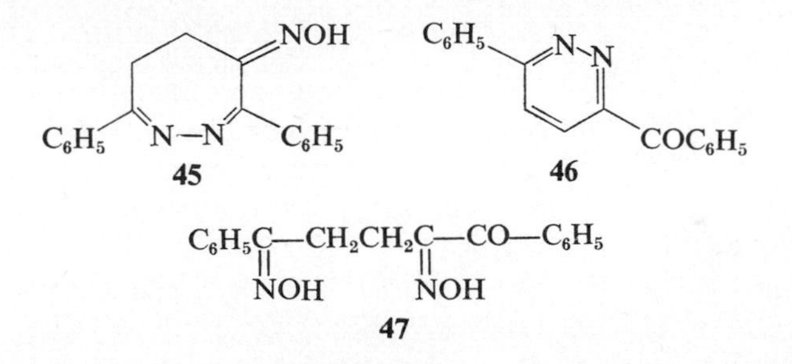

From acid hydrolysis or when heated in ethanol, **44** and **48** afforded only **43**.

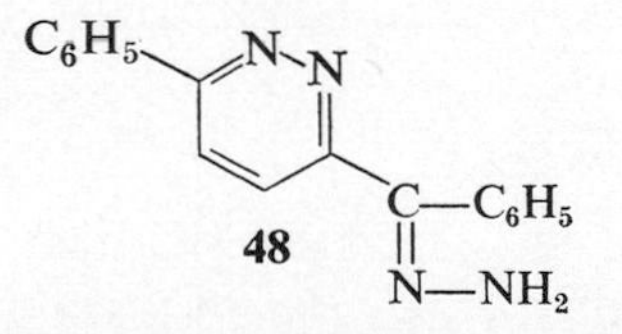

The structure of 3,6-diphenyl-*v*-triazolo[1,5-*b*]pyridazine has been substantiated by nmr spectral data, mass spectrometrically determined molecular weight, and the uv and ir spectrum have been recorded.

MO calculations (35) of total and frontier π-electron densities for this aromatic bicyclic system are presented in Table 3E-4.

VI. Tables

TABLE 3E-1. Total and Frontier π-Electron Densities of *s*-Triazolo[4,3-*b*]pyridazine

Position	1	2	3	4	5	6	7	8	8*a*
Total	1.3201	1.2210	0.9556	1.5665	1.0952	0.9522	0.9633	0.9542	0.9359
Frontier	0.2806	0.0186	0.2124	0.0162	0.1799	0.0198	0.1275	0.1203	0.0246

TABLE 3E-2. Total and Frontier π-Electron Densities of 1*H*-*v*-Triazolo[4,5-*d*]-pyridazine

Position	1	2	3	3*a*	4	5	6	7	7*a*
Total	1.6164	1.0676	1.1992	0.9997	0.9259	1.1604	1.1417	0.9493	0.9339
Frontier	0.0725	0.1814	0.2148	0.0333	0.1742	0.0214	0.1801	0.0834	0.0388

TABLE 3E-3. Total and Frontier π-Electron Densities of 1*H*-*v*-Triazolo[4,5-*c*]-pyridazine

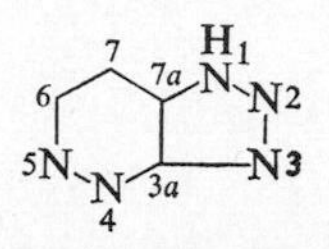

Position	1	2	3	3*a*	4	5	6	7	7*a*
Total	1.6273	1.0465	1.2149	0.9688	1.1447	1.1282	0.9725	0.9539	0.9432
Frontier	0.1137	0.0958	0.2194	0.0089	0.2464	0.0099	0.1489	0.1547	0.0023

TABLE 3E-4. Total and Frontier π-Electron Densities of *v*-Triazolo[1,5-*b*]pyridazine

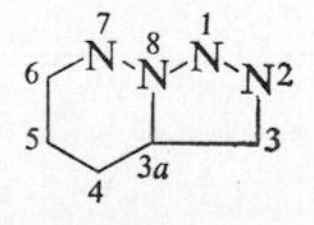

Position	1	2	3	3*a*	4	5	6	7	8
Total	1.1914	1.2051	1.0295	1.0078	0.9179	1.0072	0.9178	1.1784	1.5450
Frontier	0.0109	0.0617	0.0163	0.0271	0.3006	0.1299	0.0970	0.2851	0.0716

TABLE 3E-5. *s*-Triazolo[4,3-*b*]pyridazines

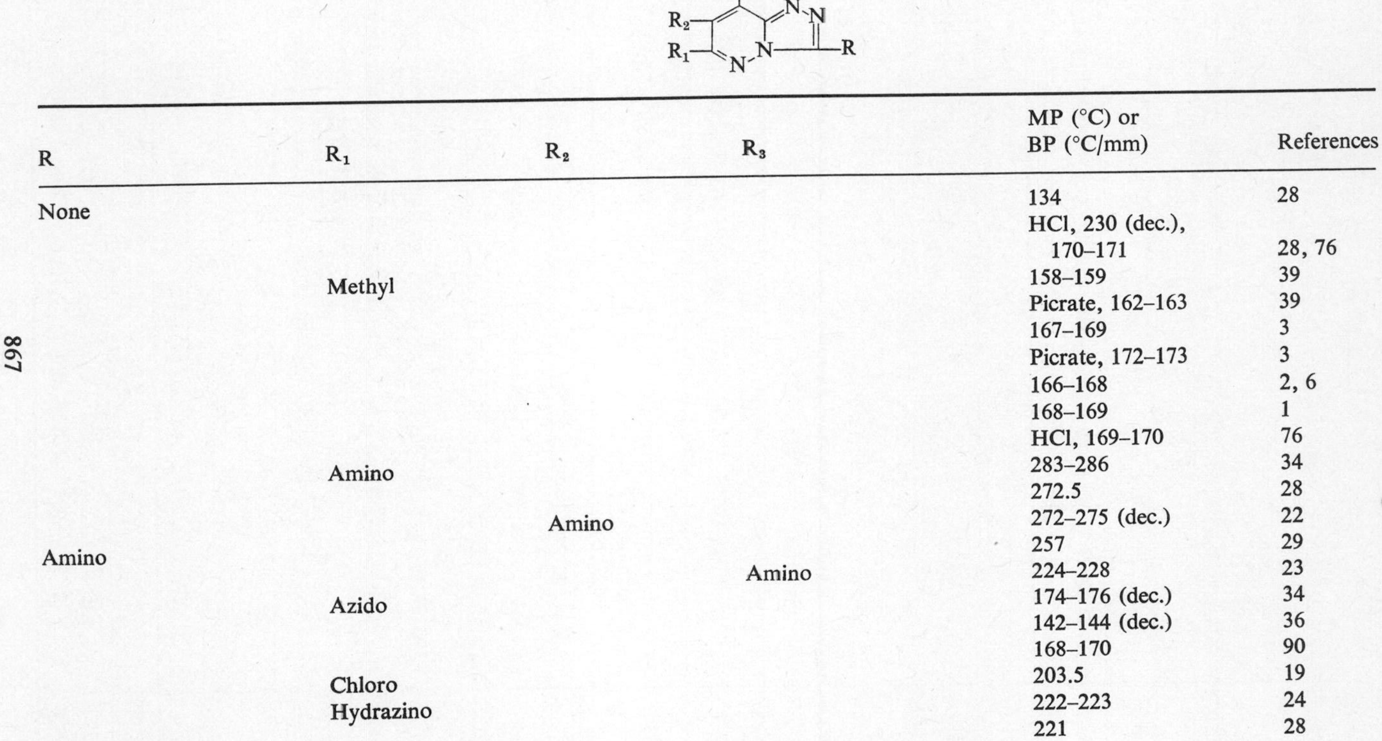

R	R_1	R_2	R_3	MP (°C) or BP (°C/mm)	References
None				134	28
				HCl, 230 (dec.), 170–171	28, 76
	Methyl			158–159	39
				Picrate, 162–163	39
				167–169	3
				Picrate, 172–173	3
				166–168	2, 6
				168–169	1
				HCl, 169–170	76
	Amino			283–286	34
				272.5	28
		Amino		272–275 (dec.)	22
Amino				257	29
			Amino	224–228	23
	Azido			174–176 (dec.)	34
				142–144 (dec.)	36
				168–170	90
	Chloro			203.5	19
	Hydrazino			222–223	24
				221	28

TABLE 3E-5 (*continued*)

R	R_1	R_2	R_3	MP (°C) or BP (°C/mm)	References
	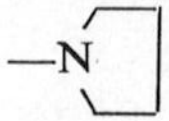 —N (pyrrolidine ring)			165–166	36
				HCl, 235–237	36
Methyl				148–149	74
				HCl, 184–186	76
	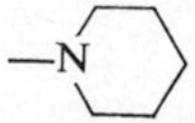 —N (piperidine ring)			123–125	36
	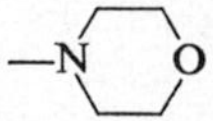 —N O (morpholine ring)			184–185	36
	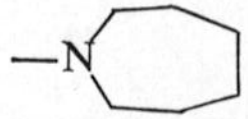 —N (seven-membered ring)			131–133	36
	$MeCOCH_2S$—			150	81
	C_6H_5CO—CH_2—S—			185	81
	Methylthio			165	20
	Ethylthio			126.5	20
	Phenylthio			140.5	20
	Phenoxy			202.5	20
	Ethoxy			176.5	20
	$MeSO_2$			181	20
	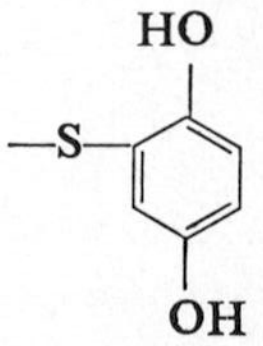 —S— (benzene ring with HO and OH)			246–247	40

—S— (2,5-dioxo-cyclohexa-3,6-dienyl; O, O)			209–210	40
—S— (1,4-dihydroxy-2-naphthyl; OH, OH)			234–235	40
—CH=CH— (5-nitro-2-furyl, NO_2)			275	41
	Methyl		168–169, HCl 188–190	74, 76
		Methyl	126–127, HCl 192–194	74, 76
NHNHCOH			252	28
Mercapto			236 (dec.)	28
			190–193 (dec.)	19
Methoxy			178	19
—S— (3-methyl-1,4-dihydroxy-2-naphthyl; OH, Me, OH)			185–186	40

TABLE 3E-5 (*continued*)

R	R_1	R_2	R_3	MP (°C) or BP (°C/mm)	References
	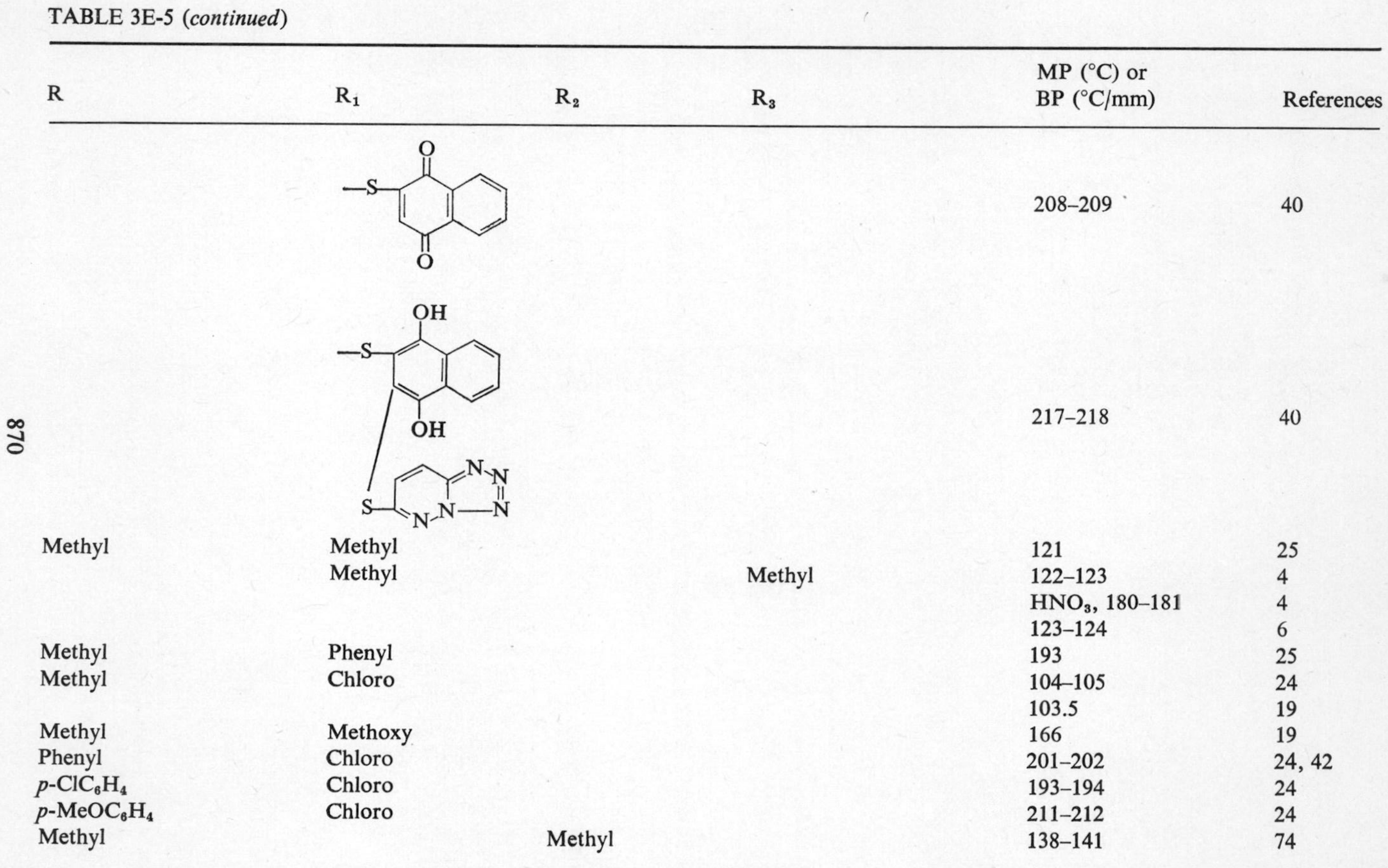			208–209	40
				217–218	40
Methyl	Methyl			121	25
	Methyl		Methyl	122–123	4
				HNO_3, 180–181	4
				123–124	6
Methyl	Phenyl			193	25
Methyl	Chloro			104–105	24
				103.5	19
Methyl	Methoxy			166	19
Phenyl	Chloro			201–202	24, 42
p-ClC_6H_4	Chloro			193–194	24
p-$MeOC_6H_4$	Chloro			211–212	24
Methyl		Methyl		138–141	74

Methyl			Methyl	130–132	74
	Methyl	Methyl		138–140	74
		Methyl	Methyl	204–205	74
NO_2–C_4H_2O–	Azido			185–188 (dec.)	82
NO_2–C_4H_2O–CH=CH–	Azido			210–212	82
NO_2–C_4H_2S–	Azido			199–200	82
NO_2–C_4H_2O–	Azido		Methyl	222–226 (dec.)	82
NO_2–C_4H_2O–	Azido	Methyl		151–152, 165–173	82
NO_2–C_4H_2O–	Azido	—N$(CH_2)_4$		260	82
NO_2–C_4H_2O–	Azido	Amino		220	82
Phenyl	Methylthio			141–142	24
Phenyl	Phenylthio			212–213	24
Phenyl	SCH_2OH			160–161	24
Phenyl	Amino			226–227	24
Phenyl	Hydrazino			261–262	24
Phenyl	—N$(CH_2)_5$			166–167	24

TABLE 3E-5 (*continued*)

R	R_1	R_2	R_3	MP (°C) or BP (°C/mm)	References
Phenyl			Amino	210–212	23
3-Pyridyl	Phenyl			188–189	27
4-Pyridyl	Phenyl			318–320	27
Amino	Chloro			>350	29
Amino	Methoxy			325 (dec.)	29
Amino	*n*-Propoxy			155–156	29
Amino	Hydrazino			293 (dec.)	29
Amino	Piperidino ($—N(CH_2)_5$)			243	29
Amino	$—NH(CH_2)_3NMe_2$			157	29
p-$ON_2C_6H_4SO_2NH$	Methoxy			255	29
p-$NH_2C_6H_4SO_2NH$	Methoxy			222–223	29
Amino	Methyl			287	29
$—SO_2NH_2$	Methyl			217	29
Amino	Amino			332–335	34
Amino	Azido			>250 (dec.)	34
	Methyl		Amino	222–223	12
				218–219	85
				214	37, 38
	Methyl		NHMe	230	37, 38

Methyl		NHEt	204	37, 38
Methyl		NH-*i*-Pr	168	37, 38
Methyl		NH-*n*-Bu	148	37, 38
Methyl		$NH(CH_2)_3NEt_2$	102	37, 38
			EtI salt, 238	37, 38
			94.5–96	12
			Oxalate, 210–211	12
			94.5–97	18
			Oxalate, 210.5–212.5	18
Methyl		$NH(CH_2)_2NEt_2$	154–155	37, 38
			151.5–152.5	18
			Oxalate, 173–174 (dec.)	12
Methyl		NMe_2	155	37, 38
Methyl		NEt_2	99	37, 38
Methyl		$N(CH_2CH_2OH)_2$	202–204	37, 38
Methyl		—N O	195	37, 38
Methyl		—N	118–119	37, 38
Chloro	Methyl		157.5–158	10, 21
Chloro		Methyl	134–134.5	10, 21
			107	19
Hydrazino		Methyl	—	21
Methyl		Chloro	185	37, 39
			187–188	3
			190–190.5	18
			190.5–191	12
Methyl		Hydrazino	~261 (dec.)	37
			223–225 (dec.)	3

TABLE 3E-5 (*continued*)

R	R_1	R_2	R_3	MP (°C) or BP (°C/mm)	References
	Methyl		$—NHCH_2CH_2C_6H_5$	230	37
	Methyl		NHC_5H_5	212	37
	Methyl		Methoxy	200	37
	Methyl		$—NHCH(CH_2)_3NEt_2$ (CH substituted by Me)	200/0.06	12, 18
				Oxalate, 217.5–218.5	12, 18
	Methyl		$—NHCH_2CHCH_2NEt$ (CH substituted by OH)	126–127	12, 18
				Oxalate, 167.5–170 (dec.)	12, 18
	Methoxy		Methyl	120, 5	19
	Methyl		$—NH(CH_2)_4NEt_2$	81–81.5	12, 18
				Oxalate, 144.5–146.5	12, 18
	Methyl		$—NH(CH_2)_4N(n\text{-}Bu)_2$	76–77	12, 18
				Oxalate, 156–160	12, 18
	Methyl		$—NHCH_2CH(CH_2)_2NEt_2$ (CH substituted by 4-chlorophenyl, C_6H_4Cl)	137–138	12, 18
				Oxalate, 199.5–201	12, 18
	Methyl		Phenyl	152–153	4
	Methyl		Methoxy	170–171	39
	Methyl		$OCOC_6H_5$	157–158	39
	Methyl		Iodo	211–212	39
				HI, 196–197	39
	Methyl		NHNHCOH	267–270 (dec.)	3

	Phenyl		Phenyl	162–163	5
	Phenyl		Chloro	214	13, 14, 1
	p-ClC_6H_4		Chloro	240.5–241.5	12, 18
	p-ClC_6H_4		$-NHCH_2CH(OH)CH_2NEt_2$	191.5–192.5	12, 18
	p-ClC_6H_4		$-NHCH(Me)CH_2CH_2CH_2NEt_2$	125.5–126.5	12, 18
	Chloro		Amino	278	23, 85
	Chloro		NHCOMe	259	23
		Amino	Chloro	285 (dec.)	22
		C_6H_5CONH	Chloro	220–222	22
		Chloro	Amino	315–317	23
$-CH{=}CH-$(furan, O)$-NO_2$	Methoxy			235–237	41
$-CH{=}CH-$(furan, O)$-NO_2$	Methyl			232–234	41
$-CH{=}CH-$(furan, O)$-NO_2$	Phenyl			252–256	41
$-CH{=}CH-$(furan, O)$-NO_2$	Amino			300	41
$-CH{=}CH-$(furan, O)$-NO_2$	$NHCOCH_3$			284–286	41

TABLE 3E-5 (*continued*)

R	R_1	R_2	R_3	MP (°C) or BP (°C/mm)	References
—CH=CH—(2,5-thienyl)—NO_2	$NHCOCH_3$			268–270	41
—CH=CH—(2,5-thienyl)—NO_2	Amino			274–276	41
	Hydrazino	Methyl		>265	10
	NHNHCOH	Methyl		>265	10
CH_2=$CHCH_2NH$	p-MeC_6H_4			184	30
C_6H_5NH—	p-MeC_6H_4			174	30
Phenyl		Amino	Chloro	295–296	22
Phenyl		Chloro	Amino	248–249	23
Phenyl	Chloro		Amino	274	23
Phenyl	Chloro		$NHCOC_6H_5$	216–223 218–222	23
Phenyl	Chloro		NHCOMe	277	23
Methyl	Chloro	Methyl		180, 183–184	21, 74
Methyl	Hydrazino	Methyl		no data	21
	Methyl	Ethyl	Chloro	112	13, 14,
	Methyl	Methyl	Chloro	153.5–154	12, 18
	Methyl	Methyl	—$NHCH_2CH_2NEt_2$	96.5–98 Oxalate, 212–215	12, 18 12, 18

	Methyl	Methyl	$—NHCH(Me)(CH_2)_3NEt_2$	97–98.5	12, 18, 85
				MeI salt 195.5–199.5	12, 18
	Methyl	Methyl	Methyl	129	4
				129–130	6
	Methyl	Ethyl	Methyl	Oil	6
Methyl	Chloro		Methyl	135	19
				170–171	74
Methyl	Chloro	Methyl	Methyl	112–113	74
	Chloro	Methyl	Methyl	121–122	74
NO_2–(C₄H₂O)–	Chloro	Methyl		136–139 (dec.)	82
Phenyl	Azido			168	84
	Chloro		$CONH_2$	247 (dec.)	85
	Chloro		Cyano	158–159	85
			$CONH_2$	257–278	85
		Bromo	Amino	281–282	85
	Methyl	Bromo	Amino	258–259	85
	Methyl	$C_6H_5—N=N—$	Chloro	142–144	86
	Methyl	$C_6H_5—N=N—$	Methoxy	135–137° (dec.)	86
	(C₄H₃S)–CONH—			224–225°	88
	Methyl		(C₄H₃S)–CONH—	227–229°	88
	Nitro			200–201	87

TABLE 3E-6. *s*-Triazolo[4,3-*b*]pyridazin-6(5*H*)ones

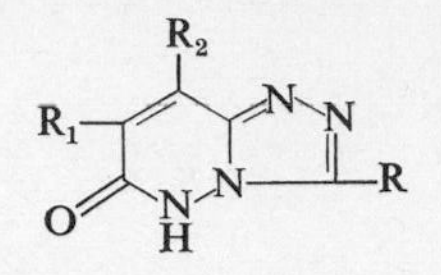

R	R_1	R_2	MP (°C) or BP (°C/mm)	References
	Methyl		>265	10
		Methyl	>265	10
None			291 (dec.)	28
			310–312	42
—CH=CH—(O)—NO_2			313	41
Phenyl			297–298	42
Methyl			263–264	42

TABLE 3E-7. *s*-Triazolo[4,3-*b*]pyridazin-8(5*H*)ones

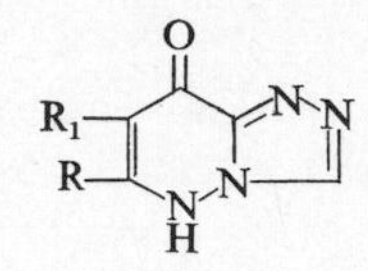

R	R_1	MP (°C) or BP (°C/mm)	References
Methyl	Ethyl	229–231	8
		229	45
		235	13, 14, 17
Methyl		295	45
		297–298 (dec.)	3
		304–306 (dec.)	12, 18
		301–303	6
		>340	7
Phenyl		283	3, 7
		280–282	6
		284–285	13, 14, 17
Benzyl		211	3
		252	6
Methyl	Methyl	252	7
		257–258	6
		264–266	12
		263.5–266.5	18

TABLE 3E-7 (*continued*)

R	R_1	MP (°C) or BP (°C/mm)	References
p-ClC_6H_4		>320	12
		320	18
Methyl	n-Butyl	206	8
Methyl	Benzyl	212 (dec.)	8
		307–308	6
n-Amyl	n-Butyl	169.5–170.5	8
Ethyl		236.5–237.5	8
Benzyl	Phenyl	286–290	8
Methyl	Heptyl	170.5–171	12, 18
Methyl	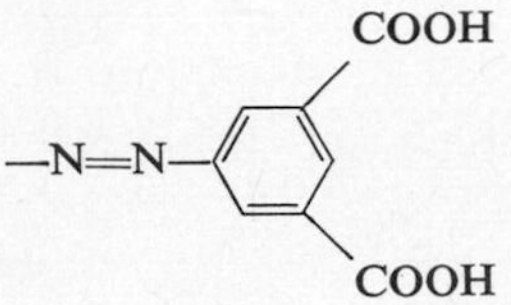 —N=N—$C_6H_3(COOH)_2$	—	15, 16
Ethoxy		274–276	83
n-Butoxy		209–211	83
Amino		>360	83
Methyl	Amino	307–308	83
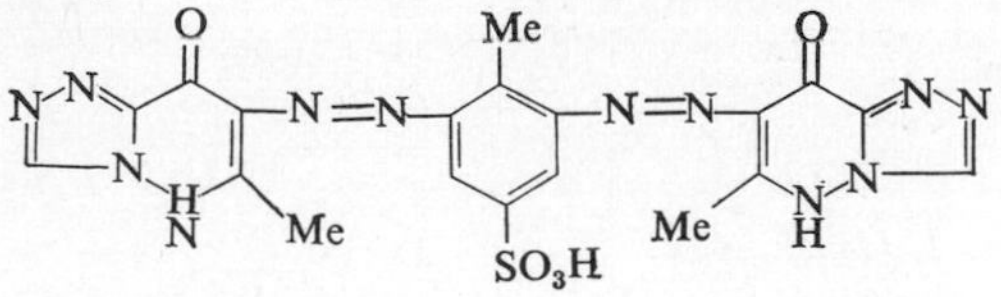		—	15
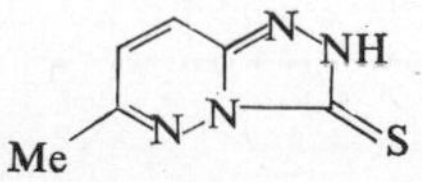		260	29

TABLE 3E-8

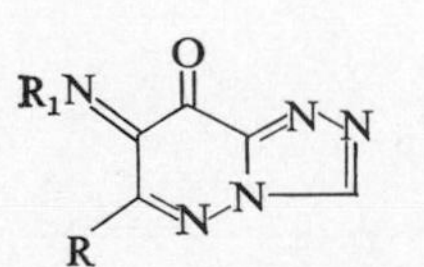

R	R_1	MP (°C) or BP (°C/mm)	References
Methyl	OH	204–205	83
Methyl	C_6H_5—NH—	269–270	86
Methyl	p-MeO—C_6H_4—NH—	265–266	86
Methyl	p-NO_2—C_6H_4—NH—	312	86
Phenyl	C_6H_5NH—	250	86
Benzyl	C_6H_5—NH—	254	86
Ethoxy	C_6H_5—NH—	248–249	86

TABLE 3E-9. *s*-Triazolo[4,3-*b*]pyridazine-6(5*H*)-thiones

R	MP (°C) or BP (°C/mm)	References
None	190–193	19
Phenyl	151–152	24

TABLE 3E-10. *s*-Triazolo[4,3-*b*]pyridazine-8(5*H*)thiones

R	R_1	MP (°C) or BP (°C/mm)	References
Methyl		228–230	3
		—	39[a]
Phenyl		173, then 250	13, 14, 17
Methyl	Ethyl	238	13, 14, 17

[a] At 150° begins to sinter, does not melt to 280° C.

TABLE 3E-11. Quaternized *s*-Triazolo[4,3-*b*]pyridazines

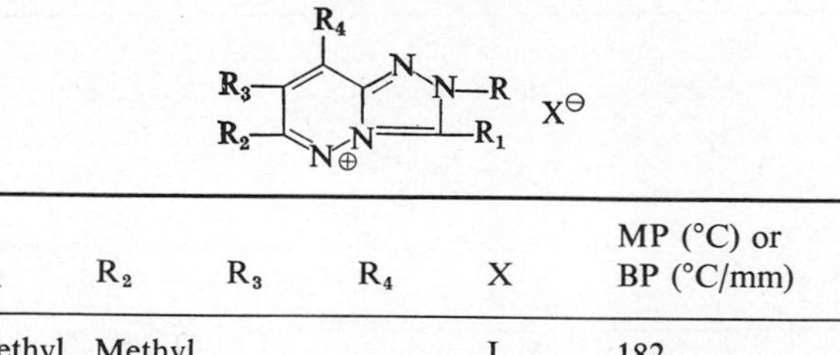

R	R_1	R_2	R_3	R_4	X	MP (°C) or BP (°C/mm)	References
Methyl	Methyl	Methyl			I	182	25
Ethyl	Methyl	Phenyl			I	153	25
Benzyl		Methyl		Methyl	Cl	198–200	5
Ethyl		Methyl		Phenyl	EtOSO	195–200	5
Benzyl		Methyl		Phenyl	Cl	223–225	5
Benzyl		Methyl	Methyl	Methyl	Cl	178–181	5
$C_6H_5COCH_2$		Methyl		Phenyl	Br	152–155	5
$MeCOCH_2$		Methyl		Phenyl	Cl	Oil	5[a]

TABLE 3E-11 (*continued*)

R	R_1	R_2	R_3	R_4	X	MP (°C) or BP (°C/mm)	References
$C_6H_5COCH_2$		Methyl		Methyl	Br	187–189	5
$MeCOCH_2$		Methyl		Methyl	Cl	Oil	5[a]
CH_2CN		Methyl		Methyl	Cl	220	5
CH_2COOEt		Methyl		Methyl	Cl	Oil	5[a]
$MeCOCH_2$		Methyl	Methyl	Methyl	Cl	Oil	5[a]
$MeCOCH_2$		Phenyl		Phenyl	Cl	Oil	5[a]
$MeCOCH_2$		Methyl	Ethyl	Methyl	Cl	Oil	5[a]
$MeCOCH_2$		Methyl	Benzyl	Methyl	Cl	Oil	5[a]
Methyl				Methyl	I	221–223	76

[a] Not analyzed.

TABLE 3E-12

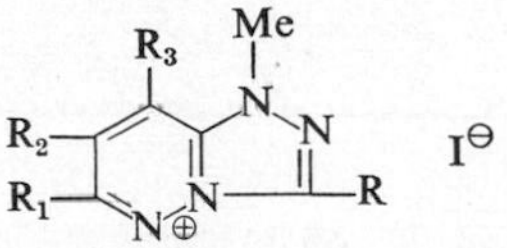

R	R_1	R_2	R_3	MP (°C) or BP (°C/mm)	References
None				269–271	76
Methyl				227–229	76
	Methyl			230–232	76
		Methyl		293–294	76
			Methyl	296–297	76
Methyl			Methyl	240–241	76

TABLE 3E-13. *s*-Triazolo[4,3-*b*]pyridazine 5-Oxides

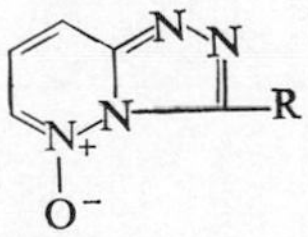

R	MP (°C) or BP (°C/mm)	References
Phenyl	154–155	42
p-$MeOC_6H_4$—	187–188	42
Methyl	199–200	42
Amino	206–208 (dec.)	42
None	243–244	42, 87

TABLE 3E-14. *s*-Triazolo[1,2-*a*]pyridazines

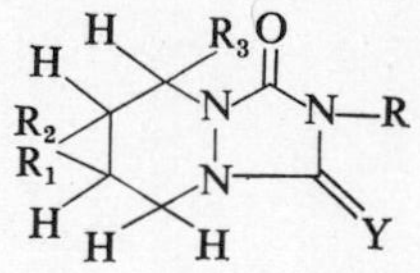

R	R_1	R_2	R_3	Y	MP (°C) or BP (°C/mm)	References
Phenyl			Methyl	S	138–139	58
	Methoxy	CH_3COOHg—		O	amorphous	56, 57
	Methoxy	Hydroxy		O	295–300 (dec.)	56, 60
	Methoxy	—SCH_2COOH		O	—	
					diNa salt, 211–223.5 (dec.)	56
					diNa salt, 219 (dec.)	57
	Methoxy	—$SCH(COOH)CH_2COOH$		O	—	
					3 Na salt, 120–125	56, 57
	Methoxy	Methylthio		O	>210 (dec.)	56
					Na salt, 178.5	57
	n-Propoxy	HgCl		O	198 (dec.)	60
	Methoxy	HgOH		O	295–300 (dec.)	57
	Hydroxy	HgOH		O	266.5 (dec.)	57
	Methoxy	HgCl		O	144–147 (dec.)	60
	Methoxy	HgBr		O	177–178 (dec.)	60
CH_2COOEt	Methoxy	CH_3COO—		O	181–183.5 (dec.)	56, 57
CH_2COOH	Methoxy	CH_3COO—		O	232–235 (dec.)	56, 57
CH_2COOH	Methoxy	Hydroxy		O	—	
					Na salt > 225 (dec.)	56, 57
Methyl	Methoxy	CH_3COO—		O	188–191 (dec.)	56, 57
Methyl	Methoxy	HgCl		O	238.5–240 (dec.)	60
n-$C_{12}H_{25}$—	Methoxy	CH_3COOHg—		O	118.5–121	60
Phenyl				O	141–142	61
n-Butyl				O	110–120/0.005	61

TABLE 3E-15. *s*-Triazolo[1,2-*a*]pyridazines

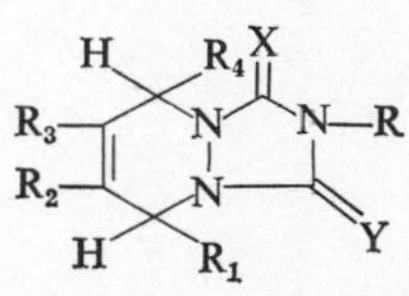

R	R_1	R_2	R_3	R_4	X	Y	MP (°C) or BP (°C/mm)	References
Phenyl				Methyl	O	S	177–178	58
Phenyl					O	S	165	59
p-BrC_6H_4					O	S	211	59
p-$Me_2NC_6H_4$					O	S	212	59
None					O	O	244–247	56, 57
CH_2COOEt					O	O	102.5–105	56, 57
CH_2COOH					O	O	160–169	56, 57
CH_2CH_2OH					O	O	162–167.5	56, 57
n-$C_{12}H_{25}$					O	O	44.5–45.5	56, 57
p-$NO_2C_6H_4CH_2$					O	O	200–202.5	56, 57
Methyl					O	O	155–159	57
Phenyl			Ethyl	CN	O	O	174–176	53
Phenyl		Methyl	Methyl		O	O	174–175.5	52
Phenyl		Methyl			O	O	110–112	52
Phenyl	Methyl				O	O	135–138	52

TABLE 3E-15 (*continued*)

R	R_1	R_2	R_3	R_4	X	Y	MP (°C) or BP (°C/mm)	References
Phenyl	Phenyl			Phenyl	O	O	160–161	52, 55
							157–158	
$HgOCOCH_3$					O	O	192–194 (dec.)	60
None					NH	NH	—	
							$HNO_3 \cdot H_2O$ 261	62
Phenyl		Chloro			O	O	157–159	55
		Methyl		Methyl	NNO_2	NH	258.5–260.5	63
		Methyl		Methyl	NH	NH	—	
							HNO_3, 196.5–196.8	63
							Picrate, 253.4–254.2	
Methyl		Chloro			O	O	134–135.5	54
Ethyl		Chloro			O	O	82.5–84	54
Isopropyl		Chloro			O	O	189–191	54
n-Butyl		Chloro			O	O	50–51.5	54
Iso-butyl		Chloro			O	O	82–83	54
Octyl		Chloro			O	O	147–149	54
C_6H_{11}		Chloro			O	O	232–233.5	54
Benzyl		Chloro			O	O	117–119	54
Phenyl					O	O	157–159	51

TABLE 3E-16. *s*-Triazolo[1,2-*a*]pyridazines

R	MP (°C) or BP (°C/mm)	References
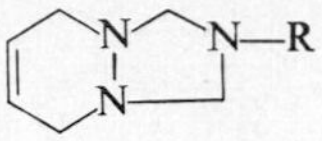 N, N, N—R		
Phenyl	145	59
p-BrC_6H_4	100	59
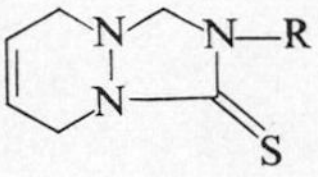 N, N, N—R, S		
p-BrC_6H_4	115	59
p-$Me_2NC_6H_4$	125	59
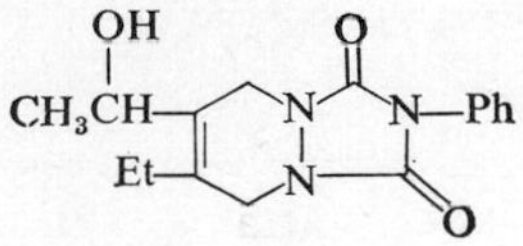 OH, CH_3CH—, Et, O, N, N, N—Ph, O	112–114	78
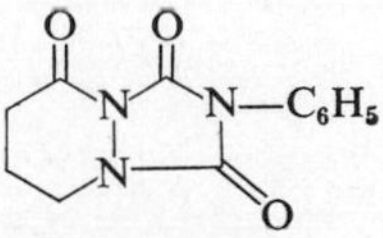 O, O, N, N, N—C_6H_5, O	195–198	61
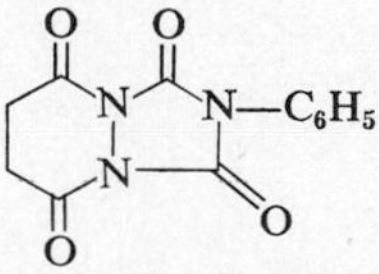 O, O, N, N, N—C_6H_5, O, O	250 (dec.)	61

TABLE 3E-17. *s*-Triazolo[1,2-*a*]pyridazines

R	R_1	R_2	R_3	MP (°C) or BP (°C/mm)	References
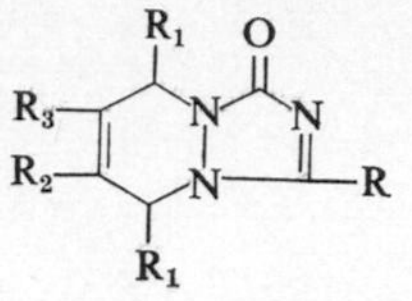 R_1 O R_3 N N R_2 N R R_1					
Phenyl		Methyl		203–205	92
Phenyl		Methyl	Methyl	198–200	92
Phenyl	Phenyl			248–250	92
p-Methoxyphenyl		Methyl	Methyl	227–230	92
p-Methoxyphenyl	Phenyl			253–255	92
p-Nitrophenyl		Methyl	Methyl	269 (dec.)	92
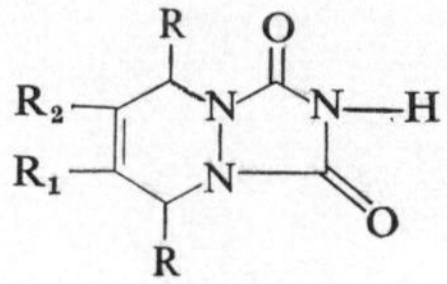 R O R_2 N N—H R_1 N O R					
	Methyl			204–206	91
Phenyl			Phenyl	284–285	91
	Methyl	Methyl		216–217	91
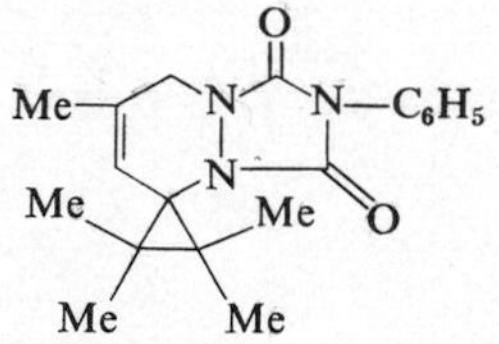 O Me N N—C_6H_5 N Me Me O Me Me				174–175	93
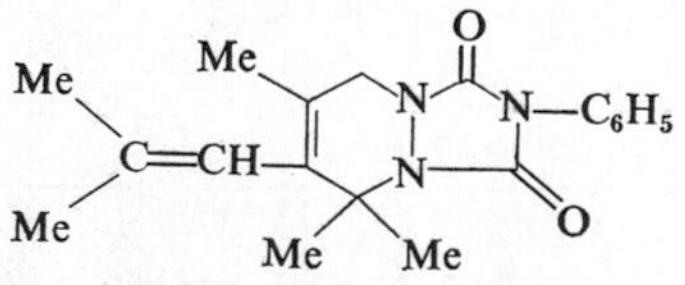 O Me Me N N—C_6H_5 C=CH N Me O Me Me				152.5–154	93
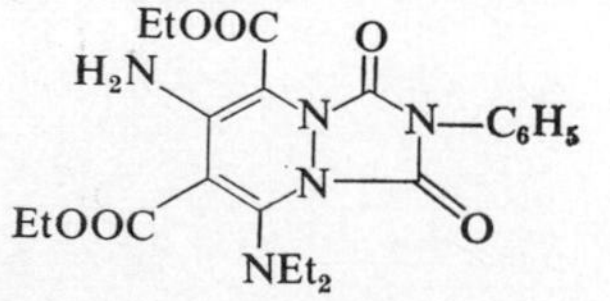 EtOOC O H_2N N N—C_6H_5 N EtOOC O NEt_2				—	94

TABLE 3E-18. 1*H*-*v*-Triazolo[4,5-*d*]pyridazines

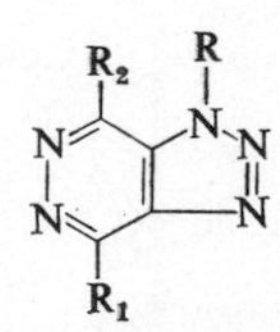

R	R_1	R_2	MP (°C) or BP (°C/mm)	References
None			>300	95
	Amino		>360, >300	79, 95
	Hydrazino		>300	79
	Methylthio		210.5–211.5 (dec.)	79
			224–225 (dec.)	95
	Benzylthio		188.5–189.5	79
	$NH(CH_2)_3NEt_2$		217–218 (dec.)	79
	$NH(CH_2)_2NMe_2$		HCl, 295 (dec.)	79
	Methoxy	Methyl	154–155[a]	69, 95
			156–157	
	Methoxy	Methoxy	183–184	68
			212–213	68
Phenyl	Chloro	Chloro	168–170	66
Phenyl	Methoxy	Methoxy	195–197	66
Phenyl	Methylthio	Methylthio	167–168	66
Phenyl	OH (or $Me_2NCH_2CH_2O$)	$Me_2NCH_2CH_2O$ (or OH)	—	
			HCl, 257–258	66
Phenyl	$Me_2NCH_2CH_2O$—	$Me_2NCH_2CH_2O$—	58–61	66
Phenyl	OH (or Cl)	Cl (or OH)	—	66
Phenyl	$NHNH_2$ (or MeS)	MeS (or $NHNH_2$)	210–211	66

TABLE 3E-18 (*continued*)

R	R_1	R_2	MP (°C) or BP (°C/mm)	References
	Methoxy		141–142	95
	Ethoxy	Methyl	172–173	95
	Amino	Chloro	>320	95
	Ethoxy	Ethoxy	155–156	95
	Amino	Methyl	>300	95
	Hydrazino	Methyl	290–291 (dec.)	95
			HCl, 250 (dec.)	95
	Benzylamino	Methyl	>300	95
	$HOCH_2CH_2NH$—	Methyl	285–287 (dec.)	95
	Cyclohexylamino	Methyl	>300	95
	Dimethylamino	Methyl	>300	95
	Hydrazino	Methoxy	177.5–178.5	95
	$HOCH_2CH_2NH$—	Methoxy	205.5–206.5	95
	$HOCH_2CH_2NH$—	$HOCH_2CH_2NH$—	252–253	95
	Benzylamino	Ethoxy	220–223°	95
	$HOCH_2CH_2NH$—	Ethoxy	227–230	95
	N—	Methyl	263–265 (dec.)	95
	O N—	O N—	>320°	95
	N—	Methoxy	211–212.5	95
	O N—	Methoxy	246.5–247 (dec.)	95

[a] As hydrate, xH_2O.

TABLE 3E-19

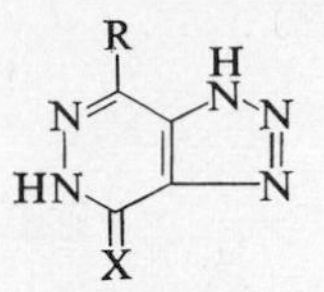

X	R	MP (°C)	References
O		>300, 305–307	79, 95
S		>300	79, 95
O	Methyl	285–285.5 (dec.)	95
O	Chloro	229–230	95
O	Ethoxy	239–241	95

TABLE 3E-20. 1*H*-*v*-Triazolo[4,5-*d*]pyridazine-4,7(5*H*,6*H*)-diones

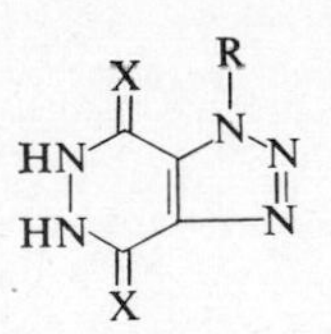

R	X	BP (°C/mm) MP (°C) or	References
Phenyl	O	310–312	66
Phenyl	S	195–198	66
p-Xylyl	O	286 (dec.)	64

TABLE 3E-21. 2*H*-*v*-Triazolo[4,5-*d*]pyridazines

Structure	MP (°C)	Reference
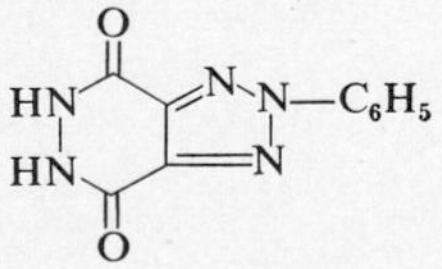	317	65

TABLE 3E-22. *v*-Triazolo[4,5-*c*]pyridazines

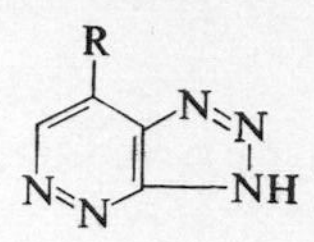

R	MP (°C) or BP (°C/mm)	References
None	227–228 (dec.)	71
Chloro	>300	71
Hydroxy	>200 (dec.)	71
Amino	>300	71
MeNH—	310–311 (dec.)	72
Me_2N	321–322 (dec.)	72
$C_6H_5CH_2NH$—	274–275	72
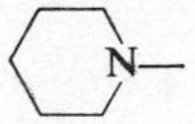 N— (piperidino ring)	324–325 (dec.)	72
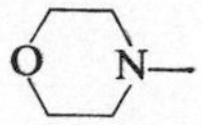 O N— (morpholino ring)	342–343 (dec.)	72
Methylthio	219–220 (dec.)	71
p-$ClC_6H_4CH_2S$—	198–198.5 (dec.)	71
2,4-di$ClC_6H_3CH_2S$—	204–205 (dec.)	71
3,4-di$ClC_6H_3CH_2S$—	197–198 (dec.)	71
2,6-di$ClC_6H_3CH_2S$—	176–177 (dec.)	71
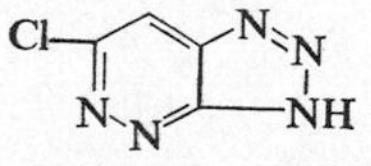 Cl— (structure: Cl on the pyridazine ring; N, N, NH)	218 (dec.)	71

TABLE 3E-23. *v*-Triazolo[1,5-*b*]pyridazines

Structure	MP (°C)	Reference
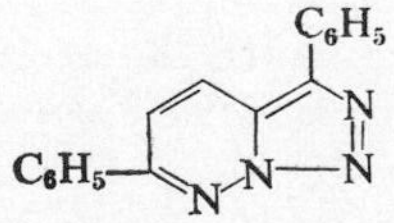 C_6H_5 (structure: C_6H_5—, N, N—N, N; C_6H_5 on the triazole ring)	254	73

References

1. C. F. H. Allen, H. R. Beilfuss, D. M. Burness, G. A. Reynolds, J. F. Tinker, and J. A. Van Allan, *J. Org. Chem.*, **24**, 796 (1959).
2. U.S. Pat. 2,837,521 (1958); *Chem. Abstr.*, **53**, 2262 (1959).
3. D. Libermann and R. Jacquier, *Bull. Soc. Chim. France*, **1962**, 355.
4. C. Bülow and F. Weber, *Ber.*, **42**, 2208 (1909).
5. H. G. O. Becker and H. Böttcher, *Tetrahedron*, **24**, 2687 (1968).
6. H. G. O. Becker and H. Böttcher, *Wiss. Z. Tech. Hochsch. Chem. Leuna-Merseburg*, **8**, 122 (1966).
7. C. Bülow and F. Weber, *Ber.*, **42**, 2594 (1909).
8. A. N. Kost and F. Genc, *Zh. Obshch. Khim.*, **28**, 2773 (1959); *Chem. Abstr.*, **53**, 9197 (1959).
9. C. R. Hauser and G. A. Reynolds, *J. Am. Chem. Soc.*, **70**, 2402 (1948).
10. S. Linholter and R. Rosenørn, *Acta Chem. Scand.*, **16**, 2389 (1962).
11. M. A. Hill, G. A. Reynolds, J. F. Tinkler, and J. A. Van Allan, *J. Org. Chem.*, **26**, 3834 (1961).
12. E. A. Steck and R. P. Brundage, *J. Am. Chem. Soc.*, **81**, 6289 (1959).
13. Belgian Pat. 566,543 (1958); *Chem. Abstr.*, **53**, 14796 (1959).
14. British Pat. 893,428 (1962); *Chem. Abstr.*, **59**, 2834 (1963).
15. U.S. Pat. 2,390,707 (1945); *Chem. Abstr.*, **40**, 2080 (1946).
16. U.S. Pat. 2,432,419 (1947); *Chem. Abstr.*, **42**, 2193 (1948).
17. U.S. Pat. 2,933,388 (1960); *Chem. Abstr.*, **55**, 1254 (1961).
18. U.S. Pat. 3,096,329 (1963); *Chem. Abstr.*, **59**, 14004 (1963).
19. N. Takahayashi, *J. Pharm. Soc. Japan*, **75**, 1242 (1955); *Chem. Abstr.*, **50**, 8655 (1956).
20. N. Takahayashi, *J. Pharm. Soc. Japan*, **76**, 1296 (1956); *Chem. Abstr.*, **51**, 6645 (1957).
21. N. Takahayashi, *Pharm. Bull. (Tokyo)*, **5**, 229 (1957); *Chem. Abstr.*, **52**, 6359 (1958).
22. T. Kuraishi and R. N. Castle, *J. Heterocyclic Chem.*, **1**, 42 (1964).
23. T. Kuraishi and R. N. Castle, *J. Heterocyclic Chem.*, **3**, 218 (1966).
24. A. Pollak and M. Tišler, *Tetrahedron*, **22**, 2073 (1966).
25. British Pat. 839,020 (1960); *Chem. Abstr.*, **55**, 2323 (1961).
26. S. Biniecki, A. Haase, J. Izdebski, E. Kesler, and L. Rylski, *Bull. acad. polon. sci., Ser. sci. chim., geol. geog.*, **6**, 227 (1958); *Chem. Abstr.*, **52**, 18424 (1958).
27. A. Haase and S. Biniecki, *Acta Polon. Pharm.*, **18**, 461 (1961); *Chem. Abstr.*, **61**, 3103 (1964).
28. N. Takahayashi, *J. Pharm. Soc. Japan*, **76**, 765 (1956); *Chem. Abstr.*, **51**, 1192 (1957).
29. N. K. Basu and F. L. Rose, *J. Chem. Soc.*, **1963**, 5660.
30. I. Zugravescu, M. Petrovanu, E. Rucinschi, and M. Caprosu, *Rev. Chim. Roum.*, **10**, 641 (1965).
31. B. Stanovnik and M. Tišler, *Tetrahedron*, **23**, 387 (1967).
32. B. Stanovnik, A. Krbavčič, and M. Tišler, *J. Org. Chem.*, **32**, 1139 (1967).
33. Unpublished results from this laboratory.
34. A. Kovačič, B. Stanovnik, and M. Tišler, *J. Hetrocyclic Chem.*, **5**, 351 (1968).
35. Unpublished data from this laboratory. Parameters for LCAO Calculations were taken from A. Streitwieser, *Molecular Orbital Theory for Organic Chemists*, *Wiley*, New York, 1961, p. 135.
36. I. B. Lundina, Ju. N. Sheinker, and I. Ja. Postovski, *Izvest. Akad. Nauk SSSR*, **1967**, 66.

37. French Pat. 1,248,409 (1960); *Chem. Abstr.*, **56,** 10160 (1962).
38. J. Sallé, M. Pesson, and H. Kornowsky, *Arch. int. pharmacodynamie*, **121,** 154 (1959); *Chem. Abstr.*, **54,** 5925 (1960).
39. C. Bülow and K. Haas, *Ber.*, **43,** 1975 (1910).
40. A. Pollak, B. Stanovnik, and M. Tišler, *Monatsh. Chem.*, **97,** 1523 (1966).
41. Neth. Pat. Appl. 6,609,264 (1967); *Chem. Abstr.*, **67,** 54151 (1967).
42. A. Pollak, B. Stanovnik, and M. Tišler, *J. Heterocyclic Chem.*, **5,** 513 (1968).
43. E. J. Birr, *Z. Wiss. Phot.*, **47,** 2 (1952).
44. Y. Kuwabara and K. Aoki, *Konishiroku Rev.*, **6,** 1 (1955); *Chem. Abstr.*, **49,** 11473 (1955).
45. K. Murobushi, Y. Kuwabara, S. Baba, and K. Aoki, *J. Chem. Soc. Japan, Ind. Chem. Sect.*, **58,** 440 (1955); *Chem. Abstr.*, **49,** 14544 (1955).
46. U.S. Pat. 3,017,270 (1962); *Chem. Abstr.*, **57,** 1791 (1962).
47. U.S. Pat. 3,053,657 (1962); *Chem. Abstr.*, **58,** 2064 (1963).
48. S. Kikuchi and K. Akiba, *J. Soc. Sci. Phot. Japan*, **18,** 20 (1955); *Chem. Abstr.*, **50,** 2331 (1956).
49. A review on azo compounds as dienophiles was published by T. Gillis in *1,4-Cycloaddition Reactions. The Diels-Alder Reaction in Heterocyclic Synthesis*, J. Hamer, Ed., Academic Press, New York-London, 1967, p. 143.
50. R. C. Cookson, S. S. H. Gilani, and I. D. R. Stevens, *Tetrahedron Lett.*, **1962,** 615.
51. R. C. Cookson, S. S. H. Gilani, and I. D. R. Stevens, *J. Chem. Soc. C*, **1967,** 1905.
52. B. T. Gillis and J. D. Hagarthy, *J. Org. Chem.*, **32,** 330 (1967).
53. S. Combrisson and E. Michel, *Bull. Soc. Chim. France*, **1968,** 787.
54. M. Furdik, S. Mikulasek, M. Livar, and S. Priehradny, *Chem. Zvesti*, **21,** 427 (1967); *Chem. Abstr.*, **67,** 116858 (1967).
55. J. Sauer and B. Schröder, *Chem. Ber.*, **100,** 678 (1967).
56. U.S. Pat. 2,813,865 (1957); *Chem. Abstr.*, **52,** 4699 (1958).
57. U.S. Pat. 2,813,866 (1957); *Chem. Abstr.*, **52,** 4698 (1958).
58. R. Ya. Levina, Yu. S. Shabarov, and M. G. Kuzmin, *Zh. Obshch. Khim.*, **30,** 2469 (1960).
59. Yu. S. Shabarov, A. R. Smirnova, and R. Ya. Levina, *Zh. Obshch. Khim.*, **34,** 390 (1964).
60. F. W. Gubitz and R. L. Clarke, *J. Org. Chem.*, **26,** 559 (1961).
61. G. Zinner and W. Deucker, *Arch. Pharm.*, **296,** 13 (1963).
62. J. C. J. MacKenzie, A. Rodgman, and G. F. Wright, *J. Org. Chem.*, **17,** 1666 (1952).
63. G. F. Wright, *Can. J. Chem.*, **30,** 62 (1952).
64. A. Bertho and F. Hölder, *J. prakt. Chem.*, **119,** 189 (1929).
65. R. Seka and H. Preissecker, *Monatsh. Chem.*, **57,** 71 (1931); *Chem. Abstr.*, **25,** 1826 (1931).
66. L. Erichomovitch and F. L. Chubb, *Can. J. Chem.*, **44,** 2095 (1966).
67. R. G. Jones, *J. Am. Chem. Soc.*, **78,** 179 (1956).
68. T. Itai and S. Suzuki, *Chem. Pharm. Bull. (Tokyo)*, **8,** 999 (1960); *Chem. Abstr.*, **55,** 24750 (1961).
69. D. L. Aldous and R. N. Castle, *Arzneimittel-Forsch.*, **13,** 878 (1963).
70. K. Dury, *Angew. Chem.*, **77,** 282 (1965).
71. G. A. Gerhardt and R. N. Castle, *J. Heterocyclic Chem.*, **1,** 247 (1964).
72. H. Murakami and R. N. Castle, *J. Heterocyclic Chem.*, **4,** 555 (1967).
73. N. A. Evans, R. B. Johns, and K. R. Markham, *Aust. J. Chem.*, **20,** 713 (1967).
74. M. Japelj, B. Stanovnik, and M. Tišler, *Monatsh.*, **100,** 671 (1969).
75. German (East) Pat. 60,042 (1968); *Chem. Abstr.*, **70,** 37825 (1969).

76. M. Japelj, B. Stanovnik, and M. Tišler, *J. Heterocyclic Chem.*, **6**, 559 (1969).
77. German (East) Pat. 60,041 (1968); *Chem. Abstr.*, **70**, 37826 (1969).
78. S. Combrisson, E. Michel, and C. Troyanowsky, *Compt. rend.*, **C267**, 326 (1968).
79. S. F. Martin and R. N. Castle, *J. Heterocyclic Chem.*, **6**, 93 (1969).
80. Y. Kuwabara, S. Irie, S. Sugita, M. Yanai, S. Takeda, and H. Sadaki, *Nippon Shashin Gakkai Kaishi*, **31**, 74 (1968); *Chem. Abstr.*, **70**, 15997 (1969).
81. B. Stanovnik, M. Tišler, and A. Vrbanič, *J. Org. Chem.*, **34**, 996 (1969).
82. S. African Pat. 06.255 (1967); *Chem. Abstr.*, **70**, 57869 (1969).
83. H. G. O. Becker, H. Böttcher, R. Ebisch, and G. Schmoz, *J. Prakt. Chem.*, **312**, 780 (1970).
84. B. Stanovnik, M. Tišler, M. Ceglar, and V. Bah, *J. Org. Chem.*, **35**, 1138 (1970).
85. M. Yanai, T. Kuraishi, T. Kinoshita, and M. Nishimura, *J. Heterocyclic Chem.*, **7**, 465 (1970).
86. H. G. O. Becker, H. Böttcher, G. Fischer, H. Rückauf, and S. Saphon, *J. Prakt. Chem.*, **312**, 591 (1970).
87. A. Pollak, B. Stanovnik, and M. Tišler, *J. Org. Chem.*, **35**, 2478 (1970).
88. M. Likar, P. Schauer, M. Japelj, M. Globokar, M. Oklobdžija, A. Povše, and V. Šunjić, *J. Med. Chem.*, **13**, 159 (1970).
89. V. Pirc, B. Stanovnik, M. Tišler, J. Marsel, and W. W. Paudler, *J. Heterocyclic Chem.*, **7**, 639 (1970).
90. T. Sasaki, K. Kanematsu, and M. Murata, *J. Org. Chem.*, **36**, 446 (1971).
91. M. G. de Amezua, M. Lora-Tamayo, and J. L. Soto, *Tetrahedron Lett.*, **1970**, 2407.
92. B. T. Gillis and J. G. Dain, *J. Org. Chem.*, **36**, 518 (1971).
93. M. L. Poutsma and P. A. Ibarbia, *J. Am. Chem. Soc.*, **93**, 440 (1971).
94. R. Gompper and G. Seybold, *Angew. Chem.*, **83**, 45 (1971).
95. M. Yanai, T. Kinoshita, S. Takeda, M. Mori, H. Sadake, and H. Watanabe, *Chem. Pharm. Bull.* (*Tokyo*), **18**, 1685 (1970).

Part F. Tetrazolo[1,5-*b*]pyridazines

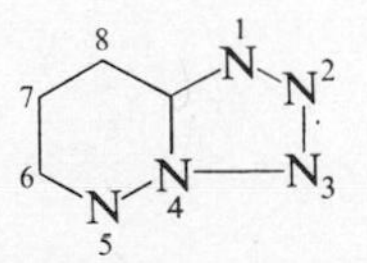

Tetrazolo[1,5-*b*]pyridazine, RRI 1034 (1,2,3-Tetrazolo-7.O-pyridazine, Tetrazolo[*b*]pyridazine)

Tetrazolo[1,5-*b*]pyridazines represent a very interesting bicyclic azaheterocycle, particularly in view of the possibility of azido-tetrazolo valence isomerization.

Tetrazolo[1,5-*b*]pyridazines (**1**) were prepared first in 1955 by Takahayashi (1) who treated 3-hydrazinopyridazines with an aqueous solution of sodium nitrite in the presence of acetic acid. The same principle has been followed in the preparation of other tetrazolo[1,5-*b*]pyridazines (2–5). That ring closure

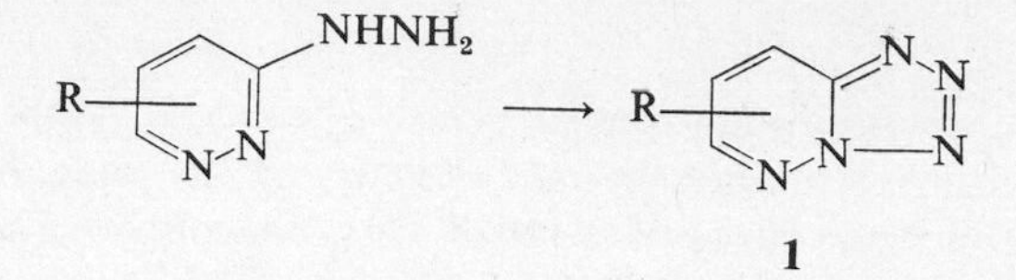

had occurred, thus excluding the tautomeric azidopyridazine structure, could be recognized from examination of the ir spectra, which show the absence of an absorption band in the region 2160–2120 cm^{-1} (2–4, 6) characteristic of the presence of an azide group.

A detailed study of 6-pyrrolidinotetrazolo[1,5-*b*]pyridazine revealed that this compound exists in solid state and in solution completely in the tetrazolo form. However, when dissolved in trifluoroacetic acid, a very weak absorption band, attributable to the azide group, appeared and attained maximum intensity after 3 days. This band disappeared after trifluoroacetic acid was removed (6).

However, treatment of 3-hydrazinopyridazine 1- or 2-oxide with nitrous acid does not give the tetrazolo derivatives, and the corresponding azides (**2**, **3**) were isolated (7) as judged from ir spectroscopic evidence.

The azido group could also be introduced by reacting 3-chloropyridazine 1-oxide with sodium azide in an aqueous solution of ethanol (**2**). That the

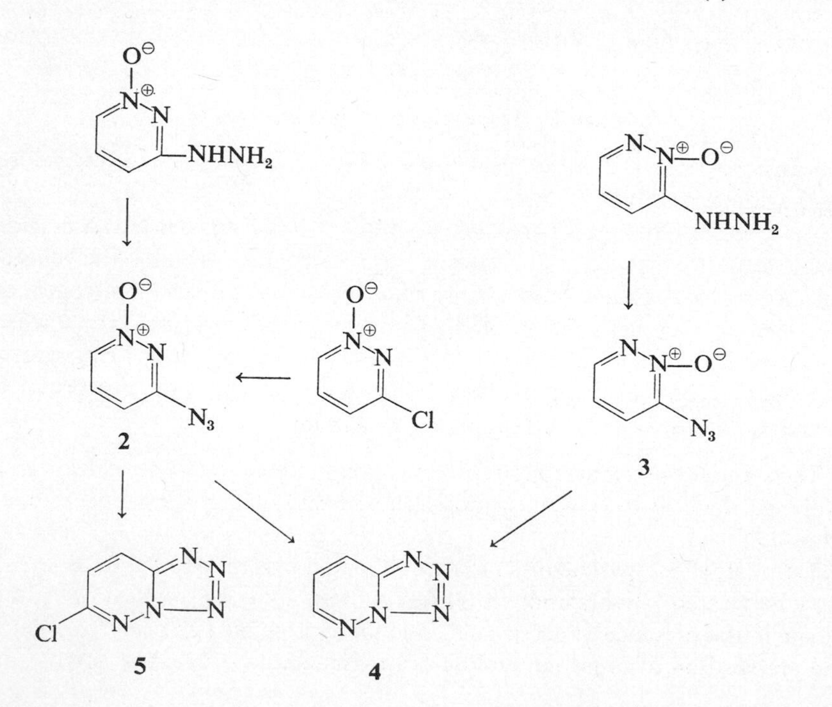

N-oxide group is responsible for the failure of cyclization of the azide group toward tetrazolo[1,5-*b*]pyridazines is clearly indicated by the observation that deoxygenation is followed by ring closure. In this manner when compounds **2** and **3** were heated under reflux with phosphorus trichloride they were converted into tetrazolo[1,5-*b*]pyridazine (**4**) (7). A similar treatment with phosphorus oxychloride yielded the 6-chloro analog **5**. On the other hand, *N*-oxidation of **4** afforded **2** by ring opening (14).

3,6-Dichloropyridazine when heated for several hours with an aqueous solution of sodium azide in ethanol was transformed into 6-azidotetrazolo-[1,5-*b*]pyridazine (**6**). Compound **6** is also obtainable from 6-hydrazinotetrazolo[1,5-*b*]pyridazine upon treatment with nitrous acid (7, 8) or from the 6-chloro analog with sodium azide (8, 9). In a like manner, 8-amino-6-

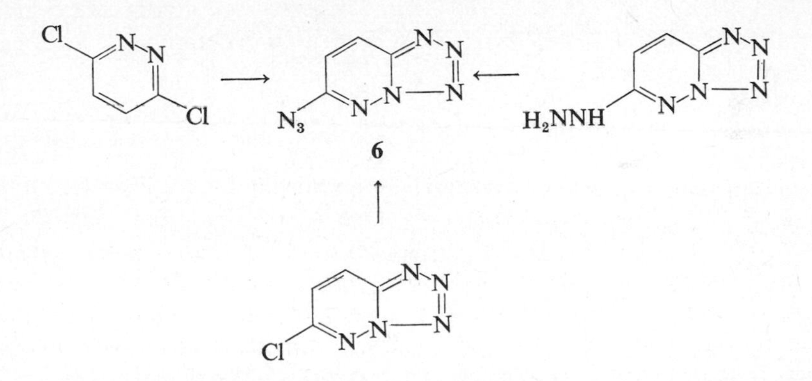

hydrazinotetrazolo[1,5-*b*]pyridazine when treated with nitrous acid yielded the 6-azido derivative (2).

It is a characteristic property of tetrazolo[1,5-*b*]pyridazines that a 6-azido group, otherwise capable of tetrazolo ring formation, remains uncyclized. This holds also for some other azolopyridazines with bridgehead nitrogen, as explained further in the text. In this manner, a 6-hydrazino group when converted by chemical reactions into a fused triazolo or triazino ring causes ring opening of the previously fused tetrazolo ring and an azido group is generated (7) (8, 13, 14, 19).

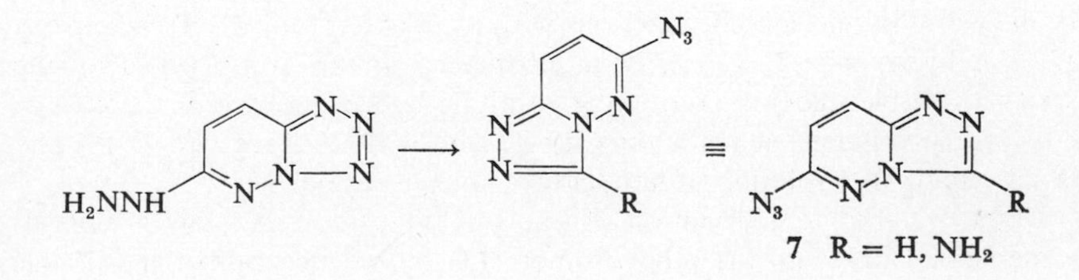

Similarly, treatment of 6-hydrazino-2-phenylimidazo[1,2-*b*]pyridazine with nitrous acid did not lead to the formation of a new fused tetrazolo cycle but to the 6-azido derivative (8). There is chemical and spectral evidence for the compound in solid state and in various solvents for the assigned structure of 6-azidotetrazolo[1,5-*b*]pyridazine.

The bis-tetrazolo structure **8** is excluded on the basis of the ir spectrum which revealed a strong absorption band, typical of an azide group, and the bis-azido form **9** is ruled out because of an easy transformation of only one azido group into the amino by means of hydrogen sulfide at room temperature (8, 12, 15, 16). The 6-azido group can also react in a cycloaddition reaction to give, for example, a 1,2,3-triazole ring (15).

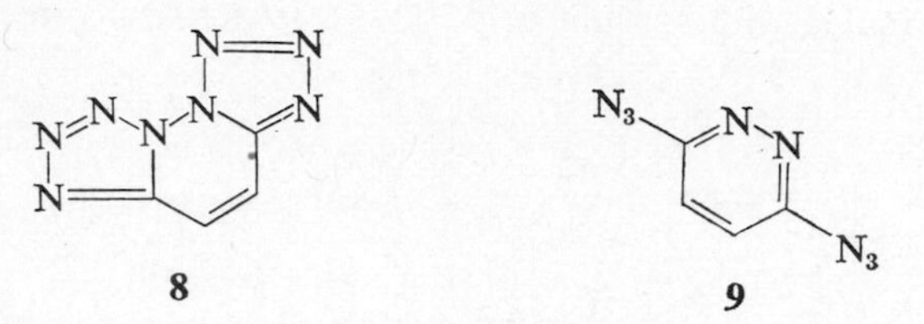

Mass spectra of several tetrazolo[1,5-*b*]pyridazines have been examined for their fragmentation patterns (17). One of the important features of fragmentation of tetrazolo[1,5-*b*]pyridazines is loss of four nitrogens, whereas the 6-azido analogs show a loss of six nitrogen atoms in what has been proposed to be a molecular ion with a bis tetrazolo structure (17). Some aliphatic fragments were detected resulting from the loss of two nitrogen atoms in the gas-phase thermolysis of tetrazolo[1,5-*b*]pyridazine (18).

MO calculations (10) of total and frontier π-electron densities for tetrazolo-[1,5-*b*]pyridazine are given in Table 3F-1.

Based on these data, it could be concluded that this system is not capable of protonation and quaternization and this has been, in fact, experimentally observed. Furthermore, *N*-oxidation of tetrazolo[1,5-*b*]pyridazine proceeds with the ring opening to give 3-azidopyridazine 1-oxide (14). However, there are no available data about direct substitutions on the pyridazine part of the molecule, which would allow conclusions about the reactivity of the parent compound.

Transformations are limited to the 6-position and the 6-chlorine atom can be displaced in nucleophilic reactions with an alkoxy group (1, 7), a hydroxy (7, 11), hydrazino (2, 8, 11), or a substituted amino group (9). The ring system is stable under conditions of catalytic hydrogenolysis of the 6,7- or 8-chloro substituent, which is thus substituted with hydrogen (2, 3, 11) and is also stable in a solution of hot acid for longer periods (9).

Recently, the transformation of one methyl azidotetrazolo[1,5-*b*]pyridazine isomer (**10**) into the other isomer (**11**) by valence isomerization was observed (12). Both isomers were prepared independently and **10** was converted at its melting point or when heated in toluene into **11**. The common method of preparing azides or tetrazoles—treatment of **13** with sodium azide in ethylene glycol—did not afford the expected product, but **12** was isolated. That an intermediate azido group formed during the reaction is transformed to an amino group was also established in other cases, such as the heating of **10** in 2-ethoxyethanol, which resulted in the formation of **12**. Apparently, isomerization of **10** to **11** must take place during this reaction. The value of $\Delta H = -5.9$ kcal/mol and the Arrhenius activation energy, $E_a = 20.5$ kcal/mol, for the process **10** → **11** have also been determined (12).

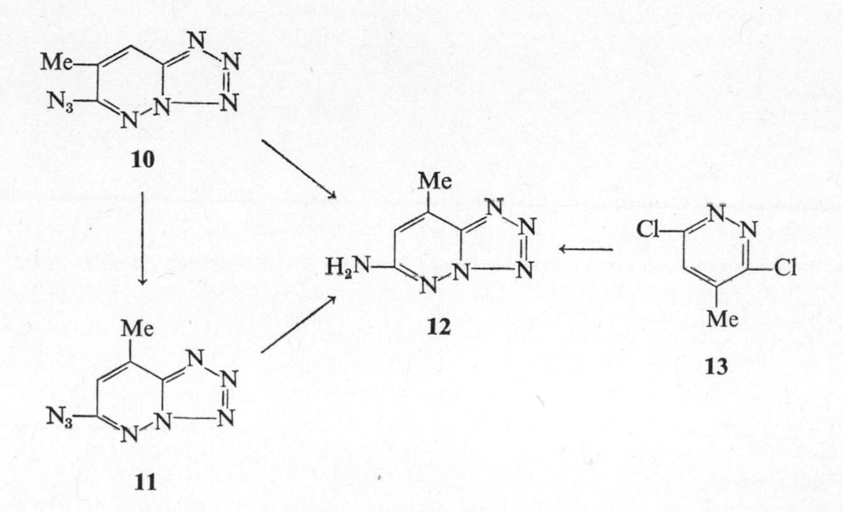

I. Tables

TABLE 3F-1. Total and Frontier π-Electron Densities of Tetrazolo[1,5-*b*]pyridazine

Position	1	2	3	4	5	6	7	8	8*a*
Total	1.2363	1.1698	1.1360	1.5468	1.1783	0.9183	1.0071	0.9185	0.8888
Frontier	0.0117	0.0665	0.0037	0.0724	0.2817	0.0969	0.1281	0.2979	0.0410

TABLE 3F-2. Tetrazolo[1,5-*b*]pyridazines

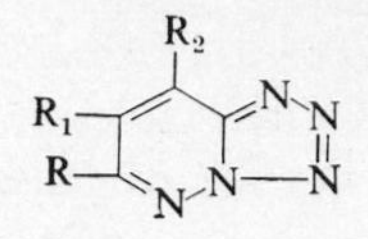

R	R_1	R_2	MP (°C) or BP (°C/mm)	Reference
None			104	11
			108–111	7
			HCl, 151	11
Chloro			107	1, 7
			104–106	
			107–109	13
Methoxy			154.5 (dec.)	1
Amino			293–295 (dec.)	7
			293	8
			313–315	8, 12
Hydrazino			238–240 (dec.)	7
			230–235	8
			230.5 (dec.)	11
Ethoxy			98	7
$C_6H_5CH_2O$			164–165	7
p-MeC_6H_4			202	5
Azido			127–129 (dec.)	9
			128–129	7
			130–131	8
EtOCH=NNH—			187–189	8
—N (pyrrolidine ring)			195–196	6, 9
—N (piperidine ring)			133–135	9
—N O (morpholine ring)			204–207	9
Mercapto			135–137	14
Methylthio			169–170	14
Phenylthio			128, 5–129, 5	14
$C_6H_5COCH_2S$—			170–172	14
CH_3COCH_2S—			124–125	14
$(MeO)_2CHCH_2$—NH—			129	14

TABLE 3F-2 (*continued*)

R	R_1	R_2	MP (°C) or BP (°C/mm)	References
HOOC—C(Me)=N—NH—			251–252	14
MeOOC—C(Me)=N—NH—			252–254	14
EtOOC—C(Me)=N—NH—			245–247	14
C_6H_5CH=N—NH—			315–317	14
C_6H_5—C(CH_2Br)=N—NH—			196–197 (dec.)	19
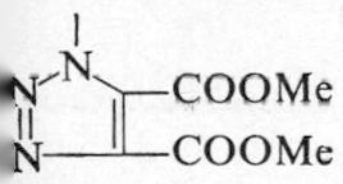 triazole ring (N—N=N) with —COOMe, —COOMe			153–155	15
Phenyl	Phenyl		166–167	15
		Methyl	126–127	17
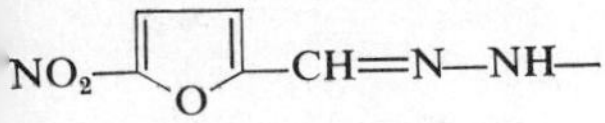 NO_2—(furan, O)—CH=N—NH—			266–270	13
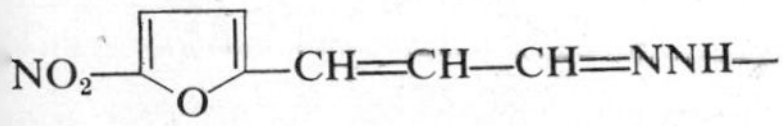 NO_2—(furan, O)—CH=CH—CH=NNH—			260	13
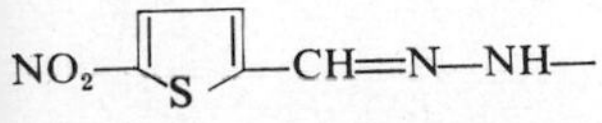 NO_2—(thiophene, S)—CH=N—NH—			—	13
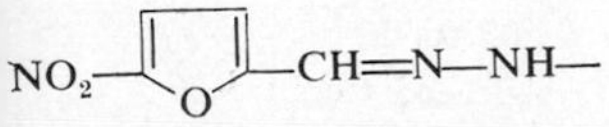 NO_2—(furan, O)—CH=N—NH—	Methyl		278–280 (dec.)	13
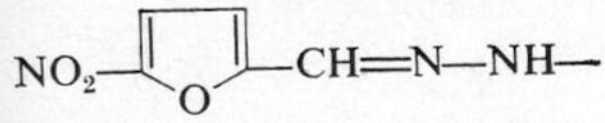 NO_2—(furan, O)—CH=N—NH—		Methyl	280–281 (dec.)	13
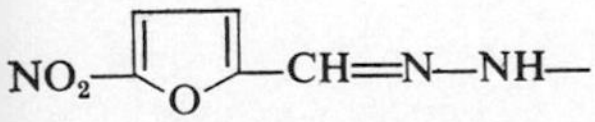 NO_2—(furan, O)—CH=N—NH—		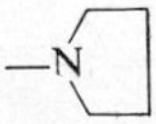 —N(pyrrolidine ring)	—	13
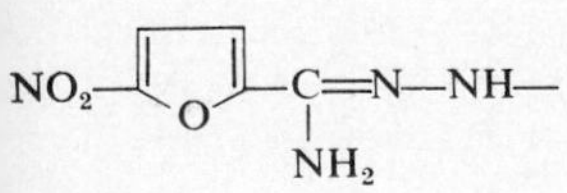 NO_2—(furan, O)—C(NH_2)=N—NH—		Amino	—	13

TABLE 3F-2 (*continued*)

R	R_1	R_2	MP (°C) or BP (°C/mm)	References
Hydrazino	Methyl		282–283	13
			285–287	12
Chloro		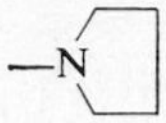 —N	151–153	13
Hydrazino		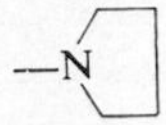 —N	223–228	13
Azido	Methyl		113–114	12
Amino	Methyl		320–323	12
Hydrazino		Methyl	247–248	12
Azido		Methyl	95	12
Amino		Methyl	302–303	12
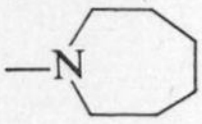 —N			192–194	9
	Amino		248 (dec.)	3
		Amino	275–277 (dec.)	2, 3
Chloro	Methyl		140.5	4
Chloro		Methyl	107	4
Chloro	H(Methyl)	Methyl(H)	97	1
Chloro		Amino	284–285	2
			270 (dec.)	13
Hydrazino		Amino	309–310	2
			291–292	13
Azido		Amino	244 (dec.)	2
Methoxy	H(Methyl)	Methyl(H)	108	1
	Chloro	Amino	292–293	3
	Amino	Chloro	270 (dec.)	3
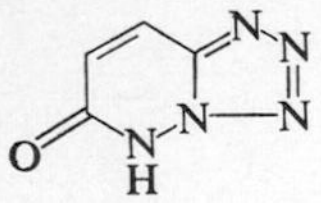 O, N–H, N, N, N, N			208 (dec.)	11
			230–231 (dec.)	7
			Na salt, 314 (dec.)	7

References

1. N. Takahayashi, *J. Pharm. Soc. Japan*, **75,** 1242 (1955); *Chem. Abstr.*, **50,** 8655 (1956).
2. W. D. Guither, D. G. Clark, and R. N. Castle, *J. Heterocyclic Chem.*, **2,** 67 (1965).
3. T. Kuraishi and R. N. Castle, *J. Heterocyclic Chem.*, **1,** 42 (1964).
4. S. Linholter and R. Rosenoern, *Acta Chem. Scand.*, **16,** 2389 (1962).
5. I. Zugravescu, M. Petrovanu, E. Rucinschi, and M. Caprosu, *Rev. Chim. Acad. Rep. Pop. Roum.*, **10,** 641 (1965).
6. N. B. Smirnova, I. J. Postovskii, N. N. Vereshchagina, I. B. Lundina, and I. I. Mudretsova, *Khim. Geterots. Soed.*, **1968,** 167.
7. T. Itai and S. Kamiya, *Chem. Pharm. Bull.* (*Tokyo*), **11,** 348 (1963); *Chem. Abstr.*, **59,** 8734 (1963).
8. A. Kovačič, B. Stanovnik, and M. Tišler, *J. Heterocyclic Chem.*, **5,** 351 (1968).
9. I. B. Lundina, J. N. Sheinker, and I. Ja. Postovskii, *Izvest. Akad. Nauk SSSR, Ser. Khim.*, **1,** 1967, 66.
10. Unpublished data from this laboratory. Parameters for LCAO Calculations were taken from A. Streitwieser, *Molecular Orbital Theory for Organic Chemists*, Wiley, New York, 1961, p. 135.
11. N. Takahayashi, *J. Pharm. Soc. Japan*, **76,** 765 (1956); *Chem. Abstr.*, **51,** 1192 (1957).
12. B. Stanovnik and M. Tišler, *Tetrahedron*, **25,** 3313 (1969).
13. S. African Pat. 06.255 (1967); *Chem. Abstr.*, **70,** 57869 (1969).
14. B. Stanovnik, M. Tišler, M. Ceglar, and V. Bah, *J. Org. Chem.*, **35,** 1138 (1970).
15. T. Sasaki, K. Kanematsu, and M. Murata, *J. Org. Chem.*, **36,** 446 (1971).
16. C. Wentrup, *Tetrahedron*, **26,** 4969 (1970).
17. V. Pirc, B. Stanovnik, M. Tišler, J. Marsel, and W. W. Paudler, *J. Heterocyclic Chem.*, **7,** 639 (1970).
18. C. Wentrup and W. D. Crow, *Tetrahedron*, **26,** 4915 (1970).
19. B. Stanovnik and M. Tišler, *Synthesis*, 180 (1970).

Part G. Furo-, Oxazolo-, Isoxazolo-, and Oxadiazolopyridazines

I. Furo[2,3-*d*]pyridazines

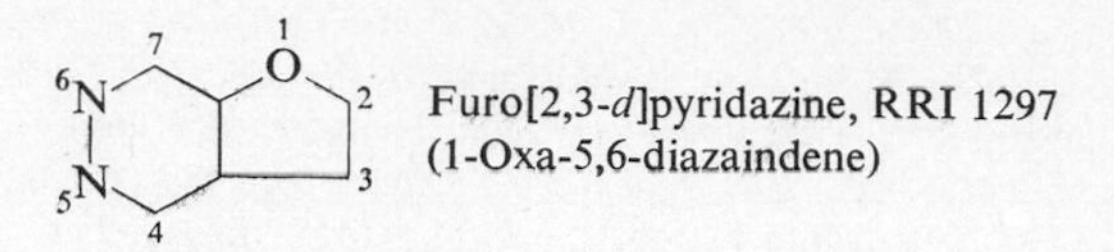

Furo[2,3-*d*]pyridazine, RRI 1297
(1-Oxa-5,6-diazaindene)

It was not until 1956 when Jones (1) described the synthesis of the first representative of this bicyclic system. 2-Methylfuro[2,3-*d*]pyridazine-4,7(5*H*,6*H*)dione (**1**) was obtained by heating 5-methylfuran-2,3-dicarboxylic acid dihydrazide with either excess hydrazine hydrate or dilute hydrochloric acid.

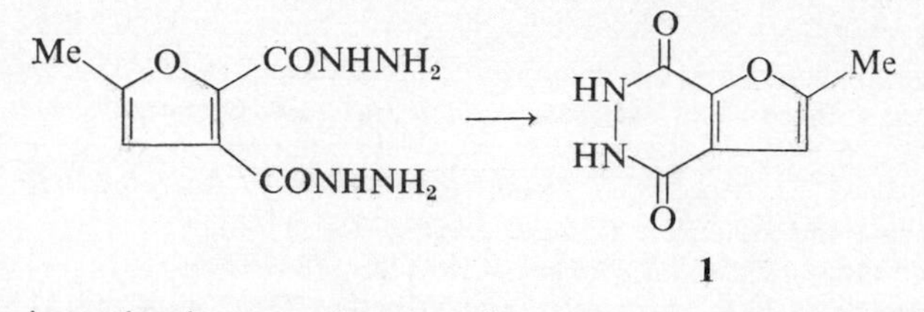

All other investigations pertinent to this system originate from two French groups. Robba, Zaluski, and Roques (2, 40) prepared the parent system **2** from 2,3-diformylfuran and hydrazine or by catalytic hydrogenolysis of 4,7-dichlorofuro[2,3-*d*]pyridazine (**3**) obtained from furan-2,3-dicarboxylic acid dihydrazide via **4**, or from 4-hydrazino derivative by elimination of the hydrazino group. The nmr spectrum of **2** [in $CDCl_3$: $\tau = 2.99$ (H_3), 2.03

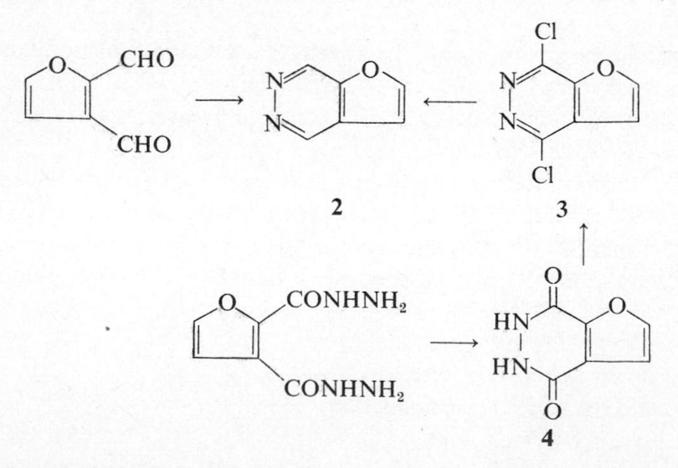

(H_2), and 0.42 (H_4, H_7)] is in accordance with the formulated structure and MO calculations.

2-Formyl-3-carboxy (or carbethoxy)furan reacted with hydrazine to form furo[2,3-*d*]pyridazin-4(5*H*)one (**5**) (3, 4, 41, 40). The 7-oxo isomer **6** was

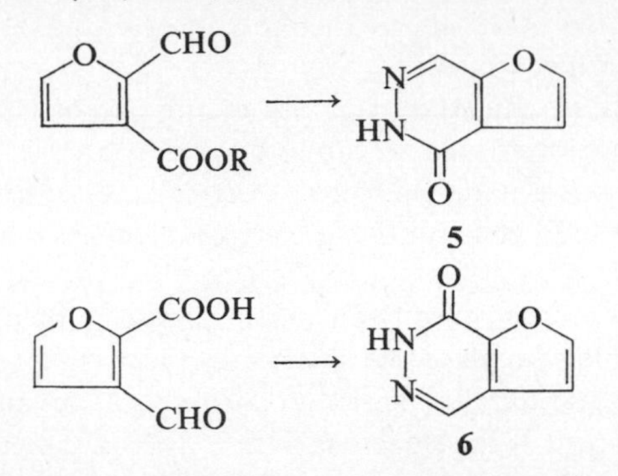

prepared similarly from 3-formylfuran-2-carboxylic acid (2, 40). From the corresponding formylacetylfurans, 4- and 7-methyl analogs of **2** were prepared (5, 40).

Moreover, it has been shown by Robba and Zaluski (5) that 4,7-diaminofuro[2,3-*d*]pyridazine (**8**) could be obtained from 2,3-dicyanofuran upon treatment with hydrazine. For the intermediate nitrile-hydrazidine structure **7** has been proposed. It is also reported that the hydrazone of 2-formyl-3,4-

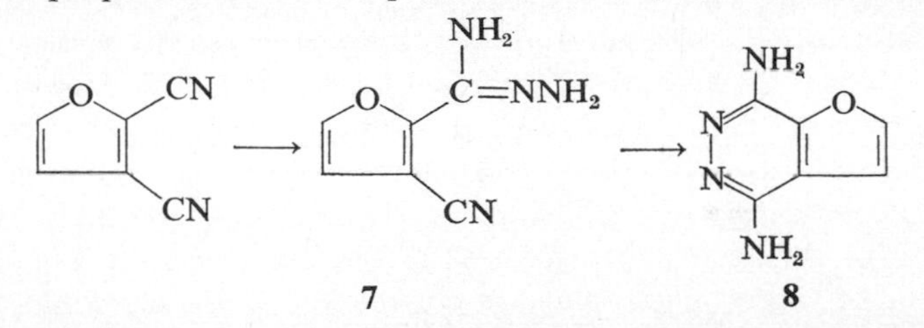

dicarbomethoxyfuran or the bis-hydrazone of the same furan derivative when heated in acetic acid is transformed into 3-carbomethoxyfuro[2,3-*d*]-pyridazin-4(5*H*)one (42).

A new approach to this system has been reported recently. It was found that diazo compounds easily add to strained double bonds of cyclopropenes and in this manner diazoethane reacted with 1,2-diphenyl-3-diacetylmethylene cyclopropene to give the furopyridazine **8a** (43). This compound is

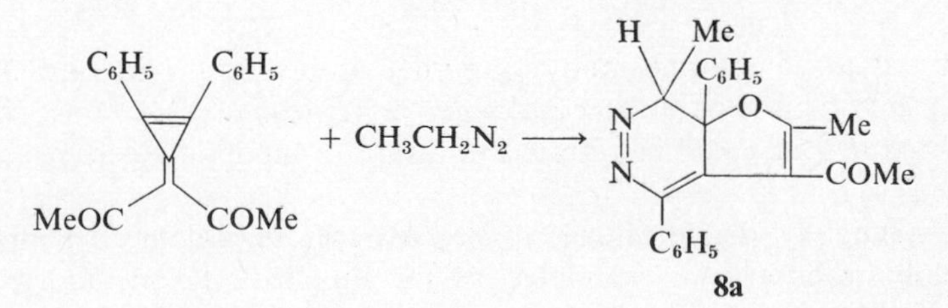

sensitive to the influence of heat, bases, or acids, adds diphenyl ketene to the —N=N— double bond, and under the influence of bases isomerizes to a pyridazine derivative. Addition of alcohols on the double bond 4, 4a also occurred upon heating.

There are no data about the reactivity of the parent bicycle, except that it forms a hydrochloride and a monoiodomethylate (2, 40). Some conclusion about the reactivity of this bicycle can be reached on the basis of MO calculations (6) of total and frontier π-electron densities which are presented in Table 3G-1.

Most of the reactions have been performed with halofuro[2,3-*d*]pyridazines. It is possible to replace the chlorine atom in 4-, or 7-chloro- or 4,7-dichlorofuro[2,3-*d*]pyridazines with most of the common nucleophiles. In this manner, furo[2,3-*d*]pyridazines with a hydroxy, mercapto, alkylthio, iodo, alkoxy, amino, substituted amino or hydrazino group attached at position 4 or 7 are accessible (3–5, 40, 44, 45). The attempted direct preparation of 4-aminofuro[2,3-*d*]pyridazine from the chloro analog failed, even at 110° when pressure was applied. However, it was possible to prepare **10** by first converting the 4-chloro compound into the 4-phenoxy derivative **9**, which yields the desired product upon heating with ammonium acetate at

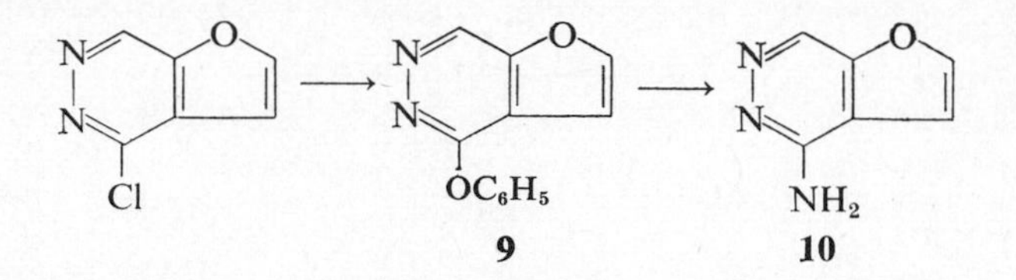

170–180° (4). A selective displacement of one chlorine atom at position 4 with the aid of various nucleophiles is possible (3, 5, 45) and compounds of the type **11** and **12** were thus obtained. This certainly indicates a greater

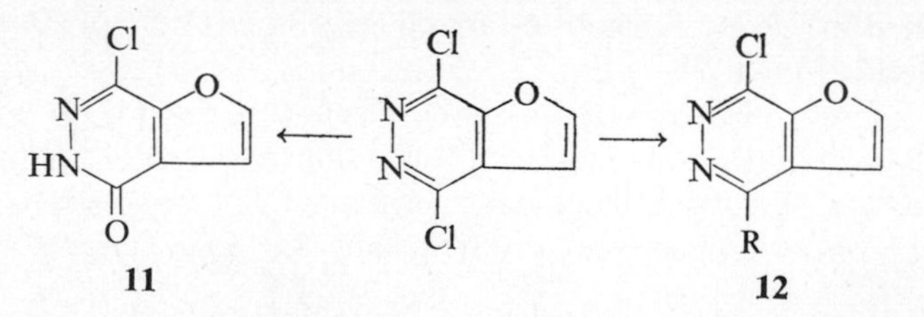

reactivity of the 4-chlorine atom than that at position 7 in nucleophilic displacements. This is also in accord with the MO calculations. Furo[2,3-*d*]pyridazin-4(5*H*)one could be nitrated to afford in moderate yield its 2-nitro derivative (41).

For 2-methyl-4,7-dichlorofuro[2,3-*d*]pyridazine, superdelocalizability for nucleophilic reactions was calculated by the simple HMO method and, as anticipated, it was found that position 4 is more reactive toward nucleophiles (44). Nevertheless, with hot alkali a mixture of isomers, 2-methyl-7-chlorofuro[2,3-*d*]pyridazin-4(5*H*)one, and 2-methyl-4-chlorofuro[2,3-*d*]pyridazin-7(6*H*)one was obtained in a ratio of 3:1. With 1 mole of sodium methoxide the 4-methoxy isomer was formed almost exclusively and similar behavior was observed for the reaction with hydrazine. With ammonia in a sealed tube only the 4-amino-7-chloro isomer is reported to be formed (44). An unusual reaction was observed when 2-methyl-4,7-dichlorofuro[2,3-*d*]pyridazine was treated with potassium cyanide in *N*,*N*-dimethylformamide or DMSO at room temperature (46). The initial addition product **12a** upon treatment with water and acid gives as the final product **12b**, whose structure is based on spectroscopic data.

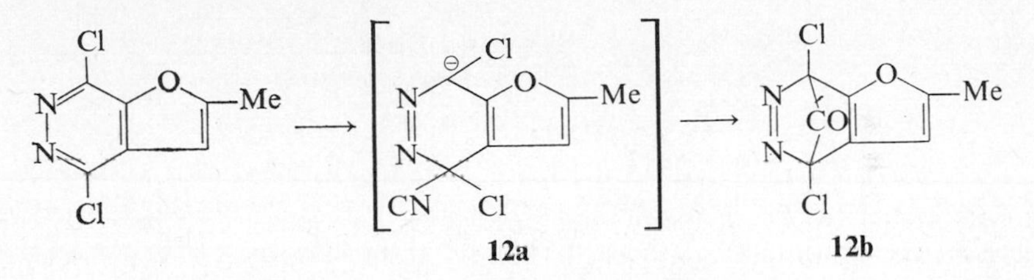

II. Furo[3,2-*c*]pyridazines

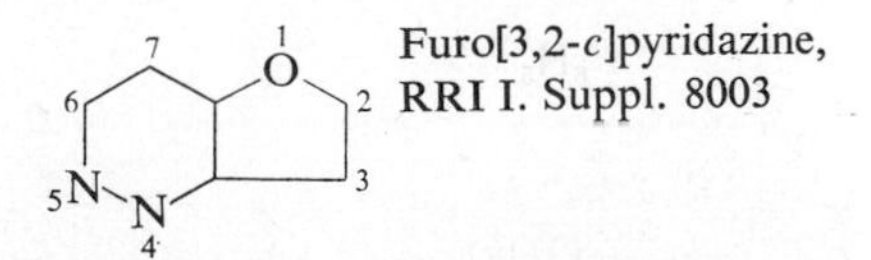

The only representatives of this system were mentioned in a communication by Hensecke, Müller, and Badicke in 1958 (7). They rejected the structural formula for dianhydrohexosazone proposed earlier by Percival (8) and proposed **13a** or **13b**.

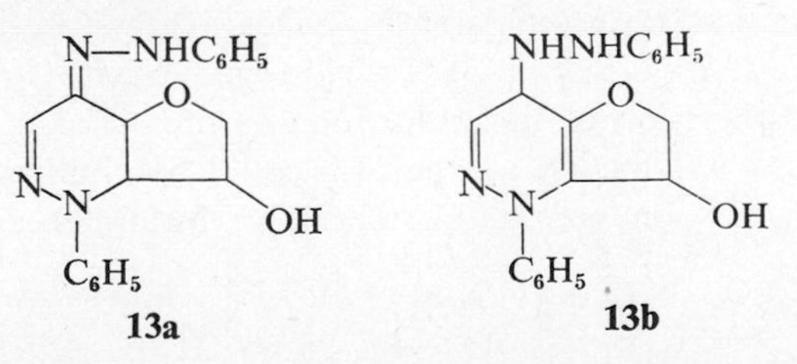

Compound **13** when treated with nitrous acid formed a product which was formulated as dianhydroosone hydrazone and for which the structure **14a** or **14b** was suggested. A more convincing structure determination of these compounds is desirable.

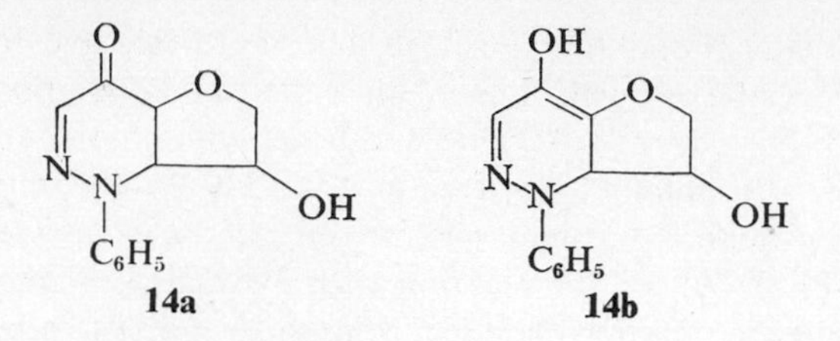

MO calculations (6) of total and frontier π-electron densities for furo-[3,2-*c*]pyridazine are presented in Table 3G-2.

III. Furo[3,4-*d*]pyridazines

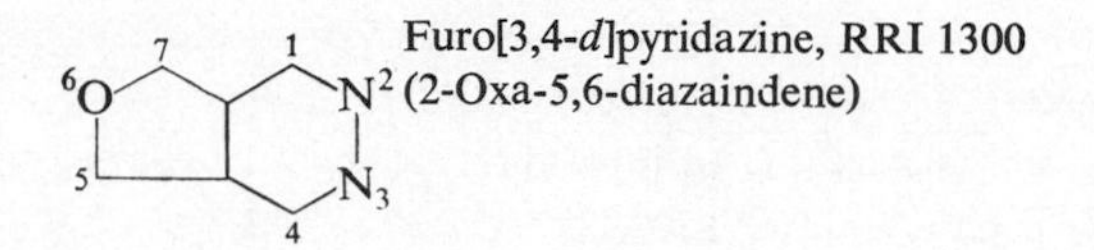

This system was first mentioned in a paper by Seka and Preissecker in 1931 (9). These authors condensed ethyl 2,5-diphenylfuran-3,4-dicarboxylate with excess of hydrazine in a solution of ethanol (100–120°, 24 hr) and obtained the bicyclic product **15**.

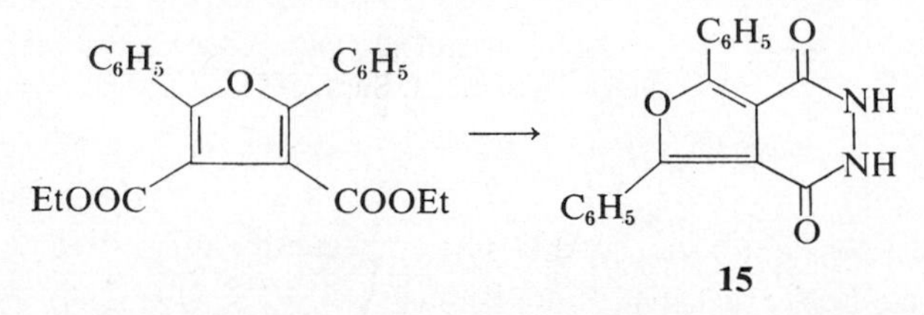

A similar treatment of ethyl 2,5-dimethylfuran-3,4-dicarboxylate is reported to cause destruction of the ester. A difficult conversion of the latter furan derivative was observed later by Jones (1), who obtained the 5,7-dimethyl bicycle (**16**) in only 35–38% yield together with 50% of unreacted material. An explanation of this behavior was presented in terms of steric hindrance, since the 5-methyl analog (**17**) could be obtained in good yield after a shorter reaction period. Furthermore, furan-3,4-dicarboxylic acid

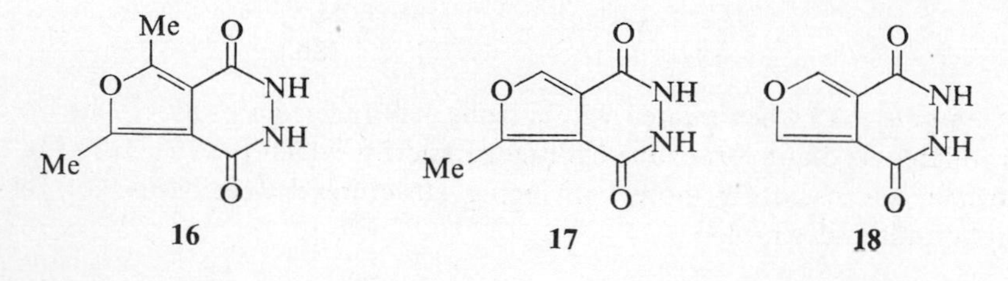

dihydrazide when heated with dilute hydrochloric acid was cyclized to **18**, whereas the otherwise successful conversion with excess hydrazine could not be used because of the furan ring opening (1).

Other furans with appropriate functional groups at positions 3 and 4 are also capable of forming furo[3,4-*d*]pyridazines. The parent compound was prepared by Robba and Zaluski (10) from 3,4-diformylfuran and anhydrous hydrazine in a solution of methanol at 0°. The recorded nmr spectroscopic data [in $CDCl_3$: $\tau = 1.69$ (H_5, H_7), 0.71 (H_1, H_4)] are in accord with the proposed structure and MO calculations.

Similarly, tetraacetyl ethylene when treated with a methanolic solution of hydrazine at room temperature afforded a furo[3,4-*d*]pyridazine derivative (47), and the same result is achieved with 2-alkoxymethyl-3,4-diacetyl-5-methylfuran.

MO calculations (6) of total and frontier π-electron densities for this symmetric aromatic system are given in Table 3G-3.

The reaction between 3,4-diacetyl-2,5-dimethylfuran and one equivalent of hydrazine yielded a product for which structure **19** was proposed (11). It is most likely that, based on similar reaction conditions and close melting point to the product of Mosby (**19**), the product described by Bradley and Watkinson (12) as the monohydrazone of 3,5-diacetyl-2,5-dimethylfuran (**20**) is in fact **19**.

Me Me O N N Me Me

19

Me O Me MeCO C—Me NNH$_2$

20

Some other compounds to which the structure of a furo[3,4-*d*]pyridazine was assigned need further confirmation of their structures. 3,4-Dibenzoyl-2,5-diphenylfuran, tetrabenzoylethane, and some of its irradiated products afforded with hydrazine products for which structures **21** and **22** were proposed (13–15).

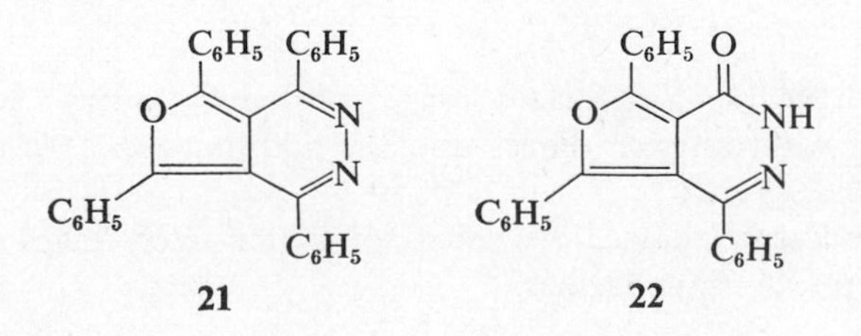

21 **22**

Furo[3,4-*d*]pyridazine represents an aromatic 10 π-electron system and its chemistry has been investigated in detail only by Robba and Zaluski (10, 16). Compound **18** when treated with phosphorus oxychloride in the presence of pyridine can be transformed into either **23** or **24**, depending on the reaction time. The dichloro derivative **24** has a low stability and is very sensitive to hydrolysis. Exposed to air for several hours it is transformed into **23**.

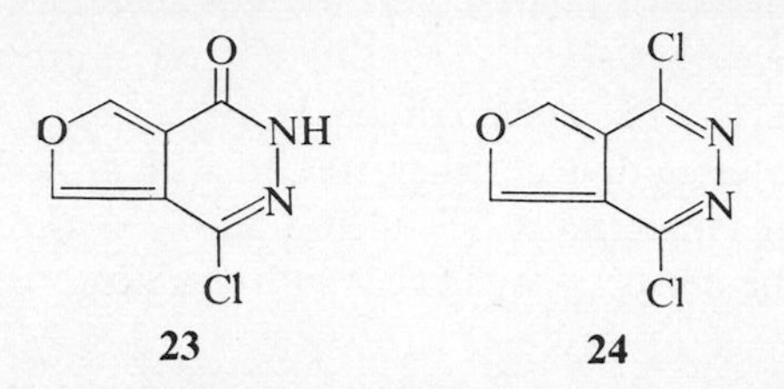

Compound **18** can be reduced with hydrogen in the presence of palladium on charcoal and under atmospheric pressure into the dihydro derivative to which structure **25** has been assigned on the basis of nmr spectroscopic data. One or both oxo groups can be converted into chloro substituents (**26**, **27**) and the monochloro derivative **26** can be catalytically dehalogenated to compound **28**, which is also obtainable from **23** by simultaneous hydrogenolysis of the chlorine atom and partial hydrogenation of the bicycle.

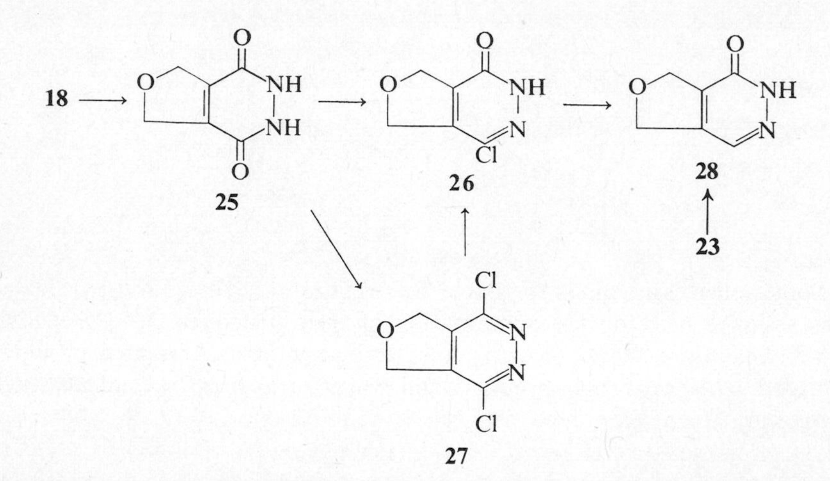

Substituted furo[3,4-*d*]pyridazines are reported to undergo a Diels-Alder type reaction with different dienes and the addition took place across the positions 5 and 7 (48).

More experimental data are needed for better interpretation of the reactivity of furo[3,4-*d*]pyridazines.

IV. Oxazolo[4,5-*d*]pyridazines

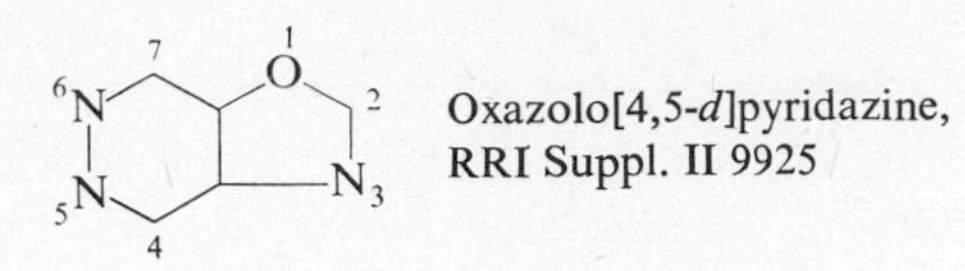

The only representative of this system was described in 1958 by Kuraishi (17). 5-Amino-3,4-dichloropyridazine or 5-amino-3-chloro-4(1*H*)pyridazinone when heated with benzoyl chloride in the absence of a solvent for 30 min afforded the same product for which, based on analytical data, the structure of 2-phenyl-7-chlorooxazolo[4,5-*d*]pyridazine (**29**) has been assigned.

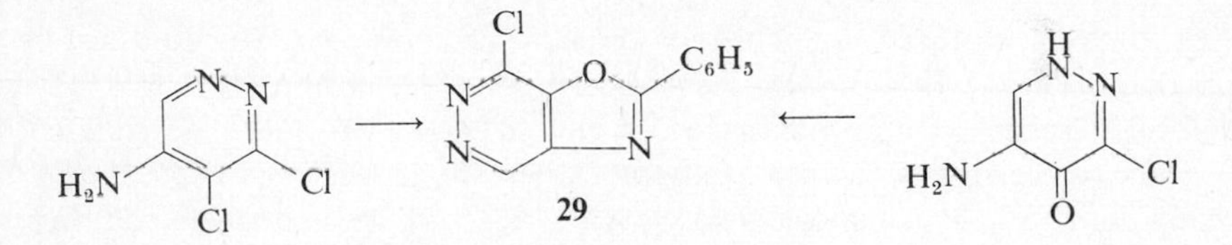

MO calculations (6) of total and frontier π-electron densities for the parent aromatic compound are presented in Table 3G-4.

V. Oxazolo[5,4-*c*]pyridazines

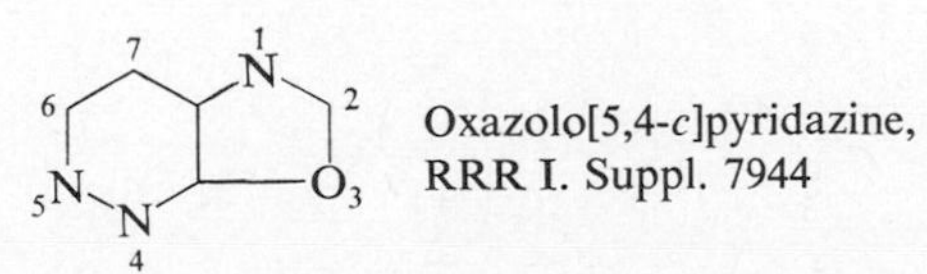

Representatives of this bicyclic system were described only in two communications by Kuraishi. In the first report in 1958 the author described the reaction between 4-amino-6-chloro-3(2*H*)pyridazinone and excess benzoyl chloride. He was able to isolate a product, insoluble in alkali, to which on the basis of analytical data the structure of 2-phenyl-6-chlorooxazolo[5,4-*c*]-pyridazine (**30**) has been assigned (18).

A product identical to compound **30** was obtained in low yield from 4-amino-3,6-dichloropyridazine under the same reaction conditions. The reaction is thought to proceed via an intermediate benzoylamino derivative.

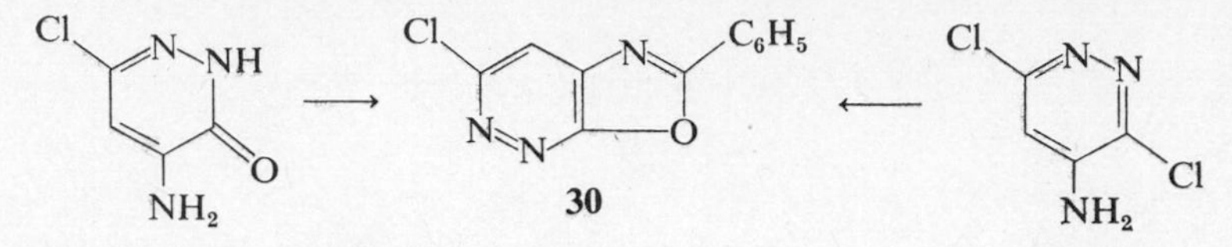

Later, Kuraishi reported the synthesis of **31**, which he was able to isolate when the reaction between 4-amino-6-chloro-3(2*H*)pyridazinone and benzoyl chloride was conducted in nitrobenzene or pyridine as solvent (19). How-

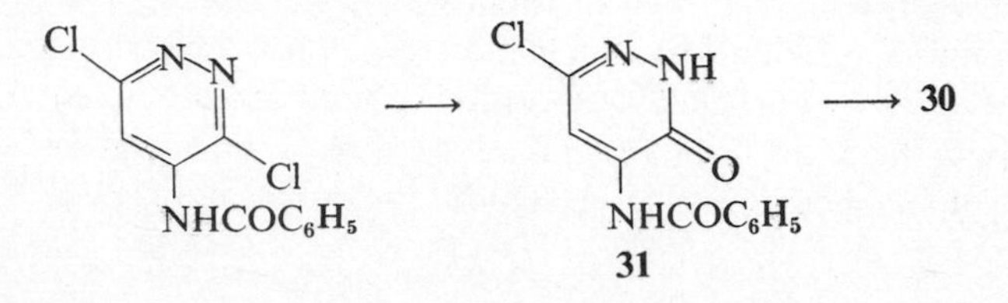

ever, under such reaction conditions the bicyclic compound **30** could not be obtained. Only when **31** was heated under reflux with phosphorus oxychloride 2-phenyl-6-chlorooxazolo[5,4-*c*]pyridazine (30) could be isolated.

The oxazole ring is cleaved easily and when the bicycle was heated under reflux with 15% hydrochloric acid or with 15% sodium hydroxide solution, **31** or 4-amino-6-chloro-3(2*H*)pyridazinone, respectively, was formed.

MO calculations (6) of total and frontier π-electron densities for oxazolo-[5,4-*c*]pyridazine are listed in Table 3G-5.

VI. Oxazolo[3,4-*b*]pyridazines

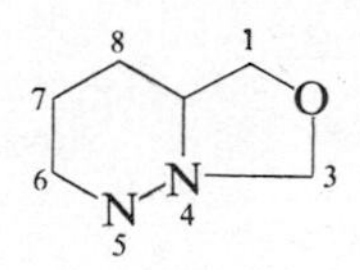

Thus far, only two compounds possessing this skeleton are reported. Molnar (20) obtained, after saponification of the diester **32** with a solution of potassium hydroxide in ethanol, a compound to which, on the basis of analytical data and ir spectroscopic evidence, he assigned the lactone structure **33**.

In a like manner, the diester **34**, obtained by catalytic hydrogenation of **32**, when saponified afforded together with a pyridazine derivative **36** the lactone **35**. This, again, can be regarded as a derivative of oxazolo[3,4-*b*]-pyridazine.

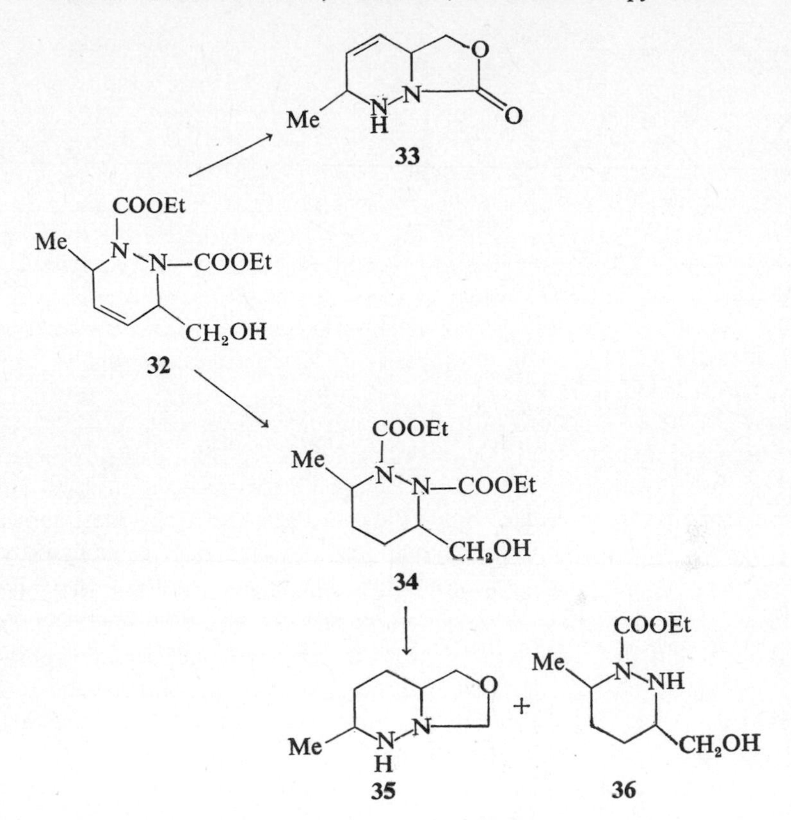

VII. Isoxazolo[4,5-*d*]pyridazines

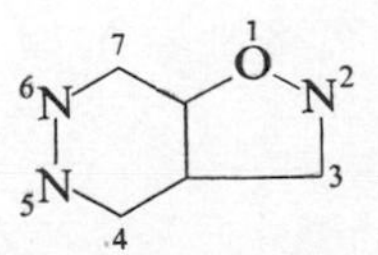

Compounds belonging to this heterocyclic system were first described in 1966 in a communication by Erichomovitch and Chubb (21). Dimethyl 3-phenylisoxazole-4,5-dicarboxylate when heated with hydrazine hydrate in a solution of absolute ethanol for 45 min afforded the isoxazolo[4,5-*d*]-pyridazine derivative **37** (50%) together with a pyrazolo[3,4-*d*]pyridazine (**38**) isolated in 40% yield from the filtrates.

A similar procedure described by Desimoni and Finzi (22) is reported to yield only **37** in somewhat better yield. It was also possible to prepare the monohydrazide of 3-phenylisoxazole-4,5-dicarboxylic acid monoester when

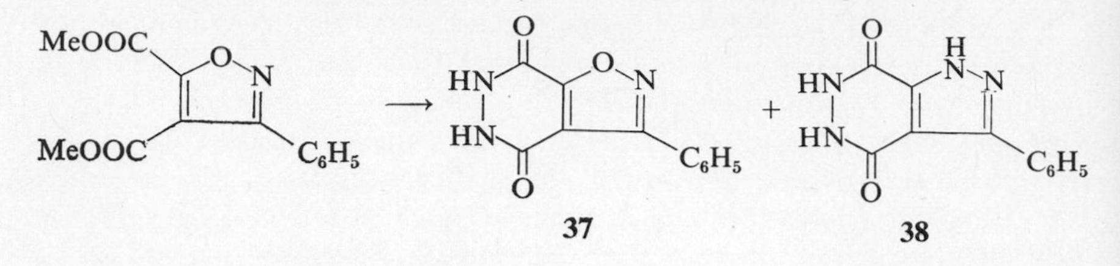

one equivalent of hydrazine hydrate was applied. The monohydrazide could be thermally cyclized to **37** (21).

The parent compound, isoxazolo[4,5-*d*]pyridazine, remains unknown. MO calculations (6) of total and frontier π-electron densities for this bicycle are given in Table 3G-6.

From **37** the 4,7-dichloro or 4,7-dibromo derivatives (**39**, X = Cl or Br) were prepared by applying the usual phosphorus oxychloride or oxybromide method. It is possible to replace both halogen atoms in reactions with different nucleophiles or to replace only, the more reactive atom. Desimoni and Finzi showed that the halogen at position 4 is more reactive for hydrolysis. Moreover, the structure of 3-phenyl-7-bromoisoxazolo[4,5-*d*]pyridazin-4(5*H*)one (**40**, X = Br), obtained from acetic acid treatment of **39**, is reported to be confirmed on the basis of X-ray analysis data.

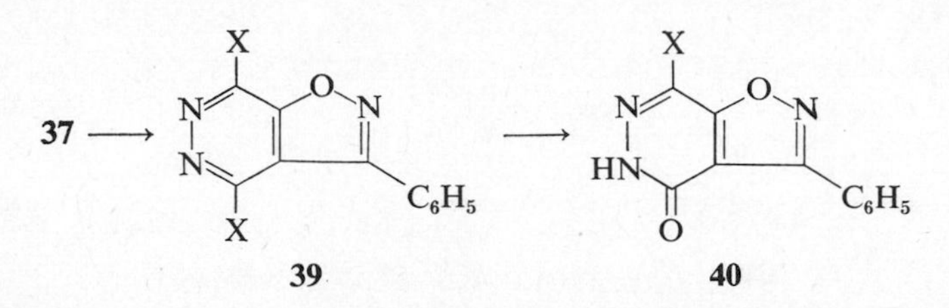

However, in the reactions with other nucleophiles the dihalo compound **39** showed no selective displaceability under the conditions employed and thus 4,7-dialkoxy, 4,7-bis-alkylthio, 4,7-bis-arylthio derivatives have been obtained (21, 22).

Moreover, 3-phenyl-4,7-dichloroisoxazolo[4,5-*d*]pyridazine when treated with only one equivalent of sodium methoxide afforded a mixture of the isomeric monomethoxy derivatives **41** and **42**. Their identification and structure assignment were based on chemical transformations (22).

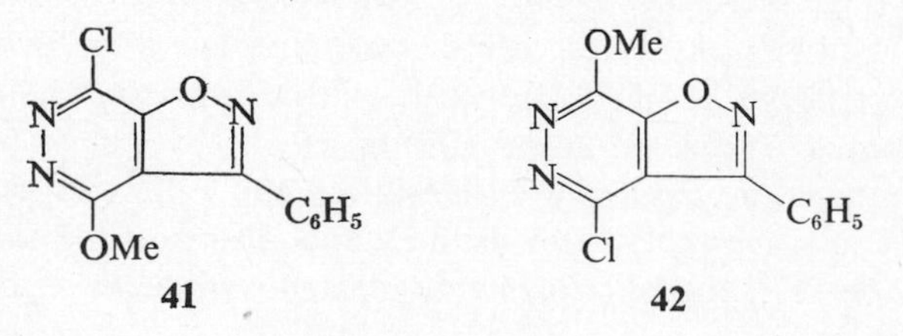

A mixture of different methyl derivatives was obtained from methylations of 3-phenylisoxazolo[4,5-*d*]pyridazine-4,7(5*H*,6*H*)dione. Two isomers were obtained when **37** was treated with dimethyl sulfate in the presence of sodium hydroxide. The structural assignments of the isomeric products **43** and **44** were based upon chemical evidence. From the reaction mixture obtained after methylation of **37** with diazomethane, **43** was isolated in a very low yield and the main product has been identified as **45** (22).

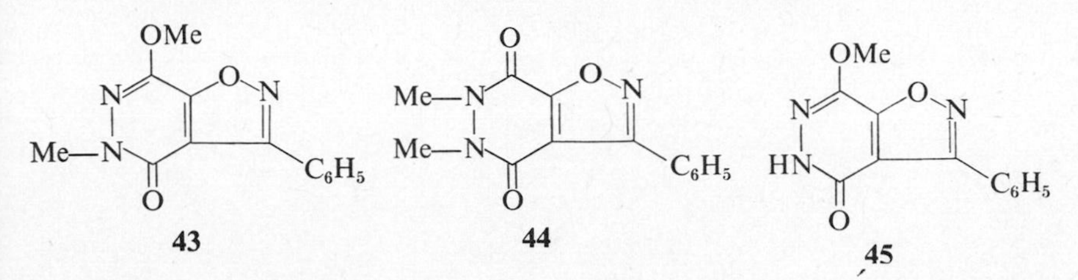

Finally, it should be mentioned that in an attempt to prepare the corresponding 4,7-dihydrazino derivative from 3-phenyl-4,7-bis-(methylthio)-isoxazolo[4,5-*d*]pyridazine (**46**) in addition to the hydrazinolysis of both substituents, the isoxazole moiety underwent ring opening and subsequent ring closure and a pyrazolo[3,4-*d*]pyridazine derivative (**47**) is claimed to have been formed (21).

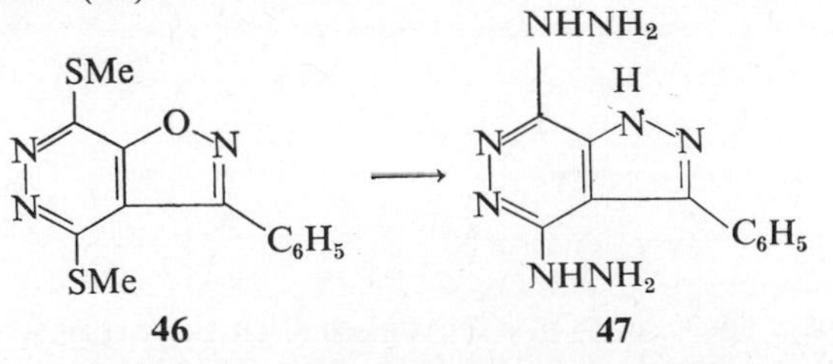

VIII. Isoxazolo[3,4-*d*]pyridazines

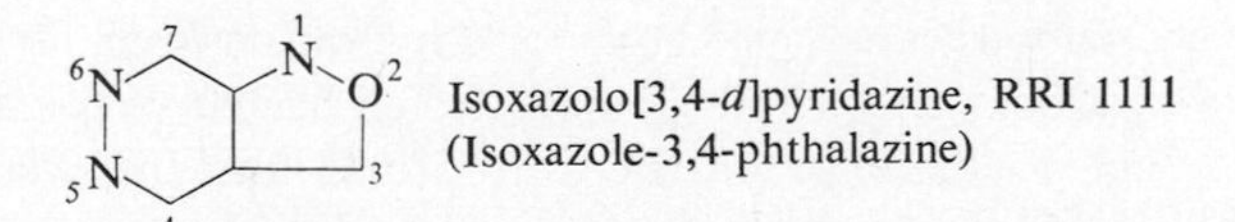

Isoxazolo[3,4-*d*]pyridazine, RRI 1111
(Isoxazole-3,4-phthalazine)

Syntheses of representatives of this bicyclic system utilize appropriate isoxazole derivatives as starting material. For example, 3,4-dibenzoyl-5-phenylisoxazole (**49**) reacted either with hydrazine hydrate in a solution of ethanol or with semicarbazide in ethanol to form a product, formulated as 3,4,7-triphenylisoxazole[3,4-*d*]pyridazine (**50**). The starting isoxazole was obtained from 4-nitroso-2,5-diphenyl-3-benzoylpyrrole when treated with a hot solution of hydrogen chloride in ethanol. Further, the pyrrole **48** could be transformed with hydrazine or semicarbazide into the bicycle **50** (23).

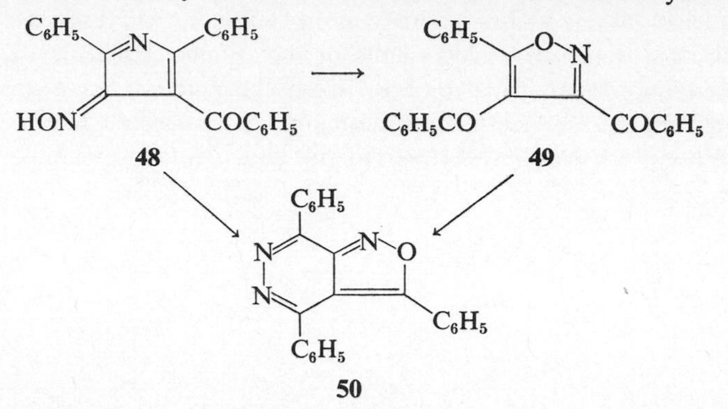

Other 3,4-diacyl isoxazoles were also used successfully for the syntheses of compounds related to **50** (24).

In a like manner, 3-carbethoxy-4-acetyl(or benzoyl)-5-methyl(or phenyl)-isoxazole when heated under reflux with a solution of hydrazine or phenylhydrazine in alcohol is reported to afford the corresponding isoxazolo [3,4-*d*]-pyridazine derivative **51** (25, 26, 49).

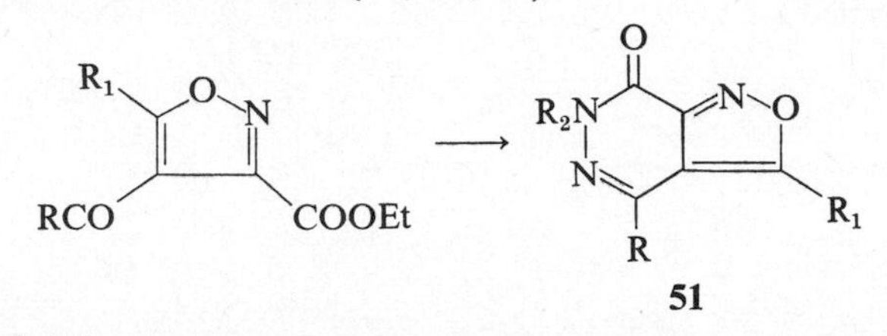

The recorded ir spectra are in accordance with the proposed structure **51**. Compound **51** ($R = R_1 = Me$, $R_2 = C_6H_5$) is claimed to be identical with a product obtained earlier by Musante (27) in a somewhat ambiguous reaction sequence. There is about 11° difference in the melting points of these compounds. Other identification data are not available.

It has been demonstrated also that a dihydrazide such as 5-methyl-isoxazolo-3,4-dicarboxylic acid dihydrazide, when heated with a solution of hydrochloric acid in ethanol could be transformed into the cyclic product **52** (26). The parent compound, isoxazolo[3,4-*d*]pyridazine, is unknown.

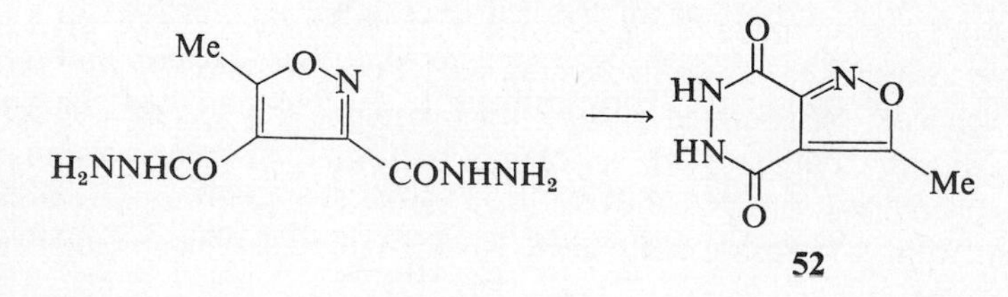

MO calculations (6) of total and frontier π-electron densities for this bicycle are presented in Table 2G-7.

The only reaction on this heterocyclic system that has been utilized by several authors is the hydrogenolytic cleavage of the isoxazole ring. Thus isoxazolo[3,4-*d*]pyridazines or pyridazinones when heated in a solution of ethanol with Raney nickel in the presence of hydrogen were transformed into the corresponding aminopyridazines (**53**, **54**, **55**) (24, 26, 28, 29).

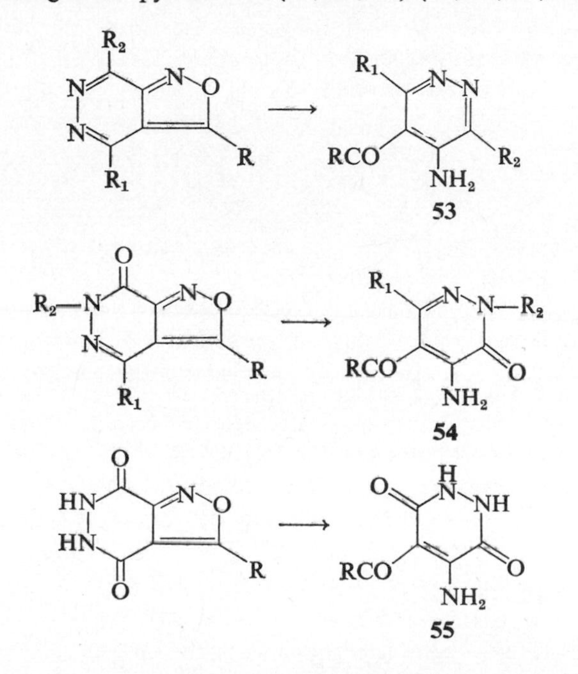

IX. 1,2,5-Oxadiazolo[3,4-*d*]pyridazines

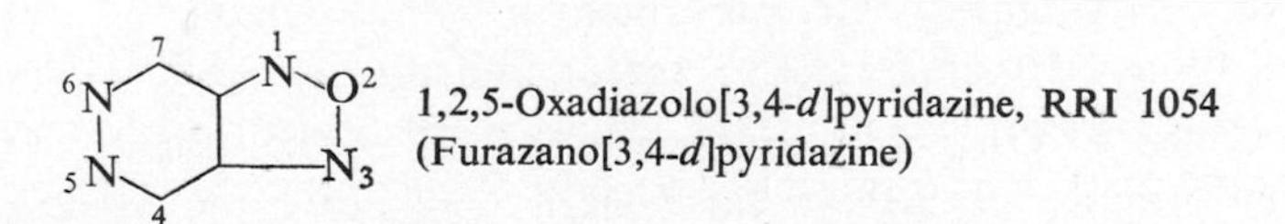

1,2,5-Oxadiazolo[3,4-*d*]pyridazine, RRI 1054 (Furazano[3,4-*d*]pyridazine)

The first representatives of this system were mentioned in 1931 in a paper by Durio (30), who treated 3,4-dibenzoyl-1,2,5-oxadiazole and related compounds with hydrazine. The products have been assigned the general formula **56**.

Boyer and Ellzey (31, 32) were able to prepare compounds that belong to this bicyclic system by deoxygenating aromatic *o*-dinitroso compounds by

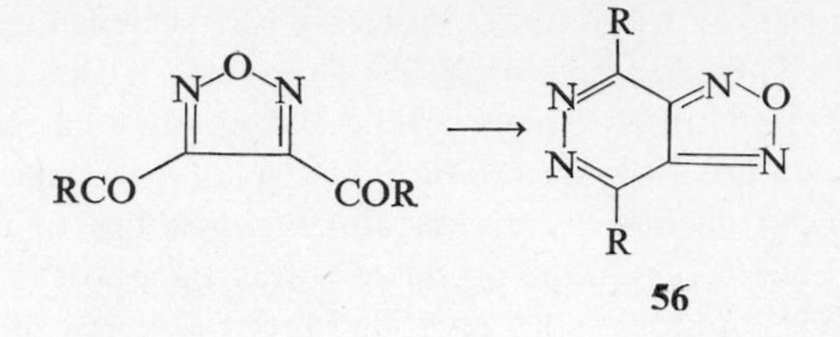

means of phosphines or sodium borohydride. Tri-*n*-butylphosphine was found to be the most powerful reagent and both agents have to be used in excess to obtain good yields of the products, for example, **57**. In the sodium boro-

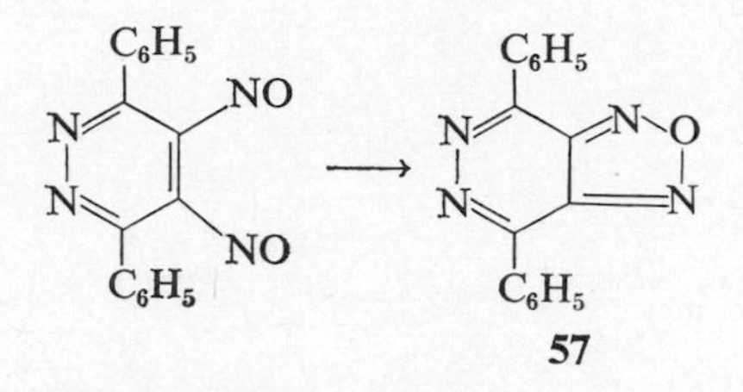

hydride reduction a red intermediate was formed. It was formulated as a dioxime and in boiling acetone it was converted into **57**. This 4,7-diphenyl-1,2,5-oxadiazolo[3,4-*d*]pyridazine is stable against attack by alkaline hypohalite. Furthermore, diacyl-1,2,5-oxadiazolo-*N*-oxides (diacylfuroxanes) when treated with hydrazine could be converted into bicyclic compounds of type **58** (33).

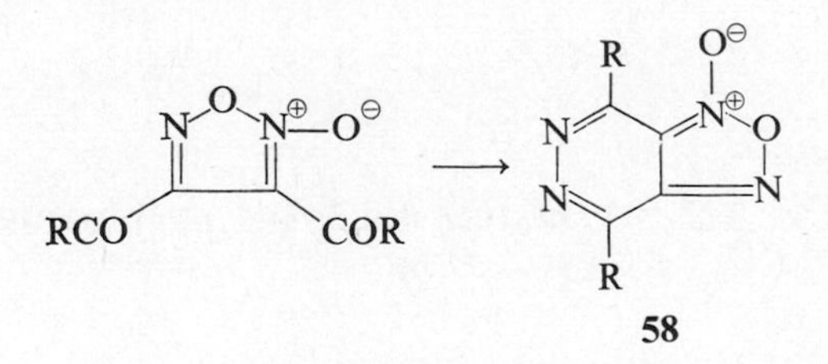

The tendency to ring closure is greatly reduced in the presence of electron-donating groups, attached to the benzene rings of dibenzoyl-1,2,5-oxadiazole *N*-oxide. A complete failure of the ring closure reaction was observed when two electron-donating *ortho* substituents were in the benzene ring. This was ascribed mainly to steric effects (33). Infrared spectra of compounds of type **58** were studied and assignments to particular vibrations have been made (34).

MO calculations (6) of total and frontier π-electron densities for this symmetric heteroaromatic system are available in Table 3G-8.

X. 1,2,4-Oxadiazolo[2,3-*b*]pyridazines

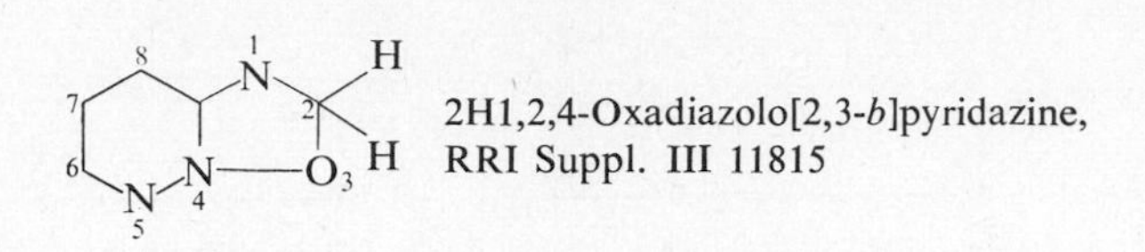

2H1,2,4-Oxadiazolo[2,3-*b*]pyridazine,
RRI Suppl. III 11815

Compounds pertinent to this system were mentioned in two communications by Itai and Nakashima (35, 36) in 1962. 3-Ethoxycarbonylaminopyridazine 2-oxide, when heated at 115° for 18 hr, was transformed in 32% yield into the bicyclic product **59**. The structure of **59** is based on analytical data and ir evidence.

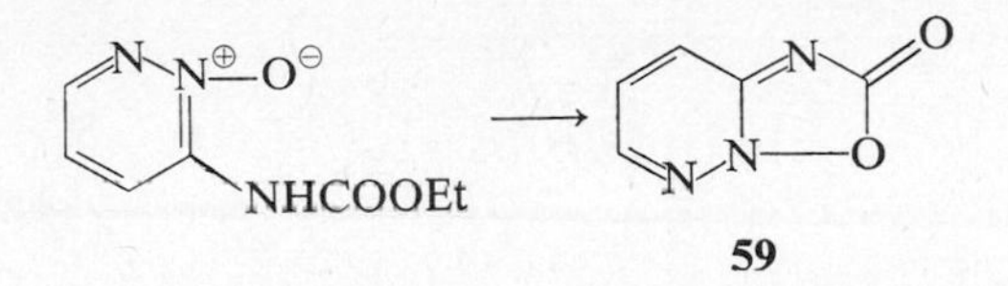

It is interesting to note that from a similar experiment with 3-ethoxycarbonylamino-6-methoxy- (or chloro) pyridazine 2-oxide it was not possible to obtain the bicyclic product under varying reaction conditions (37).

XI. 1,3,4-Oxadiazolo[3,4-*a*]pyridazines

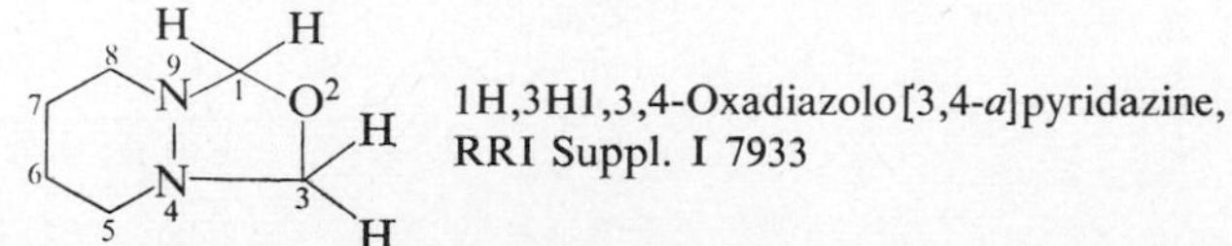

1H,3H1,3,4-Oxadiazolo[3,4-*a*]pyridazine,
RRI Suppl. I 7933

1,3,4-Oxadiazolo[3,4-*a*]pyridazines were first recorded in the literature in 1957 (38) because 1,2,3,6-tetrahydropyridazine-1,2-dicarboxylic acid anhydride has been regarded as pertaining to this heterocyclic system.

Later, a derivative of this system (**60**) was claimed to be formed by treatment of tetrahydropyridazine with 2 moles of propionaldehyde (30% yield) (39). Compound **60** is thermally unstable and can be very easily hydrolyzed in the presence of an acid.

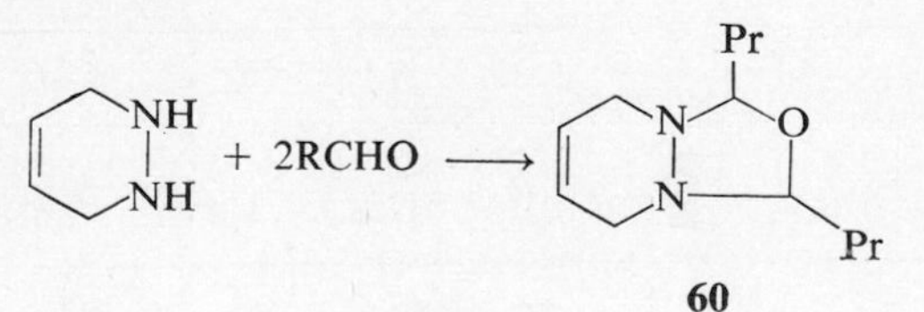

XII. Tables

TABLE 3G-1. Total and Frontier π-Electron Densities of Furo[2,3-*d*]pyridazine

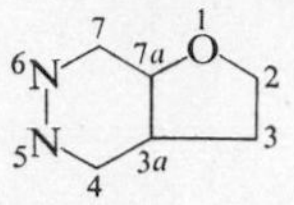

Position	1	2	3	3*a*	4	5	6	7	7*a*
Total	1.6342	1.0134	1.1080	1.0087	0.9399	1.1680	1.1700	0.9460	1.0118
Frontier	0.1645	0.1588	0.3081	0.0005	0.1117	0.0295	0.1275	0.0440	0.0555

TABLE 3G-2. Total and Frontier π-Electron Densities of Furo[3,2-*c*]pyridazine

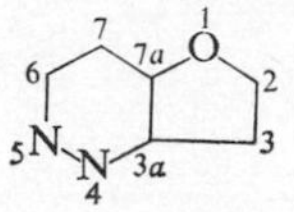

Position	1	2	3	3*a*	4	5	6	7	7*a*
Total	1.6498	1.0014	1.1187	1.0065	1.1521	1.1376	0.9852	0.9503	0.9986
Frontier	0.2061	0.1213	0.3085	0.0002	0.1577	0.0076	0.1051	0.0708	0.0227

TABLE 3G-3. Total and Frontier π-Electron Densities of Furo[3,4-*d*]pyridazine

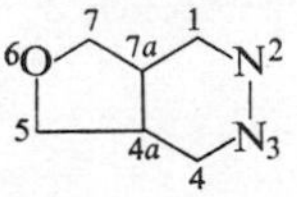

Position	1	2	3	4	4*a*	5	6	7	7*a*
Total	0.9178	1.1922	1.1922	0.9178	1.0431	1.0888	1.5162	1.0888	1.0431
Frontier	0.0612	0.0899	0.0899	0.0612	0.0429	0.3059	0.0000	0.3059	0.0492

TABLE 3G-4. Total and Frontier π-Electron Densities of Oxazolo[4,5-*d*]pyridazine

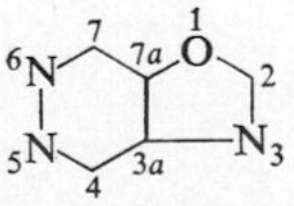

Position	1	2	3	3*a*	4	5	6	7	7*a*
Total	1.6050	0.8907	1.2955	1.0099	0.9386	1.1592	1.1594	0.9444	0.9973
Frontier	0.1525	0.0822	0.2711	0.0137	0.1629	0.0311	0.1658	0.0799	0.0408

TABLE 3G-5. Total and Frontier π-Electron Densities of Oxazolo[5,4-*c*]pyridazine

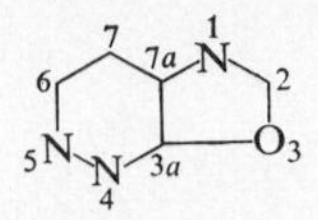

Position	1	2	3	3*a*	4	5	6	7	7*a*
Total	1.3042	0.8798	1.6244	0.9604	1.1604	1.1351	0.9842	0.9429	1.0086
Frontier	0.2618	0.0724	0.1697	0.0238	0.1242	0.0879	0.0473	0.1968	0.0162

TABLE 3G-6. Total and Frontier π-Electron Densities of Isoxazolo[4,5-*d*]pyridazine

Position	1	2	3	3*a*	4	5	6	7	7*a*
Total	1.6354	1.1875	1.0130	1.0073	0.9231	1.1665	1.1443	0.9455	0.9774
Frontier	0.3079	0.0659	0.2128	0.0035	0.1159	0.0851	0.0919	0.1115	0.0056

TABLE 3G-7. Total and Frontier π-Electron Densities of Isoxazolo[3,4-*d*]pyridazine

Position	1	2	3	3*a*	4	5	6	7	7*a*
Total	1.3081	1.5030	0.9779	1.0329	0.9056	1.1876	1.1546	0.9290	1.0013
Frontier	0.3424	0.0022	0.2392	0.0450	0.0685	0.1196	0.0691	0.1044	0.0096

TABLE 3G-8. Total and Frontier π-Electron Densities of 1,2,5-Oxadiazolo[3,4-*d*]-pyridazine

Position	1	2	3	3*a*	4	5	6	7	7*a*
Total	1.1961	1.4719	1.1961	0.9985	0.9134	1.1561	1.1561	0.9134	0.9985
Frontier	0.2697	0.0000	0.2697	0.0123	0.1157	0.1023	0.1023	0.1157	0.0123

TABLE 3G-9. Furo[2,3-*d*]pyridazines

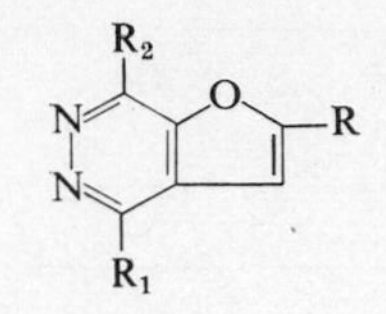

R	R_1	R_2	MP (°C) or BP (°C/mm)	References
None			108	2, 40
			205[a] HCl 212	2, 40
Methyl			90, HCl, 217, 207	5, 40
	Methyl		95	5, 40
	Methoxy		158	3
			157	45
	Hydrazino		193–195	4
			213	5, 45
			151	4
	C_6H_5NH		209	4
	m-ClC_6H_4NH		82	4
	—N (piperidine ring)		217[a]	5
			218[a]	45
	—N (morpholine ring, O)		142–143	4
	Phenoxy		114	4, 45
			115	5
	Amino		185	4
	—SCH_2COOMe		105	5, 45
	$C_6H_5CH_2S$—		115	5
	—SCH_2COOH		262	5
	—SCH_2CN		171	5
	Ethoxy		131	5, 45
	—NEt_2		96	5, 45
			129	45
	—N (pyrrolidine ring)		240[a]	5
			242[a]	45
				40
	Ethyl		78	45
	Iodo		184	40, 45
		Methyl	80	5, 40
	Chloro		108	3, 40
		Chloro	119	3, 40
		Methoxy	114	3, 45
		—SCH_2COOMe	90	5, 45
		Methylthio	114	5
		Benzylthio	80	5

TABLE 3G-9 (*continued*)

R	R_1	R_2	MP (°C) or BP (°C/mm)	References
		—SCH_2COOH	240	5
		Phenoxy	92	5, 45
		—N	145	5
	Chloro	Chloro	118	2, 40
	Bromo	Bromo	144	2, 40
	Methoxy	Chloro	116	3, 45
	—N	Chloro	130	3, 5, 45
	Ethoxy	Chloro	66	5
	—N	Chloro	125	5, 45
	—N O	Chloro	205	5, 45
	Hydrazino	Chloro	260	5, 45
	Amino	Amino	H_2O, 220	5
			HCl, 285	5
Methyl	Chloro	Chloro	152	2, 40
			151–153	44
Methyl	Methoxy	Chloro	109	5
			110	45
			113–114	44
Methyl	—N	Chloro	151	5, 45
			277[a]	5
Methyl	—N	Chloro	92	5
			236[a]	5
Methyl	—N O	Chloro	195	5, 45

[a] Iodomethylate.

TABLE 3G-10

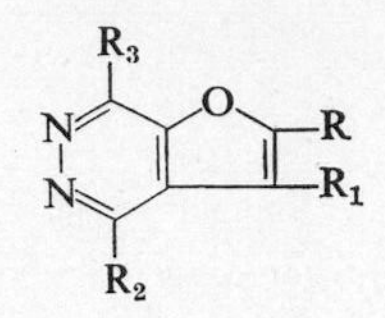

R	R_1	R_2	R_3	MP (°C) or BP (°C/mm)	References
		Ethoxy	Chloro	67	45
			NEt_2	69	45
			—N	145	45
			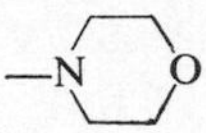 —N O	139	45
Methyl		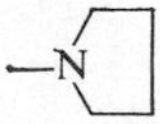 —N		MeJ, 277	45
Methyl		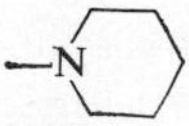 —N		MeJ, 236	45
Methyl		Chloro	Methoxy	144	44
Methyl		Amino	Chloro	208	44
Methyl		Hydrazino	Chloro	210 (dec.)	44
Methyl		Chloro	Hydrazino	192 (dec.)	44
Methyl		Bromo	Chloro	152–153	44
	COOMe	Chloro		163	40, 42
	COOMe	Chloro	Methyl	125	42
	COOMe	Hydrazino		227	45
	Cyano	Chloro		170	45
		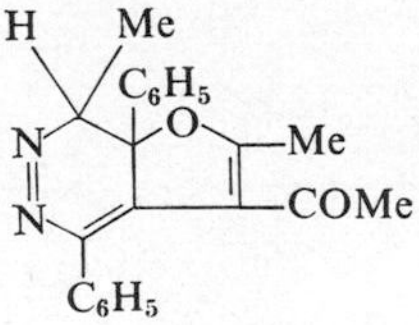 H, Me, C_6H_5, N, N, O, Me, COMe, C_6H_5		134–135	43
		H, Me, C_6H_5, HN, N, O, Me, COMe, OR, C_6H_5		R=Me, 190–191	43
				Et, 173–174	43

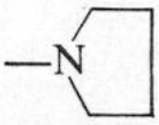

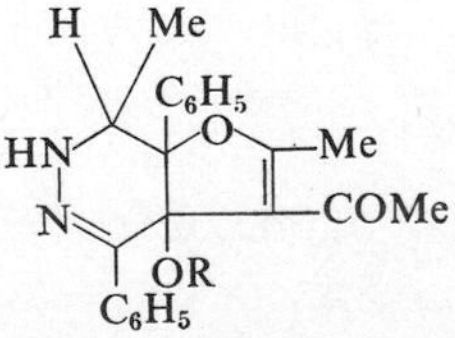

TABLE 3G-11. Furo[2,3-*d*]pyridazin-4(5*H*)ones and Furo[2,3-*d*]pyridazine-4(5*H*)thiones

R	R_1	R_2	R_3	X	MP (°C) or BP (°C/mm)	References
				O	208	4, 41
					212	3, 40
				S	184	3, 5, 45
		Methyl		O	109	5, 40
		$—CH_2CH_2CN$		O	93	5
		$—CH_2CH_2NEt_2$		O	MeJ, 218	5
			Chloro	O	281	3, 45
		CH_2OH		O	175	3
		Benzyl		O	90	3
Methyl			Chloro	O	245–246	44
Nitro				O	237–238	41
Nitro		CH_2OH		O	143[a]	41
	COOMe			O	220–222, 225	42, 40
	COOMe		Methyl	O	215–21	42
	COOMe		Ethyl	O	117	42
	$CONH_2$			O	350	45

[a] Resolidifies, melts again at 237° (loses HCHO).

TABLE 3G-12. Furo[2,3-*d*]pyridazine-7(6*H*)ones and Furo[2,3-*d*]pyridazine-7(6*H*)thiones

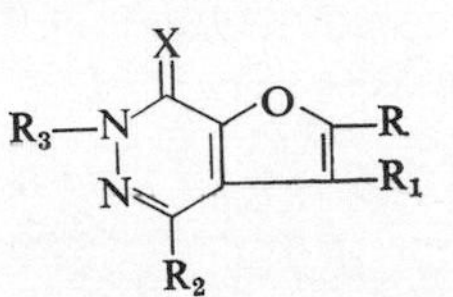

R	R_1	R_2	R_3	X	MP (°C) or BP (°C/mm)	References
			Methyl	O	128	5, 40
				O	198	2, 40
			CH_2OH	O	172	5
			Benzyl	O	114	5
			$—CH_2CH_2CN$	O	147	5
			$—CH_2COOH$	O	260	5
			$—CH_2CH_2N$	O	118	5
				S	238	5, 45

TABLE 3G-12 (*continued*)

R	R_1	R_2	R_3	X	MP (°C) or BP (°C/mm)	References
Methyl		Chloro		O	232	44
Methyl		Bromo		O	223–224	44
		$—SC(=NH)NH_2$		S	xHCl, 245	45
Methyl		$—SC(=NH)NH_2$		S	xHCl, 250	45

TABLE 3G-13. Furo[2,3-*d*]pyridazine-4,7(5*H*, 6*H*)diones

R	MP (°C) or BP (°C/mm)	References
None	295	2, 40
Methyl	290–292	1

TABLE 3G-14. Furo[3,2-*c*]pyridazines

Structure	MP (°C)	Reference
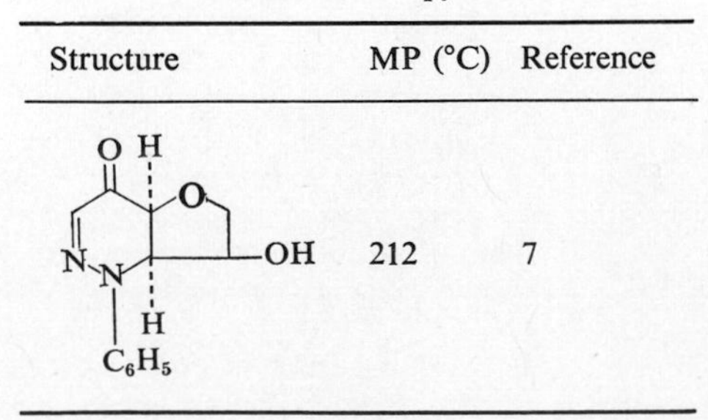	212	7

TABLE 3G-15. Furo[3,4-*d*]pyridazines

R	R_1	R_2	R_3	MP (°C) or BP (°C/mm)	References
None				161	10
Chloro	Chloro			132	10
Methyl	Methyl	Methyl	Methyl	H_2O, 144	11
				Picrate, 173–174.5	11
				136–140 (dec.)[a]	12
Phenyl	Phenyl	Phenyl	Phenyl	255	13
Methyl	Methyl	Methyl	—CH_2OH	164–166 (dec.)	47
				204–205 (dec.)	47

[a] Defined as monohydrazone.

TABLE 3G-16. Furo[3,4-*d*]pyridazine-1(2*H*)ones

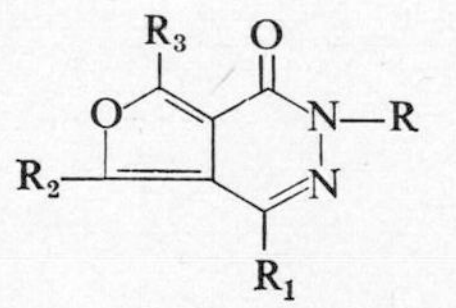

R	R_1	R_2	R_3	MP (°C) or BP (°C/mm)	References
	Chloro			232	10
	Bromo			229	10
Benzyl	Chloro			137	10
	Phenyl	Phenyl	Phenyl	252	15

TABLE 3G-17. Furo[3,4-*d*]pyridazine-1,4(2*H*, 3*H*)diones

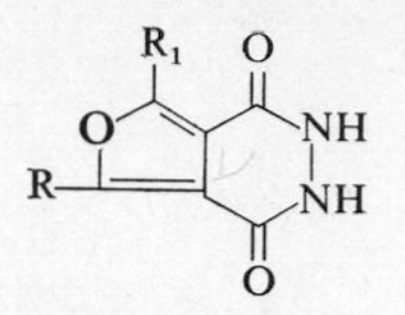

R	R_1	MP (°C) or BP (°C/mm)	References
None		>300	1
Methyl	Methyl	345 (dec.)	1
Methyl		282–283	1
Phenyl	Phenyl	—	9
$—CH{=}NNH_2$		400	42
Miscellaneous Structures			
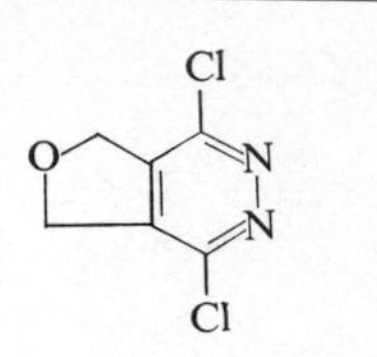		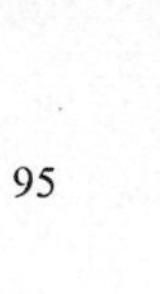 95	16
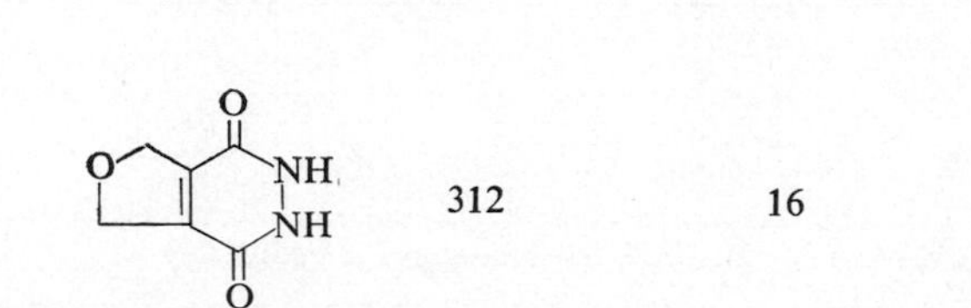		312	16

TABLE 3G-18

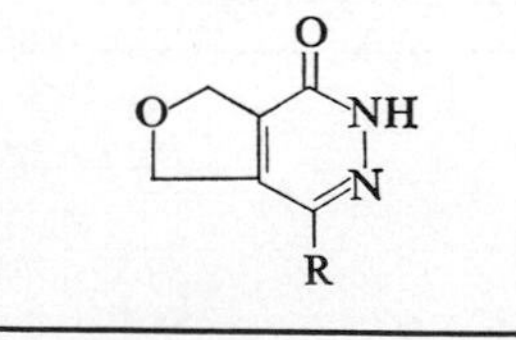

R	MP (°C) or BP (°C/mm)	References
None	197	16
Chloro	199	16

TABLE 3G-19. Oxazolo[3,4-*b*]pyridazines

Structure	MP (°C) or BP (°C/mm)	References
	105–106	20
	110–111	20

TABLE 3G-20. Oxazolo[4,5-*d*]pyridazines

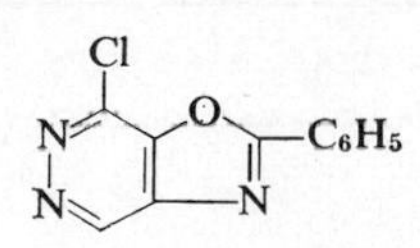	MP 188° C	Ref. 17

TABLE 3G-21. Oxazolo[5,4-*c*]pyridazines

Structure	MP (°C) or BP (°C/mm)	References
	208 209	18 19

TABLE 3G-21. Isoxazolo[4,5-*d*]pyridazines

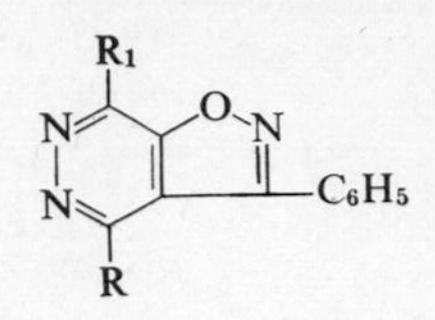

R	R_1	MP (°C) or BP (°C/mm)	References
Chloro	Chloro	158–160	21
		159–160	22
Methoxy	Methoxy	114–115	21
		116–117	22
Phenoxy	Phenoxy	183–184	22
Bromo	Bromo	189.5–191.5	22
$Me_2NCH_2CH_2O$	$Me_2NCH_2CH_2O$	52–56	21
		dipicrate 157–160	21
Methylthio	Methylthio	134–136	21
n-Butylthio	*n*-Butylthio	59–60	22
Phenylthio	Phenylthio	168	22
Chloro	Methoxy	154–155	22
Methoxy	Chloro	130.5–131	22

TABLE 3G-22. Isoxazolo[4,5-*d*]pyridazine-4(5*H*)ones

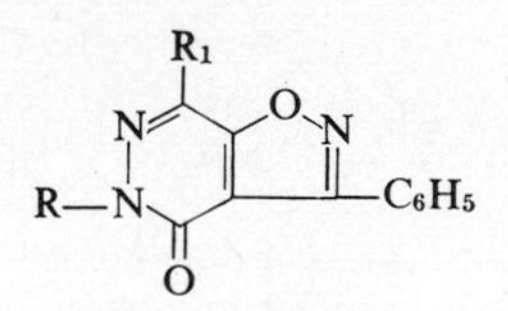

R	R_1	MP (°C) or BP (°C/mm)	References
	Chloro	244–245	22
	Bromo	236	22
	Methoxy	263–265	22
Methyl	Chloro	138.5	22
Methyl	Bromo	171–172	22
Methyl	Methoxy	147–148	22

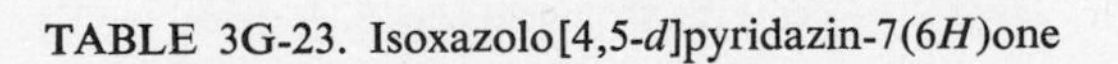

TABLE 3G-23. Isoxazolo[4,5-*d*]pyridazin-7(6*H*)one

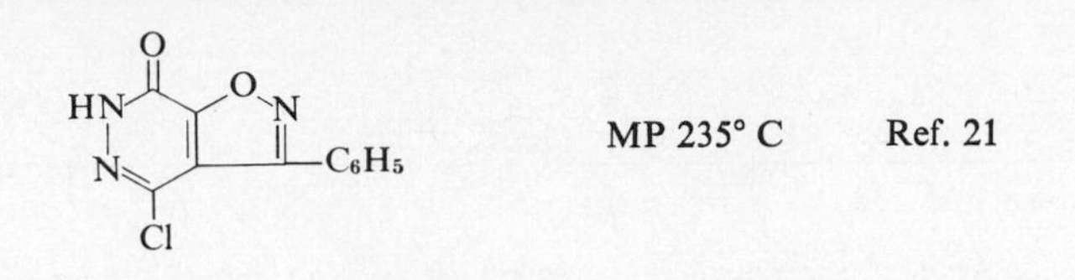

MP 235° C Ref. 21

TABLE 3G-24. Isoxazolo[4,5-*d*]pyridazine-4,7(5*H*, 6*H*)diones

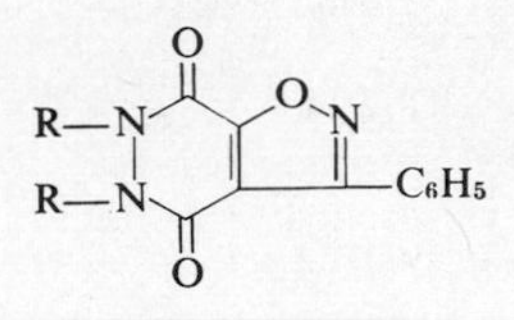

R	MP (°C) or BP (°C/mm)	References
None	280	21
	315 (dec.)	22
Methyl	148–149	22

TABLE 3G-25. Isoxazolo[3,4-*d*]pyridazines

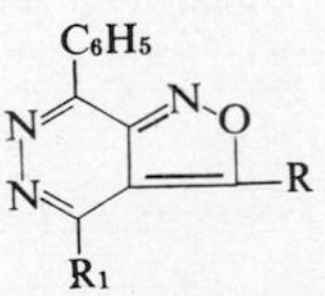

R	R_1	MP (°C) or BP (°C/mm)	References
Methyl	Methyl	185–186	24
Methyl	Phenyl	151–152	24
Phenyl	Phenyl	205	23
(structure: C_6H_5, N, O, HN, Me, O)		267–268	26

TABLE 3G-26. Isoxazolo[3,4-*d*]pyridazin-7(6*H*)ones

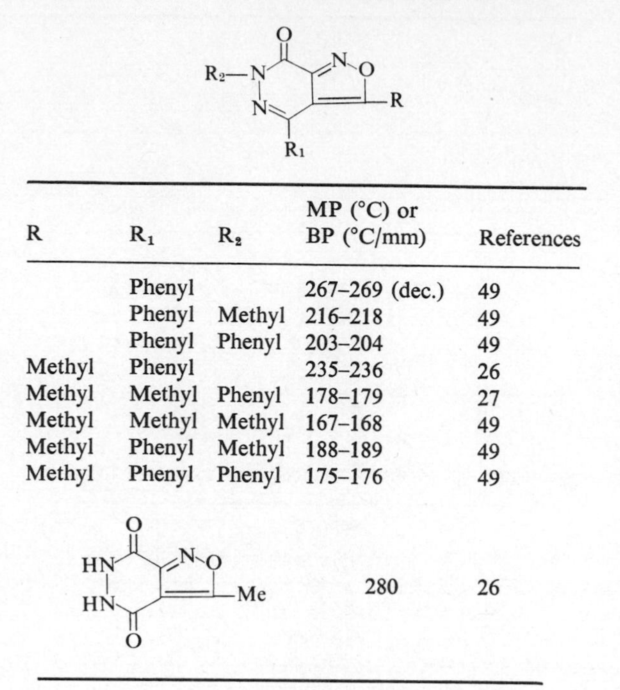

R	R_1	R_2	MP (°C) or BP (°C/mm)	References
	Phenyl		267–269 (dec.)	49
	Phenyl	Methyl	216–218	49
	Phenyl	Phenyl	203–204	49
Methyl	Phenyl		235–236	26
Methyl	Methyl	Phenyl	178–179	27
Methyl	Methyl	Methyl	167–168	49
Methyl	Phenyl	Methyl	188–189	49
Methyl	Phenyl	Phenyl	175–176	49
(structure)			280	26

TABLE 3G-27. 1,2-5-Oxadiazolo[3,4-*a*]pyridazines

R		MP (°C) or BP (°C/mm)	References
Phenyl	(structure: N-oxide-free oxadiazolopyridazine)	190 (dec.)	30
		193–195	31, 32
p-MeC_6H_4		248	30
Phenyl	(structure: N-oxide)	210	33
m-$NO_2C_6H_4$		251	33
p-$NO_2C_6H_4$		252	33
p-ClC_6H_4		235	33

TABLE 3G-28. 1,2,4-Oxadiazolo[2,3-*b*]pyridazines

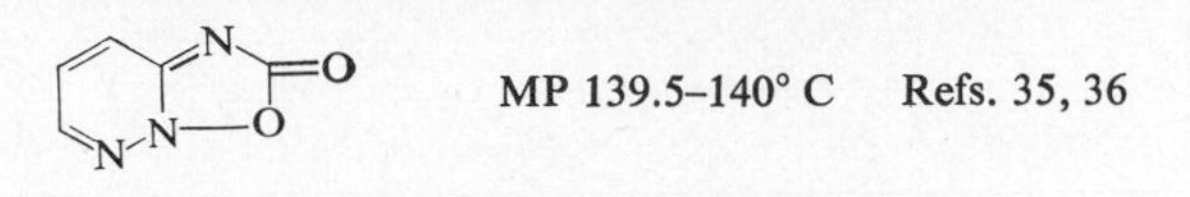

	MP 139.5–140° C	Refs. 35, 36

TABLE 3G-29. 1,3,4-Oxadiazolo[3,4-*a*]pyridazines

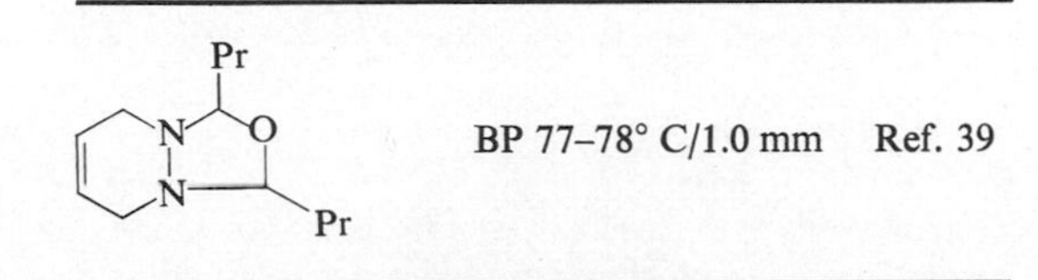

	BP 77–78° C/1.0 mm	Ref. 39

References

1. R. G. Jones, *J. Am. Chem. Soc.*, **78,** 159 (1956).
2. M. Robba, M. C. Zaluski, and B. Roques, *Compt. rend.* C, **263,** 814 (1966).
3. M. Robba, M. Zaluski, and B. Roques, *Compt. rend.* C, **264,** 413 (1967).
4. J. P. Marquet, E. Bisagni, and J. André-Louisfert, *Compt. rend.* C, **265,** 1175 (1967).
5. M. Robba and M. C. Zaluski, *Compt. rend.* C, **266,** 31 (1968).
6. Unpublished data from this laboratory. Parameters for LCAO Calculations were taken from A. Streitwieser *Molecular Orbital Theory for Organic Chemists*, Wiley, New York, 1961, p. 135.
7. G. Hensecke, U. Müller, and G. Badicke, *Chem. Ber.*, **91,** 2270 (1958).
8. E. G. V. Percival, *J. Chem. Soc.*, **1945,** 783.
9. R. Seka and H. Preissecker, *Monatsh. Chem.*, **57,** 71 (1931).
10. M. Robba and M. C. Zaluski, *Compt. rend.* C, **263,** 301 (1966).
11. W. L. Mosby, *J. Chem. Soc.*, **1957,** 3997.
12. W. Bradley and J. Watkinson, *J. Chem. Soc.*, **1956,** 319.
13. H. Keller and H. Halban, *Helv. Chim. Acta*, **27,** 1253 (1944).
14. H. Keller and H. Halban, *Helv. Chim. Acta*, **28,** 59 (1945).
15. H. Schmid, M. Hochweber, and H. Halban, *Helv. Chim. Acta*, **30,** 423 (1947).
16. M. Robba and M. C. Zaluski, *Compt. rend.* C, **263,** 429 (1966).
17. T. Kuraishi, *Chem. Pharm. Bull.* (*Tokyo*), **6,** 641 (1958); *Chem. Abstr.*, **54,** 16460 (1960).
18. T. Kuraishi, *Chem. Pharm. Bull.* (*Tokyo*), **6,** 331 (1958); *Chem. Abstr.*, **53,** 7184 (1959).
19. T. Kuraishi, *Chem. Pharm. Bull.* (*Tokyo*), **8,** 553 (1960); *Chem. Abstr.*, **55,** 13435 (1961).
20. I. Molnar, *Helv. Chim. Acta*, **49,** 586 (1966).
21. L. Erichomovich and F. L. Chubb, *Can. J. Chem.*, **44,** 2095 (1966).
22. G. Desimoni and P. V. Finzi, *Tetrahedron*, **23,** 681 (1967).
23. V. Sprio and J. Fabra, *Gazz. Chim. Ital.*, **86,** 1059 (1956).
24. V. Sprio and R. Pirisi, *Ann. Chim.* (*Rome*), **58,** 121 (1968).
25. G. Renzi and V. Dal Piaz, *Gazz. Chim. Ital.*, **95,** 1478 (1965).

26. V. Sprio, E. Ajello, and A. Massa, *Ann. Chim.* (*Rome*), **57,** 836 (1967).
27. C. Musante, *Gazz. Chim. Ital.*, **69,** 523 (1939).
28. V. Sprio and E. Ajello, *Ricerca Sci.*, *Rend. Ser. A*, **35,** 676 (1965); *Chem. Abstr.*, **64,** 5084 (1966).
29. V. Sprio, E. Ajello, and A. Mazza, *Ricerca Sci.*, **36,** 196 (1966); *Chem. Abstr.*, **65,** 2254 (1966).
30. E. Durio, *Gazz. Chim. Ital.*, **61,** 589 (1931).
31. J. H. Boyer and S. E. Ellzey, Jr., *J. Am. Chem. Soc.*, **82,** 2525 (1960).
32. J. H. Boyer and S. E. Ellzey, *J. Org. Chem.*, **26,** 4684 (1961).
33. R. H. Snyder and N. E. Boyer, *J. Am. Chem. Soc.*, **77,** 4233 (1955).
34. N. E. Boyer, G. M. Czerniak, H. S. Gutowsky, and H. R. Snyder, *J. Am. Chem. Soc.*, **77,** 4238 (1955).
35. T. Itai and T. Nakashima, *Chem. Pharm. Bull.* (*Tokyo*), **10,** 347 (1962); *Chem. Abstr.*, **58,** 4550 (1963).
36. T. Itai and T. Nakashima, *Chem. Pharm. Bull.* (*Tokyo*), **10,** 936 (1962); *Chem. Abstr.*, **59,** 8732 (1963).
37. T. Horie and T. Ueda, *Chem. Pharm. Bull.* (*Tokyo*), **11,** 114 (1963); *Chem. Abstr.*, **59,** 3917 (1963).
38. V. R. Skvarchenko, M. G. Kuz'min, and R. Ya. Levina, *Vestnik Moskov. Univ.* 12, *Ser. Mat.*, *Mekh.*, *Astron.*, *Fiz.*, *Khim.*, No. 3, 169 (1957); *Chem. Abstr.*, **52,** 6358 (1958).
39. B. Zwanenburg, W. E. Weening, and J. Strating, *Rec. Trav. Chim.*, **83,** 877 (1964).
40. M. Robba and M. C. Zaluski, *Bull. Soc. Chim. France*, **1968,** 4959.
41. J. P. Marquet, E. Bisagni, and J. Andre-Louisfert, *Chim. Ther.*, **3,** 348 (1968).
42. G. LeGuillanton and A. Daver, *Compt. rend.* C, **268,** 643 (1969).
43. T. Eicher and E. Angerer, *Chem. Ber.*, **103,** 339 (1970).
44. S. Yoshina, I. Maeba and K. Hirano, *Chem. Pharm. Bull.* (*Tokyo*), **17,** 2158 (1969).
45. M. Robba, M. C. Zaluski, B. Roques, and M. Bonhomme, *Bull. Soc. Chim. France*, **1969,** 4004.
46. S. Yoshina and I. Maeba, *Chem. Pharm. Bull.* (*Tokyo*), **18,** 379 (1970).
47. G. Adembri, F. DeSio, R. Nesi, and M. Scotton, *J. Chem. Soc. C*, 1536 (1970).
48. M. Lomme and Y. Lepage, *Bull. Soc. Chim. France*, **1969,** 4183.
49. G. Renzi and S. Pinzauti, *Farm.* (*Pavia*), *Ed. Sci.*, **24,** 885 (1969).

Part H. Thieno-, Thiazolo-, and Thiadiazolopyridazines

I. Thieno[2,3-*d*]pyridazines

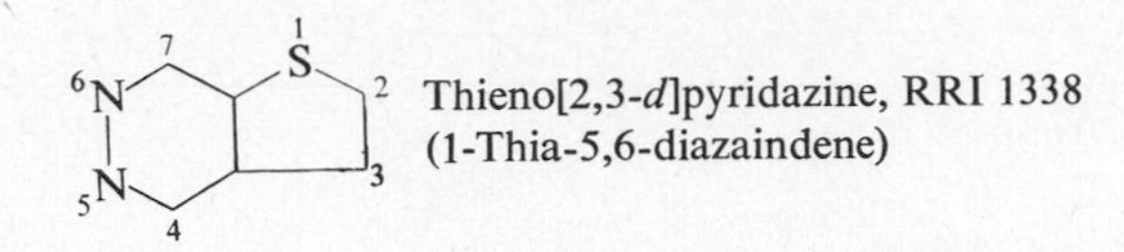

Thieno[2,3-*d*]pyridazine, RRI 1338
(1-Thia-5,6-diazaindene)

This bicyclic system was not known until 1953 when Baker et al. (1) treated 3-carbomethoxythiophene-2-carboxylic acid with hydrazine hydrate in a solution of ethanol and obtained **1**. Two years later Jones (2) synthesized

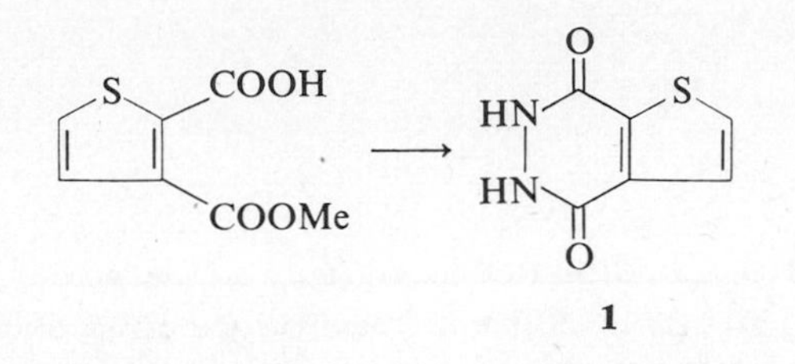

the ester **2** from 2-ethoxalyl-4-diethoxybutyronitrile and phosphorus pentasulfide; after this ester was purified a residual oil could be obtained which was believed to be dimethyl thiophene-2,3-dicarboxylate, since it reacted with hydrazine in a solution of methanol. The product thus obtained was formulated as compound **1**. There are no data other than melting points from which one could establish the identity of both products. From the corresponding

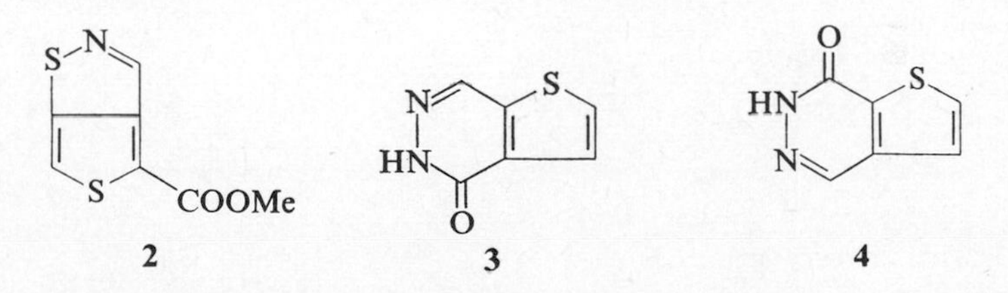

diester the 2-methyl analog of **1** has been prepared (3).

Thieno[2,3-*d*]pyridazin-4(5*H*)ones (**3**) or -7(6*H*)ones (**4**) have been prepared from 2-formyl(acetyl) thiophene-3-carboxylic acids or from 3-formyl (acyl) thiophene-2-carboxylic acids and hydrazine or substituted hydrazines (4, 5). In a like manner, 2,3-diformyl-, 2-formyl-3-acyl-, or 2-acyl-3-formylthiophenes have been transformed into the corresponding thieno[2,3-*d*]-pyridazines (**5**) (4–7). 2-Methylthiophene-4,5-dicarboxylic acid reacted with hydrazine to give the corresponding dioxothienopyridazine (36). In particular cases it was possible to isolate the intermediate hydrazones, which were

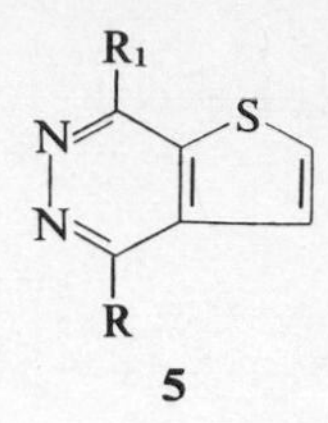

subsequently cyclized. However, it was not possible to cyclize acylhydrazones and the benzoyl hydrazone **6** was debenzoylated and then cyclized to **3** (4).

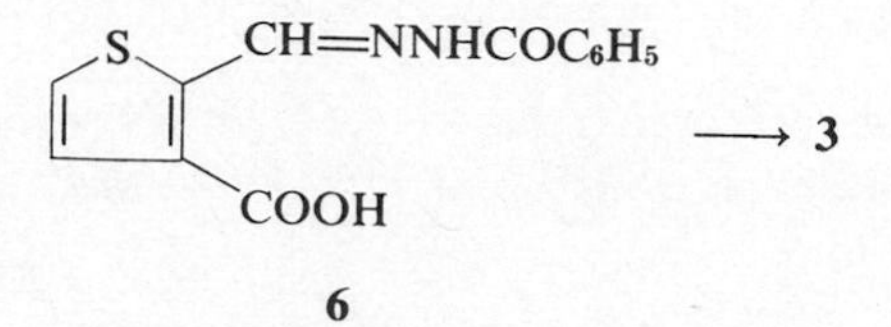

4,7-Diaminothieno[2,3-*d*]pyridazine (**7**) could be prepared in a one-step synthesis from 2,3-dicyanothiophene and hydrazine in a solution of ethanol. A mechanism for this synthesis has been outlined (8).

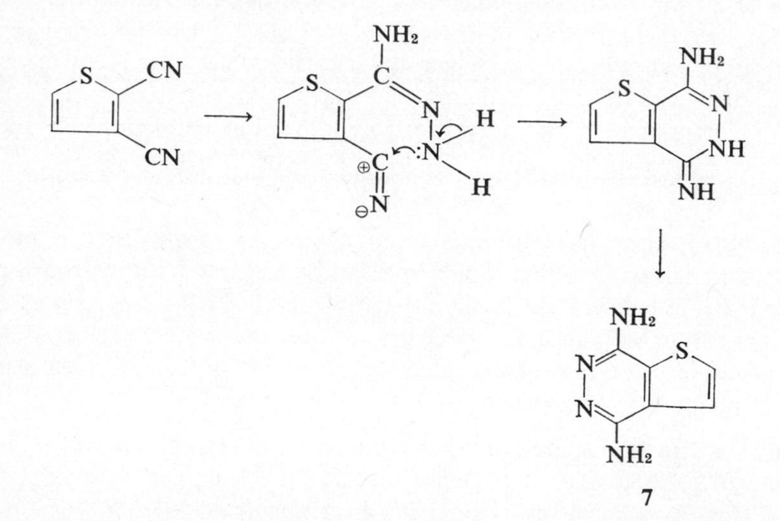

The parent compound **5** ($R = R_1 = H$) was obtained by Robba et al. (4, 7) from 2,3-diformylthiophene and hydrazine. It also resulted from decarboxylation of thieno[2,3-*d*]pyridazine-7-carboxylic acid upon sublimation at 200°/0.1 mm. The acid itself was prepared from the thiophene derivative **8**, which after being cyclized to **9** and saponified to **10** yielded the desired product (4). Finally, thieno[2,3-*d*]pyridazine and its 4- or 7-methyl

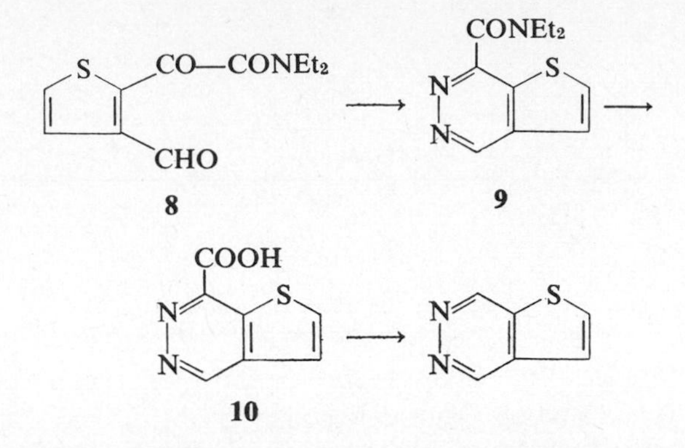

analogs were easily prepared from 4- (**11**) or 7-hydrazinothieno[2,3-*d*]-pyridazine (**12**) in a boiling solution of sodium ethoxide in ethanol (8).

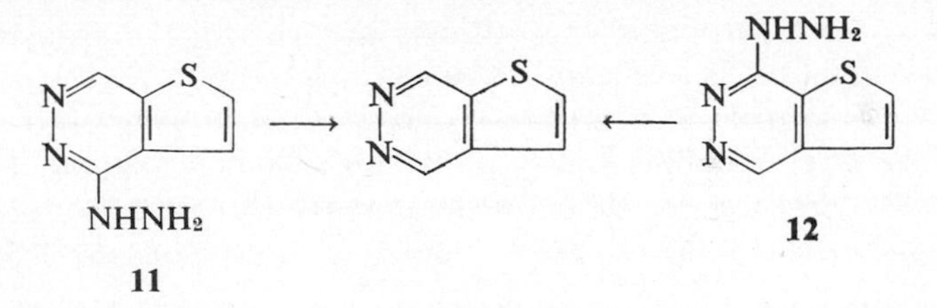

Thieno[2,3-*d*]pyridazine is soluble in water and organic solvents. Its nmr spectrum [$\tau = 2.07$ (H_2), 2.44 (H_3), 0.41 (H_4), and 0.29 (H_7)] is in accordance with the proposed structure and chemical shifts with MO calculations as well. MO calculations (10) of total and frontier π-electron densities are listed in Table 3H-1.

Thieno[2,3-*d*]pyridazine forms a monohydrochloride salt or a monoquaternary salt with methyl iodide and can be oxidized with hydrogen peroxide at room temperature to a compound formulated as an *N*-oxide.

Examination of the nmr spectrum of the quaternized thieno[2,3-*d*]-pyridazine revealed that quaternization with methyl iodide afforded a mixture of about 40% **13** and 60% **14**.

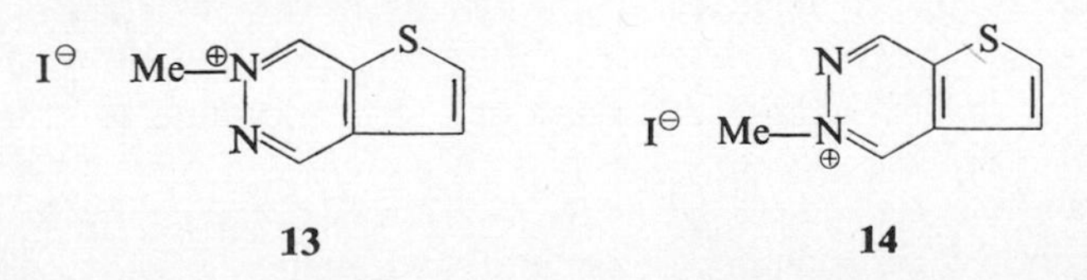

Similarly, 4-methylthieno[2,3-*d*]pyridazine afforded after quaternization a mixture of about 70% **15** and 30% **16**. However, quaternization of 7-methylthieno[2,3-*d*]pyridazine with methyl iodide resulted in the formation of only one product to which the structure of **17** was assigned (9).

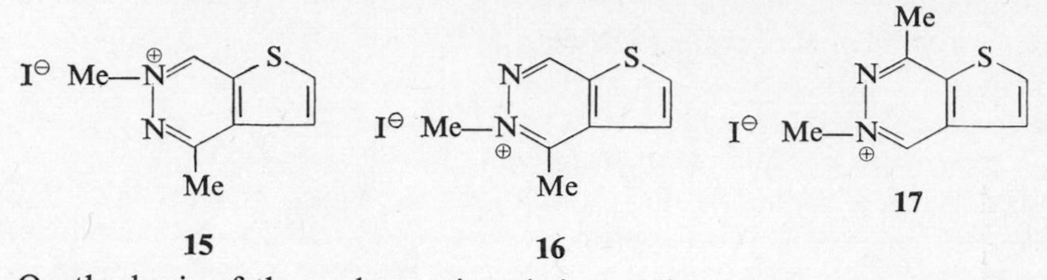

On the basis of these observations it is possible to conclude that methyl groups on the ring exert an inductive effect; however this is outweighed by steric hindrance. In addition the basicity of the respective ring nitrogen has to be taken into consideration.

Taking into account the calculated electron densities, it is possible to predict that electrophilic substitution will be favored to proceed at position 3, whereas positions 4 and 7 should be prone to nucleophilic attack. Unfortunately, no electrophilic substitutions are known and from the data about nucleophilic displacements of the chlorine atoms at position 4 or 7 it is not possible to make any conclusions about differences in reactivity.

Attempts of electrophilic substitutions such as bromination, nitration, and Friedel-Crafts reaction have failed so far (4). Other transformations of thieno[2,3-*d*]pyridazines have been extensively investigated by Robba and co-workers.

Thieno[2,3-*d*]pyridazinones **3** and **4** have been examined by ir spectroscopy. They exist in the lactam form in the solid state and are hydrogen bonded. Alkylation, as expected, yielded the *N*-alkyl derivatives (4, 8, 30). Contrary to this, **3** and **4** when heated with an alkaline solution of monochloroacetic acid formed the *O*-substituted derivatives **18** and **19** (8). The

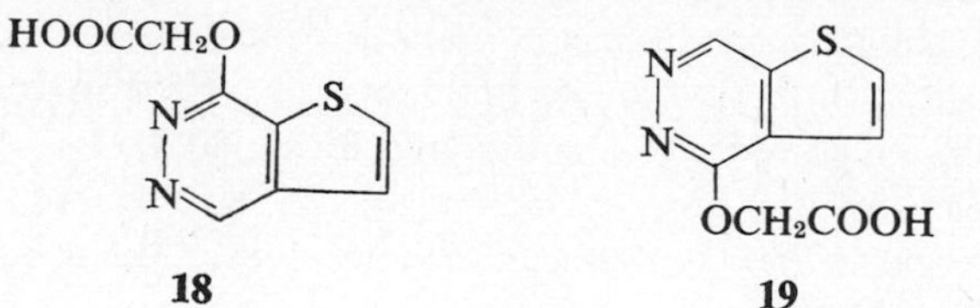

structure of both products was confirmed from the correlation of ir and nmr spectroscopic data with those of the *N*-substituted analogs **20** and **21**. These were prepared from **3** and **4** and ethyl chloroacetate and the products formed were saponified.

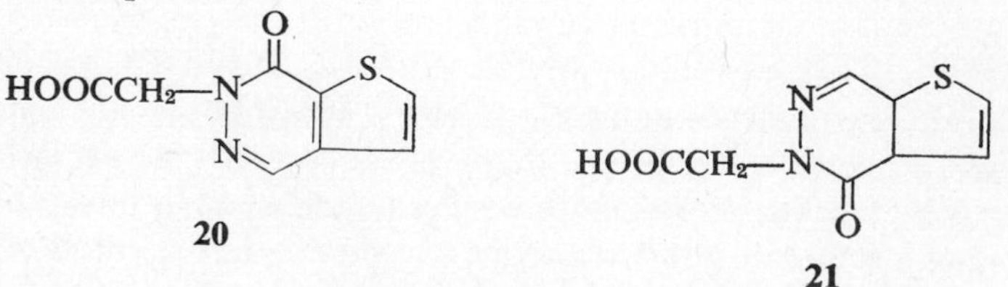

Compound **18** could be rearranged into the ester of **20** when heated in a solution of ethanol in the presence of sulfuric acid (8).

Thieno[2,3-*d*]pyridazinones **3**, **4**, and **1** have been converted into the corresponding mono- and dichloro (or bromo) derivatives by the usual procedure with phosphorus oxychloride or oxybromide (7, 8).

Halogenated thieno[2,3-*d*]pyridazines are useful as starting material for different nucleophilic displacement reactions. 4- or 7-Chlorothieno[2,3-*d*]-pyridazines could be converted into the corresponding iodo derivatives with hydriodic acid; alkoxy or aryloxy derivatives were prepared with alkoxides or phenoxides and amino derivatives with the aid of ammonia or amines. To accomplish the last mentioned reaction temperatures up to 150° and pressure have to be applied in order to obtain the desired amino derivatives. The thiourea method was used for the synthesis of thieno[2,3-*d*]pyridazine-thiones; thioethers were prepared with the aid of thiolates; and hydrazino derivatives were formed in the reaction with hydrazine (8).

In an attempt to prepare the corresponding azido derivatives, the isomeric monochloro derivatives **22** and **25** were allowed to react with sodium azide, but instead a fused tetrazole ring was formed (**23**, **26**). These isomeric polyazaheterocycles were obtained also from the corresponding hydrazino compounds **24** and **27** upon treatment with nitrous acid (8).

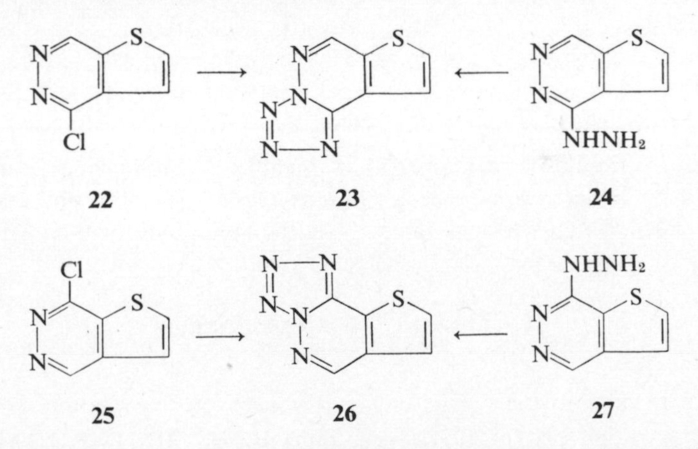

Similar displacements were successfully performed also with 4,7-dichloro-thieno[2,3-*d*]pyridazine (**28**). Although in some cases both chlorine atoms were replaced, for the most part monosubstitutions occurred. For example, with sodium β-dimethylaminoethylate a 4,7-disubstituted product was formed, whereas reactions with aniline, β-diethylaminoethylamine and hydrazine afforded the monosubstitution products. That the replacement of the halogen at position 4 took place has been established by converting 4-hy-drazino-7-chlorothieno[2,3-*d*]pyridazine (**29**) with a hot solution of sodium ethoxide in ethanol into **30**; this compound could be obtained otherwise by halogen displacement of compound **31** under similar reaction conditions (8).

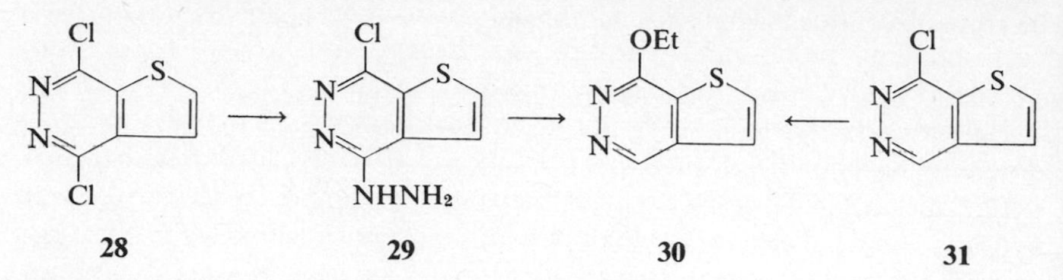

When compared to 4,7-dichlorothieno[3,4-*d*]pyridazine, 4,7-dichloro-thieno[2,3-*d*]pyridazine (**28**) is more stable and can be transformed into 7-chlorothieno[2,3-*d*]pyridazin-4(5*H*)one (**32**) only after being heated with boiling hydrochloric acid for 2 hr. The structure of the obtained product was accomplished by catalytic dehalogenation in the presence of Raney nickel and the resulting compound **3** was shown to be identical to the compound obtained by direct cyclization (8).

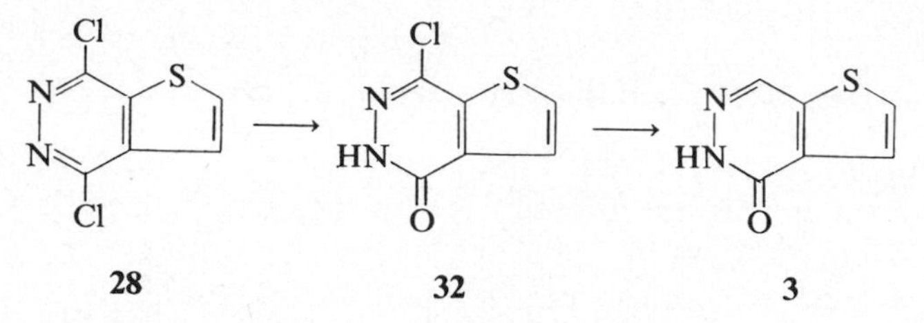

Finally, it should be mentioned that thieno[2,3-*d*]pyridazine-7-carboxylic acid and its derivatives have been examined by IR spectroscopy and it has been concluded that the acid **10** exists in the solid state in the zwitterionic form (4).

II. Thieno[3,4-*d*]pyridazines

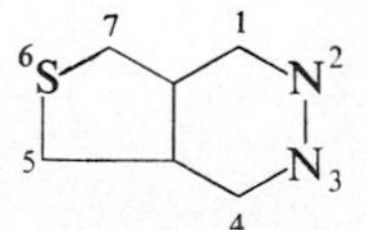

Thieno[3,4-*d*]pyridazine, RRI 1341
(2-Thia-5,6-diazaindene)

The first report on thieno[3,4-*d*]pyridazines goes back to 1947 when Baker et al. (11, 12) treated the 3,4-*cis*-anhydride of tetrahydrothiophene or its 2-*n*-propyl analog with hydrazine hydrate. The product obtained after short heating of the reaction mixture was assigned the cyclic hydrazide structure **33** on the basis of solubility in aqueous alkali and insolubility in normal hydrochloric acid or dilute solution of sodium bicarbonate. A *cis*-configuration of the bicyclic system also was postulated. If the bicyclic product **33** (R = *n*-Pr) was heated with hydrazine hydrate for 2 hr, it was transformed

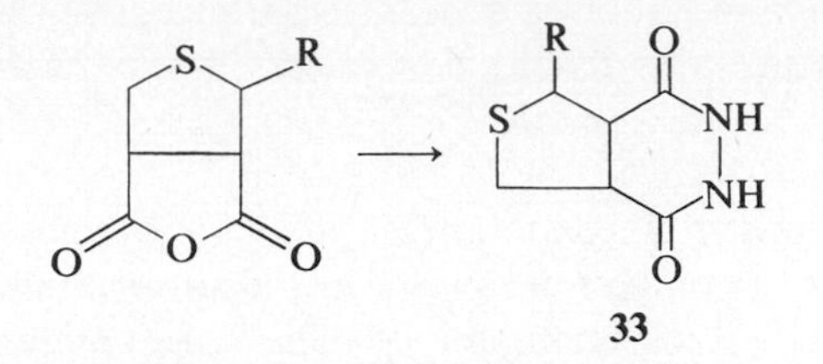

into 2-*n*-propylthiophane-3,4-*trans*-dicarbohydrazide.

Jones (3) obtained from a similar reaction from diethyl thiophene-3,4-dicarboxylate and hydrazine the dione **34**.

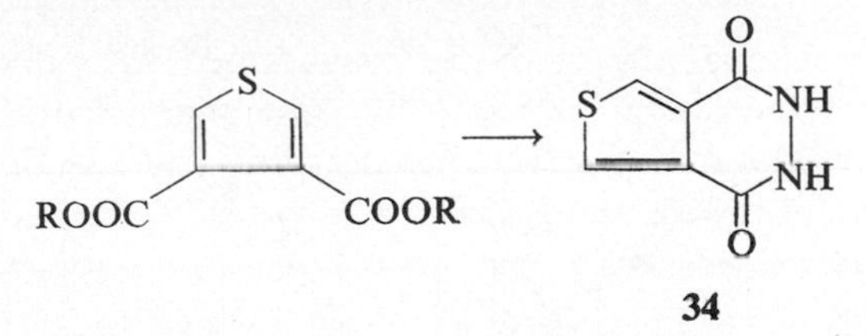

In an analogous way 3-formylthiophene-4-carboxylic acid and hydrazine or substituted hydrazines afforded the corresponding thieno[3,4-*d*]pyridazin-1(2*H*)ones (**35**) (4, 5, 13), and thieno[3,4-*d*]pyridazines (**36**) were synthesized from the corresponding 3,4-diformylthiophenes (4, 5, 13, 14).

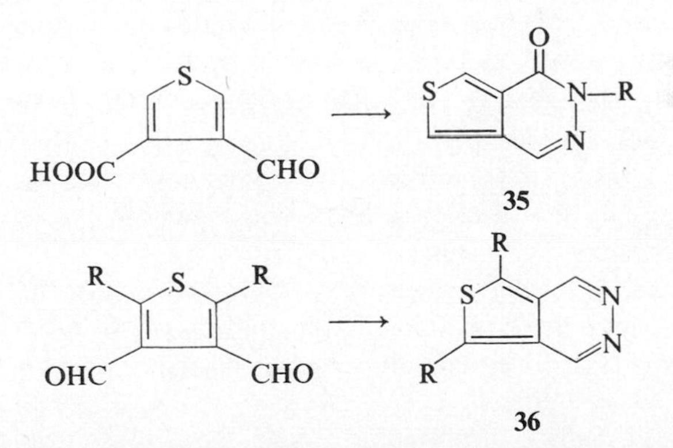

Thieno[3,4-*d*]pyridazine is a solid, soluble in water and alcohol and other organic solvents, and forms a hydrochloride salt and a monomethiodide salt. In contrast to the isomeric thieno[2,3-*d*]pyridazine which formed an *N*-oxide, thieno[3,4-*d*]pyridazine is decomposed under the same reaction conditions (4). Its nmr spectrum [in $CDCl_3$: $\tau = 0.56$ (H_1, H_4) and 1.74

(H_5, H_7)] is consistent with MO calculations. MO calculations (10) of total and frontier π-electron densities are available in Table 3H-2.

Reactivity of thieno[3,4-*d*]pyridazines has been investigated by Robba and co-workers. No reactions have been performed on the parent compound, electrophilic substitutions failed, and most of the knowledge about the chemistry of this heteroaromatic system was gained from transformations on oxo and halo analogs.

Thieno[3,4-*d*]pyridazin-1(2*H*)one (**35**, R = H) is reported to exist in the lactam form and examination of its ir spectrum indicated hydrogen-bonded-molecules. With aminoalkylhalides the corresponding *N*-substituted derivatives of the type **35** have been prepared (4, 30). It is of interest that the otherwise usual conversion of oxo into chloro compounds when applied to **35** (R = H) did not proceed under a variety of reagents employed (8).

The dioxo compound **34** could be transformed with phosphorus oxychloride in the presence of pyridine into the monochloro derivative **37** (30 min under reflux) or into the dichloro analog **38** (140°, 2 hr) (5, 8, 13). In a similar manner, with phosphorus oxybromide conversion of only one oxo group of **34** was possible.

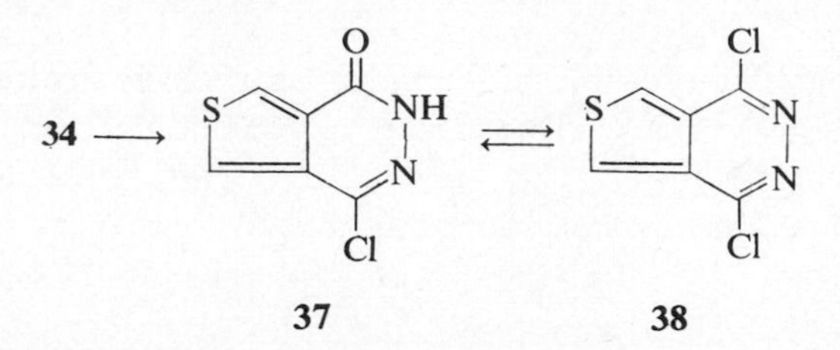

1,4-Dichlorothieno[3,4-*d*]pyridazine (**38**) is unstable. When exposed to air for a few hours or when crystallized from hydroxylic solvents **35** is transformed into **37** (8, 13). Therefore all reactions performed on this dichloro derivative have to be executed in anhydrous solvents like tetrahydrofuran and in an atmosphere of nitrogen.

With aniline, 1,4-dichlorothieno[3,4-*d*]pyridazine gave the 1,4-dianilino derivative, whereas from reactions with sodium methylate, hydrazine, or β-diethylaminoethylamine only monosubstitution products were obtained (5, 8).

It is reported that is not possible to dehalogenate **38** chemically or catalytically. The compound is either decomposed or transformed into **37** (13). However, it is possible to dehalogenate 4-bromo or 4-chlorothieno[3,4-*d*]-pyridazin-1(2*H*)one into **35** (R = H) in the presence of Raney nickel, but only under drastic conditions (100–130°, 60–100 atm). Under these reaction conditions no desulfurization could be observed (8).

III. Thiazolo[4,5-*d*]pyridazines

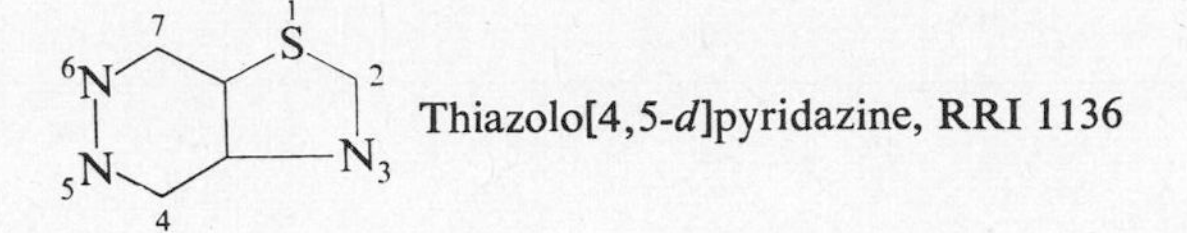

In 1943 Huntress and Pfister synthesized the first member of this series in a reaction between diethyl 2-phenylthiazole-4,5-dicarboxylate and hydrazine (15). Along with the bicyclic product **40** (R = C_6H_5), the dihydrazide **39**, (R = C_6H_5) was obtained; the relative amounts of the cyclized and uncyclized products depend on the duration of the reaction. Longer reaction time favored the formation of the bicycle **40**, and the dihydrazide **39** when heated at 200–220° was transformed into **40**.

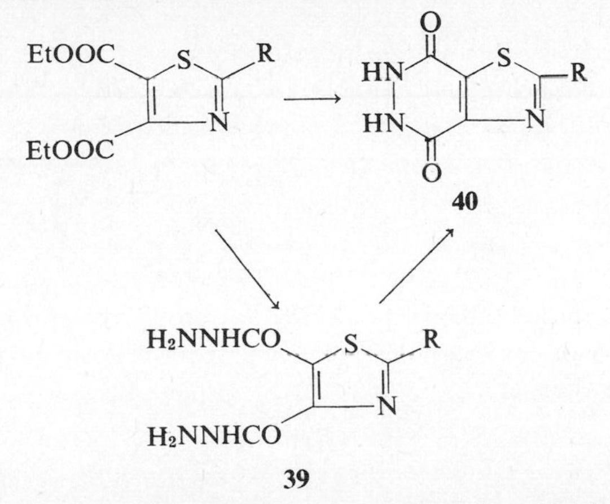

By similar methods several other thiazolo[4,5-*d*]pyridazines of the type **40** have been prepared (16–18, 31, 34). Robba and Le Guen (18, 31) applied this reaction to the synthesis of the parent compound **41**, which was obtained when 4,5-diformylthiazole was treated with a solution of hydrazine in ethanol at room temperature.

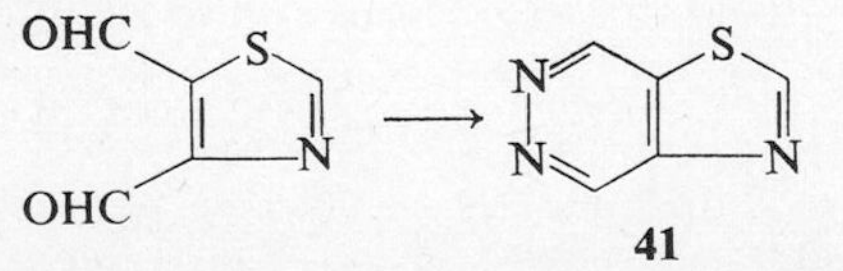

The bicyclic system could also be built up from appropriate pyridazines as shown by Kuraishi and Castle (19), who reacted the pyridazine **42** with cyanogen bromide in the presence of sodium hydroxide solution and obtained 2-aminothiazolo[4,5-*d*]pyridazine-7(6*H*)thione (**43**).

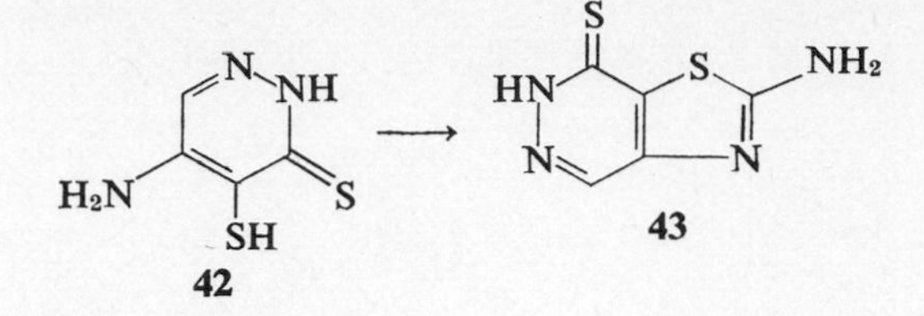

In a similar approach, 5-amino-4-bromo-2-phenyl-3(2*H*)pyridazinone (**44**) when heated under reflux (160°, 3 hr) with carbon disulfide in a solution of diethylene glycol monomethyl ether in the presence of alkoxide (from the

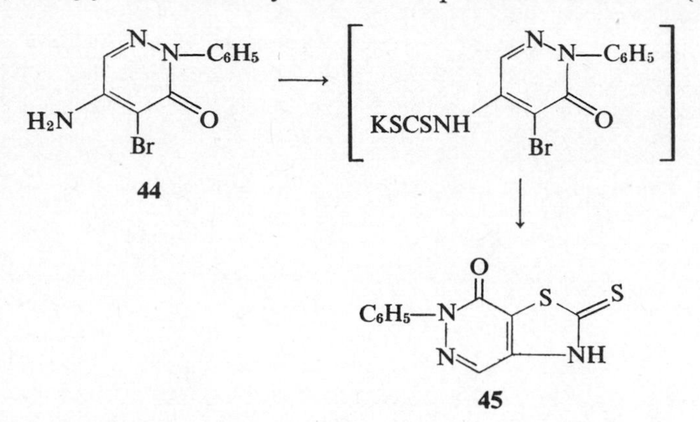

same alcohol) afforded the bicycle **45** (20). Support for the proposed structure and elimination of the possible iminothiol form followed from ir and uv spectroscopic evidence.

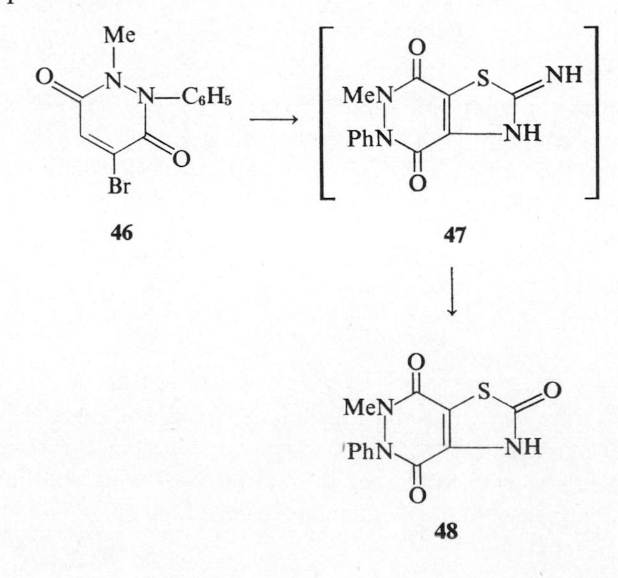

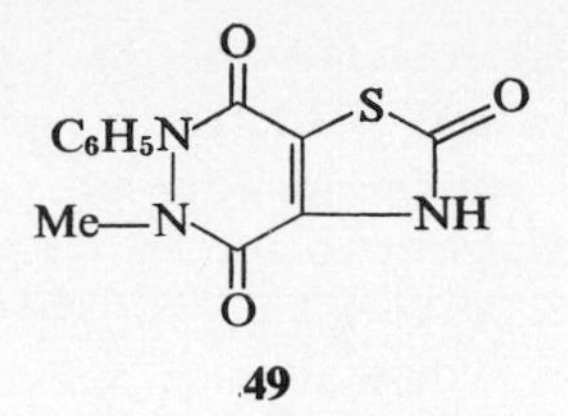

49

Druey, Meier, and Staehelin (21) treated 4-bromo-1-methyl-2-phenyl-3,6(1*H*, 2*H*)pyridazinedione (**46**) with thiourea in a solution of ethanol and the intermediate imine **47** was subsequently transformed with hot alkali into **48**. No proof for the proposed structure **48** was given and, particularly in view of the utility of thiourea in preparing mercapto pyridazines from the corresponding halopyridazines, it appears that the isomeric structure **49** is more probable.

There are no data about the behavior of thiazolo[4,5-*d*]pyridazine under the influence of different reagents and no transformations are known. MO

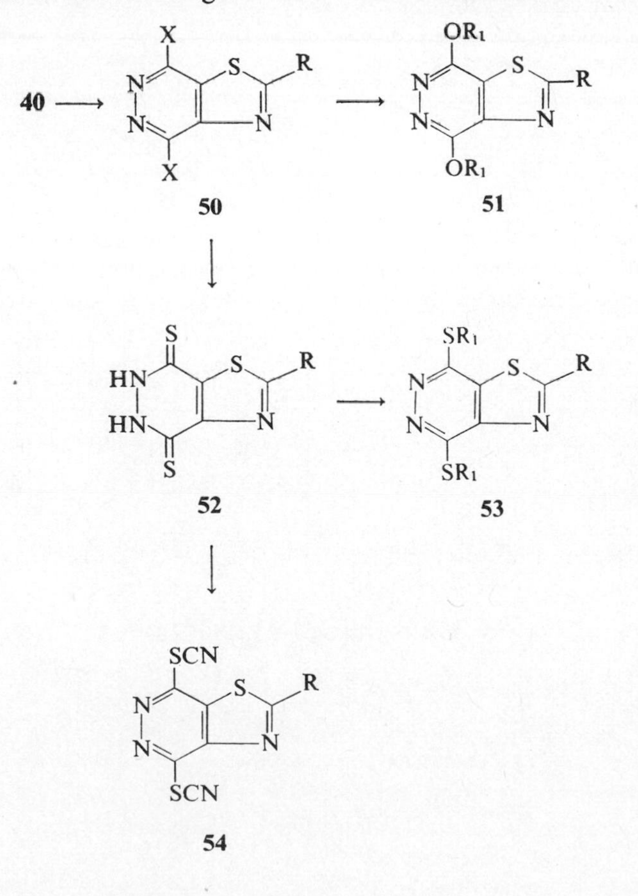

calculations (10) of total and frontier π-electron densities for thiazolo[4,5-*d*]-pyridazine are presented in Table 3H-3. From these data one can postulate that there would be no inclination for electrophilic substitutions, and nucleophilic displacements are expected to proceed preferentially at positions 4 and 7, and possibly at position 4 with somewhat greater ease.

There are no experimental data that would substantiate these suppositions and the only reported transformation is the displacement of both halogens of 4,7-dihalothiazolo[4,5-*d*]pyridazines. Compounds of the type **40** were transformed into the dihalo compounds (**50**, X = Cl or Br) in the usual way by means of phosphorus oxychloride or oxybromide, preferentially in the presence of PX_5 (18, 22).

Further transformations of **50** have led to the bis-alkoxy derivatives **51** and with the aid of thiourea the sulfur analog of **40**, **52**, could be prepared. Alkylations of **52** afforded the bis-alkylthio derivatives **53** and reaction with cyanogen bromide in the presence of alkali led to bis-thiocyanatothiazolo-[4,5-*d*]pyridazines **54** (18, 22).

IV. Thiazolo[5,4-*c*]pyridazines

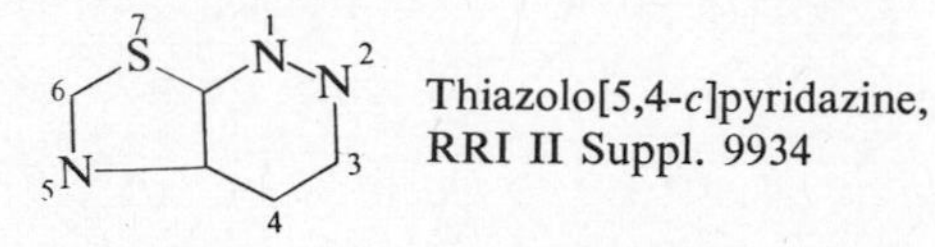

Thiazolo[5,4-*c*]pyridazine,
RRI II Suppl. 9934

The only representative of this bicyclic system was reported in 1960 by Kinugawa, Ochiai, and Yamamoto (23). 4-Amino-6-chloro-3(2*H*)pyridazinethione was treated with cyanogen bromide at 8–10° in the presence of alkali and 3-chloro-6-aminothiazolo[5,4-*c*]pyridazine (**55**) was formed.

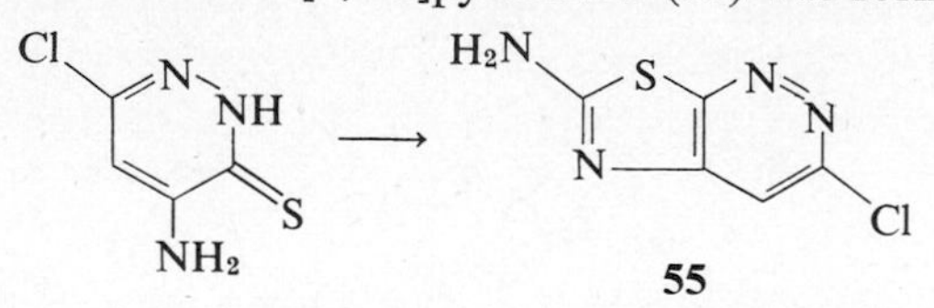

MO calculations of total and frontier π-electron densities (10) are listed in Table 3H-4.

V. Thiazolo[3,2-*b*]pyridazines

5*H*-Thiazolo[3,2-*b*]pyridazine,
RRI III Suppl. 11850

Thiazolo[3,2-*b*]pyridazin-4-ium
Cation

Most of the compounds known to belong to this heterocyclic system are mesoionic compounds. The first reported reaction in this field is due to Ohta and Kishimoto (24) in 1961. They treated 3-carboxymethylthiopyridazines (**56**) with a mixture consisting of acetic anhydride, pyridine, and trimethylamine (1:1:1) and obtained compounds of the type **57**. From a similar

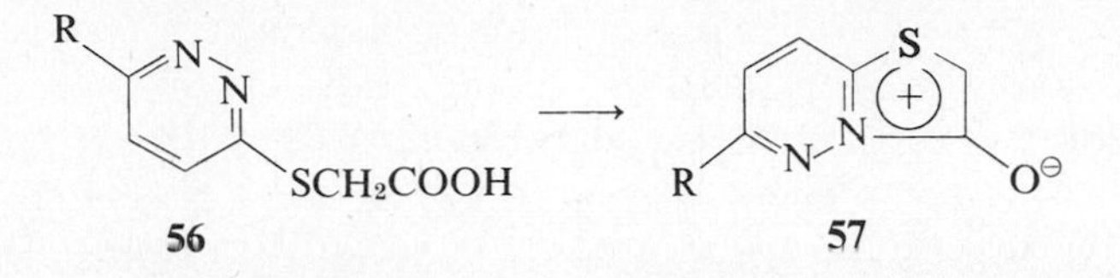

reaction the analogous **58** has been obtained (25). Thus 3(2*H*)-pyridazinethiones when heated with chloroacetonitrile in the presence of triethylamine in a solution of benzene (80°, 1.5 hr) afforded **58** (R = H) as the hydro-

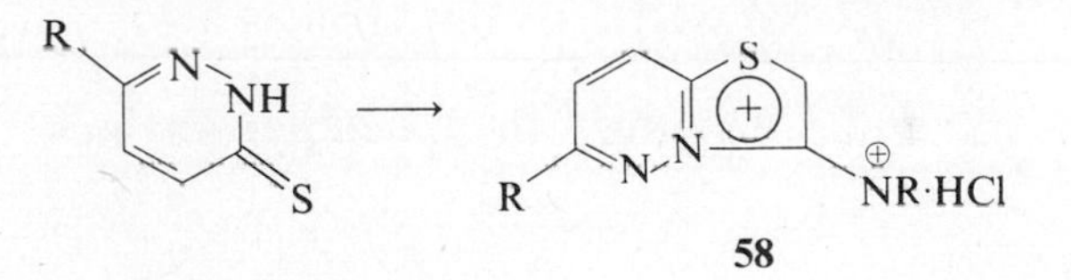

chloride salt. Cnmpounds of type **58** when boiled with excess of acetic anhydride for 6 hr formed the corresponding *N*-acetyl derivatives (**58**, R = CH_3CO) (26).

Mesoionic compounds of type **57** are soluble in water and most give yellow solutions. They are also soluble in ethanol and solutions in anhydrous benzene are red colored. They appear to be unstable to heat; with 50% sulfuric acid they are transformed back into **56** and with a boiling solution of phenylhydrazine in ethanol hydrazides of type **59** are formed (24).

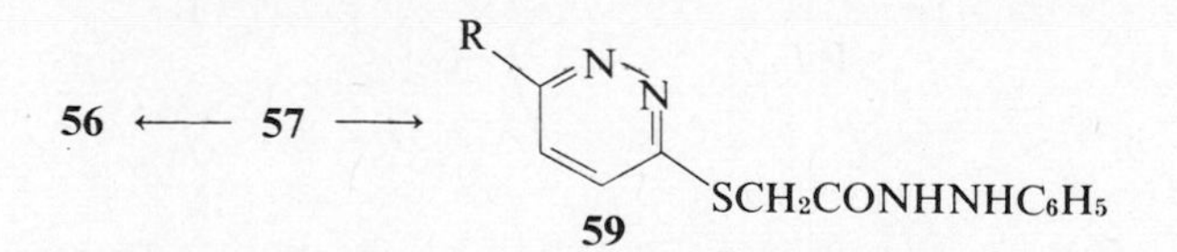

Another type of compound belonging to this heterocyclic system was discovered recently (27) during an investigation of the reaction between α-haloketones and 3(2*H*)pyridazinethiones. When the reaction was conducted in the presence of sodium alkoxide, keto sulfides (**60**) were isolated. However, in the absence of a base and if tetrahydrofuran was used as solvent it was possible to isolate the 3-hydroxy-6-chloro-2,3-dihydrothiazolo[3,2-*b*]-pyridazin-4-ium salt **61**. This salt is not very stable and is converted either

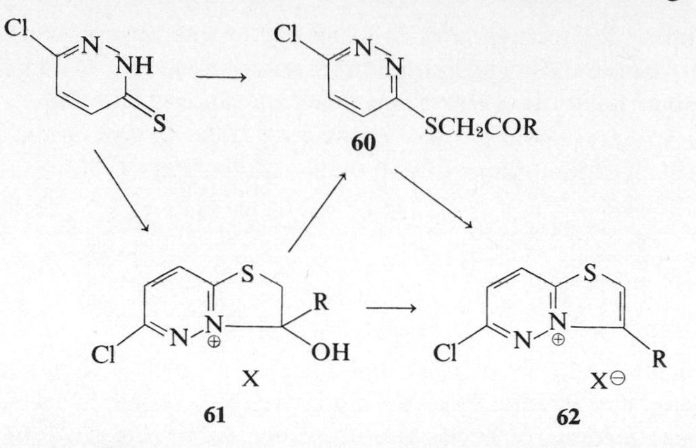

to **60** when crystallized from a mixture of ethanol and *N,N*-dimethylformamide or to **62** in the presence of concentrated sulfuric acid. On the other hand, **62** could be prepared directly from **60** with concentrated sulfuric acid.

VI. 1,2,5-Thiadiazolo[3,4-*d*]pyridazines

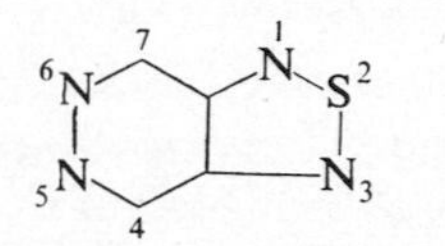

1,2,5-Thiadiazolo[3,4-*d*]pyridazine, RRI Suppl. II 9911 (Pyridazino[4,5-*d*]-2,1,3-thiadiazole)

In 1960 Sekikawa (28) reported the synthesis of 1,2,5-thiadiazolo[3,4-*d*]-pyridazine-4,7(5*H*,6*H*)dione (**64**) from diethyl 1,2,5-thiadiazole-3,4-dicarboxylate. This was first converted with hydrazine hydrate into the bis-hydrazide **63**, which when heated under reflux with dilute hydrochloric acid for 8 hr yielded the bicyclic product.

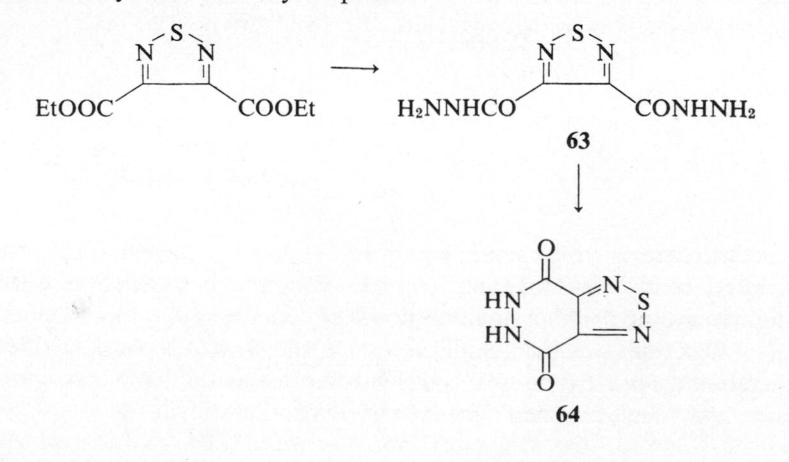

A possibility for formation of this bicycle was mentioned in a review by Dury (29); however, no experimental details were given. Accordingly, **65** can be formed from diaminopyridazinones and thionyl chloride.

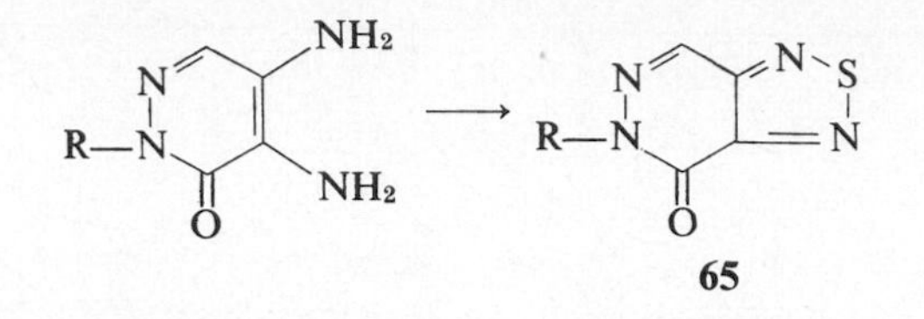

MO calculations (37) of total and frontier π-electron densities for this symmetrical heteroaromatic bicycle are presented in Table 3H-5.

Owing to the impaired resonance of the pyridazine ring, these compounds are expected to be less stable than other azolopyridazines. This has been observed, since several reactions progress with rupture of the thiadiazolo ring. Thus treatment with hot alkali afforded 4,5-diamino-3,6-dihydroxypyridazine (32) and, similarly, in attempts to convert the 4,7-bismethylthio analog to an amino derivative, it was transformed with ammonia under pressure into 4,5-diamino-3,6-bis(methylthio)pyridazine (37). Moreover, the bismethylthio derivative suffered cleavage of the thiadiazole moiety in a reaction with sodium azide in *N,N*-dimethylformamide to give 7,8-diamino-6-methylthiotetrazolo[1,5-*b*]pyridazine. Here evidently one of the methylthio groups was displaced and simultaneous ring closure of the tetrazolo ring occurred (37).

Compound **64** could be thiated or transformed into the 4,7-dichloro derivative and the 4,7-bismethylthio compound was reacted with hydrazine to give the 4-hydrazino-7-methylthio derivative (37).

VII. Thieno[3,2-*c*]pyridazines

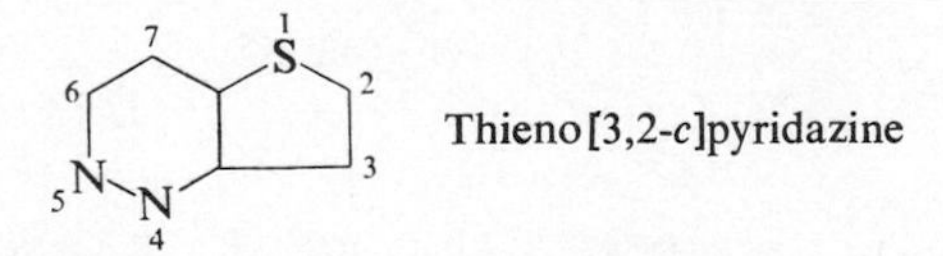

Reactions leading to the parent compound and some reduced analogs were described recently by Poole and Rose (33). The tetrahydrothiophen derivative **66** was cyclized with hydrazine to **67** (R = H or COOMe). The ester **67** (R = COOMe) was then dehydrogenated and decarboxylated to thieno-[3,2-*c*]pyridazin-6(5*H*),one (**68**), which after treatment with phosphoryl chloride and dehalogenation afforded the parent compound **69**.

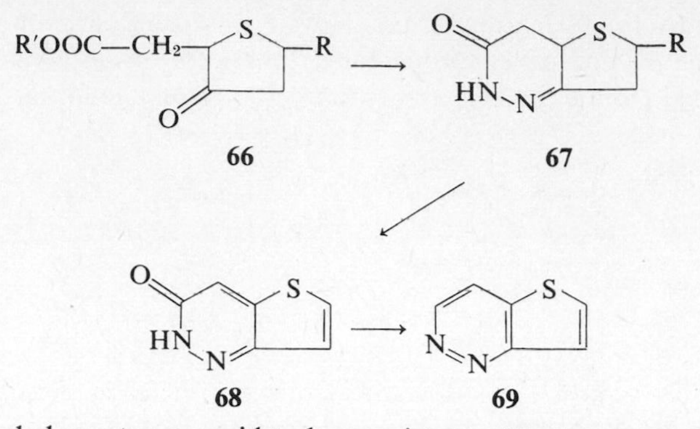

The recorded spectra are said to be consistent with the proposed structure.

VIII. Pyridazo[4,5-*d*]-1,3-dithiol

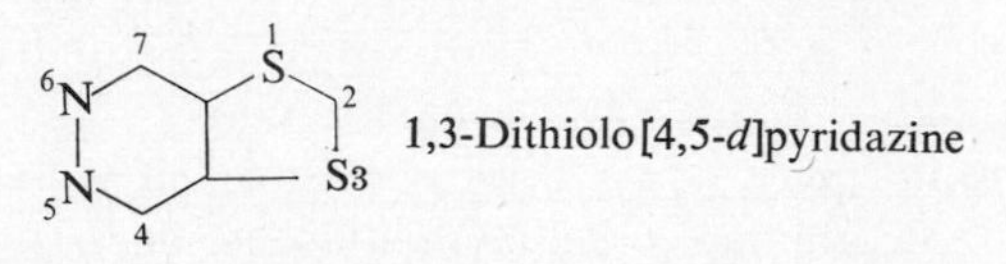

Derivatives of this new heterocyclic system were prepared recently (35) in connection with the formation of dipyridazodithiindiones. In this reaction keto-thioketo carbenes were postulated as intermediates, which could be trapped as products of type **70** when the isomeric chloro-mercapto-pyridazinones were heated with phenyl isothiocyanate in the presence of triethylamine. The same product (**70**) can be obtained from 4,5-dimercaptopyridazinones and phenylcarbamyl dichloride.

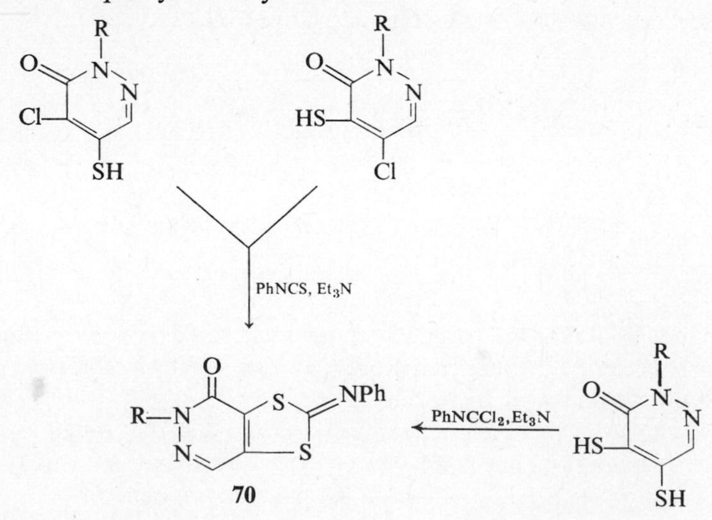

IX. Tables

TABLE 3H-1. Total and Frontier π-Electron Densities of Thieno[2,3-*d*]pyridazine

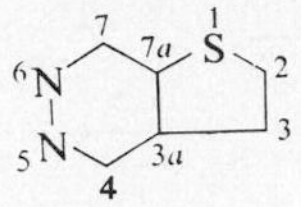

Position	1	2	3	3*a*	4	5	6	7	7*a*
Total	1.2948	1.1134	1.1606	1.0538	0.9402	1.2032	1.2084	0.9347	1.0909
Frontier	0.4060	0.0009	0.2508	0.0527	0.0408	0.0839	0.0742	0.0477	0.0429

TABLE 3H-2. Total and Frontier π-Electron Densities of Thieno[3,4-*d*]pyridazine

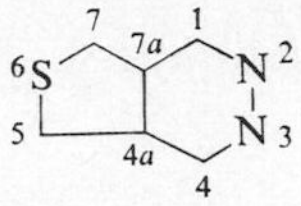

Position	1	2	3	4	4*a*	5	6	7	7*a*
Total	0.9451	1.2086	1.2086	0.9451	1.0545	1.2233	1.1370	1.2233	1.0545
Frontier	0.0612	0.0899	0.0899	0.0612	0.0429	0.3059	0.0000	0.3059	0.0429

TABLE 3H-3. Total and Frontier π-Electron densities of Thiazolo[4,5-*d*]pyridazine

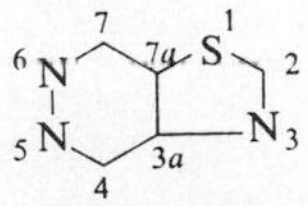

Position	1	2	3	3*a*	4	5	6	7	7*a*
Total	1.2593	1.0343	1.3478	1.0424	0.9405	1.1843	1.1912	0.9326	1.0677
Frontier	0.4344	0.0412	0.1961	0.0442	0.0464	0.0899	0.0613	0.0670	0.0196

TABLE 3H-4. Total and Frontier π-Electron Densities of Thiazolo[5,4-*c*]pyridazine

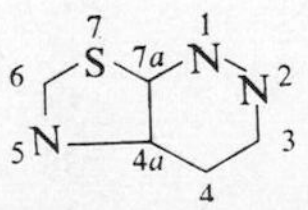

Position	1	2	3	4	4*a*	5	6	7	7*a*
Total	1.1547	1.1506	1.0035	0.9509	1.0335	1.3677	1.0183	1.2962	1.0260
Frontier	0.1034	0.0259	0.0770	0.0700	0.0231	0.2176	0.0268	0.4356	0.0210

TABLE 3H-5. Total and Frontier π-Electron Densities of 1,2,5-Thiadiazolo[3,4-*d*]-pyridazine

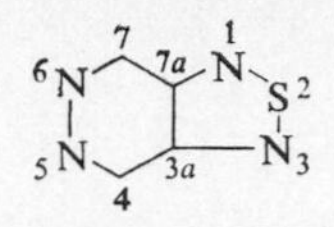

Position	1	2	3	3*a*	4	5	6	7	7*a*
Total	1.3549	0.9859	1.3549	1.0301	0.9426	1.1794	1.1794	0.9426	1.0301
Frontier	0.2697	0.0000	0.2697	0.0123	0.1157	0.1023	0.1023	0.1157	0.0123

TABLE 3H-6. Thieno[2,3-*d*]pyridazines

R	MP (°C) or BP (°C/mm	References
None	167	4, 7
	HCl, 198	
	MeI, 225	
Methyl	130	8
	131	4
	HCl, 225	
	MeI, 192	
Ethyl	58	4, 5
Iodo	204	8
	205	5
Methoxy	123	5, 8
Ethoxy	126	5, 8
n-Propoxy	79	5, 8
Iso-Propoxy	103	5, 8
n-Butoxy	84	5, 8
Phenoxy	153	8
Chloro	161	5, 7, 8
Amino	215	8
$NHCOCH_3$	238	8
$NHSO_2C_6H_5$	208	8
NHEt	175	8
$—OCH_2CH_2NEt_2$	2 MeI, 226	8
$—OCH_2CH_2N$	2 MeI, 241	8

TABLE 3H-6 (*continued*)

R	MP (°C) or BP (°C/mm)	References
$—OCH(Me)CH_2N$(pyrrolidinyl)	2 MeI, 215	5, 8
$—OCH_2COOH$	235	8
—N(pyrrolidinyl)	115	8
—N(piperidinyl)	MeI, 225	8
—N(morpholinyl) O	130	8
$—N(piperazinyl)N—CH_2C_6H_5$	155	8
Hydrazino	205	5, 8
	HCl, 282	5, 8
Methylthio	162	8
Ethylthio	74	8
Benzylthio	104	8
$—SCH_2COOH$	265	8
$—SCH_2COOMe$	130	8
$—SCH_2COOEt$	125	8
$—SCH_2CH_2CN$	194	8
$—NHN{=}CHC_6H_5$	210	5, 8
$—NHN{=}CMe_2$	211	5, 8
$—NHNHCONH_2$	321	5, 8
—NHN=CH—(4-OH-3-OMe-phenyl) OH, OMe	128	5
—NHN=CH—(2-furyl)	211	5
—NHN=CH—(2-thienyl)	217	5
$—NHN{=}C(Me)Et$	85	5
$—NHN{=}C(Me)C_6H_5$	170	5

TABLE 3H-6 (*continued*)

R	MP (°C) or BP (°C/mm)	References
—NHN=CEt $\vert$ C_6H_4OH-*p*	202	5
—NHN=C— S $\vert$ Me	270	5
m-$NO_2C_6H_4$—CH=NNH—	255	5
p-$NO_2C_6H_4$—CH=NNH—	295	5
m-OHC_6H_4—CH=NNH—	244	5

TABLE 3H-7 Thieno[2,3-*d*]pyridazines

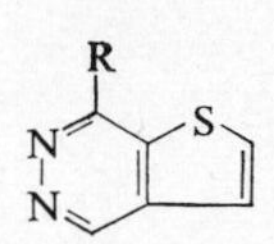

R	MP (°C) or BP (°C/mm)	References
Chloro	125	5, 7, 8
Methyl	67	4
	HCl, 225	4
	MeI, 235	4
Methoxy	130	5, 8
Ethoxy	114	5, 8
Phenoxy	121	8
Iodo	202	5, 8
COOH	265	4
COOMe	205	4
$CONH_2$	230	4
$CONEt_2$	63–64	4, 5
$CONHNH_2$	213	4
CN	157	4
$CSNH_2$	195	4
—$OCH_2CH_2NMe_2$	64	5, 8
—OCH_2COOH	270	8
Amino	218	5, 8
NHCOMe	211	8
$NHSO_2C_6H_5$	228	8

TABLE 3H-7 (*continued*)

R	MP (°C) or BP (°C/mm)	References
—N	144	8
—N	121	8
—N O	138	8
—N N—COOEt	130	8
—N N— Bu-*n*	87	8
Hydrazino	208	5, 8
	HCl 258	5, 8
$—NHN{=}CHC_6H_5$	275	5, 8
$—NHN{=}CMe_2$	157	5, 8
$—NHNHCONH_2$	250	5, 8
$—NHN(COOEt)_2$ or —N(COOEt)—NHCOOEt	165	8
$—NHN{=}C(Me)—C_6H_5$	204	5
—NHN=C(Me)— (S ring)	211	5
Methylthio	151	8
Ethylthio	90	8
Benzylthio	115	8
$—SCH_2COOH$	260	8
$—SCH_2COOMe$	114	8
$—SCH_2CONHNH_2$	202	8
$—SCH_2CH_2CN$	184	8
$—SCH_2CH_2N$ O	109	8
	2 MeI, 287	8

TABLE 3H-8. Thieno[2,3-*d*]pyridazine-4(5*H*)ones

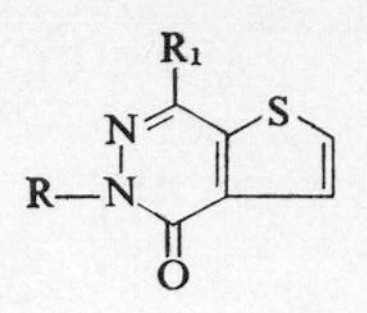

R	R_1	MP (°C) or BP (°C/mm)	References
None		239	8
		239–240	4, 5, 7
	Chloro	296	8
Methyl		135	4
Phenyl		88–89	4, 5
Benzyl		153	4, 5
p-$NO_2C_6H_4$		276	4, 5
CH_2OH		162	4
CH_2COOH		272	8
n-$PrCOOCH_2$		64	8
CH_2CH_2CN		137	4
$CH_2CH_2NMe_2$		MeI, 334	4, 30
CH_2CH_2—N		142, 119	5, 30
		MeI, 205	5
CH_2CH_2—N		142	5, 30
		MeI, 229	5
CH_2CH_2—N O		126	5, 30
—CH(Me)—CH_2—N O		140	5, 30
CH_2CH_2—N		90, 92	5, 30
$(CH_2)_3NMe_2$		MeI, 294	5, 30
$(CH_2)_3NEt_2$		MeI, 193	5, 30
—$CH_2CH_2NEt_2$		MeI, 210	30
—$CH_2CH_2N(C_4H_9)_2$		MeI, 178	30
—$(CH_2)_3$N		MeI, 205	30

TABLE 3H-8 (*continued*)

R	R_1	MP (°C) or BP (°C/mm)	References
$-(CH_2)_3N$ (six-membered ring)		MeI, 229	30
$-CH(CH_3)-CH_2N$ (five-membered ring)		MeI, 258	30
$-CH(CH_3)-CH_2N$ (seven-membered ring)		MeI, 207	30

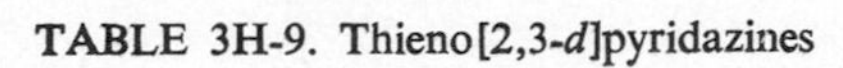

TABLE 3H-9. Thieno[2,3-*d*]pyridazines

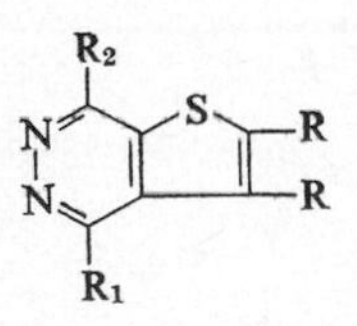

R	R_1	R_2	MP (°C) or BP (°C/mm)	References
Chloro			194–195	6
	Methyl	Chloro	149	8
	Methyl	Hydrazino	206	8
	$-OCH_2CH_2NMe_2$	$-OCH_2CH_2NMe_2$	2 HCl, 270	5, 8
	Amino	Amino	265[a]	8
	C_6H_5NH	Chloro	185	5, 8
	Hydrazino	Chloro	256	5, 8
	Hydrazino	Bromo	265	8
	$NHN{=}CMe_2$	Bromo	152	8
	$NHCH_2CH_2NEt_2$	Chloro	116	8
			2 HCl, 258	5, 8
	Chloro	Chloro	157–158	5, 7
			158	8
	Bromo	Bromo	181	5, 7, 8
		(thieno[2,3-*d*]pyridazine structure: N, N, S, —Br)	154 MeI, 251	4, 5, 7

[a] Triacetyl der. 231° C; MeI 287° C; tribenzoyl der. 270° C.

TABLE 3H-10. Thieno[2,3-*d*]pyridazine-7(6*H*)ones

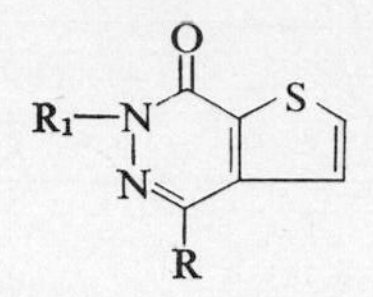

R	R_1	MP (°C) or BP (°C/mm)	References
None		220–221	4, 5, 7
Methyl		290	4, 5
Bromo		255	5
	Methyl	114	4
	Phenyl	154	4
		153–154	5
	Benzyl	106	4, 5
	CH_2OH	140	4
	CH_2Br	140	4
	CH_2COOH	215	8
	CH_2COOMe	117	8
	CH_2COOEt	93	8
	n-$PrCOOCH_2$	45	8
	CH_2CH_2OH	131	4
	$CH_2CH_2OCOCH_3$	48	4
	CH_2CH_2CN	120	4
	$CH_2CH_2CH_2$—N	MeI, 208	5, 30
	CH_2CH_2—N	98	4, 5, 30
	—$CH_2CH_2NMe_2$	MeI, 310	30
	—$CH_2CH_2NEt_2$	MeI, 252	30
	—$CH_2CH_2N(C_4H_9)_2$	MeI, 170	30
	—$(CH_2)_3NMe_2$	MeI, 256	30
	—$(CH_2)_3NEt_2$	MeI, 180	30
	—CH_2CH_2—N	MeI, 200	30
	—CH_2CH_2N	MeI, 192	30
	—CH_2CH_2N O	120	30

TABLE 3H-10 (*continued*)

R	R_1	MP (°C) or BP (°C/mm)	References
	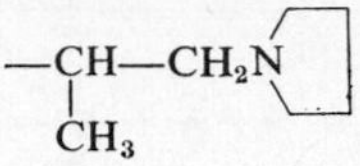 $-CH(CH_3)-CH_2N$	MeI, 229	30
	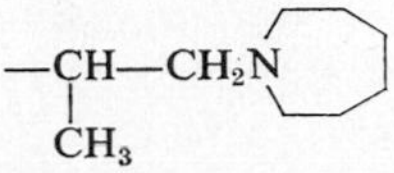 $-CH(CH_3)-CH_2N$	MeI, 186	30
	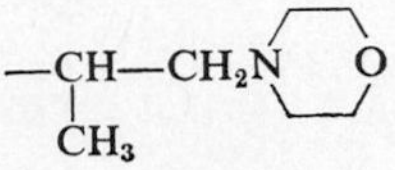 $-CH(CH_3)-CH_2N$ O	MeI, 253	30
	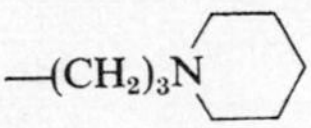 $-(CH_2)_3N$	MeI, 223	30

TABLE 3H-11

Structure		MP (°C) or BP (°C/mm)	References
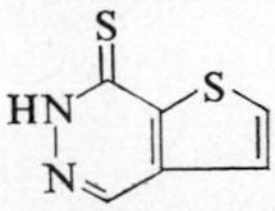 S, HN, N, S		219	8
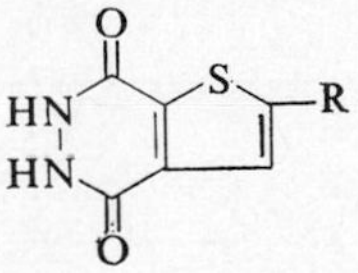 O, HN, HN, O, S, R	R = H	>300	2
		331–333 (dec.)	1
	R = Methyl	294–295	3, 36
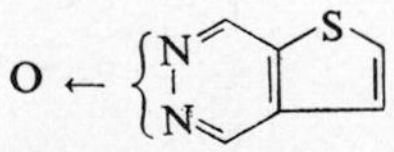 O ← {N, N}, S		183	4
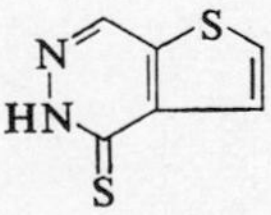 N, HN, S, S		200	8

TABLE 3H-12. Thieno[3,4-*d*]pyridazines

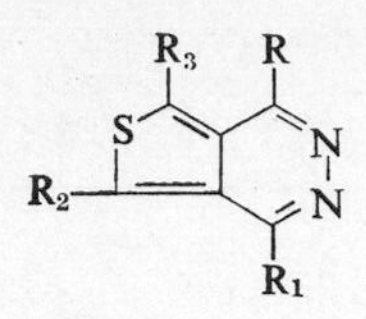

R	R_1	R_2	R_3	MP (°C) or BP (°C/mm)	References
None				136	13
				135–136	4, 5
				HCl, 175	4, 5
				MeI, 182	4, 5
		Methyl	Methyl	146–147	14
Chloro	Chloro			169	5, 8, 13
Chloro	Methoxy			139	5, 8
Chloro	Hydrazino			236	5, 8
Chloro	C_6H_5NH			186	5
Chloro	*p*-MeC_6H_4NH			216	5
C_6H_5NH	C_6H_5NH			217	5, 8
Chloro	$Et_2NCH_2CH_2NH$—			90	5, 8

TABLE 3H-13. Thieno[3,4-*d*]pyridazine-1(2*H*)ones

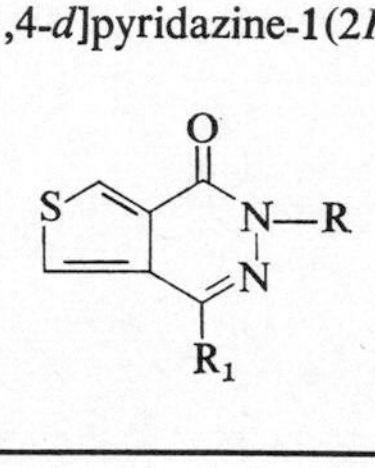

R	R_1	MP (°C) or BP (°C/mm)	References
None		252	8
		252–253	4, 5
		253	13
	Chloro	259	5, 8, 13
	Bromo	255	8, 13
Phenyl		102	4, 5
$—CH_2CH_2—N$ (ring)		221	4, 30

TABLE 3H-13 (*continued*)

R	R_1	MP (°C) or BP (°C/mm)	References
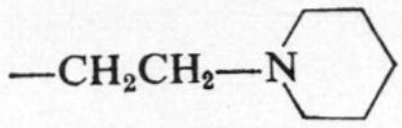 —CH_2CH_2—N		MeI, 287	5
—$(CH_2)_3NMe_2$		MeI, 261	5, 30
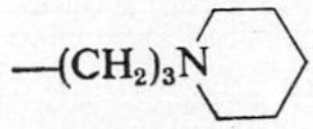 —$(CH_2)_3$N		MeI, 208	30
—$(CH_2)_2NMe_2$	Chloro	MeI, 274	30
—$(CH_2)_2NEt_2$	Chloro	MeI, 228	30
—$(CH_2)_3NMe_2$	Chloro	MeI, 239	30
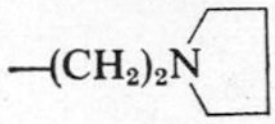 —$(CH_2)_2$N	Chloro	MeI, 202	30
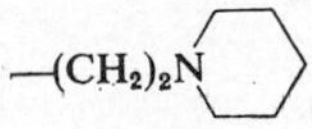 —$(CH_2)_2$N	Chloro	MeI, 287	30
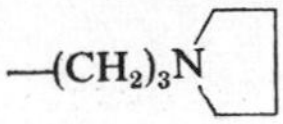 —$(CH_2)_3$N	Chloro	MeI, 155	30
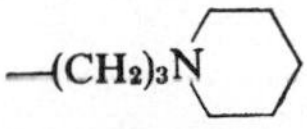 —$(CH_2)_3$N	Chloro	MeI, 225	30

TABLE 3H-14

Structure	R	MP (°C) or BP (°C/mm)	References
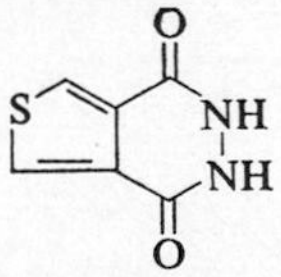 S, O, NH, NH, O		328–330	3
R, S, O, NH, NH, O	H	135–135.5	12
	Methyl	81–82	11

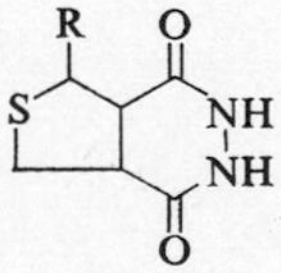

TABLE 3H-15. Thiazolo[4,5-*d*]pyridazines

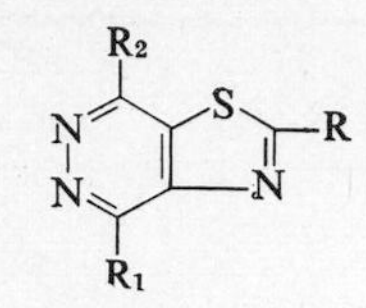

R	R_1	R_2	MP (°C) or BP (°C/mm)	References
None			202	18, 31
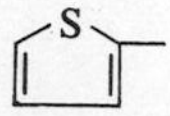			194	18, 31
	Chloro	Chloro	172	18, 31
	Bromo	Bromo	226	18, 31
Methyl	Methoxy	Methoxy	169	18, 31
Methyl	Chloro	Chloro	150–152	22
Methyl	SCN	SCN	136–138 (dec.)	22
Ethyl	Chloro	Chloro	63	18, 31
Phenyl	Chloro	Chloro	199–200	22
Phenyl	Bromo	Bromo	196	18, 31
Phenyl	Iodo	Iodo	197	18
Phenyl	Ethoxy	Ethoxy	156	18, 31
Phenyl	SCN	SCN	185–187 (dec.)	22
Phenyl	Methylthio	Methylthio	196	18, 31
Phenyl	Benzylthio	Benzylthio	159	18, 31
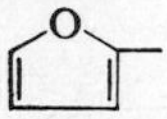	Chloro	Chloro	185	18, 31
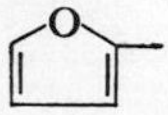	Ethoxy	Ethoxy	176	18, 31
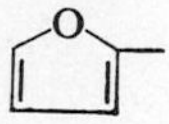	Methylthio	Methylthio	183	18, 31
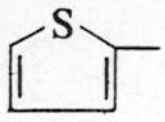	Bromo	Bromo	195	18, 31
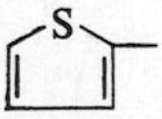	Iodo	Iodo	205	18
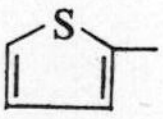	Methoxy	Methoxy	197	18, 31
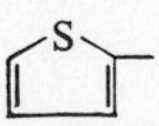	Ethylthio	Ethylthio	133	18, 31
	$—SCH_2COOH$	$—SCH_2COOH$	127	18, 31

TABLE 3H-15 (*continued*)

R	R_1	R_2	MP (°C) or BP (°C/mm)	References
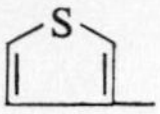	Chloro	Chloro	212	18, 31
	Iodo	Iodo	210	18
	Ethoxy	Ethoxy	171	18, 31
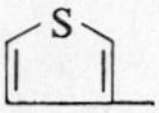	Benzylthio	Benzylthio	175	18, 31
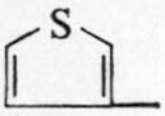	$-SCH_2CH_2CN$	$-SCH_2CH_2CN$	147	18, 31
Phenyl	Methoxy	Methoxy	184	31
	Mercapto	Mercapto	275	31
Ethyl	Mercapto	Mercapto	250	31
Phenyl	Ethylthio	Ethylthio	135	31
Phenyl	$-SCH_2COOEt$	$-SCH_2COOEt$	116	31
	Bromo	Bromo	207	31
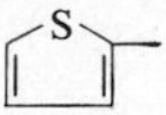	Chloro	Chloro	197	31
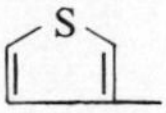	Bromo	Bromo	205	31
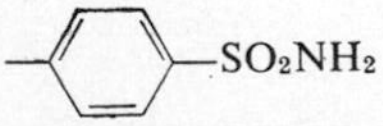 $-SO_2NH_2$	Chloro	Chloro	276	31
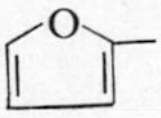	Methoxy	Methoxy	219	31
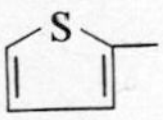	Ethoxy	Ethoxy	152	31
	Methoxy	Methoxy	212	31
Phenyl	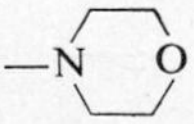 —N O	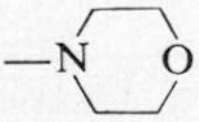 —N O	223	31
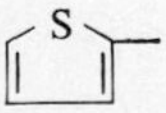	PhNH—	PhNH—	HCl, 195	31
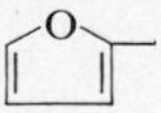	Mercapto	Mercapto	270	31

TABLE 3H-15 (*continued*)

R	R_1	R_2	MP (°C) or BP (°C/mm)	References
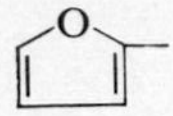	Ethylthio	Ethylthio	204	31
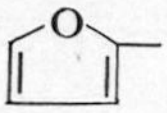	Benzylthio	Benzylthio	166	31
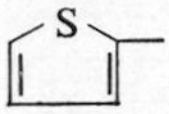	Mercapto	Mercapto	288	31
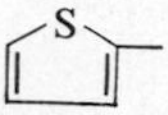	Methylthio	Methylthio	178	31
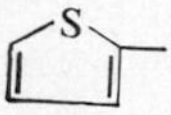	Benzylthio	Benzylthio	185	31
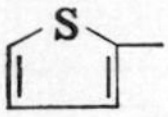	—SCH_2COOEt	—SCH_2COOEt	180	31
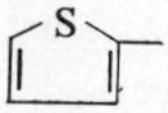	—SCH_2CH_2CN	—SCH_2CH_2CN	155	31
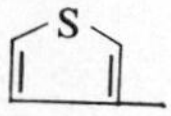	Mercapto	Mercapto	281	31
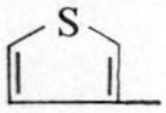	Methylthio	Methylthio	208	31
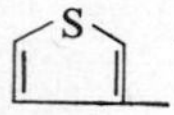	Ethylthio	Ethylthio	140	31
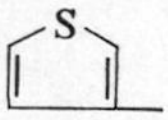	—SCH_2COOEt	—SCH_2COOEt	187	31

TABLE 3H-16. Thiazolo[4,5-*d*]pyridazine-7(6*H*)ones and Thiazolo[4,5-*d*]pyridazine-7(6*H*)thiones

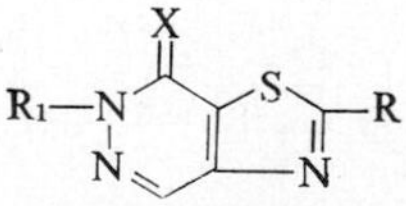

R	R_1	X	MP (°C) or BP (°C/mm)	References
Amino		S	>350 (dec.)	19
Methylthio	Phenyl	O	133–134	20

TABLE 3H-17. Thiazolo[4,5-*d*]pyridazine-4,7(5*H*, 6*H*)diones
or Thiazolo[4,5-*d*]pyridazine-4,7(5*H*. 6*H*)dithiones

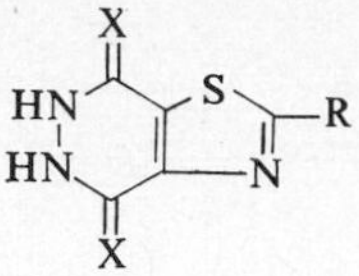

R	X	MP (°C) or BP (°C/mm)	References
	O	350	18, 31, 34
Ethyl	O	280, 283	18, 31, 34
Methyl	O	352–355 (dec.)	16
Phenyl	O	348.5–350.5, 350	15, 34
Amino	O	>330	17
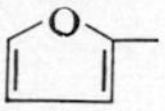	O	385	18, 31, 34
	O	350, >360	31, 34
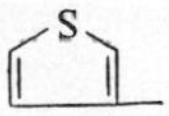	O	330	18, 31, 34
	S	275	18
Ethyl	S	250	18
Phenyl	S	250–257 (dec.)	22
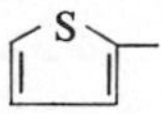	S	288	18
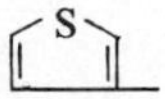	S	281	18
Methyl	S	243–245 (dec.)	22
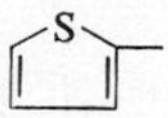	O	353	31, 34
NHCOH	O	350	34
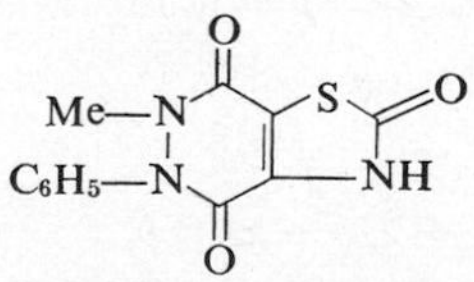		245–246	21
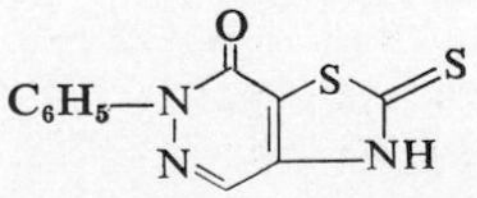		315–317	20

TABLE 3H-18. Thiazolo[5,4-*c*]pyridazine

	MP 290–295° C (dec.)	Ref. 23

TABLE 3H-19. Thiazolo[3,2-*b*]pyridazines

R	R_1	MP (°C) or BP (°C/mm)	References
$—NH^{\ominus}$	Methyl	·HCl, 160–163	25
		Picrate, 225 (dec.)	25
$—NH^{\ominus}$	Phenyl	·HCl, 217–220 (dec.)	25
$—\overset{\ominus}{N}COCH_3$	Methyl	·HCl, 270–274 (dec.)	26
$—\overset{\ominus}{N}COCH_3$	Phenyl	·HCl, 237	26
$—\overset{\ominus}{N}COCH_3$	Mercapto	·HCl, 243	26
$—O^{\ominus}$	Methyl	135 (dec.)	24
		$·H_2O$, 106 (dec.)	24
$—O^{\ominus}$	Methylthio	>300	24
$—O^{\ominus}$	Benzylthio	167–168	24
$—O^{\ominus}$	Chloro	>250	24
$—O^{\ominus}$	Phenyl	122 (dec.)	24
		$·H_2O$, 110 (dec.)	24

TABLE 3H-20

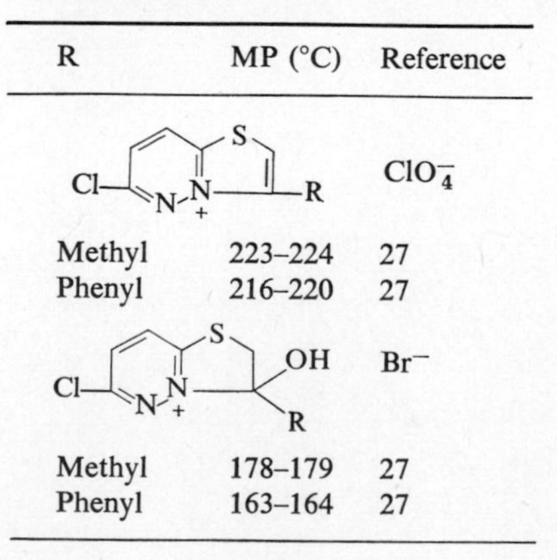

R	MP (°C)	Reference
		ClO_4^-
Methyl	223–224	27
Phenyl	216–220	27
		Br^-
Methyl	178–179	27
Phenyl	163–164	27

TABLE 3H-21. 1,2,5-Thiadiazolo[3,4-*d*]pyridazines

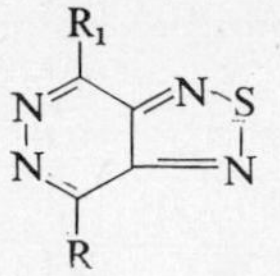

R	R$_1$	MP (°C)	References
Cl	Cl	115–130°	37
SMe	SMe	221–223, 220–221	37, 38
$NHNH_2$	SMe	245–250	37
$NHN{=}CHC_6H_5$	SMe	194	37
SMe		148–150	38
NH_2	NH_2	268 (dec.)	38
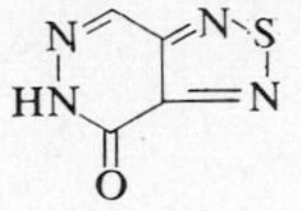		218	38
		HCl, 245, 243	37, 38
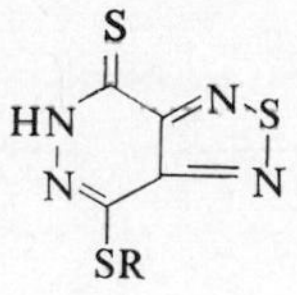	R = H	255–260	37
	$= CH_2COC_6H_5$	232–235	37
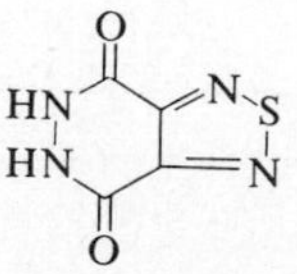		$\cdot H_2O$, 333 (dec.)	28
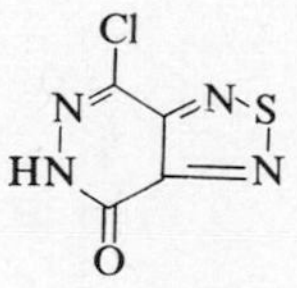		214–215	38
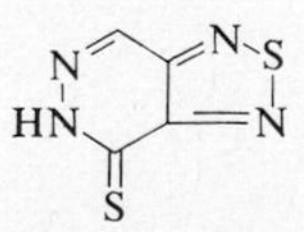		180–182°	38
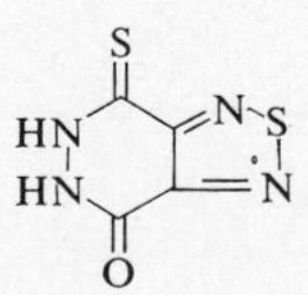		238–239°	38

TABLE 3H-21 (*continued*)

R	R_1	MP (°C)	References
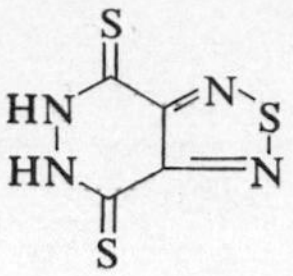 S, HN, HN, N, S, N, S		224 (dec.)	38
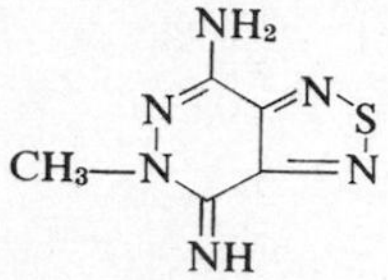 NH_2, N, CH_3—N, N, S, N, NH		209–210 (dec.)	38
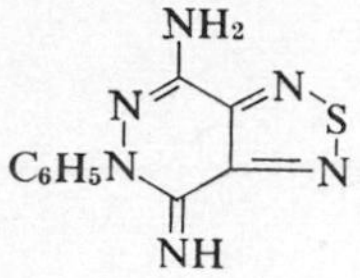 NH_2, N, C_6H_5N, N, S, N, NH		204–205 (dec.)	38

TABLE 3H-22. Thieno[3,2-*c*]pyridazines

Structure		MP (°C)	Reference
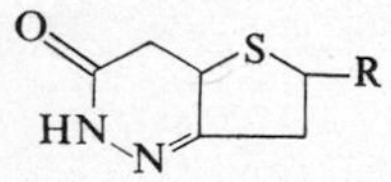 O, S, R, HN, N	R = H	136.5–137.5	33
	COOMe	119–120	33
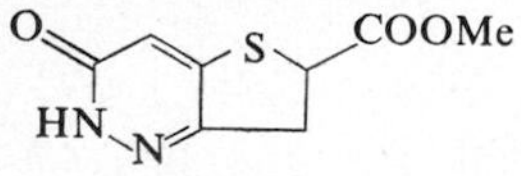 O, S, COOMe, HN, N		207–208	33
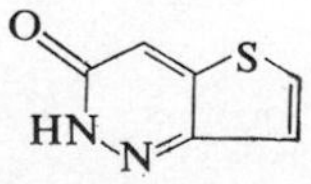 O, S, HN, N		H_2O, 219–220	33
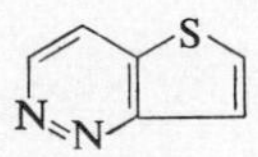 S, N, N		97.5–98.5	33

TABLE 3H-23. Pyridazo[4,5-*d*]-1,3-dithiols

R	MP (°C)	References
None	204	35
Methyl	198	35
Benzyl	145	35

References

1. B. R. Baker, J. P. Joseph, R. E. Schaub, F. J. McEvoy, and J. H. Williams, *J. Org. Chem.*, **18**, 138 (1953).
2. R. G. Jones, *J. Am. Chem. Soc.*, **77**, 4069 (1955).
3. R. G. Jones, *J. Am. Chem. Soc.*, **78**, 159 (1956).
4. M. Robba, B. Roques, and M. Bonhomme, *Bull. Soc. Chim. France*, **1967**, 2495.
5. French Pat. 1,453,897 (1966); *Chem. Abstr.*, **66**, 95066 (1967).
6. E. Profft and G. Solf, *J. Prakt. Chem.*, **24**, 38 (1964); *Chem. Abstr.*, **61**, 8254 (1964).
7. M. Robba, R. C. Moreau, and B. Roques, *Compt. rend.*, **259**, 4726 (1964).
8. M. Robba, B. Roques, and Y. Le Guen, *Bull. Soc. Chim. France*, **1967**, 4220.
9. B. Roques, Thèses, Faculté des Sciences de Paris, 1968.
10. Unpublished data from this laboratory. Parameters for LCAO Calculations were taken from A. Streitwieser *Molecular Orbital Theory for Organic Chemists*, Wiley, New York, 1961, p. 135.
11. B. R. Baker, M. V. Querry, S. R. Safir, and S. Bernstein, *J. Org. Chem.*, **12**, 138 (1947).
12. B. R. Baker, M. V. Querry, S. R. Safir, W. L. McEwen, and S. Bernstein, *J. Org. Chem.*, **12**, 174 (1947).
13. M. Robba, R. C. Moreau, and B. Roques, *Compt. rend.*, **259**, 3783 (1964).
14. A. V. Eltsov and A. A. Ginesina, *Zh. Org. Khim.*, **3**, 191 (1967).
15. E. H. Huntress and K. Pfister, *J. Am. Chem. Soc.*, **65**, 2167 (1943).
16. S. J. Childress and R. L. McKee, *J. Am. Chem. Soc.*, **73**, 3862 (1951).
17. T. S. Gardner, F. A. Smith, E. Wenis, and J. Lee, *J. Org. Chem.*, **21**, 530 (1956).
18. M. Robba and Y. Le Guen, *Compt. rend. C*, **263**, 1385 (1966).
19. T. Kuraishi and R. N. Castle, *J. Heterocyclic Chem.*, **1**, 42 (1964).
20. A. Pollak and M. Tišler, *Tetrahedron*, **21**, 1323 (1965).
21. G. Druey, K. Meier, and A. Staehelin, *Pharm. Acta Helv.*, **38**, 498 (1963).
22. J. Kinugawa, M. Ochiai, and H. Yamamoto, *Yakugaku Zasshi*, **83**, 767 (1963); *Chem. Abstr.*, **59**, 15286 (1963).
23. J. Kinugawa, M. Ochiai, and H. Yamamoto, *Yakugaku Zasshi*, **80**, 1559 (1960); *Chem. Abstr.*, **55**, 10461 (1961).
24. M. Ohta and K. Kishimoto, *Bull. Chem. Soc. Japan*, **34**, 1402 (1961); *Chem. Abstr.*, **57**, 7257 (1962).
25. Japanese Pat. 15,581 (1966); *Chem. Abstr.*, **66**, 10948 (1967).

26. Japanese Pat. 15, 582 (1966); *Chem. Abstr.*, **66,** 10949 (1967).
27. B. Stanovnik, M. Tišler, and A. Vrbanič, *J. Org. Chem.*, **34,** 996 (1969).
28. I. Sekikawa, *Bull. Chem. Soc. Japan*, **33,** 1229 (1960); *Chem. Abstr.*, **55,** 7425 (1961).
29. K. Dury, *Angew. Chem.*, **77,** 282 (1965).
30. M. Robba, B. Roques, and Y. Le Guen, *Chim. Ther.*, **3,** 413 (1968).
31. French Pat. 1,516,777 (1968).
32. I. Sekikawa, *J. Heterocyclic Chem.*, **6,** 129 (1969).
33. A. J. Poole and F. L. Rose, *J. Chem. Soc. D, Chem. Commun.*, **1969,** 281.
34. M. Robba and Y. LeGuen, *Bull. Soc. Chim. France*, **1970,** 4317.
35. K. Kaji, M. Kuzuya, and R. N. Castle, *Chem. Pharm. Bull.* (*Tokyo*), **18,** 147 (1970).
36. M. Lora-Tamayo, J. L. Soto, and E. D. Toro, *Anal. Real Soc. Espan. Fis. Quim.* (*Madrid*) *Ser. B*, **65,** 1125 (1969).
37. J. Marn, B. Stanovnik, and M. Tišler, *Croat. Chem. Acta*, **43,** 101 (1971).
38. D. Pichler and R. N. Castle, *J. Heterocyclic Chem.*, **8,** 441 (1971).

Part I. Pyridopyridazines

I. Pyrido[2,3-*d*]pyridazines

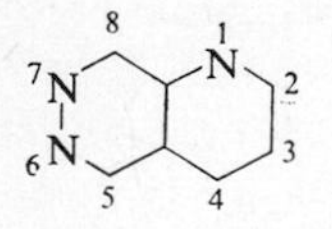

Pyrido[2,3-*d*]pyridazine, RRI 1610 (1,6,7-Triazanaphthalene, 4,6,7-Triazanaphthalene, 1,6,7-Pyridopyridazine, 5-Azaphthalazine, 1,6,7-Benzotriazine)

The preparation of the first compound belonging to this bicyclic system dates back to 1893 when Rosenheim and Tafel (1) treated 2-acetylnicotinic acid with phenylhydrazine hydrochloride. However, we are not yet certain if the product is a derivative of pyrido[2,3-*d*]pyridazine, since the meager data available (low melting point, nitrogen analysis only) do not exclude the possibility of hydrazone formation.

Nevertheless, the synthetic principle has remained useful to date and different pyridines with functional groups attached at positions 2 and 3 have been used as starting material. 2,3-Diformylpyridine reacted with hydrazine in the absence or presence of ethanol as a solvent to form the parent pyrido-[2,3-*d*]pyridazine **1** (2, 3). Similarly, the 5-deutero analog was prepared (74) and other 2,3-diacylpyridines reacted in this manner (80).

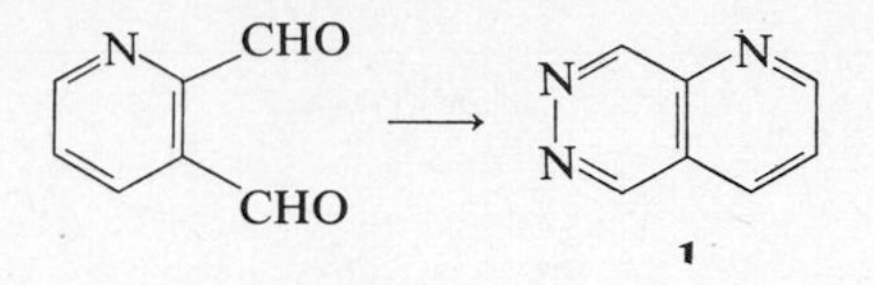

Another reaction, based on earlier observations that it is possible to transform aromatic *o*-dinitriles with hydrazine into cyclic products, was applied to pyridine-2,3-dinitrile. This compound, when heated with an equimolecular quantity of hydrazine in the presence of acetic acid for 10 min afforded 5,8-diaminopyrido[2,3-*d*]pyridazine (**2**) in 77% yield (3). A possible mechanism for this conversion has been outlined. With excess hydrazine the 5,8-di-

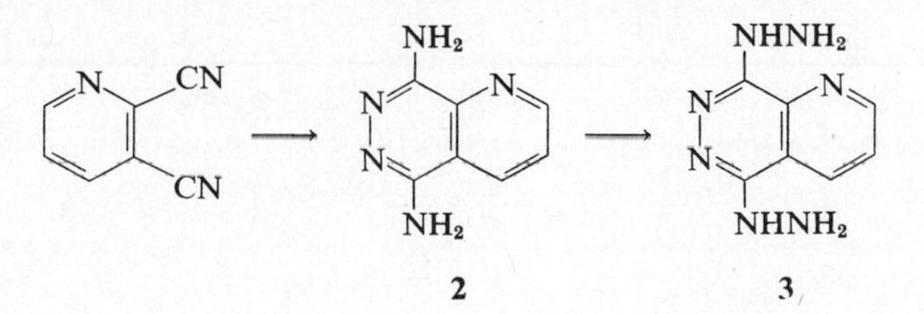

hydrazino analog **3** could be prepared (4). This reaction has been shown to be a nucleophilic displacement of the amino groups by hydrazine since in separate experiments almost quantitative conversion of **2** into **3** was demonstrated (3).

From 2-acylnicotinic acids and substituted hydrazines the corresponding pyrido[2,3-*d*]pyridazin-5(6*H*)ones are available. In addition to the reaction of Rosenheim and Tafel (1) and use of ethyl 2-formylpicolinate (81) there is another example by Wibaut and Boer (5), describing the reaction between 2-acetylnicotinic acid and *p*-nitrophenylhydrazine. It is also possible to use the lactone of 2-hydroxymethylnicotinic acid (**4**), as shown by Kakimoto and Tonooka (6). The lactone, when brominated with *N*-bromosuccinimide, yielded the bromo-lactone **5**, which was transformed into **6** upon heating with

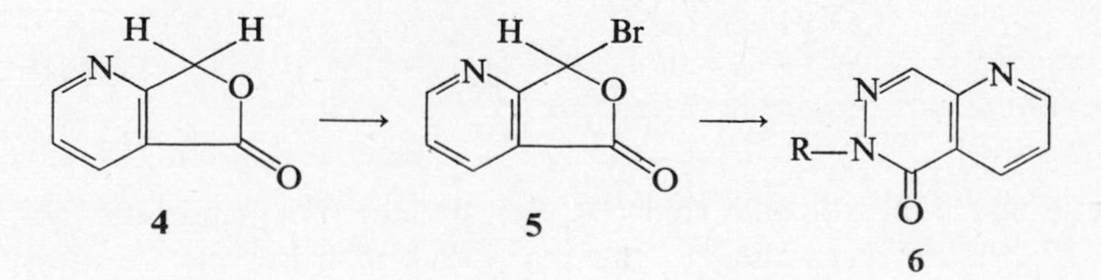

a solution of hydrazine or substituted hydrazines in ethanol. A similar reaction has been reported to occur between the lactone **7** and *p*-nitrophenylhydrazine (7).

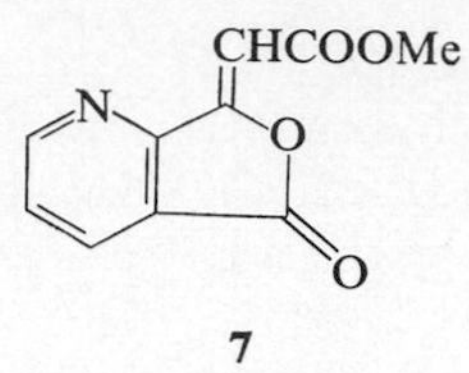

3-Acylpicolinic acids and related 2-carboxy-3-pyridineglyoxylic acid could be converted into the corresponding pyrido[2,3-*d*]pyridazin-8(7*H*)ones. 3-Benzoylpicolinic acid has been transformed with hydrazine into **8** (R = C_6H_5, R_1 = H) (8–11), and 3-formylpicolinic acid formed **8** (R = H, R_1 = H or C_6H_5) in the reaction with hydrazine or directly from its phenylhydrazone (12).

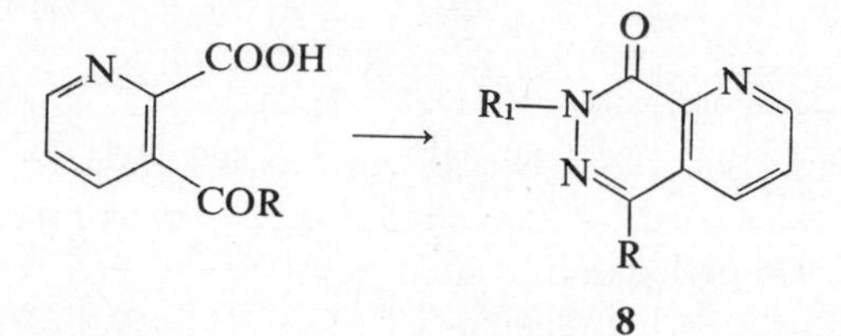

The reaction between 2-carboxy-3-pyridineglyoxylic acid and hydrazines has been investigated in detail by two Japanese groups (6, 13). The acid, when treated with hydrazine at room temperature, formed the hydrazone **9**, the same reaction at 100° afforded the cyclic product **10**. Both the hydrazide and the cyclic product **10** are interconvertible in hot water and the hydrazone could be cyclized to the methyl ester of **10** when heated under reflux in methanol and in the presence of sulfuric acid (13). In a like manner the reaction with methyl- or phenylhydrazine took place to give the 7-substituted analogs of **10** (6, 13).

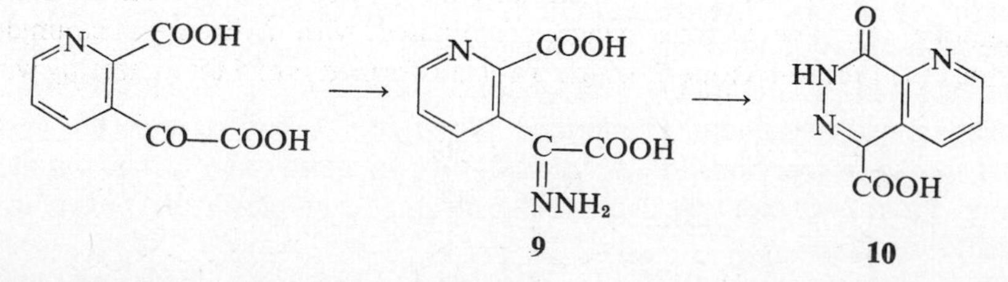

Pyridine-2,3-dicarboxylic acid (14, 82), or more frequently its anhydride (13, 15–19), has been used in the reaction with hydrazine or substituted hydrazines for the synthesis of the corresponding pyrido[2,3-*d*]pyridazine-5,8(6*H*, 7*H*)diones. In some cases esters (18, 20, 75) or the dichloride of pyridine-2,3-dicarboxylic acid (13, 21) have also been used as starting material.

It is known (22) that the anhydride of pyridine-2,3-dicarboxylic acid can form with phenylhydrazine either a monophenylhydrazide or a bis-phenylhydrazide, which on heating failed to cyclize. However, if the reaction was conducted in the presence of glacial acetic acid and the mixture heated under reflux, the cyclic hydrazide could be obtained without difficulty. Two isomers (**11** and **12**) are theoretically possible, but only one has been isolated. On chemical grounds structure **11** was proven to be correct (13).

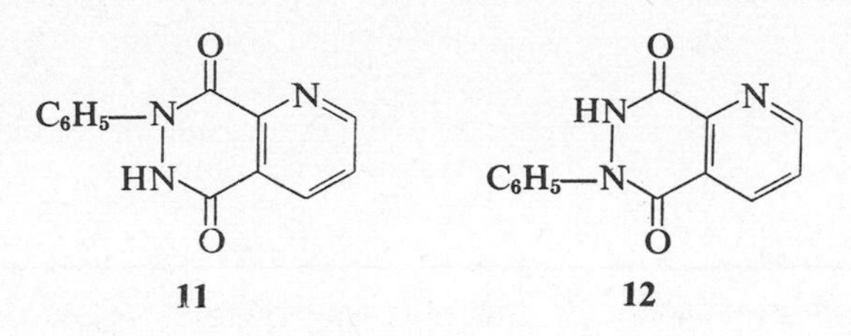

For the synthesis of 6,7-disubstituted analogs pyridine-2,3-dicarboxylic acid dichloride was used. When condensed with hydrazobenzene in the presence of *N*,*N*-dimethylaniline in a solution of benzene, this afforded 6,7-diphenylpyrido[2,3-*d*]pyridazine-5,8(6*H*, 7*H*)dione (**13**) (13, 21).

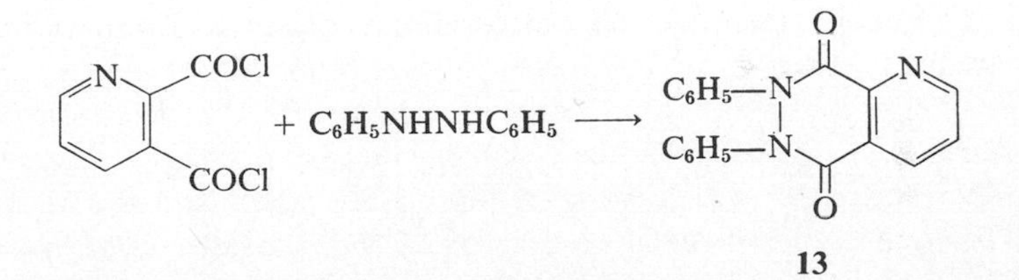

A pyridopyridazinone with the reduced pyridazine part of the molecule was obtained recently in a reaction between ethyl 2-bromomethylnicotinate hydrobromide and hydrazine (83). Here two products (**13a** and **13b**) are formed in the ratio of about 4:1. Upon treatment with alkali **13b** is converted into **13a** and this compound was obtained also from **13c** by reduction with amalgamated aluminum. Compound **13a** is easily oxidized to **6** (R = H).

It should be mentioned that pyrido[2,3-*d*]pyridazines also can be obtained from their *N*-oxides by deoxygenation, for example, with phosphorus trichloride (23).

The synthetic principles for the formation of the fused pyridazine ring of pyrido[2,3-*d*]pyridazines could be equally well applied to the synthesis of

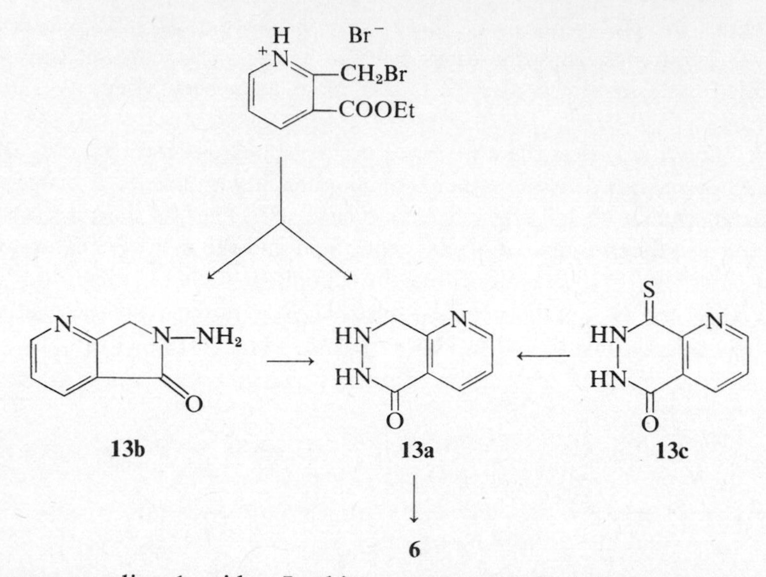

the corresponding 1-oxides. In this manner pyrido[2,3-*d*]pyridazine 1-oxide (**14**) was obtained from 2,3-diformylpyridine *N*-oxide (24) and 2-acetylnicotinic acid *N*-oxide afforded the analogous **15** (23). It is most probable that the "*p*-nitrophenylhydrazone anhydride" of 2-acetylnicotinic acid-oxide prepared by Bain and Saxton (25) is in fact the 6-*p*-nitrophenyl analog of **15**.

The parent compound, pyrido[2,3-*d*]pyridazine, was obtained either by direct cyclization from 2,3-diformylpyridine (2,3) or from substituted

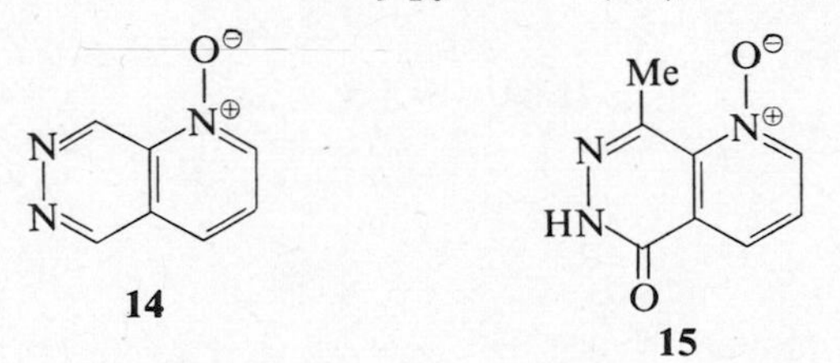

pyrido[2,3-*d*]pyridazines. Although Armarego (23) reported that 5,8-dichloropyrido[2,3-*d*]pyridazine could not be dehalogenated, Paul and Rodda (3) were able to obtain pyrido[2,3-*d*]pyridazine in 60–70% yield by dehalogenation of the 5,8-dichloro analog with hydrogen in the presence of palladized carbon and ammonium hydroxide. The presence of the latter is required for hydrogenolysis to proceed. It is also possible to obtain pyrido-[2,3-*d*]pyridazine either from its 5,8-dihydrazino analog by treating it with a suspension of yellow mercuric oxide in water, or from pyrido[2,3-*d*]pyridazine-5,8(6*H*, 7*H*)dithione by catalytic desulfurization with Raney nickel (3), although this results in poor yield.

Pyrido[2,3-*d*]pyridazine represents a stable heteroaromatic system. It is a weak base (pK_a 2.01), is monobasic, and forms salts with mineral and picric acids. It is stable to warm alkali and does not change upon heating with 65% oleum. The nmr and uv spectra in different solvents have been recorded (2, 3) and the electronic spectra have been calculated (76). The chemical shifts of ring protons in the nmr spectrum [in $CDCl_3$; $\tau = 0.67$ (H_2), 2.15 (H_3), 1.62 (H_4), 0.33 (H_5), and 0.17 (H_8)] are in accordance with MO calculations of the π-electron densities, with the exception of the C_8 position.

Although the site of protonation was not firmly established, it is probable, as indicated by calculated π-electron densities, that position 1 is favored. From the results of examination of 8-methylpyrido[2,3-*d*]pyridazine it was concluded that neither this nor the parent compound are covalently hydrated to any marked degree (3, 23).

The action of methyl iodide (at room temperature) or methyl lithium (at 0°) on pyrido[2,3-*d*]pyridazine has been investigated (26, 84). Upon hydrolysis of the reaction mixture two compounds (**16** and **17**) were isolated in a ratio of 4:1 and these were dehydrogenated into compounds **18** and **19**. The structures of both products were assigned on the basis of ir and nmr spectroscopic evidence.

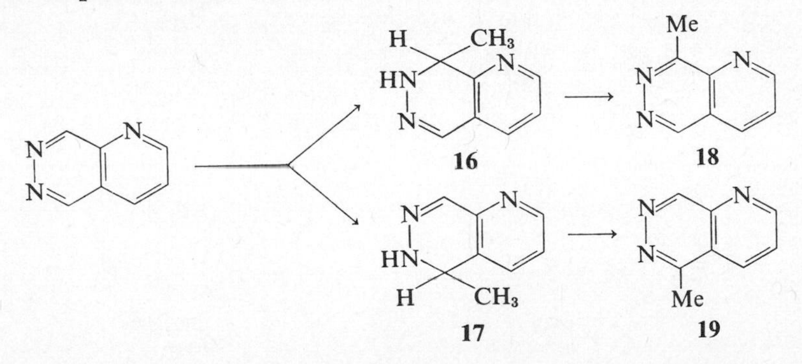

Pyrido[2,3-*d*]pyridazine was oxidized with monoperphthalic acid into a mixture of mono-*N*-oxides (26, 86). Since both products were different from the 1-oxide (**14**) prepared earlier by cyclization and because of nmr spectra they were assigned structures **20** and **21**, respectively (26, 86).

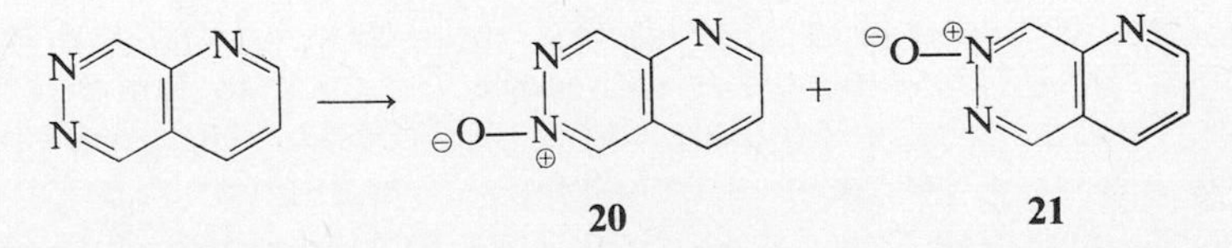

Energy levels, wave functions, and total and frontier π-electron densities for pyrido[2,3-*d*]pyridazine have been calculated by simple MO method

(27, 28). The values for total and frontier π-electron densities are listed in Table 3I-1.

This bicyclic system is not expected to be susceptible to electrophilic substitutions, although **1** afforded with bromine in CCl_4 a dibromide which on pyrolysis at 170–180° gave the 3-bromo derivative (86). Positions 2, 5, and 8 are expected to be attacked by nucleophiles. Similarly, MO calculations for 5,8-dichloropyrido[2,3-*d*]pyridazine (29) predict a greater displaceability of the 5-chlorine atom, due to resonance activation by pyridine nitrogen, and this has been verified experimentally.

The behavior of pyrido[2,3-*d*]pyridazine under electron impact has been published recently (30, 85). The compound fragments mainly by the stepwise loss of HCN, although 50% of the M-28 fragmentation mode accounts for the loss of nitrogen molecule.

For the most part, reactions performed on pyrido[2,3-*d*]pyridazines belong to displacement reactions of 5, 8-monochloro or 5,8-dichloro analogs. A direct amination at position 2 with sodium amide has been reported recently (86).

Pyrido[2,3-*d*]pyridazin-5(6*H*)ones and -8(7*H*)ones are *N*-alkylated with methyl or ethyl iodide at positions 6 and 7, respectively (8, 13). When heated with phosphorus pentasulfide in pyridine under reflux they are transformed into the corresponding pyrido[2,3-*d*]pyridazine-5(6*H*)thiones, -8(7*H*)thiones or -5,8(6*H*, 7*H*)dithione (3). 7-Phenylpyrido[2,3-*d*]pyridazine-5,8(6*H*, 7*H*)-dione was benzoylated and gave the corresponding benzoic ester (**22**) (13)

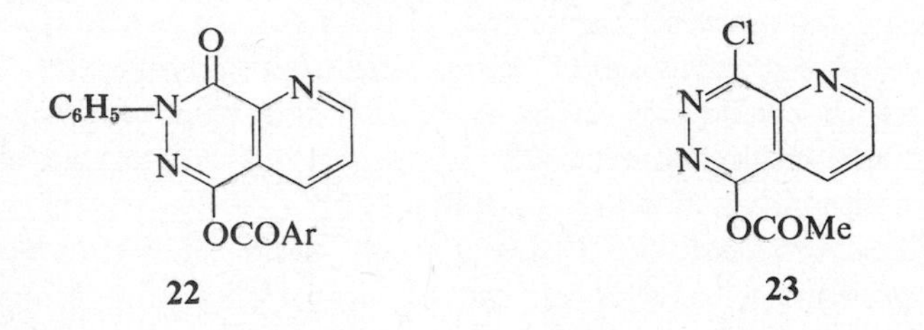

and in similar way acetylation of 8-chloropyrido[2,3-*d*]pyridazin-5(6*H*)one gave **23** (15).

Hydrolytic ring opening was observed during an attempt to hydrolyze ethyl or methyl 7-phenylpyrido[2,3-*d*]pyridazin-8(7*H*)one-5-carboxylate (13). Thus treatment of **44** (R = Et) with warm aqueous sodium hydroxide afforded a mixture of **26** (R = Et) and the known **25**. With cold aqueous sodium hydroxide only **26** (R = Et or Me) was obtained.

Several other ring-opening reactions have been reported (86). Pyrido-[2,3-*d*]pyridazine was oxidized with potassium permanganate under basic conditions into a mixture of 4-aminopyridazine-5-carboxylic acid as the

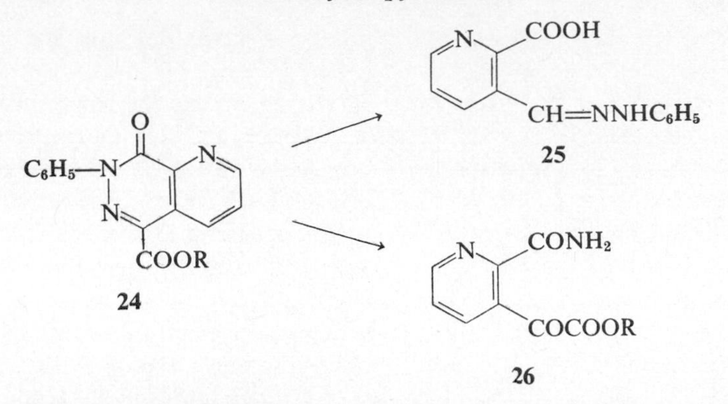

major product and quinolinic acid. Under acidic conditions a mixture of pyridopyridazinones and/or quinolinic acid could be obtained.

Pyrido[2,3-*d*]pyridazine-5,8(6*H*,7*H*)dione is slightly acidic (pK_a 6.1) and is capable of forming complexes with metal ions (31, 32). Its uv spectrum has been recorded (17, 33) and its chemoluminiscence investigated (14, 18). The cyclic hydrazide is stable against acids and alkali, it forms a diacetyl derivative, and under the influence of concentrated nitric acid it is decomposed into quinolinic acid (17). Whether one of both oxo groups in this structure may be enolized has not been firmly established. On the basis of complex formation ability, it is postulated that the 8-oxo group is enolized (16).

Substituted cyclic hydrazides of type **27** (R = *p*-nitrophenyl or 2,4-dinitrophenyl group) were transformed under the influence of alkali into the isomeric imides **28**, which could be rearranged quite easily back to **27** after treatment with acid (25% HCl) (16). The intermediate pyridine-2,3-dicarboxylic acid monohydrazide (**29**) was postulated as intermediate since its conversion into either **27** or **28** is possible.

Pyrido[2,3-*d*]pyridazin-5(6*H*)ones and -8(7*H*)ones are converted by the common phosphorus oxychloride method into the corresponding 5-chloro (23) or 8-chloro derivatives (9–11).

Whereas *N*-substituted pyrido[2,3-*d*]pyridazine-5,8(6*H*,7*H*)diones could afford only monochloro derivatives (13), the *N*-unsubstituted ones are upon treatment with phosphorus oxychloride transformed into 5,8-dichloro derivatives (23, 29, 34). The reaction is conducted preferentially in the presence of *N*,*N*-dimethylaniline since optimum yields are thus obtained.

Halopyrido[2,3-*d*]pyridazines show normal displaceability of the halogen(s) in nucleophilic substitution reactions. In this way, replacement of the 8-chlorine atom by a hydrazino (9–11, 15) or substituted amino group (35) and displacement of the 5-chlorine atom by an ethoxy (35) or hydrazino group (23) was performed. The reaction between 5-ethoxy-8-chloro- or

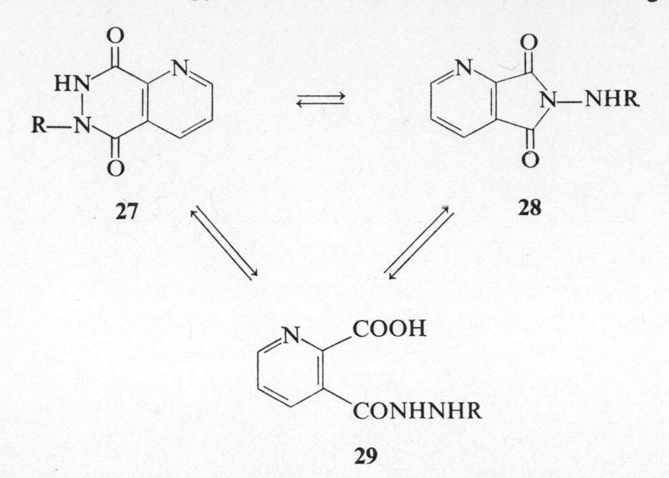

5-chloro-8-ethoxypyrido[2,3-*d*]pyridazine and 3-diethylaminopropylamine presented evidence for the preferential replacement of the ethoxy group to give **30** and **31**, respectively (35). 5- and 8-Chloropyrido[2,3-*d*]pyridazine

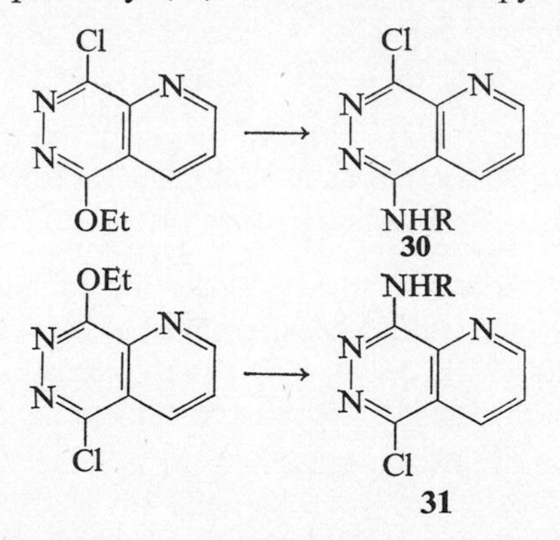

are reported to decompose on standing in air to an orange-brown solid of undetermined structure (87). Finally, it should be mentioned that attempts to replace the halogen in 5-chloropyrido[2,3-*d*]pyridazin-8(7*H*)one with ω-dialkylaminoalkylamines were unsuccessful (35).

As already mentioned, MO calculations predict a greater displaceability of the 5-chlorine atom in 5,8-dichloropyrido[2,3-*d*]pyridazine. In several cases the displacements proceeded accordingly, but, on the other hand, with a variety of nucleophiles mixtures of two isomers were formed. 5,8-Dichloropyrido[2,3-*d*]pyridazine (**32**) is stable toward water and weak acids at room temperature, but with hot alkalis or acids the hydrolytic elimination of halogen took place. Alkali hydrolysis afforded a mixture of **33** (about 70%) and **34** (about 30%) (29).

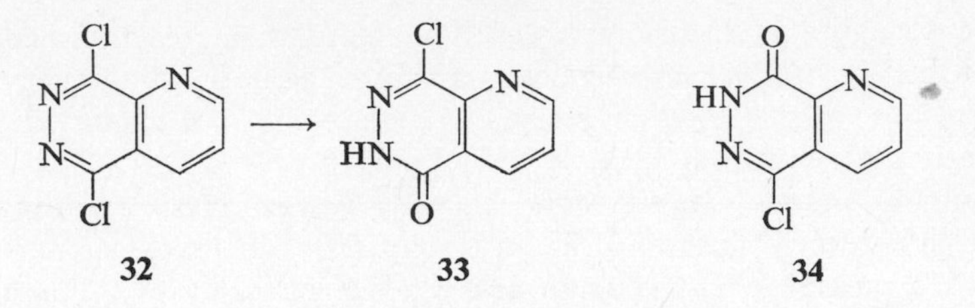

Contrary to an earlier report that **32** when heated in water only gave **33** (15), it was later found (29, 87) that both isomers, **33** and **34**, were formed in this reaction. However, these chloropyridopyridazinones possess an unreactive remaining chlorine atom which could not be displaced under the influence of alkali or hydrazine under normal conditions (87).

In the reaction between **32** and alkoxides the corresponding 5,8-dialkoxy or 5,8-diphenoxy derivatives were formed (29, 36). Furthermore, if only one equivalent of sodium ethoxide was applied, a mixturé of isomers was formed and only the 5-ethoxy derivatives could be isolated pure (29). Thiation of **32** according to the thiourea method and subsequent hydrolysis of the bis-thiuronium salt led to pyrido[2,3-*d*]pyridazine-5,8(6*H*,7*H*)-dithione (29, 37). Amination proceeded again with the formation of two isomers (**35** and **36**). With ammonia (100°, 5 hr) the 5-amino isomer was formed predominantly, whereas with an equimolar quantity of aniline the main product was the 8-anilino isomer. Similar results were observed with other amines employed for displacement reactions (29, 35, 38).

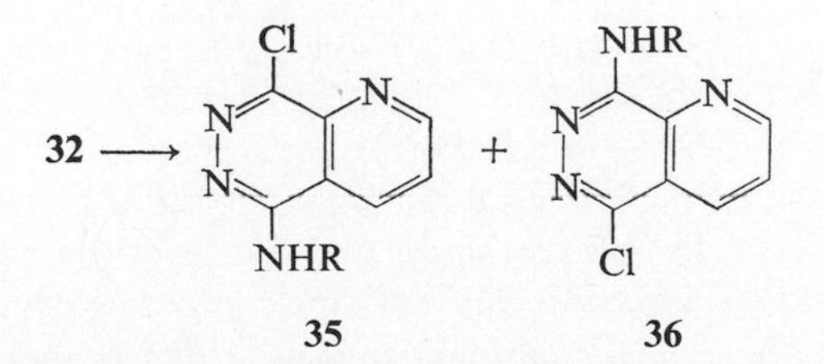

A preferential reactivity of the 5-chloro atom has been observed in the reaction between **32** and anthranilic acid where the tetracyclic compound **37** was isolated (39).

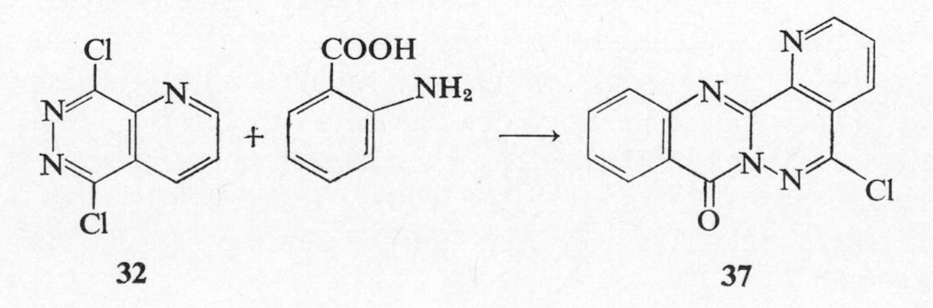

The structure of the isomers resulting from these reactions could be established by the use of chemical transformations or nmr spectroscopic correlations. With excess of amines, 5,8-diamino (or substituted amino) derivatives were formed.

A similar situation was encountered with hydrazinolysis experiments on **32**. With excess hydrazine hydrate in a solution of methanol compound **32** formed a mixture of 5-hydrazino and 8-hydrazinopyrido[2,3-*d*]pyridazine, the 5-isomer being obtained in somewhat higher yield (29, 40). Higher temperatures and prolonged reaction time (16 hr in boiling methanol) gave 5,8-dihydrazinopyrido[2,3-*d*]pyridazine (3). 5- and 8-Hydrazinopyrido-(2,3-*d*]pyridazine were prepared by hydrazinolysis of the corresponding methylthio derivatives (77).

By the action of nitrous acid, aminopyrido[2,3-*d*]pyridazines were converted into the corresponding pyrido[2,3-*d*]pyridazinones (29). The 5,8-diamino derivative could be also converted into the 5,8-dihydrazino compound with hydrazine (90°, 3 hr) and the reverse transformation took place under the influence of Raney nickel (100°, 3 hr) (3), although in moderate yield; 5- and 8-hydrazino derivatives also may be obtained from the corresponding methylthio derivatives (39). Amino and hydrazinopyrido[2,3-*d*]-pyridazines have been used as starting material for the preparation of polycyclic azaheterocycles (29, 41). Moreover, 5,8-dichloropyrido[2,3-*d*]-pyridazine reacted with sodium azide to form the more stable isomer of azidopyridotetrazolo[5,1-*b*]pyridazine (78).

Removal of the chlorine atom(s) and its replacement with hydrogen atom(s) in heterocyclic compounds is possible by several procedures, catalytic hydrogenolysis being the method of choice. Armarego (23) reported that 5-chloro-8-methylpyrido[2,3-*d*]pyridazine was resistant to catalytic dehalogenation and to the reaction with red phosphorus and hydriodic acid. A successful procedure, however, was the conversion of the chloro derivative with *p*-toluenesulfonylhydrazide (20°, 6 days) into the 5-*p*-toluenesulfonylhydrazino derivative, which could be decomposed with boiling 10% sodium carbonate solution into 8-methylpyrido[2,3-*d*]pyridazine (23). Catalytic hydrogenolysis of 5-chloropyrido[2,3-*d*]pyridazin-8(7*H*)one and its 7-phenyl analog over palladized charcoal resulted in the simultaneous reduction of the pyridine ring and the tetrahydro derivative **38** was isolated (13, 29).

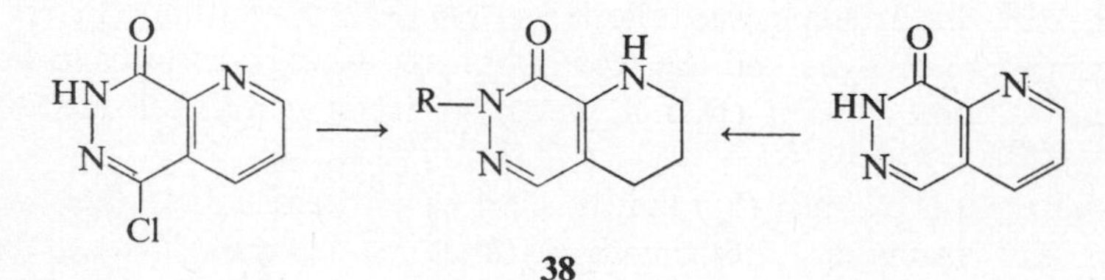

Compound **38** (R = H) could also be obtained from pyrido[2,3-*d*]pyridazin-8(7*H*)one by catalytic reduction in the presence of palladized carbon or platinum catalyst (6, 29, 87). The isomeric -5(6*H*)one and -5,8(6*H*, 7*H*)dione and the 6- or 7-substituted analogs behave similarly (6, 87). Even if the reaction was interrupted after 1 mole of hydrogen has been consumed, the tetrahydro compounds were isolated along with unreacted material. The reduced compounds could be acetylated; they form *N*-nitroso derivatives and their structure is in accord with the registered ir and nmr spectra (6).

In contrast to these observations, 5,8-dichloropyrido[2,3-*d*]pyridazine could be successfully dehalogenated over palladized carbon in the presence of ammonium hydroxide into pyrido[2,3-*d*]pyridazine (3).

Several possible applications of pyrido[2,3-*d*]pyridazines have been mentioned. Their usefulness in lowering the blood pressure (11, 10, 40, 42) as antispasmodics and muscle relaxants (21, 36), and as fungicides (19, 37, 38) is alleged. Tests for antitubercular action have also been reported (43, 44).

II. Pyrido[3,4-*d*]pyridazines

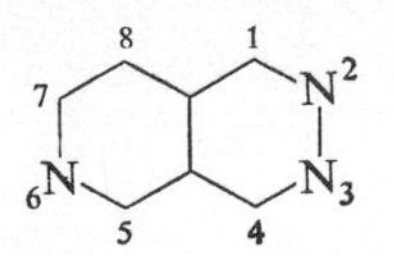

Pyrido[3,4-*d*]pyridazine, RRI 1611 (2,3,6-Triazanaphthalene, 2,6,7-Triazanaphthalene, 2,3,6-Pyridopyridazine, Pyridino-[3′;4′-4:5]pyridazine)

Although the structure is questionable, it may be possible that "monophenylhydrazide of cinchomeronic acid" obtained by Strache in 1890 (45) represents the first known representative of the pyrido[3,4-*d*]pyridazine system.

With few exceptions, pyrido[3,4-*d*]pyridazines are generally synthesized from pyridines which are substituted at positions 3 and 4 with functional groups and which are capable of cyclization with hydrazine. Most of the known procedures utilize pyridine-3,4-dicarboxylic acid (cinchomeronic acid), its esters, or anhydride (75) as starting material.

After the first report in 1912 by Meyer and Mally (46), who prepared the cyclic hydrazide **39** by heating the hydrazine salt of the acid-monohydrazide at 365–370°, this principle was followed by several other authors. Esters were preferentially employed and the reaction proceeded satisfactorily in boiling aqueous solution or using ethanol, glycol, or acetic acid as solvent (20, 33, 47–52, 82).

However, it is claimed (51) that in some cases the reaction takes another course. For example, 2,5,6-trimethylpyridine-3,4-dicarboxylic acid formed

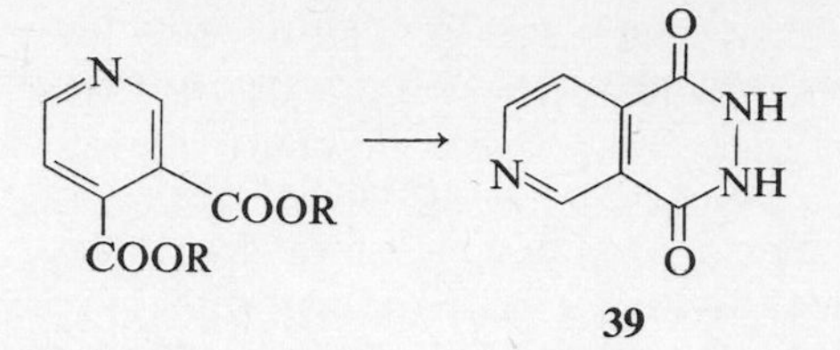

the isomeric *N*-aminoimide **40**. Prolonged heating and in particular excess hydrazine favor isomerization of **40** into **39**. Such transformations have been noted with other related imides (49, 53). Compounds of the type **39** behave

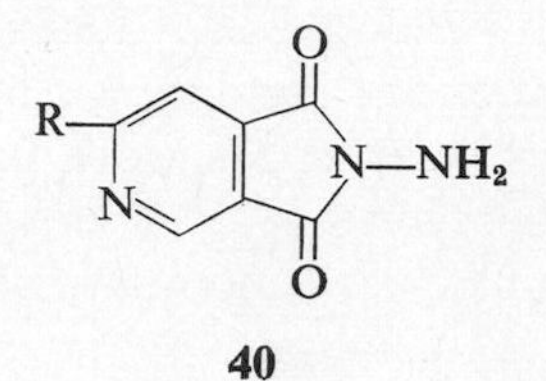

like monobasic acids, they are diacetylated (33), and their uv spectra have been recorded (33, 51).

By extending the reaction principle to 3-acylpyridine-4-carboxylic acids and their derivatives, pyrido[3,4-*d*]pyridazin-1(2*H*)ones (**41**) were obtained

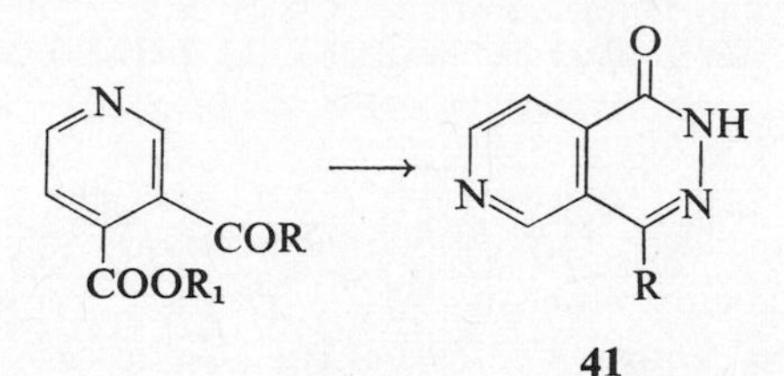

(45, 54, 55). Ethyl 3-cyanopyridine-4-carboxylate could be also used for the reaction with hydrazine and for the product structure **42** has been proposed (56–59).

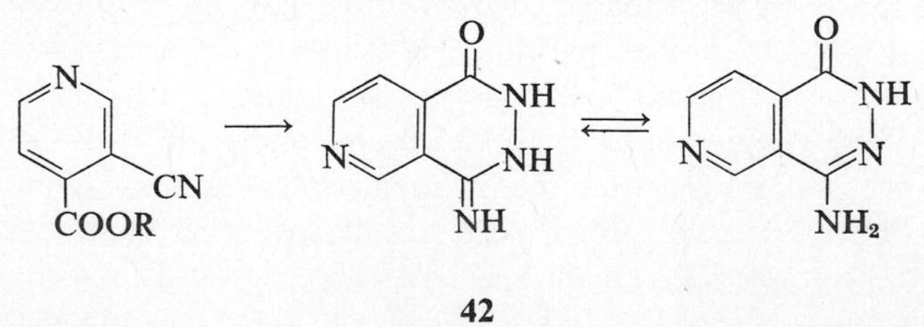

In contrast to the ready conversion of pyridine-2,3-dinitrile with hydrazine, pyridine-3,4-dinitrile could be converted in very low yield into 1,4-dihydrazinopyrido[3,4-*d*]pyridazine (3, 4).

Several pyrido[3,4-*d*]pyridazine derivatives have been reported to result from the action of hydrazine on several bicyclic or tricyclic heterocyles. It has been stated that pyrido[2,3-*d*]pyrimidines of type **43** are transformed into the pyrido[3,4-*d*]pyridazine system **44** (49, 50). Some tricyclic compounds

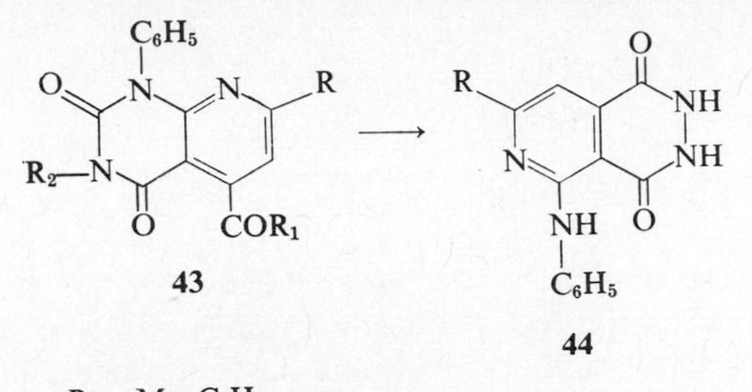

$R = Me, C_6H_5$
$R_1 = OH, OEt, NHNH_2$
$R_2 = H, Me$

may be obtained as byproducts. Alkali treatment of the tricycle **45** represents a similar reaction leading to **46** (50).

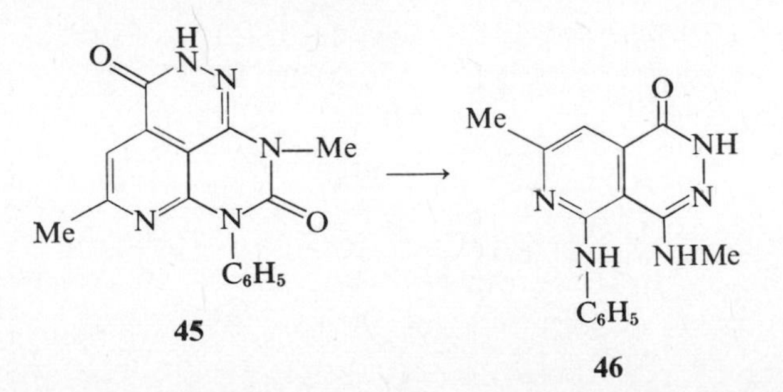

A further example are pyrrolopyridines (5-azaisoindolenines) of the type **47** or **48**, which, after treatment with hydrazine and in the presence of acetic acid, afforded 1,4-dihydrazinopyrido[3,4-*d*]pyridazine (**49**) (60, 61). More-

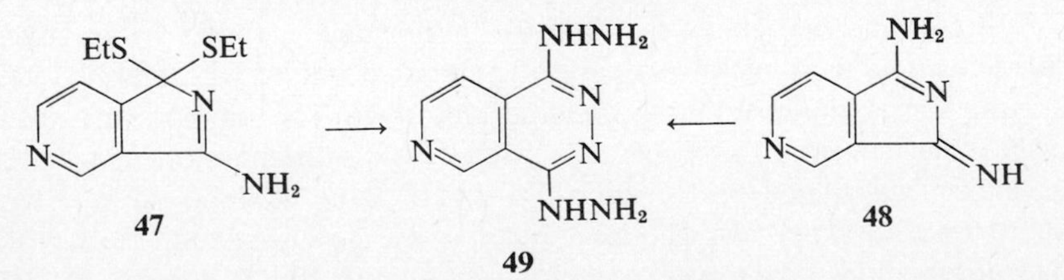

over, the pyrido[3,4-*d*]pyridazine system is capable of ring opening and recyclization as exemplified by the conversion of **50** into **51** (49).

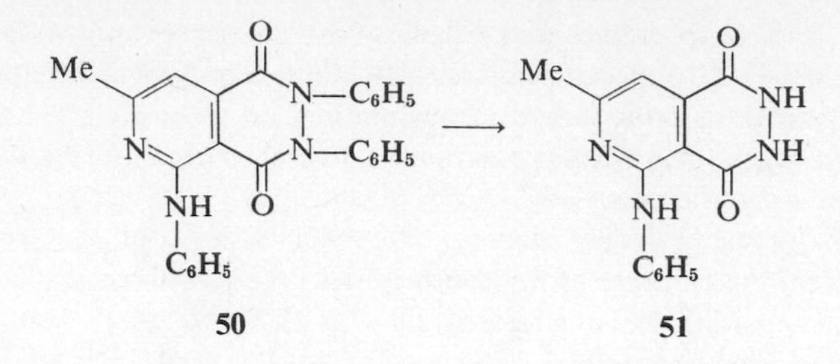

50 **51**

Pyrido[3,4-*d*]pyridazine was synthesized first by Queguiner and Pastour in 1966 (2) from 3,4-diformylpyridine and hydrazine. Paul and Rodda (3) investigated several methods for the preparation of this compound. One method consisted of treating 3,4-dimethylpyridine with *N*-bromosuccinimide in the presence of dibenzoyl peroxide and the tetrabromo compound **52** was subsequently hydrolyzed to **53**. A solution of this compound was treated with hydrazine and the parent bicyclic system **54** was obtained in an overall yield

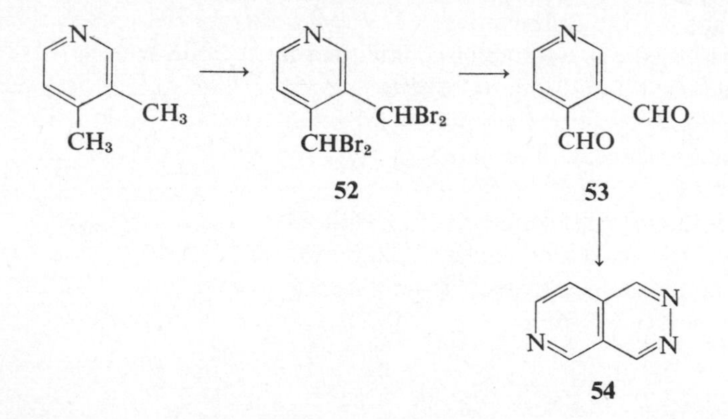

52 **53** **54**

of 13%. The parent compound **54** was also obtained from 1,4-dichloropyrido-[3,4-*d*]pyridazine by catalytic dehalogenation (62%) or by oxidation of the 1,4-dihydrazino analog with yellow mercuric oxide (58%) (3).

Pyrido[3,4-*d*]pyridazine is monobasic and a somewhat weaker base (pK_a 1.76) than the related pyrido[2,3-*d*]pyridazine. It forms a monohydrochloride and a monopicrate salt and is monoquaternized with methyl iodide, but the site of these reactions remains unknown. The nmr and uv spectra in different solvents have been recorded (2, 3) and the electronic spectra have been calculated (76). Pyrido[3,4-*d*]pyridazine is stable toward warm alkali and does not change upon heating with 65% oleum. No covalent hydration could be observed.

The values for total and frontier π-electron densities for this system obtained from MO calculations (27, 28) are listed in Table 3I-2. From these data it is possible to predict that this bicyclic heteroaromatic system should be not susceptible to electrophilic substitutions and that the most reactive position for nucleophilic substitutions should be position 4. Experimental evidence in support of these predictions is, however, very meager.

Pyrido[3,4-*d*]pyridazine-1,4(2*H*,3*H*)dione (**39**) was transformed into 1,4-dichloropyrido[3,4-*d*]pyridazine by means of phosphorus oxychloride (3, 62). Here, the chlorine atom at position 4 is expected to be more reactive than that at position 1. Thus hydrolysis with alkali afforded a mixture of **41** (R = Cl) and the isomeric 1-chloro-4-one in the ratio of 55:45, whereas acid hydrolysis gave both isomers in the reversed ratio (88). Although the dichloro compound with hydrazine at room temperature afforded an almost equal amount of both monohydrazino derivatives (88), it could be converted into the corresponding 1,4-dihydrazino derivative (16 hr, boiling methanol), which could be obtained from the 1,4-diamino derivative after treatment with hydrazine hydrate (100°, 3 hr) in almost the same yield (3). Formylation of 1,4-dihydrazinopyrido[3,4-*d*]pyridazine is reported to give a mixture of the 1- and 4-formyl derivative (63) and heating in the presence of Raney nickel afforded the 1,4-diamino compound in 31% yield (3).

Pyrido[3,4-*d*]pyridazine is oxidized slowly with potassium permanganate under basic conditions to pyridazine-4,5-dicarboxylic acid and traces of cinchomeronic acid. The latter is also formed after oxidation in acidic solution (86).

Pyrido[3,4-*d*]pyridazine-1,4(2*H*,3*H*)dione has been investigated for complex formation and luminescence (32) and some other derivatives were tested for tuberculostatic activity (64) or were claimed to be capable of reducing blood pressure (65, 66).

III. Pyrido[3,2-*c*]pyridazines

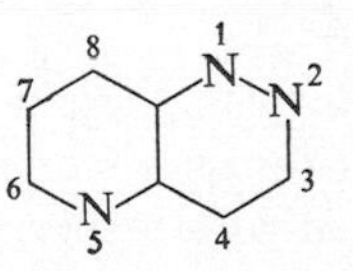

Pyrido[3,2-*c*]pyridazine (1,2,5-Triazanaphthalene)

The only communication describing a compound pertinent to this system was published in 1966 by Atkinson and Biddle (67). They applied the Widman-Stoermer reaction to the pyridine series. In this way **55**, when diazotized and left to stand 5 days in the dark in an alkaline solution, yielded a small amount of 4-methylpyrido[3,2-*c*]pyridazine (**56**), isolated as the picrate.

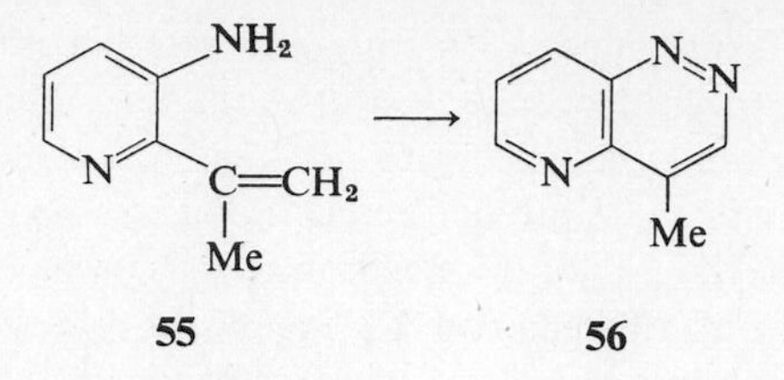

For the parent compound MO calculations of total and frontier π-electron densities have been made (27, 28) and the values are presented in Table 3I-3.

IV. Pyrido[3,4-*c*]pyridazines

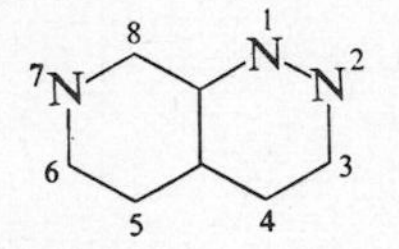

Pyrido[3,4-*c*]pyridazine (1,2,7-Triazanaphthalene)

In addition to preparing the pyrido[3,2-*c*]pyridazine system, Atkinson and Biddle (67) were able to prepare pyrido[3,4-*c*]pyridazines by applying the Widman-Stoermer reaction to pyridines. The diazotized aminopyridines (**57**) were left at room temperature for 3 days. After 2 days the reaction mixture was made alkaline and the products (**58**) were finally isolated in low yields.

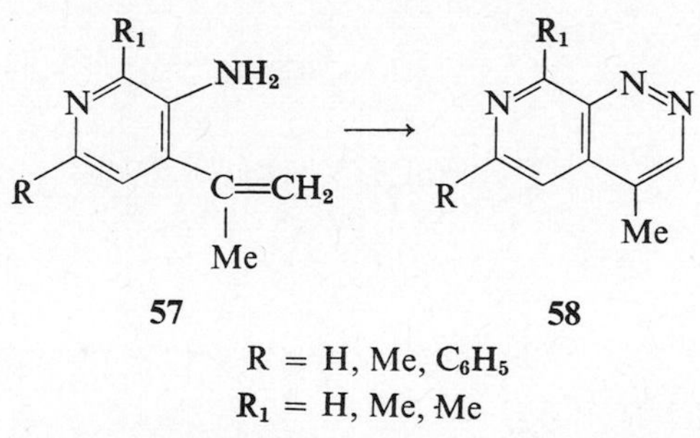

$R = H, Me, C_6H_5$

$R_1 = H, Me, Me$

In a similar reaction, 3-amino-2-chloropyridine was diazotized and coupled to methyl methylacetoacetate and the hydrazone obtained was cyclized in polyphosphoric acid to the pyrido[3,4-*c*]pyridazine derivative (89).

The methyl group of 4-methylpyrido[3,4-*c*]pyridazine could be condensed with benzaldehyde in the presence of zinc chloride and the corresponding styryl derivative was obtained.

For the parent compound MO calculations of total and frontier π-electron densities have been made (27, 28) and the results of these calculations are listed in Table 3I-4.

V. Pyrido[4,3-*c*]pyridazines

Pyrido[4,3-*c*]pyridazine (1,2,6-Triazanaphthalene)

The only compound representing this system was described by Prasad and Wermuth in 1967 (68). The condensation product from *N*-methylpiperidone and pyruvic acid (**59**) was transformed into the pseudoester **60** with methanolic hydrogen chloride. The pseudoester was then treated with hydrazine in 1-butanol and afforded the product **61** in 40% yield.

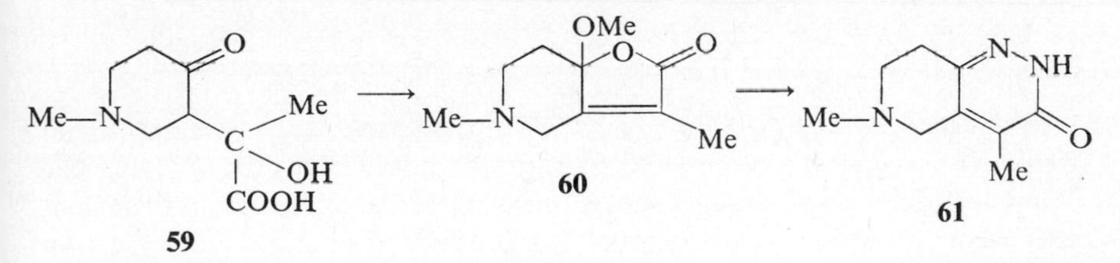

MO calculations of total and frontier π-electron densities for the fully aromatic parent compound have been performed (27, 28) and the results are presented in Table 3I-5.

VI. Pyrido[1,2-*b*]pyridazines

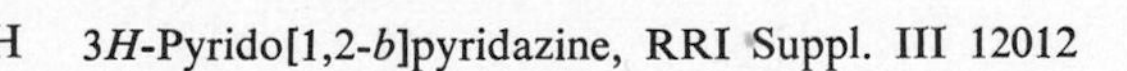

3*H*-Pyrido[1,2-*b*]pyridazine, RRI Suppl. III 12012

6*H*-Pyrido[1,2-*b*]pyridazine, RRI Suppl. III 12013

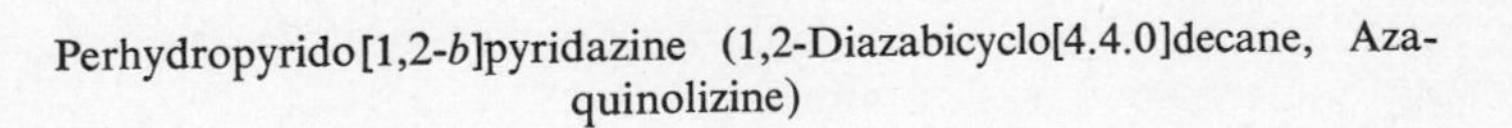

Perhydropyrido[1,2-*b*]pyridazine (1,2-Diazabicyclo[4.4.0]decane, Azaquinolizine)

Representatives of this heteroaromatic system, which would be capable of existence only in the cationized form and would be regarded as an analog of quinolizinium cation, are so far unknown. Known are different partially and fully hydrogenated compounds pertaining to this system.

The first representatives of this heterocycle were mentioned initially in 1963 in a communication by Cookson and Isaacs (69). Pyridazine, when allowed to stand for 2 days at room temperature in the presence of 2 moles of maleic anhydride in chloroform, afforded the adduct **62**, 4*aH*,5,6,7,8-tetrahydropyrido[1,2-*b*]pyridazine-5,6,7,8-tetracarboxylic acid bis-anhydride, in 85% yield. The nmr spectra of **62** and of the analogous tetradeutero adduct are in agreement with the proposed structure and the most probable stereoisomer has been presented. Compound **62** could be hydrolyzed, but an analytical pure acid could not be obtained. Reduction with 1 mole of hydrogen uptake was difficult (70°, 80 psi, 12 hr) and led to **63**. Bromination produced a compound for which the bromo-lactone structure was proposed and oxidation of **63** with concentrated nitric acid afforded pyridazine-3-carboxylic acid.

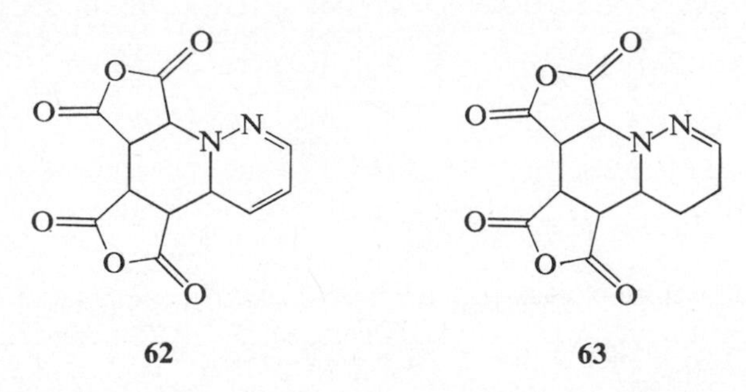

Acheson and Foxton (70), who investigated the reaction between dimethyl acetylenedicarboxylate and pyridazines, established that beside pyrrolo[1,2-*b*]pyridazines as minor products, derivatives of pyrido[1,2-*b*]-pyridazine were formed. Thus the major product from the reaction between

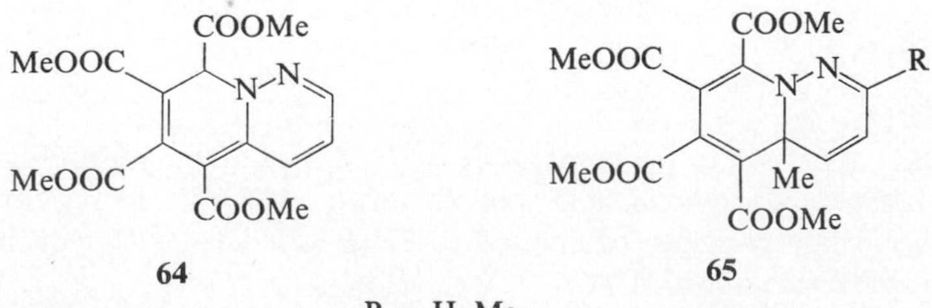

R = H, Me

pyridazine and dimethyl acetylenedicarboxylate in acetonitrile was **64**. The structure was deduced from correlation of uv, ir, and nmr spectra with those of the related quinolizine derivative, prepared in a similar reaction from pyridine.

Similarly, 1:2 molar adducts were isolated from the reaction with 3-methyl or 3,6-dimethylpyridazine (**65**). The yield in the case of the latter starting compound was low and the major product was an azepino[1,2-*b*]-pyridazine derivative. The structural assignment was made on the basis of nmr data.

Perhydropyrido[1,2-*b*]pyridazines have been prepared as follows. Yamazaki et al. (71) synthesized the perhydro compound **71** by first reducing β-(2-pyridyl)propionaldehyde dimethyl acetal (**66**), obtained from the reaction between picolyllithium and bromoacetaldehyde dimethylacetal, catalytically to **67**. After nitrosation to **68** and reduction of this compound to the *N*-amino compound **69**, the latter was cyclized to **70** and after catalytic hydrogenation the liquid perhydropyrido[1,2-*b*]pyridazine (**71**) was obtained.

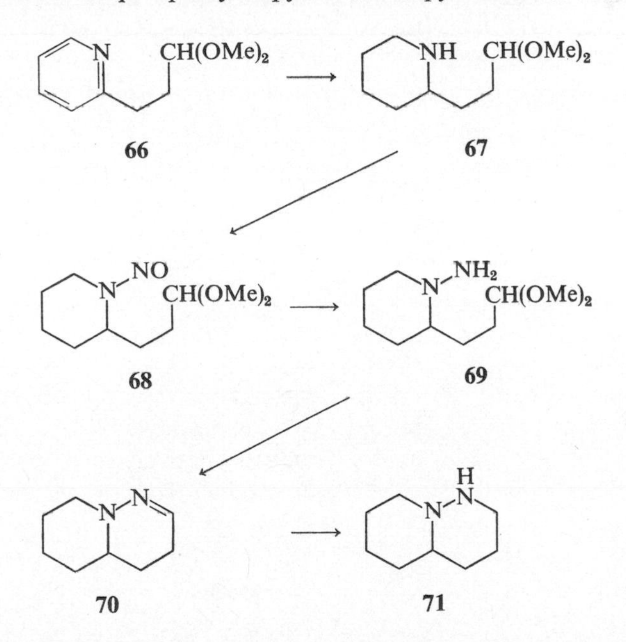

Another approach to the preparation of **71** was devised by Mikhlina et al. (72, 73). Piperidylpropionic acid was nitrosated to **72** and then cyclized via the intermediate *N*-amino compound to **73**. Reduction of **73** with lithium aluminum hydride afforded **71**.

Different transformations involving the imino group of **71** have been performed. Ring opening occurred under the influence of Raney nickel

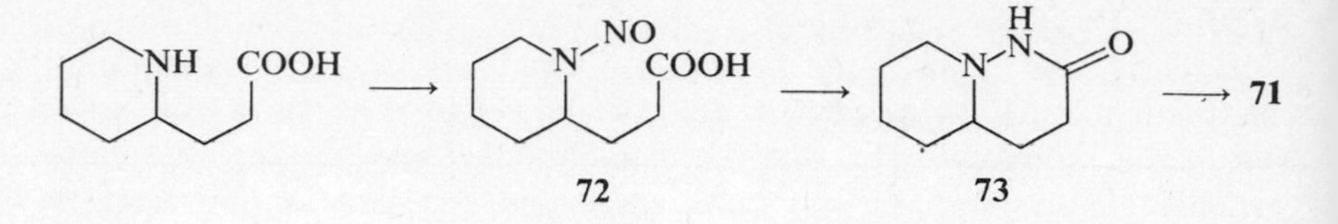

catalyst and in the presence of hydrogen (100°, 30 atm, 5 hr) to give γ-(2-piperidinyl)propylamine. Furthermore, **71** could be acylated and cyanomethylated.

In a novel synthetic approach, the Dieckmann condensation of ethyl 1-(ethoxycarbonylisopropylidene)amino-2-piperidinecarboxylate afforded a

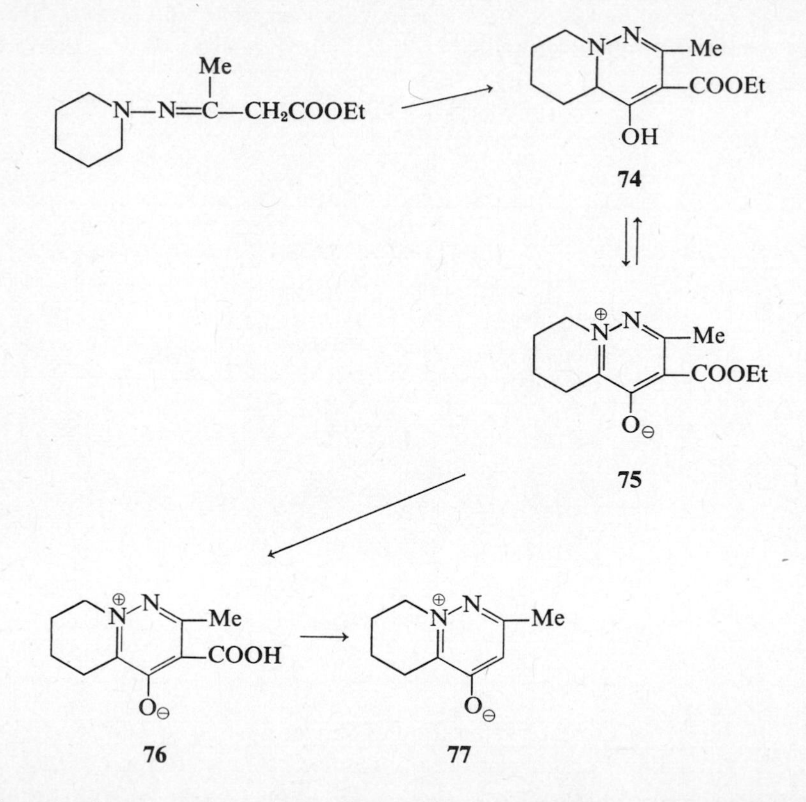

mixture of two compounds in a ratio of 2:1, to which structures **74** and **75** were assigned (90). Compound **75** was transformed into **74** by catalytic reduction and **74** spontaneously dehydrogenated in highly diluted solution back to **75**. The carbethoxy group of **75** could be saponified and the obtained acid **76** decarboxylated to **77**. The latter compound is a weak base (pK_a 2.77), but nevertheless it formed a stable hydrochloride and methiodide salt. Moreover, **77** underwent electrophilic and nucleophilic substitutions (bromination and displacement of the halogen with nucleophiles), which have been attributed to the negatively charged oxygen and positively charged bridgehead nitrogen (90). Finally, a hydroxyl group at position 3 could be introduced into **77** in the reaction with hydrogen peroxide in glacial acetic acid.

VII. Tables

TABLE 3I-1. Total and Frontier π-Electron Densities of pyrido[2,3-*d*]pyridazine

Position	1	2	3	4	4*a*	5	6	7	8	8*a*
Total (27, 28)	1.1753	0.9199	0.9737	0.9207	0.9630	0.9103	1.1412	1.1281	0.9220	0.9459
	1.203	0.881	0.992	0.920	0.981	0.890	1.152	1.134	0.912	0.935
Frontier (27)	0.1913	0.0207	0.1076	0.1815	0.0014	0.1443	0.1067	0.0953	0.1510	0.0002

TABLE 3I-2. Total and Frontier π-Electron Densities of pyrido[3,4-*d*]pyridazine

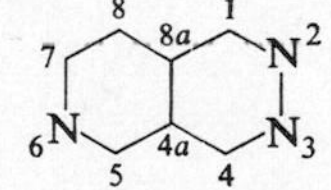

Position	1	2	3	4	4*a*	5	6	7	8	8*a*
Total (27, 28)	0.9156	1.1288	1.1386	0.9080	0.9610	0.9281	1.1486	0.9655	0.9592	0.9467
	0.902	0.135	0.150	0.885	0.983	0.883	1.183	0.929	0.995	0.955
Frontier (27)	0.1307	0.0751	0.1060	0.1113	0.0079	0.1794	0.0445	0.1284	0.2163	0.0003

TABLE 3I-3. Total and Frontier π-Electron Densities of pyrido[3,2-*c*]pyridazine

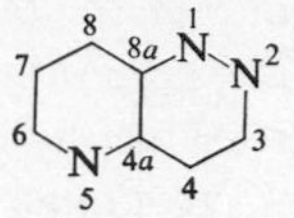

Position	1	2	3	4	4*a*	5	6	7	8	8*a*
Total (27, 28)	1.1191	1.0968	0.9633	0.9175	0.9493	1.1799	0.9168	0.9739	0.9257	0.9575
	1.107	1.103	0.935	0.950	0.934	1.208	0.882	0.992	0.929	0.958
Frontier (27)	0.1895	0.0422	0.0934	0.1794	0.0015	0.1868	0.0305	0.0946	0.1806	0.0015

TABLE 3I-4. Total and Frontier π-Electron Densities of pyrido[3,4-*c*]pyridazine

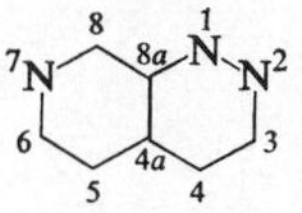

Position	1	2	3	4	4*a*	5	6	7	8	8*a*
Total (27, 28)	1.1162	1.0920	0.9638	0.9092	0.9500	0.9606	0.9737	1.1465	0.9320	0.9559
	1.102	1.099	0.936	0.941	0.954	1.000	0.934	1.182	0.891	0.961
Frontier (27)	0.1549	0.0439	0.0779	0.1561	0.0005	0.2039	0.1355	0.0428	0.1788	0.0055

TABLE 3I-5. Total and Frontier π-Electron Densities of pyrido[4,3-*c*]pyridazine

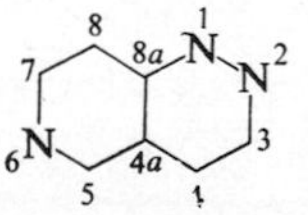

Position	1	2	3	4	4*a*	5	6	7	8	8*a*
Total (27, 28)	1.1294	1.0823	0.9704	0.8972	0.9619	0.9253	1.1545	0.9684	0.9681	0.9445
	1.122	1.084	0.950	0.921	0.980	0.885	1.186	0.930	1.007	0.934
Frontier (27)	0.1778	0.0234	0.1006	0.1335	0.0013	0.1674	0.0567	0.1190	0.2201	0.0003

TABLE 3I-6. Pyrido[2,3-*d*]pyridazines

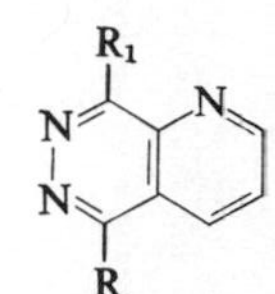

R	R_1	MP (°C) or BP (°C/mm)	References
None		152, 154–155	2, 84, 87
		Picrate, 198 (dec.), 196 (dec.)	2, 87
		154–155	3
		Picrate, 196 (dec.)	3
		HCl, 202–203 (dec.)	3
		MeI, 230	
Methyl		170	26, 84
	Methyl	115	26, 80, 84
		114–115	23
		Picrate, 173–174	23
Deuterium		152	74, 84
Methylthio		93–95	77
Hydrazino		240 (dec.)	77
		263–265 (dec.)	87
—NHN=CHPh		220	77
$PhCOCH_2S$—		160	79
	Methylthio	110	77
	Hydrazino	190–195, 232–233	77, 87
		HBr, 270	77
	—NHN=CHPh	201–202	77

TABLE 3I-6 (*continued*)

R	R_1	MP (°C) or BP (°C/mm)	References
	$MeCOCH_2S$—	146–147	79
	$PhCOCH_2S$—	196	79
Chloro		dec. over 100	87
	Chloro	no definite mp	87
Amino		Picrate, 285–286 (dec.)	87
	Amino	Picrate, 292	87
Chloro	Methyl	162–163	23
Hydrazino	Methyl	>180 (dec.)	23
Chloro	*p*-$NO_2C_6H_4NH$	>270	38
Phenyl	Chloro	—	11
		~200	9
Phenyl	Hydrazino	HCl ~ 220	11
		~200 (dec.)	9
		HCl, 259	9
Chloro	Chloro	167	3
		169	29, 34
		163–164	23
Methoxy	Methoxy	188	29
		184	36
Ethoxy	Ethoxy	158	29
		159	36
n-Propoxy	*n*-Propoxy	101	29, 36
iso-Propoxy	*iso*-Propoxy	154	29, 36
$CH_2{=}CHCH_2O$	$CH_2{=}CHCH_2O$	91	29, 36
$C_6H_5CH_2O$—	$C_6H_5CH_2O$—	168	29
C_6H_5O	C_6H_5O	178	29

Methyl	Methyl	115	80
n-Butoxy	*n*-Butoxy	95	36
Ethoxy	Chloro	155	39
Amino	Chloro	229–230 (dec.)	29
Chloro	Amino	258–261 (dec.)	29
Chloro	C_6H_5NH	180	29, 38
C_6H_5NH	Chloro	226–227 (dec.)	29, 38
Chloro	*p*-ClC_6H_4NH	220	29, 38
Chloro	*p*-BrC_6H_4NH	231–233 (dec.)	29, 38
Chloro	*p*-MeC_6H_4NH	200	29, 38
Chloro	*p*-OHC_6H_4NH	235–236 (dec.)	29, 38
Chloro	*p*-$MeOC_6H_4NH$	199–201	29, 38
Chloro	$NH(CH_2)_3NMe_2$	—	35
		2 HCl·H_2O, 243	35
Chloro	$NH(CH_2)_3NEt_2$	89–90	35
Chloro	$NHCH(CH_2)_3NEt_2$ (with Me on CH)	—	35
		2 HCl, 174–175	35
Chloro	$NH(CH_2)_2NMe_2$	112–113	35
Chloro	$NH(CH_2)_2NEt_2$	—	35
		2 HCl, 237–238	35
Chloro	$NH(CH_2)_3N/(CH_2)_3Me_2$	—	35
		2 HCl, 160–162	35
Chloro	Hydrazino	174–175, 173–175	29, 87
		175	40
p-ClC_6H_4NH	Chloro	207–208 (dec.)	29
		210 (dec.)	38
$NH(CH_2)_3NEt_2$	Chloro	120.5–121.5	35
$NH(CH_2)_3NMe_2$	Chloro	—	35
		2 HCl, 246	35

TABLE 3I-6 (*continued*)

R	R_1	MP (°C) or BP (°C/mm)	References
$NHCH(CH_2)_3NEt$ (with Me on CH)	Chloro	2 HCl, 178–181	35
$NH(CH_2)_2NMe_2$	Chloro	172–172.5	35
$NH(CH_2)_2NEt_2$	Chloro	2 HCl, 235–237	35
$NH(CH_2)_3N(n\text{-Bu})_2$	Chloro	2 HCl, 157–158	35
Hydrazino	Chloro	196–197	29, 87
		207 (dec.)	40
		HCl, 230–232	29
$NHN{=}CMe_2$	Chloro	135–136	29
NHNHCHO	Chloro	213–214	29
Hydrazino	Hydrazino	183	3
		187 (dec.)	4
Phenyl	Hydrazino	no mp	10
C_6H_5NH	C_6H_5NH	217	29, 38
$p\text{-}MeC_6H_4NH$	$p\text{-}MeC_6H_4NH$	250	29, 38
$p\text{-}MeOC_6H_4NH$	$p\text{-}MeOC_6H_4NH$	213–214	29, 38
Ethoxy	C_6H_5NH	114–116	29

p-ClC_6H_4NH	p-ClC_6H_4NH	>270	38
p-BrC_6H_4NH	p-BrC_6H_4NH	>270	38
p-$NO_2C_6H_4NH$	p-$NO_2C_6H_4NH$	>270	38
p-OHC_6H_4NH	p-OHC_6H_4NH	>270	38
Amino	Amino	251–252	3
		Picrate, 296–298	3
Ethoxy	$NH(CH_2)_3NEt_2$	—	35
		2 HCl, 130–131	35
CH_3COO	Chloro	160–162	15
CH_3COO	$NHNHCOCH_3$	201–202	15
Methylthio	Methylthio	249	29
		246–248	3
—SC(=NH)NH_2	—SC(=NH)NH_2	—	29
		2 HCl, 192	29
	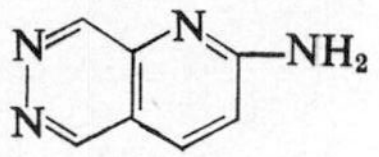 (pyrido-pyridazine structure with NH_2)	Picrate, 284–285	86
	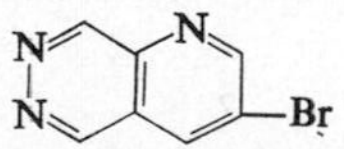 (pyrido-pyridazine structure with Br)	175 (dec.)	86

TABLE 3I-7. Pyrido[2,3-*d*]pyridazine-5(6*H*)ones and Pyrido[2,3-*d*]pyridazine-5(6*H*)thiones

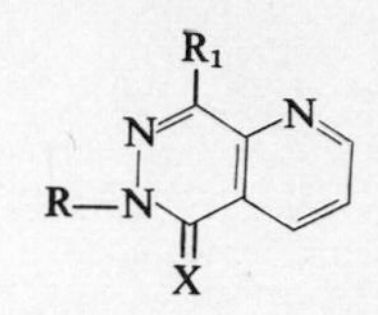

R	R_1	X	MP (°C) or BP (°C/mm)	References
None		O	261–263, 277, 262	6, 87, 81, 86
	Chloro	O	275, 296	29, 87
			284–285	15
	Hydrazino	O	280, 284	15, 87
	$NHN{=}CHC_6H_5$	O	273	15
	$NH(CH_2)_3NMe_2$	O	124–125	35
	$NH(CH_2)_3NEt_2$	O	121–123	35
Methyl		O	110–112	6
Methyl	Chloro	O	159–160	13
	Methyl	O	249–250, 249	23, 87
Phenyl	Methyl	O	121	1
p-$NO_2C_6H_4$	CH_2COOMe	O	180	7
None		S	205–207	6
p-$NO_2C_6H_4$	$OCOCH_3$	O	194–197	16
2,4-di$NO_2C_6H_3$	$OCOCH_3$	O	205–207	16

TABLE 3I-8. Pyrido[2,3-*d*]pyridazine-8(7*H*)ones and Pyrido[2,3-*d*]pyridazine-8(7*H*)thiones

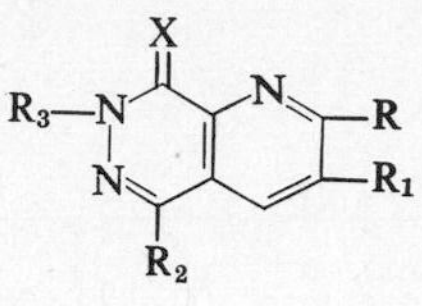

R	R_1	R_2	R_3	X	MP (°C) or BP (°C/mm)	References
None				O	305, 326–328	6, 87, 86
					300	12
None				S	212	6
		COOH		O	300–301 (dec.)	13
		COOMe		O	227–229 (dec.)	13
		$CONH_2$		O	352–357 (dec.)	13
		CONHMe		O	300–303 (dec.)	13
		CONHEt		O	270–271	13
		$CONHNH_2$		O	296 (dec.)	13
		Chloro		O	289 (dec.), 306–307	29, 87
			Phenyl	O	201–203	12
					199	6
			Methyl	O	192	6
Methyl	CN	Methyl		O	338–340	20
		COOH	Phenyl	O	258 (dec.)	13
		COOEt	Phenyl	O	138–139	13
		COOMe	Phenyl	O	178–179	13
		$CONH_2$	Phenyl	O	333 (dec.)	13
		CONHMe	Phenyl	O	199–200	13
		CONHEt	Phenyl	O	169–171	13
		$CONHNH_2$	Phenyl	O	198	13
		Chloro	Phenyl	O	184	13
		Phenyl		O	236	10, 11
					136	8
					2 HCl, 210–211	8
		Chloro	Methyl	O	194–195	13
		Phenyl	Methyl	O	173–175	8
		Phenyl	Ethyl	O	164	8
		Methoxy	Phenyl	O	162–163	13
		p-ClC_6H_4OCO	Phenyl	O	189–191	13
		p-$NO_2C_6H_4OCO$	Phenyl	O	227–230	13
		Hydrazino		O	>300	87
CH_3COO—CH_2CH_2—N (O, N, N, $OCOCH_3$)					124–125	75

TABLE 3I-9. Pyrido[2,3-*d*]pyridazine-5,8(6*H*,7*H*)diones

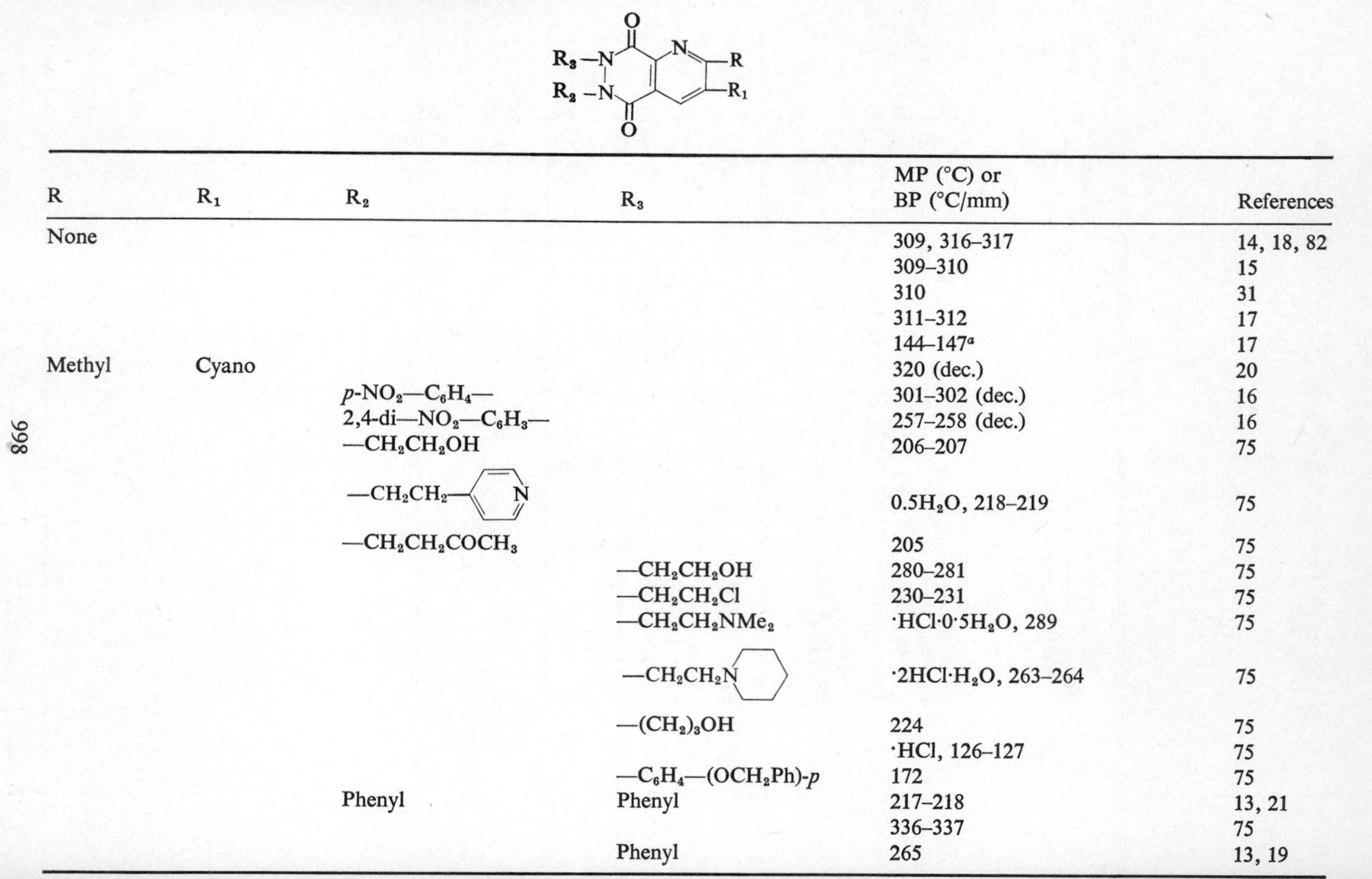

R	R_1	R_2	R_3	MP (°C) or BP (°C/mm)	References
None				309, 316–317	14, 18, 82
				309–310	15
				310	31
				311–312	17
				144–147[a]	17
Methyl	Cyano			320 (dec.)	20
		p-$NO_2—C_6H_4—$		301–302 (dec.)	16
		2,4-di—$NO_2—C_6H_3—$		257–258 (dec.)	16
		$—CH_2CH_2OH$		206–207	75
		$—CH_2CH_2—$ N		$0.5H_2O$, 218–219	75
		$—CH_2CH_2COCH_3$		205	75
			$—CH_2CH_2OH$	280–281	75
			$—CH_2CH_2Cl$	230–231	75
			$—CH_2CH_2NMe_2$	$·HCl·0·5H_2O$, 289	75
			$—CH_2CH_2N$	$·2HCl·H_2O$, 263–264	75
			$—(CH_2)_3OH$	224	75
				·HCl, 126–127	75
			$—C_6H_4—(OCH_2Ph)$-*p*	172	75
		Phenyl	Phenyl	217–218	13, 21
				336–337	75
			Phenyl	265	13, 19

TABLE 3I-10. Pyrido[2,3-*d*]pyridazine-5,8(6*H*,7*H*)dithiones

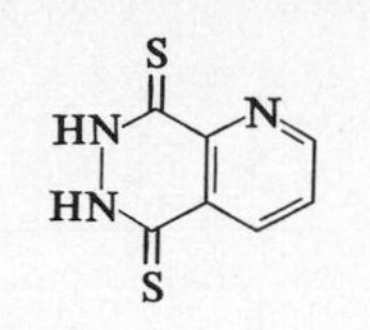

MP (°C)	References
>260	29
>300	37
270 (dec.)[a]	3

[a] Sinters above 230°.

TABLE 3I-11. Pyrido[2,3-*d*]pyridazine-2,5,8(1*H*,6*H*,7*H*)triones

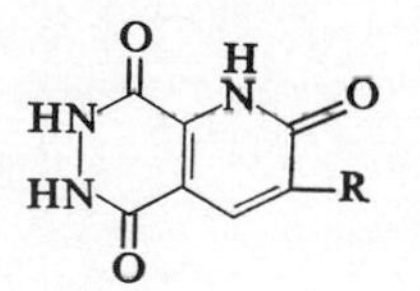

R	MP (°C) or BP (°C/mm)	References
None	>400	18
Chloro	380–400	18
Bromo	400–450	18
Iodo	420–450	18

TABLE 3I-12. 1,2,3,4-Tetrahydropyrido[2,3-*d*]pyridazines

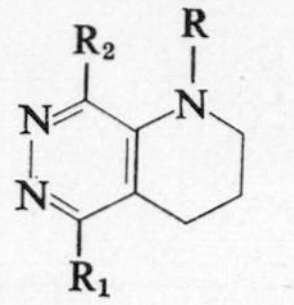

R	R_1	R_2	MP (°C) or BP (°C/mm)	References
		$OCOCH_3$	172	6
	$OCOCH_3$		204	6
CH_3CO	$OCOCH_3$		220–221	6

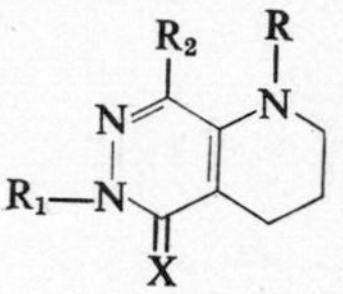

R	R_1	R_2	X	MP (°C) or BP (°C/mm)	References
None			O	241–243, 262–263	6, 87
None			S	180–182	6
	Methyl		O	172–174	6
Methyl			O	212–214	6
		$OCOCH_3$	O	196–198	6
CH_3CO		$OCOCH_3$	O	159–161	6

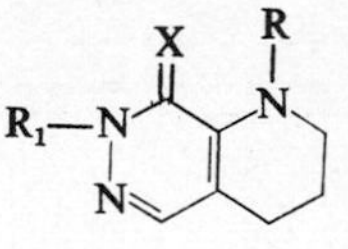

R	R_1	X	MP (°C) or BP (°C/mm)	References
None		O	223, 237–239	6, 87
			226–228	29
NO		O	159	6
None		S	182	6
	Methyl	O	115	6
	Phenyl	O	120	6
			130–131	13
Methyl		O	147	6

TABLE 3I-12 (*continued*)

Structure	MP (°C) or BP (°C/mm)	References
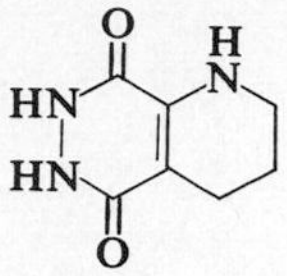 [structure]	291–293	6
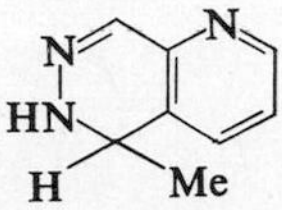 [structure]	— Oil	26 84
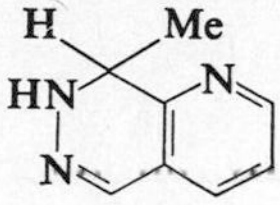 [structure]	— Oil	26 84
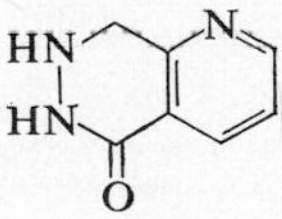 [structure]	213–215	83

TABLE 3I-13A. Pyrido[2,3-*d*]pyridazine *N*-Oxides

Structure	MP (°C) or BP (°C/mm)	References
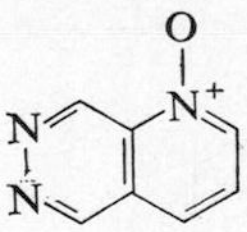 [structure]	216	24
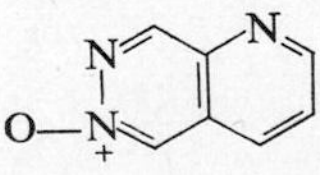 [structure]	210 209–210	26 84, 86
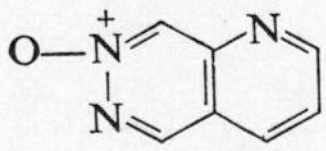 [structure]	199–200 206–207	26 84, 86

TABLE 3I-13A (*continued*)

Structure	MP (°C) or BP (°C/mm)	Reference
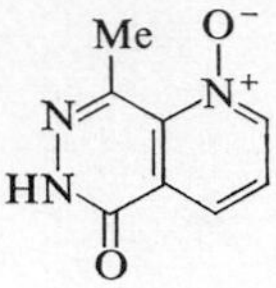	274–275	23
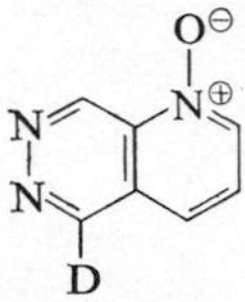	216	74

TABLE 3I-13B. Pyrido[3,4-*d*]pyridazines

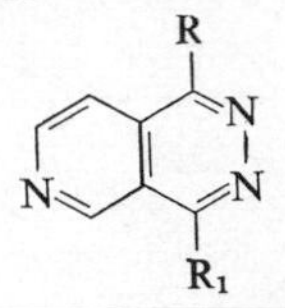

R	R_1	MP (°C) or BP (°C/mm)	References
None	Chloro	170–172, 174	3, 84
		Picrate, 184–185	3
		HCl, 217 (dec.)	3
		174	2
		Picrate, 166 (dec.)	2
Chloro	Chloro	159	62, 88
		155–156	3
Amino	Amino	Picrate, 312–313 (dec.)	3
Hydrazino	Hydrazino	198 (dec.)	60
		H_2SO_4, 203 (dec.)	60
		193–194 (dec.)	3
		200 (dec.)	61
		3 HCl, 190 (dec.)	61
		227–230[a]	63
Methoxy	Methoxy	139–140	88
Ethoxy	Ethoxy	96–97	88
$OCHMe_2$	$OCHMe_2$	80–81	88
Chloro	Hydrazino	190 (dec.)	88
Chloro	—NHN=CMe_2	198–199	88
Hydrazino	Chloro	186 (dec.)	88
—NHN=CMe_2	Chloro	201–202	88

[a] Formyl derivative (mixture of 1 and 4 formyl derivative).

TABLE 3I-14. Pyrido[3,4-*d*]pyridazine-1(2*H*)ones

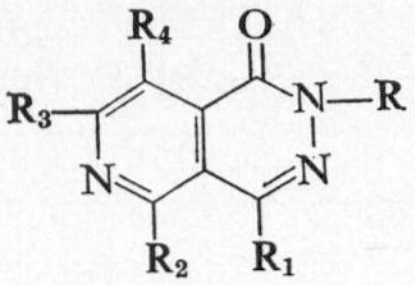

R	R_1	R_2	R_3	R_4	MP (°C) or BP (°C/mm)	References
	Methyl			MeCO—	271	55
	Methyl			MeC=NNH$_2$ (attached at the C)	>280 (dec.)	55
	Methyl	Methyl	Phenyl		261–262	54
	Methyl	Methyl	Methyl		235–237	48
	MeNH	C_6H_5NH	Methyl		275	50
	MeNH	C_6H_5NH	Phenyl		260	50
	MeNH	*p*-EtOC_6H_4NH	Methyl		235	50
Methyl	MeNH	C_6H_5NH	Methyl		170	50
Methyl	MeNH	C_6H_5NH	Phenyl		180	50
Methyl	MeNH	*p*-EtOC_6H_4NH	Methyl		—	50
MeCO	MeNH	C_6H_5NH	Methyl		185	50
MeCO	MeNH	*p*-EtOC_6H_4NH	Methyl		—	50
Phenyl	Methyl	Methyl	Phenyl		182–183	54
	Chloro				293–294	88
	Amino				>360	88
None					292–293	88

TABLE 3I-15. Pyrido[3,4-*d*]pyridazine-1,4(2*H*,3*H*)diones and Related Compounds

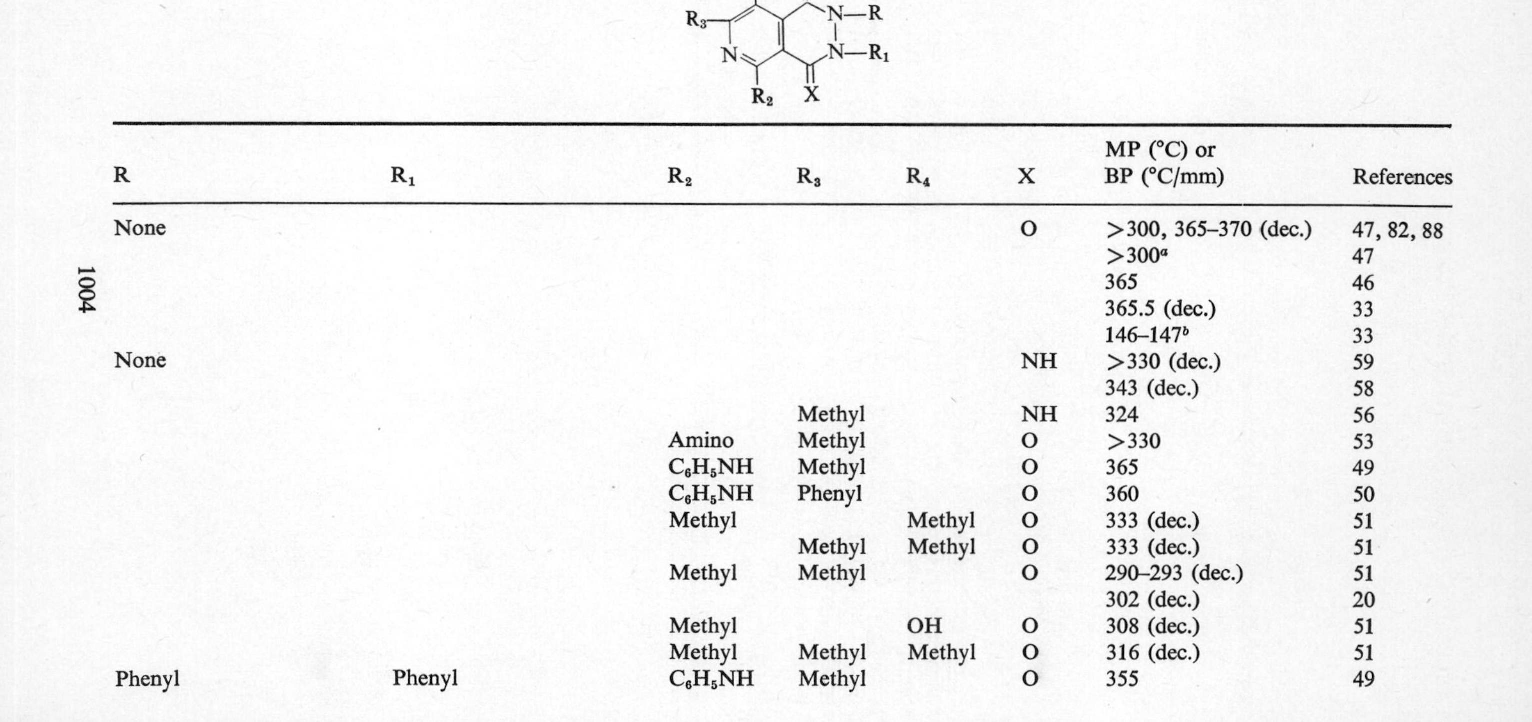

R	R_1	R_2	R_3	R_4	X	MP (°C) or BP (°C/mm)	References
None					O	>300, 365–370 (dec.)	47, 82, 88
						>300[a]	47
						365	46
						365.5 (dec.)	33
						146–147[b]	33
None					NH	>330 (dec.)	59
						343 (dec.)	58
			Methyl		NH	324	56
		Amino	Methyl		O	>330	53
		C_6H_5NH	Methyl		O	365	49
		C_6H_5NH	Phenyl		O	360	50
		Methyl		Methyl	O	333 (dec.)	51
			Methyl	Methyl	O	333 (dec.)	51
		Methyl	Methyl		O	290–293 (dec.)	51
						302 (dec.)	20
		Methyl		OH	O	308 (dec.)	51
		Methyl	Methyl	Methyl	O	316 (dec.)	51
Phenyl	Phenyl	C_6H_5NH	Methyl		O	355	49

$—CH_2CH_2COCH_3$		O	$\cdot H_2O$, 186	75
	$—CH_2CH_2OH$	O	253	75
	$—CH_2CH_2Cl$	O	·HCl, 195	75
	$—CH_2CH_2NMe_2$	O	·2HCl, 267–268	75
	$—C_6H_4—NO_2(p)$	O	181	75
	$—CH_2CH_2OH$	O	$\cdot MeI \cdot H_2O$, 209–210	75
	$—(CH_2)_3OH$	O	215	75
	$—CH_2CH(CH_3)—CH_3$	O	235	75
	$—C_6H_4—(OCH_2Ph)(p)$	O	·HCl, 175	75
Phenyl	Phenyl	O	220	75
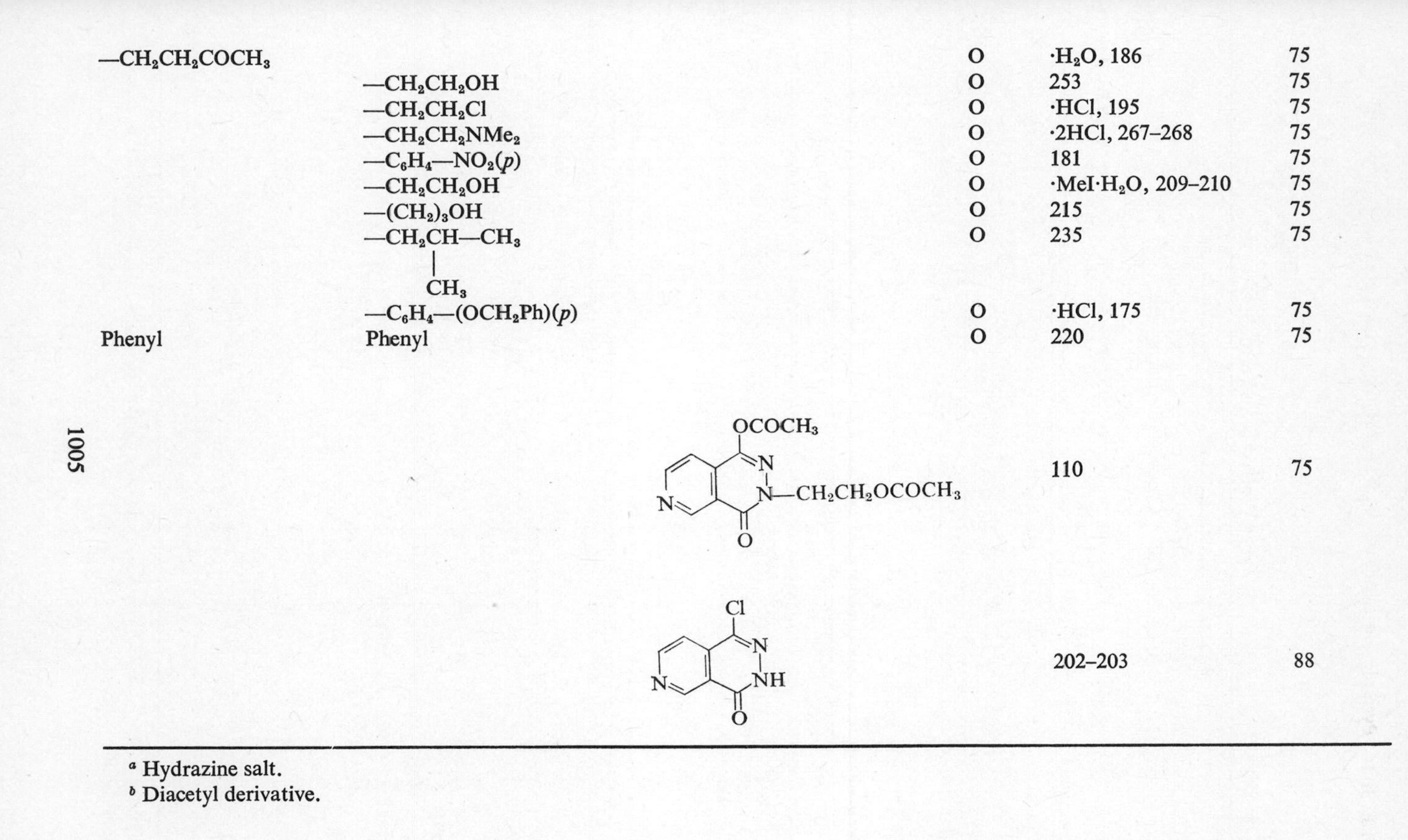			110	75
			202–203	88

[a] Hydrazine salt.
[b] Diacetyl derivative.

TABLE 3I-16. Pyrido[3,4-*d*]pyridazine-1,4(2*H*,3*H*)diones and Related Compounds

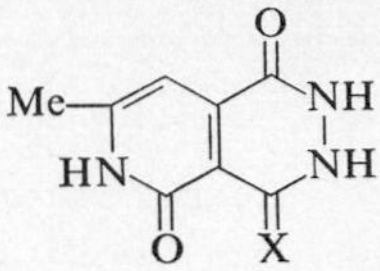

X	MP (°C) or BP (°C/mm)	References
O	>300	48
NH	>300	57

TABLE 3I-17. Pyrido[3,2-*c*]pyridazine

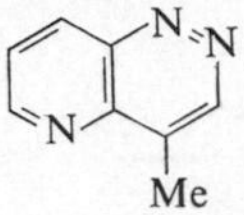 (4-Me)	MP 228° C (as picrate)	Ref. 67

TABLE 3I-18. Pyrido[3,4-*c*]pyridazines

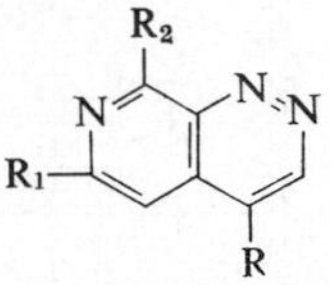

R	R_1	R_2	MP (°C) or BP (°C/mm)	References
Methyl			125	67
			255[a]	67
			163–164[b]	67
COOH			202–203	67
CH=CHC$_6$H$_5$			95–96	67
Methyl	Methyl	Methyl	142–143	67
			181–182[a]	67
Methyl	Phenyl	Methyl	217–218[a]	
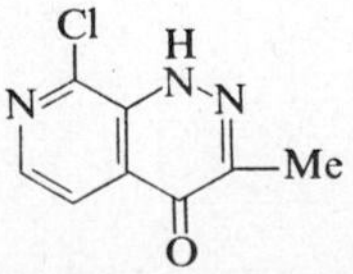 (8-Cl, 3-Me, 4-oxo, 1H compound)			198–200	89

[a] Picrate.
[b] Methiodide.

TABLE 3I-19. Pyrido[4,3-*c*]pyridazine

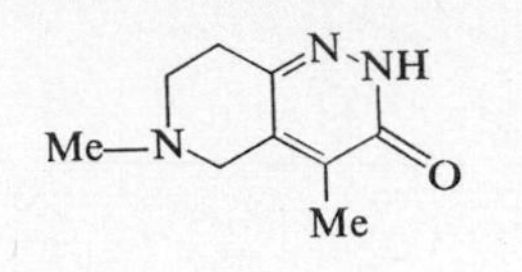

MP 204–205° C Ref. 68

TABLE 3I-20. Pyrido[1,2-*b*]pyridazines

R	MP (°C) or BP (°C/mm)	References
None	70–72/7 mm	71
	104–106/33 mm	72, 73
	HCl, 149–150	72, 73
Benzyl	162–163/7 mm	72
Ethyl	117–119/30 mm	72
$COCH_3$	145–147/11 mm	72
$CH_3CH_2CO—$	122–123/1 mm	72
$C_6H_5CO—$	162/0.7 mm	72
	60–62	72
$—CONHC_6H_5$	112–114	72
$—CH_2CN$	112–114/8 mm	72
	HCl, 154–156	72
	Picrate, 151–153 (dec.)	72
$—CH_2CH_2NH_2$	143–145/12 mm	72
$—CH_2CH_2—NHC(=NH)—NH_2$	205–207 (dec.)	72
—NHCS—	135–137	71

TABLE 3I-21. Pyrido[1,2-*b*]pyridazines

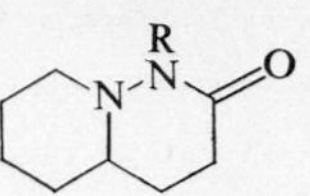

R	MP (°C) or BP (°C/mm)	References
None	147–149	73
	149–151	72
—$COCH_3$	61–63	72

Structure	MP (°C) or BP (°C/mm)	References
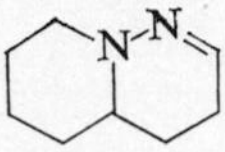	67–70/6 mm	71
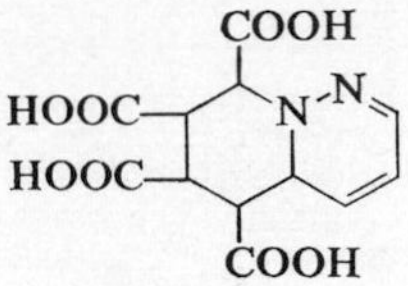	140 (dec.)	69
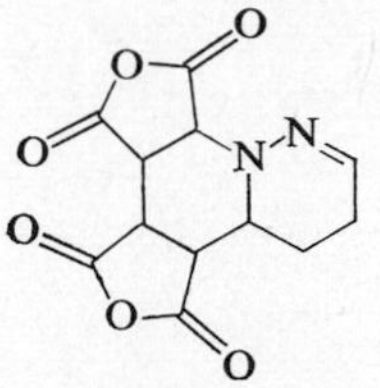	203	69
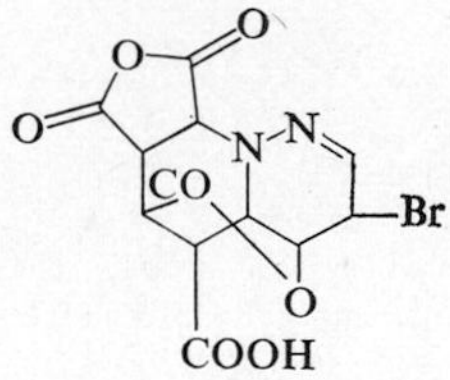	180 (dec.)	69
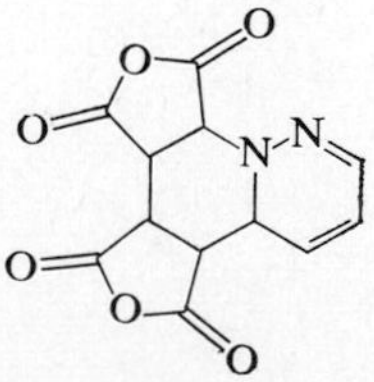	180 (dec.)	69
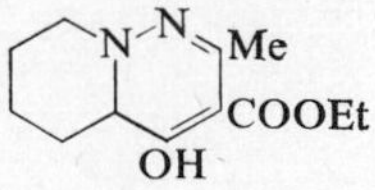	152–153	90

TABLE 3I-21 (*continued*)

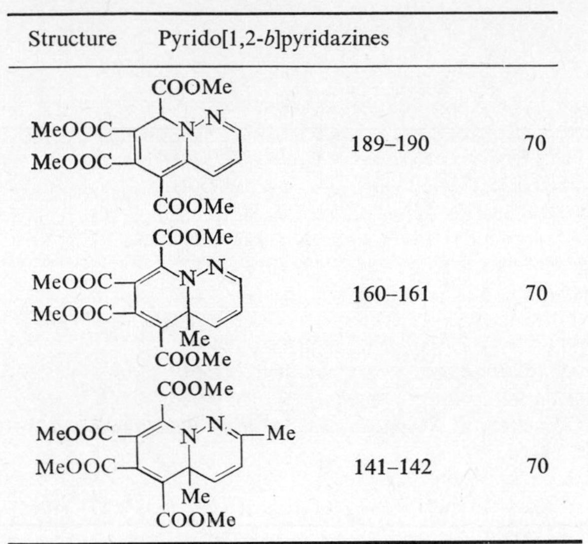

Structure	Pyrido[1,2-*b*]pyridazines	
	189–190	70
	160–161	70
	141–142	70

TABLE 3I-22

R	MP (°C)	References
None	142	90
	Hydrate, 67–68	90
	Picrate, 179	90
COOEt	135–136	90
	Hydrate, 57–58	90
COOH	190–191	90
Bromo	HBr, 206	90
C_6H_5NH	185	90
—S—C(=NH)NH_2	HBr, 152	90
Mercapto	171	90
—SHgCl	255	90
—$SCH_2C_6H_5$	96	90
	HCl, 175	90
	MeI, 175	90
Hydroxy	214	90

References

1. O. Rosenheim and J. Tafel, *Ber.*, **26,** 1501 (1893).
2. G. Queguiner and P. Pastour, *Compt. rend. C.*, **262,** 1335 (1966).
3. D. B. Paul and H. J. Rodda, *Australian J. Chem.*, **21,** 1291 (1968).
4. British Pat. 732,521 (1955); *Chem. Abstr.*, **51,** 1302 (1957).
5. J. P. Wibaut and H. Boer, *Rec. Trav. Chim.*, **74,** 241 (1955).
6. S. Kakimoto and S. Tanooka, *Bull. Chem. Soc. Japan*, **40,** 153 (1967).
7. E. Ochiai, K. Miyaki, and S. Sato, *Ber.*, **70B,** 2018 (1937).
8. B. Jeiteles, *Monatsh. Chem.*, **22,** 843 (1901).
9. J. Druey and B. H. Ringier, *Helv. Chim. Acta*, **34,** 195 (1951).
10. British Pat. 629,177 (1949); *Chem. Abstr.*, **44,** 4516 (1950).
11. U.S. Pat. 2,484,029 (1949); *Chem. Abstr.*, **44,** 4046 (1950).
12. F. Bottari and S. Carboni, *Gazz. Chim. Ital.*, **86,** 990 (1956).
13. I. Matsuura, F. Yoneda, and Y. Nitta, *Chem. Pharm. Bull. Japan*, **14,** 1010 (1966); *Chem. Abstr.*, **66,** 65438 (1967).
14. R. Wegler, *J. Prakt. Chem.*, **148,** 135 (1937).
15. E. Domagalina, I. Kurpiel, and J. Majejko, *Rozcniki Chem.*, **38,** 571 (1964).
16. E. Domagalina and I. Kurpiel, *Roczniki Chem.*, **41,** 1241 (1967).
17. G. Gheorghiu, *Bull. Soc. Chim. France*, **47,** 630 (1930).
18. K. Gleu and K. Wackernagel, *J. Prakt. Chem.*, [2], **148,** 72 (1937); *Chem. Abstr.*, **31,** 2217 (1937).
19. Japanese Pat. 192 (1967); *Chem. Abstr.*, **66,** 76026 (1967).
20. R. G. Jones, *J. Am. Chem. Soc.*, **78,** 159 (1956).
21. Japanese Pat. 191 (1967); *Chem. Abstr.*, **66,** 76025 (1967).
22. P. R. S. Gupta and A. C. Sircar, *J. Indian Chem. Soc.*, **9,** 145 (1932).
23. W. L. F. Armarego, *J. Chem. Soc.*, **1963,** 6073.
24. G. Queguiner, M. Alas, and P. Pastour, *Compt. Rend. C.*, **265,** 824 (1967).
25. B. M. Bain and J. E. Saxton, *J. Chem. Soc.*, **1961,** 5216.
26. G. Queguiner and P. Pastour, *Compt. rend. C.*, **266,** 1459 (1968).
27. Unpublished data from this laboratory. Parameters for LCAO Calculations were taken from A. Streitwieser *Molecular Orbital Theory for Organic Chemists*, Wiley, New York, 1961, p. 135.
28. S. C. Wait and J. W. Wesley, *J. Mol. Spectrosc.*, **19,** 25 (1966).
29. Y. Nitta, I. Matsuura, and F. Yoneda, *Chem. Pharm. Bull. Tokyo*, **13,** 586 (1965); *Chem. Abstr.*, **63,** 993 (1965).
30. J. H. Bowie, R. G. Cooks, P. F. Donaghue, J. A. Halleday, and H. J. Rodda, *Australian J. Chem.*, **20,** 2677 (1967).
31. P. Hemmerich and S. Fallab, *Helv. Chim. Acta*, **41,** 498 (1958).
32. T. Yoshino, *Nippon Kagaku Zasshi*, **81,** 173 (1960); *Chem. Abstr.*, **56,** 449 (1962).
33. G. Gheorghiu, *Bull. Soc. Chim. France*, **53,** 151 (1933); *Chem. Abstr.*, **27,** 3216 (1933).
34. Japanese Pat. 26,978 (1964); *Chem. Abstr.*, **62,** 11825 (1965).
35. N. R. Patel, W. M. Rich, and R. N. Castle, *J. Heterocyclic Chem.*, **5,** 13 (1968).
36. Japanese Pat. 22,266 (1965); *Chem. Abstr.*, **64,** 3567 (1966).
37. Japanese Pat. 3908 (1966); *Chem. Abstr.*, **65,** 726 (1966).
38. Japanese Pat. 18,831 (1966); *Chem. Abstr.*, **66,** 4410 (1967).
39. A. Krbavčič, B. Stanovnik, and M. Tišler, *Croat. Chem. Acta*, **40,** 181 (1968).
40. Japanese Pat. 3909 (1966); *Chem. Abstr.*, **65,** 726 (1966).

41. B. Stanovnik, A. Krbavčič, and M. Tišler, *J. Org. Chem.*, **32,** 1139 (1967).
42. German Pat. 948,976 (1956); *Chem. Abstr.*, **53,** 2263 (1959).
43. J. Shavel, F. Leonard, F. H. McMillan, and J. A. King, *J. Am. Pharm. Assoc., Sci. Ed.*, **42,** 402 (1953); *Chem. Abstr.*, **48,** 9325 (1954).
44. H. A. Offe, W. Siefken, and G. Domagk, *Z. Naturforsch.*, **76,** 462 (1952); *Chem. Abstr.*, **47,** 3929 (1953).
45. H. Stracke, *Monatsh. Chem.*, **11,** 133 (1890).
46. H. Meyer and J. Malley, *Monatsh. Chem.*, **33,** 393 (1912).
47. H. L. Yale, K. Losee, J. Martins, M. Holsing, F. M. Perry, and J. Bernstein, *J. Am. Chem. Soc.*, **75,** 1933 (1953).
48. T. S. Gardner, F. A. Smith, E. Wenis, and J. Lee, *J. Org. Chem.*, **21,** 530 (1956).
49. M. Ridi and S. Checchi, *Ann. chim. (Rome)*, **47,** 728 (1957).
50. M. Ridi, *Ann. Chim. (Rome)*, **49,** 944 (1959).
51. G. Ya. Kondrat'eva and Chin-Heng Huang, *Dokl. Akad. Nauk SSSR*, **131,** 94 (1960); *Chem. Abstr.*, **54,** 12131 (1960).
52. G. Ya. Kondrat'eva and H. Chi-Heng, *Dokl. Akad. Nauk SSSR*, **164,** 816 (1965); *Chem. Abstr.*, **64,** 2079 (1966).
53. P. Papini, M. Ridi, and S. Checchi, *Gazz. Chim. Ital.*, **90,** 1399 (1960).
54. Y. S. Kao and R. Robinson, *J. Chem. Soc.*, **1955,** 2865.
55. F. Micheel and H. Dralle, *Ann.*, **670,** 57 (1963).
56. M. J. Reider and R. C. Elderfield, *J. Org. Chem.*, **7,** 286 (1942).
57. C. Musante and S. Fatutta, *Ann. Chim. (Rome)*, **47,** 385 (1957).
58. L. Novaček, K. Palat, M. Čeladnik, and E. Matuškova, *Českoslov. Farm.*, **11,** 76 (1962); *Chem. Abstr.*, **57,** 15067 (1962).
59. K. Palat, M. Čeladnik, L. Novaček, M. Polster, R. Urbančik, and E. Matuškova, *Acta Fac. Pharm. Brun. Bratislav.*, **4,** 65 (1962); *Chem. Abstr.*, **57,** 4769 (1962).
60. German Pat. 951,995 (1956); *Chem. Abstr.*, **53,** 4313 (1959).
61. U.S. Pat. 2,786,839 (1957); *Chem. Abstr.*, **51,** 16567 (1957).
62. Japanese Pat. 19,429 (1966); *Chem. Abstr.*, **66,** 37938 (1967).
63. German Pat. 958,561 (1957); *Chem. Abstr.*, **53,** 8178 (1959).
64. E. M. Bavin, D. J. Drain, M. Seiler, and D. E. Seymour, *J. Pharm. Pharmacol.*, **4,** 844 (1952); *Chem. Abstr.*, **47,** 1841 (1953).
65. British Pat. 753,787 (1956); *Chem. Abstr.*, **51,** 7442 (1957).
66. U.S. Pat. 2,742,467 (1956); *Chem. Abstr.*, **51,** 1305 (1957).
67. C. M. Atkinson and B. N. Biddle, *J. Chem. Soc.*, **1966,** 2053.
68. M. Prasad and C. G. Wermuth, *Compt. rend. C.*, **264,** 405 (1967).
69. R. C. Cookson and N. S. Isaacs, *Tetrahedron*, **19,** 1237 (1963).
70. R. M. Acheson and M. W. Foxton, *J. Chem. Soc. C.*, **1966,** 2218.
71. T. Yamazaki, M. Nagata, H. Sugano, and N. Inoue, *Yakugaku Zasshi*, **88,** 216 (1968); *Chem. Abstr.*, **69,** 77193 (1968).
72. E. E. Mikhlina, N. A. Komarova, and M. V. Rubtsov, *Khim. Geterots. soed., Akad. Nauk Latv. SSSR*, **1966,** 91; *Chem. Abstr.*, **64,** 19602 (1966).
73. USSR Pat. 170,506 (1965); *Chem. Abstr.*, **63,** 16362 (1965).
74. G. Queguiner, M. Alas, and P. Pastour, *Compt. Rend. C*, **268,** 1531 (1969).
75. British Pat. 1,100,911 (1968); *Chem. Abstr.*, **69,** 19181 (1968).
76. G. Favini and G. Bueni, *Theor. Chim. Acta*, **13,** 79 (1969).
77. A. Krbavčič, B. Stanovnik, and M. Tišler, *Croat. Chem. Acta*, **40,** 181 (1968).
78. B. Stanovnik and M. Tišler, *Tetrahedron*, **25,** 3313 (1969).
79. B. Stanovnik, M. Tišler, and A. Vrbanič, *J. Org. Chem.*, **34,** 996 (1969).
80. G. Queguiner and P. Pastour, *Bull. Soc. Chim. France*, **1969,** 4082.

81. G. Queguiner and P. Pastour, *Bull. Soc. Chim. France*, **1969**, 3678.
82. M. Lora-Tamayo, J. L. Soto, and E. D. Toro, *Anal. Real Soc. Espan. Fis. Quim. (Madrid) Ser B.*, **65**, 1125 (1969).
83. S. Kakimoto and S. Tonooka, *Bull. Chem. Soc. Japan*, **42**, 2996 (1969).
84. G. Queguiner and P. Pastour, *Bull. Soc. Chim. France*, **1969**, 2519.
85. V. Pirc, B. Stanovnik, M. Tišler, J. Marsel, and W. W. Paudler, *J. Heterocyclic Chem.*, **7**, 639 (1970).
86. D. B. Paul and H. J. Rodda, *Aust. J. Chem.*, **22**, 1745 (1969).
87. D. B. Paul and H. J. Rodda, *Aust. J. Chem.*, **22**, 1759 (1969).
88. I. Matsuura and K. Okui, *Chem. Pharm. Bull. (Tokyo)*, **17**, 2266 (1969).
89. M. M. Kochhar, *J. Heterocyclic Chem.*, **6**, 977 (1969).
90. T. Yamazaki, M. Nagata, F. Nohara, and S. Urano, *Chem. Pharm. Bull. (Tokyo)*, **19**, 159 (1971).

Part J. Pyridazinopyridazines

I. Pyridazino[4,5-*d*]pyridazines

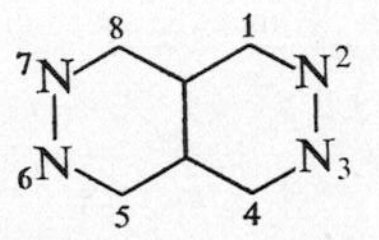

Pyridazino[4,5-d]pyridazine, RRI 1593
(2,3,6,7-Tetraazanaphthalene)

The first compound claimed to be a derivative of this bicyclic system was reported by Bülow in 1904 (1). He reinvestigated the reaction between diethyl diacetylsuccinate and one or two moles of hydrazine which, according to Curtius (2), afforded two products, formulated as **1** and **2**. Bülow showed that **1** could be converted with hydrazine into **2**, but he discarded the structure **2** and proposed instead structure **3**.

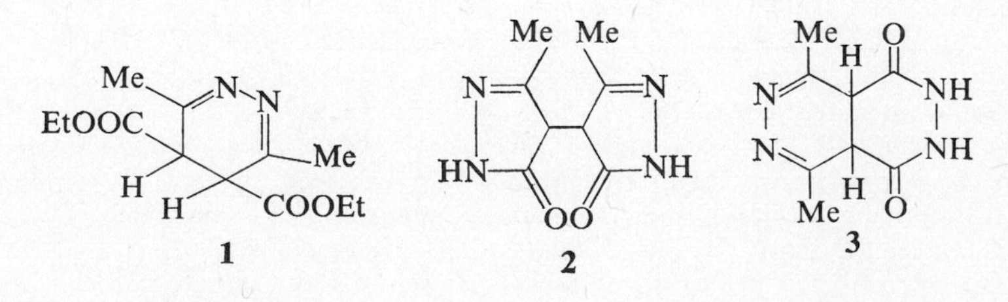

Seka and Preissecker (3) later obtained the same product either from the reaction between diethyl diacetylsuccinate and hydrazine or from diethyl 2,5-dimethylfuran-3,4-dicarboxylate and hydrazine and formulated it as **3**. Still later it was concluded by Jones (4) and shown by Mosby (5) on the basis of spectroscopic evidence that the structure of the product for which Bülow or Seka and Preissecker proposed formula **3** was not correct and that this compound was in fact **2**.

A similar structural problem was encountered for the reaction between tetrabenzoylethylene, which can react as a 1,3- or 1,4-diketone, and hydrazine. The reaction, when conducted in acetic acid, gave rise to two products for which structures **4** and **5** were proposed (6). Even with a tenfold excess of

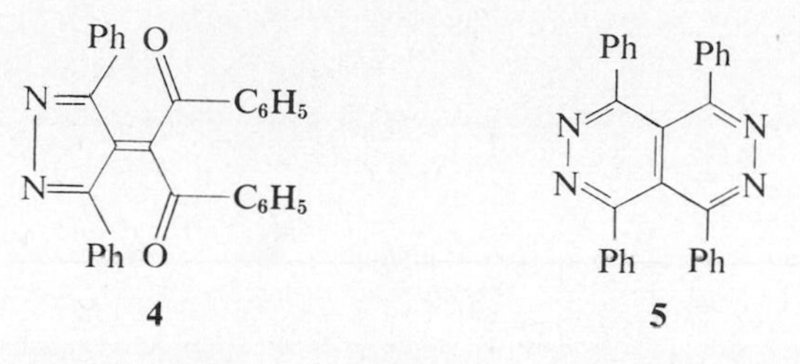

hydrazine a mixture of both compounds resulted. Structure proposal for **5** was made mainly on the basis of uv spectroscopic data; more evidence is warranted to justify this assignment.

A somewhat different behavior was observed recently for the reaction between tetraacetyl ethylene and hydrazine, which reacted upon heating to give a mixture of 1,4,5,8-tetramethylpyridazino[4,5-*d*]pyridazine and 6-amino-1,4,5,7-tetramethylpyrrolo[3,4-*d*]pyridazine (44).

It was conclusively shown that the related 3,4-diacetylhexane-2,5-dione-(tetraacetylethane) does not form with hydrazine a pyridazino[4,5-*d*]-pyridazine derivative, but that instead a bis-pyrazolo derivative resulted (5). However, Adembri, DeSio, and Nesi (7) were able to accomplish the reaction between tetraethyl ethylenetetracarboxylate and hydrazine and the structure of the product **6** is in accord with spectroscopic and chemical evidence.

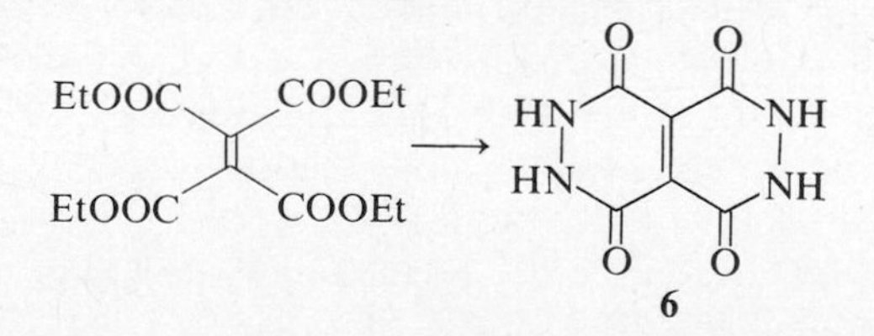

Most preparations of pyridazino[4,5-*d*]pyridazines utilize appropriate pyridazine precursors as starting material. Pyridazino[4,5-*d*]pyridazine (**8**) itself was synthesized recently (8) from diethyl pyridazine-4,5-dicarboxylate, which was reduced at -70° with lithium aluminum hydride and the resulting solution of the intermediate **7** was thereafter treated with hydrazine.

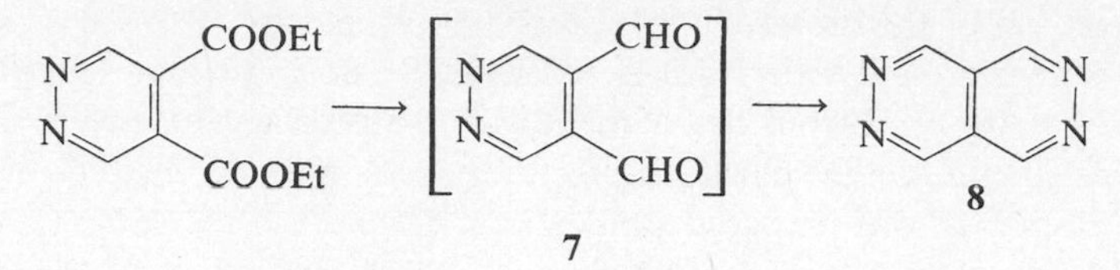

The most convenient syntheses of pyridazino[4,5-*d*]pyridazines are those from esters of pyridazine-4,5-dicarboxylic acids and hydrazine (4, 7, 9, 43). The free acid reacts equally well, but higher temperatures are required to complete the reaction. A high-boiling solvent, such as triethylene glycol, is of advantage (9, 10). The product, pyridazino[4,5-*d*]pyridazine-1,4(2*H*,3*H*)-dione, was found to exist in the solid state in the monoenol form **9** on the basis of uv and ir spectroscopic evidence (7, 43).

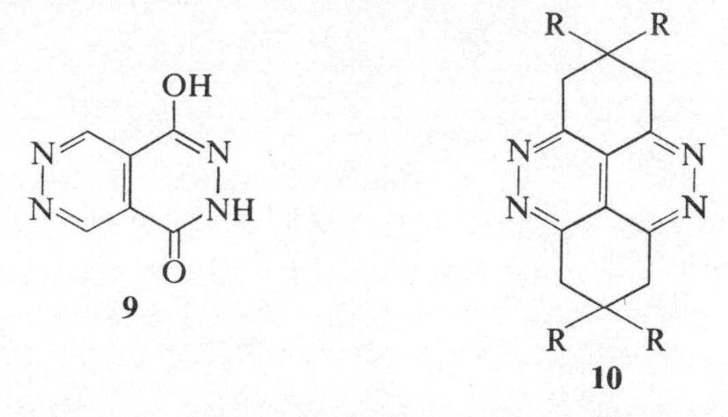

In another synthetic variant DiStefano and Castle (9) were able to cyclize 4,5-dicyanopyridazine with hydrazine at room temperature and 1,4-diamino-pyridazino[4,5-*d*]pyridazine could be obtained in moderate yield.

A derivative of pyridazino[4,5-*d*]pyridazine was formed from 1,3-cyclohexanedione and an excess of hydrazine hydrate. The structure of the product was found to be **10** (11).

Pyridazino[4,5-*d*]pyridazine is a solid, soluble in water and slightly soluble in ethanol. Its uv and ir spectra have been recorded and no covalent hydration has been observed (8).

The crystal structure of pyridazino[4,5-*d*]pyridazine has been determined and the C—C bonds are comparable to those in naphthalene, except for the C—C bonds common to both rings where a shorter distance was observed (45). MO-π-bond order-bond lengths relations for this bicycle have also been calculated (46).

MO calculations (12, 13) of total and frontier π-electron densities for this highly symmetric heterocycle have been performed and the results are presented in Table 3J-1. Considering these values, only nucleophilic displacement reactions are expected to take place on this system. Several reactions common to related heterocycles have been performed.

Methylation of the dione **11** with excess diazomethane afforded a mixture of two isomeric methyl derivatives, **12** and **13**. Both **12** and **13** could be obtained equally well from the *N*-methyl derivative **14**. Compound **14** was synthesized from diethyl pyridazine-4,5-dicarboxylate and methylhydrazine and the isomer **12** was prepared similarly from dimethylhydrazine (7, 43).

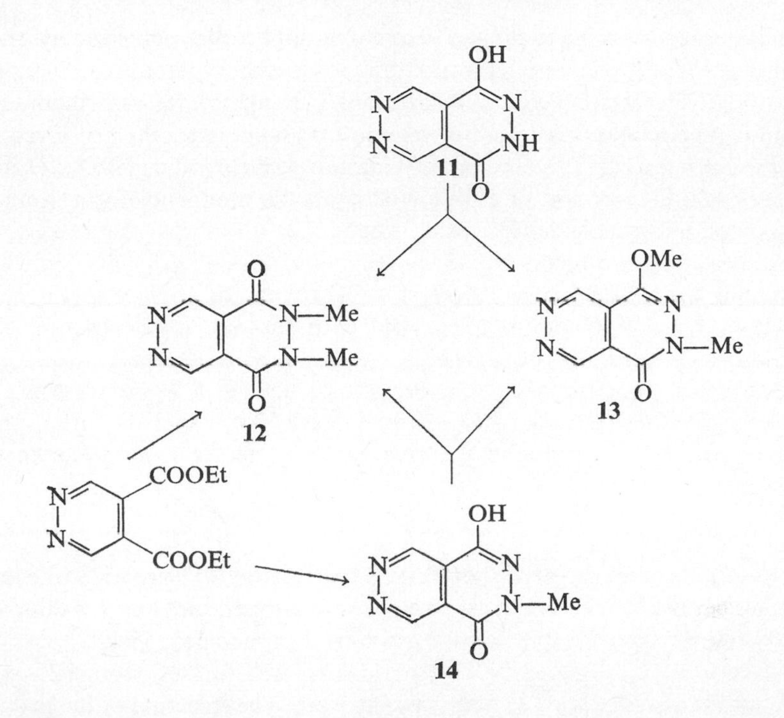

Compound **11** could be *O*-monoacylated or transformed into the corresponding sulfur analog by the phosphorus pentasulfide-pyridine method (9, 10). Subsequent treatment with methyl iodide afforded 1,4-bis(methylthio)-pyridazino[4,5-*d*]pyridazine, which was used in some displacement reactions since the 1,4-dichloro analog could not be easily formed from **11**. Only one methylthio group could be displaced in the reaction with normal sulfuric acid to produce **15**, or with amines and alkoxides to form compounds of the types **16** and **17** (10). Compounds of types **17** and **16** were used for the synthesis of substituted 1-amino-4-alkoxy-, 1,4-dialkoxypyridazino[4,5-*d*]pyridazines. Preferential replacement of the methoxyl group as compared to the methylthio group was observed. Steric, polar, and resonance effects involved in these replacements are discussed (10).

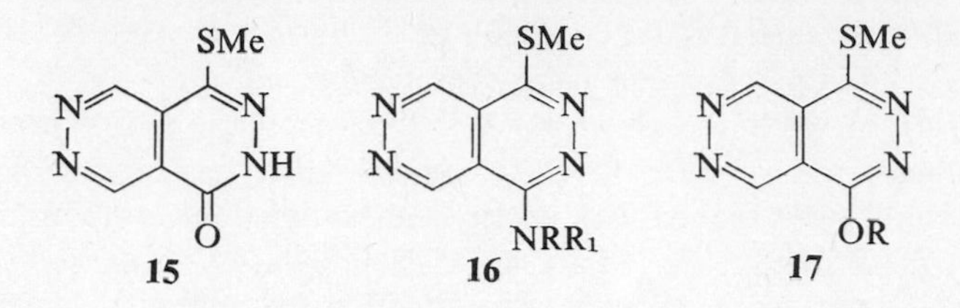

In addition to the direct synthesis of 1,4-diaminopyridazino[4,5-*d*]-pyridazine, it is possible to prepare this compound from 1,4-bis(methylthio)-pyridazino[4,5-*d*]pyridazine with a saturated solution of ammonia in ethanol under relatively drastic conditions (195–205°, 26 hr) (9). The diamino compound can be monacylated and forms a nitramino derivative.

Although it appears from the work of DiStefano and Castle (9) and Dorman (10) that the preparation of 1,4-dichloropyridazino[4,5-*d*]pyridazine under normal conditions does not proceed, Adembri et al. (43) were recently able to prepare this compound in low yield with PCl_5-$POCl_3$ (100°, 14 hr). DiStefano and Castle were successful in preparing 1-chloropyridazino-[4,5-*d*]pyridazin-4(3*H*)one in 17% yield with the aid of phosphorus oxychloride in pyridine.

It was also reported, without experimental details, that conversion of **6** into its 1,4,5,8-tetrachloro analog **18** was possible and that the latter could be converted into the tetraethoxy derivative **19** by means of sodium ethoxide (7).

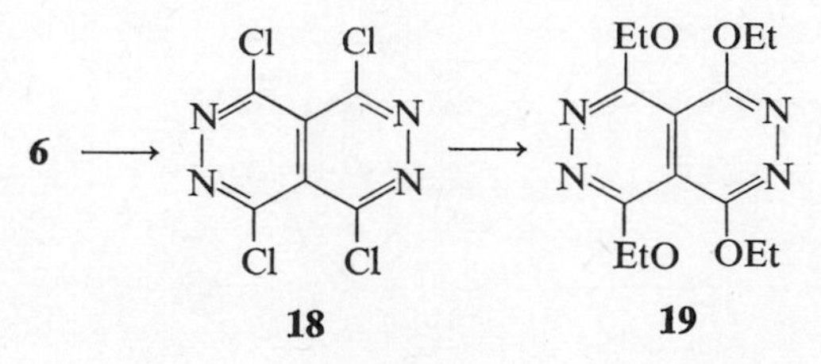

1,4,5,8-Tetramethylpyridazino[4,5-*d*]pyridazine was oxidized by means of potassium permanganate to 3,6-dimethylpyridazine-4,5-dicarboxylic acid (44).

II. Pyridazino[4,5-*c*]pyridazines

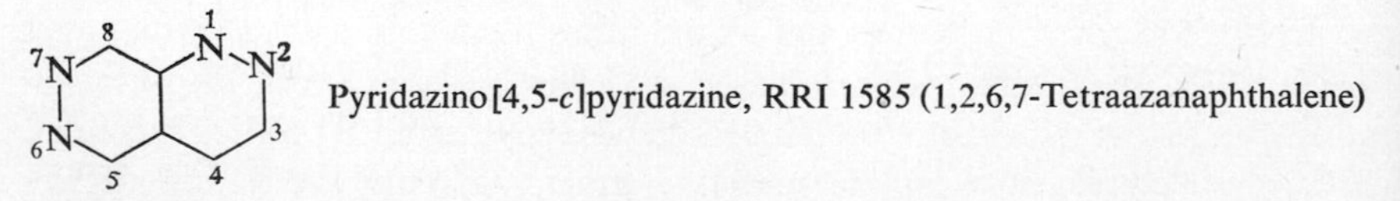

Pyridazino[4,5-*c*]pyridazine, RRI 1585 (1,2,6,7-Tetraazanaphthalene)

The first report about a representative of this heterocyclic system was published in 1954 by Gault et al. (14). They treated the reduced pyridazine

20 with hydrazine and isolated a product to which they assigned formula **21**.

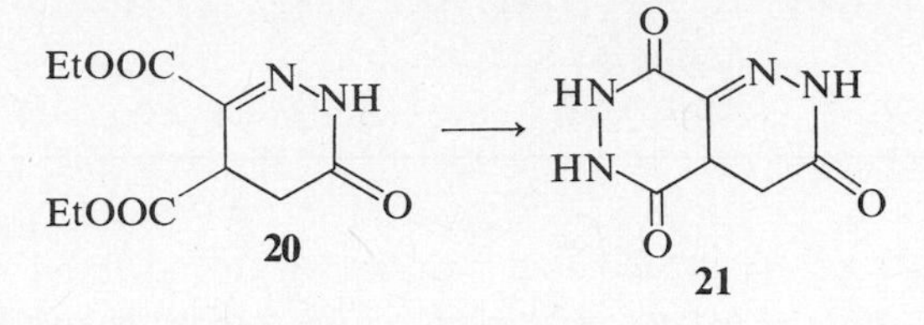

All other data about the synthesis and reactivity of this heteroaromatic system are due to Singerman and Castle (15), who presented their findings in a recent publication. They treated diethyl pyridazine-3,4-dicarboxylate and its 6-methyl analog with hydrazine and obtained the bicyclic products **22** in good yields.

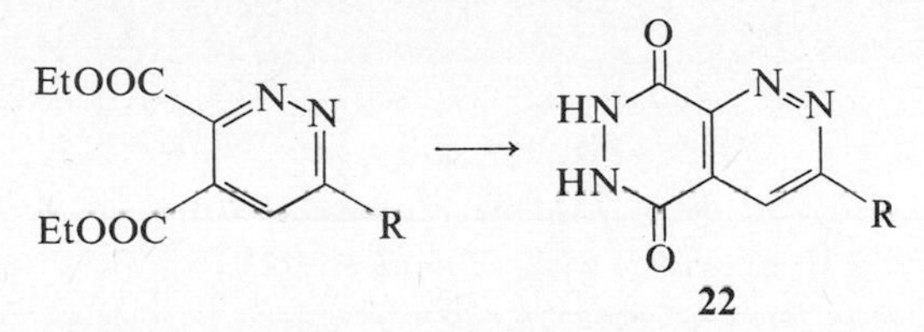

It is also possible to convert the esters via diamides into 3,4-dicyanopyridazines, which, after being treated with hydrazine, form 5,8-diaminopyridazino[4,5-*c*]pyridazine or its 3-methyl analog.

The parent compound remains unknown; nevertheless, MO calculations (12, 13) of total and frontier π-electron densities are recorded and presented in Table 3J-2.

Few reactions on this system are known. The otherwise common conversion of an oxo group into a chloro substituent failed in the case of **22** and this is comparable to low reactivity of the related pyridazino[4,5-*d*]pyridazine system. Thiation of **22** (R = Me) by means of phosphorus pentasulfide in boiling pyridine was possible under elaborated reaction conditions.

III. Pyridazino[1,2-*a*]pyridazines

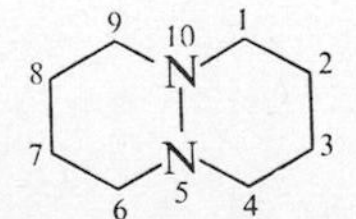

Perhydropyridazino[1,2-*a*]pyridazine, RRI 1594

Only reduced and oxo derivatives of this bicyclic system are known. Tetrahydropyridazines, containing the NH—NH group as a part of the heterocycle, are capable of condensing with appropriate aliphatic difunctional compounds to form the annelated six-membered cycle.

An intriguing structural problem represented the product obtained either in the reaction between cyclic succinhydrazide and succinoyl chloride or from the reaction of 2 moles of succinic acid in the presence of 1 mole of hydrazine. Feuer, Bachman, and White (16) obtained a compound to which they erroneously assigned structure **25** (a) by allowing either cyclic succinhydrazide (**23**) to react with diethyl succinate in nitrobenzene; or (b) by allowing **23** to react with succinoyl chloride in dioxane; or (c) by treating polysuccinhydrazide or cyclic succinhydrazide with benzenesulfonyl chloride.

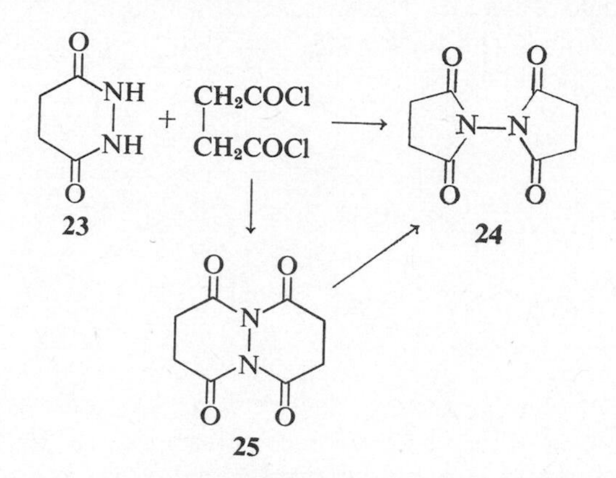

The same product can be obtained by treating with polyphosphoric acid or dehydrating thermally 2 moles of succinic acid and 1 mole of hydrazine hydrate (17). Again, the product was erroneously assigned to have structure **25**.

It was later shown that the products of Feuer et al. were in fact *N,N'*-bisuccinimide (**24**). Authentic perhydro-1,4,6,9-tetraoxopyridazino[1,2-*a*]-pyridazine (**25**) was synthesized by Hedaya et al. (18, 19) from cyclic succinhydrazide and succinoyl chloride in a solution of toluene instead of dioxane. Compound **25** can be rearranged into the imide **24** when heated under reflux in a solution of dioxane in the presence of either succinoyl chloride or hydrochloric acid or thermally, and can be reduced to perhydropyridazo[1,2-*a*]-pyridazine (**26**) with lithium aluminum hydride. The structural assignment of **26** was based on ir and nmr spectroscopic data and by establishing identity

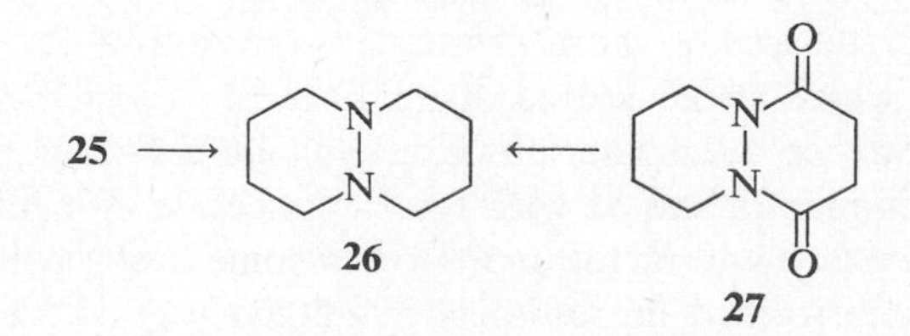

with the product obtained by reduction of **27** to **26** according to Stetter and Spangenberger (20).

Although first attempts to prepare the unsaturated analog of **25** were not successful (21), Kealy (22, 23) and later Hedaya et al. (24), who reinvestigated the reaction, were able to prepare **30** in the following manner. Maleic hydrazide (or its 4,5-difluoro analog) when treated with *t*-butyl hypochlorite at $-50°$ to $-70°$ was transformed into very unstable diazaquinone (**28**, R = H) which, as the most reactive among dienophiles, reacted instantly with dienes to form products of type **29**. However, if the solution of **28** was warmed up, nitrogen was evolved and the pyridazino[1,2-*a*]pyridazine derivative **30** was obtained in 43% yield as the principal product.

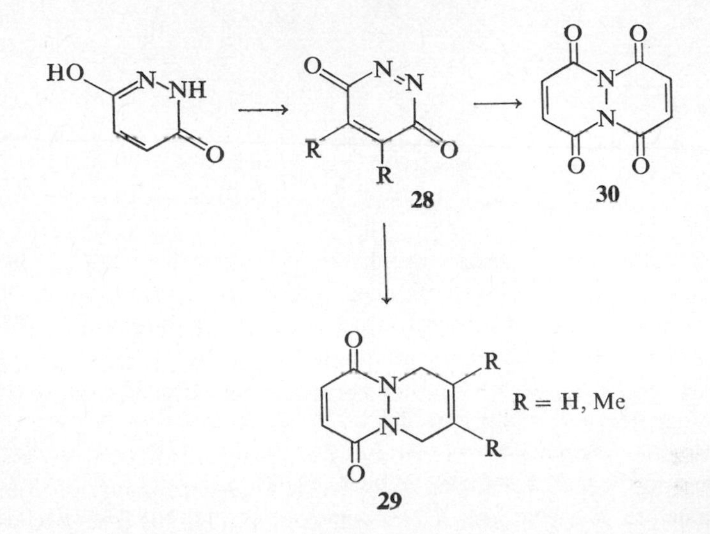

The bicycle **30** has been hydrolyzed in boiling water or with aniline in acetic acid to maleic acid and maleic hydrazide (or mixture of the dianilide of maleic and/or fumaric acid) and the structure of **29** and **30** followed from their reduction to the known compounds **27** and **25**. From the carbonyl stretching frequency in the ir spectrum it was concluded that **30** represented a strained ring system (22).

Clement (25) devised a more convenient procedure for obtaining the diazaquinones which were used as dienophiles in the Diels-Alder reaction. Maleic hydrazide or 1,2,4,5-tetrahydropyridazine-3,6-dione were oxidized to the diazaquinones **28** and **31** with lead tetraacetate in methylene chloride at room temperature and in the presence of some acetic acid and a diene. The adducts were formed immediately in relatively good yields. Oxidation with nickel peroxide has been also reported recently to generate the diazaquinone from maleic hydrazide (50).

The adduct from **31** and butadiene (25), having the formula **32**, was also obtained by either reacting succinic anhydride with 1,2,3,6-tetrahydropyridazine (26, 27) or employing succinoyl chloride in the presence of pyridine (28).

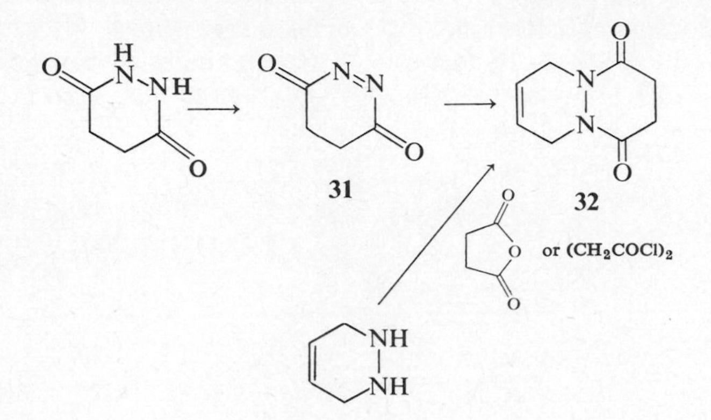

In addition to these reactions, pyridazino[1,2-*a*]pyridazines are reported to result from reactions between reduced pyridazines and different aliphatic compounds such as 1,4-dibromo- (29) or 1,4-dichlorobutane (30, 31), *meso*-1,4-dibromo-2,3-dihydroxybutane (32), maleic (28, 31) or succinic anhydride (33, 34), and 1,4-dicarboxylic acid chlorides (28, 35), derivatives of 1,4-ketoacids (49), or dibromodiacetyl (32). Compound **26** is reported to result also from the reaction between hydrazine and 1,4-dibromobutane in the presence of potassium hydroxide in low yield (36).

The reaction between *N*,*N'*-bis-(β-phenyl-γ-bromobutyryl)hydrazine (**33**) and aqueous sodium hydroxide yielded the dipyrrolidinedione **34** along with some product to which the structure of pyridazino[1,2-*a*]pyridazine (**35**) has

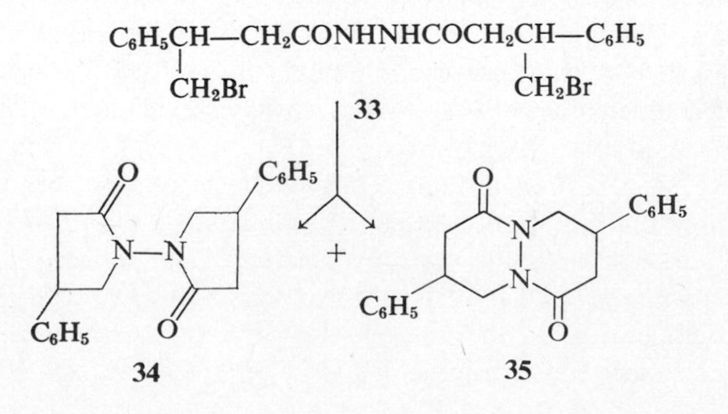

been assigned (37). In several mentioned cases there is insufficient evidence for the proposed structures and more data are needed for rigorous proof of these structural assignments.

In addition to the *trans*-diol **36**, Mayer and Gabler (38) were able to isolate from the reaction mixture resulting from the reaction between D,L-butadiene dioxide and aqueous hydrazine another compound (6%) to which the structure of **37** has been assigned. Compound **37** can be obtained in yields up to 21% from the treatment of **36** with D,L-butadiene dioxide. Compound **37**

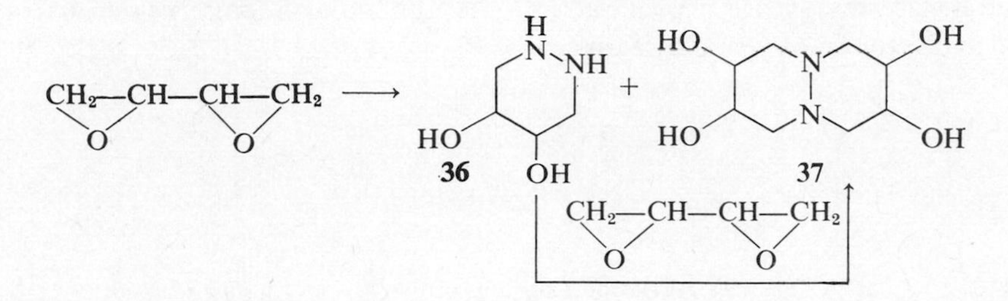

formed a tetraacetyl derivative and its nmr spectrum is in accordance with the proposed structure. A more detailed nmr spectroscopic investigation about the conformational arrangement of the tetraacetyl derivative of **37** by Koch and Zollinger (39) presented evidence for tetraaxial orientation of the acetoxy groups (38). The attempted preparation of **39** from pyridazine and either *cis* or *trans*-1,4-dibromo-2-butene failed (40).

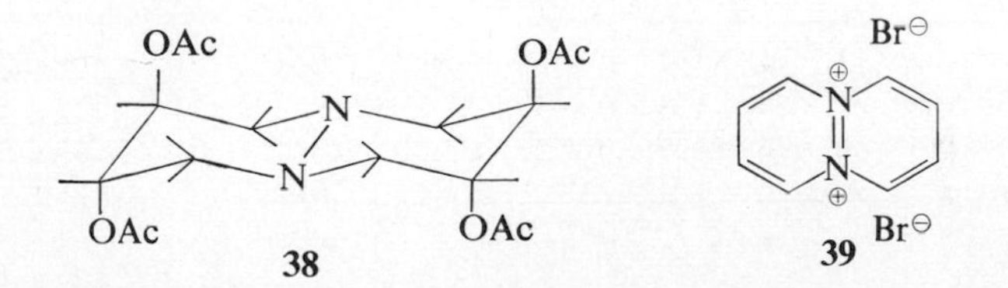

Finally, it should be mentioned that an analytical method for the determination of **25** has been devised (41) and that the pyridazino[1,2-*a*]pyridazine system has been studied by simple MO method (42). Conformational changes in **40** have been studied by nmr spectroscopy and the energy barrier to conformational inversion has been established (33).

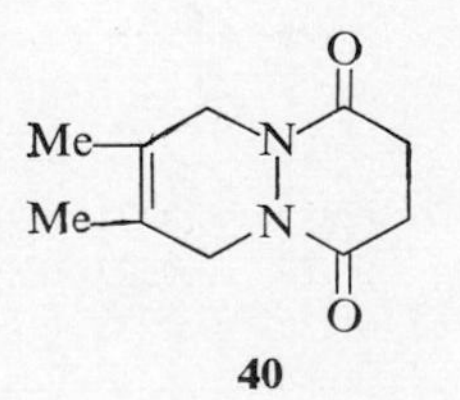

IV. Pyridazino[3,4-*c*]pyridazine

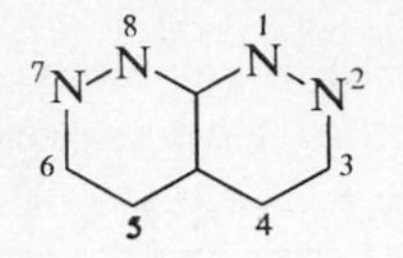

It was recently claimed (47) that a derivative of this system was obtained, but it was subsequently shown (48) that the compound is actually a derivative of pyridazino[6,1-*c*]-1,2,4-triazine.

V. Tables

TABLE 3J-1. Total and Frontier π-Electron Densities of Pyridazino[4,5-*d*]pyridazine

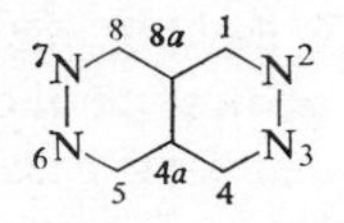

Position	1	2	3	4	4*a*	5	6	7	8	8*a*
Total (12, 13)	0.9093	1.1216	1.1216	0.9093	0.9382	0.9093	1.1216	1.1216	0.9093	0.9382
	0.888	1.133	1.133	0.888	0.958	0.888	1.133	1.133	0.888	0.958
Frontier (12)	0.1524	0.0976	0.0976	0.1524	0.0000	0.1524	0.0976	0.0976	0.1524	0.0000

TABLE 3J-2. Total and Frontier π-Electron Densities of Pyridazino[4,5-*c*]pyridazine

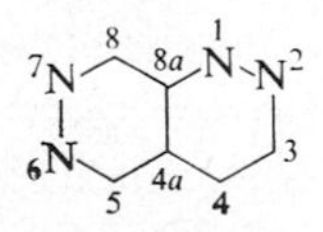

Position	1	2	3	4	4*a*	5	6	7	8	8*a*
Total (12, 13)	1.1154	1.0803	0.9565	0.8977	0.9433	0.9060	1.1297	1.1227	0.9115	0.9368
	1.105	1.084	0.934	0.924	0.957	0.891	1.138	1.134	0.897	0.937
Frontier (12)	0.1964	0.0339	0.0965	0.1762	0.0025	0.1402	0.1115	0.0897	0.1531	0.0001

TABLE 3J-3. Pyridazino[4,5-*d*]pyridazines

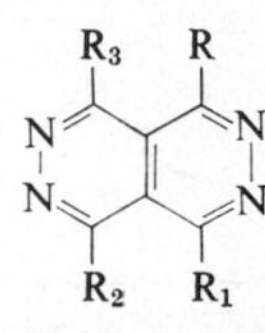

R	R_1	R_2	R_3	MP (°C) or BP (°C/mm)	References
None				~290 (dec.)	8
Chloro	Chloro			165–166	7, 43
Amino	Amino			>300	9
NHCOMe	Amino			>270	9
$NHCOC_6H_5$	Amino			237–240	9
NMe_2	Methoxy			125–126.5	10
$NHCH_2CH_2NMe_2$	Methoxy			138.5–140	10
				140.5–142	10
$OCH_2CH_2NMe_2$	NMe_2			85–86	10
$OCH_2CH_2NMe_2$	$OCH_2CH_2NMe_2$			2HBr, 248.5–250 (dec.)	10
$NHNO_2$	Amino			249–250 (dec.)	9
NMe_2	Methylthio			161–162	10
$NHCH_2CH_2NMe_2$	Methylthio			140–141.5	10

TABLE 3J-3 (*continued*)

R	R_1	R_2	R_3	MP (°C) or BP (°C/mm)	References
Hydrazino	Methylthio			~168 (dec.)	10
				HBr, 208–210	10
				$HBr \cdot H_2O$, 201–202	10
—NHN=CH—(5-nitro-2-furyl) [$-NO_2$]	Methylthio			246–246.5	10
3,5-dimethylpyrazol-1-yl (—N—N, Me, Me)	Methylthio			226–227	10
$OCH_2CH_2NMe_2$	Methylthio			HBr, 218–218.5 (dec.)	10
$OCH_2CH_2\overset{\oplus}{N}Me_3I^{\ominus}$	Methylthio			1/2 MeOH, 236–236.5	10
Methoxy	Methoxy			157–159	43
Methoxy	Methylthio			156–157	10
Methylthio	Methylthio			195	9
				192–193.5	10
Chloro	Chloro	Chloro	Chloro	200–300 (dec.)	7
Ethoxy	Ethoxy	Ethoxy	Ethoxy	259–260	7
Phenyl	Phenyl	Phenyl	Phenyl	355–360	6
Methyl	Methyl	Methyl	Methyl	262–263	44

TABLE 3J-4. Pyridazino[4,5-*d*]pyridazin-1(2*H*)ones

R	R_1	MP (°C) or BP (°C/mm)	References
	Chloro	283	9
	$OCH_2CH_2NMe_2$	168.5–169.5	10
	OCOMe	245–246	9
	OTs	225–226	9
	Methylthio	249–250	10
Methyl	Methoxy	139–140	7, 43

LE 3J-5. Pyridazino[4,5-*d*]pyridazine-1,4(2*H*,3*H*)diones

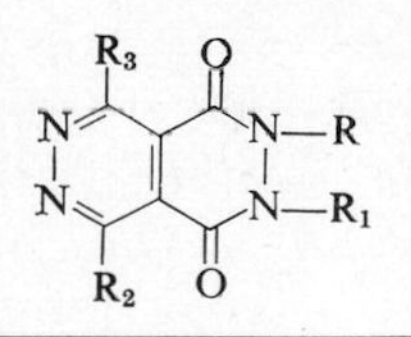

R	R_1	R_2	R_3	MP (°C) or BP (°C/mm)	References
				300 (dec.)	9
				>315	10
				>340	7, 43
	Methyl			>300 (dec.)	7, 43
yl	Methyl			239–240	7, 43
		Methyl	Methyl	320 (dec.)	4
				>300	9
				>340	7
				306–307[a]	7

etramethyl derivative (structure as bis-OMe, bis-*N*-Me-compound proposed).

TABLE 3J-6. Pyridazino[4,5-*c*]pyridazines

Structure	R	MP (°C) or BP (°C/mm)	References
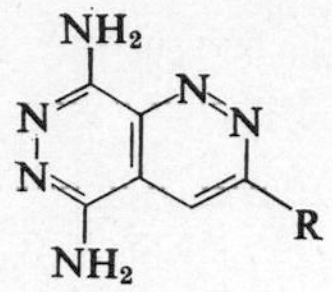	None	360[a]	15
	Methyl	—[b]	
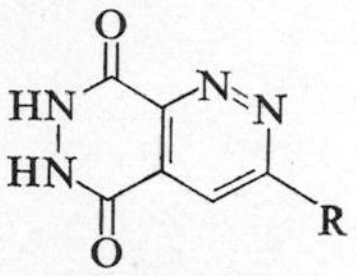	None	$\cdot H_2O$, 336–338 (dec.)	15
	Methyl	270–274 (dec.)	15
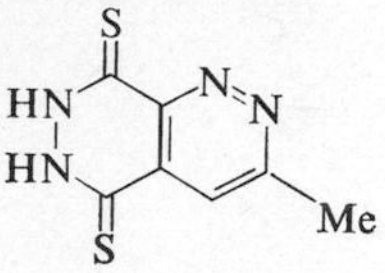		—[c]	15
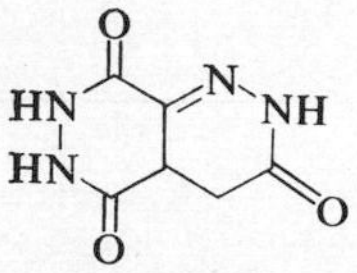		230	14

[a] Shrinks at 165° C.
[b] Gradually blackens and decomposes over 164° C.
[c] Blackens at 210° C and then gradually decomposes as temperature is raised.

TABLE 3J-7. Pyridazino[1,2-*a*]pyridazines

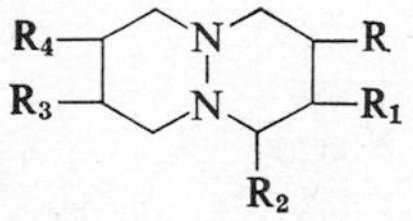

R	R_1	R_2	R_3	R_4	MP (°C) or BP (°C/mm)	References
None					oil	30
					Diperchlorate, 229	30
					66/11 mm	36
					Diperchlorate, 229	36
					MeI, 251	36
					76–77/14–15 mm	31
					Diperchlorate, 229	31
					MeI, 251	31
					79–80/18 mm	20

TABLE 3J-7 (*continued*)

R	R_1	R_2	R_3	R_4	MP (°C) or BP (°C/mm)	References
					Dipicrate, 155–158 (dec.)	20
					Dipicrate, 155	19, 18
					40/5 mm	28
					Diperchlorate, 231	28
OH	OH		OH	OH	Monohydrate, 303	32
					Monohydrate, 303–304 (dec.)	38
AcO	AcO		AcO	AcO	264	38
	OH	OH	OH	OH	302	32
					Tetraacetyl der., 267	32
		p-$MeOC_6H_4$—			66–68, 205/16	49
					HCl, 153–156	49
					MeI, 212–215	49

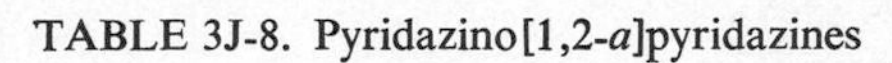

TABLE 3J-8. Pyridazino[1,2-*a*]pyridazines

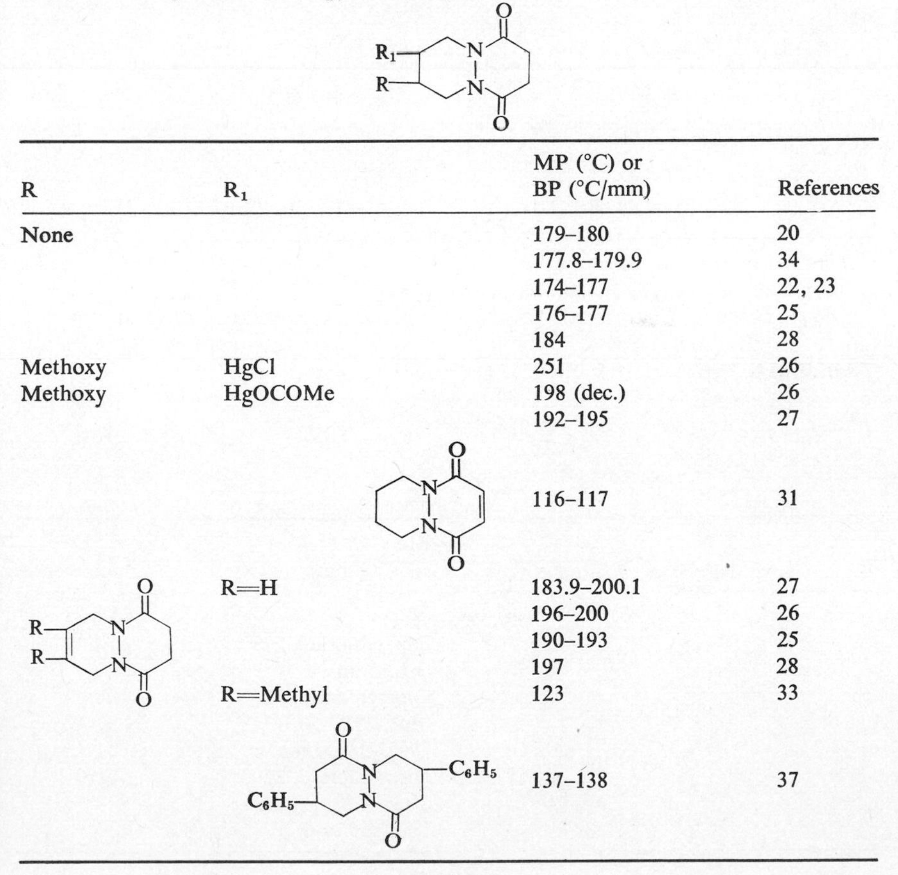

R	R_1	MP (°C) or BP (°C/mm)	References
None		179–180	20
		177.8–179.9	34
		174–177	22, 23
		176–177	25
		184	28
Methoxy	HgCl	251	26
Methoxy	HgOCOMe	198 (dec.)	26
		192–195	27
		116–117	31
	R=H	183.9–200.1	27
		196–200	26
		190–193	25
		197	28
	R=Methyl	123	33
		137–138	37

TABLE 3J-9. Pyridazino[1,2-*a*]pyridazines

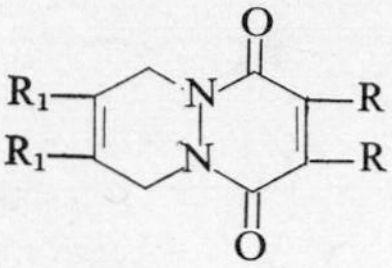

R	R_1	MP (°C) or BP (°C/mm)	References
None		156–157	25
		157–159	22, 23
		162	28
	Methyl	153–159 (dec.)	22
		161	23
Fluoro	Methyl	222.5–224 (dec.)	22
	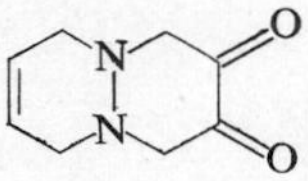	245	32
	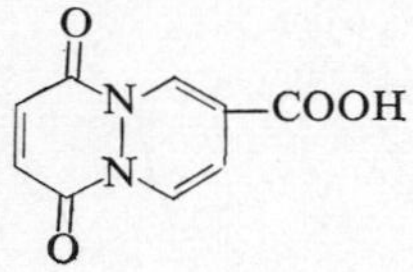	233 (dec)	22
	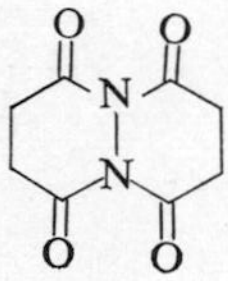	250 (dec.)	18, 24
		250–251	19
	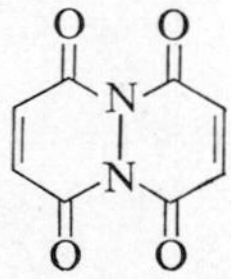	245	24
		247 (dec.)	22, 23
	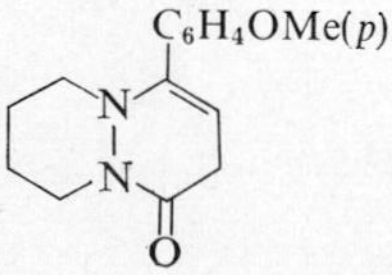	119–121	49

TABLE 3J-9 (*continued*)

R	R_1	MP (°C) or BP (°C/mm)	Reference
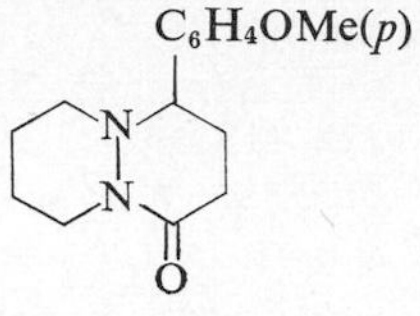		220.05 mm HCl, 174–176	49 49

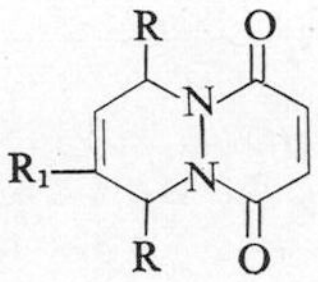

R	R_1	MP (°C)	Reference
	Methyl	117–118	50
Phenyl		225–226	50

References

1. C. Bülow, *Ber.*, **37,** 91 (1904).
2. T. Curtius, *J. Prakt. Chem.*, **50,** 508 (1894).
3. R. Seka and H. Preissecker, *Monatsh. Chem.*, **57,** 81 (1931).
4. R. G. Jones, *J. Am. Chem. Soc.*, **78,** 159 (1956).
5. W. L. Mosby, *J. Chem. Soc.*, **1957,** 3997.
6. H. Keller and H. Halban, *Helv. Chim. Acta*, **27,** 1253 (1944).
7. G. Adembri, F. DeSio, and R. Nesi, *Ricerca Sci.*, **37,** 440 (1967).
8. G. Adembri, F. DeSio, R. Nesi, and M. Scotton, *Chem. Commum.*, **1967,** 1006.
9. L. DiStefano and R. N. Castle, *J. Heterocyclic Chem.*, **5,** 53 (1968).
10. L. C. Dorman, *J. Heterocyclic Chem.*, **4,** 491 (1967).
11. J. K. Stille and R. Ertz, *J. Am. Chem. Soc.*, **86,** 661 (1964).
12. Unpublished data from this laboratory. Parameters for LCAO Calculations were taken from A. Streitwieser, *Molecular Orbital Theory for Organic Chemists*, Wiley, New York, 1961, p. 135.
13. S. C. Wait and J. W. Wesley, *J. Mol. Spectrosc.*, **19,** 25 (1966).
14. H. Gault, G. Kalopissis, N. Rist, and F. Grumbach, *Bull. Soc. Chim. France*, **1954,** 916.
15. G. M. Singermann and R. N. Castle, *J. Heterocyclic Chem.*, **4,** 393 (1967).
16. H. Feuer, G. B. Bachman, and E. H. White, *J. Am. Chem. Soc.*, **73,** 4716 (1951).
17. H. Feuer and J. E. Wyman, *Chem. Ind.* (*London*), **1956,** 577.
18. E. Hedaya, R. L. Hinman, and S. Theodoropulos, *J. Am. Chem. Soc.*, **85,** 3052 (1963).
19. E. Hedaya, R. L. Hinman, V. Schomaker, S. Theodoropulos, and L. M. Kyle, *J. Am. Chem. Soc.*, **89,** 4875 (1967).
20. H. Stetter and H. Spangenberger, *Chem. Ber.*, **91,** 1982 (1958).
21. H. Feuer and H. Rubinstein, *J. Org. Chem.*, **24,** 811 (1959).
22. T. J. Kealy, *J. Am. Chem. Soc.*, **84,** 966 (1962).
23. U.S. Pat. 3,062,820 (1962); *Chem. Abstr.*, **58,** 9101 (1963).
24. E. Hedaya, R. L. Hinman, and S. Theodoropulos, *J. Org. Chem.*, **31,** 1311 (1966).

25. R. A. Clement, *J. Org. Chem.*, **27**, 1115 (1962).
26. F. W. Gubitz and R. L. Clarke, *J. Org. Chem.*, **26**, 559 (1961).
27. U.S. Pat. 2,921,068 (1960); *Chem. Abstr.*, **54**, 8867 (1960).
28. I. Zugravescu and E. Carp, *Anal. Stiint. Univ. "Al I. Cuza" Iasi, Sect. I*, **11c**, 59 (1965); *Chem. Abstr.*, **63**, 14855 (1965).
29. R. L. Hinman and R. J. Landborg, *J. Org. Chem.*, **24**, 724 (1959).
30. M. Rink and K. Grabowski, *Naturwiss.*, **43**, 326 (1956); *Chem. Abstr.*, **52**, 18416 (1958).
31. M. Rink, S. Mehta, and K. Grabowski, *Arch. Pharm.*, **292**, 225 (1959).
32. E. Carp, M. Pahomi, and I. Zugravescu, *Anal. Stiint. Univ. "Al. I. Cuza" Iasi, Sect. IC*, **12**, 171 (1966); *Chem. Abstr.*, **68**, 49533 (1968).
33. B. Price, I. O. Sutherland, and F. G. Williamson, *Tetrahedron*, **22**, 3477 (1966).
34. U.S. Pat. 3,288,791 (1966); *Chem. Abstr.*, **66**, 65495 (1967).
35. A. Le Berre and J. Godin, *Compt. Rend.*, **260**, 5296 (1965); *Chem. Abstr.*, **63**, 6990 (1965).
36. M. Rink and S. Mehta, *Naturwiss.*, **45**, 313 (1958); *Chem. Abstr.*, **53**, 2246 (1959).
37. G. Cignarella, L. Fontanella, V. Aresi, and E. Testa, *Farm.* (Pavia), *Ed. Sci.*, **23**, 321 (1968).
38. H. R. Meyer and R. Gabler, *Helv. Chim. Acta*, **46**, 2685 (1963).
39. W. Koch and H. Zollinger, *Helv. Chim. Acta*, **46**, 2697 (1963).
40. A. E. Blood and C. R. Noller, *J. Org. Chem.*, **22**, 844 (1957).
41. P. R. Wood, *Anal. Chem.*, **25**, 1879 (1953).
42. R. D. Brown and B. A. W. Coller, *Mol. Phys.*, **2**, 158 (1959).
43. G. Adembri, F. DeSio, R. Nesi, and M. Scotton, *J. Chem. Soc. C*, **1968**, 2857.
44. G. Adembri, F. DeSio, R. Nesi, and M. Scotton, *J. Chem. Soc. C*, **1970**, 1536.
45. C. Sabelli, P. Tangocci, and P. F. Zanazzi, *Acta Crystallog.*, **B25**, 2231 (1969).
46. G. Häfelinger, *Chem. Ber.*, **103**, 3289 (1970).
47. T. LaNoce, E. Bellasio, A. Vigevani, and E. Testa, *Ann. Chim. (Roma)*, **59**, 552 (1969).
48. B. Stanovnik and M. Tišler, *Synthesis*, **1970**, 180.
49. P. Aeberli and W. J. Houlihan, *J. Org. Chem.*, **34**, 2720 (1969).
50. S. Takase and T. Motoyama, *Bull. Chem. Soc. Japan*, **43**, 3926 (1970).

Part K. Pyrimidopyridazines

I. Pyrimido[4,5-*d*]pyridazines

Pyrimido[4,5-*d*]pyridazine, RRI 1588(1,3,6,7-Tetraazanaphthalene)

There are only a few compounds known which belong to this heterocyclic system. Jones (1) in 1956 prepared the first compound representing this system by condensing diethyl 2-aminopyrimidine-4,5-dicarboxylate with hydrazine and obtained the product **1** in excellent yield.

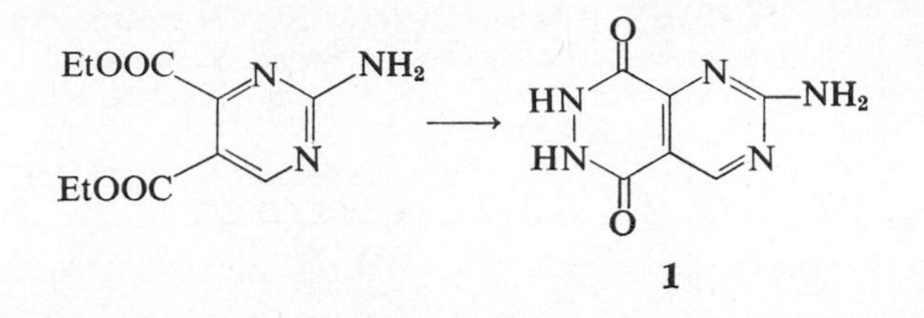

Another synthetic approach was described recently by DiStefano and Castle (2). They treated pyridazine-4,5-dicarboxamide with a solution of potassium hypobromite and the resulting pyrimido[4,5-*d*]pyridazine-2,4(1*H*, 3*H*)dione (**2**) was obtained in 87% yield.

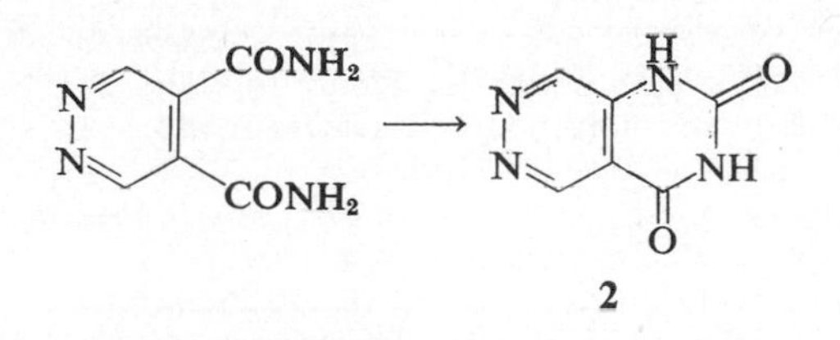

The parent compound is unknown, but from the available MO calculations (3, 4) of total and frontier π-electron densities (Table 3K-1), some conclusion about the reactivity of this bicyclic system can be drawn.

The chemistry of this system has been developed (10) mainly on the basis of the dione **2**. This compound resisted conversion into the 2,4-dichloro derivative. Upon thiation either the monothione, to which structure **3** was assigned, or the dithione **4** was obtained, depending on the reaction conditions.

The dithione **4** reacted with ethanolic ammonia under pressure to afford the 4-amino compound **5**. The dione **2** could be hydrolytically transformed into 5-aminopyridazine-4-carboxylic acid (**6**, R = H), which was converted via its ester (**6**, R = Et) upon fusion with guanidine carbonate into the 2-amino-4-one (**7**). Similarly, pyrimido[4,5-*d*]pyridazin-4-one (**8**, R = H) was prepared from the ester **6** (R = Et) via the corresponding amide. Finally, by acetylating the ester **6** and after treatment with ethanolic ammonia at room temperature the 2-methyl analog **8** (R = Me) was prepared.

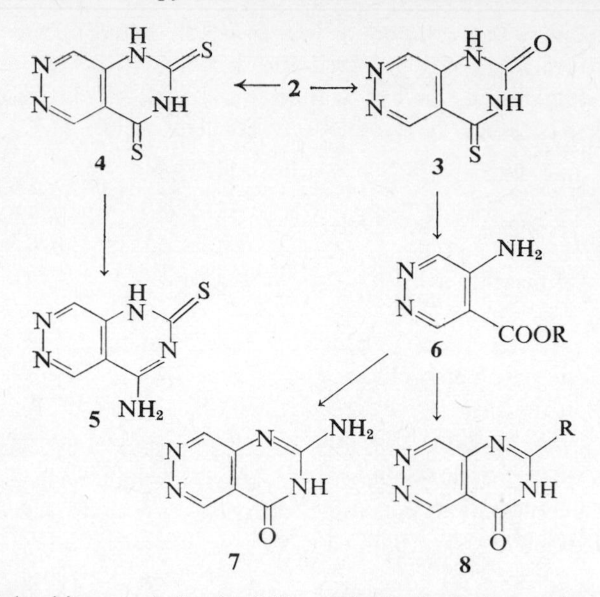

Later, Nakashima and Castle (11) synthesized the isomeric amino-one **9** and amino-thione **10**, thus indirectly confirming the structure of the previously reported **7** and **5**, obtained in a different way.

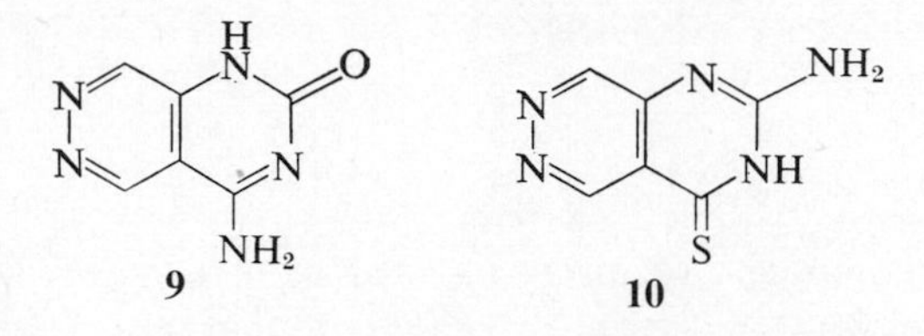

II. Pyrimido[4,5-*c*]pyridazines

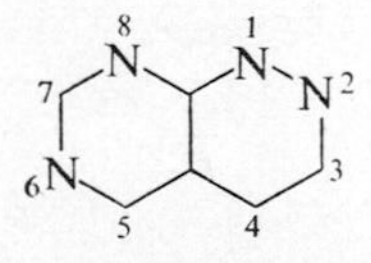

Pyrimido[4,5-*c*]pyridazine, RRI Suppl. I 8100(Pyridazino[3,4-*d*]-pyridazine,1,2,6,8-Tetraazanaphthalene,2,4,5,6-Tetraazanaphthalene)

Druey (5) in 1958 indicated in a review the possibility of building up this bicyclic system from 3-amino-5,6-dimethyl-4-carboxamidopyridazine, but no experimental details were given.

In the same year Pfleiderer and Ferch (6) showed that when 1,3-dimethyl-4-hydrazinouracil was treated with 1,2-dicarbonyl compounds, such as 1,2-diketones or α-keto esters, derivatives of pyrimido[4,5-*c*]pyridazine (**11**) could be obtained in good yields.

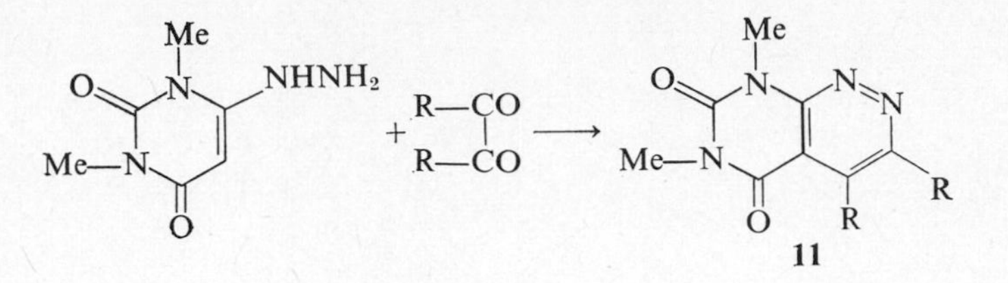

With glyoxal bisulfite a bis-hydrazone was isolated as well as the cyclic product **11**.

When α-keto esters were employed it was possible to isolate the intermediate hydrazones. Yet cyclization of these (**12**) into **13** could be accomplished only in the case of particular substituents (R = H, R = Et) and in the presence of sodium ethoxide. 1,3-Dimethyl-4-hydrazinouracil (**14**) reacted with diethyl mesoxalate to form **5** (R = COOEt).

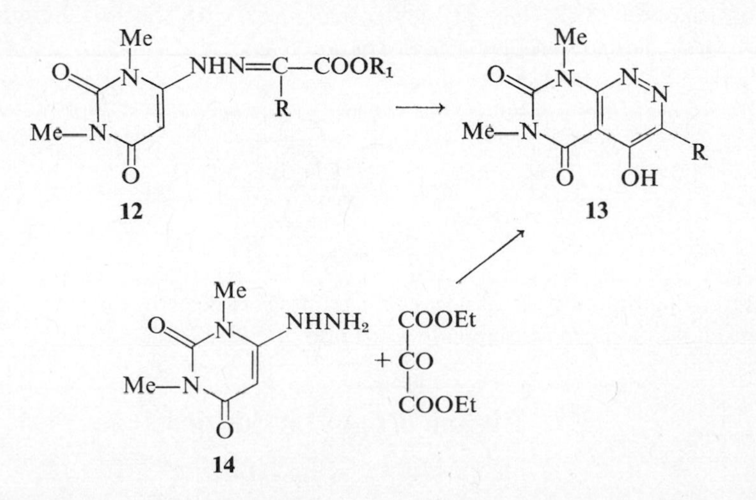

The structure of **11** (R = C_6H_5) could be confirmed by alkaline degradation, and 3,4-diphenyl-6-methylamino-5-pyridazinecarboxylic acid methylamide was obtained in almost quantitative yield.

Jones (7) applied the Hofmann reaction on 3-methylpyridazine-5,6-dicarboxamide and obtained a compound to which he tentatively assigned structure **15**. Nakagome, Castle, and Murakami (8) recently reinvestigated this reaction and obtained three different products. The major product was 3-methylpyrimido[4,5-*c*]pyridazine-5,7(6*H*,8*H*)dione (**15**); the structure of 3-methylpyrimido[5,4-*c*]pyridazine-6,8(5*H*,7*H*)dione (**16**) was ascribed to the other isomeric product on logical grounds. The third product was an acid (**17**) of still unknown structure.

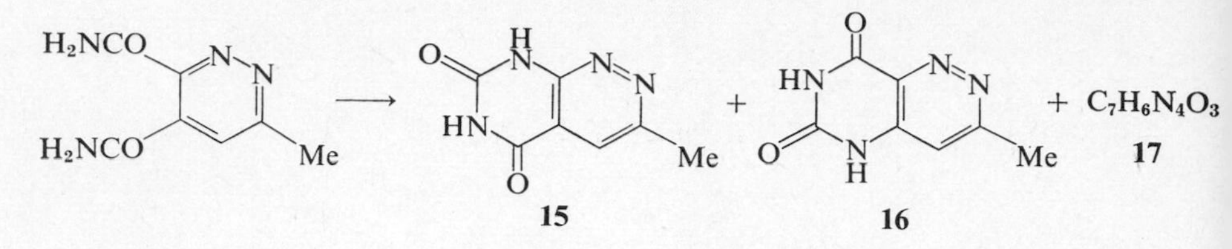

The correctness of structural assignment for **15** could be verified by ring opening with alkali or acid. The resulting 3-amino-6-methylpyridazine-4-carboxylic acid was decarboxylated to the known 3-amino-6-methylpyridazine. In a like manner, the Hofmann reaction on pyridazine-3,4-dicarboxamide gave a mixture of the demethylated analog of **15** and an acid of unknown structure.

Ring closure of the pyrimidine part of the bicycle could be also accomplished by treating 3-amino-6-methylpyridazine-4-carboxamide with ethyl orthoformate to **18** (8). The proposed structure for **18** was again confirmed on the basis of spectroscopic and degradative evidence.

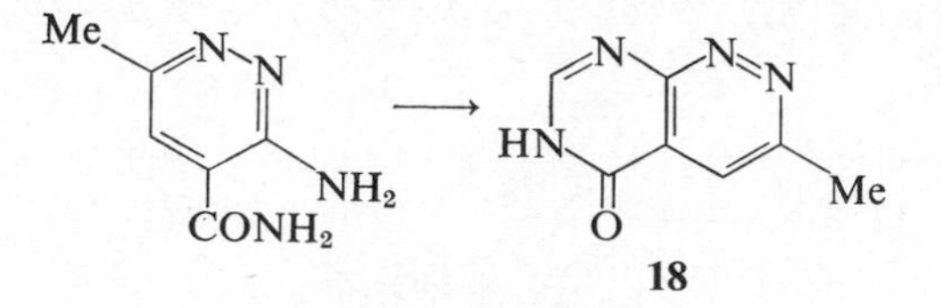

MO calculations (3, 4) of total and frontier π-electron densities for the parent compound are available and are presented in Table 3K-2.

III. Pyrimido[5,4-*c*]pyridazines

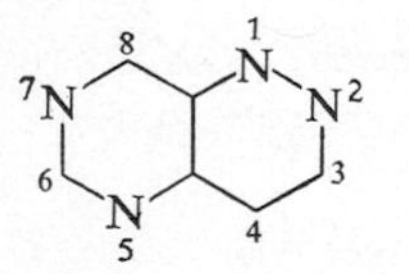

Pyrimido[5,4-*c*]pyridazine (1,2,5,7-Tetraazanaphthalene, Pyrimidino[5,6-*c*]pyridazine)

This bicyclic system eluded synthesis until recently when Nakagome, Castle, and Murakami (8) described some synthetic approaches.

As already mentioned under pyrimido[4,5-*c*]pyridazines, the Hofmann reaction on 3-methylpyridazine-5,6-dicarboxamide yielded 3-methylpyrimido-[4,5-*c*]pyridazine-5,7(6*H*,8*H*)dione (**19**) as the major product as well as 8% of the isomeric 3-methylpyrimido[5,4-*c*]pyridazine-6,8(5*H*,7*H*)dione (**20**).

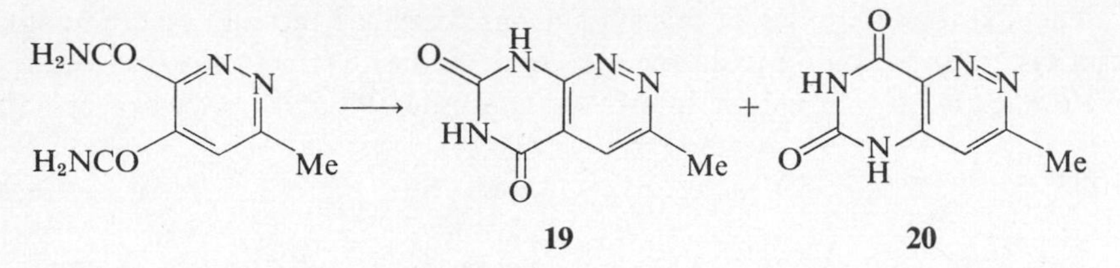

A direct synthetic approach utilized 4-amino-3-cyanopyridazine, which could be cyclized with formamide into 8-aminopyrimido[5,4-*c*]pyridazine (**21**) in moderate yield. It was also possible to convert 4-aminopyridazine-3-

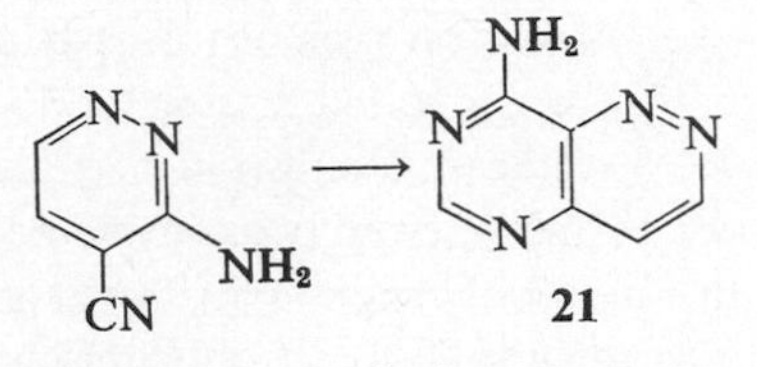

carboxamide either according a urea fusion procedure into pyrimido[5,4-*c*]-pyridazine-6,8(5*H*,7*H*)dione (**22**) or with ethyl orthoformate into the corresponding 8(7*H*)one (**23**).

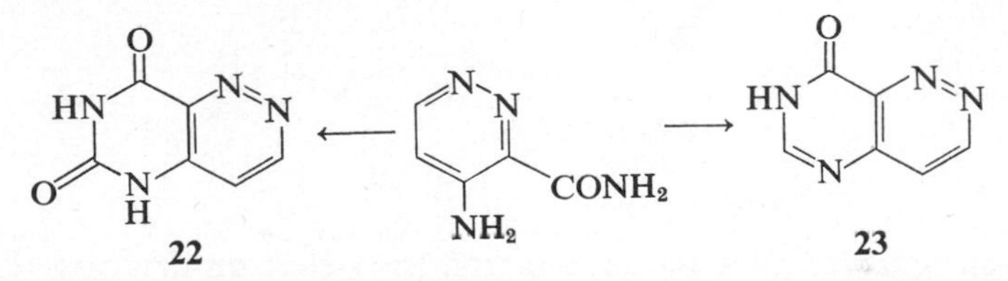

A derivative of this system was prepared from the hydrazide of uracil-6-acetic acid. After being converted into the corresponding semicarbazide derivative, this derivative could be cyclized with alkali to 1,4-dihydro-pyrimido[5,4-*c*]pyridazine-3(2*H*), 6(5*H*),8(7*H*)trione (13). The only successful reaction that was performed on this system was the thiation of **23** with phosphorus pentasulfide.

For the parent compound, MO calculations (3,4) of total and frontier π-electron densities have been made and values are listed in Table 3K-3.

IV. Pyrimido[1,2-*b*]pyridazines

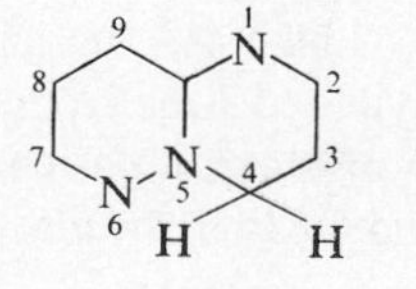

4*H*-Pyrimido[1,2-*b*]pyridazine

The first representative of this system was described recently by Stanovnik and Tišler (9). 3-Aminopyridazine was condensed with ethyl ethoxymethylenemalonate at 110° to give the intermediate product **24** and this, when heated

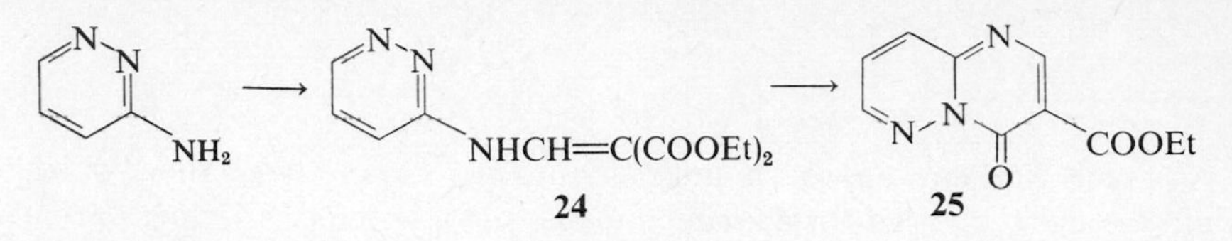

under reflux in diphenyl ether cyclized to **25**. Ultraviolet and infrared spectra of **25** have been recorded, and the nmr spectrum is in accordance with the proposed structure.

The bicyclic product **25** is stable in a solution of ethanol, but in the presence of water, acid, or alkali ring opening was observed.

Further analogs of this bicyclic system have been prepared (12). Thus 3-amino-6-chloropyridazine reacted with 1,3-dibromopropane to give **26** in low yield. Moreover, compound **26** and its free base **27** were obtained in better yield by cyclizing **28** in the presence of potassium carbonate. The structure of **27** is in accord with the recorded nmr data.

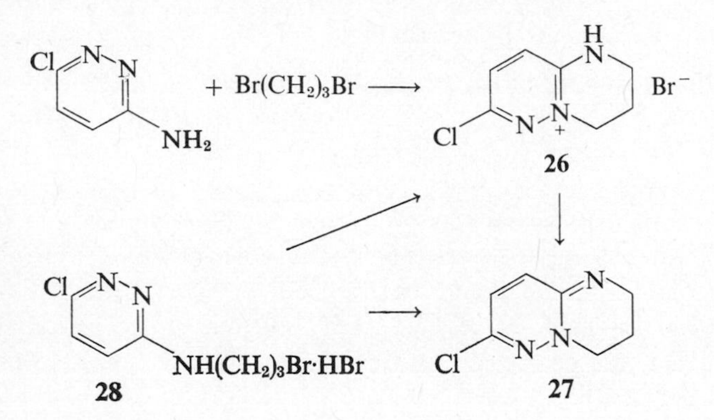

Finally, the corresponding 2-one **30** was prepared in good yield from the quaternized pyridazine **29** with the aid of polyphosphoric acid.

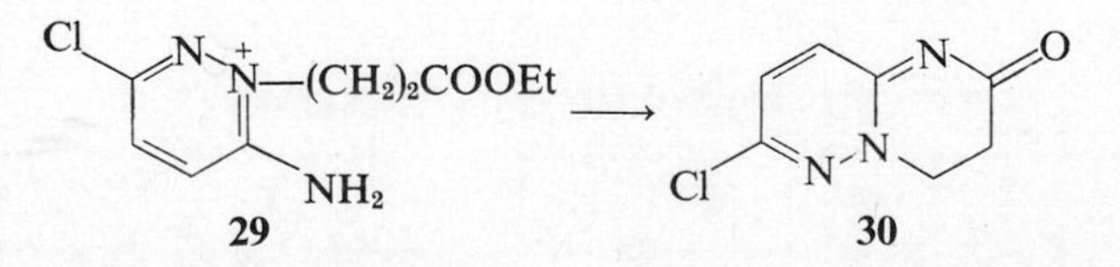

Pyrimido[1,2-*b*]pyridazinones may be obtained from 3-aminopyridazines which condense with 1,3-dicarbonyl compounds (14). In a similar reaction with 1,1,3,3-tetraethoxypropane the fully aromatic pyrimido[1,2-*b*]pyridazinium perchlorates can be prepared. The structure of 2-ones rather than 4-ones has been assigned to pyrimido[1,2-*b*]pyridazinones on the basis of chemical evidence and nmr spectroscopic correlations and π-bond orders (14).

V. Tables

ABLE 3K-1. Total and Frontier π-Electron Densities of Pyrimido[4,5-*d*]pyridazine

osition	1	2	3	4	4*a*	5	6	7	8	8*a*
otal (3, 4)	1.1783	0.8481	1.1579	0.8625	0.9596	0.9001	1.1387	1.1110	0.9236	0.9199
	1.210	0.825	1.190	0.818	0.985	0.876	1.152	1.118	0.914	0.911
rontier (3)	0.1917	0.0749	0.0423	0.1279	0.0070	0.1418	0.1506	0.0658	0.1889	0.0090

ABLE 3K-2. Total and Frontier π-Electron Densities of Pyrimido[4,5-*c*]pyridazine

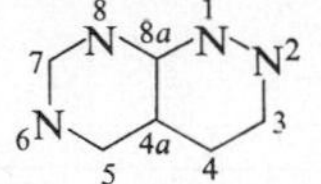

osition	1	2	3	4	4*a*	5	6	7	8	8*a*
otal (3, 4)	1.1471	1.0675	0.9721	0.8840	0.9649	0.8555	1.1656	0.8434	1.1964	0.9036
	1.135	1.071	0.951	0.910	0.982	0.818	1.194	0.824	1.222	0.892
rontier (3)	0.2416	0.0135	0.1403	0.1644	0.0004	0.1152	0.0582	0.0623	0.1971	0.0071

ABLE 3K-3. Total and Frontier π-Electron Densities of Pyrimido[5,4-*c*]pyridazine

osition	1	2	3	4	4*a*	5	6	7	8	8*a*
otal (3, 4)	1.1029	1.0941	0.9510	0.9193	0.9276	1.1830	0.8497	1.1566	0.8612	0.9547
	1.091	1.102	0.921	0.954	0.911	1.216	0.828	1.190	0.824	0.962
rontier (3)	0.1860	0.0673	0.0748	0.2180	0.0163	0.1852	0.0797	0.0419	0.1266	0.0043

TABLE 3K-4. Pyrimido[4,5-*d*]pyridazines

Structure	MP (°C) or BP (°C/mm)	References
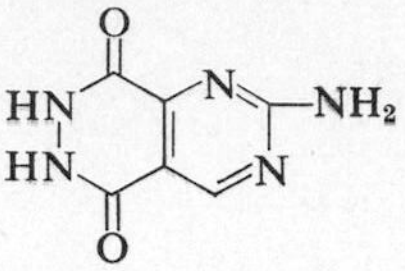	>400	1
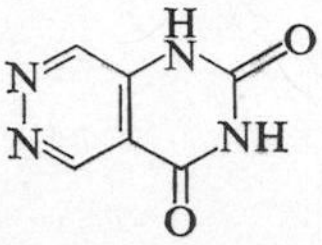	>360	2
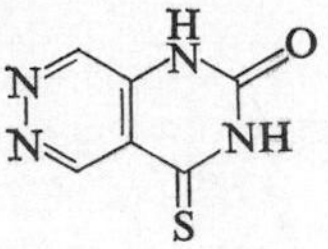	310 (dec.)	10
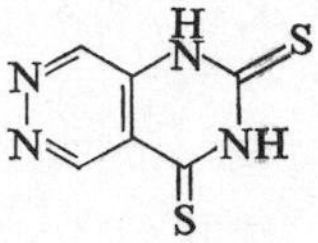	>360 (dec.)	10
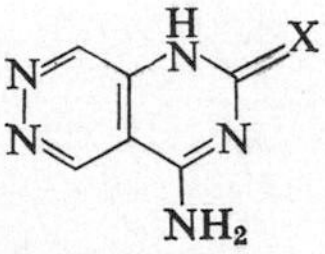	X = S >360 (dec.) O >360	10 11

TABLE 3K-5

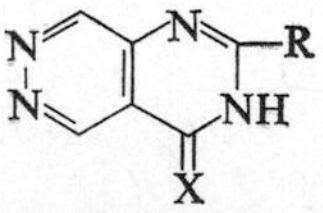

R	X	MP (°C)	References
None	O	330 (dec.)	10
Methyl	O	235 (dec.)	10
Amino	O	>350 (dec.)	10
Amino	S	>360	11

TABLE 3K-6. Pyrimido[4,5-*c*]pyridazines

Structure	MP (°C) or BP (°C/mm)	References
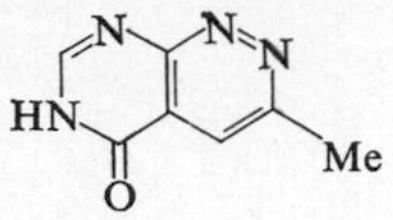	>300	8
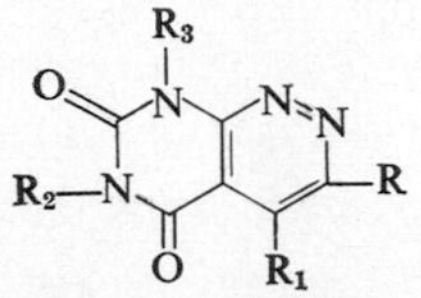		

R	R_1	R_2	R_3		
None				356 (dec.)	8
Methyl				>350	8
				—	7
Phenyl	Phenyl	Methyl	Methyl	208–209	6
Methyl	Methyl	Methyl	Methyl	146–147	6
		Methyl	Methyl	139–140	6
	OH	Methyl	Methyl	229–232	6
COOEt	OH	Methyl	Methyl	263–265	6

TABLE 3K-7. Pyrimido[5,4-*c*]pyridazines

Structure	MP (°C) or BP (°C/mm)	References
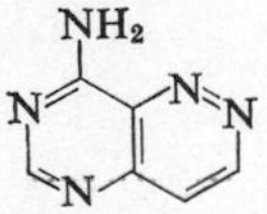	295 (dec.)	8
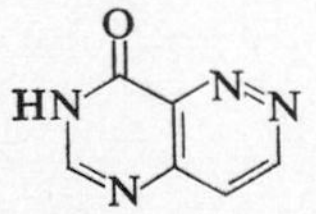	>300	8
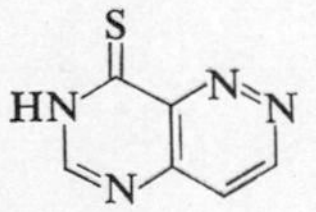	>400	8

TABLE 3K-7 (*continued*)

Structure	MP (°C) or BP (°C/mm)	References
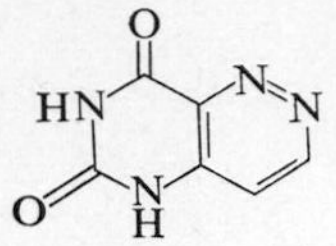	Shrinking and darking at 380° C, does not completely melt at 400° C	8
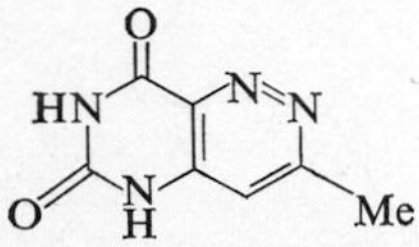	>300	8
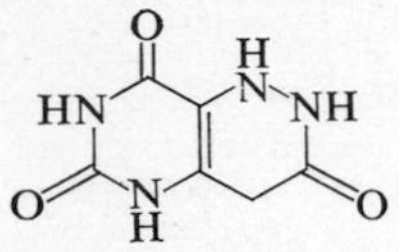	>360	13

TABLE 3K-8. Pyrimido[1,2-*b*]pyridazines

Structure	MP (°C) or BP (°C/mm)	References
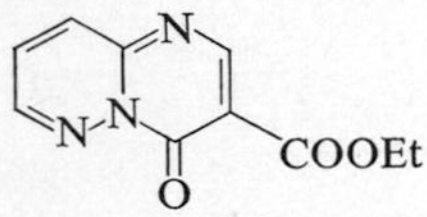	169–170	9
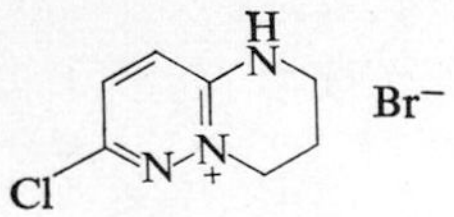 Br^-	210–211	12
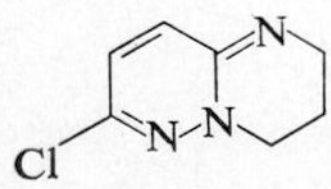	76–79	12
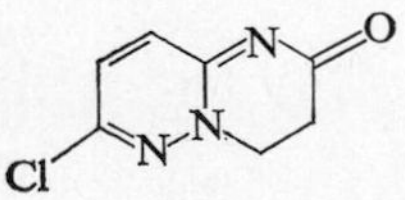	203–205	12

References

1. R. G. Jones, *J. Am. Chem. Soc.*, **78,** 159 (1956).
2. L. DiStefano and R. N. Castle, *J. Heterocyclic Chem.*, **5,** 53 (1968).
3. Unpublished data from this laboratory. Parameters for LCAO Calculations were taken from A. Streitwieser, *Molecular Orbital Theory for Organic Chemists*, Wiley, New York, 1961, p. 135.
4. S. C. Wait and J. W. Wesley, *J. Mol. Spectrosc.*, **19,** 25 (1966).
5. J. Druey, *Angew. Chem.*, **70,** 5 (1958); *Chem. Abstr.*, **52,** 11856 (1958).
6. W. Pfleiderer and H. Ferch, *Ann.*, **615,** 48 (1958).
7. R. G. Jones, *J. Org. Chem.*, **25,** 956 (1960).
8. T. Nakagome, R. N. Castle, and H. Murakami, *J. Heterocyclic Chem.*, **5,** 523 (1968).
9. B. Stanovnik and M. Tišler, *Tetrahedron Lett.*, **1968,** 33.
10. T. Kinoshita and R. N. Castle, *J. Heterocyclic Chem.*, **5,** 845 (1968).
11. T. Nakashima and R. N. Castle, *J. Heterocyclic Chem.*, **7,** 209 (1970).
12. S. Ostroveršnik, B. Stanovnik, and M. Tišler, *Croat. Chem. Acta*, **41,** 135 (1969).
13. T. Sasaki and M. Ando, *Yuki Gosei Kagaku Kyokai Shi*, **27,** 169 (1969); *Chem. Abstr.*, **71,** 22101 (1969).
14. A. Pollak, B. Stanovnik, and M. Tišler, *Chimia*, **24,** 418 (1970).

Part L. Pyrazinopyridazines

I. Pyrazino[2,3-*d*]pyridazines

Pyrazino[2,3-*d*]pyridazine, RRI 1591 (1,4,6,7-Tetraazanaphthalene)

Compounds pertaining to this heterocyclic system were unknown until 1956 when Jones (1) reported the synthesis of pyrazino[2,3-*d*]pyridazine-5,8(6*H*,7*H*)dione (**1**) from diethyl pyrazine-2,3-dicarboxylate and hydrazine.

The synthetic procedure was later improved by Patel and Castle (2). They obtained the same compound, although in somewhat lower yield, from pyrazine-2,3-dicarboxylic acid anhydride, which was used previously by Hemmerich and Fallab (3) as the starting material.

Other bifunctional pyrazines have been employed with equal success. Thus 2,3-dicanopyrazine could be converted easily into 5,8-diaminopyrazino[2,3-*d*]pyridazine with hydrazine at room temperature (2). 2,3-Di-

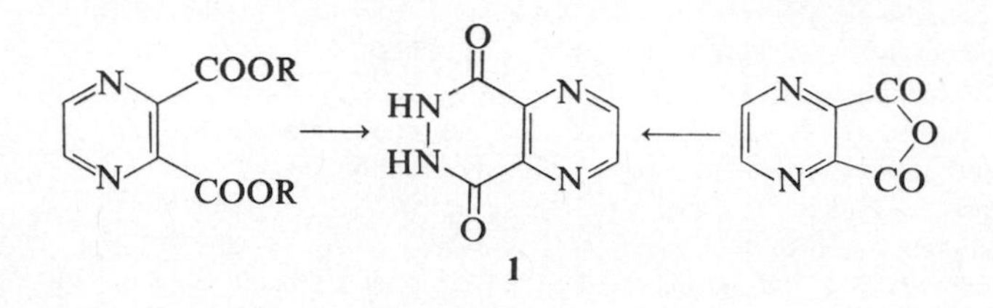

acylpyrazines were similarly transformed into 5,8-disubstituted pyrazino-[2,3-*d*]pyridazines (4, 5) and from triethylpyrazine-2,3,5-tricarboxylate and hydrazine pyrazino[2,3-*d*]pyridazine-5,8(6*H*,7*H*)dione-2-carbohydrazide was obtained (15).

Appropriate diaminopyridazines were also used in a reaction pattern first proposed in a review (6). Although no experimental details have been given, the possibility of pyrazino[2,3-*d*]pyridazine ring formation from 4,5-diamino-3(2*H*)pyridazinones and 1,2-dicarbonyl compounds or oxalic acid has been mentioned. Recently, Martin and Castle (16) described such a reaction, and they obtained from 4,5-diaminopyridazin-3(2*H*)one and glyoxal pyrazino[2,3-*d*]pyridazin-5(6*H*)one in excellent yield. Sprio, Aiello, and Fabra (5) obtained pyrazino[2,3-*d*]pyridazines (**4**) from either 4,5-diaminopyridazines (**3**) and diacetyl or from 2,3-diacylpyrazines (**6**) and hydrazine. Both starting compounds were obtained from the corresponding bicycles **2** and **5** either by hydrogenolysis in the presence of Raney nickel or oxidation with chromic acid.

In a like manner, 4,5-diaminopyridazine when condensed with pyruvaldehyde afforded 2-methylpyrazino[2,3-*d*]pyridazine in 45% yield and with an aqueous solution of glyoxal the parent compound, pyrazino[2,3-*d*] pyridazine, has been prepared in 22% yield (2). A somewhat better yield of the latter compound was attained by dechlorination of the 5,8-dichloro analog in the presence of hydrogen and palladized charcoal (2).

Pyrazino[2,3-*d*]pyridazine is a colorless solid and is expected to have reasonable stability since it is structurally similar to pteridines. Some evidence for its reactivity may be gained from the inspection of total and frontier

π-electron densities, which have been calculated by the simple MO method (Table 3L-1) (7, 8).

Reactions performed on this system have usually been nucleophilic displacements of halogen(s). Pyrazino[2,3-*d*]pyridazine-5,8(6*H*,7*H*)dione (pK_a 5.5), which has been used for the preparation of the corresponding 5,8-dihalo derivatives, could be monoacetylated only; it formed a monotosylate (2) and was also reported to form complexes with copper salts (3).

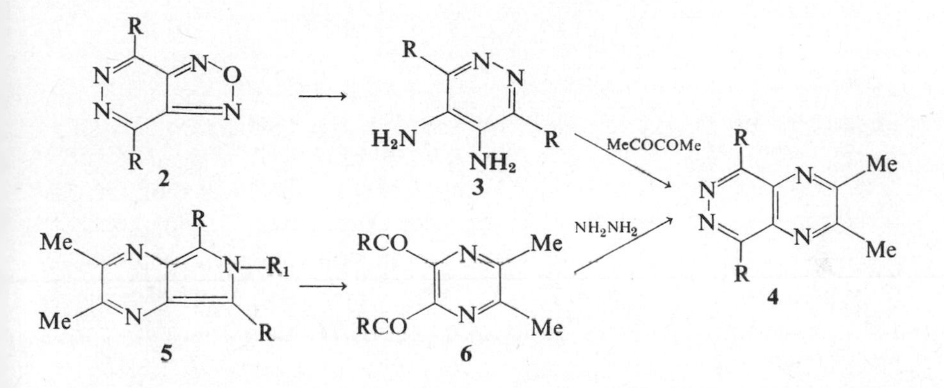

With a mixture of phosphorus oxychloride and phosphorus pentachloride it is transformed into the 5,8-dichloro compound. The 5,8-dibromo analog could be obtained in low yield from a similar treatment with phosphorus oxybromide and bromine (2).

5,8-Dichloropyrazino[2,3-*d*]pyridazine, when allowed to react with ammonia or amines of stronger basicity, afforded only the monosubstitution products (2, 9, 10). With less basic amines, such as aniline, *p*-toluidine, or *p*-nitroaniline, 5,8-disubstituted products were obtained (2, 10). This difference in reactivity has been explained in terms of a higher electron-donating capacity of an aliphatic amine substituent in comparison to an aromatic amine substituent. In this way the substitution of the remaining chlorine atom is affected.

A similar reaction course would be observed if mild alkaline hydrolysis were applied to 5,8-dichloro- or 5,8-dibromopyrazino[2,3-*d*]pyridazine. The corresponding 8-halopyrazino[2,3-*d*]pyridazin-5(6*H*)one has been obtained thus (10). With alkoxides both halogens were easily displaced (2). The reaction with sodium azide, when conducted in dimethylsulfoxide as the solvent, afforded 6-azidopyrazino[2,3-*d*]tetrazolo[4,5-*b*]pyridazine (7) (11). Spectroscopic evidence and chemical transformations are in accord with the proposed structure.

Pyrazino[2,3-*d*]pyridazine-5,8(6*H*,7*H*)dithione could not be prepared from the 5,8-dichloro compound by the thiourea method. Treatment of the 5,8-dioxo analog by the phosphorus pentasulfide-pyridine method gave an

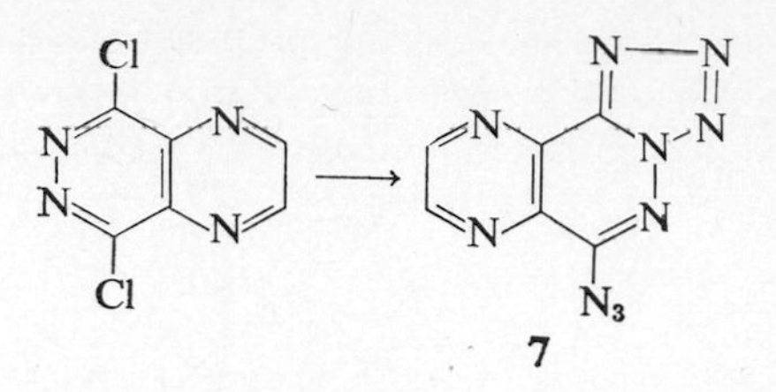

impure product in very low yield. Nevertheless, the 5,8-dichloro compound reacted easily with thiols to give the 5,8-bis(substituted thio) analogs (2).

It has been already mentioned that catalytic dehalogenation of 5,8-dichloropyrazino[2,3-*d*]pyridazine afforded the parent compound (2). However, extension of this principle to 8-chloropyrazino[2,3-*d*]pyridazin-5(6*H*)one is reported to give the reduced product, 1,2,3,4-tetrahydro-pyrazino[2,3-*d*]pyridazin-5(6*H*)one (**8**), in moderate yield. The simul-

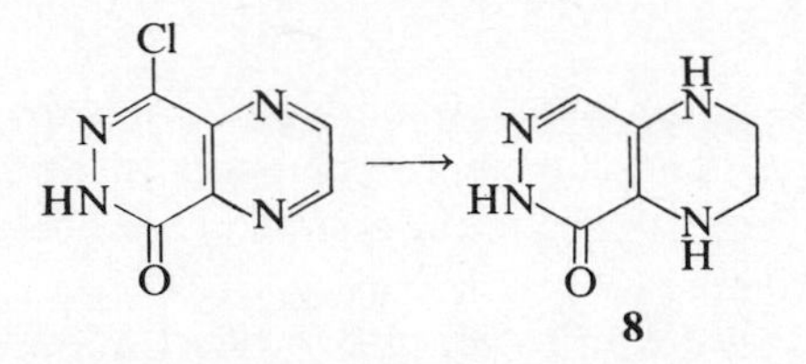

taneous dechlorination and reduction of the pyrazine part of the molecule parallels the behavior of related pyrido[2,3-*d*]pyridazines.

II. Pyrazino[2,3-*c*]pyridazines

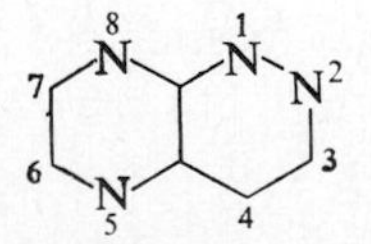

Pyrazino[2,3-*c*]pyridazine, RRI I. Suppl. 8099 (Pyrazino[*c*]pyridazine, 1,2,5,8-Tetraazanaphthalene)

This ring system and the possibilities of preparing its derivatives were mentioned first in a review by Druey in 1958 (12). The synthetic approaches to representative compounds of type **9** are outlined in Scheme 1. Additional details about these syntheses were described in a patent (13) which claims

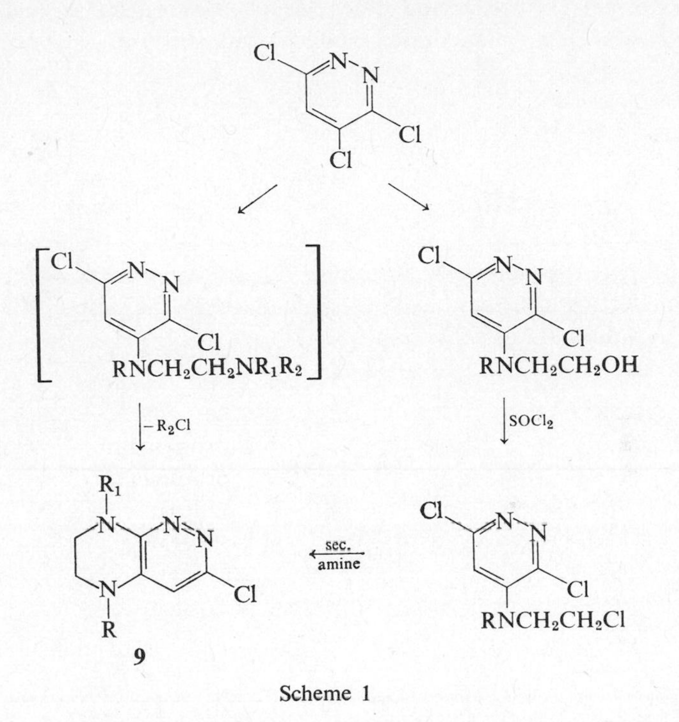

Scheme 1

that these products possess valuable properties as stimulants. However, no verification of the proposed structures was presented.

Some derivatives of the fully aromatic system were reported by Gerhardt and Castle (14). They allowed 5-chloro-3,4-diaminopyridazine to react with benzil (160–175°, 15 min) and obtained a high-melting solid for which structure **10** was proposed.

In a like manner, 6-chloro-3,4-diaminopyridazine was transformed into 3-chloro-6,7-diphenylpyrazino[2,3-*c*]pyridazine (**11**) (14). For both compounds uv and ir data are available.

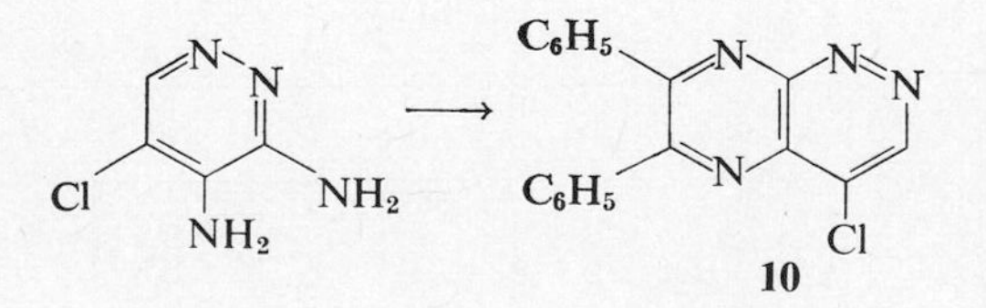

Recently, a derivative of this system was reported (17) to result from a reaction where 1-acetyl-2-carbomethoxymethylene tetrahydropyrazin-3-one was transformed into its lactim ether Meerwein's reagent and subsequent treatment with hydrazine hydrate resulted in ring closure.

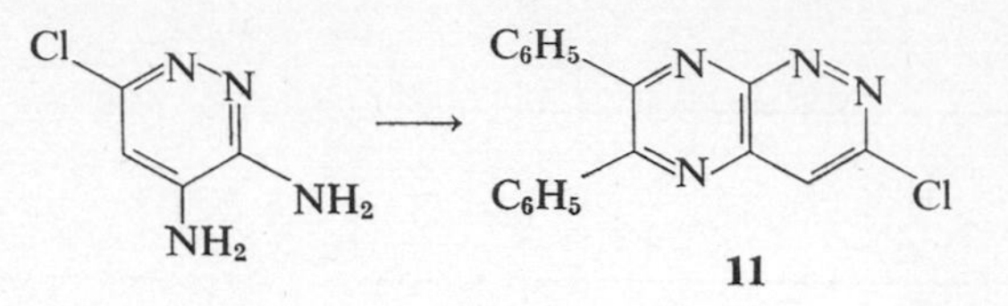

For the parent compound, which so far remains unknown, total and frontier π-electron densities have been calculated by the simple MO method (7, 8). The values are listed in Table 3L-2.

III. Tables

TABLE 3L-1. Total and Frontier π-Electron Densities of Pyrazino[2,3-*d*]pyridazine

Position	1	2	3	4	4*a*	5	6	7	8	8*a*
Total (7, 8)	1.1133	0.9112	0.9112	1.1133	0.9374	0.9139	1.1242	1.1242	0.9139	0.9374
	1.139	0.886	0.886	1.139	0.937	0.902	1.137	1.137	0.902	0.937
Frontier (7)	0.1429	0.0418	0.0418	0.1429	0.0012	0.1971	0.1169	0.1169	0.1971	0.0012

TABLE 3L-2. Total and Frontier π-Electron Densities of Pyrazino[2,3-*c*]pyridazine

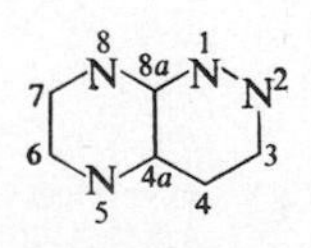

Position	1	2	3	4	4*a*	5	6	7	8	8*a*
Total (7, 8)	1.1318	1.0797	0.9628	0.9040	0.9461	1.1130	0.9137	0.9040	1.1269	0.9177
	1.121	1.088	0.938	0.940	0.935	1.216	0.828	1.190	0.824	0.962
Frontier (7)	0.2425	0.0404	0.1142	0.2226	0.0077	0.1374	0.0561	0.0327	0.1457	0.0007

TABLE 3L-3. Pyrazino[2,3-*d*]pyridazines

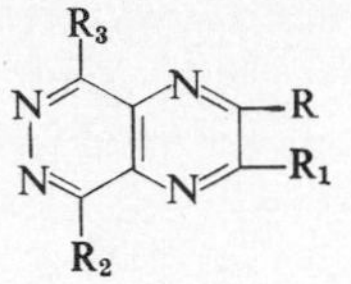

R	R_1	R_2	R_3	MP (°C) or BP (°C/mm)	References
None				157.5–158.5	2
Methyl				172.5–174	2
		Chloro	Chloro	181–182	2
		Bromo	Bromo	195	2
		Methoxy	Methoxy	278–280	2
		Ethoxy	Ethoxy	232–234	2
		$C_6H_5CH_2O$	$C_6H_5CH_2O$	239–241	2
		Chloro	$C_6H_5CH_2NH$	197–198	2
		Chloro	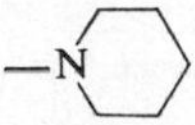 —N⟨ ⟩ (piperidine ring)	184–185	2
		Chloro	Amino	263–265	10
		Chloro	$NH(CH_2)_3NMe_2$	137–138	9
		Chloro	$NH(CH_2)_3NEt_2$	144–145	9
		Chloro	$NHCH(CH_2)_3NEt_2$ | Me	53–54	9
		Methoxy	$NH(CH_2)_3NMe_2$	157–159	9
		Methoxy	$NH(CH_2)_3NEt_2$	93–95	9
		Amino	Amino	233–234 (dec.)	2
		C_6H_5NH	C_6H_5NH	242–243	2, 10
		p-MeC_6H_4NH	*p*-MeC_6H_4NH	271–272	2
		Amino	$NHNO_2$	>300	10
		p-$NO_2C_6H_4NH$	*p*-$NO_2C_6H_4NH$	>360	10
		Phenylthio	Phenylthio	260–261	2
		$—OCOCH_3$		145–146	16
		$—NH—C_6H_2(NO_2)_3(2,4,6)$	$—NHC_6H_2(NO_2)_3(2,4,6)$	277 (dec.)	15
		Benzylthio	Benzylthio	261–262	2
		p-$ClC_6H_4CH_2S$	*p*-$ClC_6H_4CH_2S$	207–208	2
		3,4-di$ClC_6H_3CH_2S$	3,4-di$ClC_6H_3CH_2S$	199–200	2
		p-$MeOC_6H_4$	*p*-$MeOC_6H_4$	293.5–294.5	4
Methyl	Methyl	Phenyl	Phenyl	245	5

TABLE 3L-4. Pyrazino[2,3-*d*]pyridazines

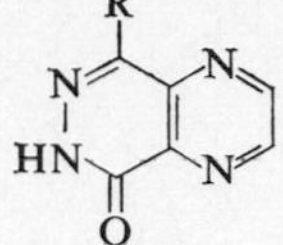

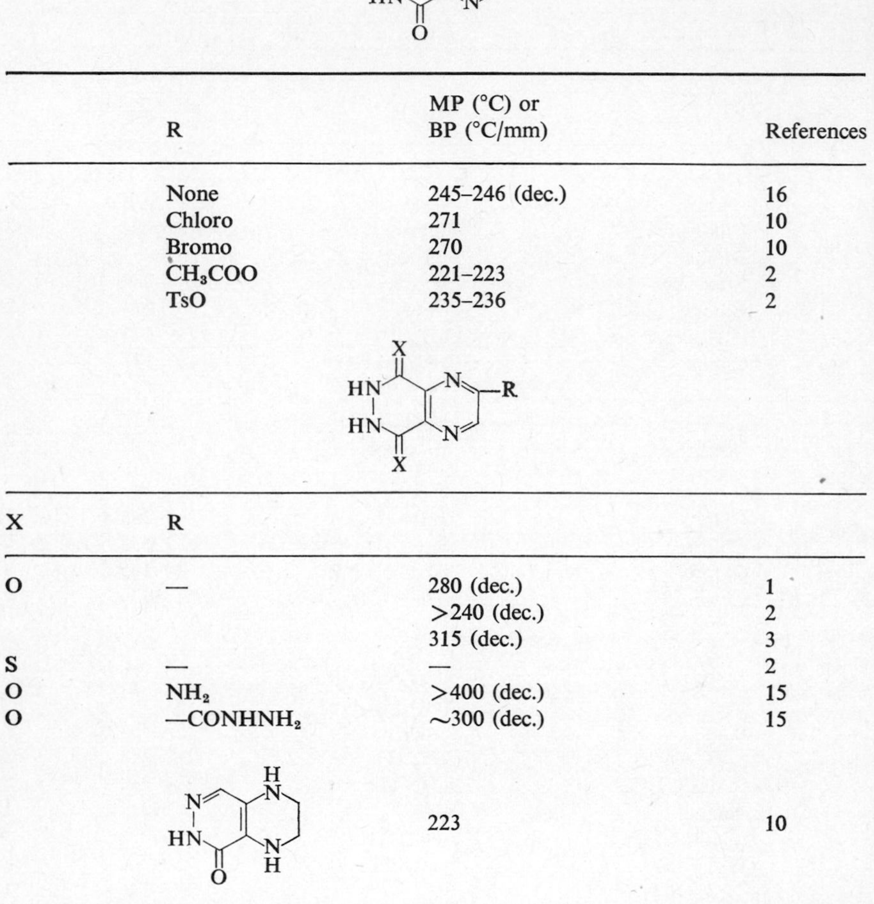

X	R	MP (°C) or BP (°C/mm)	References
	None	245–246 (dec.)	16
	Chloro	271	10
	Bromo	270	10
	CH_3COO	221–223	2
	TsO	235–236	2
O	—	280 (dec.)	1
		>240 (dec.)	2
		315 (dec.)	3
S	—	—	2
O	NH_2	>400 (dec.)	15
O	$—CONHNH_2$	~300 (dec.)	15
	(structure)	223	10

TABLE 3L-5. Pyrazino[2,3-*c*]pyridazines

R	R_1	MP (°C) or BP (°C/mm)	References
	Chloro	>300 (dec.)	14
Chloro		185 (dec.)	14

R	R_1	R_2	R_3	MP (°C) or BP (°C/mm)	References
	Methyl		Methyl	293 (dec.)	13
Chloro	Methyl		Methyl	124–124.5	13
Chloro	Ethyl		Ethyl	69–69.5	13
Chloro	Ethyl		Methyl	80–80.5	13
Chloro	$CH_2CH_2NMe_2$		Methyl	116–117	13
				HCl, 241–243	13
Chloro	Benzyl		Methyl	127–127.5	13
				HCl, 206–208	13
Chloro	*p*-$ClC_6H_4CH_2$		Ethyl	156.5–157.5	13
Chloro	*p*-$Me_2NC_6H_4CH_2$		Ethyl	134–135	13
Chloro	*m*-$MeOC_6H_4CH_2$		Ethyl	108–109	13
Methoxy	Benzyl		Ethyl	126–127	13
NMe_2	Benzyl		Methyl	168–169	13
Chloro	Benzyl	Ethyl	Methyl	145–146	13
[structure: N-COCH$_3$ hexahydropyrazino[2,3-c]pyridazinone]				214–216	17

References

1. R. G. Jones, *J. Am. Chem. Soc.*, **78,** 159 (1956).
2. N. R. Patel and R. N. Castle, *J. Heterocyclic Chem.*, **3,** 512 (1966).
3. P. Hemmerich and S. Fallab, *Helv. Chim. Acta*, **41,** 498 (1958); *Chem. Abstr.*, **52,** 20187 (1958).
4. I. Hagedorn and H. Tönjes, *Pharmazie*, **12,** 567 (1957); *Chem. Abstr.*, **52,** 6363 (1958).

5. V. Sprio, T. Aiello, and I. Fabra, *Ann. Chim. (Rome)*, **56,** 866 (1966); *Chem. Abstr.*, **66,** 1058 (1967).
6. K. Dury, *Angew. Chem.*, **77,** 282 (1965).
7. Unpublished data from this laboratory. Parameters for LCAO Calculations were taken from A. Streitwieser *Molecular Orbital Theory for Organic Chemists*, Wiley, New York, 1961, p. 135.
8. S. C. Wait and J. W. Wesley, *J. Mol. Spectrosc.*, **19,** 25 (1966).
9. N. R. Patel, W. M. Rich, and R. N. Castle, *J. Heterocyclic Chem.*, **5,** 13 (1968).
10. L. DiStefano and R. N. Castle, *J. Heterocyclic Chem.*, **5,** 53 (1968).
11. L. DiStefano and R. N. Castle, *J. Heterocyclic Chem.*, **5,** 109 (1968).
12. J. Druey, *Angew. Chem.*, **70,** 5 (1958); *Chem. Abstr.*, **52,** 11856 (1958).
13. U.S. Pat. 2,942,001 (1960); *Chem. Abstr.*, **54,** 22690 (1960).
14. G. A. Gerhardt and R. N. Castle, *J. Heterocyclic Chem.*, **1,** 247 (1964); *Chem. Abstr.*, **62,** 9126 (1965).
15. R. B. Rao and R. N. Castle, *J. Heterocyclic Chem.*, **6,** 255 (1969).
16. S. F. Martin and R. N. Castle, *J. Heterocyclic Chem.*, **6,** 93 (1969).
17. V. G. Granik and R. G. Glushkov, *Khim.-Farm. Zh.*, **1968,** 16; *Chem. Abstr.*, **69,** 59194 (1968).

Part M. Other Pyridazines with Condensed Six-Membered Heterocyclic Rings

I. Pyridazino-1,2,4-triazines

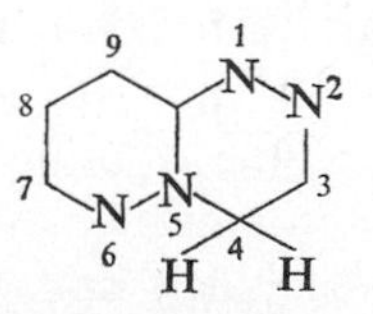

4*H*-Pyridazino[6,1-*c*]-1,2,4-triazine
(4*H*-Pyridazino[3,2-*c*]-*as*-triazine)

Compounds pertaining to this bicyclic system were synthesized first in 1969 (1, 2). Hydrazinopyridazines afforded condensation products with α-ketoesters and cyclization to **1**, for example, took place after heating in polyphosphoric acid at 140° (1). On the other hand, condensation products of hydrazinopyridazines with α-haloaldehydes or α-haloketones were similarly cyclized to **2** in boiling glacial acetic acid (2). The parent compound

2 ($R = R_1 = H$) was obtained by catalytic dehalogenation of the corresponding 7-chloro derivative.

Another synthetic approach made use of the tetrazolo-azido valence isomerization, observed earlier with certain tetrazolopyridazines. In this

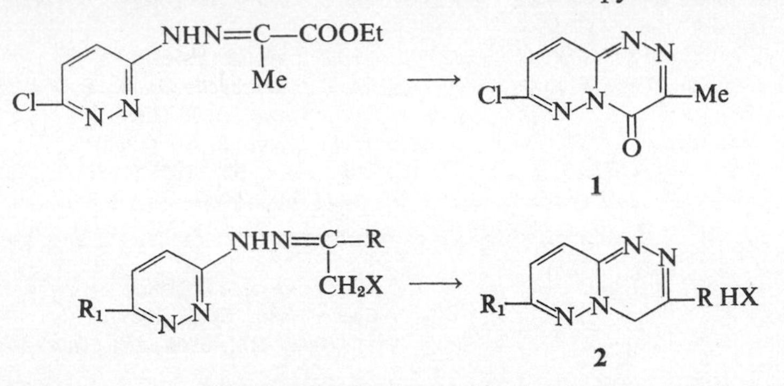

manner, hydrazones (**3**) prepared from 6-hydrazinotetrazolo[1,5-*b*]pyridazine, on heating in polyphosphoric acid, could be transformed into the corresponding pyridazinotriazinone **4** (3). These bicyclic compounds are readily hydrolyzed in the presence of mineral acid and ring opening of the fused triazino ring caused simultaneous ring closure of the azido group into the fused tetrazolo ring giving rise thus tetrazolo[1,5-*b*]pyridazines of type **3**.

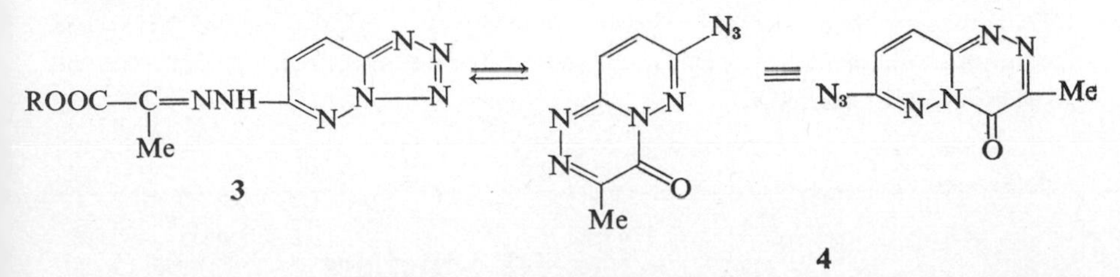

This type of interconversion was also used to demonstrate that the alleged ring opening and closure to an isomeric pyridazino[3,4-*c*]pyridazine derivative in the reaction between **2** ($R = Ph$, $R_1 = Cl$) and hydrazine (2) did not take place and that normal hydrazinolysis of the chlorine atom took place (4). The formulas **5** to **8** illustrate the reaction sequences.

There are few data about the reactivity of this bicyclic system. As already mentioned the halogen atom at position 7 is readily displaced by nucleophiles such as hydrazine; this could also be observed in the preparation of **1** since during neutralization of the reaction mixture temperature was allowed to rise above room temperature and the chlorine atom was replaced by hydroxy group (5). Under the influence of acids the condensed triazinone ring is cleaved.

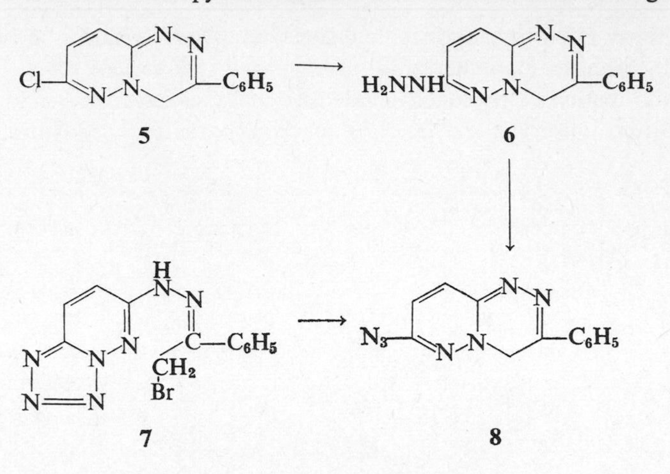

II. Pyranopyridazines

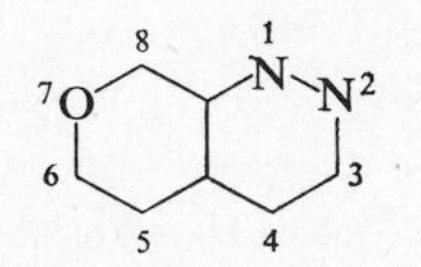

Pyrano[3,4-*c*]pyridazine

A single representative of this class of heterocycles was reported to result upon treatment of 3,6-dicarbomethoxy-5-hydroxyethyl-4-methylpyridazine in methanolic solution with hydrogen chloride and heating the mixture under reflux to give **9** (6).

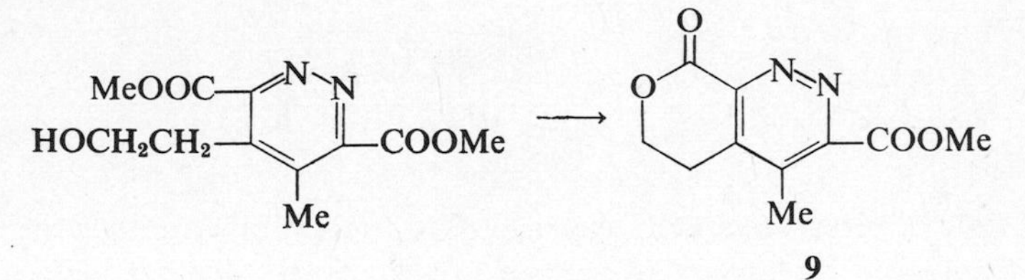

III. Dithiinopyridazines

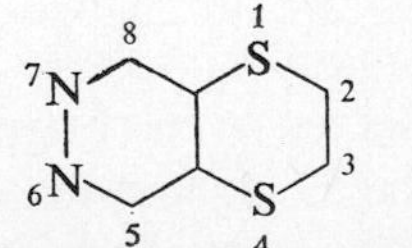

1,4-Dithiino[2,3-*d*]pyridazine
(*p*-Dithiino[2,3-*d*]pyridazine
5,8-Dithiaphthalazine),
Ring Index Suppl. II 10045

The first synthesis of **11** in this bicyclic system was mentioned in 1960 in a review article (7) that gave no details. In this instance the starting compound was dithiatetrahydrophthalic anhydride. In a later article, Schweizer (8) reported that the main reaction product with hydrazine is **10**, and the dithiinopyridazine **11** could be isolated only in small amount.

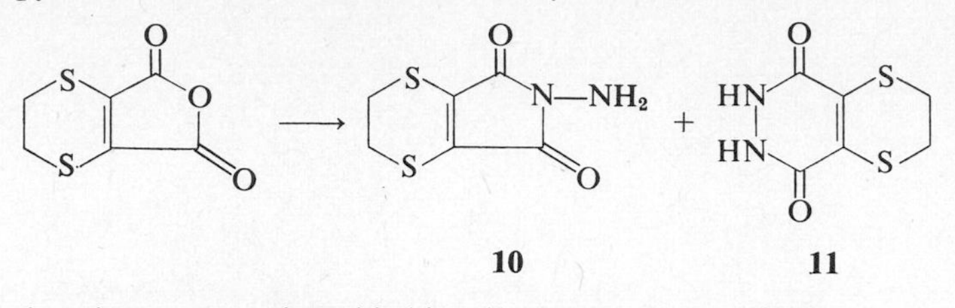

No other data concerning this bicyclic system are available.

IV. Pyridazinooxazines

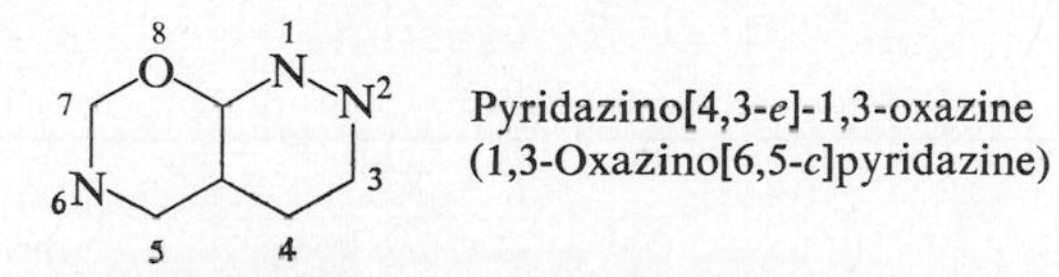

Pyridazino[4,3-e]-1,3-oxazine
(1,3-Oxazino[6,5-c]pyridazine)

It has been reported (9) that this bicyclic system was formed when the pyridazine derivative **12** was heated under reflux with 19% hydrochloric acid to give **13**. The same product was also formed when **12** was heated at 220° until evolution of ammonia was complete.

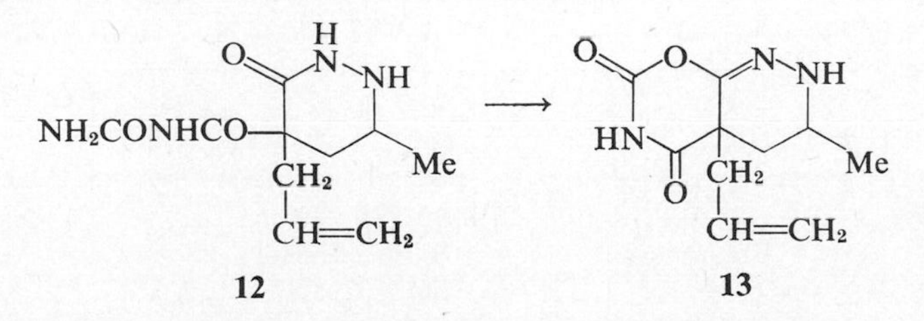

V. Pyridazinothiadiazines

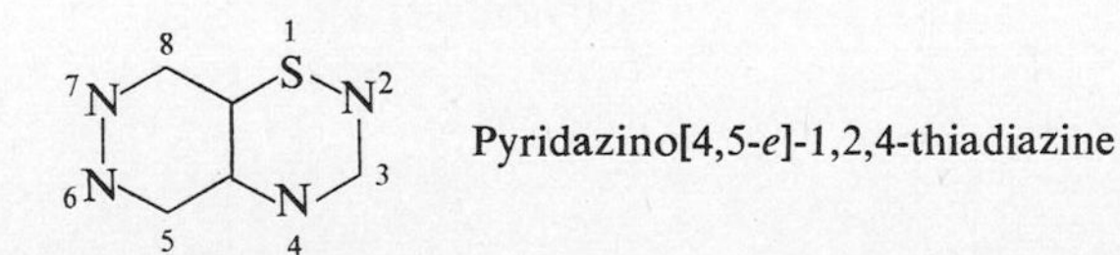

Recently, two isomeric derivatives of this bicyclic system have been reported (10) to result when the corresponding aminopyridazine sulfonamides (**14** and **15**) were cyclized with triethyl orthoformate to give compounds **16** and **17**.

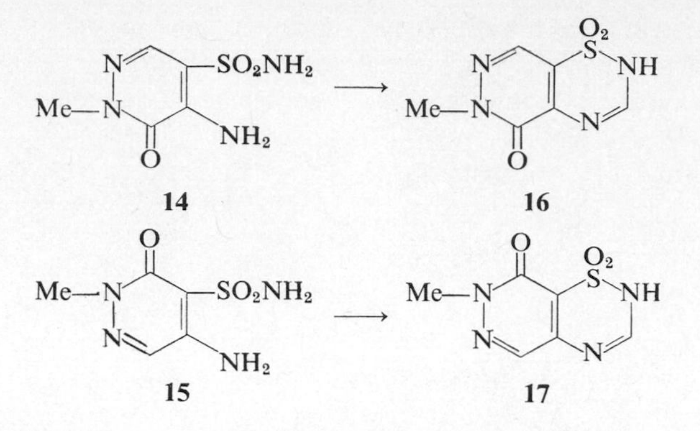

VI. Tables

TABLE 3M-1. Pyridazino[6,1-*c*]-1,2,4-triazines

R	R_1	MP (°C) or BP (°C/mm)	References
None		218	2
Phenyl	Hydrazino	239–241	4
		230–231	2
		HBr, 215–220	2
Phenyl	Chloro	165–167 (dec.)	2
		HBr, >320	
Phenyl	COOH	263	2
Phenyl	Azido	237–239	4
	Chloro	HBr, >300	2
Chloro		150–153 (dec.)	1
Azido		147–148	3
Hydroxy		285–287	5

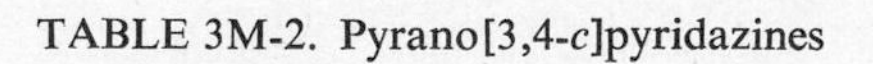

TABLE 3M-2. Pyrano[3,4-*c*]pyridazines

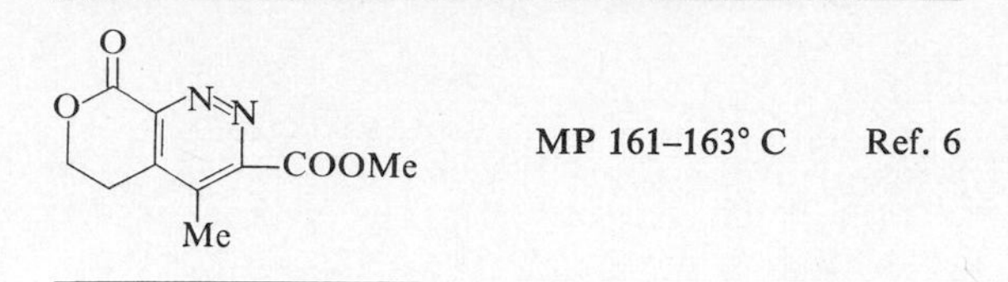

MP 161–163° C Ref. 6

TABLE 3M-3. 1,4-Dithiino[2,3-*d*]pyridazines

Structure	MP	Reference
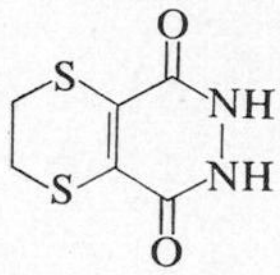	MP 340° C (dec.)	Ref. 8

TABLE 3M-4. Pyridazino[3,4-*e*]-1,3-oxazines

Structure	R	MP (°C)	Reference
	None	208	9
	Methyl	98	9

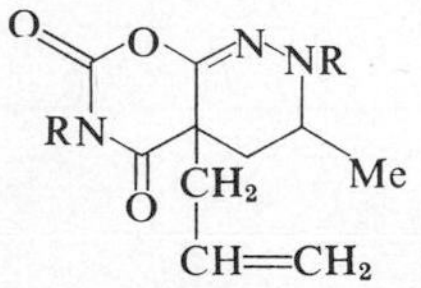

TABLE 3M-5. Pyridazino[4,5-*e*]-1,2,4-thiadiazines

Structure	MP (°C)	Reference
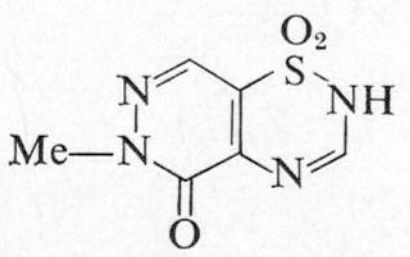	335–340 (dec.)	10
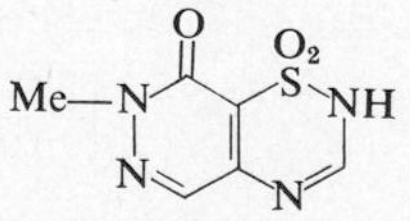	425–430	10

References

1. B. Stanovnik and M. Tišler, *J. Heterocyclic Chem.*, **6,** 413 (1969).
2. T. LaNoce, E. Bellasio, A. Vigevani, and E. Testa, *Ann. chim.* (*Roma*), **59,** 552 (1969).
3. B. Stanovnik, M. Tišler, M. Ceglar, and V. Bah, *J. Org. Chem.*, **35,** 1138 (1970).
4. B. Stanovnik and M. Tišler, *Synthesis*, **1970,** 180.
5. B. Stanovnik and M. Tišler, *Monatsh. Chem.*, **101,** 303 (1970).
6. P. Roffey and J. P. Verge, *J. Heterocyclic Chem.*, **6,** 497 (1969).
7. W. Wolf, E. Degener, and S. Petersen, *Angew. Chem.*, **72,** 963 (1960).
8. H. R. Schweizer, *Helv. Chim. Acta*, **52,** 2236 (1969).
9. M. Konieczny, *Diss. Pharm.*, **20,** 275 (1968); *Chem. Abstr.*, **70,** 11665 (1969).
10. R. F. Meyer, *J. Heterocyclic Chem.*, **6,** 407 (1969).

Author Index

Numbers in parentheses are reference numbers and show that an author's work is referred to although his name is not mentioned in the text. Numbers in *italics* indicate the pages on which the full references appear.

Subject Index

Compounds which are referred to only in tables and which can be found by examining the tables listed under the appropriate general headings covering the types of compounds are not listed individually in the index.